STUDENT SOLUTIONS MANUAL

ALLAN GUNTER

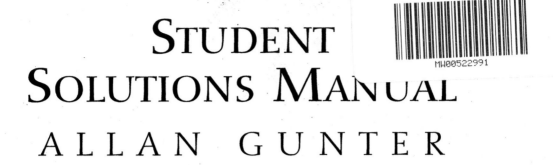

Multivariable Mathematics

Fourth Edition

WILLIAMSON • TROTTER

PEARSON

Prentice Hall

Upper Saddle River, NJ 07458

Editor-in-Chief: Sally Yagan
Executive Editor: George Lobell
Supplement Editor: Jennifer Brady
Assistant Managing Editor: John Matthews
Production Editor: Allyson Kloss
Supplement Cover Manager: Paul Gourhan
Supplement Cover Designer: Joanne Alexandris
Manufacturing Buyer: Ilene Kahn

© 2004 Pearson Education, Inc.
Pearson Prentice Hall
Pearson Education, Inc.
Upper Saddle River, NJ 07458

Pearson Prentice Hall® is a trademark of Pearson Education, Inc.

The author and publisher of this book have used their best efforts in preparing this book. These efforts include the development, research, and testing of the theories and programs to determine their effectiveness. The author and publisher make no warranty of any kind, expressed or implied, with regard to these programs or the documentation contained in this book. The author and publisher shall not be liable in any event for incidental or consequential damages in connection with, or arising out of, the furnishing, performance, or use of these programs.

Printed in the United States of America

10 9 8 7 6 5 4 3 2 1

ISBN 0-13-046192-X

Pearson Education Ltd., *London*
Pearson Education Australia Pty. Ltd., *Sydney*
Pearson Education Singapore, Pte. Ltd.
Pearson Education North Asia Ltd., *Hong Kong*
Pearson Education Canada, Inc., *Toronto*
Pearson Educación de Mexico, S.A. de C.V.
Pearson Education—Japan, *Tokyo*
Pearson Education Malaysia, Pte. Ltd.
Pearson Education, *Upper Saddle River, New Jersey*

Table of Contents

Chapter 1: Vectors

Section 1: COORDINATE VECTORS

Exercise Set 1 (pgs. 7-8)

1. Given $\mathbf{x} = (-3, 4)$ and $\mathbf{y} = (2, 2)$, we have

  (a) $\mathbf{x} + \mathbf{y} = (-3, 4) + (2, 2) = (-1, 6)$.

  (b) $2\mathbf{x} + 3\mathbf{y} = 2(-3, 4) + 3(2, 2) = (-6, 8) + (6, 6) = (0, 14)$.

  (c) $-\mathbf{x} + \mathbf{y} - (1, 4) = -(-3, 4) + (2, 2) - (1, 4) = (3, -4) + (2, 2) + (-1, -4) = (4, -6)$.

3. Given $\mathbf{x} = (3, -1, 0)$, $\mathbf{y} = (0, 1, 5)$ and $\mathbf{z} = (2, 5, -1)$, we have

  (a) $3\mathbf{x} = 3(3, -1, 0) = (9, -3, 0)$.

  (b) $4\mathbf{x} - 2\mathbf{y} + 3\mathbf{z} = 4(3, -1, 0) - 2(0, 1, 5) + 3(2, 5, -1)$
$$= (12, -4, 0) + (0, -2, -10) + (6, 15, -3) = (18, 9, -13).$$

  (c) $-\mathbf{y} + (1, 2, 1) = -(0, 1, 5) + (1, 2, 1) = (0, -1, -5) + (1, 2, 1) = (1, 1, -4)$.

5. In $\mathcal{R}^2$, $2(1, 2) - 3(-1, 4) = (2, 4) + (3, -12) = (5, -8) = 5\mathbf{i} - 8\mathbf{j}$.

7. In $\mathcal{R}^2$, $(1, 4) - (2c, d) = (1, 4) + (-2c, -d) = (1 - 2c, 4 - d) = (1 - 2c)\mathbf{i} + (4 - d)\mathbf{j}$.

9. With $\mathbf{x} = (3, -1, 0)$ and $\mathbf{y} = (0, 1, 5)$, if a and b are such that $a\mathbf{x} + b\mathbf{y} = (9, -1, 10)$ then

$$(9, -1, 10) = a(3, -1, 0) + b(0, 1, 5) = (3a, -a + b, 5b).$$

Comparing corresponding coordinates of the two outer members, gives the system

$$9 = 3a$$
$$-1 = -a + b$$
$$10 = 5b.$$

The first equation gives $a = 3$ and the third equation gives $b = 2$. Moreover, these values also satisfy the second equation. Hence, $(a, b) = (3, 2)$ solves the system. There are no other solutions because the first and third equations uniquely determine a and b, respectively.

11. With $\mathbf{x}$ and $\mathbf{y}$ as in Exercise 9, suppose a and b are real numbers such that $a\mathbf{x} + b\mathbf{y} = (3, 0, 0)$. Then $(3, 0, 0) = a(3, -1, 0) + b(0, 1, 5) = (3a, -a + b, 5b)$. Comparing coordinates shows that $3 = 3a$, $-a + b = 0$ and $5b = 0$. From the first and third equations we get $a = 1$ and $b = 0$. But these values inserted into the second equation say that $1 = a = b = 0$, a contradiction. So, no numbers a and b satisfy an equation of the form $a\mathbf{x} + b\mathbf{y} = (3, 0, 0)$, where $\mathbf{x}$ and $\mathbf{y}$ are given in Exercise 9 above.

    However, for each real number c, we can still ask "what numbers a and b satisfy $a\mathbf{x} + b\mathbf{y} = (3, 0, c)$?". If c is such that $(3, 0, c) = a(3, -1, 0) + b(0, 1, 5) = (3a, -a + b, 5b)$ for some numbers a and b then a comparison the first coordinates shows that $a = 1$, and a comparison of the second coordinates shows that $a = b$, so that $b = 1$. If there is to be a solution then the third coordinates must satisfy $c = 5b = 5$. The corresponding solution for the pair a, b is $a = b = 1$.

1

13. If $\mathbf{x} = \mathbf{i} + \mathbf{j}$, $\mathbf{y} = 2\mathbf{i} + \mathbf{j} + \mathbf{k}$ and $\mathbf{z} = -2\mathbf{i} + \mathbf{j} + 2\mathbf{k}$ then

    (a) $\mathbf{x} + 2\mathbf{y} - \mathbf{z} = -(\mathbf{i} + \mathbf{j}) + 2(2\mathbf{i} + \mathbf{j} + \mathbf{k}) - (-2\mathbf{i} + \mathbf{j} + 2\mathbf{k}) = 5\mathbf{i}$.

    (b) $6\mathbf{x} - 2\mathbf{y} + \mathbf{z} = 6(\mathbf{i} + \mathbf{j}) - 2(2\mathbf{i} + \mathbf{j} + \mathbf{k}) + (-2\mathbf{i} + \mathbf{j} + 2\mathbf{k}) = 5\mathbf{j}$.

    (c) $-4\mathbf{x} + 3\mathbf{y} + \mathbf{z} = -4(\mathbf{i} + \mathbf{j}) + 3(2\mathbf{i} + \mathbf{j} + \mathbf{k}) + (-2\mathbf{i} + \mathbf{j} + 2\mathbf{k}) = 5\mathbf{k}$.

15. Let $\mathbf{x} = (x_1, \ldots, x_n)$ be a vector in $\mathcal{R}^n$ and let r and s be real numbers. Then

$$
\begin{aligned}
r(s\mathbf{x}) &= r\big(s(x_1, \ldots, x_n)\big) \\
&= r(sx_1, \ldots, sx_n) && \text{(definition of scalar multiplication of vectors)} \\
&= \big(r(sx_1), \ldots, r(sx_n)\big) && \text{(definition of scalar mutltiplication of vector)} \\
&= \big((rs)x_1, \ldots, (rs)x_n\big) && \text{(associative property of real numbers)} \\
&= (rs)(x_1, \ldots, x_n) && \text{(definition of scalar multiplication of vectors)} \\
&= (rs)\mathbf{x}.
\end{aligned}
$$

17. We have

$$
\begin{aligned}
2(3\mathbf{x} - 2\mathbf{y} + \mathbf{z}) - 4\mathbf{x} &= 2\big(3\mathbf{x} + (-2\mathbf{y}) + \mathbf{z}\big) + (-4\mathbf{x}) && \text{(def. of subtraction)} \\
&= 2(3\mathbf{x}) + 2(-2\mathbf{y}) + 2\mathbf{z} + (-4\mathbf{x}) && \text{(formula 2 with } r = 2) \\
&= 6\mathbf{x} + (-4\mathbf{y}) + 2\mathbf{z} + (-4\mathbf{x}) && \text{(formula 3 with } r = 2, s = 3, -2) \\
&= 6\mathbf{x} + (-4\mathbf{x}) + (-4\mathbf{y}) + 2\mathbf{z} && \text{(formula 4)} \\
&= 2\mathbf{x} + (-4\mathbf{y}) + 2\mathbf{z} && \text{(formula 1 with } r = 6, s = -4) \\
&= 2\mathbf{x} - 4\mathbf{y} + 2\mathbf{z}. && \text{(def. of subtraction)}
\end{aligned}
$$

19. We have

$$
\begin{aligned}
\frac{1}{2}(\mathbf{x} + \mathbf{y}) + \mathbf{y} &= \left(\frac{1}{2}\mathbf{x} + \frac{1}{2}\mathbf{y}\right) + \mathbf{y} && \text{(formula 1 with } r = s = 1/2) \\
&= \frac{1}{2}\mathbf{x} + \left(\frac{1}{2}\mathbf{y} + \mathbf{y}\right) && \text{(formula 5)} \\
&= \frac{1}{2}\mathbf{x} + \left(\frac{1}{2}\mathbf{y} + 1\mathbf{y}\right) && \text{(formula 8)} \\
&= \frac{1}{2}\mathbf{x} + \frac{3}{2}\mathbf{y}. && \text{(formula 1 with } r = 1/2, s = 1)
\end{aligned}
$$

21. By inspection, $(-2, 3) = -2\mathbf{e}_1 + 3\mathbf{e}_2$.

23. We need to find scalars a and b such that $(2, -7) = a(1, 1) + b(1, -1) = (a + b, a - b)$. This leads to the system

$$
\begin{aligned}
a + b &= 2 \\
a - b &= -7.
\end{aligned}
$$

Adding the equations gives $2a = -5$, or $a = -\frac{5}{2}$, and subtracting the equations gives $2b = 9$, or $b = \frac{9}{2}$. Thus,

$$
(2, -7) = -\frac{5}{2}(1, 1) + \frac{9}{2}(1, -1).
$$

2

25. Let $\mathbf{x} = 2\mathbf{i}$, $\mathbf{y} = \mathbf{i} - 3\mathbf{j}$ and $\mathbf{z} = 3\mathbf{i} + 2\mathbf{j} - 2\mathbf{k}$.

    (a) The equation $\mathbf{x} = 2\mathbf{i}$ immediately gives $\mathbf{i} = \frac{1}{2}\mathbf{x}$.

    (b) By part (a), $\mathbf{y} = \mathbf{i} - 3\mathbf{j} = \frac{1}{2}\mathbf{x} - 3\mathbf{j}$. Solving for $\mathbf{j}$ gives

$$\mathbf{j} = \frac{1}{6}\mathbf{x} - \frac{1}{3}\mathbf{y}.$$

    (c) Solving $\mathbf{z} = 3\mathbf{i} + 2\mathbf{j} - 2\mathbf{k}$ for $\mathbf{k}$ gives $\mathbf{k} = \frac{3}{2}\mathbf{i} + \mathbf{j} - \frac{1}{2}\mathbf{z}$. Hence, from parts (a) and (b),

$$\mathbf{k} = \frac{3}{2}\left(\frac{1}{2}\mathbf{x}\right) + \left(\frac{1}{6}\mathbf{x} - \frac{1}{3}\mathbf{y}\right) - \frac{1}{2}\mathbf{z} = \frac{11}{12}\mathbf{x} - \frac{1}{3}\mathbf{y} - \frac{1}{2}\mathbf{z}.$$

27. Let A and B be the books represented by the vectors $\mathbf{x} = (5, 500, 10)$ and $\mathbf{y} = (4, 800, 90)$, respectively, and let $\mathbf{T}$ be the vector

$$\mathbf{T} = 100\mathbf{x} + 50\mathbf{y} = 100(5, 500, 10) + 50(4, 800, 90) = (700, 90000, 5500).$$

Then the first coordinate of $\mathbf{T}$ represents the total amount of ink (700 units) needed to produce 100 copies of A and 50 copies of B; the second coordinate of $\mathbf{T}$ represents the total amount of paper (90000 units) needed to produce 100 copies of A and 50 copies of B; and the third coordinate of $\mathbf{T}$ represents the total amount of binding material (5500 units) needed to produce 100 copies of A and 50 copies of B.

29. The temperature at the k-th site at time t is $x_k(t)$. In particular, $x_k(2)$, $x_k(8)$, $x_k(14)$ and $x_k(20)$ are the temperatures at the four times $t = 2, 8, 14, 20$. The average of these four temperatures at the k-th site is therefore $\frac{1}{4}\big(x_k(2) + x_k(8) + x_k(14) + x_k(20)\big)$. It follows that the vector that represents the average of these four temperatures at the fifty sites is

$$\left(\frac{1}{4}\big(x_1(2) + x_1(8) + x_1(14) + x_1(20)\big), \ldots, \frac{1}{4}\big(x_{50}(2) + x_{50}(8) + x_{50}(14) + x_{50}(20)\big)\right)$$

$$= \frac{1}{4}\big(x_1(2) + x_1(8) + x_1(14) + x_1(20), \ldots, (x_{50}(2) + x_{50}(8) + x_{50}(14) + x_{50}(20)\big)$$

$$= \frac{1}{4}\Big(\big(x_1(2), \ldots, x_{50}(2)\big) + \big(x_1(8), \ldots, x_{50}(8)\big) + \big(x_1(14), \ldots, x_{50}(14)\big)$$

$$+ \big(x_1(20), \ldots, x_{50}(20)\big)\Big)$$

$$= \frac{1}{4}\big(\mathbf{x}(2) + \mathbf{x}(8) + \mathbf{x}(14) + \mathbf{x}(20)\big).$$

Section 2: GEOMETRIC VECTORS

Exercise Set 2A-E (pgs. 16-17)

1. The line segment with endpoints $(1,1)$, $(-2,2)$ has length $\sqrt{\left(1-(-2)\right)^2 + (1-2)^2} = \sqrt{9+1} = \sqrt{10}$. Further, the midpoint $\mathbf{p}$ (shown in the diagram below) is located at $\mathbf{p} = \frac{1}{2}\left[(1,1) + (-2,2)\right] = \frac{1}{2}(-1,3) = (-1/2, 3/2)$.

3. The line segment with end points $(1,1,1)$, $(1,-1,1)$ has length

$$\sqrt{(1-1)^2 + \left(1-(-1)\right)^2 + (1-1)^2} = \sqrt{4} = 2.$$

The midpoint $\mathbf{p}$ (shown in the diagram below) is located at $\mathbf{p} = \frac{1}{2}\left[(1,1,1) + (1,-1,1)\right] = \frac{1}{2}(2,0,2) = (1,0,1)$.

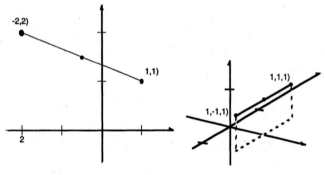
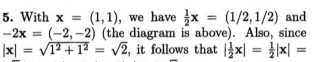
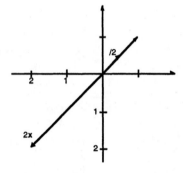

Exercise 1   Exercise 3   Exercise 5

5. With $\mathbf{x} = (1,1)$, we have $\frac{1}{2}\mathbf{x} = (1/2, 1/2)$ and $-2\mathbf{x} = (-2,-2)$ (the diagram is above). Also, since $|\mathbf{x}| = \sqrt{1^2 + 1^2} = \sqrt{2}$, it follows that $|\frac{1}{2}\mathbf{x}| = \frac{1}{2}|\mathbf{x}| = \sqrt{2}/2$ and $|-2\mathbf{x}| = 2|\mathbf{x}| = 2\sqrt{2}$.

7. With $\mathbf{x} = (1,2,2)$, we have $\frac{1}{2}\mathbf{x} = (1/2, 1, 1)$ and $-2\mathbf{x} = (-2,-4,-4)$ (the diagram is on the right). Also, since $|\mathbf{x}| = \sqrt{1^2 + 2^2 + 2^2} = 3$, $|\frac{1}{2}\mathbf{x}| = \frac{1}{2}|\mathbf{x}| = 3/2$ and $|-2\mathbf{x}| = 2|\mathbf{x}| = 6$.

9. With $\mathbf{x} = (1,1)$ and $\mathbf{y} = (1,-1)$ in $\mathcal{R}^2$, we have

$$\mathbf{x} + \mathbf{y} = (1,1) + (1,-1) = (2,0),$$
$$\mathbf{x} - \mathbf{y} = (1,1) - (1,-1) = (0,2),$$
$$\mathbf{x} + 2\mathbf{y} = (1,1) + 2(1,-1) = (3,-1).$$

(The diagram is on the next page.)

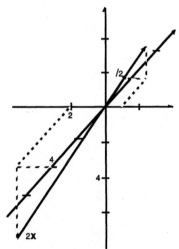

Exercise 7

4

11. With $\mathbf{x} = (1, 1, 1)$ and $\mathbf{y} = (1, 1, -1)$ in $\mathcal{R}^3$, we have

$$\mathbf{x} + \mathbf{y} = (1, 1, 1) + (1, 1, -1) = (2, 2, 0), \qquad \mathbf{x} - \mathbf{y} = (1, 1, 1) - (1, 1, -1) = (0, 0, 2),$$
$$\mathbf{x} + 2\mathbf{y} = (1, 1, 1) + 2(1, 1, -1) = (3, 3, -1).$$

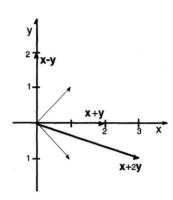

Exercise 9

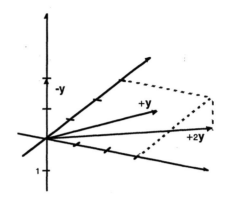

Exercise 11

13.

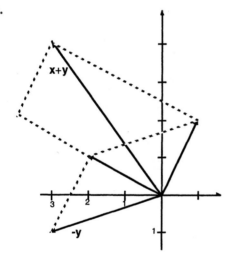

15.

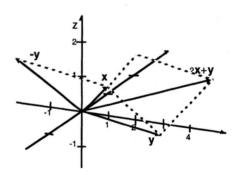

17.

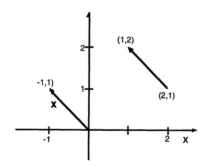

19.

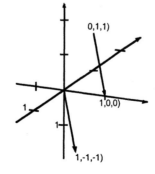

5

21.

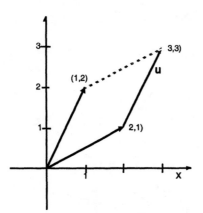

23.

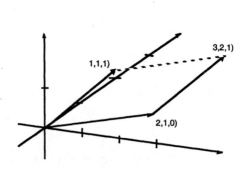

25. For **x** and **y** in $\mathcal{R}^n$,

$$\big|((1-t)\mathbf{x}+t\mathbf{y})-\mathbf{x}\big| = \big|(t\mathbf{y}+(1-t)\mathbf{x})-\mathbf{x}\big| = \big|t\mathbf{y}+((1-t)\mathbf{x}-\mathbf{x})\big|$$
$$= |t\mathbf{y}+(-t)\mathbf{x}| = |t||\mathbf{y}-\mathbf{x}|$$
$$= t|\mathbf{y}-\mathbf{x}|,$$
$$\big|((1-t)\mathbf{x}+t\mathbf{y})-\mathbf{y}\big| = |(1-t)\mathbf{x}+(t\mathbf{y}-\mathbf{y})| = |(1-t)\mathbf{x}+(t-1)\mathbf{y}|$$
$$= |1-t||\mathbf{x}-\mathbf{y}|$$
$$= (1-t)|\mathbf{y}-\mathbf{x}|,$$

where the third line holds because $t \geq 0$ and the last line holds because $1-t \geq 0$ and $|\mathbf{x}-\mathbf{y}| = |\mathbf{y}-\mathbf{x}|$.

To see why the condition $0 \leq t \leq 1$ is needed, observe that the left sides of the equations we are asked to verify are lengths of vectors and, as such, must be nonnegative. If $t < 0$ then the right side of the first equation would be $t|\mathbf{y}-\mathbf{x}| < 0$, which is impossible; and if $t > 1$ then the right side of the second equation would be $(1-t)|\mathbf{y}-\mathbf{x}| < 0$, which is also impossible. In other words, the equations we are asked to verify can simultaneously hold if, and only if, $0 \leq t \leq 1$.

27. The desired square and the four vectors $\mathbf{v}_1$, $\mathbf{v}_2$, $\mathbf{v}_3$, $\mathbf{v}_4$ are shown in the diagram at the right. In terms of coordinates, the vectors have the specific forms

$$\mathbf{v}_1 = (2,0) - (0,0) = (2,0),$$
$$\mathbf{v}_2 = (2,2) - (2,0) = (0,2),$$
$$\mathbf{v}_3 = (0,2) - (2,2) = (-2,0),$$
$$\mathbf{v}_4 = (0,0) - (0,2) = (0,-2).$$

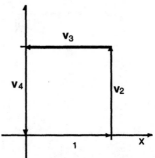

This allows computation of their sum:

$$\mathbf{v}_1 + \mathbf{v}_2 + \mathbf{v}_3 + \mathbf{v}_4 = (2,0) + (0,2) + (-2,0) + (0,-2) = (0,0).$$

29. Let $m \geq 3$ and consider the m vectors $\mathbf{v}_1, \ldots, \mathbf{v}_m$ in $\mathcal{R}^2$. Let A_1 be the arrow representing the position vector $\mathbf{v}_1$ and arrange arrows $A_2, \ldots, A_m$, representing the vectors $\mathbf{v}_2, \ldots, \mathbf{v}_m$ (resp.), in such a way that the the tail of A_k is at the tip of A_{k-1}, for $k = 2, \ldots, m$.

In the xy-plane, this is a polygonal line that starts at the origin and consists of the m directed line segments $A_1, \ldots, A_m$. We claim that the vector $\mathbf{v} = \mathbf{v}_1 + \cdots + \mathbf{v}_m$ is the zero vector if, and only if, the tip of the arrow A_m is at the origin.

In order to prove the above claim, for each k, let (x_k, y_k) be the point at the tip of the arrow A_k. By how the arrows have been arranged, the vectors $\mathbf{v}_k$ can be expressed as

$$\mathbf{v}_1 = (x_1, y_1) \quad \text{and} \quad \mathbf{v}_k = (x_k, y_k) - (x_{k-1}, y_{k-1}), \quad \text{for } k = 2, \ldots, m.$$

Hence,

$$\begin{aligned}
\mathbf{v} &= \mathbf{v}_1 + \mathbf{v}_2 + \cdots + \mathbf{v}_m \\
&= (x_1, y_1) + \big((x_2, y_2) - (x_1, y_1)\big) + \cdots + \big((x_m, y_m) - (x_{m-1}, y_{m-1})\big) \\
&= (x_m, y_m),
\end{aligned}$$

where the last equality holds because all terms cancel out except the term with the highest subscript m. It follows that $\mathbf{v}$ is the zero vector if, and only if, $(x_m, y_m) = (0, 0)$; i.e., if, and only if, the tip of A_m is at the origin. In other words, a finite sum of vectors in $\mathcal{R}^2$ is the zero vector if, and only if, the vectors can be arranged tip to tail to form a closed polygonal figure. The same condition is valid in $\mathcal{R}^3$.

31. (a) Let $\mathbf{a}$ and $\mathbf{b}$ be two different points and let $\mathbf{c} = t\mathbf{a} + (1 - t)\mathbf{b}$, where t is a scalar. The distance between $\mathbf{a}$ and $\mathbf{c}$ is $|\mathbf{c} - \mathbf{a}|$ and the distance between $\mathbf{b}$ and $\mathbf{c}$ is $|\mathbf{c} - \mathbf{b}|$, so that

$$\begin{aligned}
|\mathbf{c} - \mathbf{a}| + |\mathbf{c} - \mathbf{b}| &= \big|t\mathbf{a} + (1 - t)\mathbf{b} - \mathbf{a}\big| + \big|t\mathbf{a} + (1 - t)\mathbf{b} - \mathbf{b}\big| \\
&= |(t - 1)\mathbf{a} + (1 - t)\mathbf{b}| + |t\mathbf{a} - t\mathbf{b}| \\
&= |t - 1||\mathbf{a} - \mathbf{b}| + |t||\mathbf{a} - \mathbf{b}| \\
&= \big(|t - 1| + |t|\big)|\mathbf{a} - \mathbf{b}|.
\end{aligned}$$

Therefore, since the distance from $\mathbf{a}$ to $\mathbf{b}$ is $|\mathbf{a} - \mathbf{b}|$, $\mathbf{c}$ is *between* $\mathbf{a}$ and $\mathbf{b}$ if, and only if,

$$\big(|t - 1| + |t|\big)|\mathbf{a} - \mathbf{b}| = |\mathbf{a} - \mathbf{b}|.$$

By hypothesis, $\mathbf{a} \neq \mathbf{b}$, so we may cancel the nonzero number $|\mathbf{a} - \mathbf{b}|$ from each term to get $|t - 1| + |t| = 1$. Squaring both sides of this equation and simplifying gives $|t(t-1)| = -t(t-1)$, which implies $t(t - 1) \leq 0$. If $t < 0$ or $t > 1$ then t and $t - 1$ are the same sign and therefore their product must be positive. On the other hand, t and $t - 1$ are not of the same sign if $0 \leq t \leq 1$. In other words, $\mathbf{c}$ is between $\mathbf{a}$ and $\mathbf{b}$ if, and only if, $0 \leq t \leq 1$.

(b) We saw in part (a) that $|\mathbf{c} - \mathbf{a}| = |t - 1||\mathbf{a} - \mathbf{b}|$ and $|\mathbf{c} - \mathbf{b}| = |t||\mathbf{a} - \mathbf{b}|$. Thus, $\mathbf{a}$ is between $\mathbf{b}$ and $\mathbf{c}$ if, and only if,

$$|t - 1||\mathbf{a} - \mathbf{b}| + |\mathbf{a} - \mathbf{b}| = |t||\mathbf{a} - \mathbf{b}|.$$

Canceling the nonzero number $|\mathbf{a} - \mathbf{b}|$ from each term gives $|t - 1| + 1 = |t|$. Squaring both sides of this equation and simplifying the result gives $|t - 1| = t - 1$. Hence, $t \geq 1$. That is, $\mathbf{a}$ is between $\mathbf{b}$ and $\mathbf{c}$ if, and only if, $t \geq 1$.

(c) Taking our cue from part (b), we see that $\mathbf{b}$ is between $\mathbf{a}$ and $\mathbf{c}$ if, and only if,

$$|t||\mathbf{a} - \mathbf{b}| + |\mathbf{a} - \mathbf{b}| = |t - 1||\mathbf{a} - \mathbf{b}|.$$

Canceling the nonzero number $|\mathbf{a} - \mathbf{b}|$ from each term gives $|t| + 1 = |t - 1|$. Squaring both sides of this equation and simplifying the result gives $|t| = -t$. Hence, $t \leq 0$. That is, $\mathbf{b}$ is between $\mathbf{a}$ and $\mathbf{c}$ if, and only if, $t \leq 0$.

33. We can suppose that, in the given coordinate system, due east of the origin constitutes the positive x-axis, due north of the origin constitutes the positive y-axis, and straight up from the origin constitutes the positive z-axis. If $\mathbf{d}$ is the position vector whose tip is at the base of the flagpole then $\mathbf{d} = x_0\mathbf{i} + y_0\mathbf{j}$ for some scalars x_0, y_0. We are given that $\mathbf{d}$ makes a $20°$ angle with the positive y-axis. Assuming that $\mathbf{d}$ has length d, the component vectors $x_0\mathbf{i}$ and $y_0\mathbf{j}$ have lengths $x_0 = d\sin 20°$ and $y_0 = d\cos 20°$, respectively, so that

$$\mathbf{d} = d\sin 20°\mathbf{i} + d\cos 20°\mathbf{j}.$$

Assuming the flagpole is perpendicular to the xy-plane, if h is its height then the arrow whose tip is at the top of the flagpole and whose tail is at the bottom of the flagpole is represented by the vector $\mathbf{h} = h\mathbf{k}$. Moreover, the distance from the surveyor to the top of the flagpole is evidently $\sqrt{h^2 + d^2}$. Letting $\mathbf{T}$ be the position vector whose tip is at the top of the flagpole (so that $|\mathbf{T}| = \sqrt{h^2 + d^2}$), we see that $\mathbf{T}$ is the vector sum of $\mathbf{d}$ and $\mathbf{h}$. That is,

$$\mathbf{T} = d\sin 20°\mathbf{i} + d\cos 20°\mathbf{j} + h\mathbf{k}.$$

Since $\mathbf{T}$ makes an angle of $20°$ with the xy-plane, we also have

$$\cos 20° = \frac{d}{\sqrt{h^2 + d^2}} \quad \text{and} \quad \sin 20° = \frac{h}{\sqrt{h^2 + d^2}}.$$

Hence,

$$\mathbf{T} = \frac{dh}{\sqrt{h^2 + d^2}}\mathbf{i} + \frac{d^2}{\sqrt{h^2 + d^2}}\mathbf{j} + h\mathbf{k}.$$

If $\mathbf{t}$ denotes the unit vector pointing at the top of the flagpole then

$$\mathbf{t} = \frac{\mathbf{T}}{|\mathbf{T}|} = \frac{\mathbf{T}}{\sqrt{h^2 + d^2}} = \frac{dh}{h^2 + d^2}\mathbf{i} + \frac{d^2}{h^2 + d^2}\mathbf{j} + \frac{h}{\sqrt{h^2 + d^2}}\mathbf{k}.$$

We can eliminate h and d from the above equation by noting that $\tan 20° = h/d$, or equivalently, $h = d\tan 20°$. Using this observation, $\mathbf{t}$ can be written as

$$\begin{aligned}
\mathbf{t} &= \frac{d^2\tan 20°}{d^2\tan^2 20° + d^2}\mathbf{i} + \frac{d^2}{d^2\tan^2 20° + d^2}\mathbf{j} + \frac{d\tan 20°}{\sqrt{d^2\tan^2 20° + d^2}}\mathbf{k} \\
&= \frac{\tan 20°}{\tan^2 20° + 1}\mathbf{i} + \frac{1}{\tan^2 20° + 1}\mathbf{j} + \frac{\tan 20°}{\sqrt{\tan^2 20° + 1}}\mathbf{k} \\
&= \frac{\tan 20°}{\sec^2 20°}\mathbf{i} + \frac{1}{\sec^2 20°}\mathbf{j} + \frac{\tan 20°}{\sec 20°}\mathbf{k} \\
&= (\sin 20° \cos 20°)\mathbf{i} + (\cos^2 20°)\mathbf{j} + (\sin 20°)\mathbf{k} \\
&\approx 0.321\mathbf{i} + 0.883\mathbf{j} + 0.342\mathbf{k},
\end{aligned}$$

where we have used the identities $\tan^2 x + 1 = \sec^2 x$ and $\sec^2 x = 1/\cos^2 x$, and have rounded off the coordinates of the last member above to three significant digits.

35. Let $\mathbf{P} = (x_0, y_0, z_0)$ be the position vector for Los Angeles. Then the vector $\mathbf{P}_0 = (x_0, y_0, 0)$ is its projection onto the equatorial plane and the vector $\mathbf{P}_z = (0, 0, z_0)$ is its projection onto the earth's axis. These three vectors form a right triangle such that $\sin 34° = z_0/4000$ and $\cos 34° = |\mathbf{P}_0|/4000$, or equivalently,

(1) $$z_0 = 4000\sin 34° \quad \text{and} \quad |\mathbf{P}_0| = 4000\cos 34°.$$

8

Now observe that $\mathbf{P}_0$ makes an angle of 118° with the vector $\mathbf{i}$, so that it makes an angle of 28° with the vector $-\mathbf{j}$. The vectors $\mathbf{Q}_x = (x_0, 0, 0)$ and $\mathbf{Q}_y = (0, y_0, 0)$ lie in the equatorial plane and form the legs of a right triangle with hypotenuse $\mathbf{P}_0$. Viewing this right triangle as sitting in the third quadrant of the equatorial plane (which encompasses all points from 90° west longitude to 180° west longitude), we obtain $-\sin 28° = x_0/|\mathbf{P}_0|$ and $-\cos 28° = y_0/|\mathbf{P}_0|$, or equivalently,

$$x_0 = -|\mathbf{P}_0| \sin 28° = -4000 \cos 34° \sin 28°$$

$$\text{and} \quad y_0 = -|\mathbf{P}_0| \cos 28° = -4000 \cos 34° \cos 28°,$$

where we have used the second equation in (1) to eliminate $|\mathbf{P}_0|$. Thus,

$$\mathbf{P} = (-4000 \cos 34° \sin 28°, -4000 \cos 34° \cos 28°, 4000 \sin 34°)$$
$$\approx (-1557, -2928, 2237),$$

where the last approximation is to the nearest mile.

37. In Exercises 35 an 36 it was found that the position vectors for Los Angeles and New York are (to the nearest mile in each coordinate)

Los Angeles: $(-1557, -2928, 2237)$, New York: $(832, -2902, 2624)$.

The straight-line distance D between Los Angeles and New York is therefore the length of a vector that is the difference of these two vectors (the order in which we subtract these vectors is immaterial, since the two possible differences give vectors that are the negatives of each other). Thus,

$$D = |(832, -2902, 2624) - (-1557, -2928, 2237)| = |(2389, 26, 387)|$$
$$= \sqrt{2389^2 + 26^2 + 387^2} \approx 2420 \text{ miles},$$

where D is to the nearest mile.

Section 3: LINES AND PLANES

Exercise Set 3AB (pgs. 23-24)

1. Let $\mathbf{x} = t\mathbf{u} + \mathbf{v}$ be a parametric form representing the desired line. Since the line is parallel to the vector $(2,1)$, we can let $\mathbf{u} = (2,1)$; and since the point $(-1,2)$ is on the line, we can take $\mathbf{v} = (-1,2)$. That is,

$$\mathbf{x} = t(2,1) + (-1,2).$$

3. Let $\mathbf{x} = t\mathbf{u} + \mathbf{v}$ be a parametric form representing the desired line. The vector from the given point $\mathbf{a} = (1,2,2)$ to the given point $\mathbf{b} = (2,2,3)$ is $\mathbf{b} - \mathbf{a} = (1,0,1)$ and is parallel to the desired line. Thus, we can take $\mathbf{u} = (1,0,1)$. Since $(1,2,2)$ is on the line, we can take $\mathbf{v} = (1,2,2)$. That is,

$$\mathbf{x} = t(1,0,1) + (1,2,2).$$

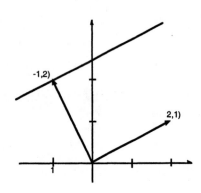

Exercise 1

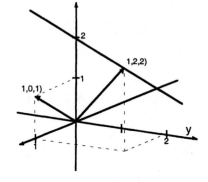

Exercise 3

5. (a) With $\mathbf{a} = (-1,1)$, $\mathbf{b} = (0,1)$, $\mathbf{c} = (2,1)$ and $\mathbf{d} = (-3,2)$, the lines $t\mathbf{a}+\mathbf{b} = t(-1,1)+(0,1)$ and $s\mathbf{c}+\mathbf{d} = s(2,1)+(-3,2)$ are shown in the diagram on the right.

(b) Suppose s and t are such that $t(-1,1) + (0,1) = s(2,1) + (-3,2)$. This can be rearranged to get $t(-1,1) - s(2,1) = (-3,2) - (0,1)$, or equivalently, $(-t-2s, t-s) = (-3,1)$. Comparing corresponding coordinates leads to the system

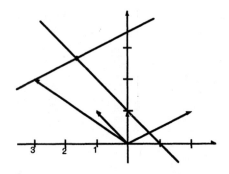

$$-t - 2s = -3$$
$$t - s = 1.$$

Adding the two equations gives $-3s = -2$, so that $s = 2/3$. With this value of s in $s(2,1) + (-3,2)$ we get $\frac{2}{3}(2,1) + (-3,2) = (4/3, 2/3) + (-3,2) = (-5/3, 8/3)$. Thus, the intersection point $\mathbf{p}$ (shown in the above diagram) is given by

$$\mathbf{p} = (-5/3, 8/3).$$

(c) If **c** is replaced by $(2, -2)$ then $s(2, -2) + (-3, 2)$ and $t(-1, 1) + (0, 1)$ have direction vectors that are scalar multiples of each other; i.e., the lines are parallel. Thus, they intersect only if they are the same line. Since $(0, 1)$ (e.g.) is on $t(-1, 1) + (0, 1)$, we check to see if it is also on $s(2, -2) + (-3, 2)$. That is, we check to see if there is a scalar s such that $(0, 1) = s(2, -2) + (-3, 2) = (2s - 3, -2s + 2)$. Comparing the first coordinates gives $s = 3/2$, while comparing the second coordinates gives $s = 1/2$. We conclude that $(0, 1)$ is not on $s(2, -2) + (-3, 2)$ and therefore the lines do not intersect.

7. Given the two lines $t(1, 2) + (2, 1)$ and $t(-2, -4) + (3, 3)$, we first notice that the direction vectors $(1, 2)$ and $(-2, -4)$ are scalar multiples of each other. Hence, the two lines are parallel.

To see if they are the same line, we check to see if $(3, 3)$ (which is on the second line) is on the first line. That is, we check to see if there is a t such that $(3, 3) = t(1, 2) + (2, 1)$. By inspection, we see that $t = 1$ is such a value. Since parallel lines with a point in common must be the same line, we conclude that the two lines are the same.

9. Given the two lines $t(2, 3, -1) + (-1, -1, 1)$ and $t(-4, -6, 2) + (1, 1, -1)$, we first notice that the direction vectors $(2, 3, -1)$ and $(-4, -6, 2)$ are related by $-2(2, 3, -1) = (-4, -6, 2)$. Hence, the lines are parallel.

To see if are the same, we check to see if $(1, 1, -1)$ (which is on the second line) is on the first line. That is, we check to see if there is a t such that $(1, 1, -1) = t(2, 3, -1) + (-1, -1, 1)$, or equivalently, $t(2, 3, -1) = (1, 1, -1) - (-1, -1, 1) = (2, 2, -2)$. Since $(2, 2, -2)$ is not a scalar multiple of $(2, 3, -1)$, there is no such t. So, the given lines are not the same line.

11. Suppose t is a scalar such that $(1, 2) = t(2, 1) = (2t, t)$. The second coordinates lead to $t = 2$, while the first coordinates lead to $t = 1/2$, a contradiction. similarly, suppose $(2, 1) = t(1, 2) = (t, 2t)$. Again, a comparison of like coordinates gives the same two contradictory equations as before. So, the given vectors are linearly independent.

13. Suppose t is a scalar such that $(3, 1, 3) = t(1, 3, 1) = (t, 3t, t)$. The first and third coordinates lead to $t = 3$, but the second coordinates lead to $t = 1/3$, a contradiction. Similarly, suppose $(1, 3, 1) = t(3, 1, 3) = (3t, t, 3t)$. The first and third coordinates lead to $t = 1/3$, but the second coordinates lead to $t = 3$, another contradiction. So, the given vectors are linearly independent.

15. Let $t_1\mathbf{u}_1 + t_2\mathbf{u}_2 + \mathbf{v}$ be a parametric representation of the desired plane. Since the plane is parallel to the linearly independent vectors $(1, 1, 0)$ and $(0, 1, 1)$, we can let $\mathbf{u}_1 = (1, 1, 0)$ and $\mathbf{u}_2 = (0, 1, 1)$; and since the origin is a point on the plane, we can let $\mathbf{v} = (0, 0, 0)$. This gives the representation

$$t_1(1, 1, 0) + t_2(0, 1, 1).$$

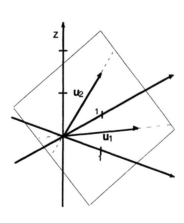

Exercise 15

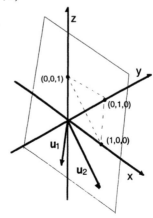

Exercise 17

17. Let $t_1\mathbf{u}_1 + t_2\mathbf{u}_2 + \mathbf{v}$ be a parametric representation of the desired plane. Since the plane contains the three points $(1,0,0)$, $(0,1,0)$ and $(0,0,1)$, the vectors $\mathbf{u}_1 = (1,0,0) - (0,1,0) = (1,-1,0)$ and $\mathbf{u}_2 = (1,0,0) - (0,0,1) = (1,0,-1)$ are parallel to the plane and, by inspection, are linearly independent. Thus, let $\mathbf{v} = (1,0,0)$ to get the representation

$$t_1(1,-1,0) + t_2(1,0,-1) + (1,0,0).$$

19. (a) Since the plane P contains the line $t(1,-1,2) + (1,2,1)$, it contains every point on the line. In particular, it contains the point $(1,2,1)$ (set $t = 0$). Moreover, P contains the point $(3,0,1)$, which is not on the line. Hence, the vector from $(1,2,1)$ to $(3,0,1)$, given by $(3,0,1) - (1,2,1) = (2,-2,0)$, can't be parallel to the direction vector $(1,-1,2)$ of the line. That is, the vectors $(2,-2,0)$ and $(1,-1,2)$ are linearly independent.

(b) Since P contains the line $t(1,-1,2) + (1,2,1)$, the direction vector $(1,-1,2)$ of the line is parallel to P. In part (a), we found a vector $(2,-2,0)$ that contains two points of P, so that it must be parallel to P. Moreover, it was shown that $(2,-2,0)$ is linearly independent of $(1,-1,2)$. Thus, a parametric representation of the plane is

$$\mathbf{x} = t_1(1,-1,2) + t_2(2,-2,0) + (1,2,1),$$

where the point $(1,2,1)$ is on P because it is on the line $t(1,-1,2) + (1,2,1)$ lying in P.

21. Let $\mathbf{a}$, $\mathbf{b}$, $\mathbf{c}$ and $\mathbf{d}$ be the vertices of a quadrilateral such that the sides are s_1 (connecting $\mathbf{a}$ and $\mathbf{b}$), s_2 (connecting $\mathbf{b}$ and $\mathbf{c}$), s_3 (connecting $\mathbf{c}$ and $\mathbf{d}$) and s_4 (connecting $\mathbf{d}$ and $\mathbf{a}$). If $\mathbf{m}_i$ denotes the midpoint of side s_i, for $i = 1, 2, 3, 4$, then

$$\mathbf{m}_1 = \frac{1}{2}(\mathbf{a} + \mathbf{b}), \qquad \mathbf{m}_2 = \frac{1}{2}(\mathbf{b} + \mathbf{c}), \qquad \mathbf{m}_3 = \frac{1}{2}(\mathbf{c} + \mathbf{d}), \qquad \mathbf{m}_4 = \frac{1}{2}(\mathbf{d} + \mathbf{a}).$$

Let Q be the quadrilateral formed by connecting $\mathbf{m}_1$ with $\mathbf{m}_2$, $\mathbf{m}_2$ with $\mathbf{m}_3$, $\mathbf{m}_3$ with $\mathbf{m}_4$, and $\mathbf{m}_4$ with $\mathbf{m}_1$. The vector $\mathbf{v}_1$ from $\mathbf{m}_1$ to $\mathbf{m}_2$ and the vector $\mathbf{v}_3$ from $\mathbf{m}_3$ to $\mathbf{m}_4$ are given by

$$\mathbf{v}_1 = \mathbf{m}_2 - \mathbf{m}_1 = \frac{1}{2}(\mathbf{b} + \mathbf{c}) - \frac{1}{2}(\mathbf{a} + \mathbf{b}) = -\frac{1}{2}\mathbf{a} + \frac{1}{2}\mathbf{c},$$

$$\mathbf{v}_3 = \mathbf{m}_4 - \mathbf{m}_3 = \frac{1}{2}(\mathbf{d} + \mathbf{a}) - \frac{1}{2}(\mathbf{c} + \mathbf{d}) = \frac{1}{2}\mathbf{a} - \frac{1}{2}\mathbf{c}.$$

Noting that $\mathbf{v}_1$ is the negative of $\mathbf{v}_3$, we conclude that $\mathbf{v}_1$ and $\mathbf{v}_3$ are parallel vectors and therefore the sides of Q representing them are parallel. Similarly, the vector $\mathbf{v}_2$ from $\mathbf{m}_2$ to $\mathbf{m}_3$ and the vector $\mathbf{v}_4$ from $\mathbf{m}_4$ to $\mathbf{m}_1$ are given by

$$\mathbf{v}_2 = \mathbf{m}_3 - \mathbf{m}_2 = \frac{1}{2}(\mathbf{c} + \mathbf{d}) - \frac{1}{2}(\mathbf{b} + \mathbf{c}) = \frac{1}{2}\mathbf{d} - \frac{1}{2}\mathbf{b},$$

$$\mathbf{v}_4 = \mathbf{m}_1 - \mathbf{m}_4 = \frac{1}{2}(\mathbf{a} + \mathbf{b}) - \frac{1}{2}(\mathbf{d} + \mathbf{a}) = \frac{1}{2}\mathbf{b} - \frac{1}{2}\mathbf{d}.$$

Again, $\mathbf{v}_2$ and $\mathbf{v}_4$ are the negatives of each other and are therefore parallel, which means that the sides of Q that they represent are parallel. In other words, Q is a parallelogram.

23. (a) Let $\mathbf{u} = (u_1, \dots, u_n)$ be a vector in $\mathcal{R}^n$ and let t be a scalar such that $t\mathbf{u} = \mathbf{0}$. Then, by the definition of scalar multiplication of a vector,

$$t\mathbf{u} = t(u_1, \dots, u_n) = (tu_1, \dots, tu_n) = \mathbf{0}.$$

Since two vectors are equal if, and only if, all corresponding coordinates are equal, $tu_i = 0$ for $i = 1, \dots, n$. If some $u_i \neq 0$ then $t = 0$. Otherwise, all $u_i = 0$ and $\mathbf{u} = (0, \dots, 0) = \mathbf{0}$.

12

(b) If the vector $\mathbf{u}_1$ is a scalar multiple of the vector $\mathbf{u}_2$, and is not zero, then $\mathbf{u}_1 = k\mathbf{u}_2$, for some scalar $k \neq 0$. Multiplying by $1/k$ gives $(1/k)\mathbf{u}_1 = \mathbf{u}_2$. Hence, $\mathbf{u}_2$ is a scalar multiple of $\mathbf{u}_1$.

(c) Let L_1 be the line $t\mathbf{u}_1 + \mathbf{v}_1$ and let L_2 be the line $s\mathbf{u}_2 + \mathbf{v}_2$. Then L_1 and L_2 are the same line if, and only if, the following two conditions are satisfied:

(i) $\mathbf{u}_1$ and $\mathbf{u_2}$ are parallel;

(ii) $\mathbf{v}_2$ is on L_1.

First, suppose L_1 and L_2 are the same line. By condition (i), one of $\mathbf{u}_1$ or $\mathbf{u}_2$ is a nonzero scalar multiple of the other. By part (b), $\mathbf{u}_2$ is a scalar multiple of $\mathbf{u}_1$, so that $\mathbf{u}_2 = k\mathbf{u}_1$ for some scalar $k \neq 0$. By condition (ii), $\mathbf{v}_2 = t_0\mathbf{u}_1 + \mathbf{v}_1$ for some $t = t_0$. This equation is equivalent to $\mathbf{v}_2 - \mathbf{v}_1 = t_0\mathbf{u}_1$, so that $\mathbf{v}_2 - \mathbf{v}_1$ is a scalar multiple of $\mathbf{u}_1$.

Conversely, suppose both $\mathbf{u}_2$ and $\mathbf{v}_2 - \mathbf{v}_1$ are scalar multiples of $\mathbf{u}_1$. Then there exist scalars k_1 and k_2 such that $\mathbf{u}_2 = k_1\mathbf{u}_1$ and $\mathbf{v}_2 - \mathbf{v}_1 = k_2\mathbf{u}_1$. The first of these equations shows that $\mathbf{u}_1$ and $\mathbf{u}_2$ are parallel, so that condition (i) is satisfied; and when the second equation is rearranged as $\mathbf{v}_2 = k_2\mathbf{u}_1 + \mathbf{v}_1$, we see that $\mathbf{v}_2$ is on L_1 (with $t = k_2$), so that condition (ii) is satisfied. Hence, L_1 and L_2 are the same line

25. Let S and T be convex subsets of $\mathcal{R}^n$ and let

$$U = S + T = \{\, \mathbf{x} + \mathbf{y} \,|\, \mathbf{x} \in S, \ \mathbf{y} \in T \,\}.$$

To show that U is convex, let $\mathbf{z}_1, \mathbf{z}_2 \in U$ be given and consider the points on the line segment $(1 - t)\mathbf{z}_1 + t\mathbf{z}_2$, where $0 \leq t \leq 1$. Arbitrarily choose such a t and fix it. By how U is defined, there exist $\mathbf{x}_1, \mathbf{x}_2 \in S$ and $\mathbf{y}_1, \mathbf{y}_2 \in T$ such that $\mathbf{z}_1 = \mathbf{x}_1 + \mathbf{y}_1$ and $\mathbf{z}_2 = \mathbf{x}_2 + \mathbf{y}_2$. For our arbitrarily chosen t we therefore have

$$(1 - t)\mathbf{z}_1 + t\mathbf{z}_2 = (1 - t)(\mathbf{x}_1 + \mathbf{y}_1) + t(\mathbf{x}_2 + \mathbf{y}_2) = \big((1 - t)\mathbf{x}_1 + t\mathbf{x}_2\big) + \big((1 - t)\mathbf{y}_1 + t\mathbf{y}_2\big).$$

Since S and T are convex, the two parenthetical terms on the right are in S and T, respectively, so that the point $(1 - t)\mathbf{z}_1 + t\mathbf{z}_2$ is in U. Since t was chosen arbitrarily, the entire line segment connecting $\mathbf{z}_1$ and $\mathbf{z}_2$ lies in U. And since $\mathbf{z}_1$ and $\mathbf{z}_2$ were chosen arbitrarily from U, the conclusion holds for any two points in U. So, U is convex.

Section 4: THE DOT PRODUCT

Exercise Set 4ABC (pgs. 31-32)

1. For $\mathbf{x} = (1, 3)$ and $\mathbf{y} = (-2, 4)$,

$$\mathbf{x} \cdot \mathbf{y} = (1, 3) \cdot (-2, 4) = (1)(-2) + (3)(4) = -2 + 12 = 10.$$

3. For $\mathbf{x} = (-1, -1, 2)$ and $\mathbf{y} = (1, 6, 1)$,

$$\mathbf{x} \cdot \mathbf{y} = (-1, -1, 2) \cdot (1, 6, 1) = (-1)(1) + (-1)(6) + (2)(1) = -1 - 6 + 2 = -5.$$

5. Let $\mathbf{u} = (1, 1)$ and $\mathbf{v} = (1, 0)$.
    (a) $\mathbf{u} \cdot \mathbf{v} = (1, 1) \cdot (1, 0) = 1 + 0 = 1$.
    (b) $|\mathbf{u}| = |(1, 1)| = \sqrt{1^2 + 1^2} = \sqrt{2}$ and $|\mathbf{v}| = \sqrt{1^2 + 0^2} = 1$.
    (c) If $0 \leq \theta \leq \pi$ is the angle between $\mathbf{u}$ and $\mathbf{v}$ then, using parts (a) and (b),

$$\cos\theta = \frac{\mathbf{u} \cdot \mathbf{v}}{|\mathbf{u}||\mathbf{v}|} = \frac{1}{\sqrt{2}}, \qquad \text{so that} \qquad \theta = \pi/4.$$

7. Let $\mathbf{u} = (2, 1, 2)$ and $\mathbf{v} = (1, 2, 2)$.
    (a) $\mathbf{u} \cdot \mathbf{v} = (2, 1, 2) \cdot (1, 2, 2) = 2 + 2 + 4 = 8$.
    (b) $|\mathbf{u}| = |(2, 1, 2)| = \sqrt{2^2 + 1^2 + 2^2} = 3$ and $|\mathbf{v}| = |(1, 2, 2)| = \sqrt{1^2 + 2^2 + 2^2} = 3$.
    (c) If $0 \leq \theta \leq \pi$ is the angle between $\mathbf{u}$ and $\mathbf{v}$ then, using parts (a) and (b),

$$\cos\theta = \frac{\mathbf{u} \cdot \mathbf{v}}{|\mathbf{u}||\mathbf{v}|} = \frac{8}{9}, \qquad \text{so that} \qquad \theta = 0.4759 \ (\approx 27.3°).$$

9. From Exercises 35 and 36 on page 19, the position vectors $\mathbf{L}$ and $\mathbf{N}$ for Los Angeles and New York, respectively, were found to be

$$\mathbf{L} = (-1557, -2928, 2237) \qquad \text{and} \qquad \mathbf{N} = (832, -2902, 2624).$$

Since $|\mathbf{L}| = |\mathbf{N}| = 4000$ (to the nearest mile), the angle θ between $\mathbf{L}$ and $\mathbf{N}$ satisfies

$$\cos\theta = \frac{\mathbf{L} \cdot \mathbf{N}}{|\mathbf{L}||\mathbf{N}|} = \frac{(-1557, -2928, 2237) \cdot (832, -2902, 2624)}{4000^2} = \frac{13,071,520}{16,000,000} = 0.8170.$$

Hence, $\theta = 0.6147$ radians ($\approx 35.2°$).
    The airline distance from Los Angeles to New York is approximately the length l of the circular arc of a great circle subtending an angle of $\theta = 0.6147$ radians over the surface of a sphere of radius $r = 4000$ miles. Hence,

$$\text{airline distance from L.A. to N.Y. is } \approx l = r\theta = 4000(0.6147) \approx 2459 \text{ miles.}$$

11. Let $\mathbf{x} = (x_1, \ldots, x_n)$, $\mathbf{y} = (y_1, \ldots, y_n)$ and $\mathbf{z} = (z_1, \ldots, z_n)$ be vectors in $\mathcal{R}^n$ and let r be a scalar.
    (a) **Positivity**: Since $\mathbf{x} \cdot \mathbf{x} = (x_1, \ldots, x_n) \cdot (x_1, \ldots, x_n) = x_1^2 + \cdots + x_n^2$ is the sum of nonnegative numbers, it's positive if at least one of the x_i is nonzero, otherwise, it's zero. That is,

$$\mathbf{x} \cdot \mathbf{x} > 0, \quad \text{except that } \mathbf{0} \cdot \mathbf{0} = 0.$$

14

(b) **Symmetry**: We have

$$
\begin{aligned}
\mathbf{x} \cdot \mathbf{y} &= (x_1, \ldots, x_n) \cdot (y_1, \ldots, y_n) \\
&= x_1 y_1 + \cdots + x_n y_n \\
&= y_1 x_1 + \cdots + y_n x_n \\
&= (y_1, \ldots, y_n) \cdot (x_1, \cdot, x_n) = \mathbf{y} \cdot \mathbf{x}.
\end{aligned}
$$

(c) **Additivity**: We have

$$
\begin{aligned}
(\mathbf{x} + \mathbf{y}) \cdot \mathbf{z} &= \big((x_1, \ldots, x_n) + (y_1, \ldots, y_n)\big) \cdot (z_1, \ldots, z_n) \\
&= (x_1 + y_1, \ldots, x_n + y_n) \cdot (z_1, \ldots, z_n) \\
&= (x_1 + y_1) z_1 + \cdots + (x_n + y_n) z_n \\
&= (x_1 z_1 + \cdots + x_n z_n) + (y_1 z_1 + \cdots + y_n z_n) \\
&= (x_1, \ldots, x_n) \cdot (z_1, \ldots, z_n) + (y_1, \ldots, y_n) \cdot (z_1, \ldots, z_n) \\
&= \mathbf{x} \cdot \mathbf{z} + \mathbf{y} \cdot \mathbf{z}.
\end{aligned}
$$

(d) **Homogeneity**: We have

$$
\begin{aligned}
(r\mathbf{x}) \cdot \mathbf{y} &= \big(r(x_1, \ldots, x_n)\big) \cdot (y_1, \ldots, y_n) \\
&= (rx_1, \ldots, rx_n) \cdot (y_1, \ldots, y_n) \\
&= (rx_1) y_1 + \cdots + (rx_n) y_n \\
&= r(x_1 y_1 + \cdots + x_n y_n) \\
&= r\big((x_1, \ldots, x_n) \cdot (y_1, \ldots, y_n)\big) \\
&= r(\mathbf{x} \cdot \mathbf{y}).
\end{aligned}
$$

13. Let $\mathbf{x} = (1, -1, 2)$ and $\mathbf{v} = (1/\sqrt{3}, 1/\sqrt{3}, 1/\sqrt{3})$ and notice that $\mathbf{v}$ is a unit vector. Therefore, the coordinate of $\mathbf{x}$ in the direction of $\mathbf{v}$ is

$$
\mathbf{v} \cdot \mathbf{x} = (1/\sqrt{3}, 1/\sqrt{3}, 1/\sqrt{3}) \cdot (1, -1, 2) = 1/\sqrt{3} - 1/\sqrt{3} + 2/\sqrt{3} = \frac{2}{\sqrt{3}}.
$$

Using this result, the component of $\mathbf{x}$ in the direction of $\mathbf{v}$ (which we denote by $\mathbf{x_v}$) is

$$
\mathbf{x_v} = (\mathbf{v} \cdot \mathbf{x})\mathbf{v} = \frac{2}{\sqrt{3}}(1/\sqrt{3}, 1/\sqrt{3}, 1/\sqrt{3}) = (2/3, 2/3, 2/3).
$$

Using this result, the component of $\mathbf{x}$ perpendicular to $\mathbf{v}$ is

$$
\mathbf{x} - \mathbf{x_v} = (1, -1, 2) - (2/3, 2/3, 2/3) = (1/3, -5/3, 4/3).
$$

15. Let $\mathbf{x} = (2, -3, 1)$ and $\mathbf{v} = (1, 3, -2)$. The unit vector $\mathbf{n}$ in the direction of $\mathbf{v}$ is

$$
\mathbf{n} = \frac{\mathbf{v}}{|\mathbf{v}|} = \frac{(1, 3, -2)}{\sqrt{1 + 9 + 4}} = \frac{(1, 3, -2)}{\sqrt{14}} = (1/\sqrt{14}, 3/\sqrt{14}, -2/\sqrt{14}).
$$

Therefore the coordinate of $\mathbf{x}$ in the direction of $\mathbf{v}$ is

$$\mathbf{n} \cdot \mathbf{x} = (1/\sqrt{14}, 3/\sqrt{14}, -2/\sqrt{14}) \cdot (2, -3, 1) = 2/\sqrt{14} - 9/\sqrt{14} - 2/\sqrt{14} = -\frac{9}{\sqrt{14}}.$$

Using this result, the component of $\mathbf{x}$ in the direction of $\mathbf{v}$ (which we denote by $\mathbf{x_v}$) is

$$\mathbf{x_v} = (\mathbf{n} \cdot \mathbf{x})\mathbf{n} = -\frac{9}{\sqrt{14}}(1/\sqrt{14}, 3/\sqrt{14}, -2/\sqrt{14}) = (-9/14, -27/14, 18/14).$$

Using this result, the component of $\mathbf{x}$ perpendicular to $\mathbf{v}$ is

$$\mathbf{x} - \mathbf{x_v} = (2, -3, 1) - (-9/14, -27/14, 18/14) = (37/14, -15/14, -4/14).$$

17. Let T be the triangle with vertices $A = (2, -3, 6)$, $B = (1, 3, -2)$ and $C = (1, 7, 1)$, and let α, β and γ be the interior angles at A, B and C, respectively. Note that the formula for the angle θ between two vectors $\mathbf{u}$ and $\mathbf{v}$ shows that $0 \leq \theta < \pi/2$ (acute angle) if $\mathbf{u} \cdot \mathbf{v} > 0$, $\pi/2 < \theta \leq \pi$ (obtuse angle) if $\mathbf{u} \cdot \mathbf{v} < 0$ and $\theta = \pi/2$ (right angle) if $\mathbf{u} \cdot \mathbf{v} = 0$. Thus, in order to determine whether the interior angles of T are acute, obtuse or right, it is only necessary to check the sign of the dot product of the various vectors representing the sides of T. However, care must be taken that the vectors representing two given sides have their tails at the same vertex.

The vectors $\mathbf{x}$ and $\mathbf{y}$ with tips at B and C (resp.) and tails at A are given by

$$\mathbf{x} = (1, 3, -2) - (2, -3, 6) = (-1, 6, -8) \quad \text{and} \quad \mathbf{y} = (1, 7, 1) - (2, -3, 6) = (-1, 10, -5).$$

So, since $\mathbf{x} \cdot \mathbf{y} = (-1, 6, -8) \cdot (-1, 10, -5) = 1 + 60 + 40 = 101 > 0$, α is acute.

The vectors $-\mathbf{x} = (1, -6, 8)$ and $\mathbf{y} - \mathbf{x} = (-1, 10, -5) - (-1, 6, -8) = (0, 4, 3)$ have their tips at A and C (resp.) and their tails at B. So, since $(-\mathbf{x}) \cdot (\mathbf{y} - \mathbf{x}) = (1, -6, 8) \cdot (0, 4, 3) = 0$, β is a right angle.

The remaining angle γ is obviously acute, since α and γ must add up to $\pi/2$. But, as a check, the vectors $-\mathbf{y} = (1, -10, 5)$ and $\mathbf{x} - \mathbf{y} = (0, -4, -3)$ have their tips at A an B (resp.) and their tails at C. So, since $(-\mathbf{y}) \cdot (\mathbf{x} - \mathbf{y}) = (1, -10, 5) \cdot (0, -4, -3) = 40 - 15 = 25 > 0$, γ is acute, as predicted.

19. (a) Let $\mathbf{x} = (x_1, x_2, x_3)$ be any vector in $\mathcal{R}^3$. Then

$$\begin{aligned}
\mathbf{x} &= (x_1, 0, 0) + (0, x_2, 0) + (0, 0, x_3) \\
&= x_1(1, 0, 0) + x_2(0, 1, 0) + x_3(0, 0, 1) \\
&= x_1 \mathbf{e}_1 + x_2 \mathbf{e}_2 + x_3 \mathbf{x}_3.
\end{aligned}$$

Since $\mathbf{x} \cdot \mathbf{e}_i = (x_1, x_2, x_3) \cdot \mathbf{e}_i = x_i$ for $i = 1, 2, 3$, equality of the first and last members above implies

$$\mathbf{x} = x_1 \mathbf{e}_1 + x_2 \mathbf{e}_2 + x_3 \mathbf{e}_3 = (\mathbf{x} \cdot \mathbf{e}_1)\mathbf{e}_1 + (\mathbf{x} \cdot \mathbf{e}_2)\mathbf{e}_2 + (\mathbf{x} \cdot \mathbf{e}_3)\mathbf{e}_3.$$

(b) If $\mathbf{u}$ is a unit vector then the angle α_i between $\mathbf{u}$ and $\mathbf{e}_i$ satisfies

$$\cos \alpha_1 = \frac{\mathbf{u} \cdot \mathbf{e}_i}{|\mathbf{u}||\mathbf{e}_i|} = \mathbf{u} \cdot \mathbf{e}_i, \quad i = 1, 2, 3,$$

where we have used the fact that $\mathbf{u}$ and $\mathbf{e}_i$ are unit vectors with $|\mathbf{u}| = 1$ and $|\mathbf{e}_i| = 1$.

16

(c) The unit vector $\mathbf{u}$ in the same direction as $(1,2,1)$ is

$$\mathbf{u} = \frac{(1,2,1)}{|(1,2,1)|} = \frac{(1,2,1)}{\sqrt{6}} = (1/\sqrt{6}, 2/\sqrt{6}, 1/\sqrt{6}).$$

Thus, the direction cosines of $(1,2,1)$ are

$$\cos\alpha_1 = \mathbf{u}\cdot\mathbf{e}_1 = \frac{1}{\sqrt{6}}, \quad \cos\alpha_2 = \mathbf{u}\cdot\mathbf{e}_2 = \frac{2}{\sqrt{6}}, \quad \cos\alpha_3 = \mathbf{u}\cdot\mathbf{e}_3 = \frac{1}{\sqrt{6}}.$$

21. For any vector $\mathbf{x}$ in $\mathcal{R}^n$ and any basis vector $\mathbf{e}_j$ for $\mathcal{R}^n$, the dot product $\mathbf{x}\cdot\mathbf{e}_j$ is the j-th coordinate of $\mathbf{x}$. Thus, if $\mathbf{e}_i$ and $\mathbf{e}_j$ are any two standard basis vectors for $\mathcal{R}^n$ then $\mathbf{e}_i \cdot \mathbf{e}_j$ is the j-th coordinate of $\mathbf{e}_i$. Since the i-th coordinate of $\mathbf{e}_i$ is the only nonzero coordinate, the j-th coordinate of $\mathbf{e}_i$ is 0 if $i \neq j$ and 1 if $i = j$. That is

$$\mathbf{e}_i \cdot \mathbf{e}_j = \begin{cases} 1, & \text{if } i = j; \\ 0, & \text{if } i \neq j. \end{cases}$$

23. The discussion preceding Example 7 says that the total amount of energy (available wattage) falling on the 15 square meter solar collector can be computed by taking the dot product of two vectors, which we'll call $\mathbf{v}$ and $\mathbf{A}$:

(i) The vector $\mathbf{v}$ should be in the direction from which the sun falls on the collector and have a magnitude equal to the sun's wattage per square meter. That is, $\mathbf{v}$ must be in the direction of $(1,1,3)$ and have a magnitude of 80 watts per square meter. Since, $|(1,1,3)| = \sqrt{1+1+9} = \sqrt{11}$, it follows that $\mathbf{v} = \frac{80}{\sqrt{11}}(1,1,3)$.

(ii) The vector $\mathbf{A}$ should be perpendicular to the solar collector and have a magnitude equal to its area. That is, $\mathbf{A}$ must be in the direction of $(4,0,3)$ and have a magnitude of 15 square meters. Since, $|(4,0,3)| = \sqrt{16+9} = 5$, it follows that $\mathbf{A} = 3(4,0,3) = (12,0,9)$.

Hence, the energy flow to the nearest watt is given by

$$\text{flow} = \mathbf{v}\cdot\mathbf{A} = \frac{80}{\sqrt{11}}(1,1,3)\cdot(12,0,9) = \frac{80}{\sqrt{11}}(12+27) = \frac{3120}{\sqrt{11}} \approx 941 \text{ watts.}$$

25. Let $\mathbf{F}$ be the force vector representing the wind. Since the wind is blowing from the northwest, $\mathbf{F}$ points southeast; i.e., in the direction of the vector $\mathbf{i} - \mathbf{j}$. Since $|\mathbf{i} - \mathbf{j}| = \sqrt{2}$, it follows that $\mathbf{F}$ is $15/\sqrt{2}$ times this vector; i.e.,

$$\mathbf{F} = \frac{15}{\sqrt{2}}\mathbf{i} - \frac{15}{\sqrt{2}}\mathbf{j}.$$

Let $\mathbf{d}_1$ be the vector representing that part of the road that is 400 feet long and let $\mathbf{d}_2$ be the vector representing that part of the road that is 500 feet long. The first part of the road points in the direction of $-\mathbf{i}$ (a unit vector) and has length 400. The second part of the road points in the direction of $(-\sqrt{3}/2)\mathbf{i} + (1/2)\mathbf{j}$ (a unit vector) and has length 500. Thus,

$$\mathbf{d}_1 = -400\mathbf{i} \quad \text{and} \quad \mathbf{d}_2 = 500\left(-\frac{\sqrt{3}}{2}\mathbf{i} + \frac{1}{2}\mathbf{j}\right) = -250\sqrt{3}\,\mathbf{i} + 250\mathbf{j}.$$

It follows that if W_1 and W_2 is the work done by the bicyclist on each part of the road then

$$W_1 = (-\mathbf{F}) \cdot \mathbf{d}_1 = \left(-\frac{15}{\sqrt{2}}\mathbf{i} + \frac{15}{\sqrt{2}}\mathbf{j}\right) \cdot (-400\mathbf{i}) = 400\frac{15}{\sqrt{2}} = \frac{6000}{\sqrt{2}} \approx 4243 \text{ foot-pounds},$$

$$W_2 = (-\mathbf{F}) \cdot \mathbf{d}_2 = \left(-\frac{15}{\sqrt{2}}\mathbf{i} + \frac{15}{\sqrt{2}}\mathbf{j}\right) \cdot (-250\sqrt{3}\,\mathbf{i} + 250\mathbf{j}) = 250\sqrt{3}\frac{15}{\sqrt{2}} + 250\frac{15}{\sqrt{2}}$$

$$= 3750\frac{\sqrt{3}+1}{\sqrt{2}} \approx 7244 \text{ foot-pounds},$$

where $-\mathbf{F}$ is used because it is the force that the bicyclist must exert to overcome the wind.

Finally, since $\mathbf{d}_1 + \mathbf{d}_2$ is the vector representing a straight road from the start to the finish, and since the dot product is distributive over addition, the total work done by the bicyclist would be

$$W = (-\mathbf{F}) \cdot (\mathbf{d}_1 + \mathbf{d}_2) = (-\mathbf{F}) \cdot \mathbf{d}_1 + (-\mathbf{F}) \cdot \mathbf{d}_2 = W_1 + W_2.$$

That is, the total work done would be the same.

27. If $\mathbf{x}$ and $\mathbf{y}$ are vectors in $\mathcal{R}^n$ then, using the triangle inequality,

$$|\mathbf{x}| = |\mathbf{x} - \mathbf{y} + \mathbf{y}| \le |\mathbf{x} - \mathbf{y}| + |\mathbf{y}|.$$

Subtracting $|\mathbf{y}|$ from the two outer members gives $|\mathbf{x}| - |\mathbf{y}| \le |\mathbf{x} - \mathbf{y}|$.

For the reversed triangle inequality, we've already shown above that the following two inequalities simultaneously hold:

$$|\mathbf{x}| - |\mathbf{y}| \le |\mathbf{x} - \mathbf{y}| \qquad \text{and} \qquad |\mathbf{y}| - |\mathbf{x}| \le |\mathbf{y} - \mathbf{x}|.$$

As $|\mathbf{x} - \mathbf{y}| = |\mathbf{y} - \mathbf{x}|$, we deduce that $|\mathbf{x} - \mathbf{y}|$ is no less than the difference of the length of $\mathbf{x}$ and $\mathbf{y}$, regardless of the order in which they're subtracted. Hence, $\big||\mathbf{x}| - |\mathbf{y}|\big| \le |\mathbf{x} - \mathbf{y}|$.

29. Let $\mathbf{x}$ and $\mathbf{y}$ be vectors in $\mathcal{R}^n$.

(a) If $\mathbf{x} = \mathbf{0}$ then all coordinates of $\mathbf{x}$ are zero. It follows that $\mathbf{x} \cdot \mathbf{y} = 0$ and therefore its absolute value is zero. Also, $|\mathbf{x}| = 0$, so that $|\mathbf{x}||\mathbf{y}| = 0$. Hence, $|\mathbf{x} \cdot \mathbf{y}| \le |\mathbf{x}||\mathbf{y}|$. The same argument shows that both sides of $|\mathbf{x} \cdot \mathbf{y}| \le |\mathbf{x}||\mathbf{y}|$ are zero if $\mathbf{y} = \mathbf{0}$. Thus, from now on we can assume $\mathbf{x} \ne \mathbf{0}$ and $\mathbf{y} \ne \mathbf{0}$.

(b) For each real number t,

$$
\begin{aligned}
|t\mathbf{x} + \mathbf{y}|^2 &= (t\mathbf{x} + \mathbf{y}) \cdot (t\mathbf{x} + \mathbf{y}) \\
&= (t\mathbf{x}) \cdot (t\mathbf{x}) + (t\mathbf{x}) \cdot \mathbf{y} + \mathbf{y} \cdot (t\mathbf{x}) + \mathbf{y} \cdot \mathbf{y} && \text{(additivity of the dot product)} \\
&= t^2(\mathbf{x} \cdot \mathbf{x}) + t(\mathbf{x} \cdot \mathbf{y}) + t(\mathbf{y} \cdot \mathbf{x}) + \mathbf{y} \cdot \mathbf{y} && \text{(homogeneity of the dot product)} \\
&= t^2|\mathbf{x}|^2 + t(\mathbf{x} \cdot \mathbf{y}) + t(\mathbf{x} \cdot \mathbf{y}) + |\mathbf{y}|^2 && \text{(symmetry of the dot product)} \\
&= t^2|\mathbf{x}|^2 + 2t(\mathbf{x} \cdot \mathbf{y}) + |\mathbf{y}|^2 \\
&\ge 0,
\end{aligned}
$$

where the last inequality holds because $|t\mathbf{x} + \mathbf{y}|^2 \ge 0$.

(c) By part (b), the quadratic function $f(t) = t^2|\mathbf{x}|^2 + 2t(\mathbf{x} \cdot \mathbf{y}) + |\mathbf{y}|^2$ is never negative. Hence, either f has a double real root and its discriminant is zero, or its roots are complex conjugates and the discriminant is negative. In either case,

$$(\text{Discriminant of } f) = 4(\mathbf{x} \cdot \mathbf{y})^2 - 4|\mathbf{x}|^2|\mathbf{y}|^2 \le 0.$$

(d) Dividing the inequality derived in part (c) by 4, adding $|\mathbf{x}|^2|\mathbf{y}|^2$ to both sides, and taking square roots gives

$$|\mathbf{x} \cdot \mathbf{y}| \le |\mathbf{x}||\mathbf{y}|,$$

which is the Cauchy-Schwarz Inequality.

Section 5: EUCLIDEAN GEOMETRY

Exercise Set 5AB (pgs. 36–37)

1. Following the procedure of Example 1, with $\mathbf{p} = \mathbf{e_2} = (0,1)$ and $\mathbf{x_0} = (2,3)$, we get

$$(0,1) \cdot (x-2, y-3) = y - 3 = 0 \quad \text{or} \quad y = 3.$$

3. Following the procedure of Example 2, with $\mathbf{p} = (1,2,4)$ and $\mathbf{x_0} = (-1,0,0)$, we get

$$(1,2,4) \cdot (x+1, y-0, z-0) = 0 \quad \text{or} \quad x+1+2y+4z = 0 \quad \text{or} \quad x+2y+4z = -1.$$

(NOTE: The vector $\mathbf{N}$ is the diagram below is parallel to the given perpendicular but is only half as long. The right triangle with $\mathbf{N}$ as one leg has its other leg in the plane and its hypotenuse parallel with the z-axis.)

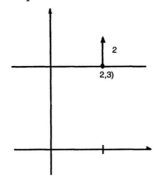

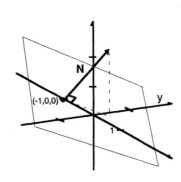

Exercise 1   Exercise 3

5. Given the line $(x,y,z) = t(1,-1,4) + (1,0,-2) = (t+1, -t, 4t-2)$, we see that $x = t+1$, $y = -t$ and $z = 4t - 2$. Inserting these into $2x + 3y - z = 2$ gives

$$2(t+1) + 3(-t) - (4t-2) = 2, \quad \text{which simplifies to} \quad -5t = -2.$$

so that $t = 2/5$. Hence, the line intersects the plane at the unique point

$$(x,y,z) = (2/5+1, -2/5, 4(2/5)-2) = (7/5, -2/5, -2/5).$$

7. Given the line $(x,y,z) = t(-1,1,1) + (-1,1,-1) = (-t-1, t+1, t-1)$, we see that $x = -t-1$, $y = t+1$ and $z = t-1$. Inserting these values into $2x + 3y - z = 2$ gives

$$2(-t-1) + 3(t+1) - (t-1) = 2, \quad \text{which simplifies to} \quad 0 = 0.$$

Since this is a true equation, every value of t works and therefore every point on the line is in the plane.

9. With the normal $(1,-1,2)$ and the point $(0,-1,0)$, the equation of the plane in question is

$$(1,-1,2) \cdot (x, y+1, z) = 0 \quad \text{or} \quad x - (y+1) + 2z = 0 \quad \text{or} \quad x - y + 2z = 1.$$

With the direction vector $(1,0,1)$ and the point $(1,1,1)$, a parametric representation of the line in question is

$$(x,y,z) = t(1,0,1) + (1,1,1) = (t+1, 1, t+1),$$

19

so that $x = t+1$, $y = 1$ and $z = t+1$. Inserting these into $x - y + 2z = 1$ gives

$$(t+1) - (1) + 2(t+1) = 1, \quad \text{which simplifies to} \quad 3t = -1,$$

so that $t = -1/3$. Hence, the line intersects the plane at the unique point

$$(x, y, z) = (-1/3 + 1, 1, -1/3 + 1) = (2/3, 1, 2/3).$$

11. Let L_1 and L_2 be the lines $x - 2y = 1$ and $2x + y = 3$, respectively. By inspection, it can be seen that the points $(1, 0)$ and $(3, 1)$ are on L_1 and the points $(1, 1)$ and $(0, 3)$ are on L_2. It follows that

$$(3, 1) - (1, 0) = (2, 1) \quad \text{and} \quad (1, 1) - (0, 3) = (1, -2)$$

are direction vectors for L_1 and L_2. Noting that $(2, 1) \cdot (1, -2) = 2 - 2 = 0$, we conclude that the lines are perpendicular to each other and therefore the angle between them is $90°$.
13. The planes $2x + y + z = 1$ and $x - y - z = -1$ have normals $(2, 1, 1)$ and $(1, -1, -1)$, respectively. Noting that $(2, 1, 1) \cdot (1, -1, -1) = 0$, we conclude that the normals are perpendicular to each other, and therefore the planes are perpendicular to each other. That is, the angle between them is $90°$.

15.

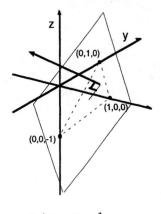

$$x + y - z = 1$$

17.

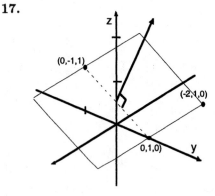

$$y + 2z = 1$$

19. Since the vector $\mathbf{p} = (3, -2, 5)$ is normal to the plane $3x - 2y + 5z = 2$, it is normal to every plane parallel to $3x - 2y + 5z = 2$. Hence, the equation of the plane we seek is of the form $3x - 2y + 5z = d$ for some scalar d. Inserting the given point $(x, y, z) = (2, 1, 1)$ into this equation gives $d = 9$. So the desired plane is $3x - 2y + 5z = 9$.
21. (a) Let (a, b, c) be a normal to the desired plane. Given that the points $(0, 0, 0)$, $(0, 1, 2)$ and $(-1, 0, 1)$ are in the plane, we have the following three vector equations for the plane:

$$(a, b, c) \cdot (x, y, z) = 0, \quad \text{or} \quad ax + by + cz = 0,$$

$$(a, b, c) \cdot (x, y - 1, z - 2) = 0, \quad \text{or} \quad ax + by + cz = b + 2c,$$

$$(a, b, c) \cdot (x + 1, y, z - 1) = 0, \quad \text{or} \quad ax + by + cz = c - a.$$

Comparing the right members of the three equations on the right shows that $b + 2c = 0$ and $c - a = 0$, so that $a = c$ and $b = -2c$. Hence, one normal to the plane can be found by setting $c = 1$. Then $(a, b, c) = (c, -2c, c) = (1, -2, 1)$ and the equation of the desired plane is

$$x - 2y + z = 0.$$

(b) Consider the given points $(1, -1, 1)$, $(1, 1, 1)$ and $(1, 2, 1)$. Note that the first coordinates of all three points are the same, as are the third coordinates. This means that the three points lie on a line parallel to the y-axis and, as such, cannot be used to determine a unique plane.

23. The vector form of the line $L : 2x + y = 2$ is given by $(2, 1) \cdot \mathbf{x} = 2$. Its normalized form is found by dividing throughout by $|(2, 1)| = \sqrt{5}$ and then using the homogeneity of the dot product to get $(2/\sqrt{5}, 1/\sqrt{5}) \cdot \mathbf{x} = 2/\sqrt{5}$. With $\mathbf{n} = (2/\sqrt{5}, 1/\sqrt{5})$ (a unit vector), $c = 2/\sqrt{5}$ and $\mathbf{x}_1 = (2, -1)$, Theorem 5.2 says that the distance from $\mathbf{x}_1$ to L is $|\delta|$, where δ is given by

$$\delta = \mathbf{n} \cdot \mathbf{x}_1 - c = (2/\sqrt{5}, 1/\sqrt{5}) \cdot (2, -1) - 2/\sqrt{5} = \frac{4}{\sqrt{5}} - \frac{1}{\sqrt{5}} - \frac{2}{\sqrt{5}} = \frac{1}{\sqrt{5}}.$$

That is, the distance is $|\delta| = 1/\sqrt{5}$. Theorem 5.2, also says that $\delta > 0$ means $(2, -1)$ is on the side of L to which $\mathbf{n}$ points. Since the y-coordinate of $\mathbf{n}$ is positive, $\mathbf{n}$ points in an upward direction, so that $(2, -1)$ is above L. Also, since $(0, 2)$ is on L, the origin is two units below L. So, $(2, -1)$ and the origin are not on the same side of L.

25. Given the plane $P : (1, 1, 1) \cdot \mathbf{x} = 1$, divide throughout by $|(1, 1, 1)| = \sqrt{3}$ and use the homogeneity of the dot product to obtain its normalization $(1/\sqrt{3}, 1/\sqrt{3}, 1/\sqrt{3}) \cdot \mathbf{x} = 1/\sqrt{3}$. With $\mathbf{n} = (1/\sqrt{3}, 1/\sqrt{3}, 1/\sqrt{3})$ (a unit vector), $c = 1/\sqrt{3}$ and $\mathbf{x}_1 = (1, 0, -1)$, Theorem 5.2 says that the distance from $\mathbf{x}_1$ to P is $|\delta|$, where δ is given by

$$\delta = \mathbf{n} \cdot \mathbf{x}_1 - c = (1/\sqrt{3}, 1/\sqrt{3}, 1/\sqrt{3}) \cdot (1, 0, -1) - 1/\sqrt{3} = \frac{1}{\sqrt{3}} - \frac{1}{\sqrt{3}} - \frac{1}{\sqrt{3}} = -\frac{1}{\sqrt{3}},$$

That is, the distance is $|\delta| = 1/\sqrt{3}$. Theorem 5.2 also says that $\delta < 0$ means $(1, 0, -1)$ is on the side of P to which $-\mathbf{n}$ points. Since the z-coordinate of $-\mathbf{n}$ is negative, it points in a downward direction. Hence, $(1, 0, -1)$ is below P.

To locate the origin relative to P, set $x = y = 0$ in the equation for P to find out where P crosses the z-axis. This gives $(1, 1, 1) \cdot (0, 0, z) = 1$, or $z = 1$. Thus, $(0, 0, 1)$ is on the plane and the origin is one unit below it; i.e., the origin is below P. It follows that $(1, 0, -1)$ and the origin are on the same side of P.

27. The vector form of the plane $P : x + 2y + 3z = 1$ is given by $(1, 2, 3) \cdot \mathbf{x} = 1$. Its normalization is obtained by dividing both sides by $|(1, 2, 3)| = \sqrt{14}$ and then using the homogeneity of the dot product to get $(1/\sqrt{14}, 2/\sqrt{14}, 3/\sqrt{14}) \cdot \mathbf{x} = 1/\sqrt{14}$. With $\mathbf{n} = (1/\sqrt{14}, 2/\sqrt{14}, 3/\sqrt{14})$ (a unit vector), $c = 1/\sqrt{14}$ and $\mathbf{x}_1 = (1, 0, -1)$, Theorem 5.2 says that the distance from $\mathbf{x}_1$ to P is $|\delta|$, where δ is given by

$$\delta = \mathbf{n} \cdot \mathbf{x}_1 - c = (1/\sqrt{14}, 2/\sqrt{14}, 3/\sqrt{14}) \cdot (1, 0, -1) - 1/\sqrt{14} = \frac{1}{\sqrt{14}} - \frac{3}{\sqrt{14}} - \frac{1}{\sqrt{14}} = -\frac{3}{\sqrt{14}}.$$

That is, the distance is $|\delta| = 3/\sqrt{14}$. Theorem 5.2 also says that $\delta < 0$ means $(1, 0, -1)$ is on the side of P to which $-\mathbf{n}$ points. Since the z-coordinate of $-\mathbf{n}$ is negative, it points in a downward direction. Hence, $(1, 0, -1)$ is below P.

To locate the origin relative to P, set $x = y = 0$ in the equation for P to find out where the plane crosses the z-axis. This gives $3z = 1$, or $z = 1/3$. Thus, $(0, 0, 1/3)$ is on the plane and $(0, 0, 0)$ is one-third unit below it; i.e., the origin is below P. It follows that $(1, 0, -1)$ and the origin are on the same side of P.

29. Let $ax + by + cz = d$ be the normalized equation of a plane in $\mathcal{R}^3$, so that $a^2 + b^2 + c^2 = 1$. Its vector form is $(a, b, c) \cdot \mathbf{x} = d$. By Theorem 5.2, the distance from the origin to the plane is $|\delta|$, where δ is given by

$$\delta = (a, b, c) \cdot (0, 0, 0) - d = -d.$$

Thus, the distance from the origin to the plane is $|d|$.

Section 6: THE CROSS PRODUCT

Exercise Set 6 (pgs. 42-44)

1. With $\mathbf{u} = \mathbf{e}_2 \,(= \mathbf{j})$ and $\mathbf{v} = \mathbf{e}_1 \,(= \mathbf{i})$, we can use the results of Example 2 to get

$$\mathbf{u} \times \mathbf{v} = \mathbf{e}_2 \times \mathbf{e}_1 = \mathbf{j} \times \mathbf{i} = -\mathbf{k} = -\mathbf{e}_3.$$

3. With $\mathbf{u} = (0, 1, 2)$ and $\mathbf{v} = (-1, 0, 1)$, we have

$$\mathbf{u} \times \mathbf{v} = (0, 1, 2) \times (-1, 0, 1) = \big((1)(1) - (2)(0), (2)(-1) - (0)(1), (0)(0) - (1)(-1)\big)$$
$$= (1, -2, 1).$$

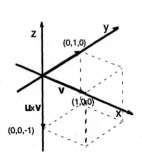

Exercise 1

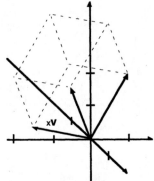

Exercise 3

5. If $\mathbf{u} = \mathbf{i}$ and $\mathbf{v} = \mathbf{j} + \mathbf{k}$ are adjacent edges of a parallelogram then its area A is given by

$$A = |\mathbf{u} \times \mathbf{v}| = |\mathbf{i} \times (\mathbf{j} + \mathbf{k})| = |(\mathbf{i} \times \mathbf{j}) + (\mathbf{i} \times \mathbf{k})| = |\mathbf{k} + (-\mathbf{j})| = |(0, -1, 1)| = \sqrt{2},$$

where the third equality follows from property (b) in Equations 6.4, and the fourth equality follows from the cross products obtained in Example 2.

7. If $\mathbf{u} = (3, -1, 2)$ and $\mathbf{v} = (-1, 0, 1)$ are adjacent edges of a triangle then its area A is given by

$$A = \frac{1}{2}|\mathbf{u} \times \mathbf{v}| = \frac{1}{2}|(3, -1, 2) \times (-1, 0, 1)|$$
$$= \frac{1}{2}\big|\big((-1)(1) - (2)(0), (2)(-1) - (3)(1), (3)(0) - (-1)(-1)\big)\big|$$
$$= \frac{1}{2}|(-1, -5, -1)| = \frac{1}{2}\sqrt{27} = \frac{3\sqrt{3}}{2}.$$

9. With $\mathbf{u} = (1, 2, 4)$ and $\mathbf{v} = (4, 2, 1)$, the vector $\mathbf{p} = \mathbf{u} \times \mathbf{v}$ is perpendicular to both $\mathbf{u}$ and $\mathbf{v}$ and therefore $\mathbf{p}$ is normal to the plane parallel to $\mathbf{u}$ and $\mathbf{v}$. Since

$$\mathbf{p} = \mathbf{u} \times \mathbf{v} = (1, 2, 4) \times (4, 2, 1)$$
$$= \big((2)(1) - (4)(2), (4)(4) - (1)(1), (1)(2) - (2)(4)\big)$$
$$= (-6, 15, -6),$$

22

the equation of the plane perpendicular to $\mathbf{p}$ and containing the point $(-1, -1, 1)$ is

$$(-6, 15, -6) \cdot (x + 1, y + 1, z - 1) = 0 \qquad \text{or} \qquad -6(x + 1) + 15(y + 1) - 6(z - 1) = 0.$$

Simplifying the last equation gives $2x - 5y + 2z = 5$.

11. With $\mathbf{i} = (1, 0, 0)$, $\mathbf{j} = (0, 1, 0)$, $\mathbf{k} = (0, 0, 1)$ and $\mathbf{u} = (a, b, c)$, we have

(a) $\quad \mathbf{i} \times \mathbf{j} = (1, 0, 0) \times (0, 1, 0)$
$$= \big((0)(0) - (0)(1), (0)(0) - (1)(0), (1)(1) - (0)(0) \big) = (0, 0, 1) = \mathbf{k};$$

(b) $\quad \mathbf{j} \times \mathbf{k} = (0, 1, 0) \times (0, 0, 1)$
$$= \big((1)(1) - (0)(0), (0)(0) - (0)(1), (0)(0) - (1)(0) \big) = (1, 0, 0) = \mathbf{i};$$

(c) $\quad \mathbf{k} \times \mathbf{i} = (0, 0, 1) \times (1, 0, 0)$
$$= \big((0)(0) - (1)(0), (1)(1) - (0)(0), (0)(0) - (0)(1) \big) = (0, 1, 0) = \mathbf{j};$$

(d) $\quad \mathbf{u} \times \mathbf{u} = (a, b, c) \times (a, b, c) = (bc - cb, ca - ac, ab - ba) = (0, 0, 0) = \mathbf{0}.$

13. The triangle with vertices $(-2, 1, 0)$, $(2, 3, 0)$, $(2, -1, 0)$ forms the base of an irregular pyramid with apex at $(0, 0, 2)$.

(a) The vectors

$$\mathbf{a} = (-2, 1, 0) - (0, 0, 2) = (-2, 1, -2),$$
$$\mathbf{b} = (2, -1, 0) - (0, 0, 2) = (2, -1, -2),$$
$$\mathbf{c} = (2, 3, 0) - (0, 0, 2) = (2, 3, -2)$$

are shown on the right with their tails at the apex.

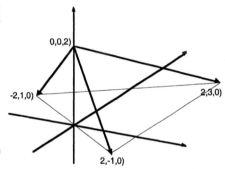

(b) Using the vectors $\mathbf{a}$, $\mathbf{b}$ and $\mathbf{c}$ derived in part (a), the vectors $\mathbf{u} = \mathbf{a} \times \mathbf{b}$, $\mathbf{v} = \mathbf{b} \times \mathbf{c}$, and $\mathbf{w} = \mathbf{c} \times \mathbf{a}$ are perpendicular to the three sides of the pyramid and, by the right-hand rule, they each point in a direction away from the pyramid. We have

$$\begin{aligned}
\mathbf{u} = \mathbf{a} \times \mathbf{b} &= (-2, 1, -2) \times (2, -1, -2) \\
&= \big((1)(-2) - (-2)(-1), (-2)(2) - (-2)(-2), (-2)(-1) - (1)(2) \big) \\
&= (-4, -8, 0),
\end{aligned}$$

$$\begin{aligned}
\mathbf{v} = \mathbf{b} \times \mathbf{c} &= (2, -1, -2) \times (2, 3, -2) \\
&= \big((-1)(-2) - (-2)(3), (-2)(2) - (2)(-2), (2)(3) - (-1)(2) \big) \\
&= (8, 0, 8),
\end{aligned}$$

$$\begin{aligned}
\mathbf{w} = \mathbf{c} \times \mathbf{a} &= (2, 3, -2) \times (-2, 1, -2) \\
&= \big((3)(-2) - (-2)(1), (-2)(-2) - (2)(-2), (2)(1) - (3)(-2) \big) \\
&= (-4, 8, 8).
\end{aligned}$$

(c) Let α, β and γ be the *interior* angles that the sides with normals $\mathbf{u}$, $\mathbf{v}$ and $\mathbf{w}$ (resp.) make with the base. The angles that the normals themselves make with the base are $\pi/2 - \alpha$, $\pi/2 - \beta$ and $\pi/2 - \gamma$, respectively, and are equal to the angles between these

23

normals and their respective projections onto the xy-plane. Since such a projection of a normal is found by setting its third coordinate equal to zero, the three projections are given by $\overline{\mathbf{u}} = (-4, -8, 0)$ (the projection of $\mathbf{u}$ onto the xy-plane); $\overline{\mathbf{v}} = (8, 0, 0)$ (the projection of $\mathbf{v}$ onto the xy-plane); and $\overline{\mathbf{w}} = (-4, 8, 0)$ (the projection of $\mathbf{w}$ onto the xy-plane). We therefore have

$$\cos(\pi/2 - \alpha) = \frac{\mathbf{u} \cdot \overline{\mathbf{u}}}{|\mathbf{u}||\overline{\mathbf{u}}|} = \frac{(-4, -8, 0) \cdot (-4 - 8, 0)}{|(-4, -8, 0)||(-4, -8, 0)|} = \frac{|(-4, -8, 0)|^2}{|(-4, -8, 0)|^2} = 1;$$

$$\cos(\pi/2 - \beta) = \frac{\mathbf{v} \cdot \overline{\mathbf{v}}}{|\mathbf{v}||\overline{\mathbf{v}}|} = \frac{(8, 0, 8) \cdot (8, 0, 0)}{|(8, 0, 8)||(8, 0, 0)|} = \frac{64}{(8\sqrt{2})(8)} = \frac{1}{\sqrt{2}};$$

$$\cos(\pi/2 - \gamma) = \frac{\mathbf{w} \cdot \overline{\mathbf{w}}}{|\mathbf{w}||\overline{\mathbf{w}}|} = \frac{(-4, 8, 8) \cdot (-4, 8, 0)}{|(-4, 8, 8)||(-4, 8, 0)|} = \frac{80}{(12)(4\sqrt{5})} = \frac{\sqrt{5}}{3}.$$

That is, $\cos(\pi/2 - \alpha) = 1$, $\cos(\pi/2 - \beta) = 1/\sqrt{2}$ and $\cos(\pi/2 - \gamma) = \sqrt{5}/3$.

The first equation implies $\pi/2 - \alpha = 0$, so that $\alpha = \pi/2$ and $\cos \alpha = 0$. The second equation implies $\pi/2 - \beta = \pi/4$, so that $\beta = \pi/4$ and $\cos \beta = 1/\sqrt{2}$. For the third equation, we use the cofunction identity $\cos(\pi/2 - x) = \sin x$ to get $\sin \gamma = \sqrt{5}/2$. Now use the pythagorean identity $\cos^2 \gamma + \sin^2 \gamma = 1$ to deduce that $\cos \gamma = 2/3$. To sum up,

$$\cos \alpha = 0, \qquad \cos \beta = \frac{1}{\sqrt{2}}, \qquad \cos \gamma = \frac{2}{3}.$$

15. (a) Let $\mathbf{u} = (u_1, u_2, u_3)$ and $\mathbf{v} = (v_1, v_2, v_3)$ and let $D = |\mathbf{u}|^2|\mathbf{v}|^2 - (\mathbf{u} \cdot \mathbf{v})^2$. The first term of D (i.e., $|\mathbf{u}|^2|\mathbf{v}|^2$) can be written as

$$\begin{aligned}
|\mathbf{u}|^2|\mathbf{v}|^2 &= (u_1^2 + u_2^2 + u_3^2)(v_1^2 + v_2^2 + v_3^2) \\
&= ((u_1v_1)^2 + (u_2v_2)^2 + u_3v_3)^2) \\
&\quad + ((u_1v_2)^2 + (u_2v_1)^2) + ((u_1v_3)^2 + (u_3v_1)^2) + ((u_2v_3)^2 + (u_3v_2)^2),
\end{aligned}$$

where we have judiciously grouped the nine terms to make the computations easier. The second term of D (i.e., $(\mathbf{u} \cdot \mathbf{v})^2$) can be written as

$$\begin{aligned}
(\mathbf{u} \cdot \mathbf{v})^2 &= (u_i v_1 + u_2 v_2 + u_3 v_3)^2 \\
&= ((u_1v_1)^2 + (u_2v_2)^2 + (u_3v_3)^2) \\
&\quad + 2(u_1v_2)(u_2v_1) + 2(u_1v_3)(u_3v_1) + 2(u_2v_3)(u_3v_2),
\end{aligned}$$

where we have again judiciously grouped the resulting terms.

Now observe that when the second term of D is subtracted from the first term of D, the first grouping of each term cancels out. The remaining nine terms can then be regrouped into three pieces: the first piece consists of the second grouping of the first term of D and the term $-2(u_1v_2)(u_2v_1)$, the second piece consists of the third grouping of the first term of D and the term $-2(u_1v_3)(u_3v_1)$, and the third piece consists of the fourth grouping of the first term of D and the term $-2(u_2v_3)(u_3v_2)$. Each of these pieces is a perfect square; namely,

$$\begin{aligned}
((u_1v_2)^2 - 2(u_1v_2)(u_2v_1) + (u_2v_1)^2) &= (u_1v_2 - u_2v_1)^2, \\
((u_1v_3)^2 - 2(u_1v_3)(u_3v_1) + (u_3v_1)^2) &= (u_1v_3 - u_3v_1)^2, \\
((u_2v_3)^2 - 2(u_2v_3)(u_3v_2) + (u_3v_2)^2) &= (u_2v_3 - u_3v_2)^2.
\end{aligned}$$

Hence,

$$D = (u_1 v_2 - u_2 v_1)^2 + (u_1 v_3 - u_3 v_1)^2 + (u_2 v_3 - u_3 v_2)^2$$
$$= |(u_3 v_2 - u_2 v_3, u_3 v_1 - u_1 v_3, u_1 v_2 - u_2 v_1)|^2$$
$$= |\mathbf{u} \times \mathbf{v}|^2.$$

That is, $|\mathbf{u} \times \mathbf{v}|^2 = |\mathbf{u}|^2 |\mathbf{v}|^2 - (\mathbf{u} \cdot \mathbf{v})^2$.

(b) Let θ be the angle between $\mathbf{u}$ and $\mathbf{v}$, where $0 \le \theta \le \pi$. Using the result of part (a), we have

$$|\mathbf{u} \times \mathbf{v}|^2 = |\mathbf{u}|^2 |\mathbf{v}|^2 - (\mathbf{u} \cdot \mathbf{v})^2 = |\mathbf{u}|^2 |\mathbf{v}|^2 - \big(|\mathbf{u}||\mathbf{v}| \cos \theta\big)^2$$
$$= |\mathbf{u}|^2 |\mathbf{v}|^2 \big(1 - \cos^2 \theta\big)$$
$$= |\mathbf{u}|^2 |\mathbf{v}|^2 \sin^2 \theta.$$

Since $\sin \theta \ge 0$ for $0 \le \theta \le \pi$, we can take square roots of the first and last members above to obtain

$$|\mathbf{u} \times \mathbf{v}| = |\mathbf{u}||\mathbf{v}| \sin \theta.$$

(c) The parallelogram with adjacent sides $\mathbf{u}$ and $\mathbf{v}$ can be thought of as two congruent triangles with adjacent sides $\mathbf{u}$ and $\mathbf{v}$ and θ as the angle at the vertex where the tails of $\mathbf{u}$ and $\mathbf{v}$ meet. It is well-known from basic trigonometric principles that the area of the triangle is one-half the product of the lengths of two sides times the sine of the angle between them; i.e., $\frac{1}{2}|\mathbf{u}||\mathbf{v}| \sin \theta$. It follows that the area $A(P)$ of the parallelogram is twice this amount. By part (b), $A(P) = |\mathbf{u} \times \mathbf{v}|$.

17. (a) Let $\mathbf{u} = (u_1, u_2, u_3)$, $\mathbf{v} = (v_1, v_2, v_3)$, $\mathbf{w} = (w_1, w_2, w_3)$ and consider the three cyclically permuted scalar triple products

$$\mathbf{u} \cdot (\mathbf{v} \times \mathbf{w}), \qquad \mathbf{v} \cdot (\mathbf{w} \times \mathbf{u}), \qquad \mathbf{w} \cdot (\mathbf{u} \times \mathbf{v}).$$

By brute force,

$$\mathbf{u} \cdot (\mathbf{v} \times \mathbf{w}) = (u_1, u_2, u_3) \cdot \big((v_1, v_2, v_3) \times (w_1, w_2, w_3)\big)$$
$$= (u_1, u_2, u_3) \cdot (v_2 w_3 - v_3 w_2, v_3 w_1 - v_1 w_3, v_1 w_2 - v_2 w_1)$$
$$= u_1 (v_2 w_3 - v_3 w_2) + u_2 (v_3 w_1 - v_1 w_3) + u_3 (v_1 w_2 - v_2 w_1)$$
$$= (u_1 v_2 w_3 + w_1 u_2 v_3 + v_1 w_2 u_3) - (u_1 w_2 v_3 + v_1 u_2 w_3 + w_1 v_2 u_3);$$

$$\mathbf{v} \cdot (\mathbf{w} \times \mathbf{u}) = (v_1, v_2, v_3) \cdot \big((w_1, w_2, w_3) \times (u_1, u_2, u_2)\big)$$
$$= (v_1, v_2, v_3) \cdot (w_2 u_3 - w_3 u_2, w_3 u_1 - w_1 u_3, w_1 u_2 - w_2 u_1)$$
$$= v_1 (w_2 u_3 - w_3 u_2) + v_2 (w_3 u_1 - w_1 u_3) + v_3 (w_1 u_2 - w_2 u_1)$$
$$= (v_1 w_2 u_3 + u_1 v_2 w_3 + w_1 u_2 v_3) - (v_1 u_2 w_3 + w_1 v_2 u_3 + u_1 w_2 v_3);$$

$$\mathbf{w} \cdot (\mathbf{u} \times \mathbf{v}) = (w_1, w_2, w_3) \cdot \big((u_1, u_2, u_3) \times (v_1, v_2, v_3)\big)$$
$$= (w_1, w_2, w_3) \cdot (u_2 v_3 - u_3 v_2, u_3 v_1 - u_1 v_3, u_1 v_2 - u_2 v_1)$$
$$= w_1 (u_2 v_3 - u_3 v_2) + w_2 (u_3 v_1 - u_1 v_3) + w_3 (u_1 v_2 - u_2 v_1)$$
$$= (w_1 u_2 v_3 + v_1 w_2 u_3 + u_1 v_2 w_3) - (w_1 v_2 u_3 + u_1 w_2 v_3 + v_1 u_2 w_3).$$

Now observe that they are all equal.

(b) There are only six ways to write a scalar triple product. Three of them are cyclically permuted and given in part (a). The remaining three are also cyclically permuted and are

$$\mathbf{u} \cdot (\mathbf{w} \times \mathbf{v}), \qquad \mathbf{v} \cdot (\mathbf{u} \times \mathbf{w}), \qquad \mathbf{w} \cdot (\mathbf{v} \times \mathbf{u}).$$

Hence, by part (a), they are equal to each other. If A is the common value of the three scalar triple products in part (a) then (e.g.)

$$\mathbf{u} \cdot (\mathbf{w} \times \mathbf{v}) = \mathbf{u} \cdot \big(- (\mathbf{v} \times \mathbf{w})\big) = -\big(\mathbf{u} \cdot (\mathbf{v} \times \mathbf{w})\big) = -A$$

implies $\mathbf{v} \cdot (\mathbf{u} \times \mathbf{w}) = -A$ and $\mathbf{w} \cdot (\mathbf{v} \times \mathbf{u}) = -A$. In other words, noncyclically permuting a scalar triple product changes the sign.

19. Let P be the parallelogram in $\mathcal{R}^3$ with adjacent edges $(1,1,0)$ and $(0,1,2)$. By Equation 6.5 in the text, the area of P is

$$A(P) = |(1,1,0) \times (0,1,2)| = \big|\big((1)(2) - (0)(1), (0)(0) - (1)(2), (1)(1) - (1)(0)\big)\big|$$
$$= |(2,-2,1)| = \sqrt{4+4+1} = 3.$$

21. Let B be the parallelepiped with edges $(1,1,0)$, $(0,1,2)$ and $(-3,5,-1)$. By Equation 6.7 in the text, the volume of B is

$$V(B) = |(-3,5,-1) \cdot \big((1,1,0) \times (0,1,2)\big)| = |(-3,5,-1) \cdot (2,-2,1)|$$
$$= |-6-10-1| = |-17| = 17,$$

where $(1,1,0) \times (0,1,2) = (2,-2,1)$ was computed in Exercise 19.

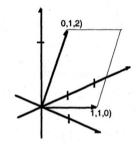

Exercise 19

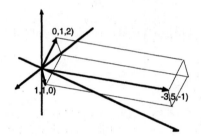

Exercise 21

23. Let $\mathbf{u} = (u_1, u_2, u_3)$ and $\mathbf{v} = (v_1, v_2, v_3)$ be nonparallel vectors in $\mathcal{R}^3$, and let $\mathbf{x} = (x, y, z)$ be such that $\mathbf{u} \cdot \mathbf{x} = \mathbf{v} \cdot \mathbf{x} = 0$, so that

$$(1) \qquad \begin{aligned} u_1 x + u_2 y + u_3 z &= 0 \\ v_1 x + v_2 y + v_3 z &= 0. \end{aligned}$$

(a) Using equations (1), subtract u_1 times the second from v_1 times the first to get

$$u_2 v_1 y - u_1 v_2 y + u_3 v_1 z - u_1 v_3 z = 0 \quad \text{or} \quad (u_2 v_1 - u_1 v_2)y = (u_1 v_3 - u_3 v_1)z.$$

(b) Using equations (1), subtract u_2 times the second from v_2 times the first to get

$$u_1 v_2 x - u_2 v_1 x + u_3 v_2 z - u_2 v_3 z = 0 \quad \text{or} \quad (u_1 v_2 - u_2 v_1)x = (u_2 v_3 - u_3 v_2)z.$$

(c) First, assume $u_2 v_1 - u_1 v_2 \neq 0$. We can then solve for x and y in the rightmost equations in parts (a) and (b) to get

$$x = \frac{u_2 v_3 - u_3 v_2}{u_1 v_2 - u_2 v_1} z \qquad \text{and} \qquad y = \frac{u_1 v_3 - u_3 v_1}{u_2 v_1 - u_1 v_2} z.$$

Choosing $z = u_1 v_2 - u_2 v_1$ gives, $x = u_2 v_3 - u_3 v_2$ and $y = u_3 v_1 - u_1 v_3$, so that

$$(x, y, z) = (u_2 v_3 - u_3 v_2, u_3 v_1 - u_1 v_3, u_1 v_2 - u_2 v_1) = \mathbf{u} \times \mathbf{v}.$$

Next, assume $u_2 v_1 - u_1 v_2 = 0$. By Exercise 24(a) in Section 3B, the projections $\overline{\mathbf{u}} = (u_1, u_2, 0)$ and $\overline{\mathbf{v}} = (v_1, v_2, 0)$ of $\mathbf{u}$ and $\mathbf{v}$ onto the xy-plane are parallel vectors. Hence, $\mathbf{u}$ and $\mathbf{v}$ both lie in the same vertical plane. It follows that, since every vector perpendicular to a vertical plane has 0 as its third coordinate, $z = 0$ (note that this means $z = u_2 v_1 - u_1 v_2$). Moreover, since $\overline{\mathbf{u}}$ and $\overline{\mathbf{v}}$ are parallel, at least one of them is a constant multiple of the other. Without loss of generality, we can suppose $\overline{\mathbf{u}} = k\overline{\mathbf{v}}$, where k is a scalar, so that $u_1 = kv_1$ and $u_2 = kv_2$. This means that the two equations in (1) are equivalent and therefore collapse into the single equation

$$(2) \qquad\qquad\qquad v_1 x + v_2 y = 0.$$

If $v_1 = v_2 = 0$ then $\overline{\mathbf{v}} = \overline{\mathbf{u}} = \mathbf{0}$ and therefore $\mathbf{u} = (0, 0, u_3)$ and $\mathbf{v} = (0, 0, v_3)$ are parallel vectors, contrary to hypothesis. So, at least one of v_1 or v_2 is nonzero. Without loss of generality, we can suppose $v_1 \neq 0$. Then equation (2) implies $x = -(v_2/v_1)y$. Setting $y = u_3 v_1 - u_1 v_3$ we get

$$x = -\frac{v_2}{v_1}y = -\frac{v_2}{v_1}(u_3 v_1 - u_1 v_3) = -u_3 v_2 + \frac{u_1 v_2 v_3}{v_1} = u_2 v_3 - u_3 v_2 + \frac{u_1 v_2 v_3}{v_1} - u_2 v_3$$

$$= u_2 v_3 - u_3 v_2 + \frac{u_1 v_2 v_3 - u_2 v_1 v_3}{v_1} = u_2 v_3 - u_3 v_2 + (u_1 v_2 - u_2 v_1)\frac{v_3}{v_1} = u_2 v_3 - u_3 v_2,$$

where the fourth equality is obtained by adding and subtracting $u_2 v_3$, and the last equality follows because the case at hand assumes $u_2 v_1 - u_1 v_2 = 0$. Thus, once again, (x, y, z) has the form

$$(x, y, z) = (u_2 v_3 - u_3 v_2, u_3 v_1 - u_1 v_3, u_2 v_1 - u_1 v_2) = \mathbf{v} \times \mathbf{v}.$$

25. (a) On one hand, $(\mathbf{i} \times \mathbf{i}) \times \mathbf{j} = \mathbf{0} \times \mathbf{j} = \mathbf{0}$. However, $\mathbf{i} \times (\mathbf{i} \times \mathbf{j}) = \mathbf{i} \times \mathbf{k} = -\mathbf{j} \neq \mathbf{0}$. So, associativity fails for this particular ordering of two copies of $\mathbf{i}$ and one copy of $\mathbf{j}$.

(b) Here, $(\mathbf{i} \times \mathbf{j}) \times \mathbf{i} = \mathbf{k} \times \mathbf{i} = -(\mathbf{i} \times \mathbf{k}) = \mathbf{i} \times (-\mathbf{k}) = \mathbf{i} \times (\mathbf{j} \times \mathbf{i})$, so that associativity holds for this particular ordering of two copies of $\mathbf{i}$ and one copy of $\mathbf{j}$.

(c) Here, $(\mathbf{i} \times \mathbf{j}) \times \mathbf{k} = \mathbf{k} \times \mathbf{k} = \mathbf{0} = \mathbf{i} \times \mathbf{i} = \mathbf{i} \times (\mathbf{j} \times \mathbf{k})$.

(d) Here, $(\mathbf{i} \times \mathbf{i}) \times \mathbf{i} = \mathbf{0} \times \mathbf{i} = \mathbf{0} = \mathbf{i} \times \mathbf{0} = \mathbf{i} \times (\mathbf{i} \times \mathbf{i})$.

27. By part (c) of Exercise 26,

$$\mathbf{u} \times (\mathbf{v} \times \mathbf{w}) = (\mathbf{u} \cdot \mathbf{w})\mathbf{v} - (\mathbf{u} \cdot \mathbf{v})\mathbf{w}$$

for any vectors $\mathbf{u}$, $\mathbf{v}$, $\mathbf{w}$ in $\mathcal{R}^3$. Using the anticommutativity and homogeneity properties of the cross product, we have

$$(\mathbf{u} \times \mathbf{v}) \times \mathbf{w} = -(\mathbf{w} \times (\mathbf{u} \times \mathbf{v})) = \mathbf{w} \times (-(\mathbf{u} \times \mathbf{v})) = \mathbf{w} \times (\mathbf{v} \times \mathbf{u}),$$

so that applying the result of Exercise 26(c) gives

$$\mathbf{w} \times (\mathbf{v} \times \mathbf{u}) = (\mathbf{w} \cdot \mathbf{u})\mathbf{v} - (\mathbf{w} \cdot \mathbf{v})\mathbf{u}.$$

It follows that if $\mathbf{u} \times (\mathbf{v} \times \mathbf{w}) = (\mathbf{u} \times \mathbf{v}) \times \mathbf{w}$ is to hold, then it is necessary that

$$(\mathbf{u} \cdot \mathbf{w})\mathbf{v} - (\mathbf{u} \cdot \mathbf{v})\mathbf{w} = (\mathbf{w} \cdot \mathbf{u})\mathbf{v} - (\mathbf{w} \cdot \mathbf{v})\mathbf{u}.$$

Since the dot product is commutative, the first terms on each side of the above equation are equal and can be subtracted off, leaving a result that can be written as

$$(\mathbf{v} \cdot \mathbf{w})\mathbf{u} - (\mathbf{v} \cdot \mathbf{u})\mathbf{w} = \mathbf{0}.$$

But, again, we use the result of Exercise 26(c) to recognize the left side of the above equation as $\mathbf{v} \times (\mathbf{u} \times \mathbf{w})$. In other words,

$$\mathbf{u} \times (\mathbf{v} \times \mathbf{w}) = (\mathbf{u} \times \mathbf{v}) \times \mathbf{w} \quad \Longrightarrow \quad \mathbf{v} \times (\mathbf{u} \times \mathbf{w}) = \mathbf{0}.$$

Since all of the above steps can be reversed, the right side of the above implication implies the left side. Finally, observe that $\mathbf{v} \times (\mathbf{u} \times \mathbf{w}) = \mathbf{0}$ if, and only if, $\mathbf{v}$ and $\mathbf{u} \times \mathbf{w}$ are parallel; i.e., if, and only if, $\mathbf{v}$ and $\mathbf{u} \times \mathbf{w}$ are linearly dependent.

Chapter Review

(pgs. 44-45)

1. Since $4\mathbf{a} = (4, 4, 4)$ and $\mathbf{c} = (2, 3, 6)$ have lengths

$$|4\mathbf{a}| = \sqrt{16 + 16 + 16} = \sqrt{48} \approx 6.93 \quad \text{and} \quad |\mathbf{c}| = \sqrt{4 + 9 + 36} = 7,$$

$\mathbf{c}$ is longer.

3. we have

$$\mathbf{b} - \mathbf{a} = (1, 2, 2) - (1, 1, 1) = (0, 1, 1) = \mathbf{j} + \mathbf{k},$$
$$\text{and} \quad \mathbf{c} - \mathbf{b} - \mathbf{a} = (2, 3, 6) - (1, 2, 2) - (1, 1, 1) = (0, 0, 3) = 3\mathbf{k}.$$

5. By inspection, $(2, -1, 3, 2) = 2\mathbf{e}_1 - \mathbf{e}_2 + 3\mathbf{e}_3 + 2\mathbf{e}_4$.

7. Suppose a and b are scalars such that

$$(4, 1, -2) = a(1, 2, 3) + b(6, 5, 4) = (a + 6b, 2a + 5b, 3a + 4b).$$

Then a, b must simultaneously satisfy the three equations

$$a + 6b = 4$$
$$2a + 5b = 1$$
$$3a + 4b = -2.$$

Subtracting the second equation from twice the first gives $7b = 7$, so that $b = 1$; and subtracting the third equation from 3 times the first gives $14b = 14$ and, again, $b = 1$. With $b = 1$ in the first equation we get $a + 6 = 4$, or $a = -2$. Thus,

$$(4, 1, -2) = -2(1, 2, 3) + (6, 5, 4).$$

9.

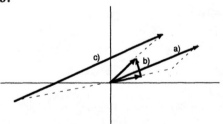

11. The vectors $(1, 2) - (2, 1) = (-1, 1)$ and $(4, 5) - (-1, -2) = (5, 7)$ are direction vectors for the lines passing through $(1, 2)$, $(2, 1)$ and $(4, 5)$, $(-1, -2)$, respectively. Thus, we consider the two lines

$$s(-1, 1) + (1, 2) \quad \text{and} \quad t(5, 7) + (4, 5).$$

They intersect if there exist scalars s and t such that $s(-1, 1) + (1, 2) = t(5, 7) + (4, 5)$, which can be written as

$$s(-1, 1) - t(5, 7) = (4, 5) - (1, 2), \quad \text{or} \quad (-s - 5t, s - 7t) = (3, 3).$$

28

This leads to the two equations $-s - 5t = 3$ and $s - 7t = 3$. Adding these equations gives $-12t = 6$, so that $t = -1/2$. Hence, the lines intersect at

$$-\frac{1}{2}(5,7) + (4,5) = (-5/2 + 4, -7/2 + 5) = (3/2, 3/2).$$

13. If f is a vector function of t that describes the motion of a particle moving in a straight line at unit speed then f has the form $f(t) = t\mathbf{u} + \mathbf{a}$, where $\mathbf{u}$ is a unit vector and $\mathbf{a}$ is a point on the line. Since $(-5, 3, 4)$ is on the line, we can take $\mathbf{a} = (-5, 3, 4)$; and since $(1, 2, 3)$ is also on the line, we can take $\mathbf{u}$ to be a unit vector in the direction of $(1, 2, 3) - (-5, 3, 4) = (6, -1, -1)$. Specifically, since $|(6, -1, -1)| = \sqrt{38}$, we can take $\mathbf{u} = (6, -1, -1)/\sqrt{38}$. Thus, a vector function that does what we want is

$$f(t) = t(6/\sqrt{38}, -1/\sqrt{38}, -1/\sqrt{38}) + (-5, 3, 4).$$

15. Let $\mathbf{p}_1(t)$ and $\mathbf{p}_2(t)$ give the positions of the two boats at time t. Since the boats maintain constant velocities $\mathbf{v}_1$ and $\mathbf{v}_2$, their directions of travel are in straight lines with direction vectors $\mathbf{v}_1$ and $\mathbf{v}_2$. And, since their initial positions are $\mathbf{u}_1$ and $\mathbf{u}_2$ (with $\mathbf{u}_1 \neq \mathbf{u}_2$),

$$\mathbf{p}_1(t) = t\mathbf{v}_1 + \mathbf{u}_1 \quad \text{and} \quad \mathbf{p}_2(t) = t\mathbf{v}_2 + \mathbf{u}_2.$$

Moreover, the displacement vector $\mathbf{d}(t)$ from the first boat to the second is given by

$$\mathbf{d}(t) = \mathbf{p}_2(t) - \mathbf{p}_1(t) = t\mathbf{v}_2 + \mathbf{u}_2 - t\mathbf{v}_1 - \mathbf{u}_1 = t(\mathbf{v}_2 - \mathbf{v}_1) + (\mathbf{u}_2 - \mathbf{u}_1).$$

(a) If the boats are on a collision course then $\mathbf{d}(t_0) = \mathbf{0}$, for some time $t = t_0$, so that

$$t_0(\mathbf{v}_2 - \mathbf{v}_1) + (\mathbf{u}_2 - \mathbf{u}_1) = \mathbf{0} \quad \text{or} \quad (\mathbf{u}_2 - \mathbf{u}_1) = -t_0(\mathbf{v}_2 - \mathbf{v}_1).$$

The displacement vector can then be written as

$$\mathbf{d}(t) = t(\mathbf{v}_2 - \mathbf{v}_1) + (\mathbf{u}_2 - \mathbf{u}_1) = t(\mathbf{v}_2 - \mathbf{v}_1) - t_0(\mathbf{v}_2 - \mathbf{v}_1) = (t - t_0)(\mathbf{v}_2 - \mathbf{v}_1),$$

so that the displacement vector is always a scalar multiple of of the constant vector $\mathbf{v}_2 - \mathbf{v}_1$, which means that the direction of the displacement does not change with time.

(b) Suppose that the direction of the displacement does not change with time. Since $\mathbf{d}(0) = \mathbf{u}_2 - \mathbf{u}_1 \ (\neq \mathbf{0})$, $\mathbf{d}(t)$ is a scalar multiple of $\mathbf{u}_2 - \mathbf{u}_1$. It follows that $\mathbf{v}_2 - \mathbf{v}_1$ is also a scalar multiple of $\mathbf{u}_2 - \mathbf{u}_1$, say $\mathbf{v}_2 - \mathbf{v}_1 = k(\mathbf{u}_2 - \mathbf{u}_1)$. Then

$$\mathbf{d}(t) = tk(\mathbf{u}_2 - \mathbf{u}_1) + (\mathbf{u}_2 - \mathbf{u}_1) = (tk + 1)(\mathbf{u}_2 - \mathbf{u}_1).$$

Thus, if a collision is to occur at some time t_0, the scalar k must be related to t_0 by the equation $t_0 k + 1 = 0$, or $t_0 = -1/k$. Since t_0 must be positive, k must be negative, and we can write

$$t_0 = -\frac{1}{k} = \frac{1}{|k|} = \frac{|\mathbf{u}_2 - \mathbf{u}_1|}{|\mathbf{v}_2 - \mathbf{v}_1|},$$

where the relation $\mathbf{v}_2 - \mathbf{v}_1 = k(\mathbf{u}_2 - \mathbf{u}_1)$ was used to obtain $|\mathbf{v}_2 - \mathbf{v}_1| = |k||(\mathbf{u}_2 - \mathbf{u}_1)|$, which implies the last equality above.

On the other hand, collision will not occur if $k \geq 0$. The case $k = 0$ implies $\mathbf{v}_1 = \mathbf{v}_2$ (i.e., the boats are traveling parallel to each other). The case $k > 0$ corresponds to the situation where the boats are initially pointed in a direction away from each other. This is saying no more than the displacement between them is increasing with time.

17. For comparison, we write the four lines in parametric form and coordinate notation. We have

$$K: \ t(5,1) + (3,4), \quad L: \ t(1,2) + (1,-3), \quad M: \ t(5,1) + (8,5), \quad N: \ t(2,4),$$

where the direction vector for K is computed from $(3,4) - (-2,3) = (5,1)$, and the direction vector for N is computed from $(2,4) - (0,0) = (2,4)$.

By inspection, K and M are parallel (same direction vector) and, further, $t = 1$ in the equation for K and $t = 0$ in the equation for M give the same point $(8,5)$, so that K and M are the same line. Also, L and M are parallel (direction vectors are scalar multiples of each other), but the point $(1,-3)$ is on L but not on N, so that L and N are not the same line.

19. By inspection, the vectors $(3,0,0) - (0,2,0) = (3,-2,0)$ and $(0,2,0) - (0,0,5) = (0,2,-5)$ are linearly independent and both are parallel to the desired plane. Using $(3,0,0)$ as our point on the plane, a parametric representation of the plane is

$$s(3,-2,0) + t(0,2,-5) + (3,0,0).$$

21. The vector $(1,0,0)$ is parallel to the x-axis and the vector $(0,1,0)$ is parallel to the y-axis. Hence, they are both parallel to a plane that is parallel to both the x-axis and the y-axis. Moreover, they are linearly independent. Using $(1,2,3)$ as our point on the plane, a parametric representation of the plane is

$$s(1,0,0) + t(0,1,0) + (1,2,3).$$

23. The vector with its tail at $(-6,-2,-4)$ and tip at $(0,0,0)$ is $(0,0,0) - (-6,-4,-2) = (6,2,4)$, and the vector with its tail at $(0,0,0)$ and tip at $(3,1,2)$ is $(3,1,2) - (0,0,0) = (3,1,2)$. Since the first vector is twice the second, they are are parallel and therefore the three given points are collinear. The line passing through these points can be given parametrically by

$$t(3,1,2).$$

(The values $t = 0, 1, -2$ give the three points.)

25. By inspection, the vectors $(1,2,3) - (3,2,1) = (-2,0,2)$ and $(1,2,3) - (4,2,2) = (-3,0,1)$ are linearly independent, and are both parallel to the plane determined by the three given points. Using $(1,2,3)$ as our point on the plane, a parametric representation of the plane is

$$s(-2,0,2) + t(-3,0,1) + (1,2,3).$$

27. Let $\mathbf{a} = (1,1,1)$, $\mathbf{b} = (1,2,2)$ and $\mathbf{c} = (2,3,6)$.

(a) If θ is the angle between $\mathbf{a}$ and $\mathbf{b}$ and ϕ is the angle between $\mathbf{a}$ and $\mathbf{c}$ then

$$\cos\theta = \frac{\mathbf{a}\cdot\mathbf{b}}{|\mathbf{a}||\mathbf{b}|} = \frac{(1,1,1)\cdot(1,2,2)}{|(1,1,1)||(1,2,2)|} = \frac{1+2+2}{\sqrt{3}\sqrt{9}} = \frac{5}{3\sqrt{3}} \approx 0.9623$$

$$\cos\phi = \frac{\mathbf{a}\cdot\mathbf{c}}{|\mathbf{a}||\mathbf{c}|} = \frac{(1,1,1)\cdot(2,3,6)}{|(1,1,1)||(2,3,6)|} = \frac{2+3+6}{\sqrt{3}\sqrt{49}} = \frac{11}{7\sqrt{3}} \approx 0.9073.$$

Since the cosine function is decreasing on $[0,\pi]$, $\phi > \theta$.

(b) Since $\mathbf{a}$ and $\mathbf{b}$ are nonparallel vectors, $\mathbf{a} \times \mathbf{b}$ is a nonzero vector perpendicular to both $\mathbf{a}$ and $\mathbf{b}$. We have

$$\mathbf{a} \times \mathbf{b} = (1,1,1) \times (1,2,2) = \big((1)(2) - (1)(2), (1)(1) - (1)(2), (1)(2) - (1)(1)\big) = (0,-1,1).$$

(c) As position vectors, **a** and **b** represent adjacent sides of a triangle T with vertices **a**, **b** and the origin. Thus, the area $A(T)$ of the triangle is

$$A(T) = \frac{1}{2}|\mathbf{a} \times \mathbf{b}| = \frac{1}{2}|(0,-1,1)| = \frac{\sqrt{2}}{2},$$

where $\mathbf{a} \times \mathbf{b}$ is computed in part (b).

(d) The vectors $\mathbf{c} - \mathbf{a} = (1,2,5)$ and $\mathbf{c} - \mathbf{b} = (1,1,4)$ represent adjacent sides of a triangle T with vertices **a**, **b** and **c**. Thus, the area $A(T)$ is

$$A(T) = \frac{1}{2}|(\mathbf{c} - \mathbf{a}) \times (\mathbf{c} - \mathbf{b})| = \frac{1}{2}|(1,2,5) \times (1,1,4)|$$

$$= \frac{1}{2}|((2)(4) - (5)(1), (5)(1) - (1)(4), (1)(1) - (2)(1))|$$

$$= \frac{1}{2}|(3,1,-1)| = \frac{\sqrt{11}}{2}.$$

29. If a force **F** moves an object from $(1,2,3)$ to $(4,5,0)$ then the object is being moved in the direction of the vector $\mathbf{v} = (4,5,0) - (1,2,3) = (3,3,-3)$. Thus, if $\mathbf{F} = (1,3,-2)$ then the work done is

$$\mathbf{F} \cdot \mathbf{v} = (1,3,-2) \cdot (3,3,-3) = 3 + 9 + 6 = 18 \text{ units.}$$

31. Let L be the line $(2,3) \cdot \mathbf{x} = 6$.

(a) We first need the normalized equation for L. So, we compute $|(2,3)| = \sqrt{13}$ and divide the given equation by this number to get $(2/\sqrt{13}, 3/\sqrt{13}) \cdot \mathbf{x} = 6/\sqrt{13}$. The distance from $(0,0)$ to L is therefore equal to $|\delta|$, where δ is given by

$$\delta = (2/\sqrt{13}, 3/\sqrt{13}) \cdot (0,0) - 6/\sqrt{13} = -\frac{6}{\sqrt{13}}.$$

So, L is $|\delta| = 6/\sqrt{13}$ units from the origin.

(b) The given equation for L is equivalent to $(2,3) \cdot (x,y) = 6$, or $2x + 3y = 6$. By inspection, $(3,0)$ and $(0,2)$ are two points on L, so that a direction vector for L is $(3,0) - (0,2) = (3,-2)$. Thus, a parametric representation for L is

$$t(3,-2) + (3,0).$$

(c) In $\mathcal{R}^2$, there are two vectors of any given positive length that are perpendicular to L. Since $(2,3)$ is perpendicular to L and has length $\sqrt{13}$, it follows $\pm 4/\sqrt{13}$ times this vector are the two vectors we seek. That is,

$$(8/\sqrt{13}, 12/\sqrt{13}) \qquad \text{and} \qquad (-8/\sqrt{13}, -12/\sqrt{13}).$$

33. Let L be the line $t(2,1) + (-2,0)$.

(a) If K is the line perpendicular to L that passes through the origin, then a direction vector (a,b) for K must be perpendicular to $(2,1)$, a direction vector for L. Thus, we want (a,b) to satisfy $(a,b) \cdot (2,1) = 0$, or $2a + b = 0$. Writing b in terms of a gives $b = -2a$, so that $(a,b) = (a,-2a) = a(1,-2)$. That is, for every nonzero scalar a, $a(1,-2)$ can be used as a direction vector for K. We choose $a = 1$. Since $(0,0)$ is a point on K, a parametric representation for K is

$$t(1, -2).$$

(b) By part (a), the vector $(1, -2)$ is perpendicular to L. The unit vector $\mathbf{n}$ in the direction of $(1, -2)$ is

$$\mathbf{n} = \frac{(1, -2)}{|(1, -2)|} = \frac{(1, -2)}{\sqrt{5}} = (1/\sqrt{5}, -2/\sqrt{5}).$$

Since $\mathbf{x}_0 = (-2, 0)$ is a point on L, the vector equation $\mathbf{n} \cdot \mathbf{x} = \mathbf{n} \cdot \mathbf{x}_0 = c$ is precisely what we want:

$$(1/\sqrt{5}, -2/\sqrt{5}) \cdot \mathbf{x} = (1/\sqrt{5}, -2/\sqrt{5}) \cdot (-2, 0) = -\frac{2}{\sqrt{5}}.$$

Here, $c = -2/\sqrt{5}$.

(c) If $\mathbf{x}$ is set equal to $(3, 5)$ in the expression $\mathbf{n} \cdot \mathbf{x} - c$, where $\mathbf{n}$ and c are identified from part (b), then the distance from $(3, 5)$ to L is $|\delta|$, where δ is given by

$$\delta = (1/\sqrt{5}, -2/\sqrt{5}) \cdot (3, 5) + 2/\sqrt{5} = 3/\sqrt{5} - 10/\sqrt{5} + 2/\sqrt{5} = -5/\sqrt{5} = -\sqrt{5}.$$

So, $|\delta| = \sqrt{5}$. Moreover, since $\delta < 0$, $(3, 5)$ is on the side of the line to which $-\mathbf{n}$ points, and since the y-coordinate of $-\mathbf{n}$ is positive, $-\mathbf{n}$ points in an upward direction. Hence, $(3, 5)$ is above L.

On the other hand, if the parametric form of L is written as $(2t - 2, t)$, we see that $(0, 1)$ is on L (just set $t = 1$). So, the origin is below L and therefore $(3, 5)$ and the origin are not on the same side of L.

35. Let P_1 and P_2 be the planes $x + y + z = 3$ and $2x - y - z = 5$, respectively, and let L be the line of intersection of P_1 and P_2. and let Q be the plane $x - y = 2$.

Since the vectors $(1, 1, 1)$ and $(2, -1, -1)$ are perpendicular to P_1 and P_2 (resp.), they are both perpendicular to their cross product and therefore their cross product is parallel to L. Thus,

$$(1, 1, 1) \times (2, -1, -1) = (1)(-1) - (1)(-1), (1)(2) - (1)(-1), (1)(-1) - (1)(2)) = (0, 3, -3)$$

is a direction vector for L.

To find a point on L, set $y = 0$ in the equations for P_1 and P_2 to get $x + z = 3$ and $2x - z = 5$, which are easily solved to get $x = 8/3$, $z = 1/3$. That is, $(8/3, 0, 1/3)$ is on both planes and therefore is on L. Hence, a parametric form for L is

$$t(0, 3, -3) + (8/3, 0, 1/3).$$

Next, to find the point where L intersects the plane $x - y = 2$, first write the equation for L as

$$(x, y, z) = (8/3, 3t, -3t + 1/3), \quad \text{so that} \quad x = 8/3, \ y = 3t.$$

Now substitute these expressions for x and y into $x - y = 2$ to get

$$\frac{8}{3} - 3t = 2 \quad \text{or} \quad t = 2/9.$$

Finally, insert this value of t into the equation for L to get the desired point of intersection:

$$(8/3, 3(2/9), -3(2/9) + 1/3) = (8/3, 2/3, -1/3).$$

Chapter 2: Equations and Matrices

Section 1: SYSTEMS OF LINEAR EQUATIONS

Exercise Set 1A (pgs. 51-52)

1. Add 1 times the first equation to the second and replace the second equation with the result. This gives the equivalent system

$$x + y = 1,$$
$$2x = 3.$$

The second equation implies $x = 3/2$. Insert this value of x into the first equation and solve for y to get $y = -1/2$. Thus, the given system has the unique solution

$$(x, y) = (3/2, -1/2).$$

The two equations of the given system are nonparallel lines in $\mathcal{R}^2$ and therefore must intersect in a unique point; namely, the point $(3/2, -1/2)$.

3. Whatever the value of $y + z$, a comparison of the first and third equations of the given system shows that $x = 0$. Since the second equation implies $x = y$, we have $y = 0$. The third equation then gives $z = 0$. That is, the system has the unique solution

$$(x, y, z) = (0, 0, 0).$$

Geometrically, the three equations of the system represent three planes, each of which contains the origin.

5. All three equations of the given system contain the sum $x + y$. Thus, if we let $X = x + y$, we can solve each equation for z and arrive at the system

$$z = -X,$$
$$z = X - 1,$$
$$z = 1 - \frac{X}{2}.$$

The first two equations implies $-X = X - 1$, so that $X = 1/2$, and the last two equations implies $X - 1 = 1 - X/2$, so that $X = 4/3$. Since $1/2 \neq 4/3$, the system is inconsistent and has no solution. Geometrically, the original system represents three mutually nonparallel planes. The reader can check that the three lines of intersection of the three possible pairs of planes are all different but all have the same direction vector $(1, -1, 0)$. Hence, these three lines no not intersect in a common point and therefore neither do the three planes.

7. Since the direction vectors of the given lines are not scalar multiples of each other, they can't be the same line. So if they intersect at all, the point of intersection is unique. Assume the two lines intersect. Then $t(1, -1, 2) + (1, 1, 1) = s(3, 2, 1) + (-2, -6, 5)$ for a unique pair of parameter values t, s. Writing each side of this equation as a single vector gives $(t + 1, -t + 1, 2t + 1) = (3s - 2, 2s - 6, s + 5)$. Equating corresponding coordinates then leads to the system

$t + 1 = 3s - 2,$	$t - 3s = -3,$
$-t + 1 = 2s - 6,$ which can be written as	$t + 2s = 7,$
$2t + 1 = s + 5,$	$2t - s = 4.$

33

Focusing on the system on the right, subtract the first equation from the second to get $5s = 10$, so that $s = 2$. Inserting this value of s into all three equations gives $t = 3$ in all three cases. That is, the system has the unique solution $(t, s) = (3, 2)$ and therefore the two lines intersect in a unique point. To find this point, one can either set $t = 3$ in the equation of the first line or set $s = 2$ in the equation of the second line. We choose the first line and find the point of intersection to be

$$(x, y, z) = \mathbf{x}(3) = 3(1, -1, 2) + (1, 1, 1) = (4, -2, 7).$$

The reader can check that setting $s = 2$ in the equation of the second line gives the same point.

9. Since the direction vectors of the given lines are not scalar multiples of each other, they can't be the same line. So if they intersect at all, the point of intersection is unique. Assume the two lines intersect. Then $t(1, 2) + (2, 1) = s(1, 3) + (3, -1)$ for a unique pair of parameter values t, s. Writing each side of this equation as a single vector gives $(t + 2, 2t + 1) = (s + 3, 3s - 1)$. Equating corresponding coordinates then leads to the system

$$\begin{aligned} t + 2 &= s + 3, \\ 2t + 1 &= 3s - 1, \end{aligned} \quad \text{which can be written as} \quad \begin{aligned} t - s &= 1, \\ 2t - 3s &= -2. \end{aligned}$$

Focusing on the system on the right, add -2 times the first equation to the second equation to get $-s = -4$, so that $s = 4$. The first equation then implies $t = 5$. That is, $(t, s) = (5, 4)$ is the unique solution of the system and therefore the two lines intersect in a unique point. To find this point, one can either set $t = 5$ in the equation of the first line or set $s = 4$ in the equation of the second line. We choose the first line and find the point of intersection to be

$$(x, y) = \mathbf{x}(5) = 5(1, 2) + (2, 1) = (7, 11).$$

The reader can check that setting $s = 4$ in the equation of the second line gives the same point.

11. (a) Referring to the system in Example 6 in the text, we immediately get $c = 1$ from the first equation. Inserting this value of c into the remaining two equations leads to the two-equation system

$$\begin{aligned} 4a + 2b &= 3, \\ 9a + 3b &= 2. \end{aligned}$$

Add -3 times the first equation to 2 times the second equation to get $6a = -5$, from which $a = -5/6$. Inserting this into the first equation above gives $4(-5/6) + 2b = 3$, from which we derive $b = 19/6$. That is,

$$(a, b, c) = (-5/6, 19/6, 1) \quad \text{and} \quad f(x) = -\frac{5}{6}x^2 + \frac{19}{6}x + 1.$$

(b) With $f(3) = y$ and $a = 0$, the system in Example 6 becomes

$$\begin{aligned} 0b + c &= 1, \\ 2b + c &= 4, \\ 3b + c &= y. \end{aligned}$$

The first equation immediately gives $c = 1$. Setting $c = 1$ in the second equation gives $2b + 1 = 4$, so that $b = 3/2$. Setting $c = 1$ and $b = 3/2$ in the third equation gives

34

$3(3/2) + 1 = y$, or equivalently, $y = 11/2$. The function f now has the form $f(x) = \frac{3}{2}x + 1$, which is a straight line and not a parabola.

13. Let S and C be the number of cubic yards of sand and cinders (resp.) in the mixture. Since sand weighs 4 tons per cubic yard, there are $4S$ tons of sand in the mixture. Similarly, since the cinders weigh 1 ton per cubic yard, there are C tons of cinders in the mixture. Given that the mixture is 10 cubic yards and weighs 34 tons, it follows that S and C satisfy the system

$$S + C = 10,$$
$$4S + C = 34.$$

Subtracting the first equation from the second gives $3S = 24$, so that $S = 8$. The first equation then gives $C = 2$. That is, the mixture consists of 8 cubic yards of sand weighing 32 tons and 2 cubic yards of cinders weighing 2 tons.

15. With $\mathbf{a}_1$, $\mathbf{a}_2$ and $\mathbf{b}$ as given, we seek scalars x_1, x_2 such that $\mathbf{b} = x_1\mathbf{a}_1 + x_2\mathbf{a}_2$. Converting to coordinate notation gives

$$\begin{pmatrix} -1 \\ 1 \end{pmatrix} = x_1 \begin{pmatrix} 1 \\ 2 \end{pmatrix} + x_2 \begin{pmatrix} 2 \\ 2 \end{pmatrix} = \begin{pmatrix} x_1 + 2x_2 \\ 2x_1 + 2x_2 \end{pmatrix}.$$

Equating corresponding coordinates leads to the system

$$-1 = x_1 + 2x_2,$$
$$1 = 2x_1 + 2x_2.$$

subtract the first equation from the second to get $2 = x_1$. Setting $x_1 = 2$ in the first equation and solving for x_2 gives $x_2 = -3/2$. The desired linear combination is therefore

$$\mathbf{b} = 2\mathbf{a}_1 - \frac{3}{2}\mathbf{a}_2.$$

17. With $\mathbf{a}_1$, $\mathbf{a}_2$, $\mathbf{a}_3$ and $\mathbf{b}$ as given, we seek scalars x_1, x_2, x_3 such that

$$\mathbf{b} = x_1\mathbf{a}_1 + x_2\mathbf{a}_2 + x_3\mathbf{a}_3.$$

Converting to coordinate notation gives

$$\begin{pmatrix} 1 \\ 0 \\ 0 \end{pmatrix} = x_1 \begin{pmatrix} 2 \\ 1 \\ 1 \end{pmatrix} + x_2 \begin{pmatrix} 2 \\ 1 \\ 2 \end{pmatrix} + x_3 \begin{pmatrix} 1 \\ 2 \\ 2 \end{pmatrix} = \begin{pmatrix} 2x_1 + 2x_2 + x_3 \\ x_1 + x_2 + 2x_3 \\ x_1 + 2x_2 + 2x_3 \end{pmatrix}.$$

Equating corresponding coordinates leads to the system

$$1 = 2x_1 + 2x_2 + x_3,$$
$$0 = x_1 + x_2 + 2x_3,$$
$$0 = x_1 + 2x_2 + 2x_3.$$

Add -2 times the second equation to the first equation and replace the second equation with the result. Add -2 times the third equation to the first equation and replace the third equation with the result. This gives the equivalent system

$$1 = 2x_1 + 2x_2 + x_3,$$
$$1 = -3x_3,$$
$$1 = -2x_2 - 3x_3.$$

The second equation gives $x_3 = -1/3$. Inserting this into the third equation and solving for x_2 gives $x_2 = 0$. Inserting $x_3 = -1/3$ and $x_2 = 0$ into the first equation and solving for x_1 gives $x_1 = 2/3$. The desired linear combination is therefore

$$\mathbf{b} = \frac{2}{3}\mathbf{a}_1 + 0\mathbf{a}_2 - \frac{1}{3}\mathbf{a}_3 = \frac{2}{3}\mathbf{a}_1 - \frac{1}{3}\mathbf{a}_3.$$

Exercise Set 1B (pg. 58)

1. Let $J_1, \ldots, J_6$ be the six junctions of the network as shown on the right and let v_i be the voltage at junction J_i ($i = 1, \ldots, 6$). The diagram shows that voltages of $v_1 = 10$ and $v_4 = 4$ are maintained at junctions J_1 and J_4 (resp.) from external power sources. The number beside each connecting wire is the number of ohms of resistance in that wire. Since junctions J_2, J_3, J_5, J_6 have no external

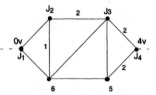

connection, the sum of the currents flowing out of each of these junctions is zero. Following the procedure of Example 7 in the text, we use equation (1) in the text to derive the four equations that describe the state of affairs:

$$\frac{1}{2}(v_2 - v_1) + \frac{1}{2}(v_2 - v_3) + \frac{1}{1}(v_2 - v_6) = 0, \qquad \text{(at } J_2)$$

$$\frac{1}{2}(v_3 - v_2) + \frac{1}{2}(v_3 - v_4) + \frac{1}{1}(v_3 - v_5) + \frac{1}{1}(v_3 - v_6) = 0, \qquad \text{(at } J_3)$$

$$\frac{1}{1}(v_5 - v_3) + \frac{1}{2}(v_5 - v_4) + \frac{1}{2}(v_5 - v_6) = 0, \qquad \text{(at } J_5)$$

$$\frac{1}{2}(v_6 - v_1) + \frac{1}{1}(v_6 - v_2) + \frac{1}{1}(v_6 - v_3) + \frac{1}{2}(v_6 - v_5) = 0. \qquad \text{(at } J_6)$$

Setting $v_1 = 10$ and $v_4 = 4$, the above system of equations can be rewritten as a system of four equations in the four unknowns v_2, v_3, v_5, v_6:

$$4v_2 - v_3 - 2v_6 = 10,$$

$$-v_2 + 6v_3 - 2v_5 - 2v_6 = 4,$$

$$-2v_3 + 4v_5 - v_6 = 4,$$

$$-2v_2 - 2v_3 - v_5 + 6v_6 = 10.$$

Solving the second equation for v_2, inserting the resulting expression for v_2 into the remaining equations and simplifying, gives the following a system of three equations in the three unknowns v_3, v_5, v_6:

$$23v_3 - 8v_5 - 10v_6 = 26,$$

$$-2v_3 + 4v_5 - v_6 = 4,$$

$$-14v_3 + 3v_5 + 10v_6 = 2.$$

Solving the second equation for v_6, inserting the resulting expression for v_6 into the remaining equations of the above three-equation system and simplifying, gives the following system of two equations in the two unknowns v_3, v_5:

$$43v_3 - 48v_5 = -14,$$

$$-34v_3 + 43v_5 = 42.$$

Solving the first equation for v_3 and inserting the result into the second equation of the above two-equation system gives a single equation in terms of v_5, which when solved for v_5 gives $v_5 = 190/31$. Inserting this value of v_5 into either of the two equations of the above two-equation system and solving for v_3 gives $v_3 = 202/31$. Inserting these values of v_3 and v_5 into the third equation of the original four-equation system and solving for v_6 gives

$v_6 = 232/31$. And finally, inserting the derived values of v_3, v_5 and v_6 into either the first or fourth equation of the original four-equation system and solving for v_2 gives $v_2 = 244/31$. To summarize:

$$v_1 = 10, \quad v_2 = \frac{244}{31}, \quad v_3 = \frac{202}{31}, \quad v_4 = 4, \quad v_5 = \frac{190}{31}, \quad v_6 = \frac{232}{31}.$$

To find the current c flowing into the network at junction A (J_1), we reason that c must equal the sum of the currents flowing out of J_1 to the two connecting junctions J_2 and J_6. That is,

$$c = c_{12} + c_{16} = \frac{1}{2}(v_1 - v_2) + \frac{1}{2}(v_1 - v_6) = \frac{1}{2}\left(10 - \frac{244}{31}\right) + \frac{1}{2}\left(10 - \frac{232}{31}\right) = \frac{72}{31} \approx 2.32 \text{ amps.}$$

As a check, a similar calculation at junction B (J_4) shows that there are $-72/31$ amps flowing into the network at junction B; i.e., $72/31$ amps are flowing out of the network at junction B.

3. Let $J_1, \ldots, J_8$ be the eight junctions of the network as shown on the right and let v_i be the voltage at junction J_i ($i = 1, \ldots, 8$). The diagram shows that voltages of $v_1 = 1$ and $v_4 = 0$ are maintained at junctions J_1 and J_4 (resp.) from external power sources. Since junctions $J_2, J_3, J_5, J_6, J_7, J_8$ have no external connection, the sum of the currents flowing out of each of these junctions is zero. As the resistance in each connecting wire is 1 ohm, we can use equation (1) in the text to derive the six equations that describe the state of affairs:

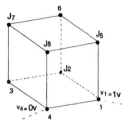

$$\frac{1}{1}(v_2 - v_1) + \frac{1}{1}(v_2 - v_3) + \frac{1}{1}(v_2 - v_6) = 0, \qquad \text{(at } J_2)$$

$$\frac{1}{1}(v_3 - v_2) + \frac{1}{1}(v_3 - v_4) + \frac{1}{1}(v_3 - v_7) = 0, \qquad \text{(at } J_3)$$

$$\frac{1}{1}(v_5 - v_1) + \frac{1}{1}(v_5 - v_6) + \frac{1}{1}(v_5 - v_8) = 0, \qquad \text{(at } J_5)$$

$$\frac{1}{1}(v_6 - v_2) + \frac{1}{1}(v_6 - v_5) + \frac{1}{1}(v_6 - v_7) = 0, \qquad \text{(at } J_6)$$

$$\frac{1}{1}(v_7 - v_3) + \frac{1}{1}(v_7 - v_6) + \frac{1}{1}(v_7 - v_8) = 0, \qquad \text{(at } J_7)$$

$$\frac{1}{1}(v_8 - v_4) + \frac{1}{1}(v_8 - v_5) + \frac{1}{1}(v_8 - v_7) = 0, \qquad \text{(at } J_8)$$

Setting $v_1 = 1$ and $v_4 = 0$, the above system of equations can be rewritten as a system of six equations in the six unknowns v_2, v_3, v_5, v_6, v_7, v_8:

$$3v_2 - v_3 - v_6 = 1,$$
$$-v_2 + 3v_3 - v_7 = 0,$$
$$3v_5 - v_6 - v_8 = 1,$$
$$-v_2 - v_5 + 3v_6 - v_7 = 0,$$
$$-v_3 - v_6 + 3v_7 - v_8 = 0,$$
$$-v_5 - v_7 + 3v_8 = 0.$$

37

Note that the second equation is in terms of v_2, v_3 and v_7, while the third equation is in terms of v_5, v_6 and v_8. Thus, solve the second equation for v_2, the third equation for v_8, insert the resulting expressions for v_2 and v_8 into the remaining four equations and simplify. This gives the following system of four equations in the four unknowns v_3, v_5, v_6, v_7:

$$8v_3 - v_6 - 3v_7 = 1,$$
$$-3v_3 - v_5 + 3v_6 = 0,$$
$$-v_3 - 3v_5 + 3v_7 = -1,$$
$$8v_5 - 3v_6 - v_7 = 3.$$

Solving the third equation for v_3, inserting the resulting expression for v_3 into the remaining equations of the above four-equation system and simplifying gives the following system of three equations in the three unknowns v_5, v_6, v_7:

$$-24v_5 - v_6 + 21v_7 = -7,$$
$$8v_5 + 3v_6 - 9v_7 = 3,$$
$$8v_5 - 3v_6 - v_7 = 3.$$

Now solve the third equation for v_7, insert the resulting expression for v_7 into the two remaining equations of the above three-equation system and simplify. This gives the following two-equation system in terms of v_5 and v_6:

$$18v_5 - 8v_6 = 7,$$
$$-32v_5 + 15v_6 = -12.$$

Solving the first equation for v_5 and inserting the resulting expression for v_5 into the second equation allows us to solve for v_6 to get $v_6 = 4/7$. Inserting this value of v_6 into either equation of the above two-equation system leads to $v_5 = 9/14$. Inserting these values of v_5 and v_6 into the third equation of the original six-equation system allows us to solve for v_8 to get $v_8 = 5/14$. Inserting the derived values of v_5 and v_8 into the sixth equation of the original six-equation system leads to $v_7 = 3/7$. Finally, inserting the appropriate derived values of v_5, v_6, v_7 and v_8 into the fourth and fifth equations of the original six-equation system allows us to solve for v_2 and v_3 to get $v_2 = 9/14$ and $v_3 = 5/14$. To summarize:

$$v_1 = 1, \quad v_2 = \frac{9}{14}, \quad v_3 = \frac{5}{14}, \quad v_4 = 0, \quad v_5 = \frac{9}{14}, \quad v_6 = \frac{4}{7}, \quad v_7 = \frac{3}{7}, \quad v_8 = \frac{5}{14}.$$

To find the current c flowing into the network at junction J_1, we reason that c must equal the sum of the currents flowing out of J_1 to the three connecting junctions J_2, J_4 and J_5. That is,

$$c = c_{12} + c_{14} + c_{15} = \frac{1}{1}(v_1 - v_2) + \frac{1}{1}(v_1 - v_4) + \frac{1}{1}(v_1 - v_5) = \frac{5}{14} + 1 + \frac{5}{14} = \frac{12}{7} \approx 1.71 \text{ amps.}$$

Similarly, letting c' be the current flowing into junction J_4 gives

$$c' = c_{41} + c_{43} + c_{48} = \frac{1}{1}(v_4 - v_1) + \frac{1}{1}(v_4 - v_3) + \frac{1}{1}(v_4 - v_8) = -1 - \frac{5}{14} - \frac{5}{14} = -\frac{12}{7} \approx -1.71 \text{ amps.}$$

The above result simply says what we should have expected; namely, 1.71 amps is flowing into the network at J_1 and 1.7 amps is flowing out of the network at J_4.

5. If $\mathbf{F}_1$, $\mathbf{F}_2$ and $\mathbf{F}_3$ are three forces in $\mathcal{R}^2$ acting parallel to the vectors $(2,1)$, $(2,2)$ and $(-3,-1)$, respectively, then they must be of the form $\mathbf{F}_1 = a(2,1)$, $\mathbf{F}_2 = b(2,2)$ and $\mathbf{F}_3 = c(-3,-1)$, where a, b, c are nonzero scalars. We want to choose a, b, c such that

$$\mathbf{F}_1 + \mathbf{F}_2 + \mathbf{F}_3 = a(2,1) + b(2,2) + c(-3,-1) = (2a + 2b - 3c, a + 2b - c) = (0,0).$$

This implies the system
$$2a + 2b - 3c = 0,$$
$$a + 2b - c = 0.$$

Add -1 times the second equation to the first equation and replace the first equation with the result to get the equivalent system

$$a - 2c = 0,$$
$$a + 2b - c = 0.$$

The first equation says $a = 2c$. Inserting this expression for a into the second equation gives $2c + 2b - c = 0$. Solving for b gives $b = -\frac{1}{2}c$. Hence, for each nonzero scalar c, the forces $\mathbf{F}_1 = 2c(2,1)$, $\mathbf{F}_2 = -\frac{1}{2}c(2,2) = -c(1,1)$ and $\mathbf{F}_3 = c(-3,-1) = -c(3,1)$ satisfy the required conditions. Their magnitudes are

$$|\mathbf{F}_1| = |2c(2,1)| = 2|c||(2,1)| = 2|c|\sqrt{5},$$
$$|\mathbf{F}_2| = |-c(1,1)| = |c||(1,1)| = |c|\sqrt{2},$$
$$|\mathbf{F}_3| = |-c(3,1)| = |c||(3,1)| = |c|\sqrt{10}.$$

7. If three forces $\mathbf{F}_1$, $\mathbf{F}_2$ and $\mathbf{F}_3$ act in the same directions as $(1,0,0)$, $(1,1,0)$ and $(1,1,1)$ (resp.) then they must be of the form $\mathbf{F}_1 = x(1,0,0)$, $\mathbf{F}_2 = y(1,1,0)$ and $\mathbf{F}_2 = z(1,1,1)$, where x, y and z are *positive* scalars. Such forces produce a resultant force $\mathbf{F}$ of the form

$$\mathbf{F} = \mathbf{F}_1 + \mathbf{F}_2 + \mathbf{F}_3 = x(1,0,0) + y(1,1,0) + z(1,1,1) = (x + y + z, y + z, z).$$

Since x, y and z are positive, so are the coordinates of $\mathbf{F}$. Hence, any force vector with at least one nonpositive coordinate cannot be the resultant force of three forces acting in the same directions as the three given vectors.

9. Referring to Fig. 2.4(a) in the text, let p_k be the probability of going from a_k to a_4 without going through a_5. It should be observed that it is not feasible to attempt to compute p_2 directly. The best strategy is to find a system of equations that relates p_1, p_2 and p_3 and then determine p_2 by eliminating the extraneous variables p_1 and p_3.

First, start at a_1 and find p_1 in terms of p_2 and p_4 $(=1)$: one can either (i) go directly to a_4 with probability $\frac{1}{3}$ or (ii) go to a_2 (with probability $\frac{1}{3}$) and then go to a_4 without going through a_5 (with probability p_2) for a combined probability of $\frac{1}{3}p_2$. Thus, p_1 is the sum of these two probabilities. That is,

(1) $$p_1 = \frac{1}{3} + \frac{1}{3}p_2, \quad \text{or equivalently,} \quad 3p_1 - p_2 = 1.$$

Next, start at a_2 and find p_2 in terms of p_1 and p_3: one can either (i) go to a_1 (with probability $\frac{1}{2}$) and then go to a_4 without going through a_5 (with probability p_1) for a combined probability of $\frac{1}{2}p_1$, or (ii) go to a_3 (with probability $\frac{1}{2}$) and then go to a_4 without going through a_5 (with probability p_3) for a combined probability of $\frac{1}{2}p_3$. Thus, p_2 is the sum of these two probabilities. That is,

$$(2) \qquad p_2 = \frac{1}{2}p_1 + \frac{1}{2}p_3, \qquad \text{or equivalently,} \qquad p_1 - 2p_2 + p_3 = 0.$$

Finally, start at a_3 and find p_3 in terms of p_2 and p_4 ($= 1$): one can either (i) go to a_2 (with probability $\frac{1}{2}$) and then go to a_4 without going through a_5 (with probability p_2) for a combined probability of $\frac{1}{2}p_2$, or (ii) go directly to a_4 with probability $\frac{1}{2}$. Thus, p_3 is the sum of these two probabilities. That is,

$$(3) \qquad p_3 = \frac{1}{2}p_2 + \frac{1}{2}, \qquad \text{or equivalently,} \qquad p_2 - 2p_3 = -1.$$

Since we only need to solve for p_2, we can solve equation (2) for p_1 and insert the resulting expression for p_1 into equation (1) and simplify to get $5p_2 - 2p_3 = 1$. This, together with equation (3), gives the two-equation system

$$5p_2 - 3p_3 = 1,$$
$$p_2 - 2p_3 = -1.$$

Adding 2 times the first equation to -3 times the second equation gives $7p_2 = 5$, so that $p_2 = \frac{5}{7}$.

11. Referring to Fig. 2.4(b) in the text, let p_k be the probability of going from b_k to b_6 without going through b_5. We will derive a system of four equations in the four unknowns p_1, p_2, p_3, p_4.

First, start at b_1 and find p_1 in terms of p_2 and p_3: one can either (i) go to b_2 (with probability $\frac{1}{2}$) and then go to b_6 without going through b_5 (with probability p_2) for a combined probability of $\frac{1}{2}p_2$, or (ii) go to b_3 (with probability $\frac{1}{2}$) and then go to b_6 without going through b_5 (with probability p_3) for a combined probability of $\frac{1}{2}p_3$. Thus, p_1 is the sum of these two probabilities. That is,

$$p_1 = \frac{1}{2}p_2 + \frac{1}{2}p_3, \qquad \text{or equivalently,} \qquad 2p_1 - p_2 - p_3 = 0.$$

Next, start at b_2 and find p_2 in terms of b_1 and b_6 ($= 1$): one can either (i) go to b_1 (with probability $\frac{1}{3}$) and then go to b_6 without going through b_5 (with probability p_1) for a combined probability of $\frac{1}{3}p_1$, or (ii) go directly to b_6 with probability $\frac{1}{3}$. Thus, p_2 is the sum of these two probabilities. That is,

$$p_2 = \frac{1}{3}p_1 + \frac{1}{3}, \qquad \text{or equivalently,} \qquad p_1 - 3p_2 = -1.$$

Next, start at b_3 and find p_3 in terms of b_1, b_4 and b_6 ($= 1$): one can either (i) go to b_1 (with probability $\frac{1}{4}$) and then go to b_6 without going through b_5 (with probability p_1) for a combined probability of $\frac{1}{4}p_1$, or (ii) go to b_4 (with probability $\frac{1}{4}$) and then go to b_6 without going through b_5 (with probability p_4) for a combined probability of $\frac{1}{4}p_4$, or (iii) go directly to b_6 with probability $\frac{1}{4}$. Thus, p_3 is the sum of these three probabilities. That is,

$$p_3 = \frac{1}{4}p_1 + \frac{1}{4}p_4 + \frac{1}{4}, \qquad \text{or equivalently,} \qquad p_1 - 4p_3 + p_4 = -1.$$

Finally, start at b_4 and find p_4 in terms of p_3. Since there is only one path leading away from b_4, one can only go to b_3 (with probability 1) and then go to b_6 without going through b_5 (with probability p_3) for a combined probability of $(1)p_3 = p_3$. Thus, $p_4 = p_3$, or equivalently, $p_3 - p_4 = 0$.

The desired system of equations is therefore

$$2p_1 - p_2 - p_3 = 0,$$
$$p_1 - 3p_2 = -1,$$
$$p_1 - 4p_3 + p_4 = -1,$$
$$p_3 - p_4 = 0.$$

Using the fourth equation, set $p_4 = p_3$ in the third equation and simplify to get the following three equation system in terms of p_1, p_2 and p_3:

$$2p_1 - p_2 - p_3 = 0,$$
$$p_1 - 3p_2 = -1,$$
$$p_1 - 3p_3 = -1.$$

Solving the second and third equations for p_2 and p_3 gives $p_2 = \frac{1}{3}(p_1+1)$ and $p_3 = \frac{1}{3}(p_1+1)$ (note that $p_2 = p_3$, so that $p_2 = p_3 = p_4$). Inserting these expressions for p_2 and p_3 into the first equation of the above three-equation system gives

$$2p_1 - \frac{1}{3}(p_1 + 1) - \frac{1}{3}(p_1 + 1) = \frac{4}{3}p_1 - \frac{2}{3} = 0, \quad \text{so that} \quad p_1 = \frac{1}{2}.$$

Hence, $p_2 = \frac{1}{3}(\frac{1}{2}+1) = \frac{1}{2}$ and therefore $p_3 = \frac{1}{2}$ and $p_4 = \frac{1}{2}$. That is, $p_k = \frac{1}{2}$ for $k = 1, 2, 3, 4$.

13. The figure on the right is Fig.2.4(b) in the text except that a new path has been introduced between b_4 and b_6. Here, we need a system of four equations in the four unknowns p_k, $k = 1, 2, 3, 4$. However, because of how the connected system of pathways has been modified, the three equations derived in Exercise 11 for p_1, p_2 and p_3 still hold. Only the last equation is different. To find this last equation, we start at b_4 and find p_4 in terms of p_3 and p_6 ($=1$): one can either (i) go to b_3 (with probability $\frac{1}{2}$) and then go to b_6 without going through b_5 (with probability p_3) for a combined probability of $\frac{1}{2}p_3$, or (ii) go directly to b_6 with probability $\frac{1}{2}$. Thus, p_4 is the sum of these two probabilities. That is

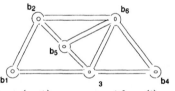

$$p_4 = \frac{1}{2}p_3 + \frac{1}{2}, \quad \text{or equivalently,} \quad p_3 - 2p_4 = -1.$$

The desired system is therefore

$$2p_1 - p_2 - p_3 = 0,$$
$$p_1 - 3p_2 = -1,$$
$$p_1 - 4p_3 + p_4 = -1,$$
$$p_3 - 2p_4 = -1.$$

41

From the fourth equation we have $p_4 = \frac{1}{2}(p_3 + 1)$. Inserting this expression for p_4 into the third equation and simplifying gives $2p_1 - 7p_3 = -3$. This equation and the first two equations constitute a system of three equations in the three unknowns p_1, p_2, p_3:

$$2p_1 - p_2 - p_3 = 0,$$
$$p_1 - 3p_2 = -1,$$
$$2p_1 - 7p_3 = -3.$$

Solving the second and third equations for p_2 and p_3 gives $p_2 = \frac{1}{3}(p_1+1)$ and $p_3 = \frac{1}{7}(2p_1+3)$. Inserting these expressions for p_2 and p_3 into the first equation gives

$$2p_1 - \frac{1}{3}(p_1 + 1) - \frac{1}{7}(2p_1 + 3) = \frac{29}{21}p_1 - \frac{16}{21} = 0,$$

from which $p_1 = \frac{16}{29}$. Hence, $p_2 = \frac{1}{3}(\frac{16}{29} + 1) = \frac{15}{29}$ and $p_3 = \frac{1}{7}[2(\frac{16}{29}) + 3] = \frac{17}{29}$. The fourth equation of the original four-equation system can now be solved for p_4 to get $p_4 = \frac{23}{29}$. To summarize:

$$p_1 = \frac{16}{29}, \qquad p_2 = \frac{15}{29}, \qquad p_3 = \frac{17}{29}, \qquad p_4 = \frac{23}{29}.$$

15. Referring to Fig. 2.5(b) in the text, we see that $s_1 = r_1 + r_2$ (from the left junction), $s_2 + r_1 = r_3$ (from the top junction), $s_3 + r_2 = r_4$ (from the bottom junction) and $r_3 + r_4 = s_4$ (from the right junction). Since we want the external flow vector $\mathbf{s} = (s_1, s_2, s_3, s_4)$ to be $\mathbf{s} = (1, 1, 2, 4)$, set $s_1 = 1$, $s_2 = 1$, $s_3 = 2$ and $s_4 = 4$ in these equations and rewrite the results in the form of the following system:

$$r_1 + r_2 = 1,$$
$$r_1 - r_3 = -1,$$
$$-r_2 + r_4 = 2,$$
$$r_3 + r_4 = 4.$$

Note that adding the first and third equations gives the same result as adding the second and fourth equations. This means that any one of the above four equations can be derived from the remaining three. We can therefore delete any one of the equations without affecting the solution set. We arbitrarily choose to delete the fourth equation and work with the system

$$r_1 + r_2 = 1,$$
$$r_1 - r_3 = -1,$$
$$-r_2 + r_4 = 2.$$

Since there are now more unknowns than equations, at least one of the r_k can be chosen arbitrarily. Indeed, letting $r_4 = a$ be arbitrarily chosen, the fourth equation uniquely determines $r_2 = a - 2$; this value of r_2 in the first equation uniquely determine $r_1 = -a + 3$; and this value of r_1 in the second equation uniquely determines $r_3 = -a + 4$. Hence, for each real number a, the vector

$$\mathbf{r} = (r_1, r_2, r_3, r_4) = (-a + 3, a - 2, -a + 4, a)$$

is a consistent internal flow vector that gives rise the the given external flow vector $\mathbf{s}$.

17. Here, we solve the system

$$A + B + C = 2h,$$
$$(a - h)A + aB + (a + h)C = 2ah,$$
$$(a^2 - 2ah + h^2)A + a^2B + (a^2 + 2ah + h^2)C = 2a^2h + \frac{2}{3}h^3.$$

The first equation implies $B = 2h - A - C$. Substituting this expression for B into the remaining equations gives

$$(a - h)A + a(2h - A - C) + (a + h)C = 2ah,$$
$$(a^2 - 2ah + h^2)A + a^2(2h - A - C) + (a^2 + 2ah + h^2)C = 2a^2h + \frac{2}{3}h^3,$$

where all terms involving a in the first equation cancel out and all terms involving a^2 in the second equation cancel out to result in a system that can be written as

$$-hA + hC = 0,$$
$$(-2ah + h^2)A + (2ah + h^2)C = \frac{2}{3}h^3,$$

or since $h \neq 0$ (because $2h$ is the length of the interval $[a - h, a + h]$),

$$-A + C = 0,$$
$$(-2a + h)A + (2a + h)C = \frac{2}{3}h^2.$$

The first equation gives $C = A$. Replacing C with A in the second equation gives the equation $(-2a + h)A + (2a + h)A = \frac{2}{3}h^2$, which simplifies to $2hA = \frac{2}{3}h^2$. Again, since $h \neq 0$, we can solve for A to get $A = \frac{1}{3}h$. So $C = \frac{1}{3}h$ as well. Inserting these expressions for A and C into the first equation of the original three-equation system gives $\frac{1}{3}h + B + \frac{1}{3}h = 2h$, from which we obtain $B = \frac{4}{3}h$. That is,

$$A = \frac{1}{3}h, \qquad B = \frac{4}{3}h, \qquad C = \frac{1}{3}h,$$

which is what we wanted to show.

Section 2: MATRIX METHODS

Exercise Set 2ABC (pgs. 69-70)

1. In matrix form, the given system is

$$\begin{pmatrix} 3 & -2 \\ 1 & -3 \end{pmatrix} \begin{pmatrix} x \\ y \end{pmatrix} = \begin{pmatrix} 1 \\ 2 \end{pmatrix}.$$

To solve the system, we first divide the first row by 1/3 to obtain a leading entry of 1. Subtracting the second row from the first and putting the result in the second row leads to

$$\begin{pmatrix} 1 & -2/3 \\ 0 & 7/3 \end{pmatrix} \begin{pmatrix} x \\ y \end{pmatrix} = \begin{pmatrix} 1/3 \\ -5/3 \end{pmatrix}.$$

Now multiply the second row by 3/7 to obtain a leading entry of 1. Adding 2/3 times the second row to the first row and putting the result in the first row gives

$$\begin{pmatrix} 1 & 0 \\ 0 & 1 \end{pmatrix} \begin{pmatrix} x \\ y \end{pmatrix} = \begin{pmatrix} -1/7 \\ -5/7 \end{pmatrix}.$$

Now that the coefficient matrix is in reduced form, write the equivalent system of equations $x = -1/7$, $y = -5/7$. This is the unique solution of the given system

3. In matrix form, the given system is

$$\begin{pmatrix} 1 & 1 & 0 \\ 0 & 1 & -1 \\ 1 & 0 & 1 \end{pmatrix} \begin{pmatrix} x \\ y \\ z \end{pmatrix} = \begin{pmatrix} 1 \\ 1 \\ 0 \end{pmatrix}.$$

To solve the system, first subtract the third row from the first row and put the result in the third row to get

$$\begin{pmatrix} 1 & 1 & 0 \\ 0 & 1 & -1 \\ 0 & 1 & -1 \end{pmatrix} \begin{pmatrix} x \\ y \\ z \end{pmatrix} = \begin{pmatrix} 1 \\ 1 \\ 1 \end{pmatrix}.$$

Now subtract the third row from the second row and put the result in the third row; then subtract the second row from the first row and put the result in the first row. This gives

$$\begin{pmatrix} 1 & 0 & 1 \\ 0 & 1 & -1 \\ 0 & 0 & 0 \end{pmatrix} \begin{pmatrix} x \\ y \\ z \end{pmatrix} = \begin{pmatrix} 0 \\ 1 \\ 0 \end{pmatrix}.$$

The coefficient matrix is now in reduced form so we write out the corresponding system of equations

$$x + z = 0,$$
$$y - z = 1,$$
$$0 = 0.$$

44

Since we have one more unknown than nontrivial equations, we let $z = t$ be arbitrary and from the first two equations we obtain $x = -t$ and $y = t + 1$. The resulting solution set is a line in $\mathcal{R}^3$;

$$(x, y, z) = (-t, t + 1, t) = t(-1, 1, 1) + (0, 1, 0).$$

5. Writing the given matrix equation as a system of equations gives

$$x + 2y = 1,$$
$$3x + y = 0.$$

Add $-1/3$ times the second equation to the first equation and replace the second equation with the result. This gives

$$x + 2y = 1,$$
$$\frac{5}{3}y = 1.$$

The second equation implies $y = 3/5$. Inserting this value of y into the first equation gives $x + 6/5 = 1$, from which $x = -1/5$. The solution of the matrix equation is therefore

$$\begin{pmatrix} x \\ y \end{pmatrix} = \begin{pmatrix} -1/5 \\ 3/5 \end{pmatrix}.$$

7. With $\mathbf{x} = \begin{pmatrix} x \\ y \\ z \end{pmatrix}$, the given matrix equation is equivalent to the system of equations

$$x + z = 0,$$
$$y = 1,$$
$$x + y = 0.$$

Since $y = 1$ from the second equation, we can insert this value into the third equation and obtain $x = -1$. With this value of x in the first equation we get $z = 1$. The unique solution of the system is therefore

$$\mathbf{x} = \begin{pmatrix} -1 \\ 1 \\ 1 \end{pmatrix}.$$

9. Add the second row to the first row and put the result in the second row. Then subtract the third row from the first row and put the result in the third row. The given matrix equation now has the form

$$\begin{pmatrix} 1 & 1 & 1 \\ 0 & 3 & -3 \\ 0 & -2 & -8 \end{pmatrix} \begin{pmatrix} x \\ y \\ z \end{pmatrix} = \begin{pmatrix} 2 \\ 4 \\ 2 \end{pmatrix}.$$

Now divide the second row by 3 and the third row by 2 to get

$$\begin{pmatrix} 1 & 1 & 1 \\ 0 & 1 & -1 \\ 0 & -1 & -4 \end{pmatrix} \begin{pmatrix} x \\ y \\ z \end{pmatrix} = \begin{pmatrix} 2 \\ 4/3 \\ 1 \end{pmatrix}.$$

Add the third row to the second row and put the result in the third row; then subtract the second row from the first row and put the result in the first row. This gives

$$\begin{pmatrix} 1 & 0 & 2 \\ 0 & 1 & -1 \\ 0 & 0 & -5 \end{pmatrix} \begin{pmatrix} x \\ y \\ z \end{pmatrix} = \begin{pmatrix} 2/3 \\ 4/3 \\ 7/3 \end{pmatrix}.$$

Now divide the last row by -5. Add this new last row to the second row and put the result in the second row; then add -2 times this new last row to the first row and put the result in the first row. The matrix equation is now

$$\begin{pmatrix} 1 & 0 & 0 \\ 0 & 1 & 0 \\ 0 & 0 & 1 \end{pmatrix} \begin{pmatrix} x \\ y \\ z \end{pmatrix} = \begin{pmatrix} 24/15 \\ 13/15 \\ -7/15 \end{pmatrix}.$$

The unique solution of the given matrix equation is therefore

$$\begin{pmatrix} x \\ y \\ z \end{pmatrix} = \begin{pmatrix} 24/15 \\ 13/15 \\ -7/15 \end{pmatrix}.$$

11. Let $A = \begin{pmatrix} 1 & -2 & 1 \\ 2 & 1 & -3 \end{pmatrix}$ and consider the matrix equation $A\mathbf{x} = \mathbf{e}_1 + \mathbf{e}_2$. We first observe that since A is a 2-by-3 matrix, the vector $\mathbf{x}$ in the expression $A\mathbf{x}$ must be a 3-by-1 column vector in order that multiplication be defined. Also, the result of the multiplication must be a 2-by-1 column vector, so that the basis vectors $\mathbf{e}_1$ and $\mathbf{e}_2$ are both in $\mathcal{R}^2$. Therefore, $\mathbf{e}_1 + \mathbf{e}_2 = \begin{pmatrix} 1 \\ 1 \end{pmatrix}$.

(a) To obtain a reduced matrix equivalent to A, we perform the following three steps:

1. subtract 2 times the second row from the first row and put the result in the second row;

2. mutiply the second row by $\dfrac{1}{5}$;

3. add 2 times the second row to the first row and put the result in the first row.

Denoting the resulting reduced matrix by A_0 we get

$$A_0 = \begin{pmatrix} 1 & 0 & -1 \\ 0 & 1 & -1 \end{pmatrix}.$$

(b) Let $\mathbf{x} = \begin{pmatrix} x \\ y \\ z \end{pmatrix}$. The solution of $A\mathbf{x} = \mathbf{0}$ (the homogeneous solution $\mathbf{x}_h$) is the same as the solution of $A_0\mathbf{x} = \mathbf{0}$. The system of equations that corresponds to $A_0\mathbf{x} = \mathbf{0}$ is

$$x - z = 0,$$
$$y - z = 0.$$

46

Setting $z = t$, we get $x = t$ and $y = t$. So,

$$\mathbf{x}_h = \begin{pmatrix} x \\ y \\ z \end{pmatrix} = \begin{pmatrix} t \\ t \\ t \end{pmatrix} = t \begin{pmatrix} 1 \\ 1 \\ 1 \end{pmatrix}.$$

(c) We can find one solution of $A\mathbf{x} = \begin{pmatrix} 1 \\ 1 \end{pmatrix}$ (a particular solution $\mathbf{x}_p$) by first applying the three steps shown in part (a) to the vector $\begin{pmatrix} 1 \\ 1 \end{pmatrix}$ to get the vector $\begin{pmatrix} 3/5 \\ -1/5 \end{pmatrix}$, and then finding one solution of $A_0\mathbf{x} = \begin{pmatrix} 3/5 \\ -1/5 \end{pmatrix}$. Since the corresponding system of equations is

$$x - z = 3/5,$$
$$y - z = -1/5,$$

and since we only need one solution, we can set $z = 0$ to get $x = 3/5$, $y = -1/5$. Thus, $\mathbf{x}_p = \begin{pmatrix} 3/5 \\ -1/5 \\ 0 \end{pmatrix}$. Using this one solution along with the homogeneous solution $\mathbf{x}_h$ found in part (b), the general solution of the system $A\mathbf{x} = \mathbf{e}_1 + \mathbf{e}_2$ can be written as

$$\mathbf{x}_h + \mathbf{x}_p = t \begin{pmatrix} 1 \\ 1 \\ 1 \end{pmatrix} + \begin{pmatrix} 3/5 \\ -1/5 \\ 0 \end{pmatrix}.$$

13. Let $A = \begin{pmatrix} 1 & 1 & 1 \\ 1 & 1 & 0 \\ 1 & 0 & 0 \end{pmatrix}$ and consider the matrix equation $A\mathbf{x} = \mathbf{e}_1 + \mathbf{e}_2$. Here, the vector $\mathbf{x}$ is a 3-by-1 column vector so that the product $A\mathbf{x}$ is also a 3-by-1 column vector. It follows that the basis vectors $\mathbf{e}_1$ and $\mathbf{e}_2$ are both in $\mathcal{R}^3$. Specifically, $\mathbf{e}_1 + \mathbf{e}_2 = \begin{pmatrix} 1 \\ 1 \\ 0 \end{pmatrix}$.

(a) To obtain a reduced matrix equivalent to A, we perform the following two steps:

1. subtract the second row from the first row and put the result in the first row;

2. subtract the third row from the second row and put the result in the second row.

Denoting the resulting reduced matrix by A_0 we get

$$A_0 = \begin{pmatrix} 0 & 0 & 1 \\ 0 & 1 & 0 \\ 1 & 0 & 0 \end{pmatrix}.$$

(b) Let $\mathbf{x} = \begin{pmatrix} x \\ y \\ z \end{pmatrix}$. The solution of $A\mathbf{x} = \mathbf{0}$ (the homogeneous solution $\mathbf{x}_h$) is the same as the solution of $A_0\mathbf{x} = \mathbf{0}$. The system of equations that corresponds to $A_0\mathbf{x} = \mathbf{0}$ is

$$z = 0,$$
$$y = 0,$$
$$x = 0.$$

So,

$$\mathbf{x}_h = \begin{pmatrix} 0 \\ 0 \\ 0 \end{pmatrix}.$$

(c) We can find one solution of $A\mathbf{x} = \begin{pmatrix} 1 \\ 1 \\ 0 \end{pmatrix}$ (a particular solution $\mathbf{x}_p$) by first applying

the two steps shown in part (a) to the vector $\begin{pmatrix} 1 \\ 1 \\ 0 \end{pmatrix}$ to get the vector $\begin{pmatrix} 0 \\ 1 \\ 0 \end{pmatrix}$. Since the

corresponding system of equations is

$$z = 0,$$
$$y = 1,$$
$$x = 0.$$

a particular solution is $\mathbf{x}_p = \begin{pmatrix} 0 \\ 1 \\ 0 \end{pmatrix}$. The general solution of the system $A\mathbf{x} = \mathbf{e}_1 + \mathbf{e}_2$ is

therefore

$$\mathbf{x}_h + \mathbf{x}_p = \begin{pmatrix} 0 \\ 0 \\ 0 \end{pmatrix} + \begin{pmatrix} 0 \\ 1 \\ 0 \end{pmatrix} = \begin{pmatrix} 0 \\ 1 \\ 0 \end{pmatrix}.$$

15. Let a_1, a_2, b_1, b_2 be scalars such that

$$\mathbf{i} = (1,0) = a_1(1,2) + b_1(2,3) = (a_1 + 2b_1, 2a_1 + 3b_1),$$
$$\mathbf{j} = (0,1) = a_2(1,2) + b_2(2,3) = (a_2 + 2b_2, 2a_2 + 3b_2).$$

The corresponding systems are

$$\begin{array}{ccc} 1 = a_1 + 2b_1, & & 0 = a_2 + 2b_2, \\ 0 = 2a_1 + 3b_1, & \text{and} & 1 = 2a_2 + 3b_2. \end{array}$$

For the first system, subtract the second equation from 2 times the first equation to get $2 = b_1$. Inserting this value into the first equation and solving for a_1 gives $a_1 = -3$. For the second system, we again subtract the second equation from 2 times the first equation to get $b_2 = -1$. Inserting this value into the first equation and solving for a_2 gives $a_2 = 2$. Hence, the desired linear combinations are

$$\mathbf{i} = -3(1,2) + 2(2,3) \qquad \text{and} \qquad \mathbf{j} = 2(1,2) - (2,3).$$

17. Let a and b be scalars such that

$$(5,0,1,2) = a(1,2,1,0) + b(2,-1,0,1) = (a+2b, 2a-b, a, b).$$

This leads to the equivalent system

$$5 = a + 2b,$$
$$0 = 2a - b,$$
$$1 = a,$$
$$2 = b.$$

The last two equations gives $a = 1$ and $b = 2$. Since these values also satisfy each of the first two equations, the desired linear combination is

$$(5, 0, 1, 2) = (1, 2, 1, 0) + 2(2, -1, 0, 1).$$

19. Let $\mathbf{x} = (x_1, x_2, x_3, x_4, x_5)$, where the coordinates x_i are to be determined. Here, we have written the vector $\mathbf{x}$ as a row vector to conserve vertical space. With this understanding, we write out the matrix equation as a system of five equations in five unknowns

$$\begin{aligned}
x_1 + 2x_2 + 3x_3 + 4x_4 + 5x_5 &= 6, \\
x_2 + 2x_3 + 3x_4 + 4x_5 &= 5, \\
x_3 + 2x_4 + 3x_5 &= 4, \\
x_4 + 2x_5 &= 3, \\
x_5 &= 2.
\end{aligned}$$

From the last equation, $x_5 = 2$. Insert this value of x_5 into the fourth equation and solve for x_4 to get $x_4 = -1$. Insert these values of x_5 and x_4 into the third equation and solve for x_3 to get $x_3 = 0$. Insert these values of x_5, x_4 and x_3 into the second equation and solve for x_2 to get $x_2 = 0$. Finally, insert the values for $x_2, \cdots, x_5$ into the first equation and solve for x_1 to get $x_1 = 0$. The unique solution of the given matrix equation is therefore

$$\mathbf{x} = \begin{pmatrix} 0 \\ 0 \\ 0 \\ -1 \\ 2 \end{pmatrix}.$$

21. Since $\frac{1}{2}(\mathbf{a} + \mathbf{b}) = 2\mathbf{i}$ and $\frac{3}{2}(\mathbf{b} - \mathbf{a}) = 3\mathbf{j}$, it follows that

$$\mathbf{v} = 2\mathbf{i} + 3\mathbf{j} = \frac{1}{2}(\mathbf{a} + \mathbf{b}) + \frac{3}{2}(\mathbf{b} - \mathbf{a}) = -\mathbf{a} + 2\mathbf{b}.$$

23. Let $\mathbf{x} = \begin{pmatrix} x \\ y \\ z \end{pmatrix}$. Writing the given vectors as column vectors, let A be the matrix whose first, second and third columns are $\mathbf{a}$, $\mathbf{b}$ and $\mathbf{c}$, respectively. Then the given problem is equivalent to determining if the matrix equation $A\mathbf{x} = \mathbf{v}$ has a solution. Thus, we will solve

$$\begin{pmatrix} 2 & 1 & 0 \\ -1 & 1 & 1 \\ 2 & -3 & 2 \end{pmatrix} \begin{pmatrix} x \\ y \\ z \end{pmatrix} = \begin{pmatrix} -1 \\ 0 \\ -1 \end{pmatrix}.$$

Add 2 times the second row to the first row and put the result in the second row; subtract the third row from the first row and put the result in the third row; then divide the first row by 2. This reduces the first column and results in

$$\begin{pmatrix} 1 & 1/2 & 0 \\ 0 & 3 & 2 \\ 0 & 4 & -2 \end{pmatrix} \begin{pmatrix} x \\ y \\ z \end{pmatrix} = \begin{pmatrix} -1/2 \\ -1 \\ 0 \end{pmatrix}.$$

Divide the second row by 3; then subtract 1/2 the second row from the first row and put the result in the first row; and subtract 1/4 the third row from the second row and put the result in the third row. This reduces the second column and results in

$$\begin{pmatrix} 1 & 0 & -1/3 \\ 0 & 1 & 2/3 \\ 0 & 0 & 7/6 \end{pmatrix} \begin{pmatrix} x \\ y \\ z \end{pmatrix} = \begin{pmatrix} -1/3 \\ -1/3 \\ -1/3 \end{pmatrix}.$$

Now multiply the last row by 6/7 to get

$$\begin{pmatrix} 1 & 0 & -1/3 \\ 0 & 1 & 2/3 \\ 0 & 0 & 1 \end{pmatrix} \begin{pmatrix} x \\ y \\ z \end{pmatrix} = \begin{pmatrix} -1/3 \\ -1/3 \\ -2/7 \end{pmatrix}.$$

The equivalent system of equations is

$$x - \frac{1}{3}z = -\frac{1}{3},$$
$$y + \frac{2}{3}z = -\frac{1}{3},$$
$$z = -\frac{2}{7}.$$

Inserting the value of z given by the last equation into the first two equations and solving for x and y gives $x = -3/7$ and $y = -1/7$. Hence, the desired linear combination is

$$\mathbf{v} = -\frac{3}{7}\mathbf{a} - \frac{1}{7}\mathbf{b} - \frac{2}{7}\mathbf{c}.$$

25. We assume there is at least one line with the desired property. Pick any one of these lines and denote it by L_3. Let $\mathbf{p}_1$ and $\mathbf{p}_2$ be the intersection points of L_1, L_3 and L_2, L_3, respectively, and let s_1 and t_1 be the corresponding parameter values so that

$$\mathbf{p}_1 = s_1(3, 2, 1) + (-1, 0, 1) = (3s_1 - 1, 2s_1, s_1 + 1),$$
$$\mathbf{p}_2 = t_1(0, 2, 1) + (-4, 2, 0) = (-4, 2t_1 + 2, t_1).$$

The vector from $\mathbf{p}_1$ to $\mathbf{p}_2$ is a direction vector for L_3 and is given by

$$\mathbf{p}_2 - \mathbf{p}_1 = (-4, 2t_1 + 2, t_1) - (3s_1 - 1, 2s_1, s_1 + 1) = (-3s_1 - 3, 2t_1 - 2s_1 + 2, t_1 - s_1 - 1).$$

Since this vector is perpendicular to both L_1 and L_2, its dot product with their direction vectors must be zero. That is,

$$(-3s_1 - 3, 2t_1 - 2s_1 + 2, t_1 - s_1 - 1) \cdot (0, 2, 1) = 5t_1 - 5s_1 + 3 = 0,$$
$$(-3s_1 - 3, 2t_1 - 2s_1 + 2, t_1 - s_1 - 1) \cdot (3, 2, 1) = 5t_1 - 14s_1 - 6 = 0.$$

This gives a system of two equations in the two unknowns s_1, t_1. Subtracting the second equation from the first gives $9s_1 + 9 = 0$, so that $s_1 = -1$. Inserting this into the first

50

equation and solving for t_1 gives $t_1 = -8/5$. Using the derived values of s_1 and t_1 in the above expressions for $\mathbf{p}_2 - \mathbf{p}_1$ and $\mathbf{p}_1$ (which is a point on L_3) we get

$$\mathbf{p}_2 - \mathbf{p}_1 = (0, 4/5, -8/5) \qquad \text{and} \qquad \mathbf{p}_1 = (-4, -2, 0).$$

This is sufficient to deduce a parametric representation for L_3. However, in order that the parametric representation be free of fractions, we use the vector $(0, 4, -8)$ as a direction vector (a constant multiple of $(0, 4/5, -8/5)$). Thus, we have

$$L_3: \quad r(0, 4, -8) + (-4, -2, 0).$$

Finally, since L_3 was chosen arbitrarily from all of the lines that may have the desired property, we conclude that there is indeed such a line (since we derived at least one) and that it is unique (since we derived exactly one).

27. Let L_1 and L_2 be two different lines in $\mathcal{R}^3$ with parametric representations

$$L_1: \quad s\mathbf{u} + \mathbf{a} \qquad \text{and} \qquad L_2: \quad t\mathbf{v} + \mathbf{b}.$$

First, if L_1 and L_2 are parallel then the fact that they are not the same line implies that they determine a plane P; namely, that plane that contains both lines. Any line in P that is perpendicular to L_1 is also perpendicular to L_2, and vice-versa. Since there are infinitely many lines in P that are perpendicular to L_1, it follows that there are infinitely many lines that intersect both L_1 and L_2 at right angles.

Second, if L_1 and L_2 intersect then $\mathbf{u} \times \mathbf{v}$ and their point of intersection $\mathbf{p}$ determines a unique plane P with vector equation $(\mathbf{u} \times \mathbf{v}) \cdot (\mathbf{x} - \mathbf{p}) = 0$. On one hand, if a line intersects L_1 and L_2 at different points then that line must lie in P. But any line in P that is perpendicular to L_1 cannot also be perpendicular to L_2, and vice-versa. It follows that any line that intersects L_1 and L_2 at right angles cannot lie in the plane P and must intersect L_1 and L_2 at the same point, which must be the point $\mathbf{p}$. There is precisely one line passing through $\mathbf{p}$ that is perpendicular to both L_1 and L_2; namely, that line with $\mathbf{u} \times \mathbf{v}$ as its direction vector.

Having disposed of these cases, we assume L_1 and L_2 are not parallel and do not intersect. Assume further, that there is at least one line with the desired property. Pick any one of these lines and denote it by L_3. Let $\mathbf{p}_1$ and $\mathbf{p}_2$ be the intersection points of L_1, L_3 and L_2, L_3, respectively, and let s_1 and t_1 be the corresponding parameter values so that

$$\mathbf{p}_1 = s_1\mathbf{u} + \mathbf{a} \qquad \text{and} \qquad \mathbf{p}_2 = t_1\mathbf{v} + \mathbf{b}.$$

The vector from $\mathbf{p}_1$ to $\mathbf{p}_2$ is a direction vector for L_3 and is given by

$$\mathbf{p}_2 - \mathbf{p}_1 = t_1\mathbf{u} - s_1\mathbf{v} + (\mathbf{b} - \mathbf{a}).$$

Since this vector is perpendicular to both L_1 and L_2, its dot product with their direction vectors must be zero. That is,

$$\left(t_1\mathbf{u} - s_1\mathbf{v} + (\mathbf{b} - \mathbf{a}) \right) \cdot \mathbf{u} = t_1(\mathbf{u} \cdot \mathbf{u}) - s_1(\mathbf{v} \cdot \mathbf{u}) + (\mathbf{b} - \mathbf{a}) \cdot \mathbf{u}$$

$$= t_1|\mathbf{u}|^2 - s_1(\mathbf{v} \cdot \mathbf{u}) + (\mathbf{b} - \mathbf{a}) \cdot \mathbf{u} = 0$$

$$\left(t_1\mathbf{u} - s_1\mathbf{v} + (\mathbf{b} - \mathbf{a}) \right) \cdot \mathbf{v} = t_1\mathbf{u} \cdot \mathbf{v} - s_1\mathbf{v} \cdot \mathbf{v} + (\mathbf{b} - \mathbf{a}) \cdot \mathbf{v}$$

$$= t_1(\mathbf{u} \cdot \mathbf{v}) - s_1|\mathbf{v}|^2 + (\mathbf{b} - \mathbf{a}) \cdot \mathbf{v} = 0.$$

Since **u** and **v** are direction vectors, their magnitudes are nonzero (that is, the dot product of each with itself is nonzero). This allows us to solve the first equation above for t_1. Without showing the algebra we get

$$t_1 = s_1 \left(\frac{\mathbf{u} \cdot \mathbf{v}}{|\mathbf{u}|^2} \right) - \frac{(\mathbf{b} - \mathbf{a}) \cdot \mathbf{u}}{|\mathbf{u}|^2}.$$

Inserting this expression for t_1 into the second equation and isolating all terms involving s_1 on one side, the result can be factored and written as

$$s_1 \left[1 - \left(\frac{\mathbf{u} \cdot \mathbf{v}}{|\mathbf{u}||\mathbf{v}|} \right)^2 \right] = c,$$

where c is a scalar. Letting θ be the angle between **u** and **v** (with $0 \leq \theta < 2\pi$), we notice that the above is equivalent to

$$s_1(1 - \cos^2 \theta) = c.$$

Since the case at hand assumes L_1 and L_2 are nonparallel, $\theta \neq 0$ and $\theta \neq \pi$, so that $1 - \cos^2 \theta \neq 0$. Hence, the above equation can be solved uniquely for s_1. The equation we solved above for t_1 then implies t_1 is also unique. Using these values of s_1 and t_1 in the above expressions for $\mathbf{p}_2 - \mathbf{p}_1$ and $\mathbf{p}_1$ (which is a point on L_3) we get a parametric representation of L_3; namely,

$$L_3: \quad r(\mathbf{p}_2 - \mathbf{p}_1) + \mathbf{p}_1.$$

Finally, since L_3 was chosen arbitrarily from all of the lines that may have the desired property, we conclude that there is indeed such a line (since we derived at least one) and that it is unique (since we derived exactly one).

29. If $\mathbf{x} = \mathbf{v}$ satisfies $A\mathbf{x} = \mathbf{b}$ then, using the linearity of matrix multiplication,

$$A(\mathbf{w} - \mathbf{v}) = A\mathbf{w} - A\mathbf{v} = A\mathbf{w} - \mathbf{b}.$$

The left member is the zero vector if, and only if, $\mathbf{x} = \mathbf{w} - \mathbf{v}$ is a solution of the homogeneous equation $A\mathbf{x} = \mathbf{0}$, and the right member is the zero vector if, and only if, $\mathbf{x} = \mathbf{w}$ is a solution of $A\mathbf{x} = \mathbf{b}$.

31. Suppose $A\mathbf{x}_1 = \mathbf{b}_1$ and $A\mathbf{x}_2 = \mathbf{b}_2$. If t_1, t_2 are any scalars then the linearity of matrix multiplication implies that the vector $\mathbf{x} = t_1\mathbf{x}_1 + t_2\mathbf{x}_2$ satisfies

$$A\mathbf{x} = A(t_1\mathbf{x}_1 + t_2\mathbf{x}_2) = t_1 A\mathbf{x}_1 + t_2 A\mathbf{x}_2 = t_1\mathbf{b}_1 + t_2\mathbf{b}_2.$$

So, $\mathbf{x} = t_1\mathbf{x}_1 + t_2\mathbf{x}_2$ satisfies $A\mathbf{x} = t_1\mathbf{b}_1 + t_1\mathbf{b}_2$.

Exercise Set 2D (pgs. 73)

1. Introduce the three parameters r, s and t and set $x = r$, $y = s$ and $z = t$. The given equation can then be solved for w to get $w = -3r + 2s - t + 3$. The vectors (x, y, z, w) satisfying the given equation can then be written as

$$(x, y, z, w) = (r, s, t, -3r + 2s - t + 3) = r(1, 0, 0, -3) + s(0, 1, 0, 2) + t(0, 0, 1, -1) + (0, 0, 0, 3).$$

This is a 3-plane in $\mathcal{R}^4$.

3. Given the matrix A in Example 15 in the text, the following three steps will reduce the first column: add -2 times the first row to the second row and put the result in the second row; add -3 times the first row to the third row and put the result in the third row; add the first row to the fourth row and put the result in the fourth row. This gives

$$\begin{pmatrix} 1 & 0 & -1 \\ 0 & 1 & 2 \\ 0 & -1 & 5 \\ 0 & 1 & -2 \end{pmatrix}.$$

The following two steps will reduce the second column: add the second row to the third row and put the result in the third row; subtract the fourth row from the second row and put the result in the fourth row. This gives

$$\begin{pmatrix} 1 & 0 & -1 \\ 0 & 1 & 2 \\ 0 & 0 & 7 \\ 0 & 0 & 4 \end{pmatrix}.$$

Now divide the third row by 7 and divide the fourth row by 4. Subtract the resulting third row from the resulting fourth row and put the result in the third row. This gives

$$\begin{pmatrix} 1 & 0 & -1 \\ 0 & 1 & 2 \\ 0 & 0 & 0 \\ 0 & 0 & 1 \end{pmatrix}.$$

Finally, the following two steps will produce the reduced matrix R shown in Example 15: add -2 times the fourth row to the second row and put the result in the second row; add the fourth row to the first row and put the result in the first row. This gives

$$R = \begin{pmatrix} 1 & 0 & 0 \\ 0 & 1 & 0 \\ 0 & 0 & 0 \\ 0 & 0 & 1 \end{pmatrix}.$$

5. (a) Consider the reduced matrix R shown in Example 16 of the text. To test the vectors $\mathbf{a}, \mathbf{b}, \mathbf{d}$, assume

$$x_1\mathbf{a} + x_2\mathbf{b} + x_4\mathbf{d} = \mathbf{0}.$$

Since the last entry in the fourth column of R is nonzero, while the last entries in the first two columns are zero, $\mathbf{d}$ is not a linear combination of $\mathbf{a}$ and $\mathbf{b}$. It follows that $x_4 = 0$ and the above equation reduces to $x_1\mathbf{a} + x_2\mathbf{b} = \mathbf{0}$, or equivalently, $x_1\mathbf{a} = (-x_2)\mathbf{b}$. Noting that

53

the first entries in the first two columns of R are 1 and 0, respectively, we see that neither one of $\mathbf{a}$ or $\mathbf{b}$ is a scalar multiple of the other, so that $x_1 = x_2 = 0$. Hence, the vectors in question are linearly independent.

To test the vectors $\mathbf{a}, \mathbf{c}, \mathbf{d}$, assume

$$x_1\mathbf{a} + x_3\mathbf{c} + x_4\mathbf{d} = \mathbf{0}.$$

Since the first entry in the first column of R is nonzero, while the first entries in the third and fourth columns of R are zero, $\mathbf{a}$ is not a linear combination of $\mathbf{c}$ and $\mathbf{d}$. It follows that $x_1 = 0$ and the above equation reduces to $x_3\mathbf{c} + x_4\mathbf{d} = \mathbf{0}$, or equivalently, $x_3\mathbf{c} = (-x_4)\mathbf{d}$. Noting that the second entries in the third and fourth columns of R are 0 and 1, respectively, we see that neither one of $\mathbf{c}$ or $\mathbf{d}$ is a scalar multiple of the other, so that $x_3 = x_4 = 0$. Hence, the vectors in question are linearly independent.

Finally, it was shown in Example 16 that $\mathbf{b} + 3\mathbf{c} + \mathbf{d} = \mathbf{0}$, so that $\mathbf{b}, \mathbf{c}, \mathbf{d}$ are linearly dependent.

(b) Given the matrix A in Example 16, subtracting the third row from the first row and then putting the result in the third row gives a matrix with a reduced first column; namely,

$$\begin{pmatrix} 1 & 0 & 1 & -3 \\ 0 & 1 & 1 & -4 \\ 0 & -1 & -1 & 4 \\ 0 & 0 & 1 & -3 \end{pmatrix}.$$

Now add the second and third rows and put the result in the third row to get

$$\begin{pmatrix} 1 & 0 & 1 & -3 \\ 0 & 1 & 1 & -4 \\ 0 & 0 & 0 & 0 \\ 0 & 0 & 1 & -3 \end{pmatrix}.$$

Finally, subtract the fourth row from the first row and put the result in the first row, and subtract the fourth row from the second row and put the result in the second row. This results in the desired reduced matrix

$$R = \begin{pmatrix} 1 & 0 & 0 & 0 \\ 0 & 1 & 0 & -1 \\ 0 & 0 & 0 & 0 \\ 0 & 0 & 1 & -3 \end{pmatrix}.$$

7. Let $\mathbf{u}$ and $\mathbf{v}$ be nonzero vectors in $\mathcal{R}^3$.

If $\mathbf{u}$ and $\mathbf{v}$ are linearly dependent then each is a scalar multiple of the other and we can write $\mathbf{u} = k\mathbf{v}$, for some scalar k. This shows that the vector $\mathbf{u}$ is on the line through the origin with parametric representation $s\mathbf{v}$, where $s = k$ is the parameter value corresponding to the point $\mathbf{u}$. Since $\mathbf{v}$ is obviously also on this line, they are both on the line $s\mathbf{v}$.

Conversely, suppose $\mathbf{u}$ and $\mathbf{v}$ both lie on the same line through the origin. Specifically, suppose they both lie on the line with parametric representation $s\mathbf{w}$. Then there exist nonzero parameter values s_1, s_2 such that $\mathbf{u} = s_1\mathbf{w}$ and $\mathbf{v} = s_2\mathbf{w}$. Multiplying the first equation by s_2 gives

$$s_2\mathbf{u} = s_1 s_2 \mathbf{w} = s_1\mathbf{v}, \qquad \text{or equivalently,} \qquad s_2\mathbf{u} + (-s_1)\mathbf{v} = \mathbf{0}.$$

Since s_1 and $-s_2$ are nonzero, the equation on the right, together with Definition 2.7, implies that $\mathbf{u}$ and $\mathbf{v}$ are linearly dependent.

9. Given $A = \begin{pmatrix} 1 & 2 & -1 \\ 0 & 1 & 3 \end{pmatrix}$, we can obtain a reduced form for A by adding -2 times the second row to the first row and putting the result in the first row. This gives

$$R = \begin{pmatrix} 1 & 0 & -7 \\ 0 & 1 & 3 \end{pmatrix}.$$

The solutions of the system $A\mathbf{x} = \mathbf{0}$ and the reduced system $R\mathbf{x} = \mathbf{0}$ are the same. Thus, with $\mathbf{x} = \begin{pmatrix} x \\ y \\ z \end{pmatrix}$, we solve the system of equations corresponding to $R\mathbf{x} = \mathbf{0}$:

$$x - 7z = 0,$$
$$y + 3z = 0.$$

Since z is the only nonleading variable, we set $z = t$ in the above equations and find that $x = 7t$ and $y = -3t$. The general solution of the system $A\mathbf{x} = \mathbf{0}$ is therefore

$$\mathbf{x} = \begin{pmatrix} 7t \\ -3t \\ t \end{pmatrix} = t \begin{pmatrix} 7 \\ -3 \\ 1 \end{pmatrix},$$

which is a line through the origin in $\mathcal{R}^3$.

11. Subtract the third row of the given matrix equation from the first row and put the result in the third row. Then subtract the fourth row from the first row and put the result in the fourth row. This results in the matrix equation

$$\begin{pmatrix} 1 & 0 & 0 & 1 \\ 0 & 0 & 1 & 1 \\ 0 & 0 & -1 & -1 \\ 0 & 0 & -2 & -2 \end{pmatrix} \begin{pmatrix} x \\ y \\ z \\ w \end{pmatrix} = \begin{pmatrix} 0 \\ 0 \\ 0 \\ 0 \end{pmatrix}.$$

Now add the second row to the third row and put the result in the third row, and add 2 times the second row to the fourth row and put the result in the fourth row. This gives

$$\begin{pmatrix} 1 & 0 & 0 & 1 \\ 0 & 0 & 1 & 1 \\ 0 & 0 & 0 & 0 \\ 0 & 0 & 0 & 0 \end{pmatrix} \begin{pmatrix} x \\ y \\ z \\ w \end{pmatrix} = \begin{pmatrix} 0 \\ 0 \\ 0 \\ 0 \end{pmatrix},$$

where we observe that the coefficient matrix is reduced. The corresponding system of equations is

$$x + w = 0,$$
$$z + w = 0,$$
$$0 = 0,$$
$$0 = 0.$$

55

Since y and w are the only nonleading variables, we set $y = t$ and $w = s$ in the above equations and find that $x = -s$ and $z = -s$. The general solution of the given system is therefore

$$\begin{pmatrix} x \\ y \\ z \\ w \end{pmatrix} = \begin{pmatrix} -s \\ t \\ -s \\ s \end{pmatrix} = s \begin{pmatrix} -1 \\ 0 \\ -1 \\ 1 \end{pmatrix} + t \begin{pmatrix} 0 \\ 1 \\ 0 \\ 0 \end{pmatrix},$$

which is a 2-plane through the origin in $\mathcal{R}^4$.

13. Let E be an n-by-n matrix in echelon form. If E has zero rows then we are done. Therefore, assume E has no zero rows. To fix ideas, let $\mathbf{r}_1, \ldots, \mathbf{r}_n$ be the rows of E and let $\mathbf{r}_k$ be fixed. We will show that the first nonzero entry of $\mathbf{r}_k$ is in the kth row. We will then show that this nonzero entry is the only nonzero entry in the kth column. It will then follow that the entries on the main diagonal of E are all 1 (because E is reduced, the first nonzero entry in any row is 1) and all entries off the main diagonal are 0. This says that $\mathbf{r}_k = \mathbf{e}_k$ is the kth standard basis vector in $\mathcal{R}^n$.

Let the first nonzero entry of $\mathbf{r}_k$ be in the jth column. Because E is in echelon form, the first nonzero entry of $\mathbf{r}_k$ is farther to the right than the first nonzero entry of $\mathbf{r}_{k-1}$. Since this is true of any k, and since there are $k - 1$ rows above $\mathbf{r}_k$, it follows that the first nonzero entry of $\mathbf{r}_k$ is at least as far to the right as the kth column; i.e., $j \geq k$. Also, the first nonzero entry of $\mathbf{r}_k$ is farther to the left than the first nonzero entry of $\mathbf{r}_{k+1}$; and since this is true of any k, and since there are $n - k$ rows below $\mathbf{r}_k$, it follows that the first nonzero entry of $\mathbf{r}_k$ is at least as far to the left as the $n - (n - k) = k$th column; i.e., $j \leq k$. Hence, $j = k$. As $\mathbf{r}_k$ was arbitrary, the main diagonal entries are all 1's and all entries below the main diagonal are 0's.

Given the same arbitrarily chosen $\mathbf{r}_k$, suppose there are nonzero entries in the kth column that are not on the main diagonal. By what we have already shown, all such entries are *above* the main diagonal. For each such entry a, add $-a\mathbf{r}_k$ to the row with this entry and put the result in that row. The new row now has a zero where before there was a nonzero entry. The entries in this new row that are to the left of the kth column are unaffected because the first $k - 1$ entries of $\mathbf{r}_k$ are all zero. Once this process has been done for each such entry, the resulting kth column will be reduced. But this implies that the kth column was not reduced to begin with, which is contrary to the definition of a matrix in echelon form. It follows that there are no nonzero entries in the kth column except the entry 1 on the main diagonal.

The result now follows.

Section 3: MATRIX ALGEBRA

Exercise Set 3A-D (pgs. 80-81)

1. $3A - B = 3\begin{pmatrix} -1 & 2 \\ 0 & 1 \end{pmatrix} - \begin{pmatrix} 1 & 2 \\ 1 & 4 \end{pmatrix} = \begin{pmatrix} -3 & 6 \\ 0 & 3 \end{pmatrix} + \begin{pmatrix} -1 & -2 \\ -1 & -4 \end{pmatrix} = \begin{pmatrix} -4 & 4 \\ -1 & -1 \end{pmatrix}.$

3. $2A + B + C = 2\begin{pmatrix} -1 & 2 \\ 0 & 1 \end{pmatrix} + \begin{pmatrix} 1 & 2 \\ 1 & 4 \end{pmatrix} + \begin{pmatrix} 1 & 1 \\ 1 & 2 \end{pmatrix} = \begin{pmatrix} -2 & 4 \\ 0 & 2 \end{pmatrix} + \begin{pmatrix} 2 & 3 \\ 2 & 6 \end{pmatrix} = \begin{pmatrix} 0 & 7 \\ 2 & 8 \end{pmatrix}.$

5. $AB = \begin{pmatrix} -1 & 2 \\ 0 & 1 \end{pmatrix}\begin{pmatrix} 1 & 2 \\ 1 & 4 \end{pmatrix} = \begin{pmatrix} -1+2 & -2+8 \\ 0+1 & 0+4 \end{pmatrix} = \begin{pmatrix} 1 & 6 \\ 1 & 4 \end{pmatrix}.$

7. Using the product $AB = \begin{pmatrix} 1 & 6 \\ 1 & 4 \end{pmatrix}$ computed in Exercise 5 above, we have

$$ABC = \begin{pmatrix} 1 & 6 \\ 1 & 4 \end{pmatrix}\begin{pmatrix} 1 & 1 \\ 1 & 2 \end{pmatrix} = \begin{pmatrix} 1+6 & 1+12 \\ 1+4 & 1+8 \end{pmatrix} = \begin{pmatrix} 7 & 13 \\ 5 & 9 \end{pmatrix}.$$

9. $C + AC = \begin{pmatrix} 1 & 1 \\ 1 & 2 \end{pmatrix} + \begin{pmatrix} -1 & 2 \\ 0 & 1 \end{pmatrix}\begin{pmatrix} 1 & 1 \\ 1 & 2 \end{pmatrix} = \begin{pmatrix} 1 & 1 \\ 1 & 2 \end{pmatrix} + \begin{pmatrix} -1+2 & -1+4 \\ 0+1 & 0+2 \end{pmatrix}$

$= \begin{pmatrix} 1 & 1 \\ 1 & 2 \end{pmatrix} + \begin{pmatrix} 1 & 3 \\ 1 & 2 \end{pmatrix} = \begin{pmatrix} 2 & 4 \\ 2 & 4 \end{pmatrix}.$

11. Since B and G are both 2-by-3 matrices, so is any linear combination of them. That is, the given expression is defined and

$$2B - 3G = 2\begin{pmatrix} 0 & -2 & 1 \\ -1 & 3 & 0 \end{pmatrix} - 3\begin{pmatrix} 1 & -1 & 2 \\ 1 & 0 & 3 \end{pmatrix} = \begin{pmatrix} 0 & -4 & 2 \\ -2 & 6 & 0 \end{pmatrix} + \begin{pmatrix} -3 & 3 & -6 \\ -3 & 0 & -9 \end{pmatrix}$$

$$= \begin{pmatrix} -3 & -1 & -4 \\ -5 & 6 & -9 \end{pmatrix}.$$

13. Since B has three columns and A has two rows, the product BA is not defined.

15. The product DB is defined and is a 3-by-3 matrix. Specifically,

$$DB = \begin{pmatrix} 2 & -4 \\ 0 & 0 \\ 3 & 3 \end{pmatrix}\begin{pmatrix} 0 & -2 & 1 \\ -1 & 3 & 0 \end{pmatrix} = \begin{pmatrix} 0+4 & -4-12 & 2+0 \\ 0+0 & 0+0 & 0+0 \\ 0-3 & -6+9 & 3+0 \end{pmatrix} = \begin{pmatrix} 4 & -16 & 2 \\ 0 & 0 & 0 \\ -3 & 3 & 3 \end{pmatrix}.$$

17. The product AB is defined and is a 2-by-3 matrix. Since G is also a 2-by-3 matrix, every linear combination of AB and G is defined and is a 2-by-3 matrix. In particular,

$$2AB - 5G = 2\begin{pmatrix} 1 & 3 \\ -4 & 2 \end{pmatrix}\begin{pmatrix} 0 & -2 & 1 \\ -1 & 3 & 0 \end{pmatrix} - 5\begin{pmatrix} 1 & -1 & 2 \\ 1 & 0 & 3 \end{pmatrix}$$

$$= 2\begin{pmatrix} 0-3 & -2+9 & 1+0 \\ 0-2 & 8+6 & -4+0 \end{pmatrix} + \begin{pmatrix} -5 & 5 & -10 \\ -5 & 0 & -15 \end{pmatrix}$$

$$= \begin{pmatrix} -6 & 14 & 2 \\ -4 & 28 & -8 \end{pmatrix} + \begin{pmatrix} -5 & 5 & -10 \\ -5 & 0 & -15 \end{pmatrix} = \begin{pmatrix} -11 & 19 & -8 \\ -9 & 28 & -23 \end{pmatrix}.$$

19. Since D has two columns while C has three rows, the product DC is not defined. Hence, neither is the product CDC.

21. Assume X and Y are matrices such that the equation

$$AX = B + Y$$

makes sense. Since A is a 2-by-2 matrix, the product AX on the left side is defined only if X is a 2-by-m matrix for some positive integer m. The product will then be a 2-by-m matrix. The sum $B + Y$ on the right side is defined only if Y has the same dimensions as B; i.e., only if Y is a 2-by-3 matrix. The resulting sum will then be a 2-by-3 matrix. Since the matrices on each side of the above equation must have the same dimensions, $m = 3$. That is, the equation makes sense if, and only if, X and Y are both 2-by-3 matrices.

23. Assume X and Y are matrices such that the equation

$$AX = YD$$

makes sense. Since A is a 2-by-2 matrix, the matrix AX must have two rows, and since D is a 3-by-2 matrix, the matrix YD must have two columns. Since each side of the equation is a matrix of the same dimensions, it follows that each side is a matrix with two rows and two columns; i.e., each side is a 2-by-2 matrix. In particular, X has two columns and Y has two rows. Moreover, since AX is defined, X must have two rows, so that X is a 2-by-2 matrix. Similarly, since YD is defined, Y must have three columns, so that Y is a 2-by-3 matrix.

25. Assume X and Y are matrices such that the equation

$$AX = YC$$

makes sense. Since A is a 2-by-2 matrix, the matrix AX must have two rows, and since C is a 3-by-3 matrix, the matrix YC must have three columns. Since each side of the equation is a matrix of the same dimensions, it follows that each side is a matrix with two rows and three columns; i.e., each side is a 2-by-3 matrix. In particular, X has three columns and Y has two rows. Moreover, since AX is defined, X must have two rows, so that X is a 2-by-3 matrix. Similarly, since YC is defined, Y must have three columns, so that Y is a 2-by-3 matrix.

27. We have

$$C\mathbf{i} = \begin{pmatrix} -2 & 0 & 1 \\ 0 & 3 & 0 \\ 2 & 3 & -1 \end{pmatrix} \begin{pmatrix} 1 \\ 0 \\ 0 \end{pmatrix} = \begin{pmatrix} -2+0+0 \\ 0+0+0 \\ 2+0+0 \end{pmatrix} = \begin{pmatrix} -2 \\ 0 \\ 2 \end{pmatrix}$$

$$C\mathbf{j} = \begin{pmatrix} -2 & 0 & 1 \\ 0 & 3 & 0 \\ 2 & 3 & -1 \end{pmatrix} \begin{pmatrix} 0 \\ 1 \\ 0 \end{pmatrix} = \begin{pmatrix} 0+0+0 \\ 0+3+0 \\ 0+3+0 \end{pmatrix} = \begin{pmatrix} 0 \\ 3 \\ 3 \end{pmatrix}$$

$$C\mathbf{k} = \begin{pmatrix} -2 & 0 & 1 \\ 0 & 3 & 0 \\ 2 & 3 & -1 \end{pmatrix} \begin{pmatrix} 0 \\ 0 \\ 1 \end{pmatrix} = \begin{pmatrix} 0+0+1 \\ 0+0+0 \\ 0+0-1 \end{pmatrix} = \begin{pmatrix} 1 \\ 0 \\ -1 \end{pmatrix}$$

29. $\begin{pmatrix} 1 & 2 & 3 \\ 4 & 5 & 6 \\ 7 & 8 & 9 \end{pmatrix} \begin{pmatrix} 0 \\ 1 \\ 0 \end{pmatrix} = \begin{pmatrix} 0+2+0 \\ 0+5+0 \\ 0+8+0 \end{pmatrix} = \begin{pmatrix} 2 \\ 5 \\ 8 \end{pmatrix}.$

31. $(2 \quad 1 \quad 4) \begin{pmatrix} 3 \\ 5 \\ 7 \end{pmatrix} = (6+5+28) = (39).$

33. $\begin{pmatrix} 2 & 1 \\ 5 & 6 \\ 3 & 4 \end{pmatrix} \begin{pmatrix} 1 & -1 & 1 \\ -1 & 1 & -1 \end{pmatrix} = \begin{pmatrix} 2-1 & -2+1 & 2-1 \\ 5-6 & -5+6 & 5-6 \\ 3-4 & -3+4 & 3-4 \end{pmatrix} = \begin{pmatrix} 1 & -1 & 1 \\ -1 & 1 & -1 \\ -1 & 1 & -1 \end{pmatrix}.$

35. Let A be an m-by-n matrix with a_{kj} as its kjth entry ($k = 1\ldots,m$, $j = 1,\ldots,n$) and suppose O is a zero matrix such that AO is defined. Then O must have n rows and therefore must have dimensions n-by-p for some positive integer p. Since all entries of O are zero, we use the summation notation and find that the ijth entry of AO is

$$\sum_{k=1}^{m} a_{ik}0 = 0.$$

That is, all entries of AO are zero, so that AO is a zero matrix of dimensions m-by-p.

Similarly, suppose O is a zero matrix such that OA is defined. Then O must have m columns and therefore must have dimensions q-by-m for some positive integer q. Again, since all entries of O are zero, we use the summation notation and find that the ijth entry of OA is

$$\sum_{k=1}^{n} 0a_{kj} = 0.$$

That is, all entries of OA are zero, so that OA is a zero matrix of dimensions q-by-n.

37. (a) Let r_k and c_k be the kth entries of the 1-by-n row vector $\mathbf{r}$ and the n-by-1 column vector $\mathbf{c}$, respectively. Then the matrix product $\mathbf{rc}$ is a 1-by-1 matrix whose sole entry is the dot product of the single row in $\mathbf{r}$ and the single column in $\mathbf{c}$. Specifically,

$$\mathbf{rc} = (r_1 \quad \cdots \quad r_n) \begin{pmatrix} c_1 \\ \vdots \\ c_n \end{pmatrix} = (r_1c_1 + \cdots + r_nc_n).$$

Notice that the formal definition of a matrix A is "a rectangular array of numbers". The parentheses that usually accompany the matrix notation are used only for clarity as a way to delineate the boundaries of A and are not a part of the definition. Since a single scalar is "a rectangular array of numbers" written without the usual parentheses, it follows that every scalar is actually a matrix of dimension 1-by-1. Hence, the rightmost expression displayed above can be written without the parentheses to get

$$\mathbf{rc} = r_1c_1 + \cdots + r_nc_n = \mathbf{r} \cdot \mathbf{c}.$$

(b) The matrix product $\mathbf{cr}$ is an n-by-n square matrix given by

$$\mathbf{cr} = \begin{pmatrix} c_1 \\ \vdots \\ c_n \end{pmatrix} (r_1 \quad \cdots \quad r_n) = \begin{pmatrix} c_1r_1 & \cdots & c_1r_n \\ \vdots & \ddots & \vdots \\ c_nr_1 & \cdots & c_nr_n \end{pmatrix},$$

where the ith row is the 1-by-n vector $c_i\mathbf{r}$ and the jth column is the n-by-1 vector $r_j\mathbf{c}$.

39. With A as given, we compute

$$A^2 = AA = \begin{pmatrix} 1 & 1 \\ -2 & 3 \end{pmatrix} \begin{pmatrix} 1 & 1 \\ -2 & 3 \end{pmatrix} = \begin{pmatrix} -1 & 4 \\ -8 & 7 \end{pmatrix},$$

$$A^3 = AA^2 = \begin{pmatrix} 1 & 1 \\ -2 & 3 \end{pmatrix} \begin{pmatrix} -1 & 4 \\ -8 & 7 \end{pmatrix} = \begin{pmatrix} -9 & 11 \\ -22 & 13 \end{pmatrix}.$$

With $p(x) = 2x^2 - 3x + 3$, we set $x = A$ and replace the constant term 3 with the scalar product $3I$, where I is the 2-by-2 identity matrix. This gives

$$p(A) = 2A^2 - 3A + 3I = 2\begin{pmatrix} -1 & 4 \\ -8 & 7 \end{pmatrix} - 3\begin{pmatrix} 1 & 1 \\ -2 & 3 \end{pmatrix} + 3\begin{pmatrix} 1 & 0 \\ 0 & 1 \end{pmatrix}$$

$$= \begin{pmatrix} -2 & 8 \\ -16 & 14 \end{pmatrix} + \begin{pmatrix} -3 & -3 \\ 6 & -9 \end{pmatrix} + \begin{pmatrix} 3 & 0 \\ 0 & 3 \end{pmatrix} = \begin{pmatrix} -2 & 5 \\ -10 & 8 \end{pmatrix}.$$

41. With A as given, we compute

$$A^2 = AA = \begin{pmatrix} -1 & 0 & 0 \\ 0 & 2 & 0 \\ 0 & 0 & 3 \end{pmatrix} \begin{pmatrix} -1 & 0 & 0 \\ 0 & 2 & 0 \\ 0 & 0 & 3 \end{pmatrix} = \begin{pmatrix} 1 & 0 & 0 \\ 0 & 4 & 0 \\ 0 & 0 & 9 \end{pmatrix},$$

$$A^3 = AA^2 = \begin{pmatrix} -1 & 0 & 0 \\ 0 & 2 & 0 \\ 0 & 0 & 3 \end{pmatrix} \begin{pmatrix} 1 & 0 & 0 \\ 0 & 4 & 0 \\ 0 & 0 & 9 \end{pmatrix} = \begin{pmatrix} -1 & 0 & 0 \\ 0 & 8 & 0 \\ 0 & 0 & 27 \end{pmatrix}.$$

With $p(x) = 2x^2 - 3x + 3$, we set $x = A$ and replace the constant term 3 with the scalar product $3I$, where I is the 3-by-3 identity matrix. This gives

$$p(A) = 2A^2 - 3A + 3I$$

$$= 2 \begin{pmatrix} 1 & 0 & 0 \\ 0 & 4 & 0 \\ 0 & 0 & 9 \end{pmatrix} - 3 \begin{pmatrix} -1 & 0 & 0 \\ 0 & 2 & 0 \\ 0 & 0 & 3 \end{pmatrix} + 3 \begin{pmatrix} 1 & 0 & 0 \\ 0 & 1 & 0 \\ 0 & 0 & 1 \end{pmatrix}$$

$$= \begin{pmatrix} 2 & 0 & 0 \\ 0 & 8 & 0 \\ 0 & 0 & 18 \end{pmatrix} + \begin{pmatrix} 3 & 0 & 0 \\ 0 & -6 & 0 \\ 0 & 0 & -9 \end{pmatrix} + \begin{pmatrix} 3 & 0 & 0 \\ 0 & 3 & 0 \\ 0 & 0 & 3 \end{pmatrix} = \begin{pmatrix} 8 & 0 & 0 \\ 0 & 5 & 0 \\ 0 & 0 & 12 \end{pmatrix}.$$

43. With U and V as given, we compute

$$UV = \begin{pmatrix} -1 & 2 \\ 2 & -4 \end{pmatrix} \begin{pmatrix} 2 & 6 \\ 1 & 3 \end{pmatrix} = \begin{pmatrix} 0 & 0 \\ 0 & 0 \end{pmatrix},$$

$$VU = \begin{pmatrix} 2 & 6 \\ 1 & 3 \end{pmatrix} \begin{pmatrix} -1 & 2 \\ 2 & -4 \end{pmatrix} = \begin{pmatrix} 10 & -20 \\ 5 & -10 \end{pmatrix}.$$

Since the corresponding entries in the two products are not the same, the products are different matrices. As the first product shows, it is certainly possible for the product of two nonzero matrices to be a zero matrix.

45. Let A be an n-by-n matrix and let I be the n-by-n identity matrix. In the following string of equalities, we use the basic property of the identity matrix without explicitly stating so (i.e., $XI = IX = X$ for any n-by-n matrix X). We have

$$\begin{aligned} (A + I)^2 &= (A + I)(A + I) && \text{definition of the square of a matrix} \\ &= (A + I)A + (A + I)I && \text{left distributive law} \\ &= AA + IA + AI + II && \text{right distributive law} \\ &= A^2 + A + A + I && \text{definition of } A^2 \\ &= A^2 + 2A + I. \end{aligned}$$

(b) Let a be a nonzero scalar and let $A = \begin{pmatrix} a & 0 \\ 0 & 0 \end{pmatrix}$, $B = \begin{pmatrix} 0 & a \\ 0 & 0 \end{pmatrix}$. The reader can verify that $A^2 = \begin{pmatrix} a^2 & 0 \\ 0 & 0 \end{pmatrix}$, $B^2 = \begin{pmatrix} 0 & a^2 \\ 0 & 0 \end{pmatrix}$, $AB = \begin{pmatrix} 0 & a^2 \\ 0 & 0 \end{pmatrix}$ and $A + B = \begin{pmatrix} a & a \\ 0 & 0 \end{pmatrix}$. Hence,

$$(A + B)^2 = (A + B)(A + B) = \begin{pmatrix} a & a \\ 0 & 0 \end{pmatrix} \begin{pmatrix} a & a \\ 0 & 0 \end{pmatrix} = \begin{pmatrix} a^2 & a^2 \\ 0 & 0 \end{pmatrix},$$

$$A^2 + 2AB + B^2 = \begin{pmatrix} a^2 & 0 \\ 0 & 0 \end{pmatrix} + 2 \begin{pmatrix} 0 & a^2 \\ 0 & 0 \end{pmatrix} + \begin{pmatrix} 0 & a^2 \\ 0 & 0 \end{pmatrix} = \begin{pmatrix} a^2 & 3a^2 \\ 0 & 0 \end{pmatrix}.$$

Since $a \neq 0$, the upper right entries of $(A + B)^2$ and $A^2 + 2AB + B^2$ are not equal. Hence, $(A + B)^2 \neq A^2 + 2AB + B^2$. This gives infinitely many examples of the fact that $(A + B)^2 = A^2 + 2AB + B^2$ does not always hold. These are perhaps the simplest such examples but are by no means the only ones.

47. Let $X = \begin{pmatrix} 2 & a \\ 0 & 1 \end{pmatrix}$, where a is a scalar, and let $I = \begin{pmatrix} 1 & 0 \\ 0 & 1 \end{pmatrix}$. Then

$$X^2 - 3X + 2I = XX - 3X + 2I$$

$$= \begin{pmatrix} 2 & a \\ 0 & 1 \end{pmatrix}\begin{pmatrix} 2 & a \\ 0 & 1 \end{pmatrix} - 3\begin{pmatrix} 2 & a \\ 0 & 1 \end{pmatrix} + 2\begin{pmatrix} 1 & 0 \\ 0 & 1 \end{pmatrix}$$

$$= \begin{pmatrix} 4 & 3a \\ 0 & 1 \end{pmatrix} + \begin{pmatrix} -6 & -3a \\ 0 & -3 \end{pmatrix} + \begin{pmatrix} 2 & 0 \\ 0 & 2 \end{pmatrix} = \begin{pmatrix} 0 & 0 \\ 0 & 0 \end{pmatrix}.$$

49. (a) Setting $a = 1$, $b = -3$, $c = 1$, $d - 1$ in the given polynomial yields $p(x) = x^2 + 2$. These same values in the matrix A gives $A = \begin{pmatrix} 1 & -3 \\ 1 & -1 \end{pmatrix}$. In computing $p(A)$, we replace the constant term 2 with the scalar product $2I$, where I is the 2-by-2 identity matrix. We get

$$p(A) = A^2 + 2I = AA + 2I = \begin{pmatrix} 1 & -3 \\ 1 & -1 \end{pmatrix}\begin{pmatrix} 1 & -3 \\ 1 & -1 \end{pmatrix} + 2\begin{pmatrix} 1 & 0 \\ 0 & 1 \end{pmatrix}$$

$$= \begin{pmatrix} -2 & 0 \\ 0 & -2 \end{pmatrix} + \begin{pmatrix} 2 & 0 \\ 0 & 2 \end{pmatrix} = \begin{pmatrix} 0 & 0 \\ 0 & 0 \end{pmatrix} = O.$$

(b) Here, we replace the constant term $ad - bc$ in the given polynomial with the scalar product $(ad - bc)I$. The computation of $p(A)$ then gives

$$p(A) = A^2 - (a + d)A + (ad - bc)I = AA - (a + d)A + (ad - bc)I$$

$$= \begin{pmatrix} a & b \\ c & d \end{pmatrix}\begin{pmatrix} a & b \\ c & d \end{pmatrix} - (a + d)\begin{pmatrix} a & b \\ c & d \end{pmatrix} + (ad - bd)\begin{pmatrix} 1 & 0 \\ 0 & 1 \end{pmatrix}$$

$$= \begin{pmatrix} a^2 + bc & ab + bd \\ ca + dc & cb + d^2 \end{pmatrix} + \begin{pmatrix} -(a + d)a & -(a + d)b \\ -(a + d)c & -(a + d)d \end{pmatrix} + \begin{pmatrix} ad - bc & 0 \\ 0 & ad - bc \end{pmatrix}$$

$$= \begin{pmatrix} a^2 + bc - (a + d)a + ad - bc & ab + bd - (a + d)b \\ ca + dc - (a + d)c & cb + d^2 - (a + d)d + ad - bc \end{pmatrix} = \begin{pmatrix} 0 & 0 \\ 0 & 0 \end{pmatrix} = O.$$

Section 4: INVERSE MATRICES

Exercise Set 4ABC (pgs. 86-88)

1. By Formula 4.1,

$$\begin{pmatrix} 1 & 1 \\ 1 & 2 \end{pmatrix}^{-1} = \frac{1}{(1)(2)-(1)(1)} \begin{pmatrix} 2 & -1 \\ -1 & 1 \end{pmatrix} = \frac{1}{1}\begin{pmatrix} 2 & -1 \\ -1 & 1 \end{pmatrix} = \begin{pmatrix} 2 & -1 \\ -1 & 1 \end{pmatrix}.$$

3. By Formula 4.1,

$$\begin{pmatrix} 1/2 & 1/4 \\ 1/4 & 1/5 \end{pmatrix}^{-1} = \frac{1}{(1/2)(1/5)-(1/4)(1/4)} \begin{pmatrix} 1/5 & -1/4 \\ -1/4 & 1/2 \end{pmatrix} = \frac{80}{3}\begin{pmatrix} 1/5 & -1/4 \\ -1/4 & 1/2 \end{pmatrix}$$
$$= \begin{pmatrix} 16/3 & -20/3 \\ -20/3 & 40/3 \end{pmatrix}.$$

5. Applying Formula 4.1 to the given matrix A yields

$$A^{-1} = \frac{1}{8-(-3)} \begin{pmatrix} 4 & 1 \\ -3 & 2 \end{pmatrix} = \begin{pmatrix} 4/11 & 1/11 \\ -3/11 & 2/11 \end{pmatrix}.$$

With **b** as given, use statement 4.2 in the text to conclude that the solution of the matrix equation $A\mathbf{x} = \mathbf{b}$ is

$$\mathbf{x} = A^{-1}\mathbf{b} = \begin{pmatrix} 4/11 & 1/11 \\ -3/11 & 2/11 \end{pmatrix}\begin{pmatrix} 1 \\ 1 \end{pmatrix} = \begin{pmatrix} 4/11+1/11 \\ -3/11+2/11 \end{pmatrix} = \begin{pmatrix} 5/11 \\ -1/11 \end{pmatrix}.$$

7. We begin with the given matrix A on the left and the identity matrix on the right. The following three row operations result in the identity matrix on the left and A^{-1} on the right (note: a designation such as $-3r_1 + r_2 \to r_2$ indicates a row operation and means "add -3 times the first row to the second row and put the result in the second row"):

$$\begin{pmatrix} 1 & 0 & 0 & \vdots & 1 & 0 & 0 \\ 3 & 1 & 5 & \vdots & 0 & 1 & 0 \\ -2 & 0 & 1 & \vdots & 0 & 0 & 1 \end{pmatrix} \quad -3r_1+r_2 \to r_2 \quad \begin{pmatrix} 1 & 0 & 0 & \vdots & 1 & 0 & 0 \\ 0 & 1 & 5 & \vdots & -3 & 1 & 0 \\ -2 & 0 & 1 & \vdots & 0 & 0 & 1 \end{pmatrix}$$

$$2r_1+r_3 \to r_3 \quad \begin{pmatrix} 1 & 0 & 0 & \vdots & 1 & 0 & 0 \\ 0 & 1 & 5 & \vdots & -3 & 1 & 0 \\ 0 & 0 & 1 & \vdots & 2 & 0 & 1 \end{pmatrix}$$

$$-5r_3+r_2 \to r_2 \quad \begin{pmatrix} 1 & 0 & 0 & \vdots & 1 & 0 & 0 \\ 0 & 1 & 0 & \vdots & -13 & 1 & -5 \\ 0 & 0 & 1 & \vdots & 2 & 0 & 1 \end{pmatrix}.$$

Hence, $A^{-1} = \begin{pmatrix} 1 & 0 & 0 \\ -13 & 1 & -5 \\ 2 & 0 & 1 \end{pmatrix}$. The reader should check that the matrix we have called A^{-1} actually satisfies $AA^{-1} = I$.

9. We begin with the given matrix A on the left and the identity matrix on the right. The following four row operations result in a zero row, from which we conclude that A is not invertible (note: a designation such as $r_1 - r_2 \to r_2$ indicates a row operation and means "subtract the second row from the first row and put the result in the second row"):

$$\left(\begin{array}{ccc:ccc} 2 & 6 & 10 & 1 & 0 & 0 \\ 1 & 1 & 1 & 0 & 1 & 0 \\ 1 & -3 & -7 & 0 & 0 & 1 \end{array}\right) \quad \tfrac{1}{2}r_1 \to r_1 \quad \left(\begin{array}{ccc:ccc} 1 & 3 & 5 & 1/2 & 0 & 0 \\ 1 & 1 & 1 & 0 & 1 & 0 \\ 1 & -3 & -7 & 0 & 0 & 1 \end{array}\right)$$

$$r_1 - r_2 \to r_2 \quad \left(\begin{array}{ccc:ccc} 1 & 3 & 5 & 1/2 & 0 & 0 \\ 0 & 2 & 4 & 1/2 & -1 & 0 \\ 1 & -3 & -7 & 0 & 0 & 1 \end{array}\right)$$

$$r_1 - r_3 \to r_3 \quad \left(\begin{array}{ccc:ccc} 1 & 3 & 5 & 1/2 & 0 & 0 \\ 0 & 2 & 4 & 1/2 & -1 & 0 \\ 0 & 6 & 12 & 1/2 & 0 & -1 \end{array}\right)$$

$$-3r_2 + r_3 \to r_3 \quad \left(\begin{array}{ccc:ccc} 1 & 3 & 5 & 1/2 & 0 & 0 \\ 0 & 2 & 4 & 1/2 & -1 & 0 \\ 0 & 0 & 0 & -1 & 3 & -1 \end{array}\right).$$

11. By Formula 4.1, the inverse of the matrix multiplying X is

$$\begin{pmatrix} 1 & 2 \\ 5 & 6 \end{pmatrix}^{-1} = -\frac{1}{4}\begin{pmatrix} 6 & -2 \\ -5 & 1 \end{pmatrix} = \begin{pmatrix} -3/2 & 1/2 \\ 5/4 & -1/4 \end{pmatrix}.$$

Solve for X by left-multiplication of both sides of the given matrix equation by the above inverse matrix. This gives

$$X = \begin{pmatrix} 1 & 2 \\ 5 & 6 \end{pmatrix}^{-1}\begin{pmatrix} 0 & -3 & 4 \\ 1 & 2 & 0 \end{pmatrix} = \begin{pmatrix} -3/2 & 1/2 \\ 5/4 & -1/4 \end{pmatrix}\begin{pmatrix} 0 & -3 & 4 \\ 1 & 2 & 0 \end{pmatrix}$$
$$= \begin{pmatrix} 1/2 & 11/2 & -6 \\ -1/4 & -17/4 & 5 \end{pmatrix}.$$

13. Let A and B be invertible matrices of dimensions n-by-n. Then B^{-1} and A^{-1} both exist and have dimensions n-by-n. Hence, the matrix $M = B^{-1}A^{-1}$ is defined and has dimensions n-by-n. Also, AB is defined and has dimensions n-by-n, so that the product $(AB)M$ is defined. Using $AA^{-1} = I$ and $BB^{-1} = I$, associativity of matrix multiplication gives

$$(AB)M = (AB)B^{-1}A^{-1} = A(BB^{-1})A^{-1} = AIA^{-1} = AA^{-1} = I.$$

Therefore, by definition, AB is invertible and, by Theorem 4.5, M is *the* inverse of AB. That is, $(AB)^{-1} = B^{-1}A^{-1}$.

15. If A is a square matrix and I is the identity matrix of the same dimensions then $I + A + A^2$ and $I - A$ are both defined and have the same dimension, so that their product is defined. Using left and right distributivity of matrix multiplication gives

$$
\begin{aligned}
(I + A + A^2)(I - A) &= (I + A + A^2)I - (I + A + A^2)A \\
&= II + AI + A^2I - IA - AA - A^2A \\
&= I + A + A^2 - A - A^2 - A^3 = I - A^3.
\end{aligned}
$$

Hence, if $A^3 = O$ then $(I + A + A^2)(I - A) = I$ and, By Theorem 4.5, $I + A + A^2$ is the inverse of $I - A$.

17. The given matrix is a diagonal matrix which we can denote by $\mathrm{diag}(1, 2, 1)$. By Equation 4.6 in the text,

$$
\left(\mathrm{diag}(1, 2, 1)\right)^{-1} = \mathrm{diag}(1, 1/2, 1), \quad \text{or equivalently} \quad
\begin{pmatrix} 1 & 0 & 0 \\ 0 & 2 & 0 \\ 0 & 0 & 1 \end{pmatrix}^{-1} =
\begin{pmatrix} 1 & 0 & 0 \\ 0 & 1/2 & 0 \\ 0 & 0 & 1 \end{pmatrix}.
$$

19. We apply the Inversion Process to the given triangular matrix, which we shall call T. Starting with T on the left and the identity matrix on the right, we first multiply the second row by $1/2$ and then multiply the fourth row by $1/4$. This gives

$$
\left(\begin{array}{cccc:cccc}
1 & 2 & -1 & 3 & 1 & 0 & 0 & 0 \\
0 & 2 & 0 & 1 & 0 & 1 & 0 & 0 \\
0 & 0 & 1 & 0 & 0 & 0 & 1 & 0 \\
0 & 0 & 0 & 4 & 0 & 0 & 0 & 1
\end{array}\right) \cdot
\quad \rightarrow \quad
\left(\begin{array}{cccc:cccc}
1 & 2 & -1 & 3 & 1 & 0 & 0 & 0 \\
0 & 1 & 0 & 1/2 & 0 & 1/2 & 0 & 0 \\
0 & 0 & 1 & 0 & 0 & 0 & 1 & 0 \\
0 & 0 & 0 & 1 & 0 & 0 & 0 & 1/4
\end{array}\right).
$$

Now add $-1/2$ times the fourth row to the second row and put the result in the second row. Then add -3 times the fourth row to the first row and put the result in the first row. These two steps give

$$
\left(\begin{array}{cccc:cccc}
1 & 2 & -1 & 0 & 1 & 0 & 0 & -3/4 \\
0 & 1 & 0 & 0 & 0 & 1/2 & 0 & -1/8 \\
0 & 0 & 1 & 0 & 0 & 0 & 1 & 0 \\
0 & 0 & 0 & 1 & 0 & 0 & 0 & 1/4
\end{array}\right).
$$

Finally, add the third row to the first row and put the result in the first row; then add -2 times the second row to the first row and put the result in the first row. These two steps give

$$
\left(\begin{array}{cccc:cccc}
1 & 0 & 0 & 0 & 1 & -1 & 1 & -1/2 \\
0 & 1 & 0 & 0 & 0 & 1/2 & 0 & -1/8 \\
0 & 0 & 1 & 0 & 0 & 0 & 1 & 0 \\
0 & 0 & 0 & 1 & 0 & 0 & 0 & 1/4
\end{array}\right).
$$

Since the identity matrix is on the left, the matrix on the right, call it S, is the inverse of the given matrix. That is,

$$T^{-1} = \begin{pmatrix} 1 & 2 & -1 & 3 \\ 0 & 2 & 0 & 1 \\ 0 & 0 & 1 & 0 \\ 0 & 0 & 0 & 4 \end{pmatrix}^{-1} = \begin{pmatrix} 1 & -1 & 1 & -1/2 \\ 0 & 1/2 & 0 & -1/8 \\ 0 & 0 & 1 & 0 \\ 0 & 0 & 0 & 1/4 \end{pmatrix} = S.$$

The reader should check that $TS = I$.

21. Let $A = (a_{ij})$. Then

$$(A^t)^t = \left((a_{ij})^t\right)^t = (a_{ji})^t = (a_{ij}) = A.$$

23. Let $A = (a_{ij})$ and $B = (b_{ij})$. Further, assume they have the same dimensions so that $A + B$ is defined. In particular, if $A + B = C$ and $C = (c_{ij})$ then $a_{ij} + b_{ij} = c_{ij}$. Thus,

$$(A+B)^t = \left((a_{ij}) + (b_{ij})\right)^t = (a_{ij} + b_{ij})^t = (c_{ij})^t = (c_{ji}) = (a_{ji} + b_{ji}) = (a_{ji}) + (b_{ji}) = A^t + B^t.$$

25. Suppose A and B are square matrices such that $BA = I$. If A is not invertible then there is a vector $\mathbf{x} \neq \mathbf{0}$ such that $A\mathbf{x} = \mathbf{0}$. But then $BA\mathbf{x} = I\mathbf{x} = \mathbf{x} = \mathbf{0}$, a contradiction. Hence, A is invertible and A^{-1} exists. So

$$B = BI = B(AA^{-1}) = (BA)A^{-1} = IA^{-1} = A^{-1}.$$

27. The rows of the given matrix are $\mathbf{r}_1 = (\cos\theta, -\sin\theta)$ and $\mathbf{r}_2 = (\sin\theta, \cos\theta)$. First,

$$|\mathbf{r}_1| = \sqrt{\cos^2\theta + \sin^2\theta} = 1 \qquad \text{and} \qquad |\mathbf{r}_2| = \sqrt{\sin^2\theta + \cos^2\theta} = 1,$$

so that both rows are vectors of length 1. Also,

$$\mathbf{r}_1 \cdot \mathbf{r}_2 = (\cos\theta, -\sin\theta) \cdot (\sin\theta, \cos\theta) = \cos\theta\sin\theta - \sin\theta\cos\theta = 0,$$

so that the rows are perpendicular. The given matrix is therefore orthogonal.

29. The rows of the given matrix are $\mathbf{r}_1 = (1/\sqrt{3}, -1/\sqrt{2}, 1/\sqrt{6})$, $\mathbf{r}_2 = (1/\sqrt{3}, 1/\sqrt{2}, 1/\sqrt{6})$ and $\mathbf{r}_3 = (1/\sqrt{3}, 0, -2/\sqrt{6})$. First,

$$|\mathbf{r}_1| = \sqrt{(1/\sqrt{3})^2 + (-1/\sqrt{2})^2 + (1/\sqrt{6})^2} = \sqrt{1/3 + 1/2 + 1/6} = \sqrt{1} = 1,$$

$$|\mathbf{r}_2| = \sqrt{(1/\sqrt{3})^2 + (1/\sqrt{2})^2 + (1/\sqrt{6})^2} = \sqrt{1/3 + 1/2 + 1/6} = \sqrt{1} = 1,$$

$$|\mathbf{r}_3| = \sqrt{(1/\sqrt{3})^2 + (0)^2 + (-2/\sqrt{6})^2} = \sqrt{1/3 + 4/6} = \sqrt{1} = 1,$$

so that the rows are vectors of length 1. Also,

$$\mathbf{r}_1 \cdot \mathbf{r}_2 = (1/\sqrt{3}, -1/\sqrt{2}, 1/\sqrt{6}) \cdot (1/\sqrt{3}, 1/\sqrt{2}, 1/\sqrt{6}) = 1/3 - 1/2 + 1/6 = 0,$$

$$\mathbf{r}_1 \cdot \mathbf{r}_3 = (1/\sqrt{3}, -1/\sqrt{2}, 1/\sqrt{6}) \cdot (1/\sqrt{3}, 0, -2/\sqrt{6}) = 1/3 - 0 - 2/6 = 0,$$

$$\mathbf{r}_2 \cdot \mathbf{r}_3 = (1/\sqrt{3}, 1/\sqrt{2}, 1/\sqrt{6}) \cdot (1/\sqrt{3}, 0, -2/\sqrt{6}) = 1/3 + 0 - 2/6 = 0,$$

so that the rows are mutually perpendicular. The given matrix is therefore orthogonal.

31. (a) If Q is an orthogonal matrix of dimensions n-by-n and its columns $\mathbf{r}_1, \ldots, \mathbf{r}_n$ are mutually perpendicular vectors of length 1 then

$$\mathbf{r}_i \cdot \mathbf{r}_j = \begin{cases} 1, & i = j, \\ 0, & i \neq j. \end{cases}$$

Since $\mathbf{r}_i$ is the ith row of Q^t, the ijth entry of the matrix $Q^t Q$ is $\mathbf{r}_i \cdot \mathbf{r}_j$. Hence, the ijth entry of $Q^t Q$ is 1 if $i = j$ (i.e., if the entry is on the main diagonal) and is zero otherwise (i.e., if the entry is off the main diagonal). In other words, $Q^t Q = I$. By Theorem 4.5, $Q^t = Q^{-1}$.

If Q is an orthogonal matrix whose *rows* $\mathbf{r}_1, \ldots, \mathbf{r}_n$ are mutually perpendicular vectors of length 1 then the above argument can still be used with the roles of columns and rows interchanged to conclude that $QQ^t = I$. By Theorem 4.5, $Q = (Q^t)^{-1}$. Taking the transpose of both sides gives

$$Q^t = \left((Q^t)^{-1}\right)^t = \left((Q^t)^t\right)^{-1} = Q^{-1},$$

where we have used the result of Exercise 24 to obtain the third member and the result of Exercise 21 to obtain the last member.

(b) Let Q be a square matrix. If the columns of Q are mutually perpendicular and have length 1 then Q is orthogonal. Since the columns of Q^t are the rows of Q, and since the ijth entry of QQ^t is the dot product of the ith row of Q and the jth column of Q^t, it follows that the ijth entry of QQ^t is the dot product of the ith and jth rows of Q. The equation $QQ^t = I$ (from part (a)) then shows that such an entry is 1 if $i = j$ and is 0 if $i \neq j$. So the rows of Q are mutually perpendicular and have length 1.

Conversely, if the rows of Q are mutually perpendicular and have length 1 then the columns of Q^t are mutually perpendicular and have length 1. By what we have just shown, the rows of Q^t are mutually perpendicular and have length 1. So the columns of Q are mutually perpendicular and have length 1.

33. (a) Let A be a square matrix. Then A^t has the same dimensions as A and $A + A^t$ and $A - A^t$ are both defined. Using the properties of the transpose shown in Exercises 21 and 23 above,

$$(A + A^t)^t = A^t + (A^t)^t = A^t + A.$$

So $A + A^t$ is symmetric. To show that $A - A^t$ is skew-symmetric, note first that if k is a scalar then $(kA)^t = \left(k(a_{ij})\right)^t = (ka_{ij})^t = (ka_{ji}) = k(a_{ji}) = kA^t$. Applying this property with $k = -1$ and using Exercises 21 and 23 above,

$$(A - A^t)^t = (A + (-1)A^t)^t = A^t + \left((-1)A^t\right)^t = A^t + (-1)(A^t)^t = A^t + (-1)A$$
$$= A^t - A = -(A - A^t).$$

So $A - A^t$ is skew-symmetric.

(b) Let A be a square matrix. By part (a), $A + A^t$ is symmetric, $A - A^t$ is skew-symmetric. Moreover, it was shown in part (a) that the transpose is linear with respect to scalar multiplication; i.e., $(kA)^t = kA^t$ for scalars k. This property directly shows that kA is symmetric (skew-symmetric) if A is symmetric (skew-symmetric). In particular, $\frac{1}{2}(A + A^t)$ and $\frac{1}{2}(A - A^t)$ are symmetric and skew-symmetric, respectively. Therefore, since

$$\frac{1}{2}(A + A^t) + \frac{1}{2}(A - A^t) = A,$$

every square matrix is the sum of a symmetric matrix and a skew-symmetric matrix.

(c) Let A be invertible and symmetric. Then, using the property of the transpose shown in Exercise 24 in this section, $(A^{-1})^t = (A^t)^{-1} = A^{-1}$. That is, A^{-1} is symmetric. If A is invertible and skew-symmetric then $(A^{-1})^t = (A^t)^{-1} = (-A)^{-1} = -A^{-1}$, where the last equality follows from the fact that the inverse of the negative of a matrix is the negative of the inverse of the matrix. Hence, A^{-1} is skew-symmetric.

35. Let $b_0, \ldots, b_n$ be scalars, let $x_0, \ldots, x_n$ be distinct scalars and let A be the matrix

$$A = \begin{pmatrix} 1 & x_0 & \cdots & x_0^n \\ \vdots & \vdots & \ddots & \vdots \\ 1 & x_n & \cdots & x_n^n \end{pmatrix}.$$

If A is not invertible then, by Theorem 4.4, there exists a vector $\mathbf{c} \neq \mathbf{0}$ such that $A\mathbf{c} = \mathbf{0}$. Letting c_k be the kth coordinate of $\mathbf{c}$ we have

$$A\mathbf{c} = \begin{pmatrix} 1 & x_0 & \cdots & x_0^n \\ \vdots & \vdots & \ddots & \vdots \\ 1 & x_n & \cdots & x_n^n \end{pmatrix} \begin{pmatrix} c_1 \\ \vdots \\ c_n \end{pmatrix} = \begin{pmatrix} 0 \\ \vdots \\ 0 \end{pmatrix}.$$

This gives rise to the system of scalar equations

$$c_0 + c_1 x_0 + \cdots + c_n x_0^n = 0,$$

$$\vdots$$

$$c_0 + c_1 x_n + \cdots + c_n x_n^n = 0,$$

This says that the polynomial $f(x) = c_0 + c_1 x + \cdots + c_n x^n$ has the property that $f(x_k) = 0$ for $k = 0, \ldots, n$. The algebra theorem alluded to in the statement of the problem then implies that all coefficients $c_k = 0$. But this says that $\mathbf{c} = \mathbf{0}$, a contradiction. We conclude from Theorem 4.4 that A is invertible.

Now let $\mathbf{b} = \begin{pmatrix} b_0 \\ \vdots \\ b_n \end{pmatrix}$ and consider solutions of the matrix equation $A\mathbf{a} = \mathbf{b}$, where

$\mathbf{a} = \begin{pmatrix} a_0 \\ \vdots \\ a_n \end{pmatrix}$ is to be determined. Since A^{-1} exists, we can multiply this matrix equation on the left by A^{-1} to obtain the unique solution $\mathbf{a} = A^{-1}\mathbf{b}$. In other words, there is exactly one vector $\mathbf{a}$ such that $A\mathbf{a} = \mathbf{b}$. In coordinate form, there is exactly one set of real numbers $a_0, \ldots, a_n$ such that

$$\begin{pmatrix} 1 & x_0 & \cdots & x_0^n \\ \vdots & \vdots & \ddots & \vdots \\ 1 & x_n & \cdots & x_n^n \end{pmatrix} \begin{pmatrix} a_1 \\ \vdots \\ a_n \end{pmatrix} = \begin{pmatrix} b_0 \\ \vdots \\ b_n \end{pmatrix}.$$

In terms of scalar equations, there is exactly one collection of coefficients $a_0, \ldots, a_n$ such that

$$a_0 + a_1 x_0 + \cdots + a_n x_0^n = b_0,$$

$$\vdots$$

$$a_0 + a_1 x_n + \cdots + a_n x_n^n = b_n.$$

This is equivalent to saying that the polynomial $p(x) = a_0 + a_1 x + \cdots + a_n x^n$ is the only polynomial of degree at most n such that $p(x_k) = b_k$ for $k = 0, \ldots, n$.

Section 5: DETERMINANTS

Exercise Set 5A-E (pgs. 98-99)

1. Expansion of A about the first row gives

$$\det A = \begin{vmatrix} 1 & -2 & 3 \\ 3 & 1 & 4 \\ 5 & 6 & 7 \end{vmatrix} = (1)\begin{vmatrix} 1 & 4 \\ 6 & 7 \end{vmatrix} - (-2)\begin{vmatrix} 3 & 4 \\ 5 & 7 \end{vmatrix} + (3)\begin{vmatrix} 3 & 1 \\ 5 & 6 \end{vmatrix}$$

$$= (7 - 24) + 2(21 - 20) + 3(18 - 5) = -17 + 2 + 39 = 24.$$

Since $2A$ is a matrix with each of its three rows multiplied by 2, $\det(2A) = 2^3 \det A$, where the number of factors of 2 multiplying $\det A$ is the number of rows of A. Hence, $\det(2A) = 8(24) = 192$.

3. If A is an n-by-n matrix then $\det(2A) = 2^n \det A$. The reason is that we can view $2A$ as a matrix with each of its n rows multiplied by 2. For each of these rows, an extra factor of 2 multiplies $\det A$.

5. Letting A be the given matrix, we add the third row to the first row and then add -2 times the third row to the fourth row. This leaves the determinant unchanged so that

$$\det A = \begin{vmatrix} 0 & 2 & 0 & 2 \\ 0 & 1 & 2 & -1 \\ 1 & 2 & -1 & 0 \\ 0 & -5 & 2 & 1 \end{vmatrix}.$$

Since the resulting matrix has only one nonzero entry in its first column, we expand about the first column to get

$$\det A = (1)\begin{vmatrix} 2 & 0 & 2 \\ 1 & 2 & -1 \\ -5 & 2 & 1 \end{vmatrix} = \begin{vmatrix} 2 & 0 & 2 \\ 1 & 2 & -1 \\ -5 & 2 & 1 \end{vmatrix}.$$

Adding -1 times the first column to the third column leaves the determinant unchanged. This gives

$$\det A = \begin{vmatrix} 2 & 0 & 0 \\ 1 & 2 & -2 \\ -5 & 2 & 6 \end{vmatrix}.$$

Since the above matrix has only one nonzero entry in its first row, we expand about the first row to get

$$\det A = (2)\begin{vmatrix} 2 & -2 \\ 2 & 6 \end{vmatrix} = 2(12 + 4) = 32.$$

7. With A and B as given, we first compute

$$AB = \begin{pmatrix} 1 & -2 \\ 3 & 1 \end{pmatrix}\begin{pmatrix} 0 & 1 \\ 2 & -3 \end{pmatrix} = \begin{pmatrix} -4 & 7 \\ 2 & 0 \end{pmatrix},$$

$$BA = \begin{pmatrix} 0 & 1 \\ 2 & -3 \end{pmatrix}\begin{pmatrix} 1 & -2 \\ 3 & 1 \end{pmatrix} = \begin{pmatrix} 3 & 1 \\ -7 & -7 \end{pmatrix}.$$

Therefore,

$$\det A = \begin{vmatrix} 1 & -2 \\ 3 & 1 \end{vmatrix} = 1 + 6 = 7, \qquad \det B = \begin{vmatrix} 0 & 1 \\ 2 & -3 \end{vmatrix} = 0 - 2 = 2,$$

$$\det(AB) = \begin{vmatrix} -4 & 7 \\ 2 & 0 \end{vmatrix} = 0 - 14 = 14, \qquad \det(BA) = \begin{vmatrix} 3 & 1 \\ -7 & -7 \end{vmatrix} = -21 + 7 = 14.$$

Since $\det(A)\det(B) = (7)(2) = 14 = \det(AB) = \det(BA)$, the product rule for determinants holds in these cases.

9. Let I_n denote the n-by-n identity matrix. Then, by definition, $\det I_1 = \det(1) = 1$, and for $n \geq 2$, expansion about the first row gives

$$\det I_n = (1)\det I_{n-1} = \det I_{n-1}.$$

This establishes a proof by induction that every identity matrix I has determinant 1. Now let A be an invertible matrix so that A^{-1} exists. Then, with the help of the product rule for determinants, we can write

$$1 = \det I = \det(AA^{-1}) = \det A \det(A^{-1}).$$

Equality of the two outer members shows that $\det A \neq 0$, so we can divide throughout by $\det A$ to get

$$\det(A^{-1}) = \frac{1}{\det A} = (\det A)^{-1}.$$

11. Let A be the given matrix. Expansion about the first column gives

$$\det A = \begin{vmatrix} 1 & 2 & 3 & 4 \\ 0 & -1 & 5 & 6 \\ 0 & 0 & 3 & -1 \\ 0 & 0 & 0 & 4 \end{vmatrix} = (1) \begin{vmatrix} -1 & 5 & 6 \\ 0 & 3 & -1 \\ 0 & 0 & 4 \end{vmatrix} = \begin{vmatrix} -1 & 5 & 6 \\ 0 & 3 & -1 \\ 0 & 0 & 4 \end{vmatrix}.$$

Again, expansion about the first column gives

$$\det A = (-1) \begin{vmatrix} 3 & -1 \\ 0 & 4 \end{vmatrix} = -(12 - 0) = -12.$$

13. Let A be the given matrix. Expansion about the first row gives

$$\det A = \begin{vmatrix} 1 & 0 & 0 \\ 3 & 1 & 5 \\ -2 & 0 & 1 \end{vmatrix} = (1) \begin{vmatrix} 1 & 5 \\ 0 & 1 \end{vmatrix} = 1 \neq 0.$$

By Theorem 5.7, A is invertible. To compute A^{-1} using Theorem 5.8, we first compute the entries $\tilde{a}_{ij}$ of $\tilde{A}$. We have

$$\tilde{a}_{11} = \begin{vmatrix} 1 & 5 \\ 0 & 1 \end{vmatrix} = 1, \qquad \tilde{a}_{12} = -\begin{vmatrix} 3 & 5 \\ -2 & 1 \end{vmatrix} = -13, \qquad \tilde{a}_{13} = \begin{vmatrix} 3 & 1 \\ -2 & 0 \end{vmatrix} = 2,$$

$$\tilde{a}_{21} = -\begin{vmatrix} 0 & 0 \\ 0 & 1 \end{vmatrix} = 0, \qquad \tilde{a}_{22} = \begin{vmatrix} 1 & 0 \\ -2 & 1 \end{vmatrix} = 1, \qquad \tilde{a}_{23} = -\begin{vmatrix} 1 & 0 \\ -2 & 0 \end{vmatrix} = 0,$$

$$\tilde{a}_{31} = \begin{vmatrix} 0 & 0 \\ 1 & 5 \end{vmatrix} = 0, \qquad \tilde{a}_{32} = -\begin{vmatrix} 1 & 0 \\ 3 & 5 \end{vmatrix} = -5, \qquad \tilde{a}_{33} = \begin{vmatrix} 1 & 0 \\ 3 & 1 \end{vmatrix} = 1.$$

Therefore,

$$\tilde{A} = \begin{pmatrix} \tilde{a}_{11} & \tilde{a}_{12} & \tilde{a}_{13} \\ \tilde{a}_{21} & \tilde{a}_{22} & \tilde{a}_{23} \\ \tilde{a}_{31} & \tilde{a}_{32} & \tilde{a}_{33} \end{pmatrix} = \begin{pmatrix} 1 & -13 & 2 \\ 0 & 1 & 0 \\ 0 & -5 & 1 \end{pmatrix}, \qquad \text{so that} \qquad \tilde{A}^t = \begin{pmatrix} 1 & 0 & 0 \\ -13 & 1 & -5 \\ 2 & 0 & 1 \end{pmatrix}.$$

Theorem 5.8 then gives

$$A^{-1} = \frac{1}{\det A}\,\tilde{A}^t = \frac{1}{1}\begin{pmatrix} 1 & 0 & 0 \\ -13 & 1 & -5 \\ 2 & 0 & 1 \end{pmatrix} = \begin{pmatrix} 1 & 0 & 0 \\ -13 & 1 & -5 \\ 2 & 0 & 1 \end{pmatrix}.$$

The reader should check that A times the matrix identified above as A^{-1} is actually the identity matrix.

15. Let A be the given matrix. Expansion about the second row gives

$$\det A = \begin{vmatrix} 2 & 4 & 8 \\ 1 & 0 & 0 \\ 1 & -3 & -7 \end{vmatrix} = -(1)\begin{vmatrix} 4 & 8 \\ -3 & -7 \end{vmatrix} = -(-28 + 24) = 4 \neq 0.$$

By Theorem 5.7, A is invertible. To compute A^{-1} using Theorem 5.8, we first compute the entries $\tilde{a}_{ij}$ of $\tilde{A}$. We have

$$\tilde{a}_{11} = \begin{vmatrix} 0 & 0 \\ -3 & -7 \end{vmatrix} = 0, \qquad \tilde{a}_{12} = -\begin{vmatrix} 1 & 0 \\ 1 & -7 \end{vmatrix} = 7, \qquad \tilde{a}_{13} = \begin{vmatrix} 1 & 0 \\ 1 & -3 \end{vmatrix} = -3,$$

$$\tilde{a}_{21} = -\begin{vmatrix} 4 & 8 \\ -3 & -7 \end{vmatrix} = 4, \qquad \tilde{a}_{22} = \begin{vmatrix} 2 & 8 \\ 1 & -7 \end{vmatrix} = -22, \qquad \tilde{a}_{23} = -\begin{vmatrix} 2 & 4 \\ 1 & -3 \end{vmatrix} = 10,$$

$$\tilde{a}_{31} = \begin{vmatrix} 4 & 8 \\ 0 & 0 \end{vmatrix} = 0, \qquad \tilde{a}_{32} = -\begin{vmatrix} 2 & 8 \\ 1 & 0 \end{vmatrix} = 8, \qquad \tilde{a}_{33} = \begin{vmatrix} 2 & 4 \\ 1 & 0 \end{vmatrix} = -4.$$

Therefore,

$$\tilde{A} = \begin{pmatrix} \tilde{a}_{11} & \tilde{a}_{12} & \tilde{a}_{13} \\ \tilde{a}_{21} & \tilde{a}_{22} & \tilde{a}_{23} \\ \tilde{a}_{31} & \tilde{a}_{32} & \tilde{a}_{33} \end{pmatrix} = \begin{pmatrix} 0 & 7 & -3 \\ 4 & -22 & 10 \\ 0 & 8 & -4 \end{pmatrix}, \qquad \text{so that} \qquad \tilde{A}^t = \begin{pmatrix} 0 & 4 & 0 \\ 7 & -22 & 8 \\ -3 & 10 & -4 \end{pmatrix}.$$

Theorem 5.8 then gives

$$A^{-1} = \frac{1}{\det A}\,\tilde{A}^t = \frac{1}{4}\begin{pmatrix} 0 & 4 & 0 \\ 7 & -22 & 8 \\ -3 & 10 & -4 \end{pmatrix} = \begin{pmatrix} 0 & 1 & 0 \\ 7/4 & -11/2 & 2 \\ -3/4 & 5/2 & -1 \end{pmatrix}.$$

The reader should check that A times the matrix identified above as A^{-1} is actually the identity matrix.

17. Let A be the given matrix and observe that A is upper triangular. By Exercise 12 of this section, $\det A$ is the product of its diagonal entries. So $\det A = (1)(0)(3) = 0$. By Theorem 5.5, A is not invertible.

19. Let A be the given matrix and observe that A is upper triangular. By Exercise 12 of this section, $\det A$ is the product of its diagonal entries. So $\det A = (1)(2)(1)(4) = 8 \neq 0$. By Theorem 5.7, A is invertible.

To compute A^{-1} using Theorem 5.8, we first compute the entries $\tilde{a}_{ij}$ of $\tilde{A}$. In order to facilitate these computations we will use the fact that thirteen of the sixteen entries of $\tilde{A}$ are determinants of upper triangular matrices. The three exceptional cases are $\tilde{a}_{31}$, $\tilde{a}_{41}$ and $\tilde{a}_{42}$, which we compute first.

For $\tilde{a}_{31}$, interchange the first two columns and then observe that the resulting matrix is upper triangular. Hence,

$$\tilde{a}_{31} = \begin{vmatrix} 2 & -1 & 3 \\ 2 & 0 & 1 \\ 0 & 0 & 4 \end{vmatrix} = -\begin{vmatrix} -1 & 2 & 3 \\ 0 & 2 & 1 \\ 0 & 0 & 4 \end{vmatrix} = -(-1)(2)(4) = 8.$$

For $\tilde{a}_{41}$, add -1 times the first row to the second row and then expand about the first column. This gives

$$\tilde{a}_{41} = -\begin{vmatrix} 2 & -1 & 3 \\ 2 & 0 & 1 \\ 0 & 1 & 1 \end{vmatrix} = -\begin{vmatrix} 2 & -1 & 3 \\ 0 & 1 & -2 \\ 0 & 1 & 1 \end{vmatrix} = -(2)\begin{vmatrix} 1 & -2 \\ 1 & 1 \end{vmatrix} = -2(1+2) = -6.$$

For $\tilde{a}_{42}$, expand about the first column to get

$$\tilde{a}_{42} = \begin{vmatrix} 1 & -1 & 3 \\ 0 & 0 & 1 \\ 0 & 1 & 1 \end{vmatrix} = (1)\begin{vmatrix} 0 & 1 \\ 1 & 1 \end{vmatrix} = -1.$$

The remaining thirteen entries are as follows:

$$\tilde{a}_{11} = \begin{vmatrix} 2 & 0 & 1 \\ 0 & 1 & 1 \\ 0 & 0 & 4 \end{vmatrix} = 8, \qquad \tilde{a}_{12} = -\begin{vmatrix} 0 & 0 & 1 \\ 0 & 1 & 1 \\ 0 & 0 & 4 \end{vmatrix} = 0, \qquad \tilde{a}_{13} = \begin{vmatrix} 0 & 2 & 1 \\ 0 & 0 & 1 \\ 0 & 0 & 4 \end{vmatrix} = 0,$$

$$\tilde{a}_{14} = -\begin{vmatrix} 0 & 2 & 0 \\ 0 & 0 & 1 \\ 0 & 0 & 0 \end{vmatrix} = 0, \qquad \tilde{a}_{21} = -\begin{vmatrix} 2 & -1 & 3 \\ 0 & 1 & 1 \\ 0 & 0 & 4 \end{vmatrix} = -8, \qquad \tilde{a}_{22} = \begin{vmatrix} 1 & -1 & 3 \\ 0 & 1 & 1 \\ 0 & 0 & 4 \end{vmatrix} = 4,$$

$$\tilde{a}_{23} = -\begin{vmatrix} 1 & 2 & 3 \\ 0 & 0 & 1 \\ 0 & 0 & 4 \end{vmatrix} = 0, \qquad \tilde{a}_{24} = \begin{vmatrix} 1 & 2 & -1 \\ 0 & 0 & 1 \\ 0 & 0 & 0 \end{vmatrix} = 0, \qquad \tilde{a}_{32} = -\begin{vmatrix} 1 & -1 & 3 \\ 0 & 0 & 1 \\ 0 & 0 & 4 \end{vmatrix} = 0,$$

$$\tilde{a}_{33} = \begin{vmatrix} 1 & 2 & 3 \\ 0 & 2 & 1 \\ 0 & 0 & 4 \end{vmatrix} = 8, \qquad \tilde{a}_{34} = -\begin{vmatrix} 1 & 2 & -1 \\ 0 & 2 & 0 \\ 0 & 0 & 0 \end{vmatrix} = 0, \qquad \tilde{a}_{43} = -\begin{vmatrix} 1 & 2 & 3 \\ 0 & 2 & 1 \\ 0 & 0 & 1 \end{vmatrix} = -2,$$

$$\tilde{a}_{44} = \begin{vmatrix} 1 & 2 & -1 \\ 0 & 2 & 0 \\ 0 & 0 & 1 \end{vmatrix} = 2.$$

Therefore,

$$\tilde{A} = \begin{pmatrix} \tilde{a}_{11} & \tilde{a}_{12} & \tilde{a}_{13} & \tilde{a}_{14} \\ \tilde{a}_{21} & \tilde{a}_{22} & \tilde{a}_{23} & \tilde{a}_{24} \\ \tilde{a}_{31} & \tilde{a}_{32} & \tilde{a}_{33} & \tilde{a}_{34} \\ \tilde{a}_{41} & \tilde{a}_{42} & \tilde{a}_{43} & \tilde{a}_{44} \end{pmatrix} = \begin{pmatrix} 8 & 0 & 0 & 0 \\ -8 & 4 & 0 & 0 \\ 8 & 0 & 8 & 0 \\ -6 & -1 & -2 & 2 \end{pmatrix} \quad \text{and} \quad \tilde{A}^t = \begin{pmatrix} 8 & -8 & 8 & -6 \\ 0 & 4 & 0 & -1 \\ 0 & 0 & 8 & -2 \\ 0 & 0 & 0 & 2 \end{pmatrix}.$$

Theorem 5.8 then gives

$$A^{-1} = \frac{1}{\det A}\,\tilde{A}^t = \frac{1}{8}\begin{pmatrix} 8 & -8 & 8 & -6 \\ 0 & 4 & 0 & -1 \\ 0 & 0 & 8 & -2 \\ 0 & 0 & 0 & 2 \end{pmatrix} = \begin{pmatrix} 1 & -1 & 1 & -3/4 \\ 0 & 1/2 & 0 & -1/8 \\ 0 & 0 & 1 & -1/4 \\ 0 & 0 & 0 & 1/4 \end{pmatrix}.$$

The reader should check that A times the matrix identified above as A^{-1} is actually the identity matrix.

21. For each scalar t, define the matrix A_t by

$$A_t = \begin{pmatrix} 1-t & 2 & 0 \\ 0 & 2-t & 5 \\ 0 & 0 & 3-t \end{pmatrix}.$$

This is an upper triangular matrix so that the determinant of A_t is the product of its diagonal entries. That is,

$$\det A_t = (1-t)(2-t)(3-t).$$

Hence, $\det A_t = 0$ only if $t = 1$, $t = 2$ or $t = 3$. By Theorem 5.7, A fails to have an inverse only for these values of t.

23. For each scalar t, define the matrix A_t by

$$A_t = \begin{pmatrix} 2 & 4 & 8 \\ 1 & 0 & 0 \\ 1 & 2 & t \end{pmatrix}.$$

Expansion about the second row gives

$$\det A_t = -(1)\begin{vmatrix} 4 & 8 \\ 2 & t \end{vmatrix} = -(4t - 16) = -4(t-4).$$

Hence, $\det A_t = 0$ only if $t = 4$. By Theorem 5.7, A fails to have an inverse only for $t = 4$.

25. Let $\mathbf{u} = (u_1, u_2, u_3)$ and $\mathbf{v} = (v_1, v_2, v_3)$ be vectors in $\mathcal{R}^3$ and let A be the array $A = \begin{pmatrix} \mathbf{i} & \mathbf{j} & \mathbf{k} \\ u_1 & u_2 & u_3 \\ v_1 & v_2 & v_3 \end{pmatrix}$. Even though the entries making up the array defining A are not all the same kind of object, we can still regard A as a matrix and formally compute $\det A$ by using all of the theorems in this section. In particular, we can formally expand A about the first row to get

$$\det A = \begin{vmatrix} \mathbf{i} & \mathbf{j} & \mathbf{k} \\ u_1 & u_2 & u_3 \\ v_1 & v_2 & v_3 \end{vmatrix} = \mathbf{i}\begin{vmatrix} u_2 & u_3 \\ v_2 & v_3 \end{vmatrix} - \mathbf{j}\begin{vmatrix} u_1 & u_3 \\ v_1 & v_3 \end{vmatrix} + \mathbf{k}\begin{vmatrix} u_1 & u_2 \\ v_1 & v_2 \end{vmatrix}$$

$$= \mathbf{i}(u_2v_3 - u_3v_2) - \mathbf{j}(u_1v_3 - u_3v_1) + \mathbf{k}(u_1v_2 - u_2v_1)$$

$$= (u_2v_3 - u_3v_2)\mathbf{i} + (u_3v_1 - u_1v_3)\mathbf{j} + (u_1v_2 - u_2v_1)\mathbf{k}$$

$$= (u_2v_3 - u_3v_2,\ u_3v_1 - u_1v_3,\ u_1v_2 - u_2v_1).$$

The last vector above is the right side of Equation 6.1 in Chapter 1, which is the definition of the cross product of $\mathbf{u}$ and $\mathbf{v}$. Hence,

$$\det \begin{pmatrix} \mathbf{i} & \mathbf{j} & \mathbf{k} \\ u_1 & u_2 & u_3 \\ v_1 & v_2 & v_3 \end{pmatrix} = \mathbf{u} \times \mathbf{v}.$$

27. Let A be an m-by-m matrix and let B be an n-by-n matrix. For each ordered pair i, j of positive integers, let $O_{i,j}$ be the zero matrix of dimensions i-by-j, and define the $(m+n)$-by-$(m+n)$ matrix C by

$$C = \begin{pmatrix} A & O_{m,n} \\ O_{n,m} & B \end{pmatrix}.$$

Further, for each positive integer k, let I_k be the identity matrix of dimensions k-by-k and define the $(m+n)$-by-$(m+n)$ matrices A_1 and B_1 by

$$A_1 = \begin{pmatrix} A & O_{m,n} \\ O_{n,m} & I_n \end{pmatrix} \quad \text{and} \quad B_1 = \begin{pmatrix} I_m & O_{m,n} \\ O_{n,m} & B \end{pmatrix}.$$

Taking care to attend to the order of multiplication of matrices, we find that

$$A_1 B_1 = \begin{pmatrix} A & O_{m,n} \\ O_{n,m} & I_n \end{pmatrix} \begin{pmatrix} I_m & O_{m,n} \\ O_{n,m} & B \end{pmatrix} = \begin{pmatrix} AI_m + O_{m,n}O_{n,m} & AO_{m,n} + O_{m,n}B \\ O_{n,m}I_m + I_nO_{n,m} & O_{n,m}O_{m,n} + I_nB \end{pmatrix}$$
$$= \begin{pmatrix} A + O_{m,m} & O_{m,n} + O_{m,n} \\ O_{n,m} + O_{n,m} & O_{n,n} + B \end{pmatrix} = \begin{pmatrix} A & O_{m,n} \\ O_{n,m} & B \end{pmatrix} = C.$$

Therefore, the product rule for determinants gives

$$\det C = \det(A_1 B_1) = (\det A_1)(\det B_1).$$

We can find $\det A_1$ by n repeated expansions about the last row, and we can find $\det B_1$ by m repeated expansions about the first row. Specifically,

$$\det A_1 = \begin{vmatrix} A & O_{m,n} \\ O_{n,m} & I_n \end{vmatrix} = \begin{vmatrix} A & O_{m,n-1} \\ O_{n-1,m} & I_{n-1} \end{vmatrix} = \cdots = \begin{vmatrix} A & O_{m,1} \\ O_{1,m} & 1 \end{vmatrix} = |A| = \det A$$

and

$$\det B_1 = \begin{vmatrix} I_m & O_{m,n} \\ O_{n,m} & B \end{vmatrix} = \begin{vmatrix} I_{m-1} & O_{m-1,n} \\ O_{n,m-1} & B \end{vmatrix} = \cdots = \begin{vmatrix} 1 & O_{1,n} \\ O_{n,1} & B \end{vmatrix} = |B| = \det B.$$

Hence, $\det C = (\det A_1)(\det B_1) = (\det A)(\det B)$.

(pgs. 99-101)

1. Since A and B have different dimensions, their sum $A + B$ is not defined.
3. We have

$$B + 2C = \begin{pmatrix} 4 & 1 & 0 \\ -2 & -1 & 2 \end{pmatrix} + 2 \begin{pmatrix} 1 & -3 & -1 \\ 2 & 1 & 0 \end{pmatrix} = \begin{pmatrix} 4 & 1 & 0 \\ -2 & -1 & 2 \end{pmatrix} + \begin{pmatrix} 2 & -6 & -2 \\ 4 & 2 & 0 \end{pmatrix}$$
$$= \begin{pmatrix} 6 & -5 & -2 \\ 2 & 1 & 2 \end{pmatrix}.$$

5. Since E has dimensions 3-by-2 and B has dimensions 2-by-3, their product EB is defined and has dimensions 3-by-3. But A has dimensions 2-by-2, so that the sum $A + EB$ is not defined.
7. The second column of D is a zero column. Consequently, D is not invertible and D^{-1} does not exist. So, the product CD^{-1} does not exist.
9. Subtracting AX from both sides of the given equation gives $2B = CX - AX$ which, by right distributivity of matrix multiplication, can be written as $2B = (C - A)X$. For this to have a unique solution, $C - A$ must be invertible. In that case, left-multiply both sides of the equation by $(C - A)^{-1}$ and use commutativity of scalar-matrix multiplication to obtain

$$X = (C - A)^{-1}(2B) = 2(C - A)^{-1}B.$$

11. In order for $AX = 3A$ to have a unique solution, A must be invertible. In that case, left-multiply both sides of given equation by A^{-1} and use commutativity of scalar-matrix multiplication to obtain
$$X = A^{-1}(3A) = 3AA^{-1} = 3I.$$

13. The given statement is sometimes true and sometimes false. To see that it is sometimes true, let a, b and c be scalars with $ac \neq 0$ and define the 2-by-2 matrices A and B by $A = \begin{pmatrix} a & b \\ c & bc/a \end{pmatrix}$ and $B = \begin{pmatrix} -b/a & b/c \\ 1 & -a/c \end{pmatrix}$. The reader can check that $AB = BA = O$.
To see that it is sometimes false, let a and b be nonzero scalars and define A and B by $A = \begin{pmatrix} a & 1/(2b) \\ b & 1/(2a) \end{pmatrix}$, $B = \begin{pmatrix} -1/(2a) & -1/(2b) \\ b & a \end{pmatrix}$. The reader can check that

$$AB = O \qquad \text{and} \qquad BA = \begin{pmatrix} -1 & -1/(4ab) \\ 2ab & 1 \end{pmatrix} \neq O.$$

15. The given statement is sometimes true and sometimes false. To see that it is sometimes true, let A be invertible. Then $AB = A$ can be left-multiplied by A^{-1} to get $B = I$. Hence, $BA = IA = A$. To see that it is sometimes false, let a be a nonzero scalar and define the 2-by-2 matrices A and B by $A = \begin{pmatrix} a & 0 \\ 0 & 0 \end{pmatrix}$ and $B = \begin{pmatrix} 1 & 0 \\ a & a \end{pmatrix}$. The reader can check that

$$AB = A \qquad \text{and} \qquad BA = \begin{pmatrix} a & 0 \\ a^2 & 0 \end{pmatrix} \neq A.$$

17. If $A = O$ then every entry of A is 0 and therefore the ijth entry of A^2 is the dot product of two zero vectors. Hence, $A^2 = O$ and the given statement is always true.

19. The given statement is sometimes true and sometimes false. To see that it is sometimes true, take $A = I$. To see that it is sometimes false, let b be a scalar and let A be the nonidentity matrix $A = \begin{pmatrix} -1 & b \\ 0 & 1 \end{pmatrix}$. The reader can check that $A^2 = I$, so that $A^2 = I$ does not imply $A = I$.

21. Let $A = \begin{pmatrix} 1 & 0 & 1 \\ 0 & 1 & 1 \\ 0 & 0 & 0 \end{pmatrix}$ and let $\mathbf{x} = \begin{pmatrix} x \\ y \\ z \end{pmatrix}$. Notice that A is reduced.

(a) The homogeneous system $A\mathbf{x} = \mathbf{0}$ is equivalent to the system of equations

$$x + z = 0,$$
$$y + z = 0,$$
$$0 = 0.$$

Since z is the only nonleading variable, set $z = t$ and deduce from the above system that $x = y = -t$. The desired solution set is therefore

$$\mathbf{x}_h = \begin{pmatrix} -t \\ -t \\ t \end{pmatrix} = t \begin{pmatrix} -1 \\ -1 \\ 1 \end{pmatrix},$$

where the subscript h on the solution vector denotes the homogeneous solution.

(b) The system $A\mathbf{x} = \begin{pmatrix} 1 \\ 2 \\ 0 \end{pmatrix}$ is equivalent to the system of equations

$$x + z = 1,$$
$$y + z = 2,$$
$$0 = 0.$$

One solution of this system is $\mathbf{x}_p = \begin{pmatrix} 1 \\ 2 \\ 0 \end{pmatrix}$ (found by setting $z = 0$ in the above system).

We can use the solution from part (a) to conclude that the general solution of the above system is

$$\mathbf{x} = \mathbf{x}_h + \mathbf{x}_p = t \begin{pmatrix} -1 \\ -1 \\ 1 \end{pmatrix} + \begin{pmatrix} 1 \\ 2 \\ 0 \end{pmatrix}.$$

(c) The system $A\mathbf{x} = \begin{pmatrix} 0 \\ 1 \\ 2 \end{pmatrix}$ is equivalent to the system of equations

$$x + z = 0,$$
$$y + z = 1,$$
$$0 = 2,$$

which we observe is inconsistent because the last equation is false. That is, the system has no solution.

In general, if $\mathbf{b} = \begin{pmatrix} b_1 \\ b_2 \\ b_3 \end{pmatrix}$ then the equation $A\mathbf{x} = \mathbf{b}$ has solutions precisely when the system of equations

$$x + z = b_1,$$
$$y + z = b_2,$$
$$0 = b_3$$

has solutions. The third equation forces $b_3 = 0$. Setting $z = 0$ in the remaining two equations gives $x = b_1$ and $y = b_2$, so that one solution of the above system is $\mathbf{x}_p = \begin{pmatrix} b_1 \\ b_2 \\ 0 \end{pmatrix}$.

That is, the equation $A\mathbf{x} = \mathbf{b}$ has a solution if, and only if, the third coordinate of $\mathbf{b}$ is zero.

23. The given system in matrix form is

$$\begin{pmatrix} 2 & 1 & -1 \\ 0 & 3 & -1 \\ 1 & -1 & 0 \end{pmatrix} \begin{pmatrix} x \\ y \\ z \end{pmatrix} = \begin{pmatrix} 0 \\ b \\ 1 \end{pmatrix}.$$

Add -2 times the third row to the first row and put the result in the first row to get

$$\begin{pmatrix} 0 & 3 & -1 \\ 0 & 3 & -1 \\ 1 & -1 & 0 \end{pmatrix} \begin{pmatrix} x \\ y \\ z \end{pmatrix} = \begin{pmatrix} -2 \\ b \\ 1 \end{pmatrix}.$$

Add -1 times the first row to the second row and put the result in the second row; then divide the first row by 3. This results in the system

$$\begin{pmatrix} 0 & 1 & -1/3 \\ 0 & 0 & 0 \\ 1 & -1 & 0 \end{pmatrix} \begin{pmatrix} x \\ y \\ z \end{pmatrix} = \begin{pmatrix} -2/3 \\ b+2 \\ 1 \end{pmatrix},$$

where the coefficient matrix is reduced. The equivalent system of equations is

$$y - \frac{1}{3}z = -\frac{2}{3},$$
$$0 = b + 2,$$
$$x - y = 1.$$

The second equation is false unless $b = -2$. In this case, y is the only nonleading variable and therefore one solution can be found by setting $y = 0$ in the above equations. The corresponding values of x and y are $x = 1$ and $z = 2$. In other words, the given system has a solution only if $b = -2$, and in this case, one solution is $x = 1$, $y = 0$, $z = 2$.

25. With $\mathbf{v}$, $\mathbf{a}$ and $\mathbf{b}$ as given, we seek scalars x and y such that $\mathbf{v} = x\mathbf{a} + y\mathbf{b}$. That is, we want x and y to satisfy

$$2\mathbf{i} + 3\mathbf{j} = x(2\mathbf{i} - \mathbf{j}) + y(2\mathbf{i} + \mathbf{j}) = (2x + 2y)\mathbf{i} + (-x + y)\mathbf{j}.$$

Since **i** and **j** are linearly independent, the coefficient of **i** on the right must equal to the coefficient of **i** on the left, and the coefficient of **j** on the right must equal the coefficient of **j** on the left. This gives the system

$$2x + 2y = 2,$$
$$-x + y = 3.$$

Adding 1/2 times the first equation to the second equation gives $2y = 4$, so that $y = 2$. Inserting this into the second equation gives $-x + 2 = 3$, so that $x = -1$. Hence,

$$\mathbf{v} = -\mathbf{a} + 2\mathbf{b}, \qquad \text{or} \qquad 2\mathbf{i} + 3\mathbf{j} = -(2\mathbf{i} - \mathbf{j}) + 2(2\mathbf{i} + \mathbf{j}).$$

27. With **v**, **a**, **b**, **c** as given, we seek scalars x, y and z such that $\mathbf{v} = x\mathbf{a} + y\mathbf{b} + z\mathbf{c}$. That is, we want x, y and z to satisfy

$$(3, -1, 0, -1) = x(2, -1, 3, 2) + y(-1, 1, 1, -3) + z(1, 1, 9, -5)$$
$$= (2x - y + z, -x + y + z, 3x + y + 9z, 2x - 3y - 5z).$$

This implies the system

$$2x - y + z = 3,$$
$$-x + y + z = -1,$$
$$3x + y + 9z = 0,$$
$$2x - 3y - 5z = -1.$$

Adding the first equation to the second and third equations gives $x + 2z = 2$ and $5x + 10z = 3$. Multiplying the first of these equations by 5 gives $5x + 10z = 10$, which is inconsistent with $5x + 10z = 3$. It follows that the above four-equation system is also inconsistent. Therefore, **v** is not a linear combination of **a**, **b** and **c**.

29. We have $\begin{pmatrix} 3 & 5 \\ -2 & 1 \end{pmatrix}^{-1} = \frac{1}{13} \begin{pmatrix} 1 & -5 \\ 2 & 3 \end{pmatrix} = \begin{pmatrix} 1/13 & -5/13 \\ 2/13 & 3/13 \end{pmatrix}.$

31. Let A be the given matrix. We first compute $\det A$ by expansion about the first column. We have

$$\det A = \begin{vmatrix} 5 & -2 & 3 \\ 4 & -3 & 2 \\ -3 & 4 & -1 \end{vmatrix} = (5) \begin{vmatrix} -3 & 2 \\ 4 & -1 \end{vmatrix} - (4) \begin{vmatrix} -2 & 3 \\ 4 & -1 \end{vmatrix} + (-3) \begin{vmatrix} -2 & 3 \\ -3 & 2 \end{vmatrix}$$
$$= 5(3 - 8) - 4(2 - 12) - 3(-4 + 9) = -25 + 40 - 15 = 0.$$

Hence, A is not invertible.

33. Let A be the given matrix. We first compute $\det A$ by expanding about the third row to get

$$\det A = \begin{vmatrix} -1 & 0 & 3 & 2 \\ -2 & 1 & 6 & 4 \\ 0 & 1 & 0 & 0 \\ 3 & -2 & 1 & 3 \end{vmatrix} = -(1) \begin{vmatrix} -1 & 3 & 2 \\ -2 & 6 & 4 \\ 3 & 1 & 3 \end{vmatrix} = 0,$$

where the last equality follows because the first and second rows of the rightmost matrix are proportional. Hence, A is not invertible.

35. Let A be the given matrix. We will use Theorem 5.8 to compute A^{-1}. First, we compute $\det A$. The following four elementary operations leave the determinant unchanged and produce an upper triangular matrix: add -1 times the first row to the second row; add the second row to the third row; add -1 times the fifth column to the fourth column; and add $-1/3$ times the third row to the fourth row. This gives

$$\det A = \begin{vmatrix} 1 & 2 & 0 & 0 & 0 \\ 1 & 1 & 2 & 0 & 0 \\ 0 & 1 & 1 & 2 & 0 \\ 0 & 0 & 1 & 1 & 2 \\ 0 & 0 & 0 & 1 & 1 \end{vmatrix} = \begin{vmatrix} 1 & 2 & 0 & 0 & 0 \\ 0 & -1 & 2 & 0 & 0 \\ 0 & 0 & 3 & 2 & 0 \\ 0 & 0 & 0 & -5/3 & 2 \\ 0 & 0 & 0 & 0 & 1 \end{vmatrix}.$$

The desired determinant is the product of the diagonal entries of the matrix on the right. That is, $\det A = (1)(-1)(3)(-5/3)(1) = 5$ and therefore A is invertible.

Next, we compute the twenty-five entries $\tilde{a}_{ij}$ of $\tilde{A}$. These computations are straightforward but involve the evaluation of the determinants of twenty-five 4-by-4 matrices. As such, we will leave the details to the reader and present only the resulting values of the $\tilde{a}_{ij}$s:

$$\tilde{a}_{11} = -1, \quad \tilde{a}_{12} = 3, \quad \tilde{a}_{13} = -1, \quad \tilde{a}_{14} = -1, \quad \tilde{a}_{15} = 1,$$
$$\tilde{a}_{21} = 6, \quad \tilde{a}_{22} = -3, \quad \tilde{a}_{23} = 1, \quad \tilde{a}_{24} = 1, \quad \tilde{a}_{25} = -1,$$
$$\tilde{a}_{31} = -4, \quad \tilde{a}_{32} = 2, \quad \tilde{a}_{33} = 1, \quad \tilde{a}_{34} = 1, \quad \tilde{a}_{35} = -1,$$
$$\tilde{a}_{41} = -8, \quad \tilde{a}_{42} = 4, \quad \tilde{a}_{43} = 2, \quad \tilde{a}_{44} = -3, \quad \tilde{a}_{45} = 3,$$
$$\tilde{a}_{51} = 16, \quad \tilde{a}_{52} = -8, \quad \tilde{a}_{53} = -4, \quad \tilde{a}_{54} = 6, \quad \tilde{a}_{55} = -1.$$

Therefore,

$$\tilde{A} = \begin{pmatrix} -1 & 3 & -1 & -1 & 1 \\ 6 & -3 & 1 & 1 & -1 \\ -4 & 2 & 1 & 1 & -1 \\ -8 & 4 & 2 & -3 & 3 \\ 16 & -8 & -4 & 6 & -1 \end{pmatrix}, \quad \text{so that} \quad \tilde{A}^t = \begin{pmatrix} -1 & 6 & -4 & -8 & 16 \\ 3 & -3 & 2 & 4 & -8 \\ -1 & 1 & 1 & 2 & -4 \\ -1 & 1 & 1 & -3 & 6 \\ 1 & -1 & -1 & 3 & -1 \end{pmatrix}.$$

Hence, by Theorem 5.8,

$$A^{-1} = \frac{1}{\det A} \tilde{A}^t = \begin{pmatrix} -1/5 & 6/5 & -4/5 & -8/5 & 16/5 \\ 3/5 & -3/5 & 2/5 & 4/5 & -8/5 \\ -1/5 & 1/5 & 1/5 & 2/5 & -4/5 \\ -1/5 & 1/5 & 1/5 & -3/5 & 6/5 \\ 1/5 & -1/5 & -1/5 & 3/5 & -1/5 \end{pmatrix}.$$

The reader should check that A times the matrix identified above as A^{-1} is actually the identity matrix.

37. Let $E = \text{diag}(e_1, \ldots, e_n)$ and $F = \text{diag}(f_1, \ldots, f_m)$ be diagonal matrices of the same size. The equation $EF = FE$ follows directly from the commutative property of the scalar entries of E and F; namely

$$EF = \big(\text{diag}(e_1, \ldots, e_n)\big)\big(\text{diag}(f_1, \ldots, f_m)\big) = \text{diag}(e_1 f_1, \ldots, e_n f_n)$$
$$= \text{diag}(f_1 e_1, \ldots, f_n e_n) = \big(\text{diag}(f_1, \ldots, f_m)\big)\big(\text{diag}(e_1, \ldots, e_n)\big) = FE.$$

(b) With A as given in Exercise 31 in this section and $D = \text{diag}(a, b, c)$, we have

$$DA = \big(\text{diag}(a, b, c)\big)A = \begin{pmatrix} a & 0 & 0 \\ 0 & b & 0 \\ 0 & 0 & c \end{pmatrix} \begin{pmatrix} 5 & -2 & 3 \\ 4 & -3 & 2 \\ -3 & 4 & -1 \end{pmatrix} = \begin{pmatrix} 5a & -2a & 3a \\ 4b & -3b & 2b \\ -3c & 4c & -c \end{pmatrix},$$

$$AD = A\big(\text{diag}(a, b, c)\big) = \begin{pmatrix} 5 & -2 & 3 \\ 4 & -3 & 2 \\ -3 & 4 & -1 \end{pmatrix} \begin{pmatrix} a & 0 & 0 \\ 0 & b & 0 \\ 0 & 0 & c \end{pmatrix} = \begin{pmatrix} 5a & -2b & 3c \\ 4a & -3b & 2c \\ -3a & 4b & -c \end{pmatrix}.$$

In words, multiplying a matrix A on the left by a diagonal matrix D results in a matrix DA whose jth column is the jth column of A multiplied by the jth diagonal entry of D; and multiplying a matrix A on the right by a diagonal matrix D results in a matrix AD whose ith row is the ith row of A multiplied by the ith diagonal entry of D.

(c) Let $D = \text{diag}(a, b, c)$ be given and suppose $B = \begin{pmatrix} r & s & t \\ u & v & w \\ x & y & z \end{pmatrix}$ is a matrix such that $DB = BD$. Using the description given in part (b) we have

$$DB = \begin{pmatrix} ar & bs & ct \\ au & bv & cw \\ ax & by & cz \end{pmatrix} = \begin{pmatrix} ra & sa & ta \\ ub & rb & wb \\ xc & yc & zc \end{pmatrix} = BD.$$

Since the diagonal entries of the left matrix are the same as the diagonal entries of the right matrix (e.g., $ar = ra$ for all r), equality of the two matrices does nothing to restrict the values of the diagonal entries r, v, z of B. However, the above equation does give rise to the following six equation system that relates the corresponding nondiagonal entries:

$$bs = sa, \qquad ct = ta, \qquad au = ub, \qquad cw = wb, \qquad ax = xc, \qquad by = yc.$$

which can be written as

$$s(b - a) = 0, \quad t(c - a) = 0, \quad u(a - b) = 0, \quad w(c - b) = 0, \quad x(a - c) = 0, \quad y(b - c) = 0.$$

The first thing we notice is that if a, b, c are all different then none of the parenthetical factors are zero, so that each equation implies that the nonparenthetical factor is zero. That is $s = t = u = w = x = y = 0$. Hence,

$$B = \begin{pmatrix} r & 0 & 0 \\ 0 & v & 0 \\ 0 & 0 & z \end{pmatrix} = \text{diag}(r, v, z).$$

The second thing we notice is that if $a = b = c$ then each of the above equations leaves its corresponding nonparenthetical factor undetermined. That is, the nondiagonal entries of B are arbitrary (as well as the diagonal entries) and therefore B itself is arbitrary. This can also be seen by writing D as $D = \text{diag}(a, a, a) = a\,\text{diag}(1, 1, 1) = aI$ and observing that if B is any 3-by-3 matrix then $DB = (aI)B = a(IB) = a(BI) = B(aI) = BD$.

The third thing we notice is that if $a = b \neq c$ then the first and third equation leave s and u undetermined while the nonparenthetical factors of the remaining four equations are all zero. That is, $t = w = x = y = 0$ and

$$B = \begin{pmatrix} r & s & 0 \\ u & v & 0 \\ 0 & 0 & z \end{pmatrix}.$$

39. With $f(x) = a \sin x + b \cos x + c$, we have

$$f'(x) = a \cos x - b \sin x \qquad \text{and} \qquad f''(x) = -a \sin x - b \cos x.$$

Evaluating f, f' and f'' at $x = 0$ and using the conditions $f(0) = 1$, $f'(0) = 2$, $f''(0) = 3$ gives rise to the system

$$b + c = 1,$$
$$a = 2,$$
$$-b = 3.$$

The last two equations give $a = 2$, $b = -3$, and substituting $b = -3$ into the first equation yields $c = 4$. Hence,

$$f(x) = 2 \sin x - 3 \cos x + 4.$$

41. Let $A = \begin{pmatrix} 1 & 2 & 1 & 2 \\ 1 & 2 & 3 & 4 \\ 1 & 0 & 0 & 1 \end{pmatrix}$ (the rows of A are the three given vectors). To say that a

vector $\mathbf{x} = (x, y, z, w)$ in $\mathcal{R}^4$ is perpendicular to each of the given vectors is to say that the dot product of $\mathbf{x}$ with each of the given vectors is zero. It follows from the definition of the ijth entry of a product matrix that $\mathbf{x}$ must satisfy the matrix equation $A\mathbf{x} = \mathbf{0}$, where $\mathbf{x}$ is considered as a column vector. We therefore find a reduced form R for A and solve the reduced system $R\mathbf{x} = \mathbf{0}$.

Starting with A, subtract the last row from the first row and put the result in the first row. Then subtract the last row from the second row and put the result in the second row. This gives the matrix

$$\begin{pmatrix} 0 & 2 & 1 & 1 \\ 0 & 2 & 3 & 3 \\ 1 & 0 & 0 & 1 \end{pmatrix}.$$

Now subtract the first row from the second row and put the result in the second row, and then divide the new second row by 2 to obtain

$$\begin{pmatrix} 0 & 2 & 1 & 1 \\ 0 & 0 & 1 & 1 \\ 1 & 0 & 0 & 1 \end{pmatrix}.$$

Finally, subtract the second row from the first row and put the result in the first row, and then divide the new first row by 2. This gives the reduced matrix

$$R = \begin{pmatrix} 0 & 1 & 0 & 0 \\ 0 & 0 & 1 & 1 \\ 1 & 0 & 0 & 1 \end{pmatrix}.$$

The reduced system $R\mathbf{x} = \mathbf{0}$ is therefore equivalent to the system of equations

$$y = 0,$$
$$z + w = 0,$$
$$x + w = 0.$$

Since w is the only nonleading variable, set $w = t$ and deduce from the remaining equations that $y = 0$, $z = -t$, $x = -t$. Hence,

$$(x, y, z, w) = (-t, 0, -t, t) = t(-1, 0, -1, 1).$$

43. Let A be the given matrix. Expanding about the first row gives

$$\det A = \begin{vmatrix} 1 & 2 & 0 \\ 0 & 1 & 2 \\ 2 & 0 & 1 \end{vmatrix} = (1)\begin{vmatrix} 1 & 2 \\ 0 & 1 \end{vmatrix} - (2)\begin{vmatrix} 0 & 2 \\ 2 & 1 \end{vmatrix} = (1-0) - 2(0-4) = 9.$$

45. Let A be the given matrix. Add -2 times the first column to the second column and then expand about the first row to get

$$\det A = \begin{vmatrix} 1 & 2 & 0 & 0 \\ 0 & 1 & 2 & 0 \\ -2 & -3 & 0 & 2 \\ 2 & 0 & -1 & 3 \end{vmatrix} = \begin{vmatrix} 1 & 0 & 0 & 0 \\ 0 & 1 & 2 & 0 \\ -2 & 1 & 0 & 2 \\ 2 & -4 & -1 & 3 \end{vmatrix} = (1)\begin{vmatrix} 1 & 2 & 0 \\ 1 & 0 & 2 \\ -4 & -1 & 3 \end{vmatrix} = \begin{vmatrix} 1 & 2 & 0 \\ 1 & 0 & 2 \\ -4 & -1 & 3 \end{vmatrix}.$$

Now add -2 times the first column to the second column and then expand about the first row to obtain

$$\det A = \begin{vmatrix} 1 & 0 & 0 \\ 1 & -2 & 2 \\ -4 & 7 & 3 \end{vmatrix} = (1)\begin{vmatrix} -2 & 2 \\ 7 & 3 \end{vmatrix} = -6 - 14 = -20.$$

81

Chapter 3: Vector Spaces & Linearity

Section 1: LINEAR FUNCTIONS ON $\mathcal{R}^n$

Exercise Set 1ABC (pgs. 110-112)

1. We have $f : \mathcal{R}^2 \longrightarrow \mathcal{R}^2$, so that the matrix A representing f has dimensions 2-by-2. The first and second columns of A are $f(\mathbf{e}_1) = \begin{pmatrix} 1 \\ 2 \end{pmatrix}$ and $f(\mathbf{e}_2) = \begin{pmatrix} 2 \\ 4 \end{pmatrix}$, respectively. Hence,

$$A = \begin{pmatrix} 1 & 2 \\ 2 & 4 \end{pmatrix} \text{ and}$$

$$f(\mathbf{x}) = \begin{pmatrix} 1 & 2 \\ 2 & 4 \end{pmatrix} \mathbf{x}, \qquad \text{for } \mathbf{x} \in \mathcal{R}^2.$$

Since the second column of A is twice the first column, the columns of A are not linearly independent vectors and therefore f is not one-to-one.

3. We have $f : \mathcal{R}^2 \longrightarrow \mathcal{R}^3$, so that the matrix A representing f has dimensions 3-by-2. The first and second columns of A are $f(\mathbf{e}_1) = \begin{pmatrix} 1 \\ 2 \\ 3 \end{pmatrix}$ and $f(\mathbf{e}_2) = \begin{pmatrix} 3 \\ 2 \\ 1 \end{pmatrix}$, respectively. Hence,

$$A = \begin{pmatrix} 1 & 3 \\ 2 & 2 \\ 3 & 1 \end{pmatrix} \text{ and}$$

$$f(\mathbf{x}) = \begin{pmatrix} 1 & 3 \\ 2 & 2 \\ 3 & 1 \end{pmatrix} \mathbf{x}, \qquad \text{for } \mathbf{x} \in \mathcal{R}^2.$$

By inspection, the columns of A are linearly independent vectors, so that f is one-to-one.

5. We first find scalars a_1, b_1, a_2, b_2 such that

$$\mathbf{e}_1 = \begin{pmatrix} 1 \\ 0 \end{pmatrix} = a_1 \begin{pmatrix} 1 \\ 1 \end{pmatrix} + b_1 \begin{pmatrix} -1 \\ 1 \end{pmatrix} = \begin{pmatrix} a_1 - b_1 \\ a_1 + b_1 \end{pmatrix},$$

$$\mathbf{e}_2 = \begin{pmatrix} 0 \\ 1 \end{pmatrix} = a_2 \begin{pmatrix} 1 \\ 1 \end{pmatrix} + b_2 \begin{pmatrix} -1 \\ 1 \end{pmatrix} = \begin{pmatrix} a_2 - b_2 \\ a_2 + b_2 \end{pmatrix}.$$

The first equation implies the system $a_1 - b_1 = 1$, $a_1 + b_1 = 0$, with solution $a_1 = 1/2$, $b_1 = -1/2$. The second equation implies the system $a_2 - b_2 = 0$, $a_2 + b_2 = 1$, with solution $a_2 = 1/2$, $b_2 = 1/2$. Hence,

$$\mathbf{e}_1 = \frac{1}{2} \begin{pmatrix} 1 \\ 1 \end{pmatrix} - \frac{1}{2} \begin{pmatrix} -1 \\ 1 \end{pmatrix} \quad \text{and} \quad \mathbf{e}_2 = \frac{1}{2} \begin{pmatrix} 1 \\ 1 \end{pmatrix} + \frac{1}{2} \begin{pmatrix} -1 \\ 1 \end{pmatrix}.$$

Using the linearity of f along with the given values of $f \begin{pmatrix} 1 \\ 1 \end{pmatrix}$ and $f \begin{pmatrix} -1 \\ 1 \end{pmatrix}$ gives

$$f(\mathbf{e}_1) = f \left[\frac{1}{2} \begin{pmatrix} 1 \\ 1 \end{pmatrix} - \frac{1}{2} \begin{pmatrix} -1 \\ 1 \end{pmatrix} \right] = \frac{1}{2} f \begin{pmatrix} 1 \\ 1 \end{pmatrix} - \frac{1}{2} f \begin{pmatrix} -1 \\ 1 \end{pmatrix}$$

$$= \frac{1}{2} \begin{pmatrix} 2 \\ 1 \end{pmatrix} - \frac{1}{2} \begin{pmatrix} 1 \\ -1 \end{pmatrix} = \begin{pmatrix} 1/2 \\ 1 \end{pmatrix}$$

and

$$f(\mathbf{e}_2) = f\left[\frac{1}{2}\begin{pmatrix}1\\1\end{pmatrix} + \frac{1}{2}\begin{pmatrix}-1\\1\end{pmatrix}\right] = \frac{1}{2}f\begin{pmatrix}1\\1\end{pmatrix} + \frac{1}{2}f\begin{pmatrix}-1\\1\end{pmatrix}$$
$$= \frac{1}{2}\begin{pmatrix}2\\1\end{pmatrix} + \frac{1}{2}\begin{pmatrix}1\\-1\end{pmatrix} = \begin{pmatrix}3/2\\0\end{pmatrix}.$$

7. By inspection, we see that

$$\mathbf{e}_1 = \begin{pmatrix}1\\0\\0\end{pmatrix}, \quad \mathbf{e}_2 = \begin{pmatrix}1\\1\\0\end{pmatrix} - \begin{pmatrix}1\\0\\0\end{pmatrix}, \quad \mathbf{e}_3 = \begin{pmatrix}1\\1\\1\end{pmatrix} - \begin{pmatrix}1\\1\\0\end{pmatrix}.$$

We are already given that $f(\mathbf{e}_1) = \begin{pmatrix}2\\3\end{pmatrix}$. Using the linearity of f along with the given

values of $f\begin{pmatrix}1\\1\\1\end{pmatrix}$, $f\begin{pmatrix}1\\1\\0\end{pmatrix}$ and $f\begin{pmatrix}1\\0\\0\end{pmatrix}$, we obtain the values of $f(\mathbf{e}_2)$ and $f(\mathbf{e}_3)$ as follows:

$$f(\mathbf{e}_2) = f\left[\begin{pmatrix}1\\1\\0\end{pmatrix} - \begin{pmatrix}1\\0\\0\end{pmatrix}\right] = f\begin{pmatrix}1\\1\\0\end{pmatrix} - f\begin{pmatrix}1\\0\\0\end{pmatrix} = \begin{pmatrix}2\\1\end{pmatrix} - \begin{pmatrix}2\\3\end{pmatrix} = \begin{pmatrix}0\\-2\end{pmatrix},$$

$$f(\mathbf{e}_3) = f\left[\begin{pmatrix}1\\1\\1\end{pmatrix} - \begin{pmatrix}1\\1\\0\end{pmatrix}\right] = f\begin{pmatrix}1\\1\\1\end{pmatrix} - f\begin{pmatrix}1\\1\\0\end{pmatrix} = \begin{pmatrix}1\\2\end{pmatrix} - \begin{pmatrix}2\\1\end{pmatrix} = \begin{pmatrix}-1\\1\end{pmatrix}.$$

9. Let $\mathbf{x} = \begin{pmatrix}x\\y\end{pmatrix}$ and let R_θ be as given. Then

$$|R_\theta\mathbf{x}| = \left|\begin{pmatrix}\cos\theta & -\sin\theta\\\sin\theta & \cos\theta\end{pmatrix}\begin{pmatrix}x\\y\end{pmatrix}\right| = \left|\begin{pmatrix}x\cos\theta - y\sin\theta\\x\sin\theta + y\cos\theta\end{pmatrix}\right|$$
$$= \sqrt{(x\cos\theta - y\sin\theta)^2 + (x\sin\theta + y\cos\theta)^2}$$
$$= \sqrt{(x^2\cos^2\theta - 2xy\cos\theta\sin\theta + y^2\sin^2\theta) + (x^2\sin^2\theta + 2xy\sin\theta\cos\theta + y^2\cos^2\theta)}$$
$$= \sqrt{x^2(\cos^2\theta + \sin^2\theta) + y^2(\sin^2\theta + \cos^2\theta)} = \sqrt{x^2 + y^2} = \left|\begin{pmatrix}x\\y\end{pmatrix}\right| = |\mathbf{x}|.$$

Hence, R_θ preserves length.

11. With f and g as defined, we have $f : \mathcal{R}^2 \longrightarrow \mathcal{R}^2$ and $g : \mathcal{R}^2 \longrightarrow \mathcal{R}^2$, so that

$$g \circ f : \mathcal{R}^2 \longrightarrow \mathcal{R}^2.$$

That is, the domain of $g \circ f$ is $\mathcal{R}^2$ and the range of $g \circ f$ is $\mathcal{R}^2$. Furthermore, the matrices representing f and g are $A = \begin{pmatrix}2 & -1\\3 & 1\end{pmatrix}$ and $B = \begin{pmatrix}1 & 0\\2 & 1\end{pmatrix}$, respectively. Therefore, by Theorem 1.5, the matrix BA represents $g \circ f$. That is,

$$(g \circ f)(\mathbf{x}) = (BA)\mathbf{x} = \begin{pmatrix}1 & 0\\2 & 1\end{pmatrix}\begin{pmatrix}2 & -1\\3 & 1\end{pmatrix}\mathbf{x} = \begin{pmatrix}2 & -1\\7 & -1\end{pmatrix}\mathbf{x}, \quad \text{for } \mathbf{x} \in \mathcal{R}^2.$$

13. With f and g as defined, we have $f : \mathcal{R}^3 \longrightarrow \mathcal{R}^3$ and $g : \mathcal{R}^3 \longrightarrow \mathcal{R}^3$, so that

$$g \circ f : \mathcal{R}^3 \longrightarrow \mathcal{R}^3.$$

That is, the domain of $g \circ f$ is $\mathcal{R}^3$ and the range of $g \circ f$ is $\mathcal{R}^3$. Furthermore, the matrices representing f and g are $A = \begin{pmatrix} 1 & 0 & 2 \\ -1 & 1 & 3 \\ 2 & 1 & 0 \end{pmatrix}$ and $B = \begin{pmatrix} 1 & -1 & 1 \\ -1 & 1 & 1 \\ 1 & 1 & -1 \end{pmatrix}$, respectively.

Thus, by Theorem 1.5, the matrix BA represents $g \circ f$. That is,

$$(g \circ f)(\mathbf{x} = (BA)\mathbf{x} = \begin{pmatrix} 1 & -1 & 1 \\ -1 & 1 & 1 \\ 1 & 1 & -1 \end{pmatrix} \begin{pmatrix} 1 & 0 & 2 \\ -1 & 1 & 3 \\ 2 & 1 & 0 \end{pmatrix} \mathbf{x} = \begin{pmatrix} 4 & 0 & -1 \\ 0 & 2 & 1 \\ -2 & 0 & 5 \end{pmatrix} \mathbf{x}.$$

15. (a) Define the function $f : \mathcal{R}^2 \longrightarrow \mathcal{R}^2$ by

$$f(\mathbf{x}) = \begin{pmatrix} 0 & 1 \\ 1 & 0 \end{pmatrix} \mathbf{x}, \qquad \text{for } \mathbf{x} \in \mathcal{R}^2.$$

If $\mathbf{x} = \begin{pmatrix} x \\ y \end{pmatrix}$ then

$$f\begin{pmatrix} x \\ y \end{pmatrix} = \begin{pmatrix} 0 & 1 \\ 1 & 0 \end{pmatrix} \begin{pmatrix} x \\ y \end{pmatrix} = \begin{pmatrix} y \\ x \end{pmatrix}.$$

In other words, f interchanges the coordinates of vectors in $\mathcal{R}^2$. This is equivalent to saying that f reflects a point in $\mathcal{R}^2$ about the line $y = x$ (the line through the origin 45° counterclockwise from the horizontal).

(b) The line through the origin 135° counterclockwise form the horizontal is the line $y = -x$. When a point (x, y) in $\mathcal{R}^2$ is reflected about this line, the coordinates of the reflection are $(-y, -x)$. That is, the coordinates are interchanged and then multiplied by -1. Since $\begin{pmatrix} 0 & 1 \\ 1 & 0 \end{pmatrix}$ interchanges coordinates, it follows that $-\begin{pmatrix} 0 & 1 \\ 1 & 0 \end{pmatrix} = \begin{pmatrix} 0 & -1 \\ -1 & 0 \end{pmatrix}$ not only interchanges coordinates but multiplies them by -1 as well.

(c) When the matrix given in part (a) is multiplied on the right by the matrix found in part (b) the result is

$$\begin{pmatrix} 0 & 1 \\ 1 & 0 \end{pmatrix}\begin{pmatrix} 0 & -1 \\ -1 & 0 \end{pmatrix} = \begin{pmatrix} -1 & 0 \\ 0 & -1 \end{pmatrix} = -\begin{pmatrix} 1 & 0 \\ 0 & 1 \end{pmatrix} = -I.$$

The same result is obtained by reversing the order of multiplication. The resulting matrix $-I$ represents a linear function from $\mathcal{R}^2$ to $\mathcal{R}^2$ that sends a vector to its negative. Another way of saying this is that $-I$ reflects a point in $\mathcal{R}^2$ through the origin.

17. Let P_x be the perspective in $\mathcal{R}_3$ that looks down the x-axis with the positive y-axis pointing to the right and the positive z-axis pointing up; let P_y be the perspective that looks down the y-axis with the positive x-axis pointing to the left and the positive z-axis pointing up; and let P_z be the perspective that looks down the z-axis with the positive x-axis pointing to the right and the positive y-axis pointing up.

(a) The effect that the matrix $U = \begin{pmatrix} 1 & 0 & 0 \\ 0 & 0 & -1 \\ 0 & 1 & 0 \end{pmatrix}$ has on a point in $\mathcal{R}^3$ can be determined by the effect it has on the standard basis vectors $\mathbf{e}_1$, $\mathbf{e}_2$, $\mathbf{e}_3$. Since $U\mathbf{e}_k$ is the kth column of U, we have

$$U\mathbf{e}_1 = \begin{pmatrix} 1 \\ 0 \\ 0 \end{pmatrix} = \mathbf{e}_1, \qquad U\mathbf{e}_2 = \begin{pmatrix} 0 \\ 0 \\ 1 \end{pmatrix} = \mathbf{e}_3, \qquad U\mathbf{e}_3 = \begin{pmatrix} 0 \\ -1 \\ 0 \end{pmatrix} = -\mathbf{e}_2.$$

The first equation says that the unit vector pointing up the positive x-axis remains fixed; the second equation says that the unit vector pointing up the positive y-axis becomes the unit vector pointing up the positive z-axis; and the third equation says that the unit vector pointing up the positive z-axis becomes the unit vector pointing down the negative y-axis. From the perspective P_x, the effect of U depicts a 90° counterclockwise rotation about the x-axis.

The computation of U^{-1} can be done in many ways. However, choosing to use the same conceptual method just employed, we use the perspective P_x and note that a 90° *clockwise* rotation about the x-axis leaves $\mathbf{e}_1$ fixed, sends $\mathbf{e}_2$ to $-\mathbf{e}_3$, and sends $\mathbf{e}_3$ to $\mathbf{e}_2$. That is, $\mathbf{e}_1$, $-\mathbf{e}_3$, $\mathbf{e}_2$ are the columns of U^{-1} and we have

$$U^{-1} = \begin{pmatrix} 1 & 0 & 0 \\ 0 & 0 & 1 \\ 0 & -1 & 0 \end{pmatrix}.$$

The reader can check that the product of U and the above matrix is the identity matrix.

Similarly, the matrix $V = \begin{pmatrix} 0 & 0 & 1 \\ 0 & 1 & 0 \\ -1 & 0 & 0 \end{pmatrix}$ has the following effect on the standard basis vectors:

$$V\mathbf{e}_1 = \begin{pmatrix} 0 \\ 0 \\ -1 \end{pmatrix} = -\mathbf{e}_3, \qquad V\mathbf{e}_2 = \begin{pmatrix} 0 \\ 1 \\ 0 \end{pmatrix} = \mathbf{e}_2, \qquad V\mathbf{e}_3 = \begin{pmatrix} 1 \\ 0 \\ 0 \end{pmatrix} = \mathbf{e}_1.$$

The first equation says that the unit vector pointing up the positive x-axis becomes the unit vector pointing down the negative z-axis; the second equation says the the unit vector pointing up the positive y-axis remains fixed; and the third equation says that the unit vector pointing up the positive z-axis becomes the unit vector pointing up the positive x-axis. From the perspective P_y, the effect of V depicts a 90° counterclockwise rotation about the y-axis.

To find V^{-1}, we use the perspective P_y and note that a 90° clockwise rotation about the y-axis sends $\mathbf{e}_1$ to $\mathbf{e}_3$, leaves $\mathbf{e}_2$ fixed, and sends $\mathbf{e}_3$ to $-\mathbf{e}_1$. That is, $\mathbf{e}_3$, $\mathbf{e}_2$, $-\mathbf{e}_1$ are the columns of V^{-1} and we have

$$V^{-1} = \begin{pmatrix} 0 & 0 & -1 \\ 0 & 1 & 0 \\ 1 & 0 & 0 \end{pmatrix}.$$

The reader can check that the product of V and the above matrix is the identity matrix.

From the perspective P_z let W be the matrix that effects a 90° counterclockwise rotation about the z-axis. Then W sends the unit vector that points up the positive x-axis to the unit vector that points up the positive y-axis; it sends the unit vector that points up the positive y-axis to the unit vector that points down the negative x axis; and it leaves the unit vector that points up the positive z-axis unchanged. In other words,

$$W\mathbf{e}_1 = \mathbf{e}_2 = \begin{pmatrix} 0 \\ 1 \\ 0 \end{pmatrix}, \qquad W\mathbf{e}_2 = -\mathbf{e}_1 = \begin{pmatrix} -1 \\ 0 \\ 0 \end{pmatrix}, \qquad W\mathbf{e}_3 = \mathbf{e}_3 = \begin{pmatrix} 0 \\ 0 \\ 1 \end{pmatrix}.$$

Since $W\mathbf{e}_k$ is the kth column of W, it follows that

$$W = \begin{pmatrix} 0 & -1 & 0 \\ 1 & 0 & 0 \\ 0 & 0 & 1 \end{pmatrix}.$$

(b) Focusing first on UVU^{-1}, we can use the results of part (a) to get

$$\begin{aligned}
UVU^{-1}\mathbf{e}_1 &= UV\mathbf{e}_1 = U(-\mathbf{e}_3) = -U\mathbf{e}_3 = -(-\mathbf{e}_2) = \mathbf{e}_2, \\
UVU^{-1}\mathbf{e}_2 &= UV(-\mathbf{e}_3) = -UV\mathbf{e}_3 = -U\mathbf{e}_1 = -\mathbf{e}_1, \\
UVU^{-1}\mathbf{e}_3 &= UV\mathbf{e}_2 = U\mathbf{e}_2 = \mathbf{e}_3.
\end{aligned}$$

From the perspective P_z, this is a $90°$ counterclockwise rotation about the z-axis. Moreover, since $UVU^{-1}\mathbf{e}_k$ is the kth column of UVU^{-1}, we have

$$UVU^{-1} = \begin{pmatrix} 0 & -1 & 0 \\ 1 & 0 & 0 \\ 0 & 0 & 1 \end{pmatrix}.$$

Note that this is the matrix W that was derived in part (a).

Focusing next on VUV^{-1}, we again use the results from part (a) to get

$$\begin{aligned}
VUV^{-1}\mathbf{e}_1 &= VU\mathbf{e}_3 = V(-\mathbf{e}_2) = -V\mathbf{e}_2 = -\mathbf{e}_2, \\
VUV^{-1}\mathbf{e}_2 &= VU\mathbf{e}_2 = V\mathbf{e}_3 = \mathbf{e}_1, \\
VUV^{-1}\mathbf{e}_3 &= VU(-\mathbf{e}_1) = -VU\mathbf{e}_1 = -V\mathbf{e}_1 = -(-\mathbf{e}_3) = \mathbf{e}_3.
\end{aligned}$$

From the perspective P_z, this is a $90°$ clockwise rotation about the z-axis. Moreover, since $VUV^{-1}\mathbf{e}_k$ is the kth column of VUV^{-1}, we have

$$VUV^{-1} = \begin{pmatrix} 0 & 1 & 0 \\ -1 & 0 & 0 \\ 0 & 0 & 1 \end{pmatrix}.$$

From the geometric interpretation, we conclude that $VUV^{-1} = W^{-1}$.

19. If $\mathbf{n}$ is a unit vector in $\mathcal{R}^3$ then the associated projection function $P_{\mathbf{n}} : \mathcal{R}^3 \longrightarrow \mathcal{R}^3$ is given by $P_{\mathbf{n}}(\mathbf{x}) = (\mathbf{x} \cdot \mathbf{n})\mathbf{n}$. In particular, if $\mathbf{n} = (\frac{3}{7}, \frac{6}{7}, \frac{2}{7})$ then, for $\mathbf{x} = \mathbf{e}_k$ $(k = 1, 2, 3)$ we have

$$P_{\mathbf{n}}(\mathbf{e}_1) = \left[\mathbf{e}_1 \cdot \left(\frac{3}{7}, \frac{6}{7}, \frac{2}{7} \right) \right] \left(\frac{3}{7}, \frac{6}{7}, \frac{2}{7} \right) = \frac{3}{7} \left(\frac{3}{7}, \frac{6}{7}, \frac{2}{7} \right) = \left(\frac{9}{49}, \frac{18}{49}, \frac{6}{49} \right),$$

$$P_n(\mathbf{e}_2) = \left[\mathbf{e}_2 \cdot \left(\frac{3}{7}, \frac{6}{7}, \frac{2}{7}\right)\right]\left(\frac{3}{7}, \frac{6}{7}, \frac{2}{7}\right) = \frac{6}{7}\left(\frac{3}{7}, \frac{6}{7}, \frac{2}{7}\right) = \left(\frac{18}{49}, \frac{36}{49}, \frac{12}{49}\right),$$

$$P_n(\mathbf{e}_3) = \left[\mathbf{e}_3 \cdot \left(\frac{3}{7}, \frac{6}{7}, \frac{2}{7}\right)\right]\left(\frac{3}{7}, \frac{6}{7}, \frac{2}{7}\right) = \frac{2}{7}\left(\frac{3}{7}, \frac{6}{7}, \frac{2}{7}\right) = \left(\frac{6}{49}, \frac{12}{49}, \frac{4}{49}\right).$$

Since these are the columns of the matrix representing $P_{\mathbf{n}}$, the matrix of $P_{\mathbf{n}}$ is

$$\begin{pmatrix} 9/49 & 18/49 & 6/49 \\ 18/49 & 36/49 & 12/49 \\ 6/49 & 12/49 & 4/49 \end{pmatrix}.$$

21. For the given function f we have $f : \mathcal{R}^2 \longrightarrow \mathcal{R}^2$ and

$$f(x, y) = (2x - y, 6x - 3y) = (2x - y)(1, 3).$$

As (x, y) ranges over $\mathcal{R}^2$, the scalar $2x - y$ ranges over $\mathcal{R}$. Hence, the above equation show that the image of f is a line through the origin in $\mathcal{R}^2$ in the direction of $(1, 3)$. This has the scalar form $y = 3x$.

23. For the given function f we have $f : \mathcal{R}^3 \longrightarrow \mathcal{R}^2$ and

$$f(x, y, z) = (x + y - z, -x - y + z) = (x + y - z)(1, -1).$$

As (x, y, z) ranges over $\mathcal{R}^3$, the scalar $x + y - z$ ranges over $\mathcal{R}$. Hence, the above equation shows that image of f is a line through the origin in $\mathcal{R}^2$ in the direction of $(1, -1)$. This has the scalar form $y = -x$

25. For the given function f we have $f : \mathcal{R}^3 \longrightarrow \mathcal{R}^3$ and

$$(*) \quad f(x, y, z) = (x - 2y + z, -2x - y - z, -5y + z) = x(1, -2, 0) + y(-2, -1, -5) + z(1, -1, 1).$$

The vectors $(1, -2, 0)$, $(-2, -1, -5)$ and $(1, -1, 1)$ are not linearly independent. For example, $(1, -1, 1)$ is a linear combination of $(1, -2, 0)$ and $(-2, -1, -5)$. To see this, suppose a and b are scalars such that

$$(1, -1, 1) = a(1, -2, 0) + b(-2, -1, -5) = (a - 2b, -2a - b, -5b).$$

This implies the three equations $1 = a - 2b$, $-1 = -2a - b$ and $1 = -5b$. The last equation says $b = -1/5$. Using this value of b in each of the first two equations gives $a = 3/5$ in both cases. Therefore $(1, -1, 1) = \frac{3}{5}(1, -2, 0) - \frac{1}{5}(-2, -1, -5)$ and we can write $(*)$ as

$$(**) \quad f(x, y, z) = x(1, -2, 0) + y(-2, -1, -5) + z\left(\frac{3}{5}(1, -2, 0) - \frac{1}{5}(-2, -1, -5)\right)$$

$$= \left(x + \frac{3}{5}z\right)(1, -2, 0) + \left(y - \frac{1}{5}z\right)(-2, -1, -5).$$

The vectors $(1, -2, 0)$ and $(-2, -1, -5)$ are linearly independent (since neither is a scalar multiple of the other). Moreover, the scalars $x + \frac{3}{5}z$ and $y - \frac{1}{5}z$ range independently over $\mathcal{R}$ as (x, y, z) ranges over $\mathcal{R}^3$. Hence, equation $(**)$ shows that the image of f is the plane through the origin in $\mathcal{R}^3$ determined by the vectors $(1, -2, 0)$ and $(-2, -1, -5)$. The reader can check that the scalar equation of this plane is $2x + y - z = 0$.

27. Let $f : \mathcal{R} \longrightarrow \mathcal{R}^2$ and $g : \mathcal{R}^2 \longrightarrow \mathcal{R}$. Since the image of f (a subset of $\mathcal{R}^2$) is contained in the domain of g $(\mathcal{R}^2)$, $g \circ f$ makes sense. Further, the image of g (a subset of $\mathcal{R}$) is contained in the domain of f $(\mathcal{R}^2)$, so that $f \circ g$ also makes sense.

29. Let $f : \mathcal{R}^2 \longrightarrow \mathcal{R}^2$ and $g : \mathcal{R}^2 \longrightarrow \mathcal{R}^3$. Since the image of f (a subset of $\mathcal{R}^2$) is contained in the domain of g $(\mathcal{R}^2)$, $g \circ f$ makes sense. However, the image of g (a subset of $\mathcal{R}^3$) is not contained in the domain of f $(\mathcal{R}^2)$, so that $f \circ g$ does not make sense.

31. Let $f : \mathcal{R}^n \longrightarrow \mathcal{R}$ be a linear function and let $\mathbf{e}_1, \ldots, \mathbf{e}_n$ be the standard basis vectors in $\mathcal{R}^n$. Consider the constant vector $\mathbf{a} = \big(f(\mathbf{e}_1), \ldots, f(\mathbf{e}_n)\big)$. With $\mathbf{x} = (x_1, \ldots, x_n)$, the linearity of f implies

$$f(\mathbf{x}) = f(x_1\mathbf{e}_1 + \cdots + x_n\mathbf{e}_n) = x_1 f(\mathbf{e}_1) + \cdots + x_n f(\mathbf{e}_n)$$
$$= (x_1, \ldots, x_n) \cdot \big(f(\mathbf{e}_1), \ldots, f(\mathbf{e}_n)\big) = \mathbf{x} \cdot \mathbf{a}.$$

That is, $f(\mathbf{x}) = \mathbf{a} \cdot \mathbf{x}$ for every $\mathbf{x}$ in $\mathcal{R}^n$.

33. (a) Since $f_a(\mathbf{e}_1) = f_a(1,0) = (1,0)$ and $f_a(\mathbf{e}_2) = f_a(0,1) = (a,1)$, we use the fact that f_a is linear to conclude from Theorem 1.3 that the matrix of f_a is $\begin{pmatrix} 1 & a \\ 0 & 1 \end{pmatrix}$.

(b) The points $f_a(1,0)$ and $f_a(0,1)$ are given in part (a). For the other points we have

$$f_a(-1,0) = (-1,0), \qquad f_a(0,-1) = (-a,-1).$$

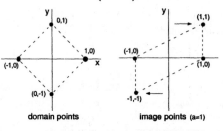

domain points   image points (a=1)

The domain points are shown in the first diagram on the right and the corresponding image points for $a = 1$ are shown in the second diagram on the right. The arrows in the right diagram indicate "shifts" (as discussed in part (c) below).

(c) First, the defining equation $f_a(x,y) = (x + ay, y)$ shows that the y-coordinate remains the same, so that no vertical shift occurs under f_a, regardless of the sign of a or where the point (x,y) lies. With this in mind, let $a > 0$ and let (x,y) be above the x-axis so that $y > 0$. Then $ay > 0$ and the defining equation shows that the x-coordinate *increases* by the amount ay. That is, f_a shifts the domain point ay units to the right. On the other hand, if (x,y) is below the x-axis (so that $ay < 0$) then the x-coordinate *decreases* by the amount $|ay|$. That is, f_a shifts the domain point $|ay|$ units to the left. If $a < 0$ then the shifts are reversed, with points above the x-axis shifted $|ay|$ units to the left, and points below the x-axis shifted ay units to the right.

(d) A fixed point (x,y) is one that satisfies the equation $f_a(x,y) = (x,y)$, or equivalently, $(x + ay, y) = (x,y)$. This holds only if $ay = 0$. Since $a \neq 0$ by hypothesis, we must have $y = 0$. Thus, the points that remain fixed under f_a are the points on the x-axis.

(e) Let L be a line in the plane such that $f(L) = L$. If (x,y) is any point on L then $(x + ay, y)$ must also be a point on L. Since two such points have the same y-coordinate, it follows that L is a horizontal line and therefore has an equation of the form $y = b$ for some constant b. Moreover, the equation $f_a(x - ab, b) = (x - ab + ab, b) = (x, b)$ shows that the point (x,b) on L is the image of the point $(x - ab, b)$ on L. This is sufficient to conclude that a line in the plane is its own image under f_a if, and only if, it is a horizontal line.

(f) Let f_a and f_b be two shear transformations. Then, for (x,y) in $\mathcal{R}^2$,

$$(f_a \circ f_b)(x,y) = f_a\big(f_b(x,y)\big) = f_a(x+by, y) = (x+by+ay, y) = \big(x+(b+a)y, y\big) = f_{b+a}(x,y).$$

That is, $f_a \circ f_b = f_{b+a}$. Interchanging the roles of a and b shows that we also have $f_b \circ f_a = f_{a+b}$. Since $a + b = b + a$, we obtain

$$f_a \circ f_b = f_b \circ f_a = f_{a+b}.$$

88

Section 2: VECTOR SPACES

Exercise Set 2AB (pgs. 118-119)

1. Let S be the collection of all vectors (x, y, z) in $\mathcal{R}^3$ such that $x + 2y = 0$. Thus, if $\mathbf{x} = (x_1, y_1, z_1)$ and $\mathbf{y} = (x_2, y_2, z_2)$ are in S then $x_1 + 2y_1 = 0$ and $x_2 + 2y_2 = 0$. Since the coordinates of $\mathbf{x} + \mathbf{y} = (x_1 + x_2, y_1 + y_2, z_1 + z_2)$ and $r\mathbf{x} = (rx_1, ry_1, rz_1)$ satisfy the respective equations

$$(x_1 + y_1) + 2(y_1 + y_2) = (x_1 + 2y_1) + (x_2 + 2y_2) = 0 + 0 = 0,$$
$$rx_1 + 2(ry_1) = r(x_1 + 2y_1) = r(0) = 0,$$

$\mathbf{x} + \mathbf{y}$ and $r\mathbf{x}$ are both in S, so that S is closed with respect to both vector addition and scalar multiplication. Hence, S is a subspace of $\mathcal{R}^3$.

3. Let S be the collection of all vectors (x, y, z) in $\mathcal{R}^3$ such that $x + y = 0$ and $z = 0$. Since the vectors in S are those of the form $(x, -x, 0) = x(1, -1, 0)$ with x is arbitrary, the vectors in S are those that are parallel to the line through the origin in $\mathcal{R}^3$ in the direction of $(1, -1, 0)$. Since the sum of any two vectors of this form is also a vector parallel to $(1, -1, 0)$, and since every scalar multiple of a vector of this form is a vector parallel to $(1, -1, 0)$, S is closed with respect to vector addition and scalar multiplication. Hence, S is a subspace of $\mathcal{R}^3$.

5. If S is the collection of all vectors (x, y, z) in $\mathcal{R}^3$ such that $x = y^3$ then $(1, 1, 0)$ is in S. However, since $2 \neq 2^3$, the scalar multiple $2(1, 1, 0) = (2, 2, 0)$ is not in S. That is, S is not closed with respect to scalar multiplication and therefore S is not a subspace of $\mathcal{R}^3$.

7. Let S be the set of all 2-by-2 matrices of the form $\begin{pmatrix} x & y \\ z & x \end{pmatrix}$. If $A_1 = \begin{pmatrix} x_1 & y_1 \\ z_1 & x_1 \end{pmatrix}$ and $A_2 = \begin{pmatrix} x_2 & y_2 \\ z_2 & x_2 \end{pmatrix}$ are in S and r is a scalar then

$$A_1 + A_2 = \begin{pmatrix} x_1 & y_1 \\ z_1 & x_1 \end{pmatrix} + \begin{pmatrix} x_2 & y_2 \\ z_2 & x_2 \end{pmatrix} = \begin{pmatrix} x_1 + x_2 & y_1 + y_2 \\ z_1 + z_2 & x_1 + x_2 \end{pmatrix}$$
$$rA_1 = r\begin{pmatrix} x_1 & y_1 \\ z_1 & x_1 \end{pmatrix} = \begin{pmatrix} rx_1 & ry_1 \\ rz_1 & rx_1 \end{pmatrix}.$$

So $A_1 + A_2$ and rA_1 are both in S. Thus, S is closed with respect to addition and scalar multiplication and is therefore a subspace of $M_{2,2}$.

9. The set S of all 2-by-2 matrices of the form $A = \begin{pmatrix} x & 1 \\ 1 & w \end{pmatrix}$ is not a subspace of $M_{2,2}$. It is not closed with respect to addition and it is not closed with respect to scalar multiplication. For example, if A is in S then

$$2\begin{pmatrix} x & 1 \\ 1 & w \end{pmatrix} = \begin{pmatrix} 2x & 2 \\ 2 & 2w \end{pmatrix},$$

so that the scalar multiple $2A$ is not in S.

11. Let S be the set of all vectors (x, y, z) in $\mathcal{R}^3$ such that $x + 2y - z = 0$.

(a) Let $\mathbf{x} = (x_1, y_1, z_1)$ and $\mathbf{y} = (x_2, y_2, z_2)$ be two vectors in S and let r be an arbitrary scalar. Then $\mathbf{x} + \mathbf{y} = (x_1 + x_2, y_1 + y_2, z_1 + z_2)$ and $r\mathbf{x} = (rx_1, ry_1, rz_1)$. Since $x_1 - 2y_1 - z_1 = 0$ and $x_1 + 2y_2 - z_2 = 0$, we have

$$(x_1 + x_1) + 2(y_1 + y_2) - (z_1 + z_2) = (x_1 + 2y_1 - z_1) + (x_2 + 2y_2 - z_2) = 0 + 0 = 0,$$
$$(rx_1) + 2(ry_1) - (rz_1) = r(x_1 + 2y_1 - z_1) = r(0) = 0,$$

so that $\mathbf{x} + \mathbf{y}$ and $r\mathbf{x}$ are both in S. Thus, S is closed with respect to both vector addition and scalar multiplication and is therefore a subspace of $\mathcal{R}^3$.

(b) The defining condition that must hold for a vector to be in S can be solved for z to get $z = x + 2y$. This shows two things. First, if x and y are selected arbitrarily then z is uniquely determined and the triple x, y, z determines a vector in S; and second, every vector in S can be written as

$$(x, y, z) = (x, y, x + 2y) = x(1, 0, 1) + y(0, 1, 2).$$

It follows that the vectors in S are precisely the linear combinations of $(1, 0, 1)$ and $(0, 1, 2)$. That is, the pair of vectors $(1, 0, 1)$, $(0, 1, 2)$ span the subspace in part (a) and a parametric representation for the solutions of $x + 2y - z = 0$ is given by the above equation.

13. Let S be any nonempty subset of $\mathcal{R}^n$ and let

$$S^\perp = \{\mathbf{p} \in \mathcal{R}^n \mid \mathbf{p} \cdot \mathbf{s} = 0, \text{ for all } \mathbf{s} \in S\}.$$

To show that $S^\perp$ is closed with respect to vector addition, let $\mathbf{p}_1, \mathbf{p}_2 \in S^\perp$. Then, for each $\mathbf{s} \in S$, additivity of the dot product gives

$$(\mathbf{p}_1 + \mathbf{p}_2) \cdot \mathbf{s} = \mathbf{p}_1 \cdot \mathbf{s} + \mathbf{p}_2 \cdot \mathbf{s} = 0 + 0 = 0,$$

so that $\mathbf{p}_1 + \mathbf{p}_2 \in S^\perp$. To show that $S^\perp$ is closed with respect to scalar multiplication, let $\mathbf{p} \in S^\perp$ and let r be a scalar. Then, for each $\mathbf{s} \in S$, homogeneity of the dot product gives

$$(r\mathbf{p}) \cdot \mathbf{s} = r(\mathbf{p} \cdot \mathbf{s}) = r(0) = 0,$$

so that $r\mathbf{p} \in S^\perp$. Since $S^\perp$ is closed with respect to vector addition and scalar multiplication, $S^\perp$ is a subspace of $\mathcal{R}^n$.

15. Let $S_\infty = \{\mathbf{e}_i \mid i = 1, 2, 3, \dots\}$, where $\mathbf{e}_i$ is the vector in $\mathcal{R}^\infty$ with 1 in the ith position and 0 elsewhere, and for each n let $S_n = \{\mathbf{e}_1, \dots, \mathbf{e}_n\}$.

First, if $\mathbf{x}$ is in the span of S_n then there exist scalars $a_1, \dots, a_n$ such that

$$\mathbf{x} = a_1\mathbf{e}_1 + a_2\mathbf{e}_2 + \cdots + a_n\mathbf{e}_n = a_1(1, 0, \dots) + a_2(0, 1, 0, \dots) + \cdots + a_n(0, \dots, 0, 1, 0, \dots)$$
$$= (a_1, \dots, a_n, 0, 0, \dots).$$

That is, the vectors in the span of S_n are those with 0 in every position after the nth position.

In considering the span of S_∞, which we denote by $\text{span}(S_\infty)$, we are only allowed to form linear combinations of *finitely many* vectors from S_∞. This means that a vector in the span of S_∞ has 0 in every position from some point on. So,

$$\text{span}(S_\infty) = \{\mathbf{x} \in \mathcal{R}^\infty \mid \mathbf{x} \text{ in the span of } S_n, \text{ for some } n\}.$$

The vectors in $\mathcal{R}^\infty$ that are not in span(S_∞) are not in the span of any S_n; i.e., those that have infinitely many nonzero entries. An example of this is the vector with 1 in every position.

17. Let S be the set of all polynomials p in P_3 such that $p(1) = 0$. Then, for $p_1, p_2 \in S$ and each scalar r, the definition of the sum of two functions and the definition of a scalar multiple of a function gives

$$(p_1 + p_2)(1) = p_1(1) + p_2(1) = 0 + 0 = 0,$$
$$(2p_1)(1) = rp_1(1) = r(0) = 0,$$

so that $p_1 + p_2$ and rp_1 are both in S. That is, S is closed with respect to polynomial addition and scalar multiplication and is therefore a subspace of P_3.

19. Let S be the set of all polynomials p in P_3 such that $p(1) = p'(2)$. Let $p_1, p_2 \in S$ and let r be a scalar. Using the definitions of the sum of two functions and the scalar multiple of a function, along with the linearity of the derivative, we obtain

$$(p_1 + p_2)(1) = p_1(1) + p_2(1) = p_1'(2) + p_2'(2) = (p_1' + p_2')(2) = (p_1 + p_2)'(2),$$
$$(rp_1)(1) = rp_1(1) = rp_1'(2) = (rp_1')(2) = (rp_1)'(2).$$

Thus, $p_1 + p_2$ and rp_1 are both in S, so that S is closed with respect to polynomial addition and scalar multiplication. That is, S is a subspace of P_3.

21. Let S be the set of all f in $C^{(1)}(-\infty, \infty)$ such that $f'(0)$ exists and let $f \in C^{(1)}(-\infty, \infty)$. Then, by definition, f has a continuous derivative. Therefore, $f'(x)$ is defined everywhere and $f'(0)$ exists. So $C^{(1)}(-\infty, \infty)$ is contained in S. Since S is also contained in $C^{(1)}(-\infty, \infty)$, it follows that $S = C^{(1)}(-\infty, \infty)$. Hence, S is a subspace of $C^{(1)}(-\infty, \infty)$.

23. Let S be the set of all f in $C^{(1)}(-\infty, \infty)$ such that $f'(0) = f(2)$. Let $f_1, f_2 \in S$ and let r be a scalar. Then

$$(f_1 + f_2)'(2) = (f_1' + f_2')(2) = f_1'(2) + f_2'(2) = f_1(0) + f_2(0) = (f_1 + f_2)(0),$$
$$(rf_1)'(2) = rf_1'(2) = rf_1(0) = (rf_1)(0),$$

so that $f_1 + f_2$ and rf_1 are both in S. Thus, S is closed with respect to function addition and scalar multiplication and is therefore a subspace of $C^{(1)}(-\infty, \infty)$.

25. Let $f_1, f_2 \in C_0[a, b]$ and let r be a scalar. Then

$$(f_1 + f_2)(a) = f_1(a) + f_2(a) = f_1(b) + f_2(b) = (f_1 + f_2)(b) = f_1(b) + f_2(b) = 0 + 0 = 0,$$
$$(rf_1)(a) = rf_1(a) = rf_1(b) = (rf_1)(b) = rf_1(b) = r(0) = 0,$$

so that $f_1 + f_2$ and rf_1 are both in $C_0[a, b]$. Thus, $C_0[a, b]$ is closed with respect to function addition and scalar multiplication and is therefore a subspace of $C[a, b]$.

27. Let $S = \{(2, 3, 1), (1, 2, 3)\}$ and $T = \{(3, 5, 4), (1, 1, -2)\}$ and let U and V be the spans of S and T, respectively. To show that $U = V$, we show that each is a subset of the other. This will follow if each is a subspace of the other.

To show that U is a subspace of V, we first show that S in contained in V. This is equivalent to showing that there exist scalars a, b, c, d such that

$$a(3, 5, 4) + b(1, 1, -2) = (2, 3, 1) \qquad \text{and} \qquad c(3, 5, 4) + d(1, 1, -2) = (1, 2, 3).$$

These can be written as $(3a+b, 5a+b, 4a-2b) = (2, 3, 1)$ and $(3c+d, 5c+d, 4c-2d) = (1, 2, 3)$. These in turn imply the systems

$$
\begin{array}{ccc}
3a + b = 2, & & 3c + d = 1, \\
5a + b = 3, & \text{and} & 5c + d = 2, \\
4a - 2b = 1, & & 4c - 2d = 3.
\end{array}
$$

In each system, subtract the first equation from the second equation to get $2a = 1$ and $2c = 1$, from which $a = 1/2$ and $c = 1/2$. The first equations of the systems can then be written as $3(1/2) + b = 2$ and $3(1/2) + d = 1$, from which $b = 1/2$ and $d = -1/2$. By inspection, we notice that these values of a and b and of c and d also satisfy the third equations of their respective systems. Hence, the required scalars exist and S is contained in V. By Theorem 2.4, the vector space spanned by S is a subspace of V; i.e., U is a subspace of V.

We use the same procedure to show that V is a subspace of U. However, here it is a little more straightforward because we simply observe that the sum of the vectors in S is $(3, 5, 4)$ and the difference of the vectors in S is $(1, 1, -2)$. So T is contained in U. Again, by Theorem 2,4, V is a subspace of U. As has already been observed, this is sufficient to conclude that $U = V$.

29. Let S be a subset of a vector space W and let V be any subspace of W that contains S. Since V is a subspace of W, it is closed under both addition and scalar multiplication. By Theorem 2.1, it is a vector space. Since S is a subset of V, Theorem 2.4 say that the span of S is a subspace of V. Hence, the span of S is contained in V.

31. Let V be a vector space and let A and B be subspaces of V. First of all, if A is contained in B then $B = A \cup B$. By hypothesis, B is a subspace of V. Hence, so is $A \cup B$. A similar result is obtained if we assume that B is contained in A.

Conversely, suppose $A \cup B$ is a subspace of V and that neither A nor B is contained in the other. Then there exists a vector $\mathbf{x} \in A$ that is not in B and a vector $\mathbf{y} \in B$ that is not in A. Since $A \cup B$ is a subspace, the vector $\mathbf{z} = \mathbf{x} + \mathbf{y}$ is in $A \cup B$ and is therefore either in A or B. If $\mathbf{z} \in A$ then the linear combination $\mathbf{z} - \mathbf{x} = \mathbf{y}$ is in A (because A is a subspace), contrary to assumption; and if $\mathbf{z} \in B$ then the linear combination $\mathbf{z} - \mathbf{y} = \mathbf{x}$ is in B (because B is a subspace), which is also contrary to assumption. We conclude that either A is contained in B or B is contained in A.

33. We make the preliminary remark that if $l \geq k$ and f has l continuous derivatives on $[a, b]$ then f also has k continuous derivatives on $[a, b]$. Thus, $C^{(l)}[a, b]$ is a subset of $C^{(k)}[a, b]$. Since $C^{(l)}[a, b]$ is a vector space, $C^{(l)}[a, b]$ and the span of $C^{(l)}[a, b]$ are the same set, so that Theorem 2.4 implies $C^{(l)}[a, b]$ is a subspace of the vector space $C^{(k)}[a, b]$. Thus, we need only show that $C^{(l)}[a, b]$ is a *proper* subset of $C^{(k)}[a, b]$ when $l > k$.

(a) Consider the function $f(x) = (x-a)^{1/2}$ with derivative $f'(x) = \frac{1}{2}(x-a)^{-1/2}$. Since f is continuous on $[a, b]$ but $f'(a)$ does not exist, $f \in C[a, b]$ but $f \notin C^{(1)}[a, b]$. Hence, $C^{(1)}[a, b]$ is properly contained in $C[a, b]$ and is therefore a *proper* subspace of $C[a, b]$.

(b) Using the function f from part (a), the fundamental theorem of calculus says that the function f_1 defined by

$$f_1(x) = \int_a^x f(t)\, dt = \int_a^x (x-a)^{1/2} dt = \frac{2}{3}(x-a)^{3/2}$$

satisfies $f_1'(x) = f(x) = (x-a)^{1/2}$ for all $x \in [a, b]$. Hence, f_1' is continuous on $[a, b]$ and therefore $f_1 \in C^{(1)}[a, b]$. However, $f_1''(x) = f'(x)$ is not continuous on $[a, b]$, so that $f_1 \notin C^{(2)}[a, b]$.

(c) Define the sequence $f_0, f_1, f_2, \ldots$ of continuous functions on $[a, b]$ by

$$f_k(x) = \frac{2^k}{(1)(3)\cdots(2k+1)}(x-a)^{(2k+1)/2} \quad \text{for } k \geq 0.$$

Note that f_k is differentiable on $[a, b]$ for $k \geq 1$ and that

$$f_k'(x) = \frac{2^{k-1}}{(1)(3)\cdots(2(k-1)+1)}(x-a)^{[2(k-1)+1]/2} = f_{k-1}(x).$$

By induction, $f_k^{(k)}(x) = f_0(x)$ for all $x \in [a, b]$ and all $k \geq 1$. Hence, the first k derivatives of f_k are continuous on $[a, b]$, so that $f_k \in C^{(k)}[a, b]$. However, $f_k^{(k+1)}(x) = \frac{1}{2}(x - a)^{-1/2}$, which is not continuous at $x = a$, so that $f_k \notin C^{(k+1)}[a, b]$. Since $C^{(l)}[a, b]$ is contained in $C^{(k+1)}[a, b]$ for $l > k$, it follows that f_k is also not in $C^{(l)}[a, b]$. Thus, $C^{(l)}[a, b]$ is a proper subset of $C^{(k)}[a, b]$ when $l > k$. By our preliminary remarks, we are done.

35. Let $f : \mathcal{R}^3 \longrightarrow \mathcal{R}$ be the linear function such that $f(\mathbf{e}_1) = 1$, $f(\mathbf{e}_2) = 2$ and $f(\mathbf{e}_3) = 1$. Then the matrix representing f is the 1-by-3 matrix whose kth column is $f(\mathbf{e}_k)$. That is,

$$f(\mathbf{x}) = (1 \quad 2 \quad 1)\,\mathbf{x}, \quad \text{for all } \mathbf{x} \in \mathcal{R}^3.$$

In order to solve the equation $f(\mathbf{x}) = 1$, set $\mathbf{x} = \begin{pmatrix} x \\ y \\ z \end{pmatrix}$ and write

$$f(\mathbf{x}) = f\begin{pmatrix} x \\ y \\ z \end{pmatrix} = (1 \quad 2 \quad 1)\begin{pmatrix} x \\ y \\ z \end{pmatrix} = x + 2y + z = 1,$$

where we have removed the unnecessary parentheses from the 1-by-1 matrix $(x+2y+z)$. The solutions of the equation $x + 2y + z = 1$ are precisely the points in the plane $x + 2y + z = 1$. Noting that $(1, 2, 1)$ is normal to this plane, and that $(1, 0, 0)$ is a point on this plane, we conclude that the solutions of $f(\mathbf{x}) = 1$ are precisely the vectors in the plane perpendicular to $(1, 2, 1)$ and passing through $(1, 0, 0)$.

37. The given statement is always true. To see this, suppose S is a subspace of the vector space W and that $\mathbf{x}$ is a vector such that $\mathbf{x} \in W$ but $\mathbf{x} \notin S$. Further, suppose the set

$$V = \{\mathbf{x} + \mathbf{y} \mid \mathbf{y} \in S\}$$

is a subspace of W. If $\mathbf{y}_1$ is a fixed vector in S then $\mathbf{x} + \mathbf{y}_1$ is in V and therefore, since V is closed with respect to vector addition, $2(\mathbf{x} + \mathbf{y}_1)$ is in V. Hence, $2(\mathbf{x} + \mathbf{y}_1) = \mathbf{x} + \mathbf{y}_2$ for some $\mathbf{y}_2 \in S$, or equivalently, $\mathbf{x} = \mathbf{y}_2 - 2\mathbf{y}_1$ for some $\mathbf{y}_2 \in S$. Since S is a subspace of W, and since $\mathbf{y}_1$ and $\mathbf{y}_2$ are both in in S, it follows that the linear combination $\mathbf{y}_1 - 2\mathbf{y}_2$ is also in S. That is, $\mathbf{x} \in S$, contrary to supposition. We conclude that V can't be a subspace of W.

39. The given statement is sometimes false. For example, let $W = \mathcal{R}^2$ and let S_1 be the subspace consisting of all scalar multiples of $(1, 0)$ and let S_2 be the subspace consisting of all scalar multiples of $(0, 1)$. Any subspace of W that contains both S_1 and S_2 must contain $(1, 0)$ and $(0, 1)$ and therefore must contain every linear combination of $(1, 0)$ and $(0, 1)$. That is, any such subspace must contain all vectors of the form $x(1, 0) + y(0, 1) = (x, y)$, where x and y are any two scalars. Since this is all of $\mathcal{R}^2$, we conclude that there is no *proper* subspace of $\mathcal{R}^2$ that contains both S_1 and S_2.

41. Since every subspace of $\mathcal{R}^3$ contains the zero vector, and since the zero vector does not satisfy the equation $\mathbf{x} \cdot (1, 2, 1) = 1$, it follows that there is no subspace S of $\mathcal{R}^3$ such that $\mathbf{x} \cdot (1, 2, 1) = 1$ for all $\mathbf{x}$ in S. So the given statement is true.

Section 3: LINEAR FUNCTIONS

Exercise Set 3ABC (pgs. 125-126)

1. With f as given, the columns of the matrix A such that $f(\mathbf{x}) = A\mathbf{x}$ for all $\mathbf{x} \in \mathcal{R}^2$ are the vectors $f(\mathbf{e}_1) = (1,2)$ and $f(\mathbf{e}_2) = (2,1)$. That is,

$$A = \begin{pmatrix} 1 & 2 \\ 2 & 1 \end{pmatrix}.$$

3. With f as given, the columns of the matrix A such that $f(\mathbf{x}) = A\mathbf{x}$ for all $\mathbf{x} \in \mathcal{R}^2$ are the the vectors $f(\mathbf{e}_1)$ and $f(\mathbf{e}_2)$. To find these image vectors, we first note that the domain vectors $(1,1)$ and $(2,1)$ can be written as $(1,1) = \mathbf{e}_1 + \mathbf{e}_2$ and $(2,1) = 2\mathbf{e}_1 + \mathbf{e}_2$. We then use the linearity of f and the given images of these domain vectors to obtain the system of equations

$$f(1,1) = f(\mathbf{e}_1 + \mathbf{e}_2) = f(\mathbf{e}_1) + f(\mathbf{e}_2) = (1,2),$$
$$f(2,1) = f(2\mathbf{e}_1 + \mathbf{e}_2) = 2f(\mathbf{e}_1) + f(\mathbf{e}_2) = (2,1).$$

Subtracting the first equation from the second gives $f(\mathbf{e}_1) = (2,1) - (1,2) = (1,-1)$. Substituting this into the first equation gives $(1,-1) + f(\mathbf{e}_2) = (1,2)$, or equivalently, $f(\mathbf{e}_2) = (1,2) - (1,-1) = (0,3)$. Hence,

$$A = \begin{pmatrix} 1 & 0 \\ -1 & 3 \end{pmatrix}.$$

5. Let $(x_1, x_2, x_3, \ldots)$ and $(y_1, y_2, y_3, \ldots)$ be in $\mathcal{R}^\infty$ and let r and s be scalars. Then

$$
\begin{aligned}
f\big(r(x_1, x_2, x_3 \ldots) + s(y_1, y_2, y_3, \ldots)\big) &= f(rx_1 + sy_1, rx_2 + sy_2, rx_3 + sy_3, \ldots) \\
&= 2(rx_1 + sy_1, rx_2 + sy_2, rx_3 + sy_3, \ldots) \\
&= 2\big(r(x_1, x_2, x_3, \ldots) + s(y_1, y_2, y_3, \ldots)\big) \\
&= r\big(2(x_1, x_2, x_3, \ldots)\big) + s\big(2(y_1, y_2, y_3, \ldots)\big) \\
&= rf(x_1, x_2, x_3, \ldots) + sf(y_1, y_2, y_3, \ldots).
\end{aligned}
$$

So f is linear. Also, $f(x_1, x_2, x_3, \ldots) = (0,0,0,\ldots)$ only if $2(x_1, x_2, x_3, \ldots) = (0,0,0,\ldots)$, which holds only if $x_k = 0$ for $k = 1, 2, 3 \ldots$. That is, $f(x_1, x_2, x_3, \ldots) = (0,0,0,\ldots)$ only if $(x_1, x_2, x_3, \ldots)$ is the zero sequence, so that f is one-to-one.

To describe f^{-1}, divide both sides of the equation defining f by 2 and use the linearity of f to get $(x_1, x_2, x_3, \ldots) = \frac{1}{2}f(x_1, x_2, x_3, \ldots) = f(\frac{1}{2}(x_1, x_2, x_3, \ldots))$. Now apply f^{-1} to both sides to get

$$f^{-1}(x_1, x_2, x_3, \ldots) = f^{-1}f\left(\frac{1}{2}(x_1, x_2, x_3, \ldots)\right) = \frac{1}{2}(x_1, x_2, x_3, \ldots).$$

Since this holds for all $(x_1, x_2, x_3, \ldots)$ in $\mathcal{R}^\infty$, the domain of f^{-1} is all of $\mathcal{R}^\infty$.

7. Let $(x_1, x_2, x_3, \ldots)$ and $(y_1, y_2, y_3, \ldots)$ be in $\mathcal{R}^\infty$ and let r and s be scalars. Then

$$
\begin{aligned}
h\big(r(x_1, x_2, x_3 \ldots) + s(y_1, y_2, y_3, \ldots)\big) &= h(rx_1 + sy_1, rx_2 + sy_2, rx_3 + sy_3, \ldots) \\
&= (rx_2 + sy_2, rx_3 + sy_3, rx_4 + sy_4, \ldots) \\
&= r(x_2, x_3, x_4, \ldots) + s(y_2, y_3, y_4, \ldots) \\
&= rh(x_1, x_2, x_3, \ldots) + sh(y_1, y_2, y_3, \ldots).
\end{aligned}
$$

So h is linear. However, h is not one-to-one because $h(x_1, 0, 0, \ldots) = (0, 0, 0, \ldots)$ for each of the infinitely many sequences $(x_1, 0, 0, \ldots)$ in $\mathcal{R}^\infty$.

9. Let f and g be as in Exercises 5 and 6 in this exercise set. Then, for $(x_1, x_2, x_3, \ldots)$ in $\mathcal{R}^\infty$,

$$
\begin{aligned}
(f \circ g)(x_1, x_2, x_3, \ldots) &= f\big(g(x_1, x_2, x_3, \ldots)\big) = f(x_1, 2x_2, 3x_3, \ldots) = 2(x_1, 2x_2, 3x_3, \ldots) \\
&= (2x_1, 4x_2, 6x_3, \ldots), \\
(g \circ f)(x_1, x_2, x_3, \ldots) &= g\big(f(x_1, x_2, x_3, \ldots)\big) = g\big(2(x_1, x_2, x_3, \ldots)\big) = g(2x_1, 2x_2, 2x_3, \ldots) \\
&= (2x_1, 4x_2, 6x_3, \ldots).
\end{aligned}
$$

11. Let g and p be as in Exercises 6 and 8 in this exercise set. Then, for $(x_1, x_2, x_3, \ldots)$ in $\mathcal{R}^\infty$,

$$
\begin{aligned}
(g \circ p)(x_1, x_2, x_3, \ldots) &= g\big(p(x_1, x_2, x_3, \ldots)\big) = g(0, x_1, x_2, x_3, \ldots) \\
&= (0, 2x_1, 3x_2, 4x_3, \ldots), \\
(p \circ g)(x_1, x_2, x_3, \ldots) &= p\big(g(x_1, x_2, x_3, \ldots)\big) = p(x_1, 2x_2, 3x_3, \ldots) \\
&= (0, x_1, 2x_2, 3x_3, \ldots).
\end{aligned}
$$

13. Let $M_{q,m}$ be the vector space of all q-by-m matrices, let $M_{q,n}$ be the vector space of all q-by-n matrices, let B be a fixed m-by-n matrix and let $g_B : M_{q,m} \longrightarrow M_{q,n}$ be defined by $g_B(X) = XB$ for $X \in M_{q,m}$. Let X and Y be matrices in $M_{q,m}$ and let r and s be scalars. Using the right distributive law and the scalar commutativity law of Theorem 3.2 in Chapter 2 (basic properties of matrix products), we have

$$
g_B(rX + sY) = (rX + sY)B = (rX)B + (sY)B = r(XB) + s(YB) = rg_B(X) + sg_B(Y).
$$

So g_B is linear.

15. Let $D = d/dx$ and $u(x) = 2x^3 - 4x$. Then

$$
Du(x) = \frac{du}{dx} = \frac{d(2x^3 - 4x)}{dx} = 6x^2 - 4, \qquad xu(x) = x(2x^3 - 4x) = 2x^4 - 4x^2,
$$

$$
D(xu(x)) = \frac{d(xu)}{dx} = \frac{d(2x^4 - 4x^2)}{dx} = 8x^3 - 8x, \qquad xDu(x) = x(6x^2 - 4) = 6x^3 - 4x.
$$

17. Let $D = d/dx$. (CAUTION: In parts (a) and (b), the transformation Dx is not to be interpreted as D acting on the polynomial x, but rather as the transformation that first multiplies by x and then differentiates with respect to x.)

(a) If $u(x) = 2x^3$ then

$$
\begin{aligned}
(Dx - xD)u(x) &= (Dx - xD)(2x^3) = Dx(2x^3) - xD(2x^3) \\
&= D(2x^4) - x(6x^2) = 8x^3 - 6x^3 = 2x^3.
\end{aligned}
$$

(b) For $u = u(x)$ in $C^{(1)}(-\infty, \infty)$ we have

$$(Dx - xD)u = (Dx)u - (xD)u = D(ux) - x(Du) = (xu' + u) - xu' = u.$$

Hence, the transformation acts on a function by doing nothing to it. That is, it acts as the identity transformation $Dx - xD = I$.

(c) The transformation $D^2 - x^2$ and $(D + x)(D - x)$ are not the same transformation. To see this, let $u = u(x)$ be any function in $C^{(2)}(-\infty, \infty)$. On one hand,

$$(D^2 - x^2)u = D^2u - x^2u = u'' - x^2u;$$

and on the other hand,

$$(D + x)(D - x)u = (D + x)(Du - xu) = (D + x)(u' - xu) = D(u' - xu) + x(u' - xu)$$
$$= (u'' - xu' - u) + xu' - x^2u = u'' - (1 + x^2)u.$$

Since $u'' - x^2y \neq u'' - (1 + x^2)u$ for all $u \in C^{(2)}(-\infty, \infty)$, $D^2 - x^2$ and $(D + x)(D - x)$ can't be the same transformation.

19. Define the function $S : C[0, 1] \longrightarrow C[0, 1]$ by $Su(x) = \int_0^x u(t)\, dt$ for $u(x) \in C[0, 1]$. Let $u(x)$ and $v(x)$ be in $C[0, 1]$ and let a and b be scalars (real numbers). Then linearity of the integral gives

$$S(au(x) + bv(x)) = \int_0^x (au(t) + bv(t))\, dt = a \int_0^x u(t)\, dt + b \int_0^x v(t)\, dt = aSu(x) + bSv(x).$$

So S is linear.

21. Let $L : \mathcal{R}^2 \longrightarrow \mathcal{R}$ be a linear function and let $\mathbf{u}_0$, $\mathbf{u}_1$ be in $\mathcal{R}^2$. Given that $L(\mathbf{u}_0) = 1$ and $L(\mathbf{u}_1) = -2$,

$$L(3\mathbf{u}_0 - 4\mathbf{u}_1) = 3L(\mathbf{u}_0) - 4L(\mathbf{u}_1) = 3(1) - 4(-2) = 11.$$

23. Let $L : \mathcal{R}^3 \longrightarrow \mathcal{R}^2$ with $L(\mathbf{e}_1) = (1, 2)$, $L(\mathbf{e}_2) = (-1, 0)$ and $L(\mathbf{e}_3) = (2, 2)$. Given that L is linear,

$$L(-1, 3, 2) = L(-\mathbf{e}_1 + 3\mathbf{e}_2 + 2\mathbf{e}_3) = -L(\mathbf{e}_1) + 3L(\mathbf{e}_2) + 2L(\mathbf{e}_3)$$
$$= -(1, 2) + 3(-1, 0) + 2(2, 2) = (0, 2).$$

25. Let $L : C[0, 1] \longrightarrow C[0, 1]$ be a linear function such that $L(1) = x$ and $L(x) = x^2$. Then

$$L(2x + 3) = 2L(x) + 3L(1) = 2x^2 + 3x.$$

27. Let $L : C^{(1)}[0, 1] \longrightarrow C[0, 1]$ be defined by $Lf = f' - 2f$. In order that a function f be in $C^{(1)}[0, 1]$ it is first of all necessary that f' be defined at each point in the closed interval $[0, 1]$. Since $f'(0)$ is not defined for the function $f(x) = |x|$, $|x|$ is not in $C^{(1)}[0, 1]$, the domain of L.

29. Let $L : C[0, 1] \longrightarrow C[0, 1]$ be defined by $Lf(x) = xf(x)$. In order that there be a function f in $C[0, 1]$ such that $Lf(x) = x^2 + 1$ it is necessary and sufficient that $xf(x) = x^2 + 1$. Thus, we can find such an f if this equation can be solved for $f(x)$ on the closed interval $[0, 1]$. Since this can't be done (we can't divide by zero), there is no such f.

31. Let $f : \mathcal{R}^2 \longrightarrow \mathcal{R}^2$ be defined by $f(\mathbf{x}) = A\mathbf{x}$ for $\mathbf{x} \in \mathcal{R}^2$, where A is the square matrix $A = \begin{pmatrix} 6 & 3 \\ 4 & 2 \end{pmatrix}$. Since $\det A = 0$, A^{-1} does not exist. By Theorem 1.6 (Chapter 3), f^{-1} can't exist either.

33. Let V be the subspace of $\mathcal{R}^3$ consisting of all linear combinations of $(1,1,1)$ and $(1,2,3)$. Writing the vectors in V as column vectors we see that the vectors $\mathbf{x} \in V$ can be written as

$$(1) \qquad \mathbf{x} = a \begin{pmatrix} 1 \\ 1 \\ 1 \end{pmatrix} + b \begin{pmatrix} 1 \\ 2 \\ 3 \end{pmatrix} = \begin{pmatrix} a+b \\ a+2b \\ a+3b \end{pmatrix} = \begin{pmatrix} 1 & 1 \\ 1 & 2 \\ 1 & 3 \end{pmatrix} \begin{pmatrix} a \\ b \end{pmatrix}.$$

Thus, if $f : V \longrightarrow \mathcal{R}^2$ is defined by $f(\mathbf{x}) = \begin{pmatrix} 1 & 1 & 3 \\ -1 & 2 & 2 \end{pmatrix} \mathbf{x}$ for $\mathbf{x} \in V$ then

$$(2) \quad f(\mathbf{x}) = f \begin{pmatrix} a+b \\ a+2b \\ a+3b \end{pmatrix} = \begin{pmatrix} 1 & 1 & 3 \\ -1 & 2 & 2 \end{pmatrix} \begin{pmatrix} a+b \\ a+2b \\ a+3b \end{pmatrix}$$

$$= \begin{pmatrix} a+b+a+2b+3(a+3b) \\ -(a+b)+2(a+2b)+2(a+3b) \end{pmatrix} = \begin{pmatrix} 5a+12b \\ 3a+9b \end{pmatrix} = \begin{pmatrix} 5 & 12 \\ 3 & 9 \end{pmatrix} \begin{pmatrix} a \\ b \end{pmatrix}.$$

Letting $A = \begin{pmatrix} 5 & 12 \\ 3 & 9 \end{pmatrix}$ and $\mathbf{a} = \begin{pmatrix} a \\ b \end{pmatrix}$, we see that the columns of A are linearly independent. Therefore, the only solution of $A\mathbf{a} = \mathbf{0}$ is $\mathbf{a} = \mathbf{0}$, so that $a = b = 0$. Hence, by equation (1) above, $\mathbf{x} = \mathbf{0}$. That is, f is one-to-one and has an inverse f^{-1}. The second thing we can deduce from equation (2) is that the image of f (the domain of f^{-1}) is all of $\mathcal{R}^2$.

We can find an explicit formula for $f^{-1}(\mathbf{y})$ for $\mathbf{y}$ in $\mathcal{R}^2$ in the following way. Equation (2) shows that f sends $\mathbf{x}$ in V (where $\mathbf{x}$ is given by (1)) to the point $\mathbf{y} = A\mathbf{a}$. Therefore, f^{-1} sends the point $\mathbf{y} = A\mathbf{a}$ in $\mathcal{R}^2$ to the point $\mathbf{x}$. Left-multiplying both sides of the equation $A\mathbf{a} = \mathbf{y}$ by $A^{-1} = \begin{pmatrix} 1 & -4/3 \\ -1/3 & 5/9 \end{pmatrix}$ gives

$$\mathbf{a} = \begin{pmatrix} 1 & -4/3 \\ -1/3 & 5/9 \end{pmatrix} \mathbf{y}.$$

Therefore, by equation (1) above,

$$\mathbf{x} = \begin{pmatrix} 1 & 1 \\ 1 & 2 \\ 1 & 3 \end{pmatrix} \begin{pmatrix} 1 & -4/3 \\ -1/3 & 5/9 \end{pmatrix} \mathbf{y} = \begin{pmatrix} 2/3 & -7/9 \\ 1/3 & -2/9 \\ 0 & 1/3 \end{pmatrix} \mathbf{y}.$$

That is,

$$f^{-1}(\mathbf{y}) = \begin{pmatrix} 2/3 & -7/9 \\ 1/3 & -2/9 \\ 0 & 1/3 \end{pmatrix} \mathbf{y}, \quad \text{for } \mathbf{y} \in \mathcal{R}^2.$$

35. Let $D : V \longrightarrow C(-\infty, \infty)$ be the differential transformation $D = d/dx$ and let V be the subspace of $C^{(1)}(-\infty, \infty)$ consisting of the continuously differentiable functions with $u(0) = u(1)$. Note that V contains all constant functions (they all have the same value at $x = 0$ and $x = 1$) and that their derivatives are all identically zero. That is, $Du = 0$ for infinitely many functions in V and therefore D is not one-to-one. So D doesn't have an inverse.

Section 4: IMAGE AND NULL-SPACE

Exercise Set 4ABC (pgs. 130-131)

1. By inspection, the given function f maps a vector in $\mathcal{R}^2$ to a vector in $\mathcal{R}^3$, so that $\mathcal{R}^3$ is the range of f. Using column vectors, f can be written as

$$f\begin{pmatrix} x \\ y \end{pmatrix} = \begin{pmatrix} x \\ y \\ x+y \end{pmatrix} = x\begin{pmatrix} 1 \\ 0 \\ 1 \end{pmatrix} + y\begin{pmatrix} 0 \\ 1 \\ 1 \end{pmatrix} = \begin{pmatrix} 1 & 0 \\ 0 & 1 \\ 1 & 1 \end{pmatrix}\begin{pmatrix} x \\ y \end{pmatrix},$$

which shows that $f(\mathcal{R}^2)$, the image of f, is the plane through the origin in $\mathcal{R}^3$ parallel to $(1,0,1)$ and $(0,1,1)$. Hence, $f(\mathcal{R}^2)$ is a subspace of $\mathcal{R}^3$. The above equation also displays a matrix representation for f, so that f is linear. The fact that the columns of the matrix of f are linearly independent implies f is one-to-one, so the null-space is $N = \{\mathbf{0}\}$.

3. By inspection, the given function f maps a vector in $\mathcal{R}^2$ to a vector in $\mathcal{R}^3$, so that $\mathcal{R}^3$ is the range of f. To find $f(\mathcal{R}^2)$, the image of f, write

$$f(u,v) = (u, v, 2u + v + 1) = u(1,0,2) + v(0,1,1) + (0,0,1),$$

and then observe that this is the equation of the plane containing the point $(0,0,1)$ in $\mathcal{R}^3$ parallel to $(1,0,2)$ and $(0,1,1)$. However, it does not contain the origin and is therefore not a subspace of $\mathcal{R}^3$. For linearity, note that

$$f\big(2(1,1)\big) = f(2,2) = (2,2,7) \neq (2,2,8) = 2(1,1,4) = 2f(1,1),$$

so that f is not linear.

5. By inspection, the given function f maps a vector in $\mathcal{R}^3$ to a vector in $\mathcal{R}^2$, so that $\mathcal{R}^2$ is the range of f. To find $f(\mathcal{R}^3)$, the image of f, write

$$\begin{aligned} f(x,y,z) &= (x + 4y + 3z, 2x + 5y + 4z) = x(1,2) + y(4,5) + z(3,4) \\ &= x(1,2) + y(4,5) + z\left(\frac{1}{3}(1,2) + \frac{2}{3}(4,5)\right) \\ &= \left(x + \frac{1}{3}z\right)(1,2) + \left(y + \frac{2}{3}z\right)(4,5), \end{aligned}$$

and observe that $x + \frac{1}{3}z$ and $y + \frac{2}{3}z$ range independently over the real numbers as (x,y,z) ranges over $\mathcal{R}^3$. Therefore, since $(1,2)$ and $(4,5)$ are linearly independent, $f(\mathcal{R}^3)$ is all of $\mathcal{R}^2$. Hence, $f(\mathcal{R}^3)$ is a subspace of $\mathcal{R}^2$. For linearity, use column vectors to write f as

$$f\begin{pmatrix} x \\ y \\ z \end{pmatrix} = \begin{pmatrix} x + 4y + 3z \\ 2x + 5y + 4z \end{pmatrix} = \begin{pmatrix} 1 & 4 & 3 \\ 2 & 5 & 4 \end{pmatrix}\begin{pmatrix} x \\ y \\ z \end{pmatrix}.$$

This shows that f has a matrix representation and is therefore linear.

The null-space N is the solution set of the homogeneous matrix equation

$$\begin{pmatrix} 1 & 4 & 3 \\ 2 & 5 & 4 \end{pmatrix}\mathbf{x} = \mathbf{0}.$$

Adding -2 times the first row to the second row; dividing the new second row by -3; and then adding -4 times the new second row to the first row results in the reduced matrix equation

$$\begin{pmatrix} 1 & 0 & 1/3 \\ 0 & 1 & 2/3 \end{pmatrix} \mathbf{x} = \mathbf{0},$$

which is equivalent to the scalar system

$$x + \frac{1}{3}z = 0,$$

$$y + \frac{2}{3}z = 0.$$

Since z is the only nonleading variable, let $z = t$ and deduce from the first equation that $x = -\frac{1}{3}t$ and from the second equation that $y = -\frac{2}{3}t$. Hence, $f(\mathbf{x}) = \mathbf{0}$ if, and only if, $\mathbf{x} = (-\frac{1}{3}t, -\frac{2}{3}t, t) = -\frac{1}{3}t(1,2,-3)$, which is a line through the origin in $\mathcal{R}^3$ in the direction of $(1,2,-3)$. Therefore, letting $s = -\frac{1}{3}t$, the null-space can be written as

$$N = \{\, s(1,2,-3) \,|\, s \text{ is a scalar } \}.$$

7. Let $F : C(-\infty,\infty) \longrightarrow C(-\infty,\infty)$ be defined by $\big(F(u)\big)(x) = u(x) + x$ for $u = u(x)$ in $C(-\infty,\infty)$. Observe that if $u = u(x)$ is continuous on $(-\infty,\infty)$ then so is $u - x$. Hence, since

$$\big(F(u-x)\big)(x) = \big(u(x) - x\big) + x = u(x),$$

every function in $C(-\infty,\infty)$ is in the image of F. Hence, $F\big(C(-\infty,\infty)\big)$, the image of F, is all of $C(-\infty,\infty)$. As for linearity, note that F maps a constant function $u(x) = c$ to $c + x$, so that

$$\big(F(2)\big)(x) = 2 + x, \quad \text{and} \quad 2\big(F(1)\big)(x) = 2(1 + x) = 2 + 2x.$$

Since $F(2) \neq 2F(1)$, F is not linear.

9. Let $F : C(-\infty,\infty) \longrightarrow C^{(1)}(-\infty,\infty)$ be defined by $\big(F(u)\big)(x) = \int_0^x e^{-t}u(t)\,dt$ for $u = u(x)$ in $C(-\infty,\infty)$. The first thing we observe is that if $v(x)$ is in the image of F then $v(x) = \int_0^x e^{-t}u(t)\,dt$ must hold for all x and some $u(x) \in C(-\infty,\infty)$. In particular, $v(0) = 0$ is a necessary condition for $v(x)$ to be in the image of F. On the other hand, if $v(x)$ is continuously differentiable on $(-\infty,\infty)$ with $v(0) = 0$, then the function $e^x v'(x)$ is continuous on $(-\infty,\infty)$ (i.e., is in $C(-\infty,\infty)$) and

$$\Big(F\big(e^x v'(x)\big)\Big)(x) = \int_0^x e^{-t}\big(e^t v'(t)\big)\,dt = \int_0^x v'(t)\,dt = v(x) - v(0) = v(x),$$

so that $v(x)$ is in the image of F. That is, $v(0) = 0$ is a sufficient condition for a function in $C^{(1)}(-\infty,\infty)$ to be in the image of F. It follows that $F\big(C(-\infty,\infty)\big)$, the image of F, is

$$F\big(C(-\infty,\infty)\big) = \{\, v(x) \in C^{(1)}(-\infty,\infty) \,|\, v(0) = 0 \,\}.$$

Moreover, F is linear because linearity of the integral implies

$$\big(F(au_1 + bu_2)\big)(x) = \int_0^x e^{-t}\big(au_1(t) + bu_2(t)\big)\,dt = a\int_0^x e^{-t}u_1(t)\,cr + a\int_0^x e^{-t}u_2(t)\,dt$$
$$= a\big(F(u_1)\big)(x) + b\big(F(u_2)\big)(x),$$

for all $u_1, u_2 \in C(-\infty, \infty)$ and scalars a, b. And finally, if $(F(u))(x)$ is identically zero then $\int_0^x e^{-t} u(t)\, dt = 0$, identically. Differentiating this with respect to x shows that $e^{-x} u(x) = 0$, identically. Since e^{-x} is never zero, $u(x)$ is identically zero. That is, the null-space of F is $N = \{$ the zero function $\}$.

11. First of all, given that f is defined by $f(\mathbf{x}) = A\mathbf{x}$, where $A = \begin{pmatrix} 1 & 1 \\ 0 & 1 \end{pmatrix}$ and $\mathbf{x}$ is a vector, we conclude that $\mathbf{x}$ must be in $\mathcal{R}^2$ in order for the product $A\mathbf{x}$ to be defined. That is, the domain of f is $\mathcal{R}^2$. Further, for such vectors, $A\mathbf{x}$ is in $\mathcal{R}^2$, so that $\mathcal{R}^2$ is also the range of f. In other words, $f : \mathcal{R}^2 \longrightarrow \mathcal{R}^2$.

Next, the fact that A is a square matrix with linearly independent columns implies A is invertible. Therefore, for each $\mathbf{b}$ in $\mathcal{R}^2$ the equation $f(\mathbf{x}) = \mathbf{b}$ has the unique solution $\mathbf{x} = A^{-1}\mathbf{b}$. This means two things: (i) $f(A^{-1}\mathbf{b}) = \mathbf{b}$ and the image of f is all of $\mathcal{R}^2$, and (ii) the null-space N of f is $N = \{\mathbf{0}\}$ (only the zero solution when $\mathbf{b}$ is the zero vector).

13. Given that f is defined by $f(\mathbf{x}) = A\mathbf{x}$, where $A = \begin{pmatrix} 2 & 4 & 1 \\ 0 & 1 & 0 \\ 2 & 1 & 1 \end{pmatrix}$ and $\mathbf{x}$ is a vector, we conclude that $\mathbf{x}$ must be in $\mathcal{R}^3$ in order for the product $A\mathbf{x}$ to be defined. That is, the domain of f is $\mathcal{R}^3$. Further, for such vectors, $A\mathbf{x}$ is in $\mathcal{R}^3$, so that $\mathcal{R}^3$ is also the range of f. In other words, $f : \mathcal{R}^3 \longrightarrow \mathcal{R}^3$.

To find the image of f and the null-space of f, let $\mathbf{x} = \begin{pmatrix} x \\ y \\ z \end{pmatrix}$ and note that

$$f(\mathbf{x}) = \begin{pmatrix} 2 & 4 & 1 \\ 0 & 1 & 0 \\ 2 & 1 & 1 \end{pmatrix} \begin{pmatrix} x \\ y \\ z \end{pmatrix} = \begin{pmatrix} 2x + 4y + z \\ y \\ 2x + y + z \end{pmatrix} = (2x + z) \begin{pmatrix} 1 \\ 0 \\ 1 \end{pmatrix} + y \begin{pmatrix} 4 \\ 1 \\ 1 \end{pmatrix}.$$

Since $2x + z$ and y vary independently as (x, y, z) ranges over $\mathcal{R}^3$, and since $(1, 0, 1)$, $(4, 1, 1)$ are linearly independent, the above equation shows that $f(\mathcal{R}^3)$, the image of f, is the plane through the origin in $\mathcal{R}^3$ parallel to $(1, 0, 1)$ and $(4, 1, 1)$ and is formally given by

$$f(\mathcal{R}^3) = \{\, s(1, 0, 1) + t(4, 1, 1) \,|\, s \text{ and } t \text{ are scalars} \,\}.$$

It also shows that $f(\mathbf{x}) = \mathbf{0}$ only if $\mathbf{x} = (x, y, z)$ is such that $2x + z = 0$ and $y = 0$. The first of these equations says $z = -2x$, so that $\mathbf{x} = (x, y, z) = (x, 0, -2x) = x(1, 0, -2)$. That is, the null-space N of f is the line through the origin in $\mathcal{R}^3$ in the direction of $(1, 0, -2)$ and is formally given by

$$N = \{\, t(1, 0, -2) \,|\, t \text{ is a scalar} \,\}.$$

15. This exercise deals with the function $f : \mathcal{R}^2 \longrightarrow \mathcal{R}$ defined by $f(x, y) = 2x - 5y$ for (x, y) in $\mathcal{R}^2$. Since $2x - 5y = (2, -5) \cdot (x, y)$, f has a matrix representation with matrix $(2 \quad -5)$ and is therefore linear.

(a) A solution pair x, y satisfies the homogeneous equation $2x - 5y = 0$ if, and only if, $y = \frac{2}{5}x$. In order to eliminate fractions, we can introduce the parameter t by setting $x = 5t$, which then allows us to write $y = 2t$. That is, all solution pairs x, y of $2x - 5y = 0$ are of the form $x = 5t$, $y = 2t$, where t is a real number. In terms of f, a vector $\mathbf{x}$ is in the null-space N of f (i,e, $f(\mathbf{x}) = \mathbf{0}$) if, and only if, $\mathbf{x} = (5t, 2t)$, where t is a scalar.

(b) We have already shown that f is linear. By Theorem 4.2, N is a subspace of $\mathcal{R}^2$. Hence, every finite linear combination of vectors in N is also in N. That is, $f(a\mathbf{x}_1 + b\mathbf{x}_2) = \mathbf{0}$

for all $\mathbf{x}_1, \mathbf{x}_2$ in N and all scalars a, b. This is equivalent to saying that if x_1, y_1 and x_2, y_2 are any two solution pairs of the homogeneous equation $2x - 5y = 0$ and a, b are real numbers then the pair $ax_1 + bx_2$, $ay_1 + by_2$ is also a solution.

(c) By inspection, $x = 1$, $y = -1$ is a solution pair of the nonhomogeneous equation $2x - 5y = 7$. That is, $f(1, -1) = 7$. Using this particular solution and the solution found in part (a), Theorem 4.4 says that all solutions of the nonhomogeneous equation $f(\mathbf{x}) = 7$ are given by $\mathbf{x} = (5t, 2t) + (1, -1) = (5t + 1, 2t - 1)$, where t is a scalar. That is, the solution pairs x, y of the nonhomogeneous equation $2x - 5y = 7$ are of the form

$$x = 5t + 1, \ y = 2t - 1, \quad t \text{ is a real number.}$$

17. Let $f : \mathcal{R}^n \longrightarrow \mathcal{R}$ be linear. Let $\mathbf{e}_k$ be the kth standard basis vector in $\mathcal{R}^n$ and let $f(\mathbf{e}_k) = c_k$ for $k = 1, \dots, n$. Define $\mathbf{x}_0 \in \mathcal{R}^n$ by $\mathbf{x}_0 = (c_1, \dots, c_n)$. Then, since f is linear, f sends the vector $\mathbf{x} = (x_1, \dots, x_n)$ in $\mathcal{R}^n$ to the real number

$$f(\mathbf{x}) = f(x_1, \dots, x_n) = f(x_1\mathbf{e}_1 + \dots + x_n\mathbf{e}_n) = x_1 f(\mathbf{e}_1) + \dots + x_n f(\mathbf{e}_n)$$
$$= x_1 c_1 + \dots + x_n c_n = (x_1, \dots, x_n) \cdot (c_1, \dots, c_n) = \mathbf{x} \cdot \mathbf{x}_0.$$

Thus, $f(\mathbf{x}) = 0$ if, and only if, $\mathbf{x} \cdot \mathbf{x}_0 = 0$. This is equivalent to

$$N = \{ \mathbf{x} \in \mathcal{R}^n \,|\, \mathbf{x} \text{ is orthogonal to } \mathbf{x}_0 \}.$$

19. If $D = d/dx$ then the transformation $D - 1$ acts on functions $y = y(x)$ in the vector space $C^{(1)}(-\infty, \infty)$ by the rule $(D - 1)y = dy/dx - y$. Hence, the image of $D - 1$ lies in the vector space $C(-\infty, \infty)$.

(a) Let $y_1 = y_1(x)$ and $y_2 = y_2(x)$ be two solutions of $(D - 1)y = 0$ and let a, b be real numbers. Then $(D - 1)y_1 = 0$ and $(D - 1)y_2 = 0$ implies

$$(D - 1)(ay_1 + by_2) = \frac{d(ay_1 + by_2)}{dx} - (ay_1 + by_2) = a\frac{dy_1}{dx} + b\frac{dy_2}{dx} - ay_1 - by_2$$
$$= a\left(\frac{dy_1}{dx} - y_1\right) + b\left(\frac{dy_2}{dx} - y_2\right) = a(D - 1)y_1 + b(D - 1)y_2 = 0.$$

So $ay_1(x) + by_2(x)$ is a solution of the homogeneous equation $(D - 1)y = 0$.

(b) If $y(x) = x + 1$ then

$$(D - 1)y = (D - 1)(x + 1) = \frac{d(x + 1)}{dx} - (x + 1) = 1 - (x + 1) = -x,$$

so that $y(x) = x + 1$ is a solution of $(D - 1)y = -x$.

To find all solutions of the nonhomogeneous equation $(D - 1)y = -x$, we note that $D - 1$ is the difference of two linear functions and is therefore linear by Theorem 3.5 (this is also shown in the computations in part (a)). Moreover, we are given that all solutions of the homogeneous equation $(D - 1)y = 0$ are represented by $y(x) = ce^x$, where c is a constant. Using these facts, along with the nonhomogeneous solutions found above, Theorem 4.4 says that all solutions of the nonhomogeneous equation $(D - 1)y = -x$ are given by

$$y(x) = ce^x + x + 1, \quad c \text{ is a constant.}$$

21. Let $C = C(-\infty, \infty)$ and define the function $G : C \longrightarrow C$ by

$$Gu(x) = \int_0^x tu(t)\, dt.$$

(a) To show that G is linear, let $u_1(x)$ and $u_2(x)$ be in C and let a, b be scalars. Then, using the linearity of the integral,

$$G\big(au_1(x) + bu_2(x)\big) = \int_0^x t\big(au_1(t) + bu_2(t)\big)\, dt = a\int_0^x tu_1(t)\, dt + b\int_0^x tu_2(t)\, dt$$
$$= aGu_1(x) + bGu_2(x).$$

So G is linear.

(b) To show that G is one-to-one, we first show that $Gu(x) = 0$ for all x only if $u(x) = 0$ for all x. To this end, suppose $u(x)$ satisfies $Gu(x) = 0$ for all x. Then $\int_0^x tu(t)\, dt = 0$ for all x. Differentiating this equation with respect to x gives $0 = xu(x)$ for all x. Hence, $u(x) = 0$ for all nonzero x. But since $u(x)$ is continuous at $x = 0$, $u(0) = 0$ as well. That is, $u(x)$ is the zero function. Now use Theorem 3.3 to conclude that G is one-to-one.

(c) The vector space P_n of all polynomials in x of degree $\leq n$ is a subspace of P, the set of all polynomials in x. Since P is a subspace of C, so is P_n. It therefore makes sense to consider $G(P_n)$, the image of P_n under G.

Let $p(x) = a_0 + a_1 x + \cdots + a_n x^n$ be in P_n. Then

$$Gp(x) = \int_0^x tp(t)\, dt = \int_0^x (a_0 t + a_1 t^2 + \cdots + a_n t^{n+1})dt$$
$$= \left[\frac{a_0}{2}t^2 + \frac{a_1}{3}t^3 + \cdots + \frac{a_n}{n+2}t^{n+2}\right]_0^x$$
$$= \frac{a_0}{2}x^2 + \frac{a_1}{3}x^3 + \cdots + \frac{a_n}{n+2}x^{n+2} = x^2\left(\frac{a_0}{2} + \frac{a_1}{3}x + \cdots + \frac{a_n}{n+2}x^n\right).$$

Thus, every function in $G(P_n)$ is a polynomial in P_n multiplied by x^2. The above computation suggests a way to show that every polynomial that is x^2 times a polynomial in P_n is in $G(P_n)$. To see this, take the polynomial $p(x)$ given above and define the polynomial $q(x)$ by $q(x) = 2a_0 + 3a_1 x + \cdots + (n+2)x^n$. Then $q(x)$ is in P_n and, by analogy to what we have computed above,

$$Gq(x) = x^2\left(\frac{2a_0}{2} + \frac{3a_1}{3}x + \cdots + \frac{(n+2)a_n}{n+2}x^n\right) = x^2(a_0 + a_1 x + \cdots + a_n x^n) = x^2 p(x).$$

It follows that
$$G(P_n) = \{x^2 p(x) \mid p(x) \text{ is in } P_n\}.$$

(d) To find $G(P)$, the image of P under G, we first observe that if $p(x)$ is a polynomial of degree $n \geq 0$ or if $p(x)$ is the zero polynomial then $p(x)$ is in P_n. That is, P is the union of all P_n. Also note that the result of part (c) holds for any nonnegative integer n. It follows that $G(P)$ is the union of all $G(P_n)$ and therefore can be written as

$$G(P) = \{x^2 p(x) \mid p(x) \text{ is in } P\}.$$

(e) To find G^{-1}, we first need to find the domain of G^{-1}, which is the same as the image of G. To this end, let $y(x)$ be in the image of G. Then $Gu(x) = y(x)$ for a unique $u(x)$ in C and $y(x) = \int_0^x tu(t)\, dt$. This shows several things. First, it shows that $y(0) = 0$. Second, since $tu(t)$ is continuous on every closed interval with endpoints 0 and x, the fundamental theorem of calculus implies $y(x)$ is differentiable on $(-\infty, \infty)$ and $y'(x) = xu(x)$ for all x, or equivalently, $u(x) = y'(x)/x$ for all x. This last condition implies $y'(x)/x$ is in C. These observations show that

$$\text{domain of } G^{-1} = \{\, y(x) \in C^{(1)}(-\infty, \infty) \,|\, y(0) = 0 \text{ and } y'(x)/x \text{ is in } C \,\}.$$

And, furthermore,

$$G^{-1}y(x) = \frac{1}{x}y'(x), \quad \text{for all } y(x) \text{ in the domain of } G^{-1}.$$

(f) It was shown in part (e) that all functions $y(x)$ in the image of G satisfy $y(0) = 0$. It follows that one function in the range of G that is not in the image of G is the constant function $y(x) = 1$.

23. Let $C = C(-\infty, \infty)$ and define the operator $R : C \longrightarrow C$ by $Ru(x) = u(-x)$. Also, let I be the identity operator $Iu(x) = u(x)$.

(a) Let $u(x)$ be in C and consider the graphs of both Ru and u in the xy-plane. Replacing x with $-x$ in the definition of $Ru(x)$ gives $Ru(-x) = u(x)$ for all x. Therefore, a typical point on the graph of Ru can be written as $\big(-x, u(x)\big)$. But this is also the reflection in the y-axis of the point $\big(x, u(x)\big)$, a point on the graph of u. Since this holds for all x, the graph of Ru is the reflection of the graph of u in the y-axis.

(b) To show that the operators R and I are linear, let $u_1(x)$ and $u_2(x)$ be in C and let a, b be scalars. Then

$$\big(R(au_1 + bu_2)\big)(x) = (au_1 + bu_2)(-x) = (au_1)(-x) + (bu_2)(-x)$$
$$= au_1(-x) + bu_2(-x) = aRu_1(x) + bRu_2(x),$$
$$\big(I(au_1 + bu_2)\big)(x) = (au_1 + bu_2)(x) = (au_1)(x) + (bu_2)(x)$$
$$= au_1(x) + bu_2(x) = aIu_1(x) + bIu_2(x).$$

So R and I are both linear. Moreover, for each u in C,

$$R^2u(x) = R(Ru)(x) = Ru(-x) = u(x),$$

so that R^2 and I have the same action on each u in C. That is, $R^2 = I$.

(c) Let $F_e = \frac{1}{2}(I + R)$. Then for an arbitrary u in C,

(1) $$F_e u(x) = \frac{1}{2}(I + R)u(x) = \frac{1}{2}\big(Iu(x) + Ru(x)\big) = \frac{1}{2}\big(u(x) + u(-x)\big).$$

We first show that the image $F_e(C)$ of F_e is the set of all even functions in C. To this end, observe that replacing x with $-x$ in the expression on the right of (1) doesn't alter the value of the expression (it just changes the order in which the terms are added). That is, every function in $F_e(C)$ is an even function. To show that $F_e(C)$ contains every even function in C, note that if u is an even function (so that $u(x) = u(-x)$ for all x) then (1) becomes $F_e u(x) = \frac{1}{2}\big(u(x) + u(x)\big) = u(x)$. That is, F_e maps an even function to itself and therefore u is in $F_e(C)$. It follows that $F_e(C)$ is precisely the set of even functions in C.

For the the null-space N of F, observe that the right side of (1) is identically zero if, and only if, $u(x) = -u(-x)$ for all x; i.e., if, and only if, u is an odd function in C. So, N is precisely the set of all odd functions in C.

103

(d) Let $F_o = I - F_e = \frac{1}{2}(I - R)$. Then for an arbitrary u in C,

$$(2) \qquad F_o u(x) = \frac{1}{2}(I - R)u(x) = \frac{1}{2}\big(Iu(x) - Ru(x)\big) = \frac{1}{2}\big(u(x) - u(-x)\big).$$

We first find the image $F_o(C)$ of F_o. To this end, observe that replacing x with $-x$ in the expression on the right of (2) changes the order in which the terms are subtracted. Hence, $F_o u(-x) = -F_o u(x)$ and $F_o u(x)$ is an odd function. That is, every function in $F_o(C)$ is an odd function. To show that $F_o(C)$ contains every odd function in C, note that if u is an odd function (so that $u(x) = -u(-x)$ for all x) then (2) becomes $F_o u(x) = \frac{1}{2}\big(u(x) + u(x)\big) = u(x)$. That is, F_o maps an odd function to itself and therefore u is in $F_o(C)$. It follows that $F_o(C)$ is precisely the set of odd functions in C.

For the null-space N of F_o, observe that the right side of (2) is identically zero if, and only if, $u(x) = u(-x)$ for all x; i.e., if, and only if, u is an even function in C. So, N is precisely the set of all even functions in C.

(e) Let $u(x)$ be in C. It was shown in part (c) that $F_e u(x)$ is an even function in C. It was also shown that F_e maps an even function to itself. Hence, F_e applied to $F_e u(x)$ gives the function $F_e u(x)$. That is,

$$F_e^2 u(x) = F_e(F_e u)(x) = F_e u(x) \quad \text{for all } u(x) \in C.$$

This means $F_e^2 = F_e$.

Similarly, it was shown in part (d) that $F_o u(x)$ is an odd function for all $u(x)$ in C, and that F_o maps an odd function to itself. Hence, F_o applied to $F_o u(x)$ gives the function $F_o u(x)$. That is,

$$F_o^2 u(x) = F_o(F_o u)(x) = F_o u(x) \quad \text{for all } u(x) \in C.$$

This means $F_o^2 = F_o$.

Section 5: COORDINATES AND DIMENSION

Exercise Set 5AB (pgs. 137-138)

1. Let $B = \{(-1,1),(1,1)\}$. To show that B spans $\mathcal{R}^2$, we arbitrarily choose (x,y) in $\mathcal{R}^2$ and attempt to find scalars a,b such that

$$(x,y) = a(-1,1) + b(1,1) = (-a+b, a+b).$$

This leads to the two equations $x = -a + b$ and $y = a + b$. Adding them gives $x + y = 2b$ and subtracting them gives $x - y = -2a$. Hence, $a = -\frac{1}{2}(x-y)$ and $b = \frac{1}{2}(x+y)$. So (x,y) is in the span of B. Since (x,y) was an arbitrary vector in $\mathcal{R}^2$, B spans $\mathcal{R}^2$.

By inspection, neither vector in B is a scalar multiple of the other and therefore they are linearly independent. It follows that B is a basis for $\mathcal{R}^2$.

3. Let $B = \{(1,0,0),(1,1,0),(1,1,1)\}$. To show that B spans $\mathcal{R}^3$, we arbitrarily choose (x,y,z) in $\mathcal{R}^3$ and attempt to find scalars a,b,c such that

$$(x,y,z) = a(1,0,0) + b(1,1,0) + c(1,1,1) = (a+b+c, b+c, c).$$

This leads to the three equations $a + b + c = x$, $b + c = y$, $c = z$. Setting $c = z$ in the first two equations gives $a + b + z = x$ and $b + z = y$, or equivalently, $a + b = x - z$ and $b = y - z$. Setting $b = y - z$ in the first of these equations gives $a + y - z = x - z$, so that $a = x - y$. So (x,y,z) is in the span of B. Since (x,y,z) was an arbitrary vector in $\mathcal{R}^3$, B spans $\mathcal{R}^3$.

To show that the vectors in B are linearly independent, we use what we have just shown; namely, if (x,y,z) is in $\mathcal{R}^3$ then

$$(x,y,z) = (x-y)(1,0,0) + (y-z)(1,1,0) + z(1,1,1),$$

where the coefficients $x-y$, $y-z$ and z are uniquely determined by x, y and z. In particular, if $(x,y,z) = \mathbf{0}$, so that $x = y = z = 0$, then each coefficient is zero. That is, B is a linearly independent set. It follows that B is a basis for $\mathcal{R}^3$.

5. Let $S = \{(-1,1),(1,-1)\}$ and note that the vectors in S are the negatives of each other. So they are not linearly independent. Thus, the singleton set $\{(-1,1)\}$ is a basis for the subspace of $\mathcal{R}^2$ spanned by S. By definition, the dimension of this subspace is 1.

7. Let $S = \{(1,0,1),(0,0,1),(1,0,2)\}$ and note that $(1,0,1) + (0,0,1) = (1,0,2)$. That is, $(1,0,2)$ is in the span of the vectors $(1,0,1)$ and $(0,0,1)$. Moreover, these two vectors are linearly independent because neither is a scalar multiple of the other. Thus, the set $\{(1,0,1),(0,0,1)\}$ is a basis for the subspace of $\mathcal{R}^3$ spanned by S. By definition, the dimension of this subspace is 2.

9. Let $S = \{e^x, e^{2x}, e^{3x}\}$ and suppose a,b,c are scalars such that

$$ae^x + be^{2x} + ce^{3x} = 0 \quad \text{for all } x \in \mathcal{R}.$$

Then we can set $x = 0$, $x = \ln 2$ and $x = \ln 3$ (e.g.) in the above equation and obtain the system

$$\begin{aligned} a + b + c &= 0, \\ 2a + 4b + 8c &= 0, \\ 3a + 9b + 27c &= 0, \end{aligned} \qquad \text{or in matrix form} \qquad \begin{pmatrix} 1 & 1 & 1 \\ 2 & 4 & 8 \\ 3 & 9 & 27 \end{pmatrix} \begin{pmatrix} a \\ b \\ c \end{pmatrix} = \begin{pmatrix} 0 \\ 0 \\ 0 \end{pmatrix}.$$

Focusing on the matrix equation, add -2 times the first row to the second row and then add -3 times the first row to the third row. This gives

$$\begin{pmatrix} 1 & 1 & 1 \\ 0 & 2 & 6 \\ 0 & 6 & 24 \end{pmatrix} \begin{pmatrix} a \\ b \\ c \end{pmatrix} = \begin{pmatrix} 0 \\ 0 \\ 0 \end{pmatrix}.$$

Now divide the second row by 2; add -6 times this new second row to the third row; and then divide the new third row by 6. This gives

$$\begin{pmatrix} 1 & 1 & 1 \\ 0 & 1 & 3 \\ 0 & 0 & 1 \end{pmatrix} \begin{pmatrix} a \\ b \\ c \end{pmatrix} = \begin{pmatrix} 0 \\ 0 \\ 0 \end{pmatrix}, \qquad \text{or in scalar form} \qquad \begin{aligned} a+b+c &= 0, \\ b+3c &= 0, \\ c &= 0. \end{aligned}$$

The last equation on the right directly gives $c = 0$. Setting $c = 0$ in the second equation gives $b = 0$. Setting $b = c = 0$ in the first equation gives $a = 0$. Hence, the above proposed linear combination holds only if all coefficients are zero. So the functions in S are linearly independent.

11. Let $S = \{\cos x, \sin x\}$ and suppose a, b are scalars such that

$$a \cos x + b \sin x = 0 \quad \text{for all } x \in \mathcal{R}.$$

Then we can set $x = 0$ and $x = \pi/2$ (e.g.) in the above equation to directly get $a = 0$ and $b = 0$, respectively. So the functions in S are linearly independent.

13. Let S be the subspace of $C(-\infty, \infty)$ with basis $B = \{e^x, e^{-x}\}$.

(a) Let $B' = \{\cosh x, \sinh x\}$ and let S' be the subspace generated by B'. Because $\cosh x$ and $\sinh x$ are linear combinations of e^x and e^{-x}, B' is contained in S and therefore S' is a subspace of S. And because $\cosh x + \sinh x = e^x$ and $\cosh x - \sinh x = e^{-x}$, B is contained in S' and therefore S is a subspace of S'. It follows that $S' = S$. Moreover, since neither $\cosh x$ nor $\sinh x$ is a scalar multiple of the other, they are linearly independent. So, B' is a basis for S.

(b) As was shown in part (a), $e^x = \cosh x + \sinh x$ and $e^{-x} = \cosh x - \sinh x$. Hence, the coordinates of e^x relative to the basis B' is the ordered pair $(1, 1)$ and the coordinates of e^{-x} relative to the basis B' is the ordered pair $(1, -1)$.

15. (a) Let $B = \{1, x, x^2\}$ be the "natural" basis for P_2 and let $B' = \{1, x+1, (x+1)^2\}$ be the proposed basis. First, note that every linear combination of the functions in B' is a polynomial in P_2, so that the span of B' is a subspace of P_2.

We now find the change of coordinate matrix A that sends the vector in $\mathcal{R}^3$ with coordinates relative to B to the vector in $\mathcal{R}^3$ with coordinates relative to B'. To do this, we observe that the three equations

$$1 = 1(1) + 0(x+1) + 0(x+1)^2, \quad x = -1(1) + 1(x+1) + 0(x+1)^2, \quad x^2 = 1(1) - 2(x+1) + 1(x+1)^2$$

exhibit the basis functions in B in terms of the basis functions in B'. The corresponding coefficients are the coordinates relative to B' and constitute the columns of A. That is,

$$A = \begin{pmatrix} 1 & -1 & 1 \\ 0 & 1 & -2 \\ 0 & 0 & 1 \end{pmatrix}.$$

Thus, the vector $\mathbf{a} = \begin{pmatrix} a \\ b \\ c \end{pmatrix}$ with coordinates relative to B corresponds to the vector

$$A\mathbf{a} = \begin{pmatrix} 1 & -1 & 1 \\ 0 & 1 & -2 \\ 0 & 0 & 1 \end{pmatrix} \begin{pmatrix} a \\ b \\ c \end{pmatrix} = \begin{pmatrix} a - b + c \\ b - 2c \\ c \end{pmatrix}$$

with coordinates relative to B'. In other words,

$$a + 2b + cx^2 = (a - b + c) + (b - 2c)(x + 1) + c(x + 1)^2.$$

Hence, every polynomial in P_2 is a linear combination of the functions in B' and therefore P_2 is a subspace of the span of B'. We conclude that the B' spans P_2.

(b) The coordinates of $x^2 + x + 1$ are the coordinates of the vector $\begin{pmatrix} 1 \\ 1 \\ 1 \end{pmatrix}$. With $a = b = c = 1$ in part (a), we use the result of part (a) to conclude that the coordinate vector relative to B' is

$$\begin{pmatrix} a - b + c \\ b - 2c \\ c \end{pmatrix} = \begin{pmatrix} 1 - 1 + 1 \\ 1 - 2 \\ 1 \end{pmatrix} = \begin{pmatrix} 1 \\ -1 \\ 1 \end{pmatrix}.$$

That is, $x^2 + x + 1 = (x + 1)^2 - (x + 1) + 1$.

17. Let $B_2 = \{1, \cos x, \sin x, \cos 2x, \sin 2x\}$ and let T_2 be the subspace of $C(-\infty, \infty)$ spanned by B_2. The identities

$$\cos^2 x = \frac{1}{2} + \frac{1}{2} \cos 2x \qquad \text{and} \qquad \sin^2 x = \frac{1}{2} - \frac{1}{2} \cos 2x.$$

show $\cos^2 x$ and $\sin^2 x$ as linear combinations of functions in B_2. Hence, $\cos^2 x$ and $\sin^2 x$ are both in T_2.

In order to find their coordinates relative to the basis B_2, we make the convention that the coordinates of the 5-tuple $(a_0, a_1, b_1, b_2, b_3)$ in $\mathcal{R}^5$ are the coefficients of the linear combination

$$a_0 + a_1 \cos x + b_1 \sin x + a_2 \cos 2x + b_2 \sin 2x.$$

With this convention, the coordinates of $\cos^2 x$ and $\sin^2 x$ are

$$(1/2, 0, 0, 1/2, 0) \quad (\text{for } \cos^2 x), \qquad \text{and} \qquad (1/2, 0, 0, -1/2, 0) \quad (\text{for } \sin^2 x).$$

19. Let p and q be nonnegative integers. Each $f(x)$ in T_p is a linear combination of functions in B_p, and each $g(x)$ in T_q is a linear combination of functions in B_q. It follows that the product $f(x)g(x)$ is a linear combination of three types of functions: $\cos kx \sin jx$, $\cos kx \cos jx$ and $\sin kx \sin jx$, where k and j run over all permissible indices such that $k + j \leq p + q$. Thus, $f(x)g(x)$ will be a linear combination of functions in B_{p+q} if each of these three types of products are in T_{p+q}. To show that these three types of products are in T_{p+q}, we will use the following four trigonometric identities

(1a) $\qquad\qquad\qquad \cos(a + b) = \cos a \cos b - \sin a \sin b,$

(1b) $\qquad\qquad\qquad \cos(a - b) = \cos a \cos b + \sin a \sin b,$

(1c) $\qquad\qquad\qquad \sin(a + b) = \sin a \cos b + \cos a \sin b,$

(1d) $\qquad\qquad\qquad \sin(a - b) = \sin a \cos b - \cos a \sin b.$

By selectively adding or subtracting two of these equations, the following identities are obtained

(2a)
$$\cos a \cos b = \frac{1}{2}\cos(a-b) + \frac{1}{2}\cos(a+b),$$

(2b)
$$\sin a \sin b = \frac{1}{2}\cos(a-b) - \frac{1}{2}\cos(a+b),$$

(2c)
$$\cos a \sin b = \frac{1}{2}\sin(a+b) - \frac{1}{2}\sin(a-b),$$

(2d)
$$\cos b \sin a = \frac{1}{2}\sin(a+b) + \frac{1}{2}\sin(a-b).$$

(For example, adding equations (1a) and (1b) allows one to solve for $\cos a \cos b$ to obtain equation (2a)). With $a = kx$ and $b = jx$, equations (2a)-(2d) show that each of the products on the left of these equations is a linear combination of functions from B_{p+q}, where (2c) is used if $k \geq j$ and (2d) is used if $j \geq k$ (note that (2c) and (2d) are the same identity if $k = j$). Hence, $f(x)g(x)$ is in T_{p+q}.

21. Let $S = \{e^x, e^{-x}, \cosh x, \sinh x\}$ and let V be the vector space spanned by S. First observe that $\cosh x$ and $\sinh x$ are both linear combinations of e^x and e^{-x}. Hence, we may delete these functions and deduce that V is also spanned by the smaller set $S = \{e^x, e^{-x}\}$. Moreover, since neither function in S' is a multiple of the other, S' is a linearly independent set. It is therefore a basis for V. (See the solution for Exercise 13 in this section.)

23. Let $S = \{\cos x, \sin x, \sin 2x\}$ and let V be the vector space spanned by S. The functions in S are linearly independent. To see this, suppose a, b, c are scalars such that

$$a\cos x + b\sin x + c\sin 2x = 0 \quad \text{for all } x \in \mathcal{R}.$$

Setting $x = 0$ and $x = \pi/2$ directly gives $a = 0$ and $b = 0$, respectively. The above linear combination then reduces to $c\sin 2x = 0$ for all x. Therefore we can choose any x for which $\sin 2x \neq 0$ and conclude that $c = 0$ as well. So S is a linearly independent set. It follows that S is a basis for V.

25. Let $f : \mathcal{R}^2 \longrightarrow \mathcal{R}^2$ be defined by $f(\mathbf{x}) = A\mathbf{x}$, where $A = \begin{pmatrix} 2 & 1 \\ 1 & 2 \end{pmatrix}$. The image of f is spanned by the columns of A, which are linearly independent since neither is a scalar multiple of the other. Therefore, $\{(2,1),(1,2)\}$ is a basis for the image of f. Moreover, since the columns of A are linearly independent, f is one-to-one, which implies that the null-space of f is the zero subspace. Hence, the null-space of f has no basis.

27. Let $f : \mathcal{R}^3 \longrightarrow \mathcal{R}^3$ be defined by $f(\mathbf{x}) = A\mathbf{x}$, where $A = \begin{pmatrix} 2 & 4 & 2 \\ 0 & 1 & 1 \\ 1 & 3 & 2 \end{pmatrix}$. The image of f is spanned by the columns of A. But note that the third column is the difference of the first and second columns and is therefore automatically in the span of the first two columns. Moreover, the first two columns are linearly independent since neither is a scalar multiple of the other. It follows that $\{(2,0,1),(4,1,3)\}$ is a basis for the image of f.

For a basis for the null-space of f, we first find all solutions of $A\mathbf{x} = \mathbf{0}$. These solutions are precisely the solutions of the system $R\mathbf{x} = \mathbf{0}$, where R is a reduced form for A. To find such a matrix R, add the first row of A to -2 times the third row and then add 2 times the second row to the new third row to get

$$\begin{pmatrix} 2 & 4 & 2 \\ 0 & 1 & 1 \\ 0 & 0 & 0 \end{pmatrix}.$$

Now add -4 times the second row to the first row and divide the new first row by 2 to get the reduced matrix
$$R = \begin{pmatrix} 1 & 0 & -1 \\ 0 & 1 & 1 \\ 0 & 0 & 0 \end{pmatrix}.$$

The reduced system $R\mathbf{x} = \mathbf{0}$ is equivalent to the scalar system

$$x - z = 0,$$
$$y + z = 0.$$

The nonleading variable is z, so we set $z = t$ and deduce from the above system that $x = t$ and $y = -t$. Hence, the null-space of f consists of all vectors of the form

$$(x, y, z) = (t, -t, t) = t(1, -1, 1).$$

Therefore, $\{(1, -1, 1)\}$ is a basis for the null-space of f.

29. Let S be the solution set in $\mathcal{R}^3$ of the system

$$x + y - z = 0,$$
$$2x + y = 0.$$

Since these equations are equations of planes in $\mathcal{R}^3$, S is precisely the set of points in the intersection of the two planes. Note that the two planes are not parallel (since their normals $(1, 1, -1)$ and $(2, 1, 0)$ are not parallel), so that S must be a line in $\mathcal{R}^3$. Further, this line contains the origin because $(0, 0, 0)$ is a solution of the above system. Hence, S is a subspace of dimension 1.

31. Let k be a positive integer. If $p_1(x)$ is a nonzero polynomial in P then the singleton set $\{p_1\}$ is a automatically a linearly independent set. This establishes the case $k = 1$ for a proof by mathematical induction that if $p_1(x), \ldots, p_k(x)$ are k nonzero polynomials in P whose degrees are all different then $\{p_1, \ldots, p_k\}$ is a linearly independent set. For our induction hypothesis, we assume that the assertion has been proven for all sets of size $k \leq k_0$ for some $k_0 \geq 1$. Let $p_1(x), \ldots, p_{k_0}(x), p_{k_0+1}(x)$ be $k_0 + 1$ polynomials in P whose degrees are all different. Specifically, we can suppose that p_j has degree n_j and that $n_1 < n_2 < \cdots < n_{k_0+1}$. To show that $\{p_1, \ldots, p_{k_0+1}\}$ is a linearly independent set, suppose $c_1, \ldots, c_{k_0}, c_{k_0+1}$ are scalars such that

(1) $\qquad c_1 p_1(x) + \cdots + c_{k_0} p_{k_0}(x) + c_{k_0+1} p_{k_0+1}(x) = 0 \quad$ for all $x \in \mathcal{R}.$

Since the degrees of $p_1, \ldots, p_{k_0}$ are all no greater than n_{k_0}, $c_1 p_1(x) + \cdots + c_{k_0} p_{k_0}(x)$ is a polynomial whose degree is no greater than n_{k_0}. If $c_{k_0+1} \neq 0$ then $c_{k_0+1} p_{k_0+1}(x)$ has degree n_{k_0+1} and therefore the left side of equation (1) is a polynomial of degree n_{k_0+1}. However, the right side is the zero polynomial whose degree is not defined. It follows that $c_{k_0+1} = 0$ and (1) can be written as

$$c_1 p_1(x) + \cdots + c_{k_0} p_{k_0}(x) = 0 \quad \text{for all } x \in \mathcal{R}.$$

By the induction hypothesis, $\{p_1, \ldots, p_{k_0}\}$ is a linearly independent set, so that $c_k = 0$ for $k = 1, \ldots, k_0$. Hence, equation (1) can hold only if all coefficients are zero. That is, $\{p_1, \ldots, p_{k_0+1}\}$ is a linearly independent set. By induction, $\{p_1, \ldots, p_k\}$ is a linearly independent set for each integer $k \geq 1$.

33. From Exercise 32 in this section, the set $S = \{x, x^3, x^5\}$ was shown to be a basis for O. Thus, in order that the set $S' = \{x - x^3, x^3 + x^5, p(x)\}$ be a basis for O, it is necessary that $p(x)$ be a linear combination of functions in S. Thus, we let $p(x) = a_1 x + a_3 x^3 + a_5 x^5$, and seek to discover the relation(s) among the coefficients a_1, a_3, a_5 (if any) that must hold in order that S' be a basis for O.

To do this, we assume that $p(x)$ is such that S' is a basis for O and let $f : O \longrightarrow O$ be such that

$$f(x) = x - x^3, \qquad f(x^3) = x^3 + x^5, \qquad f(x^5) = a_1 x + a_3 x^3 + a_5 x^5.$$

That is, f maps the basis functions in S to the basis functions in S'. Requiring f to be linear, uniquely defines f on all of O. If A is the matrix of f then, by Theorem 5.4, the columns of A are the coordinates of $f(x)$, $f(x^3)$ and $f(x^5)$ relative to S'. That is,

$$A = \begin{pmatrix} 1 & 0 & a_1 \\ -1 & 1 & a_3 \\ 0 & 1 & a_5 \end{pmatrix}.$$

Since S' is a basis, the columns of A are linearly independent, so that A is invertible. This forces a_1, a_3, a_5 to be such that $\det A \neq 0$. That is, a_1, a_3, a_5 must satisfy

$$\det A = \begin{vmatrix} 1 & 0 & a_1 \\ -1 & 1 & a_3 \\ 0 & 1 & a_5 \end{vmatrix} = \begin{vmatrix} 0 & 1 & a_1 + a_3 \\ -1 & 1 & a_3 \\ 0 & 1 & a_5 \end{vmatrix} = -(-1) \begin{vmatrix} 1 & a_1 + a_3 \\ 1 & a_5 \end{vmatrix} = a_5 - (a_1 + a_3) \neq 0,$$

where the second row was added to the first row to obtain the third member, and the fourth member is expansion about the first column. As can be seen, there are infinitely many triples a_1, a_3, a_5 such that $a_5 - (a_1 + a_3) \neq 0$. For example, we choose $a_1 = 1$, $a_3 = a_5 = 0$, so that $S' = \{x - x^3, x^3 + x^5, x\}$ is a basis for O.

35. $\mathbf{x} = (3, -4, 5, 2)$ is in the span of $S = \{(1, -2, 1, 1), (2, 1, -2, 1), (3, 1, 1, 1)\}$ if there exist scalars a, b, c such that

$$(3, -4, 5, 2) = a(1, -2, 1, 1) + b(2, 1, -2, 1) + c(3, 1, 1, 1)$$
$$= (a + 2b + 3c, -2a + b + c, a - 2b + c, a + b + c).$$

This is equivalent to the scalar system

$$a + 2b + 3c = 3,$$
$$-2a + b + c = -4,$$
$$a - 2b + c = 5,$$
$$a + b + c = 2.$$

Subtracting the second and third equations from the fourth equation gives $3a = 6$ and $3b = -3$, respectively, from which $b = -1$ and $a = 2$. Substituting these values of a and b into each of the above four equations gives the four equations $2 - 2 + 3c = 3$, $-4 - 1 + c = -4$, $2 + 2 + c = 5$ and $2 - 1 + c = 2$, all of which have the solution $c = 1$. That is,

$$(3, -4, 5, 2) = 2(1, -2, 1, 1) - (2, 1, -2, 1) + (3, 1, 1, 1)$$

and $\mathbf{x}$ is in the span of S.

110

37. One form of the double-angle identity for the cosine is $\cos 2x = 2\cos^2 x - 1$. Hence, $\mathbf{x} = \cos 2x$ is in the span of $S = \{1, \cos x, \cos^2 x\}$.

39. Let B_2 be the natural basis for $\mathcal{R}^2$ and let $S_2 = \{(1,1),(1,2)\}$ be the other basis for $\mathcal{R}^2$ given in the statement of the problem. Similarly, let B_3 be the natural basis for $\mathcal{R}^3$ and let and $S_3 = \{(1,0,0),(1,1,0),(1,1,1)\}$ be the other basis given for $\mathcal{R}^3$ in the statement of the problem. We can compute the desired matrix by forming the product of three matrices as follows.

By Theorem 5.4, the columns of the matrix P sending a coordinate vector relative to the basis S_2 to its image coordinate vector relative to the basis B_2 are the vectors in S_2; i.e., $P = \begin{pmatrix} 1 & 1 \\ 1 & 2 \end{pmatrix}$. By Theorem 1.3 (the special case of Theorem 5.4 given in Section 1A in this chapter), the matrix sending a coordinate vector relative to the basis B_2 to its image coordinate vector relative to the basis B_3 is the given matrix $A = \begin{pmatrix} 2 & -1 \\ 1 & 2 \\ -2 & 2 \end{pmatrix}$. And the matrix sending a coordinate vector relative to the basis B_3 to its image coordinate vector relative to the basis S_3 is the inverse of the matrix $Q = \begin{pmatrix} 1 & 1 & 1 \\ 0 & 1 & 1 \\ 0 & 0 & 1 \end{pmatrix}$ (whose columns are the vectors in S_3); i.e., $Q^{-1} = \begin{pmatrix} 1 & -1 & 0 \\ 0 & 1 & -1 \\ 0 & 0 & 1 \end{pmatrix}$ (we omit the details of the computation of Q^{-1}). Therefore, the matrix of g sending a coordinate vector relative to the basis S_2 to its image coordinate vector relative to S_3 is given by

$$Q^{-1}AP = \begin{pmatrix} 1 & -1 & 0 \\ 0 & 1 & -1 \\ 0 & 0 & 1 \end{pmatrix}\begin{pmatrix} 2 & -1 \\ 1 & 2 \\ -2 & 2 \end{pmatrix}\begin{pmatrix} 1 & 1 \\ 1 & 2 \end{pmatrix} = \begin{pmatrix} -2 & -5 \\ 3 & 3 \\ 0 & 2 \end{pmatrix}.$$

Note that multiplying a vector $\mathbf{x} = \begin{pmatrix} x \\ y \end{pmatrix}$ by $Q^{-1}AP$ produces the vector $\begin{pmatrix} -2x - 5y \\ 3x + 3y \\ 2y \end{pmatrix}$.

We leave it to the reader to verify that

$$g\big(x(1,1) + y(1,2)\big) = (-2x - 5y)(1,0,0) + (3x + 3y)(1,1,0) + 2y(1,1,1).$$

41. Let P_2 be the vector space of polynomials of degree ≤ 2 and let $S : P_2 \longrightarrow P_2$ be the shift operator defined by $Sp(x) = p(x+1)$ for $p(x)$ in P_2.

(a) Let $p(x) = a_0 + a_1 x + a_2 x^2$ and $q(x) = b_0 + b_1 x + b_2 x^2$ be in P_2 and let r be a scalar. Then

$$\begin{aligned} S\big(p(x) + q(x)\big) &= S\big((a_0 + a_1 x + a_2 x^2) + (b_0 + b_1 x + b_2 x^2)\big) \\ &= S\big((a_0 + b_0) + (a_1 + b_1)x + (a_2 + b_2)x^2\big) \\ &= (a_0 + b_0) + (a_1 + b_1)(x+1) + (a_2 + b_2)(x+1)^2 \\ &= a_0 + a_1(x+1) + a_2(x+1)^2 + b_0 + b_1(x+1) + b_2(x+1)^2 \\ &= p(x+1) + q(x+1) = Sp(x) + Sq(x), \end{aligned}$$

and

$$S\big(rp(x)\big) = S(rp)(x) = (rp)(x+1) = rp(x+1) = rSp(x).$$

So S is linear.

(b) Let $D = d/dx$ and $D^2 = d^2/dx^2$ be the first and second derivative operators and let I be the identity operator. Then $I + D + \frac{1}{2}D^2$ is an operator on P_2. Specifically, if $p(x) = a_0 + a_1 x + a_2 x^2$ is in P_2 then

$$\left(I + D + \frac{1}{2}D^2\right) p(x) = p(x) + p'(x) + \frac{1}{2}p''(x)$$

$$= (a_0 + a_1 x + a_2 x^2) + (a_1 + 2a_2 x) + \frac{1}{2}(2a_2)$$

$$= a_0 + a_1(x+1) + a_2(x^2 + 2x + 1)$$

$$= a_0 + a_1(x+1) + a_2(x+1)^2$$

$$= p(x+1) = Sp(x).$$

That is, $I + D + \frac{1}{2}D^2$ and S have the same effect on polynomials in P_2 and therefore, as operators on P_2, $S = I + D + \frac{1}{2}D^2$.

(c) This part of the exercise generalizes the result of part (b). The shift operator S can be applied to *any* polynomial. In particular, if P_n is the vector space of polynomials in x of degree $\leq n$, then $Sp(x)$ has degree n for each $p(x)$ in P_n. Hence, S can be viewed as an operator on P_n.

Let $p(x)$ be in P_n. Note that p has derivatives of all orders and that the kth derivative of p is identically zero for all $k \geq n+1$. Therefore, for any number a, Taylor's formula applied to p takes the form

$$p(x) = p(a) + p'(a)(x-a) + \frac{1}{2!}p''(a)(x-a)^2 + \cdots + \frac{1}{n!}p^{(n)}(a)(x-a)^n \quad \text{for all } x.$$

In particular, setting $x = a+1$ and then replacing a with x gives

$$p(x+1) = p(x) + p'(x) + \frac{1}{2!}p''(x) + \cdots + \frac{1}{n!}p^{(n)}(x) \quad \text{for all } x.$$

Since the kth derivative operator $D^k = d^k/dx^k$ satisfies $D^k p(x) = p^{(k)}(x)$, the above formula can be written as

$$p(x+1) = Ip(x) + Dp(x) + \frac{1}{2!}D^2 p(x) + \cdots + \frac{1}{n!}D^n p(x)$$

$$= \left(I + D + \frac{1}{2!}D^2 + \cdots + \frac{1}{n!}D^n\right) p(x),$$

where I is the identity operator. Since the leftmost member is $Sp(x)$, the above equation shows that S and the operator $I + D + \frac{1}{2!}D^2 + \cdots + \frac{1}{n!}D^n$ have the same effect on functions in P_n. Thus, as operators on P_n, $S = I + D + \frac{1}{2!}D^2 + \cdots + \frac{1}{n!}D^n$.

Exercise Set 5C (pgs. 142-143)

1. (a) Let $M_{m,n}$ be the vector space of all m-by-n matrices. Let E_{ij} be the m-by-n matrix with 1 in the ijth position and 0 elsewhere and let $B_{m,n}$ be the collection of all such matrices. First, note that if a is any scalar then aE_{ij} is the m-by-n matrix with a in the ijth position and 0 elsewhere. It follows that if $A = (a_{ij})$ is in $M_{m,n}$ then

$$A = a_{11}E_{11} + \cdots + a_{1n}E_{1n} + a_{21}E_{21} + \cdots + a_{2n}E_{2n} + \cdots + a_{m1}E_{m1} + \cdots + a_{mn}E_{mn}.$$

Hence, $B_{m,n}$ spans $M_{m,n}$. Further, if E_{ij} is any fixed member of $B_{m,n}$ then the ijth entry of any linear combination of the remaining members of $B_{m,n}$ is zero. Hence, the matrix defined by such a linear combination cannot be E_{ij}. That is, $B_{m,n}$ is a linearly independent set and therefore $B_{m,n}$ is a basis for $M_{m,n}$. Since $B_{m,n}$ has mn members, $\dim(M_{m,n}) = mn$.

(b) Let $B_{n,n}$ be as in part (a). Let D_n be the subspace of $M_{n,n}$ consisting of all diagonal matrices in $M_{n,n}$ and let $B'_{n,n}$ be the subset of $B_{n,n}$ consisting of $E_{11}, \ldots, E_{nn}$. Since $B_{n,n}$ is a linearly independent set (by part (a)), so is $B'_{n,n}$. Moreover, D_n is spanned by $B'_{n,n}$. Hence, $B'_{n,n}$ is a basis for D_n. Since $B'_{n,n}$ has n members, $\dim(D_n) = n$.

3. Let f be a one-to-one linear function. Let $S = \{x_1, \ldots, x_k\}$ be a finite subset of the domain of f and let $f(S) = \{f(x_1), \ldots, f(x_k)\}$.

First, suppose $f(S)$ is a linearly independent set and let $a_1, \ldots, a_n$ be scalars such that

$$a_1 x_1 + \cdots + a_k x_k = 0.$$

Applying f to both sides of this equation and using the linearity of f gives

$$f(a_1 x_1 + \cdots + a_k x_k) = a_1 f(x_1) + \cdots + a_k f(x_k) = 0.$$

Since the vectors in $f(S)$ are linearly independent, equality of the two rightmost members of the above equation implies $a_1 = \cdots = a_k = 0$. So S is a linearly independent set.

Conversely, suppose S is a linearly independent set and let $a_1, \ldots, a_k$ be scalars such that

$$a_1 f(x_1) + \cdots + a_k f(x_k) = 0.$$

By the linearity of f, this equation implies $f(a_1 x_1 + \cdots + a_k x_k) = 0$. Hence, $a_1 x_1 + \cdots + a_k x_k$ is in the null-space of f. But since f is one-to-one, the null-space of f contains only the zero vector, so that

$$a_1 x_1 + \cdots + a_k x_k = 0.$$

As the vectors in S are linearly independent, $a_1 = \cdots = a_k = 0$. So $f(S)$ is a linearly independent set.

We can analyze the above result as follows. Let V be the domain of f and let W be the image of f. If V is infinite dimensional and B is a basis for V then the above result implies $f(B)$ is a linearly independent set that contains infinitely many vectors, so that W is infinite dimensional. If $\dim(V) = n$ is finite then the above result implies $f(B)$ contains n linearly independent vectors, so that every basis of W contains at least n vectors. This means that $\dim(W) \geq \dim(V)$. But since f is one-to-one and linear on V, f^{-1} exists and is one-to-one and linear on W. The same arguments that were just made with respect to f can therefore be made with respect to f^{-1} to conclude that W is finite dimensional and $\dim(V) \geq \dim(W)$. Hence, $\dim(V) = \dim(W)$.

5. Let P_1 and P_2 be two planes through the origin in $\mathcal{R}^3$ and let $K = P_1 \cap P_2$. As subspaces of $\mathcal{R}^3$ of dimension 2, we can let $B_1 = \{u_1, u_1\}$ and $B_2 = \{v_1, v_2\}$ be bases for P_1 and P_2, respectively. Since $\mathcal{R}^3$ has dimension 3, the four member set $B_1 \cup B_2$ is linearly dependent. Hence, there exist coefficients a_1, a_2, b_1, b_2, not all zero, such that

$$a_1 u_1 + a_2 u_2 + b_1 v_1 + b_2 v_2 = 0, \quad \text{or equivalently,} \quad a_1 u_1 + a_2 u_2 = -b_1 v_1 - b_2 v_2.$$

113

Since B_1 and B_2 are linearly independent sets, and since at least one of the coefficients is nonzero, it follows that each side of the equation on the above right is nonzero. Moreover, the equation also shows that the point $a_1\mathbf{u}_1 + a_2\mathbf{u}_2$ in P_1 is the same as the point $-b_1\mathbf{v}_1 - b_2\mathbf{v}_2$ in P_2. It follows that K contains nonzero vectors. Hence, as K is a subspace of P_1, its dimension is at least 1. On the other hand, by Theorem 5.5, no subset of P_1 can contain more than two linearly independent vectors. In particular, a basis for K can have no more than two members. We conclude that $1 \leq \dim(K) \leq 2$.

If $\dim(K) = 1$ then K is a line. If $\dim(K) = 2$ then K is a plane. But since the only plane in P_1 is P_1 itself, $K = P_1$. This means $P_1 \subseteq P_2$. Since the exact same argument can be be made with respect to P_2, we also have $P_2 \subseteq P_1$. So $P_1 = P_2$.

7. Let W be a vector space with $\dim(W) = n$. By Theorem 5.6, every basis of W contains exactly n elements. Hence, there cannot be more than n linearly independent vectors in W. In particular, every basis of every subspace of W has no more than n elements. Thus, if V is a subspace of W then $\dim(V) \leq n$.

9. (a) Let S be a k-dimensional subspace of $\mathcal{R}^n$ and let $B_k = \{\mathbf{x}_1, \ldots, \mathbf{x}_k\}$ be a basis for S. By Theorem 5.8, we can extend B_k to a basis $B_n = B_k \cup \{\mathbf{x}_{k+1}, \ldots, \mathbf{x}_n\}$ for $\mathcal{R}^n$. Define the function $f : \mathcal{R}^n \longrightarrow \mathcal{R}^{n-k}$ by

$$f(a_1\mathbf{x}_1 + \cdots + a_n\mathbf{x}_n) = (a_{k+1}, \ldots, a_n) \quad \text{for } \mathbf{x} \in \mathcal{R}^n.$$

By definition, f acts on the basis vectors in B_n by

$$f(\mathbf{x}_i) = \begin{cases} \mathbf{e}_j, & \text{if } i = k + j, \\ \mathbf{0}, & \text{otherwise,} \end{cases}$$

where $\mathbf{e}_j$ is the jth standard basis vector for $\mathcal{R}^{n-k}$. Hence,

$$a_1 f(\mathbf{x}_1) + \cdots + a_k f(\mathbf{x}_k) + a_{k+1} f(\mathbf{x}_{k+1}) + \cdots + a_n f(\mathbf{x}_n)$$
$$= a_1(\mathbf{0}) + \cdots + a_k(\mathbf{0}) + a_{k+1}\mathbf{e}_1 + \cdots + a_n\mathbf{e}_{n-k} = a_{k+1}\mathbf{e}_1 + \cdots + a_n\mathbf{e}_{n-k}$$
$$= (a_{k+1}, \ldots, a_n) = f(a_1\mathbf{x}_1 + \cdots + a_n\mathbf{x}_n).$$

So f is linear. Furthermore, if $\mathbf{x} = a_1\mathbf{x}_1 + \cdots + a_n\mathbf{x}_n$ then $f(\mathbf{x})$ is the zero vector if, and only if, $a_{k+1} = \cdots = a_n = 0$; i.e., if, and only if, $\mathbf{x}$ is in S. Hence, the null-space of f is precisely S.

(b) Let S and B_n be as in part (a). For $j = 1, \ldots, n - k$, let H_j be the hyperplane of $\mathcal{R}^n$ spanned by the set $B_n \backslash \{\mathbf{x}_{k+j}\}$ of all vectors in B_n *except* the vector $\mathbf{x}_{k+j}$. Then, a vector $\mathbf{x} = a_1\mathbf{x}_1 + \cdots + a_n\mathbf{x}_n$ in $\mathcal{R}^n$ is in H_j if, and only if, the basis vector $\mathbf{x}_{k+j}$ does not appear in the linear combination for $\mathbf{x}$. It follows that $\mathbf{x}$ is in every H_j if, and only if, none of the basis vectors $\mathbf{x}_{k+1}, \ldots, \mathbf{x}_n$ appears in the linear combination for $\mathbf{x}$; i.e., if, and only if, $\mathbf{x}$ is in S. Hence,

$$S = H_1 \cap \cdots \cap H_{n-k}.$$

11. Let V and W be finite-dimensional vector spaces with $\dim(V) = n$ and $\dim(W) = m$. Let $f : V \longrightarrow W$ be linear and denote the null-space and image of f by N and $f(V)$, respectively. Pick bases for V and W and let A be the matrix of f with respect to these bases. Then A has dimension m-by-n. By Theorem 5.10, if R is a reduced form of A then $\dim(N)$ is the number of nonleading variables associated with R and $\dim\big(f(V)\big)$ is the number of leading variables associated with R. Since there are n variables associated with R (the number of columns of R) and since every variable is either leading or nonleading, it follows that

$$\dim(N) + \dim\big(f(V)\big) = n = \dim(V).$$

114

13. (a) Let V and W be vector spaces and let $f : V \longrightarrow W$ be linear. Let N be the null-space of f and let $f(V)$ be the image of f. By Theorem 4.1 in this chapter, $f(V)$ is a subspace of W and therefore, by Exercise 7 in this section, $\dim(f(V)) \leq \dim(W)$. Thus,

$$\dim(V) > \dim(W) \quad \Longrightarrow \quad \dim(V) > \dim(f(V)).$$

By Exercise 11 in this section, $\dim(V) = \dim(N) + \dim(f(V))$. It follows from this and the above displayed implication that $\dim(N) > 0$.

(b) Each m-by-n matrix A with real entries defines a linear function $f : \mathcal{R}^n \longrightarrow \mathcal{R}^m$ by the formula $f(\mathbf{x}) = A\mathbf{x}$ for $\mathbf{x} \in \mathcal{R}^n$, where A is relative to the standard bases in $\mathcal{R}^n$ and $\mathcal{R}^m$. Similarly, each n-by-m matrix B defines a linear function $g : \mathcal{R}^m \longrightarrow \mathcal{R}^n$. The matrix BA is the matrix of the linear composite function $g \circ f : \mathcal{R}^n \longrightarrow \mathcal{R}^n$ with respect to the standard bases in $\mathcal{R}^n$.

If N is the null-space of f and $m < n$ then, by part (a), $\dim(N) > 0$. Moreover, if $\mathbf{x}$ is in N then

$$(g \circ f)(\mathbf{x}) = g(f(\mathbf{x})) = g(\mathbf{0}) = \mathbf{0},$$

so that $\mathbf{x}$ is in the null-space, call it N', of $g \circ f$. That is, N is a subspace of N'. Therefore, by Exercise 7 in this section $\dim(N) \leq \dim(N')$, from which we conclude that $\dim(N') > 0$. By Exercise 11 in this section,

$$\dim(N') + \dim((g \circ f)(\mathcal{R}^n)) = \dim(\mathcal{R}^n) = n,$$

where $(g \circ f)(\mathcal{R}^n)$ is the image of $g \circ f$. Hence, $\dim((g \circ f)(\mathcal{R}^n)) < n$. Since the image of the identity function on $\mathcal{R}^n$ is all of $\mathcal{R}^n$, it follows that $g \circ f$ is *not* the identity function. And since the n-by-n identity matrix I is the matrix of the identity function with respect to the standard basis in $\mathcal{R}^n$, the matrix of $g \circ f$ with respect to the standard basis in $\mathcal{R}^n$ can't be the identity matrix. That is $BA \neq I$.

Section 6: EIGENVALUES & EIGENVECTORS

Exercise Set 6A (pg. 148)

1. Let L be the linear operator with matrix $A = \begin{pmatrix} 1 & 12 \\ 3 & 1 \end{pmatrix}$. Given that L has the eigenvalues 7 and -5, we compute $A - \lambda I$ for each eigenvalue λ:

$$A - 7I = \begin{pmatrix} 1-7 & 12 \\ 3 & 1-7 \end{pmatrix} = \begin{pmatrix} -6 & 12 \\ 3 & -6 \end{pmatrix}, \qquad A + 5I = \begin{pmatrix} 1+5 & 12 \\ 3 & 1+5 \end{pmatrix} = \begin{pmatrix} 6 & 12 \\ 3 & 6 \end{pmatrix}.$$

Let $\mathbf{u}_1, \ldots, \mathbf{u}_5$ be the five test vectors in the order in which they are given. We multiply each $\mathbf{u}_k$ on the left by each of the above matrices. If the result of any such multiplication is the zero vector then the corresponding $\mathbf{u}_k$ is an eigenvector of the associated eigenvalue. Otherwise, we discard $\mathbf{u}_k$ and move on to $\mathbf{u}_{k+1}$. The multiplications are straightforward. We immediately find that

$$\begin{pmatrix} -6 & 12 \\ 3 & -6 \end{pmatrix} \mathbf{u}_1 = \begin{pmatrix} 0 \\ 0 \end{pmatrix} \qquad \text{and} \qquad \begin{pmatrix} 6 & 12 \\ 3 & 6 \end{pmatrix} \mathbf{u}_2 = \begin{pmatrix} 0 \\ 0 \end{pmatrix}.$$

Since the eigenvectors associated with a given eigenvalue are the nonzero scalar multiples of any one eigenvector, a cursory examination of the given vectors reveals that $\mathbf{u}_3 = -2\mathbf{u}_1$ and that neither $\mathbf{u}_4$ nor $\mathbf{u}_5$ are scalar multiples of either $\mathbf{u}_1$ or $\mathbf{u}_2$. To summarize:

$\mathbf{u}_1 = \begin{pmatrix} 2 \\ 1 \end{pmatrix}$ and $\mathbf{u}_3 = \begin{pmatrix} -4 \\ -2 \end{pmatrix}$ are the only eigenvectors of L associated with $\lambda = 7$,

$\mathbf{u}_2 = \begin{pmatrix} -2 \\ 1 \end{pmatrix}$ is the only eigenvector of L associated with $\lambda = -5$.

(NOTE: Had the list of given vectors been more extensive, it would have perhaps been more practical to actually construct an eigenvector for each of the eigenvalues as was done in Example 3 of the text. Once this is done, one need only scan the list to see which of the given vectors is a scalar multiple of the two constructed eigenvectors. However, in this case, the list was small enough to use the above method.)

3. Let L be the operator defined by $A = \begin{pmatrix} 0 & 4 \\ 1 & 0 \end{pmatrix}$. The eigenvalues of L are the solutions λ of the characteristic equation

$$\det(A - \lambda I) = \begin{vmatrix} -\lambda & 4 \\ 1 & -\lambda \end{vmatrix} = \lambda^2 - 4 = (\lambda - 2)(\lambda + 2) = 0.$$

Hence, the eigenvalues of L are $\lambda_1 = 2$ and $\lambda_2 = -2$.

Each eigenvector $\mathbf{u}_1 = \begin{pmatrix} u_1 \\ v_1 \end{pmatrix}$ associated with $\lambda_1 = 2$ must satisfy $(A - \lambda_1 I)\mathbf{u}_1 = \mathbf{0}$. So we compute

$$(A - \lambda_1 I)\mathbf{u}_1 = \begin{pmatrix} -2 & 4 \\ 1 & -2 \end{pmatrix} \begin{pmatrix} u_1 \\ v_1 \end{pmatrix} = \begin{pmatrix} -2u_1 + 4v_1 \\ u_1 - 2v_1 \end{pmatrix} = (u_1 - 2v_1) \begin{pmatrix} -2 \\ 1 \end{pmatrix} = \mathbf{0}.$$

116

Hence, the coordinates of $\mathbf{u}_1$ must satisfy $u_1 - 2v_1 = 0$. Choosing $v_1 = 1$, $u_1 = 2$ we get the eigenvector $\mathbf{u}_1 = \begin{pmatrix} 2 \\ 1 \end{pmatrix}$.

Each eigenvector $\mathbf{u}_2 = \begin{pmatrix} u_2 \\ v_2 \end{pmatrix}$ associated with $\lambda_2 = -2$ must satisfy $(A - \lambda_2 I)\mathbf{u}_2 = \mathbf{0}$. So we compute

$$(A - \lambda_2 I)\mathbf{u}_2 = \begin{pmatrix} 2 & 4 \\ 1 & 2 \end{pmatrix} \begin{pmatrix} u_2 \\ v_2 \end{pmatrix} = \begin{pmatrix} 2u_2 + 4v_2 \\ u_2 + 2v_2 \end{pmatrix} = (u_2 + 2v_2) \begin{pmatrix} 2 \\ 1 \end{pmatrix} = \mathbf{0}.$$

Hence, the coordinates of $\mathbf{u}_2$ must satisfy $u_2 + 2v_2 = 0$. Choosing $v_2 = 1$, $u_2 = -2$ we get the eigenvector $\mathbf{u}_2 = \begin{pmatrix} -2 \\ 1 \end{pmatrix}$.

5. Let L be the operator defined by $A = \begin{pmatrix} 2 & 4 \\ 1 & 2 \end{pmatrix}$. The eigenvalues of L are the solutions λ of the characteristic equation

$$\det(A - \lambda I) = \begin{vmatrix} 2 - \lambda & 4 \\ 1 & 2 - \lambda \end{vmatrix} = (2 - \lambda)^2 - 4 = \lambda(\lambda - 4) = 0.$$

Hence, the eigenvalues of L are $\lambda_1 = 0$ and $\lambda_2 = 4$.

Each eigenvector $\mathbf{u}_1 = \begin{pmatrix} u_1 \\ v_1 \end{pmatrix}$ associated with $\lambda_1 = 0$ must satisfy $(A - \lambda_1 I)\mathbf{u}_1 = \mathbf{0}$. So we compute

$$(A - \lambda_1 I)\mathbf{u}_1 = \begin{pmatrix} 2 & 4 \\ 1 & 2 \end{pmatrix} \begin{pmatrix} u_1 \\ v_1 \end{pmatrix} = \begin{pmatrix} 2u_1 + 4v_1 \\ u_1 + 2v_1 \end{pmatrix} = (u_1 + 2v_1) \begin{pmatrix} 2 \\ 1 \end{pmatrix} = \mathbf{0}.$$

Hence, the coordinates of $\mathbf{u}_1$ must satisfy $u_1 + 2v_1 = 0$. Choosing $v_1 = 1$, $u_1 = -2$ we get the eigenvector $\mathbf{u}_1 = \begin{pmatrix} -2 \\ 1 \end{pmatrix}$.

Each eigenvector $\mathbf{u}_2 = \begin{pmatrix} u_2 \\ v_2 \end{pmatrix}$ associated with $\lambda_2 = 4$ must satisfy $(A - \lambda_2 I)\mathbf{u}_2 = \mathbf{0}$. So we compute

$$(A - \lambda_2 I)\mathbf{u}_2 = \begin{pmatrix} -2 & 4 \\ 1 & -2 \end{pmatrix} \begin{pmatrix} u_2 \\ v_2 \end{pmatrix} = \begin{pmatrix} -2u_2 + 4v_2 \\ u_2 - 2v_2 \end{pmatrix} = (u_2 - 2v_2) \begin{pmatrix} -2 \\ 1 \end{pmatrix} = \mathbf{0}.$$

Hence, the coordinates of $\mathbf{u}_2$ must satisfy $u_2 - 2v_2 = 0$. Choosing $v_2 = 1$, $u_2 = 2$ we get the eigenvector $\mathbf{u}_2 = \begin{pmatrix} 2 \\ 1 \end{pmatrix}$.

7. Let L be the operator defined by $A = \begin{pmatrix} 0 & 0 & 1 \\ 0 & 1 & 0 \\ 0 & 0 & 2 \end{pmatrix}$. The eigenvalues of L are the solutions λ of the characteristic equation

$$\det(A - \lambda I) = \begin{vmatrix} -\lambda & 0 & 1 \\ 0 & 1 - \lambda & 0 \\ 0 & 0 & 2 - \lambda \end{vmatrix} = -\lambda(1 - \lambda)(2 - \lambda) = 0.$$

117

Hence, the eigenvalues of L are $\lambda_1 = 0$, $\lambda_2 = 1$ and $\lambda_3 = 2$.

Each eigenvector $\mathbf{u}_1 = \begin{pmatrix} u_1 \\ v_1 \\ w_1 \end{pmatrix}$ associated with $\lambda_1 = 0$ must satisfy $(A - \lambda_1 I)\mathbf{u}_1 = \mathbf{0}$.

So we compute

$$(A - \lambda_1 I)\mathbf{u}_1 = \begin{pmatrix} 0 & 0 & 1 \\ 0 & 1 & 0 \\ 0 & 0 & 2 \end{pmatrix} \begin{pmatrix} u_1 \\ v_1 \\ w_1 \end{pmatrix} = \begin{pmatrix} w_1 \\ v_1 \\ 2w_1 \end{pmatrix} = w_1 \begin{pmatrix} 1 \\ 0 \\ 2 \end{pmatrix} + v_1 \begin{pmatrix} 0 \\ 1 \\ 0 \end{pmatrix} = \mathbf{0}.$$

Since the two vectors forming the linear combination in the fourth member are linearly independent, their coefficients w_1 and v_1 must be zero while u_1 remains arbitrary. Choosing $u_1 = 1$ we get the eigenvector $\mathbf{u}_1 = \begin{pmatrix} 1 \\ 0 \\ 0 \end{pmatrix}$.

Each eigenvector $\mathbf{u}_2 = \begin{pmatrix} u_2 \\ v_2 \\ w_2 \end{pmatrix}$ associated with $\lambda_2 = 1$ must satisfy $(A - \lambda_2 I)\mathbf{u}_2 = \mathbf{0}$.

So we compute

$$(A - \lambda_2 I)\mathbf{u}_2 = \begin{pmatrix} -1 & 0 & 1 \\ 0 & 0 & 0 \\ 0 & 0 & 1 \end{pmatrix} \begin{pmatrix} u_2 \\ v_2 \\ w_2 \end{pmatrix} = \begin{pmatrix} -u_2 + w_2 \\ 0 \\ w_2 \end{pmatrix} = -u_2 \begin{pmatrix} 1 \\ 0 \\ 0 \end{pmatrix} + w_2 \begin{pmatrix} 1 \\ 0 \\ 1 \end{pmatrix} = \mathbf{0}.$$

Since the two vectors forming the linear combination in the fourth member are linearly independent, their coefficients $-u_2$ and w_2 must be zero while v_1 remains arbitrary. Choosing $v_2 = 1$ we get the eigenvector $\mathbf{u}_2 = \begin{pmatrix} 0 \\ 1 \\ 0 \end{pmatrix}$.

Each eigenvector $\mathbf{u}_3 = \begin{pmatrix} u_3 \\ v_3 \\ w_3 \end{pmatrix}$ associated with $\lambda_3 = 2$ must satisfy $(A - \lambda_3 I)\mathbf{u}_3 = \mathbf{0}$.

So we compute

$$(A - \lambda_3 I)\mathbf{u}_3 = \begin{pmatrix} -2 & 0 & 1 \\ 0 & -1 & 0 \\ 0 & 0 & 0 \end{pmatrix} \begin{pmatrix} u_3 \\ v_3 \\ w_3 \end{pmatrix} = \begin{pmatrix} -2u_2 + w_3 \\ -v_3 \\ 0 \end{pmatrix}$$

$$= (2u_3 + w_3) \begin{pmatrix} 1 \\ 0 \\ 0 \end{pmatrix} + v_3 \begin{pmatrix} 0 \\ 1 \\ 0 \end{pmatrix} = \mathbf{0}.$$

Since the two vectors forming the linear combination in the bottom row are linearly independent, their coefficients $-2u_3 + w_3$ and $-v_3$ must be zero. Choosing $u_3 = 1$, $w_3 = 2$ we get the eigenvector $\mathbf{u}_3 = \begin{pmatrix} 1 \\ 0 \\ 2 \end{pmatrix}$.

9. Let f be a one-to-one linear operator on the vector space V, let λ be an eigenvalue of f and let $\mathbf{u}$ be an associated eigenvector. Then f^{-1} exists and is a linear operator on the image of f. Therefore, since $f^{-1} \circ f$ is the identity operator on V, the definitions of eigenvalue and eigenvector imply that

$$\mathbf{u} = (f^{-1} \circ f)(\mathbf{u}) = f^{-1}\big(f(\mathbf{u})\big) = f^{-1}(\lambda \mathbf{u}) = \lambda f^{-1}(\mathbf{u}).$$

That is, $\lambda f^{-1}(\mathbf{u}) = \mathbf{u}$. As $\mathbf{u} \neq 0$, neither is $\lambda f^{-1}(\mathbf{u})$. So $\lambda \neq 0$ and we can divide by λ to get

$$f^{-1}(\mathbf{u}) = \frac{1}{\lambda}\mathbf{u}.$$

This is precisely the condition that must hold in order that $1/\lambda$ be an eigenvalue of f^{-1}.

11. Let $C^{(\infty)}(\mathcal{R})$ be the vector space of infinitely often differentiable functions $f(x)$ for $x \in \mathcal{R}$ and let $D^2 = d^2/dx^2$ act on functions in $C^{(\infty)}(\mathcal{R})$,

(a) Let $\lambda > 0$, let $k = \sqrt{\lambda}$ and consider the linear combination $c_1 e^{kx} + c_2 e^{-kx}$. We have

$$D^2(c_1 e^{kx} + c_2 e^{-kx}) = D\big(D(c_1 e^{kx} + c_2 e^{-kx})\big) = D(kc_1 e^{kx} - kc_2 e^{-kx})$$
$$= k^2 c_2 e^{kx} + k^2 c_2 e^{-kx} = k^2(c_1 e^{kx} + c_2 e^{-kx}) = \lambda(c_1 e^{kx} + c_2 e^{-kx}).$$

Hence, $c_1 e^{kx} + c_2 e^{-kx}$ is an eigenvector associated with $\lambda > 0$.

(b) Let $\lambda < 0$, let $k = \sqrt{-\lambda}$ and consider the linear combination $c_1 \cos kx + c_2 \sin kx$. We have

$$D^2(c_1 \cos kx + c_2 \sin kx) = D\big(D(c_1 \cos kx + c_2 \sin kx)\big) = D(-kc_1 \sin kx + kc_2 \cos kx)$$
$$= -k^2 c_1 \cos kx - k^2 c_2 \sin kx = -k^2(c_1 \cos kx + c_2 \sin kx)$$
$$= -(-\lambda)(c_1 \cos kx + c_2 \sin kx) = \lambda(c_1 \cos kx + c_2 \sin kx).$$

Hence, $c_1 \cos kx + c_2 \sin kx$ is an eigenvector for $\lambda < 0$.

(c) If $\lambda = 0$ is an eigenvalue then a function f in $C^{(\infty)}(\mathcal{R})$ is an eigenvector if $D^2 f = 0$. The functions in $C^{(\infty)}(\mathcal{R})$ whose second derivative is identically zero are the polynomials of the form $f(x) = c_1 + c_2 x$. Two linearly independent functions of this form can then be found by choosing $c_1 = 1, c_2 = 0$ and $c_1 = 0, c_2 = 1$ to get $f_1(x) = 1$ and $f_2(x) = x$.

(d) Given that the only eigenvectors for the operator D^2 are those found in parts (a), (b) and (c), we set $x = 0$ and $x = \pi$ in each of the three linear combinations to see which of them is zero in both cases.

For the linear combination $c_1 e^{kx} + c_2 e^{-kx}$ we obtain the two equations $c_1 + c_2 = 0$ and $c_1 e^{k\pi} + c_2 e^{-k\pi} = 0$. The first equation gives $c_2 = -c_1$, which we substitute into the second equation to get $c_1 e^{k\pi} - c_1 e^{-k\pi} = 0$, or equivalently,

$$c_1 e^{-k\pi}(e^{2k\pi} - 1) = 0.$$

Since the exponential function is never zero, we must have $c_1 = 0$ or $e^{2k\pi} = 1$. The only value of k that satisfies the second equation is $k = 0$, which is contrary to the condition $k = \sqrt{\lambda} > 0$. Hence, $c_1 = 0$ and, since $c_2 = -c_1$, $c_2 = 0$ as well. This says that $c_1 e^{kx} + c_2 e^{-kx}$ is the zero function and is therefore not an eigenvector.

For the linear combination $c_1 + c_2 x$ we obtain the two equation $c_1 = 0$ and $c_1 + c_2 \pi = 0$, with the unique solution $c_1 = c_2 = 0$. Again, this is the zero function and is therefore not an eigenvector.

For the linear combination $c_1 \cos kx + c_2 \sin kx$ we obtain $c_1 = 0$ and $c_1 \cos k\pi = 0$. These determine $c_1 = 0$ but leave c_2 arbitrary. Thus, the the only eigenvectors of D^2 that are zero when $x = 0$ and $x = \pi$ are the nonzero constant multiples of $\sin kx$. By part (b), the associated eigenvalue is $\lambda < 0$ with $k = \sqrt{-\lambda}$. In terms of k, $\lambda = -k^2$.

13. Let G be an operator on $\mathcal{R}^2$ with matrix $A = \begin{pmatrix} 1 & 2 \\ 1 & 1 \end{pmatrix}$. The eigenvalues of G are the solutions λ of the equation

$$\det(A - \lambda I) = \begin{vmatrix} 1 - \lambda & 2 \\ 1 & 1 - \lambda \end{vmatrix} = (1 - \lambda)^2 - 2 = (1 + \sqrt{2} - \lambda)(1 - \sqrt{2} - \lambda) = 0.$$

So the eigenvalues of G are $\lambda_1 = 1 + \sqrt{2}$ and $\lambda_2 = 1 - \sqrt{2}$.

Each eigenvector $\mathbf{u}_1 = \begin{pmatrix} u_1 \\ v_1 \end{pmatrix}$ associated with $\lambda_1 = 1 + \sqrt{2}$ must satisfy $(A - \lambda_1 I)\mathbf{u}_1 = \mathbf{0}$. So we compute

$$(A - \lambda_1 I)\mathbf{u}_1 = \begin{pmatrix} -\sqrt{2} & 2 \\ 1 & -\sqrt{2} \end{pmatrix} \begin{pmatrix} u_1 \\ v_1 \end{pmatrix} = \begin{pmatrix} -\sqrt{2}u_1 + 2v_1 \\ u_1 - \sqrt{2}v_1 \end{pmatrix} = (u_1 - \sqrt{2}v_1)\begin{pmatrix} -\sqrt{2} \\ 1 \end{pmatrix} = \mathbf{0}.$$

Thus, the coordinates of $\mathbf{u}_1$ must satisfy $u_1 - \sqrt{2}v_1 = 0$. Choosing $v_1 = 1$, $u_1 = \sqrt{2}$ gives the eigenvector $\mathbf{u}_1 = \begin{pmatrix} \sqrt{2} \\ 1 \end{pmatrix}$.

Each eigenvector $\mathbf{u}_2 = \begin{pmatrix} u_2 \\ v_2 \end{pmatrix}$ associated with $\lambda_2 = 1 - \sqrt{2}$ must satisfy $(A - \lambda_2 I)\mathbf{u}_2 = \mathbf{0}$. So we compute

$$(A - \lambda_2 I)\mathbf{u}_2 = \begin{pmatrix} \sqrt{2} & 2 \\ 1 & \sqrt{2} \end{pmatrix} \begin{pmatrix} u_2 \\ v_2 \end{pmatrix} = \begin{pmatrix} \sqrt{2}u_2 + 2v_2 \\ u_2 + \sqrt{2}v_2 \end{pmatrix} = (u_2 + \sqrt{2}v_2)\begin{pmatrix} \sqrt{2} \\ 1 \end{pmatrix} = \mathbf{0}.$$

Thus, the coordinates of $\mathbf{u}_2$ must satisfy $u_2 + \sqrt{2}v_2 = 0$. Choosing $v_2 = 1$, $u_2 = -\sqrt{2}$ gives the eigenvector $\mathbf{u}_2 = \begin{pmatrix} -\sqrt{2} \\ 1 \end{pmatrix}$.

To see that $B = \{\mathbf{u}_1, \mathbf{u}_2\}$ is a basis for $\mathcal{R}^2$, let (x, y) be an arbitrary vector in $\mathcal{R}^2$ and observe that

$$(x, y) = \left(\frac{1}{2\sqrt{2}}x + \frac{1}{2}y\right)(\sqrt{2}, 1) + \left(-\frac{1}{2\sqrt{2}}x + \frac{1}{2}y\right)(-\sqrt{2}, 1)$$

$$= \left(\frac{1}{2\sqrt{2}}x + \frac{1}{2}y\right)\mathbf{u}_1 + \left(-\frac{1}{2\sqrt{2}}x + \frac{1}{2}y\right)\mathbf{u}_2.$$

That is, (x, y) is in the span of B and therefore B spans $\mathcal{R}^2$. Since B contains exactly two vectors and since $\mathcal{R}^2$ has dimension 2, Theorem 5.9 in this chapter says that B is a basis for $\mathcal{R}^2$.

As in Example 2, the eigenvalue matrix describing G in terms of the above eigenvector basis is

$$\begin{pmatrix} 1 + \sqrt{2} & 0 \\ 0 & 1 - \sqrt{2} \end{pmatrix}.$$

This is a composition of two actions: the first is a stretch by a factor of $1 + \sqrt{2}$ along the line in the direction of $\mathbf{u}_1 = \begin{pmatrix} \sqrt{2} \\ 1 \end{pmatrix}$ with coordinate matrix $\begin{pmatrix} 1 + \sqrt{2} & 0 \\ 0 & 1 \end{pmatrix}$, and the second is a stretch by a factor of $\sqrt{2} - 1$ in the direction of $-\mathbf{u}_2 = \begin{pmatrix} \sqrt{2} \\ -1 \end{pmatrix}$ with coordinate matrix $\begin{pmatrix} 1 & 0 \\ 0 & 1 - \sqrt{2} \end{pmatrix}$.

15. Let L be the linear operator defined by

$$(1) \qquad L(\mathbf{x}) = \begin{pmatrix} 0 & 4 \\ 1 & 0 \end{pmatrix} \mathbf{x}.$$

Here, $\mathbf{x} = \begin{pmatrix} x \\ y \end{pmatrix}$ with x and y as coordinates with respect to the standard basis vectors in $\mathcal{R}^2$. It was found in Exercise 3 in this section that 2 and -2 are the eigenvalues of L and that $\begin{pmatrix} 2 \\ 1 \end{pmatrix}$ and $\begin{pmatrix} -2 \\ 1 \end{pmatrix}$ are respective associated eigenvectors. Hence, if u and v are the eigenvector coordinates for $\mathbf{x}$ then L has the diagonal matrix representation

$$(2) \qquad L(\mathbf{x}) = \begin{pmatrix} 2 & 0 \\ 0 & -2 \end{pmatrix} \begin{pmatrix} u \\ v \end{pmatrix}.$$

Now let $\mathbf{x}$ be a solution of the given system of differential equations, where $x = x(t)$ and $y = y(t)$. In terms of eigenvector coordinates,

$$\mathbf{x} = \begin{pmatrix} u \\ v \end{pmatrix},$$

where $u = u(t)$ and $v = v(t)$ are functions of t. Differentiating both sides of this equation with respect to t gives

$$\frac{d\mathbf{x}}{dt} = \begin{pmatrix} u' \\ v' \end{pmatrix}.$$

Since the right side of equations (1) and (2) are equal, it follows that the given system of differential equations can be written as

$$\begin{pmatrix} u' \\ v' \end{pmatrix} = \begin{pmatrix} 2 & 0 \\ 0 & -2 \end{pmatrix} \begin{pmatrix} u \\ v \end{pmatrix}, \qquad \text{or equivalently,} \qquad \begin{matrix} u' = 2u, \\ v' = -2v. \end{matrix}$$

By Example 4 in the text, the solutions of the two scalar equations are $u = c_1 e^{2t}$ and $v = c_2 e^{-2t}$. Since u and v are the eigenvalue coordinates of $\mathbf{x}$,

$$\mathbf{x} = c_1 e^{2t} \begin{pmatrix} 2 \\ 1 \end{pmatrix} + c_2 e^{-2t} \begin{pmatrix} -2 \\ 1 \end{pmatrix}.$$

The solution in scalar form is therefore

$$x(t) = 2c_1 e^{2t} - 2c_2 e^{-2t},$$
$$y(t) = c_1 e^{2t} + c_2 e^{-2t}.$$

17. Let L be the linear operator defined by

$$(1) \qquad L(\mathbf{x}) = \begin{pmatrix} 2 & 4 \\ 1 & 2 \end{pmatrix} \mathbf{x}.$$

Here, $\mathbf{x} = \begin{pmatrix} x \\ y \end{pmatrix}$ with x and y as coordinates with respect to the standard basis vectors in $\mathcal{R}^2$. It was found in Exercise 5 in this section that 0 and 4 are the eigenvalues of L and that $\begin{pmatrix} -2 \\ 1 \end{pmatrix}$ and $\begin{pmatrix} 2 \\ 1 \end{pmatrix}$ are respective associated eigenvectors. Hence, if u and v are the eigenvector coordinates for $\mathbf{x}$ then L has the diagonal matrix representation

$$(2) \qquad L(\mathbf{x}) = \begin{pmatrix} 0 & 0 \\ 0 & 4 \end{pmatrix} \begin{pmatrix} u \\ v \end{pmatrix}.$$

Now let $\mathbf{x}$ be a solution of the given system of differential equations, where $x = x(t)$ and $y = y(t)$. In terms of eigenvector coordinates,

$$\mathbf{x} = \begin{pmatrix} u \\ v \end{pmatrix},$$

where $u = u(t)$ and $v = v(t)$ are functions of t. Differentiating both sides of this equation with respect to t gives

$$\frac{d\mathbf{x}}{dt} = \begin{pmatrix} u' \\ v' \end{pmatrix}.$$

Since the right side of equations (1) and (2) are equal, it follows that the given system of differential equations can be written as

$$\begin{pmatrix} u' \\ v' \end{pmatrix} = \begin{pmatrix} 0 & 0 \\ 0 & 4 \end{pmatrix} \begin{pmatrix} u \\ v \end{pmatrix}, \qquad \text{or equivalently,} \qquad \begin{aligned} u' &= 0, \\ v' &= 4v. \end{aligned}$$

Integrating the equation for u gives $u = c_1$ and, by Example 4 in the text, the solution for v is $v = c_2 e^{4t}$. Since u and v are the eigenvalue coordinates of $\mathbf{x}$,

$$\mathbf{x} = c_1 \begin{pmatrix} -2 \\ 1 \end{pmatrix} + c_2 e^{4t} \begin{pmatrix} 2 \\ 1 \end{pmatrix}.$$

The solution in scalar form is therefore

$$\begin{aligned} x(t) &= -2c_1 + 2c_2 e^{4t}, \\ y(t) &= c_1 + c_2 e^{4t}. \end{aligned}$$

Exercise Set 6BC (pgs. 154-155)

1. Let L be the linear operator on $\mathcal{R}^2$ with matrix $A = \begin{pmatrix} 0 & 1 \\ 1 & 1 \end{pmatrix}$. Its characteristic equation is $\det(A - \lambda I) = \begin{vmatrix} -\lambda & 1 \\ 1 & 1-\lambda \end{vmatrix} = -\lambda(1-\lambda) - 1 = \lambda^2 - \lambda - 1 = 0$. The quadratic formula gives the two distinct *real* roots $\lambda_{1,2} = \frac{1}{2}(1 \pm \sqrt{5})$. By Theorem 6.7, $\mathcal{R}^2$ has a basis consisting of eigenvectors of L.

3. Let L be the linear operator on $\mathcal{R}^2$ with matrix $A = \begin{pmatrix} 0 & 1 \\ -1 & 0 \end{pmatrix}$. Its characteristic equation is $\det(A - \lambda I) = \begin{vmatrix} -\lambda & 1 \\ -1 & -\lambda \end{vmatrix} = \lambda^2 + 1 = 0$, which has no real roots. Hence, $\mathcal{R}^2$ does not have a basis consisting of eigenvectors of L. However, the equation does have two distinct complex roots. So Theorem 6.7 implies that the coordinate space $\mathcal{C}^2$ has a basis of eigenvectors of L if we view L as an operator on $\mathcal{C}^2$.

5. Let L be the linear operator on $\mathcal{R}^3$ with matrix $A = \begin{pmatrix} 3 & -2 & -2 \\ -2 & -2 & 1 \\ 2 & 2 & -2 \end{pmatrix}$. Its characteristic equation is

$$\det(A - \lambda I) = \begin{vmatrix} 3-\lambda & -2 & -2 \\ -2 & -2-\lambda & 1 \\ 2 & 2 & -2-\lambda \end{vmatrix}$$

$$= (3-\lambda)\begin{vmatrix} -2-\lambda & 1 \\ 2 & -2-\lambda \end{vmatrix} - (-2)\begin{vmatrix} -2 & -2 \\ 2 & -2-\lambda \end{vmatrix} + (2)\begin{vmatrix} -2 & -2 \\ -2-\lambda & 1 \end{vmatrix}$$

$$= (3-\lambda)(\lambda^2 + 4\lambda + 2)) + 2(2\lambda + 8) + 2(-2\lambda - 6) = -\lambda^3 - \lambda^2 + 10\lambda + 10$$

$$= (-\lambda^2 + 10)(\lambda + 1) = 0,$$

where the second line is obtained by expansion about the first column. This has the three distinct *real* roots $\lambda_1 = \sqrt{10}$, $\lambda_1 = -\sqrt{10}$ and $\lambda_3 = -1$. By Theorem 6.7, $\mathcal{R}^3$ has a basis consisting of eigenvectors of L.

7. For each θ satisfying $0 \le \theta < 2\pi$, let L_θ be the operator on $\mathcal{R}^2$ given by the matrix $R_\theta = \begin{pmatrix} \cos\theta & -\sin\theta \\ \sin\theta & \cos\theta \end{pmatrix}$.

(a) The eigenvalues of L_θ are the solutions λ of

$$\det(R_\theta - \lambda I) = \begin{vmatrix} \cos\theta - \lambda & -\sin\theta \\ \sin\theta & \cos\theta - \lambda \end{vmatrix} = (\cos\theta - \lambda)^2 + \sin^2\theta$$

$$= \cos^2\theta - 2\lambda\cos\theta + \lambda^2 + \sin^2\theta = \lambda^2 - 2\lambda\cos\theta + 1 = 0.$$

By the quadratic formula,

$$\lambda = \frac{2\cos\theta \pm \sqrt{4\cos^2\theta - 4}}{2} = \cos\theta \pm \sqrt{\cos^2\theta - 1} = \cos\theta \pm \sqrt{-\sin^2\theta} = \cos\theta \pm i\sin\theta.$$

Hence, the eigenvalues of L_θ are real if, and only if, $\sin\theta = 0$. Since $\theta \in [0, 2\pi)$, we must have $\theta = 0$ or $\theta = \pi$.

If $\theta = 0$ then R_0 is the identity matrix and L_0 is the identity operator. Hence, the equations $L_0(\mathbf{e}_1) = \mathbf{e}_1$ and $L_0(\mathbf{e}_2) = \mathbf{e}_2$ show that $\mathbf{e}_1$ and $\mathbf{e}_2$ are linearly independent eigenvectors associated with the eigenvalue $\lambda = 1$.

In a similar fashion, if $\theta = \pi$ then R_π is the negative of the identity matrix and L_π maps a vector to its negative. Hence, the equations $L_\pi(\mathbf{e}_1) = -\mathbf{e}_1$ and $L_\pi(\mathbf{e}_2) = -\mathbf{e}_2$ show that $\mathbf{e}_1$ and $\mathbf{e}_2$ are linearly independent eigenvectors associated with the eigenvalue $\lambda = -1$.

(b) Note that $\det R_\theta = 1$ for all θ so that R_θ is one-to-one. By Exercise 8 in Section 6A in this chapter, 0 is not an eigenvalue of R_θ for all θ. Thus, if $\mathbf{u}$ is a real eigenvector of R_θ and λ is its associated eigenvalue then $\mathbf{u}$ and $R_\theta \mathbf{u} = \lambda \mathbf{u}$ must be parallel vectors. So the angle θ between $\mathbf{u}$ and $R_\theta \mathbf{u}$ must be a multiple of π. Since $0 \le \theta < 2\pi$, the only possibilities are $\theta = 0$ and $\theta = \pi$.

9. Let $A = \begin{pmatrix} 0 & 2 \\ -1 & 3 \end{pmatrix}$. First, straightforward computation gives

$$A^2 = AA = \begin{pmatrix} 0 & 2 \\ -1 & 3 \end{pmatrix}\begin{pmatrix} 0 & 2 \\ -1 & 3 \end{pmatrix} = \begin{pmatrix} -2 & 6 \\ -3 & 7 \end{pmatrix},$$

$$A^4 = A^2 A^2 = \begin{pmatrix} -2 & 6 \\ -3 & 7 \end{pmatrix}\begin{pmatrix} -2 & 6 \\ -3 & 7 \end{pmatrix} = \begin{pmatrix} -14 & 30 \\ -15 & 31 \end{pmatrix},$$

$$A^8 = A^4 A^4 = \begin{pmatrix} -14 & 30 \\ -15 & 31 \end{pmatrix}\begin{pmatrix} -14 & 30 \\ -15 & 31 \end{pmatrix} = \begin{pmatrix} -254 & 510 \\ -255 & 511 \end{pmatrix},$$

$$A^{10} = A^2 A^8 = \begin{pmatrix} 0 & 2 \\ -1 & 3 \end{pmatrix}\begin{pmatrix} -254 & 510 \\ -255 & 511 \end{pmatrix} = \begin{pmatrix} -1022 & 2046 \\ -1023 & 2047 \end{pmatrix}.$$

The above result can also be obtained by following the hint given in the statement of the problem. The matrix Λ in the matrix equation $A = U\Lambda U^{-1}$ is a diagonal matrix whose diagonal entries are the eigenvalues of A, and the columns of U are associated eigenvectors. Doing the calculation, one finds that 1 and 2 are the eigenvalues of A and that $\mathbf{u}_1 = \begin{pmatrix} 2 \\ 1 \end{pmatrix}$ and $\mathbf{u}_2 = \begin{pmatrix} 1 \\ 1 \end{pmatrix}$ are corresponding eigenvectors. So

$$A = U\Lambda U^{-1} = \begin{pmatrix} 2 & 1 \\ 1 & 1 \end{pmatrix}\begin{pmatrix} 1 & 0 \\ 0 & 2 \end{pmatrix}\begin{pmatrix} 1 & -1 \\ -1 & 2 \end{pmatrix}.$$

The reason the above form for A is helpful in finding powers of A is that (i) $A^n = U\Lambda^n U^{-1}$ for $n \ge 0$ (which can easily be proven by mathematical induction) and (ii) the nth power of a diagonal matrix can be computed by simply raising each of its diagonal entries to the nth power. So

$$A^{10} = U\lambda^{10}U^{-1} = \begin{pmatrix} 2 & 1 \\ 1 & 1 \end{pmatrix}\begin{pmatrix} 1 & 0 \\ 0 & 2 \end{pmatrix}^{10}\begin{pmatrix} 1 & -1 \\ -1 & 2 \end{pmatrix}$$

$$= \begin{pmatrix} 2 & 1 \\ 1 & 1 \end{pmatrix}\begin{pmatrix} 1 & 0 \\ 0 & 2^{10} \end{pmatrix}\begin{pmatrix} 1 & -1 \\ -1 & 2 \end{pmatrix} = \begin{pmatrix} -1022 & 2046 \\ -1023 & 2047 \end{pmatrix},$$

which is the same answer obtained by the first method.

(NOTE: While the second method is perhaps not the easiest method to apply to this particular problem, it is quite instructive to learn and, in many cases, greatly reduces the computations involved. Of course, the method can only be used if the matrix U is invertible; i.e., if the eigenvectors form a basis for the underlying vector space.)

11. Let L be a linear operator on $\mathcal{R}^2$ and let its matrix relative to the standard basis be given by $A = \begin{pmatrix} 4 & -3 \\ 2 & -1 \end{pmatrix}$. The eigenvalues of L are the solutions λ of the equation

$$\det(A - \lambda I) = \begin{vmatrix} 4 - \lambda & -3 \\ 2 & -1 - \lambda \end{vmatrix} = (4 - \lambda)(-1 - \lambda) + 6 = \lambda^2 - 3\lambda + 2$$

$$= (\lambda - 1)(\lambda - 2) = 0.$$

Hence, the eigenvalues of L are $\lambda_1 = 1$ and $\lambda_2 = 2$.

Let $\mathbf{u}_1 = \begin{pmatrix} u_1 \\ v_1 \end{pmatrix}$, $\mathbf{u}_2 = \begin{pmatrix} u_2 \\ v_2 \end{pmatrix}$ be eigenvectors associated with λ_1 and λ_2, respectively. Then $\mathbf{u}_1$ and $\mathbf{u}_2$ satisfy

$$(A - \lambda_1 I)\mathbf{u}_1 = \begin{pmatrix} 3 & -3 \\ 2 & -2 \end{pmatrix} \begin{pmatrix} u_1 \\ v_1 \end{pmatrix} = \begin{pmatrix} 3u_1 - 3v_1 \\ 2u_1 - 2v_1 \end{pmatrix} = (u_1 - v_1) \begin{pmatrix} 3 \\ 2 \end{pmatrix} = \mathbf{0},$$

and

$$(A - \lambda_2 I)\mathbf{u}_2 = \begin{pmatrix} 2 & -3 \\ 2 & -3 \end{pmatrix} \begin{pmatrix} u_2 \\ v_2 \end{pmatrix} = \begin{pmatrix} 2u_2 - 3v_2 \\ 2u_2 - 3v_2 \end{pmatrix} = (2u_2 - 3v_2) \begin{pmatrix} 1 \\ 1 \end{pmatrix} = \mathbf{0}.$$

The first equation shows that the coordinates of $\mathbf{u}_1$ must be related by $u_1 - v_1 = 0$, and the second equation shows that the coordinates of $\mathbf{u}_2$ must be related by $2u_2 - 3v_2 = 0$. Choosing $u_1 = 1, v_1 = 1$ and $u_2 = 3, v_2 = 2$ gives the two eigenvectors $\mathbf{u}_1 = \begin{pmatrix} 1 \\ 1 \end{pmatrix}$ and $\mathbf{u}_2 = \begin{pmatrix} 3 \\ 2 \end{pmatrix}$. This is a basis for $\mathcal{R}^2$. Listed in this order, the diagonal matrix that represents L with respect to this basis is

$$\Lambda = \begin{pmatrix} 1 & 0 \\ 0 & 2 \end{pmatrix}.$$

13. Let L be a linear operator on $\mathcal{R}^3$ and let its matrix relative to the standard basis be given by $A = \begin{pmatrix} -1 & 0 & 0 \\ -1 & 0 & 0 \\ -1 & -1 & 1 \end{pmatrix}$. The eigenvalues of L are the solutions λ of the equation

$$\det(A - \lambda I) = \begin{vmatrix} -1 - \lambda & 0 & 0 \\ -1 & -\lambda & 0 \\ -1 & -1 & 1 - \lambda \end{vmatrix} = (-1 - \lambda)(-\lambda)(1 - \lambda) = 0.$$

Hence, the eigenvalues of L are $\lambda_1 = -1$, $\lambda_2 = 0$ and $\lambda_3 = 1$.

For $k = 1, 2, 3$, let $\mathbf{u}_k = \begin{pmatrix} u_k \\ v_k \\ w_k \end{pmatrix}$ be an eigenvector associated with λ_k. Then

$$(A - \lambda_1 I)\mathbf{u}_1 = \begin{pmatrix} 0 & 0 & 0 \\ -1 & 1 & 0 \\ -1 & -1 & 2 \end{pmatrix} \begin{pmatrix} u_1 \\ v_1 \\ w_1 \end{pmatrix} = \begin{pmatrix} 0 \\ -u_1 + v_1 \\ -u_1 - v_1 + 2w_1 \end{pmatrix} = \mathbf{0},$$

$$(A - \lambda_2 I)\mathbf{u}_2 = \begin{pmatrix} -1 & 0 & 0 \\ -1 & 0 & 0 \\ -1 & -1 & 1 \end{pmatrix} \begin{pmatrix} u_2 \\ v_2 \\ w_2 \end{pmatrix} = \begin{pmatrix} -u_2 \\ -u_2 \\ -u_2 - v_2 + w_2 \end{pmatrix} = \mathbf{0},$$

$$(A - \lambda_3 I)\mathbf{u}_3 = \begin{pmatrix} -2 & 0 & 0 \\ -1 & -1 & 0 \\ -1 & -1 & 0 \end{pmatrix} \begin{pmatrix} u_3 \\ v_3 \\ w_3 \end{pmatrix} = \begin{pmatrix} -2u_3 \\ -u_3 - v_3 \\ -u_3 - v_3 \end{pmatrix} = \mathbf{0}.$$

The first equation gives the two-equation system $-u_1 + v_1 = 0$, $-u_1 - v_1 + 2w_1 = 0$. The first equation is satisfied by the choice $u_1 = v_1 = 1$. Inserting these values into the second equation gives $w_1 = 1$. The corresponding eigenvector is $\mathbf{u}_1 = \begin{pmatrix} 1 \\ 1 \\ 1 \end{pmatrix}$. The second equation above gives the equivalent two-equation system $u_2 = 0$, $u_2 + v_2 - w_2 = 0$. Inserting $u_2 = 0$ into the second equation and choosing $v_2 = w_2 = 1$ gives the second eigenvector $\mathbf{u}_2 = \begin{pmatrix} 0 \\ 1 \\ 1 \end{pmatrix}$.

The third equation above leads to the equivalent two-equation system $2u_3 = 0$, $u_3 + v_3 = 0$. These imply that $u_3 = v_3 = 0$, while w_3 remains arbitrary. Choosing $w_3 = 1$ gives the third eigenvector $\mathbf{u}_3 = \begin{pmatrix} 0 \\ 0 \\ 1 \end{pmatrix}$.

Thus, a basis for $\mathcal{R}^3$ consisting of eigenvectors of L is given by

$$\left\{ \begin{pmatrix} 1 \\ 1 \\ 1 \end{pmatrix}, \begin{pmatrix} 0 \\ 1 \\ 1 \end{pmatrix}, \begin{pmatrix} 0 \\ 0 \\ 1 \end{pmatrix} \right\}.$$

With the basis vectors listed in this order, the corresponding diagonal matrix that represents L with respect to this basis is

$$\Lambda = \begin{pmatrix} -1 & 0 & 0 \\ 0 & 0 & 0 \\ 0 & 0 & 1 \end{pmatrix}.$$

15. Let L be a linear operator and let its matrix relative to the standard basis be given by $A = \begin{pmatrix} 0 & 0 \\ -4 & 2 \end{pmatrix}$. The eigenvalues of L are the solutions λ of the equation

$$\det(A - \lambda I) = \begin{vmatrix} -\lambda & 0 \\ -4 & 2 - \lambda \end{vmatrix} = -\lambda(2 - \lambda) = 0.$$

Hence, the eigenvalues of L are $\lambda_1 = 0$ and $\lambda_2 = 2$. With the eigenvalues listed in this order, the eigenvalue matrix Λ is

$$\Lambda = \begin{pmatrix} 0 & 0 \\ 0 & 2 \end{pmatrix}.$$

To find an associated eigenvector matrix U, let $\mathbf{u}_1 = \begin{pmatrix} u_1 \\ v_1 \end{pmatrix}$ and $\mathbf{u}_2 = \begin{pmatrix} u_2 \\ v_2 \end{pmatrix}$ be eigenvectors associated with λ_1 and λ_2, respectively. Then $\mathbf{u}_1$ and $\mathbf{u}_2$ must satisfy

$$(A - \lambda_1 I)\mathbf{u}_1 = \begin{pmatrix} 0 & 0 \\ -4 & 2 \end{pmatrix} \begin{pmatrix} u_1 \\ v_1 \end{pmatrix} = \begin{pmatrix} 0 \\ -4u_1 + 2v_1 \end{pmatrix} = (-4u_1 + 2v_1) \begin{pmatrix} 0 \\ 1 \end{pmatrix} = \mathbf{0},$$

$$(A - \lambda_2 I)\mathbf{u}_2 = \begin{pmatrix} -2 & 0 \\ -4 & 0 \end{pmatrix} \begin{pmatrix} u_2 \\ v_2 \end{pmatrix} = \begin{pmatrix} -2u_2 \\ -4u_2 \end{pmatrix} = -2u_2 \begin{pmatrix} 1 \\ 2 \end{pmatrix} = \mathbf{0}.$$

The first equation shows that $-4u_1 + 2v_1 = 0$, and the second equation shows that $u_2 = 0$ and v_2 remains arbitrary. The choices $u_1 = 1, v_1 = 2$ and $v_2 = 1$ give the eigenvectors

$\mathbf{u}_1 = \begin{pmatrix} 1 \\ 2 \end{pmatrix}$ and $\mathbf{u}_2 = \begin{pmatrix} 0 \\ 1 \end{pmatrix}$. Hence, $U = \begin{pmatrix} 1 & 0 \\ 2 & 1 \end{pmatrix}$ with inverse $U^{-1} = \begin{pmatrix} 1 & 0 \\ -2 & 1 \end{pmatrix}$. So $A = U\Lambda U^{-1}$ takes the form

$$\begin{pmatrix} 0 & 0 \\ -4 & 2 \end{pmatrix} = \begin{pmatrix} 1 & 0 \\ 2 & 1 \end{pmatrix} \begin{pmatrix} 0 & 0 \\ 0 & 2 \end{pmatrix} \begin{pmatrix} 1 & 0 \\ -2 & 1 \end{pmatrix}.$$

The reader should verify that the product on the right is equal to the given matrix A on the left.

17. Let L be a linear operator and let its matrix relative to the standard basis be given by $A = \begin{pmatrix} 0 & 1 \\ -4 & 2 \end{pmatrix}$. The eigenvalues of L are the solutions λ of the equation

$$\det(A - \lambda I) = \begin{vmatrix} -\lambda & 1 \\ -4 & 2-\lambda \end{vmatrix} = -\lambda(2 - \lambda) + 4 = \lambda^2 - 2\lambda + 4 = 0.$$

The quadratic formula gives complex eigenvalues for L: $\lambda_1 = 1 + i\sqrt{3}$ and $\lambda_2 = 1 - i\sqrt{3}$. With the eigenvalues listed in this order, the eigenvalue matrix Λ is

$$\Lambda = \begin{pmatrix} 1 + i\sqrt{3} & 0 \\ 0 & 1 - i\sqrt{3} \end{pmatrix}.$$

Note that this forces us to view L as an operator on $\mathcal{C}^2$.

To find an associated eigenvector matrix U, let $\mathbf{u}_1 = \begin{pmatrix} u_1 \\ v_1 \end{pmatrix}$ and $\mathbf{u}_2 = \begin{pmatrix} u_2 \\ v_2 \end{pmatrix}$ be eigenvectors associated with λ_1 and λ_2, respectively. Then $\mathbf{u}_1$ and $\mathbf{u}_2$ must satisfy

$$(A - \lambda_1 I)\mathbf{u}_1 = \begin{pmatrix} -1 - i\sqrt{3} & 1 \\ -4 & 1 - i\sqrt{3} \end{pmatrix} \begin{pmatrix} u_1 \\ v_1 \end{pmatrix} = \begin{pmatrix} (-1 - i\sqrt{3})u_1 + v_1 \\ -4u_1 + (1 - i\sqrt{3})v_1 \end{pmatrix}$$

$$= ((-1 - i\sqrt{3})u_1 + v_1) \begin{pmatrix} 1 \\ 1 - i\sqrt{3} \end{pmatrix} = \mathbf{0},$$

$$(A - \lambda_2 I)\mathbf{u}_2 = \begin{pmatrix} -1 + i\sqrt{3} & 1 \\ -4 & 1 + i\sqrt{3} \end{pmatrix} \begin{pmatrix} u_2 \\ v_2 \end{pmatrix} = \begin{pmatrix} (-1 + i\sqrt{3})u_2 + v_2 \\ -4u_2 + (1 + i\sqrt{3})v_2 \end{pmatrix}$$

$$= ((-1 + i\sqrt{3})u_2 + v_2) \begin{pmatrix} 1 \\ 1 + i\sqrt{3} \end{pmatrix} = \mathbf{0}.$$

The first equation shows that $(-1 - i\sqrt{3})u_1 + v_1 = 0$, and the second equation shows that $(-1 + i\sqrt{3})u_2 + v_2 = 0$. The choices $u_1 = 1, v_1 = 1 + i\sqrt{3}$ and $u_2 = 1, v_2 = 1 - i\sqrt{3}$ give the eigenvectors $\mathbf{u}_1 = \begin{pmatrix} 1 \\ 1 + i\sqrt{3} \end{pmatrix}$ and $\mathbf{u}_2 = \begin{pmatrix} 1 \\ 1 - i\sqrt{3} \end{pmatrix}$. Hence,

$$U = \begin{pmatrix} 1 & 1 \\ 1 + i\sqrt{3} & 1 - i\sqrt{3} \end{pmatrix} \quad \text{with inverse} \quad U^{-1} = \begin{pmatrix} (3 + i\sqrt{3})/6 & -i\sqrt{3}/6 \\ (3 - i\sqrt{3})/6 & i\sqrt{3}/6 \end{pmatrix}.$$

So $A = U\Lambda U^{-1}$ takes the form

$$\begin{pmatrix} 0 & 1 \\ -4 & 2 \end{pmatrix} = \begin{pmatrix} 1 & 1 \\ 1 + i\sqrt{3} & 1 - i\sqrt{3} \end{pmatrix} \begin{pmatrix} 1 + i\sqrt{3} & 0 \\ 0 & 1 - i\sqrt{3} \end{pmatrix} \begin{pmatrix} (3 + i\sqrt{3})/6 & -i\sqrt{3}/6 \\ (3 - i\sqrt{3})/6 & i\sqrt{3}/6 \end{pmatrix}.$$

The reader should verify that the product on the right is equal to the given matrix A on the left.

19. Let D be the differentiation operator on the vector space $C^{(1)}(-\infty, \infty)$ and let S be the subspace spanned by the set $B = \{e^x, e^{-x}\}$. If $c_1 e^x + c_2 e^{-x}$ is in S then the equation $D(c_1 e^x + c_2 e^{-x}) = c_1 e^x - c_2 e^{-x}$ shows that the image of S under D is contained in S. Thus, D restricted to S is an operator on S.

The matrix A of D with respect to the basis B is found by noting that

$$De^x = e^x = (1)e^x + (0)e^{-x} \qquad \text{and} \qquad De^{-x} = -e^{-x} = (0)e^x + (-1)e^{-x}$$

and then using the coefficients shown on the right of each equation to get

$$A = \begin{pmatrix} 1 & 0 \\ 0 & -1 \end{pmatrix}.$$

Moreover, the equations $De^x = e^x$ and $De^{-x} = -e^{-x}$ directly show that the eigenvalues of D are 1 and -1 with corresponding eigenvectors e^x and e^{-x}. Thus, B itself is a basis of eigenvectors for S.

21. Let S be the subspace of the vector space $C^{(1)}(-\infty, \infty)$ spanned by the set $B = \{\sin x, \cos x\}$. If D is the differentiation operator and $c_1 \sin x + c_2 \cos x$ is in S then the equation $D(c_1 \sin x + c_2 \cos x) = -c_2 \sin x + c_1 \cos x$ shows that the image of S under D is contained in S. Thus, D restricted to S is an operator on S.

To find the matrix A of D relative to the basis B, we compute

$$D\sin x = \cos x = (0)\sin x + (1)\cos x, \qquad D\cos x = -\sin x = (-1)\sin x + (0)\cos x.$$

The columns of A are the vectors whose coordinates are the above shown coefficients. That is,

$$A = \begin{pmatrix} 0 & -1 \\ 1 & 0 \end{pmatrix}.$$

The eigenvalues of D are the solutions λ of

$$\det(A - \lambda I) = \begin{vmatrix} -\lambda & -1 \\ 1 & -\lambda \end{vmatrix} = \lambda^2 + 1 = 0.$$

Hence, the eigenvalues of D are $\lambda_1 = i$ and $\lambda_2 = -i$. By Theorem 6.7, the associated coordinate space C^2 has a basis $\{\mathbf{u}, \mathbf{v}\}$ of eigenvectors of the linear operator on C^2 with matrix A. The two linear combinations of $\sin x$ and $\cos x$ with coefficients equal to the coordinates of $\mathbf{u}$ and $\mathbf{v}$ then constitute a basis of eigenvectors for S. The reader can verify that $\{\cos x + i\sin x, \cos x - i\sin x\}$ is such a basis.

Section 7: INNER PRODUCTS

Exercise Set 7A (pg. 158)

1. With $\mathbf{x} = (x_1, x_2)$ and $\mathbf{y} = (y_1, y_2)$, the formula $\langle \mathbf{x}, \mathbf{y} \rangle = x_1 y_1 + 2 x_2 y_2$ defines an inner product on $\mathcal{R}^2$ because

Positivity : $\quad \langle \mathbf{x}, \mathbf{x} \rangle = x_1 x_1 + 2 x_2 x_2 = x_1^2 + 2 x_2^2 > 0,$ except $\langle \mathbf{0}, \mathbf{0} \rangle = (0)(0) + 2(0)(0) = 0;$

Symmetry : $\quad \langle \mathbf{x}, \mathbf{y} \rangle = x_1 y_1 + 2 x_2 y_2 = y_1 x_1 + 2 y_2 x_2 = \langle \mathbf{y}, \mathbf{x} \rangle;$

Additivity : $\quad \langle \mathbf{x} + \mathbf{y}, \mathbf{z} \rangle = (x_1 + y_1) z_1 + 2(x_2 + y_2) z_2 = x_1 z_1 + y_1 z_1 + 2 x_2 z_2 + 2 y_2 z_2$
$$= x_1 z_1 + 2 x_2 z_2 + y_1 z_1 + 2 y_2 z_2 = \langle \mathbf{x}, \mathbf{z} \rangle + \langle \mathbf{y}, \mathbf{z} \rangle;$$

Homogeneity : $\quad \langle r\mathbf{x}, \mathbf{y} \rangle = (r x_1) y_1 + 2(r x_2) y_2 = r(x_1 y_1 + 2 x_2 y_2) = r \langle \mathbf{x}, \mathbf{y} \rangle.$

3. With $\mathbf{x} = (x_1, x_2)$ and $\mathbf{y} = (y_1, y_2)$, the formula $\langle \mathbf{x}, \mathbf{y} \rangle = x_1 y_1 + x_1 y_2 + x_2 y_1 + 2 x_2 y_2$ defines an inner product on $\mathcal{R}^2$. First of all,

$$\langle \mathbf{x}, \mathbf{x} \rangle = x_1 x_1 + x_1 x_2 + x_2 x_1 + 2 x_2 x_2 = x_1^2 + 2 x_1 x_2 + 2 x_2^2$$
$$= (x_1^2 + 2 x_1 x_2 + x_2^2) + x_2^2 = (x_1 + x_2)^2 + x_2^2.$$

The last expression in the second line is never negative because it is the sum of two squares of real numbers. Furthermore, it is zero only if $x_1 + x_2 = 0$ and $x_2 = 0$. This happens only if both x_1 and x_2 are zero. That is, $\langle \mathbf{x}, \mathbf{x} \rangle > 0$, except that $\langle \mathbf{0}, \mathbf{0} \rangle = 0$. So positivity holds. The remaining three properties are shown as follows:

Symmetry : $\quad \langle \mathbf{x}, \mathbf{y} \rangle = x_1 y_1 + x_1 y_2 + x_2 y_1 + 2 x_2 y_2$
$$= y_1 x_1 + y_1 x_2 + y_2 x_1 + 2 y_2 x_2 = \langle \mathbf{y}, \mathbf{x} \rangle;$$

Additivity : $\quad \langle \mathbf{x} + \mathbf{y}, \mathbf{z} \rangle = (x_1 + y_1) z_1 + (x_1 + y_1) z_2 + (x_2 + y_2) z_1 + 2(x_2 + y_2) z_2$
$$= x_1 z_1 + y_1 z_1 + x_1 z_2 + y_1 z_2 + x_2 z_1 + y_2 z_1 + 2 x_2 z_2 + 2 y_2 z_2$$
$$= (x_1 z_1 + x_1 z_2 + x_2 z_1 + 2 x_2 z_2) + (y_1 z_1 + y_1 z_2 + y_2 z_1 + 2 y_2 z_2)$$
$$= \langle \mathbf{x}, \mathbf{z} \rangle + \langle \mathbf{y}, \mathbf{z} \rangle;$$

Homogeneity : $\quad \langle r\mathbf{x}, \mathbf{y} \rangle = (r x_1) y_1 + (r x_1) y_2 + (r x_2) y_1 + 2(r x_2) y_2$
$$= r(x_1 y_1 + x_1 y_2 + x_2 y_1 + 2 x_2 y_2) = r \langle \mathbf{x}, \mathbf{y} \rangle.$$

5. The norm $\|\mathbf{x}\|$ under the inner product $\langle \mathbf{x}, \mathbf{y} \rangle = \langle (x_1, x_2), (y_1, y_2) \rangle = 3 x_1 y_1 + 2 y_1 y_2$ is given by

$$\|\mathbf{x}\| = \sqrt{\langle \mathbf{x}, \mathbf{x} \rangle} = \sqrt{3 x_1 x_1 + 2 x_2 x_2} = \sqrt{3 x_1^2 + 2 x_2^2}.$$

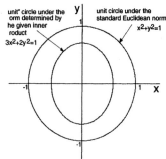

Thus, the set of points (x, y) such that $\|(x, y)\| = 1$ satisfy the equation $\sqrt{3 x^3 + 2 y^2} = 1$, or equivalently, $3 x^2 + 2 y^2 = 1$. This is the "unit" circle under this inner product and is shown on the right. The usual unit circle under the standard Euclidean norm defined by the dot product is also shown for comparison.

7. Letting $a = kx$ and $b = lx$ in the identities $\cos a \cos b = \frac{1}{2}(\cos(a+b) + \cos(a-b))$ and $\sin a \sin b = \frac{1}{2}(\cos(a+b) + \cos(a-b))$ allows us to write the first two integrals in Example 2 of the text as

$$\frac{1}{\pi}\int_{-\pi}^{\pi}\cos kx\cos lx\,dx = \frac{2}{\pi}\int_0^\pi \cos kx\cos lx\,dx = \frac{2}{\pi}\int_0^\pi\left[\frac{1}{2}\cos(k+l)x + \frac{1}{2}\cos(k-l)x\right]dx$$

$$= \frac{1}{\pi}\int_0^\pi\cos(k+l)x\,dx + \frac{1}{\pi}\int_0^\pi\cos(k-l)x\,dx$$

$$\frac{1}{\pi}\int_{-\pi}^{\pi}\sin kx\sin lx\,dx = \frac{2}{\pi}\int_0^\pi \sin kx\sin lx\,dx = \frac{2}{\pi}\int_0^\pi\left[\frac{1}{2}\cos(k-l)x - \frac{1}{2}\cos(k+l)x\right]dx$$

$$= \frac{1}{\pi}\int_0^\pi\cos(k-l)x\,dx - \frac{1}{\pi}\int_0^\pi\cos(k+l)x\,dx,$$

where the second integral in the first and third lines follow because $\cos kx\cos lx$ and $\sin kx\sin lx$ are even functions for all k and l. If $k \neq l$ then $\cos(k+l)x$ and $\cos(k-l)x$ are symmetric about the midpoint of the interval $[0,\pi]$. Hence, both integrals in the second line are zero and both integrals in the fourth line are zero. If $k = l \neq 0$ then the first integral in the second line and the second integral in the fourth line are zero, while the remaining integral in these lines reduces to $\frac{1}{\pi}\int_0^\pi 1\,dx = 1$. This proves the first two orthogonality relations given in Example 2

The last orthogonality relation in Example 2 is most easily shown by observing that $\cos kx\sin lx$ is an odd function for all k and l, so that its integral over any symmetric interval is zero.

9. Let e_1, e_2, e_3 be the standard basis vectors for $\mathcal{R}^3$ and suppose $\langle \mathbf{x}, \mathbf{y}\rangle$ is an inner product on $\mathcal{R}^3$ such that

$$\langle e_1, e_1\rangle = 1, \qquad \langle e_2, e_2\rangle = 1, \qquad \langle e_3, e_3\rangle = 5,$$
$$\langle e_1, e_2\rangle = 0, \qquad \langle e_1, e_3\rangle = 2, \qquad \langle e_2, e_3\rangle = 2.$$

A typical vector in $\mathcal{R}^3$ has the form $(a,b,c) = ae_1 + be_2 + ce_3$. We can find the length of this vector by using additivity to get

$$\langle (a,b,c),(a,b,c)\rangle = \langle ae_1 + be_2 + ce_3, ae_1 + be_2 + ce_3\rangle$$
$$= \langle ae_1, ae_1\rangle + \langle ae_1, be_2\rangle + \langle ae_1, ce_3\rangle + \langle be_2, ae_1\rangle + \langle be_2, be_2\rangle + \langle be_2, ce_3\rangle$$
$$+ \langle ce_3, ae_1\rangle + \langle ce_3, be_2\rangle + \langle ce_3, ce_3\rangle.$$

Now use symmetry to reverse the order of the first product in the third row and the first and second products in the fourth row. This gives

$$\langle (a,b,c),(a,b,c)\rangle = \langle ae_1, ae_1\rangle + \langle ae_1, be_2\rangle + \langle ae_1, ce_3\rangle$$
$$+ \langle ae_1, be_2\rangle + \langle be_2, be_2\rangle + \langle be_2, ce_3\rangle + \langle ae_1, ce_3\rangle + \langle be_2, ce_3\rangle + \langle ce_3, ce_3\rangle.$$

Now use homogeneity to get

$$\langle (a,b,c),(a,b,c)\rangle = a^2\langle e_1, e_1\rangle + ab\langle e_1, e_2\rangle + ac\langle e_1, e_3\rangle$$
$$+ ab\langle e_1, e_2\rangle + b^2\langle e_2, e_2\rangle + bc\langle e_2, e_3\rangle + ac\langle e_1, e_3\rangle + bc\langle e_2, e_3\rangle + c^2\langle e_3, e_3\rangle$$
$$= a^2\langle e_1, e_1\rangle + 2ab\langle e_1, e_2\rangle + 2ac\langle e_1, e_3\rangle + b^2\langle e_2, e_2\rangle + 2bc\langle e_2, e_3\rangle + c^2\langle e_3, e_3\rangle$$
$$= a^2(1) + 2ab(0) + 2ac(2) + b^2(1) + 2bc(2) + c^2(5)$$
$$= a^2 + 4ac + b^2 + 4bc + 5c^2$$
$$= (a^2 + 4ac + 4c^2) + (b^2 + 4bc + 4c^2) - 3c^2$$
$$= (a + 2c)^2 + (b + 2c)^2 - 3c^2.$$

The last expression above is negative whenever $c \neq 0$ and a and b are chosen sufficiently close to $-2c$. For example, if $c = -1$ and $a = b = 2$ then $(a, b, c) = (2, 2, -1)$ and the above becomes $\langle(2, 2, -1), (2, 2, -1)\rangle = -3 < 0$. So positivity fails. It follows that the assumption that such an inner product exists on $\mathcal{R}^3$ was false.

11. (a) Let $C[-\pi, \pi]$ be the vector space of real-valued continuous functions on $[-\pi, \pi]$. Define the product of f and g in $C[-\pi, \pi]$ by

$$\langle f, g \rangle = \int_{-\pi}^{\pi} f(x)g(x)\,dx.$$

We check that this product satisfies Properties 7.1 in the text. First observe that

$$\langle f, f \rangle = \int_{-\pi}^{\pi} f(x)f(x)\,dx = \int_{-\pi}^{\pi} f^2(x)\,dx.$$

Note that the integrand is nonnegative. Thus, if $f(x)$ is not identically zero on $[-\pi, \pi]$ then the continuity of f implies that the above integral is positive. Otherwise, its value is zero. So positivity holds. Further, symmetry holds because of commutativity of real numbers, and additivity and homogeneity follow from the linearity of the integral. Specifically,

$$\langle f, g \rangle = \int_{-\pi}^{\pi} f(x)g(x)\,dx = \int_{-\pi}^{\pi} g(x)f(x)\,dx = \langle g, f \rangle,$$

$$\langle f + g, h \rangle = \int_{-\pi}^{\pi} (f + g)(x)h(x)\,dx = \int_{-\pi}^{\pi} \big(f(x) + g(x)\big)h(x)\,dx$$

$$= \int_{-\pi}^{\pi} \big(f(x)h(x) + g(x)h(x)\big)\,dx = \int_{-\pi}^{\pi} f(x)h(x)\,dx + \int_{-\pi}^{\pi} g(x)h(x)\,dx$$

$$= \langle f, h \rangle + \langle g, h \rangle,$$

$$\langle rf, g \rangle = \int_{-\pi}^{\pi} (rf)(x)g(x)\,dx = \int_{-\pi}^{\pi} rf(x)g(x)\,dx = r\int_{-\pi}^{\pi} f(x)g(x)\,dx = r\langle f, g \rangle.$$

So the product is an inner product on $C[-\pi, \pi]$.

(b) In terms of the Cauchy-Schwarz inequality, the inner product defined in part (a) says

$$\left| \int_{-\pi}^{\pi} f(x)g(x)\,dx \right| \leq \left(\int_{-\pi}^{\pi} f^2(x)\,dx \right)^{1/2} \left(\int_{-\pi}^{\pi} g^2(x)\,dx \right)^{1/2}.$$

13. Let $\mathbf{x}$ and $\mathbf{y}$ be given vectors in a vector space V and let $\langle \mathbf{x}, \mathbf{y} \rangle$ be an inner product defined on V. Then symmetry, homogeneity and additivity of the inner product gives

$$\|\mathbf{x} - \mathbf{y}\|^2 = \langle \mathbf{x} - \mathbf{y}, \mathbf{x} - \mathbf{y} \rangle = \langle \mathbf{x}, \mathbf{x} \rangle - 2\langle \mathbf{x}, \mathbf{y} \rangle + \langle \mathbf{y}, \mathbf{y} \rangle = \|\mathbf{x}\|^2 + \|\mathbf{y}\|^2 - 2\langle \mathbf{x}, \mathbf{y} \rangle.$$

Hence,

$$\|\mathbf{x} - \mathbf{y}\|^2 = \|\mathbf{x}\|^2 + \|\mathbf{y}\|^2$$

if, and only if, $\langle \mathbf{x}, \mathbf{y} \rangle = 0$. That is, the proposed identity holds if, and only if, $\mathbf{x}$ and $\mathbf{y}$ are orthogonal.

131

Exercise Set 7B (pg. 167)

1. Let $\mathbf{x}_1 = (1,1,1)$, $\mathbf{x}_2 = (-1, \frac{1}{2}, \frac{1}{2})$, $\mathbf{x}_3 = (x,y,z)$ be a basis for $\mathcal{R}^3$. First note that $\mathbf{x}_1 \cdot \mathbf{x}_2 = (1,1,1) \cdot (-1, 1/2, 1/2) = -1 + 1/2 + 1/2 = 0$, so that $\mathbf{x}_3$ can be chosen so that $\{\mathbf{x}_1, \mathbf{x}_2, \mathbf{x}_3\}$ is an orthogonal set. For this to be the case, $\mathbf{x}_3$ must be chosen so that

$$\mathbf{x}_1 \cdot \mathbf{x}_3 = (1,1,1) \cdot (x,y,z) = x + y + z = 0,$$
$$\mathbf{x}_2 \cdot \mathbf{x}_3 = (-1, 1/2, 1/2) \cdot (x,y,z) = -x + y/2 + z/2 = 0.$$

The second equation can be solved for z to get $z = 2x - y$. Inserting this expression for z into the first equation gives $x + y + (2x - y) = 3x = 0$. So $x = 0$. The equation $z = 2x - y$ then gives $z = -y$. Hence, $\mathbf{x}_3 = (x,y,z) = (0, y, -y) = y(0,1,-1)$. Choosing $y = 1$ gives the orthogonal set $\{(1,1,1), (-1, 1/2, 1/2), (0,1,-1)\}$, which must be a basis for $\mathcal{R}^3$ because the vectors in an orthogonal set are automatically independent. For an orthonormal basis, we compute the unit vectors $\mathbf{x}_i / \|\mathbf{x}_i\|$ for $i = 1, 2, 3$:

$$\frac{\mathbf{x}_1}{\|\mathbf{x}_1\|} = \frac{(1,1,1)}{\sqrt{3}} = (1/\sqrt{3}, 1/\sqrt{3}, 1/\sqrt{3}),$$
$$\frac{\mathbf{x}_2}{\|\mathbf{x}_2\|} = \frac{(-1, 1/2, 1/2)}{\sqrt{3/2}} = ((-1/\sqrt{3/2}, 1/(2\sqrt{3/2}), 1/(2\sqrt{3/2})),$$
$$\frac{\mathbf{x}_3}{\|\mathbf{x}_3\|} = \frac{(0, 1, -1)}{\sqrt{2}} = (0, 1/\sqrt{2}, -1/\sqrt{2}).$$

Rationalizing all denominators in the coordinates of these unit vectors gives the orthonormal basis

$$\left\{ \left(\frac{\sqrt{3}}{3}, \frac{\sqrt{3}}{3}, \frac{\sqrt{3}}{3} \right), \left(-\frac{\sqrt{6}}{3}, \frac{\sqrt{6}}{6}, \frac{\sqrt{6}}{6} \right), \left(0, \frac{\sqrt{2}}{2}, -\frac{\sqrt{2}}{2} \right) \right\}.$$

3. Let P be the subspace spanned by $\mathbf{x}_1 = (1,2,1,1)$, $\mathbf{x}_2 = (-1,0,1,0)$, $\mathbf{x}_3 = (0,1,0,2)$. We first find a vector $\mathbf{x}_4 = (x,y,z,w)$ independent of these three by forming the matrix

$$A = \begin{pmatrix} 1 & -1 & 0 & x \\ 2 & 0 & 1 & y \\ 1 & 1 & 0 & z \\ 1 & 0 & 2 & w \end{pmatrix},$$

with $\mathbf{x}_1$, $\mathbf{x}_2$, $\mathbf{x}_3$, $\mathbf{x}_4$ as its columns. By expanding about the second column, we find that $\det A = -3x + 4y - 3z - 2w$. Since the columns of A will be linearly independent if, and only if, $\det A \neq 0$, we choose x, y, z, w such that $\det A \neq 0$. For simplicity, let $x = 1$ and $y = z = w = 0$ to get $\det A = -3 \neq 0$ and $\mathbf{x}_4 = (1,0,0,0)$. Hence,

$$\{\mathbf{x}_1, \mathbf{x}_2, \mathbf{x}_3, \mathbf{x}_4\} = \{(1,2,1,1), (-1,0,1,0), (0,1,0,2), (1,0,0,0)\}$$

is a basis for $\mathcal{R}^4$.

Using only the first three vectors, we choose $\mathbf{y}_1 = \mathbf{x}_1 = (1,2,1,1)$ and happen to notice that $\mathbf{x}_1 \cdot \mathbf{x}_2 = 0$, so that, with the dot product relative to the above basis, the Gram-Schmidt process directly gives $\mathbf{y}_2 = \mathbf{x}_2 = (-1,0,1,0)$. For $\mathbf{y}_3$, we first compute $\|\mathbf{y}_1\|^2 = 7$

and $\|\mathbf{y}_2\|^2 = 2$ and then apply the Gram-Schmidt process to get

$$
\begin{aligned}
\mathbf{y}_3 &= \mathbf{x}_3 - \frac{(\mathbf{x}_3 \cdot \mathbf{y}_1)}{\|\mathbf{y}_1\|^2}\mathbf{y}_1 - \frac{(\mathbf{x}_3 \cdot \mathbf{y}_2)}{\|\mathbf{y}_2\|^2}\mathbf{y}_2 \\
&= (0,1,0,2) - \frac{(0,1,0,2) \cdot (1,2,1,1)}{7}(1,2,1,1) - \frac{(0,1,0,2) \cdot (-1,0,1,0)}{2}(-1,0,1,0) \\
&= (-4/7, -1/7, -4/7, 10/7).
\end{aligned}
$$

Hence, $\{\mathbf{y}_1, \mathbf{y}_2, \mathbf{y}_3\}$ is an orthogonal basis for P. Computing $\|\mathbf{y}_3\|^2 = 133/49$, we apply the Gram-Schmidt process one more time to get

$$
\begin{aligned}
\mathbf{y}_4 &= \mathbf{x}_4 - \frac{(\mathbf{x}_4 \cdot \mathbf{y}_1)}{\|\mathbf{y}_1\|^2}\mathbf{y}_1 - \frac{(\mathbf{x}_4 \cdot \mathbf{y}_2)}{\|\mathbf{y}_2\|^2}\mathbf{y}_2 - \frac{(\mathbf{x}_4 \cdot \mathbf{y}_3)}{\|\mathbf{y}_3\|^2}\mathbf{y}_3 \\
&= (1,0,0,0) - \frac{(1,0,0,0) \cdot (1,2,1,1)}{7}(1,2,1,1) - \frac{(1,0,0,0) \cdot (-1,0,1,0)}{2}(-1,0,1,0) \\
&\quad - \frac{(1,0,0,0) \cdot (-4/7,-1/7,-4/7,10/7)}{133/49}(-4/7,-1/7,-4/7,10/7) \\
&= (1,0,0,0) - \frac{1}{7}(1,2,1,1) + \frac{1}{2}(-1,0,1,0) - \frac{4}{133}(4,1,4,-10) \\
&= = (9/38, -6/19, 9/38, 3/19).
\end{aligned}
$$

Thus, $\{\mathbf{y}_1, \mathbf{y}_2, \mathbf{y}_3, \mathbf{y}_4\}$ is an orthogonal basis for $\mathcal{R}^4$ in which the first three vectors are an orthogonal basis for P. However, because of homogeneity of the dot product, the set $\{a\mathbf{y}_1, b\mathbf{y}_2, c\mathbf{y}_3, d\mathbf{y}_4\}$ is also an orthogonal basis for $\mathcal{R}^4$ for any nonzero scalars a, b, c, d. We can therefore eliminate the fractions in the basis we have constructed by choosing $a = 1$, $b = 1$, $c = -7$ and $d = 38/3$ to get the orthogonal basis

$$
\{(1,2,1,1), (-1,0,1,0), (4,1,4,-10), (3,-4,3,2)\}.
$$

5. The vectors $\mathbf{x}_1 = (3,0,0)$, $\mathbf{x}_2 = (1,1,0)$ and $\mathbf{x}_3 = (1,1,1)$ are linearly independent and therefore form a basis for $\mathcal{R}^3$. Let $\mathbf{y}_1 = \mathbf{x}_1$. With the inner product on $\mathcal{R}^3$ given by the dot product relative to the basis $\{\mathbf{x}_1, \mathbf{x}_2, \mathbf{x}_3\}$, we normalize $\mathbf{y}_1$ to get $\mathbf{y}_1/\|\mathbf{y}_1\| = (1,0,0) = \mathbf{e}_1$. Applying the Gram-Schmidt process gives

$$
\mathbf{y}_2 = \mathbf{x}_2 - (\mathbf{x}_2 \cdot \mathbf{e}_1)\mathbf{e}_1 = (1,1,0) - \big((1,1,0) \cdot (1,0,0)\big)(1,0,0) = (0,1,0) = \mathbf{e}_2.
$$

Applying the Gram-Schmidt process again gives

$$
\begin{aligned}
\mathbf{y}_3 &= \mathbf{x}_3 - (\mathbf{x}_3 \cdot \mathbf{e}_1)\mathbf{e}_1 - (\mathbf{x}_3 \cdot \mathbf{e}_2)\mathbf{e}_2 \\
&= (1,1,1) - \big((1,1,1) \cdot (1,0,0)\big)(1,0,0) - \big((1,1,1) \cdot (0,1,0)\big)(0,1,0) \\
&= (1,1,1) - (1,0,0) - (0,1,0) = (0,0,1) = \mathbf{e}_3.
\end{aligned}
$$

That is, $\{\mathbf{e}_1, \mathbf{e}_2, \mathbf{e}_3\}$ is the resulting orthonormal basis.

7. Let V be a finite dimensional vector space with $\dim(V) = n$. Let $\{\mathbf{u}_1, \ldots, \mathbf{u}_n\}$ and $\{\mathbf{v}_1, \ldots, \mathbf{v}_n\}$ be two orthonormal bases for V. Express each $\mathbf{u}_k$ as a linear combination of $\mathbf{v}_1, \ldots, \mathbf{v}_n$;

$$
\mathbf{u}_1 = a_{11}\mathbf{v}_1 + \cdots + a_{1n}\mathbf{v}_n,
$$

$$
\vdots
$$

$$
\mathbf{u}_n = a_{n1}\mathbf{v}_1 + \cdots + a_{nn}\mathbf{v}_n,
$$

where the vectors $(a_{11}, \ldots, a_{1n}), \ldots, (a_{n1}, \ldots, a_{nn})$ are the columns of the matrix M that is used to change from coordinates in terms of $\mathbf{u}_1, \ldots, \mathbf{u}_n$ to coordinates in terms of $\mathbf{v}_1, \ldots, \mathbf{v}_n$. If V is endowed with an inner product then Theorem 7.6. says that

$$\langle \mathbf{u}_j, \mathbf{u}_k \rangle = (a_{j1}, \ldots, a_{jn}) \cdot (a_{k1}, \ldots, a_{kn}).$$

However, since $\{\mathbf{u}_1, \ldots, \mathbf{u}_n\}$ is an orthonormal set, $\langle \mathbf{u}_j, \mathbf{u}_k \rangle = 1$ if $j = k$ and is zero otherwise. That is,

$$(a_{j1}, \ldots, a_{jn}) \cdot (a_{k1}, \ldots, a_{kn}) = \begin{cases} 1, & \text{if } j = k, \\ 0, & \text{otherwise.} \end{cases}$$

This is precisely the statement that the columns of M are an orthonormal set.

9. First, the ordinary dot product is an inner product on $\mathcal{R}^3$ and its associated norm satisfies $\|\mathbf{x}\| = |\mathbf{x}|$. Now let P be the plane parametrized by $s(1,1,1) + t(1,-1,1)$. The vectors $\mathbf{x}_1 = (1,1,1)$, $\mathbf{x}_2 = (1,-1,1)$ form a basis for P. To find an orthogonal basis for P, let $\mathbf{y}_1 = \mathbf{x}_1$ and use the Gram-Schmidt process to get

$$\mathbf{y}_2 = \mathbf{x}_2 - \frac{\mathbf{x}_2 \cdot \mathbf{x}_1}{\|\mathbf{x}_1\|^2}\mathbf{x}_1 = (1,-1,1) - \frac{(1,-1,1) \cdot (1,1,1)}{\|(1,1,1)\|^2}(1,1,1) = \left(\frac{2}{3}, -\frac{4}{3}, \frac{2}{3}\right).$$

Next compute $\mathbf{y}_i/\|\mathbf{y}_i\| = \mathbf{u}_i$ for $i = 1, 2$ to get the orthonormal basis $\{\mathbf{u}_1, \mathbf{u}_2\}$, where

$$\mathbf{u}_1 = \left(\frac{1}{\sqrt{3}}, \frac{1}{\sqrt{3}}, \frac{1}{\sqrt{3}}\right) \quad \text{and} \quad \mathbf{u}_2 = \left(\frac{1}{\sqrt{6}}, -\frac{2}{\sqrt{6}}, \frac{1}{\sqrt{6}}\right).$$

Since $(2,3,4) \cdot \mathbf{u}_1 = 9/\sqrt{3}$ and $(2,3,4) \cdot \mathbf{u}_2 = 0$, Theorem 7.5 says that the point $\mathbf{y}$ on P closest to $(2,3,4)$ is

$$\mathbf{y} = ((2,3,4) \cdot \mathbf{u}_1)\mathbf{u}_1 + ((2,3,4) \cdot \mathbf{u}_2)\mathbf{u}_2 = \frac{9}{\sqrt{3}}\left(\frac{1}{\sqrt{3}}, \frac{1}{\sqrt{3}}, \frac{1}{\sqrt{3}}\right) + 0\left(\frac{1}{\sqrt{6}}, -\frac{2}{\sqrt{6}}, \frac{1}{\sqrt{6}}\right)$$

$$= (3,3,3).$$

Hence, the distance between $(2,3,4)$ and P is

$$d = \|\mathbf{x} - \mathbf{y}\| = |\mathbf{x} - \mathbf{y}| = |(2,3,4) - (3,3,3)| = |(-1,0,1)| = \sqrt{2}.$$

11. Let $\langle \mathbf{x}, \mathbf{y} \rangle$ be an inner product on $\mathcal{R}^n$ and let A be the n-by-n matrix whose ijth entry is $a_{ij} = \langle \mathbf{e}_i, \mathbf{e}_j \rangle$, where $\mathbf{e}_1, \ldots, \mathbf{e}_n$ are the standard basis vectors for $\mathcal{R}^n$. If $\mathbf{x}$ and $\mathbf{y}$ are in $\mathcal{R}^n$ then there exist scalars $x_1, \ldots, x_n$ and $y_1, \ldots, y_n$ such that $\mathbf{x} = x_1\mathbf{e}_1 + \cdots + x_n\mathbf{e}_n$ and $\mathbf{y} = y_1\mathbf{e}_1 + \cdots + y_n\mathbf{e}_n$. Hence, additivity and homogeneity of the inner product gives

$$\langle \mathbf{x}, \mathbf{y} \rangle = \langle x_1\mathbf{e}_1 + \cdots + x_n\mathbf{e}_n, y_1\mathbf{e}_1 + \cdots + y_n\mathbf{e}_n \rangle = \sum_{i=1}^{n}\sum_{j=1}^{n} x_i y_j \langle \mathbf{e}_i, \mathbf{e}_j \rangle$$

$$= \sum_{i=1}^{n}\sum_{j=1}^{n} x_i y_j a_{ij} = \sum_{i=1}^{n} x_i \left(\sum_{j=1}^{n} y_j a_{ij}\right).$$

134

Note that the parenthetical sum in the last line is the dot product of the ith row of A and $\mathbf{y}$; i.e., the ith entry in the vector $A\mathbf{y}$. Thus, multiplication by x_i and then summing over i is just the dot product of $\mathbf{x}$ and $A\mathbf{y}$. That is, $\langle \mathbf{x}, \mathbf{y} \rangle = \mathbf{x} \cdot A\mathbf{y}$.

13. Let $\langle \mathbf{x}, \mathbf{y} \rangle = \mathbf{x} \cdot A\mathbf{y}$ for all $\mathbf{x}, \mathbf{y} \in \mathcal{R}^n$. Exercise 12 in this section establishes the symmetry property of an inner product on $\mathcal{R}^n$. For additivity, we use distributivity of the dot product to get

$$\langle \mathbf{x} + \mathbf{y}, \mathbf{z} \rangle = (\mathbf{x} + \mathbf{y}) \cdot A\mathbf{z} = \mathbf{x} \cdot A\mathbf{z} + \mathbf{y} \cdot A\mathbf{z} = \langle \mathbf{x}, \mathbf{z} \rangle + \langle \mathbf{y}, \mathbf{z} \rangle.$$

Also, homogeneity of the dot product gives

$$\langle r\mathbf{x}, \mathbf{y} \rangle = (r\mathbf{x}) \cdot A\mathbf{y} = r(\mathbf{x} \cdot A\mathbf{y}) = r\langle \mathbf{x}, \mathbf{y} \rangle.$$

However, positivity fails. To see this, note that the n-by-n zero matrix A is symmetric. Using this matrix we have, for all $\mathbf{x} \in \mathcal{R}^n$,

$$\langle \mathbf{x}, \mathbf{x} \rangle = \mathbf{x} \cdot A\mathbf{x} = \mathbf{x} \cdot \mathbf{0} = 0.$$

That is, $\langle \mathbf{x}, \mathbf{x} \rangle = 0$ for all $\mathbf{x} \neq \mathbf{0}$.

15. Let $A = \begin{pmatrix} a & b \\ b & c \end{pmatrix}$ and let $\mathbf{x} = (x_1, x_2)$ represent an arbitrary vector in $\mathcal{R}^2$. The definition $\langle \mathbf{x}, \mathbf{y} \rangle = \mathbf{x} \cdot A\mathbf{y}$ gives

$$\langle \mathbf{x}, \mathbf{x} \rangle = \mathbf{x} \cdot A\mathbf{x} = (x_1, x_2) \cdot \begin{pmatrix} ax_1 + bx_2 \\ bx_1 + cx_2 \end{pmatrix} = ax_1^2 + bx_1x_2 + bx_2x_1 + cx_2^2$$

$$= ax_1^2 + 2bx_1x_2 + cx_2^2.$$

Note that if $a = 0$ then the last expression can be written as $x_2(2bx_1 + x_2)$, which is zero when $x_2 = 0$. That is, $\langle \mathbf{x}, \mathbf{x} \rangle = 0$ for all nonzero $\mathbf{x} = (x_1, 0)$ and A is not positive definite. Thus, in showing that A is positive definite if, and only if, $a > 0$ and $ac - b^2 > 0$, it will always be the case that $a \neq 0$. This allows us to be able to view the last expression above as a quadratic expression in x_1. Completing the square on $ax_1^2 + 2bx_2x_1$, equality of the two outer members of the above equation can be written as

$(*)$
$$\langle \mathbf{x}, \mathbf{x} \rangle = a \left(x_1 + \frac{b}{a} x_2 \right)^2 + \frac{ac - b^2}{a} x_2^2.$$

If A is positive definite then the right member of $(*)$ is positive whenever $\mathbf{x} \neq \mathbf{0}$. By inspection, the choice $\mathbf{x} = (1, 0)$ gives $\langle \mathbf{x}, \mathbf{x} \rangle = a > 0$, and the choice $\mathbf{x} = (-b, a)$ gives $\langle \mathbf{x}, \mathbf{x} \rangle = (ac - b^2)a > 0$. That is, $a > 0$ and $ac - b^2 > 0$.

Conversely, if $a > 0$ and $ac - b^2 > 0$ then the right member of $(*)$ is a linear combination of squares of real numbers with positive coefficients. Thus, the right side of $(*)$ is positive if at least one of $x_1 + (b/a)x_2$ or x_2 is nonzero. This is equivalent to saying that $\langle \mathbf{x}, \mathbf{x} \rangle > 0$ if $\mathbf{x} \neq \mathbf{0}$. Further, the right side of $(*)$ is zero if $\mathbf{x} = \mathbf{0}$. So $\langle \mathbf{x}, \mathbf{y} \rangle = \mathbf{x} \cdot A\mathbf{y}$ has the positivity property of an inner product and A is therefore positive definite.

17. Let A be an n-by-n symmetric matrix with real entries, let λ be an eigenvalue of A and let $\mathbf{u} = (u_1, \ldots, u_n)$ be an eigenvector associated with λ so that $A\mathbf{u} = \lambda\mathbf{u}$ (here, we leave open the possibility that some of the entries of $\mathbf{u}$ may be complex numbers).

By definition, the conjugate of a vector is the result of conjugating each entry of the vector. With this in mind, we show that $\overline{\lambda}$ (the conjugate of λ) is an eigenvalue of A and that $\overline{\mathbf{u}}$ (the conjugate of $\mathbf{u}$) is an associated eigenvector. To do this, let $u_k = r_k + is_k$ for $k = 1, \ldots, n$ and write $\mathbf{u}$ as

$$\mathbf{u} = (r_1 + is_1, \ldots, r_n + is_n) = (r_1, \ldots, r_n) + i(s_1, \ldots, s_n) = \mathbf{r} + i\mathbf{s},$$

where $\mathbf{r} = (r_1, \ldots, r_n)$ and $\mathbf{s} = (s_1, \ldots, s_n)$ are in $\mathcal{R}^n$. The fact that matrix multiplication distributes over vector addition gives

$$\overline{A\mathbf{u}} = \overline{A(\mathbf{r} + i\mathbf{s})} = \overline{A\mathbf{r} + iA\mathbf{s}} = A\mathbf{r} - iA\mathbf{s} = A(\mathbf{r} - i\mathbf{s}) = A\overline{\mathbf{u}},$$

Moreover, the conjugate of a scalar multiple of a vector is the product of the conjugate of the scalar and the conjugate of the vector (we leave the proof of this to the reader). This implies $\overline{\lambda \mathbf{u}} = \overline{\lambda}\overline{\mathbf{u}}$. Thus, conjugating both sides of $A\mathbf{u} = \lambda\mathbf{u}$ results in $A\overline{\mathbf{u}} = \overline{\lambda}\overline{\mathbf{u}}$. That is, $\overline{\lambda}$ is an eigenvalue of A and $\overline{\mathbf{u}}$ is an associated eigenvector.

Next, the proof of the identity $\mathbf{x} \cdot A\mathbf{y} = A\mathbf{x} \cdot \mathbf{y}$ that was given in Exercises 12 did not depend on whether the entries of A, $\mathbf{x}$ and $\mathbf{y}$ were real or complex. This allows us to use the result of Exercise 12 even if λ and $\mathbf{u}$ are complex. Thus, $\mathbf{u} \cdot A\overline{\mathbf{u}} = A\mathbf{u} \cdot \overline{\mathbf{u}}$. The left side of this equation can be written as $\mathbf{u} \cdot (\overline{\lambda}\overline{\mathbf{u}}) = \overline{\lambda}(\mathbf{u} \cdot \overline{\mathbf{u}})$ and the right side can be written as $(\lambda\mathbf{u}) \cdot \overline{\mathbf{u}} = \lambda(\mathbf{u} \cdot \overline{\mathbf{u}})$. Hence,

$$\overline{\lambda}(\mathbf{u} \cdot \overline{\mathbf{u}}) = \lambda(\mathbf{u} \cdot \overline{\mathbf{u}}).$$

Note that the dot product shown on both sides of the above equation can be written as

$$\mathbf{u} \cdot \overline{\mathbf{u}} = (u_1, \ldots, u_n) \cdot (\overline{u_1}, \ldots, \overline{u_n}) = u_1\overline{u_1} + \cdots + u_n\overline{u_n} = |u_1|^2 + \cdots + |u_n|^2.$$

Since $\mathbf{u} \neq \mathbf{0}$, at least one of the u_k is nonzero, so that $\mathbf{u} \cdot \overline{\mathbf{u}} \neq 0$. We can therefore cancel this dot product from both sides to get $\lambda = \overline{\lambda}$. So λ is real.

Exercise Set 7C (pgs. 170-171)

1. Let R be the rotation in $\mathcal{R}^3$ with matrix A as given in Example 14 of the text. Define
the matrix $U = \begin{pmatrix} 3/\sqrt{14} & 0 & 5/\sqrt{70} \\ 2/\sqrt{14} & -1/\sqrt{5} & -6/\sqrt{70} \\ 1/\sqrt{14} & 2/\sqrt{5} & -3/\sqrt{70} \end{pmatrix}$ and let $\mathbf{u}_1, \mathbf{u}_2, \mathbf{u}_3$ be the respective columns
of U. We have,

$$\mathbf{u}_1 \cdot \mathbf{u}_2 = 0 - \frac{2}{\sqrt{70}} + \frac{2}{\sqrt{70}} = 0; \qquad\qquad \mathbf{u}_1 \cdot \mathbf{u}_1 = \frac{9}{14} + \frac{4}{14} + \frac{1}{14} = 1;$$

$$\mathbf{u}_1 \cdot \mathbf{u}_3 = \frac{15}{14\sqrt{5}} - \frac{12}{14\sqrt{5}} - \frac{3}{14\sqrt{5}} = 0; \qquad\qquad \mathbf{u}_2 \cdot \mathbf{u}_2 = 0 + \frac{1}{5} + \frac{4}{5} = 1;$$

$$\mathbf{u}_2 \cdot \mathbf{u}_3 = 0 + \frac{6}{5\sqrt{14}} - \frac{6}{5\sqrt{14}} = 0; \qquad\qquad \mathbf{u}_3 \cdot \mathbf{u}_3 = \frac{25}{70} + \frac{36}{70} + \frac{9}{70} = 1.$$

Hence, the columns of U are an orthonormal basis for $\mathcal{R}^3$. Moreover, it was shown in
Example 14 that the axis of the rotation is in the direction of $\mathbf{a} = (3, 2, 1)$. Since $\mathbf{a} = \sqrt{14}\mathbf{u}_1$
(i.e., $\mathbf{a}$ is a *positive* scalar multiple of $\mathbf{u}_1$), the vectors $\mathbf{a}$ and $\mathbf{u}_1$ have the same direction.
That is, the rotation is in the direction of $\mathbf{u}_1$.

The matrix M of the rotation R relative to the basis $\{\mathbf{u}_1, \mathbf{u}_2, \mathbf{u}_3\}$ is given by

$$M = U^t A U$$

$$= \begin{pmatrix} 3/\sqrt{14} & 2/\sqrt{14} & 1/\sqrt{14} \\ 0 & -1/\sqrt{5} & 2/\sqrt{5} \\ 5/\sqrt{70} & -6/\sqrt{70} & -3/\sqrt{7} \end{pmatrix} \begin{pmatrix} 2/7 & 6/7 & 3/7 \\ 6/7 & -3/7 & 2/7 \\ 3/7 & 2/7 & -6/7 \end{pmatrix} \begin{pmatrix} 3/\sqrt{14} & 0 & 5/\sqrt{70} \\ 2/\sqrt{14} & -1/\sqrt{5} & -6/\sqrt{70} \\ 1/\sqrt{14} & 2/\sqrt{5} & -3/\sqrt{70} \end{pmatrix}$$

$$= \begin{pmatrix} 3/\sqrt{14} & 2/\sqrt{14} & 1/\sqrt{14} \\ 0 & 1/\sqrt{5} & -2/\sqrt{5} \\ -5/\sqrt{70} & 6/\sqrt{70} & 3/\sqrt{70} \end{pmatrix} \begin{pmatrix} 3/\sqrt{14} & 0 & 5/\sqrt{70} \\ 2/\sqrt{14} & -1/\sqrt{5} & -6/\sqrt{70} \\ 1/\sqrt{14} & 2/\sqrt{5} & -3/\sqrt{70} \end{pmatrix} = \begin{pmatrix} 1 & 0 & 0 \\ 0 & -1 & 0 \\ 0 & 0 & -1 \end{pmatrix}.$$

Comparing M with the matrix M_θ of Example 12 in the text, we see that $\cos\theta = -1$ and
$\sin\theta = 0$. Hence, the angle of rotation is $\theta = \pi$.

3. Since the axes of rotation are the x_1-axis and the x_2-axis, they are in the direction of $\mathbf{e}_1$
and $\mathbf{e}_2$, respectively, so that the matrices R and S are with respect to the standard basis
$\mathbf{e}_1, \mathbf{e}_2, \mathbf{e}_3$. Thus, the matrix R for the rotation through $90°$ about the x_1-axis is the matrix
given in Example 12 in the text with $\theta = \pi/2$; namely,

$$R = \begin{pmatrix} 1 & 0 & 0 \\ 0 & 0 & -1 \\ 0 & 1 & 0 \end{pmatrix}.$$

Note that the columns of R are $R(\mathbf{e}_1) = \mathbf{e}_1$, $R(\mathbf{e}_2) = \mathbf{e}_3$ and $R(\mathbf{e}_3) = -\mathbf{e}_2$, which depicts
a counterclockwise rotation of $90°$ in the $x_2 x_3$-plane. By analogy, the matrix S for the
rotation through $90°$ about the x_2-axis is the matrix $\begin{pmatrix} \cos\theta & 0 & -\sin\theta \\ 0 & 1 & 0 \\ \sin\theta & 0 & \cos\theta \end{pmatrix}$ with $\theta = \pi/2$;
namely,

$$S = \begin{pmatrix} 0 & 0 & -1 \\ 0 & 1 & 0 \\ 1 & 0 & 0 \end{pmatrix}.$$

Again, note that the columns of S are $S(\mathbf{e}_1) = \mathbf{e}_3$, $S(\mathbf{e}_2) = \mathbf{e}_2$ and $S(\mathbf{e}_3) = -\mathbf{e}_2$, which
depicts a counterclockwise rotation of $90°$ in the $x_1 x_3$-plane.

5. Let θ be the rotation angle satisfying $\cos\theta = \frac{3}{5}$, $\sin\theta = \frac{4}{5}$ and let $(1,1,0)$ be the axis of rotation. We first need to find an orthogonal basis for $\mathcal{R}^3$ that contains $(1,1,0)$. By inspection, we see that $(1,-1,0)$ is orthogonal to $(1,1,0)$ (their dot product is zero) and both vectors lie in the xy-plane. Moreover, since $(0,0,1)$ is perpendicular to the xy-plane, it is perpendicular to both $(1,1,0)$ and $(1,-1,0)$. Thus, $\{(1,1,0),(1,-1,0),(0,0,1)\}$ is an orthogonal basis containing $(1,1,0)$. Normalizing this set we obtain the orthonormal basis $\{(1/\sqrt{2},1/\sqrt{2},0),(1/\sqrt{2},-1/\sqrt{2},0),(0,0,1)\}$. Hence, the matrix of the given rotation relative to the above basis is the matrix M given in Example 12 in the text; namely,

$$M = \begin{pmatrix} 1 & 0 & 0 \\ 0 & 3/5 & -4/5 \\ 0 & 4/5 & 3/5 \end{pmatrix}.$$

Using Theorem 6.8 in this chapter, the matrix A of the given rotation relative to the standard basis is $A = UMU^{-1}$, where the respective columns of U are the above orthonormal basis vectors. That is, $U = \begin{pmatrix} 1/\sqrt{2} & 1/\sqrt{2} & 0 \\ 1/\sqrt{2} & -1/\sqrt{2} & 0 \\ 0 & 0 & 1 \end{pmatrix}$. Note that U is a symmetric matrix so that $U = U^t$. Moreover, since $U^{-1} = U^t$, we have $U^{-1} = U$. Hence,

$$A = UMU = \begin{pmatrix} 1/\sqrt{2} & 1/\sqrt{2} & 0 \\ 1/\sqrt{2} & -1/\sqrt{2} & 0 \\ 0 & 0 & 1 \end{pmatrix} \begin{pmatrix} 1 & 0 & 0 \\ 0 & 3/5 & -4/5 \\ 0 & 4/5 & 3/5 \end{pmatrix} \begin{pmatrix} 1/\sqrt{2} & 1/\sqrt{2} & 0 \\ 1/\sqrt{2} & -1/\sqrt{2} & 0 \\ 0 & 0 & 1 \end{pmatrix}$$

$$= \begin{pmatrix} 4/5 & 1/5 & -2\sqrt{2}/5 \\ 1/5 & 4/5 & 2\sqrt{2}/5 \\ 2\sqrt{2}/5 & -2\sqrt{2}/5 & 3/5 \end{pmatrix}.$$

As a partial check on the calculations, the reader can verify that if $\mathbf{a}$ is a vector parallel to $(1,1,0)$ then $A\mathbf{a} = \mathbf{a}$, as it should.

7. Let $M = \begin{pmatrix} a & b \\ c & d \end{pmatrix}$ be a matrix whose columns form an orthonormal set.

    (a) First, the orthonormal condition on the column vectors $\begin{pmatrix} a \\ c \end{pmatrix}$, $\begin{pmatrix} b \\ d \end{pmatrix}$ implies the three equations

$$(*) \quad (a,c)\cdot(a,c) = a^2 + c^2 = 1, \qquad (b,d)\cdot(b,d) = b^2 + d^2 = 1, \qquad (a,c)\cdot(b,d) = ab + cd = 0.$$

Writing the third equation as $ab = -cd$ and squaring gives $a^2b^2 = c^2d^2$. Use the first two equations of $(*)$ to replace a^2 and d^2 with $1 - c^2$ and $1 - b^2$, respectively, then simplify the result to get $c^2 = b^2$. Hence, $c = \pm b$ and the third equation of $(*)$ can be written as

$$0 = ab + cd = ab + (\pm b)d = (a \pm d)b.$$

So $a \pm d = 0$ or $b = 0$. If $a \pm d = 0$ then $d = \mp a$, so that $(c,d) = (\pm b, \mp a) = \pm(b,-a)$. And if $b = 0$ then $c = 0$ and the first two equations of $(*)$ become $a^2 = 1$ and $d^2 = 1$. Hence, $(a,d) = (\pm 1, \pm 1)$, where the signs of the coordinates can be independently chosen. The reader can check that each of the four possible choices is equivalent to either $(c,b) = (-b,a)$ or $(c,b) = (b,-a)$ (e.g., if $a = d = 1$ then $(c,d) = (0,1) = (-0,1) = (-b,a)$). In any case, $(c,d) = \pm(b,-a)$.

138

(b) Assume $(c, d) = (-b, a)$. The third equation of $(*)$ in part (a) is equivalent to $ab = -cd$, which implies $a^2 b^2 = c^2 d^2$. Hence, the first two equations of $(*)$ in part (a) implies $a^2 b^2 = (1 - a^2)(1 - b^2)$, which simplifies to $a^2 + b^2 = 1$. This relation implies the existence of a unique real number $0 \le \theta < 2\pi$ such that $\cos\theta = a$ and $\sin\theta = b$. The corresponding matrix M is therefore of the form

$$M = \begin{pmatrix} a & -b \\ b & a \end{pmatrix} = \begin{pmatrix} \cos\theta & -\sin\theta \\ \sin\theta & \cos\theta \end{pmatrix},$$

which we recognize as representing a rotation in $\mathcal{R}^2$ through an angle θ.

(c) Assume $(c, d) = (b, -a)$. The corresponding matrix is of the form $M = \begin{pmatrix} a & b \\ b & -a \end{pmatrix}$.

Using the fact that $a^2 + b^2 = 1$ (from part (b)), the eigenvalues λ of M satisfy

$$\det(A - \lambda I) = \begin{vmatrix} a - \lambda & b \\ b & -a - \lambda \end{vmatrix} = -(a - \lambda)(a + \lambda) - b^2 = \lambda^2 - (a^2 + b^2) = \lambda^2 - 1 = 0.$$

So the eigenvalues of M are 1 and -1. The reader can verify by direct calculation that the vectors $\begin{pmatrix} a + 1 \\ b \end{pmatrix}$ and $\begin{pmatrix} a - 1 \\ b \end{pmatrix}$ are eigenvectors associated with 1 and -1, respectively. Moreover, $(a + 1, b) \cdot (a - 1, b) = a^2 - 1 + b^2 = 0$, so that the eigenvectors are orthogonal. Thus, if we choose eigenvectors of length 1 then they form an orthonormal basis for $\mathcal{R}^2$. With this in mind, let $\mathbf{u}_1$ be an eigenvector of length 1 associated with the eigenvalue 1 and let $\mathbf{u}_2$ be an eigenvector of length 1 associated with the eigenvalue -1. Then $M\mathbf{u}_1 = \mathbf{u}_1$ and $M\mathbf{u}_2 = -\mathbf{u}_2$.

Since $\{\mathbf{u}_1, \mathbf{u}_2\}$ is a basis for $\mathcal{R}^2$, if $\mathbf{x}$ is in $\mathcal{R}^2$ then $\mathbf{x} = x_1 \mathbf{u}_1 + x_2 \mathbf{u}_2$ for some scalars x_1, x_2. Hence,

$$M\mathbf{x} = M(x_1 \mathbf{u}_1 + x_2 \mathbf{u}_2) = x_1 M\mathbf{u}_1 + x_2 M\mathbf{u}_2 = x_1 \mathbf{u}_1 - x_2 \mathbf{u}_2.$$

By Example 15 in the text, M is the matrix of a reflection in the subspace generated by the vector $\mathbf{u}_1$. Since this subspace is a line through the origin in the direction of the eigenvector $\mathbf{u}_1$, M is the matrix of a reflection in the line through the origin in the direction of the eigenvector $\mathbf{u}_1$.

9. First note that the columns of M give the coordinates of $f(\mathbf{u}_1)$, $f(\mathbf{u}_2)$, $f(\mathbf{u}_3)$ with respect to the basis $\{\mathbf{u}_1, \mathbf{u}_2, \mathbf{u}_3\}$, and since $\{\mathbf{u}_1, \mathbf{u}_2, \mathbf{u}_3\}$ is an orthonormal set, Theorem 7.6 implies that the dot products and lengths of these columns are the same as the dot products and lengths of $f(\mathbf{u}_1)$, $f(\mathbf{u}_2)$, $f(\mathbf{u}_3)$. That is, $\{f(\mathbf{u}_1), f(\mathbf{u}_2), f(\mathbf{u}_3)\}$ is an orthonormal set.

(a) Since $\mathbf{u}_1$ is an eigenvector associated with the real eigenvalue λ, $f(\mathbf{u}_1) = \lambda \mathbf{u}_1$ and the coordinate vector of $f(\mathbf{u}_1)$ is $(\lambda, 0, 0)$. Hence, the first column of M is as shown. The vectors $f(\mathbf{u}_2)$ and $f(\mathbf{u}_3)$ are orthogonal to $f(\mathbf{u}_1)$ so the second and third columns of M have dot product 0 with the first column, and therefore their first entries are 0. Hence, they have the form $(0, a, c)$ and $(0, b, d)$ for some scalars a, b, c, d, so the second and third columns of M are as shown. Moreover, the second and third columns of M are an orthonormal set in $\mathcal{R}^3$, so that the columns of the submatrix $S = \begin{pmatrix} a & b \\ c & d \end{pmatrix}$ are an orthonormal set in $\mathcal{R}^2$.

(b) Apply the results of Exercise 7 in this section to the submatrix S. It is either a rotation about the origin in the plane spanned by $\mathbf{u}_2, \mathbf{u}_3$, or it is the matrix of a reflection in the same plane, and therefore there are orthonormal vectors $\mathbf{a}$ and $\mathbf{b}$ in the plane such that $f(\mathbf{a}) = \mathbf{a}$ and $f(\mathbf{b}) = -\mathbf{b}$. In the first case, f is a rotation about the axis $\mathbf{u}_1$. In the second case, $\{\mathbf{u}_1, \mathbf{a}, \mathbf{b}\}$ is an orthonormal basis with $f(\mathbf{u}_1) = \mathbf{u}_1$, $f(\mathbf{a}) = \mathbf{a}$ and $f(\mathbf{b}) = -\mathbf{b}$, and therefore f is a reflection in the plane spanned by $\mathbf{u}_1, \mathbf{a}$.

(c) Again, consider the submatrix S as in part (b). If it gives a rotation then f is the composition of a rotation with axis $\mathbf{u}_1$ composed with a reflection in the plane perpendicular to $\mathbf{u}_1$. In the second case, f is a reflection in the line with direction $\mathbf{a}$.

139

Chapter Review

(pg. 171)

1. Since every basis for $\mathcal{R}^2$ has two vectors, every subset of $\mathcal{R}^2$ with more that two vectors is linearly dependent (Theorem 5.5). Hence, the three given vectors from $\mathcal{R}^2$ are linearly dependent.

3. With $\mathbf{x} = (-1, 2, 0, 3)$, $\mathbf{y} = (0, 1, -1, 4)$ and $\mathbf{z} = (-1, 3, -1, 1)$, let a, b, c be scalars such that $a\mathbf{x} + b\mathbf{y} + c\mathbf{y} = \mathbf{0}$. Then

$$a(-1, 2, 0, 3) + b(0, 1, -1, 4) + c(-1, 3, -1, 1)$$
$$= (-a - c, 2a + b + 3c, -b - c, 3a + 4b + c) = (0, 0, 0, 0).$$

This is equivalent to the four-equation scalar system

$$-a - c = 0, \qquad 2a + b + 3c = 0, \qquad -b - c = 0, \qquad 3a + 4b + c = 0.$$

The first and third equations give $a = -c$ and $b = -c$. When these expressions for a and b are inserted into the fourth equation the result reduces to $-6c = 0$, from which $c = 0$. This implies $a = b = 0$ as well. So the three given vectors are linearly independent.

5. Let $f : \mathcal{R}^3 \longrightarrow \mathcal{R}^3$ be a linear function with null-space spanned by $(1, 1, 1)$ and $(1, -1, 1)$. Since the null-space is parametrized by $\mathbf{x} = r(1, 1, 1) + s(1, -1, 1)$, for each $\mathbf{x}$ in the null-space we can use the linearity of f to obtain

$$f(\mathbf{x}) = f\big(r(1, 1, 1) + s(1, -1, 1)\big) = f\big((r + s)\mathbf{e}_1 + (r - s)\mathbf{e}_2 + (r + s)\mathbf{e}_3\big)$$
$$= (r + s)f(\mathbf{e}_1) + (r - s)f(\mathbf{e}_2) + (r + s)f(\mathbf{e}_3) = \mathbf{0}.$$

In particular, $r = s = 1/2$ gives $f(\mathbf{e}_1) + f(\mathbf{e}_3) = \mathbf{0}$, or equivalently, $f(\mathbf{e}_1) = -f(\mathbf{e}_3)$. Given that $f(\mathbf{e}_3) = \mathbf{e}_3$, we get $f(\mathbf{e}_1) = -\mathbf{e}_3$. Furthermore, $r = -s = 1/2$ gives $f(\mathbf{e}_2) = \mathbf{0}$. Since the columns of the matrix A of f with respect to the standard basis are $f(\mathbf{e}_1)$, $f(\mathbf{e}_2)$, $f(\mathbf{e}_3)$, we have

$$A = \begin{pmatrix} 0 & 0 & 0 \\ 0 & 0 & 0 \\ -1 & 0 & 1 \end{pmatrix}.$$

7. If $f : \mathcal{R}^3 \longrightarrow \mathcal{R}^2$ is a linear function whose null-space is the line $\mathbf{x} = t(1, 1, 1)$ then, in particular, $(1, 1, 1)$ is in the null-space of f and the linearity of f gives

$$f\big((1, 1, 1)\big) = f(\mathbf{e}_1 + \mathbf{e}_2 + \mathbf{e}_2) = f(\mathbf{e}_1) + f(\mathbf{e}_2) + f(\mathbf{e}_3) = \mathbf{0}.$$

Since we are also given that $f(\mathbf{e}_1) = (1, 1)$ and $f(\mathbf{e}_2) = (1, 2)$, equality of the last two members above gives $(1, 1) + (1, 2) + f(\mathbf{e}_3) = \mathbf{0}$, or equivalently, $f(\mathbf{e}_3) = (-2, -3)$. Hence, since the columns of the matrix of f relative to the standard basis are $f(\mathbf{e}_1)$, $f(\mathbf{e}_2)$, $f(\mathbf{e}_3)$, we have

$$A = \begin{pmatrix} 1 & 1 & -2 \\ 1 & 2 & -3 \end{pmatrix}.$$

9. Let $f : \mathcal{R}^3 \longrightarrow \mathcal{R}^3$ be a linear function with $f(2, 2, 2) = (1, 0, 1)$ and $f(\mathbf{x}) = \mathbf{x}$ for $\mathbf{x}$ in the plane spanned by $(0, 1, 1)$ and $(1, 1, 0)$. If A is the matrix of f relative to the standard basis then the columns of A are $f(\mathbf{e}_1)$, $f(\mathbf{e}_2)$, $f(\mathbf{e}_3)$. We will construct a system of three

140

vector equations in the three unknowns $f(\mathbf{e}_1)$, $f(\mathbf{e}_2)$, $f(\mathbf{e}_3)$ and then solve the system. The resulting solution will be the columns of A.

Since $f(\mathbf{x}) = \mathbf{x}$ for $\mathbf{x}$ in the plane spanned by $(0,1,1) = \mathbf{e}_2 + \mathbf{e}_3$ and $(1,1,0) = \mathbf{e}_1 + \mathbf{e}_2$, this condition holds for the spanning vectors themselves. That is, linearity allows us to obtain the two vector equations

(1) $$f(\mathbf{e}_2 + \mathbf{e}_3) = f(\mathbf{e}_2) + f(\mathbf{e}_3) = \mathbf{e}_2 + \mathbf{e}_3;$$
(2) $$f(\mathbf{e}_1 + \mathbf{e}_2) = f(\mathbf{e}_1) + f(\mathbf{e}_2) = \mathbf{e}_1 + \mathbf{e}_2.$$

Further, the condition $f(2,2,2) = (1,0,1) = \mathbf{e}_1 + \mathbf{e}_3$ is equivalent to

(3) $$f(2\mathbf{e}_1 + 2\mathbf{e}_2 + 2\mathbf{e}_3) = 2f(\mathbf{e}_1) + 2f(\mathbf{e}_2) + 2f(\mathbf{e}_3) = \mathbf{e}_1 + \mathbf{e}_3.$$

Equality of the two rightmost members of each of equations (1) and (2) gives

(4) $$f(\mathbf{e}_3) = \mathbf{e}_2 + \mathbf{e}_3 - f(\mathbf{e}_2) \qquad \text{and} \qquad f(\mathbf{e}_1) = \mathbf{e}_1 + \mathbf{e}_2 - f(\mathbf{e}_2).$$

Inserting these values of $f(\mathbf{e}_3)$ and $f(\mathbf{e}_1)$ into equation (3) gives

$$2\big(\mathbf{e}_1 + \mathbf{e}_2 - f(\mathbf{e}_2)\big) + 2f(\mathbf{e}_2) + 2\big(\mathbf{e}_2 + \mathbf{e}_3 - f(\mathbf{e}_2)\big) = \mathbf{e}_1 + \mathbf{e}_3.$$

Solving for $f(\mathbf{e}_2)$ gives

$$f(\mathbf{e}_2) = \frac{1}{2}\mathbf{e}_1 + 2\mathbf{e}_2 + \frac{1}{2}\mathbf{e}_3.$$

Using this value of $f(\mathbf{e}_2)$ in equations (4) and simplifying gives

$$f(\mathbf{e}_3) = -\frac{1}{2}\mathbf{e}_1 - \mathbf{e}_2 + \frac{1}{2}\mathbf{e}_3 \qquad \text{and} \qquad f(\mathbf{e}_1) = \frac{1}{2}\mathbf{e}_1 - \mathbf{e}_2 - \frac{1}{2}\mathbf{e}_3.$$

Hence, $f(\mathbf{e}_1) = (1/2, -1, -1/2)$, $f(\mathbf{e}_2) = (1/2, 2, 1/2)$ and $f(\mathbf{e}_3) = (-1/2, -1, 1/2)$. So

$$A = \begin{pmatrix} 1/2 & 1/2 & -1/2 \\ -1 & 2 & -1 \\ -1/2 & 1/2 & 1/2 \end{pmatrix}.$$

11. Since $M_{2,2}$ has dimension 4 and $M_{2,3}$ has dimension 6, every matrix of the linear function $f : M_{2,2} \longrightarrow M_{2,3}$ defined by $f(X) = X \begin{pmatrix} -1 & 0 & 2 \\ 3 & 1 & 2 \end{pmatrix}$ for X in $M_{2,2}$ is a 6-by-4 matrix, regardless of which bases for $M_{2,2}$ and $M_{2,3}$ are under consideration. In this case, we are considering the bases

$$B_{2,2} = \{E_{11}, E_{12}, E_{21}, E_{22}\} \qquad \text{and} \qquad B_{2,3} = \{E'_{11}, E'_{12}, E'_{13}, E'_{21}, E'_{22}, E'_{23}\}.$$

We make the convention that a matrix X in $M_{2,2}$ relative to the basis $B_{2,2}$ will be written as

$$X = x_1 E_{11} + x_2 E_{12} + x_3 E_{21} + x_4 E_{22}$$

so that the coordinate vector of X is (x_1, x_2, x_3, x_4). Similarly, a matrix Y in $M_{2,3}$ relative to the basis $B_{2,3}$ will be written as

$$Y = y_1 E'_{11} + y_2 E'_{12} + y_3 E'_{13} + y_4 E'_{21} + y_5 E'_{22} + y_6 E'_{23}$$

141

so that the coordinate vector for Y is $(y_1, y_2, y_3, y_4, y_5, y_6)$.

Let A be the desired matrix. The columns of A (from left to right) are the coordinate vectors of $f(E_{11})$, $f(E_{12})$, $f(E_{21})$ and $f(E_{22})$, respectively. Since

$$f(E_{11}) = \begin{pmatrix} 1 & 0 \\ 0 & 0 \end{pmatrix} \begin{pmatrix} -1 & 0 & 2 \\ 3 & 1 & 2 \end{pmatrix} = \begin{pmatrix} -1 & 0 & 2 \\ 0 & 0 & 0 \end{pmatrix} = -E'_{11} + 2E'_{13};$$

$$f(E_{12}) = \begin{pmatrix} 0 & 1 \\ 0 & 0 \end{pmatrix} \begin{pmatrix} -1 & 0 & 2 \\ 3 & 1 & 2 \end{pmatrix} = \begin{pmatrix} 3 & 1 & 2 \\ 0 & 0 & 0 \end{pmatrix} = 3E'_{11} + E'_{12} + 2e'_{13};$$

$$f(E_{21}) = \begin{pmatrix} 0 & 0 \\ 1 & 0 \end{pmatrix} \begin{pmatrix} -1 & 0 & 2 \\ 3 & 1 & 2 \end{pmatrix} = \begin{pmatrix} 0 & 0 & 0 \\ -1 & 0 & 2 \end{pmatrix} = -E'_{21} + 2E'_{23};$$

$$f(E_{22}) = \begin{pmatrix} 0 & 0 \\ 0 & 1 \end{pmatrix} \begin{pmatrix} -1 & 0 & 2 \\ 3 & 1 & 2 \end{pmatrix} = \begin{pmatrix} 0 & 0 & 0 \\ 3 & 1 & 2 \end{pmatrix} = 3E'_{21} + E'_{22} + 2E'_{23},$$

the columns of A are $(-1, 0, 2, 0, 0, 0)$, $(3, 1, 2, 0, 0, 0)$, $(0, 0, 0, -1, 0, 2)$ and $(0, 0, 0, 3, 1, 2)$. That is,

$$A = \begin{pmatrix} -1 & 3 & 0 & 0 \\ 0 & 1 & 0 & 0 \\ 2 & 2 & 0 & 0 \\ 0 & 0 & -1 & 3 \\ 0 & 0 & 0 & 1 \\ 0 & 0 & 2 & 2 \end{pmatrix}.$$

13. Since $M_{2,2}$ has dimension 4, every matrix of the linear function $f : M_{2,2} \longrightarrow M_{2,2}$ defined by $f(X) = \begin{pmatrix} 1 & 1 \\ 1 & -2 \end{pmatrix} X \begin{pmatrix} -1 & 1 \\ 1 & 1 \end{pmatrix}$ for X in $M_{2,2}$ is a 4-by-4 matrix, regardless of which basis for $M_{2,2}$ is under consideration. In this case, we are considering the basis $B_{2,2}$. Relative to this basis, we make the convention that (a, b, c, d) is the coordinate vector of the matrix $\begin{pmatrix} a & b \\ c & d \end{pmatrix} = aE_{11} + bE_{12} + cE_{21} + dE_{22}$.

Let A be the desired matrix. The columns of A (from left to right) are the coordinate vectors of $f(E_{11})$, $f(E_{12})$, $f(E_{21})$ and $f(E_{22})$, respectively. Since

$$f(E_{11}) = \begin{pmatrix} 1 & 1 \\ 1 & -2 \end{pmatrix} E_{11} \begin{pmatrix} -1 & 1 \\ 1 & 1 \end{pmatrix} = \begin{pmatrix} -1 & 1 \\ -1 & 1 \end{pmatrix} = -E_{11} + E_{12} - E_{21} + E_{22};$$

$$f(E_{12}) = \begin{pmatrix} 1 & 1 \\ 1 & -2 \end{pmatrix} E_{12} \begin{pmatrix} -1 & 1 \\ 1 & 1 \end{pmatrix} = \begin{pmatrix} 1 & -1 \\ 1 & -1 \end{pmatrix} = E_{11} - E_{12} + E_{21} - E_{22};$$

$$f(E_{21}) = \begin{pmatrix} 1 & 1 \\ 1 & -2 \end{pmatrix} E_{21} \begin{pmatrix} -1 & 1 \\ 1 & 1 \end{pmatrix} = \begin{pmatrix} -1 & 1 \\ 2 & -2 \end{pmatrix} = -E_{11} + E_{12} + 2E_{21} - 2E_{22};$$

$$f(E_{22}) = \begin{pmatrix} 1 & 1 \\ 1 & -2 \end{pmatrix} E_{22} \begin{pmatrix} -1 & 1 \\ 1 & 1 \end{pmatrix} = \begin{pmatrix} 1 & -1 \\ -2 & 2 \end{pmatrix} = E_{11} - E_{12} - 2E_{21} + 2E_{22},$$

the columns of A are $(-1, 1, -1, 1)$, $(1, -1, 1, -1)$, $(-1, 1, 2, -2)$ and $(1, -1, -2, 2)$. That is,

$$A = \begin{pmatrix} -1 & 1 & -1 & 1 \\ 1 & -1 & 1 & -1 \\ -1 & 1 & 2 & -2 \\ 1 & -1 & -2 & 2 \end{pmatrix}.$$

15. Imagine looking down the positive x-axis toward the origin in $\mathcal{R}^3$ with the positive y-axis pointing to the right and the positive z-axis pointing straight up. Relative to the standard basis, the matrix R that represents a counterclockwise rotation through a 90° angle about the x-axis would send $\mathbf{e}_1$ to itself, send $\mathbf{e}_2$ to $\mathbf{e}_3$, and send $\mathbf{e}_3$ to $-\mathbf{e}_2$. That is,

$\mathbf{e}_1$, $\mathbf{e}_3$, $-\mathbf{e}_2$ are the columns of R so that $R = \begin{pmatrix} 1 & 0 & 0 \\ 0 & 0 & -1 \\ 0 & 1 & 0 \end{pmatrix}$. (This is the matrix M given

in Example 12 in Section 7C with $\theta = \pi/2$.)

Now imagine looking down the negative y-axis toward the origin in $\mathcal{R}^3$ with the positive x-axis pointing to the right and the positive z-axis pointing straight up. Relative to the standard basis, the matrix S that represents a counterclockwise rotation of 90° about the y-axis would send $\mathbf{e}_1$ to $\mathbf{e}_3$, send $\mathbf{e}_2$ to itself, and send $\mathbf{e}_3$ to $-\mathbf{e}_1$. That is, $\mathbf{e}_3$, $\mathbf{e}_2$, $-\mathbf{e}_1$ are

the columns of S so that $S = \begin{pmatrix} 0 & 0 & -1 \\ 0 & 1 & 0 \\ 1 & 0 & 0 \end{pmatrix}$.

(b) With R and S as in part (a), we find that $R^{-1} = \begin{pmatrix} 1 & 0 & 0 \\ 0 & 0 & 1 \\ 0 & -1 & 0 \end{pmatrix}$ and therefore

$$R^{-1}SR = \begin{pmatrix} 1 & 0 & 0 \\ 0 & 0 & 1 \\ 0 & -1 & 0 \end{pmatrix} \begin{pmatrix} 0 & 0 & -1 \\ 0 & 1 & 0 \\ 1 & 0 & 0 \end{pmatrix} \begin{pmatrix} 1 & 0 & 0 \\ 0 & 0 & -1 \\ 0 & 1 & 0 \end{pmatrix} = \begin{pmatrix} 0 & -1 & 0 \\ 1 & 0 & 0 \\ 0 & 0 & 1 \end{pmatrix}.$$

Note that this matrix sends $\mathbf{e}_1$ to $\mathbf{e}_2$, sends $\mathbf{e}_2$ to $-\mathbf{e}_1$, and sends $\mathbf{e}_3$ to itself. If we image ourselves looking down the positive z-axis toward the origin in $\mathcal{R}^3$ with the positive x-axis pointing to the right and the positive y-axis pointing up, we recognize the matrix $R^{-1}SR$ as effecting a 90° counterclockwise rotation about the z-axis.

17. Let $A = \begin{pmatrix} 1 & -3 \\ -2 & -4 \end{pmatrix}$. The eigenvalues of A are the solutions λ of the characteristic equation

$$\det(A - \lambda I) = \begin{vmatrix} 1 - \lambda & -3 \\ -2 & -4 - \lambda \end{vmatrix} = (1 - \lambda)(-4 - \lambda) - 6 = (\lambda + 5)(\lambda - 2) = 0.$$

Hence, the eigenvalues of A are $\lambda_1 = -5$ and $\lambda_2 = 2$. By Theorem 6.7, $\mathcal{R}^2$ has a basis of eigenvectors of A.

For the eigenvalue $\lambda_1 = -5$, if $\mathbf{u} = \begin{pmatrix} u_1 \\ u_2 \end{pmatrix}$ is an associated eigenvector then $\mathbf{u}$ must satisfy $(A - \lambda_1 I)\mathbf{u} = \mathbf{0}$. Since

$$(A - \lambda_1 I)\mathbf{u} = \begin{pmatrix} 6 & -3 \\ -2 & 1 \end{pmatrix} \begin{pmatrix} u_1 \\ u_2 \end{pmatrix} = \begin{pmatrix} 6u_1 - 3u_2 \\ -2u_1 + u_2 \end{pmatrix} = (-2u_1 + u_2) \begin{pmatrix} -3 \\ 1 \end{pmatrix} = \mathbf{0},$$

it follows that the coordinates of $\mathbf{u}$ must satisfy $-2u_1 + u_2 = 0$. Choosing $u_1 = 1$ gives $u_2 = 2$ and the resulting eigenvector is $\mathbf{u} = \begin{pmatrix} 1 \\ 2 \end{pmatrix}$.

For the eigenvalue $\lambda_2 = 2$, if $\mathbf{v} = \begin{pmatrix} v_1 \\ v_2 \end{pmatrix}$ is an associated eigenvector then $\mathbf{v}$ must satisfy $(A - \lambda_2 I)\mathbf{v} = \mathbf{0}$. Since

$$(A - \lambda_2 I)\mathbf{v} = \begin{pmatrix} -1 & -3 \\ -2 & -6 \end{pmatrix} \begin{pmatrix} v_1 \\ v_2 \end{pmatrix} = \begin{pmatrix} -v_1 - 3v_2 \\ -2v_1 - 6v_2 \end{pmatrix} = (-v_1 - 3v_2) \begin{pmatrix} 1 \\ 2 \end{pmatrix} = \mathbf{0},$$

it follows that the coordinates of $\mathbf{v}$ must satisfy $-v_1 - 3v_2 = 0$. Choosing $v_2 = 1$ gives $v_1 = -3$ and the resulting eigenvector is $\mathbf{v} = \begin{pmatrix} -3 \\ 1 \end{pmatrix}$.

19. Let $A = \begin{pmatrix} 1 & -2 & 4 \\ 0 & 2 & 0 \\ 0 & -1 & 3 \end{pmatrix}$. The eigenvalues of A are the solutions λ of the characteristic equation

$$\det(A - \lambda I) = \begin{vmatrix} 1-\lambda & -2 & 4 \\ 0 & 2-\lambda & 0 \\ 0 & -1 & 3-\lambda \end{vmatrix} = (1-\lambda)\begin{vmatrix} 2-\lambda & 0 \\ -1 & 3-\lambda \end{vmatrix} = (1-\lambda)(2-\lambda)(3-\lambda) = 0.$$

Hence, there are three real eigenvalues: $\lambda_1 = 1$, $\lambda_2 = 2$ and $\lambda_3 = 3$. By Theorem 6.7, $\mathcal{R}^3$ has a basis of eigenvectors of A.

For the eigenvalue $\lambda_1 = 1$, if $\mathbf{u} = \begin{pmatrix} u_1 \\ u_2 \\ u_3 \end{pmatrix}$ is an associated eigenvector then $\mathbf{u}$ must satisfy $(A - \lambda_1 I)\mathbf{u} = \mathbf{0}$. Since

$$(A - \lambda_1 I)\mathbf{u} = \begin{pmatrix} 0 & -2 & 4 \\ 0 & 1 & 0 \\ 0 & -1 & 2 \end{pmatrix}\begin{pmatrix} u_1 \\ u_2 \\ u_3 \end{pmatrix} = \begin{pmatrix} -2u_2 + 4u_3 \\ u_2 \\ -u_2 + 2u_3 \end{pmatrix} = u_2\begin{pmatrix} -2 \\ 1 \\ -1 \end{pmatrix} + u_3\begin{pmatrix} 4 \\ 0 \\ 2 \end{pmatrix} = \mathbf{0},$$

and since the two vectors in the linear combination in the fourth member are linearly independent, it follows that $u_2 = u_3 = 0$ while $u_1 \ (\neq 0)$ is arbitrary. Choosing $u_1 = 1$ gives the eigenvector $\mathbf{u} = \begin{pmatrix} 1 \\ 0 \\ 0 \end{pmatrix}$.

For the eigenvalue $\lambda_2 = 2$, if $\mathbf{v} = \begin{pmatrix} v_1 \\ v_2 \\ v_3 \end{pmatrix}$ is an associated eigenvector then $\mathbf{v}$ must satisfy $(A - \lambda_2 I)\mathbf{v} = \mathbf{0}$. Since

$$(A - \lambda_2 I)\mathbf{v} = \begin{pmatrix} -1 & -2 & 4 \\ 0 & 0 & 0 \\ 0 & -1 & 1 \end{pmatrix}\begin{pmatrix} v_1 \\ v_2 \\ v_3 \end{pmatrix} = \begin{pmatrix} -v_1 - 2v_2 + 4v_3 \\ 0 \\ -v_2 + v_3 \end{pmatrix}$$

$$= (-v_1 + 2v_2)\begin{pmatrix} 1 \\ 0 \\ 0 \end{pmatrix} + (v_2 - v_3)\begin{pmatrix} 4 \\ 0 \\ -1 \end{pmatrix} = \mathbf{0},$$

and since the two vectors in the linear combination in the second line are linearly independent, it follows that $-v_1 + 2v_2 = 0$ and $v_2 - v_3 = 0$. Writing each coordinate in terms of v_2 gives $v_1 = 2v_2$, $v_2 = v_2$, $v_3 = v_2$. Choosing $v_2 = 1$ gives the eigenvector $\mathbf{v} = \begin{pmatrix} 2 \\ 1 \\ 1 \end{pmatrix}$.

For the eigenvalue $\lambda_3 = 3$, if $\mathbf{w} = \begin{pmatrix} w_1 \\ w_2 \\ w_3 \end{pmatrix}$ is an associated eigenvector then $\mathbf{w}$ must satisfy $(A - \lambda_3 I)\mathbf{w} = \mathbf{0}$. Since

$$(A - \lambda_3 I)\mathbf{w} = \begin{pmatrix} -2 & -2 & 4 \\ 0 & -1 & 0 \\ 0 & -1 & 0 \end{pmatrix} \begin{pmatrix} w_1 \\ w_2 \\ w_3 \end{pmatrix} = \begin{pmatrix} -2w_1 - 2w_2 + 4w_3 \\ -w_2 \\ -w_2 \end{pmatrix}$$

$$= (-2w_1 + 4w_3) \begin{pmatrix} 1 \\ 0 \\ 0 \end{pmatrix} - w_2 \begin{pmatrix} 2 \\ 1 \\ 1 \end{pmatrix} = \mathbf{0},$$

and since the two vectors in the linear combination in the second line are linearly independent, it follows that $-2w_1 - 4w_3 = 0$ and $w_2 = 0$. Choosing $w_3 = 1$ gives $w_1 = -2$ and results in the eigenvector $\mathbf{w} = \begin{pmatrix} -2 \\ 0 \\ 1 \end{pmatrix}$.

21. (a) The eigenvalues of the given matrix A are the solutions λ of the characteristic equation

$$\det(A - \lambda I) = \begin{vmatrix} -\lambda & a & -b \\ -a & -\lambda & c \\ b & -c & -\lambda \end{vmatrix} = -\lambda \begin{vmatrix} -\lambda & c \\ -c & -\lambda \end{vmatrix} - (-a) \begin{vmatrix} a & -b \\ -c & -\lambda \end{vmatrix} + b \begin{vmatrix} a & -b \\ -\lambda & c \end{vmatrix}$$

$$= -\lambda(\lambda^2 + c^2) + a(-\lambda a - bc) + b(ac - \lambda b) = -\lambda(\lambda^2 + a^2 + b^2 + c^2) = 0.$$

Hence, $\lambda_1 = 0$ and $\lambda_{2,3} = \pm i\sqrt{a^2 + b^2 + c^2}$. If $a = b = c = 0$ then $\lambda_2 = \lambda_3 = 0$ so that 0 is the only eigenvalue and has multiplicity 3. Otherwise, 0 has multiplicity 1 and $\lambda_{2,3}$ are nonreal complex conjugates.

(b) First, assume that at least one of a, b, c is nonzero. By our comments in part (a), $\lambda_1 = 0$ is the real eigenvalue of A and has multiplicity 1. If $\mathbf{u} = \begin{pmatrix} u_1 \\ u_2 \\ u_3 \end{pmatrix}$ is an eigenvector associated with $\lambda_1 = 0$ then $\mathbf{u}$ satisfies $(A - \lambda_1 I)\mathbf{u} = A\mathbf{u} = \mathbf{0}$. That is,

$$A\mathbf{u} = \begin{pmatrix} 0 & a & -b \\ -a & 0 & c \\ b & -c & 0 \end{pmatrix} \begin{pmatrix} u_1 \\ u_2 \\ u_3 \end{pmatrix} = \begin{pmatrix} au_2 - bu_3 \\ -au_1 + cu_3 \\ bu_1 - cu_2 \end{pmatrix} = \mathbf{0}.$$

This is equivalent to the three-equation scalar system

$$au_2 = bu_3, \qquad au_1 = cu_3, \qquad bu_1 = cu_2.$$

Note that the first equation holds if $u_2 = b$ and $u_3 = a$. Inserting these values for u_2 and u_3 into the second and third equations gives $au_1 = ca$ and $bu_1 = cb$, or equivalently, $a(u_1 - c) = 0$ and $b(u_1 - c) = 0$. If at least one of a or b is nonzero then $u_1 = c$ and therefore

$$\mathbf{u} = \begin{pmatrix} c \\ b \\ a \end{pmatrix}$$

is an eigenvalue. If $a = b = 0$ then $u_2 = u_3 = 0$ and $u_1 \ (\neq 0)$ remains arbitrary. As $c \neq 0$, we can choose $u_1 = c$ and, again, obtain the above above displayed eigenvector. In either case, the above displayed vector is an eigenvector associated with the real eigenvalue 0.

145

Finally, if $a = b = c = 0$ then A is the 3-by-3 zero matrix with 0 as an eigenvalue of multiplicity 3. In this case, $A\mathbf{x} = \mathbf{0}$ for all $\mathbf{x}$ in $\mathcal{R}^3$ and therefore for any three linearly independent vectors. Thus, the eigenvectors associated with the triple eigenvalue 0 can be chosen to be the vectors in any basis for $\mathcal{R}^3$.

(c) Setting $a = 1$, $b = 2$ and $c = -2$ in the result of part (a) shows that $\lambda_2 = 3i$ and $\lambda_3 = -3i$ are the two nonreal complex eigenvalues.

For the eigenvalue $\lambda_2 = 3i$ we observe that if $\mathbf{v}$ is an associated eigenvector then the equation $(A - \lambda_2 I)\mathbf{v} = \mathbf{0}$ shows that $\mathbf{v}$ must have at least one nonreal complex entry. Thus, we set $\mathbf{v} = \begin{pmatrix} r_1 + s_1 i \\ r_2 + s_2 i \\ r_3 + s_3 i \end{pmatrix}$, where $r_1, r_2, r_3, s_1, s_2, s_3$ are real numbers. Without showing the tedious algebra, we write the entries of the vector $(A - \lambda_2 I)\mathbf{v}$ in terms of their real and imaginary parts and obtain

$$(A - \lambda_2 I)\mathbf{v} = \begin{pmatrix} -3i & 1 & -2 \\ -1 & -3i & -2 \\ 2 & 2 & -3i \end{pmatrix} \begin{pmatrix} r_1 + s_1 i \\ r_2 + s_2 i \\ r_3 + s_3 i \end{pmatrix}$$
$$= \begin{pmatrix} (r_2 - 2r_3 + 3s_1) + i(s_2 - 2s_3 - 3r_1) \\ (-r_1 - 2r_3 + 3s_2) + i(-s_1 - 2s_3 - 3r_2) \\ (2r_1 + 2r_2 + 3s_3) + i(2s_1 + 2s_2 - 3r_3) \end{pmatrix} = \mathbf{0}.$$

This leads to the three-equation system

$$(r_2 - 2r_3 + 3s_1) + i(s_2 - 2s_3 - 3r_1) = 0;$$
$$(-r_1 - 2r_3 + 3s_2) + i(-s_1 - 2s_3 - 3r_2) = 0;$$
$$(2r_1 + 2r_2 + 3s_3) + i(2s_1 + 2s_2 - 3r_3) = 0.$$

This shows that the real parts of the left members are all zero and the the imaginary parts of the left members are all zero. In particular, the imaginary parts give the three-equation system $s_2 - 2s_3 - 3r_1 = 0$, $-s_1 - 2s_3 - 3r_2 = 0$, $2s_1 + 2s_2 - 3r_3 = 0$, which allows us to write r_1, r_2 and r_3 in terms of s_1, s_2 and s_3; namely,

(*) $\qquad r_1 = \dfrac{1}{3}s_2 - \dfrac{2}{3}s_3, \qquad r_2 = -\dfrac{1}{3}s_1 - \dfrac{2}{3}s_3, \qquad r_3 = \dfrac{2}{3}s_1 + \dfrac{2}{3}s_2.$

The real parts give the three-equation system $r_2 - 2r_3 + 3s_1 = 0$, $-r_1 - 2r_3 + 3s_2 = 0$, $2r_1 + 2r_2 + 3s_3 = 0$. Inserting the above expressions for r_1, r_2, r_3 into these equations and simplifying gives the three-equation system in three unknowns:

$$\frac{4}{3}s_1 - \frac{4}{3}s_2 - \frac{2}{3}s_3 = 0;$$
$$-\frac{4}{3}s_1 + \frac{4}{3}s_2 + \frac{2}{3}s_3 = 0;$$
$$-\frac{2}{3}s_1 + \frac{2}{3}s_2 + \frac{4}{3}s_3 = 0.$$

Note that -2 times the third equation added to the second equation gives $-6s_3 = 0$, from which $s_3 = 0$. Also, since the first two equations are equivalent, the entire system implies the single equation $\frac{4}{3}s_1 - \frac{4}{3}s_2 = 0$, or equivalently, $s_1 = s_2$. Setting $s_1 = s_2 = 3$ and $s_3 = 0$

in equations $(*)$ gives $r_1 = 1$, $r_2 = -1$ and $r_3 = 4$. Hence, an eigenvector associated with the eigenvalue $3i$ is $\mathbf{v} = \begin{pmatrix} 1 + 3i \\ -1 + 3i \\ 4 \end{pmatrix}$.

An eigenvector for $\lambda_3 = -3i$ can be found by the same strategy as was used above to find an eigenvector for $3i$. However, here is a simpler method: the fact that A has real entries implies that conjugation of the equation $A\mathbf{v} = 3i\mathbf{v}$ leads to $A\overline{\mathbf{v}} = -3i\overline{\mathbf{v}}$, which directly shows that the conjugate of an eigenvector associated with $3i$ is an eigenvector associated with $-3i$. In other words, $\mathbf{w} = \overline{\mathbf{v}} = \begin{pmatrix} 1 - 3i \\ -1 - 3i \\ 4 \end{pmatrix}$ is an eigenvector associated with $\lambda_3 = -3i$.

Finally, we set $a = 1$, $b = 2$ and $c = -2$ in the eigenvector found in part (b) to get an eigenvector associated with $\lambda_1 = 0$; namely, $\mathbf{u} = \begin{pmatrix} -2 \\ 2 \\ 1 \end{pmatrix}$. Therefore, one set of three linearly independent eigenvectors of A when $a = 1$, $b = 2$, $c = -2$ is

$$\left\{ \begin{pmatrix} -2 \\ 2 \\ 1 \end{pmatrix}, \begin{pmatrix} 1 + 3i \\ -1 + 3i \\ 4 \end{pmatrix}, \begin{pmatrix} 1 - 3i \\ -1 - 3i \\ 4 \end{pmatrix} \right\}.$$

23. Let M be an n-by-n matrix such that $M^2 = M$, let $\{\mathbf{v}_1, \ldots, \mathbf{v}_r\}$ be a basis for the null-space of M and let $\{\mathbf{w}_1, \ldots, \mathbf{w}_s\}$ be a basis for the image of M. We want to show that the set $B = \{\mathbf{v}_1, \ldots, \mathbf{v}_r, \mathbf{w}_1, \ldots, \mathbf{w}_s\}$ is a basis for $\mathcal{R}^n$.

Let $\mathbf{x}$ be an arbitrary vector in $\mathcal{R}^n$ and let $\mathbf{v} = \mathbf{x} - M\mathbf{x}$. Left-multiplication by M of both sides of this equation gives

$$M\mathbf{v} = M(\mathbf{x} - M\mathbf{x}) = M\mathbf{x} - M^2\mathbf{x} = \mathbf{0},$$

where the last equality follows because M^2 and M are assumed to be the same matrix. So $\mathbf{v}$ is in the null-space of M and is therefore a linear combination of $\mathbf{v}_1, \ldots, \mathbf{v}_r$. On the other hand, since $\mathbf{w} = M\mathbf{x}$ is in the image of M, $\mathbf{w}$ is a linear combination of $\mathbf{w}_1, \ldots, \mathbf{w}_s$. Hence, $\mathbf{v} + \mathbf{w}$ is a linear combination of the vectors in B. But since

$$\mathbf{v} + \mathbf{w} = M\mathbf{x} + (\mathbf{x} - M\mathbf{x}) = \mathbf{x},$$

we see that $\mathbf{x}$ itself is a linear combination of vectors in B. Since $\mathbf{x}$ was an arbitrary vector in $\mathcal{R}^n$, B spans $\mathcal{R}^n$. Furthermore, Exercise 11 in Section 5C shows that $r + s = n$ so that B contains n vectors. Since $\dim(\mathcal{R}^n) = n$, we now use Theorem 5.9 to conclude that B is a basis for $\mathcal{R}^n$.

Chapter 4: Derivatives

Section 1: FUNCTIONS OF ONE VARIABLE

Exercise Set 1A-D (pgs. 182-184)

1. With $f(t) = (1 + t^2, 1 + t^3)$, we have

$$f'(t) = (2t, 3t^2) \qquad \text{and} \qquad f''(t) = (2, 6t).$$

So, $f(2) = (5, 9)$, $f'(2) = (4, 12)$ and $f''(2) = (2, 12)$. A parametric representation of the tangent line at $t = 2$ is therefore $\mathbf{t}(s) = sf'(2) + f(2) = s(4, 12) + (5, 9)$.

3. With $f(t) = \begin{pmatrix} e^t \\ e^{2t} \end{pmatrix}$, we have

$$f'(t) = \begin{pmatrix} e^t \\ 2e^{2t} \end{pmatrix} \qquad \text{and} \qquad f''(t) = \begin{pmatrix} e^t \\ 4e^{2t} \end{pmatrix}.$$

So, $f(-1) = \begin{pmatrix} e^{-1} \\ e^{-2} \end{pmatrix}$, $f'(-1) = \begin{pmatrix} e^{-1} \\ 2e^{-2} \end{pmatrix}$, and $f''(-1) = \begin{pmatrix} e^{-1} \\ 4e^{-2} \end{pmatrix}$. A parametric representation of the tangent line at $t = -1$ is therefore

$$\mathbf{t}(s) = sf'(-1) + f(-1) = s \begin{pmatrix} e^{-1} \\ 2e^{-2} \end{pmatrix} + \begin{pmatrix} e^{-1} \\ e^{-2} \end{pmatrix}.$$

5. With $f(t) = (\cos t, \cos 2t, \cos 3t, \cos 4t)$ we have

$$f'(t) = (-\sin t, -2\sin 2t, -3\sin 3t, -4\sin 4t)$$
$$\text{and} \qquad f''(t) = (-\cos t, -4\cos 2t, -9\cos 3t, -16\cos 4t).$$

So, $f(\pi/2) = (0, -1, 0, 1)$, $f'(\pi/2) = (-1, 0, 3, 0)$, and $f''(\pi/2) = (0, 4, 0, -16)$. A parametric representation of the tangent line at $t = \pi/2$ is therefore

$$\mathbf{t}(s) = sf'(\pi/2) + f(\pi/2) = s(-1, 0, 3, 0) + (0, -1, 0, 1).$$

7. $f(t) = t(1, 2, 0) + (1, 1, 1)$ **9.** $f(t) = (2t, t)$ **11.** $f(t) = (2t, |t|)$

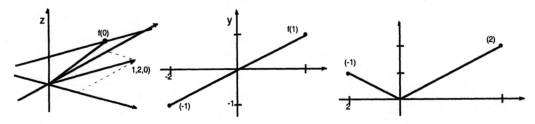

148

13. Consider the path of a particle defined by $(x, y, z) = (t, t^2, t^3)$. Given the temperature function $T(x, y, z) = x^2 + y^2 + z^2$, the temperature at each point on the path can be viewed as a function of the single real variable t and can be written as $T(t) = t^2 + t^4 + t^6$. In particular, at $t = 1/2$,

$$T(1/2) = (1/2)^2 + (1/2)^4 + (1/2)^6 = 21/64.$$

Moreover, $T'(t) = dT/dt = 2t + 4t^3 + 6t^5$ is the rate of change of the temperature of the point occupied by the particle at time t. So,

$$T'(1/2) = 2(1/2) + 4(1/2)^3 + 6(1/2)^5 = 27/16.$$

15. If $\mathbf{x}$ in the formula proven in Exercise 14 is replaced with $\dot{\mathbf{x}}$, the result is

$$d(\dot{\mathbf{x}} \cdot \dot{\mathbf{x}})/dt = 2\dot{\mathbf{x}} \cdot \ddot{\mathbf{x}}.$$

However, $\dot{\mathbf{x}} \cdot \dot{\mathbf{x}} = |\dot{\mathbf{x}}|^2$ is the square of the speed, which is assumed to be constant. Hence, the left side of the above displayed equation is 0. Therefore the right side is also 0. That is, the velocity and acceleration are perpendicular vectors.

17. The curve traced out by $g(t)$ is the graph of $y = \ln x$, with $t = 0$ and $t = 1$ corresponding to the points $(1, 0)$ and $(e, 1)$, respectively. Also, $g'(t) = (e^t, 1)$, so that $g'(0) = (1, 1)$ and $g'(1) = (e, 1)$. The pertinent curve and tangent vectors are shown below.

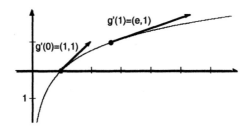

Exer. 17 $\quad g(t) = (e^t, t)$

19. With $f(t) = (t, t^2, t^3)$, the tangent vector is $f'(t) = (1, 2t, 3t^2)$. This vector is parallel to the vector $(4, 4, 3)$ if, and only if, $(4, 4, 3) = kf'(t) = (k, 2kt, 3kt^2)$ for some constant k. Comparing the first coordinates shows that $k = 4$. Thus, the second and third coordinates of $f'(t)$ must simultaneously satisfy $4 = 8t$ and $3 = 12t^2$. By inspection, $t = 1/2$ is the only solution. Hence, there is only one point on the curve satisfying the desired condition: $f(1/2) = (1/2, 1/4, 1/8)$. Further, since the dot product

$$f'(t) \cdot (4, 4, 3) = (1, 2t, 3t^2) \cdot (4, 4, 3) = (4, 8t, 9t^2)$$

is non-zero for all t, the tangent vector is never perpendicular to $(4, 4, 3)$.

21. Using the identity $\sin 2t = 2 \sin t \cos t$, we can write $g(t) = (2 \sin t \cos t, 2 \sin^2 t, 2 \cos t)$. Then the length of the position vector is given by

$$|g(t)| = \sqrt{4 \sin^2 t \cos^2 t + 4 \sin^4 t + 4 \cos^2 t} = 2\sqrt{\sin^2 t(\cos^2 t + \sin^2 t) + \cos^2 t}$$

$$= 2\sqrt{\sin^2 t + \cos^2 t} = 2.$$

That is, all points on the curve are two units from the origin and therefore lie on a sphere in $\mathcal{R}^3$ of radius 2 centered at the origin.

The velocity vector is $g'(t) = (2\cos 2t, 4\sin t\cos t, -4\sin t) = (2\cos 2t, 2\sin 2t, -4\sin t)$, where we have again used the identity $\sin 2t = 2\sin t\cos t$. The projection of this vector onto the xy-plane is found by setting the third coordinate equal to 0. The length of this projection is therefore $\sqrt{4\cos^2 2t + 4\sin^2 2t + 0} = 2$, a constant.

23. Let γ be the curve traced out by $f(t) = (3t, 4t)$ for $0 \le t \le 4$. The velocity at which γ is traced out is $v(t) = f'(t) = (3, 4)$ and the speed at which it is traced out is $|v(t)| = \sqrt{3^2 + 4^2} = 5$. Since γ is traced out exactly once over the given time interval, the distance traveled over the given time interval is exactly the arc length of γ; namely,

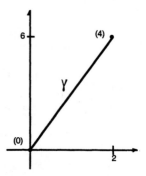

$$l(\gamma) = \int_0^4 |v(t)|dt = \int_0^4 5dt = 5t\Big|_0^4 = 20.$$

25. Let γ be the curve traced out by $h(t) = (t, 2t^{3/2})$ for $0 \le t \le 5/3$. The velocity at which γ is traced out is $v(t) = h'(t) = (1, 3t^{1/2})$ and the speed at which it is traced out is $|v(t)| = \sqrt{1 + 9t}$. Since γ is traced out exactly once over the given time interval, the distance traveled over the given time interval is exactly the arc length of γ; namely,

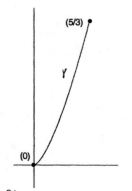

$$l(\gamma) = \int_0^{5/3} |v(t)|dt = \int_0^{5/3} \sqrt{1+9t}\,dt = \frac{1}{9}\int_1^{16} \sqrt{u}\,du$$
$$= \frac{2}{27}u^{3/2}\Big|_1^{16} = \frac{14}{3},$$

where the third equality is obtained using the substitution $u = 1 + 9t$.

27. Since the orbiting moon follows a circular path of radius b and constant angular speed δ, the orbital motion can be parametrized by $\mathbf{y}(t) = (b\cos\delta t, b\sin\delta t)$ (this parametrization does not attempt to characterize the position of the moon relative to the star, but rather describes only its orbital motion with respect to the planet). Since the orbital motion of the planet around the sun is parametrized by $\mathbf{x}(t) = (a\cos\omega t, a\sin\omega t)$, an obvious parametrization, call it $\mathbf{z}(t)$, of the motion of the moon *relative to the star* can be found by the parallelogram rule of vector addition: simply add the two vector parametrizations. This gives
$$\mathbf{z}(t) = \mathbf{x}(t) + \mathbf{y}(t) = (a\cos\omega t + b\cos\delta t, a\sin\omega t + b\sin\delta t)$$

as a parametric representation for the path of the moon relative to the fixed star. Moreover, the vectors $\mathbf{x}(0) = (a, 0)$ and $\mathbf{y}(0) = (b, 0)$ are pointing in the same direction and are therefore parallel. In other words, this parametric representation has the three bodies lined up at time $t = 0$.

29. Let $\mathbf{x} = \mathbf{x}(t) = (x_1, \ldots, x_n)$ and $\mathbf{y} = \mathbf{y}(t) = (y_1, \ldots, y_n)$ be vector-valued functions of the real variable t, where their coordinate functions $x_1, \ldots, x_n$ and $y_1, \ldots, y_n$ are all differentiable on a common interval $a < t < b$. Then, for $t \in (a, b)$,

$$\frac{d}{dt}(\mathbf{x} + \mathbf{y}) = \frac{d}{dt}\Big((x_1, \ldots, x_n) + (y_1, \ldots, y_n)\Big) = \frac{d}{dt}(x_1 + y_1, \ldots, x_n + y_n)$$
$$= (x_1' + y_1', \ldots, x_n' + y_n') = (x_1', \ldots, x_n') + (y_1', \ldots, y_n') = \dot{\mathbf{x}} + \dot{\mathbf{y}}.$$

150

Also, for any constant c,

$$\frac{d}{dt}(c\mathbf{x}) = \frac{d}{dt}\Big(c(x_1,\ldots,x_n)\Big) = \frac{d}{dt}(cx_1,\ldots,cx_n) = (cx'_1,\ldots,cx'_n) = c(x'_1,\ldots,x'_n) = c\dot{\mathbf{x}}.$$

31. Let $\mathbf{x} = \mathbf{x}(t) = (x_1,\ldots,x_n)$ and $\mathbf{y} = \mathbf{y}(t) = (y_1,\ldots,y_n)$ be vector-valued functions of the real variable t, where their coordinate functions $x_1,\ldots,x_n$ and $y_1,\ldots,y_n$ are all differentiable on a common interval $a < t < b$. Then, for $t \in (a,b)$,

$$\frac{d}{dt}(\mathbf{x}\cdot\mathbf{y}) = \frac{d}{dt}\Big((x_1,\ldots,x_n)\cdot(y_1,\ldots,y_n)\Big) = \frac{d}{dt}(x_1 y_1 + \cdots + x_n y_n)$$

$$= x_1 y'_1 + x'_1 y_1 + \cdots + x_n y'_n + x'_n y_n$$

$$= (x_1 y'_1 + \cdots + x_n y'_n) + (x'_1 y_1 + \cdots + x'_n y_n)$$

$$= (x_1,\ldots,x_n)\cdot(y'_1,\ldots,y'_n) + (x'_1,\ldots,x'_n)\cdot(y_1,\ldots,y_n)$$

$$= \mathbf{x}\cdot\dot{\mathbf{y}} + \dot{\mathbf{x}}\cdot\mathbf{y}.$$

33. Suppose $g : \mathcal{R} \longrightarrow \mathcal{R}^n$ and that $g'(t) = \mathbf{0}$ on $a < t < b$. If $g_1,\ldots,g_n$ are the corresponding coordinate functions then each g_k is differentiable on $a < t < b$ and $g'(t) = (g'_1(t),\ldots,g'_n(t))$. The condition $g'(t) = \mathbf{0}$ on $a < t < b$ then implies $g'_k(t) = 0$ on $a < t < b$, for $k = 1,\ldots,n$. A direct application of the mean-value theorem for real-valued functions of a real variable shows that each $g_k(t)$ is constant on $a < t < b$. It follows that $g(t)$ is constant on $a < t < b$.

35. We have

$$\mathbf{L}'(t) = \frac{d}{dt}\big(\mathbf{L}(t)\big) = \frac{d}{dt}\big(\mathbf{g}(t) \times \mathbf{P}(t)\big) = \mathbf{g}'(t) \times \mathbf{P}(t) + \mathbf{g}(t) \times \mathbf{P}'(t)$$

$$= \mathbf{g}'(t) \times \mathbf{P}(t) + \mathbf{g}(t) \times \mathbf{F}(t)$$

$$= \mathbf{g}'(t) \times \mathbf{P}(t) + \mathbf{N}(t)$$

$$= \mathbf{g}'(t) \times m(t)\mathbf{v}(t) + \mathbf{N}(t)$$

$$= \mathbf{v}(t) \times m(t)\mathbf{v}(t) + \mathbf{N}(t) = \mathbf{N}(t),$$

where the last equality follows from the fact that $\mathbf{v}(t)$ and $m(t)\mathbf{v}(t)$ are parallel vectors whose cross product is therefore $\mathbf{0}$. The above result shows that if the torque $\mathbf{N}(t)$ is identically zero, then the angular momentum $\mathbf{L}(t)$ is constant. (see Exercise 33)

37. Let $\mathbf{x}(t) = (a\cos\omega t, b\sin\omega t)$, where a, b and ω are positive constants. As a curve in $\mathcal{R}^2$, $x = a\cos\omega t$ and $y = b\sin\omega t$. Eliminating t from these two equations gives $x^2/a^2 + y^2/b^2 = \cos^2\omega t + \sin^2\omega t = 1$. This is an equation of an ellipse and is independent of both t and ω.

39. $\mathbf{x}(t) = (t, t, t^2)$, $0 \le t \le 1$ **41.** $\mathbf{x}(t) = (t, t^2, t^3)$, $0 \le t \le 1$

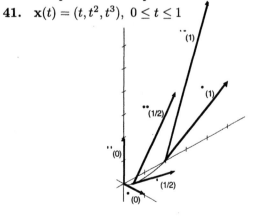

43. While the projectile is in the air, only the force of gravity $g = 32$ ft/sec^2 is acting on it. We arbitrarily choose upward motion as positive, so that the acceleration of the projectile is given by $a(t) = -32$ (the minus sign indicates that the force of gravity is acting in a downward direction). To find the velocity of the projectile, integrate this equation with respect to t and use the initial condition $v(0) = 300$ to get $v(t) = -32t + 300$. To find the height $h(t)$ of the projectile, integrate this equation with respect to t and use the initial condition $h(0) = 0$ to get $h(t) = -16t^2 + 300t$.

Now let $T(t)$ be the temperature of the air around the projectile t seconds after it is fired. Since $T(t)$ decreases 3° F every 1000 feet, and since $T(0) = 32$, it follows that $T(t)$ is given by

$$T(t) = 32 - \frac{3}{1000}h(t) = 32 - \frac{3}{1000}(-16t^2 + 300t),$$

which holds for all times t that the projectile is in the air. Once the projectile hits the ground, $T(t) = 32$ from that time on.

The minimum temperature is attained when the projectile reaches its maximum height. This occurs when $v(t) = 0$; i.e., when $-32t + 300 = 0$, or at $t = 75/8$ seconds. The minimum temperature attained is therefore

$$T(75/8) = 32 - \frac{3}{1000}\left(-16(75/8)^2 + 300(75/8)\right) \approx 27.8° \text{ F.}$$

1. $g(t) = (t \cos t, t \sin t),\ 0 \le t \le 2\pi$

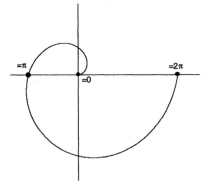

3. $g(t) = (\sin 2t, 2 \sin^2 t, 1 \cos t),\ 0 \le t \le 2\pi$

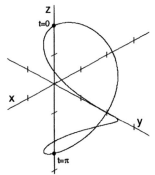

5. The figure on the right shows the image curve of $g(t) = (\sin 2t, 2 \sin^2 t, 2 \cos t)$ on the surface of a sphere of radius 2 centered at the origin. It is important to realize that this graphical representation does not "prove" that the curve lies on the surface of the sphere, only that it is reasonable to believe that it does. It is still possible that the curve rises off the surface at some points and goes inside the sphere at other points, but due to the particular perspective presented, this departure from what we appear to see cannot be observed. A mathematical proof that the curve does indeed lie on the surface of the sphere is given in the solution to Exercise 21, Section 1D in this Chapter.

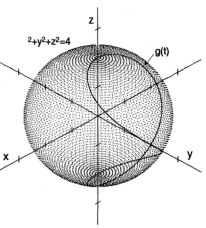

7. $\mathbf{x} = (2 \cos t, 3 \sin t, e^t),\ 1 \le z \le 2$

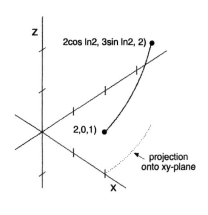

9. $\mathbf{x} = (t^2, t^3, t^4),\ |y| \le 5$

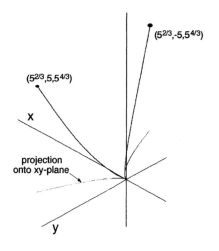

Exercise Set 1F (pgs. 187-188)

1. $f(t) = (t^2 + 1, t^3 - 1)$. Integrating each coordinate function gives

$$F(t) = \left(\frac{1}{3}t^3 + t + c_1, \frac{1}{4}t^4 - t + c_2 \right).$$

Given $F(1) = \left(\frac{4}{3} + c_1, -\frac{3}{4} + c_2 \right) = (2, 2)$ implies $c_1 = \frac{2}{3}$ and $c_2 = \frac{11}{4}$. Therefore,

$$F(t) = \left(\frac{1}{3}t^3 + t + \frac{2}{3}, \frac{1}{4}t^4 - t + \frac{11}{4} \right).$$

3. $f(t) = (t \cos t, t \sin t)$. Integrating each coordinate function by parts gives

$$F(t) = (t \sin t + \cos t + c_1, -t \cos t + \sin t + c_2).$$

Given $F(0) = (1 + c_1, c_2) = (1, 1)$ implies $c_1 = 0$ and $c_2 = 1$. Therefore

$$F(t) = (t \sin t + \cos t, -t \cos t + \sin t + 1).$$

5. $f(t) = (1, t^2, -1, t^2)$. Integrating each coordinate function gives

$$F(t) = \left(t + c_1, \frac{1}{3}t^3 + c_2, -t + c_3, \frac{1}{3}t^3 + c_4 \right).$$

Given $F(1) = \left(1 + c_1, \frac{1}{3} + c_2, -1 + c_3, \frac{1}{3} + c_4 \right) = (2, 2, 2, 2)$ implies $c_1 = 1$, $c_2 = \frac{5}{3}$, $c_3 = 3$ and $c_4 = \frac{5}{3}$. Therefore,

$$F(t) = \left(t + 1, \frac{1}{3}t^3 + \frac{5}{3}, -t + 3, \frac{1}{3}t^3 + \frac{5}{3} \right).$$

7. $\dot{x}(t) = (t, -t^2)$. Integrating each coordinate function gives

$$x(t) = \left(\frac{1}{2}t^2 + c_1, -\frac{1}{3}t^3 + c_2 \right).$$

Given $x(0) = (c_1, c_2) = (2, 1)$ implies $c_1 = 2$ and $c_2 = 1$. Therefore,

$$x(t) = \left(\frac{1}{2}t^2 + 2, -\frac{1}{3}t^3 + 1 \right).$$

9. $\dot{x}(t) = (\cos t, \sin 2t)$. Integrating each coordinate function gives

$$x(t) = \left(\sin t + c_1, -\frac{1}{2}\cos 2t + c_2 \right).$$

Given $x(\pi/2) = (1 + c_1, 1/2 + c_2) = (-1, 1)$ implies $c_1 = -2$ and $c_2 = \frac{1}{2}$. Therefore,

$$x(t) = \left(\sin t - 2, -\frac{1}{2}\cos 2t + \frac{1}{2} \right).$$

11. $\dot{\mathbf{x}}(t) = (t, t, t^2)$. Integrating each coordinate function gives

$$\mathbf{x}(t) = \left(\frac{1}{2}t^2 + c_1, \frac{1}{2}t^2 + c_2, \frac{1}{3}t^3 + c_3\right).$$

Given $\mathbf{x}(1) = (1/2 + c_1, 1/2 + c_2, 1/3 + c_3) = (1, -1, 1)$ implies $c_1 = \frac{1}{2}$, $c_2 = -\frac{3}{2}$ and $c_3 = \frac{2}{3}$. Therefore,

$$\mathbf{x}(t) = \left(\frac{1}{2}t^2 + \frac{1}{2}, \frac{1}{2}t^2 - \frac{3}{2}, \frac{1}{3}t^3 + \frac{2}{3}\right).$$

13. $\ddot{\mathbf{x}}(t) = (t, -t^2)$. Integrating each coordinate function gives have

$$\dot{\mathbf{x}}(t) = \left(\frac{1}{2}t^2 + c_1, -\frac{1}{3}t^3 + c_2\right).$$

Given $\dot{\mathbf{x}}(0) = (c_1, c_2) = (1, 1)$ implies $c_1 = c_2 = 1$. Therefore,

$$\dot{\mathbf{x}}(t) = \left(\frac{1}{2}t^2 + 1, -\frac{1}{3}t^3 + 1\right).$$

Integrating each coordinate function again gives

$$\mathbf{x}(t) = \left(\frac{1}{6}t^3 + t + d_1, -\frac{1}{12}t^4 + t + d_2\right).$$

Given $\mathbf{x}(0) = (d_1, d_2) = (2, 1)$ implies $d_1 = 2$ and $d_2 = 1$. Therefore,

$$\mathbf{x}(t) = \left(\frac{1}{6}t^3 + t + 2, -\frac{1}{12}t^4 + t + 1\right).$$

15. Let $\mathbf{x} = \mathbf{x}(t)$ be the vector position of the ball at time t. Imposing an xy-coordinate system (so that $\mathbf{x}(t) = (x, y)$), we shall assume that the ball is kicked from the origin and that it travels in the direction of increasing x values. The initial conditions are therefore $\mathbf{x}(0) = (0, 0)$ and $\dot{\mathbf{x}}(0) = (v_0 \cos\theta, v_0 \sin\theta)$, where v_0 is the initial speed of the ball and $0 < \theta < \pi/2$ is the angle at which the ball is kicked.

Neglecting air resistance, the acceleration vector is given by $\ddot{\mathbf{x}} = (0, -g)$, where g is the acceleration of gravity. One integration gives the velocity vector $\dot{\mathbf{x}}(t) = (c_1, -gt + c_2)$. Given $\dot{\mathbf{x}}(0) = (c_1, c_2) = (v_0 \cos\theta, v_0, \sin\theta)$, we get $c_1 = v_1 \cos\theta$ and $c_2 = v_0 \sin\theta$, so that

$$\dot{\mathbf{x}}(t) = (v_0 \cos\theta, -gt + v_0 \sin\theta).$$

One more integration gives $\mathbf{x}(t) = ((v_0 \cos\theta)t + d_1, -(g/2)t^2 + (v_0 \sin\theta)t + d_2)$. Given $\mathbf{x}(0) = (d_1, d_2) = (0, 0)$, we get $d_1 = d_2 = 0$, so that

$$\mathbf{x}(t) = ((v_0 \cos\theta)t, -(g/2)t^2 + (v_0 \sin\theta)t).$$

In terms of x and y, $x = (v_0 \cos\theta)t$ and $y = -(g/2)t^2 + (v_0 \sin\theta)t$. Eliminating t from these two equations gives an equation of a parabola which describes the curve taken by the ball in its flight; namely,

$$y = -\frac{g}{2v_0^2 \cos^2\theta}x^2 + (\tan\theta)x \qquad \text{for } 0 \le x \le a + b.$$

155

Letting $\alpha = -g/(2v_0^2 \cos^2\theta)$ and $\beta = \tan\theta$, use the fact that (a, h) and $(a + b, 0)$ are both on this curve to obtain the two equations

$$h = \alpha a^2 + \beta a \qquad \text{and} \qquad 0 = \alpha(a + b)^2 + \beta(a + b).$$

This is a linear system of equations in the two unknowns α and β. Solving for α and β (by the elimination method, Cramer's Rule, matrix inversion, or some other technique), gives

$$\alpha = -\frac{g}{2v_0^2 \cos^2\theta} = -\frac{h}{ab} \qquad \text{and} \qquad \beta = \tan\theta = \frac{h(a + b)}{ab}.$$

(NOTE: The bounds on θ imply $\theta = \arctan\big(h(a + b)/ab\big)$. This can be inserted into the first equation and the resulting expression can then be solved for v_0. But the answers get rather messy. So it is quite all right to leave the second equation as it is and simply solve the first equation for v_0^2.)

17. Impose an xy-coordinate system on the scene, with the origin at the base of the building on which Clark stands. Let $\mathbf{x}(t)$ be the position vector of Clark's position at time t. Since Clark's acceleration vector is $\ddot{\mathbf{x}}(t) = (0, -32)$ (where the acceleration of gravity is taken to be 32 ft/sec^2), two integrations and the use of the initial conditions $\dot{\mathbf{x}}(0) = (v_x, v_y)$ and $\mathbf{x}(0) = (0, 200)$ gives

$$\mathbf{x}(t) = (v_x t, -16t^2 + v_y t + 200).$$

Let $\mathbf{y}(t)$ be the victim's position vector at time t. Since the victim's acceleration vector is $\ddot{\mathbf{y}}(t) = (0, -32)$, two integrations and the use of the initial conditions $\dot{\mathbf{y}}(0) = (0, 0)$ and $\mathbf{y}(0) = (50, 100)$ gives

$$\mathbf{y}(t) = (50, -16t^2 + 100).$$

Note that $(50, 0)$ is the point where the victim will hit the ground. Letting $t = t_0$ be the time at which this occurs, the second coordinate of $\mathbf{y}(t_0)$ satisfies $-16t_0^2 + 100 = 0$. Solving for t_0 gives $t_0 = 5/2$. Since Clark wishes to also be at $(50, 0)$ at time t_0, we must have

$$\mathbf{x}(5/2) = (5v_x/2, 100 + 5v_y/2) = (50, 0),$$

from which we obtain $v_x = 20$ and $v_y = -40$. Hence,

$$\dot{\mathbf{x}}(0) = (v_x, v_y) = (20, -40) \qquad \text{and} \qquad v_0 = \sqrt{20^2 + (-40)^2} = 20\sqrt{5} \approx 45 \text{ ft/sec.}$$

Also, $\dot{\mathbf{x}}(t) = (v_x, -32t + v_y) = (20, -32t - 40)$ and $\dot{\mathbf{y}}(t) = (0, -32t)$ are the velocity vectors for Clark and the victim (resp.), so that Clark's and the victim's speeds at $t_0 = 5/2$ are given by

$$|\dot{\mathbf{x}}(5/2)| = |(20, -120)| = 20\sqrt{37} \approx 122 \text{ ft/sec} \qquad \text{and} \qquad |\dot{\mathbf{y}}(5/2)| = |(0, -80)| = 80 \text{ ft/sec.}$$

19. We impose an xy-coordinate system onto the vertical plane containing the ball's trajectory, with the origin $(0, 0)$ as the point where the ball is kicked. Then, assuming the ball's flight is from left to right, the points (a, h) and $(a + b, h)$ are where the ball just grazes the edge of the building and where the ball lands on the flat roof of the building (resp.). Since the ball's trajectory is a parabolic arc that touches the x-axis at $x = 0$, it is a segment of the graph of some quadratic function f of the form $f(x) = \alpha x^2 + \beta x$, where α and β are constants. These constants can be determined by using the fact that $f(a) = h$

and $f(a + b) = h$. When this is done, we get $\alpha = -h/a(a + b)$ and $\beta = h/a + h/(a + b)$, so that

$$f(x) = -\frac{h}{a(a + b)}x^2 + \left(\frac{h}{a} + \frac{h}{a + b}\right)x, \qquad 0 \leq x \leq a + b.$$

On the other hand, if $\mathbf{x}(t) = (x(t), y(t))$ is the position vector of the ball's trajectory and θ is the angle at which the ball is kicked then $\ddot{\mathbf{x}}(t) = (0, -g)$, with the initial conditions $\dot{\mathbf{x}}(0) = (v_0 \cos \theta, v_0 \sin \theta)$ and $\mathbf{x}(0) = (0, 0)$. Two integrations gives $\mathbf{x}(t) = ((v_0 \cos \theta)t, -\frac{1}{2}gt^2 + (v_0 \sin \theta)t)$. Setting $x = x(t)$, $y = y(t)$, we see that the x and y coordinates of the ball's trajectory are related by the equation

$$y = -\frac{g}{2v_0^2 \cos^2 \theta}x^2 + (\tan \theta)x, \qquad 0 \leq x \leq a + b.$$

Since $f(x)$ and $y(x)$ are two quadratics describing the same trajectory, they must have the same coefficients, so that equating coefficients of like powers of x gives

$$-\frac{h}{a(a + b)} = -\frac{g}{2v_0^2 \cos^2 \theta} \qquad \text{and} \qquad \frac{h}{a} + \frac{h}{a + b} = \tan \theta.$$

From the first equation, the initial speed is $v_0 = \sqrt{ga(a + b)/(2h \cos \theta)}$; and from the second equation $\theta = \arctan(h/a + h/(a + b))$. (Compare this with Exercise 15 above)

21. Let θ be the angle at which the gun is fired, let v_0 be the initial speed of the shell, let $h_{\max}$ be the maximum height of the trajectory, and let $\mathbf{x}(t)$ be the position vector for the trajectory. With initial conditions $\dot{\mathbf{x}}(0) = (v_0 \cos \theta, v_0 \sin \theta)$ and $\mathbf{x}(0) = (0, 0)$, the acceleration vector is $\ddot{\mathbf{x}}(t) = (0, -32)$ ($g \approx 32$ ft/sec^2). Two integrations then gives

$$\mathbf{x}(t) = ((v_0 \cos \theta)t, -16t^2 + (v_0 \sin \theta)t).$$

Given that the shell's flight time is 186 seconds, we use 396000 feet = 75 miles to get

$$\mathbf{x}(186) = (186v_0 \cos \theta, -553536 + 186v_0 \sin \theta) = (396000, 0).$$

Therefore, $v_0 \cos \theta = 2129$ and $v_0 \sin \theta = 2976$, so that $\tan \theta = 2976/2129 = 1.3978$. Hence, $\theta \approx 54.4°$. Finally, with no air resistance, $h_{\max}$ is reached half way through the flight; i.e., at $t = 93$ seconds. Since the second coordinate of $\mathbf{x}(93)$ is $h_{\max}$, we have

$$h_{\max} = -16(93)^2 + 2976(93) = 138384 \text{ feet} \approx 26.2 \text{ miles}.$$

23. Suppose that the momentum is constant at $\mathbf{p}_0 = m\dot{\mathbf{c}}$. If $\mathbf{p}_0 = \mathbf{0}$ then each coordinate of $m\dot{\mathbf{c}}$ is zero. Integrating each coordinate, results in each coordinate of $m\mathbf{c}$ being constant. Since m is constant, $\mathbf{c}$ is a constant vector. If $\mathbf{p}_0 \neq \mathbf{0}$ then integrating both sides of $\mathbf{p}_0 = m\dot{\mathbf{c}}$ with respect to t gives

$$m\mathbf{c} = \mathbf{p}_0 t + \mathbf{c}_1, \qquad \text{or equivalently,} \qquad \mathbf{c} = \frac{1}{m}\mathbf{p}_0 t + \mathbf{c}_2$$

where $\mathbf{c}_2 = \frac{1}{m}\mathbf{c}_1$ are constant vectors. The right side of this equation is a straight line parallel to the vector $\frac{1}{m}\mathbf{p}_0$, and therefore parallel to $\mathbf{p}_0$ itself. That is, $\mathbf{c}$ moves at a constant speed $\frac{1}{m}|\mathbf{p}_0|$ along a fixed line parallel to $\mathbf{p}_0$.

25. (a) With $f(t) = (\cos t, \sin t)$ on $[0, \pi/2]$, we have

$$\int_a^b f(t)\, dt = \int_0^{\pi/2} (\cos t, \sin t)\, dt = \left(\int_0^{\pi/2} \cos t\, dt, \int_0^{\pi/2} \sin t\, dt \right)$$

$$= \left([\sin t]_0^{\pi/2}, [-\cos t]_0^{\pi/2} \right) = (1, 1).$$

(b) With $g(t) = (t, t^2, t^3)$ on $[0, 1]$, we have

$$\int_a^b g(t)\, dt = \int_0^1 (t, t^2, t^3)\, dt = \left(\int_0^1 t\, dt, \int_0^1 t^2\, dt, \int_0^1 t^3\, dt \right)$$

$$= \left(\left[\frac{1}{2}t^2\right]_0^1, \left[\frac{1}{3}t^3\right]_0^1, \left[\frac{1}{4}t^4\right]_0^1 \right) = \left(\frac{1}{2}, \frac{1}{3}, \frac{1}{4} \right).$$

27. If f is a vector-valued function of a real variable such that f' is continuous on $[a, b]$ then each of the coordinate functions of f has a continuous derivative on $[a, b]$, so that the derivative of each coordinate function is integrable on $[a, b]$. Therefore, by definition, f' is integrable on $[a, b]$. Thus, let $f(t) = (f_1(t), \ldots, f_n(t))$, so that $f'(t) = (f_1'(t), \ldots, f_n'(t))$. Then $\int_a^b f'(t)\, dt$ exists and we have

$$\int_a^b f'(t)\, dt = \left(\int_a^b f_1'(t)\, dt, \ldots, \int_a^b f_n'(t)\, dt \right)$$

$$= \big(f_1(b) - f_1(a), \ldots, f_n(b) - f_n(a) \big)$$

$$= \big(f_1(b), \ldots, f_n(b) \big) - \big(f_1(a), \ldots, f_n(a) \big)$$

$$= f(b) - f(a),$$

where the second equality follows from the fundamental theorem of calculus for real-valued functions of a real variable.

29. (a) The definition $h(t) = \int_a^t g(u)\, du$ is equivalent to $h'(t) = g(t)$, $h(a) = 0$, so that

$$\frac{d}{dt}\big(e^{-h(t)} \dot{\mathbf{x}}(t) \big) = \mathbf{0} \quad \Longleftrightarrow \quad e^{-h(t)} \ddot{\mathbf{x}}(t) - h'(t) e^{-h(t)} \dot{\mathbf{x}}(t) = \mathbf{0}$$

$$\Longleftrightarrow \quad \ddot{\mathbf{x}}(t) - g(t) \dot{\mathbf{x}}(t) = \mathbf{0}$$

$$\Longleftrightarrow \quad \ddot{\mathbf{x}}(t) = g(t) \dot{\mathbf{x}}(t).$$

(b) If $f(t)$ is a vector-valued function such that $f'(t) = \mathbf{0}$ on an interval then $f(t)$ is constant on that interval (see Exercise 33, Ch.4, Sec.1D). The result of part (a) therefore implies $e^{-h(t)} \dot{\mathbf{x}}(t) = \mathbf{c}$, for some constant vector $\mathbf{c}$. Thus, $\dot{\mathbf{x}}(t) = e^{h(t)} \mathbf{c}$. (Note: The equation $\ddot{\mathbf{x}}(t) = g(t) \dot{\mathbf{x}}(t)$ together with the assumption that $\ddot{\mathbf{x}}(t)$ is nonzero on $[a, b]$ implies that $\dot{\mathbf{x}}(t)$ is nonzero on $[a, b]$, so that $\mathbf{c} \neq \mathbf{0}$.)

(c) Let $H(t) = \int_a^t e^{h(u)}\, du$ for $a \leq t \leq b$, so that $H'(t) = e^{h(t)}$. Integrating both sides of $\dot{\mathbf{x}}(t) = e^{h(t)} \mathbf{c}$ from a to t, the fundamental theorem of calculus gives

$$\mathbf{x}(t) - \mathbf{x}(a) = \left(\int_a^t e^{h(u)}\, du \right) \mathbf{c} = H(t)\mathbf{c} \quad \text{or equivalently,} \quad \mathbf{x}(t) = H(t)\mathbf{c} + \mathbf{d},$$

where $\mathbf{d} = \mathbf{x}(a)$ is a constant vector. Since $\mathbf{c} \neq \mathbf{0}$ (see "Note" in part (b)), the equation on the right is the vector equation of some portion of the parametrized line $s\mathbf{c} + \mathbf{d}$. That is, $\mathbf{x}(t)$ stays on a line.

158

Section 2: SEVERAL INDEPENDENT VARIABLES

Exercise Set 2AB (pgs. 192-193)

1 (a) The domain of $f(x,y) = \sqrt{4 - x^2 - y^2}$ is the set of points (x,y) in $\mathcal{R}^2$ such that $4 - x^2 - y^2 \geq 0$, or equivalently, $x^2 + y^2 \leq 4$. This is a closed disk of radius 2 centered at $(0,0)$ in the xy-plane, and is the solid base of the figure shown on the right.

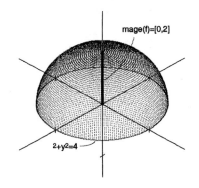

(b) Setting $f(x,y) = z$, the graph of f is the set of points (x,y,z) in $\mathcal{R}^3$ such that $z = \sqrt{4 - x^2 - y^2}$, or equivalently, $x^2 + y^2 + z^2 = 4$, with $z \geq 0$. This is the top half of a sphere of radius 2 centered at $(0,0,0)$ in $\mathcal{R}^3$.

(c) The image of f is the set of all possible values of $f(x,y) = z$; i.e., the interval $[0,2]$. This is depicted in the figure on the right as the vertical black "axis" of the hemisphere.

3. The graph of the function $f(x,y) = 2 - x^2 - y^2$ is depicted on the right. The domain of f is the xy-plane. Setting $z = f(x,y)$, we have $z = 2 - x^2 - y^2$, which can be written as $x^2 + y^2 = 2 - z$. Since the left side of this equation is never negative, $z \leq 2$ (the image of f). The level sets of f are therefore circles of the form $x^2 + y^2 = 2 - k$ (k constant) on the plane $z = k$ (in particular, the figure on the right shows the level $k = 0$). Note too that the trace of the figure in the xz-plane and yz-plane are the parabolas $z = 2 - x^2$ and $z = 2 - y^2$, respectively. The overall figure is an inverted paraboloid with its vertex at $(0,0,2)$ and its axis coincident with the z-axis.

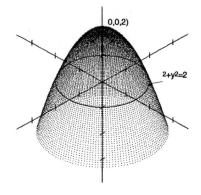

5. The graph of the function $g(x,y) = \sin x$ is depicted on the right. The domain of g is the xy-plane. The fact that g is independent of y makes its graph particularly easy to envision. First set $y = 0$ in the equation $z = g(x,y)$ and graph the trace of g in the xz-plane (shown in the figure as the curve $g(x,0) = \sin x$), then extend this curve parallel to the y-axis infinitely far in both directions. Also, the image of g is the closed interval $[-1,1]$, the range of the sine function.

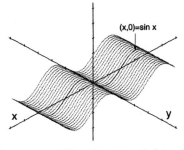

7. The graph of the function $f(x,y) = e^{x+y}$ is on the next page. The domain of f is the xy-plane. Since the exponential function is always positive, the image of f is the interval $(0,+\infty)$. Setting $z = f(x,y)$ and solving for y gives $y = -x + \ln z$, which puts the relation among x, y and z in a form to examine level curves in a familiar way. Holding z fixed at the positive constant value $z = k$, gives $y = -x + \ln k$, a line parallel to the line $y = -x$, and since the range of the log function is all real numbers, the level curves consist of *all* lines parallel to $y = -x$ (which is shown in the accompanying figure as a dotted line in the xy-plane). As a further aid in sketching the graph of f, note that its trace in the xz-plane and yz-plane are the exponentials $z = e^x$ and $z = e^y$, respectively. The overall figure looks like an infinite sheet curving exponentially upward along the line $y = x$.

9. With $z = f(x, y)$, we keep z fixed at $z = 1$, so that the equation of the desired level curve is $x + y = 1$, a line in $\mathcal{R}^2$. The line is sketched below on the right.

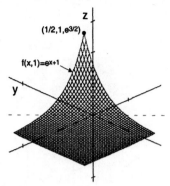

Exer. 7: $f(x, y) = e^{x+y}$

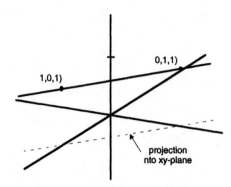

Exercise 9

11. With $z = f(x, y)$, keep z fixed at $z = 0$, so that $(x^2 + y^2 + 1)^2 - 4x^2 = 0$. Observe that the left side of this equation can be factored to get

$$(x^2 + y^2 + 1 - 2x)(x^2 + y^2 + 1 + 2x)$$
$$= \big((x-1)^2 + y^2\big)\big((x+1)^2 + y^2\big) = 0.$$

Hence, the desired level curve is the set of points such that either $(x-1)^2 + y^2 = 0$ or $(x+1)^2 + y^2 = 0$. The first equation is satisfied only by $x = 1$, $y = 0$, and the second equation is satisfied only by $x = -1$, $y = 0$. Thus, the desired level curve lies on the xy-plane and consists of the two points $(1, 0)$ and $(-1, 0)$.

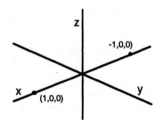

13. With $w = f(x, y, z)$, keep w fixed at $w = 0$, so that $xyz = 0$. This equation is satisfied by $x = 0$ (with y and z arbitrary), $y = 0$ (with x and z arbitrary) and $z = 0$ (with x and y arbitrary). These are the three coordinate planes in $\mathcal{R}^3$ (while each plane is in $\mathcal{R}^2$, the level curve consists of all three planes and therefore lies in $\mathcal{R}^3$). The level set is shown on the right.

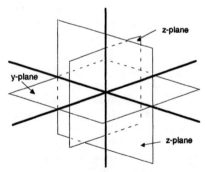

15. With $w = f(x, y, z)$, we hold w fixed at $k = 0, 1, 2$ to get the three level curves $x + y = 0$, $x + y = 1$ and $x + y = 2$, respectively. These are parallel vertical planes in $\mathcal{R}^3$, and are shown in the figure on the right. Note that for each plane in the figure, a solid line runs from left to right across the plane. This line is the intersection of that plane with the xy-plane, and is drawn to give the viewer the right perspective as to how the planes are situated with respect to the z-axis.

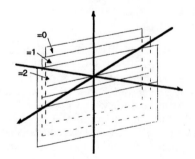

17. With $w = f(x, y, z)$, the level curve in $\mathcal{R}^3$ of level $k \geq 0$ is the graph of $k = \sqrt{x^2 + y^2 + z^2}$, or equivalently, $x^2 + y^2 + z^2 = k^2$, which is a sphere of radius k centered at the origin. In particular, for $k = 0, 1$, we have the level curves $x^2 + y^2 + z^2 = 0$ and $x^2 + y^2 + z^2 = 1$. The first level curve is the single point $(0, 0, 0)$, and the second level curve is a sphere of radius 1 centered at $(0, 0, 0)$. The level curves in question are shown on the right.

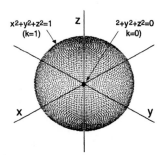

19. With $f(x, y, z) = \begin{pmatrix} x - y \\ y + z \end{pmatrix} = \begin{pmatrix} 0 \\ 0 \end{pmatrix} = \mathbf{k}$, we compare corresponding coordinates and obtain the two equations $x - y = 0$ and $y + z = 0$, which are planes in $\mathcal{R}^3$. The line defined by their intersection is the desired level set. Letting $y = t$, we see that $x = t$ and $z = -t$. Thus, the desired line is $(x, y, z) = (t, t, -t) = t(1, 1, -1)$. The figure is shown below.

21. With $f(x, y, z) = \begin{pmatrix} x^2 + y^2 + z^2 \\ x - z \end{pmatrix} = \begin{pmatrix} 1 \\ 0 \end{pmatrix} = \mathbf{k}$, we compare corresponding coordinates and obtain the two equations $x^2 + y^2 + z^2 = 1$ and $x - z = 0$. The first equation describes a sphere of radius 1 centered at $(0, 0, 0)$, and the second equation is a plane passing through the origin. The intersection of these two level sets is therefore a circle of radius 1 in $\mathcal{R}^3$. The figure is shown below.

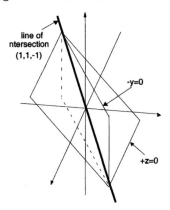

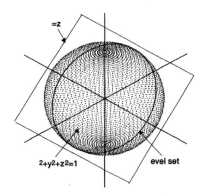

Exercise 19   Exercise 21

23. With $d(x, y) = x^2 + 2y^2 - x + 1$, we need only graph the curve $x^2 + 2y^2 - x + 1 = 7/4$. Completing the square in x gives $(x - 1/2)^2 + 2y^2 = 1$, which is an ellipse centered at $(1/2, 0)$, with its major axis in the x direction and of length 2 and minor axis of length $\sqrt{2}$. The only part of the ellipse that is not part of the level set is that part to the right of the line $x = 1$.

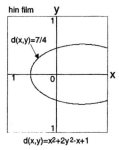

25. (a) The region D is the set of points inside and on the right circular cylinder of radius 2 and height 5, with its base in the xy-plane. (see figure on next page)

    (b) With the temperature of the form $T(x, y, z) = x^2 + y^2 - z$, we want all points (x, y, z) in D such that $T(x, y, z) = -1$. That is, we want to sketch the set of points in D for which $x^2 + y^2 - z = -1$, or equivalently, $x^2 + y^2 = z - 1$. Since the temperature only applies to points in D, $1 \le z \le 5$.

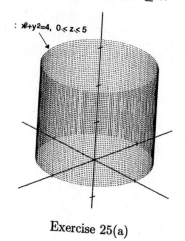

: $x^2+y^2=4$, $0 \le z \le 5$

Exercise 25(a)

$T(x,y,z)=x^2+y^2-z=-1$

$1 \le z \le 5$

$D \rightarrow$

2

y

Exercise 25(b)

Exercise Set 2C (pgs. 195-196)

1. $f(x,y) = x^2 - y^2$, for $|x| \leq 2$, $|y| \leq 2$

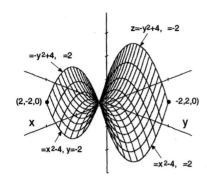

3. $z = 1 - x - y$, for $0 \leq x \leq 2$, $0 \leq y \leq 2$

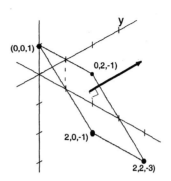

5. $f(x,y) = x^2 + y^3$, for $|x| \leq 3$, $|y| \leq 3$

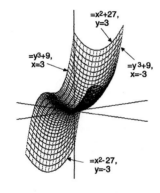

7. $f(x,y) = x + y$, for $0 \leq x \leq 1$, $0 \leq y \leq 2$

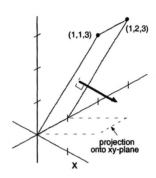

9. $f(x,y) = \cos x \sin y$, $0 \leq x, y \leq 2\pi$

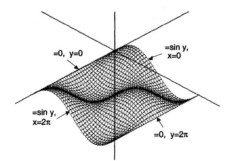

11. $f(x,y) = \frac{xy(x^3+y^3)}{x^2+y^2}$, $|x| \leq 1$, $|y| \leq 1$

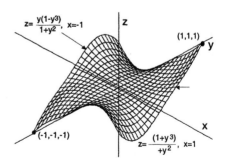

(COMMENT: The difficulty at $(0,0)$ in Exercise 11 is that $f(0,0)$ is not defined. There are several ways to get around this. The graph above avoided $(0,0)$ by simply graphing from $x = -1$ to $x = -0.01$ and then from $x = 0.01$ to $x = 1$, all the while allowing y to vary freely between -1 and 1. The grid-like pattern easily masked the fact that the x-interval $(-0.01, 0.01)$ was ignored.

13. $P(x,y) = H(x)H(y)H(1-y)H(y-x)$

15. $P(x,y) = H(1-x^2-y^2)H(x-y)$

17. $P(x,y) = H(x)H(1-x)H(y)H(1-y)$

19. $f(x,y) = y^2 - x^2, 0 \le x \le y \le 2$

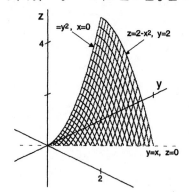

21. $f(x,y) = \frac{\cos(x^2+y^2)}{1+x^2+y^2}, x^2 + y^2 \le 2$

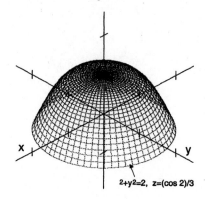

23.

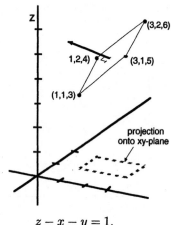

$z - x - y = 1,$
$1 \le x \le 3, 1 \le y \le 2$

25.

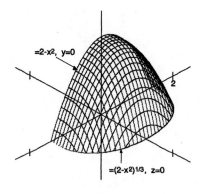

$f(x,y) = 2 - x^2 - y^3,$
$y > 0, f(x,y) > 0$

164

Exercise Set 2D (pg. 198)

1.

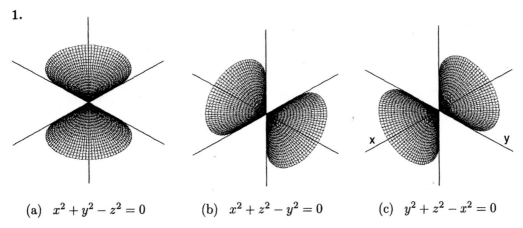

(a) $x^2 + y^2 - z^2 = 0$ (b) $x^2 + z^2 - y^2 = 0$ (c) $y^2 + z^2 - x^2 = 0$

3. For the ellipsoid of level 2, the pertinent equation is $x^2/a^2 + y^2/b^2 + z^2/c^2 = 2$, or equivalently, $x^2/(a\sqrt{2})^2 + y^2/(b\sqrt{2})^2 + z^2/(c\sqrt{2})^2 = 1$. So, the length of each axis increases by a factor of $\sqrt{2}$.

For the elliptic paraboloid of level 1, the pertinent equation is $x^2/a^2 + y^2/b^2 - cz = 1$, or equivalently, $x^2/a^2 + y^2/b^2 = c(z + 1/c)$. Since the general form of an elliptic paraboloid with vertex at (x_0, y_0, z_0) is $(x - x_0)^2/a^2 + (y - y_0)^2/b^2 = c(z - z_0)$, if follows that the basic dimensions and orientation of the surface stay the same for any level. Only the vertex changes. In this case the level 1 vertex is $(0, 0, -1/c)$. Similar comments hold for the generic hyperbolic paraboloid of level 1, except we then speak of a saddle point and not a vertex.

5. (a) A plane perpendicular to the z-axis has an equation of the form $z = k$ for some constant k. Setting $z = k$ in the equation for H_2 gives $(x/a)^2 + (y/b)^2 - (k/c)^2 = -1$, or equivalently, $(x/a)^2 + (y/b)^2 = (k/c)^2 - 1$. This equation has no real solutions if $|k| < c$ (i.e., the plane does not intersect the quadric). And if $|k| = c$ then $(x/a)^2 + (y/b)^2 = 0$, which can be considered to be a *degenerate* ellipse in $\mathcal{R}^2$. Otherwise, $|k| > c$ and the equation can be written as

$$\frac{x^2}{(a\sqrt{(k/c)^2 - 1})^2} + \frac{y^2}{(b\sqrt{(k/c)^2 - 1})^2} = 1,$$

which is the standard form for an ellipse in $\mathcal{R}^2$.

(b) A plane perpendicular to the x-axis has an equation of the form $x = k$ for some constant k. Setting $x = k$ in the equation for H_2 gives $(k/a)^2 + (y/b)^2 - (z/c)^2 = -1$, which can be written as

$$\frac{z^2}{(c\sqrt{(k/a)^2 + 1})^2} - \frac{y^2}{(b\sqrt{(k/a)^2 + 1})^2} = 1,$$

which is the standard form for an hyperbola in $\mathcal{R}^2$ that opens up and down.

A similar analysis (with a similar result) can be done for the case where the plane is perpendicular the y-axis.

7. The distance $d_1(x, y, z)$ from the point (x, y, z) to the plane $z = -1$ is the length of the line segment perpendicular to the plane with endpoints (x, y, z) and $(x, y, -1)$. This distance is

$$d_1(x, y, z) = \sqrt{(x - x)^2 + (y - y)^2 + (z + 1)^2} = |z + 1|.$$

On the other hand, the distance $d_2(x,y,z)$ from the point (x,y,z) to the point $(0,0,1)$ is

$$d_2(x,y,z) = \sqrt{(x-0)^2 + (y-0)^2 + (z-1)^2}$$
$$= \sqrt{x^2 + y^2 + (z-1)^2}.$$

Since (x,y,z) must satisfy $d_1(x,y,z) = d_2(x,y,z)$, $|z+1| = \sqrt{x^2 + y^2 + (z-1)^2}$. Squaring both sides, expanding and rearranging, gives

$$x^2 + y^2 = 4z,$$

an elliptic paraboloid. This is the desired quadric surface Q and is shown in the figure on the right.

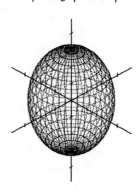

9. $x^2/4 - y^2/16 + z^2/4 = 1$ **11.** $x^2/4 - y^2/4 + z^2/9 = 1$ **13.** $x^2/4 + y^2/4 + z^2/9 = 1$

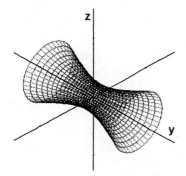
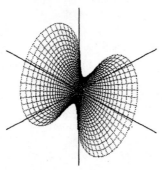

hyperboloid of one sheet   hyperboloid of one sheet   ellipsoid

15. $x^2 - y^2/4 = z/4$ **17.** $x^2/4 - y^2/4 = -z + 1$

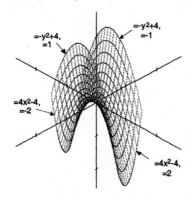
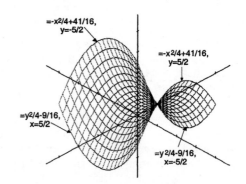

hyperbolic paraboloid   hyperbolic paraboloid

19. $x^2/4 + z^2/9 = y$

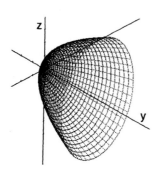

elliptic paraboloid

21. $x^2 + y = 0$

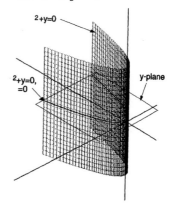

parabolic cylinder

23. $y^2 + z^2 = 1$

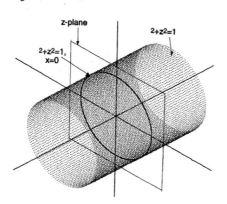

circular cylinder

25. $x^2 + z = 1$

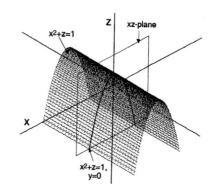

parabolic cylinder

27. $x^2 + 4z^2 = 4$

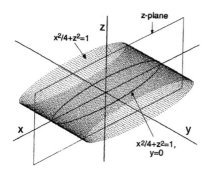

elliptic cylinder

167

Section 3: Partial Derivatives

Exercise Set 3ABC (pgs. 203-204)

1. If $f(x,y) = x^2 + x\sin(x+y)$ then

$$\frac{\partial f}{\partial x} = 2x + x\cos(x+y) + \sin(x+y) \qquad \text{and} \qquad \frac{\partial f}{\partial y} = x\cos(x+y).$$

3. If $f(x,y) = e^{x+y+1}$ then

$$\frac{\partial f}{\partial x} = e^{x+y+1} \qquad \text{and} \qquad \frac{\partial f}{\partial y} = e^{x+y+1}.$$

5. If $f(x,y) = x^y$ then

$$\frac{\partial f}{\partial x} = yx^{y-1} \qquad \text{and} \qquad \frac{\partial f}{\partial y} = x^y \ln x.$$

7. If $f(x,y) = x^2y + xy^2$ then $f_x = \partial f/\partial x = 2xy + y^2$ and $f_y = \partial f/\partial y = x^2 + 2xy$. At the point $(a,b) = (1,-1)$, we have $f(1,-1) = 0$, $f_x(1,-1) = -1$ and $f_y(1,-1) = -1$. By Eq. 3.2, the equation of the tangent plane to the graph of f at the point $(1,-1)$ is given by

$$z = f(1,-1) + (x-1)f_x(1,-1) + (y+1)f_y(1,-1) = -(x-1) - (y+1) = -x - y,$$

or equivalently, $x + y + z = 0$.

9. If $f(x,y) = 1/(x^2 + y^2)$ then $f_x = \partial f/\partial x = -2x/(x^2+y^2)^2$ and $f_y = \partial f/\partial y = -2y/(x^2+y^2)^2$. At the point $(1,1)$, we have $f(1,1) = \frac{1}{2}$, $f_x(1,1) = -\frac{1}{2}$ and $f_y(1,1) = -\frac{1}{2}$. By Eq. 3.2, the equation of the tangent plane to the graph of f at the point $(1,1)$ is given by

$$z = f(1,1) + (x-1)f_x(1,1) + (y-1)f_y(1,1) = \frac{1}{2} - \frac{1}{2}(x-1) - \frac{1}{2}(y-1) = -\frac{1}{2}x - \frac{1}{2}y + \frac{3}{2},$$

or equivalently, $x + y + 2z = 3$.

11. if $f(x,y) = xy + x^2y^3$ then $\partial f/\partial x = y + 2xy^3$ and $\partial f/\partial y = x + 3x^2y^2$, so that

$$\frac{\partial^2 f}{\partial y \partial x} = \frac{\partial}{\partial y}\left(\frac{\partial f}{\partial x}\right) = 1 + 6xy^2 \qquad \text{and} \qquad \frac{\partial^2 f}{\partial x \partial y} = \frac{\partial}{\partial x}\left(\frac{\partial f}{\partial y}\right) = 1 + 6xy^2.$$

13. If $f(x,y) = 1/(x^2 + y^2)$ then $\partial f/\partial x = -2x/(x^2+y^2)^2$ and $\partial f/\partial yx = -2y/(x^2+y^2)^2$, so that

$$\frac{\partial^2 f}{\partial y \partial x} = \frac{\partial}{\partial y}\left(\frac{\partial f}{\partial x}\right) = \frac{8xy}{(x^2+y^2)^3} \qquad \text{and} \qquad \frac{\partial^2 f}{\partial x \partial y} = \frac{\partial}{\partial x}\left(\frac{\partial f}{\partial y}\right) = \frac{8xy}{(x^2+y^2)^3}.$$

15. With $f(x,y,z) = x^2e^{x+y+z}\cos y$, there are three first-order partial derivatives:

$$\frac{\partial f}{\partial x} = 2xe^{x+y+z}\cos y + x^2e^{x+y+z}\cos y = (2x + x^2)e^{x+y+z}\cos y,$$

$$\frac{\partial f}{\partial y} = x^2e^{x+y+z}\cos y - x^2e^{x+y+z}\sin y = x^2e^{x+y+z}(\cos y - \sin y),$$

$$\frac{\partial f}{\partial z} = x^2e^{x+y+z}\cos y.$$

168

17. With $f(x, y, z, w) = (x^2 - y^2)/(z^2 + w^2)$, there are four first-order partial derivatives:

$$\frac{\partial f}{\partial x} = \frac{2x}{z^2 + w^2}, \qquad \frac{\partial f}{\partial y} = -\frac{2y}{z^2 + w^2}, \qquad \frac{\partial f}{\partial z} = \frac{2z(y^2 - x^2)}{(z^2 + w^2)^2}, \qquad \frac{\partial f}{\partial w} = \frac{2w(y^2 - x^2)}{(z^2 + w^2)^2}.$$

19. With $f(x, y, z) = x + 2yz$, there are three first-order partial derivatives:

$$\frac{\partial f}{\partial x} = 1, \qquad \frac{\partial f}{\partial y} = 2z, \qquad \frac{\partial f}{\partial z} = 2y.$$

21. If $f(x, y) = \log(x + y)$ then

$$\frac{\partial^3 f}{\partial x^2 \partial y} = \frac{\partial}{\partial x^2}\left(\frac{\partial f}{\partial y}\right) = \frac{\partial}{\partial x^2}\left(\frac{1}{x + y}\right)$$
$$= \frac{\partial}{\partial x}\left(\frac{\partial}{\partial x}\left(\frac{1}{x + y}\right)\right) = \frac{\partial}{\partial x}\left(-\frac{1}{(x + y)^2}\right) = \frac{2}{(x + y)^3}.$$

23. If $f(x, y) = x^3 - 3xy^2$ then

$$f_{xx} = \frac{\partial}{\partial x}\left(\frac{\partial f}{\partial x}\right) = \frac{\partial}{\partial x}(3x^2 - 3y^2) = 6x, \qquad f_{yy} = \frac{\partial}{\partial y}\left(\frac{\partial f}{\partial y}\right) = \frac{\partial}{\partial y}(-6xy) = -6x.$$

Hence, $f_{xx} + f_{yy} = 0$.

25. If $f(x, y) = e^x \cos y$ then

$$f_{xx} = \frac{\partial}{\partial x}\left(\frac{\partial f}{\partial x}\right) = \frac{\partial}{\partial x}(e^x \cos y) = e^x \cos y$$

and

$$f_{yy} = \frac{\partial}{\partial y}\left(\frac{\partial f}{\partial y}\right) = \frac{\partial}{\partial y}(-e^x \sin y) = -e^x \cos y.$$

Hence, $f_{xx} + f_{yy} = 0$.

27. If $f(x_1, \ldots, x_n) = 1/(x_1^2 + \cdots + x_n^2)^{(n-2)/2}$ then, for $1 \leq k \leq n$,

$$f_{x_k x_k} = \frac{\partial}{\partial x_k}\left(\frac{\partial f}{\partial x_k}\right) = \frac{\partial}{\partial x_k}\left(-\frac{(n-2)x_k}{(x_1^2 + \cdots + x_n^2)^{n/2}}\right)$$
$$= \frac{-(n-2)x_1^2 - \cdots + (n-1)(n-2)x_k^2 - \cdots - (n-2)x_n^2}{(x_1^2 + \cdots + x_n^2)^{(n+2)/2}}.$$

Hence, for each k, the numerator in the sum $f_{x_1 x_1} + \cdots + f_{x_n x_n}$ has $n-1$ copies of $-(n-2)x_k^2$, and one copy of $(n-1)(n-2)x_k^2$, so that the term involving x_k^2 vanishes. It follows that $f_{x_1 x_1} + \cdots + f_{x_n x_n} = 0$.

29. If $f(x, y) = \sqrt{1 - x^2 - y^2}$ then $f_x = -x/\sqrt{1 - x^2 - y^2}$ and $f_y = -y/\sqrt{1 - x^2 - y^2}$. Thus, $f_x(1/2, 1/2) = -\sqrt{2}/2$ and $f_y(1/2, 1/2) = -\sqrt{2}/2$. By Eq. 3.2, the tangent plane to the graph of f at the point $(1/2, 1/2, \sqrt{2}/2)$ is given by

$$z = f(1/2, 1/2) + (x - 1/2)f_x(1/2, 1/2) + (y - 1/2)f_y(1/2, 1/2)$$
$$= \frac{\sqrt{2}}{2} - \frac{\sqrt{2}}{2}(x - 1/2) - \frac{\sqrt{2}}{2}(y - 1/2) = -\frac{\sqrt{2}}{2}x - \frac{\sqrt{2}}{2}y + \sqrt{2}.$$

Thus, we let $\Gamma(x,y) = z$ be the function defined by $\Gamma(x,y) = -\frac{\sqrt{2}}{2}x - \frac{\sqrt{2}}{2}y + \sqrt{2}$, and observe that the graph of Γ is the desired tangent plane. The graphs of f and Γ are shown below.

31. Let $f(x,y) = e^{-x^2-y^2}$, so that $f_x = -2xe^{-x^2-y^2}$ and $f_y = -2ye^{-x^2-y^2}$. Then $f_x(0,0) = 0$ and $f_y(0,0) = 0$. By Eq. 3.2, the tangent plane to the graph of f at the point $(0,0,1)$ is given by

$$z = f(0,0) + (x-0)f_x(0,0) + (y-0)f_y(0,0) = 1 + (x-0)(0) + (y-0)(0) = 1.$$

Thus, we let $\Gamma(x,y) = z$ be the constant function $\Gamma(x,y) = 1$ and observe that the graph of Γ is the desired tangent plane. The graphs of f and Γ are shown below.

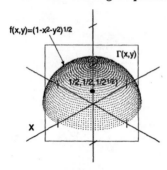

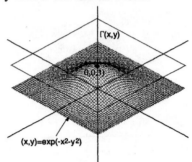

Exercise 29   Exercise 31

33. From Exercise 29, $z = -(\sqrt{2}/2)x - (\sqrt{2}/2)y + \sqrt{2}$ is the equation of the relevant tangent plane, and $(1/2, 1/2, \sqrt{2}/2)$ is the relevant point of tangency. Writing the equation of the tangent plane as $x + y + \sqrt{2}\,z = 2$, we see that the vector $(1,1,\sqrt{2})$ is normal to the plane. It follows that, since $(1/2, 1/2, \sqrt{2}/2)$ must be on the desired line, a parametric representation for the desired line is
$$g(t) = (1,1,\sqrt{2})t + (1/2, 1/2, \sqrt{2}/2).$$

35. For $u(x,t) = t^{-1/2}e^{-x^2/t}$, we have

$$u_{xx} = \frac{\partial}{\partial x}\left(\frac{\partial}{\partial x}\left(t^{-1/2}e^{-x^2/t}\right)\right) = \frac{\partial}{\partial x}\left(-2t^{-3/2}xe^{-x^2/t}\right) = 4t^{-5/2}x^2e^{-x^2/t} - 2t^{-3/2}e^{-x^2/t}$$

$$u_t = \frac{\partial}{\partial t}\left(t^{-1/2}e^{-x^2/t}\right) = t^{-5/2}x^2e^{-x^2/t} - \frac{1}{2}t^{-3/2}e^{-x^2/t}.$$

Therefore, $u_{xx} = 4u_t$ and $u(x,y)$ satisfies the diffusion equation.

37. With $y(x,t) = \cosh(x+t)$, we easily get $y_{xx} = \cosh(x+t)$ and $y_{tt} = \cosh(x+t)$. That is, $y_{xx} = y_{tt}$ and $u(x,y)$ satisfies the wave equation.

39. With $u(x,y) = e^x\cos y$ and $v(x,y) = e^x\sin y$, we have four first-order partials

$$u_x(x,y) = e^x\cos y, \quad u_y(x,y) = -e^x\sin y, \quad v_x(x,y) = e^x\sin y, \quad v_y(x,y) = e^x\cos y.$$

Hence, $u_x(x,y) = v_y(x,y)$ and $u_y(x,y) = -v_x(x,y)$, and the two given functions satisfy the Cauchy-Riemann equations.

41. The domain D of $u(x,y) = x^3 - y^3$ is the xy-plane. Moreover, $u_{xx} = 6x$ on D and $u_{yy} = -6y$ on D. Since $u_{xx}(x,y) + u_{yy}(x,y) = 6x - 6y \neq 0$ on D, u is not harmonic on D.

43. The domain D of $u(x,y) = x^3 - 3xy^2$ is the xy-plane. Moreover, $u_{xx} = 6x$ on D and $u_{yy} = -6x$ on D, so that $u_{xx}(x,y) + u_{yy}(x,y) = 0$ on D. That is, u is harmonic on D.

45. The domain D of $u(x,y) = \sin(x-y)$ is the xy-plane. Moreover, $u_{xx} = -\sin(x-y)$ on D and $u_{yy} = -\sin(x-y)$ on D, so that $u_{xx}(x,y) + u_{yy}(x,y) = -2\sin(x-y) \neq 0$ on D. That is, u is not harmonic on D.

47. The domain D of $u(x,y) = \arctan(y/x)$ is the xy-plane with the y-axis deleted (since x can't be zero). Moreover, $u_{xx} = 2xy/(x^2+y^2)^2$ on D and $u_{yy} = -2xy/(x^2+y^2)^2$ on D, so that $u_{xx}(x,y) + u_{yy}(x,y) = 0$ on D. That is, u is harmonic on D.

49. The domain D of $u(x,y) = \sin(x+y)$ is the xy-plane. Moreover, $u_{xx} = -\sin(x+y)$ on D and $u_{yy} = -\sin(x+y)$ on D. Since $u_{xx}(x,y) + u_{yy}(x,y) = -2\sin(x+y) \neq 0$ on D, u is not harmonic on D.

51. Let $u(x,y)$ be harmonic on a domain D, and let $(x_0, y_0) \in D$ be arbitrarily chosen and fixed. The definition of "harmonic" then says that,

$$u_{xx}(x_0, y_0) + u_{yy}(x_0, y_0) = 0.$$

Evidently, if either double partial is zero then the other is also zero. Thus, assuming that neither is zero, the above displayed equation shows that the real numbers $u_{xx}(x_0, y_0)$ and $u_{yy}(x_0, y_0)$, which measure the concavities of the coordinate curves $u(x, y_0)$ and $u(x_0, y)$ in the planes $y = y_0$ and $x = x_0$ (resp.), must be the negatives of each other at the point $(x_0, y_0, u(x_0, y_0))$ on the graph of u where they intersect. That is, the concavities of $u(x, y_0)$ and $u(x_0, y)$ have opposite directions at $(x_0, y_0, u(x_0, y_0))$. Since $(x_0, y_0) \in D$ is arbitrary, at each point $(x_0, y_0, u(x_0, y_0))$ on the graph of u, either both double partials are zero or the concavities of the corresponding coordinate curves have opposite directions.

Section 4: Parametrized Surfaces

Exercise Set 4AB (pgs. 210-212)

1. If $f(x,y) = \begin{pmatrix} x+y \\ x-y \\ x^2+y^2 \end{pmatrix}$ then $f_x(x,y) = \begin{pmatrix} 1 \\ 1 \\ 2x \end{pmatrix}$ and $f_y(x,y) = \begin{pmatrix} 1 \\ -1 \\ 2y \end{pmatrix}$.

3. If $f(x,y) = \begin{pmatrix} xy \\ x+y \end{pmatrix}$ then $f_x(x,y) = \begin{pmatrix} y \\ 1 \end{pmatrix}$ and $f_y(x,y) = \begin{pmatrix} x \\ 1 \end{pmatrix}$.

5. If $f(x,y) = (e^x, e^y, e^{x+y})$ then $f_x(x,y) = (e^x, 0, e^{x+y})$ and $f_y(x,y) = (0, e^y, e^{x+y})$.

7. Let $g(u,v) = (u, v, u^2+v^2)$, so that $g_u(u,v) = (1,0,2u)$ and $g_v(u,v) = (0,1,2v)$. In particular, at the point $(u_0,v_0) = (1,1)$, we have $g_u(1,1) = (1,0,2)$ and $g_v(1,1) = (0,1,2)$. Setting $(x,y,z) = g(u,v)$, we see that $x = u$ and $y = v$. Therefore, the coordinate curve defined by $g(u,1)$ is the graph of $z = x^2+1$ and lies in the plane $y = 1$; and the curve defined by $g(1,v)$ is the graph of $z = y^2+1$ and lies in the plane $x = 1$. The overall surface defined by g is the elliptic paraboloid shown on the right.

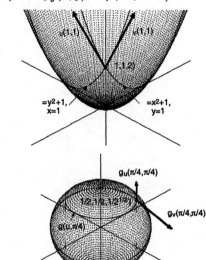

9. If $g(u,v) = (\cos u \sin v, \sin u \sin v, \cos v)$ then

$$g_u(u,v) = (-\sin u \sin v, \cos u \sin v, 0),$$

$$g_v(u,v) = (\cos u \cos v, \sin u \cos v, -\sin v).$$

In particular, at the point $(u_0,v_0) = (\pi/4, \pi/4)$,

$$g_u(\pi/4, \pi/4) = (-1/2, 1/2, 0)$$

$$\text{and} \quad g_v(\pi/4, \pi/4) = (1/2, 1/2, -\sqrt{2}/2).$$

Setting $(x,y,z) = g(u,v)$, we see that $x = \cos u \sin v$, $y = \sin u \sin v$ and $z = \cos v$. Therefore, the coordinate curve defined by $g(u, \pi/4)$ is the circle $x^2+y^2 = 1/2$ and lies in the plane $z = 1/\sqrt{2}$; and the coordinate curve defined by $g(\pi/4, v)$ is a circle of radius 1, centered at the origin, and lies in the plane $x = y$. The overall surface is the sphere of radius 1 shown above on the right.

11. The given function g and the given point of tangency are the same as in Exercise 7 above. Using that solution, the tangent plane at $(1,1)$ can be given parametrically as

$$(x,y,z) = sg_u(1,1) + tg_v(1,1) + g(1,1) = s(1,0,2) + t(0,1,2) + (1,1,2)$$

$$= (s+1, t+1, 2s+2t+2).$$

In terms of x,y,z, we have $z = 2s+2t+2 = 2(x-1) + 2(y-1) + 2 = 2x+2y-2$, or equivalently, $2x+2y-z = 2$.

13. The given function g and the given point of tangency are the same as in Exercise 9 above. Using that solution, the tangent plane at $(\pi/4, \pi/4)$ can be given parametrically as

$$(x,y,z) = sg_u(\pi/4, \pi/4) + tg_v(\pi/4, \pi/4) + g(\pi/4, \pi/4)$$

$$= s(-1/2, 1/2, 0) + t(1/2, 1/2, -\sqrt{2}/2) + (1/2, 1/2, \sqrt{2}/2)$$

$$= s(-1,1,0) + t(1,1,-\sqrt{2}) + (1/2, 1/2, \sqrt{2}/2)$$

$$= (-s+t+1/2, s+t+1/2, -\sqrt{2}t + \sqrt{2}/2),$$

where the third equality is obtained by absorbing $1/2$ into the scalars s and t. In terms of x,y,z, the equations $x = -s + t + 1/2$ and $y = s + t + 1/2$ implies $s = (y - x)/2$ and $t = (x+y-1)/2$. So, $z = -\sqrt{2}t + \sqrt{2}/2 = -\sqrt{2}(x+y-1)/2 + \sqrt{2}/2 = -\sqrt{2}(x+y)/2 + 2\sqrt{2}$, or equivalently, $x + y + \sqrt{2}\,z = 2$.

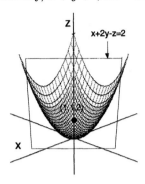

Exercise 11

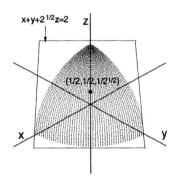

Exercise 13

15. (a) For (x_0, y_0) in $\mathcal{R}^2$, $t \geq 0$, the parametrization $(x, y) = f(x_0, y_0, t) = (x_0 + t, y_0 + t^2)$ is the right half of the parabola $y = (x - x_0)^2 + y_0$, where $t = 0$ corresponds to the vertex (x_0, y_0). The paths starting at $(0, 0)$, $(1, 0)$ and $(1, 1)$ are show below on the left.

(b) In general, the vector partial derivative $f_t(-1, 0, t)$ can be interpreted geometrically as the velocity at time t of the moving point on the path starting at $(-1, 0)$ in the xy-plane. The direction of this vector is parallel to the line tangent to the path at time t, and the length of this vector is the slope of the tangent line. The path starting at $(-1, 0)$, as well as the vector derivatives $f_t(-1, 0, 0)$ and $f_t(-1, 0, 1)$, is shown below on the right.

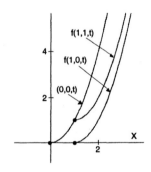

Exercise 15(a)

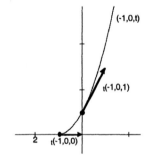

Exercise 15(b)

17. Let $f(x, y) = (u, v) = \left(x, (x + y)^2/(4x)\right)$, $x \neq 0$.

(a) The coordinate functions are $f_1(x, y) = x$, $x \neq 0$, and $f_2(x, y) = (x + y)^2/(4x)$, $x \neq 0$.

(b) We have

$$f_x(x, y) = \left(1, \frac{2(x + y) - (x + y)^2}{4x^2}\right) \qquad \text{and} \qquad f_y(x, y) = \left(0, \frac{x + y}{2x}\right),$$

both of which are defined for $x \neq 0$.

173

(c) The given region (call it R) is a square with vertices $(0,0)$, $(4,4)$, $(8,0)$ and $(4,-4)$. To simplify matters, if $y = mx + b$ is a line then $f(x,y) = \left(x, ((m+1)x+b)^2/(4x)\right)$, so that $u = x$ and $v = ((m+1)x+b)^2/(4x)$ implies $v = ((m+1)u+b)^2/(4u)$, $u \neq 0$. That is,

f maps the line $y = mx + b$ in the xy-plane

to the curve $v = \dfrac{((m+1)u+b)^2}{4u}$ in the uv-plane.

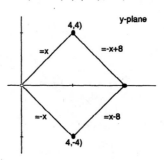

On the boundary piece associated with $y = x$, we have $m = 1$, $b = 0$, and x satisfies $0 < x \leq 4$. The image curve in the uv-plane (call it A) is then $v = u$, for $0 < u \leq 4$. On the boundary piece associated with $y = 8 - x$, we have $m = -1$, $b = 8$, and x satisfies $4 \leq x \leq 8$. The image curve in the uv-plane (call it B) is then $v = 16/u$, for $4 \leq u \leq 8$. On the boundary piece associated with $y = x - 8$, we have $m = 1$, $b = -8$, and x satisfies $4 \leq x \leq 8$. The image curve in the uv-plane (call it C) is then $v = u - 8 + 16/u$, for $4 \leq u \leq 8$. And finally, on the boundary piece associated with $y = -x$, we have $m = -1$, $b = 0$, and x satisfies $0 < x \leq 4$. The image curve in the uv-plane (call it D) is then $v = 0$, for $0 < u \leq 4$. The image curves A, B, C, D bound the desired region $f(R)$. For clarity, the region R and its image $f(R)$ are shown in the accompanying sketch.

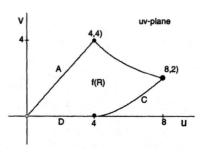

19. Given the parametrization $\mathbf{x}(u,v) = (au\cos v, bu\sin v, u)$ of the elliptic cone in Example 10, we have

$$\mathbf{x}_u(x,y) = (a\cos v, b\sin v, 1) \qquad \text{and} \qquad \mathbf{x}_v(u,v) = (-au\sin v, bu\cos v, 0),$$

where we observe that the six coordinate functions of $\mathbf{x}_u$ and $\mathbf{x}_v$ are continuous on the entire uv-plane. Further, we have

$$\mathbf{x}_u \times \mathbf{x}_v = (a\cos v, b\sin v, 1) \times (-au\sin v, bu\cos v, 0) = (-bu\cos v, -au\sin v, abu)$$
$$= u(-b\cos v, -a\sin v, ab).$$

At points on the cone that are different from the pointed tip, $u \neq 0$. Thus, the cross product is non-zero. This means that neither of the vector partial derivatives is the zero vector and they are not parallel to each other; i.e., neither is a scalar multiple of the other and therefore they are linearly independent. So, the definition of "smooth" implies that the cone is smooth at all points different from the pointed tip. However, at the pointed tip, $\mathbf{x}(u,v)$ is the zero vector, which implies $u = 0$. Hence, $\mathbf{x}_v(u,v)$ is the zero vector and is therefore linearly dependent with $\mathbf{x}_u(u,v)$. Again, by the definition of "smooth", the cone is not smooth at its pointed tip

21. Let $g(u,v) = \left(u, v, \sqrt{u^2+v^2}\right)$.

(a) Setting $x = u$, $y = v$ and $z = \sqrt{u^2+v^2}$, we have

$$x^2 + y^2 = u^2 + v^2 = \left(\sqrt{u^2+v^2}\right)^2 = z^2, \qquad z \geq 0,$$

which is the top half of a right circular cone.

174

(b) For the given parametrization g, we have

$$g_u(u,v) = \left(1, 0, u/\sqrt{u^2 + v^2}\right) \qquad \text{and} \qquad g_v(u,v) = \left(0, 1, v/\sqrt{u^2 + v^2}\right),$$

and note that the third coordinate function of each partial derivative is not continuous at $(u, v) = (0, 0)$. So, the pointed end of the cone is singular with respect to the parametrization g because of a discontinuity at that point.

On the other hand, it was shown in Exercise 19 above that the pointed tip of the cone is singular with respect to the parametrization $\mathbf{x}(u, v)$. However, there it was singular because one of the partial derivatives was the zero vector.

(c) It was shown in part (b) that the tip of the cone is singular with respect to the parametrization g. Thus, it suffices to show that the surface of the cone is smooth with respect to g at all other points. To do this, let (u_0, v_0) be different from $(0, 0)$, and note (from part (b)) that $g_u(u_0, v_0)$ and $g_v(u_0, v_0)$ are nonzero vectors. Therefore, since

$$g_u(u_0, v_0) \times g_v(u_0, v_0) = \left(-\frac{u_0}{\sqrt{u_0^2 + v_0^2}}, -\frac{v_0}{\sqrt{u_0^2 + v_0^2}}, 1 \right)$$

is not the zero vector, $g_u(u_0, v_0)$ and $g_v(u_0, v_0)$ are non-parallel nonzero vectors; i.e., they are linearly independent. So, the surface of the cone is smooth with respect to g at all points different from $(0, 0, 0)$.

23. Let $a > b > 0$, and let T be the surface parametrized by

$$\mathbf{x}(u, v) = \begin{pmatrix} x \\ y \\ z \end{pmatrix} = \begin{pmatrix} (a + b\cos v)\cos u \\ (a + b\cos v)\sin u \\ b\sin v \end{pmatrix}, \qquad 0 \leq u \leq 2\pi, \qquad 0 \leq v \leq 2\pi.$$

(a) For fixed $v = v_0$, $x = (a + b\cos v_0)\cos u$, $y = (a + b\cos v_0)\sin u$ and $z = b\sin v_0$. Eliminating u from the first two equations gives the curve traced out by $\mathbf{x}(u, v_0)$; namely, the circle $x^2 + y^2 = (a + b\cos v_0)^2$ in the plane $z = b\sin v_0$ parallel to the xy-plane.

For fixed $u = u_0$, $x = (a + b\cos v)\cos u_0$, $y = (a + b\cos v)\sin u_0$ and $z = b\sin v$. Eliminating v from the first two equations gives the plane on which the curve $\mathbf{x}(u_0, v)$ is traced; namely, $x\sin u_0 - y\cos u_0 = 0$, which contains the z-axis. Moreover, the position vector $\mathbf{x}(u_0, v)$ can be written as the sum of two vectors; namely,

$$\mathbf{x}(u_0, v) = \begin{pmatrix} a\cos u_0 \\ a\sin u_0 \\ 0 \end{pmatrix} + \begin{pmatrix} (b\cos v)\cos v_0 \\ (b\cos v)\sin u_0 \\ b\sin v \end{pmatrix},$$

where we note that, as a position vector, the first vector on the right side above (call it $\mathbf{a}$) corresponds to a point (also denoted by $\mathbf{a}$) in the plane $x\sin u_0 - y\cos u_0 = 0$. Thus, for each $0 \leq v \leq 2\pi$, the points $\mathbf{a}$ and $\mathbf{x}(u_0, v)$ both lie in the plane $x\sin u_0 - y\cos u_0 = 0$, and the distance between them is

$$|\mathbf{x}(u_0, v) - \mathbf{a}| = \sqrt{b^2\cos^2 v\cos^2 u_0 + b^2\cos^2 v\sin^2 u_0 + b^2\sin^2 v} = b.$$

It follows that, the curve traced out by $\mathbf{x}(u_0, v)$ is a circle of radius b, centered at $\mathbf{a}$, that lies in the plane $x\sin u_0 - y\cos u_0 = 0$. Note that as $u = u_0$ takes on different fixed values, the radius of the circles remains the same; namely, b.

The two families of circles obtained above allow us to conclude that T is a torus, with outer radius $a + b$ (set $v = 0$ in the first family of circles), and inner radius $a - b$ (set $v = \pi$ in the first family of circles).

(b) Eliminating u and v, we get

$$\left((x^2 + y^2 + z^2) - (a^2 + b^2)\right)^2$$
$$= \left((a + b\cos v)^2 \cos^2 u + (a + b\cos v)^2 \sin^2 u + b^2 \sin^2 v - (a^2 + b^2)\right)^2$$
$$= \left((a + b\cos v)^2 + b^2 \sin^2 v - (a^2 + b^2)\right)^2$$
$$= \left(2ab\cos v + b^2 \cos^2 v + b^2 \sin^2 v - b^2\right)^2$$
$$= \left(2ab\cos v\right)^2$$
$$= 4a^2 b^2 \cos^2 v$$
$$= 4a^2 b^2 (1 - sin^2 v)$$
$$= 4a^2 (b^2 - b^2 \sin^2 v)$$
$$= 4a^2 (b^2 - z^2),$$

25. (a) The given parametrizations $\mathbf{z}(u, v)$ and $\mathbf{x}(u, v)$ are obviously different (their first two coordinates are interchanged). As a surface in $\mathcal{R}^3$, we have the two designations

$$x = a\cos v, \quad y = b\cos u \sin v, \quad z = c\sin u \sin v$$
$$\text{and} \quad x = a\cos u \sin v, \quad y = b\cos v, \quad z = c\sin u \sin v.$$

The first parametrization implies

$$(x/a)^2 + (y/b)^2 + (z/c)^2 = \cos^2 v + \cos^2 u \sin^2 v + \sin^2 u \sin^2 v$$
$$= \cos^2 v + (\cos^2 u + \sin^2 u) \sin^2 v = 1,$$

and the second parametrization implies

$$(x/a)^2 + (y/b)^2 + (z/c)^2 = \cos^2 u \sin^2 v + \cos^2 v + \sin^2 u \sin^2 v$$
$$= (\cos^2 u + \sin^2 u) \sin^2 v + \cos^2 v = 1.$$

Hence, both parametrizations coincide with the ellipsoid $(x/a)^2 + (y/b)^2 + (z/c)^2 = 1$

(b) To show that the surface has elliptic cross sections perpendicular to the x-axis, we must fix a plane parallel to the yz-plane. The parametrization $\mathbf{z}(u, v)$ is the logical choice since its first coordinate is in terms of v only. Set $v = v_0$, so that

$$\mathbf{z}(u, v_0) = (a\cos v_0, b\cos u \sin v_0, c\sin u \sin v_0).$$

Using only the last two coordinates, the cross section of the ellipsoid in the plane $x = a\cos v_0$ satisfies

$$(y/b)^2 + (z/c)^2 = \cos^2 u \sin^2 v_0 + \sin^2 u \sin^2 v_0 = (\cos^2 u + \sin^2 u) \sin^2 v_0 = \sin^2 v_0.$$

In standard form, $\left(y/(b\sin v_0)\right)^2 + \left(z/(c\sin v_0)\right)^2 = 1$ is an ellipse in the plane $x = a\cos v_0$ (unless v_0 is an integer multiple of π, in which case the curve is a degenerate ellipse).

Similarly, to show that the surface has elliptic cross sections perpendicular to the y-axis, we use the parametrization $\mathbf{x}(u, v)$. Set $v = v_0$, so that

$$\mathbf{x}(u, v_0) = (a \cos u \sin v_0, b \cos v_0, c \sin u \sin v_0).$$

Using only the first and third coordinates, the cross section of the ellipsoid in the plane $y = b \cos v_0$ satisfies

$$(x/a)^2 + (z/c)^2 = \cos^2 u \sin^2 v_0 + sin^2 u \sin^2 v_0 = (\cos^2 u + \sin^2 u) \sin^2 v_0 = \sin^2 v_0.$$

In standard form, $\left(x/(a \sin v_0)\right)^2 + \left(z/(c \sin v_0)\right)^2 = 1$ is an ellipse in the plane $y = b \cos v_0$ (unless v_0 is an integer multiple of π, in which case the curve is a degenerate ellipse)

27. Suppose $(y, z) = \left(g(v), h(v)\right)$ parametrizes a curve in the yz-plane of $\mathcal{R}^3$, and let S be the surface of revolution obtained by rotating the curve about the z-axis. In order to show that S is parametrized by $\mathbf{x}(u, v) = \left(g(v) \cos u, g(v) \sin u, h(v)\right)$, it suffices to show that, for each $v = v_0$, $\mathbf{x}(u, v_0)$ traces out a circle of radius $g(v_0)$ about the z-axis on the plane $z = h(v_0)$.

To this end, fix $v = v_0$, so that $\mathbf{x}(u, v_0) = \left(g(v_0) \cos u, g(v_0) \sin u, h(v_0)\right)$. We now write $\mathbf{x}(u, v_0)$ as a linear combination of two mutually perpendicular unit vectors; namely,

$$\mathbf{x}(u, v_0) = g(v_0)(\cos u, \sin u, 0) + h(v_0)(0, 0, 1).$$

The first term $g(v_0)(\cos u, \sin u, 0)$ parametrizes a circle in the xy-plane with radius $g(v_0)$ centered on the z-axis. The second term $h(v_0)(0, 0, 1)$ is a constant vector of length $|h(v_0)|$ parallel to the z-axis. Thus, $\mathbf{x}(u, v_0)$ traces out a circle of radius $g(v_0)$ in the plane $z = h(v_0)$.

Exercise Set 4C (pg. 213)

1. The parametrization f (without the restrictions on u and v) gives a sphere of radius 1 centered at $(0,0,0)$. The restrictions on u and v give that portion of the surface lying in the first octant. The figure is shown below.

3. When the surface defined by the parametrization $f(u,v) = (e^{u-v}, u, v)$, $0 \le u \le 2$, $0 \le v \le 2$, is projected onto the yz-plane, one obtains a square with vertices $(0,0,0)$, $(0,2,0)$, $(0,2,2)$ and $(0,0,2)$. In the figure below, a few points on the surface are indicated to give the viewer a sense of its orientation.

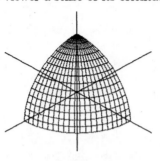

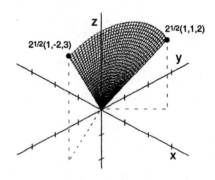

Exercise 1   Exercise 3

5. Without the restrictions on u and v, the given parametrization f defines a plane that contains the origin. In fact, it is not hard to show that the plane is given by $7x - y - 3z = 0$. The main difficulty in graphing the surface lies in the fact that u and v are restricted to the closed quarter-disk $u^2 + y^2 \le 2$, $0 \le u$, $0 \le v$. The figure at the right was done by converting to polar coordinates by letting $u = r\cos\theta$ and $v = r\sin\theta$. The pair of conditions $0 \le u$ and $0 \le v$ then translates to $0 \le \theta \le \pi/2$; and the condition $u^2 + v^2 \le 2$ translates to $0 \le r \le \sqrt{2}$. This makes the surface amenable to being easily rendered via the plotting program presented in this section.

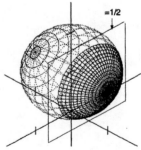

7. In the parametrizations of Exer. 1 and 2, u is the nonnegative angle made by the positive x-axis and the projection of the vector f onto the xy-plane; and v is the nonnegative angle made by the positive z-axis and the vector f. Thus, if $u = u_0$ is held fixed while v varies then $f(u_0, v)$ traces out a circle of radius 1 in a plane containing the z-axis; and if $v = v_0$ is held fixed while u varies then $f(u, v_0)$ traces out a circle in the plane $z = \cos v_0$ (for Exercise 1) and $z = 1 + \cos v_0$ (for Exercise 2).

9. The given parametrization (without the restriction on y) is a sphere of radius 1 centered at $(0,0,1)$ in $\mathcal{R}^3$. The "end-piece" with $1/2 \le y \le 1$ is shown on the right. In order to give the viewer a better perspective, the entire sphere is also shown, as well as the plane $y = 1/2$.

11. It was observed in Exercise 5 that the given parametrized surface (without the restrictions on x, y, and z) is the plane $7x - y - 3z = 0$. That portion of the plane inside the rectangular box $|x| \le 1$, $|y| \le 2$, $|z| \le 3$ is shown on the next page.

13. The parametrization $(x, y, z) = \big(a \cos u \cosh v, b \sin u \cosh .v, c \sinh v\big)$ coincides with the hyperboloid of one sheet $(x/a)^2 + (y/b)^2 - (z/c)^2 = 1$ (see Exercise 26(b), Ch.4, Sec.4B). It follows that the hyperboloid of one sheet $x^2 + y^2 - 2z^2 = 1$ is parametrized by

$$(x, y, z) = \big(\cos u \cosh v, \sin u \cosh .v, (1/\sqrt{2}) \sinh v\big).$$

15. If $c < 0$, the parametrization $(x, y, z) = (a \cos u \sinh v, b \sin u \sinh v, c \cosh v)$ coincides with one sheet $(z \leq c)$ of the hyperboloid of two sheets $(z/c)^2 - (x/a)^2 - (y/b)^2 = 1$ (see Exercise 26(c), Ch.4, Sec.4B). It follows that the sheet below the xy-plane of the hyperboloid of two sheets $z^2 - x^2 - y^2 = 1$ is parametrized by

$$(x, y, z) = (\cos u \sinh v, \sin u \sinh v, -\cosh v).$$

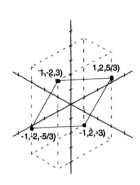

Exercise 11

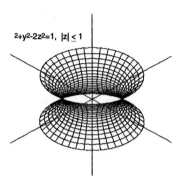

Exercise 13:
hyperboloid of one sheet, $|z| \leq 1$

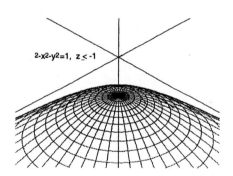

Exercise 15:
hyperboloid of two sheets, $z \leq -1$

179

Chapter Review

1. With $\mathbf{x}(t) = (e^t \cos t)\mathbf{i} + (e^t \sin t)\mathbf{j} + e^t\mathbf{k}$, we have

$$\dot{\mathbf{x}}(t) = e^t(\cos t - \sin t)\mathbf{i} + e^t(\cos t + \sin t)\mathbf{j} + e^t\mathbf{k} \quad \text{and} \quad \ddot{\mathbf{x}}(t) = (-2e^t \sin t)\mathbf{i} + (2e^t \cos t)\mathbf{j} + e^t\mathbf{k}.$$

The unit vector $\mathbf{t}(t)$ that points in the same direction as $\dot{\mathbf{x}}(t)$ is

$$\mathbf{t}(t) = \frac{\dot{\mathbf{x}}(t)}{|\dot{\mathbf{x}}(t)|} = \frac{e^t(\cos t - \sin t)\mathbf{i} + e^t(\cos t + \sin t)\mathbf{j} + e^t\mathbf{k}}{\sqrt{e^{2t}(\cos t - \sin t)^2 + e^{2t}(\cos t + \sin t)^2 + e^{2t}}}$$

$$= \frac{e^t(\cos t - \sin t)\mathbf{i} + e^t(\cos t + \sin t)\mathbf{j} + e^t\mathbf{k}}{e^t\sqrt{3}}$$

$$= \frac{1}{\sqrt{3}}(\cos t - \sin t)\mathbf{i} + \frac{1}{\sqrt{3}}(\cos t + \sin t)\mathbf{j} + \frac{1}{\sqrt{3}}\mathbf{k}.$$

3. Write $\mathbf{x}(t) = ((5/\sqrt{2})\sin t, (5/\sqrt{2})\sin t, 5\cos t)$ as the position of the particle at time t.

(a) We have

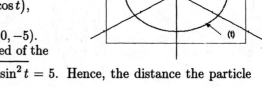

$$\dot{\mathbf{x}}(t) = ((5/\sqrt{2})\cos t, (5/\sqrt{2})\cos t, -5\sin t),$$

$$\ddot{\mathbf{x}}(t) = (-(5/\sqrt{2})\sin t, -(5/\sqrt{2})\sin t, -5\cos t),$$

so that $\dot{\mathbf{x}}(0) = (5/\sqrt{2}, 5/\sqrt{2}, 0)$ and $\ddot{\mathbf{x}}(0) = (0, 0, -5)$.

(b) With $\dot{\mathbf{x}}(t)$ as found in part (a), the speed of the particle is $|\dot{\mathbf{x}}(t)| = \sqrt{\frac{25}{2}\cos^2 t + \frac{25}{2}\cos^2 t + 25\sin^2 t} = 5$. Hence, the distance the particle travels from $t = 0$ to $t = 2\pi$ is

$$\int_0^{2\pi} |\dot{\mathbf{x}}(t)|dt = \int_0^{2\pi} 5\,dt = 5t\Big|_0^{2\pi} = 10\pi.$$

(c) With $x = (5/\sqrt{2})\sin t$, $y = (5/\sqrt{2})\sin t$ and $z = 5\cos t$, we have $x^2 + y^2 + z^2 = 25$, so that all points on the given curve lie on the sphere of radius 5 centered at the origin.

(d) Let t_1 and t_2 be any two times such that $t_1 - t_2$ is not an integer multiple of π. Then

$$\mathbf{x}(t_1) \times \mathbf{x}(t_2) = \left(\frac{25}{\sqrt{2}}(\sin t_1 \cos t_2 - \cos t_1 \sin t_2), \frac{25}{\sqrt{2}}(\sin t_2 \cos t_1 - \cos t_2 \sin t_1), 0\right)$$

$$= \left(\frac{25}{\sqrt{2}}\sin(t_1 - t_2), \frac{25}{\sqrt{2}}\sin(t_2 - t_1), 0\right)$$

$$= \left(\frac{25}{\sqrt{2}}\sin(t_1 - t_2)\right)(1, -1, 0).$$

The restriction placed on $t_1 - t_2$ shows that any two position vectors satisfying the restriction lie in a plane with $(1, -1, 0)$ as a normal vector. It follows that the curve lies in a plane in $\mathcal{R}^3$ and that an equation for this plane is $x - y = 0$. All of the details are shown in the above figure.

5. With $\mathbf{x}(t) = (a\cos t, b\sin t, ct)$, where a, b and c are constants and $a > b > 0$, the speed of the curve traced by $\mathbf{x}(t)$ is the length of the vector $\dot{\mathbf{x}}(t) = (-a\sin t, b\cos t, c)$. That is,

$$\text{speed} = |\dot{\mathbf{x}}(t)| = |(-a\sin t, b\cos t, c)| = \sqrt{a^2\sin^2 t + b^2\cos^2 t + c^2}.$$

The maximum and minimum values of the speed occur at those times t where the first derivative of the speed is zero, or at times when the first derivative of the speed does not exist. Since

$$\frac{d}{dt}|\dot{\mathbf{x}}(t)| = \frac{(a^2 - b^2)\sin t\cos t}{\sqrt{a^2\sin^2 t + b^2\cos^2 t + c^2}} = \frac{(a^2 - b^2)\sin 2t}{2\sqrt{a^2\sin^2 t + b^2\cos^2 t + c^2}}$$

is continuous everywhere, the maximum and minimum speeds will occur when $d|\dot{\mathbf{x}}(t)|/dt = 0$. This occurs if, and only if, $t = n\pi/2$ for some integer n. If $n = 2k$ is even then $\sin t = \sin k\pi = 0$ and $\cos t = \cos k\pi = (-1)^k$, so that $|\dot{\mathbf{x}}(k\pi)| = \sqrt{b^2 + c^2}$. Similarly, if $n = 2k + 1$ is odd then $\sin t = \sin((2k+1)\pi/2) = (-1)^k$ and $\cos t = \cos((2k+1)\pi/2) = 0$, so that $|\dot{\mathbf{x}}((2k+1)\pi/2)| = \sqrt{a^2 + c^2}$. Since $a > b$, it follows that, regardless of the value of c,

$$\max\{|\dot{\mathbf{x}}(t)|\} = \sqrt{a^2 + c^2} \quad \text{and} \quad \min\{|\dot{\mathbf{x}}(t)|\} = \sqrt{b^2 + c^2}.$$

7. Consider the curve parametrized by $\mathbf{x}(t) = (e^t\cos t, e^t\sin t)$.

(a) That portion of the curve for the parameter interval $0 \le t \le \pi/2$ is shown below on the right and is denoted by λ_1.

(b) With λ_1 as given in part (a), the length $l(\lambda_1)$ of λ_1 is given by

$$l(\lambda_1) = \int_0^{\pi/2} |\dot{\mathbf{x}}(t)|dt = \int_0^{\pi/2} |(e^t(\cos t - \sin t), e^t(\cos t + \sin t))|dt$$

$$= \int_0^{\pi/2} \sqrt{e^{2t}(\cos t - \sin t)^2 + e^{2t}(\cos t + \sin t)^2}\, dt$$

$$= \int_0^{\pi/2} e^t\sqrt{2}\, dt = e^t\sqrt{2}\Big|_0^{\pi/2} = (e^{\pi/2} - 1)\sqrt{2}.$$

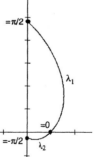

(c) Let λ_2 be that portion of the curve for the parameter interval $-\pi/2 \le t \le 0$. Then the length $l(\lambda_2)$ of λ_2 is given by

$$l(\lambda_2) = \int_{-\pi/2}^0 |\dot{\mathbf{x}}(t)|dt = e^t\sqrt{2}\Big|_{-\pi/2}^0 = (1 - e^{-\pi/2})\sqrt{2},$$

where the second equality follows from part (b). In order that their relative sizes can be more easily seen, λ_1 and λ_2 are shown on the same set of axes in the figure on the right.

9. Let $f(x, y) = x^2 + y^3$. Then $f_x(x, y) = 2x$ and $f_y(x, y) = 3y^2$, and an equation for the tangent plane at the point $(1, 2, 9)$ is given by

$$z = f(1, 2) + (x - 1)f_x(1, 2) + (y - 2)f_y(1, 2) = 9 + (x - 1)2 + (y - 2)12 = 2x + 12y - 17.$$

11. With $u(x,y) = x^2 - y^2$, we have $u_x = 2x$, $u_{xx} = 2$, $u_y = -2y$ and $u_{yy} = -2$. Thus, $u_{xx} + u_{yy} = 2 - 2 = 0$.

13. With $u(x,y) = x^3y - xy^3$, we have $u_x = 3x^2y - y^3$, $u_{xx} = 6xy$, $u_y = x^3 - 3xy^2$ and $u_{yy} = -6xy$. Thus, $u_{xx} + u_{yy} = 6xy - 6xy = 0$.

15. With $u(x,y) = \arctan(y/x)$ and $x \neq 0$, we have

$$u_x = -\frac{y}{x^2 + y^2}, \quad u_{xx} = \frac{2xy}{(x^2 + y^2)^2}, \quad \text{and} \quad u_y = \frac{x}{x^2 + y^2}, \quad u_{yy} = -\frac{2xy}{(x^2 + y^2)^2}.$$

Thus, $u_{xx} + u_{yy} = 2xy/(x^2 + y^2)^2 - 2xy/(x^2 + y^2)^2 = 0$.

17. With $u(x,y,z) = xyz + 2x^2 - y^2 - z^2$, we have $u_x = yz + 4x$, $u_{xx} = 4$, $u_y = xz - 2y$, $u_{yy} = -2$, $u_z = xy - 2z$ and $u_{zz} = -2$. Thus, $u_{xx} + u_{yy} + u_{zz} = 4 - 2 - 2 = 0$.

19. Let $u(x,y,z) = (x^2 + y^2 + z^2)^\alpha$, where α is a constant (to be determined) such that $u(x,y,z)$ satisfies the Laplace equation in three dimensions. We have

$$u_x = 2x\alpha(x^2 + y^2 + z^2)^{\alpha-1}, \quad u_{xx} = 2\alpha\Big((2\alpha - 1)x^2 + y^2 + z^2\Big)(x^2 + y^2 + z^2)^{\alpha-2},$$

$$u_y = 2y\alpha(x^2 + y^2 + z^2)^{\alpha-1}, \quad u_{yy} = 2\alpha\Big(x^2 + (2\alpha - 1)y^2 + z^2\Big)(x^2 + y^2 + z^2)^{\alpha-2},$$

$$u_z = 2z\alpha(x^2 + y^2 + z^2)^{\alpha-1}, \quad u_{zz} = 2\alpha\Big(x^2 + y^2 + (2\alpha - 1)z^2\Big)(x^2 + y^2 + z^2)^{\alpha-2}.$$

Thus,

$$\begin{aligned}
u_{xx} + u_{yy} + u_{zz} &= 2\alpha\Big((2\alpha + 1)x^2 + (2\alpha + 1)y^2 + (2\alpha + 1)z^2\Big)(x^2 + y^2 + z^2)^{\alpha-2} \\
&= 2\alpha(2\alpha + 1)(x^2 + y^2 + z^2)(x^2 + y^2 + z^2)^{\alpha-2} \\
&= 2\alpha(2\alpha + 1)(x^2 + y^2 + z^2)^{\alpha-1}
\end{aligned}$$

The last expression above can be identically zero for all (x,y,z) in the domain of u if, and only if, $2\alpha(2\alpha + 1) = 0$. Hence, $u(x,y,z) = (x^2 + y^2 + z^2)^\alpha$ satisfies the Laplace equation in three dimensions if, and only if, $\alpha = 0$ or $\alpha = -1/2$.

21. (a) Let $z = k$ be a plane perpendicular to the z-axis. The intersection of the elliptic cone $(x/a)^2 + (y/b)^2 - z^2 = 0$ and this plane is found by setting $z = k$ in the elliptic cone equation. This gives

$$(x/a)^2 + (y/b)^2 - k^2 = 0, \quad \text{or equivalently,} \quad \big(x/(ak)\big)^2 + \big(y/(bk)\big)^2 = 1,$$

which holds for $k \neq 0$. In this case, we have an ellipse in the plane $z = k$ with axes of lengths $2a|k|$ and $2b|k|$. In case $k = 0$, the resulting curve is a single point $(0,0,0)$, which can be regarded as a degenerate ellipse.

(b) Let $x = k$ be a plane perpendicular to the x-axis. The intersection of the cone with this plane is found by setting $x = k$ in the elliptic cone equation. This gives

$$(k/a)^2 + (y/b)^2 - z^2 = 0, \quad \text{or equivalently,} \quad \big(z/(k/a)\big)^2 - \big(y/(bk/a)\big)^2 = 1,$$

which holds for $k \neq 0$. In this case, we have an hyperbola in the plane $x = k$ that opens up and down with vertex at $(k,0,0)$. In case $k = 0$, the resulting curve consists of the two lines $z = \pm y/b$ (which can be regarded as a degenerate hyperbola). The curves obtained by intersecting the cone with a plane perpendicular to the y-axis are similar, except that the roles of x and y are interchanged, as are the roles of a and b.

23. (a) Consider the curve of intersection of the elliptic paraboloid $(x/a)^2 + (y/b)^2 = z$ and the plane $z = k$. Setting $z = k$ in the equation of the elliptic paraboloid gives

$$(x/a)^2 + (y/b)^2 = k.$$

If $k < 0$ then the plane does not intersect the elliptic paraboloid; and if $k = 0$, the curve reduces to a single point $(0, 0, 0)$. But if $k > 0$ then the curve has the equation $(x/ka)^2 + (y/bk)^2 = 1$, which is an ellipse in the plane $z = k$, with axes of lengths $2ka$ and $2kb$, and centered at the point $(0, 0, k)$ in $\mathcal{R}^3$.

(b) Let $x = k$ be a plane perpendicular to the x-axis. Its intersection with the given elliptic paraboloid can be found by setting $x = k$ in the equation of the paraboloid. This gives

$$(k/a)^2 + (y/b)^2 = z,$$

which is the equation of a parabola in the plane $x = k$, opening upward, with vertex at the point $(k, 0, k^2/a^2)$. A similar result is obtained by considering planes perpendicular to the y-axis.

25. With $y = f(x)$, the parabola P can be written as $y - x^2 = 0$, which we recognize as a level set (level 0) of the function $F(x, y) = y - x^2$. So, F is an implicit representation for P in $\mathcal{R}^2$.

27. The function $h : \mathcal{R}^1 \longrightarrow \mathcal{R}^2$, defined by $h(x) = (x, x^2)$, has the parabola P as its graph in $\mathcal{R}^2$. So h is an explicit representation of P in $\mathcal{R}^2$. The graph of P is shown below.

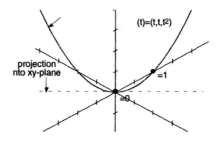

Exercise 27

183

Chapter 5: Differentiability

Section 1: LIMITS AND CONTINUITY

Exercise Set 1ABC (pgs. 224-225)

1.

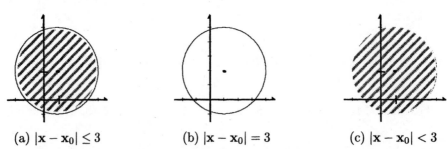

(a) $|\mathbf{x} - \mathbf{x_0}| \leq 3$ (b) $|\mathbf{x} - \mathbf{x_0}| = 3$ (c) $|\mathbf{x} - \mathbf{x_0}| < 3$

3. The given set $S = \{\, \mathbf{x} \in \mathcal{R}^2 \,|\, |\mathbf{x} - (1,2)| < 0.5 \,\}$ is the set of points lying inside the circle $\mathcal{C} : (x-1)^2 + (y-2)^2 = 0.25$. The interior of S is S and, therefore, S is open; the boundary of S is $\mathcal{C}$.

5. Let $S = \{\, (x,y) \,|\, 0 < x < 3, \; 0 < y < 2 \,\}$. The interior of S is S and, therefore, S is open. The boundary of S consists of the four line segments l_1, l_2, l_3, l_4, where l_1 has endpoints $(0,0)$ and $(3,0)$; l_2 has endpoints $(3,0)$ and $(3,2)$; l_3 has endpoints $(3,2)$ and $(0,2)$; and l_4 has endpoints $(0,2)$ and $(0,0)$.

7. The set $S = \{\, (x,y) \,|\, x^2 + 2y^2 < 1 \,\}$ contains all points inside the ellipse $\mathcal{E} : x^2 + 2y^2 = 1$. The interior of S is S and, therefore, S is open. The boundary of S is $\mathcal{E}$.

9. The set $S = \{\, (x,y) \,|\, x^2 + y^2 > 0 \,\}$ is the xy-plane with the origin deleted. The interior of S is S and, therefore, S is open. The boundary of S consists of the single point $(0,0)$.

11. The given set $S = \{\, (x,y) \,|\, x > y \,\}$ is the region below the line $y = x$ in the xy-plane. The interior of S is S and, therefore, S is open. The boundary of S consists of all points on the line $y = x$.

13. Lines and planes in $\mathcal{R}^3$ are not open subsets of $\mathcal{R}^3$ because no point on a line or a plane is an interior point. For example, in the case of a line, if $\mathbf{x_0}$ is a point on the line then every neighborhood of $\mathbf{x_0}$ contains points not on the line. Hence, no point on a line is an interior point (in fact, they are all boundary points). A similar observation holds for planes.

15. Since $\begin{pmatrix} 1 & 3 \\ 0 & 2 \end{pmatrix} \begin{pmatrix} x \\ y \end{pmatrix} = \begin{pmatrix} x + 3y \\ 2y \end{pmatrix}$, the given function can be written as $f(x,y) = \begin{pmatrix} x + 3y \\ 2y \end{pmatrix}$. Thus, the domain space and the range space are both of dimension 2 (i.e., $n = m = 2$). The real-valued coordinate functions of f are $f_1(x,y) = x + 3y$ and $f_1(x,y) = 2y$.

17. The domain space of $f(t) = (t, t^2, t^3, t^4)$ has dimension 1, and the range space has dimension 4 (i.e., $n = 1$, $m = 4$). The real-valued coordinate functions of f are $f_1(t) = t$, $f_2(t) = t^2$, $f_3(t) = t^3$ and $f_4(t) = t^4$.

19. The domain space and range space of $f(x,y,z) = (2x, 2y, 2z)$ are both of dimension 3 (i.e., $n = m = 3$). The real valued coordinate functions of f are $f_1(x,y,z) = 2x$, $f_2(x,y,z) = 2y$ and $f_3(x,y,z) = 2z$.

21. The coordinate functions of the given function f are

$$f_1(x,y) = \frac{y}{x^2 + 1} \qquad \text{and} \qquad f_2(x,y) = \frac{x}{y^2 - 1}.$$

184

f_1 is continuous everywhere on the xy-plane. f_2 is continuous on the xy-plane except on the horizontal lines $y = \pm 1$, where $\lim_{\mathbf{x} \to \mathbf{x}_0} f_2(\mathbf{x})$ fails to exist. It follows that $\lim_{\mathbf{x} \to \mathbf{x}_0} f(\mathbf{x})$ fails to exist for $\mathbf{x}_0$ on the horizontal lines $y = \pm 1$. (NOTE: For points of the form $\mathbf{x}_0 = (x_0, \pm 1)$, where $x_0 \neq 0$, $\lim_{\mathbf{x} \to \mathbf{x}_0} f_2(\mathbf{x})$ fails to exist because $f_2(\mathbf{x}_0)$ is infinitely large. But for points of the form $\mathbf{x}_0 = (0, \pm 1)$, $\lim_{\mathbf{x} \to \mathbf{x}_0} f_2(\mathbf{x})$ fails to exist because its value depends on the direction from which we approach $\mathbf{x}_0$.)

23. There is only one coordinate function of the given function f; namely,

$$f(x, y) = \begin{cases} x/\sin x + y, & \text{if } x \neq 0; \\ 2 + y, & \text{if } x = 0. \end{cases}$$

For points $\mathbf{x}_0$ on the vertical lines $l_n : x = n\pi$ (n an integer), $\lim_{\mathbf{x} \to \mathbf{x}_0} f(\mathbf{x})$ fails to exist, but this limit does exist at all other points in the xy-plane.

25. The coordinate functions of the given function f are

$$f_1(u, v) = \frac{uv}{1 - u^2 - v^2} \qquad \text{and} \qquad f_2(u, v) = \frac{1}{2 - u^2 - v^2}.$$

f_1 is continuous everywhere on the uv-plane except on the circle $\mathcal{C}_1 : u^2 + v^2 = 1$. For points $\mathbf{x}_0$ on $\mathcal{C}_1$, $\lim_{\mathbf{x} \to \mathbf{x}_0} f(\mathbf{x})$ fails to exist. f_2 is continuous everywhere on the uv-plane except on the circle $\mathcal{C} : u^2 + v^2 = 2$. For points $\mathbf{x}_0$ on $\mathcal{C}_2$, $\lim_{\mathbf{x} \to \mathbf{x}_0} f(\mathbf{x})$ fails to exist. It follows that f is continuous everywhere on the uv-plane except on the two circles $\mathcal{C}_1$ and $\mathcal{C}_2$, and $\lim_{\mathbf{x} \to \mathbf{x}_0} f(\mathbf{x})$ fails to exist for points $\mathbf{x}_0$ on these two circles.

27. The coordinate functions of the given function f are $f_1(u, v) = 3u - 4v$ and $f_2(u, v) = u + 8$, both of which are continuous on the uv-plane. Thus, f is continuous on the uv-plane.

29. The given function f is continuous on the xy-plane *except* possibly at $(0, 0)$. However, it was shown in Example 6 that $\lim_{\mathbf{x} \to (0,0)} f(x, y)$ fails to exist, so that f can't be continuous at $(0, 0)$ (regardless of how $f(0, 0)$ is defined).

31. The function $f(\mathbf{x}) = |\mathbf{x}|/(1 - |\mathbf{x}|^2)$ is continuous everywhere on $\mathcal{R}^n$ except at points $\mathbf{x}$ with $|\mathbf{x}| = 1$. Specifically, $\lim_{\mathbf{x} \to \mathbf{x}_0} f(\mathbf{x})$ does not exist whenever $|\mathbf{x}_0| = 1$.

33. (a) The translation $T : \mathcal{R}^2 \longrightarrow \mathcal{R}^2$ defined by $T(x, y) = (x, y) + (1, 1)$ for (x, y) in $\mathcal{R}^2$ moves a point in the xy-plane a distance of $|(1, 1)| = \sqrt{2}$ units in the direction of the vector $(1, 1)$.

(b) Let $\mathbf{y}_0 = (y_1, \ldots, y_n)$ be a constant vector in $\mathcal{R}^n$ and let $T(\mathbf{x}) = \mathbf{x} + \mathbf{y}_0$ be a translation of $\mathcal{R}^n$ by $\mathbf{y}_0$. If $\mathbf{x} = (x_1, \ldots, x_n)$ then $T(\mathbf{x}) = (x_1 + y_1, \ldots, x_n + y_n)$, so that the coordinate functions of T are

$$T_k(\mathbf{x}) = x_k + y_k, \qquad k = 1, \ldots, n.$$

The functions x_k are coordinate projections and are therefore continuous by Theorem 1.3; and the functions y_k are continuous because they're constant functions. Hence, each T_k is the sum of two continuous functions and is therefore continuous. By Theorem 1.2, T is continuous.

35. Let $O_1, \ldots, O_m$ be open subsets of $\mathcal{R}^n$ and let $S = \bigcap_{k=1}^m O_k$. If $\mathbf{x}_0 \in S$ then $\mathbf{x}_0 \in O_k$, for $k = 1, \ldots, m$. Hence, for each k, there exists a neighborhood $\mathcal{N}_k$ of $\mathbf{x}_0$ such that $\mathcal{N}_k \subset O_k$. Let ϵ_k be the radius of the k-th neighborhood, and let $\epsilon = \min\{\epsilon_k\}$. Letting $\mathcal{N}$ denote the neighborhood associated with ϵ, we have $\mathcal{N} \subset \mathcal{N}_k$, for each k. Hence, $\mathcal{N} \subset O_k$ for each k, so that $\mathcal{N} \subset S$. Thus, $\mathbf{x}_0$ is an interior point of S. As $\mathbf{x}_0$ was an arbitrary point of S, all points of S are interior points. Hence, S is open.

37. Let $P_k : \mathcal{R}^n \longrightarrow \mathcal{R}$ be the k-th coordinate projection and pick $\mathbf{x}_0 = (\overline{x}_1, \ldots, \overline{x}_n) \in \mathcal{R}^n$. Let $\epsilon > 0$ be given. If $\mathbf{x} = (x_1, \ldots, x_n) \in \mathcal{R}^n$ satisfies $|\mathbf{x} - \mathbf{x}_0| < \epsilon$. Then

$$
\begin{aligned}
|P_k(\mathbf{x}) - P_k(\mathbf{x}_0)| &= |P_k(x_1, \ldots, x_n) - P_k(\overline{x}_1, \ldots, \overline{x}_n)| \\
&= |x_k - \overline{x}_k| = \sqrt{(x_k - \overline{x}_k)^2} \leq \sqrt{(x_1 - \overline{x}_1)^2 + \cdots (x_n - \overline{x}_n)^2} \\
&= |(x_1 - \overline{x}_1, \ldots, x_n - \overline{x}_n)| = |(x_1, \ldots, x_n) - (\overline{x}_1, \ldots, \overline{x}_n)| \\
&= |\mathbf{x} - \mathbf{x}_0| < \epsilon.
\end{aligned}
$$

Thus, given $\epsilon > 0$, the choice $\delta = \epsilon$ is such that $|P_k(\mathbf{x}) - P_k(\mathbf{x}_0)| < \epsilon$ whenever $|\mathbf{x} - \mathbf{x}_0| < \delta$. So, P_k is continuous at $\mathbf{x}_0$. Since $\mathbf{x}_0$ was an arbitrary point in $\mathcal{R}^n$, P_k is continuous.

39. Let f and g be vector functions with the same domain and same range space, and let $\mathbf{x}_0$ be a point in their common domain such that $\lim_{\mathbf{x} \to \mathbf{x}_0} f(\mathbf{x})$ and $\lim_{\mathbf{x} \to \mathbf{x}_0} g(\mathbf{x})$ exist. Then there exist vectors $\mathbf{y}_1$ and $\mathbf{y}_2$ in their common range space such that

$$
\lim_{\mathbf{x} \to \mathbf{x}_0} f(\mathbf{x}) = \mathbf{y}_1 \qquad \text{and} \qquad \lim_{\mathbf{x} \to \mathbf{x}_0} g(\mathbf{x}) = \mathbf{y}_2.
$$

Then, given $\epsilon > 0$, there exists $\delta > 0$ such that $|f(\mathbf{x}) - \mathbf{y}_1| < \epsilon/2$ and $|g(\mathbf{x}) - \mathbf{y}_2| < \epsilon/2$ whenever $|\mathbf{x} - \mathbf{x}_0| < \delta$. Therefore, whenever $|\mathbf{x} - \mathbf{x}_0| < \delta$,

$$
\begin{aligned}
|f(\mathbf{x}) + g(\mathbf{x}) - (\mathbf{y}_1 - \mathbf{y}_2)| &= |f(\mathbf{x}) - \mathbf{y}_1 + g(\mathbf{x}) - \mathbf{y}_2| \\
&\leq |f(\mathbf{x}) - \mathbf{y}_1| + |g(\mathbf{x}) - \mathbf{y}_2| \\
&< \epsilon/2 + \epsilon/2 = \epsilon.
\end{aligned}
$$

Hence, $\lim_{\mathbf{x} - \mathbf{x}_0} (f(\mathbf{x}) + g(\mathbf{x}))$ exists and equals $\mathbf{y}_1 + \mathbf{y}_2$. That is,

$$
\lim_{\mathbf{x} - \mathbf{x}_0} (f(\mathbf{x}) + g(\mathbf{x})) = \mathbf{y}_1 + \mathbf{y}_2 = \lim_{\mathbf{x} \to \mathbf{x}_0} f(\mathbf{x}) + \lim_{\mathbf{x} \to \mathbf{x}_0} g(\mathbf{x}).
$$

41. Let S be an open subset of $\mathcal{R}^n$ and let S^c be its complement.

If S^c has no boundary points then, by default, it contains all of its boundary points and is therefore closed.

Otherwise, let $\mathbf{x}_0$ be a boundary point of S^c. If $\mathbf{x}_0$ is in S^c then there is nothing to show. Thus, assume $\mathbf{x}_0$ is not in S^c; i.e., assume $\mathbf{x}_0$ is in S. On one hand, since S is open, there is a neighborhood $\mathcal{N}$ of $\mathbf{x}_0$ that contains only points of S. On the other hand, since $\mathbf{x}_0$ is a boundary point of S^c, $\mathcal{N}$ must contain points of S^c, a contradiction. It follows that $\mathbf{x}_0$ is in S^c. That is, S^c contains all of its boundary points and is therefore closed.

Section 2: REAL-VALUED FUNCTIONS

Exercise Set 2AB (pgs. 232)

1. For $f(x, y) = x^2 - y^2$ we have $\nabla f(\mathbf{x}) = (f_x, f_y) = (2x, -2y)$.

3. For $f(x, y) = x + 2y$ we have $\nabla f(\mathbf{x}) = (f_x, f_y) = (1, 2)$.

5. For $f(x, y, z) = x + y - z^2$ we have $\nabla f(\mathbf{x}) = (f_x, f_y, f_z) = (1, 1, -2z)$.

7. For $f(x_1, x_2) = x_1^2 + 2x_2^4$ we have $\nabla f(\mathbf{x}) = (f_{x_1}, f_{x_2}) = (2x_1, 8x_2^3)$.

9. For $f(x, y) = x^3 - y^3$ we have $\nabla f(\mathbf{x}) = (3x^2, -3y^2)$. Thus, $\nabla f(1, 1) = (3 - 3)$ and $f(1, 1) = 0$. The tangent approximation $T(x, y)$ at $\mathbf{x}_0 = (1, 1)$ is then

$$T(x, y) = f(1, 1) + \nabla f(1, 1) \cdot \big((x, y) - (1, 1)\big) = 0 + (3, -3) \cdot (x - 1, y - 1)$$
$$= 3(x - 1) - 3(y - 1) = 3x - 3y.$$

Introducing the new variable z, the equation for the tangent plane at $\mathbf{x}_0 = (1, 1)$ is $z = 3x - 3y$.

11. For $f(x, y) = x + 2y$ we have $\nabla f(\mathbf{x}) = (1, 2)$. Thus, $\nabla f(1, 2) = (1, 2)$ and $f(1, 2) = 5$. The tangent approximation $T(x, y)$ at $\mathbf{x}_0 = (1, 2)$ is then

$$T(x, y) = f(1, 2) + \nabla f(1, 2) \cdot \big((x, y) - (1, 2)\big) = 5 + (1, 2) \cdot (x - 1, y - 2)$$
$$= 5 + (x - 1) + 2(y - 2) = x + 2y.$$

Introducing the variable z, the equation of the tangent plane at $\mathbf{x}_0 = (1, 2)$ is $z = x + 2y$.

13. For $f(x, y, z)) = x + y - z^2$ we have $\nabla f(\mathbf{x}) = (1, 1, -2z)$. Thus, $\nabla f(0, 0, 1) = (1, 1, -2)$ and $f(0, 0, 1) = -1$. The tangent approximation at $\mathbf{x}_0 = (0, 0, 1)$ is then

$$T(x, y, z) = f(0, 0, 1) + \nabla f(0, 0, 1) \cdot \big((x, y, z) - (0, 0, 1)\big) = -1 + (1, 1, -2) \cdot (x, y, z - 1)$$
$$= -1 + x + y - 2(z - 1) = x + y - 2z + 1.$$

Introducing the variable w, the equation of the tangent plane at $\mathbf{x}_0$ is therefore $w = x + y - 2z + 1$.

15. For $f(x, y, z) = x + 2yz$ we have $\nabla f(\mathbf{x}) = (1, 2z, 2y)$. Thus, $\nabla f(1, 1, 1) = (1, 2, 2)$ and $f(1, 1, 1) = 3$. The tangent approximation at $\mathbf{x}_0 = (1, 1, 1)$ is then

$$T(x, y, z) = f(1, 1, 1) + \nabla f(1, 1, 1) \cdot \big((x, y, z) - (1, 1, 1)\big)$$
$$= 3 + (1, 2, 2) \cdot (x - 1, y - 1, z - 1)$$
$$= 3 + (x - 1) + 2(y - 1) + 2(z - 1)$$
$$= x + 2y + 2z - 2.$$

Introducing the variable w, the equation of the tangent plane at $\mathbf{x}_0 = (1, 1, 1)$ is $w = x + 2y + 2z - 2$.

17. The domain D of the function $f(x, y) = x^{-2} + y^{-2}$ is the xy-plane with the axes deleted. This is an open set in $\mathcal{R}^2$. Thus, $\mathbf{x}_0 \in D$ implies $\mathbf{x}_0$ is an interior point of the domain of f and therefore condition (i) of Definition 2.1 in the text holds for all points in D. Also, $\nabla f(x, y) = (-2x^{-3}, -2y^{-3})$ exists on all of D, so that if $\mathbf{x}_0$ is in D then the vector $\mathbf{a} = \nabla f(\mathbf{x}_0)$ satisfies the limit equation in the definition and f is differentiable at $\mathbf{x}_0$. We conclude that there are no points in the domain of f at which f fails to be differentiable.

19. The domain D of the function $f(x, y) = |x + y|$ is the xy-plane, so that all points are interior points of the domain of f. That is, condition (i) of Definition 2.1 in the text is satisfied. Moreover, if $x + y \neq 0$ then

$$f(x, y)) = \begin{cases} x + y, & x + y > 0; \\ -x - y, & x + y < 0 \end{cases} \quad \text{implies} \quad \nabla f(x, y) = \begin{cases} (1, 1), & x + y > 0; \\ -(1, 1), & x + y < 0. \end{cases}$$

This shows that $\nabla f(x, y)$ exists at all points (x, y) that are *not* on the line $x + y = 0$. So, if $\mathbf{x}_0$ is such a point then the vector $\mathbf{a} = \nabla f(\mathbf{x}_0)$ satisfies the limit equation in Definition 2.1 and therefore f is differentiable at $\mathbf{x}_0$.

However, f fails to be differentiable on the line $x + y = 0$. To see this, let $\mathbf{x}_0 = (x_0, -x_0)$ be such a point, and consider approaching $\mathbf{x}_0$ along the horizontal line $y = -x_0$ first from the right and then from the left. The value of $\mathbf{a}$ that works in the limit in Definition 2.1 is $\mathbf{a} = (1, 1)$ when approaching from the right and is $\mathbf{a} = -(1, 1)$ when approaching from the left. Thus, there is no single vector $\mathbf{a}$ for which the limit in Definition 2.1 can hold. So condition (ii) does not hold for points on the line $x + y = 0$.

21. Let $f : \mathcal{R}^n \longrightarrow \mathcal{R}$ be defined by $f(\mathbf{x}) = |\mathbf{x}|^2$. With $\mathbf{x} = (x_1, \ldots, x_n)$, we can write $f(\mathbf{x}) = |(x_1, \ldots, x_n)|^2 = x_1^2 + \cdots + x_n^2$, so that

$$\nabla f(\mathbf{x}) = (f_{x_1}, \ldots, f_{x_n}) = (2x_1, \ldots, 2x_n) = 2(x_1, \ldots, x_n) = 2\mathbf{x} \quad \text{for all } \mathbf{x} \in \mathcal{R}^n.$$

23. If $\mathbf{x}$ is the zero vector then both sides of the proposed equation hold because both sides are equal to zero. Thus, assume $\mathbf{x}$ is not the zero vector. Since f is differentiable at $\mathbf{x}_0$,

$$\lim_{t \to 0} \frac{f(\mathbf{x}_0 + t\mathbf{x}) - f(\mathbf{x}_0) - \nabla f(\mathbf{x}_0) \cdot (t\mathbf{x})}{|t\mathbf{x}|} = 0.$$

Writing the denominator as $|t||\mathbf{x}|$ and removing the absolute value sign from t (the sign of t is irrelevant since the numerator goes to zero) gives the equivalent equation

$$\lim_{t \to 0} \frac{1}{|\mathbf{x}|} \left(\frac{f(\mathbf{x}_0 + t\mathbf{x}) - f(\mathbf{x}_0)}{t} - \nabla f(\mathbf{x}_0) \cdot \mathbf{x} \right) = 0.$$

Multiplying throughout by $|\mathbf{x}|$ produces a limit equation that is equivalent to

$$\lim_{t \to 0} \frac{f(\mathbf{x}_0 + t\mathbf{x}) - f(\mathbf{x}_0)}{t} = \nabla f(\mathbf{x}_0) \cdot \mathbf{x}.$$

Thus, the proposed equation holds for nonzero $\mathbf{x}$. We conclude that the proposed equation holds for all vectors $\mathbf{x}$.

25. Let the function f be defined by $f(x) = x^2 \sin(1/x)$, for $x \neq 0$, and $f(0) = 0$. For $x \neq 0$, we can directly compute

$$f'(x) = \frac{d}{dx} \left(x^2 \sin(1/x) \right) = -\cos(1/x) + 2x \sin(1/x).$$

For $x = 0$, we use the definition of a derivative of a real-valued function of a real variable to get

$$f'(0) = \lim_{t \to 0} \frac{f(0 + t) - f(0)}{t} = \lim_{t \to 0} \frac{t^2 \sin(1/t) - 0}{t} = \lim_{t \to 0} t \sin(1/t) = 0.$$

Thus, f is differentiable for all x. However,

$$\lim_{x \to 0} f'(x) = \lim_{x \to 0} \left(-\cos(1/x) + 2x \sin(1/x) \right) = \lim_{x \to 0} \left(-\cos(1/x) \right) + \lim_{x \to 0} \left(2x \sin(1/x) \right).$$

The second limit of the rightmost member exists and equals 0, but the first limit of the rightmost member does not exist. That is, $\lim_{x \to 0} f'(x) \neq f'(0)$, so that f' is not continuous at $x = 0$ and therefore f is not continuously differentiable at $x = 0$.

Section 3: DIRECTIONAL DERIVATIVES

Exercise Set 3AB (pgs. 236-237)

1. For $f(x,y,z) = x^2 + y^2 + z^2$ and $\mathbf{u} = (u_1, u_2, u_3) = (1/\sqrt{3}, 1/\sqrt{3}, 1/\sqrt{3})$ we have

$$\frac{\partial f}{\partial \mathbf{u}}(\mathbf{x}) = u_1 \frac{\partial f}{\partial x}(\mathbf{x}) + u_2 \frac{\partial f}{\partial y}(\mathbf{x}) + u_3 \frac{\partial f}{\partial z}(\mathbf{x}) = \frac{1}{\sqrt{3}} 2x + \frac{1}{\sqrt{3}} 3y + \frac{1}{\sqrt{3}} 2z.$$

Hence,

$$\frac{\partial f}{\partial \mathbf{u}}(1,0,1) = \frac{1}{\sqrt{3}} 2 + \frac{1}{\sqrt{3}} 2 = \frac{4}{\sqrt{3}}.$$

3. For $f(x,y) = x + y$ and $\mathbf{u} = (u_1, u_2) = (1,0)$ we have

$$\frac{\partial f}{\partial \mathbf{u}}(\mathbf{x}) = u_1 \frac{\partial f}{\partial x}(\mathbf{x}) + u_2 \frac{\partial f}{\partial y}(\mathbf{x}) = 1(1) + 0(1) = 1.$$

Hence, $\partial f / \partial \mathbf{u}(2,3) = 1$.

5. Let $f(x,y) = x^2 - y^2$ and let $\mathbf{u} = (1/\sqrt{5}, 2/\sqrt{5})$. Then

$$\frac{\partial f}{\partial \mathbf{u}}(\mathbf{x}) = \nabla f(\mathbf{x}) \cdot \mathbf{u} = (2x, -2y) \cdot (1/\sqrt{5}, 2/\sqrt{5}) = \frac{2x}{\sqrt{5}} - \frac{4y}{\sqrt{5}}.$$

Hence, $\partial f / \partial \mathbf{u}(1,1) = -2/\sqrt{5}$.

7. The curve defined by $g(t) = (t^2, t^3)$ has derivative $g'(t) = (2t, 3t^2)$, so that at $t = 2$ the velocity vector is $g'(2) = (4, 12)$. Since $|(4,12)| = 4\sqrt{10}$, the unit vector pointing in the direction of $g'(2)$ is $\mathbf{u} = (1/\sqrt{10}, 3/\sqrt{10})$. So, for $f(x,y) = e^{x+y}$ we have

$$\frac{\partial f}{\partial \mathbf{u}}(\mathbf{x}) = \nabla f(\mathbf{x}) \cdot \mathbf{u} = (e^{x+y}, e^{x+y}) \cdot (1/\sqrt{10}, 3/\sqrt{10}) = e^{x+y} \frac{4}{\sqrt{10}}.$$

Hence, $\partial f / \partial \mathbf{u}(1,1) = 4e^2/\sqrt{10}$.

9. The curve defined by $g(t) = (3t^2 + t + 1, 2t, t^2)$ has derivative $g'(t) = (6t + 1, 2, 2t)$, so that the tangent vector at $t = 0$ is $g'(0) = (1, 2, 0)$. Since $|(1,2,0)| = \sqrt{5}$, the unit vector pointing in this direction of $g'(0)$ is $\mathbf{u} = (1/\sqrt{5}, 2/\sqrt{5}, 0)$. So, for $f(x,y,z) = x^2 + ye^z$ we have

$$\frac{\partial f}{\partial \mathbf{u}}(\mathbf{x}) = \nabla f(\mathbf{x}) \cdot \mathbf{u} = (2x, e^z, ye^z) \cdot (1/\sqrt{5}, 2/\sqrt{5}, 0) = \frac{2x}{\sqrt{5}} + \frac{2e^z}{\sqrt{5}}.$$

Hence, $\partial f / \partial \mathbf{u}(1,0,0) = 4/\sqrt{5}$.

11. Let $f(x,y,z) = 4x^2y + y^2z$. The unit vector pointing in the direction of the vector $(1,1,1)$ is $\mathbf{u} = (1,1,1)/|(1,1,1)| = (1/\sqrt{3}, 1/\sqrt{3}, 1/\sqrt{3})$, so that

$$\frac{\partial f}{\partial \mathbf{u}}(\mathbf{x}) = \nabla f(\mathbf{x}) \cdot \mathbf{u} = (8xy, 4x^2 + 2yz, y^2) \cdot (1/\sqrt{3}, 1/\sqrt{3}, 1/\sqrt{3}) = \frac{8xy}{\sqrt{3}} + \frac{4x^2 + 2yz}{\sqrt{3}} + \frac{y^2}{\sqrt{3}}.$$

Hence, $\partial f / \partial \mathbf{u}(1,0,1) = 4/\sqrt{3}$.

13. Let S be the surface defined parametrically by $(x, y, z) = g(u, v) = (u^2v, u + v, u)$. If (x_0, y_0, z_0) is a point on S and (u, v) satisfies $g(u, v) = (x_0, y_0, z_0)$ then the vector

$$g_u(u, v) \times g_v(u, v) = (2uv, 1, 1) \times (u^2, 1, 0) = (-1, u^2, 2uv - u^2)$$

is perpendicular to S at (x_0, y_0, z_0). In particular, if $(x_0, y_0, z_0) = (1, 2, 1)$ then comparing coordinates in the equation $g(u, v) = (u^2v, u + v, u) = (1, 2, 1)$ shows that $u = v = 1$. That is, the vector $g_u(, 1, 1) \times g_v(1, 1) = (-1, 1, 1)$ is perpendicular to S at $(1, 2, 1)$. The unit vector pointing in the direction of $(-1, 1, 1)$ is $\mathbf{u} = (-1/\sqrt{3}, 1/\sqrt{3}, 1/\sqrt{3})$. So, for $f(x, y, z) = x^3 + y^2 + z$ we have

$$\frac{\partial f}{\partial \mathbf{u}}(\mathbf{x}) = \nabla f(\mathbf{x}) \cdot \mathbf{u} = (3x^2, 2y, 1) \cdot (-1/\sqrt{3}, 1/\sqrt{3}, 1/\sqrt{3}) = -\frac{3x^2}{\sqrt{3}} + \frac{2y}{\sqrt{3}} + \frac{z}{\sqrt{3}}.$$

Hence, $\partial f / \partial \mathbf{u}(1, 2, 1) = 2/\sqrt{3}$.

15. If $f : \mathcal{R}^2 \to \mathcal{R}$ is differentiable then $\nabla f(\mathbf{x})$ exists for all $\mathbf{x}$ in the domain of f. Thus, for the unit vector $\mathbf{u} = (\cos \alpha, \sin \alpha)$,

$$\frac{\partial f}{\partial \mathbf{u}}(\mathbf{x}) = \nabla f(\mathbf{x}) \cdot \mathbf{u} = \left(\frac{\partial f}{\partial x}, \frac{\partial f}{\partial y} \right) \cdot (\cos \alpha, \sin \alpha) = \cos \alpha \frac{\partial f}{\partial x} + \sin \alpha \frac{\partial f}{\partial y}.$$

17. If $f : \mathcal{R}^n \longrightarrow \mathcal{R}$ is differentiable then $\nabla f(\mathbf{x})$ exists for all $\mathbf{x}$ in the domain of f. Thus, for the unit vector $\mathbf{u} = (\cos \alpha_1, \ldots, \cos \alpha_n)$,

$$\frac{\partial f}{\partial \mathbf{u}}(\mathbf{x}) = \nabla f(\mathbf{x}) \cdot \mathbf{u} = \left(\frac{\partial f}{\partial x_1}, \ldots, \frac{\partial f}{\partial x_n} \right) \cdot (\cos \alpha_1, \ldots, \cos \alpha_n)$$

$$= \cos \alpha_1 \frac{\partial f}{\partial x_1} + \cdots + \cos \alpha_n \frac{\partial f}{\partial x_n}.$$

19. The mean-value formula of Theorem 3.3 is trivial if $\mathbf{x} = \mathbf{y}$. Hence, assume $\mathbf{x} \neq \mathbf{y}$, so that $|\mathbf{y} - \mathbf{x}| \neq 0$. Dividing each side of the mean-value formula of Theorem 3.3 by the scalar $|\mathbf{y} - \mathbf{x}|$ gives

$$\frac{f(\mathbf{y}) - f(\mathbf{x})}{|\mathbf{y} - \mathbf{x}|} = \frac{\nabla f(\mathbf{x}_0) \cdot (\mathbf{y} - \mathbf{x})}{|\mathbf{y} - \mathbf{x}|} = \nabla f(\mathbf{x}_0) \cdot \frac{\mathbf{y} - \mathbf{x}}{|\mathbf{y} - \mathbf{x}|} = \nabla f(\mathbf{x}_0) \cdot \mathbf{u} = \frac{\partial f}{\partial \mathbf{u}}(\mathbf{x}_0),$$

where the second equality follows from the homogeneity of the dot product, and the third equality is obtained by setting $\mathbf{u} = (\mathbf{y} - \mathbf{x})/|\mathbf{y} - \mathbf{x}|$.

21. Let $f(x) = (\sin x, \sin 2x)$ and note that

$$\frac{f(\pi) - f(0)}{\pi - 0} = \frac{(0, 0) - (0, 0)}{\pi} = (0, 0),$$

while $f'(x) = (\cos x, 2 \cos 2x)$ is never the zero vector. In particular,

$$\frac{f(\pi) - f(0)}{\pi - 0} \neq f'(x_0) \quad \text{for any } x_0 \text{ between } x = 0 \text{ and } x = \pi.$$

23. If $f(x, y, z) = e^{x^2+y^2-z^2}$, $\mathbf{x} = (0, 0, 0)$ and $\mathbf{h} = (h, k, l)$ then the second-degree Taylor approximation to $f(\mathbf{x} + \mathbf{h}) = f(h, k, l)$ is given by

$$f(h, k, l) \approx f(0, 0, 0) + \frac{\partial f}{\partial \mathbf{h}}(0, 0, 0) + \frac{1}{2!}\frac{\partial^2 f}{\partial \mathbf{h}^2}(0, 0, 0).$$

By Theorem 3.1,

$$\frac{\partial f}{\partial \mathbf{h}}(\mathbf{x}) = \nabla f(\mathbf{x}) \cdot \mathbf{h} = (2xe^{x^2+y^2-z^2}, 2ye^{x^2+y^2-z^2}, -2ze^{x^2+y^2-z^2}) \cdot (h, k, l)$$

$$= (2hx + 2ky - 2lz)e^{x^2+y^2-z^2};$$

and, using the given definition, the second-order partial of f with respect to $\mathbf{h}$ is

$$\frac{\partial^2 f}{\partial \mathbf{h}^2}(\mathbf{x}) = \frac{\partial}{\partial \mathbf{h}}\left(\frac{\partial f}{\partial \mathbf{h}}\right)(\mathbf{x}) = \nabla\left((2hx + 2ky - 2lz)e^{x^2+y^2-z^2}\right)(\mathbf{x}) \cdot \mathbf{h}$$

$$= \Big((4hx^2 + 4kxy - 4lxz + 2h)e^{x^2+y^2-z^2}, (4hxy + 4ky^2 - 4lyz + 2k)e^{x^2+y^2-z^2},$$

$$-(4hxz + 4kyz - 4lz^2 + 2l)e^{x^2+y^2-z^2}\Big) \cdot (h, k, l).$$

Hence, at $\mathbf{x} = (0, 0, 0)$

$$\frac{\partial f}{\partial \mathbf{h}}(0, 0, 0) = 0 \qquad \text{and} \qquad \frac{\partial^2 f}{\partial \mathbf{h}^2}(0, 0, 0) = (2h, 2k, -2l) \cdot (h, k, l) = 2h^2 + 2k^2 - 2l^2.$$

Since $f(0, 0, 0) = 1$, the desired approximation is

$$f(h, k, l) \approx 1 + 0 + \frac{1}{2!}(2h^2 + 2k^2 - 2l^2) = 1 + h^2 + k^2 - l^2.$$

Section 4: VECTOR-VALUED FUNCTIONS

Exercise Set 4ABC (pgs. 243-245)

1. The coordinate functions of $f(x,y) = \begin{pmatrix} xy \\ x+y \end{pmatrix}$ are

$$f_1(x,y) = xy \qquad \text{and} \qquad f_2(x,y) = x + y.$$

Therefore, the derivative matrix of f is the 2×2 matrix

$$f'(\mathbf{x}) = \begin{pmatrix} \frac{\partial f_1}{\partial x}(\mathbf{x}) & \frac{\partial f_1}{\partial y}(\mathbf{x}) \\ \frac{\partial f_2}{\partial x}(\mathbf{x}) & \frac{\partial f_2}{\partial y}(\mathbf{x}) \end{pmatrix} = \begin{pmatrix} y & x \\ 1 & 1 \end{pmatrix},$$

for each $\mathbf{x} = (x,y)$ in the domain of f.

3. The coordinate functions of $f(x,y,z) = \begin{pmatrix} x + \sin y \\ y + \cos z \\ x + y + z \end{pmatrix}$ are

$$f_1(x,y,z) = x + \sin y, \qquad f_2(x,y,z) = y + \cos z, \qquad f_3(x,y,z) = x + y + z.$$

Therefore, derivative matrix of f is the 3×3 matrix

$$f'(\mathbf{x}) = \begin{pmatrix} \frac{\partial f_1}{\partial x}(\mathbf{x}) & \frac{\partial f_1}{\partial y}(\mathbf{x}) & \frac{\partial f_1}{\partial z}(\mathbf{x}) \\ \frac{\partial f_2}{\partial x}(\mathbf{x}) & \frac{\partial f_2}{\partial y}(\mathbf{x}) & \frac{\partial f_2}{\partial z}(\mathbf{x}) \\ \frac{\partial f_3}{\partial x}(\mathbf{x}) & \frac{\partial f_3}{\partial y}(\mathbf{x}) & \frac{\partial f_3}{\partial z}(\mathbf{x}) \end{pmatrix} = \begin{pmatrix} 1 & \cos y & 0 \\ 0 & 1 & -\sin z \\ 1 & 1 & 1 \end{pmatrix},$$

for each $\mathbf{x} = (x,y,z)$ in the domain of f.

5. The derivative matrix of the function $f(x) = x^2 e^x$ is the 1×1 matrix whose single entry is the ordinary derivative of f with respect to x. Therefore, since the matrix parentheses are unnecessary in writing a 1-by-1 matrix, the derivative matrix of f can be written as

$$f'(x) = \frac{df}{dx} = (x^2 + 2x)e^x = x^2 e^x + 2xe^x.$$

7. The coordinate functions of $f(u,v,w) = \begin{pmatrix} uv \\ vw \\ wu \end{pmatrix}$ are

$$f_1(u,v,w) = uv, \qquad f_2(u,v,w) = vw, \qquad f_3(u,v,w) = wu.$$

Therefore, the derivative matrix of f is the 3×3 matrix

$$f'(\mathbf{x}) = \begin{pmatrix} \frac{\partial f_1}{\partial u}(\mathbf{x}) & \frac{\partial f_1}{\partial v}(\mathbf{x}) & \frac{\partial f_1}{\partial w}(\mathbf{x}) \\ \frac{\partial f_2}{\partial u}(\mathbf{x}) & \frac{\partial f_2}{\partial v}(\mathbf{x}) & \frac{\partial f_2}{\partial w}(\mathbf{x}) \\ \frac{\partial f_3}{\partial u}(\mathbf{x}) & \frac{\partial f_3}{\partial v}(\mathbf{x}) & \frac{\partial f_3}{\partial w}(\mathbf{x}) \end{pmatrix} = \begin{pmatrix} v & u & 0 \\ 0 & w & v \\ w & 0 & u \end{pmatrix},$$

for each $\mathbf{x} = (u,v,w)$ in the domain of f.

9. The coordinate functions of $f(x,y) = \begin{pmatrix} x^2 + y^2 \\ x^2 - y^2 \\ xy \end{pmatrix}$ are

$$f_1(x,y) = x^2 + y^2, \qquad f_2(x,y) = x^2 - y^2, \qquad f_3(x,y) = xy.$$

Therefore, the derivative matrix of f is the 3×2 matrix

$$f'(\mathbf{x}) = \begin{pmatrix} \frac{\partial f_1}{\partial x}(\mathbf{x}) & \frac{\partial f_1}{\partial y}(\mathbf{x}) \\ \frac{\partial f_2}{\partial x}(\mathbf{x}) & \frac{\partial f_2}{\partial y}(\mathbf{x}) \\ \frac{\partial f_3}{\partial x}(\mathbf{x}) & \frac{\partial f_3}{\partial y}(\mathbf{x}) \end{pmatrix} = \begin{pmatrix} 2x & 2y \\ 2x & -2y \\ y & x \end{pmatrix},$$

for each $\mathbf{x} = (x, y)$ in the domain of f.

11. The derivative matrix of the vector function $f\begin{pmatrix} x \\ y \end{pmatrix} = \begin{pmatrix} x^2 - y^2 \\ 2xy \end{pmatrix}$ is given by

$$f'\begin{pmatrix} x \\ y \end{pmatrix} = \begin{pmatrix} 2x & -2y \\ 2y & 2x \end{pmatrix}.$$

This matrix is already evaluated at the point $\begin{pmatrix} x \\ y \end{pmatrix}$.

13. With $f\begin{pmatrix} x \\ y \end{pmatrix}$ and $f'\begin{pmatrix} x \\ y \end{pmatrix}$ as given in Exercise 11 above, we have

$$f'\begin{pmatrix} 1 \\ 0 \end{pmatrix} = \begin{pmatrix} 2 & 0 \\ 0 & 2 \end{pmatrix}.$$

15. The derivative matrix of the function $f\begin{pmatrix} x \\ y \end{pmatrix} = x^2 + y^2$ is the 1×2 matrix

$$f'\begin{pmatrix} x \\ y \end{pmatrix} = (2x, 2y), \quad \text{so that} \quad f'\begin{pmatrix} 1 \\ 1 \end{pmatrix} = (2, 2).$$

17. The derivative matrix of the function $f(t) = \begin{pmatrix} \sin t \\ \cos t \end{pmatrix}$ is the 2×1 matrix

$$f'(t) = \begin{pmatrix} \cos t \\ -\sin t \end{pmatrix}, \quad \text{so that} \quad f'(\pi/4) = \begin{pmatrix} \sqrt{2}/2 \\ -\sqrt{2}/2 \end{pmatrix}.$$

19. The derivative matrix of the function $g(x,y) = \begin{pmatrix} x + y \\ x^2 + y^2 \end{pmatrix}$ is the 2×2 matrix

$$g'(x,y) = \begin{pmatrix} 1 & 1 \\ 2x & 2y \end{pmatrix}, \quad \text{so that} \quad g'(1,2) = \begin{pmatrix} 1 & 1 \\ 2 & 4 \end{pmatrix}.$$

193

21. The derivative matrix of the function $T\begin{pmatrix} u \\ v \end{pmatrix} = \begin{pmatrix} u\cos v \\ u\sin v \\ v \end{pmatrix}$ is the 3×2 matrix

$$T'\begin{pmatrix} u \\ v \end{pmatrix} = \begin{pmatrix} \cos v & -u\sin v \\ \sin v & u\cos v \\ 0 & 1 \end{pmatrix}, \quad \text{so that} \quad T'\begin{pmatrix} 1 \\ \pi \end{pmatrix} = \begin{pmatrix} -1 & 0 \\ 0 & -1 \\ 0 & 1 \end{pmatrix}.$$

23. Let $P(x,y,z) = (x,y)$

(a) P is the projection of the vector (x,y,z) in $\mathcal{R}^3$ onto the xy-plane.

(b) The coordinate functions of P are $P_1(x,y,z) = x$ and $P_2(x,y,z) = y$. For each $\mathbf{x} \in \mathcal{R}^3$, we have $\nabla P_1(\mathbf{x}) = (1,0,0)$ and $\nabla P_2(\mathbf{x}) = (0,1,0)$, so that the coordinate functions of P are differentiable on all of $\mathcal{R}^3$. Therefore, so is P. Perhaps an easier way to show that P is differentiable is to simply compute its 2×3 derivative matrix:

$$P'(\mathbf{x}) = \begin{pmatrix} 1 & 0 & 0 \\ 0 & 1 & 0 \end{pmatrix},$$

which holds for all $\mathbf{x} \in \mathcal{R}^3$. In particular, for $\mathbf{x} = (1,1,1)$, $P'(1,1,1) = \begin{pmatrix} 1 & 0 & 0 \\ 0 & 1 & 0 \end{pmatrix}$.

25. Let $f(x,y) = \begin{pmatrix} xy \\ x+y \end{pmatrix}$. From Exercise 1, $f'(x,y) = \begin{pmatrix} y & x \\ 1 & 1 \end{pmatrix}$

(a) With $\mathbf{x}_0 = \begin{pmatrix} 1 \\ 0 \end{pmatrix}$, $\mathbf{y}_1 = \begin{pmatrix} 0.1 \\ 0 \end{pmatrix}$, $\mathbf{y}_2 = \begin{pmatrix} 0 \\ 0.1 \end{pmatrix}$ and $\mathbf{y}_3 = \begin{pmatrix} 0.1 \\ 0.1 \end{pmatrix}$, we have

$$f(\mathbf{x}_0 + \mathbf{y}_1) = f(1.1, 0) = \begin{pmatrix} 0 \\ 1.1 \end{pmatrix},$$

$$f(\mathbf{x}_0 + \mathbf{y}_2) = f(1, 0.1) = \begin{pmatrix} 0.1 \\ 1.1 \end{pmatrix},$$

$$f(\mathbf{x}_0 + \mathbf{y}_3) = f(1.1, 0.1) = \begin{pmatrix} 0.11 \\ 1.2 \end{pmatrix}.$$

(b) From part (a), $f(1,0) = \begin{pmatrix} 0 \\ 1 \end{pmatrix}$ and $f'(1,0) = \begin{pmatrix} 0 & 1 \\ 1 & 1 \end{pmatrix}$. Hence, by Definition 4.1, the first-degree approximation at $\mathbf{x}_0 = \begin{pmatrix} 1 \\ 0 \end{pmatrix}$ is

$$T(x,y) = f(1,0) + f'(1,0)\left(\begin{pmatrix} x \\ y \end{pmatrix} - \begin{pmatrix} 1 \\ 0 \end{pmatrix}\right) = \begin{pmatrix} 0 \\ 1 \end{pmatrix} + \begin{pmatrix} 0 & 1 \\ 1 & 1 \end{pmatrix}\begin{pmatrix} x-1 \\ y \end{pmatrix}$$

$$= \begin{pmatrix} 0 \\ 1 \end{pmatrix} + \begin{pmatrix} y \\ x-1+y \end{pmatrix} = \begin{pmatrix} y \\ x+y \end{pmatrix}.$$

(c) To approximate the vectors $f(\mathbf{x}_0 + \mathbf{y}_i)$, we use the result of part (b) to get

$$f(\mathbf{x}_0 + \mathbf{y}_1) = f(1.1, 0) \approx T(1.1, 0) = \begin{pmatrix} 0 \\ 1.1 \end{pmatrix},$$

$$f(\mathbf{x}_0 + \mathbf{y}_2) = f(1, 0.1) \approx T(1, 0.1) = \begin{pmatrix} 0.1 \\ 1.1 \end{pmatrix},$$

$$f(\mathbf{x}_0 + \mathbf{y}_3) = f(1.1, 0.1) \approx T(1.1, 0.1) = \begin{pmatrix} 0.1 \\ 1.2 \end{pmatrix}.$$

27. Write the given function f as a single 3×1 column vector:

$$f(x, y, z) = \begin{pmatrix} a_1 x + a_2 y + a_3 z \\ b_1 x + b_2 y + b_3 z \\ c_1 x + c_2 y + c_3 z \end{pmatrix} + \begin{pmatrix} a_0 \\ b_0 \\ c_0 \end{pmatrix} = \begin{pmatrix} a_1 x + a_2 y + a_3 z + a_0 \\ b_1 x + b_2 y + b_3 z + b_0 \\ c_1 x + c_2 y + c_3 z + c_0 \end{pmatrix}$$

and identify the derivative matrix of f to be the 3×3 matrix

$$f'(x, y, z) = \begin{pmatrix} a_1 & a_2 & a_3 \\ b_1 & b_2 & b_3 \\ c_1 & c_2 & c_3 \end{pmatrix}.$$

(NOTE: If $g : \mathcal{R}^n \longrightarrow \mathcal{R}^m$ is of the form $g(\mathbf{x}) = A\mathbf{x} + \mathbf{b}$, where A is an $m \times n$ constant matrix and $\mathbf{b}$ is a constant vector in $\mathcal{R}^m$, then $g'(\mathbf{x}) = A$. The above exercise proves this for the case $m = n = 3$.)

29. First, since f and g are assumed to be differentiable at $\mathbf{x}_0$, the domains of f and g are subsets of $\mathcal{R}^n$ whose intersection contains a neighborhood $\mathcal{N}$ of $\mathbf{x}_0$. Second, we must assume throughout that the ranges of f and g are both subsets of $\mathcal{R}^m$ for some m, otherwise, the expression $f + g$ makes no sense.

With the above understanding, let $f_1, \ldots, f_m$ and $g_1, \ldots, g_m$ be the coordinate functions of f and g (resp.) and note that, for $i = 1, \ldots, m$, both f_i and g_i are differentiable on $\mathcal{N}$. In particular, for the point $\mathbf{x}_0$, Definition 2.1 guarantees the existence of vectors $\mathbf{a}_i$, $\mathbf{b}_i$ such that

$$\lim_{\mathbf{x} \to \mathbf{x}_0} \frac{f_i(\mathbf{x}) - f_i(\mathbf{x}_0) - \mathbf{a}_i \cdot (\mathbf{x} - \mathbf{x}_0)}{|\mathbf{x} - \mathbf{x}_0|} = 0 \quad \text{and} \quad \lim_{\mathbf{x} \to \mathbf{x}_0} \frac{g_i(\mathbf{x}) - g_i(\mathbf{x}_0) - \mathbf{b}_i \cdot (\mathbf{x} - \mathbf{x}_0)}{|\mathbf{x} - \mathbf{x}_0|} = 0.$$

Adding these limit equations member by member and combining the resulting terms into a single limit, we use the fact that the dot product distributes over addition to obtain

$$\lim_{\mathbf{x} \to \mathbf{x}_0} \frac{\big(f_i(\mathbf{x}) + g_i(\mathbf{x})\big) - \big(f_i(\mathbf{x}_0) + g_i(\mathbf{x}_0)\big) - \big(\mathbf{a}_i + \mathbf{b}_i\big) \cdot (\mathbf{x} - \mathbf{x}_0)}{|\mathbf{x} - \mathbf{x}_0|} = 0.$$

Since $f_i(\mathbf{x}) + g_i(\mathbf{x}) = (f_i + g_i)(\mathbf{x})$ for all $\mathbf{x} \in \mathcal{N}$, the above limit equation is equivalent to

$$\lim_{\mathbf{x} \to \mathbf{x}_0} \frac{(f_i + g_i)(\mathbf{x}) - (f_i + g_i)(\mathbf{x}_0) - (\mathbf{a}_i + \mathbf{b}_i) \cdot (\mathbf{x} - \mathbf{x}_0)}{|\mathbf{x} - \mathbf{x}_0|} = 0.$$

Hence, with $\mathbf{a}_i + \mathbf{b}_i$ playing the role of $\mathbf{a}$ in Definition 2.1, $f_i + g_i$ is differentiable at $\mathbf{x}_0$. Since $f_i + g_i$ is the i-th coordinate function of $f + g$, it follows from the definition of differentiability of a vector-valued function that $f + g$ is differentiable at $\mathbf{x}_0$.

Having observed that the i-th coordinate function f_i of f satisfies the limit equation of Definition 2.1, we can multiply both sides of that limit equation by an arbitrarily chosen real number a to obtain

$$\lim_{\mathbf{x} \to \mathbf{x}_0} \frac{a f_i(\mathbf{x}) - a f_i(\mathbf{x}_0) - a \mathbf{a}_i \cdot (\mathbf{x} - \mathbf{x}_0)}{|\mathbf{x} - \mathbf{x}_0|} = 0.$$

Since $a f_i(\mathbf{x}) = (a f_i)(\mathbf{x})$ for all $\mathbf{x} \in \mathcal{N}$, the above limit equation is equivalent to

$$\lim_{\mathbf{x} \to \mathbf{x}_0} \frac{(a f_i)(\mathbf{x}) - (a f_i)(\mathbf{x}_0) - (a \mathbf{a}_i) \cdot (\mathbf{x} - \mathbf{x}_0)}{|\mathbf{x} - \mathbf{x}_0|} = 0.$$

Hence, with $a\mathbf{a}_i$ playing the role of $\mathbf{a}$ in Definition 2.1, af_i is differentiable at $\mathbf{x}_0$. Since af_i is the i-th coordinate function of af, it follows from the definition of differentiability of a vector-valued function that af is differentiable at $\mathbf{x}_0$.

For parts (a) and (b) below, let $\mathbf{x} = (x_1, \ldots, x_n)$ for each $\mathbf{x}$ in the domain of f and g.

(a) By what we have shown above, the derivative matrix of $f + g$ exist at $\mathbf{x}_0$ and is of size $m \times n$. Moreover, the derivative matrices of f and g are both $m \times n$ matrices, so that their sum exists and is of size $m \times n$. Thus, to show that $(f + g)'(\mathbf{x}_0) = f'(\mathbf{x}_0) + g'(\mathbf{x}_0)$, it suffices to show that the matrices on each side of this equation have the same corresponding entries. To this end, we observe that the entry in the i-th row, j-th column of the sum $f'(\mathbf{x}_0) + g'(\mathbf{x}_0)$ is

$$\frac{\partial f_i}{\partial x_j}(\mathbf{x}_0) + \frac{\partial g_i}{\partial x_j}(\mathbf{x}_0) = \left(\frac{\partial f_i}{\partial x_j} + \frac{\partial g_i}{\partial x_j}\right)(\mathbf{x}_0) = \frac{\partial(f_i + g_i)}{\partial x_j}(\mathbf{x}_0),$$

where the first equality follows from the definition of the sum of two functions and the second equality follows from the additive property of the derivatives of real-valued functions. Recognizing the rightmost member above as the entry in the i-th row, j-th column of the derivative matrix of $f + g$, we conclude that $f'(\mathbf{x}_0) + g'(\mathbf{x}_0) = (f + g)'(\mathbf{x}_0)$.

(b) By what we have shown above, the derivative matrix of af exists at $\mathbf{x}_0$ and is of size $m \times n$. Since a scalar multiple of the derivative matrix of f is also of size $m \times n$, to show that $(af)'(\mathbf{x}_0) = af'(\mathbf{x}_0)$, it suffices to show that the matrices on each side of this equation have the same corresponding entries. To this end, we observe that multiplication of a matrix by a scalar a results in a matrix with each entry multiplied by a. Thus, the entry in the i-th row, j-th column of $af'(\mathbf{x}_0)$ is

$$a\frac{\partial f_i}{\partial x_j}(\mathbf{x}_0) = \frac{\partial(af_i)}{\partial x_j}(\mathbf{x}_0),$$

where we have used the homogeneity of the derivative of a real-valued function to move the scalar a past the derivative sign. Recognizing the right member above as the entry in the i-th row, j-th column of $(af)'(\mathbf{x}_0)$, we conclude that $af'(\mathbf{x}_0) = (af)'(\mathbf{x}_0)$.

Section 5: NEWTON'S METHOD

Exercise Set 5 (pg. 250)

1. (a) The graph of $f(x) = x^{1/3} - x$, $-2 < x < 2$ is shown on the right.

(b) The tangent lines l_1, l_2 and l_3 to the graph of f for $x_0 = 3/4$, $x_0 = -3/4$ and $x_0 = -1/4$ (resp.) are shown on the graph given in part (a). Their respective points of tangency p_1, p_2 and p_3 are also shown.

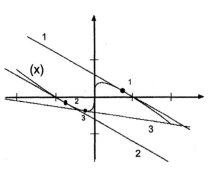

(c) The equation $x^{1/3} - x = 0$ can be factored as $x^{1/3}(1 - x^{1/3})(1 + x^{1/3}) = 0$, from which we obtain the three solutions 1, -1 and 0.

With $f(x)$ as in part (a), we have $f'(x) = \frac{1}{3}x^{-2/3} - 1$, so that Newton's formula for x_{k+1} is

$$x_{k+1} = x_k - \frac{f(x_k)}{f'(x_k)} = x_k - \frac{x_k^{1/3} - x_k}{(1/3)x^{-2/3} - 1}$$

$$= \frac{x_k\big((1/3)x^{-2/3} - 1\big) - (x^{1/3} - x_k)}{(1/3)x^{-2/3} - 1} = \frac{2x_k}{3x_k^{2/3} - 1}.$$

Here are the first four approximations for each of the starting choices $x_0 = 3/4, -3/4, -1/4$:

$x_0 = 0.75000$	$x_0 = -0.75000$	$x_0 = -0.25000$
$x_1 = 1.01595$	$x_1 = -1.01595$	$x_1 = -2.62397$
$x_2 = 1.00004$	$x_2 = -1.00004$	$x_2 = -1.11486$
$x_3 = 1.00000$	$x_3 = -1.00000$	$x_3 = -1.00188.$

Hence, we expect the Newton approximations for the choice $x_0 = 3/4$ to converge to 1; and we expect the Newton approximations for the choices $x_0 = -3/4$ and $x_0 = -1/4$ to converge to -1. Note that x_3 is accurate to five decimal places for $x_0 = \pm 3/4$, while x_3 is accurate to only two decimal places for $x_0 = -1/4$. However, one more iteration in the latter case gives $x_4 = -1.00000$.

(d) If we begin with the initial guess $x_0 = \frac{1}{9}3^{1/3} \approx 0.16025$, the first four Newton approximations are

$$x_1 = -2.78910, \quad x_2 = -1.12822, \quad x_3 = -1.00231, \quad x_4 = -1.00000.$$

Hence, the Newton approximations appear to converge to -1. What sets this apart from the three choices examined in part (c) is that here the initial choice produces approximations that converge to that solution farthest from x_0. This suggests that the solution approximated by a given initial choice x_0 cannot always be predicted by how near x_0 is from that solution.

3. (a) Let $f(x) = \cos x - x$. A rough sketch of the intersection points of the graphs of $y = x$ and $y = \cos x$ shows that there is exactly one solution of $f(x) = 0$. We want to use Newton's

method to approximate this solution. Let x_0 be chosen. Then, since, $f'(x) = -\sin x - 1$, Newton's formula for x_{x+1} becomes

$$x_{x+1} = x_k - \frac{f(x_k)}{f'(x_k)} = x_k - \frac{\cos x_k - x_k}{-\sin x_k - 1} = \frac{x_k(-\sin x_k - 1) - (\cos x_k - x_k)}{-\sin x_k - 1}$$
$$= \frac{-x_k \sin x_k - \cos x_k}{-\sin x_k - 1} = \frac{x_k \sin x_k + \cos x_k}{\sin x_k + 1}.$$

(b) With $x_0 = 1$ as our initial approximation to the solution of $f(x) = \cos x - x = 0$, we use the formula derived in part (a) to get the following approximations:

$$x_1 = 0.750363868, \quad x_2 = 0.739112891, \quad x_3 = 0.739085133, \quad x_4 = 0.739085133.$$

The fact that two consecutive approximations agree to nine decimal places means that all subsequent approximations will agree to nine decimal places.

5. Let $f(x, y, z) = \begin{pmatrix} x + y + z \\ x^2 + y^2 + z^2 \\ x^3 + y^3 + z^3 \end{pmatrix}$ and let $\mathbf{a} = (2.1, 5.7, 8.2)$. We want a solution of $f(x, y, z) = \mathbf{a}$, which can be written as

$$f(x, y, z) - \mathbf{a} = \begin{pmatrix} x + y + z - 2.1 \\ x^2 + y^2 + z^2 - 5.7 \\ x^3 + y^3 + z^3 - 8.2 \end{pmatrix} = \mathbf{0},$$

where $\mathbf{a}$ is thought of as a column vector. Thus, let $f_*(x, y, z) = f(x, y, z) - \mathbf{a}$ and observe that $f_*(x, y, z) = \mathbf{0}$ if, and only if, $f(x, y, z) = \mathbf{a}$. Moreover, $f(1, 2, -1) = \begin{pmatrix} 2 \\ 6 \\ 8 \end{pmatrix}$, so

that $f_*(1, 2, -1) = \begin{pmatrix} -0.1 \\ 0.3 \\ -0.2 \end{pmatrix}$. This suggests that $\mathbf{x}_0 = \begin{pmatrix} 1 \\ 2 \\ -1 \end{pmatrix}$ is a reasonable initial

approximation to a solution of $f_*(x, y, z) = \mathbf{0}$. We will therefore apply the modified Newton Formula 5.2 to approximate a solution of $f_*(x, y, z) = \mathbf{0}$, and we will use $\mathbf{x}_0 = (1, 2, -1)$ as our initial approximation.

We have

$$f_*'(x, y, z) = \begin{pmatrix} 1 & 1 & 1 \\ 2x & 2y & 2z \\ 3x^2 & 3y^2 & 3z^2 \end{pmatrix}, \qquad f_*'(1, 2, -1) = \begin{pmatrix} 1 & 1 & 1 \\ 2 & 4 & -2 \\ 3 & 12 & 3 \end{pmatrix},$$

$$[f_*'(1, 2, -1)]^{-1} = \frac{1}{36} \begin{pmatrix} 36 & 9 & -6 \\ -12 & 0 & 4 \\ 12 & -9 & 2 \end{pmatrix}.$$

Hence, Formula 5.2 becomes

$$\begin{pmatrix} x_{k+1} \\ y_{k+1} \\ z_{k+1} \end{pmatrix} = \begin{pmatrix} x_k \\ y_k \\ z_k \end{pmatrix} - \frac{1}{36} \begin{pmatrix} 36 & 9 & -6 \\ -12 & 0 & 4 \\ 12 & -9 & 2 \end{pmatrix} \begin{pmatrix} x_k + y_k + z_k - 2.1 \\ x_k^2 + y_k^2 + z_k^2 - 5.7 \\ x_k^3 + y_k^3 + z_k^3 - 8.2 \end{pmatrix},$$

198

where, for ease of computation, the fraction 1/36 has been left outside of the above matrix. Using a hand held calculator, we get

$$\mathbf{x}_1 = \begin{pmatrix} x_1 \\ y_1 \\ z_1 \end{pmatrix} = \begin{pmatrix} 0.991666667 \\ 1.988888889 \\ -0.880555556 \end{pmatrix}, \qquad \mathbf{x}_6 = \begin{pmatrix} x_6 \\ y_6 \\ z_6 \end{pmatrix} = \begin{pmatrix} 0.980222889 \\ 1.993801547 \\ -0.874024437 \end{pmatrix},$$

$$\mathbf{x}_2 = \begin{pmatrix} x_2 \\ y_2 \\ z_2 \end{pmatrix} = \begin{pmatrix} 0.981360082 \\ 1.993349966 \\ -0.874710048 \end{pmatrix}, \qquad \mathbf{x}_7 = \begin{pmatrix} x_7 \\ y_7 \\ z_7 \end{pmatrix} = \begin{pmatrix} 0.980222757 \\ 1.993801596 \\ -0.874024353 \end{pmatrix},$$

$$\mathbf{x}_3 = \begin{pmatrix} x_3 \\ y_3 \\ z_3 \end{pmatrix} = \begin{pmatrix} 0.980340179 \\ 1.993758340 \\ -0.874098519 \end{pmatrix}, \qquad \mathbf{x}_8 = \begin{pmatrix} x_8 \\ y_8 \\ z_8 \end{pmatrix} = \begin{pmatrix} 0.980222742 \\ 1.993801601 \\ -0.874024344 \end{pmatrix},$$

$$\mathbf{x}_4 = \begin{pmatrix} x_4 \\ y_4 \\ z_4 \end{pmatrix} = \begin{pmatrix} 0.980235433 \\ 1.993796936 \\ -0.874032369 \end{pmatrix}, \qquad \mathbf{x}_9 = \begin{pmatrix} x_9 \\ y_9 \\ z_9 \end{pmatrix} = \begin{pmatrix} 0.980222741 \\ 1.993801602 \\ -0.874024343 \end{pmatrix},$$

$$\mathbf{x}_5 = \begin{pmatrix} x_5 \\ y_5 \\ z_5 \end{pmatrix} = \begin{pmatrix} 0.980224114 \\ 1.993801097 \\ -0.874025211 \end{pmatrix}, \qquad \mathbf{x}_{10} = \begin{pmatrix} x_{10} \\ y_{10} \\ z_{10} \end{pmatrix} = \begin{pmatrix} 0.980222741 \\ 1.993801602 \\ -0.874024343 \end{pmatrix}.$$

Since the coordinates of $\mathbf{x}_9$ and $\mathbf{x}_{10}$ agree to nine decimal places, all coordinates of $\mathbf{x}_9$ are accurate to nine decimal places.

Chapter Review

(pgs. 250-251)

1. Let Q_1 be the first quadrant of $\mathcal{R}^2$. Then every point of Q_1 has positive coordinates and therefore is the center of some ϵ-ball lying entirely inside Q_1. That is, every point of Q_1 is an interior point and Q_1 is therefore open. Moreover, the boundary of Q_1 is the set of points on the nonnegative x-axis together with the points on the nonnegative y-axis. This says that Q_1 is not closed.

3. Let H be the set of points (x, y, z) in $\mathcal{R}^3$ such that $x^2 + y^2 + z^2 \leq 1$ and $x < 0$. The set of points with $x^2 + y^2 + z^2 \leq 1$ is the solid unit sphere S_3 in $\mathcal{R}^3$ and includes the "skin". Restricting $x < 0$ simply identifies that hemisphere of S_3 lying to the negative x-axis side of the yz-plane. The boundary of H is that part of the "skin" of S_3 lying in H, together with the closed unit disk in the yz-plane. So, H is not open (it contains some of its boundary points), and H is not closed (points interior to the unit disk in the yz-plane are not in H). Moreover, the interior of H is H without its "skin".

5. In $\mathcal{R}^3$, ϵ-balls are spheres. Hence, every ϵ-ball of a point on the xy-plane contains points on the plane and points not on the plane. That is, every point on the xy-plane is a boundary point in $\mathcal{R}^3$, so that the interior of the xy-plane is the empty set and the xy-plane is not open in $\mathcal{R}^3$. And, since every point not on the xy-plane is the center of some ϵ-ball that does not intersect the xy-plane, the boundary of the xy-plane is itself. Hence, the xy-plane is closed in $\mathcal{R}^3$.

7. Let O_1 be the set of points in $\mathcal{R}^3$ such that $x > 0$, $y > 1$ and $z > 3$. Then O_1 is essentially the first octant of $\mathcal{R}^3$ displaced one unit in the direction of the positive y-axis and three units in the direction of the positive z-axis. Every point in O_1 is the center of some ϵ-ball lying entirely inside O_1. Hence, O_1 is open. The boundary of O_1 is the union of the sets B_1, B_2, B_3 defined by

$$B_1 = \{\, (x, y, z) \in \mathcal{R}^3 \mid x = 0, y \geq 1, z \geq 3 \,\},$$
$$B_2 = \{\, (x, y, z) \in \mathcal{R}^3 \mid x \geq 0, y = 1, z \geq 3 \,\},$$
$$B_3 = \{\, (x, y, z) \in \mathcal{R}^3 \mid x \geq 0, y \geq 1, z = 3 \,\},$$

no point of which is in O_1. Hence, O_1 is not closed.

9. The set S is the solid unit sphere in $\mathcal{R}^3$, where the hemisphere with $y \leq 0$ has its "skin" and the hemisphere with $y > 0$ does not. The interior of S is the solid unit sphere in $\mathcal{R}^3$ without its "skin", and the boundary of S is its "skin". It follows that the smallest closed set containing S is the solid unit sphere in $\mathcal{R}^3$ together with its "skin".

11. The domain D of $H \circ f$ is the same as the domain of f. With $f(x, y, z) = x^2 + y^2 + z^2$, D is all of $\mathcal{R}^3$. However, in this case, since $f(x, y, z) \geq 0$ on D, $H\big(f(x, y)\big) = 1$ on D, so that $H \circ f$ has no points of discontinuity.

13. The domain D of $H \circ f$ is the same as the domain of $f(x, y) = (x - y)/(1 + (x - y)^2)$, i.e., D is the xy-plane. If D^+ is the set of points below the line $y = x$ and D^- is the set of points above the line $y = x$ then $f(x, y) > 0$ on D^+ and $f(x, y) < 0$ on D^-. Hence,

$$(H \circ f)(x, y) = \begin{cases} 1, & \text{if } (x, y) \in D^+; \\ 0, & \text{if } (x, y) \in D^-. \end{cases}$$

The points of discontinuity of $H \circ f$ are therefore the points on the curve separating D^+ and D^-; i.e., all points on the line $y = x$.

15. The domain D of $H \circ f$ is the same as the domain of f. With $f(x,y) = 1/(1+x^2+y^2)$, D is the xy-plane. Since $0 < f(x,y) \le 1$ on D, $(H \circ f)(x,y) = 1$ on D and therefore $H \circ f$ has no points of discontinuity.

17. For $f(x,y) = (x^2 + y^2)^{-1}$ and $\mathbf{x} = (x,y)$ in the domain of f,

$$\nabla f(\mathbf{x}) = \left(\frac{\partial f}{\partial x}(\mathbf{x}), \frac{\partial f}{\partial y}(\mathbf{x}) \right) = \left(-2x(x^2+y^2)^{-2}, -2y(x^2+y^2)^{-2} \right).$$

19. For $f(x,y,z) = x - y$ and $\mathbf{x} = (x,y,z)$ in the domain of f,

$$\nabla f(\mathbf{x}) = \left(\frac{\partial f}{\partial x}(\mathbf{x}), \frac{\partial f}{\partial y}(\mathbf{x}), \frac{\partial f}{\partial z}(\mathbf{x}) \right) = (1, -1, 0).$$

21. For $f(x,y,z,w) = xy + yz + zw + wx$ and $\mathbf{x} = (x,y,z,w)$ in the domain of f,

$$\nabla f(\mathbf{x}) = \left(\frac{\partial f}{\partial x}(\mathbf{x}), \frac{\partial f}{\partial y}(\mathbf{x}), \frac{\partial f}{\partial z}(\mathbf{x}), \frac{\partial f}{\partial w}(\mathbf{x}) \right) = (y+w, x+z, y+w, z+x).$$

23. For $f(x,y) = e^x \sin y$, there are four second-order partial derivatives, two of which are the same:

$$\frac{\partial^2 f}{\partial x^2} = \frac{\partial}{\partial x} \left(\frac{\partial f}{\partial x} \right) = \frac{\partial}{\partial x} (e^x \sin y) = e^x \sin y,$$

$$\frac{\partial^2 f}{\partial y^2} = \frac{\partial}{\partial y} \left(\frac{\partial f}{\partial y} \right) = \frac{\partial}{\partial y} (e^x \cos y) = -e^x \sin y,$$

$$\frac{\partial^2 f}{\partial x \partial y} = \frac{\partial}{\partial x} \left(\frac{\partial f}{\partial y} \right) = \frac{\partial}{\partial x} (e^x \cos y) = e^x \cos y,$$

$$\frac{\partial^2 f}{\partial y \partial x} = \frac{\partial}{\partial y} \left(\frac{\partial f}{\partial x} \right) = \frac{\partial}{\partial y} (e^x \sin y) = e^x \cos y.$$

25. For $f(x,y,z) = yze^x$, there are nine second-order partial derivatives:

$$\frac{\partial^2 f}{\partial x^2} = \frac{\partial}{\partial x} \left(\frac{\partial f}{\partial x} \right) = \frac{\partial}{\partial x} (yze^x) = yze^x, \quad \frac{\partial^2 f}{\partial x \partial z} = \frac{\partial}{\partial x} \left(\frac{\partial f}{\partial z} \right) = \frac{\partial}{\partial x} (ye^x) = ye^x,$$

$$\frac{\partial^2 f}{\partial y^2} = \frac{\partial}{\partial y} \left(\frac{\partial f}{\partial y} \right) = \frac{\partial}{\partial y} (ze^x) = 0, \quad \frac{\partial^2 f}{\partial z \partial x} = \frac{\partial}{\partial z} \left(\frac{\partial f}{\partial x} \right) = \frac{\partial}{\partial z} (yze^x) = ye^x,$$

$$\frac{\partial^2 f}{\partial z^2} = \frac{\partial}{\partial z} \left(\frac{\partial f}{\partial z} \right) = \frac{\partial}{\partial z} (ye^x) = 0, \quad \frac{\partial^2 f}{\partial y \partial z} = \frac{\partial}{\partial y} \left(\frac{\partial f}{\partial z} \right) = \frac{\partial}{\partial y} (ye^x) = e^x,$$

$$\frac{\partial^2 f}{\partial x \partial y} = \frac{\partial}{\partial x} \left(\frac{\partial f}{\partial y} \right) = \frac{\partial}{\partial x} (ze^x) = ze^x, \quad \frac{\partial^2 f}{\partial z \partial y} = \frac{\partial}{\partial z} \left(\frac{\partial f}{\partial y} \right) = \frac{\partial}{\partial z} (ze^x) = e^x.$$

$$\frac{\partial^2 f}{\partial y \partial x} = \frac{\partial}{\partial y} \left(\frac{\partial f}{\partial x} \right) = \frac{\partial}{\partial y} (yze^x) = ze^x,$$

27. For $f(x,y,z) = x^4 + y^3 + z^2$, there are nine second-order partial derivatives. However, f_x is a function of x only, f_y is a function of y only, and f_z is a function of z only. This means that all mixed partials are zero. We therefore have

$$f_{xx} = 12x^2, \quad f_{yy} = 6y, \quad f_{zz} = 2, \quad f_{xy} = f_{yx} = f_{xz} = f_{zx} = f_{yz} = f_{zy} = 0.$$

29. For $f(u,v) = (u, v, u+v, u-1)$, there are four coordinate functions: $f_1(u,v) = u$, $f_2(u,v) = v$, $f_3(u,v) = u+v$ and $f_4(u,v) = u-1$. Hence, the derivative matrix for f is the 4×2 matrix

$$f'(u,v) = \begin{pmatrix} (f_1)_u & (f_1)_v \\ (f_2)_u & (f_2)_v \\ (f_3)_u &)f_3)_v \\ (f_4)_u & (f_4)_v \end{pmatrix} = \begin{pmatrix} 1 & 0 \\ 0 & 1 \\ 1 & 1 \\ 1 & 0 \end{pmatrix}.$$

31. For $f(x,y) = (\frac{1}{2}x^2 - \frac{1}{2}y^2, xy)$, there are two coordinate functions: $f_1(x,y) = \frac{1}{2}x^2 - \frac{1}{2}y^2$ and $f_2(x,y) = xy$. Hence, the derivative matrix for f is the 2×2 matrix

$$f'(x,y) = \begin{pmatrix} (f_1)_x & (f_1)_y \\ (f_2)_x & (f_2)_y \end{pmatrix} = \begin{pmatrix} x & -y \\ y & x \end{pmatrix}.$$

33. For $f(u,v,w) = (uv, vw, wu, uvw)$, there are four coordinate functions: $f_1(u,v,w) = uv$, $f_2(u,v,w) = vw$, $f_3(u,v,w) = wu$ and $f_4(u,v,w) = uvw$. Hence, the derivative matrix for f is the 4×3 matrix

$$f'(u,v,w) = \begin{pmatrix} (f_1)_u & (f_1)_v & (f_1)_w \\ (f_2)_u & (f_2)_v & (f_2)_w \\ (f_3)_u & (f_3)_v & (f_3)_w \\ (f_4)_u & (f_4)_v & (f_4)_w \end{pmatrix} = \begin{pmatrix} v & u & 0 \\ 0 & w & v \\ w & 0 & u \\ vw & uw & uv \end{pmatrix}.$$

35. Let $f(x,y) = xy^2$ and let $\mathbf{u} = (\cos\theta, \sin\theta)$.

(a) Since $\nabla f(\mathbf{x}) = (f_x(\mathbf{x}), f_y(\mathbf{x})) = (y^2, 2xy)$, $\nabla f(-1,2) = (4, -4)$, so that

$$\frac{\partial f}{\partial \mathbf{u}}(-1,2) = \nabla f(-1,2) \cdot \mathbf{u} = (4, -4) \cdot (\cos\theta, \sin\theta) = 4\cos\theta - 4\sin\theta.$$

(b) In order to find the direction that f increases most rapidly from $f(-1,2)$, we must maximize the directional derivative of f at $(-1,2)$ found in part (a). There, we found that the directional derivative is a real-valued function of a single real variable θ. Hence, to simplify notation, we let $g(\theta) = 4\cos\theta - 4\sin\theta$ and use the techniques of elementary calculus to find the global maximum of g.

Now, $g'(\theta) = -4\sin\theta - 4\cos\theta$, so that $g'(\theta) = 0$ implies $\tan\theta = -1$, from which we conclude that $\theta = -\pi/4 + n\pi$, for some integer n. Since g has period 2π, we can restrict θ to the interval $(-\pi, \pi]$, which determines two possible values for θ; namely, $\theta = -\pi/4$ or $\theta = 3\pi/4$. Since

$$g(-\pi/4) = 4\cos(-\pi/4) - 4\sin(-\pi/4) = 2\sqrt{2} + 2\sqrt{2} = 4\sqrt{2}$$

and $\quad g(3\pi/4) = 4\cos(3\pi/4) - 4\sin(3\pi/4) = -2\sqrt{2} - 2\sqrt{2} = -4\sqrt{2},$

it follows that $\theta = -\pi/4$ maximizes g. That is, f increases most rapidly in the direction of

$$\mathbf{u} = (\cos(-\pi/4), \sin(-\pi/4)) = (1/\sqrt{2}, -1/\sqrt{2}).$$

(NOTE: It is no accident that the vector $\mathbf{u}$ derived above points in the same direction as $\nabla f(-1,2) = (4,-4)$. This result turns out to be a special case of a more general concept; namely, if $f : \mathcal{R}^n \longrightarrow \mathcal{R}$ is differentiable on an open set D in $\mathcal{R}^n$ then, at each point $\mathbf{x}$ in D for which $\nabla f(\mathbf{x}) \neq 0$, the vector $\nabla f(\mathbf{x})$ points in the direction of maximum increase for f.)

Chapter 6: Vector Differential Calculus

Section 1: GRADIENT FIELDS

Exercise Set 1ABC (pgs. 257-259)

1. With $f(x,y) = x^2 - y^2$ and $\mathbf{x} = (x,y)$, we have

$$\nabla f(\mathbf{x}) = (f_x(\mathbf{x}), f_y(\mathbf{x})) = (2x, -2y).$$

3. With $f(x,y) = x + 2y$ and $\mathbf{x} = (x,y)$, we have

$$\nabla f(\mathbf{x}) = (f_x(\mathbf{x}), f_y(\mathbf{x})) = (1, 2).$$

5. With $f(x,y,z) = x + y - z^2$ and $\mathbf{x} = (x,y,z)$, we have

$$\nabla f(\mathbf{x}) = (f_x(\mathbf{x}), f_y(\mathbf{x}), f_z(\mathbf{x})) = (1, 1, -2z).$$

7. With $f(x,y) = x^2 - y^3$ and $\mathbf{x} = (x,y)$, we have $\nabla f(\mathbf{x}) = (2x, -3y^2)$, so that the rate of maximum increase at $\mathbf{x}$ is

$$|\nabla f(\mathbf{x})| = |(2x, -3y^2)| = \sqrt{4x^2 + 9y^4}.$$

For the point $\mathbf{x}_0 = (x_0, y_0) = (1,1)$ we have $\nabla f(1,1) = (2, -3)$ and $|\nabla f(1,1)| = \sqrt{13}$. Hence, the unit vector $\mathbf{u}$ in the direction of $\nabla f(1,1)$ is

$$\mathbf{u} = \frac{\nabla f(1,1)}{|\nabla f(1,1)|} = \frac{(2,-3)}{\sqrt{13}} = (2/\sqrt{13}, -3/\sqrt{13}).$$

9. With $h(x,y,z) = xy \sin z$ and $\mathbf{x} = (x,y,z)$, we have $\nabla h(\mathbf{x}) = (y \sin z, x \sin z, xy \cos z)$, so that the rate of maximum increase at $\mathbf{x}$ is

$$|\nabla h(\mathbf{x})| = |(y \sin z, x \sin z, xy \cos z)| = \sqrt{y^2 \sin^2 z + x^2 \sin^2 z + x^2 y^2 \cos^2 z}.$$

For $\mathbf{x}_0 = (x_0, y_0, z_0) = (1, 2, \pi)$, $\nabla h(1, 2, \pi) = (0, 0, -2)$. By inspection, $\mathbf{u} = (0, 0, -1)$ is the unit vector in the direction of $\nabla h(1, 2, \pi)$.

11. $\mathbf{F}(x,y) = (1, x),$
$-1 \le x \le 2, 0 \le y \le 2$

13. $\mathbf{F}(x,y) = (y, x)$, $x^2 + y^2 \le 4$

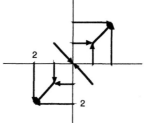

15. If $f(x, y) = xy + y^2$ then $\mathbf{F}(x, y) = \nabla f(x, y) = (y, x + 2y)$. The vector field $\mathbf{F}$ is shown below on the left for $|x| \le 2$, $|y| \le 2$. The tails of the arrows are spaced 0.5 units vertically and horizontally and have been scaled by a factor of $3/10$ in order to avoid overlapping arrows.

17. If $f(x, y, z) = x^2 + y^2$ then $\mathbf{F}(x, y, z) = \nabla f(x, y, z) = (2x, 2y, 0)$. The vector field $\mathbf{F}$ is shown below on the right.

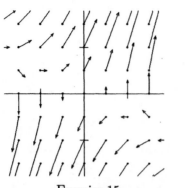

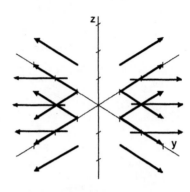

Exercise 15   Exercise 17

19. The level set S of level 2 of the function $f(x, y, z) = x^2 + y^2 - z^2$ is a surface in $\mathcal{R}^3$ consisting of all points satisfying $x^2 + y^2 - z^2 = 2$ (an hyperboloid of one sheet). Moreover, the point $(1, 1, 0)$ is on S. To find the tangent plane at $(1, 1, 0)$, we compute $\nabla f(x, y, z) = (2x, 2y, -2z)$ and find that $\nabla f(1, 1, 0) = (2, 2, 0)$. By Theorem 1.4, $(2, 2, 0)$ is a normal vector to S at $(1, 1, 0)$ and the tangent plane at $(1, 1, 0)$ has the equation

$$(2, 2, 0) \cdot (x - 1, y - 1, z - 0) = 0, \quad \text{or equivalently,} \quad x + y = 2.$$

21. Let $f(\mathbf{x}) = |\mathbf{x}|$, for $\mathbf{x} \in \mathcal{R}^n$. The level set S of level 1 of f consists of all points $\mathbf{x}$ in $\mathcal{R}^n$ such that $|\mathbf{x}| = 1$.

First, we compute $\nabla f(\mathbf{x})$. With $\mathbf{x} = (x_1, \ldots, x_n)$ $(\ne \mathbf{0})$ we see that $\partial f / \partial x_i(\mathbf{x}) = x_1 / \sqrt{x_1^2 + \cdots + x_n^2} = x_i / f(\mathbf{x})$, for $i = 1, \ldots, n$. Hence,

$$\nabla f(\mathbf{x}) = \left(\frac{\partial f}{\partial x_1}(\mathbf{x}), \ldots, \frac{\partial f}{\partial x_n}(\mathbf{x}) \right) = \left(x_1 / f(\mathbf{x}), \ldots, x_n / f(\mathbf{x}) \right) = \frac{1}{f(\mathbf{x})} \mathbf{x}.$$

Since $f(\mathbf{e}_1) = 1$, $\nabla f(\mathbf{e}_1) = \mathbf{e}_1$. By Theorem 1.4, $\mathbf{e}_1$ is a normal to S at $\mathbf{e}_1$ and the equation of the tangent hyperplane at $\mathbf{e}_1$ is

$$0 = \mathbf{e}_1 \cdot (\mathbf{x} - \mathbf{e}_1) = \mathbf{e}_1 \cdot \mathbf{x} - \mathbf{e}_1 \cdot \mathbf{e}_1 = x_1 - 1,$$

where $\mathbf{x} = (x_1, \ldots, x_n)$ is a point on the hyperplane. That is, $x_1 = 1$ is the hyperplane perpendicular to the x_1-axis that contains the point $(1, 0, \ldots, 0)$.

23. The level set S of level 1 of the function $f(x, y, z) = xyz$ consists of all points (x, y, z) satisfying $xyz = 1$. Moreover, the point $(1, 1, 1)$ is on S. To find the tangent plane at $(1, 1, 1)$, we compute $\nabla f(x, y, z) = (yz, xz, xy)$ and find that $\nabla f(1, 1, 1) = (1, 1, 1)$. By Theorem 1.4, $(1, 1, 1)$ is a normal to S at $(1, 1, 1)$ and the tangent plane at $(1, 1, 1)$ has the equation

$$0 = (1, 1, 1) \cdot (x - 1, y - 1, z - 1) = x + y + z - 3, \quad \text{or equivalently,} \quad x + y + z = 3.$$

25. Let $f : \mathcal{R}^2 \longrightarrow \mathcal{R}$ be continuously differentiable. With $z = f(x, y)$, let $F(x, y, z) = z - f(x, y)$, so that the level surface S of the function F of level 0 is precisely the set of points (x, y, z) satisfying $z - f(x, y) = 0$; i.e., S is the graph of f.

(a) With $F(x, y, z)$ defined as above and $\mathbf{x} = (x, y, z)$, note that

$$F_x(\mathbf{x}) = \frac{\partial\big(z - f(x, y)\big)}{\partial x}(\mathbf{x}) = -\frac{\partial f(x, y)}{\partial x}(\mathbf{x})$$

$$F_y(\mathbf{x}) = \frac{\partial\big(z - f(x, y)\big)}{\partial y}(\mathbf{x}) = -\frac{\partial f(x, y)}{\partial y}(\mathbf{x})$$

$$F_x(\mathbf{x}) = \frac{\partial\big(z - f(x, y)\big)}{\partial z}(\mathbf{x}) = 1.$$

The first two partials evaluated at $\mathbf{x} = (x, y, z)$ is equivalent to them being evaluated at (x, y) because the third coordinate does not enter into the evaluation (the expression for $f(x, y)$ is in terms of x and y only and therefore so are the partials $f_x(x, y)$ and $f_y(x, y)$). In short,

$$\nabla F(x, y, z) = (-f_x(x, y), -f_y(x, y), 1).$$

Since f is continuously differentiable, its partials are continuous and therefore the partials of F are continuous at $\mathbf{x} = (x, y, z)$ for all (x, y) in the domain of f. That is, F is continuously differentiable. Note too that $\nabla F(x, y, z)$ is never the zero vector since its third coordinate is a nonzero constant.

(b) The function $f(x, y) = xy + ye^x$ has partials $f_x = y + ye^x$ and $f_y = x + e^x$, which are continuous on the xy-plane (i.e., f is continuously differentiable). If $F : \mathcal{R}^3 \longrightarrow \mathcal{R}$ is defined by $F(x, y, z) = z - f(x, y)$ then, by part (a), F is continuously differentiable and

$$\nabla F(x, y, z) = (-f_x(x, y), -f_y(x, y), 1) = (-y - ye^e, -x - e^x, 1)$$

is never the zero vector. The level set S of level 0 of F contains the point $(1, 1, f(1, 1)) = (1, 1, 1 + e)$. Since $\nabla F(1, 1, 1 + e) = (-f_x(1, 1), -f_y(1, 1), 1) = (-1 - e, -1 - e, 1)$, by Theorem 1.4, $(-1 - e, -1 - e, 1)$ is a normal to S at $(1, 1, 1 + e)$ and the equation of the tangent plane at $(1, 1, 1 + e)$ is

$$0 = (-1 - e, -1 - e, 1) \cdot (x - 1, y - 1, z - 1 - e)$$
$$= -(1 + e)(x - 1) - (1 + e)(y - 1) + (z - 1 - e)$$
$$= -(1 + e)x - (1 + e)y + z + (1 + e),$$

or equivalently, $x + y - z/(1 + e) = 1$. (NOTE: The same tangent plane is obtained by using Equation 3.2 in Section 3B of Chapter 4).

27. Let $g(x, y) = e^{x+y}$, let $f(t)$ be differentiable in some neighborhood of $t = 0$, and let F be the composite function $F(t) = g(f(t))$. Further, suppose $f(0) = (1, -1)$ and $f'(0) = (1, 2)$. Since $\nabla g(x, y) = (e^{x+y}, e^{x+y})$, we have

$$\nabla g(f(0)) = \nabla g(1, -1) = (e^{1-1}, e^{1-1}) = (1, 1).$$

Hence, by the chain rule,

$$F'(0) = \nabla g(f(0)) \cdot f'(0) = (1, 1) \cdot (1, 2) = 3.$$

29. Let $f(x, y, z) = \sin z$, $F(t) = (\cos t, \sin t, t)$ and $g(t) = f(F(t))$. Since the gradient of f is $\nabla f(x, y, z) = (0, 0, \cos z)$ and $F(\pi) = (-1, 0, \pi)$, we have

$$\nabla f(F(\pi)) = \nabla f(-1, 0, \pi) = (0, 0, -1).$$

Also, $F'(t) = (-\sin t, \cos t, 1)$ implies $F'(\pi) = (0, -1, 1)$. Hence, by the chain rule,

$$g'(\pi) = \nabla f(F(\pi)) \cdot F'(\pi) = (0, 0, -1) \cdot (0, -1, 1) = -1.$$

31. The position of the spaceship at time t is given parametrically by $g(t) = (3t^2, t^3)$; and the intensity of gamma radiation at the point (x, y) is $I(x, y) = x^2 - y^2$.

(a) Since the spaceship is at the point $(3t^2, t^3)$ at time t, it follows that the gamma radiation experienced by the spaceship at time t is represented by the composite function

$$I\big(g(t)\big) = I(3t^2, t^3) = (3t^2)^2 - (t^3)^2 = 9t^4 - t^6.$$

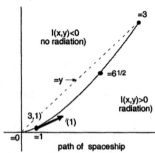

In particular, at time $t = 1$, $I\big(g(1)\big) = 9 - 1 = 8$. Since a level curve of I is just the set of points (x, y) that satisfy $I(x, y) = k$, where k is some fixed constant, it follows that at time $t = 1$ the spaceship is on the level curve of I with equation $x^2 - y^2 = 8$.

(b) The rectangular coordinates of the spaceship's position at time t satisfy $x = 3t^2$, $y = t^3$. For $t \geq 0$, we have $t = (x/3)^{1/2}$, so that $y = (x/3)^{3/2}$. The graph of this relationship is the path of the spaceship, with $t = 0$ corresponding to $(x, y) = (0, 0)$, and is shown on the right.

(c) Since $\nabla I(x, y) = (2x, -2y)$, we use $g(1) = (3, 1)$ to get

$$\nabla I\big(g(1)\big) = \nabla I(3, 1) = (6, -2).$$

(d) Since $g'(t) = (6t, 3t^2)$, the spaceship's velocity at time $t = 1$ is $g'(1) = (6, 3)$. The velocity vector of the spaceship is shown in the diagram above.

(e) By part (a), $I\big(g(t)\big) = 9t^4 - t^6 = t^4(9 - t^2)$ is the radiation experienced by the spaceship at time t. Notice that $I\big(g(t)\big) < 0$ when $t < 0$ and when $t > 3$. Since negative radiation makes no sense in this context, t will be restricted to the interval $[0, 3]$.

The rate at which the radiation is changing is

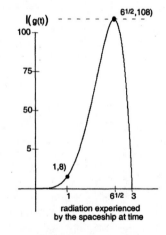

$$\frac{d}{dt}\big(I\big(g(t)\big)\big) = 36t^3 - 6t^5 = 6t^3(6 - t^2).$$

This is zero when $t = 0$ and $t = \sqrt{6}$. Since $I(g(0)) = 0$, $I(g(\sqrt{6})) = 108$ and $I(g(3)) = 0$, it follows that the radiation is increasing on the interval $[0, \sqrt{6}]$ (reaching a maximum at $t = \sqrt{6}$) and is decreasing on the interval $[\sqrt{6}, 3]$ (reaching a minimum at $t = 3$). Thus, the spaceship begins its race to safety at $t = \sqrt{6}$, from which time onward it experiences less and less radiation until it drops to zero at $t = 3$, never again to bother the ship or its crew. The details are shown in the corresponding graph on the right.

33. The function $\mathbf{F}(x, y) = (-y, x)$ defines a vector field on $\mathcal{R}^2$. Suppose the vector field is also a gradient field; i.e., suppose there exists a function $f : \mathcal{R}^2 \longrightarrow \mathcal{R}$ such that $\nabla f(x, y) = \mathbf{F}(x, y)$ for $(x, y) \in \mathcal{R}^2$. Then the first-order partial derivatives of f are given by

$$\frac{\partial f}{\partial x}(x, y) = -y \quad \text{and} \quad \frac{\partial f}{\partial y}(x, y) = x,$$

206

both of which are continuous on $\mathcal{R}^2$. By Theorem 2.3 of Sec.2A in Ch.5, f is differentiable on $\mathcal{R}^2$, so that f is necessarily continuous on $\mathcal{R}^2$. Differentiating the first equation with respect to y and the second equation with respect to x gives

$$\frac{\partial^2 f}{\partial y \partial x}(x,y) = -1 \quad \text{and} \quad \frac{\partial^2 f}{\partial x \partial y}(x,y) = 1,$$

which are also continuous on $\mathcal{R}^2$. Under these continuity conditions, Clairaut's Theorem says that the mixed second-order partials of f must be equal. Since they are not equal, we conclude that $\mathbf{F}$ is not a gradient field.

35. If $f(x,y,z)$ is such that $\nabla f(x,y,z) = \mathbf{F}(x,y,z) = (x,y,z)$ then $\partial f/\partial x = x$, $\partial f/\partial y = y$ and $\partial f/\partial z = z$. The form of the first-order partials and their dependence on only one variable at a time, suggests that f could be the sum of three functions, where one is a function of x only, one is a function of y only and one is a function of z only. So, we try the obvious:

$$f(x,y,z) = \frac{1}{2}x^2 + \frac{1}{2}y^2 + \frac{1}{2}z^2$$

and find that is does what we want. Of course, any function whose gradient is $\mathbf{F}$ must differ from this by at most a constant.
(NOTE: We will see that there is a relatively simple algorithm that constructs the desired functions.)

37. If $f(x,y)$ is such that $\nabla f(x,y) = \mathbf{F}(x,y) = (x^2, y^2)$ then $\partial f/\partial x = x^2$ and $\partial f/\partial y = y^2$. As in Exercise 35 above, the form of the first-order partials suggests that f may be the sum of two functions, one of which is a function of x only and the other a function of y only. This basic set up leads us to the guess

$$f(x,y) = \frac{1}{3}x^3 + \frac{1}{3}y^3,$$

which does what we want. (See the NOTE following Exercise 35 above).

39. (a) Given the vector equation $\ddot{\mathbf{x}} = -k|\mathbf{x}|^{-3}\mathbf{x}$, where k is a positive constant, the magnitude of the acceleration vector $\ddot{\mathbf{x}}$ satisfies

$$|\ddot{\mathbf{x}}| = \left| -k|\mathbf{x}|^{-3}\mathbf{x} \right| = |k|\left||\mathbf{x}|^{-3}\right||\mathbf{x}| = k|\mathbf{x}|^{-3}|\mathbf{x}| = k/|\mathbf{x}|^2.$$

(b) Since $\mathbf{x} = (x,y)$ is a function of t, so are x and y, and therefore $\ddot{\mathbf{x}} = (\ddot{x}, \ddot{y})$, where the derivatives $\ddot{x}$ and $\ddot{y}$ are with respect to t. Furthermore, $|\mathbf{x}|^{-3} = \left(\sqrt{x^2+y^2}\right)^{-3} = (x^2+y^2)^{-3/2}$. Hence, the given vector equation can be written as

$$(\ddot{x}, \ddot{y}) = -k(x^2+y^2)^{-3/2}(x,y) = \left(-k(x^2+y^2)^{-3/2}x, -k(x^2+y^2)^{-3/2}y\right).$$

Equating like coordinates of each side leads to the two equation system

$$\ddot{x} = -kx(x^2+y^2)^{-3/2}, \qquad \ddot{y} = -ky(x^2+y^2)^{-3/2}.$$

(c) Let $f(x,y) = k(x^2+y^2)^{-1/2}$. Then the first-order partials of f are

$$\frac{\partial f}{\partial x}(x,y) = -kx(x^2+y^2)^{-3/2}, \qquad \frac{\partial f}{\partial y}(x,y) = -ky(x^2+y^2)^{-3/2}.$$

Hence, $\nabla f(x,y) = \left(-kx(x^2+y^2)^{-3/2}, -ky(x^2+y^2)^{-3/2}\right)$, which is the given vector field $\mathbf{F}(x,y)$.

Exercise Set 1D (pg. 261)

1. The vector field $\mathbf{F}(x,y) = (x,y)$ is shown below for $|x| \leq 2$, $|y| \leq 2$, as well as three typical flow lines. The spacing between the initial points of the arrows is 0.7 units. No scaling was used.

3. The vector field $\mathbf{F}(x,y) = (2x, x)$ is shown below for $|x| \leq 4$, $|y| \leq 4$, as well as three typical flow lines. The spacing between the initial points of the arrows is 1.25 units. No scaling was used.

5. The vector field $\mathbf{F}(x,y) = \nabla e^{x+y} = (e^{x+y}, e^{x+y})$ is shown below for $|x| \leq 4$, $|y| \leq 4$, as well as two typical flow lines. The spacing between the initial points of the arrows is 0.75 units. No scaling was used.

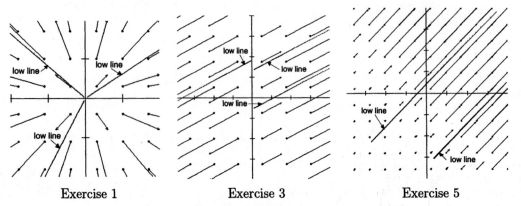

Exercise 1   Exercise 3   Exercise 5

7. (a) Let a and b be constants, not both zero, and let $g(t) = (a\cos t + b\sin t, b\cos t - a\sin t)$. If $(x,y) = g(t)$, so that $x = a\cos t + b\sin t$ and $y = b\cos t - a\sin t$, then

$$g'(t) = (-a\sin t + b\cos t, -b\sin t - a\cos t) = (y, -x).$$

That is, the velocity vector $g'(t)$ at a point (x,y) on the curve parametrized by g coincides with the vector $(y, -x)$. Hence, the image of the curve parametrized by g is a flow line of the vector field $\mathbf{F}(x,y) = (y, -x)$.

(b) By part (a), the x and y coordinates of a point (x,y) on the flow line parametrized by g satisfy

$$
\begin{aligned}
x^2 + y^2 &= (a\cos t + b\sin t)^2 + (b\cos t - a\sin t)^2 \\
&= (a^2\cos^2 t + 2ab\cos t\sin t + b^2\sin^2 t) + (b^2\cos^2 t - 2ab\cos t\sin t + a^2\sin^2 t) \\
&= (a^2 + b^2)(\cos^2 t + \sin^2 t) = a^2 + b^2.
\end{aligned}
$$

Since a and b are not both zero, the right side of the above equation is positive, so that the equation is that of a circle of radius $\sqrt{a^2 + b^2}$ centered at the origin.

Section 2: THE CHAIN RULE

Exercise Set 2A (pgs. 269-271)

1. Let $f\begin{pmatrix} x \\ y \end{pmatrix} = \begin{pmatrix} x^2 + xy + 1 \\ y^2 + 2 \end{pmatrix}$ and $g\begin{pmatrix} u \\ v \end{pmatrix} = \begin{pmatrix} u+v \\ 2u \\ v^2 \end{pmatrix}$.

(a) The derivative matrices of $f(x,y)$ and $g(u,v)$ are

$$f'(x,y) = \begin{pmatrix} 2x+y & x \\ 0 & 2y \end{pmatrix} \quad \text{and} \quad g'(u,v) = \begin{pmatrix} 1 & 1 \\ 2 & 0 \\ 0 & 2v \end{pmatrix}.$$

To find $g'\big(f(x,y)\big)$, we compute

$$g'\big(f(x,y)\big) = g'(x^2 + xy + 1, y^2 + 2) = \begin{pmatrix} 1 & 1 \\ 2 & 0 \\ 0 & 2y^2 + 4 \end{pmatrix}.$$

(b) With $g'\big(f(x,y)\big)$ and $f'(x,y)$ as found in part (a), the chain rule gives

$$(g \circ f)'(x,y) = g'\big(f(x,y)\big) f'(x,y)$$

$$= \begin{pmatrix} 1 & 1 \\ 2 & 0 \\ 0 & 2y^2 + 4 \end{pmatrix} \begin{pmatrix} 2x+y & x \\ 0 & 2y \end{pmatrix} = \begin{pmatrix} 2x+y & x+2y \\ 4x+2y & 2x \\ 0 & 4y^3 + 8y \end{pmatrix},$$

so that at $(x,y) = (1,1)$ and $(x,y) = (0,0)$ we have

$$(g \circ f)'(1,1) = \begin{pmatrix} 3 & 3 \\ 6 & 2 \\ 0 & 12 \end{pmatrix} \quad \text{and} \quad (g \circ f)'(0,0) = \begin{pmatrix} 0 & 0 \\ 0 & 0 \\ 0 & 0 \end{pmatrix}.$$

3. If g is a real-valued differentiable function with domain $\mathcal{R}^3$ then the derivative matrix of g is the 1×3 matrix

$$g'(\mathbf{x}) = \left(\frac{\partial g}{\partial x}(\mathbf{x}), \frac{\partial g}{\partial y}(\mathbf{x}), \frac{\partial g}{\partial z}(\mathbf{x}) \right),$$

where $\mathbf{x} = (x,y,z)$. When $\mathbf{x} = \mathbf{x}_0 = (2,0,1)$, we are given that

$$\frac{\partial g}{\partial x}(\mathbf{x}_0) = 4, \quad \frac{\partial g}{\partial y}(\mathbf{x}_0) = 2, \quad \frac{\partial g}{\partial z}(\mathbf{x}_0) = 2,$$

so that $g'(\mathbf{x}_0) = (4,2,2)$.

Now let $f(t) = \begin{pmatrix} t \\ t^2 - 4 \\ e^{t-2} \end{pmatrix}$ and note that $f(2) = \mathbf{x}_0$, so that $g'(\mathbf{x}_0) = g'\big(f(2)\big)$. Also,

the derivative matrix of f is $f'(t) = \begin{pmatrix} 1 \\ 2t \\ e^{t-2} \end{pmatrix}$, so that $f'(2) = \begin{pmatrix} 1 \\ 4 \\ 1 \end{pmatrix}$. Therefore,

$$\frac{d(g \circ f)}{dt}(2) = g'\big(f(2)\big) f'(2) = (4,2,2) \begin{pmatrix} 1 \\ 4 \\ 1 \end{pmatrix} = 4 + 8 + 2 = 14.$$

5. Let $F(x, y, z) = x^2 + y^2 + z^2$ and $f\begin{pmatrix} u \\ v \end{pmatrix} = \begin{pmatrix} u + v \\ u - v \\ u^2 - v^2 \end{pmatrix} = \begin{pmatrix} x \\ y \\ z \end{pmatrix}$.

(a) The composite function $(F \circ f)(u, v)$ is given by

$$(F \circ f)(u, v) = F\big(f(u, v)\big) = F(u + v, u - v, u^2 - v^2)$$
$$= (u + v)^2 + (u - v)^2 + (u^2 - v^2)^2 = 2u^2 + 2v^2 + u^4 - 2u^2v^2 + v^4.$$

Hence, the derivative matrix of $(F \circ f)(u, v)$ is the 1×2 matrix

$$(F \circ f)'(u, v) = \left(\frac{\partial(F \circ f)}{\partial u}, \frac{\partial(F \circ f)}{\partial v} \right) = (4u + 4u^3 - 4uv^2, 4v - 4vu^2 + 4v^3).$$

(b) If $w = F(x, y, z) = F\big(f(u, v)\big) = (F \circ f)(u, v)$ then the partial derivatives of w with respect to u and v are the coordinate functions of the derivative matrix of $(F \circ f)(u, v)$, which we have already computed in part (a). That is,

$$\frac{\partial w}{\partial u} = \frac{\partial(F \circ f)}{\partial u} = 4u + 4u^3 - 4uv^2 \quad \text{and} \quad \frac{\partial w}{\partial v} = \frac{\partial(F \circ f)}{\partial v} = 4v - 4vu^2 + 4v^3.$$

7. Let $w = w(x, y, z) = \sqrt{x^2 + y^2 + z^2}$ and let $f(r, \theta) = (r\cos\theta, r\sin\theta, r) = \begin{pmatrix} x \\ y \\ z \end{pmatrix}$. The derivative matrix of $w(x, y, z)$ is given by

$$w'(x, y, z) = \left(\frac{x}{\sqrt{x^2 + y^2 + z^2}}, \frac{y}{\sqrt{x^2 + y^2 + z^2}}, \frac{z}{\sqrt{x^2 + y^2 + z^2}} \right).$$

Since $x = r\cos\theta$, $y = \sin\theta$ and $z = r$, we get

$$\sqrt{x^2 + y^2 + z^2} = \sqrt{r^2 \cos^2\theta + r^2 \sin^2\theta + r^2} = \sqrt{2r^2} = r\sqrt{2},$$

so that

$$w'\big(f(r, \theta)\big) = \left(\frac{r\cos\theta}{r\sqrt{2}}, \frac{r\sin\theta}{r\sqrt{2}}, \frac{r}{r\sqrt{2}} \right) = \frac{1}{\sqrt{2}}(\cos\theta, \sin\theta, 1).$$

Also, the derivative matrix of $f(r, \theta)$ is given by

$$f'(r, \theta) = \begin{pmatrix} \cos\theta & -r\sin\theta \\ \sin\theta & r\cos\theta \\ 1 & 0 \end{pmatrix}.$$

Thus, by the chain rule,

$$(w \circ f)'(r, \theta) = w'\big(f(r, \theta)\big) f'(r, \theta) = \frac{1}{\sqrt{2}}(\cos\theta, \sin\theta, 1) \begin{pmatrix} \cos\theta & -r\sin\theta \\ \sin\theta & r\cos\theta \\ 1 & 0 \end{pmatrix}$$

$$= \frac{1}{\sqrt{2}}\left(\cos^2\theta + \sin^2\theta + 1, -r\cos\theta\sin\theta + r\cos\theta\sin\theta + 0 \right) = \frac{1}{\sqrt{2}}(2, 0).$$

Finally, since the entries of the derivative matrix of $(w \circ f)(r, \theta)$ are the partials $\partial w / \partial r(r, \theta)$ and $\partial w / \partial \theta(r, \theta)$, we have

$$\frac{\partial w}{\partial r}(r, \theta) = \frac{2}{\sqrt{2}} = \sqrt{2}, \qquad \text{and} \qquad \frac{\partial w}{\partial \theta}(r, \theta) = 0.$$

9. Let C_g be the curve parametrized by a differentiable vector function g. Let $\mathbf{x}_0$ be a point on C_g and let $\mathbf{u}_0$ be a point in the domain of g such that $g(\mathbf{u}_0) = \mathbf{x}_0$. To say that a vector $\mathbf{v}$ is tangent to C_g at $\mathbf{x}_0$ is to assert the existence of a nonzero scalar k such that $g'(\mathbf{u}_0) = k\mathbf{v}$.

Now let $\mathbf{x}_0$ be in the domain of a differentiable vector function F and let C_F be the curve defined parametrically by $F \circ g$. Assuming $(F \circ g)(\mathbf{u}_0) \neq \mathbf{0}$, the chain rule gives

$$(F \circ g)'(\mathbf{u}_0) = F'\big(g(\mathbf{u}_0)\big)g'(\mathbf{u}_0) = F'(\mathbf{x}_0)(k\mathbf{v}) = kF'(\mathbf{x}_0)\mathbf{v} \neq \mathbf{0}.$$

Hence, the vector $(F \circ g)'(\mathbf{u}_0)$ is a nonzero scalar multiple of $F'(\mathbf{x}_0)\mathbf{v}$. But $(F \circ g)(\mathbf{u}_0) = F(\mathbf{x}_0)$ implies $(F \circ g)'(\mathbf{u}_0)$ is a tangent vector to C_F at $F(\mathbf{x}_0)$. So, $F'(\mathbf{x}_0)\mathbf{v}$ is also a tangent vector to C_F at $F(\mathbf{x}_0)$.

11. Let a be a constant and let $y = f(x - at) + g(x + at)$. Setting $u = x - at$ and $v = x + at$, we note that $u_x = v_x = 1$, $u_t = -a$ and $v_t = a$. Also note that $f(u)$ and $\partial f / \partial u = df / du$ are independent of v and that $g(v)$ and $\partial g / \partial v = dg / dv$ are independent of u. Therefore, $f_v = g_u = f_{uv} = g_{vu} = 0$. So, by the chain rule,

$$\frac{\partial^2 y}{\partial x^2} = y_{xx} = (y_x)_x = (f_u u_x + f_v v_x + g_u u_x + g_v v_x)_x = (f_u + g_v)_x$$

$$= (f_{uu} u_x + f_{uv} v_x + g_{vu} u_x + g_{vv} v_x) = f_{uu} + g_{vv},$$

$$\frac{\partial^2 y}{\partial t^2} = y_{tt} = (y_t)_t = (f_u u_t + f_v v_t + g_u u_t + g_v v_t)_t = (-a f_u + a g_v)_t = -a(f_u)_t + a(g_v)_t$$

$$= -a(f_{uu} u_t + f_{uv} v_t) + a(g_{vu} u_t + g_{vv} v_t) = -a(-a f_{uu}) + a(a g_{vv})$$

$$= a^2 f_{uu} + a^2 g_{vv}$$

Hence,

$$a^2 \frac{\partial^2 y}{\partial x^2} = \frac{\partial^2 y}{\partial t^2}.$$

13. Suppose $f : \mathcal{R}^2 \longrightarrow \mathcal{R}$ is a differentiable function such that, for some integer n, $f(tx, ty) = t^n f(x, y)$ for all x, y, t. Let $F(x, y, t) = f(tx, ty)$ and note that $F_t(x, y, t)$ can be computed in two ways. On one hand, $F(x, y, t) = t^n f(x, y)$, so that

$$F_t(x, y, t) = nt^{n-1} f(x, y).$$

On the other hand, the chain rule gives

$$F_t(x, y, t) = f_x(tx, ty)(tx)_t + f_y(tx, ty)(ty)_t = x f_x(tx, ty) + y f_y(tx, ty).$$

Hence, $x f_x(tx, ty) + y f_y(tx, ty) = nt^{n-1} f(x, y)$ for all x, y, t. In particular, when $t = 1$,

$$x f_x(x, y) + y f_y(x, y) = nf(x, y),$$

which is what we wanted to show.

15. Let $z = f(x, y)$ and $(x, y) = g(u, v) = (u + v, uv)$, so that $f \circ g = z$. Noting that $x_v = x_u = 1$, $y_u = v$ and $y_v = u$, the chain rule gives $z_u = f_x x_u + f_y y_u = f_x + v f_y$, so that

$$
\begin{aligned}
z_{uv} = (f_x + v f_y)_v &= (f_x)_v + (v f_y)_v \\
&= (f_{xx} x_v + f_{xy} y_v) + (v(f_{yx} x_v + f_{yy} y_v) + f_y) \\
&= f_{xx} + u f_{xy} + v f_{yx} + uv f_{yy} + f_y.
\end{aligned}
$$

Since $x = 2$ and $y = 1$ when $u = v = 1$, we use $f_{xx}(2, 1) = 0$, $f_{xy}(2, 1) = f_{yx}(2, 1) = 1$, $f_{yy}(2, 1) = 2$ and $f_y(2, 1) = -2$ to get

$$
\begin{aligned}
\frac{\partial^2 (f \circ g)}{\partial v \partial u}(1, 1) &= z_{uv}(1, 1) \\
&= f_{xx}(2, 1) + (1) f_{xy}(2, 1) + (1) f_{yx}(2, 1) + (1)(1) f_{yy}(2, 1) + f_y(2, 1) \\
&= 0 + (1)(1) + (1)(1) + (1)(1)(2) - 2 = 2.
\end{aligned}
$$

17. With $u(x, y) = f(xy)$, let $t = xy$. By the chain rule,

$$
\frac{\partial u}{\partial x} = \frac{df}{dt} \frac{\partial t}{\partial x} = \frac{df}{dt} y \qquad \text{and} \qquad \frac{\partial u}{\partial y} = \frac{df}{dt} \frac{\partial t}{\partial y} = \frac{df}{dt} x.
$$

Multiplying the first equation by x, the second by y, and subtracting, gives

$$
x \frac{\partial u}{\partial x} - y \frac{\partial u}{\partial y} = 0, \qquad \text{or} \qquad x \frac{\partial u}{\partial x} = y \frac{\partial u}{\partial y}.
$$

19. With $u(x, y) = f(x^2 + y^2)$, let $t = x^2 + y^2$. By the chain rule,

$$
\frac{\partial u}{\partial x} = \frac{df}{dt} \frac{\partial t}{\partial x} = \frac{df}{dt}(2x) \qquad \text{and} \qquad \frac{\partial u}{\partial y} = \frac{df}{dt} \frac{\partial t}{\partial y} = \frac{df}{dt}(2y).
$$

Multiplying the first equation by y, the second by x, and subtracting, gives

$$
y \frac{\partial u}{\partial x} - x \frac{\partial u}{\partial y} = 0, \qquad \text{or} \qquad y \frac{\partial u}{\partial x} = x \frac{\partial u}{\partial y}.
$$

21. If $f : \mathcal{R}^3 \longrightarrow \mathcal{R}^2$ and $g : \mathcal{R}^2 \longrightarrow \mathcal{R}$ then the range values of f are of the same dimension as the domain values of g (namely, dimension 2). Thus, in terms of dimension, it may be possible to define $g \circ f : \mathcal{R}^3 \longrightarrow \mathcal{R}$. However, the range values of g are of a different dimension from the domain values of f. So, based on dimension alone, $f \circ g$ cannot be defined.

23. If $f : \mathcal{R}^3 \longrightarrow \mathcal{R}^2$ and $g : \mathcal{R}^3 \longrightarrow \mathcal{R}^3$ then the range values of g are of the same dimension as the domain values of f (namely, dimension 3). Thus, in terms of dimension, it may be possible to define $f \circ g : \mathcal{R}^3 \longrightarrow \mathcal{R}^2$. However, the range values of f are of a different dimension from the domain values of g. So, based on dimension alone, $g \circ f$ cannot be defined.

25. If $f : \mathcal{R} \longrightarrow \mathcal{R}^3$ and $g : \mathcal{R}^3 \longrightarrow \mathcal{R}^3$ then the range values of f are of the same dimension as the domain values of g (namely, dimension 3). Thus, in terms of dimension, it may be possible to define $g \circ f : \mathcal{R} \longrightarrow \mathcal{R}^3$. However, the range values of g are of a different dimensions from the domain values of f. So, based on dimension alone, $f \circ g$ cannot be defined.

212

Exercise Set 2B (pgs. 274-275)

1. Let $u(x,y)$ be differentiable for all (x,y) in $\mathcal{R}^2$. Let $x = s+t$, $y = s-t$, and define $\overline{u}$ by $\overline{u}(s,t) = u(s+t, s-t)$. Noting that $x_s = x_t = y_s = 1$ and $y_t = -1$, we use the chain rule to compute $\overline{u}_s$ and $\overline{u}_t$:

$$\overline{u}_s = u_x x_s + u_y y_s = u_x + u_y \qquad \text{and} \qquad \overline{u}_t = u_x x_t + u_y y_t = u_x - u_y.$$

Squaring both equations and adding the results gives

$$\overline{u}_s^2 + \overline{u}_y^2 = (u_x^2 + 2u_x u_y + u_y^2) + (u_x^2 - 2u_x u_y + u_y^2) = 2u_x^2 + 2u_y^2.$$

3. Let $z = f(x,y)$ be differentiable, and let $x = u\cos v$ and $y = u\sin v$. Noting that $x_u = \cos v$, $x_v = -u\sin v$, $y_u = \sin v$ and $y_v = u\cos v$, we use the chain rule to compute z_u and z_v:

$$z_u = z_x x_u + z_y y_u = z_x(\cos v) + z_y(\sin v),$$
$$z_v = z_x x_v + z_y y_v = z_x(-u\sin v) + z_y(u\cos v).$$

For $u \neq 0$, divide the second equation by u. Now square both equations and add the results to get

$$z_u^2 + \frac{1}{u^2}z_v^2 = (z_x^2 \cos^2 v + 2z_x z_y \cos v \sin v + z_y^2 \sin^2 v)$$
$$+ (z_x^2 \sin^2 v - 2z_x z_y \sin v \cos v + z_y^2 \cos^2 v)$$
$$= z_x^2(\cos^2 v + \sin^2 v) + z_y^2(\sin^2 v + \cos^2 v) = z_x^2 + z_y^2.$$

5. Let $f : \mathcal{R}^2 \longrightarrow \mathcal{R}$ be continuously differentiable. Let $x(u,v) = u+v$, $y(u,v) = u^2 - v^2$, and set $z = f\big(x(u,v), y(u,v)\big)$. With $x = x(u,v)$ and $y = y(u,v)$, we can dispense with the cumbersome notation and write $x = u+v$, $y = u^2 - v^2$ and $z = f(x,y)$. Noting that $x_u = x_v = 1$, $y_u = 2u$ and $y_v = -2v$, we use the chain rule to compute z_u and z_v:

$$z_u = z_x x_u + z_y y_u = z_x + z_y(2u), \qquad z_v = z_x x_v + z_y y_v = z_x + z_y(-2v).$$

Squaring both equations and subtracting the results gives

$$z_u^2 - z_v^2 = (z_x^2 + 4u z_x z_y + 4u^2 z_y^2) - (z_x^2 - 4v z_x z_y + 4v^2 z_y^2)$$
$$= 4(u+v)z_x z_y + 4(u^2 - v^2)z_y^2 = 4x z_x z_y + 4y z_y^2.$$

Since $z_x = f_x$ and $z_y = f_y$, we obtain

$$z_u^2 - z_v^2 = 4x f_x f_y + 4y f_y^2.$$

7. Let $u(x,y)$ be differentiable and real-valued. Let $z = (x+y)/\sqrt{2}$, $w = (x-y)/\sqrt{2}$, and define $\overline{u}(z,w) = u(x,y)$. Noting that $z_x = z_y = w_x = 1/\sqrt{2}$ and $w_y = -1/\sqrt{2}$, we use the chain rule to compute u_x and u_y:

$$u_x = \overline{u}_z z_x + \overline{u}_w w_x = \overline{u}_z(1/\sqrt{2}) + \overline{u}_w(1/\sqrt{2}) = \frac{1}{\sqrt{2}}(\overline{u}_z + \overline{u}_w)$$

$$u_y = \overline{u}_z z_y + \overline{u}_w w_y = \overline{u}_z(1/\sqrt{2}) + \overline{u}_w(-1/\sqrt{2}) = \frac{1}{\sqrt{2}}(\overline{u}_z - \overline{u}_w)$$

Applying the chain rule again gives

$$u_{xx} = \frac{1}{\sqrt{2}}\big((\overline{u}_z)_x + (\overline{u}_w)_x\big) = \frac{1}{\sqrt{2}}\big(\overline{u}_{zz}z_x + \overline{u}_{zw}w_x\big) + (\overline{u}_{wz}z_x + \overline{u}_{ww}w_x\big)\big)$$

$$= \frac{1}{\sqrt{2}}\big(\overline{u}_{zz}(1/\sqrt{2}) + \overline{u}_{zw}(1/\sqrt{2}) + \overline{u}_{wz}(1/\sqrt{2}) + \overline{u}_{ww}(1/\sqrt{2})\big)$$

$$= \frac{1}{2}(\overline{u}_{zz} + \overline{u}_{zw} + \overline{u}_{wz} + \overline{u}_{ww}),$$

$$u_{yy} = \frac{1}{\sqrt{2}}\big((\overline{u}_z)_y - (\overline{u}_w)_y\big) = \frac{1}{\sqrt{2}}\big(\overline{u}_{zz}z_y + \overline{u}_{zw}w_y\big) - (\overline{u}_{wz}z_y + \overline{u}_{ww}w_y\big)\big)$$

$$= \frac{1}{\sqrt{2}}\big(\overline{u}_{zz}(1/\sqrt{2}) + \overline{u}_{zw}(-1/\sqrt{2}) - \overline{u}_{wz}(1/\sqrt{2}) - \overline{u}_{ww}(-1/\sqrt{2})\big)$$

$$= \frac{1}{2}(\overline{u}_{zz} - \overline{u}_{zw} - \overline{u}_{wz} + \overline{u}_{ww}).$$

That is, $u_{xx} = \frac{1}{2}(\overline{u}_{zz} + \overline{u}_{zw} + \overline{u}_{wz} + \overline{u}_{ww})$ and $u_{yy} = \frac{1}{2}(\overline{u}_{zz} - \overline{u}_{zw} - \overline{u}_{wz} + \overline{u}_{ww})$. Adding these equations gives

$$u_{xx} + u_{yy} = \overline{u}_{zz} + \overline{u}_{ww}.$$

9. Let $f : \mathcal{R}^2 \longrightarrow \mathcal{R}^2$ be defined by $f(u,v) = (2uv, u^2 - v^2) = (x, y)$.

(a) To each angle $0 \le \theta < \pi$ there corresponds a unique line L_θ in the uv-plane that passes through the origin. A parametric representation of L_θ is given by

$$(u, v) = (t\cos\theta, t\sin\theta), \quad -\infty < t < \infty.$$

The image of L_θ under f is then

$$f(t\cos\theta, t\sin\theta) = (2t^2\cos\theta\sin\theta, t^2\cos^2\theta - t^2\sin^2\theta) = t^2(\sin 2\theta, \cos 2\theta), \quad -\infty < t < \infty,$$

which is the collection of all nonnegative scalar multiples of the unit vector $(\sin 2\theta, \cos 2\theta)$ and therefore constitutes a ray in the xy-plane that emanates from the origin at an angle of $\pi/2 - 2\theta$ radians (use the identities $\sin 2\theta = \cos(\pi/2 - 2\theta)$ and $\cos 2\theta = \sin(\pi/2 - 2\theta)$). Clearly, every such ray is the image of a unique line L_θ and all points in the xy-plane except the origin is on a unique such ray.

Observe that if (x_0, y_0) is any point except the origin in the xy-plane then there is a unique angle $0 \le \theta_0 < \pi$ such that $(x_0, y_0) = \big(\sqrt{x_0^2 + y_0^2}\,\sin 2\theta_0, \sqrt{x_0^2 + y_0^2}\,\cos 2\theta_0\big)$, and there are exactly two real numbers t such that $t^2 = \sqrt{x_0^2 + y_0^2}$; namely, $t_1 = (x_0^2 + y_0^2)^{1/4}$ and $t_2 = -(x_0^2 + y_0^2)^{1/4}$. Therefore, (x_0, y_0) is the image of exactly two points in the uv-plane and both are on the line L_{θ_0}. Specifically,

$$f(t_1\cos\theta_0, t_1\sin\theta_0) = f(t_2\cos\theta_0, t_2\sin\theta_0) = (x_0, y_0).$$

(b) Let $\mathcal{C} : u^2 + v^2 = a^2$ $(a > 0)$ be a circle in the uv-plane. Then $\mathcal{C}$ can be parametrized by $(u, v) = (a\cos\theta, a\sin\theta)$, where $0 \le \theta < 2\pi$. The image of $\mathcal{C}$ under f is therefore

$$f(a\cos\theta, a\sin\theta) = (2a^2\cos\theta\sin\theta, a^2\cos^2\theta - a^2\sin^2\theta) = a^2(\sin 2\theta, \cos 2\theta), \quad 0 \le \theta < 2\pi,$$

which is a parametrization of the circle $x^2 + y^2 = a^4\sin^2 2\theta + a^4\cos^2 2\theta = a^4$ in the xy-plane.

(c) The derivative matrix of f is given by $f'(u,v) = \begin{pmatrix} 2v & 2u \\ 2u & -2v \end{pmatrix}$ with Jacobian determinant

$$\det\left(f'(u,v)\right) = -2v^2 - 2u^2 = -2(u^2 + v^2),$$

which is nonzero for $(u,v) \neq (0,0)$. Note too that the entries of $f'(u,v)$ are continuous on the uv-plane so that f is continuously differentiable on all of $\mathcal{R}^2$. By the inverse function theorem, if (u_0, v_0) is a nonzero point then there is a neighborhood N of (u_0, v_0) such that f has a continuously differentiable local inverse function f^{-1} defined on the image set $f(N)$.

11. Suppose $f : \mathcal{R} \longrightarrow \mathcal{R}$ is continuously differentiable and let x_0 be a point in the domain of f such that $f'(x_0) \neq 0$. To fix ideas, assume $f'(x_0) > 0$.

Let ϵ be chosen to satisfy $0 < \epsilon < f'(x_0)$. Since f' is continuous, there exist $\delta > 0$ such that

$$|f'(x) - f'(x_0)| < \epsilon \quad \text{whenever} \quad |x - x_0| < \delta.$$

For such δ, the left inequality above is equivalent to $-\epsilon < f'(x) - f'(x_0) < \epsilon$. Adding $f'(x_0)$ to each member gives

$$f'(x_0) - \epsilon < f'(x) < f'(x_0) + \epsilon.$$

Because of how ϵ was chosen, the leftmost member of this string of inequalities is positive. Hence,

(1) $\qquad\qquad 0 < f'(x)$ for all x in the interval $(x_0 - \delta, x_0 + \delta)$.

Let $I = (x_0 - \delta, x_0 + \delta)$. By the mean-value theorem for derivatives, if $a, b \in I$ and $a < b$ then there exists $x_{a,b} \in (a,b)$ such that

(2) $\qquad\qquad f(b) - f(a) = f'(x_{a,b})(b - a).$

Since $x_{a,b} \in I$, the statement given in (1) implies $f'(x_{a,b}) > 0$. As $b - a$ is positive, equation (2) implies $f(b) - f(a)$ is positive. That is, $f(b) > f(a)$. Since a and b were chosen arbitrarily from I, $f(b) > f(a)$ whenever $a, b \in I$ and $b > a$. So, f is strictly increasing on I. A slight modification of the above argument works if we begin by assuming $f'(x_0) < 0$. In this case, choose ϵ to satisfy $0 < \epsilon < -f'(x_0)$.

13. Suppose $F : \mathcal{R}^n \longrightarrow \mathcal{R}^n$ satisfies the assumptions of the inverse function theorem at some point $\mathbf{x}_0$ in some neighborhood N. By the definition of inverse function,

$$(F^{-1} \circ F)(\mathbf{x}) = \mathbf{x}, \quad \text{for all } \mathbf{x} \in N,$$

so that the derivative matrix of $(F^{-1} \circ F)$ at $\mathbf{x}$ is the $n \times n$ identity matrix, which we will denote by I_n. Therefore, the chain rule gives

$$I_n = \left(F^{-1}\right)'(F(\mathbf{x}))F'(\mathbf{x}), \quad \text{for all } \mathbf{x} \in N.$$

This shows that the $n \times n$ matrix $F'(\mathbf{x})$ is invertible for all $\mathbf{x}$ in N. In particular, $\left(F'(\mathbf{x}_0)\right)^{-1}$ exists. Evaluating the above displayed equation at $\mathbf{x} = \mathbf{x}_0$ and then right-multiplying the resulting equation by $\left(F'(\mathbf{x}_0)\right)^{-1}$ gives

$$\left(F'(\mathbf{x}_0)\right)^{-1} = \left(F^{-1}\right)'(F(\mathbf{x}_0)).$$

215

Section 3: IMPLICIT DIFFERENTIATION

Exercise Set 3 (pgs. 281-283)

1. Consider the equation $x^2 + y^2 = 1$.
   (a) Viewing y as a function of x, implicit differentiation gives

$$2x + 2y\frac{dy}{dx} = 0, \quad \text{so that} \quad \frac{dy}{dx} = -x/y, \quad y \neq 0.$$

Hence, $dy/dx(1/\sqrt{2}, 1/\sqrt{2}) = -1$. Further, it makes sense to evaluate dy/dx at $(0,1)$ but not at $(-1,1)$. The reason is that $(0,1)$ satisfies the given equation and $y \neq 0$, but $(-1,1)$ does not satisfy the given equation.
   (b) Viewing x as a function of y, implicit differentiation gives

$$2x\frac{dx}{dy} + 2y = 0, \quad \text{so that} \quad \frac{dx}{dy} = -y/x, \quad x \neq 0.$$

Hence, $dx/dy(1/\sqrt{2}, -1/\sqrt{2}) = 1$. Further, it makes sense to evaluate dx/dy at $(1,0)$ but not at $(0,1)$. The reason is that while both points satisfy the given equation, dx/dy is not defined for $x = 0$. (NOTE: The graph of the given equation is the unit circle, with vertical tangent line $x = 1$ at the point $(1,0)$. However, $dx/dy(1,0) = 0$, which we normally take to mean that the tangent line is horizontal. But remember, we are viewing x as a function of y. Orienting the coordinate axes with the x-axis vertical and the y-axis horizontal should convince the reader that $dx/dy(1,0) = 0$ is exactly correct!)

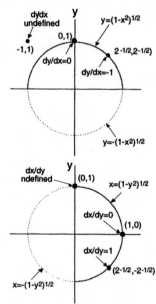

   (c) When the given equation is solved explicitly for y, one obtains $y = \pm\sqrt{1-x^2}$, where "+" is chosen if we are interested in the top half of the circle, and "−" is chosen if we are interested in the bottom half of the circle. The points $(1/\sqrt{2}, 1/\sqrt{2})$ and $(0,1)$ suggest that we are dealing with the function $y(x) = \sqrt{1-x^2}$. It should be noted that had we been asked if it makes sense to evaluate dy/dx at both of the points $(1/\sqrt{2}, 1/\sqrt{2})$ and, say $(0,-1)$ (which is on the bottom half of the circle), then the answer would still have been "yes", but with the qualification that the function y in the two instances are different. The results are shown in the graph on the right.
   (d) When the given equation is solved explicitly for x, one obtains $x = \pm\sqrt{1-y^2}$, where "+" is chosen if we are interested in the right half of the circle, and "−" is chosen if we are interested in the left half of the circle. The points $(1/\sqrt{2}, -1/\sqrt{2})$ and $(1,0)$ suggest that we are dealing with the function $x(y) = \sqrt{1-y^2}$. The same note made in part (c) above can be made here as well. The results are shown in the graph on the right.

3. Given $xy + 1 = 0$, implicit differentiation, first with respect to x and then with respect to y, gives the two equations

$$x\frac{dy}{dx} + y = 0 \qquad \text{and} \qquad x + \frac{dx}{dy}y = 0,$$

216

so that $dy/dx = -y/x$, $x \neq 0$, and $dx/dy = -x/y$, $y \neq 0$. Hence,

$$\frac{dy}{dx}(-1,1) = 1 \qquad \text{and} \qquad \frac{dx}{dy}(-1,1) = 1.$$

5. Given $x + y(x^2 + 1) + \frac{1}{2} = 0$, implicit differentiation, first with respect to x and then with respect to y, gives the two equations

$$1 + 2xy + \frac{dy}{dx}(x^2 + 1) = 0 \qquad \text{and} \qquad \frac{dx}{dy} + 2xy\frac{dx}{dy} + (x^2 + 1) = 0,$$

so that $dy/dx = -(2xy+1)/(x^2+1)$, and $dx/dy = -(x^2+1)/(2xy+1)$, $2xy+1 \neq 0$. Hence,

$$\frac{dy}{dx}(-1,1/4) = -\frac{1}{4} \qquad \text{and} \qquad \frac{dx}{dy}(-1,1/4) = -4.$$

7. Suppose $x^2y + yz = 0$ and $xyz + 1 = 0$.

(a) Since (x, y, z) must satisfy both equations, the second equation shows that $x \neq 0$, $y \neq 0$ and $z \neq 0$. Hence, we can divide the first equation throughout by y to obtain $x^2 + z = 0$. Viewing x as a function of z, implicit differentiation with respect to z gives $2xdx/dz = -1$, so that $dx/dz = -(1/2)x^{-1}$. Replacing z with $-x^2$ in the second equation and solving for y gives $y = x^{-3}$. Viewing y as a function of z, implicit differentiation gives $dy/dz = -3x^{-4}dx/dz = -3x^{-4}(-(1/2)x^{-1}) = (3/2)x^{-5}$. Hence,

$$\frac{dx}{dz}(1,1,-1) = -\frac{1}{2} \qquad \text{and} \qquad \frac{dy}{dz}(1,1,-1) = \frac{3}{2}.$$

(b) From part (a),

$$\frac{dz}{dx}(1,1,-1) = \left(\frac{dx}{dz}(1,1,-1)\right)^{-1} = -2,$$

$$\frac{dy}{dx}(1,1,-1) = \left(\frac{dy}{dz}(1,1,-1)\right)\left(\frac{dx}{dz}(1,1,-1)\right)^{-1} = \left(\frac{3}{2}\right)(-2) = -3.$$

(c) From parts (a) and (b),

$$\frac{dx}{dy}(1,1,-1) = \left(\frac{dy}{dx}(1,1,-1)\right)^{-1} = -\frac{1}{3}, \qquad \frac{dz}{dy}(1,1,-1) = \left(\frac{dy}{dz}(1,1,-2)\right)^{-1} = \frac{2}{3}.$$

(NOTE: The above manipulations may seem a bit too simplistic or even unjustified. To see why they are permissible, suppose x and y are continuous variables such that each is a differentiable function of the other. Since (e.g.) $dy/dy = 1$, the chain rule implies $1 = dy/dy = (dy/dx)(dx/dy)$. Therefore, at any point $\mathbf{x}$ where $dy/dx(\mathbf{x})$ and $dx/dy(\mathbf{x})$ are both defined, $dy/dx(\mathbf{x}) = 1/\big(dx/dy(\mathbf{x})\big) = \big(dx/dy(\mathbf{x})\big)^{-1}$.)

9. Referring to Exercise 7 above, the function $F : \mathcal{R}^3 \longrightarrow \mathcal{R}^2$ is the same for all three parts; namely,

$$F(x,y,z) = \begin{pmatrix} x^2y + yz \\ xyz + 1 \end{pmatrix}.$$

For part (a), $z = \mathbf{x}$ is the independent vector variable and $\begin{pmatrix} x \\ y \end{pmatrix} = \mathbf{y}$ is the dependent vector variable. The function $G(\mathbf{x}) = \mathbf{y}$ is then implicitly defined by the equation

$$(1) \qquad F(\mathbf{x}, G(\mathbf{x})) = F(\mathbf{x}, \mathbf{y}) = \begin{pmatrix} x^2 y + yz \\ xyz + 1 \end{pmatrix} = \begin{pmatrix} 0 \\ 0 \end{pmatrix}.$$

$F_\mathbf{x}$ is computed by holding $\mathbf{y}$ fixed (i.e., holding x and y fixed) in the expression for F and differentiating the coordinate function of F with respect to $\mathbf{x}$ (i.e., with respect to z) to get the 2×1 matrix

$$F_\mathbf{x}(\mathbf{x}, G(\mathbf{x})) = \begin{pmatrix} y \\ xy \end{pmatrix}.$$

Similarly, $F_\mathbf{y}$ is computed by holding $\mathbf{x}$ fixed (i.e., holding z fixed) in the expression for F and differentiating the coordinate functions of F with respect to $\mathbf{y}$ (i.e., with respect to x and y) to get the 2×2 matrix

$$F_\mathbf{y}(\mathbf{x}, G(\mathbf{x})) = \begin{pmatrix} 2xy & x^2 + z \\ yz & xz \end{pmatrix}.$$

For part (b), $x = \mathbf{x}$ is the independent vector variable and $\begin{pmatrix} y \\ z \end{pmatrix} = \mathbf{y}$ is the dependent vector variable. The function $G(\mathbf{x}) = \begin{pmatrix} y \\ z \end{pmatrix} = \mathbf{y}$ is then implicitly defined by equation (1). The functions $F_\mathbf{x}$ and $F_\mathbf{y}$ are computed in the same manner as described in part (a) (keeping in mind the various variable have changed places) and are given by

$$F_\mathbf{x}(\mathbf{x}, G(\mathbf{x})) = \begin{pmatrix} 2xy \\ yz \end{pmatrix}, \qquad F_\mathbf{y}(\mathbf{x}, G(\mathbf{x})) = \begin{pmatrix} x^2 + z & y \\ xz & xy \end{pmatrix}.$$

For part (c), $y = \mathbf{x}$ is the independent vector variable and $\begin{pmatrix} x \\ z \end{pmatrix} = \mathbf{y}$ is the dependent vector variable. The function $G(\mathbf{x})$ is then implicitly defined by equation (1). Computing $F_\mathbf{x}$ an $F_\mathbf{y}$ in the same manner as described in parts (a) and (b), gives

$$F_\mathbf{x}(\mathbf{x}, G(\mathbf{x})) = \begin{pmatrix} x^2 + z \\ xz \end{pmatrix}, \qquad F_\mathbf{y}(\mathbf{x}, G(\mathbf{x})) = \begin{pmatrix} 2xy & y \\ yz & xy \end{pmatrix}.$$

11. Assuming x and y are functions of u, v and w, we implicitly differentiate the given equations $x^2 + yu + xv + w = 0$ and $x + y + uvw + 1 = 0$ with respect to u to get

$$2x\frac{\partial x}{\partial u} + y + \frac{\partial y}{\partial u}u + \frac{\partial x}{\partial u}v = 0, \qquad \frac{\partial x}{\partial u} + \frac{\partial y}{\partial u} + vw = 0.$$

Setting $x = u = v = 1$ and $y = w = -1$, then simplifying gives the two equations

$$3\frac{\partial x}{\partial u}(1, -1, 1, 1, -1) + \frac{\partial y}{\partial u}(1, -1, 1, 1, -1) = 1,$$

$$\frac{\partial x}{\partial u}(1, -1, 1, 1, -1) + \frac{\partial y}{\partial u}(1, -1, 1, 1, -1) = 1,$$

218

which is easily solved to get

$$\frac{\partial x}{\partial u}(1, -1, 1, 1, -1) = 0 \qquad \text{and} \qquad \frac{\partial y}{\partial u}(1, -1, 1, 1, -1) = 1.$$

13. Let the equation $x^2/4 + y^2 + z^2/9 - 1 = 0$ define z implicitly as a function $z = f(x, y)$ near the point $x = 1$, $y = \sqrt{11}/6$, $z = 2$. Implicit differentiation of the given equation with respect to x and y gives

$$\frac{x}{2} + \frac{2z}{9} f_x(x, y) = 0, \qquad 2y + \frac{2z}{9} f_y(x, y) = 0.$$

Setting $x = 1$, $y = \sqrt{11}/6$ and $z = 2$ gives

$$1/2 + (4/9) f_x(1, \sqrt{11}/6) = 0, \qquad \sqrt{11}/3 + (4/9) f_y(1, \sqrt{11}/6) = 0,$$

from which we obtain $f_x(1, \sqrt{11}/6) = -9/8$ and $f_y(1, \sqrt{11}/6) = -3\sqrt{11}/4$. Hence, the equation of the desired tangent plane is

$$z = f(1, \sqrt{11}/6) + (x - 1) f_x(1, \sqrt{11}/6) + (y - \sqrt{11}/6) f_y(1, \sqrt{11}/6)$$
$$= 2 + (x - 1)(-9/8) + (y - \sqrt{11}/6)(-3\sqrt{11}/6) = 9/2 - (9/8)x - (3\sqrt{11}/4)y.$$

Clearing fractions and simplifying gives $9x + (6\sqrt{11})y + 8z = 36$.

15. Let $F : \mathcal{R}^5 \longrightarrow \mathcal{R}^3$ be defined by

$$F(x, y, z, u, v) = \begin{pmatrix} 2x + y + 2z + u - v - 1 \\ xy + z - u + 2v - 1 \\ yz + xz + u^2 + v \end{pmatrix}.$$

The equation $F(x, y, z, u, v) = \mathbf{0}$ implicitly defines x, y and z as functions of u and v near $(x, y, z, u, v) = (1, 1, -1, 1, 1)$.

    (a) By Theorem 3.3, the derivative matrix $f'(u, v)$ of the given implicitly defined function $f(u, v) = (x, y, z)$ is given by

$$f'(\mathbf{x}) = -F_{\mathbf{y}}^{-1}(\mathbf{x}, f(\mathbf{x})) F_{\mathbf{x}}(\mathbf{x}, f(\mathbf{x})) = -\begin{pmatrix} 2 & 1 & 2 \\ y & x & 1 \\ z & z & x+y \end{pmatrix}^{-1} \begin{pmatrix} 1 & -1 \\ -1 & 2 \\ 2u & 1 \end{pmatrix},$$

where $\mathbf{x} = (u, v)$, $\mathbf{y} = (x, y, z)$ and the entries of $F_{\mathbf{x}}$ are found by differentiating the coordinate functions of F while keeping x, y and z fixed; and the entries of $F_{\mathbf{y}}$ are found by differentiating the coordinate functions of F while keeping u and v fixed. Setting $x = y = u = v = 1$, $z = -1$, the inverse matrix $F_{\mathbf{y}}^{-1}$ is found to be

$$F_{\mathbf{y}}^{-1}((1, 1), f(1, 1)) = \begin{pmatrix} 1 & -4/3 & -1/3 \\ -1 & 2 & 0 \\ 0 & 1/3 & 1/3 \end{pmatrix},$$

so that

$$f'(1,1) = -\begin{pmatrix} 2 & 1 & 2 \\ 1 & 1 & 1 \\ -1 & -1 & 2 \end{pmatrix}^{-1} \begin{pmatrix} 1 & -1 \\ -1 & 2 \\ 2 & 1 \end{pmatrix}$$

$$= -\begin{pmatrix} 1 & -4/3 & -1/3 \\ -1 & 2 & 0 \\ 0 & 1/3 & 1/3 \end{pmatrix} \begin{pmatrix} 1 & -1 \\ -1 & 2 \\ 2 & 1 \end{pmatrix} = \begin{pmatrix} -5/3 & 4 \\ 3 & -5 \\ -1/3 & -1 \end{pmatrix}.$$

(b) The two column vectors of $f'(1,1)$ found in part (a) are $f_u(1,1)$ and $f_v(1,1)$, which we write as row vectors: $f_u(1,1) = (-5/3, 3, -1/3)$ and $f_v(1,1) = (4, -5, -1)$. We also have $f(1,1) = (1,1,-1)$. Hence, a parametric representation of the tangent plane at $(1,1,-1)$ is

$$\Gamma(s,t) = sf_u(1,1) + tf_v(1,1) + f(1,1) = s(-5/3, 3, -1/3) + t(4,-5,-1) + (1,1,-1).$$

(NOTE: The equation of the plane in rectangular coordinates is $14x + 9y + 11z = 12$.)

17. (a) Let $F : \mathcal{R}^3 \longrightarrow \mathcal{R}$ be defined by $F(x,y,z) = x^2 + y^2 + z^2 - a^2$. Since $F_x = 2x$, $F_y = 2y$ and $F_z = 2z$ are all continuous, F is continuously differentiable. Let $\mathbf{p_0} = (x_0, y_0, z_0)$ be a point on the sphere $x^2 + y^2 + z^2 = a^2$. Then $F(\mathbf{p_0}) = F(\mathbf{x_0}, \mathbf{y_0}) = 0$ and condition (i) of the implicit function theorem is satisfied, regardless of how the variables x, y and z are related to $\mathbf{x_0}$ and $\mathbf{y_0}$. Now, observe that at least one of x_0, y_0, z_0 must be nonzero (since the origin is not a point on the sphere).

If $x_0 \neq 0$, set $\mathbf{x} = (y,z)$, $\mathbf{y} = x$. Then $F_{\mathbf{y}} = \partial F / \partial x = 2x$ and $F_{\mathbf{y}}(\mathbf{x_0}, \mathbf{y_0}) = 2x_0 \neq 0$ and condition (ii) of the implicit function theorem is satisfied. So there exists a continuously differentiable function $G : \mathcal{R}^2 \longrightarrow \mathcal{R}$ defined on a neighborhood N of $\mathbf{x_0} = (y_0, z_0)$ in $\mathcal{R}^2$ such that $F(\mathbf{x}, G(\mathbf{x})) = 0$ for all $\mathbf{x}$ in N. So, the sphere is smooth at the point $\mathbf{p_0}$.

Similar arguments can be made if $y_0 \neq 0$ or if $z_0 \neq 0$. In the first case, $\mathbf{x} = (x,z)$, $\mathbf{y} = y$ and $F_{\mathbf{y}}(\mathbf{x}, \mathbf{y}) = \partial F / \partial y = 2y$, so that $F_{\mathbf{y}}(\mathbf{x_0}, \mathbf{y_0}) = 2y_0 \neq 0$; and in the second case, $\mathbf{x} = (x,y)$, $\mathbf{y} = z$ and $F_{\mathbf{y}}(\mathbf{x}, \mathbf{y}) = \partial F / \partial z = 2z$, so that $F_{\mathbf{y}}(\mathbf{x_0}, \mathbf{y_0}) = 2z_0 \neq 0$. That is, condition (ii) is satisfied in both cases and, as above, we can conclude that the sphere is smooth at $\mathbf{p_0}$.

(b) Solving $x^2 + y^2 + z^2 = a^2$ for x, y and z gives the six functions

$$x_1(y,z) = \sqrt{a^2 - y^2 - z^2}, \quad x_2(y,z) = -\sqrt{a^2 - y^2 - z^2};$$
$$y_1(x,z) = \sqrt{a^2 - x^2 - z^2}, \quad y_2(x,z) = -\sqrt{a^2 - x^2 - z^2};$$
$$z_1(x,y) = \sqrt{a^2 - x^2 - y^2}, \quad z_2(x,y) = -\sqrt{a^2 - x^2 - y^2}.$$

The twelve partial derivatives of these six functions can be written as

$$\frac{\partial x_k}{\partial y} = (-1)^k y / x_k, \quad \frac{\partial y_k}{\partial x} = (-1)^k x / y_k, \quad \frac{\partial z_k}{\partial x} = (-1)^k x / z_k,$$

$$\frac{\partial x_k}{\partial z} = (-1)^k z / x_k, \quad \frac{\partial y_k}{\partial z} = (-1)^k z / y_k, \quad \frac{\partial z_k}{\partial y} = (-1)^k y / z_k,$$

which all hold for $k = 1, 2$. Note that each of these functions is continuous at all points where its denominator is nonzero. As pointed out in part (a), letting $\mathbf{p_0} = (x_0, y_0, z_0)$ be a point on the sphere, at least one of x_0, y_0, z_0 is nonzero. As an example of how the sphere can be shown to be smooth at $\mathbf{p_0}$, suppose $y_0 \neq 0$. Then the second two partials can be used to conclude that y_1 is continuously differentiable on some neighborhood of (x_0, z_0) if $y_0 > 0$, or that y_2 is continuously differentiable on some neighborhood of (x_0, z_0) if $y_0 < 0$. The remaining five possibilities are done in a similar manner.

19. The equation $xyz - yz^2 + x^2y = 1$ is satisfied by the points on the level 1 set S in $\mathcal{R}^3$ of the function $F(x, y, z) = xyz - yz^2 + x^2y$.

(a) We have $F_x(x, y, z) = yz + 2xy = y(z + 2x)$, so that $F_x(x, y, z) = 0$ when $y = 0$ and when $z + 2x = 0$.

(b) Using the points found in part (a), we first note that y is never zero in the given equation, so that even though $F_x(x, 0, z) = 0$, such points are not in S. For points such that $z + 2x = 0$, we check to see if they are in S by substituting $x = -z/2$ into the given equation. This gives

$$(-z/2)yz - yz^2 + (-z/2)^2 y = 1, \quad \text{or equivalently,} \quad 5yz^2 + 4 = 0.$$

In other words, if (x, y, z) is a point such that $z + 2x = 0$ and $5yz^2 + 4 = 0$ then (x, y, z) is in S and $F_x(x, y, z) = 0$. This is equivalent to saying that if the coordinates of (x_0, y_0, z_0) satisfy both of these equations then $F_y(x_0, y_0, z_0) = F_x(x_0, y_0, z_0) = 0$, so that condition (ii) of the implicit function theorem does not hold.

(c) The given equation is a quadratic equation in x and, as such, we can solve for x using the quadratic formula:

$$x = \frac{-yz \pm \sqrt{y^2z^2 + 4y(yz^2 + 1)}}{2y} = \frac{-yz \pm \sqrt{y(5yz^2 + 4)}}{2y}.$$

It is perhaps revealing that this defines x as two different functions of y and z (one for the "+" sign and one for the "$-$" sign), and that the graphs of these two functions intersect precisely when $y(5yz^2 + 4) = 0$ or, since $y \neq 0$, precisely when $5yz^2 + 4 = 0$. We also have

$$\frac{\partial x}{\partial y} = \mp \frac{1}{y\sqrt{y(5yz^2 + 4)}} \quad \text{and} \quad \frac{\partial x}{\partial z} = -\frac{1}{2} \pm \frac{5yz}{2\sqrt{y(5yz^2 + 4)}},$$

so that x fails to be continuously differentiable only when $5yz^2 + 4 = 0$. The formula for x reduces to $x = -z/2$ when $5yz^2 + 4 = 0$, so that $z + 2x = 0$. Note that the points in S whose coordinates are related by $5yz^2 + 4 = 0$ and $z + 2x = 0$ are the same ones found in part (b).

21. Let $f : \mathcal{R}^n \longrightarrow \mathcal{R}^2$ be continuously differentiable on an open subset S of $\mathcal{R}^n$, and let $\mathbf{y}_0$ be a point in S such that the derivative matrix $f'(\mathbf{y}_0)$ is invertible.

If $F : \mathcal{R}^{2n} \longrightarrow \mathcal{R}^n$ is defined by $F(\mathbf{x}, \mathbf{y}) = \mathbf{x} - f(\mathbf{y})$, for all $\mathbf{x}, \mathbf{y}$ in $\mathcal{R}^n$, then F is continuously differentiable (being the difference of two continuously differentiable functions). Letting $f(\mathbf{y}_0) = \mathbf{x}_0$, we see that $F(\mathbf{x}_0, \mathbf{y}_0) = \mathbf{0}$. Furthermore, the matrix $F_\mathbf{y}$ is found by holding $\mathbf{x}$ fixed and differentiating $\mathbf{x} - f(\mathbf{y})$ with respect to $\mathbf{y}$. In other words, $F_\mathbf{y}(\mathbf{x}_0, \mathbf{y}_0)$ is just the derivative matrix $f'(\mathbf{y}_0)$, which is assumed to be invertible. Hence, the hypotheses of the implicit function theorem hold and, consequently, there exists a continuously differentiable function $G : \mathcal{R}^n \longrightarrow \mathcal{R}^n$ defined on a neighborhood N of $\mathbf{x}_0$ ($= f(\mathbf{y}_0)$) such that

$$F(\mathbf{x}, G(\mathbf{x})) = \mathbf{x} - f(G(\mathbf{x})) = \mathbf{0}, \quad \text{for all } \mathbf{x} \in N.$$

That is, $(f \circ G)(\mathbf{x}) = \mathbf{x}$ for all $\mathbf{x} \in N$. This says that $G = f^{-1}$ on N, which is the conclusion of the inverse function theorem.

Section 4: EXTREME VALUES

Exercise Set 4A-D (pgs. 292-293)

1. Let $f(x,y) = x^2 + 4xy - y^2 - 8x - 6y$. The critical points of f are those for which

$$\nabla f(x,y) = (2x + 4y - 8, 4x - 2y - 6) = (0,0).$$

This gives the linear system $2x + 4y = 8$, $4x - 2y = 6$, with solution $x = 2$, $y = 1$. Thus, the only critical point of f is $(x,y) = (2,1)$.

3. Let $f(x,y) = x^2 - y^2 - 2xy - y$. The critical points of f are those for which

$$\nabla f(x,y) = (2x - 2y, -2y - 2x - 1) = (0,0).$$

This gives the linear system $2x - 2y = 0$, $-2y - 2x = 1$, with solution $x = y = -1/4$. Thus, the only critical point of f is $(x,y) = (-1/4, -1/4)$.

5. Let $f(x,y,z) = x^2 + y^2 - z^2 - xz + x$. The critical points of f are those for which $\nabla f(x,y,z) = (2x - z + 1, 2y, -2z - x) = (0,0,0)$. This gives the linear system

$$2x - z = -1$$
$$2y = 0$$
$$-2z - x = 0,$$

with solution $x = -2/5$, $y = 0$, $z = 1/5$. Thus, $(x,y,z) = (-2/5, 0, 1/5)$ is the only critical point of f.

7. Let $f(x,y,z,w) = xy + yz + zw + wx + x + y - z + w$. The critical points of f are those for which $\nabla f(x,y,z,w) = (y + w + 1, x + z + 1, y + w - 1, z + x + 1) = (0,0,0,0)$. Examination of the first and third coordinates reveals that $y + w + 1 = 0$ and $y + w - 1 = 0$, which are inconsistent. Thus, f has no critical points.

9. Let $f(x,y) = x + y$ and let R be the square with corners $(\pm 1, \pm 1)$. Observe that $f(x,y)$ will be maximum on R when both x and y are maximal on R; i.e., at the corner $(1,1)$. Similarly, $f(x,y)$ will be minimal on R when both x and y are minimal on R; i.e., at the corner $(-1,-1)$.

11. Let $f(x,y) = x^2 + 24xy + 8y^2$ and let R be the disk $x^2 + y^2 \leq 25$. We first look for critical points on the interior of R. Setting $\nabla f(x,y) = (2x + 24y, 16y + 24x) = (0,0)$ gives the equations $2x + 24y = 0$ and $16y + 24x = 0$. The obvious solution is $x = y = 0$, so that $f(0,0) = 0$.

Next, we look for extreme values on the circle $x^2 + y^2 = 25$ (the boundary of R), which has the parametrization $g(t) = (5\cos t, 5\sin t)$, $0 \leq t \leq 2\pi$. The values of f on the boundary are $(f \circ g)(t) = 25(\cos^2 t + 24\cos t \sin t + 8\sin^2 t)$ and the critical points of $f \circ g$ therefore satisfy

$$\frac{d(f \circ g)}{dt}(t) = 25(-2\cos t \sin t + 24\cos^2 t - 24\sin^2 t + 16\sin t \cos t)$$

$$= 25\cos^2 t(24 + 14\tan t - 24\tan^2 t) = 0.$$

By the quadratic formula, $\tan t = 4/3$ and $\tan t = -3/4$. (It should be noted that $\cos^2 t = 0$ when $t = \pi/2$ and $3\pi/2$, and that these values of t do not satisfy the above equation as it is written in the first line. This fact allows us to factor out $\cos^2 t$ as is shown in the second line above.) These two values of t lead to

$$g(t) = (5\cos t, 5\sin t) = (\pm 3, \pm 4), \quad \text{for } \tan t = 4/3,$$
$$g(t) = (5\cos t, 5\sin t) = (\pm 4, \mp 3), \quad \text{for } \tan t = -3/4.$$

The values of f at these points are therefore

$$f(\pm 3, \pm 4) = 425 \qquad \text{and} \qquad f(\pm 4, \mp 3) = -200.$$

Hence, f attains its maximum on R at $(\pm 3, \pm 4)$, and its minimum on R at $(\pm 4, \mp 3)$.

13. Let $f(x,y) = x^2 + y^2 + (2\sqrt{2}/3)xy$ and let R be the elliptic region $x^2 + 2y^2 \leq 1$. We first look for critical points on the interior of R. Setting

$$\nabla f(x,y) = \left(2x + (2\sqrt{2}/3)y, 2y + (2\sqrt{2}/3)x\right) = (0,0)$$

gives the equations $2x + (2\sqrt{2}/3)y = 0$ and $2y + (2\sqrt{2}/3)x = 0$. The obvious solution is $x = y = 0$, so that $f(0,0) = 0$.

Next, we look for extreme values on the ellipse $x^2 + 2y^2 = 1$ (the boundary of R), which has the parametrization $g(t) = \left(\cos t, (1/\sqrt{2})\sin t\right)$, $0 \leq t \leq 2\pi$. The values of f on the boundary are $(f \circ g)(t) = \cos^2 t + \frac{1}{2}\sin^2 t + \frac{2}{3}\cos t \sin t$ and the critical points of $f \circ g$ therefore satisfy

$$\frac{d(f \circ g)}{dt}(t) = -2\cos t \sin t + \cos t \sin t + \frac{2}{3}\cos^2 t - \frac{2}{3}\sin^2 t$$

$$= \frac{1}{3}\cos^2 t(2 - 3\tan t - 2\tan^2 t) = 0.$$

By the quadratic formula, $\tan t = -2$ and $\tan t = 1/2$. (It should be noted that $\cos^2 t = 0$ when $t = \pi/2$ and $3\pi/2$, and that these values of t do not satisfy the above equation as it is written in the first line. This fact allows us to factor out $\cos^2 t$ as is shown in the second line above.) These two values of t lead to

$$g(t) = \left(\cos t, (1/\sqrt{2})\sin t\right) = (\pm 1/\sqrt{5}, \mp 2/\sqrt{10}), \quad \text{for } \tan t = -2,$$
$$g(t) = \left(\cos t, (1/\sqrt{2})\sin t\right) = (\pm 2/\sqrt{5}, \pm 1/\sqrt{10}), \quad \text{for } \tan t = 1/2.$$

The values of f at these points are therefore

$$f(\pm 1/\sqrt{5}, \mp 2/\sqrt{10}) = 1/3 \qquad \text{and} \qquad f(\pm 2/\sqrt{5}, \pm 1/\sqrt{10}) = 7/6.$$

Hence, f attains its maximum of $7/6$ on R at $(\pm 2/\sqrt{5}, \pm 1/\sqrt{10})$, and its minimum of 0 on R at $(0,0)$.

15. Consider the closed curve $\mathcal{C}$ in $\mathcal{R}^3$ parametrized by $g(t) = \left(\cos t, \sin t, \sin(t/2)\right)$. Note that $g(t)$ is periodic with fundamental period 4π. Thus, we can restrict $0 \leq t \leq 4\pi$.

Maximizing the distance from the origin to the points on $\mathcal{C}$ is the same as maximizing the square of the distance. Thus, letting $f(t)$ represent the square of the distance from the origin to the point $g(t)$, we have

$$f(t) = \cos^2 t + \sin^2 t + \sin^2(t/2) = 1 + \sin^2(t/2).$$

Its derivative is $f'(t) = \sin(t/2)\cos(t/2) = \frac{1}{2}\sin t$, so that $f'(t) = 0$ when $t = 0, \pi, 2\pi, 3\pi, 4\pi$. Since

$$f(0) = f(2\pi) = f(4\pi) = 1 \qquad \text{and} \qquad f(\pi) = f(3\pi) = 2,$$

there are two points on $\mathcal{C}$ that are farthest from the origin; namely, $g(\pi) = (-1, 0, 1)$ and $g(3\pi) = (-1, 0, -1)$.

17. The critical points of the function $xy - xz$ that lie inside or on the three-dimensional cube bounded by $|x| \leq 1$, $|y| \leq 1$, $|z| \leq 1$ satisfy $\nabla(xy - yz) = (y - z, x, -x) = (0, 0, 0)$. So $x = 0$ and $y = z$. These points are of the form $(0, t, t)$, where $|t| \leq \sqrt{2}$.

19. The critical points of the function $x^3 - y^3 + z^3 - x + y - z$ that lie in the region R bounded by $0 \leq x \leq 1$, $0 \leq y \leq 1$, $0 \leq z \leq 1$ satisfy

$$\nabla(x^3 - y^3 + z^3 - x + y - z) = (3x^2 - 1, -3y^2 + 1, 3z^2 - 1) = (0, 0, 0).$$

So, $x = \pm 1/\sqrt{3}$, $y = \pm 1/\sqrt{3}$, $z = \pm 1/\sqrt{3}$, where the signs are chosen independently. However, of these eight points, only one is in R; namely, $(1/\sqrt{3}, 1/\sqrt{3}, 1/\sqrt{3})$.

21. Let $z = f(x, y) = x + y^2$. To minimize f subject to the condition $2x^2 + y^2 = 1$, it is easiest to simply write the restricting condition as $y^2 = 1 - 2x^2$ and substitute back into f to get $z = x + 1 - 2x^2$. This is a parabola opening downward. Thus, its minimum value occurs either when x is maximal or when x is minimal. Since x must satisfy $2x^2 + y^2 = 1$, we see that its square can be as large as $1/2$, but no larger. That is, $-1/\sqrt{2} \leq x \leq 1/\sqrt{2}$. The y value on the ellipse at these values of x is $y = 0$. So, the minimum value of f on the ellipse is the smaller of the two values

$$z = f(\pm 1/\sqrt{2}, 0) = \pm 1/\sqrt{2} + 0^2 = \pm 1/\sqrt{2};$$

i.e., $z = -1/\sqrt{2}$. (NOTE: This problem can also be solved by parametrizing the ellipse by some function $g(t)$, forming the composite function $(f \circ g)(t)$ and then finding the critical points of $f \circ g$. It also lends itself to the Lagrange multiplier method. However, the above method uses no differentiation, only an understanding of the geometry of parabolas.)

23. Let x be the length and width of the square base of the box and let y be the height of the box. The surface area is then given by the function $S(x, y) = 4xy + x^2$. We want to minimize S subject to the condition that the volume of the box is to be 108 cubic units; i.e., subject to the condition $x^2 y = 108$. Using the Lagrange multiplier λ we find the critical points of the function

$$4xy + x^2 + \lambda(x^2 y - 108).$$

Setting its gradient equal to zero leads to the two equations

$$4y + 2x + 2xy\lambda = 0, \qquad 4x + \lambda x^2 = 0.$$

The second equation holds when $x = 0$ and when $4 + \lambda x = 0$. The choice $x = 0$ must be rejected as it does not satisfy the condition $x^2 y = 108$. So, $4 + \lambda x = 0$, or $x = -4/\lambda$. Inserting this expression for x into the first equation and solving for y gives $y = -2/\lambda$. It follows that $2y = x$. The condition $x^2 y = 108$ is therefore equivalent to $4y^3 = 108$, from which we obtain $y = 3$. So, $x = 6$. Thus, if $S(x, y)$ can be minimized subject to the given volume condition then it must be minimized for $x = 6$, $y = 3$.

To find out whether $(x, y) = (6, 3)$ gives a minimum, a maximum, or neither, observe that the volume condition can be written as $y = 108/x^2$, so that the surface area can be written as a function of x only; namely,

$$S_1(x) = S(x, 108/x^2) = 432/x + x^2.$$

Hence, $S_1'(x) = -432/x^2 + 2x$, $S_1''(x) = 864/x^3 + 2$, so that $S_1''(6) = 6 > 0$. By the second-derivative test from one-variable calculus, $S_1(6)$ is a local minimum. Thus, the dimensions of the box that will minimize surface area are

$$\text{side length of the square base} = 6, \quad \text{height} = 3.$$

(NOTE: The reader might have noticed that the surface area could have been written in terms of x from the beginning, allowing us to use the usual one-variable calculus methods of solving extreme-value problems. Unfortunately, in practical situations, this is the exception and not the rule. The Langrange multiplier method, while not being definitive, can be a valuable tool for finding critical points *if they exist*. In light of this, we need all the practice we can get.)

25. Let x be the length and width of the square base of the box and let y be the height of the box. If $C > 0$ is the per square unit area cost of material for the least expensive sides then the total cost of materials for the box is

$$P(x,y) = 3Cxy + 2Cxy + 3Cx^2 = 5Cxy + 3Cx^2,$$

where $3Cxy$ is the total cost of the three least expensive sides, $2Cxy$ is the cost of the fourth side, and $3Cx^2$ is the cost of the base. We want to minimize P subject to the constraint $x^2y = V$, where V ($\neq 0$) is some unspecified but constant volume. Using the Lagrange multiplier λ, we find the critical points of the function

$$5Cxy + 3Cx^2 + \lambda(x^2y - V).$$

Setting its gradient equal to zero leads to the two equations

$$5Cy + 6Cx + 2\lambda xy = 0, \qquad 5Cx + \lambda x^2 = 0.$$

The second equation is equivalent to $x(5C + \lambda x) = 0$, so either $x = 0$ or $5C + \lambda x = 0$. Since $x = 0$ is inconsistent with the condition $x^2y = V \neq 0$, we reject $x = 0$ and conclude that $5C + \lambda x = 0$. Since $C \neq 0$, neither is λ, and we can write $x = -5C/\lambda$. Inserting this into the first equation and solving for y gives $y = -6C/\lambda$. The two equations $x = -5C/\lambda$, $y = -6C/\lambda$ implies $6x = 5y$. The condition $x^2y = V$ is therefore equivalent to $x^2(6x/5) = V$, or $x = (5V/6)^{1/3}$. So, $y = (36V/25)^{1/3}$.

To find out whether $(x,y) = \big((5V/6)^{1/3}, (36V/25)^{1/3}\big)$ gives a minimum, a maximum, or neither, observe that the volume condition can be written as $y = V/x^2$, so that the surface area can be written as a function of x only; namely,

$$P_1(x) = P(x, V/x^2) = 5CV/x + 3Cx^2.$$

Hence, $P_1'(x) = -5CV/x^2 + 6Cx$, $P_1''(x) = 10CV/x^3 + 6C$, so that $P_1''\big((5V/6)^{1/3}\big) = 18C > 0$. By the second-derivative test from one-variable calculus, $P_1\big((5V/6)^{1/3}\big)$ is a local minimum. Thus, the dimensions of the box that will minimize cost are

$$\text{side length of the square base} = (5V/6)^{1/3}, \quad \text{height} = (36V/25)^{1/3}.$$

(See the NOTE following Exercise 23 above.)

27. Let $f(x,y,z) = x^2 + xy + y^2 + yz + z^2$. In part (a) below, we are asked to maximize f given the restriction $x^2 + y^2 + z^2 = 1$; i.e., the points in the domain of f that are under consideration are those on the unit sphere in $\mathcal{R}^3$. Since the unit sphere is closed and bounded, Theorem 4.1 guarantees that f will attain its maximum value there.

(a) To maximize f subject to the condition $x^2 + y^2 + z^2 = 1$, we use the Lagrange multiplier λ to find the critical points of the function

$$x^2 + xy + y^2 + yz + z^2 + \lambda(x^2 + y^2 + z^2 - 1).$$

Setting its gradient equal to zero leads to the three equations

$$2x + y + 2\lambda x = 0, \qquad x + 2y + z + 2\lambda y = 0, \qquad y + 2z + 2\lambda z = 0,$$

which can be more suggestively written as

$$y = -2x(1 + \lambda), \qquad x + z = -2y(1 + \lambda), \qquad y = -2z(1 + \lambda).$$

If $\lambda = -1$ then $y = 0$ and $x + z = 0$. The condition $x^2 + y^2 + z^2 = 1$ is then equivalent to $2x^2 = 1$, so that $x = \pm 1/\sqrt{2}$ and $z = -x = \mp 1/\sqrt{2}$. Thus, we have the two critical points

$$(1/\sqrt{2}, 0, -1/\sqrt{2}) \qquad \text{and} \qquad (-1/\sqrt{2}, 0, 1/\sqrt{2}).$$

If $\lambda \neq -1$, we add the first and third equations and use the second equation to get

$$(1) \qquad 2y = -2x(1 + \lambda) - 2z(1 + \lambda) = -2(x + z)(1 + \lambda) = 4y(1 + \lambda)^2.$$

If $y = 0$ then the first and third equations, along with $1 + \lambda \neq 0$, gives $x = z = 0$, which violates $x^2 + y^2 + z^2 = 1$. So $y \neq 0$ and equation (1) is therefore equivalent to $2 = 4(1 + \lambda)^2$, from which $1 + \lambda = \pm 1/\sqrt{2}$. With these values of $1 + \lambda$, the first and third equations reduce to $y = \pm x\sqrt{2}$ and $y = \pm z\sqrt{2}$, so that $x = z = \pm y/\sqrt{2}$. The equation $x^2 + y^2 + z^2 = 1$ then gives $(\pm y/\sqrt{2})^2 + y^2 + (\pm y/\sqrt{2})^2 = 1$, so that $2y^2 = 1$, or $y = \pm 1/\sqrt{2}$. From this we get $x = z = \pm 1/2$, where the sign on x and z can be chosen independently from the sign on y. This gives the four critical points:

$$(1/2, -1/\sqrt{2}, 1/2), \quad (-1/2, 1/\sqrt{2}, -1/2), \quad (1/2, 1/\sqrt{2}, 1/2), \quad (-1/2, -1/\sqrt{2}, -1/2).$$

Evaluating f at the six critical points gives

$$f(\pm 1/\sqrt{2}, 0, \mp 1/\sqrt{2}) = 1,$$
$$f(\pm 1/2, \pm 1/\sqrt{2}, \pm 1/2) = 1 + 1/\sqrt{2},$$
$$f(\pm 1/2, \mp 1/\sqrt{2}, \pm 1/2) = 1 - 1/\sqrt{2}.$$

Hence, f attains its maximum value of $1 + 1/\sqrt{2}$ at the two points $\pm(1/2, 1/\sqrt{2}, 1/2)$.

(b) To maximize f subject to the two conditions $x^2 + y^2 + z^2 = 1$ and $x + \sqrt{2}y + z = 0$, we first observe that f can be written as $f(x, y, z) = x^2 + y^2 + z^2 + (x + z)y$, so that

$$f(x, y, z) = 1 + (-\sqrt{2}\,y)y = 1 - \sqrt{2}\,y^2.$$

Hence, $f(x, y, z) \leq 1$ for all (x, y, z) satisfying the two conditions. In part (a), we saw that $f(\pm 1/\sqrt{2}, 0, \mp 1/\sqrt{2}) = 1$. We also note that these two points are on the plane. It follows that the maximum value of f is 1, attained at the two points $(\pm 1/\sqrt{2}, 0, \mp 1/\sqrt{2})$.

29. Let $f(x, y, z, w) = x^2 + y^2 + z^2 + w^2$ and note that f is the square of the distance from the point (x, y, z, w) in $\mathcal{R}^4$ to the origin. We therefore want to minimize f subject to the two conditions $x + y - z - 2w = 1$ and $x - y + z + w = 2$. Using the Lagrange multipliers λ and μ, we find the critical points of the function

$$x^2 + y^2 + z^2 + w^2 + \lambda(x + y - z - 2w - 1) + \mu(x - y + z + w - 2).$$

Setting its gradient equal to zero leads to the four equations

$$2x + \lambda + \mu = 0, \qquad 2y + \lambda - \mu = 0, \qquad 2z - \lambda + \mu = 0, \qquad 2w - 2\lambda + \mu = 0.$$

These four equations, together with the equations of the two given planes, is a linear system of six equations in six unknowns. This solver chose to use the elimination method. The solution is found to be

$$x = 27/19, \quad y = -7/19, \quad z = 7/19, \quad w = -3/19, \quad \lambda = -20/19, \quad \mu = -34/19.$$

Although the values for λ and μ are not needed to determine the desired solution, they are nevertheless included here as a check for the reader.

Now, the intersection of the given planes, which the authors denote by $\mathcal{F}$, is a plane in $\mathcal{R}^3$. As such, there is a unique point on $\mathcal{F}$ that is nearest the origin, so that f attains a minimum value on $\mathcal{F}$; and, since the Lagrange multiplier method must give all points where f attains its minimum on $\mathcal{F}$, the point we have found must be the desired point.

31. We proceed by induction on n. The given inequality clearly holds for $n = 1$. Suppose it holds for $n = N$ and all collections of N positive real numbers $a_1, \ldots, a_N$. Pick any collection $\mathcal{C}_0$ of N positive real numbers and fix it, say $\mathcal{C}_0 = \{a_1, \ldots, a_N\}$, so that

$$(1) \qquad (a_1 \cdots a_N)^{1/N} \leq \frac{a_1 + \cdots + a_N}{N};$$

and define the function $f(x)$ by

$$f(x) = \frac{a_1 + \cdots + a_N + x}{N+1} - (a_1 \cdots a_N x)^{1/(N+1)}, \qquad x > 0.$$

We have

$$f'(x) = \frac{1}{N+1} - \frac{1}{N+1}(a_1 \cdots a_N x)^{1/(N+1)-1}(a_1 \cdots a_N)$$
$$= \frac{1}{N+1}\left(1 - (a_1 \cdots a_N)^{1/(N+1)} x^{-N/(N+1)}\right),$$

so that $f'(x) = 0$ if, and only if, $x = (a_1 \cdots a_N)^{1/N}$. Moreover,

$$f''(x) = \frac{N}{(N+1)^2}(a_1 \cdots a_N)^{1/(N+1)} x^{-N/(N+1)-1} > 0, \qquad \text{for all } x > 0.$$

In particular, $f''\left((a_1 \cdots a_N)^{1/N}\right) > 0$, so that $f\left((a_1 \cdots a_N)^{1/N}\right)$ is a global minimum for f. Now,

$$f\left((a_1 \cdots a_N)^{1/N}\right) = \frac{a_1 + \cdots + a_N + (a_1 \cdots a_N)^{1/N}}{N+1} - \left(a_1 \cdots a_N (a_1 \cdots a_N)^{1/N}\right)^{1/(N+1)}$$
$$= \frac{a_1 + \cdots + a_N + (a_1 \cdots a_N)^{1/N}}{N+1} - (a_1 \cdots a_N)^{1/N}$$
$$= \frac{N}{N+1}\left(\frac{a_1 + \cdots + a_N}{N} - (a_1 \cdots a_N)^{1/N}\right) \geq 0,$$

227

where the last inequality follows from the induction hypothesis given by equation (1). It follows that $f(x) \geq 0$ for all $x > 0$. Thus, for our specific collection $\mathcal{C}_0$,

$$(a_1 \cdots a_N a_{N+1})^{1/(N+1)} \leq \frac{a_1 + \cdots + a_N + a_{N+1}}{N+1}, \qquad \text{for all } a_{N+1} > 0.$$

Since $\mathcal{C}_0$ was an arbitrary collection of N positive real numbers it follows that the given inequality holds for every collection of $N+1$ positive real numbers. We have shown that if the given inequality holds for some $n = N$ then it holds for $n = N+1$. By induction, it holds for all n. This completes the proof.

33. (a) The set C_h is the curve of intersection of the unit sphere $x^2 + y^2 + z^2 = 1$ in $\mathcal{R}^3$ and the plane $z = h$. If $|h| > 1$ then the plane does not intersect the sphere and $C_h = \emptyset$; but if $|h| < 1$ then the curve of intersection is a circle or radius $\sqrt{1 - h^2}$ in the plane $z = h$. Hence, C_h is closed and bounded if, and only if, $|h| < 1$.

   (b) Let $f(x, y, z) = x$. To apply the Lagrange multiplier method to the function f restricted to C_h ($|h| < 1$) means that we look for the critical points of the function $x + \lambda(x^2 + y^2 + z^2 - 1) + \mu(z - h)$. Setting its gradient equal to zero gives the three equations

$$1 + 2\lambda x = 0, \qquad 2\lambda y = 0, \qquad 2\lambda z + \mu = 0.$$

The first equation says $\lambda \neq 0$, so the first two equations imply $x = -1/(2\lambda)$ and $y = 0$, respectively. The equations $z = h$ and $x^2 + y^2 + z^2 = 1$ then gives $1 = 1/(2\lambda)^2 + h^2$, from which $\lambda = \pm 1/(2\sqrt{1 - h^2})$. So, $x = \pm\sqrt{1 - h^2}$ and there are two critical points; namely, $(\pm\sqrt{1 - h^2}, 0, h)$.

   By part (a), C_h is closed and bounded. Therefore, by Theorem 4.1, f attains its absolute maximum and minimum values on C_h. Further, C_h is a level set of the differentiable function $x^2 + y^2$. Since f is differentiable as well, the Lagrange multiplier method must give the points at which these extreme values are attained, it follows that $\sqrt{1 - h^2}$ is the absolute maximum of f on C_h and $-\sqrt{1 - h^2}$ is the absolute minimum of f on C_h.

   (c) The set C_1 is the set of points on the intersection of the unit sphere and the plane $z = 1$, which consists of a single point; namely, $(0, 0, 1)$. It follows that f restricted to C_1 has only one value, namely, $f(0, 0, 1) = 0$. So, 0 is both the maximum and minimum value on C_1.

35. We seek to find the extreme values of $f(x, y, z) = x^2 - y^2 + z^2$, subject to the conditions $x^2 + y^2 + z^2 = 3$ and $x + y + z = 3$. We are told by the authors that there is a "peculiar reason" why there is a unique point that gives an extreme value for f. With our suspicions aroused, we examine a little more closely the relationship between the given sphere and the given plane and find that the plane is tangent to the sphere! To convince ourselves that this is the case, we first write the vector equation of the plane $x + y + z = 3$ as $(1, 1, 1) \cdot \mathbf{x} = 3$, so that the normalized equation has the form $(1/\sqrt{3}, 1/\sqrt{3}, 1/\sqrt{3}) \cdot \mathbf{x} = 3/\sqrt{3} = \sqrt{3}$. The distance from this plane to the origin is the absolute value of the number

$$(1/\sqrt{3}, 1/\sqrt{3}, 1/\sqrt{3}) \cdot (0, 0, 0) - \sqrt{3} = -\sqrt{3}.$$

That is, the plane $x + y + z = 3$ is $\sqrt{3}$ units from the origin. Since the sphere $x^2 + y^2 + z^2 = 3$ has radius $\sqrt{3}$, it follows that the plane is tangent to the sphere. Moreover, the point of tangency is on the sphere in the direction of the normal vector $(1, 1, 1)$; i.e., at the point $(1, 1, 1)$. Since this is the only point at which f can be evaluated under the given conditions, $f(1, 1, 1) = 1$ is both the absolute maximum and absolute minimum value attained by f subject to the given conditions.

37. (a) If C is the square unit cost of the side material then $2C$ is the square unit cost of the roof materials. If l, w and h denote the length, width and height of the shed then the cost of roof materials is $2Clw$ and the total cost of the side material is $2Clh + Cwh$. Hence, the total cost of materials to construct the shed is

$$P(l, w, h) = 2Clw + 2Clh + Cwh.$$

We want to minimize P subject to the condition that the volume of the shed is to be 108 cubic feet; i.e., subject to the condition $lwh = 108$. We first find the critical points of P by using the Lagrange multiplier method to find the critical points of the function

$$2Clw + 2Clh + Cwh + \lambda(lwh - 108).$$

Setting its gradient equal to zero leads to the three equations

$$2Cw + 2Ch + \lambda wh = 0, \qquad 2Cl + Ch + \lambda lh = 0, \qquad 2Cl + Cw + \lambda lw = 0.$$

Subtracting the third equation from the second and factoring gives $(h - w)(C + \lambda l) = 0$, so that either $h = w$ or $C = -\lambda l$. However, inserting $C = -\lambda l$ into the second equation leads to $2Cl = 0$, which is impossible since $C \neq 0$ and $l = 0$ is inconsistent with $lwh = 108$. So we must have $h = w$. Setting $h = w$ in the first equation, simplifying, and factoring, gives $w(4C + \lambda w) = 0$. Since $w = 0$ is inconsistent with $lwh = 108$, we must have $4C + \lambda w = 0$. Since $C \neq 0$, $\lambda \neq 0$ and we can write $w = -4C/\lambda$. Inserting this into the third equation and solving for l gives $l = -2C/\lambda$. The three equations $h = w$, $w = -4C/\lambda$, $l = -2C/\lambda$ implies $h = w = 2l$. The condition $lwh = 108$ then leads to $l(2l)(2l) = 108$, or equivalently, $4l^3 = 108$, from which $l = 3$. So, $h = w = 6$. So there is at most one critical point of P; namely, $(l, w, h) = (3, 6, 6)$.

To show that the above obtained critical point minimizes P, observe that if we fix $l = 1$ then the condition $lwh = 108$ becomes $wh = 108$, so that w (e.g.) can be made arbitrarily large by choosing h sufficiently close to zero. Therefore, $P(1, w, h) = C(2w + 2h + wh)$ has no maximum value. It follows that the above critical point minimizes P. That is, the dimensions that will minimize cost are

$$\text{length} = 3, \qquad \text{width} = 6, \qquad \text{height} = 6.$$

(b) As in part (a), let C be the unit cost of the side material and suppose that C is also the cost of the roof material. Using the same reasoning as in part (a), the new cost function we want to minimize is

$$Q(l, w, h) = Clw + 2Clh + Cwh,$$

where this minimization is still subject to the condition $lwh = 108$. Again, we use a Lagrange multiplier and find the critical points of the function

$$Clw + 2Clh + Cwh + \lambda(lwh - 108).$$

Setting its gradient equal to zero leads to the three equations

$$Cw + 2Ch + \lambda wh = 0, \qquad Cl + Ch + \lambda lh = 0, \qquad 2Cl + Cw + \lambda lw = 0.$$

Subtract the third equation from the first equation and factor to get $(2C + \lambda w)(h - l) = 0$. Hence, either $2C + \lambda w = 0$ or $h = l$. If $2C + \lambda w = 0$ then $\lambda w = -2C$. When this is inserted into the first equation, the result is $Cw = 0$. As $C \neq 0$ and $w \neq 0$, the choice $2C + \lambda w = 0$ is rejected. We conclude that $h = l$. Inserting $h = l$ into the second equation and factoring gives $l(2C + \lambda l) = 0$. Since $l \neq 0$, we must have $2C + \lambda l = 0$. This equation shows that since C and l are nonzero, so is λ and we may write $l = -2C/\lambda$. Using this expression for l in the third equation and then solving for w gives $w = -4C/\lambda$. The three equations $h = l$, $l = -2C/\lambda$, $w = -4C/\lambda$ implies $w = 2h = 2l$. The condition $lwh = 108$ can then be written as $l(2l)l = 108$, from which $l = 3(2)^{1/3}$. Arguing in the same way as in part (a), this critical point must minimize the cost. That is, the dimensions that will minimize the cost are

$$\text{length} = 3(2)^{1/3}, \qquad \text{width} = 6(2)^{1/3}, \qquad \text{height} = 3(2)^{1/3}.$$

Exercise Set 4E (pgs. 297-298)

1. (a) The functions $f(x) = x^3$ and $g(x) = x^4$ both have the single critical point $x = 0$. The function f is positive for $x > 0$ and negative for $x < 0$, so that $f(0) = 0$ is neither a maximum nor a minimum for f. On the other hand, g satisfies $g(x) \geq 0$ for all x, with equality if, and only if, $x = 0$, so that $g(0) = 0$ is a strict global minimum. However, $f''(0) = 0$ and $g''(0) = 0$, so that the second derivative test fails to distinguish between them.

(b) Let $f(x, y) = (x + y)^3$. We have $f_x(x, y) = f_y(x, y) = 3(x + y)^2$, so that these are both zero when $x + y = 0$. That is, $(a, -a)$ is a critical point of f for every real number a. Note that $f(a, -a) = 0$ for each critical point, while for any $\epsilon > 0$,

$$f\big(a, -(a + \epsilon)\big) = -\epsilon^3 < 0 < \epsilon^3 = f(a + \epsilon, -a),$$

which shows that $f(a, -a) = 0$ is neither a local maximum nor a local minimum value of f. Moreover, by inspection, we see that all second-order partials of f are equal and have the value zero at each critical point. Thus, although each critical point of f is a saddle point, Theorem 4.4 gives no information as to the behavior of f at any of its critical points.

For the function $g(x, y) = (x + y)^4$, we have $g_x(x, y) = g_y(x, y) = 4(x + y)^3$, so that $(a, -a)$ is a critical point of g for every real number a. Note that $g(x, y) \geq 0$ for all (x, y), with equality if, and only if, (x, y) is a critical point of g. So $g(a, -a) = 0$ is a global minimum value of g but it is not a *strict* global minimum because each neighborhood of a critical point contains infinitely many critical points. However, as in the case with f, all second-order partials of g are equal and have the value zero at each critical point. Thus, although each critical point of g globally minimizes g, Theorem 4.4 gives no information as to the behavior of g at any of its critical points.

3. Let $f(x, y) = x^2 - 2x + y^2 + 4y$. Then $f_x(x, y) = 2x - 2$ and $f_y(x, y) = 2y + 4$. Setting these equal to zero leads to $x = 1$, $y = -2$, so that $(1, -2)$ is the only critical point of f. Since $f_{xx}(x, y) = 2$, $f_{xy}(x, y) = 0$ and $f_{yy}(x, y) = 2$, we also have $f_{xx}(1, -2) = f_{yy}(1, -2) = 2$, $f_{xy}(1, -2) = 0$. Thus, by Theorem 4.5, if $\mathbf{u} = (u, v)$ is a unit vector then

$$\frac{\partial^2 f}{\partial \mathbf{u}^2}(1, -2) = f_{xx}(1, -2)u^2 + 2f_{xy}(1, -2)uv + f_{yy}(1, -2)v^2 = 2(u^2 + v^2) > 0.$$

By Theorem 4.4, $f(1, -2) = -5$ is a strict local minimum value of f.

5. Let $f(x, y) = x^2 - xy - y^2 + 5y$. Then $f_x(x, y) = 2x - y$ and $f_y(x, y) = -x - 2y + 5$. Setting these equal to zero leads to $x = 1$, $y = 2$, so that $(1, 2)$ is the only critical point of f. Since $f_{xx}(x, y) = 2$, $f_{xy}(x, y) = -1$ and $f_{yy}(x, y) = -2$, we also have $f_{xx}(1, 2) = 2$, $f_{xy}(1, 2) = -1$, $f_{yy}(1, 2) = -2$. Since

$$f_{xx}(1, 2)f_{yy}(1, 2) - f_{xy}^2(1, 2) = (2)(-2) - (-1)^2 = -5 < 0,$$

Theorem 4.6 shows that $(1, 2)$ is a saddle point of f.

7. Let $f(x, y) = x^4 + y^4$, so that $f_x(x, y) = 4x^3$ and $f_y(x, y) = 4y^3$. Setting these equal to zero easily gives $x = 0$, $y = 0$, so that $(0, 0)$ is the only critical point of f. Without appealing to any further tests, we simply observe that $f(x, y) \geq 0$, for all (x, y), with equality if, and only if, $(x, y) = (0, 0)$. It follows that $f(0, 0) = 0$ is a strict global minimum value of f.

9. Let $f(x, y) = x^2 + 2xy$, so that $f_x(x, y) = 2x + 2y$ and $f_y(x, y) = 2x$. Setting these equal to zero gives the solution $x = y = 0$, so that $(0, 0)$ is the only critical point of f. Note that $f(x, x) = 3x^2$ and $f(x, -x) = -x^2$, so that every neighborhood of $(0, 0)$ contains points for which f is positive and points for which f is negative. Hence, $(0, 0)$ is a saddle point of f.

11. Let $f(x,y) = x^{-1} + xy - 8y^{-1}$, so that $f_x(x,y) = -x^{-2} + y$ and $f_y(x,y) = x + 8y^{-2}$. Setting these equal to zero gives the system $-x^{-2} + y = 0$, $x + 8y^{-2} = 0$. The first equation gives $y = x^{-2}$, so that the second equation is equivalent to $x + 8x^4 = 0$, from which $x = 0$ or $x = -1/2$. Since points on the coordinate axes are not in the domain of f, we reject $x = 0$ and keep $x = -1/2$, from which we get $y = x^{-2} = 4$. Thus, $(-1/2, 4)$ is the only critical point of f. Also, $f_{xx}(x,y) = 2x^{-3}$, $f_{xy}(x,y) = 1$ and $f_{yy}(x,y) = -16y^{-3}$, so that $f_{xx}(-1/2, 4) = -16$, $f_{xy}(-1/2, 4) = 1$, $f_{yy}(-1/2, 4) = -1/4$. Since $f_{xx}(-1/2, 4) < 0$ and

$$f_{xx}(-1/2, 4)f_{yy}(-1/2, 4) - f_{xy}^2(-1/2, 4) = (-16)(-1/4) - 1 = 3 > 0,$$

Theorem 4.6 (ii) says that $f(-1/2, 4) = -6$ is a strict local maximum value of f.

13. Let $f(x,y) = e^{x^2 - y^2}$, so that $f_x(x,y) = 2xe^{x^2 - y^2}$ and $f_y(x,y) = -2ye^{x^2 - y^2}$. Setting these equal to zero gives $x = 0$, $y = 0$, so that $(0,0)$ is the only critical point of f. Note that $f(0,y) = e^{-y^2} < 1$ for $y \neq 0$, and $f(x,0) = e^{x^2} > 1$ for $x \neq 0$. It follows that $f(0,0) = 1$ is neither a local minimum nor a local maximum. Hence, $(0,0)$ is a saddle point of f.

15. (a) Let $p(x,y) = ax^2 + bxy + cy^2$. Then $p_x(x,y) = 2ax + by$, $p_{xx}(x,y) = 2a$, $p_{xy}(x,y) = b$, $p_y(x,y) = bx + 2cy$, $p_{yy}(x,y) = 2c$. So, if $\mathbf{u} = (u, v)$ is a unit vector then, by Theorem 4.5,

$$\frac{1}{2}\frac{\partial^2 p}{\partial \mathbf{u}^2}(x,y) = \frac{1}{2}\left(p_{xx}(x,y)u^2 + 2p_{xy}(x,y)uv + p_{yy}(x,y)v^2\right)$$

$$= \frac{1}{2}(2au^2 + 2buv + 2cv^2) = au^2 + buv + cv^2 = p(u,v),$$

which is independent of (x,y).

(b) Let $q(x,y,z) = ax^2 + by^2 + cz^2 + lyz + mxz + nxy$. Then

$$q_x(x,y,z) = 2ax + mz + ny, \qquad q_{xx}(x.y.z) = 2a, \qquad q_{xz}(x,y) = m,$$
$$q_y(x,y,z) = 2by + lz + nx, \qquad q_{yy}(x,y,z) = 2b, \qquad q_{yz}(x,y) = l,$$
$$q_z(x,y,z) = 2cz + ly + mx, \qquad q_{zz}(x,y,z) = 2c, \qquad q_{xy}(x,y) = n.$$

By definition, if $\mathbf{u} = (u, v, w)$ is a unit vector then

$$\frac{\partial^2 q}{\partial \mathbf{u}^2} = \frac{\partial}{\partial \mathbf{u}}\left(\frac{\partial q}{\partial \mathbf{u}}\right) = \frac{\partial}{\partial \mathbf{u}}(\nabla q \cdot \mathbf{u}) = \frac{\partial}{\partial \mathbf{u}}(q_x u + q_y v + q_z w)$$

$$= \nabla(q_x u + q_y v + q_z w) \cdot (u, v, w)$$

$$= (q_x u + q_y v + q_z w)_x u + (q_x u + q_y v + q_z w)_y v + (q_x u + q_y v + q_z w)_z w$$

$$= (q_{xx} u + q_{yx} v + q_{zx} w)u + (q_{xy} u + q_{yy} v + q_{zy} w)v + (q_{xz} u + q_{yz} v + q_{zz} w)w$$

$$= q_{xx} v^2 + q_{yy} v^2 + q_{zz} w^2 + 2q_{yz} vw + 2q_{xz} uw + 2q_{xy} uv$$

$$= 2au^2 + 2bv^2 + 2cw^2 + 2lvw + 2muw + 2nuv = 2q(u,v,w),$$

which is independent of (x,y,z). Dividing throughout by $1/2$ gives the desired result.

17. Let $f(x,y)$ be a real-valued twice continuously differentiable function. The 2×2 Hessian matrix of f is given by

$$H_f(\mathbf{x}) = \begin{pmatrix} f_{xx}(\mathbf{x}) & f_{xy}(\mathbf{x}) \\ f_{yx}(\mathbf{x}) & f_{yy}(\mathbf{x}) \end{pmatrix}.$$

Hence, if I is the 2×2 identity matrix then, for a critical point $\mathbf{x}_0$ of f, the eigenvalues λ of $H_f(\mathbf{x}_0)$ satisfy

$$
\begin{aligned}
0 = \det\big(H_f(\mathbf{x}_0) - \lambda I\big) &= \det\left\{ \begin{pmatrix} f_{xx}(\mathbf{x}_0) & f_{xy}(\mathbf{x}_0) \\ f_{yx}(\mathbf{x}_0) & f_{yy}(\mathbf{x}_0) \end{pmatrix} - \begin{pmatrix} \lambda & 0 \\ 0 & \lambda \end{pmatrix} \right\} \\
&= \det\begin{pmatrix} f_{xx}(\mathbf{x}_0) - \lambda & f_{xy}(\mathbf{x}_0) \\ f_{yx}(\mathbf{x}_0) & f_{yy}(\mathbf{x}_0) - \lambda \end{pmatrix} \\
&= \lambda^2 - \big(f_{xx}(\mathbf{x}_0) + f_{yy}(\mathbf{x}_0)\big)\lambda + \big(f_{xx}(\mathbf{x}_0) f_{yy}(\mathbf{x}_0) - f_{xy}^2(\mathbf{x}_0)\big).
\end{aligned}
$$

Now, if a quadratic equation with leading coefficient 1 has roots r_1, r_2, then the equation can be factored as $0 = (x - r_1)(x - r_2)$, so that $0 = x^2 - (r_1 + r_2)x + r_1 r_2$. That is, the coefficient of x is the negative of the sum of the roots and the constant term is the product of the roots. Since the eigenvalues, call them λ_1, λ_2, of $H_f(\mathbf{x}_0)$ are the roots of the above quadratic equation, it follows that

$$
\lambda_1 + \lambda_2 = f_{xx}(\mathbf{x}_0) + f_{yy}(\mathbf{x}_0) \qquad \text{and} \qquad \lambda_1 \lambda_2 = D,
$$

where $D = f_{xx}(\mathbf{x}_0) f_{yy}(\mathbf{x}_0) - f_{xy}^2(\mathbf{x}_0)$ is the discriminant appearing in Theorem 4.6. If the eigenvalues are both positive then the first equation implies that at least one of $f_{xx}(\mathbf{x}_0)$ or $f_{yy}(\mathbf{x}_0)$ is positive, and the second equation implies $D > 0$. In this case, criterion (i) of Theorem 4.6 is satisfied. If the eigenvalues are both negative then the first equation implies that at least one of $f_{xx}(\mathbf{x}_0)$ or $f_{yy}(\mathbf{x}_0)$ is negative, and the second equation implies $D > 0$, so that criterion (ii) of Theorem 4.6 is satisfied. And if the eigenvalues have opposite signs then the second equation implies $D < 0$, and criterion (iii) of Theorem 4.6 is satisfied.

(b) Let $f(x, y, z) = x^2 + y^2 + z^2 - 2xy - 4yz - 6xz$, so that

$$
\nabla f(x, y, z) = \big(f_x(x, y, z), f_y(x.y.z), f_z(x.y.z)\big) = (2x - 2y - 6z, 2y - 2x - 4z, 2z - 4y - 6x).
$$

Setting this equal to zero leads to the linear system $2x - 2y - 6z = 0$, $2y - 2x - 4z = 0$, $2z - 4y - 6x = 0$, with the unique solution $(0, 0, 0)$. Moreover,

$$
H_f(0, 0, 0) = \begin{pmatrix} f_{xx} & f_{xy} & f_{xz} \\ f_{yx} & f_{yy} & f_{yz} \\ f_{zx} & f_{zy} & f_{zz} \end{pmatrix} = \begin{pmatrix} 2 & -2 & -6 \\ -2 & 2 & -4 \\ -6 & -4 & 2 \end{pmatrix},
$$

so that if I is the 3×3 identity matrix then the eigenvalues λ_1, λ_2, λ_3 of $H_f(0, 0, 0)$ satisfy

$$
\det\big(H_f(0, 0, 0) - \lambda I\big) = \det\begin{pmatrix} 2 - \lambda & -2 & -6 \\ -2 & 2 - \lambda & -4 \\ -6 & -4 & 2 - \lambda \end{pmatrix} = \lambda^3 - 6\lambda^2 - 44\lambda + 200 = 0.
$$

The reader can check that the three solutions of this cubic are $\lambda_1 \approx -6.2$, $\lambda_2 \approx 3.8$ and $\lambda_3 \approx 8.4$. Since both signs occur among the eigenvalues, $(0, 0, 0)$ is a saddle point.

Exercise Set 4F (pgs. 302-303)

1. Let $f(x, y) = \sin^2 x + \sin^2 y$. Notice that $f(x, y)$ has maximum value 2 when x and y are such that $\sin x = \pm 1$ and $\sin y = \pm 1$ both hold. These are points of the form $(x, y) = \big((n + 1/2)\pi, (m + 1/2)\pi\big)$, where n and m are integers. Similarly, $f(x, y)$ has minimum value 0 when x and y are such that $\sin x = 0$ and $\sin y = 0$ both hold. These are the points of the form $(x, y) = (n\pi, m\pi)$, where n and m are integers. On the square R defined by $0 \leq x \leq 5$, $0 \leq y \leq 5$, there are eight extreme points; namely.

$(0, 0)$, $(\pi, 0)$, (π, π), $(0, \pi)$ and $(\pi/2, \pi/2)$, $(3\pi/2, \pi/2)$, $(3\pi/2, 3\pi/2)$, $(\pi/2, 3\pi/2)$,

where the first four give $f(x, y) = 0$ and the second four give $f(x, y) = 2$. Since we are asked to find numerical approximations for the above coordinates, and since all coordinates are rational multiples of π, it suffices to find a numerical approximation for π.

With $f_x(x, y) = \sin 2x$, $f_y(x, y) = \sin 2y$, $f_{xx}(x, y) = 2\cos 2x$, $f_{xy}(x, y) = 0$, and $f_{yy}(x, y) = 2\cos 2y$, the modified program with the variable step size

$$h = -\frac{\sin^2 2x + \sin^2 2y}{2(\cos 2x \sin^2 2x + \cos 2y \sin^2 2y)}$$

was used along with the initial guess $(x_0, y_0) = (3, 3)$. The result $(3.14159, 3.14159)$ occurred at the third iteration.

3. Let $f(x, y) = (x + 2y)e^{-x^2 - 2y^2}$ for (x, y) in $\mathcal{R}^2$. We have

$$f_x(x, y) = (-2x^2 - 4xy + 1)e^{-x^2 - y^2}, \qquad f_y(x, y) = (-8y^2 - 4xy + 2)e^{-x^2 - 2y^2}.$$

It so happens that we can find the exact coordinates of the local extreme points by setting each of the above equations equal to zero and solving the resulting system. To do this, we see that the above equations are simultaneously zero precisely when

$$2x^2 + 4xy - 1 = 0 \qquad \text{and} \qquad 8y^2 + 4xy - 2 = 0.$$

Solve the first equation for y to get $y = (1 - 2x^2)/(4x)$ (notice that $x = 0$ is not possible in the first equation so that division by x does not eliminate any possible solutions). Inserting this expression for y into the second equation results in a quadratic equation in x; namely, $-6x^2 + 1 = 0$. This gives, $x = \pm 1/\sqrt{6}$, so that $y = (1 - 2x^2)/(4x) = \pm 1/\sqrt{6}$. That is, the only critical points of f are

$$\pm\big(1/\sqrt{6}, 1/\sqrt{6}\big) \approx \pm(0.40824829, 0.40824829),$$

which is accurate to eight decimal places. We have

$$f\big(1/\sqrt{6}, 1/\sqrt{6}\big) = \frac{3}{\sqrt{6}}e^{-1/2} \approx 0.742845315$$

$$f\big(-1/\sqrt{6}, -1/\sqrt{6}\big) = -\frac{3}{\sqrt{6}}e^{-1/2} \approx -0.742845315.$$

Since $f(x, y) \to 0$ as $x^2 + y^2 \to \pm\infty$, it follows that the critical points give the extreme values of f on $\mathcal{R}^2$.

5. Let $f(x,y) = \left[(x-1)^2 + y^2\right]^{-1} + \left[(x+1)^2 + y^2\right]^{-1}$. We observe that $f(x,y)$ is always positive, except at $(1,0)$ and $(-1,0)$ where it becomes infinite. Moreover,

$$f_x(x,y) = -\left(\frac{2(x-1)}{\left[(x-1)^2 + y^2\right]^2} + \frac{2(x+1)}{\left[(x+1)^2 + y^2\right]^2}\right),$$

$$f_y(x,y) = -2y\left(\frac{1}{\left[(x-1)^2 + y^2\right]^2} + \frac{1}{\left[(x+1)^2 + y^2\right]^2}\right).$$

First observe that the parenthetical factor on the right in the second line is always positive. It follows that $f_y(x,y) = 0$ if, and only if, $y = 0$. Inserting this value of y into the formula for $f_x(x,y)$ gives

$$f_x(x,0) = -\left(\frac{2}{(x-1)^3} + \frac{2}{(x+1)^3}\right) = -\frac{2(x+1)^3 + 2(x-1)^3}{(x^2-1)^3} = -\frac{2x(x^2+3)}{(x^2-1)^3}.$$

Hence, $f_x(x,0) = 0$ only when $x = 0$. The point $(0,0)$ is therefore the only critical point of f. The corresponding value of f is $f(0,0) = 2$.

We claim that $(0,0)$ is a saddle pont of f. To see this, let $\epsilon > 0$. A simple calculation gives

$$f(\epsilon,0) = 2\left(\frac{1}{\epsilon^3 - \epsilon^2 - \epsilon + 1}\right) \qquad \text{and} \qquad f(0,\epsilon) = 2\left(\frac{1}{1 + \epsilon^2}\right).$$

On the left, consider the cubic $p(x) = x^3 - x^2 - x + 1$, with derivative $p'(x) = 3x^2 - 2x - 1 = (3x+1)(x-1)$. This is negative for $0 < x < 1$ and therefore $p(x)$ is strictly decreasing on $(0,1)$. Since $p(0) = 1$, it follows that $p(\epsilon) < 1$ for $0 < \epsilon < 1$. So, the corresponding parenthetical fraction is greater than 1 for $0 < \epsilon < 1$. That is, $f(\epsilon,0) > 2$ for $0 < \epsilon < 1$ and therefore $f(0,0)$ is not a local maximum.

On the right, it is evident that the parenthetical fraction is less than 1 for $\epsilon > 0$. Hence, $f(0,\epsilon) < 2$ for $\epsilon > 0$ and therefore $f(0,0)$ is not a local minimum.

We conclude that f has no maximum value and no minimum value.

7. Let $f(x) = e^{-x^2}\sin x$, for $0 \le x \le 4$, with derivative $f'(x) = e^{-x^2}(\cos x - 2x\sin x)$. Note that $f(x) \ge 0$ for $0 \le x \le \pi$ with equality if, and only if, $x = 0$ or $x = \pi$; and that $f(x) < 0$ for $\pi < x \le 4$. Also note that $f'(0) = 1 > 0$, $f'(\pi) = -e^{-\pi^2} < 0$ and $f'(4) = e^{-16}(\cos 4 - 8\sin 4) > 0$. Since f' is continuous on $(0,4)$, the intermediate value theorem says that $f'(r_1) = 0$ for some $r_1 \in (0,\pi)$ and $f'(r_2) = 0$ for some $r_2 \in (\pi,4)$. Furthermore, a relatively accurate sketch of the graph of $y = \cos x - 2x\sin x$ shows that it crosses the x-axis exactly twice in the interval $[0,4]$. It follows that r_1 is the only critical point on $(0,\pi)$, so that $f(r_1)$ is the maximum of f on $[0,4]$, and that r_2 is the only critical point on $(\pi,4)$, so that $f(r_2)$ is the minimum of f on $[0,4]$.

To approximate the extreme values of f, the successive approximations $x_1, x_2, \ldots$ were computed by means of the equation

$$x_{n+1} = x_n + he^{-x^2}(\cos x - 2x\sin x), \quad n = 0,1,2,\ldots.$$

For the maximum value of f, we chose the initial guess $x_0 = 0.5$ and the step size $h = 0.5$; and for the minimum value of f, we chose the initial guess of $x_0 = 3.25$ and the step size $h = -5,000$. The iterations were done on a TI-36X hand calculator and were continued until nine place accuracy was achieved for the initial point $x_0 = 0.5$ and seven place accuracy was achieved for the initial point $x_0 = 3.25$. The approximations obtained were as follows:

$$x_0 = 0.500000000, \qquad x_0 = 3.250000000,$$
$$x_1 = 0.655042501, \qquad x_1 = 3.287620001,$$
$$x_2 = 0.653391128, \qquad x_2 = 3.290917245,$$
$$x_3 = 0.653279169, \qquad x_3 = 3.291876306,$$
$$x_4 = 0.653271718, \qquad x_4 = 3.292173162,$$
$$x_5 = 0.653271222, \qquad x_5 = 3.292266660,$$
$$x_6 = 0.653271189, \qquad x_6 = 3.292296266,$$
$$x_7 = 0.653271187, \qquad x_7 = 3.292305656,$$
$$x_8 = 3.292308636,$$
$$x_9 = 3.292309581,$$
$$x_{10} = 3.292309882,$$
$$x_{11} = 3.292309977,$$
$$x_{12} = 3.292310007.$$

The corresponding approximations for the extreme values for f on $[0, 4]$ are

$$f(0.653271187) \approx 0.396652961 \qquad \text{and} \qquad f(3.292310007) \approx -0.000002945,$$

both of which are accurate to nine decimal places.

(NOTE: (i) It is absolutely essential to use a positive step size when seeking a critical point that gives a maximum, and a negative step size when seeking a critical point that gives a minimum. (ii) The step size $h = -5000$ may seem unduly large but was chosen to compensate for the extremely small values of $\exp(-x_n^2)$. This would not have been necessary had we simply sought to find the zeros of the function $\cos x - 2x \sin x$, which are the same as the zeros we sought in the above computation. (iii) The graph of f is extremely flat around the extreme point to the right of $x = \pi$. For this reason it is necessary to use an initial approximation that is relatively close to the actual value.)

9. (a) The conditions of the problem allow us to suppose that the first two coordinates of each of the vertices of the box are nonnegative. Thus, x and y are the *length* and *width* of the box. We have the following three facts regarding the vertices of the box:

(i) The vertices on the z-axis have the form

$$\left(0, 0, e^{-0.1x^2 - 0.2y^2}\right) \qquad \text{and} \qquad \left(0, 0, -e^{-0.4x^2 - 0.1y^2}\right),$$

(ii) The vertices farthest from the origin have the form

$$\left(x, y, e^{-0.1x^2 - 0.2y^2}\right) \qquad \text{and} \qquad \left(x, y, -e^{-0.4x^2 - 0.1y^2}\right),$$

(iii) The vertical distance between the vertices in (a) equals the vertical distance between the vertices in (b) and is the *height* of the box. Its value is

$$e^{-0.1x^2 - 0.2y^2} + e^{-0.4x^2 - 0.1y^2}.$$

Given this information, the volume of the box is of the form

$$V(x,y) = xy(e^{-0.1x^2-0.2y^2} + e^{-0.4x^2-0.1y^2}), \quad x > 0, \ y > 0.$$

The first and second-order partials of V are

$$V_x(x,y) = (y - 0.2x^2y)e^{-0.1x^2-0.2y^2} + (y - 0.8x^2y)e^{-0.4x^2-0.1y^2},$$

$$V_y(x,y) = (x - 0.4xy^2)e^{-0.1x^2-0.2y^2} + (x - 0.2xy^2)e^{-0.4x^2-0.1y^2},$$

$$V_{xx}(x,y) = (0.04x^3y - 0.6xy)e^{-0.1x^2-0.2y^2} + (0.64x^3y - 2.4xy)e^{-0.4x^2-0.1y^2},$$

$$V_{yy}(x,y) = (0.16xy^3 - 1.2xy)e^{-0.1x^2-0.2y^2} + (0.04xy^3 - 0.6xy)e^{-0.4x^2-0.1y^2},$$

$$V_{xy}(x,y) = (0.08x^2y^2 - 0.4y^2 - 0.2x^2 + 1)e^{-0.1x^2-0.2y^2}$$
$$+ (0.16x^2y^2 - 0.2y^2 - 0.8x^2 + 1)e^{-0.4x^2-0.1y^2}.$$

In order to get a feel for the approximate coordinates that maximize V, the graph of $V(x,y)$ was drawn and the result is the picture on the right. The grid lines are 0.25 units apart in both the x and y directions. The solid arrow pointing upward represents a normal to the tangent plane at the highest point on the surface. The values of x and y where the surface appears to be at its highest point are approximately $x_0 = 1.5$ and $y_0 = 1.75$, which are depicted by the line segments forming the rectangle shown on the xy-plane. Using an ASCENT+ program with the variable step size h as given in the text, and the initial values $x_0 = 1.5$, $y_0 = 1.75$, it took only five iterations to arrive at $x = 1.49546$, $y = 1.77438$. The corresponding height of the box was then computed by using the expression given in (iii) above and was found to be 0.72437. That is, the maximum volume of the box is approximately

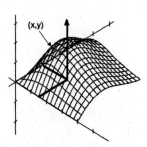

$$V_{\max} \approx (1.49546)(1.77438)(0.72437) = 1.92213 \text{ cubic units.}$$

(b) Replacing $e^{-0.1x^2-0.2y^2}$ and $-e^{-0.4x^2-0,1y^2}$ given in part (a) with $e^{-x^2-y^2}$ and $-e^{-x^2-y^2}$, the generic dimensions of the box are x, y and $e^{-x^2-y^2} + e^{-x^2-y^2} = 2e^{-x^2-y^2}$. The corresponding volume function is therefore

$$V(x,y) = 2xye^{-x^2-y^2}, \quad x > 0, \ y > 0,$$

with partial derivatives

$$V_x(x,y) = 2y(1 - 2x^2)e^{-x^2-y^2} \quad \text{and} \quad V_y(x,y) = 2x(1 - 2y^2)e^{-x^2-y^2}.$$

Since $x > 0$ and $y > 0$, these partials are simultaneously zero only when $1 - 2x^2 = 0$ and $1 - 2y^2 = 0$: i.e., only when $x = 1/\sqrt{2}$ and $y = 1/\sqrt{2}$. The height of the box is then $2e^{-1/2-1/2} = 2e^{-1}$. Therefore, the maximum volume of the box is

$$V_{\max} = (1/\sqrt{2})(1/\sqrt{2})(2e^{-1}) = e^{-1}.$$

236

Section 5: CURVILINEAR COORDINATES

Exercise 5A-D (pgs. 308-309)

1. $r = 1$, $\pi \leq \theta \leq 3\pi/2$

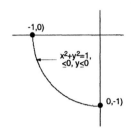

3. $r(\sin\theta - \cos\theta) = \pi/2$

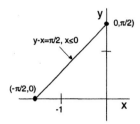

5. $r = 2$, $0 \leq \theta \leq \pi/4$, $\pi/4 \leq \phi \leq \pi/2$

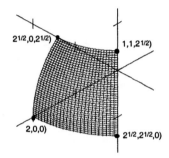

7. $0 \leq r \leq 1$, $0 \leq \theta \leq \pi/2$, $\phi = \pi/4$

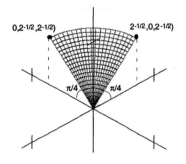

9. If R is the region to be "described" then

$$R = \{\,(r\cos\theta, r\sin\theta, z)\,|\,0 \leq r \leq 1,\ -\pi/2 \leq \theta \leq \pi/2\,\}.$$

11. Let (r, ϕ, θ) be spherical coordinates in $\mathcal{R}^3$ and consider the curve defined parametrically by $(r, \phi, \theta) = (1, t, t)$, for $0 \leq t \leq \pi/2$. The image of this curve under the spherical coordinate transformation is given parametrically in rectangular coordinates by

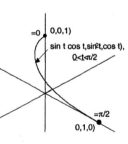

$$S(1, t, t) = (x, y, z) = (\sin t \cos t, \sin t \sin t, \cos t)$$
$$= (\sin t \cos t, \sin^2 t, \cos t), \quad 0 \leq t \leq \pi/2.$$

Its graph is shown on the right.

13. Consider the *ellipsoidal coordinate transformation* defined by

$$(x, y, z) = (r\sin\phi\cos\theta, 2r\sin\phi\sin\theta, 2r\cos\phi), \quad 0 < r < \infty,\ 0 < \phi < \pi,\ 0 \leq \theta < 2\pi.$$

A coordinate surface is found by fixing one of the variables r, ϕ, θ, while allowing the other two to vary. Hence, the three surfaces are given by

(*i*) $\qquad (x, y, z) = (r_0\sin\phi\cos\theta, 2r_0\sin\phi\sin\theta, 2r_0\cos\phi), \quad r = r_0$ fixed;

(*ii*) $\qquad (x, y, z) = (r\sin\phi_0\cos\theta, 2r\sin\phi_0\sin\theta, 2r\cos\phi_0), \quad \phi = \phi_0$ fixed;

(*iii*) $\qquad (x, y, z) = (r\sin\phi\cos\theta_0, 2r\sin\phi\sin\theta_0, 2r\cos\phi), \quad \theta = \theta_0$ fixed.

237

The "degenerate" case for (ii) occurs for $\phi_0 = \pi/2$. Here, $(x, y, z) = (r\cos\theta, 2r\sin\theta, 0)$, which is the xy-plane with the origin deleted. Other than this case, the first coordinate curve is an ellipsoid (whose radii depend of the choice of r_0), the second is either the top half of a cone or the bottom half of a cone (depending on whether $\phi_0 < \pi/2$ ar $\phi_0 > \pi/2$, respectively), and the third one is a vertical halfplane $2(\sin\theta_0)x - (\cos\theta_0)y = 0$. The three coordinate surfaces shown below correspond to $r_0 = 1$, $\phi_0 = \pi/3$ and $\theta_0 = 2\pi/3$.

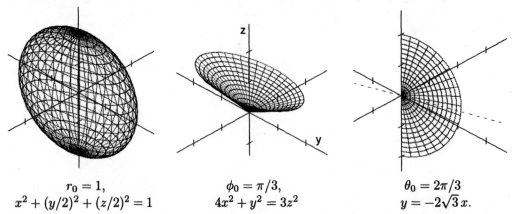

$$r_0 = 1, \qquad\qquad \phi_0 = \pi/3, \qquad\qquad \theta_0 = 2\pi/3$$
$$x^2 + (y/2)^2 + (z/2)^2 = 1 \qquad 4x^2 + y^2 = 3z^2 \qquad y = -2\sqrt{3}\,x.$$

15. Let $(r, \phi, \theta) = (1, t, t^2)$ be a parametrization in spherical coordinates in $\mathcal{R}^3$. Under the spherical coordinate transformation, the rectangular coordinates of this parametrization are given by

$$S(1, t, t^2) = (x, y, z) = (\sin t \cos t^2, \sin t \sin t^2, \cos t).$$

Setting $f(t) = (\sin t \cos t^2, \sin t \sin t^2, \cos t)$ we have

$$f'(t) = (-2t \sin t \sin t^2 + \cos t \cos t^2, 2t \sin t \cos t^2 + \cos t \sin t^2, -\sin t).$$

This is the tangent vector for the curve defined by f.

(pgs. 309-311)

1. With $f(x,y,z) = x^2 + y^2 - z^2$, we have $f'(x,y,z) = (2x, 2y, -2z)$.

3. With $g(u,v) = (\cos u, \sin u, v)$, we have

$$g'(u,v) = \begin{pmatrix} -\sin u & 0 \\ \cos u & 0 \\ 0 & 1 \end{pmatrix}, \quad \text{so that} \quad g'(\pi/3, \pi^2/36) = \begin{pmatrix} -\sqrt{3}/2 & 0 \\ 1/2 & 0 \\ 0 & 1 \end{pmatrix}.$$

5. With $K(x,y) = xy$, we have $\nabla K(x,y) = (y,x)$, so that $\nabla K(1, \sqrt{3}) = (\sqrt{3}, 1)$. The unit vector $\mathbf{u}$ in the direction of $(\sqrt{3}, 1)$ is $\mathbf{u} = (\sqrt{3}/2, 1/2)$. Thus,

$$\frac{\partial K}{\partial \mathbf{u}}(1, \sqrt{3}) = \nabla K(1, \sqrt{3}) \cdot \mathbf{u} = (\sqrt{3}, 1) \cdot (\sqrt{3}/2, 1/2) = 2.$$

7. Let the temperature at a point (x,y,z) of a solid ball of radius 3 centered at $(0,0,0)$ be given by $T(x,y,z) = yz + zx + xy$. Since $\nabla T(x,y,z) = (z+y, z+x, y+x)$, the direction in which T is increasing most rapidly at $(1,1,2)$ is in the direction of the vector $\nabla T(1,1,2) = (3,3,2)$.

9. Let $\mathbf{F}(x,y) = (x/\sqrt{x^2 + y^2}, y/\sqrt{x^2 + y^2})$, for $(x,y) \neq (0,0)$.

(a) The vector field $\mathbf{F}$ is shown on the right. The vectors have been scaled by a factor of 0.3 to avoid overlapping arrows

(b) First observe that $\mathbf{F}(x,y)$ is a unit vector for all $(x,y) \neq (0,0)$. Then note that $\mathbf{F}$ is a gradient field of the function $f(x,y) = \sqrt{x^2 + y^2}$ $((x,y) \neq (0,0))$; i.e., $\nabla f = \mathbf{F}$. Hence, the vector $\mathbf{F}(x,y)$ is a unit vector pointing in the direction of maximum increase for f. It follows that this rate is given by

$$\frac{\partial f}{\partial \mathbf{F}}(x,y) = \nabla f(x,y) \cdot \mathbf{F}(x,y) = \mathbf{F}(x,y) \cdot \mathbf{F}(x,y) = |\mathbf{F}(x,y)|^2 = 1.$$

11. Let $f : \mathcal{R}^n \longrightarrow \mathcal{R}$ be a differentiable function that is homogeneous of degree m. For each positive real number t and each $x = (x_1, \ldots, x_n) \in \mathcal{R}^n$, define $g : \mathcal{R}^{n+1} \longrightarrow \mathcal{R}^n$, by $g(t, \mathbf{x}) = t\mathbf{x}$, where $(t, \mathbf{x}) = (t, x_1, \ldots, x_n)$. The composite function $f \circ g$ then satisfies

$$(f \circ g)(t, \mathbf{x}) = f\big(g(t, \mathbf{x})\big) = f(t\mathbf{x}) = t^m f(\mathbf{x}),$$

so that $\partial (f \circ g)/\partial t(t, \mathbf{x}) = mt^{m-1} f(\mathbf{x})$, which is the first entry in $(f \circ g)'(t, \mathbf{x})$, the derivative matrix of $f \circ g$. Now, by the chain rule,

$$(f \circ g)'(t, \mathbf{x}) = f'\big(g(t, \mathbf{x})\big) g'(t, \mathbf{x}) = \nabla f(t\mathbf{x}) g'(t, \mathbf{x}).$$

The first entry in the matrix on the right is the dot product of the first column of $g'(t, \mathbf{x})$ with $\nabla f(t\mathbf{x})$. Since $\big(\partial (tx_1)/\partial t(t, \mathbf{x}), \ldots, \partial (tx_n)/\partial t(t, \mathbf{x})\big) = (x_1, \ldots, x_n) = \mathbf{x}$ is the first column of $g'(t, \mathbf{x})$ (written as a row vector), we must have $mt^{m-1}f(\mathbf{x}) = \nabla f(t\mathbf{x}) \cdot \mathbf{x}$. Setting $t = 1$ and then dividing both sides by $m \neq 0$, gives

$$f(\mathbf{x}) = \frac{1}{m} \nabla f(\mathbf{x}) \cdot \mathbf{x}.$$

13. Let $h(z)$ be a real-valued function, differentiable for all z. Define $u(x,t) = h(x - at)$, where a is a constant. Let $z = z(x,t) = x - at$ and note that $z_x = 1$ and $z_t = -a$. Then, by the chain rule,

$$u_{xx} = (u_x)_x = (h_z z_x)_x = (h_z)_x = h_{zz}z_x = h_{zz},$$
$$u_{tt} = (u_t)_t = (h_z z_t)_t = (-ah_z)_t = -ah_{zz}z_t = a^2 h_{zz}.$$

Hence, $a^2 u_{xx} = u_{tt}$, or $a^2 u_{xx} - u_{tt} = 0$, for all pairs (x, t).

15. Let $f(u,v) = u^2 v + uv^3$ and suppose u and v are differentiable functions of s and t with $u(2,1) = -2$, $u_s(2,1) = 3$, $v(2,1) = 2$ and $v_s(2,1) = -4$. At the point $(s,t) = (2,1)$ we have $(u,v) = \big(u(2,1), v(2,1)\big) = (-2, 2)$. Therefore, the values of $f_u(u,v) = 2uv + v^3$ and $f_v(u,v) = u^2 + 3uv^2$ when $(s,t) = (2,1)$ are $f_u(-2,2) = 2(-2)(2) + (2)^3 = 0$ and $f_v(-2,2) = (-2)^2 + 3(-2)(2)^2 = -20$. The chain rule then gives

$$f_s(2,1) = f_u(-2,2)u_s(2,1) + f_v(-2,2)v_s(2,1) = (0)(3) + (-20)(-4) = 80.$$

17. Let $P = P(t)$, $V = V(t)$, $T = T(t)$ be the pressure, volume and temperature (resp.) of a given gas at time $t \geq 0$. and suppose they are related by $PV = 6T$.

(a) Suppose P, V and T all increase simultaneously at their own constant rates. Then each must be a linear function of t with a positive constant derivative. It follows that P, V and T have the form $P(t) = a_1 t + b_1$, $V(t) = a_2 t + b_2$ and $T(t) = a_3 t + b_3$, where $a_1, a_2, a_3, b_1, b_2, b_3$ are constant with $a_1 > 0$, $a_2 >$, $a_3 > 0$. Then, for all $t \geq 0$,

$$(a_1 t + b_1)(a_2 t + b_2) = 6(a_3 t + b_3), \quad \text{or} \quad a_1 a_2 t^2 + (a_1 b_2 + a_2 b_1 - 6a_3)t + (b_1 b_2 - 6b_3) = 0.$$

The only way for the right equation to hold for all $t \geq 0$ is for the three coefficients of the polynomial in t on the left side to be equal to zero. Since, $a_1 a_2 > 0$, we have a contradiction. Hence, P, V, and T cannot all simultaneously increase at their own constant rates.

(b) The given descriptions of how V and T are changing with respect to time t implies that $V(t) = 2t + V(0) = 2t + 10$ and $T(t) = 3t + T(0) = 3t + 8$. Hence, $P(t)V(t) = 6T(t)$ implies

$$P(t) = 6\frac{T(t)}{V(t)} = 6\frac{2t + 10}{3t + 8}, \qquad \text{so that} \qquad \frac{dP}{dt}(t) = -\frac{84}{(3t + 8)^2}.$$

Hence, $dP/dt(0) = -84/64 = -21/16 = -1.3125$.

(c) Using the same data as in part (b), $dP/dt(30) = -84/98^2 = -3/7^3 \approx -0.009$.

19. Let $z = f(x,y)$ be a function defined implicitly by $x + y^2 + z^2 = 3z$. Implicit differentiation with respect to x and then y gives $1 + 2zz_x = 3z_x$ and $2y + 2zz_y = 3z_y$. Evaluating these at $(x,y,z) = (1,1,1)$, we then solve for $z_x(1,1,1)$ and $z_y(1,1,1)$ to get

$$z_x(1,1,1) = 1 \qquad \text{and} \qquad z_y(1,1,1) = 2.$$

21. Let γ be the curve of intersection of the two level surfaces

$$x^2 + 2y^2 + 3z^2 = 6,$$
$$x^2 + y^2 - z^2 = 1.$$

240

Observe that the point $(1, 1, 1)$ is on γ

(a) Subtracting the second equation from the first equation and solving for y gives $y = \pm\sqrt{5 - 4z^2}$. Since we are interested in the point $(1, 1, 1)$, where y is positive, we choose the plus sign: $y(z) = \sqrt{5 - 4z^2}$. Similarly, subtracting the first equation from twice the second and solving for x gives $x(z) = \sqrt{5z^2 - 4}$, where we have again chosen the plus sign because x is positive at the point $(1, 1, 1)$. This defines a parametric representation (with parameter z) of that part of γ where the first two coordinates are positive; namely.

$$(x, y, z) = (x(z), y(z), z) = \left(\sqrt{5z^2 - 4}, \sqrt{5 - 4z^2}, z\right).$$

A tangent vector to γ at such points is therefore

$$\frac{d\left(\sqrt{5z^2 - 4}, \sqrt{5 - 4z^2}, z\right)}{dz}(z) = \left(\frac{5z}{\sqrt{5z^2 - 4}}, -\frac{4z}{\sqrt{5 - 4z^2}}, 1\right).$$

Hence, at $(1, 1, 1)$ (i.e., $z = 1$), we have $(5, -4, 1)$.

(b) Viewing x and y as functions of z, we implicitly differentiate the two given equations with respect to z to get

$$2x\frac{dx}{dz} + 4y\frac{dy}{dz} + 6z = 0 \qquad \text{and} \qquad 2x\frac{dx}{dz} + 2y\frac{dy}{dz} - 2z = 0.$$

Solving the system for the two derivatives we get $dx/dz = 5z/x$ and $dy/dz = -4z/y$. In particular, $dx/dz(1, 1, 1) = 5$ and $dy/dz(1, 1, 1) = -4$. Hence, a tangent vector to γ at $(1, 1, 1)$ is given by

$$\frac{d(x, y, z)}{dz}(1, 1, 1) = \left(\frac{dx}{dz}(1, 1, 1), \frac{dy}{dz}(1, 1, 1), \frac{dz}{dz}(1, 1, 1)\right) = (5, -4, 1).$$

Note that this is the same tangent vector obtained in part (a).

23. Let $f(x, y)$ be differentiable, and let $\mathbf{u} = (\cos\theta, \sin\theta)$.

(a) Let $\mathbf{x}_0$ be a point in the domain of f and let $\theta = \theta_0$ be such that $\nabla f(\mathbf{x}_0)$ and $\mathbf{u} = \mathbf{u}_0$ are parallel. Then there exists a nonzero constant k such that $\nabla f(\mathbf{x}_0) = k\mathbf{u}_0$. Hence,

$$\frac{\partial f}{\partial \mathbf{u}}(\mathbf{x}_0) = \nabla f(\mathbf{x}_0) \cdot \mathbf{u} = k(\cos\theta_0, \sin\theta_0) \cdot (\cos\theta, \sin\theta) = k(\cos\theta_0 \cos\theta + \sin\theta_0 \sin\theta)$$

Thus, as a function of θ, the derivative of the directional derivative of f at $\mathbf{x}_0$ satisfies

$$\frac{d}{d\theta}\left(\frac{\partial f}{\partial \mathbf{u}}(\mathbf{x}_0)\right)(\theta) = k(-\cos\theta_0 \sin\theta + \sin\theta_0 \cos\theta).$$

This is zero for $\theta = \theta_0$, so that θ_0 is a critical point of $\partial f/\partial \mathbf{u}(\mathbf{x}_0)$.

(b) If $\mathbf{x}_0$ is a point in the domain of f where $\nabla f(\mathbf{x}_0)$ is parallel to *some* unit vector, it must be the case that $\nabla f(\mathbf{x}_0) \neq \mathbf{0}$. The fact that θ_0 is a critical point of the directional derivative of f at $\mathbf{x}_0$ means that the directional derivative evaluated at θ_0 must be an extreme value. That is,

$$\frac{\partial f}{\partial \mathbf{u}_0}(\mathbf{x}_0) = \nabla f(\mathbf{x}_0) \cdot \mathbf{u}_0 = k(\cos\theta_0, \sin\theta_0) \cdot (\cos\theta_0, \sin\theta_0) = k(\cos^2\theta + \sin^2\theta_0) = k$$

is either a local maximum or a local minimum value for the rate of increase of f at $\mathbf{x}_0$. Since $|\nabla f(\mathbf{x}_0)| = |k|$, it follows that $\nabla f(\mathbf{x}_0)$ points in the direction of maximum rate of increase of f (if $k > 0$), or in the direction of minimum rate of increase of f (if $k < 0$). Thus, part (a) is a partial proof of Theorem 1.2 in Section 1. All that needs to be shown to complete the proof is to observe that we can always choose the parallel unit vector $\mathbf{u}_0 = (\cos\theta_0, \sin\theta_0)$ to be going in the same direction as $\nabla f(\mathbf{x}_0)$, forcing $k > 0$.

25. Since the given equations implicitly define x and y as functions of u and v, we implicitly differentiate the two equations with respect to u and get the system

$$(x_u + 1)\cos(x + u) + y_u\sin(y + v) + xy_u + x_uy - 1 + v = 0,$$

$$1 + x_u + y_u - ux_u - x - vy_u = 0.$$

When these equation are evaluated at the point $(u, v, x, y) = (2, 1, -2, -1)$, we find that x_u cancels out of the first equation, leaving $y_u = 1/2$, and that y_u cancels out of the second equation, leaving $x_u = 3$.

27. Since the given equations implicitly define x and z as functions of y and t near the point $(x, y, z, t) = (-2, 1, -1, 2)$, we implicitly differentiate the two equations with respect to y and get

$$x + yx_y + tz_y = 0, \qquad 2xx_y + 1 + z_y = 0.$$

When these equations are evaluated at the point $(x, y, z, t) = (-2, 1, -1, 2)$, we get the system $x_y + 2z_y = 2$, $4x_y - z_y = 1$, with solution $x_y = 4/9$, $z_y = 7/9$.

29. Let $f(x, y) = x^2 + y$ and let R be the region on or inside the ellipse $x^2 + 2y^2 = 1$. Since f is continuous on R, a closed and bounded set, f attains its maximum and minimum values on R. Moreover, the fact that $\nabla f(x, y) = (2x, 1)$ is never zero shows that f does not attain either of its extreme values on the interior of R and therefore must attain them on the ellipse $x^2 + 2y^2 = 1$, the boundary of R.

The x coordinate of a point (x, y) on the ellipse satisfies $x^2 = -2y^2 + 1$, so that the value of f at such a point is $z = -2y^2 + y + 1$, which is a parabola opening downward with vertex at $(y, z) = (1/4, 9/8)$. Hence, the maximum value of $f(x, y) = z$ on the ellipse occurs at $y = 1/4$. The corresponding x value satisfies $f(x, 1/4) = 9/8$, or $9/8 = x^2 + 1/4$, from which $x = \pm\sqrt{7/8} = \pm\sqrt{14}/4$. That is, the two points $(x, y) = (\pm\sqrt{14}/4, 1/4)$ give the maximum value of f. And since the parabola is opening downward, the lowest point on the parabola (the minimum value of z) will occur when y has maximal absolute value in the equation $x^2 + 2y^2 = 1$; i.e., when $y = \pm 1/\sqrt{2}$. Since the two corresponding x values are zero, $f(x, y) = z$ is minimal when $(x, y) = (0, \pm 1/\sqrt{2})$.

31. Letting x, y and z denote the length, width and height (resp.) of the rectangular box, its volume is $V(x, y, z) = xyz$. The length (3 units) of the internal diagonal of the box is related to x, y and z by the equation $3 = \sqrt{x^2 + y^2 + z^2}$ or, $x^2 + y^2 + z^2 = 9$. Thus, we want to maximize V subject to the condition $x^2 + y^2 + z^2 = 9$, where x, y and z are all nonnegative (it will be logistically easier to allow for boxes with zero volume).

The region R defined by these conditions is the portion of the sphere $x^2 + y^2 + z^2 = 9$ that lies in the first octant, together with its boundary, where either x, y or z is zero. Therefore, R is closed and bounded and V attains both maximum and minimum values on R. Clearly, V is never negative and $V(x, y, z) = 0$ if, and only if, at least one of x, y or z is zero. It follows that $V = 0$ is the minimum value attained by V on R and that the maximum value of V occurs on the interior of R. Further, since V is continuously differentiable, the points that give these maximal values are critical points of V, and are therefore among the critical points of the function $xyz + \lambda(x^2 + y^2 + z^2 - 9)$, for some constant λ; i.e., they are

242

among the solutions of the system

$$yz + 2x\lambda = 0,$$
$$xz + 2y\lambda = 0,$$
$$xy + 2z\lambda = 0,$$
$$x^2 + y^2 + z^2 = 9.$$

Multiplying the first equation by x, the second equation by y, and the third equation by z, we add the resulting equations and get $3xyz + 2\lambda(x^2 + y^2 + z^2) = 0$, which the fourth equation allows us to write as $3xyz + 18\lambda = 0$. Solving this for λ and inserting into the first three equations produces three equations that can be written as

$$yz\left(1 - \frac{x^2}{3}\right) = 0, \quad xz\left(1 - \frac{y^2}{3}\right) = 0, \quad xy\left(1 - \frac{z^2}{3}\right) = 0.$$

By our previous remarks, we need only look for solutions with x, y, z positive. In this case, we immediately see that $(x, y, z) = (\sqrt{3}, \sqrt{3}, \sqrt{3})$ is the only solution. It follows that $V(\sqrt{3}, \sqrt{3}, \sqrt{3}) = 3\sqrt{3}$ is the maximum volume of a rectangular box with an internal diagonal of 3 units.

33. Let $u = f(r)$ and $r = \sqrt{x^2 + y^2}$. Then, by the chain rule,

$$u_{xx} = (u_x)_x = \left(\frac{df}{dr}r_x\right)_x = \frac{df}{dr}r_{xx} + \frac{d^2f}{dr^2}r_x^2, \quad u_{yy} = (u_y)_y = \left(\frac{df}{dr}r_y\right)_y = \frac{df}{dr}r_{yy} + \frac{d^2f}{dr^2}r_y^2,$$

so that

$$u_{xx} + u_{yy} = \frac{df}{dr}(r_{xx} + r_{yy}) + \frac{d^2f}{dr^2}(r_x^2 + r_y^2)$$
$$= \frac{df}{dr}\left(\frac{x^2}{(x^2+y^2)^{3/2}} + \frac{y^2}{(x^2+y^2)^{3/2}}\right) + \frac{d^2f}{dr^2}\left(\left(\frac{x}{\sqrt{x^2+y^2}}\right)^2 + \left(\frac{y}{\sqrt{x^2+y^2}}\right)^2\right)$$
$$= \frac{d^2f}{dr^2} + \frac{1}{r}\frac{df}{dr}.$$

35. Let $f(x, y, z) = xy + yz + xz$, so that $\nabla f(x, y, z) = (y + z, x + z, x + y) = (0, 0, 0)$ when $y + z = x + z = x + y = 0$. The first equality implies $x = y$, the second equality implies $z = y$, and the third equality implies $x = -y$. Taken together, deduce $x = y = z = 0$ and $(0, 0, 0)$ is the only critical point of f.

37. First observe that the function $f(x, y) = 2xy + y^2 + 4y + 2x$ is real-valued and twice continuously differentiable on $\mathcal{R}^2$. Since $\nabla f(x, y) = (2y + 2, 2x + 2y + 4)$, the critical points of f are those with x and y satisfying $2y + 2 = 0$, $2x + 2y + 4 = 0$. This gives the single critical point $(-1, -1)$.

Next, $f_{xx}(x, y) = 0$, $f_{yy}(x, y) = 2$ and $f_{xy}(x, y) = 2$, from which $f_{xx}(-1, -1) = 0$, $f_{yy}(-1, -1) = 2$ and $f_{xy}(-1, -1) = 2$. Hence,

$$f_{xx}(-1, -1)f_{yy}(-1, -1) - f_{xy}^2(-1, -1) = (0)(2) - (2)^2 = -2 < 0.$$

By Theorem 4.6, $(-1, -1)$ is a saddle point.

39. For the purposes of this exercise, let U_n be the set of all unit vectors in $\mathcal{R}^n$ and let $\overline{U}_n$ be the set of all vectors $\mathbf{x}$ in $\mathcal{R}^n$ such that $|\mathbf{x}| \leq 1$.

Let $\mathbf{x}_0$ be a nonzero vector in $\mathcal{R}^n$ and define $f(\mathbf{x}) = \mathbf{x}_0 \cdot \mathbf{x}$. Suppose $\mathbf{x}_1$ is in U_n such that $f(\mathbf{x}_1) = \mathbf{x}_0 \cdot \mathbf{x}_1 > |\mathbf{x}_0|$. Then the angle θ between $\mathbf{x}_0$ and $\mathbf{x}_1$ satisfies

$$\cos\theta = \frac{\mathbf{x}_0 \cdot \mathbf{x}_1}{|\mathbf{x}_0||\mathbf{x}_1|} = \frac{\mathbf{x}_0 \cdot \mathbf{x}_1}{|\mathbf{x}_0|} > \frac{|\mathbf{x}_0|}{|\mathbf{x}_0|} = 1,$$

where the second equality follows from $|\mathbf{x}_1| = 1$. But this asserts $\cos\theta > 1$, a contradiction. We conclude that $f(\mathbf{x}) \leq |\mathbf{x}_0|$ for all $\mathbf{x}$ in U_n. To show that f actually attains this value, observe that $\mathbf{x}_0/|\mathbf{x}_0|$ is a unit vector and

$$(*) \qquad\qquad f\left(\mathbf{x}_0/|\mathbf{x}_0|\right) = \mathbf{x}_0 \cdot \frac{\mathbf{x}_0}{|\mathbf{x}_0|} = \frac{|\mathbf{x}_0|^2}{|\mathbf{x}_0|} = |\mathbf{x}_0|.$$

Thus, $|\mathbf{x}_0|$ is the maximum value of f on U_n. Moreover, an almost identical argument shows that $-|\mathbf{x}_0|$ is the minimum value of f on U_n.

We now consider the maximum and minimum values of f on $\overline{U}_n$. Assume $\mathbf{x}_1$ is a nonzero vector in $\overline{U}_n$ such that $f(\mathbf{x}_1) = \mathbf{x}_0 \cdot \mathbf{x}_1 > |\mathbf{x}_0|$. Then the angle θ between $\mathbf{x}_0$ and $\mathbf{x}_1$ satisfies

$$\cos\theta = \frac{\mathbf{x}_0 \cdot \mathbf{x}_1}{|\mathbf{x}_0||\mathbf{x}_1|} > \frac{|\mathbf{x}_0|}{|\mathbf{x}_0||\mathbf{x}_1|} = \frac{1}{|\mathbf{x}_1|} > 1,$$

a contradiction. Thus, $f(\mathbf{x}) \leq |\mathbf{x}_0|$ for all $\mathbf{x}$ in $\overline{U}_n$. Since equation $(*)$ is still valid, we conclude that $|\mathbf{x}_0|$ is the maximum value of f on $\overline{U}_n$. A similar argument shows that $-|\mathbf{x}_0|$ is the minimum value of f on $\overline{U}_n$.

41. Let $f(x, y) = x^2 y - 2x - y$.

(a) We have $\nabla f(x, y) = (2xy - 2, x^2 - 1) = (0, 0)$ when $x = y = \pm 1$. Thus, $(\pm 1, \pm 1)$ are the only two critical points in $\mathcal{R}^2$.

(b) The points (x, y) an the line segment joining $(0, 1)$ and $(1, 0)$ satisfy $y = -x + 1$, with $0 \leq x \leq 1$. Setting $F(x) = f(x, -x + 1)$, gives

$$F(x) = -x^3 + x^2 - x - 1 \quad \text{with derivative} \quad F'(x) = -3x^2 + 2x - 1.$$

Since $-3x^2 + 2x - 1$ cannot be factored over the reals, $F'(x) \neq 0$ for all x. It follows that the maximum and minimum values of f on the line segment must occur at the endpoints $(0, 1)$ and $(1, 0)$. Since $F(0) = -1$ and $F(1) = -2$, we conclude that $-2 \leq f(x, y) \leq -1$, with equality on the left only if $(x, y) = (1, 0)$ and equality on the right only if $(x, y) = (0, 1)$.

43. Let $f(x, y) = x^3 + 3xy + y^2$.

(a) First, $\nabla f(x, y) = (3x^2 + 3y, 3x + 2y)$, so that $\nabla f(x, y) = \mathbf{0}$ implies the system $x^2 + y = 0$, $3x + 2y = 0$. The second equation subtracted from twice the first equation results in $2x^2 - 3x = 0$, or $2x(x - 3/2) = 0$. This has solutions $x = 0$ and $x = 3/2$. The corresponding y values are $y = 0$ and $y = -9/4$, respectively. The critical points of f are therefore $(0, 0)$ and $(3/2, -9/4)$.

(b) Of the two critical points found in part (a), only $(0, 0)$ is interior to the square $|x| \leq 1$, $|y| \leq 1$ and, further, $f(0, 0) = 0$. Thus, to find the extreme values of f on the square, it is only necessary to find the extreme values of f on the boundary of the square and compare them to $f(0, 0)$.

On the upper boundary of the square, $y = 1$ and we have $f(x, 1) = x^3 + 3x + 1$. Its derivative with respect to x is $3x^2 + 3$ and is always positive. It follows that f is strictly monotonic across the top of the square.

On the right boundary of the square, $x = 1$ and we have $f(1, y) = y^2 + 3y + 1$. This is a parabola opening to the right with vertex at $(-3/2, -5/4)$, so that $f(1, y)$ is strictly increasing on $-1 \leq y \leq 1$. That is, f is strictly monotonic on the right side of the square.

On the lower boundary of the square, $y = -1$ and we have $f(x, -1) = x^3 - 3x + 1$. Its derivative with respect to x is $3x^2 - 3$, which is zero if $x = \pm 1$. These correspond to the "humps" of the graph of the cubic, between which f is strictly decreasing and is therefore strictly monotonic across the bottom of the square.

On the left boundary of the square, $x = -1$ and we have $f(-1, y) = y^2 - 3y + 1$. This is a parabola opening to the right with vertex at $(3/2, -5/4)$, so that $f(-1, y)$ is strictly increasing on $-1 \leq y \leq 1$ That is, f is strictly monotonic on the left side of the square.

The above results implies that f must attain its maximum and minimum values either at the corners of the square or at the origin. Since,

$$f(1, 1) = 5, \quad f(1, -1) = -1, \quad f(-1, -1) = 3, \quad f(-1, 1) = -3, \quad f(0, 0) = 0,$$

the maximum and minimum values of f on the closed square are 5 and -3, respectively.

45. The circle of radius a centered at (x_0, y_0) has the form $(x - x_0)^2 + (y - y_0)^2 = a^2$. In terms of polar coordinates, we set $x_0 = r_0 \cos \theta_0$, $y_0 = r_0 \sin \theta_0$, $x = r \cos \theta$ and $y = r \sin \theta$ and compute:

$$
\begin{aligned}
a^2 &= (r \cos \theta - r_0 \cos \theta_0)^2 + (r \sin \theta - r_0 \sin \theta_0)^2 \\
&= r^2 \cos^2 \theta - 2 r r_0 \cos \theta \cos \theta_0 + r_0^2 \cos^2 \theta_0 + r^2 \sin^2 \theta - 2 r r_0 \sin \theta \sin \theta_0 + r_0^2 \sin^2 \theta_0 \\
&= r^2 - 2 r r_0 (\cos \theta \cos \theta_0 + \sin \theta \sin \theta_0) + r_0^2 \\
&= r^2 - 2 r r_0 \cos(\theta - \theta_0) + r_0^2,
\end{aligned}
$$

where the last equality follows from the identity $\cos(\alpha - \beta) = \cos \alpha \cos \beta - \sin \alpha \sin \beta$.

Chapter 7: Multiple Integration

Section 1: ITERATED INTEGRALS

Exercise Set 1A-D (pgs. 321-322)

1.

$$\int_{-1}^{0} \left[\int_{1}^{2} (x^2 y^2 + xy^3) dy \right] dx = \int_{-1}^{0} \left[\frac{1}{3} x^2 y^3 + \frac{1}{4} xy^4 \right]_{1}^{2} dx$$

$$= \int_{-1}^{0} \left(\frac{7}{3} x^2 + \frac{15}{4} x \right) dx = \left[\frac{7}{9} x^3 + \frac{15}{8} x^2 \right]_{-1}^{0} = -\frac{79}{72}.$$

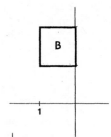

3.

$$\int_{1}^{0} \left[\int_{2}^{0} (x + y^2) dy \right] dx = \int_{1}^{0} \left[xy + \frac{1}{3} y^3 \right]_{2}^{0} dx$$

$$= \int_{1}^{0} \left(-2x - \frac{8}{3} \right) dx = \left[-x^2 - \frac{8}{3} x \right]_{1}^{0} = \frac{11}{3}.$$

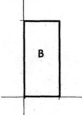

5.

$$\int_{-2}^{1} dy \int_{0}^{y^2} (x^2 + y) dx = \int_{-2}^{1} \left[\frac{1}{3} x^3 + xy \right]_{0}^{y^2} dy$$

$$= \int_{-2}^{1} \left(\frac{1}{3} y^6 + y^3 \right) dy = \left[\frac{1}{21} y^7 + \frac{1}{4} y^4 \right]_{-2}^{1} = \frac{67}{28}.$$

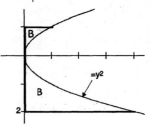

7.

$$\int_{0}^{1} dx \int_{0}^{\sqrt{1-x}} dy = \int_{0}^{1} [y]_{0}^{\sqrt{1-x}} dx = \int_{0}^{1} \sqrt{1-x} \, dx$$

$$= \left[-\frac{2}{3} (1-x)^{3/2} \right]_{0}^{1} = \frac{2}{3}.$$

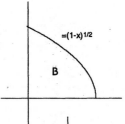

9.

$$\int_{0}^{\pi/2} dy \int_{0}^{\cos y} x \sin y \, dx = \int_{0}^{\pi/2} \left[\frac{1}{2} x^2 \sin y \right]_{0}^{\cos y} dx$$

$$= \int_{0}^{\pi/2} \left(\frac{1}{2} \cos^2 y \sin y \right) dy = \left[-\frac{1}{6} \cos^3 y \right]_{0}^{\pi/2} = \frac{1}{6}.$$

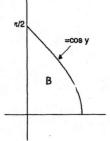

246

11.

$$\int_0^1\left[\int_0^z\left[\int_0^y dx\right]dy\right]dz = \int_0^1\left[\int_0^z\left[x\right]_0^y dy\right]dz$$

$$= \int_0^1\left[\int_0^z y\,dy\right]dz = \int_0^1\left[\frac{1}{2}y^2\right]_0^z dz$$

$$= \int_0^1 \frac{1}{2}z^2\,dz = \left[\frac{1}{6}z^3\right]_0^1 = \frac{1}{6}.$$

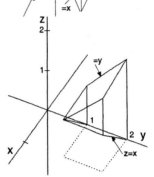

13.

$$\int_1^2 dy\int_0^1 dx\int_x^y dz = \int_1^2 dy\int_0^1 \left[z\right]_x^y dx$$

$$= \int_1^2 dy\int_0^1 (y-x)dx = \int_1^2\left[yx-\frac{1}{2}x^2\right]_0^1 dy$$

$$= \int_1^2\left(y-\frac{1}{2}\right)dy = \left[\frac{1}{2}y^2-\frac{1}{2}y\right]_1^2 = 1.$$

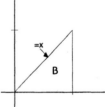

15.

$$\int_0^\pi \sin x\,dx\int_0^1 dy\int_0^2 (x+y+z)dz = \int_0^\pi \sin x\,dx\int_0^1 \left[xz+yz+\frac{1}{2}z^2\right]_0^2 dy$$

$$\int_0^\pi \sin x\,dx\int_0^1 (2x+2y+2)dy = \int_0^\pi \sin x\left[2xy+y^2+2y\right]_0^1 dx$$

$$\int_0^\pi (2x+3)\sin x\,dx = \left[-(2x+3)\cos x+2\sin x\right]_0^\pi = 2\pi+6.$$

17. The region B is the shaded triangular region shown on the right. The two integrals are

$$\int_0^1\left[\int_0^x x\sin y\,dy\right]dx \quad \text{and} \quad \int_0^1\left[\int_y^1 x\sin y\,dx\right]dy.$$

For the first integral we have

$$\int_0^1\left[\int_0^x x\sin y\,dy\right]dx = \int_0^1\left[-x\cos y\right]_0^x dx = \int_0^1(-x\cos x+x)dx$$

$$= \left[-x\sin x-\cos x+\frac{1}{2}x^2\right]_0^1 = -\sin 1-\cos 1+\frac{3}{2}.$$

And for the second integral we have

$$\int_0^1\left[\int_y^1 x\sin y\,dx\right]dy = \int_0^1\left[\frac{1}{2}x^2\sin y\right]_y^1 dy = \int_0^1\left(\frac{1}{2}\sin y-\frac{1}{2}y^2\sin y\right)dy$$

$$= \left[\frac{1}{2}y^2\cos y-y\sin y-\frac{3}{2}\cos y\right]_0^1 = -\sin 1-\cos 1+\frac{3}{2}.$$

247

19. (a) Let $c(x) = \sin \pi x$ and $d(x) = 4x - 2x^2$. On the x-interval $0 \le x \le 2$, the graph of $d(x)$ lies above the graph of $c(x)$. The region B in the xy-plane lying between $c(x)$ and $d(x)$ for $0 \le x \le 2$ is shown on the right.

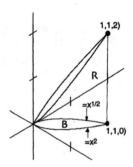

(b) Let $A(B)$ be the area of B. The volume of the solid lying above B and below the plane $z = 1$ is found by simply multiplying the area of its base by its height. In this case, the volume is numerically equal to $A(B)$. Thus, $A(B)$ is given by

$$A(B) = \int_0^2 \left[\int_{\sin \pi x}^{4x - 2x^2} 1 \, dy \right] dx.$$

(c) An iterated integral of $f(x, y)$ over B can be set up the same way as in part (b). That is,

$$\int_0^2 \left[\int_{\sin \pi x}^{4x - 2x^2} f(x, y) \, dy \right] dx.$$

Note that the integral in part (b) is a special case of the above iterated integral with $f(x, y) = 1$ over B.

21. Let B be the region in the xy-plane defined by $0 \le x \le 1$, $x^2 \le y \le \sqrt{x}$, and let R be the solid region with the added condition $0 \le z \le x + y$. The condition $0 \le z$ says that each point in R is either on or above B, and the conditions $z \le x + y$ says the each point is on or below the plane $z = x + y$. The region B and the solid region R are shown on the right.

Now let $f(x, y, z) = x + y + z$. Only two of the six possible iterated integrals of f over R can be written without having to break it into at least two pieces; namely, those that are computed first with respect to z. In terms of complexity of computation, the order of integration for the remaining variables is immaterial. Thus, we choose

$$\int_0^1 \left[\int_{x^2}^{\sqrt{x}} \left[\int_0^{x+y} (x + y + z) dz \right] dy \right] dx = \int_0^1 \left[\int_{x^2}^{\sqrt{x}} \left[(x+y)z + \frac{1}{2}z^2 \right]_0^{x+y} dy \right] dx$$

$$= \int_0^1 \left[\int_{x^2}^{\sqrt{x}} \frac{3}{2}(x + y)^2 dy \right] dx = \int_0^1 \left[\frac{1}{2}(x + y)^3 \right]_{x^2}^{\sqrt{x}} dx$$

$$= \frac{1}{2} \int_0^1 \left(3x^{5/2} + 3x^2 + x^{3/2} - 3x^4 - 3x^5 - x^6 \right) dx$$

$$= \frac{1}{2} \left[\frac{6}{7}x^{7/2} + x^3 + \frac{2}{5}x^{5/2} - \frac{3}{5}x^5 - \frac{1}{2}x^6 - \frac{1}{7}x^7 \right]_0^1 = \frac{71}{140}.$$

23. The solid region B is shown in the figure on the next page. A typical cross-sectional square S_y is shown with dashed borders. Note that the coordinates of the indicated corner of S_y shows that the area of S_y is given by $A(y) = 4y^2$. By symmetry, the volume of that

248

part of B that lies to the negative x-axis side of B is the same as the volume of that part of B that lies to the positive x-axis side of B. Thus, $V(B)$ can be found by integrating $A(y)$ from $y = 0$ to $y = a$ and then multiplying the result by 2. That is,

$$V(B) = 2 \int_0^a A(y)dy = 2 \int_0^a 4y^2 dy$$
$$= \left[\frac{8}{3}y^3\right]_0^a = \frac{8}{3}a^3.$$

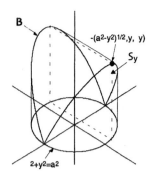

25. (a) Let the z-axis be the central axis of the cylinder and let the xy-plane be one of the planes. Let the second plane P be the plane with the following two properties: (i) the line of intersection l of P and the xy-plane has the equation $y = -a$; and (ii) the angle between P and the xy-plane is $\pi/4$. Property (i) implies that P is parallel to the x-axis and therefore the vector $(0, 1, k)$ is normal to P for some constant k; and property (ii) says $k = -1$. Hence, $(0, 1, -1)$ is a normal to P. Property (i) also says that $(0, -a, 0)$ is on P. Thus, a vector equation for the plane P is $(0, 1, -1) \cdot (x, y + a, z) = 0$, or $y - z = -a$. The wedge shaped solid W bounded by the planes and the cylinder are shown on the right.

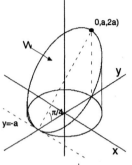

(b) A cross-section of W perpendicular to the axis of the cylinder is a segment of the circle $x^2 + y^2 = a^2$ and lies in a plane parallel to the xy-plane. A typical such cross-section is shown on the right. As can be seen, there is a one-to-one correspondence between these segments and the values of y between $-a$ and a. Letting $A(y)$ denote the area of the segment corresponding to y we have

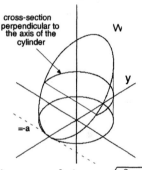

cross-section perpendicular to the axis of the cylinder

$$A(y) = \frac{\pi a^2}{2} - a^2 \arcsin(y/a) - y\sqrt{a^2 - y^2}, \quad -a \le y \le a.$$

Since the integral of an odd function over a symmetric interval is zero, and since $y\sqrt{a^2 - y^2}$ and the arcsine function are odd functions, the volume $V(W)$ of W is

$$V(W) = \int_{-a}^a A(y)dy = \int_{-a}^a \left(\frac{\pi a^2}{2} - a^2 \arcsin(y/a) - y\sqrt{a^2 - y^2}\right) dy = \frac{\pi a^2}{2} \int_{-a}^a dy = \pi a^3.$$

(c) A cross-section of W perpendicular to the line $y = -a$ is a plane polygonal figure. Two typical such cross-sections are shown on the right. The cross-section lying in the plane $x = 0$ is a 45° right triangle with legs of length $2a$, so that its area is $2a^2$. All of the other cross-sections (one for each nonzero x between $-a$ and a) are quadrilaterals with two vertical sides and a base of length $2\sqrt{a^2 - x^2}$. If such a quadrilateral is cut vertically down the middle then the two pieces can be rearranged to form a rectangle of height $2a$ and width $\sqrt{a^2 - x^2}$, so that its area is $B(x) = 2a\sqrt{a^2 - x^2}$ (which holds for $x = 0$).

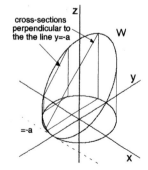

cross-sections perpendicular to the the line $y=-a$

249

Hence, the volume of W is

$$V(W) = \int_{-a}^{a} B(x)dx = 2a \int_{-a}^{a} \sqrt{a^2 - x^2}\, dx = 2a\left(\frac{\pi a^2}{2}\right) = \pi a^3,$$

where the third equality holds because we can interpret the second integral as half the area of a circle of radius a.

27. Let F_n be the value of the expression for $n \geq 2$. Note the two special cases

$$F_2 = \int_0^1 dx_1 \int_0^{x_1} (x_1 + x_2)dx_2 = \frac{3}{2}\int_0^1 x_1^2 dx_1 = \frac{1}{2},$$

$$F_3 = \int_0^1 dx_1 \int_0^1 dx_2 \int_0^{x_1} (x_1 + x_2)dx_3$$

$$= \int_0^1 dx_1 \int_0^1 (x_1 + x_2)x_1\, dx_2 = \int_0^1 \left(x_1^2 + \frac{1}{2}x_1\right)dx_1 = \frac{7}{12}.$$

For $n \geq 4$, we have

$$F_n = \int_0^1 dx_1 \int_0^1 dx_2 \cdots \int_0^1 d_{n-1} \int_0^{x_1}(x_1 + x_2)dx_n$$

$$= \int_0^1 dx_1 \int_0^1 dx_2 \cdots \int_0^1 (x_1 + x_2)x_1 d_{n-1}$$

$$= \int_0^1 dx_1 \int_0^1 dx_2 \cdots \int_0^1 (x_1 + x_2)x_1 d_{n-2}$$

$$\vdots$$

$$= \int_0^1 dx_1 \int_0^1 (x_1 + x_2)x_1 dx_2 = \frac{7}{12},$$

where the last integral has already been evaluated above as the second iterated integral in the equations showing the stages in computing F_3. In summary, $F_2 = 1/2$ and $F_n = 7/12$ for $n \geq 3$.

29. (a) Evidently, x and y are positive, so that $1/\sqrt{xy} = x^{-1/2}y^{-1/2}$. Hence,

$$J_\delta = \int_\delta^1 dy \int_\delta^1 x^{-1/2}y^{-1/2}dx = \int_\delta^1 y^{-1/2}dy \int_\delta^1 x^{-1/2}dx = \left(2(1 - \delta^{1/2})\right)^2 = 4(1 - \delta^{1/2})^2.$$

(b) By part (a), $\lim_{\delta \to 0+} J_\delta = \lim_{\delta \to 0+} 4(1 - \delta^{1/2})^2 = 4$. On the other hand,

$$\lim_{\delta \to 0+} J_\delta = \lim_{\delta \to 0+} \int_\delta^1 y^{-1/2}dy \int_\delta^1 x^{-1/2}dx = \left(\lim_{\delta \to 0+}\int_\delta^1 y^{-1/2}dy\right)\left(\lim_{\delta \to 0+}\int_\delta^1 x^{-1/2}dx\right)$$

$$= \int_0^1 y^{-1/2}dy \int_0^1 x^{-1/2}dx = \int_0^1 dy \int_0^1 (xy)^{-1/2}dx,$$

where the distribution of the limit to each integral in the first line can be done because all limits exist, and the first member of the last line is the result of using the definition of the value of an improper integral where the integrand becomes unbounded at one of the limits of integration. Therefore, $\int_0^1 dy \int_0^1 \frac{1}{\sqrt{xy}}dx = 4$.

Section 2: MULTIPLE INTEGRALS

Exercise Set 2A-E (pgs. 332-333)

1. Let B be the unit disk shown on the right. Then

$$\int_B (x^2 + 3y^2)\,dx\,dy = \int_{-1}^{1} dx \int_{-\sqrt{1-x^2}}^{\sqrt{1-x^2}} (x^2 + 3y^2)\,dy$$

$$= \int_{-1}^{1} \left[x^2 y + y^3 \right]_{-\sqrt{1-x^2}}^{\sqrt{1-x^2}} dx = \int_{-1}^{1} 2\sqrt{1 - x^2}\,dx$$

$$= 2 \int_{-1}^{1} \sqrt{1 - x^2}\,dx = \pi,$$

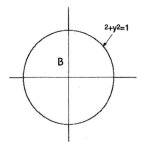

where we recognize the last integral to be one-half the area of a circle of radius 1.

3. Let B be the rectangle shown on the right. Then

$$\int_B x \sin xy\,dx\,dy = \int_0^{\pi} dx \int_0^1 x \sin xy\,dy$$

$$= \int_0^{\pi} \left[-\cos xy \right]_0^1 dx = \int_0^{\pi} (1 - \cos x)\,dx$$

$$= \left[x - \sin x \right]_0^{\pi} = \pi.$$

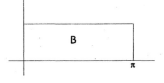

5. Let B be the rectangle $0 \le x \le 2$, $0 \le y \le 1$. For each positive integer n, consider the grid consisting of the vertical lines $x = i/n$, $i = 0, \ldots, 2n$, and the horizontal lines $y = j/n$, $j = 1, \ldots, n$. Each rectangle R_{ij} is a square with side length $1/n$ and therefore has an area of $1/n^2$. Setting $(x_i, y_j) = (i/n, j/n)$, the corresponding Riemann sum for the function $f(x,y) = x + 4y$ is

$$\sum_{i=1}^{2n} \sum_{j=1}^{n} (x_i + 4y_j) A(R_{ij}) = \sum_{i=1}^{2n} \sum_{j=1}^{n} \left(\frac{i}{n} + \frac{4j}{n} \right) \frac{1}{n^2} = \frac{1}{n^2} \sum_{i=1}^{2n} \left(\sum_{j=1}^{n} \frac{i}{n} + \sum_{j=1}^{n} \frac{4j}{n} \right)$$

$$= \frac{1}{n^2} \sum_{i=1}^{2n} (i + 2(n+1)) = \frac{1}{n^2} \left(n(2n+1) + 4n(n+1) \right)$$

$$= \frac{1}{n^2} (6n^2 + 5n) = 6 + \frac{5}{n}.$$

Hence, $\int_B (x + 4y)\,dx\,dy = \lim_{n \to \infty} (6 + 5/n) = 6$. We verify this answer by computing $\int_B (x + 4y)\,dx\,dy$ directly as an iterated integral. We have

$$\int_B (x + 4y)\,dx\,dy = \int_0^1 dy \int_0^2 (x + 4y)\,dx = \int_0^1 \left[\frac{1}{2} x^2 + 4xy \right]_0^2 dy$$

$$= \int_0^1 (2 + 8y)\,dy = \left[2y + 4y^2 \right]_0^1 = 2 + 4 = 6.$$

251

7. Let B be the rectangle with corners $(1,1)$, $(1,3)$, $(2,3)$ and $(2,1)$. Here, B has the limits $1 \le x \le 2$, $1 \le y \le 3$. With $f(x,y) = x + y^2$, the volume under the graph of f and above B is

$$\int_B (x+y^2)\,dA = \int_1^3 dy \int_1^2 (x+y^2)\,dx = \int_1^3 \left[\frac{1}{2}x^2 + y^2 x\right]_1^2$$

$$= \int_1^3 \left(\frac{3}{2} + y^2\right) dy = \left[\frac{3}{2}y + \frac{1}{3}y^3\right]_1^3 = \frac{35}{3}.$$

9. Let B be the unit disk. With $f(x,y) = |x+y|$, the volume under the graph of f and above B is

$$\int_B |x+y|\,dA = \int_{-1}^1 dx \int_{-\sqrt{1-x^2}}^{\sqrt{1-x^2}} |x+y|\,dy$$

$$= \int_{-1}^1 dx \left(\int_{-\sqrt{1-x^2}}^{-x} (-x-y)\,dy + \int_{-x}^{\sqrt{1-x^2}} (x+y)\,dy\right)$$

$$= \int_{-1}^1 \left(\left[-xy - \frac{1}{2}y^2\right]_{-\sqrt{1-x^2}}^{-x} + \left[xy + \frac{1}{2}y^2\right]_{-x}^{\sqrt{1-x^2}}\right) dx$$

$$= \int_{-1}^1 dx = 2.$$

11. If B is the region bounded by the curve $x^2 - 2x + 4y^2 - 8y + 1 = 0$ then its area is the content of B and is given by $A(B) = \int_B dA$. To get a handle on the region B, we complete the square on each variable in the given equation to get $(x-1)^2 + 4(y-1)^2 = 4$ or $\left(\frac{x-1}{2}\right)^2 + (y-1)^2 = 1$, which is an ellipse centered at $(1,1)$ with major axis of length 4 in the x direction and minor axis of length 2 in the y direction. We arbitrarily choose to integrate first with respect to y and then with respect to x. The upper limits of integration with respect to y are found by solving the given equation for y to get $y = 1 \pm \frac{1}{2}\sqrt{4 - (x-1)^2}$. We therefore have

$$\int_B dA = \int_{-1}^3 dx \int_{1-\frac{1}{2}\sqrt{4-(x-1)^2}}^{1+\frac{1}{2}\sqrt{4-(x-1)^2}} dy = \int_{-1}^3 \sqrt{4-(x-1)^2}\,dx = 2\pi,$$

where the last integral is done by recognizing it as the area of the top half of the circle of radius 2 centered at $(1,0)$. Note that we can check our answer by recalling that the area of the ellipse $(x-x_0)^2/a^2 + (y-y_0)^2/b^2 = 1$ is πab. Here, $a = 2$ and $b = 1$, so that the area 2π is in agreement with our result

13. Let B be the region bounded by the surface $z = 4 - 4x^2 - y^2$ and the xy-plane. The equation for the surface can be written as $4x^2 + y^2 = 4 - z$, which we recognize as an elliptic paraboloid opening downward with the z-axis as its central axis. The region R in the xy-plane lying under the surface is the ellipse $x^2 + (y/2)^2 = 1$, found by setting $z = 0$ in the given equation. The region B is shown on the next page.

As a triple integral, we integrate the function $f(x,y,z) = 1$ over B to get

$$V(B) = \iiint_B dV = \int_{-1}^1 dx \int_{-2\sqrt{1-x^2}}^{2\sqrt{1-x^2}} dy \int_0^{4-4x^2-y^2} dz.$$

As a double integral, we integrate the function $f(x,y) = 4 - 4x^2 - y^2$ over R to get

$$V(B) = \iint_R (4 - 4x^2 - y^2)\, dA$$

$$= \int_{-1}^{1} dx \int_{-2\sqrt{1-x^2}}^{2\sqrt{1-x^2}} (4 - 4x^2 - y^2)\, dy.$$

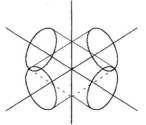

However, it is perhaps easiest to integrate an area over a fixed interval by observing that the intersection of the plane $z = z_0$ $(0 \le z_0 \le 4)$ with the solid B is an elliptic region whose boundary is an equation of the form

$$x^2/\left(\sqrt{4 - z_0}/2\right)^2 + y^2/\left(\sqrt{4 - z_0}\right)^2 = 1.$$

In Exercise 11 above, it was noted that the area of such an ellipse is $\frac{\pi}{2}(4 - z_0)$. Hence, if we let $A(z) = \frac{\pi}{2}(4 - z)$, we can integrate $A(z)$ from $z = 0$ to $z = 4$ to get

$$V(B) = \int_0^4 A(z)\, dz = \int_0^4 \frac{\pi}{2}(4 - z)\, dz = \frac{\pi}{2}\left[4z - \frac{1}{2}z^2\right]_0^4 = 4\pi.$$

The reader can check that the same answer is obtained if $V(B)$ is computed using either of the multiple integrals given above.

15. Let B be the region of intersection of the two cylindrical solids shown on the right. Notice that cross-sections perpendicular to the z-axis are squares. For a given value of z between -1 and 1, the side lengths of the corresponding square are $2x = 2\sqrt{1 - z^2}$ and $2y = 2\sqrt{1 - z^2}$, so that the area of the cross-section for such a value of z is $4(1 - z^2)$. Thus, we can find the volume of B by integrating the area function $A(z) = 4(1 - z^2)$ from $z = -1$ to $z = 1$. This gives

$$V(B) = \int_{-1}^{1} A(z)\, dz = \int_{-1}^{1} 4(1 - z^2)\, dz = 4\left[z - \frac{1}{3}z^3\right]_{-1}^{1} = \frac{16}{3}.$$

The volume of B can also be found with multiple integrals but is much more complicated.

17. Let B be the solid represented by the shape of the liquid in the hemispherical bowl. We view the bowl as the bottom half of the sphere $x^2 + y^2 + z^2 = a^2$. Notice that cross-sections perpendicular to the z-axis are circles. In particular, if a cross-section is at height z (where $-a \le z \le h - a$) then the radius of the cross-section can be found by setting $y = 0$ in the equation of the sphere and solving for x (one can also set $x = 0$ and solve for y). This gives the radius of the cross-sectional circle as $|x| = \sqrt{a^2 - z^2}$, so that its area is $A(z) = \pi(a^2 - z^2)$. Since the liquid is at heights z only for $-a \le z \le h - a$, the volume of the liquid can be found by integrating the area function $A(z)$ from $z = -a$ to $z = h - a$. This gives

$$V(B) = \int_{-a}^{h-a} A(z)\, dz = \int_{-a}^{h-a} \pi(a^2 - z^2)\, dz = \pi\left[a^2 z - \frac{1}{3}z^3\right]_{-a}^{h-a} = \pi h^2\left(a - \frac{1}{3}h\right).$$

19. Orient the steel plate in the upper xy-plane so that its straight edge lies on the x-axis and is symmetrical with the y-axis. If A is the semicircular region defined by $x^2 + y^2 \le 4$,

$y \geq 0$, and B is the semicircular region defined by $x^2 + y^2 \leq 1/4$, $y \geq 0$, then the region covered by the plate is $A - B$ (i.e., the region A with the region B removed). Given that the plate has uniform density $\mu(x, y) = 12$, the mass of the plate is given by

$$\iint_{A-B} 12 \, dx \, dy = \iint_A 12 \, dx \, dy - \iint_B 12 \, dx \, dy$$

$$= \int_{-2}^{2} dx \int_0^{\sqrt{4-x^2}} 12 \, dy - \int_{-1/2}^{1/2} dx \int_0^{\sqrt{1/4-x^2}} 12 \, dy.$$

The computations are unnecessary because the entire problem can be done without integration. The area of a semicircular plate of radius 2 is 2π and the area of a semicircular plate of radius $1/2$ is $\pi/8$. Thus, the area of the semicircular plate with the semicircular piece removed is $2\pi - \pi/8 = 15\pi/8$. Since the density is uniform at 12 pounds per square foot, it follows that the total mass of the plate is $(12)(15\pi/8) = 45\pi/2$ pounds.

21. Orient the column with its base on the xy-plane and its axis coincident with the z-axis. Let C_z be the circular cross-section perpendicular to the z-axis at height z above the xy-plane.

Now, imagine extending the column upward until it becomes an inverted cone. Since the column loses 4 inches in diameter for every 10 feet in height, it follows that the cone is 30 feet high. A side view of this cone is a triangle which is similar to the triangle formed by the diameter of C_z, the axis of the column, and the side of the column. By similar triangles,

$$\frac{\text{height of the smaller triangle}}{\text{height of the cone}} = \frac{\text{radius of the base of the smaller triangle}}{\text{radius of the base of the cone}}.$$

By definition, C_z is at height z, so that the vertical distance from C_z to the top of the cone (height of the smaller triangle) is $30 - z$. Since the radius of the base of the cone is 6 inches $= 1/2$ feet, we have

$$\frac{30 - z}{30} = \frac{r(z)}{1/2} \qquad \text{so that} \qquad r(z) = \frac{30 - z}{60}.$$

Hence,

$$A(C_z) = \pi r^2(z) = \pi \left(\frac{30 - z}{60} \right)^2, \quad 0 \leq z \leq 10.$$

From the information given, the point mass density of the material depends only on the height z of the cross-section on which the point mass of material is situated. Thus, for each $0 \leq z \leq 10$, we let $\mu(z)$ denote the uniform density of the material on C_z. Since $\mu(z)$ is a linear function of z, it must have the form $\mu(z) = az + b$ for some constants a, b. Using $\mu(0) = 50$ and $\mu(10) = 40$, we find that $a = -1$, $b = 50$, so that $\mu(z) = -z + 50$. It follows that the total mass of C_z, which we will denote by $m(z)$, is

$$m(z) = \mu(z) A(C_z) = (-z + 50)\pi \left(\frac{30 - z}{60} \right)^2 = \frac{\pi}{3600}(-z^3 + 110z^2 - 3900z + 45000).$$

Therefore, integrating $m(z)$ from $z = 0$ to $z = 10$ gives the total mass M of the column; namely,

$$M = \int_0^{10} m(z) \, dz = \int_0^{10} \frac{\pi}{3600}(-z^3 + 110z^2 - 3900z + 45000) \, dz$$

$$= \frac{\pi}{3600} \left[-\frac{1}{4}z^4 + \frac{110}{3}z^3 - 1950z^2 + 45000z \right]_0^{10} \approx 252.3 \text{ pounds.}$$

Section 3: INTEGRATION THEOREMS

Exercise Set 3 (pgs. 336-337)

1. Regions B_1 and B_2 are disjoint. Moreover, $f(x,y) = 2x - y$ if (x,y) is in B_1, and $f(x,y) = x^2 + y$ if (x,y) is in B_2. So, by Theorem 3.6,

$$\int_{B_1 \cup B_2} f(x,y)\, dx\, dy = \int_{B_1} (2x - y)\, dx\, dy + \int_{B_2} (x^2 + y)\, dx\, dy$$

$$= \int_0^1 dy \int_0^1 (2x - y)\, dx + \int_{-1}^1 dy \int_1^2 (x^2 + y)\, dx.$$

The first iterated integral evaluates to

$$\int_0^1 dy \int_0^1 (2x - y)\, dx = \int_0^1 \left[x^2 - xy \right]_0^1 dy = \int_0^1 (1 - y)\, dy = \left[y - \frac{1}{2} y^2 \right]_0^1 = \frac{1}{2}.$$

And the second iterated integral evaluates to

$$\int_{-1}^1 dy \int_1^2 (x^2 + y)\, dx = \int_{-1}^1 \left[\frac{1}{3} x^3 + xy \right]_1^2 dy = \int_{-1}^1 \left(\frac{7}{3} + y \right) dy = \left[\frac{7}{3} y + \frac{1}{2} y^2 \right]_{-1}^1 = \frac{14}{3}.$$

Hence,

$$\int_{B_1 \cup B_2} f(x,y)\, dx\, dy = \frac{1}{2} + \frac{14}{3} = \frac{31}{6}.$$

3. Let $f : \mathcal{R}^n \longrightarrow \mathcal{R}$ be continuous on a set B and let $\mathbf{x}_0$ be interior to B. If B_r is the ball of radius r centered at $\mathbf{x}_0$ then $B_r \subset B$ for all sufficiently small r. Let $\epsilon > 0$ be given. Then the continuity of f at $\mathbf{x}_0$ implies the existence of $r > 0$ such that

$$|f(\mathbf{x}) - f(\mathbf{x}_0)| < \epsilon, \quad \text{for all } \mathbf{x} \in B_r.$$

Therefore, for such r, we can use the result of Exercise 2 in this section, along with Theorems 3.1 and 3.5, to get

$$(*) \quad \left| \int_{B_r} \left(f(\mathbf{x}) - f(\mathbf{x}_0) \right) dV \right| \leq \int_{B_r} |f(\mathbf{x}) - f(\mathbf{x}_0)|\, dV \leq \int_{B_r} \epsilon\, dV \leq \epsilon \int_{B_r} dV = \epsilon V(B_r).$$

Now observe that linearity allows us to write the leftmost member above as

$$\left| \int_{B_r} \left(f(\mathbf{x}) - f(\mathbf{x}_0) \right) dV \right| = \left| \int_{B_r} f(\mathbf{x})\, dV - f(\mathbf{x}_0) \int_{B_r} dV \right| = \left| \int_{B_r} f(\mathbf{x})\, dV - f(\mathbf{x}_0) V(B_r) \right|.$$

Thus, when the two outer members of $(*)$ are divided by the *positive* number $V(B_r)$, the result can be written as

$$\left| \frac{1}{V(B_r)} \int_{B_r} f(\mathbf{x})\, dV - f(\mathbf{x}_0) \right| < \epsilon, \quad \text{so that} \quad \lim_{r \to 0^+} \frac{1}{V(B_r)} \int_{B_r} f(\mathbf{x})\, dV = f(\mathbf{x}_0).$$

5. Let B be the rectangle $0 \leq x \leq 1$, $0 \leq y \leq 1$ and let $f(x,y) = 1$ if x is rational and $f(x,y) = 2y$ if x is irrational. By direct computation, we find that $\int_0^1 f(x,y)\,dy = 1$, regardless of whether x is rational or irrational. It follows that

$$\int_0^1 dx \int_0^1 f(x,y)\,dy = \int_0^1 1\,dx = 1.$$

Now, if f is integrable over B then $\int_B f\,dV$ exists and is equal to the above value of the corresponding iterated integral. That is, the corresponding Riemann sums must tend to 1 as the mesh size of the grids tends to zero. We show that this does not happen by assuming that f is integrable over B and deriving a contradiction.

First, let B_1 be the rectangle $0 \leq x \leq 1$, $0 \leq y \leq 3/4$. Let G be any grid covering B with mesh size less than $1/16$ (i.e., $M(G) < 1/16$), and let $R_1, \ldots, R_n$ be the corresponding rectangles. We choose a point (x_i, y_i) in R_i by the rule:

$$x_i \text{ is rational if } R_i \subseteq B_1 \qquad \text{and} \qquad x_i \text{ is irrational and } y_i > 3/4 \text{ if } R_i \not\subseteq B_1.$$

Since $M(G) < 1/16$, there will always be rectangles that are subsets of B_1 and rectangles that are not subsets of B_1. Thus, we can write the corresponding Riemann sum as

$$\sum_{i=1}^n f(x_i, y_i) A(R_i) = \sum_{R_i \subseteq B_1} 1\, A(R_i) + \sum_{R_i \not\subseteq B_1} 2y_i\, A(R_i).$$

Again, since the side lengths of the R_i are all less than $1/16$, the sum of the areas of all rectangles contained in B_1 is no less than $3/4 - 1/16 = 11/16$, which is a lower bound for the first sum on the right. Further, we notice that each y_i counted in the second sum on the right is greater than $3/4$ and that the sum of the areas of the rectangles that are not subsets of B_1 is at least $1/4$, so that a lower bound for the second sum on the right is $2(3/4)(1/4) = 6/16$. Hence,

$$\sum_{i=1}^n f(x_i, y_i) A(R_i) \geq \frac{11}{16} + \frac{6}{16} = \frac{17}{16}.$$

In other words, all Riemann sums with sufficiently small mesh size are bounded away from 1 and we conclude that

$$\lim_{M(G) \to 0} \sum_{i=1}^n f(x_i, y_i) A(R_i) \neq 1,$$

the contradiction we predicted. Therefore, f is not integrable over B.

7. In parts (a), (b) and (c) below, let $f : \mathcal{R}^n \to \mathcal{R}^m$ and $g : \mathcal{R}^n \to \mathcal{R}^m$ be defined on a subset B in $\mathcal{R}^n$. Further, let $f_1, \ldots, f_m$ and $g_1, \ldots, g_m$ be the coordinate function of f and g, respectively. We shall say that f and g are integrable over B if, and only if, each f_i and g_i is integrable over B, in which case, $\int_B f\,dV$ (e.g.) is defined to be

$$\int_B f\,dV = \left(\int_B f_1\,dV \ldots, \int_B f_m\,dV \right).$$

(a) Let f and g be integrable over B and let a and b be real constants. Then $af_i + bg_i$ are the coordinate functions of $af + bg$ and each is integrable over B. Therefore, $af + bg$ is integrable over B and

$$\int_B (af + bg)\, dV = \left(\int_B (af_1 + bg_1)\, dV, \ldots, \int_B (af_m + bg_m)\, dV \right)$$

$$= \left(a \int_B f_1\, dV + b \int_B g_1\, dV, \ldots, a \int_B f_m\, dV + b \int_B g_m\, dV \right)$$

$$= a \left(\int_B f_1\, dV, \ldots, \int_B f_m\, dV \right) + b \left(\int_B g_1\, dV, \ldots, \int_B g_m\, dV \right)$$

$$= a \int_B f\, dV + b \int_B g\, dV,$$

where the second equality follows from the linearity of the multiple integral of real-valued functions.

(b) Let $\mathbf{k} = (k_1, \ldots, k_m)$ be a fixed vector in $\mathcal{R}^m$ and let f be integrable over B. Then each $k_i f_i$ is integrable over B, so that $k_1 f_1 + \cdots + k_m f_m = \mathbf{k} \cdot f$ is integrable over B. So,

$$\int_B \mathbf{k} \cdot f\, dV = \int_B [(k_1, \ldots, k_m) \cdot (f_1, \ldots, f_m)]\, dV$$

$$= \int_B (k_i f_1 + \cdots + k_m f_m)\, dV = k_1 \int_B f_1\, dV + \cdots + k_m \int_B f_m\, dV$$

$$= (k_1, \ldots, k_m) \cdot \left(\int_B f_1\, dV, \ldots, \int_B f_m\, dV \right)$$

$$= \mathbf{k} \cdot \int_B f\, dV,$$

where the last expression in the second line follows from the linearity of the multiple integral of real-valued functions.

(c) Let f and g be integrable over B. The inequality we are asked to establish clearly holds if $\int_B f\, dV = \mathbf{0}$. Thus, we will assume that $\int_B f\, dV \neq \mathbf{0}$. By the Cauchy-Schwarz inequality,

$$f(\mathbf{x}) \cdot \int_B f dV \leq |f(\mathbf{x})| \left| \int_B f\, dV \right|, \qquad \text{for all } \mathbf{x} \in B.$$

Note that each side is a real-valued function of $\mathbf{x}$, so that Theorem 3.5 says we can integrate both sides with respect to $\mathbf{x}$ and still retain the same direction on the inequality. This gives

$$\int_B \left(f(\mathbf{x}) \cdot \int_B f dV \right) dV \leq \int_B \left(|f(\mathbf{x})| \left| \int_B f dV \right| \right) dV.$$

By part (b) (with $\mathbf{k} = \int_B f dv$), the left member of this inequality becomes

$$\int_B \left(f(\mathbf{x}) \cdot \int_B f dV \right) dV = \int_B f dV \cdot \int_B f(\mathbf{x})\, dV = \left| \int_B f dV \right|^2.$$

In the right member of the inequality, we note that $\left| \int_B f dV \right|$ is a real constant multiplying the real-valued function $|f(\mathbf{x})|$, so that the linearity of the multiple integral of real-valued functions allows us to write the right member of the inequality as

$$\int_B \left(|f(\mathbf{x})| \left| \int_B f dV \right| \right) dV = \left| \int_B f dV \right| \int_B |f(\mathbf{x})|\, dV.$$

257

Hence,

$$\left| \int_B f \, dV \right|^2 \le \left| \int_B f \, dV \right| \int_B |f(\mathbf{x})| \, dV.$$

Under our assumption that $\int_B f \, dV \ne 0$, we can divide both sides by $\left| \int_B f \, dV \right|$ to get

$$\left| \int_B f \, dV \right| \le \int_B |f(\mathbf{x})| \, dV.$$

9. Applying the Leibniz rule to the function $f(y) = \int_0^1 (y^2 + t^2) \, dt$, we get

$$f'(y) = \frac{d}{dy} \int_0^1 (y^2 + t^2) \, dt = \int_0^1 \frac{\partial}{\partial y} (y^2 + t^2) \, dt = \int_0^1 2y \, dt = \left[2yt \right]_0^1 = 2y.$$

11. Let $h(x) = \int_0^x (x - u) e^{u^2} \, du$. Since the upper limit of integration is not a constant with respect to h, we cannot apply the Leibniz rule given in the text. Rather, we apply the more general Leibniz rule established in Exercise 8 in this section. We get

$$h'(x) = \frac{d}{dx} \int_0^x (x - u) e^{u^2} \, du$$

$$= \int_0^x \frac{\partial}{\partial x} (x - u) e^{u^2} \, du + \frac{dx}{dx} (x - x) e^{x^2} - \frac{d(0)}{dx} (x - 0) e^{0^2} = \int_0^x e^{u^2} \, du.$$

To find $h''(x)$, we simply recognize that $h'(x)$ is an antiderivative of e^{x^2} and apply one version of the fundamental theorem of calculus to get $h''(x) = e^{x^2}$.

13. Let $f : \mathcal{R}^2 \longrightarrow \mathcal{R}$ be such that f_x, f_y, f_{yx} and f_{xy} are continuous. By the fundamental theorem of calculus,

$$f(x, y) - f(a, y) = \int_a^x f_x(t, y) \, dt.$$

By the Leibniz rule, differentiation of this equation with respect to y gives

$$f_y(x, y) - f_y(a, y) = \frac{\partial}{\partial y} \int_a^x f_x(t, y) \, dt = \int_a^x f_{xy}(t, y) \, dt.$$

We can then use the fundamental theorem of calculus to conclude that differentiation of the rightmost member of the above equation with respect to x gives $f_{xy}(x, y)$, while differentiation of the leftmost member with respect to x gives $f_{yx}(x, y)$. Hence,

$$f_{yx}(x, y) = f_{xy}(x, y).$$

258

Section 4: CHANGE OF VARIABLE

Exercise Set 4A-D (pgs. 346-348)

1. Let $x = \sqrt{u}$ in $\int_0^2 x e^{x^2} dx$. Then $u = x^2$ and $\frac{1}{2} du = x dx$. Also, $u = x$ when $x = 0$ and $u = 4$ when $x = 2$. So,

$$\int_0^2 x e^{x^2} dx = \frac{1}{2} \int_0^4 e^u du = \left[\frac{1}{2} e^u\right]_0^4 = \frac{1}{2}(e^4 - 1).$$

3. Let B be the region in $\mathcal{R}^2$ described by $0 \le x$, $0 \le y$, and $x^2 + y^2 \le 4$.

(a) The region B is the area in the first quadrant of the xy-plane bounded by the circle of radius 2 centered at the origin. Its description in terms of polar coordinates is $0 \le r \le 2$, $0 \le \theta \le \pi/2$.

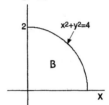

(b) If $f(x,y) = \sqrt{x^2 + y^2}$ then $f(r \cos \theta, r \sin \theta) = r$. Using the limits for r and θ found in part (a), Equation 4.1 gives

$$\int_B \sqrt{x^2 + y^2}\, dx\, dy = \int_0^{\pi/2} d\theta \int_0^2 r\, r\, dr = \int_0^{\pi/2} d\theta \int_0^2 r^2\, dr = \left[\theta\right]_0^{\pi/2} \left[\frac{1}{3} r^3\right]_0^2 d\theta = \frac{4\pi}{3}.$$

(c) With B as given in part (a), if $f(x,y) = x$ then $f(r \cos \theta, r \sin \theta) = r \cos \theta$. Equation 4.1 then gives

$$\int_B x\, dx\, dy = \int_0^{\pi/2} d\theta \int_0^2 r \cos \theta\, r\, dr = \int_0^{\pi/2} \cos \theta\, d\theta \int_0^2 r^2\, dr = \left[\sin \theta\right]_0^{\pi/2} \left[\frac{1}{3} r^3\right]_0^2 = \frac{8}{3}.$$

5. The annular region A is the area between the concentric circles about the origin of radii 1 and 2. In terms of polar coordinates, A is described by $1 \le r \le 2$, $0 \le \theta \le 2\pi$. Let $\mu(x,y)$ be the density at the point (x,y) on the corresponding plastic ring. To say that $\mu(x,y)$ varies inversely as the distance from (x,y) to the center of the ring means that $\mu(x,y) = k/\sqrt{x^2 + y^2}$ for some positive constant k. Changing to polar coordinates, $\mu(r \cos \theta, r \sin \theta) = k/r$ and the mass is given by a double integral of μ over A. Specifically, by Equation 4.1,

$$M(A) = \int_A \frac{k}{\sqrt{x^2 + y^2}} dx\, dy = \int_0^{2\pi} d\theta \int_1^2 \frac{k}{r} r\, dr = \int_0^{2\pi} d\theta \int_1^2 k\, dr = \left[\theta\right]_0^{2\pi} \left[kr\right]_1^2 = 2k\pi.$$

For points (x,y) at the inner edge of the ring, $\sqrt{x^2 + y^2} = 1$, so that $\mu(x,y) = k$. Since we are given that $\mu(x,y) = 10$ on the inner edge of the ring, $k = 10$. Hence, the mass of the plastic ring is $M(A) = 20\pi$ grams.

7. Let D be the disk of radius $\sqrt{\pi/2}$ centered at $(0,0)$. In terms of polar coordinates, D is described by $0 \le r \le \sqrt{\pi/2}$, $0 \le \theta \le 2\pi$. When a change of variable to polar coordinates is made in the integral to be evaluated, the integrand $\cos(x^2 + y^2)$ becomes $\cos r^2$ and Equation 4.1 gives

$$\int_D \cos(x^2 + y^2)\, dA = \int_0^{2\pi} d\theta \int_0^{\sqrt{\pi/2}} \cos r^2\, r\, dr = \left[\theta\right]_0^{2\pi} \left[\frac{1}{2} \sin r^2\right]_0^{\sqrt{\pi/2}} = \pi.$$

259

9. Consider the curve C defined by $(x^2 + y^2)^2 = 2a^2(x^2 - y^2)$. A quick sketch of C reveals that C has the shape of an infinity symbol and is symmetric with respect to both axes. Thus, the region bounded by C, call it B, can be found by computing the area of that part of B that lies in the first quadrant and then multiplying the result by 4. Changing to polar coordinates we get $r^4 = 2a^2r^2(\cos^2\theta - \sin^2\theta) = 2a^2r^2\cos 2\theta$ or, on solving for r,

$$r = \sqrt{2a^2\cos 2\theta}.$$

This shows $\cos 2\theta \geq 0$, which happens for points in the first quadrant if, and only if, $0 \leq \theta \leq \pi/4$. Hence, the area of B is given by

$$A(B) = 4\int_0^{\pi/4} d\theta \int_0^{\sqrt{2a^2\cos 2\theta}} r\,dr = 4\int_0^{\pi/4}\left[\frac{1}{2}r^2\right]_0^{\sqrt{2a^2\cos 2\theta}} d\theta$$

$$= 4a^2\int_0^{\pi/4}\cos 2\theta\,d\theta = 4a^2\left[\frac{1}{2}\sin 2\theta\right]_0^{\pi/4} = 2a^2.$$

11. Let $f(x,y,z) = \sqrt{x^2 + y^2}$ and let B be the solid ball of radius 1 centered at the origin in $\mathcal{R}^3$. In terms of spherical coordinates,

$$\overline{f}(r,\phi,\theta) = f(r\sin\phi\cos\theta, r\sin\phi\sin\theta, r\cos\phi) = r\sin\phi.$$

Also, B is described by $0 \leq r \leq 1$, $0 \leq \phi \leq \pi$, $0 \leq \theta \leq 2\pi$. Hence, Equation 4.2 gives

$$\int_B \sqrt{x^2 + y^2}\,dx\,dy\,dz = \int_0^{2\pi} d\theta \int_0^{\pi} d\phi \int_0^1 r\sin\phi\, r^2\sin\phi\,dr$$

$$= \int_0^{2\pi} d\theta \int_0^{\pi}\sin^2\phi\,d\phi \int_0^1 r^3\,dr = \int_0^{2\pi} d\theta \int_0^{\pi}\frac{1}{2}(1 - \cos 2\phi)\,d\phi \int_0^1 r^3\,dr$$

$$= [\theta]_0^{2\pi}\left[\frac{1}{2}\phi - \frac{1}{4}\sin 2\phi\right]_0^{\pi}\left[\frac{1}{4}r^4\right]_0^1 = \frac{\pi^2}{4}.$$

13. We view the solid ball B of radius a as the solid sphere of radius a centered at the origin in $\mathcal{R}^3$. The density of B at the point (x,y,z) is $\mu(x,y,z) = \sqrt{x^2 + y^2 + z^2}$, which is numerically equal to the distance from (x,y,z) to the origin (the center of B). The mass of B is therefore given by the triple integral

$$M(B) = \int_B \sqrt{x^2 + y^2 + z^2}\,dx\,dy\,dz.$$

Changing to spherical coordinates, we find that B is described by $0 \leq r \leq a$, $0 \leq \phi \leq \pi$, $0 \leq \theta \leq 2\pi$, and that the density function becomes

$$\overline{\mu}(r,\phi,\theta) = \mu(r\sin\phi\cos\theta, r\sin\phi\sin\theta, r\cos\phi) = r.$$

By Equation 4.2,

$$\int_B \sqrt{x^2 + y^2 + z^2}\,dx\,dy\,dz = \int_0^{2\pi} d\theta \int_0^{\pi} d\phi \int_0^a r\,r^2\sin\phi\,dr$$

$$= \int_0^{2\pi} d\theta \int_0^{\pi}\sin\phi\,d\phi \int_0^a r^3\,dr = [\theta]_0^{2\pi}\big[-\cos\phi\big]_0^{\pi}\left[\frac{1}{4}r^4\right]_0^a = \pi a^4.$$

15. The region C in $\mathcal{R}^3$ described by $0 \leq x$, $0 \leq y$, $0 \leq z \leq 1$, $x^2 + y^2 \leq 2$ is the "quarter" wedge one unit high in the first octant cut from the vertical circular cylinder of radius $\sqrt{2}$ centered at the origin. Its description in cylindrical coordinates is $0 \leq r \leq \sqrt{2}$, $0 \leq \theta \leq \pi/2$, $0 \leq z \leq 1$. Moreover, the given integrand $x^2 + y^2 + z^2$ reduces to $r^2 + z^2$ in cylindrical coordinates. Thus,

$$\int_C (x^2 + y^2 + z^2)\, dx\, dy\, dz = \int_0^1 dz \int_0^{\pi/2} d\theta \int_0^{\sqrt{2}} (r^2 + z^2) r\, dr$$

$$= \int_0^1 dz \int_0^{\pi/2} d\theta \int_0^{\sqrt{2}} (r^3 + z^2 r)\, dr = \int_0^1 dz \int_0^{\pi/2} \left[\frac{1}{4} r^4 + \frac{1}{2} z^2 r^2\right]_0^{\sqrt{2}} d\theta$$

$$= \int_0^1 dz \int_0^{\pi/2} (1 + z^2)\, d\theta = \int_0^1 (1 + z^2)\, dz \int_0^{\pi/2} d\theta = \left[z + \frac{1}{3} z^3\right]_0^1 [\theta]_0^{\pi/2} = \frac{2\pi}{3}.$$

17. Let S be a solid ball of radius a. If S is oriented with its center at the origin in $\mathcal{R}^3$, then S is the set of points (x, y, z) satisfying $x^2 + y^2 + z^2 \leq a^2$. In terms of spherical coordinates, S is described by $0 \leq r \leq a$, $0 \leq \phi \leq \pi$, $0 \leq \theta \leq 2\pi$. Hence, the volume of S is

$$V(S) = \int_S dx\, dy\, dz = \int_0^{2\pi} d\theta \int_0^{\pi} d\phi \int_0^a r^2 \sin \phi\, dr$$

$$= \int_0^{2\pi} d\theta \int_0^{\pi} \sin \phi\, d\phi \int_0^a r^2\, dr = [\theta]_0^{2\pi} \left[-\cos \phi\right]_0^{\pi} \left[\frac{1}{3} r^3\right]_0^a = \frac{4}{3} \pi a^3.$$

19. Let L be a solid right circular cylinder of height k and radius a. If L is oriented so that it sits on the xy-plane with its central axis coincident with the z-axis, then L is the set of points (x, y, z) satisfying $x^2 + y^2 \leq a^2$, $0 \leq z \leq k$. In terms of cylindrical coordinates, L is described by $0 \leq r \leq a$, $0 \leq \theta \leq 2\pi$, $0 \leq z \leq k$. Hence, the volume of L is

$$V(L) = \int_S dx\, dy\, dz = \int_0^k dz \int_0^{2\pi} d\theta \int_0^a r\, dr = [z]_0^k [\theta]_0^{2\pi} \left[\frac{1}{2} r^2\right]_0^a = \pi a^2 k.$$

21. Let $T : \mathcal{U}^2 \longrightarrow \mathcal{R}^2$ be defined by $T(u, v) = (x, y) = (u^2 - v^2, 2uv)$ and note that T is continuously differentiable. Let R_{uv} be the region in the uv-plane defined by $1 \leq u^2 + v^2 \leq 4$, $u \geq 0$, $v \geq 0$, and note that the boundary of R_{uv} consists of four smooth curves and R_{uv} is contained in the domain of T. In order to use Jacobi's theorem to compute what is needed in part (b) below, we must first establish the conditions (i) and (ii) in the statement of the theorem.

To see that T is one-to-one on the interior of R_{uv}, suppose $T(u_1, v_1) = T(u_2, v_2)$ for (u_1, v_1) and (u_2, v_2) in the interior of R_{uv}. Then $u_1^2 - v_1^2 = u_2^2 - v_2^2$ and $2u_1 v_1 = 2u_2 v_2$. Since u_1, v_1, u_2, v_2 are all positive, the second equation is equivalent to $v_2 = u_1 v_1 / u_2$. When this expression for v_2 is inserted into the first equation, the result can be written as $(u_1^2 - u_2^2)(u_2^2 + v_1^2) = 0$. As $u_2^2 + v_1^2 \neq 0$, we must have $u_1^2 = u_2^2$, so that $u_1 = u_2$ (since u_1 and u_2 are positive). It follows from $v_2 = u_1 v_1 / u_2$ that $v_1 = v_2$. So, T is one-to-one on R_{uv} and condition (i) is satisfied.

To see that condition (ii) holds, we compute the derivative matrix of T to get

$$T'(u, v) = \begin{pmatrix} 2u & -2v \\ 2v & 2u \end{pmatrix},$$

so that $\det T' = 4(u^2 + v^2)$, which is nonzero for (u, v) in the interior of R_{uv}. That is, condition (ii) holds.

(a) By definition, $x = u^2 - v^2$ and $y = 2uv$. First, note that $y \geq 0$ for (u, v) in R_{uv}, with equality if, and only if, (u, v) is on one of the coordinate axes. This means that T maps points on the straight line segments of the boundary of R_{uv} to the x-axis. Second, note that x can take on both positive and negative values, depending on whether $u > v$ or $u < v$, respectively, and that $-4 \leq x \leq 4$, with equality on the left for $(u, v) = (0, 2)$ and equality on the right for $(u, v) = (2, 0)$. Third, squaring each equation then adding them gives an equation that can be written as $\sqrt{x^2 + y^2} = u^2 + v^2$. Points on the boundary piece $u^2 + v^2 = 4$ are therefore mapped to the upper half of the circle $x^2 + y^2 = 16$ and points on the boundary piece $u^2 + v^2 = 1$ are mapped to the upper half of the circle $x^2 + y^2 = 1$. We conclude that the image of R_{uv} under T is the region R_{xy} in the xy-plane lying between the circles $x^2 + y^2 = 1$ and $x^2 + y^2 = 16$ and above the x-axis. The image region R_{xy} is shown on the right.

(b) The function $f(x, y) = 1/\sqrt{x^2 + y^2}$ is continuous and bounded on R_{xy} and, by what we have shown in part (a), $f(T(\mathbf{u})) = 1/(u^2 + v^2)$. Since $\det T' = 4(u^2 + v^2)$, Jacobi's theorem gives

$$\int_{R_{xy}} \frac{1}{\sqrt{x^2 + y^2}} \, dx \, dy = \int_{R_{uv}} \frac{1}{u^2 + v^2} \, |4(u^2 + v^2)| \, du \, dv$$

$$= 4 \int_{R_{uv}} du \, dv = 4A(R_{uv}) = 4\left(\pi - \frac{1}{4}\pi\right) = 3\pi,$$

where $A(R_{uv})$ is the area of R_{uv} and is the difference of the areas of the quarter circle $u^2 + v^2 = 4$, $0 \leq u$, $0 \leq v$ and the quarter circle $u^2 + v^2 = 1$, $0 \leq u$, $0 \leq v$.

23. Let T be the transformation $T(u, v) = (x, y) = (u, v(1 + u^2))$, and let R_{uv} be the rectangle $0 \leq u \leq 3$, $0 \leq 2 \leq v$ in $\mathcal{U}^2$. Note that the boundary of R_{uv} consists of four smooth curves (line segments) and that R_{uv} and its boundary are contained in the interior of the domain of T.

(a) To find $T(R_{uv}) = R_{xy}$, we will determine the images of the four lines segments that comprise the boundary of T. The four images of these line segments will then comprise the boundary of R_{xy}.

For the vertical line segment $u = 0$, $0 \leq v \leq 2$, we have $T(0, v) = (0, v) = (x, y)$, so that $x = 0$, $0 \leq y \leq 2$, which is a vertical line segment on the y-axis. For the vertical line segment $u = 3$, $0 \leq v \leq 2$, we have $T(3, v) = (3, 20v) = (x, y)$, so that $x = 3$, $0 \leq y \leq 20$, which is also a vertical line segment. For the horizontal line segment $v = 0$, $0 \leq u \leq 3$, we have $T(u, 0) = (u, 0) = (x, y)$, so that $y = 0$, $0 \leq x \leq 3$, which is a horizontal line segment on the x-axis. For the horizontal line segment $v = 2$, $0 \leq u \leq 3$, we have $T(u, 2) = (u, 2(1 + u^2)) = (x, y)$, so that

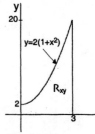

$y = 2(1 + x^2)$, $0 \leq x \leq 3$, which is a portion of a parabola. The region R_{xy} is shown on the right.

(b) We have,

$$T'(u, v) = \begin{pmatrix} 1 & 0 \\ 2uv & 1 + u^2 \end{pmatrix} \qquad \text{so that} \qquad \frac{\partial(x, y)}{\partial(u, v)} = \det T' = 1 + u^2.$$

262

(c) The result of part (b) shows that $\det T' \neq 0$ on the interior of R_{uv}, so that condition (ii) in Jacobi's theorem holds. By inspection, we can see T is one-to-one on its entire domain and is therefore one-to-one on the interior of R_{uv}, so that condition (i) in Jacobi's theorem holds.

The function $f(x, y) = x$ is bounded and continuous on every bounded subset of $\mathcal{R}^2$. In particular, f is bounded and continuous on R_{xy}. Also, it's clear that $f\big(T(u, v)\big) = u$. Hence, by Jacobi's theorem,

$$\int_{R_{xy}} x \, dx, dy = \int_{R_{uv}} u|1 + u^2| \, du \, dv = \int_0^2 dv \int_0^3 (u + u^3) \, du = \big[v\big]_0^2 \left[\frac{1}{2}u^2 + \frac{1}{4}u^4\right]_0^3 = \frac{99}{2}.$$

25. Let D be the circular disk in the xz-plane defined by $(x-1)^2 + z^2 = 4$. The z-axis cuts the disk into two pieces. If the piece with $x \geq 0$ is revolved about the z-axis then the resulting solid is the same as the solid generated by revolving the entire disk D about the z-axis. Let B be the solid generated by revolving D about the z-axis. The figure on the right shows B and a typical cylindrical shell inside B. For $0 \leq x \leq 3$, the height of the cylindrical shell of radius x in B is $2\sqrt{4 - (x-1)^2}$. By Equation 4.3, the volume of B is given by

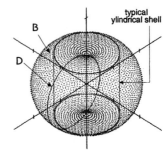

$$V(B) = \int_0^3 4\pi x\sqrt{4 - (x-1)^2} \, dx = 4\pi \int_0^3 x\sqrt{4 - (x-1)^2} \, dx.$$

The change of variable $x - 1 = 2\sin u$, where $-\pi/6 \leq u \leq \pi/2$, results in

$$V(B) = 4\pi \int_{-\pi/6}^{\pi/2} (4\cos^2 u + 8\sin u \cos^2 u) \, du = 8\pi \int_{-\pi/6}^{\pi/4} (1 + \cos 2u + 4\sin u \cos^2 u) \, du$$

$$= 8\pi \left[u + \frac{1}{2}\sin 2u - \frac{4}{3}\cos^3 u\right]_{-\pi/6}^{\pi/2} = \frac{16\pi^2}{3} + 6\pi\sqrt{3}.$$

27. Consider the sphere $x^2 + y^2 + z^2 = a^2$. Its equation in cylindrical coordinates is $r^2 + z^2 = a^2$. Let B be what remains of the solid region bounded by the sphere when a circular hole of radius b is bored out through the center of the sphere. Further, assume that the z-axis is the axis of the bore hole. For $b \leq r \leq a$, the height if the cylindrical shell in B of radius r is $2\sqrt{a^2 - r^2}$, so that Equation 4.3 gives the volume of B as

$$V(B) = \int_b^a 4\pi r\sqrt{a^2 - r^2} \, dr = 2\pi \int_0^{a^2 - b^2} u^{1/2} du = 2\pi \left[\frac{2}{3}u^{3/2}\right]_0^{a^2 - b^2} = \frac{4\pi}{3}(a^2 - b^2)^{3/2},$$

where the second equality is obtained by making the substitution $u = a^2 - r^2$. Since the volume V of the entire sphere is $V = 4\pi a^3/3$, the proportion of the sphere that remains is $k = V(B)/V$, which simplifies to

$$k = \frac{\frac{4\pi}{3}(a^2 - b^2)^{3/2}}{\frac{4\pi}{3}a^3} = \frac{(a^2 - b^2)^{3/2}}{a^3} = (1 - b^2/a^2)^{3/2}.$$

Solving for b gives

$$b = a\sqrt{1 - k^{2/3}}.$$

Section 5: CENTROIDS AND MOMENTS

Exercise Set 5 (pgs. 352-353)

1. Since the sum of the masses is $M = 6$, Equation 5.1 gives

$$\bar{x} = \frac{1}{M} \sum_{k=1}^{3} m_k x_k = \frac{1}{6} [(1)(1) + (3)(2) + (2)(-4)] = -\frac{1}{6}.$$

The three discrete masses and their center of mass are shown on the right.

3. Since the sum of the masses is $M = 6$, Equation 5.1 gives

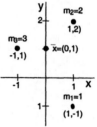

$$\bar{x} = \frac{1}{M} \sum_{k=1}^{3} m_k \mathbf{x}_k = \frac{1}{6} [1(1,-1) + 2(1,2) + 3(-1,1)]$$

$$= \frac{1}{6}(0,6) = (0,1).$$

The three discrete masses and their center of mass are shown on the right.

5. With $I = [0,2]$ and $\mu(x) = 1 - \frac{1}{2}x$, the mass of I is

$$M(I) = \int_0^2 \left(1 - \frac{1}{2}x\right) dx = \left[x - \frac{1}{4}x^2\right]_0^2 = 1.$$

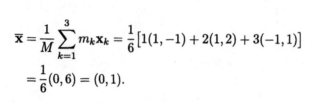

By Equation 5.3, the center of mass is given by

$$\bar{x} = \frac{1}{1} \int_0^2 \left(1 - \frac{1}{2}x\right) x \, dx = \int_0^2 \left(x - \frac{1}{2}x^2\right) dx = \left[\frac{1}{2}x^2 - \frac{1}{6}x^3\right]_0^2 = \frac{2}{3}.$$

The interval I and the center of mass are shown on the right.

7. Let Q be the region bounded by the unit circle and the first quadrant of the xy-plane and let $\mu(x,y) = x + y$. For each $0 \le y \le 1$, x is bounded by $0 \le x \le \sqrt{1 - y^2}$. Hence, the mass of Q is given by

$$M = \int_Q (x+y) \, dx \, dy = \int_0^1 dy \int_0^{\sqrt{1-y^2}} (x+y) \, dx = \int_0^1 \left[\frac{1}{2}x^2 + xy\right]_0^{\sqrt{1-y^2}} dy$$

$$= \int_0^1 \left(\frac{1}{2}(1 - y^2) + y\sqrt{1 - y^2}\right) dy = \left[\frac{1}{2}y - \frac{1}{6}y^3 - \frac{1}{3}(1 - y^2)^{3/2}\right]_0^1 = \frac{2}{3}.$$

Observe that if (x, y) is in Q then so is (y, x) and $\mu(x, y) = \mu(y, x)$. This implies that the moment of Q about the vertical plane containing the line of symmetry $y = x$ is zero, so that the center of mass must lie on the line $y = x$. This means that we need only compute the x-coordinate of $\bar{\mathbf{x}}$. We will use the first equation of Equation 5.5, but in polar coordinates.

We have $\mu(r\cos\theta, r\sin\theta) = \bar\mu(r,\theta) = r(\cos\theta + \sin\theta)$ and the limits on r and θ are $0 \le r \le 1$, $0 \le \theta \le \pi/2$. Hence,

$$\bar x = \frac{1}{M}\int_Q x\mu(x,y)\,dx\,dy = \frac{3}{2}\int_Q (r\cos\theta)\,r(\cos\theta + \sin\theta)\,r\,dr\,d\theta$$

$$= \frac{3}{2}\int_0^{\pi/2} d\theta \int_0^1 r^3(\cos^2\theta + \sin\theta\cos\theta)\,dr = \frac{3}{2}\int_0^{\pi/2}(\cos^2\theta + \sin\theta\cos\theta)\,d\theta \int_0^1 r^3\,dr$$

$$= \frac{3}{2}\int_0^{\pi/2}\left(\frac{1}{2}(1 + \cos 2\theta) + \frac{1}{2}\sin 2\theta\right)d\theta \int_0^1 r^3\,dr$$

$$= \frac{3}{4}\left[\theta + \frac{1}{2}\sin 2\theta - \frac{1}{2}\cos 2\theta\right]_0^{\pi/2}\left[\frac{1}{4}r^4\right]_0^1 = \frac{3(\pi + 2)}{32} \approx 0.48.$$

By our previous remarks, $\bar y = 3(\pi + 2)/32$, so that

$$\bar{\mathbf{x}} = (\bar x, \bar y) = \left(\frac{3(\pi + 2)}{32}, \frac{3(\pi + 2)}{32}\right).$$

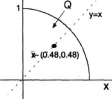

The region Q and its center of mass $\bar{\mathbf{x}}$ are shown on the right.

9. Let R be the region in the first quadrant of the xy-plane bounded by the curve $y = x^2$, the line $y = 4$ and the y-axis, and let the density μ be constant on R. For each $0 \le x \le 2$, y satisfies $x^2 \le y \le 4$. Thus, the mass of R is given by

$$M(R) = \int_R \mu\,dV = \mu\int_0^2 dx \int_{x^2}^4 dy = \mu\int_0^2 (4 - x^2)\,dx = \mu\left[4x - \frac{1}{3}x^3\right]_0^2 = \frac{16\mu}{3},$$

so that the center of mass is

$$\bar{\mathbf{x}} = \frac{1}{M(R)}\int_R \mu\mathbf{x}\,dV = \frac{3}{16}\int_0^2 dx \int_{x^2}^4 (x,y)\,dy = \frac{3}{16}\int_0^2\left[\left(xy, \frac{1}{2}y^2\right)\right]_{x^2}^4 dx$$

$$= \frac{3}{16}\int_0^2\left(4x - x^3, 8 - \frac{1}{2}x^4\right)dx = \frac{3}{16}\left[\left(2x^2 - \frac{1}{4}x^4, 8x - \frac{1}{10}x^5\right)\right]_0^2 = \left(\frac{3}{4}, \frac{12}{5}\right).$$

11. Consider the homogeneous solid region B bounded by $x^2 + y^2 + z^2 = a^2$ and let μ_0 be its constant density. In terms of spherical coordinates, B is described by the inequalities $0 \le r \le a$, $0 \le \phi \le \pi$, $0 \le \theta \le 2\pi$. Letting M be the mass of B, the three coordinates of $\bar{\mathbf{x}}$ are given by Equations 5.5 as

$$\bar x = \frac{1}{M}\int_B (r\sin\phi\cos\theta)\mu_0\,r^2\sin\phi\,dr\,d\phi\,d\theta = \frac{\mu_0}{M}\int_0^{2\pi}\cos\theta\,d\theta \int_0^\pi \sin^2\phi\,d\phi \int_0^a r^3\,dr,$$

$$\bar y = \frac{1}{M}\int_B (r\sin\phi\sin\theta)\mu_0\,r^2\sin\phi\,dr\,d\phi\,d\theta = \frac{\mu_0}{M}\int_0^{2\pi}\sin\theta\,d\theta \int_0^\pi \sin^2\phi\,d\phi \int_0^a r^3\,dr,$$

$$\bar z = \frac{1}{M}\int_B (r\cos\phi)\mu_0\,r^2\sin\phi\,dr\,d\phi\,d\theta = \frac{\mu_0}{M}\int_0^{2\pi} d\theta \int_0^\pi \cos\phi\sin\phi\,d\phi \int_0^a r^3\,dr.$$

Note that the integrands of the integrals $\int_0^{2\pi}\cos\theta\,d\theta$, $\int_0^{2\pi}\sin\theta\,d\theta$ and $\int_0^\pi \cos\phi\sin\phi\,d\phi$ appearing on the right of the above equations are symmetric with respect to the center of their respective intervals of integration. That is, each integral evaluates to zero. It follows that $\bar{\mathbf{x}} = (\bar{x}, \bar{y}, \bar{z}) = (0, 0, 0)$.

13. (a) Let A be the region in the first quadrant of $\mathcal{R}^2$ bounded by $x^2 + y^2 = a^2$ and $x^2 + y^2 = b^2$, where $a > b$. In polar coordinates, this is described by $b \le r \le a$, $0 \le \theta \le \pi/2$. With density constant at $\mu = 1$, its mass is

$$M = \int_A r \, dr \, d\theta = \int_0^{\pi/2} d\theta \int_b^a r \, dr = [\theta]_0^{\pi/2} \left[\frac{1}{2}r^2\right]_b^a = \frac{(a^2 - b^2)\pi}{4}.$$

By symmetry, the centroid is on the line $y = x$, so that we need only compute the first coordinate of $\bar{\mathbf{x}}$. We have

$$\bar{x} = \frac{4}{(a^2 - b^2)\pi} \int_A (r\cos\theta) r \, dr \, d\theta = \frac{4}{(a^2 - b^2)\pi} \int_0^{\pi/2} \cos\theta \, d\theta \int_b^a r^2 \, dr$$

$$= \frac{4}{(a^2 - b^2)\pi} [\sin\theta]_0^{\pi/2} \left[\frac{1}{3}r^3\right]_b^a = \frac{4(a^3 - b^3)}{3\pi(a^2 - b^2)}.$$

Hence, the centroid of A is

$$\bar{\mathbf{x}} = (\bar{x}, \bar{y}) \left(\frac{4(a^3 - b^3)}{3\pi(a^2 - b^2)}, \frac{4(a^3 - b^3)}{3\pi(a^2 - b^2)} \right).$$

(b) The case $b = 0$ for the annulus described in part (a) corresponds to a quarter circle. Setting $b = 0$ in the formula derived in part (a), gives the centroid of that part of the circle $x^2 + y^2 = a^2$ that lies in the first quadrant; namely,

$$\bar{\mathbf{x}} = \left(\frac{4a}{3\pi}, \frac{4a}{3\pi} \right).$$

15. A plane semicircular region R of radius a has area $A(R) = \frac{1}{2}\pi a^2$. If R is rotated about its flat edge, the resulting solid B is a solid ball of radius a with volume $V(B) = \frac{4}{3}\pi a^3$. The centroid of R lies on the line of symmetry of R at a distance $\bar{r}$ from its flat edges. By Pappus's theorem, $A(R)$, $V(B)$ and $\bar{r}$ are related by the equation

$$\frac{4}{3}\pi a^3 = 2\pi\bar{r}\frac{1}{2}\pi a^2, \quad \text{so that} \quad \bar{r} = \frac{\frac{4}{3}\pi a^3}{2\pi\frac{1}{2}\pi a^2} = \frac{4a}{3\pi}.$$

Thus, the centroid of R lies on the line of symmetry of R at a distance $4a/3\pi$ units from its flat edge.

17. Let B be a set in $\mathcal{R}^n$ and $\mu(\mathbf{x})$ the density at $\mathbf{x}$ in B. Let P be a plane containing the point $\mathbf{x}_0$ and let $\mathbf{n} \cdot (\mathbf{x} - \mathbf{x}_0) = 0$ be its normalized vector equation.

(a) If $\mathbf{x}_1$ is any point on P then $\mathbf{n} \cdot (\mathbf{x}_1 - \mathbf{x}_0) = 0$. Hence, $\int_B \mu(\mathbf{x}) \mathbf{n} \cdot (\mathbf{x}_1 - \mathbf{x}_0) \, dV = 0$. Given the definition of $M_P(B)$, we have

$$M_P(B) = \int_B \mu(\mathbf{x}) \mathbf{n} \cdot (\mathbf{x} - \mathbf{x}_0) \, dV + \int_B \mu(\mathbf{x}) \mathbf{n} \cdot (\mathbf{x}_1 - \mathbf{x}_0) \, dV$$

$$= \int_B [\mu(\mathbf{x}) \mathbf{n} \cdot (\mathbf{x} - \mathbf{x}_0) + \mu(\mathbf{x}) \mathbf{n} \cdot (\mathbf{x}_1 - \mathbf{x}_0)] \, dV$$

$$= \int_B \mu(\mathbf{x}) \left(\mathbf{n} \cdot (\mathbf{x} - \mathbf{x}_0 - \mathbf{x}_1 + \mathbf{x}_0) \right) dV = \int_B \mu(\mathbf{x}) \mathbf{n} \cdot (\mathbf{x} - \mathbf{x}_1) \, dV.$$

266

This shows that the point $\mathbf{x}_0$ can be replaced by any point in P without affecting the value of $M_P(B)$. In other words, $M_P(B)$ is independent of the point $\mathbf{x}_0$ as long as $\mathbf{x}_0$ is in P.

(b) Suppose P passes through the center of mass $\overline{\mathbf{x}}$ of B. Replacing $\mathbf{x}_0$ with $\overline{\mathbf{x}}$ in the equation defining $M_P(B)$ gives

$$M_P(B) = \int_B \mu(\mathbf{x})\,\mathbf{n} \cdot (\mathbf{x} - \overline{\mathbf{x}})\,dV = \int_B \mu(\mathbf{x})\left[\mathbf{n} \cdot \mathbf{x} - \mathbf{n} \cdot \overline{\mathbf{x}}\right] dV$$

$$= \int_B \mu(\mathbf{x})\,\mathbf{n} \cdot \mathbf{x}\,dV - \int_B \mu(\mathbf{x})\,\mathbf{n} \cdot \overline{\mathbf{x}}\,dV = \mathbf{n} \cdot \int_B \mu(\mathbf{x})\,\mathbf{x}\,dV - \mathbf{n} \cdot \overline{\mathbf{x}}\int_B \mu(\mathbf{x})\,dV,$$

where we have used the homogeneity of the dot product, the linearity of the integral and the result of Exercise 7(b) in Section 3 of this chapter. Note that the very last integral is the mass $M(B)$ of B, which we assume is finite. Dividing throughout by this mass gives

$$\frac{M_P(B)}{M(B)} = \mathbf{n} \cdot \frac{1}{M(B)}\int_B \mu(\mathbf{x})\,\mathbf{x}\,dV - \mathbf{n} \cdot \overline{\mathbf{x}} = \mathbf{n} \cdot \overline{\mathbf{x}} - \mathbf{n} \cdot \overline{\mathbf{x}} = 0,$$

where the second equality follows from the definition of $\overline{\mathbf{x}}$ given by Equation 5.3. Hence, $M_P(B) = 0$.

19. Let S be the square with corners at $(0,0)$, $(b,0)$, (b,b), $(0,b)$ and let μ be its constant density. With $\mathbf{z}_0 = (0,0)$, the moment of inertia of S about $(0,0)$ is

$$I(0,0) = \int_S |\mathbf{x}|^2 \mu\,dx\,dy = \mu \int_0^b dy \int_0^b (x^2 + y^2)\,dx = \mu \int_0^b \left[\frac{1}{3}x^3 + xy^2\right]_0^b dy$$

$$= \mu \int_0^b \left(\frac{b^3}{3} + by^2\right) dy = \mu \left[\frac{b^3}{3}y + \frac{b}{3}y^3\right]_0^b = \frac{2\mu b^4}{3}.$$

21. Let R be a region in $\mathcal{R}^2$, let $\mu(\mathbf{x})$ be the density at $\mathbf{x}$ in R, let $\overline{\mathbf{x}}$ be the center of mass of R, and let $\mathbf{z}_0$ be a fixed point in $\mathcal{R}^2$. Using the formula for the moment of inertia of R about $\mathbf{z}_0$, together with the hint, gives

$$I(\mathbf{z}_0) = \int_R |\mathbf{x} - \mathbf{z}_0|^2 \mu(\mathbf{x})\,dV = \int_R |(\mathbf{x} - \overline{\mathbf{x}}) + (\overline{\mathbf{x}} - \mathbf{z}_0)|^2 \mu(\mathbf{x})\,dV$$

$$= \int_B \left[(\mathbf{x} - \overline{\mathbf{x}}) + (\overline{\mathbf{x}} - \mathbf{z}_0)\right] \cdot \left[(\mathbf{x} - \overline{\mathbf{x}}) + (\overline{\mathbf{x}} - \mathbf{z}_0)\right]\mu(\mathbf{x})\,dV$$

$$= \int_R \left[|\mathbf{x} - \overline{\mathbf{x}}|^2 + 2(\mathbf{x} - \overline{\mathbf{x}}) \cdot (\overline{\mathbf{x}} - \mathbf{z}_0) + |\overline{\mathbf{x}} - \mathbf{z}_0|^2\right]\mu(\mathbf{x})\,dV$$

$$= \int_R |\mathbf{x} - \overline{\mathbf{x}}|^2 \mu(\mathbf{x})\,dV + 2\int_R (\mathbf{x} - \overline{\mathbf{x}}) \cdot (\overline{\mathbf{x}} - \mathbf{z}_0)\mu(\mathbf{x})\,dV + \int_R |\overline{\mathbf{x}} - \mathbf{z}_0|^2 \mu(\mathbf{x})\,dV$$

$$= I(\overline{\mathbf{x}}) + 2(\overline{\mathbf{x}} - \mathbf{z}_0) \cdot \int_R (\mathbf{x} - \overline{\mathbf{x}})\mu(\mathbf{x})\,dV + |\overline{\mathbf{x}} - \mathbf{z}_0|^2 \int_R \mu(\mathbf{x})\,dV.$$

The last integral is the mass of R. Further, since the integral in the middle term in the last line can be written as

$$\int_R \mathbf{x}\mu(\mathbf{x})\,dV - \int_R \overline{\mathbf{x}}\mu(\mathbf{x})\,dV = \int_R \mathbf{x}\mu(\mathbf{x})\,dV - \overline{\mathbf{x}}\int_R \mu(\mathbf{x})\,dV = \int_R \mathbf{x}\mu(\mathbf{x})\,dV - \overline{\mathbf{x}}M(R)$$

$$= M(R)\left(\frac{1}{M(R)}\int_R \mathbf{x}\mu(\mathbf{x})\,dV - \overline{\mathbf{x}}\right) = M(R)(\overline{\mathbf{x}} - \overline{\mathbf{x}}) = \mathbf{0},$$

267

the middle term in the last line is zero. Hence,

$$I(\mathbf{z}_0) = I(\overline{\mathbf{x}}) + |\overline{\mathbf{x}} - \mathbf{z}_0|^2 M(R),$$

which is what we were asked to show.

23. Let $B_1, \ldots, B_N$ be nonoverlapping regions with union B, having respective masses $M(B_k)$, $M(B)$ and centers of mass $\overline{\mathbf{x}}_k$, $\overline{\mathbf{x}}$.

(a) Let $\mu(\mathbf{x})$ be the density at the point $\mathbf{x}$ in B. Then, for each k, $\mu(\mathbf{x})$ is the density at the point $\mathbf{x}$ in B_k. Since the B_k are disjoint and their union is B, the definition of $\overline{\mathbf{x}}_k$ and $\overline{\mathbf{x}}$ gives,

$$\overline{\mathbf{x}} = \frac{1}{M(B)} \int_B \mu(\mathbf{x})\mathbf{x}\, dV = \frac{1}{M(B)} \int_{B_1 \cup \cdots \cup B_N} \mu(\mathbf{x})\mathbf{x}\, dV = \frac{1}{M(B)} \sum_{k=1}^{N} \int_{B_k} \mu(\mathbf{x})\mathbf{x}\, dV$$

$$= \frac{1}{M(B)} \sum_{k=1}^{N} \left[M(B_k) \frac{1}{M(B_k)} \int_{B_k} \mu(\mathbf{x})\mathbf{x}\, dV \right] = \frac{1}{M(B)} \sum_{k=1}^{N} M(B_k)\overline{\mathbf{x}}_k.$$

(b) The rectangles R_1 and R_2 with respective corners at $(1,3)$, $(3,3)$, $(3,2)$, $(1,2)$ and $(3,-1)$, $(4,-1)$, $(4,-2)$, $(3,-2)$ are disjoint. Note that the area of R_1 is 2 and the area of R_2 is 1. Since the density on each rectangle is constant at $\mu = 1$, their masses are numerically equal to their corresponding areas; i.e., $M(R_1) = 2$ and $M(R_2) = 1$. Moreover, their centers of mass are $\overline{\mathbf{x}}_1 = (2, 5/2)$ and $\overline{\mathbf{x}}_2 = (7/2, -3/2)$, respectively. Treating each center of mass as a point-mass, the discrete point-mass formula (Equation 5.1) gives the center of mass $\overline{\mathbf{y}}$ of the two-point system as

$$\overline{\mathbf{y}} = \frac{1}{M(R_1) + M(R_2)} \sum_{k=1}^{2} M(R_k)\overline{\mathbf{x}}_k = \frac{1}{3}\left(2(2, 5/2) + 1(7/2, -3/2) \right) = \left(\frac{5}{2}, \frac{7}{6} \right).$$

On the other hand, if $R = R_1 \cup R_2$ then the density of R is constant at $\mu = 1$ and its mass is numerically equal to its area, which is the sum of the areas of R_1 and R_2; i.e., $M(R) = 3$. Thus, the center of mass $\overline{\mathbf{x}}$ of R is

$$\overline{\mathbf{x}} = \frac{1}{M(B)} \int_B \mathbf{x}\, dV = \frac{1}{3} \int_{R_1 \cup R_2} \mathbf{x}\, dV = \frac{1}{3}\left(\int_{R_1} \mathbf{x}\, dV + \int_{R_2} \mathbf{x}\, dV \right)$$

$$= \frac{1}{3}\left(\int_2^3 dy \int_1^3 (x,y)\, dx + \int_{-2}^{-1} dy \int_3^4 (x,y)\, dx \right)$$

$$= \frac{1}{3}\left(\int_2^3 \left[\left(\frac{1}{2}x^2, xy \right) \right]_1^3 dy + \int_{-2}^{-1} \left[\left(\frac{1}{2}x^2, xy \right) \right]_3^4 dy \right)$$

$$= \frac{1}{3}\left(\int_2^3 (4, 2y)\, dy + \int_{-2}^{-1} \left(\frac{7}{2}, y \right) dy \right) = \frac{1}{3}\left([(4y, y^2)]_2^3 + \left[\left(\frac{7}{2}y, \frac{1}{2}y^2 \right) \right]_{-2}^{-1} \right)$$

$$= \frac{1}{3}\left((4, 5) + (7/2, -3/2) \right) = \left(\frac{5}{2}, \frac{7}{6} \right).$$

Hence, $\overline{\mathbf{x}} = \overline{\mathbf{y}}$. Using the result of part (a) is clearly easier when the masses and the centers of mass of the individual regions can be done by inspection.

Section 6: IMPROPER INTEGRALS

Exercise Set 6 (pgs. 358-359)

1.

$$\int_0^\infty (1+x)^{-4}dx = \lim_{b\to\infty}\int_0^b (1+x)^{-4}dx$$

$$= \lim_{b\to\infty}\left[\frac{(1+x)^{-3}}{-3}\right]_0^b = -\frac{1}{3}\lim_{b\to\infty}\left((1+b)^{-3}-1\right) = \frac{1}{3}.$$

3. Let R be the infinite rectangle $0 \le x \le 1$, $2 \le y$ in $\mathcal{R}^2$. The region R is unbounded so we integrate over the finite rectangle R_b defined by $0 \le x \le 1$, $2 \le y \le b$. We have

$$\int_{R_b}\frac{x}{y^2}\,dx\,dy = \int_2^b dy\int_0^1\frac{x}{y^2}\,dx = \int_2^b\left[\frac{x^2}{2y^2}\right]_0^1 dy = \int_2^b\frac{1}{2y^2}\,dy = \left[-\frac{1}{2y}\right]_2^b = \left(-\frac{1}{2b}+\frac{1}{4}\right).$$

Hence,

$$\int_R\frac{x}{y^2}\,dx\,dy = \lim_{b\to\infty}\int_{R_b}\frac{x}{y^2}\,dx\,dy = \lim_{b\to\infty}\left(-\frac{1}{2b}+\frac{1}{4}\right) = \frac{1}{4}.$$

5. Let C be the infinite rectangle $0 \le x \le 1$, $0 \le y \le 1$, $0 \le z$ in $\mathcal{R}^3$. The region C is unbounded so we integrate over the finite rectangle C_b defined by $0 \le x \le 1$, $0 \le y \le 1$, $0 \le z \le b$. We have

$$\int_{C_b}e^{-x-y-z}dx\,dy\,dz = \int_0^b dz\int_0^1 dy\int_0^1 e^{-x-y-z}dx = \int_0^b e^{-z}dz\int_0^1 e^{-y}dy\int_0^1 e^{-x}dx$$

$$= \left[-e^{-z}\right]_0^b\left[-e^{-y}\right]_0^1\left[-e^{-x}\right]_0^1 = (1-e^{-b})(1-e^{-1})(1-e^{-1}).$$

Hence,

$$\int_C e^{-x-y-z}dx\,dy\,dz = \lim_{b\to\infty}\int_{C_b}e^{-x-y-z}dx\,dy\,dz$$

$$= \lim_{b\to\infty}(1-e^{-b})(1-e^{-1})(1-e^{-1}) = (1-e^{-1})^2.$$

7. Let D be the unit disk in $\mathcal{R}^2$. Since $\ln(x^2+y^2)$ tends to $-\infty$ as x^2+y^2 tends to zero, we integrate over the annulus D_δ in $\mathcal{R}^2$ defined by $0 < \delta \le x^2+y^2 \le 1$. In polar coordinates, D_δ is described by $\delta \le r \le 1$, $0 \le \theta \le 2\pi$. Moreover, $\ln(x^2+y^2) = \ln r^2 = 2\ln r$ and $dx\,dy = r\,dr\,d\theta$. We have

$$\int_{D_\delta}\ln(x^2+y^2)\,dx\,dy = \int_{D_\delta}2r\ln r\,dr\,d\theta = 2\int_0^{2\pi}d\theta\int_\delta^1 r\ln r\,dr$$

$$= 2\left[\theta\right]_0^{2\pi}\left[\frac{1}{2}r^2\ln r - \frac{1}{4}r^2\right]_\delta^1 = \pi\left(-1 - 2\delta^2\ln\delta + \delta^2\right),$$

where the integral with respect to r is done by parts (let $u = \ln r$, $dv = r\,dr$). Hence,

$$\int_D\ln(x^2+y^2)\,dx\,dy = \lim_{\delta\to 0^+}\int_{D_\delta}\ln(x^2+y^2)\,dx\,dy = \lim_{\delta\to 0^+}\left(\pi\left(-1 - 2\delta^2\ln\delta + \delta^2\right)\right) = -\pi,$$

where l'Hôpital's rule is used to get $\lim_{\delta\to 0^+} r^2\ln r = 0$.

269

9. Let Q be the quarter disk in $\mathcal{R}^2$ defined by $x^2 + y^2 \leq 1$, $0 \leq x$, $0 \leq y$. The integrand $1/\sqrt{x^2 + y^2}$ is unbounded near the origin. However, if we change to polar coordinates, with $0 \leq r \leq 1$ and $0 \leq \theta \leq \pi/2$, then

$$\int_Q \frac{1}{\sqrt{x^2 + y^2}}\, dx\, dy = \int_0^{\pi/2} d\theta \int_0^1 \frac{1}{r} r\, dr = \int_0^{\pi/2} d\theta \int_0^1 dr = [\theta]_0^{\pi/2} [r]_0^1 = \frac{\pi}{2}.$$

11. The integral $\int_1^\infty x^\alpha dx$ is improper for all values of α since the interval of integration is unbounded. The case $\alpha = -1$ is exceptional and we dispose of it now. We have

$$\int_1^\infty x^{-1} dx = \lim_{b \to \infty} \int_1^b x^{-1} dx = \lim_{b \to \infty} \left[\ln x\right]_1^b = \lim_{b \to \infty} \ln b = +\infty.$$

For $\alpha \neq -1$, we integrate over the bounded interval $[1, b]$ to get

$$\int_1^b x^\alpha dx = \left[\frac{1}{\alpha + 1} x^{\alpha + 1}\right]_1^b = \frac{1}{\alpha + 1}(b^{\alpha + 1} - 1).$$

Hence,

$$\int_1^\infty x^\alpha dx = \lim_{b \to \infty} \int_1^b x^\alpha dx = \lim_{b \to \infty} \frac{1}{\alpha + 1}(b^{\alpha + 1} - 1) = \begin{cases} +\infty, & \alpha > -1 \\ -1/(\alpha + 1), & \alpha < -1. \end{cases}$$

13. Let D be the unit disk in $\mathcal{R}^2$. In polar coordinates, D is described by $0 \leq r \leq 1$, $0 \leq \theta \leq 2\pi$.

(a) In order that $p(x, y) = k(1 - x^2 - y^2)$ be a probability density on D, we must have

$$1 = \int_D p(x, y)\, dx\, dy = \int_D k(1 - x^2 - y^2)\, dx\, dy = \int_0^{2\pi} d\theta \int_0^1 k(1 - r^2) r\, dr$$

$$= k \int_0^{2\pi} d\theta \int_0^1 (r - r^3)\, dr = k [\theta]_0^{2\pi} \left[\frac{1}{2}r^2 - \frac{1}{4}r^4\right]_0^1 = k\frac{\pi}{2}.$$

So, $k = 2/\pi$ and $p(x, y) = 2(1 - x^2 - y^2)/\pi$.

(b) Suppose the outcomes (x, y) of a probability experiment E are distributed on D according to the probability density $p(x, y)$ found in part (a). We want to compute $\Pr[x > 1/2]$. These are the outcomes on D lying in the segment $D_{1/2}$ to the right of the vertical line $x = 1/2$. Inserting $x = 1/2$ into $x^2 + y^2 \leq 1$, we see that $-\sqrt{3}/2 \leq y \leq \sqrt{3}/2$. For each such y, x is bounded by $1/2 \leq x \leq \sqrt{1 - y^2}$. Thus,

$$\Pr[x > 1/2] = \int_{D_{1/2}} p(x, y)\, dx\, dy = \frac{2}{\pi} \int_{-\sqrt{3}/2}^{\sqrt{3}/2} dy \int_{1/2}^{\sqrt{1 - y^2}} (1 - x^2 - y^2)\, dx$$

$$= \frac{2}{\pi} \int_{-\sqrt{3}/2}^{\sqrt{3}/2} \left[(1 - y^2)x - \frac{1}{3}x^3\right]_{1/2}^{\sqrt{1 - y^2}} dy = \frac{2}{\pi} \int_{-\sqrt{3}/2}^{\sqrt{3}/2} \left(\frac{2}{3}(1 - y^2)^{3/2} + \frac{1}{2}y^2 - \frac{11}{24}\right) dy$$

$$= \frac{2}{\pi} \left[\frac{1}{6}y(1 - y^2)^{3/2} + \frac{1}{4}y\sqrt{1 - y^2} + \frac{1}{4}\arcsin y + \frac{1}{6}y^3 - \frac{11}{24}y\right]_{-\sqrt{3}/2}^{\sqrt{3}/2} = \frac{1}{3} - \frac{3\sqrt{3}}{8\pi} \approx 0.13$$

(c) The mean of $p(x, y)$ is the vector $\mathbf{m}[p]$ in $\mathcal{R}^2$ defined by

$$\mathbf{m}[p] = \int_D \mathbf{x}\, p(\mathbf{x})\, dV = \frac{2}{\pi} \int_D (x, y)(1 - x^2 - y^2)\, dx\, dy$$

$$= \frac{2}{\pi} \left(\int_D x(1 - x^2 - y^2)\, dx\, dy, \int_D y(1 - x^2 - y^2)\, dx\, dy \right)$$

$$= \frac{2}{\pi} \left(\int_0^{2\pi} d\theta \int_0^1 r\cos\theta(1 - r^2)\, r\, dr, \int_0^{2\pi} d\theta \int_0^1 r\sin\theta(1 - r^2)\, r\, dr \right)$$

$$= \frac{2}{\pi} \left(\int_0^{2\pi} \cos\theta\, d\theta \int_0^1 (r^2 - r^4)\, dr, \int_0^{2\pi} \sin\theta\, d\theta \int_0^1 (r^2 - r^4)\, dr \right),$$

$$= (0, 0)$$

where the third line is a conversion to polar coordinates and the last equality follows from the fact that the cosine function is being integrated over one period in the first coordinate and the sine function is being integrated over one period in the second coordinate.

15. (a) We can write the result of Example 6 in the text as

$$\frac{1}{2\pi\sigma^2} \int_{-\infty}^{\infty} dy \int_{-\infty}^{\infty} e^{-(x^2+y^2)/(2\sigma^2)}\, dx = \frac{1}{2\pi\sigma^2} \int_{-\infty}^{\infty} e^{-y^2/(2\sigma^2)}\, dy \int_{-\infty}^{\infty} e^{-x^2/(2\sigma^2)}\, dx$$

$$= \left(\frac{1}{\sqrt{2\pi\sigma^2}} \int_{-\infty}^{\infty} e^{-y^2/(2\sigma^2)}\, dy \right) \left(\frac{1}{\sqrt{2\pi\sigma^2}} \int_{-\infty}^{\infty} e^{-x^2/(2\sigma^2)}\, dx \right)$$

$$= \left(\frac{1}{\sqrt{2\pi\sigma^2}} \int_{-\infty}^{\infty} e^{-x^2/(2\sigma^2)}\, dx \right)^2 = 1.$$

Taking square roots gives

$$\frac{1}{\sqrt{2\pi\sigma^2}} \int_{-\infty}^{\infty} e^{-x^2/(2\sigma^2)}\, dx = 1.$$

Note that we took the *positive* square root. This is because the integrand is positive and therefore so is the integral.

(b) The function $N(x) = \frac{1}{\sigma\sqrt{2\pi}} e^{-(x-m)^2/(2\sigma^2)}$ is a probability density on the interval $-\infty < x < \infty$ if, and only if, its integral over this interval is 1. We have

$$\int_{-\infty}^{\infty} N(x)\, dx = \frac{1}{\sigma\sqrt{2\pi}} \int_{-\infty}^{\infty} e^{-(x-m)^2/(2\sigma^2)}\, dx = \frac{1}{\sigma\sqrt{2\pi}} \int_{-\infty}^{\infty} e^{-u^2/(2\sigma^2)}\, du = 1,$$

where the second equality follows from the change of variable $u = x - m$, and the last equality follows from part (a).

(c) The mean of $N(x)$ is a vector quantity with dimension equal to the dimension of the region of integration. In this case, the mean has dimension one and is therefore a real number. As defined in the text, the mean is given by

$$\mathbf{m}[N] = \int_{-\infty}^{\infty} x N(x)\, dx = \frac{1}{\sigma\sqrt{2\pi}} \int_{-\infty}^{\infty} x\, e^{-(x-m)^2/(2\sigma^2)}\, dx$$

$$= \frac{1}{\sigma\sqrt{2\pi}} \int_{-\infty}^{\infty} (u + m)\, e^{-u^2/(2\sigma^2)}\, du$$

$$= \frac{1}{\sigma\sqrt{2\pi}} \int_{-\infty}^{\infty} u e^{-u^2/(2\sigma^2)}\, du + \frac{m}{\sigma\sqrt{2\pi}} \int_{-\infty}^{\infty} e^{-u^2/(2\sigma^2)}\, du = m,$$

where the second line is obtained by making the change of variable $u = x - m$ and the last equality follows from part (b) and the fact that the first integral in the last line is an odd function being integrated over a symmetric interval and its value is therefore zero.

(d) With $A = 1/(\sigma\sqrt{2\pi})$, the variance of $N(x)$ is given by

$$\int_{-\infty}^{\infty} (x-m)^2 N(x)\, dx = A \int_{-\infty}^{\infty} (x-m)^2 e^{-(x-m)^2/(2\sigma^2)}\, dx = A \int_{-\infty}^{\infty} u^2 e^{-u^2/(2\sigma^2)}\, du,$$

where the last integral is obtained by the change of variable $u = x - m$. For $b > 0$, integrate over the symmetric interval $[-b, b]$ to get

$$A \int_{-b}^{b} u^2 e^{-u^2/(2\sigma^2)}\, du = A \left[-\sigma^2 u e^{-u^2/(2\sigma^2)} \right]_{-b}^{b} + \sigma^2 A \int_{-b}^{b} e^{-u^2/(2\sigma^2)}\, du$$

$$= -2\sigma^2 A e^{-b^2/(2\sigma^2)} + \sigma^2 A \int_{-b}^{b} e^{-u^2/(2\sigma^2)}\, du,$$

where the right member of the first line is the result of integration by parts with $v = u$ and $dw = u e^{-u^2/(2\sigma^2)}\, du$. Letting $b \to \infty$, the left member of the first line tends to the variance. In the second line, the first term tends to zero and the second term tends to σ^2 (since the integral tends to $A \int_{-\infty}^{\infty} e^{-u^2/(2\sigma^2)}\, du$, which has value 1 as shown in part (b)). In other words,

$$\int_{-\infty}^{\infty} (x-m)^2 N(x)\, dx = \sigma^2.$$

(NOTE: An expression of the form $\lim_{b\to\infty} \int_{-b}^{b} f(x)\, dx$ is called a *symmetric limit* and asserts that both limits of integration simultaneously increase without bound. One does not necessarily obtain the same value, if any, as alternately holding each limit of integration fixed while allowing the other to increase without bound. However, an equation of the form

$$\lim_{b\to\infty} \int_{-b}^{b} f(x)\, dx = \int_{-\infty}^{\infty} f(x)\, dx$$

will hold if the integrals $I_1 = \int_{-\infty}^{a} f(x)\, dx$ and $I_2 = \int_{a}^{\infty} f(x)\, dx$ both exist. In this case, the symmetric limit is the sum $I_1 + I_1$. It is therefore important to observe that $\int_{-\infty}^{0} e^{-u^2/(2\sigma^2)}\, du$ and $\int_{0}^{\infty} u^2 e^{-u^2/(2\sigma^2)}\, du$ both exist.)

17. (a) Since the range of speeds of a gas molecule satisfies $0 \le v \le \infty$, the gist of what we are asked to do is to verify that the function

$$F(v) = 4\pi \left(\frac{m}{2\pi kT} \right)^{3/2} v^2 e^{-mv^2/(2kT)}$$

is a probability density function by showing that the integral of F over the interval $I = [0, \infty)$ has value 1. Since such an integral is improper, we let $I_b = [0, b]$ and compute

$$\int_{I_b} F(v)\, dv = 4\pi \left(\frac{m}{2\pi kT} \right)^{3/2} \int_{0}^{b} v^2 e^{-mv^2/(2kT)}\, dv$$

$$= \sqrt{\frac{2m}{\pi kT}} \left(-b e^{-mb^2/(2kT)} + \int_{0}^{b} e^{-mv^2/(2kT)}\, dv \right),$$

where the last integral follows by integration by parts and simplification of the expressions involving the constants. Letting $b \to \infty$ gives

(∗)
$$\int_0^\infty F(v)\,dv = \sqrt{\frac{2m}{\pi kT}} \int_0^\infty e^{-mv^2/(2kT)}\,dv.$$

In part (a) of Exercise 15 above, it was shown that

$$\int_{-\infty}^\infty e^{-x^2/(2\sigma^2)}\,dx = \sigma\sqrt{2\pi}.$$

As the integrand of the above integral is an even function, we can write

$$\int_0^\infty e^{-x^2/(2\sigma^2)}\,dx = \frac{1}{2}\int_{-\infty}^\infty e^{-x^2/(2\sigma^2)}\,dx = \frac{1}{2}\sigma\sqrt{2\pi} = \sigma\sqrt{\frac{\pi}{2}}.$$

Setting $2\sigma^2 = 2kT/m$ and solving for σ gives $\sigma = \sqrt{kT/m}$, so that

$$\int_0^\infty e^{-mv^2/(2kT)}\,dv = \sqrt{\frac{kT}{m}}\sqrt{\frac{\pi}{2}} = \sqrt{\frac{\pi kT}{2m}}.$$

Hence, equation (∗) becomes

$$\int_0^\infty F(v)\,dv = \sqrt{\frac{2m}{\pi kT}}\sqrt{\frac{\pi kT}{2m}} = 1,$$

which is what we wanted to show.
(NOTE: Theoretically, no gas molecule can exceed the speed of light. Therefore, letting c be the speed of light, we have $\Pr[0 \le v \le c] = 1$ and we conclude from the above result that $\Pr[v > c] = 0$. But clearly, $\int_c^\infty F(v)\,dv \ne 0$. However, the value of this integral is infinitesimal and can, for all practical purposes, be ignored.)
   (b) With $F(v)$ as defined in part (a), the mean of F is given by

$$\mathbf{m}[F] = \int_0^\infty vF(v)\,dv = 4\pi\left(\frac{m}{2\pi kT}\right)^{3/2}\int_0^\infty v^3 e^{-mv^2/(2kT)}\,dv.$$

Rather than drag around all of the messy constants and present the reader with line after line of complicated looking transformations, we will evaluate the integral $\int_0^b v^3 e^{-\alpha v^3}\,dv$, where b and α are positive constants, and then apply the result to the above integral on the right. We have

$$\int_0^b v^3 e^{-\alpha v^3}\,dv = \frac{1}{2\alpha^2}\int_0^{\alpha b^2} te^{-t}\,dt = \frac{1}{2\alpha^2}\big[-te^{-t} - e^{-t}\big]_0^{\alpha b^2} = \frac{1}{2\alpha^2}\big(1 - \alpha b^2 e^{-\alpha b^2} - e^{-\alpha b^2}\big),$$

where the first equality is obtained by the change of variable $t = \alpha v^2$ and the second integral is evaluated using integration by parts. Letting $b \to \infty$ gives

$$\int_0^\infty v^3 e^{-\alpha v^2}\,dv = \lim_{b\to\infty}\int_0^b v^3 e^{-\alpha v^2}\,dv = \lim_{b\to\infty}\frac{1}{2\alpha^2}\big(1 - \alpha b^2 e^{-\alpha b^2} - e^{-\alpha b^2}\big) = \frac{1}{2\alpha^2}.$$

With $\alpha = m/(2kT)$, the mean we are interested in becomes

$$\mathbf{M}[F] = 4\pi\left(\frac{m}{2\pi kT}\right)^{3/2}\frac{1}{2\left(\frac{m}{2kT}\right)^2} = \frac{2}{\sqrt{\pi}}\left(\frac{m}{2kT}\right)^{3/2}\left(\frac{2kT}{m}\right)^2 = \sqrt{\frac{8kT}{\pi m}}.$$

Section 7: NUMERICAL INTEGRATION

Exercise Set 7AB (pgs. 362-363)

1. With $f(x, y) = x^2 + y^4$, let R be the rectangle $0 \le x \le 1$, $0 \le y \le 1$. Then the exact value of $\int_R f\, dV$ is

$$\int_0^1 dy \int_0^1 (x^2 + y^4)\, dx = \int_0^1 \left[\frac{1}{3}x^3 + xy^4\right]_0^1 dy = \int_0^1 \left(\frac{1}{3} + y^4\right) dy = \left[\frac{1}{3}y + \frac{1}{5}y^5\right]_0^1 = \frac{8}{15}.$$

Applying the midpoint approximation to the integral with $a = 0$, $b = 1$, $c = 0$, $d = 1$, required $p = 86$, $q = 87$ to achieve the value

$$\int_R (x^2 + y^4)\, dV \approx 0.5333000.$$

Since $8/15 = 0.5\overline{3}$, the above approximation is accurate to four decimal places.

3. Let R be the rectangle $0 \le x \le 2$, $0 \le y \le 3$. With $f(x, y) = 1 - x^2 - y^2$, the integral of f over R is

$$\int_R (1 - x^2 - y^2)\, dx\, dy = \int_0^3 dy \int_0^2 (1 - x^2 - y^2)\, dx = \int_0^3 \left[x - \frac{1}{3}x^3 - xy^2\right]_0^2 dy$$

$$= \int_0^3 \left(-\frac{2}{3} - 2y^2\right) dy = \left[-\frac{2}{3}y - \frac{2}{3}y^3\right]_0^3 = -20.$$

Using this as a guide, Simpson's rule was applied with $p = q = 2$ and the exact value of -20 was obtained for the integral. The midpoint approximation does not give the exact value but rather approaches it from below as p and q increase without bound.

5. Let R be the unit disk in $\mathcal{R}^2$ and let $f(x, y) = 1 - x^2 - y^2$. In terms of polar coordinates, R is described by $0 \le r \le 1$, $0 \le \theta \le 2\pi$ and $f(x, y) = 1 - r^2$. The exact value of the integral of f over R is therefore

$$\int_R f(x, y)\, dx\, dy = \int_R (1 - r^2) r\, dr\, d\theta = \int_0^{2\pi} d\theta \int_0^1 (r - r^3)\, dr = [\theta]_0^{2\pi} \left[\frac{1}{2}r^2 - \frac{1}{4}r^4\right]_0^1 = \frac{\pi}{2}.$$

Because the given region R is not a rectangle, the function that was used for the approximation was $F(x, y) = (1 - x^2 - y^2)H(1 - x^2 - y^2)$, where H is the *Heaviside unit step function* introduced in Ch.4. Thus, $F(x, y) = f(x, y)$ inside the unit disk and is zero elsewhere. The Simpson approximation and the midpoint approximation were then applied to $F(x, y)$ on the rectangle $-1 \le x \le 1$, $-1 \le y \le 1$ for $p = q = 150$. Both methods produced the value 1.5708. Since $\pi/2 = 1.570796327\ldots$, the approximation is accurate to three decimal places.

7. Since the region R given here is the quarter unit disk in the first quadrant of the xy-plane, and since the integrand is the same here as it was in Exercise 5 above, we deduce that the exact value of $\int_R (1 - x^2 - y^2)\, dx\, dy$ is one-fourth the value found in Exercise 5; namely $\pi/8 \approx 0.392699082$. As the region R is not a rectangle, we applied the Simpson and midpoint approximations to the function $F(x, y) = (1 - x^2 - y^2)H(1 - x^2 - y^2)$ over the rectangle $0 \le x \le 1$, $0 \le y \le 1$. Six decimal place accuracy was first achieved with the Simpson approximation at $p = q = 136$, while the same accuracy was first achieved with the midpoint approximation at $p = q = 70$.

9. (a) For $R > 0$, let D_R be the disk $x^2 + y^2 \leq R^2$ in $\mathcal{R}^2$. In polar coordinates, D_R is described by the inequalities $0 \leq r \leq R$, $0 \leq \theta \leq 2\pi$. When the given integral is written in terms of polar coordinates, the integrand becomes $e^{-x^2-y^2} = e^{-r^2}$ and the area element becomes $dx\,dy = r\,dr\,d\theta$. Hence,

$$\int_{D_R} e^{-x^2-y^2}\,dx\,dy = \int_{D_R} re^{-r^2}\,dr\,d\theta = \int_0^{2\pi} d\theta \int_0^R re^{-r^2}\,dr = [\theta]_0^{2\pi}\left[-\frac{1}{2}e^{-r^2}\right]_0^R = \pi(1 - e^{-R^2}).$$

Letting $R \to \infty$ gives

$$\int_{\mathcal{R}^2} e^{-x^2-y^2}\,dx\,dy = \lim_{R\to\infty}\int_{D_R} e^{-x^2-y^2}\,dx\,dy = \lim_{R\to\infty}\pi(1 - e^{-R^2}) = \pi.$$

(b) For each $a > 0$, let S_a be the square in $\mathcal{R}^2$ with corners at $(\pm a, \pm a)$. The strategy is to let $p = q$ be large enough so that applying Simpson's rule over S_a with $p = q$ gives the same value for $4G(0, a, 0, a)$ (to eight-place accuracy) as applying Simpson's rule over S_a with $2p = 2q$. This produces the following table:

$$4G(0, 2.600, 0, 2.600) \approx 4(0.78502744) = 3.140110976$$

$$4G(0, 3.050, 0, 3.050) \approx 4(0.78537291) = 3.14149164$$

$$4G(0, 3.075, 0, 3.075) \approx 4(0.78537665) = 3.1415066$$

$$4G(0, 3.575, 0, 3.575) \approx 4(0.78539749) = 3.14158996$$

$$4G(0, 3.600, 0, 3.600) \approx 4(0.78539760) = 3.1415904$$

(c) The table given in part (b) suggests that $4G(0, a, 0, a)$ first approximates π to four-decimal places at about $a = 3.07$.

11. The region R is a hemisphere of radius 1 centered at the origin in $\mathcal{R}^3$ with its flat base on the xy-plane. We first evaluate the integral directly so as to have a check on the accuracy of our approximations. In spherical coordinates the region of integration is described by $0 \leq r \leq 1$, $0 \leq \phi \leq \pi/2$, $0 \leq \theta \leq 2\pi$. The integrand becomes $\sqrt{x^2 + y^2 + z^2} = r$ and the volume element becomes $dx\,dy\,dz = r^2\sin\phi\,dr\,d\phi\,d\theta$. Hence,

$$\int_R \sqrt{x^2 + y^2 + z^2}\,dx\,dy\,dz = \int_R r\,r^2\sin\phi\,dr\,d\phi\,d\theta$$

$$= \int_0^{2\pi} d\theta \int_0^{\pi/2}\sin\phi\,d\phi \int_0^1 r^3\,dr = [\theta]_0^{2\pi}\left[-\cos\phi\right]_0^{\pi/2}\left[\frac{1}{4}r^4\right]_0^1 = \frac{\pi}{2} \approx 1.570796327.$$

Although the given region is not a rectangle in xyz-space, it *is* a rectangle in $r\phi\theta$-space. Since the names of the variables is irrelevant, we applied Simpson's rule in three-dimensions to the function $x^2\sin y$ over the rectangle $0 \leq x \leq 1$, $0 \leq y \leq \pi/2$, $0 \leq z \leq 2\pi$. For $p = q = r = 20$, the approximation was 1.5707960690, which is accurate to six decimal places.

13. Letting R be the rectangle $0 \leq x \leq 1$, $0 \leq y \leq 1$, $0 \leq z \leq 1$ in $\mathcal{R}^3$, we have

$$\int_R (x + y + z)\,dx\,dy\,dz = \int_0^1 dz \int_0^1 dy \int_0^1 (x + y + z)\,dx$$

$$= \int_0^1 dz \int_0^1 \left[\frac{1}{2}x^2 + (y + z)x\right]_0^1 dy = \int_0^1 dz \int_0^1 \left(\frac{1}{2} + y + z\right) dy$$

$$= \int_0^1 \left[\frac{1}{2}y + \frac{1}{2}y^2 + yz\right]_0^1 dz = \int_0^1 (1 + z)\,dz = \left[z + \frac{1}{2}z^2\right]_0^1 = \frac{3}{2}.$$

275

Using Simpson's formula for a triple integral over the given three-dimensional rectangle and letting $p = q = r = 2$, gives

$$\int_R (x + y + z)\, dx\, dy\, dz \approx 1.5000000.$$

Clearly, these values of p, q and r are minimal so that the above approximation holds for any p, q and r.

15. (a) The region R in $\mathcal{R}^3$ bounded by the five planes $x = 0$, $y = 0$, $z = 1$, $z = 2$, $x + y + z = 3$ is the polyhedron shown on the right.

(b) As the figure on the right shows, the region R is the frustum of a tetrahedron whose cross sections perpendicular to the z-axis are isosceles right triangles. The triangle forming the base of R has legs of length 2 so that its area is $A = 2$, and the triangle forming the top of R has legs of length 1 so that its area is $B = 1/2$. The height of R (i.e., the perpendicular distance between the base and its top) is $h = 1$. From geometry, the volume of R is

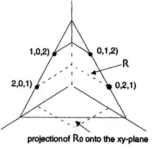

projection of R_0 onto the xy-plane

$$V(R) = \frac{1}{3} h \left(A + B + \sqrt{AB}\right) = \frac{1}{3}(1)\left(2 + \frac{1}{2} + 1\right) = \frac{7}{6}.$$

One can also compute the volume of R with a triple integral. Specifically,

$$V(R) = \int_R dx\, dy\, dz = \int_1^2 dz \int_0^{3-z} dy \int_0^{3-y-z} dx = \int_1^2 dz \int_0^{3-z} (3 - y - z)\, dy$$

$$= \int_1^2 \left[(3-z)y - \frac{1}{2}y^2\right]_0^{3-z} dz = \int_0^1 \frac{1}{2}(z-3)^2 dz = \left[\frac{1}{6}(z-3)^3\right]_1^2 = \frac{7}{6}.$$

In order to use Simpson's rule in three dimensions, we first observe that the smallest rectangle, call it R_0, containing R is $0 \le x \le 2$, $0 \le y \le 2$, $1 \le z \le 2$ (see figure). Second, the function we need to integrate is the Heaviside unit step function $H(3 - x - y - z)$, which has the value 1 for points below the plane $x + y + z = 3$ and is zero elsewhere. When Simpson's rule in three dimensions was applied to this function over R_0, the approximation for $p = q = r = 50$ was 1.176576, which is accurate to only one decimal place. When the midpoint approximation in three dimensions was applied with $p = q = r = 50$, the result was 1.166400, which is accurate to three decimal places.

(c) The apparent superiority of the midpoint approximation suggested by the result of part (b), compelled us to forego the use of the Simpson approximation in favor of the midpoint approximation, which was applied to the function $(x^4 + y^4 + z^4)H(3 - x - y - z)$ over the rectangle R_0 described in part (b). Using $p = q = r = 50$, the result was 6.679043. In order to check this answer, one can directly compute

$$\int_R (x^4 + y^4 + z^4)\, dx\, dy\, dz = \int_1^2 dz \int_0^{3-z} dy \int_0^{3-y-z} (x^4 + y^4 + z^4)\, dx = \frac{1403}{210} \approx 6.680952381.$$

We see that our midpoint approximation is accurate to only one decimal place (although rounding to two places produces two-place accuracy). We leave the verification of the above exact value to the reader.

(NOTE: In previous exercises, values of p and q were tried for much larger values than were tried here. The reason is that the number of operations required to carry out both the Simpson approximation and the midpoint approximation is roughly proportional to the cube of the dimension. Thus, all things being equal, computation time for the computer is over three times longer in three dimensions than it is in two dimensions.)

Chapter Review

(pgs. 363-366)

1. The region of integration is the rectangle $0 \leq x \leq 2$, $0 \leq y \leq 1$ in $\mathcal{R}^2$ and is shown on the right. Integrating first with respect to x and then y gives

$$\int_0^1 \left[\int_0^2 xy^2 \, dx \right] dy = \int_0^1 \left[\frac{1}{2} x^2 y^2 \right]_0^2 dy = \int_0^1 2y^2 dy = \left[\frac{2}{3} y^3 \right]_0^1 = \frac{2}{3}.$$

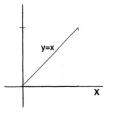

As a check, we reverse the order of integration to get

$$\int_0^2 \left[\int_0^1 xy^2 dy \right] dx = \int_0^2 \left[\frac{1}{3} xy^3 \right]_0^1 dx = \int_0^2 \frac{1}{3} x \, dx = \left[\frac{1}{6} x^2 \right]_0^2 = \frac{2}{3}.$$

3. The region of integration (shown on the right) is the triangular region in $\mathcal{R}^2$ lying below the line $y = x$, to the left of the vertical line $x = 2$ and above the x-axis. We integrate first with respect to y and then x:

$$\int_0^2 \left[\int_0^x e^{x-y} dy \right] dx = \int_0^2 \left[-e^{x-y} \right]_0^x dx = \int_0^2 (-1 + e^x) \, dx$$
$$= \left[-x + e^x \right]_0^2 = -3 + e^2.$$

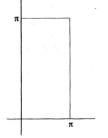

As a check, we reverse the order of integration. For each fixed y satisfying $0 \leq y \leq 2$, the x values satisfy $y \leq x \leq 2$. We have

$$\int_0^2 \left[\int_y^2 e^{x-y} dx \right] dy = \int_0^2 \left[e^{x-y} \right]_y^2 dy = \int_0^2 (e^{2-y} - 1) \, dy = \left[-e^{2-y} - y \right]_0^2 = -3 + e^2.$$

5. The region of integration is the rectangle $0 \leq x \leq \pi$, $0 \leq y \leq 2\pi$ in $\mathcal{R}^2$ and is shown on the right. We integrate first with respect to y and then x:

$$\int_0^\pi \left[\int_0^{2\pi} \sin(x-y) \, dy \right] dx = \int_0^\pi \left[\cos(x-y) \right]_0^{2\pi} dx$$
$$= \int_0^\pi (\cos(x - 2\pi) - \cos x) dx = \int_0^\pi (\cos x - \cos x) \, dx = 0,$$

As a check, we now integrate in the reverse order:

$$\int_0^{2\pi} \left[\int_0^\pi \sin(x-y) \, dx \right] dy = \int_0^{2\pi} \left[-\cos(x-y) \right]_0^\pi dy$$
$$= \int_0^{2\pi} \left(-\cos(\pi - y) + \cos(-y) \right) dy = \int_0^{2\pi} 2\cos y \, dy = 0,$$

where the last equality follows because the cosine function has period 2π and therefore any integral of the cosine over an interval of length 2π has value zero.

7. The figure on the right shows the triangle T in $\mathcal{R}^2$ with corners at $(0,0)$, $(1,0)$, $(1,1)$. The given double integral over T is perhaps slightly easier to compute if we integrate first with respect to y and then x. The reason is that both lower limits of integration on the corresponding iterated integral are zero. Specifically, For each x satisfying $0 \le x \le 1$, the y values satisfy $0 \le y \le x$. We therefore have

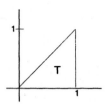

$$\int_T (x+y)\,dx\,dy = \int_0^1 dx \int_0^x (x+y)\,dy = \int_0^1 \left[xy + \frac{1}{2}y^2\right]_0^x dx = \int_0^1 \frac{3}{2}x^2\,dx = \left[\frac{1}{2}x^3\right]_0^1 = \frac{1}{2}.$$

9. The region Q (shown below on the left) is the part of the unit disk that lies in the first quadrant of $\mathcal{R}^2$. The circular nature of the region Q suggests we convert to polar coordinates. For the limits on r and θ we have $0 \le r \le 1$, $0 \le \theta \le \pi/2$. The integrand becomes $x = r\cos\theta$ and the area element becomes $dx\,dy = r\,dr\,d\theta$. Hence,

$$\int_Q x\,dx\,dy = \int_0^{\pi/2} d\theta \int_0^1 (r\cos\theta)\,r\,dr = \int_0^{\pi/2}\cos\theta\,d\theta \int_0^1 r^2 dr = \left[\sin\theta\right]_0^{\pi/2}\left[\frac{1}{3}r^3\right]_0^1 = \frac{1}{3}.$$

11. Let D be the unit disk in $\mathcal{R}^2$ as shown in the middle figure below. In terms of polar coordinates, D is described by $0 \le r \le 1$, $0 \le \theta \le 2\pi$. The integrand becomes

$$(x^2 - y^2)^2 = (r^2\cos^2\theta - r^2\sin^2\theta)^2 = r^4(\cos^2\theta - \sin^2\theta)^2 = r^4\cos^2 2\theta = \frac{1}{2}r^4(1+\cos 4\theta)$$

and the area element is $r\,dr\,d\theta$. We therefore have

$$\int_D (x^2-y^2)^2 dx\,dy = \int_0^{2\pi} d\theta \int_0^1 \left(\frac{1}{2}r^4(1+\cos 4\theta)\right) r\,dr$$

$$= \frac{1}{2}\int_0^{2\pi}(1+\cos 4\theta)\int_0^1 r^5 dr = \frac{1}{2}\left[\theta + \frac{1}{4}\sin 4\theta\right]_0^{2\pi}\left[\frac{1}{6}r^6\right]_0^1 = \frac{\pi}{6}.$$

13. The square S in $\mathcal{R}^2$ of side length 2 centered at $(1,0)$ is shown below on the right. In the evaluation of the given double integral over S using an iterated integral, there is no advantage in using any one order of integration over the other, so we arbitrarily choose to integrate first with respect to x and then y. We have

$$\int_S (x^2 + y^2)\,dx\,dy = \int_{-1}^1 dy \int_0^2 (x^2+y^2)\,dx$$

$$= \int_{-1}^1 \left[\frac{1}{3}x^3 + xy^2\right]_0^2 dy = \int_{-1}^1 \left(\frac{8}{3} + 2y^2\right) dy = \left[\frac{8}{3}y + \frac{2}{3}y^3\right]_{-1}^1 = \frac{20}{3}.$$

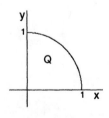

Exercise 9

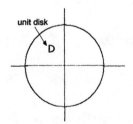

Exercise 11

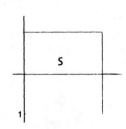

Exercise 13

278

15. The region of integration is the rectangle $0 \le x \le 1$, $1 \le y \le 2$, $2 \le z \le 3$ in $\mathcal{R}^3$ and is shown on the right. The integration is straightforward:

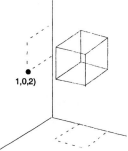

$$\int_0^1 \left[\int_1^2 \left[\int_2^3 xyz\, dz \right] dy \right] dx = \int_0^1 x\, dx \int_1^2 y\, dy \int_2^3 z\, dz$$

$$= \left[\frac{1}{2} x^2 \right]_0^1 \left[\frac{1}{2} y^2 \right]_1^2 \left[\frac{1}{2} z^2 \right]_2^3 = \frac{15}{8}.$$

17. The solid cylinder C in $\mathcal{R}^3$ defined by $x^2 + y^2 \le 1$, $0 \le z \le 2$ is shown on the right. The integrand and the region C suggests a change to cylindrical coordinates. The region C is described by the inequalities $0 \le r \le 1$, $0 \le \theta \le 2\pi$, $0 \le z \le 2$. The integrand becomes $z(x^2 + y^2) = zr^2$ and the volume element is $dV = r\, dr\, d\theta\, dz$. Hence,

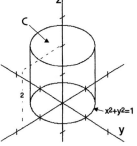

$$\int_C z(x^2 + y^2)\, dV = \int_0^{2\pi} d\theta \int_0^2 dz \int_0^1 zr^2 r\, dr$$

$$= \int_0^{2\pi} d\theta \int_0^2 z\, dz \int_0^1 r^3 dr = [\theta]_0^{2\pi} \left[\frac{1}{2} z^2 \right]_0^2 \left[\frac{1}{4} r^4 \right]_0^1 = \pi.$$

19. The solid ball B defined by $x^2 + y^2 + z^2 \le 1$ is shown on the right. In spherical coordinates, its description is $0 \le r \le 1$, $0 \le \phi \le \pi$, $0 \le \theta \le 2\pi$. The integrand becomes $x^2 + y^2 + z^2 = r^2$, the volume element is $dV = r^2 \sin\phi\, dr\, d\phi\, d\theta$, and we get

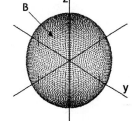

$$\int_B (x^2 + y^2 + z^2)\, dV = \int_0^{2\pi} d\theta \int_0^\pi d\phi \int_0^1 r^2 r^2 \sin\phi\, dr$$

$$= \int_0^{2\pi} d\theta \int_0^\pi \sin\phi\, d\phi \int_0^1 r^4\, dr = [\theta]_0^{2\pi} \left[-\cos\phi \right]_0^\pi \left[\frac{1}{5} r^5 \right]_0^1 = \frac{4\pi}{5}.$$

NOTE: *The region of integration in Exercises 21 and 23 below are truncated cones. The one for Exercise 21 is 1 unit high and the one for Exercise 23 is 2 units high. Perceptually, they look identical. As such, the drawings for these two exercises is done only once and is shown to the right of the computations evaluating the integral in Exercise 21.*

21. The solid cone K defined by $\sqrt{x^2 + y^2} \le z \le 1$ is shown on the right. Its description in cylindrical coordinates is $0 \le z \le 1$, $0 \le \theta \le 2\pi$, $0 \le r \le z$. The integrand is $x^2 + y^2 = r^2$ and the volume element is $dV = r\, dr\, d\theta\, dz$. Hence,

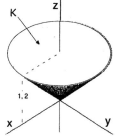

$$\int_K (x^2 + y^2)\, dV = \int_0^{2\pi} d\theta \int_0^1 dz \int_0^z r^2 r\, dr$$

$$= \int_0^{2\pi} d\theta \int_0^1 dz \int_0^z r^3 dr = \int_0^{2\pi} d\theta \int_0^1 \left[\frac{1}{4} r^4 \right]_0^z dz$$

$$= \int_0^{2\pi} d\theta \int_0^1 \frac{1}{4} z^4 dz = [\theta]_0^{2\pi} \left[\frac{1}{20} z^5 \right]_0^1 = \frac{\pi}{10}.$$

23. The drawing of the solid cone K defined by $\sqrt{x^2 + y^2} \leq z \leq 2$ is shown to the right of the computations in Exercise 21. Its description in cylindrical coordinates is $0 \leq z \leq 2$, $0 \leq \theta \leq 2\pi$, $0 \leq r \leq z$. The integrand is $\sqrt{x^2 + y^2} = r$ and the volume element is $dV = r\,dr\,d\theta\,dz$. Hence,

$$\int_K \sqrt{x^2 + y^2}\,dV = \int_0^{2\pi} d\theta \int_0^2 dz \int_0^z r\,r\,dr = \int_0^{2\pi} d\theta \int_0^2 dz \int_0^z r^2\,dr$$

$$= \int_0^{2\pi} d\theta \int_0^2 \left[\frac{1}{3}r^3\right]_0^z dz = \int_0^{2\pi} d\theta \int_0^2 \frac{1}{3}z^3\,dz = \left[\theta\right]_0^{2\pi}\left[\frac{1}{12}z^4\right]_0^2 = \frac{8\pi}{3}.$$

25. Let R be the region in $\mathcal{R}^2$ bounded by the parabola $y = x^2$ and the line $y = 2x + 3$. A simple calculation shows that the two boundary curves intersect at the points $(3, 9)$ and $(-1, 1)$. The region R and all of the information necessary to do this exercise are shown in the diagram on the right. As can be seen, for each x satisfying $-1 \leq x \leq 3$, y satisfies $x^2 \leq y \leq 2x + 3$. Thus, by integrating first with respect to y and then x, we get

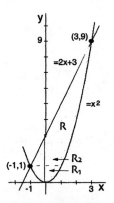

$$\int_R x^2 y\,dA = \int_{-1}^3 dx \int_{x^2}^{2x+3} x^2 y\,dy = \int_{-1}^3 \left[\frac{1}{2}x^2 y^2\right]_{x^2}^{2x+3} dx$$

$$= \frac{1}{2}\int_{-1}^3 \left(x^2(2x+3)^2 - x^6\right) dx = \frac{1}{2}\int_{-1}^3 (4x^4 + 12x^3 + 9x^2 - x^6)\,dx$$

$$= \frac{1}{2}\left[\frac{4}{5}x^5 + 3x^4 + 3x^3 - \frac{1}{7}x^7\right]_{-1}^3 = \frac{3616}{35}.$$

The other iterated integral must be done as two separate iterated integrals, one over the region R_1 and one over the region R_2 (see diagram). On R_1, for each y satisfying $0 \leq y \leq 1$, x satisfies $-\sqrt{y} \leq x \leq \sqrt{y}$; and on R_2, for each y satisfying $1 \leq y \leq 9$, x satisfies $\frac{1}{2}y - \frac{3}{2} \leq x \leq \sqrt{y}$. The result is

$$\int_R x^2 y\,dA = \int_{R_1} x^2 y\,dA + \int_{R_2} x^2 y\,dA = \int_0^1 dy \int_{-\sqrt{y}}^{\sqrt{y}} x^2 y\,dx + \int_1^9 dy \int_{\frac{1}{2}y - \frac{3}{2}}^{\sqrt{y}} x^2 y\,dx.$$

27. The solid B with its base on the elliptic region E given in Exercise 26 is shown on the right. Given that cross-sections perpendicular to the x-axis are squares, for $-a \leq x \leq a$, let S_x be the square cross-section and let $A(x)$ be its area. Then

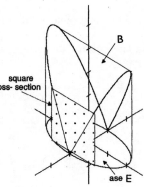

$$V(B) = \int_{-a}^a A(x)\,dx = 2\int_0^a A(x)\,dx,$$

where the second equality follows because the solid is symmetrical with respect to the cross-section S_0. To find $A(x)$, note that one-half the side length of S_x is the positive y value in the equation $x^2/a^2 + y^2/b^2 = 1$; i.e. $y = b\sqrt{1 - x^2/a^2}$. It follows that $A(x) = 4b^2(1 - x^2/a^2)$. Therefore,

$$V(B) = 2\int_0^a A(x)\,dx = 2\int_0^a 4b^2\left(1 - \frac{x^2}{a^2}\right)dx = 8b^2\left[x - \frac{x^3}{3a^2}\right]_0^a = \frac{16}{3}ab^2.$$

29. Let B be the ellipsoidal region $x^2/a^2 + y^2/b^2 + z^2/c^2 \leq 1$ (see figure on the right). The observation that cross-sections perpendicular to any of the coordinate planes are ellipses suggests that the volume of B can be found by cross-sectional slicing. We arbitrarily choose to use those cross-sections parallel to the yz-plane. So, for each fixed $-a < x < a$, we subtract x^2/a^2 from both sides of the equation of the ellipsoid and write the result as

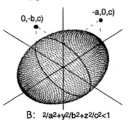

$$\frac{y^2}{b^2} + \frac{z^2}{c^2} = \left(\sqrt{1 - \frac{x^2}{a^2}}\right)^2, \qquad \text{or} \qquad \frac{y^2}{\left(b\sqrt{1 - \frac{x^2}{a^2}}\right)^2} + \frac{z^2}{\left(c\sqrt{1 - \frac{x^2}{a^2}}\right)^2} = 1.$$

The equation on the right is the equation of the elliptical boundary of the cross-section corresponding to x. Its major and minor semi-axes are the two parenthetical denominators. Using the fact that the area of an ellipse is equal to π times the product of its two semi-axes, it follows that the area $A(x)$ of the elliptical cross-section corresponding to x is

$$A(x) = \pi\left(b\sqrt{1 - \frac{x^2}{a^2}}\right)\left(c\sqrt{1 - \frac{x^2}{a^2}}\right) = \pi bc\left(1 - \frac{x^2}{a^2}\right).$$

The volume of B is then $V(B) = \int_{-a}^a A(x)\,dx$. However, because of the symmetry of B about the plane $x = 0$, we need only integrate over the interval $[0, a]$ and then multiply the result by 2. Hence, the volume of B is

$$V(B) = 2\int_0^a A(x)\,dx = 2\pi bc\int_0^a \left(1 - \frac{x^2}{a^2}\right)dx = 2\pi bc\left[x - \frac{x^3}{3a^2}\right]_0^a = \frac{4}{3}\pi abc.$$

31. Let H be the region in $\mathcal{R}^2$ bounded by the two branches of the hyperbola $x^2 - y^2 = 1$ and the lines $y = \pm 1$, and let B be the solid generated by rotating H about the y-axis. The way in which the generation of B is described might lead one to think that the volume of B could be found by using cylindrical shells. While this is true, it's not as straightforward as it may first appear. The drawing on the right may be helpful.

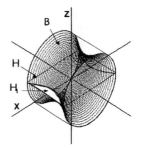

Let R be the rectangle in $\mathcal{R}^2$ with corners at $(\pm\sqrt{2}, \pm 1)$ and note that H is contained in R. Let H_1 be the part of R that lies outside of H. The solid cylinder of radius $\sqrt{2}$ and height 2 generated by rotating R about the y-axis has volume 4π (area of its base times its height). This is the sum of the volumes of two disjoint regions; namely, B and the solid B_1 generated by rotating H_1 about the y-axis. It follows that

$$V(B) = 4\pi - V(B_1).$$

We will apply the method of cylindrical shells to find $V(B_1)$.

281

For $1 \leq x \leq \sqrt{2}$, the associated cylindrical shell has height $2y = 2x\sqrt{x^2 - 1}$ and circumference $2\pi x$. The method of cylindrical shells then gives the volume of B_1 as

$$V(B_1) = 4\pi \int_1^{\sqrt{2}} x\sqrt{x^2 - 1}\, dx = 4\pi \left[\frac{1}{3}(x^2 - 1)^{3/2} \right]_1^{\sqrt{2}} = \frac{4\pi}{3}.$$

Hence,

$$V(B) = 4\pi - \frac{4\pi}{3}\pi = \frac{8\pi}{3}.$$

33. Let a, b and c be positive constants and let B be the region in $\mathcal{R}^3$ bounded by the three coordinate planes and the plane $ax + by + cz = 1$. Then B is the solid lying under the graph of $f(x,y) = (1 - ax - by)/c$ and above the triangle T with vertices $(0,0,0)$, $(1/a,0,0)$, $(0,1/b,0)$ in the xy-plane. The line in the xy-plane containing the points $(1/a,0,0)$ and $(0,1/b,0)$ has the equation $ax + by = 1$. Hence, for each $0 \leq x \leq 1/a$, the y values satisfy $0 \leq y \leq (1 - ax)/b$. So, the volume of B is

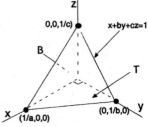

$$V(B) = \int_T f(x,y)\, dy\, dx = \frac{1}{c} \int_0^{1/a} dx \int_0^{(1-ax)/b} (1 - ax - by)\, dy$$

$$= \frac{1}{c} \int_0^{1/a} \left[(1 - ax)y - \frac{b}{2}y^2 \right]_0^{(1-ax)/b} dx = \frac{1}{2bc} \int_0^{1/a} (1 - ax)^2 dx$$

$$= \frac{1}{2bc} \left[-\frac{1}{3a}(1 - ax)^3 \right]_0^{1/a} = \frac{1}{6abc}.$$

35. Consider the iterated integral

$$\int_0^1 \int_0^{\sqrt{2 - 2x^2}} 2x\, dy\, dx.$$

    (a) The first thing to observe about the above iterated integral is that integration is first done with respect to y. Thus, letting R be the region of integration, the limits of integration show that for each $0 \leq x \leq 1$, the y values range from the curve defined by $y = 0$ to the curve defined by $y = \sqrt{2 - 2x^2}$. The first rquation describes the x-axis and the second equation is the ellipse $x^2 + y^2/2 = 1$. It follows that R is the region in the first quadrant of $\mathcal{R}^2$ bounded by the ellipse $x^2 + y^2/2 = 1$. The region is shown on the right.

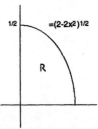

    (b) Now that we know the region R, we can reverse the order of integration to obtain the iterated integral with integration first with respect to x. Here, we fix y between 0 and $\sqrt{2}$ and allow x to vary from the curve $x = 0$ to the curve $x = \sqrt{1 - y^2/2}$. Hence, the equivalent iterated integral to the one given is

$$\int_0^{\sqrt{2}} \int_0^{\sqrt{1 - y^2/2}} 2x\, dx\, dy.$$

(c) The given iterated integral evaluates as follows:

$$\int_0^1 \int_0^{\sqrt{2-2x^2}} 2x\, dy\, dx = \int_0^1 \left[2xy\right]_0^{\sqrt{2-2x^2}} dx$$

$$= 2\int_0^1 x\sqrt{2-2x^2}\, dx = 2\left[-\frac{1}{6}(2-2x^2)^{3/2}\right]_0^1 = \frac{2\sqrt{2}}{3}.$$

The iterated integral derived in part (b) evaluates as follows:

$$\int_0^{\sqrt{2}} \int_0^{\sqrt{1-y^2/2}} 2x\, dx\, dy = \int_0^{\sqrt{2}} \left[x^2\right]_0^{\sqrt{1-y^2/2}} dy$$

$$= \int_0^{\sqrt{2}} \left(1 - \frac{1}{2}y^2\right) dy = \left[y - \frac{1}{6}y^3\right]_0^{\sqrt{2}} = \frac{2\sqrt{2}}{3}.$$

37. We are asked to evaluate the double integral $\int_R (3x^2 + 2y)\, dA$, where R is the region in $\mathcal{R}^2$ bounded by the parabola $y = x^2$ and the line $y = 2 - x$. A simple calculation shows that the two boundary curves intersect at the points $(-2, 4)$ and $(1, 1)$. Observe that if y is fixed between zero and 4 then the upper limit for x will be $x = \sqrt{y}$ if $0 \le y \le 1$ but will be $x = 2 - y$ if $1 \le y \le 4$. This would require that the associated iterated integral be split into two pieces. Instead of doing this, integrate first with respect to y. Then, for each $-2 \le x \le 1$, the y values vary from $y = x^2$ to $y = 2 - x$ and the associated iterated integral is

$$\int_R (3x^2 + 2y)\, dA = \int_{-2}^1 dx \int_{x^2}^{2-x} (3x^2 + 2y)\, dy = \int_{-2}^1 \left[3x^2 y + y^2\right]_{x^2}^{2-x} dx$$

$$= \int_{-2}^1 (3x^2(2-x) + (2-x)^2 - 4x^4)\, dx = \int_{-2}^1 (-4x^4 - 3x^3 + 7x^2 - 4x + 4)\, dx$$

$$= \left[-\frac{4}{5}x^5 - \frac{3}{4}x^4 + \frac{7}{3}x^3 - 2x^2 + 4x\right]_{-2}^1 = \frac{477}{20}.$$

39. When the given polar equation for the boundary of R is multiplied by r and the result is converted to rectangular coordinates, one obtains the equation $(x - 1/2)^2 + y^2 = 1/4$, which is a circle of radius $1/2$ centered at $(1/2, 0)$ in the xy-plane. Moreover, as θ varies from $-\pi/2$ to $\pi/2$, the entire circle is traced out once. In terms of polar coordinates, the integrand becomes $\sqrt{x^2 + y^2} = r$ and the area element becomes $dx\, dy = r\, dr\, d\theta$. Hence,

$$\int_R \sqrt{x^2 + y^2}\, dx\, dy = \int_{-\pi/2}^{\pi/2} d\theta \int_0^{\cos\theta} r^2\, dr = \int_{-\pi/2}^{\pi/2} \left[\frac{1}{3}r^3\right]_0^{\cos\theta} d\theta$$

$$= \int_{-\pi/2}^{\pi/2} \frac{1}{3}\cos^3\theta\, d\theta = \frac{1}{3}\int_{-\pi/2}^{\pi/2} (\cos\theta - \sin^2\theta \cos\theta)\, d\theta$$

$$= \frac{1}{3}\left[\sin\theta - \frac{1}{3}\sin^3\theta\right]_{-\pi/2}^{\pi/2} = \frac{4}{9}.$$

41. (a) A second characterization of W is the following: if B is the plane region in $\mathcal{R}^2$ defined by $x^2 + y^2 \le 1$, $x \ge 0$ then the wedge shaped region W in $\mathcal{R}^3$ is the region above B and below the plane $z = x$. Thus, with the point (x, y) fixed in B, we allow z to vary from $z = 0$ to $z = x$. This traces out a vertical line segment l with endpoints $(x, y, 0)$, (x, y, x). Still regarding x as fixed, we now allow y to vary from $y = -\sqrt{1 - x^2}$ to $y = \sqrt{1 - x^2}$. This sweeps out the vertical rectangular plane region inside W parallel to the yz-plane and containing the line segment l. Finally, we allow x to vary from $x = 0$ to $x = 1$, which sums the areas of these rectangles over W. The corresponding iterated integral is

$$\int_W z\, dV = \int_0^1 \left[\int_{-\sqrt{1-x^2}}^{\sqrt{1-x^2}} \left[\int_0^x z\, dz \right] dy \right] dx.$$

(b) Notice first that if (x, y, z) is in W then it lies below the plane $z = x$, so that $z \le x$. Since $x = r\cos\theta$ in cylindrical coordinates, it follows that $z \le r\cos\theta$ and, since $r \le 1$, $z \le \cos\theta$. Thus, for each $0 \le z \le 1$, the bounds on r depend on z and θ and the bounds on θ depend on z. Specifically, $z\sec\theta \le r \le 1$ and $-\arccos z \le \theta \le \arccos z$, where r takes on the single value $r = 1$ in case $|\theta| = \pi/2$. The corresponding iterated integral is therefore

$$\int_W z\, dV = \int_0^1 \left[\int_{-\arccos z}^{\arccos z} \left[\int_{z\sec\theta}^1 zr\, dr \right] d\theta \right] dz.$$

(c) We will evaluate the iterated integral derived in part (a). We have

$$\int_W z\, dV = \int_0^1 \left[\int_{-\sqrt{1-x^2}}^{\sqrt{1-x^2}} \left[\int_0^x z\, dz \right] dy \right] dx = \int_0^1 \left[\int_{-\sqrt{1-x^2}}^{\sqrt{1-x^2}} \left[\frac{1}{2}z^2 \right]_0^x dy \right] dx$$

$$= \int_0^1 \left[\int_{-\sqrt{1-x^2}}^{\sqrt{1-x^2}} \frac{1}{2}x^2 dy \right] dx = \int_0^1 \left[\frac{1}{2}x^2 y \right]_{-\sqrt{1-x^2}}^{\sqrt{1-x^2}} dx = \int_0^1 x^2\sqrt{1 - x^2}\, dx$$

$$= \left[-\frac{1}{4}x(1 - x^2)^{3/2} + \frac{1}{8}x\sqrt{1 - x^2} + \frac{1}{8}\arcsin x \right]_0^1 = \frac{\pi}{16}.$$

43. The region R in $\mathcal{R}^3$ determined by the inequalities $x \ge 0$, $y \ge 0$, $x^2 + y^2 \le 1$, $0 \le z \le x^2 + y^2$ is the region above the quarter unit disk in the first quadrant and below the circular paraboloid $z = x^2 + y^2$. In terms of cylindrical coordinates $0 \le \theta \le \pi/2$, $0 \le r \le 1$, $0 \le z \le r^2$. The integrand becomes $xyz = (r\cos\theta)(r\sin\theta)z = r^2 z\cos\theta\sin\theta$ and the volume element is $dV = r\, dr\, d\theta\, dz$. Hence,

$$\int_R xyz\, dV = \int_0^{\pi/2} d\theta \int_0^1 dr \int_0^{r^2} (r^2 z\cos\theta\sin\theta)\, r\, dz$$

$$= \int_0^{\pi/2} \cos\theta\sin\theta\, d\theta \int_0^1 r^3 dr \int_0^{r^2} z\, dz = \int_0^{\pi/2} \frac{1}{2}\sin 2\theta\, d\theta \int_0^1 r^3 \left[\frac{1}{2}z^2 \right]_0^{r^2} dr$$

$$= \int_0^{\pi/2} \frac{1}{2}\sin 2\theta\, d\theta \int_0^1 \frac{1}{2}r^7 dr = \left[-\frac{1}{4}\cos 2\theta \right]_0^{\pi/2} \left[\frac{1}{16}r^8 \right]_0^1 = \frac{1}{32}.$$

45. The region of integration is the region bounded by the circle $(x - 3)^2 + y^2 = 4$. Since the given iterated integral requires that integration first be done with respect to x, it follows

284

that the limits on the outside integral are the maximum and minimum values that y can attain on the circle. So, $a = -2$ and $b = 2$. For each y in this range, x varies from the lowest x value satisfying the equation $(x - 3)^2 + y^2 = 4$ to the largest x value satisfying the equation. That is, x varies from $x = 3 - \sqrt{4 - y^2}$ to $x = 3 + \sqrt{4 - y^2}$. So, $c = 3 - \sqrt{4 - y^2}$ and $d = 3 + \sqrt{4 - y^2}$. The appropriate integral is therefore

$$\int_{-2}^{2} \left[\int_{3-\sqrt{4-y^2}}^{3+\sqrt{4-y^2}} f(x, y) \, dx \right] dy.$$

47. Let B be the rectangle in $\mathcal{U}^2$ with corners at $(0, 0)$, $(2, 0)$, $(2, 1)$ and $(0, 1)$. The given transformation can be formally defined as $T : \mathcal{U}^2 \longrightarrow \mathcal{R}^2$, where

$$T(u, v) = (x, y) = (u + v, u + 3v).$$

The derivative matrix of T satisfies

$$T'(u, v) = \begin{pmatrix} 1 & 1 \\ 1 & 3 \end{pmatrix} \qquad \text{and} \qquad \det T'(u, v) = 2.$$

Note too that T is one-to-one on B (and, in fact, on all of $\mathcal{U}^2$). Because T is linear, it maps bounded regions to bounded regions, so that the image of B under T, denoted by $T(B) = R$, is bounded. Hence, the function $f(x, y) = y$ is bounded and continuous on R. Moreover, $f(T(u, v)) = f(u + v, u + 3v) = u + 3v$. By Jacobi's theorem,

$$\int_R y \, dV_{\mathbf{x}} = \int_B (u + 3v)(2) \, dV_{\mathbf{u}} = 2 \int_B (u + 3v) \, dV_{\mathbf{u}}.$$

Since B is a rectangle with $0 \le u \le 2$ and $0 \le v \le 1$, we have

$$2 \int_B (u + 3v) \, dV_{\mathbf{u}} = 2 \int_0^1 \left[\int_0^2 (u + 3v) \, du \right] dv = 2 \int_0^1 \left[\frac{1}{2} u^2 + 3uv \right]_0^2 dv$$

$$= 2 \int_0^1 (2 + 6v) \, dv = 2 \left[2v + 3v^2 \right]_0^1 = 10.$$

Since the context is clear, we can let $dV_{\mathbf{x}} = dV$ to obtain

$$\int_R y \, dV = 10.$$

49. Let B be the semicircular region described by $x^2 + y^2 \le 9$, $x \ge 0$. The region C described in the statement of this exercise is then the region above B and below the surface defined by $f(x, y) = 1 + x + y^2$. Moreover, its volume is given by

$$V(C) = \int_B f(x, y) \, dV = \int_B (1 + x + y^2) \, dV.$$

The iterated integral we will use to evaluate this double integral will be done first with respect to x and then y. To find the limits of integration, observe that for each $-3 \le y \le 3$,

x varies from $x = 0$ to $x = \sqrt{9 - y^2}$. Also observe that the integrand and the base B are symmetric with respect to the x axis, so that it is only necessary to let y range from $y = 0$ to $y = 3$ and then multiply the result by 2. Hence,

$$V(C) = 2\int_0^3 dy \int_0^{\sqrt{9-y^2}} (1 + x + y^2)\, dx = 2\int_0^3 \left[x + \frac{1}{2}x^2 + xy^2 \right]_0^{\sqrt{9-y^2}} dy$$

$$= 2\int_0^3 \left(\sqrt{9-y^2} + \frac{1}{2}(9-y^2) + y^2\sqrt{9-y^2} \right) dy$$

$$= 2\int_0^3 \sqrt{9-y^2}\, dy + \int_0^3 (9-y^2)\, dy + 2\int_0^3 y^2\sqrt{9-y^2}\, dy.$$

The first term in the last line is the area of B, which is $9\pi/2$ (one-half the area of a circle of radius 3). The second term can be evaluated directly as

$$\int_0^3 (9-y^2)\, dy = \left[9y - \frac{1}{3}y^3 \right]_0^3 = 18.$$

The third term is best done with standard integral tables. We have

$$2\int_0^3 y^2\sqrt{9-y^2}\, dy = 2\left[-\frac{1}{4}y(9-y^2)^{3/2} + \frac{9}{8}y\sqrt{9-y^2} + \frac{81}{8}\arcsin(y/3) \right]_0^3 = \frac{81\pi}{8}.$$

Hence,

$$V(C) = \frac{9\pi}{2} + 18 + \frac{81\pi}{8} = \frac{117\pi}{8} + 18.$$

51. The given integral is improper because the region of integration is unbounded. Thus, let D_b be the disk of radius b. In terms of polar coordinates, D_b is described by $0 \le r \le b$, $0 \le \theta \le 2\pi$. The integrand becomes $e^{-(x^2+y^2)} = e^{-r^2}$ and the area element becomes $dx\, dy = r\, dr\, d\theta$. Thus,

$$\int_{D_b} e^{-(x^2+y^2)}\, dx\, dy = \int_{D_b} e^{-r^2} r\, dr\, d\theta = \int_0^{2\pi} d\theta \int_0^b re^{-r^2}\, dr$$

$$= [\theta]_0^{2\pi} \left[-\frac{1}{2}e^{-r^2} \right]_0^b = \pi(1 - e^{-b^2}).$$

Therefore,

$$\int_{\mathcal{R}^2} e^{-(x^2+y^2)}\, dx\, dy = \lim_{b\to\infty} \int_{D_b} e^{-(x^2+y^2)}\, dx\, dy = \lim_{b\to\infty} \pi(1 - e^{-b^2}) = \pi.$$

53. The given integral is improper because the integrand is unbounded near $(0,0)$. Thus, for $0 < \delta < 1$, let A_δ be the annulus $\delta \le x^2 + y^2 \le 1$. In terms of polar coordinates, A_δ is described by $\delta \le r \le 1, 0 \le \theta \le 2\pi$. The integrand becomes $(x^2+y^2)^{-1/3} = r^{-2/3}$ and the area element is $dA = r\, dr\, d\theta$. We then have

$$\int_{A_\delta} (x^2+y^2)^{-1/3} dA = \int_{A_\delta} r^{-2/3} r\, dr\, d\theta = \int_0^{2\pi} d\theta \int_\delta^1 r^{1/3} dr$$

$$= [\theta]_0^{2\pi} \left[\frac{3}{4}r^{4/3} \right]_\delta^1 = \frac{3\pi}{2}(1 - \delta^{4/3}).$$

Therefore,

$$\int_{x^2+y^2\leq 1} (x^2+y^2)^{-1/3} dA = \lim_{\delta\to 0^+}\int_{A_\delta}(x^2+y^2)^{-1/3}dA = \lim_{\delta\to 0^+}\frac{3\pi}{2}(1-\delta^{4/3}) = \frac{3\pi}{2}.$$

55. The given integral is improper because the region of integration is unbounded. Thus, for $a > 0$, let S_a be the sphere of radius a in $\mathcal{R}^3$. In spherical coordinates, S_a is described by $0 \leq r \leq a$, $0 \leq \phi \leq \pi$, $0 \leq \theta \leq 2\pi$. The integrand becomes $e^{-(x^2+y^2+z^2)^{3/2}} = e^{-r^3}$ and the volume element becomes $dx\,dy\,dz = r^2\sin\phi\,dr\,d\phi\,d\theta$. We then have

$$\int_{S_a} e^{-(x^2+y^2+z^2)^{3/2}} dx\,dy\,dz$$

$$= \int_{S_a} e^{-r^3} r^2 \sin\phi\,dr\,d\phi\,d\theta = \int_0^{2\pi} d\theta \int_0^\pi \sin\phi\,d\phi \int_0^a r^2 e^{-r^3}\,dr$$

$$= [\theta]_0^{2\pi} [-\cos\phi]_0^\pi \left[-\frac{1}{3}e^{-r^3}\right]_0^a = \frac{4\pi}{3}(1-e^{-a^3}).$$

Therefore,

$$\int_{\mathcal{R}^3} e^{-(x^2+y^2+z^2)^{3/2}} dx\,dy\,dz = \lim_{a\to\infty}\int_{S_a} e^{-(x^2+y^2+z^2)^{3/2}} dx\,dy\,dz = \lim_{a\to\infty}\frac{4\pi}{3}(1-e^{-a^3}) = \frac{4\pi}{3}.$$

57. Since the region of integration on the given integral is bounded, the given integral is proper if, and only if, $\alpha \leq 0$. In this case, the integral has a finite value. Otherwise, the integrand is unbounded near $(0,0)$ and the integral is improper. In this case, for $0 < \delta < 1$, let A_δ be the annulus $\delta^2 \leq x^2+y^2 \leq 1$. In polar coordinates, A_δ is described by $\delta \leq r \leq 1$, $0 \leq \theta \leq 2\pi$. The integrand becomes $(x^2+y^2)^{-\alpha} = r^{-2\alpha}$ and the area element is $dA = r\,dr\,d\theta$. We then have

$$\int_{A_\delta} \frac{1}{(x^2+y^2)^\alpha}\,dA = \int_{A_\delta} r^{-2\alpha} r\,dr\,d\theta = \int_0^{2\pi} d\theta \int_\delta^1 r^{1-2\alpha}dr.$$

The case $\alpha = 1$ is exceptional and we dispose of it now:

$$\int_0^{2\pi} d\theta \int_\delta^1 r^{-1}dr = [\theta]_0^{2\pi} [\ln r]_\delta^1 = -2\pi\ln\delta.$$

Since the expression on the right tends to $+\infty$ as $\delta \to 0^+$, the given integral fails to converge for $\alpha = 1$.

For $\alpha \neq 1$, we have

$$\int_0^{2\pi} d\theta \int_\delta^1 r^{1-2\alpha}dr = [\theta]_0^{2\pi} \left[\frac{r^{2-2\alpha}}{2-2\alpha}\right]_\delta^1 = \frac{\pi}{1-\alpha}(1-\delta^{2(1-\alpha)}).$$

If $0 < \alpha < 1$ then, in the expression on the right, δ is in the numerator, so that the right side tends to the finite value $\pi/(1-\alpha)$ as $\delta \to 0^+$; and if $\alpha > 1$ then δ is in the denominator and the right side tends to $+\infty$ as $\delta \to 0^+$. Thus, the given integral has a finite value if, and only it, $\alpha < 1$.

59. Since the region of integration on the given integral is bounded, the given integral is proper if, and only if, $\alpha \leq 0$. In this case, the integral has a finite value. Otherwise, the integrand is unbounded near $(0,0,0)$ and the integral is improper. In this case, for $0 < \delta < 1$, let A_δ be the annulus $\delta^2 \leq x^2 + y^2 + z^2 \leq 1$. In spherical coordinates, A_δ is described by $\delta \leq r \leq 1$, $0 \leq \phi \leq \pi$, $0 \leq \theta \leq 2\pi$. The integrand becomes $(x^2 + y^2 + z^2)^{-\alpha} = r^{-2\alpha}$ and the volume element is $dV = r^2 \sin\phi \, dr \, d\phi \, d\theta$. We then have

$$\int_{A_\delta} \frac{1}{(x^2 + y^2 + z^2)^\alpha} \, dV = \int_{A_\delta} r^{-2\alpha} r^2 \sin\phi \, dr \, d\phi \, d\theta = \int_0^{2\pi} d\theta \int_0^\pi \sin\phi \, d\phi \int_\delta^1 r^{2-2\alpha} dr.$$

The case $\alpha = 3/2$ is exceptional and we dispose of it now:

$$\int_0^{2\pi} d\theta \int_0^\pi \sin\phi \, d\phi \int_\delta^1 r^{-1} dr = \big[\theta\big]_0^{2\pi} \big[-\cos\phi\big]_0^\pi \big[\ln r\big]_\delta^1 = -4\pi \ln\delta.$$

Since the expression on the right tends to $+\infty$ as $\delta \to 0^+$, the given integral fails to converge for $\alpha = 3/2$.

For $\alpha \neq 3/2$, we have

$$\int_0^{2\pi} d\theta \int_0^\pi \sin\phi \, d\phi \int_\delta^1 r^{2-2\alpha} dr = \big[\theta\big]_0^{2\pi} \big[-\cos\phi\big]_0^\pi \left[\frac{r^{3-2\alpha}}{3-2\alpha}\right]_\delta^1 = \frac{4\pi}{3-2\alpha}(1 - \delta^{3-2\alpha}).$$

If $0 < \alpha < 3/2$ then, in the expression on the right, δ is in the numerator, so that the right side tends to the finite value $4\pi/(3-2\alpha)$ as $\delta \to 0^+$; and if $\alpha > 3/2$ then δ is in the denominator and the right side tends to $+\infty$ as $\delta \to 0^+$. Thus, the given integral has a finite value if, and only if, $\alpha < 3/2$.

61. (a) Let H be the region of integration of the given integral. There are two key ideas to focus on in trying to determine H: (i) the given integral is written in terms of cylindrical coordinates, and (ii) integration is done first with respect to z, then r, and then θ. The limits of integration with respect to r and θ show that we can think of choosing a direction θ $(0 \leq \theta \leq 2\pi)$ and a length r $(0 \leq r \leq 1)$ and temporarily holding them fixed. We then allow z to vary from the plane $z = 0$ to the surface defined by $z = \sqrt{1 - r^2}$, which is a sphere of radius 1 centered at the origin in $\mathcal{R}^3$. This traces out a vertical line segment l. Still thinking of θ as fixed, we allow r to vary from $r = 0$ to $r = 1$. This sweeps out a vertical quarter disk of radius 1 that contains l. Finally, as θ varies 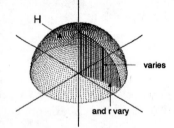 from $\theta = 0$ to $\theta = 2\pi$, the quarter circle sweeps out a hemispherical region of radius 1 with base on the xy-plane and axis of symmetry at the origin. The region H is shown on the right.

(b) The result of part (a) implies that we must set up an integral in terms of rectangular coordinates whose value is the volume of a hemispherical solid H of radius 1, which we position with its flat base coincident with the unit disk on the xy-plane. We will integrate first with respect to z, then y and then x. So we think of choosing a point (x, y) on the unit disk base and fixing it. Allowing z to vary from $z = 0$ to the inside surface of the hemisphere gives the inside integral; namely, $\int_0^{\sqrt{1-x^2-y^2}} dz$. Still thinking of x as fixed, we allow y to vary from one side of the unit disk to the other; i.e., from $y = -\sqrt{1 - x^2}$ to $y = \sqrt{1 - x^2}$. And finally, we allow x to vary from $x = -1$ to $x = 1$. This gives the iterated integral

$$\int_{-1}^1 dx \int_{-\sqrt{1-x^2}}^{\sqrt{1-x^2}} dy \int_0^{\sqrt{1-x^2-y^2}} dz.$$

288

(c) In spherical coordinates, H is described by $0 \le r \le 1$, $0 \le \phi \le \pi/2$, $0 \le \theta \le 2\pi$. With the volume element equal to $r^2 \sin \phi \, dr \, d\phi \, d\theta$, we obtain the equivalent integral

$$\int_0^{2\pi} d\theta \int_0^{\pi/2} \sin \phi \, d\phi \int_0^2 r^2 \, dr.$$

(d) We formally compute the value of the given integral:

$$
\begin{aligned}
V(H) &= \int_0^{2\pi} d\theta \int_0^1 dr \int_0^{\sqrt{1-r^2}} r \, dz = \int_0^{2\pi} d\theta \int_0^1 [rz]_0^{\sqrt{1-r^2}} dr \\
&= \int_0^{2\pi} d\theta \int_0^1 r\sqrt{1-r^2} \, dr = [\theta]_0^{2\pi} \left[-\frac{1}{3}(1-r^2)^{3/2} \right]_0^1 = \frac{2\pi}{3}.
\end{aligned}
$$

Note that the volume of a sphere of radius 1 is $4\pi/3$, so that the above result for $V(H)$ is consistent with our identification of H derived in part (a).

Chapter 8: Integrals and Derivatives on Curves

Section 1: LINE INTEGRALS

Exercise Set 1AB (pgs. 376-377)

1. With L parametrized by $g(t) = (t, t, t)$, for $0 \leq t \leq 1$, and $\mathbf{F}(x, y, z) = (x, x^2, y)$, we have $\mathbf{F}\big(g(t)\big) = \mathbf{F}(t, t, t) = (t, t^2, t)$ and $g'(t)\, dt = (1, 1, 1)\, dt = (dx, dy, dz)$, so that

$$\int_L x\, dx + x^2 dy + y\, dz = \int_L \mathbf{F}\big(g(t)\big) \cdot g'(t)\, dt = \int_0^1 (t, t^2, t) \cdot (1, 1, 1)\, dt$$

$$= \int_0^1 (2t + t^2)\, dt = \left[t^2 + \frac{1}{3} t^3 \right]_0^1 = \frac{4}{3}.$$

3. Let $\mathbf{F}(x, y) = (0, x)$ and let γ_1 be parametrized by $g(t) = (\cos t, \sin t)$, for $0 \leq t \leq 2\pi$. Then $\mathbf{F}\big(g(t)\big) = \mathbf{F}(\cos t, \sin t) = (0, \cos t)$ and $g'(t)\, dt = (-\sin t, \cos t)\, dt = (dx, dy)$, so that

$$\int_{\gamma_1} x\, dy = \int_{\gamma_1} (0, x) \cdot (dx, dy) = \int_0^{2\pi} (0, \cos t) \cdot (-\sin t, \cos t)\, dt$$

$$= \int_0^{2\pi} \cos^2 t\, dt = \int_0^{2\pi} \frac{1}{2}(1 + \cos 2t)\, dt = \frac{1}{2}\left[t + \frac{1}{2} \sin 2t \right]_0^{2\pi} = \pi.$$

The same integral over γ_2 parametrized by $h(t) = (\cos t, \sin t)$, for $0 \leq t \leq 4\pi$, uses the same $\mathbf{F}$ and the same form for h as was used for g above. The only difference is that the upper limit of integration is 4π instead of 2π. It follows that

$$\int_{\gamma_2} x\, dy = \int_0^{4\pi} \cos^2 t\, dt = \frac{1}{2}\left[t + \frac{1}{2} \sin 2t \right]_0^{4\pi} = 2\pi.$$

5. With $\mathbf{F}(x, y) = \big(1/(x^2 + y^2), 1/(x^2 + y^2)\big)$ and γ_1 parametrized by $(x, y) = (\cos t, \sin t)$, for $0 \leq t \leq 2\pi$, we have $\mathbf{F}\big(g(t)\big) = \mathbf{F}(\cos t, \sin t) = (1, 1)$ and $(-\sin t, \cos t)\, dt = (dx, dy)$. Hence,

$$\int_{\gamma_1} \frac{dx + dy}{x^2 + y^2} = \int_{\gamma_1} \left(\frac{1}{x^2 + y^2}, \frac{1}{x^2 + y^2} \right) \cdot (dx, dy)$$

$$= \int_0^{2\pi} (1, 1) \cdot (-\sin t, \cos t)\, dt = \int_0^{2\pi} (-\sin t + \cos t)\, dt = 0,$$

where the last equality follows because the cosine and sine functions have period 2π.

7. With $\mathbf{F}(x, y, z) = (z, x, y)$ and γ parametrized by $(x, y, z) = (\cos t, \sin t, t)$, $0 \leq t \leq 2\pi$, we have (on γ) $\mathbf{F}(x, y, z) = (t, \cos t, \sin t)$ and $d\mathbf{x} = (dx, dy, dz) = (-\sin t, \cos t, 1)\, dt$. Hence,

$$\int_\gamma \mathbf{F} \cdot d\mathbf{x} = \int_0^{2\pi} (t, \cos t, \sin t) \cdot (-\sin t, \cos t, 1)\, dt$$

$$= \int_0^{2\pi} (-t \sin t + \cos^2 t + \sin t)\, dt = \int_0^{2\pi} \left(-t \sin t + \frac{1}{2}(1 + \cos 2t) + \sin t \right) dt$$

$$= \left[t \cos t - \sin t + \frac{t}{2} + \frac{1}{4} \sin 2t - \cos t \right]_0^{2\pi} = 3\pi.$$

Note: *In Exercises 9 and 11 below, γ_1 is given by $(x,y) = (\cos t, \sin t)$, for $0 \le t \le \pi/2$, and γ_2 is given by $(x,y) = (1-u, u)$, for $0 \le u \le 1$. Hence, $(dx, dy) = (-\sin t, \cos t)\,dt$ on γ_1 and $(dx, dy) = (-1,1)\,du$ on γ_2. Further, for the given functions f and g, we let $\mathbf{F}(x,y) = \big(f(x,y), g(x,y)\big)$, so that*

$$f\,dx + g\,dy = (f,g)\cdot(dx,dy) = \mathbf{F}\cdot(-\sin t, \cos t)\,dt \qquad \text{(on } \gamma_1\text{)},$$
$$f\,dx + g\,dy = (f,g)\cdot(dx,dy) = \mathbf{F}\cdot(-1,1)\,du \qquad \text{(on } \gamma_2\text{)}.$$

9. Here $\mathbf{F}(x,y) = (x, x+1)$, so that $\mathbf{F}(x,y) = (\cos t, \cos t + 1)$ on γ_1. Hence, on γ_1,

$$\int_{\gamma_1} (f\,dx + g\,dy) = \int_0^{\pi/2} (\cos t, \cos t + 1)\cdot(-\sin t, \cos t)\,dt$$

$$= \int_0^{\pi/2} (-\sin t \cos t + \cos^2 t + \cos t)\,dt$$

$$= \int_0^{\pi/2} \left(-\sin t \cos t + \frac{1}{2}(1 + \cos 2t) + \cos t\right) dt$$

$$= \left[\frac{1}{2}\cos^2 t + \frac{t}{2} + \frac{1}{4}\sin 2t + \sin t\right]_0^{\pi/2} = \frac{1}{2} + \frac{\pi}{4}.$$

On γ_2 we have $\mathbf{F}(x,y) = (1-u, 2-u)$, so that

$$\int_{\gamma_2} (f\,dx + g\,dy) = \int_0^1 (1-u, 2-u)\cdot(-1,1)\,du = \int_0^1 1\,du = 1.$$

11. Here, $\mathbf{F}(x,y) = \big(1/(x^2+y^2), 1/(x^2+y^2)\big)$, so that $\mathbf{F}(x,y) = (1,1)$ on γ_1. Hence, on γ_1,

$$\int_{\gamma_1} (f\,dx + g\,dy) = \int_0^{\pi/2} (1,1)\cdot(-\sin t, \cos t)\,dt = \int_0^{\pi/2} (-\sin t + \cos t)\,dt$$

$$= \big[\cos t + \sin t\big]_0^{\pi/2} = 0.$$

On γ_2 we have $\mathbf{F}(x,y) = \big(1/(2u^2 - 2u + 1), 1/(2u^2 - 2u + 1)\big)$, so that

$$\int_{\gamma_1} (f\,dx + g\,dy) = \int_0^1 \left(\frac{1}{2u^2 - 2u + 1}, \frac{1}{2u^2 - 2u + 1}\right)\cdot(-1,1)\,du = \int_0^1 0\,du = 0.$$

13. Letting $\mathbf{F}(x,y) = \big(F_1(x,y), F_2(x,y)\big) = (x^2, y^2)$, we observe that the gradient of the function $f(x,y) = \frac{1}{3}x^3 + \frac{1}{3}y^3$ satisfies $\nabla f = \mathbf{F}$. The curve γ_2 given by $(x,y) = (1-u, u)$, $0 \le u \le 1$, has initial and terminal points $\mathbf{a} = (1,0)$ and $\mathbf{b} = (0,1)$, respectively, so that $f(\mathbf{a}) = f(\mathbf{b}) = \frac{1}{3}$. By Theorem 1.3,

$$\int_{\gamma_2} (F_1\,dx + F_2\,dy) = \int_{\gamma_2} \mathbf{F}\cdot d\mathbf{x} = \int_{\mathbf{a}}^{\mathbf{b}} \nabla f \cdot d\mathbf{x} = f(\mathbf{b}) - f(\mathbf{a}) = 0.$$

15. Letting $\mathbf{F}(x,y) = \big(F_1(x,y), F_2(x,y)\big) = (\sin y, x \cos y)$, we observe that the gradient of the function $f(x,y) = x \sin y$ satisfies $\nabla f = \mathbf{F}$. The curve γ_2 given by $(x,y) = (1-u, u)$,

$0 \le u \le 1$, has initial and terminal points $\mathbf{a} = (1,0)$ and $\mathbf{b} = (0,1)$, respectively, so that $f(\mathbf{a}) = f(\mathbf{b}) = 0$. By Theorem 1.3,

$$\int_{\gamma_2} (F_1\, dx + F_2\, dy) = \int_{\gamma_2} \mathbf{F} \cdot d\mathbf{x} = \int_{\mathbf{a}}^{\mathbf{b}} \nabla f \cdot d\mathbf{x} = f(\mathbf{b}) - f(\mathbf{a}) = 0.$$

17. Consider moving a particle along a curve γ defined by $(x, y, z) = (t, t^2, t^3)$, $0 \le t \le 2$, under the influence of the force field $\mathbf{F}(x, y, z) = (x+y, y, y)$. On γ, $\mathbf{F}(x, y, z) = (t+t^2, t^2, t^2)$. Moreover, $(dx, dy, dz) = (1, 2t, 3t^2)\, dt$. Hence, the work done in moving the particle along γ is

$$W = \int_{\gamma} \mathbf{F} \cdot d\mathbf{x} = \int_0^2 (t + t^2, t^2, t^2) \cdot (1, 2t, 3t^2)\, dt = \int_0^2 (t + t^2 + 2t^3 + 3t^4)\, dt$$

$$= \left[\frac{1}{2}t^2 + \frac{1}{3}t^3 + \frac{1}{2}t^4 + \frac{3}{5}t^5 \right]_0^2 = \frac{478}{15}.$$

19. (a) The vector field $\mathbf{F}(x, y) = (y, x)$ and the curve γ defined by $g(t) = (e^t, e^{-t})$, for $0 \le t \le 1$, are shown on the right.
 (b) We have

$$\int_{\gamma} \mathbf{F}\big(g(t)\big) \cdot g'(t)\, dt = \int_0^1 (e^{-t}, e^t) \cdot (e^t, -e^{-t})\, dt$$

$$= \int_0^1 (1 - 1)\, dt = \int_0^1 0\, dt = 0.$$

Note that this result could have been anticipated because the sketch on the right suggests that the curve parametrized by g appears to be perpendicular to the flow of the vector field.

21. Consider the two parametrizations

$$f(t) = (t^{1/2}, t^{3/2}), \quad 1 \le t \le 2, \qquad \text{and} \qquad g(u) = (u, u^3), \quad 1 \le u \le \sqrt{2}.$$

In both parametrizations, the second coordinate is the cube of the first coordinate so that each is a parametrization of some part of the graph of the cubic equation $y = x^3$.

To show that the parametrizations are equivalent, let $\phi(u) = u^2$ and note that ϕ is continuously differentiable. It is straightforward to observe that $\phi(1) = 1$, $\phi(\sqrt{2}) = 2$ and that $\phi'(u) = 2u > 0$, for $1 < u < \sqrt{2}$. Finally,

$$f\big(\phi(u)\big) = f(u^2) = \big((u^2)^{1/2}, (u^2)^{3/2}\big) = (u, u^3) = g(u),$$

Hence, f and g are equivalent parametrizations.

23. (a) Suppose $g(t)$ is the position of a particle in $\mathcal{R}^3$ at time t and that its mass is given by the differentiable real-valued function $m(t)$. Then its velocity vector is given by $\mathbf{v}(t) = g'(t)$. Given that $\mathbf{F}\big(g(t)\big) = \big[m(t)\mathbf{v}(t)\big]'$, we have

$$\mathbf{F}\big(g(t)\big) = \frac{d}{dt}\big(m(t)\mathbf{v}(t)\big) = m'(t)\mathbf{v}(t) + m(t)\mathbf{v}'(t).$$

Using the fact that $\mathbf{v}(t) = g'(t)$, the dot product of the left hand member with $g'(t)$ is equal to the dot product of the righthand member with $\mathbf{v}(t)$. So,

$$\mathbf{F}\big(g(t)\big) \cdot g'(t) = \Big(m'(t)\mathbf{v}(t) + m(t)\mathbf{v}'(t)\Big) \cdot \mathbf{v}(t) = m'(t)\Big(\mathbf{v}(t) \cdot \mathbf{v}(t)\Big) + m(t)\Big(\mathbf{v}'(t) \cdot \mathbf{v}(t)\Big).$$

If $v(t)$ denotes the speed of the particle at time t then the parenthetical factor of the first term on the right is $v^2(t)$. Also, $v'(t)$ is the acceleration of the particle in the direction of motion; i.e., in the direction of the unit tangent vector $\mathbf{v}(t)/|\mathbf{v}(t)|$. This can be stated in vector form as

$$v'(t) = \mathbf{v}'(t) \cdot \frac{\mathbf{v}(t)}{|\mathbf{v}(t)|}, \quad \text{so that} \quad \mathbf{v}'(t) \cdot \mathbf{v}(t) = v'(t)\,|\mathbf{v}(t)| = v'(t)v(t).$$

In other words, the parenthetical factor of the second term above is $v'(t)v(t)$. So,

$$\mathbf{F}\big(g(t)\big) \cdot g'(t) = m'(t)v^2(t) + m(t)v'(t)v(t).$$

(b) If $m(t) = m$ is constant then $m'(t) = 0$ and the result of part (a) becomes

$$\mathbf{F}\big(g(t)\big) \cdot g'(t) = m\,v'(t)v(t).$$

By definition, the work done in moving the particle along γ between the times $t = a$ and $t = b$ is

$$W = \int_a^b \mathbf{F}\big(g(t) \cdot g'(t)\big)\, dt = \int_a^b m\,v'(t)v(t)\, dt$$

$$= m \int_{v(a)}^{v(b)} u\, du = m \left[\frac{1}{2}u^2\right]_{v(a)}^{v(b)} = \frac{m}{2}\big(v^2(b) - v^2(a)\big),$$

where the first integral in the second line is obtained via the substitution $u = v(t)$.

25. The vector field $\mathbf{F}(x, y) = (-y, x)$ is shown on the right. The first thing to notice is that the the flow lines are concentric circles about the origin. Thus, if γ is any path in the field and (x, y) is a point on γ then rotating γ about the origin to a new fixed position will not affect the geometric relationship between the direction in which the path is traced out at (x, y) and the direction in which the field acts at the point (x, y). With this in mind, let γ be an ellipse with center at the origin. By what we have just said, we can assume that the axes of the ellipse are parallel to the coordinate axes (as shown in the drawing).

(x,y)=(-y,x)

For a particle moving counterclockwise around γ, we can see from the drawing that the angle θ between the velocity vector of the particle's motion and the associated field vector always satisfies $0 \le \theta < \pi/2$. The significance of this is that if $g(t)$ is any continuously differentiable parametrization of γ then the dot product $\mathbf{F}\big(g(t)\big) \cdot g'(t)$ is positive for all t. It follows that the integral of $\mathbf{F}$ along γ is also positive. If the parametrization moves the particle in the opposite direction then the dot product is always negative, so that the integral of $\mathbf{F}$ along γ is negative. In any case, $\int_\gamma \mathbf{F} \cdot d\mathbf{x} \ne 0$, for all ellipses γ centered at the origin.

27. (a) The vector field $\mathbf{G}(x, y) = (-\frac{1}{2}y, \frac{1}{2}x)$ is a scaled version of the vector field $\mathbf{F}(x, y) = (-y, x)$ given in Exercise 25 above. The drawing of $\mathbf{F}$ provided in Exercise 25 was scaled in order to display it more clearly. Consequently, there is no need to provide another scaled version here. That is, for visual aid purposes, we can take the field shown in Exercise 25 to be the field $\mathbf{G}$.

(b) If c is a circular path of radius a centered at (α, β) then the parametrization $g(t) = (\alpha + a\cos t, \beta + a\sin t)$, for $0 \le t \le 2\pi$, traces c once in a counterclockwise direction. Since $\mathbf{G}\big(g(t)\big) = \left(-\frac{\beta}{2} - \frac{a}{2}\sin t, \frac{\alpha}{2} + \frac{a}{2}\cos t\right)$ and $g'(t) = (-a\sin t, a\cos t)$, we have

$$\int_c \mathbf{F} \cdot d\mathbf{x} = \int_0^{2\pi} \left(-\frac{\beta}{2} - \frac{a}{2}\sin t, \frac{\alpha}{2} + \frac{a}{2}\cos t\right) \cdot (-a\sin t, a\cos t)\, dt$$

$$= \int_0^{2\pi} \left(\frac{a\beta}{2}\sin t + \frac{a^2}{2}\sin^2 t + \frac{a\alpha}{2}\cos t + \frac{a^2}{2}\cos^2 t\right) dt = \int_0^{2\pi} \frac{a^2}{2}\, dt = \left[\frac{a^2}{2}t\right]_0^{2\pi} = \pi a^2,$$

where the fourth integral is obtained by noting that the integral of any linear combination of the sine and cosine functions over an interval of length 2π is zero.

(c) Let r be a rectangle with sides parallel to the coordinate axes. For definiteness, we shall suppose that the four sides of r are as follows: r_1 is the line segment with endpoints (α, β) and $(a+\alpha, \beta)$; r_2 is the line segment with endpoints $(a+\alpha, \beta)$ and $(a+\alpha, b+\beta)$; r_3 is the line segment with endpoints $(a+\alpha, b+\beta)$ and $(\alpha, b+\beta)$; and r_4 is the line segment with endpoints $(\alpha, b + \beta)$ and (α, β). Hence, the dimensions of r are a units in the x direction and b units in the y direction. Consider the following four parametrizations:

$$g_1(t) = (t + \alpha, \beta), \quad 0 \le t \le a; \qquad g_3(t) = (a + \alpha - t, b + \beta), \quad 0 \le t \le a;$$
$$g_2(t) = (a + \alpha, t + \beta), \quad 0 \le t \le b; \qquad g_4(t) = (\alpha, b + \beta - t), \quad 0 \le t \le b.$$

Notice that g_1 traces out r_1 from left to right, g_2 traces out r_2 from bottom to top, g_3 traces out r_3 from right to left, and g_4 traces out r_4 from top to bottom. In other words, taken in order, the four parametrizations trace out r once in a counterclockwise direction. Moreover, on the four sides we have

$$\mathbf{G}\big(g_1(t)\big) = \left(-\frac{\beta}{2}, \frac{t}{2} + \frac{\alpha}{2}\right) \quad \text{and} \quad g_1'(t) = (1, 0);$$

$$\mathbf{G}\big(g_2(t)\big) = \left(-\frac{t}{2} - \frac{\beta}{2}, \frac{1}{2}(a + \alpha)\right) \quad \text{and} \quad g_2'(t) = (0, 1);$$

$$\mathbf{G}\big(g_3(t)\big) = \left(-\frac{1}{2}(b + \beta), \frac{1}{2}(a + \alpha) - \frac{t}{2}\right) \quad \text{and} \quad g_3'(t) = (-1, 0);$$

$$\mathbf{G}\big(g_4(t)\big) = \left(-\frac{1}{2}(b + \beta) + \frac{t}{2}, \frac{\alpha}{2}\right) \quad \text{and} \quad g_4'(t) = (0, -1).$$

Therefore,

$$\mathbf{G}\big(g_1(t)\big) \cdot g_1'(t) = -\frac{\beta}{2}; \qquad \mathbf{G}\big(g_2(t)\big) \cdot g_2'(t) = \frac{1}{2}(a + \alpha);$$
$$\mathbf{G}\big(g_3(t)\big) \cdot g_3'(t) = \frac{1}{2}(b + \beta); \qquad \mathbf{G}\big(g_4(t)\big) \cdot g_4'(t) = -\frac{\alpha}{2}.$$

So,

$$\int_{r_1} \mathbf{G} \cdot d\mathbf{x} = \int_0^a -\frac{\beta}{2}\, dt = \left[-\frac{\beta}{2}t\right]_0^a = -\frac{\beta a}{2};$$

$$\int_{r_2} \mathbf{G} \cdot d\mathbf{x} = \int_0^b \frac{1}{2}(a+\alpha)\, dt = \left[\frac{1}{2}(a+\alpha)t\right]_0^b = \frac{b}{2}(a+\alpha);$$

$$\int_{r_3} \mathbf{G} \cdot d\mathbf{x} = \int_0^a \frac{1}{2}(b+\beta)\, dt = \left[\frac{1}{2}(b+\beta)t\right]_0^a = \frac{a}{2}(b+\beta);$$

$$\int_{r_4} \mathbf{G} \cdot d\mathbf{x} = \int_0^b -\frac{\alpha}{2}\, dt = \left[-\frac{\alpha}{2}t\right]_0^b = -\frac{\alpha b}{2}.$$

And finally, since r_1, r_2, r_3, r_4 are essentially disjoint curves (except at the corners of r),

$$\int_r \mathbf{G} \cdot d\mathbf{x} = \int_{r_1} \mathbf{G} \cdot d\mathbf{x} + \int_{r_2} \mathbf{G} \cdot d\mathbf{x} + \int_{r_3} \mathbf{G} \cdot d\mathbf{x} + \int_{r_4} \mathbf{G} \cdot d\mathbf{x}$$

$$= -\frac{\beta a}{2} + \frac{b}{2}(a+\alpha) + \frac{a}{2}(b+\beta) - \frac{\alpha b}{2} = ab.$$

(d) Let a and b be positive constants and let γ be the triangle with vertices $(0,0)$, $(a,0)$ and $(0,b)$. The segments of γ on the x-axis and the y-axis are perpendicular to the vector field $\mathbf{G}$, so that the integrals of $\mathbf{G}$ over these parts of γ are zero. Thus, the integral of $\mathbf{G}$ around γ in a counterclockwise direction is equal to the integral of $\mathbf{G}$ along that part of γ (call it γ_1) traced out from $(a,0)$ to $(0,b)$. Note that the function $g(t) = (a-t, bt/a)$, $0 \le t \le a$, parametrizes γ_1 in a counterclockwise direction. Moreover,

$$\mathbf{G}\big(g(t)\big) = \left(-\frac{bt}{2a}, \frac{1}{2}(a-t)\right) \qquad \text{and} \qquad d\mathbf{x} = g'(t)\, dt = \left(-1, \frac{b}{a}\right) dt.$$

Hence,

$$\int_\gamma \mathbf{G} \cdot d\mathbf{x} = \int_{\gamma_1} \mathbf{G} \cdot d\mathbf{x} = \int_0^a \left(-\frac{bt}{2a}, \frac{1}{2}(a-t)\right) \cdot \left(-1, \frac{b}{a}\right) dt = \int_0^a \frac{b}{2}\, dt = \left[\frac{b}{2}t\right]_0^a = \frac{1}{2}ab.$$

In each case above, the curve in question is a closed curve and the integral of $\mathbf{G}$ along the curve in a counterclockwise direction is the area of the region enclosed by the curve.
29. Let $f(x,y,z)) = x - y^2 + z$ and let γ be parametrized by $g(t) = (t, t^2, -t^2)$, $0 \le t \le 2$. With $\mathbf{a} = g(0) = (0,0,0)$ and $\mathbf{b} = g(2) = (2,4,-4)$, Theorem 1.3 gives

$$\int_\gamma \nabla f \cdot d\mathbf{x} = \int_{(0,0,0)}^{(2,4,-4)} \nabla f \cdot d\mathbf{x} = f(2,4,-4) - f(0,0,0) = -18 - 0 = -18.$$

31. Let $f(x,y,z) = (x-y+z)^2$ and let γ be one turn of the helix from $(1,0,0)$ to $(1,0,4)$. With $\mathbf{a} = (1,0,0)$ and $\mathbf{b} = (1,0,4)$, Theorem 1.3 gives

$$\int_\gamma \nabla f \cdot d\mathbf{x} = \int_{(1,0,0)}^{(1,0,4)} \nabla f \cdot d\mathbf{x} = f(1,0,4) - f(1,0,0) = 25 - 1 = 24.$$

33. (a) The vector field $\mathbf{F}(x,y) = (-y, x)$ is shown on the right.

(b) Let $\mathbf{a} = (1,0)$ and $\mathbf{b} = (-1,0)$. Let γ_1 be the line segment joining $\mathbf{a}$ and $\mathbf{b}$ and let γ_2 be the half circle of radius 1 with endpoints $\mathbf{a}$ and $\mathbf{b}$ $(y \geq 0)$. Since the path γ_1 is perpendicular to the vector field $\mathbf{F}$,

$$\int_{\gamma_1} \mathbf{F} \cdot d\mathbf{x} = 0.$$

Alternatively, γ_2 is traced out in a counterclockwise direction by the parametrization $g(t) = (\cos t, \sin t)$, $0 \leq t \leq \pi$. Since $\mathbf{F}\big(g(t)\big) = (-\sin t, \cos t) = g'(t)$, we have

$$\int_{\gamma_2} \mathbf{F} \cdot d\mathbf{x} = \int_0^\pi (-\sin t, \cos t) \cdot (-\sin t, \cos t)\, dt = \int_0^\pi 1\, dt = \pi.$$

If $\mathbf{F}$ were a gradient field then the value of the integrals of $\mathbf{F}$ along γ_1 and γ_2 would be the same. Since this is not the case, $\mathbf{F}$ can't be the gradient of a real-valued function f.

(c) Let $\mathbf{a}$ and $\mathbf{b}$ be as in part (a) and let γ be the closed path that starts at $\mathbf{a}$, proceeds to $\mathbf{b}$ along the half-circle of radius 1 in the upper half-plane, and then goes back to $\mathbf{a}$ along the x-axis. By part (a), the integral of $\mathbf{F}$ along the half-circle is π and the integral of $\mathbf{F}$ along the x-axis is zero. Therefore, $\int_\gamma \mathbf{F} \cdot d\mathbf{x} = \pi$. Since the integral of a gradient field over a closed curve is zero, it follows that $\mathbf{F}$ is not a gradient field; i.e., $\mathbf{F}$ is not the gradient of a real-valued function f.

Section 2: WEIGHTED CURVES AND SURFACES OF REVOLUTION

Exercise Set 2 (pgs. 382-383)

1. Let γ be parametrized by $g(t) = (t, \ln \cos t)$, $0 \le t \le 1$. Since $g'(t) = (1, -\tan t)$, we have

$$|g'(t)| = |(1, -\tan t)| = \sqrt{1 + \tan^2 t} = \sqrt{\sec^2 t} = |\sec t| = \sec t,$$

where the absolute value sign on the secant function is not needed because $\sec t > 0$ for $0 \le t \le 1$. Hence, the arc length of γ is

$$l(\gamma) = \int_0^1 |g'(t)| \, dt = \int_0^1 \sec t \, dt = \left[\ln |\sec t + \tan t| \right]_0^1 = \ln(\sec 1 + \tan 1).$$

3. Let γ be the curve defined by $y = x^{3/2}$, $0 \le x \le 5$. A parametrization of γ is given by $g(t) = (t, t^{3/2})$, $0 \le t \le 5$. Since, $g'(t) = \left(1, \frac{3}{2} t^{1/2}\right)$, we have

$$|g'(t)| = \left| \left(1, \frac{3}{2} t^{1/2}\right) \right| = \sqrt{1 + \frac{9}{4} t}.$$

The arc length of γ is then

$$l(\gamma) = \int_0^5 |g'(t)| \, dt = \int_0^5 \sqrt{1 + \frac{9}{4} t} \, dt = \left[\frac{8}{27} \left(1 + \frac{9}{4} t\right)^{3/2} \right]_0^5 = \frac{335}{27}.$$

5. Let γ be a curve defined in polar coordinates by a function $r = f(\theta)$ for $a \le \theta \le b$. In rectangular coordinates, we have

$$(x, y) = (r \cos \theta, r \sin \theta) = (f(\theta) \cos \theta, f(\theta) \sin \theta), \quad a \le \theta \le b.$$

(a) Let $g(\theta) = (f(\theta) \cos \theta, f(\theta) \sin \theta)$, for $a \le \theta \le b$. Then

$$g'(\theta) = (-f(\theta) \sin \theta + f'(\theta) \cos \theta, f(\theta) \cos \theta + f'(\theta) \sin \theta)$$

and

$$|g'(\theta)| = \left| (-f(\theta) \sin \theta + f'(\theta) \cos \theta, f(\theta) \cos \theta + f'(\theta) \sin \theta) \right|$$
$$= \sqrt{(-f(\theta) \sin \theta + f'(\theta) \cos \theta)^2 + (f(\theta) \cos \theta + f'(\theta) \sin \theta)^2}.$$

When the terms under the radical are multiplied out, the two terms involving the product $f(\theta) f'(\theta)$ cancel each other. The identity $\cos^2 \theta + \sin^2 \theta = 1$ can then be used to reduce the remaining four terms to get the two terms $\left(f(\theta)\right)^2$ and $\left(f'(\theta)\right)^2$. That is,

$$|g'(\theta)| = \sqrt{\left(f(\theta)\right)^2 + \left(f'(\theta)\right)^2}.$$

(b) Let γ be the curve defined by $r = f(\theta) = 1 + \cos \theta$, $0 \le \theta \le \pi$. Since $f'(\theta) = -\sin \theta$, we have

$$\sqrt{\left(f(\theta)\right)^2 + \left(f'(\theta)\right)^2} = \sqrt{(1 + \cos \theta)^2 + (-\sin \theta)^2} = \sqrt{2 + 2 \cos \theta} = 2 \sqrt{\frac{1 + \cos \theta}{2}}$$
$$= 2\sqrt{\cos^2(\theta/2)} = 2|\cos(\theta/2)| = 2\cos(\theta/2),$$

297

where the first expression in the second line follows from the half-angle identity for the cosine, and the absolute value sign has been removed from the cosine in the last member because $\cos(\theta/2) \geq 0$ for $0 \leq \theta \leq \pi$. Hence, the result of part (a) says that the arc length of γ is

$$l(\gamma) = \int_0^\pi \sqrt{(f(\theta))^2 + (f'(\theta))^2}\, d\theta$$

$$= 2\int_0^\pi \cos(\theta/2)\, d\theta = 4\big[\sin(\theta/2)\big]_0^\pi = 4.$$

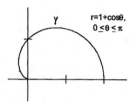

A sketch of γ is shown on the right.

7. Consider the spiral parametrized by $g(t) = (a\cos t, a\sin t, bt)$, $0 \leq t \leq 2\pi$, where a and b are constants. Let $\mu(x, y, z) = x^2 + y^2 + z^2$ be the density per unit length at the point (x, y, z) on the spiral. In terms of the parameter t, the density is given by

$$\mu\big(g(t)\big) = \mu(a\cos t, a\sin t, bt) = a^2\cos^2 t + a^2\sin^2 t + b^2 t^2 = a^2 + b^2 t^2.$$

Moreover, $g'(t) = (-a\sin t, a\cos t, b)$, so that $|g'(t)| = \sqrt{a^2 + b^2}$. The total mass of the spiral is then

$$M = \int_0^{2\pi} \mu\big(g(t)\big)|g'(t)|\, dt = \sqrt{a^2 + b^2}\int_0^{2\pi} (a^2 + b^2 t^2)\, dt$$

$$= \sqrt{a^2 + b^2}\left[a^2 t + \frac{b^2}{3}t^3\right]_0^{2\pi} = \sqrt{a^2 + b^2}\left(2\pi a^2 + \frac{8b^2\pi^3}{3}\right).$$

9. Let γ be the curve defined by $g(t) = (x, y, z) = (6t^2, 4\sqrt{2}\,t^3, 3t^4)$, $0 \leq t \leq 1$. We then have $g'(t) = (12t, 12\sqrt{2}\,t^2, 12t^3)$ and $|g'(t)| = 12t(1 + t^2)$. Suppose that a wire is bent into the shape of γ.

(a) If the density of the wire at the point corresponding to t is $\mu(t) = t^2$ then the mass of the wire is

$$M = \int_0^1 \mu(t)|g'(t)|\, dt = \int_0^1 (t^2)(12t(1 + t^2))\, dt = 12\int_0^1 (t^3 + t^5)\, dt = 5.$$

(b) Suppose the density $\mu(x, y, z)$ of the wire at the point (x, y, z) is the square of the distance to the yz-plane. For a given point $(x, y, z) = (6t^2, 4\sqrt{2}\,t^3, 3t^4)$ on the wire, the point on the yz-plane that is closest to (x, y, z) is $(0, y, z) = (0, 4\sqrt{2}\,y^3, 3t^4)$. Hence,

$$\mu\big(g(t)\big) = \mu(x, y, z) = |(x, y, z) - (0, y, z)|^2 = |(x, 0, 0)|^2 = |(6t^2, 0, 0)|^2 = 36t^4$$

and the mass of the wire is

$$M = \int_0^1 \mu\big(g(t)\big)|g'(t)|\, dt = \int_0^1 (36t^4)(12t(1 + t^2))\, dt = 432\int_0^1 (t^5 + t^7)\, dt = 126.$$

11. Let γ_a be the semicircle of radius a centered at the origin in $\mathcal{R}^2$ with $y \geq 0$. Rotating γ_a about the x-axis generates a sphere S_a of radius a centered at the origin in $\mathcal{R}^3$.

(a) Parametrize γ_a by $g(t) = (a\cos t, a\sin t)$, for $0 \leq t \leq \pi$. Since the arc length differential is $ds = |g'(t)|\,dt = |(-a\sin t, a\cos t)|\,dt = a\,dt$, and since $r(s(t)) = a\sin t$ (the y-coordinate of $g(t)$), the integral in Equation 2.2 can be written in terms of t to get

$$\sigma(S_a) = \int_0^\pi 2\pi\,(a\sin t)\,a\,dt = 2\pi a^2 \big[-\cos t\big]_0^\pi = 4\pi a^2.$$

(b) Parametrize γ_a by $g(t) = \left(t, \sqrt{a^2 - t^2}\right)$, $-a \leq t \leq a$. Since the arc length differential is given by

$$ds = |g'(t)|\,dt = \left|\left(1, \frac{-t}{\sqrt{a^2 - t^2}}\right)\right|\,dt = \sqrt{1 + \frac{t^2}{a^2 - t^2}}\,dt = \sqrt{\frac{a^2}{a^2 - t^2}}\,dt = \frac{a}{\sqrt{a^2 - t^2}}\,dt,$$

and since $r(s(t)) = \sqrt{a^2 - t^2}$ (the y-coordinate of $g(t)$), the integral in Equation 2.2, can be written in terms of t to get

$$\sigma(S_a) = \int_{-a}^a 2\pi\sqrt{a^2 - t^2}\,\frac{a}{\sqrt{a^2 - t^2}}\,dt = 2\pi a \int_{-a}^a dt = 2\pi a\big[t\big]_{-a}^a = 4\pi a^2.$$

13. (a) For positive constants a and b, let γ be the ellipse in $\mathcal{R}^2$ parametrized by

$$(x, y) = (a\cos t, b\sin t), \quad 0 \leq t \leq 2\pi.$$

By definition, the arc length of γ is

$$l(\gamma) = \int_0^{2\pi} |(a\cos t, b\sin t)|\,dt = \int_0^{2\pi} \sqrt{a^2\cos^2 t + b^2\sin^2 t}\,dt.$$

(b) Assuming $a > b$, the integrand of the arc length integral derived in part (a) can be written as

$$\sqrt{a^2\cos^2 t + b^2\sin^2 t} = a\sqrt{\cos^2 + (b^2/a^2)\sin^2 t}$$
$$= a\sqrt{1 - \sin^2 t + (b^2/a^2)\sin^2 t}$$
$$= a\sqrt{1 - (1 - b^2/a^2)\sin^2 t}$$
$$= a\sqrt{1 - k^2\sin^2 t}, \quad \text{where } k^2 = (1 - b^2/a^2).$$

Further, we observe that the ellipse is symmetric with respect to the origin, so that the arc length of the part of the ellipse in the first quadrant is the same as the arc length of the part of the ellipse in any other quadrant. This means that the we need only compute the arc length for $0 \leq t \leq \pi/2$ and multiply the result by 4. Putting these two ideas together gives

$$l(\gamma) = 4a \int_0^{\pi/2} \sqrt{1 - k^2\sin^2 t}\,dt, \quad \text{where } k^2 = (1 - b^2/a^2).$$

(c) With $a = 2$ and $b = 1$, we have $k^2 = 3/4$ and

$$l(\gamma) = 8 \int_0^{\pi/2} \sqrt{1 - (3/4)\sin^2 t}\,dt.$$

299

From a CRC <u>Standard Mathematical Tables</u>, in the section on "Elliptic integrals of the second kind" one finds that the above integral evaluates to 1.2111 (to four decimal places). Multiplying this result by 8 gives $l(\gamma) \approx 9.6888$.

15. Consider the helix $g(t) = (\cos t, \sin t, t)$. We have

$$g'(t) = (-\sin t, \cos t, 1), \quad \text{so that} \quad |g'(t)| = \sqrt{\sin^2 t + \cos^2 t + 1} = \sqrt{2}.$$

For each $t \geq 0$, the arc length parameter s then satisfies

$$s = \int_0^t |g'(t)|\, dt = \int_0^t \sqrt{2}\, dt = \sqrt{2}\, t.$$

Solving for t gives $t = s/\sqrt{2}$. Therefore an arc length parametrization of the helix is

$$h(s) = g\big(s/\sqrt{2}\big) = \big(\cos(s/\sqrt{2}), \sin(s/\sqrt{2}), s/\sqrt{2}\big), \quad 0 \leq s.$$

17. (a) Let γ be an arc of a circle of radius a such that the ends of the arc subtend angle θ, where $0 \leq \theta \leq 2\pi$. We can suppose that the arc is parametrized by $g(t) = (a \cos t, a \sin t)$, for $0 \leq t \leq \theta$. Since we will be using the second formulation for the centroid of γ, we first compute

$$g(t)\,|g'(t)| = (a \cos t, a \sin t)|(-a \sin t, a \cos t)| = (a \cos t, a \sin t)a = (a^2 \cos t, a^2 \sin t).$$

Using the fact that the length of an arc of a circle is the product of the radius and the angle (in radians) it subtends, we have $l(\gamma) = a\theta$. Hence, the centroid of γ is given by

$$\mathbf{p}_0 = \frac{1}{l(\gamma)} \int_0^\theta g(t)\,|g'(t)|\, dt = \frac{1}{a\theta} \int_0^\theta (a^2 \cos t, a^2 \sin t)\, dt = \frac{a}{\theta}\left(\int_0^\theta \cos t\, dt, \int_0^\theta \sin t\, dt \right)$$

$$= \frac{a}{\theta}\left(\big[\sin t\big]_0^\theta, \big[-\cos t\big]_0^\theta \right) = \frac{a}{\theta}(\sin\theta, 1 - \cos\theta) = \left(\frac{a}{\theta}\sin\theta, \frac{a}{\theta}(1 - \cos\theta) \right).$$

The distance from $\mathbf{p}_0$ to the origin is therefore the length of the vector $\mathbf{p}_0$; namely,

$$\left| \left(\frac{a}{\theta}\sin\theta, \frac{a}{\theta}(1 - \cos\theta) \right) \right| = \sqrt{ \frac{a^2}{\theta^2}\sin^2\theta + \frac{a^2}{\theta^2}(1 - 2\cos\theta + \cos^2\theta) }$$

$$= \frac{a}{\theta}\sqrt{2 - 2\cos\theta} = \frac{2a}{\theta}\sqrt{\frac{1 - \cos\theta}{2}} = \frac{2a}{\theta}\sin(\theta/2),$$

where the last equality follows from the half-angle identity for the sine. Note that the absolute value sign is not needed on $\sin(\theta/2)$ since this is nonnegative for $0 \leq \theta \leq 2\pi$. Finally, by symmetry, the centroid must lie on the line from the center of the circle to the midpoint of the arc.

(b) Let γ be a half-turn of a helix and let γ be parametrized by $g(t) = (a \cos t, a \sin t, bt)$, $0 \leq t \leq \pi$, To find the centroid of γ, we will need its arc length. We have $g'(t) = (-a \sin t, a \cos t, b)$, so that $|g'(t)| = \sqrt{a^2 \sin^2 t + a^2 \cos^2 t + b^2} = \sqrt{a^2 + b^2}$. Hence,

$$l(\gamma) = \int_0^\pi |g'(t)|\, dt = \int_0^\pi \sqrt{a^2 + b^2}\, dt = \big[\sqrt{a^2 + b^2}\, t\big]_0^\pi = \pi\sqrt{a^2 + b^2}.$$

The centroid of γ is therefore

$$\mathbf{p}_0 = \frac{1}{l(\gamma)} \int_0^\pi g(t)\,|g'(t)|\, dt = \frac{1}{\pi\sqrt{a^2 + b^2}} \int_0^\pi (a \cos t, a \sin t, bt)\sqrt{a^2 + b^2}\, dt$$

$$= \frac{1}{\pi}\left(\int_0^\pi a \cos t\, dt, \int_0^\pi a \sin t\, dt, \int_0^\pi bt\, dt \right) = \frac{1}{\pi}\left(\big[a \sin t\big]_0^\pi, \big[-a \cos t\big]_0^\pi, \left[\frac{b}{2}t^2\right]_0^\pi \right)$$

$$= (0, 2a/\pi, b\pi/2).$$

300

Section 3: NORMAL VECTORS AND CURVATURE

Exercise Set 3 (pgs. 385-386)

1. Let γ be the circle of radius $a > 0$ centered at the origin in $\mathcal{R}^2$, and let γ be parametrized by $g(t) = (a\cos t, a\sin t)$, $0 \leq t \leq 2\pi$. To find its curvature, we need $\dot{\mathbf{x}}(t)$ and its length $\dot{s}$. We have

$$\dot{\mathbf{x}}(t) = (-a\sin t, a\cos t), \qquad \text{and} \qquad \dot{s} = |\dot{\mathbf{x}}(t)| = \sqrt{a^2\cos^2 t + a^2\sin^2 t} = a.$$

Hence, the unit tangent vector $\mathbf{t}(t)$ is

$$\mathbf{t}(t) = \frac{1}{\dot{s}}\dot{\mathbf{x}}(t) = \frac{1}{a}(-a\sin t, a\cos t) = (-\sin t, \cos t).$$

Therefore, $\dot{\mathbf{t}}(t) = (-\cos t, -\sin t)$. This immediately tells us that $|\dot{\mathbf{t}}(t)| = 1$, so that

$$\kappa = \frac{1}{\dot{s}}|\dot{\mathbf{t}}(t)| = \frac{1}{a}.$$

3. Theorem 3.4 shows that the centripetal component of acceleration is

$$\mathbf{a_n} = \dot{s}^2\kappa\mathbf{n}.$$

Therefore, if $\dot{s}$ is increased by a factor $\rho > 1$ the new speed would be $\rho\dot{s}$, and if curvature κ is increased by the factor ρ then the new curvature would be $\rho\kappa$. The resulting expressions for centripetal acceleration become

$$(\rho\dot{s})^2\kappa\mathbf{n} = \rho^2(\dot{s}^2\kappa\mathbf{n}) \qquad \text{and} \qquad \dot{s}^2(\rho\kappa)\mathbf{n} = \rho(\dot{s}^2\kappa\mathbf{n}).$$

Since $\rho > 1$, $\rho^2 > \rho$, so that

$$\rho^2(\dot{s}^2\kappa\mathbf{n}) > \rho(\dot{s}^2\kappa\mathbf{n}).$$

Hence, an increase in speed by a factor greater than 1 does more to increase centripetal acceleration than does an increase in curvature by the same factor.

5. By the Cauchy-Schwarz inequality $|\dot{\mathbf{x}} \cdot \ddot{\mathbf{x}}| \leq |\dot{\mathbf{x}}|\,|\ddot{\mathbf{x}}|$. Since both sides of this inequality are nonnegative, we can square both sides and still retain the same direction on the inequality. Hence,

$$|\dot{\mathbf{x}} \cdot \ddot{\mathbf{x}}|^2 \leq |\dot{\mathbf{x}}|^2|\ddot{\mathbf{x}}|^2, \qquad \text{or equivalently,} \qquad |\dot{\mathbf{x}}|^2|\ddot{\mathbf{x}}|^2 - |\dot{\mathbf{x}} \cdot \ddot{\mathbf{x}}|^2 \geq 0.$$

Since the absolute value sign on the dot product on the right is just the ordinary absolute value of a real number, we can (because the dot product is squared) replace the absolute value with parentheses and still retain a true inequality. Thus, the expression under the radical in Equation 3.5 is always nonnegative.

7. Suppose $\mathbf{x}(t)$ describes motion along a path such that the centripetal component of acceleration is identically zero but $\dot{s} \neq 0$.

(a) By Equation 3.1, the centripetal component of acceleration is $\dot{s}\dot{\mathbf{t}}$. If this expression is identically zero and $\dot{s} \neq 0$ then $\dot{\mathbf{t}} = \mathbf{0}$. It follows that the unit tangent vector $\mathbf{t}(t)$ is a constant, say $\mathbf{t}(t) = \mathbf{c}$, where $|\mathbf{c}| = 1$ (and therefore $\mathbf{c} \neq \mathbf{0}$). So,

$$\dot{\mathbf{x}}(t) = |\dot{\mathbf{x}}(t)|\frac{\dot{\mathbf{x}}(t)}{|\dot{\mathbf{x}}(t)|} = \dot{s}(t)\mathbf{t}(t) = \dot{s}(t)\mathbf{c}.$$

301

(b) Using the fundamental theorem of calculus, we integrate both sides of the equation $\dot{\mathbf{x}}(t) = \dot{s}(t)\mathbf{c}$ from t_0 to t to get $\mathbf{x}(t) - \mathbf{x}(t_0)$ on the left and $s(t)\mathbf{c} - s(t_0)\mathbf{c}$ on the right. Hence,

$$\mathbf{x}(t) - \mathbf{x}(t_0) = s(t)\mathbf{c} - s(t_0)\mathbf{c}, \qquad \text{or equivalently,} \qquad \mathbf{x}(t) = s(t)\mathbf{c} + \mathbf{d},$$

where $\mathbf{d} = \mathbf{x}(t_0) - s(t_0)\mathbf{c}$ is a constant. That is, $\mathbf{x}(t)$ describes motion along a straight line.

9. The curve γ defined by $(x, y) = (\cos s, \sin s)$, $0 \le s \le 2\pi$, is the unit circle. Differentiating this parametrization with respect to s gives the velocity vector $(-\sin s, \cos s)$. By inspection, we see that its length is 1 for $0 \le s \le 2\pi$, so that this is an arc length parametrization. Moreover, differentiating a second time with respect to s gives the acceleration vector $(-\cos s, -\sin s)$. Evaluating both the velocity vector and the acceleration vector at $s = \pi/2$ gives the vectors $(-1, 0)$ and $(0, -1)$, respectively. The details are shown in the sketch on the right.

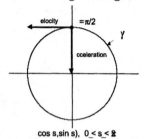

11. We view the particle as moving in $\mathcal{R}^3$, where its motion along its path γ at time t is given by $\mathbf{x}(t)$. This motion gives rise to a natural force field defined on γ. Specifically, if m is the constant mass of the particle then Newton's second law of motion says that the force exerted on the particle at time t is $m\ddot{\mathbf{x}}(t)$. It follows that the work done in moving the particle along the path between the times $t = a$ and $t = b$ is

$$W = \int_a^b m\ddot{\mathbf{x}}(t) \cdot \dot{\mathbf{x}}(t)\, dt.$$

Using Theorem 3.2, the integrand can be written as

$$m\ddot{\mathbf{x}}(t) \cdot \dot{\mathbf{x}}(t) = m\Big(\ddot{s}(t)\mathbf{t}(t) + \dot{s}(t)|\dot{\mathbf{t}}(t)|\mathbf{n}(t)\Big) \cdot \dot{\mathbf{x}}(t)$$
$$= m\ddot{s}(t)\big(\mathbf{t}(t) \cdot \dot{\mathbf{x}}(t)\big) + m\dot{s}(t)|\dot{\mathbf{t}}(t)|\big(\mathbf{n}(t) \cdot \dot{\mathbf{x}}(t)\big).$$

Since $\dot{\mathbf{x}}(t)$ and $\mathbf{n}(t)$ are perpendicular, $\mathbf{n}(t) \cdot \dot{\mathbf{x}}(t) = 0$ and the second term of the last member above vanishes. Also, since $|\mathbf{t}(t)| = 1$, we have

$$\mathbf{t}(t) \cdot \dot{\mathbf{x}}(t) = \mathbf{t}(t) \cdot \dot{s}(t)\mathbf{t}(t) = \dot{s}(t)\big(\mathbf{t}(t) \cdot \mathbf{t}(t)\big) = \dot{s}(t)|\mathbf{t}(t)|^2 = \dot{s}(t).$$

So, the work integral can be written as

$$W = \int_a^b m\ddot{s}(t)\dot{s}(t)\, dt.$$

Changing to the arc length variable $s = s(t)$, we have $ds = \dot{s}(t)\, dt$ and $d^2s/dt^2 = \ddot{s}(t)$. Also, since s_0 is the length of the path, the new limits of integration are $s(a) = 0$ and $s(b) = s_0$. Hence,

$$W = \int_0^{s_0} m\frac{d^2s}{dt^2}\, ds.$$

13. (a) Let $f(x)$ be a real-valued function that is twice differentiable on $a < x < b$ and whose graph is a smooth curve in $\mathcal{R}^2$. We can view the graph of f as a curve parametrized by $\mathbf{x}(x) = \big(x, f(x)\big)$, for $a \le x \le b$. Then

$$\dot{\mathbf{x}} = \big(1, f'(x)\big) \qquad \text{and} \qquad \ddot{\mathbf{x}} = \big(0, f''(x)\big).$$

302

These imply

$$|\dot{\mathbf{x}}|^2 = 1 + \left(f'(x)\right)^2, \qquad |\ddot{\mathbf{x}}|^2 = \left(f''(x)\right)^2, \qquad (\dot{\mathbf{x}} \cdot \ddot{\mathbf{x}})^2 = \left(f'(x)\right)^2\left(f''(x)\right)^2.$$

Inserting these expressions into Equation 3.5, we get

$$\kappa(x) = \frac{\sqrt{\left[1 + \left(f'(x)\right)^2\right]\left(f''(x)\right)^2 - \left(f'(x)\right)^2\left(f''(x)\right)^2}}{\left[1 + \left(f'(x)\right)^2\right]^{3/2}} = \frac{\sqrt{\left(f''(x)\right)^2}}{\left[1 + \left(f'(x)\right)^2\right]^{3/2}}$$

$$= \frac{|f''(x)|}{\left[1 + \left(f'(x)\right)^2\right]^{3/2}}.$$

(b) Let $f(x) = \cos x$, for $-\pi/2 \leq x \leq \pi/2$. Then $f'(x) = -\sin x$, $f''(x) = -\cos x$ and the result of part (a) becomes

$$\kappa(x) = \frac{|\cos x|}{(1 + \sin^2 x)^{3/2}} = \frac{\cos x}{(1 + \sin^2 x)^{3/2}},$$

where the absolute value sign is not needed on the cosine because $\cos x \geq 0$ on $[-\pi/2, \pi/2]$. Since differentiating $\kappa(x)$ to find its critical points is somewhat messy, we take a more direct approach. Observe that the numerator is maximum when $x = 0$ and the denominator is minimum when $x = 0$. This means that $\kappa(0) = 1$ is the maximum curvature on $[-\pi/2, \pi/2]$. Also observe that on the interval $[0, \pi/2]$ the numerator is strictly decreasing and the denominator is strictly increasing. It follows that $\kappa(x)$ is strictly decreasing on $[0, \pi/2]$, so that its minimum value on this interval is $\kappa(\pi/2) = 0$. Since curvature can never be negative, this must be the minimum value of $\kappa(x)$ on $[-\pi/2, \pi/2]$.

(c) Let $f(x) = x^4$, for $-\infty < x < \infty$. Then $f'(x) = 4x^3$, $f''(x) = 12x^2$ and the result of part (a) gives

$$\kappa(x) = \frac{12x^2}{(1 + 16x^6)^{3/2}} \quad \text{so that} \quad \kappa'(x) = \frac{24x(1 - 56x^6)}{(1 + 16x^6)^{5/2}}.$$

First note that κ is an even function of x, so that if $x_0 \geq 0$ is a critical point of κ then so is $-x_0$. Thus, we can concentrate only on the x-interval $[0, \infty)$. Setting $\kappa'(x) = 0$, we find that the critical points of κ on $[0, \infty)$ are $x = 0$ and $x = 56^{-1/6}$. Since $\kappa(0) = 0$, $x = 0$ is point of minimum curvature of f. Moreover, $\kappa'(x) > 0$ on $[0, 56^{-1/6})$ and $\kappa(x) < 0$ on $(56^{-1/6}, \infty)$. It follows that $\kappa(56^{-1/6})$ is a strict maximum of κ on $[0, \infty)$ and therefore $x = 56^{-1/6}$ is a point of maximum curvature of f on $[0, \infty)$. By our previous remarks, $\kappa(-56^{-1/6}) = \kappa(56^{-1/6})$ and $x = -56^{-1/6}$ is a point of maximum curvature of f on $(-\infty, 0]$. To summarize: $x = 0$ gives a strict global minimum curvature for f and the two values $x = \pm 56^{-1/6}$ give a strict global maximum curvature for f.

15. (a) Consider a curve in $\mathcal{R}^2$ parametrized by $\mathbf{x} = \mathbf{x}(t) = \left(x(t), y(t)\right) = (x, y)$. Assuming $x(t)$ and $y(t)$ are twice differentiable, we have $\dot{\mathbf{x}} = (\dot{x}, \dot{y})$ and $\ddot{\mathbf{x}} = (\ddot{x}, \ddot{y})$, so that

$$|\dot{\mathbf{x}}|^2|\ddot{\mathbf{x}}|^2 = (\dot{x}^2 + \dot{y}^2)(\ddot{x}^2 + \ddot{y}^2) = \dot{x}^2\ddot{x}^2 + \dot{x}^2\ddot{y}^2 + \ddot{x}^2\dot{y}^2 + \dot{y}^2\ddot{y}^2,$$

$$(\dot{\mathbf{x}} \cdot \ddot{\mathbf{x}})^2 = \left[(\dot{x}, \dot{y}) \cdot (\ddot{x}, \ddot{y})\right]^2 = (\dot{x}\ddot{x} + \dot{y}\ddot{y})^2 = \dot{x}^2\ddot{x}^2 + 2\dot{x}\ddot{x}\dot{y}\ddot{y} + \dot{y}^2\ddot{y}^2.$$

When these two expressions are subtracted from each other the terms involving only the derivatives of x cancel out and the terms involving only the derivatives of y cancel out. The difference of the two expressions is therefore

$$|\dot{\mathbf{x}}|^2|\ddot{\mathbf{x}}|^2 - (\dot{\mathbf{x}} \cdot \ddot{\mathbf{x}})^2 = \dot{x}^2\ddot{y}^2 + \ddot{x}^2\dot{y}^2 - 2\dot{x}\ddot{x}\dot{y}\ddot{y} = (\dot{x}\ddot{y} - \ddot{x}\dot{y})^2.$$

It follows that Equation 3.5 can be written as

$$\kappa(t) = \frac{\sqrt{(\dot{x}\ddot{y} - \ddot{x}\dot{y})^2}}{(\dot{x}^2 + \dot{y}^2)^{3/2}} = \frac{|\dot{x}\ddot{y} - \ddot{x}\dot{y}|}{(\dot{x}^2 + \dot{y}^2)^{3/2}}.$$

(b) Let $x = t^2$, $y = t^3$ so that $\dot{x} = 2t$, $\ddot{x} = 2$, $\dot{y} = 3t^2$, $\ddot{y} = 6t$. For $t \neq 0$, we can use the result of part (a) to get

$$\kappa = \frac{|(2t)(6t) - (3t^2)(2)|}{((2t)^2 + (3t^2)^2)^{3/2}} = \frac{|6t^2|}{(4t^2 + 9t^4)^{3/2}} = \frac{6t^2}{|t|^3(4 + 9t^2)^{3/2}} = \frac{6}{|t|(4 + 9t^2)^{3/2}}.$$

Hence, as a function of t, $\lim_{t \to 0} \kappa(t) = \infty$.

17. For an arc length parametrization, we write the unit tangent vector as $\mathbf{t}(s)$. Throughout, we consider s to be fixed at a value where $\mathbf{t}(s)$ is differentiable with respect to arc length in a neighborhood N of s. This guarantees that the curvature $\kappa(s)$ exists on N. For each real number h such that $s + h$ is in N, let $\theta(s, h)$ be the angle between $\mathbf{t}(s + h)$ and $\mathbf{t}(s)$. Keeping in mind that $|\mathbf{t}(s)|$ is identically 1 on N, we compute

$$
\begin{aligned}
|\mathbf{t}(s + h) - \mathbf{t}(s))|^2 &= (\mathbf{t}(s + h) - \mathbf{t}(s)) \cdot (\mathbf{t}(s + h) - \mathbf{t}(s)) \\
&= \mathbf{t}(s + h) \cdot \mathbf{t}(s + h) - 2\mathbf{t}(s + h) \cdot \mathbf{t}(s) + \mathbf{t}(s) \cdot \mathbf{t}(s) \\
&= |\mathbf{t}(s + h)|^2 - 2\mathbf{t}(s + h) \cdot \mathbf{t}(s) + |\mathbf{t}(s)|^2 \\
&= 2 - 2|\mathbf{t}(s + h)|\,|\mathbf{t}(s)| \cos\theta(s, h) \\
&= 2(1 - \cos\theta(s, h)) \\
&= 4\sin^2(\theta(s, h)/2),
\end{aligned}
$$

where the last line follows from the half-angle identity for the sine. Hence, assuming for the moment that $\theta(s, h) \neq 0$ for all sufficiently small nonzero h (note that this is equivalent to assuming that the curvature is nonzero in some punctured neighborhood of s), we can write

$$
\left| \frac{\mathbf{t}(s + h) - \mathbf{t}(s)}{h} \right| = \left| \frac{2\,|\sin(\theta(s, h)/2)|}{h} \right| = \left| \frac{\sin(\theta(s, h)/2)}{\theta(s, h)/2} \frac{\theta(s, h)}{h} \right|
$$

$$
= \left| \frac{\sin(\theta(s, h)/2)}{\theta(s, h)/2} \right| \left| \frac{\theta(s, h)}{h} \right|.
$$

The definition of curvature and the definition of the derivative of a vector-valued function of a real variable gives

$$
\kappa(s) = \left| \frac{d}{ds}\mathbf{t}(s) \right| = \left| \lim_{h \to 0} \frac{\mathbf{t}(s + h) - \mathbf{t}(s)}{h} \right| = \lim_{h \to 0} \left| \frac{\mathbf{t}(s + h) - \mathbf{t}(s)}{h} \right|
$$

$$
= \lim_{h \to 0} \left| \frac{\sin(\theta(s, h)/2)}{\theta(s, h)/2} \right| \left| \frac{\theta(s, h)}{h} \right| = \lim_{h \to 0} \left| \frac{\sin(\theta(s, h)/2)}{\theta(s, h)/2} \right| \lim_{h \to 0} \left| \frac{\theta(s, h)}{h} \right|
$$

$$
= \lim_{h \to 0} \left| \frac{\theta(s, h)}{h} \right|,
$$

where the second member of the second line is justified by our assumption that the curvature exists at $\kappa(s)$ and by the fact that $\theta(s, h)$ tends to zero as h tends to zero implies

$$\lim_{h \to 0} \frac{\sin\big(\theta(s, h)/2\big)}{\theta(s, h)/2} = \lim_{\theta(s,h) \to 0} \frac{\sin\big(\theta(s, h)/2\big)}{\theta(s, h)/2} = 1.$$

Finally, since $\theta(s, h)$ is a continuous function of h on N (because $\mathbf{t}(s + h)$ exists on N), it can't equal zero for infinitely many h around s unless it is identically zero on some neighborhood of s. In this case, the curve is straight around s and therefore $\kappa(s) = 0$. Since the limit shown at the bottom of the previous page is also zero (because $\theta(s, h) = 0$ for all sufficiently small h), the equation

$$\kappa(s) = \lim_{h \to 0} \left| \frac{\theta(s, h)}{h} \right|$$

holds in this case as well. We conclude that it holds in every case and for each s for which $\kappa(s)$ exists.

Section 4: FLOW LINES, DIVERGENCE AND CURL

Exercise Set 4 (pgs. 394-395)

1. With $\mathbf{F}(x, y) = \cos(xy)\,\mathbf{i} + \sin(xy)\,\mathbf{j}$, we have

$$\mathrm{div}\mathbf{F} = \frac{\partial\big(\cos(xy)\big)}{\partial x} + \frac{\partial\big(\sin(xy)\big)}{\partial y} = -y\sin(xy) + x\cos(xy),$$

$$\mathrm{curl}\mathbf{F} = \frac{\partial\big(\sin(xy)\big)}{\partial x} - \frac{\partial\big(\cos(xy)\big)}{\partial y} = y\cos(xy) + x\sin(xy).$$

3. With $\mathbf{F}(x, y) = (2x - y)\,\mathbf{i} + (x - 3y)\,\mathbf{j}$, we have

$$\mathrm{div}\mathbf{F} = \frac{\partial(2x - y)}{\partial x} + \frac{\partial(x - 3y)}{\partial y} = 2 - 3 = -1,$$

$$\mathrm{curl}\mathbf{F} = \frac{\partial(x - 3y)}{\partial x} - \frac{\partial(2x - y)}{\partial y} = 1 - (-1) = 2.$$

5. With $\mathbf{F}(x, y) = (x^2 + y^2)^2\mathbf{i} + (x^2 - y^2)^2\mathbf{j}$, we have

$$\mathrm{div}\mathbf{F} = \frac{\partial\big((x^2 + y^2)^2\big)}{\partial x} + \frac{\partial\big((x^2 - y^2)^2\big)}{\partial y}$$
$$= 4x(x^2 + y^2) - 4y(x^2 - y^2) = 4x^3 - 4xy(x - y) + 4y^3,$$

$$\mathrm{curl}\mathbf{F} = \frac{\partial\big((x^2 - y^2)^2\big)}{\partial x} - \frac{\partial\big((x^2 + y^2)^2\big)}{\partial y}$$
$$= 4x(x^2 - y^2) - 4y(x^2 + y^2) = 4x^3 - 4xy(x + y) - 4y^3.$$

7. Let $\mathbf{F}(x, y) = (x^2 + y^2)\,\mathbf{i} + (x^2 - y^2)\,\mathbf{j}$, so that

$$\mathrm{div}\mathbf{F} = \frac{\partial(x^2 + y^2)}{\partial x} + \frac{\partial(x^2 - y^2)}{\partial y} = 2x - 2y.$$

Therefore, the flow is expanding when $2x - 2y > 0$, or equivalently, when $x > y$. These are the points in $\mathcal{R}^2$ that lie below the line $y = x$.

9. Let $\mathbf{F}(x, y) = e^{x+y}\mathbf{i} + e^{x-y}\mathbf{j}$, so that

$$\mathrm{div}\mathbf{F} = \frac{\partial\big(e^{x+y}\big)}{\partial x} + \frac{\partial\big(e^{x-y}\big)}{\partial y} = e^{x+y} - e^{x-y}.$$

Therefore, the flow is expanding when $e^{x+y} - e^{x-y} > 0$, or equivalently, when $e^{x+y} > e^{x-y}$. Since the exponential function is strictly increasing everywhere, this inequality is equivalent to $x + y > x - y$ or $y > 0$. This is the set of points above the x-axis in $\mathcal{R}^2$.

11. (a) Let $\mathbf{F}$ be the gradient field given in Example 3 of the text and let γ be the curve parametrized by $\mathbf{x}(t) = c_1 e^{t/4}\,\mathbf{i} + c_2 e^{t/2}\mathbf{j}$, where c_1 and c_2 are real constants. The coordinates of a point (x, y) on γ are differentiable functions of t and can be written as

$$x = x(t) = c_1 e^{t/4} \qquad \text{and} \qquad y = y(t) = c_2 e^{t/2}.$$

Differentiating these with respect to t gives

$$\dot{x} = \frac{1}{4}c_1 e^{t/4} = \frac{1}{4}x \qquad \text{and} \qquad \dot{y} = \frac{1}{2}c_2 e^{t/2} = \frac{1}{2}y.$$

Hence, $\dot{\mathbf{x}} = \dot{\mathbf{x}}(t)$ is given by

$$\dot{\mathbf{x}} = \frac{1}{4}x\,\mathbf{i} + \frac{1}{2}y\,\mathbf{j} = \mathbf{F}(x,y).$$

That is, the velocity vector at each point (x,y) on γ coincides with the vector field $\mathbf{F}$. This is precisely what it means for γ to be a flow line of $\mathbf{F}$.

(b) With $\mathbf{x}(t)$ as given in part (a), if $c_1 \neq 0$ and $c_2 \neq 0$ then

$$y = c_2 e^{t/2} = \frac{c_2}{c_1^2}\left(c_1 e^{t/4}\right)^2 = \frac{c_2}{c_1^2}x^2,$$

which is some portion of a parabola with its vertex at the origin. Specifically, if $c_2 > 0$ then the parabola opens upward and the flow line is the right half if $c_1 > 0$ (because $x = c_1 e^{t/4} > 0$) and the left half if $c_1 < 0$ (because $x = c_1 e^{t/4} < 0$). A similar result holds if $c_2 < 0$, except the parabola opens downward. If $c_1 = 0$ and $c_2 \neq 0$ then $\mathbf{x}(t) = (0, c_2 e^{t/2})$, which is the positive y-axis if $c_2 > 0$ and the negative y-axis if $c_2 < 0$. If $c_1 \neq 0$ and $c_2 = 0$ then $\mathbf{x}(t) = (c_1 e^{t/4}, 0)$, which is the positive x-axis if $c_1 > 0$ and the negative x-axis if $c_1 < 0$. The case where $c_1 = c_2 = 0$ degenerates to the single point $\mathbf{x}(t) = (0,0)$ for all t.

13. (a) Let $A > 0$ and α be real constants and let γ be the curve defined by

$$\mathbf{x}(t) = A\cos\big(t/(4A) + \alpha\big)\,\mathbf{i} + A\sin\big(t/(4A) + \alpha\big)\,\mathbf{j}, \quad -\infty < t < \infty.$$

For $(x,y) = \big(x(t), y(t)\big)$ on γ we have $x = A\cos\big(t/(4A) + \alpha\big)$ and $y = A\sin\big(t/(4A) + \alpha\big)$. Differentiating these with respect to t gives

$$\dot{x} = -\frac{1}{4A}A\sin\big(t/(4A) + \alpha\big) = -\frac{1}{4A}y \qquad \text{and} \qquad \dot{y} = \frac{1}{4A}A\cos\big(t/(4A) + \alpha\big) = \frac{1}{4A}x.$$

Noting that $\sqrt{x^2 + y^2} = \sqrt{A^2\cos^2\big(t/(4A) + \alpha\big) + A^2\sin^2\big(t/(4A) + \alpha\big)} = A$, the velocity vector to γ at (x,y) can be written as

$$\dot{\mathbf{x}} = -\frac{y}{4\sqrt{x^2 + y^2}}\,\mathbf{i} + \frac{x}{4\sqrt{x^2 + y^2}}\,\mathbf{j} = \mathbf{G}(x,y).$$

That is, the velocity vector at each point on γ coincides with the vector field $\mathbf{G}$. Hence, γ is a flow line of $\mathbf{F}$.

(b) For a specific flow line as given in part (a), the position vector has the constant length $|\mathbf{x}(t)| = A$, so that all points on the flow line lie on a circle of radius A. Moreover, the entire circle is traced out exactly once on every t-interval of length $8\pi A$ (the period of the associated sine and cosine) and, since $A > 0$, the direction in which the circle is traced out is counterclockwise.

15. Let $\mathbf{F}$ be a continuously differentiable 3-dimensional vector field such that $\mathbf{F} \cdot \text{curl}\mathbf{F}$ is identically zero. If γ is a flow line of $\mathbf{F}$ that is parametrized by a differentiable function $g(t)$, then the velocity vector $g'(t)$ coincides with $\mathbf{F}\big(g(t)\big)$ for all t, i.e., $g'(t) = \mathbf{F}\big(g(t)\big)$ for all t. The line integral of $\text{curl}\mathbf{F}$ along γ is therefore equal to

$$\int_\gamma \text{curl}\mathbf{F}\big(g(t)\big) \cdot g'(t)\,dt = \int_\gamma \text{curl}\mathbf{F}\big(g(t)\big) \cdot \mathbf{F}\big(g(t)\big)\,dt = \int_\gamma 0\,dt = 0.$$

17. (a) Let $f : \mathcal{R}^3 \longrightarrow \mathcal{R}$ be twice continuously differentiable and let $\mathbf{F}$ be the gradient field defined by $\mathbf{F} = \nabla f$. Then the coordinate functions of $\mathbf{F}$ are $F_1 = \partial f/\partial x$, $F_2 = \partial f/\partial y$ and $F_3 = \partial f/\partial z$. So,

$$
\begin{aligned}
\operatorname{curl}\mathbf{F} &= \left(\frac{\partial F_3}{\partial y} - \frac{\partial F_2}{\partial z}\right)\mathbf{i} + \left(\frac{\partial F_1}{\partial z} - \frac{\partial F_3}{\partial x}\right)\mathbf{j} + \left(\frac{\partial F_2}{\partial x} - \frac{\partial F_1}{\partial y}\right)\mathbf{k} \\
&= \left[\frac{\partial}{\partial y}\left(\frac{\partial f}{\partial z}\right) - \frac{\partial}{\partial z}\left(\frac{\partial f}{\partial y}\right)\right]\mathbf{i} + \left[\frac{\partial}{\partial z}\left(\frac{\partial f}{\partial x}\right) - \frac{\partial}{\partial x}\left(\frac{\partial f}{\partial z}\right)\right]\mathbf{j} \\
&\qquad\qquad + \left[\frac{\partial}{\partial x}\left(\frac{\partial f}{\partial y}\right) - \frac{\partial}{\partial y}\left(\frac{\partial f}{\partial x}\right)\right]\mathbf{k} \\
&= \left(\frac{\partial^2 f}{\partial y \partial z} - \frac{\partial^2 f}{\partial z \partial y}\right)\mathbf{i} + \left(\frac{\partial^2 f}{\partial z \partial x} - \frac{\partial^2 f}{\partial x \partial z}\right)\mathbf{j} + \left(\frac{\partial^2 f}{\partial x \partial y} - \frac{\partial^2 f}{\partial y \partial x}\right)\mathbf{k}.
\end{aligned}
$$

Since f is twice continuously differentiable, all second-order mixed partials of f are continuous. By a straightforward extension of Theorem 3.3 in Chapter 4, Section 3C (Clairaut's Theorem), the pair of mixed partials in each parenthetical term above are equal, so that each term is identically zero. That is, $\operatorname{curl}\mathbf{F}$ is identically zero.

(b) Pick any twice continuously differentiable function f such that its gradient is nonzero (this solver picks $f(x,y,z) = x$ and observes that $\nabla f = (1,0,0)$). Then the vector field $\mathbf{F} = \nabla f$ is a nonzero gradient field. By part (a), $\operatorname{curl}\mathbf{F}$ is identically zero, so that $\mathbf{F} \cdot \operatorname{curl}\mathbf{F}$ is identically zero. Hence, for every differentiable real-valued function g defined on the domain of $\mathbf{F}$, the right side of the identity given in Exercise 16 is identically zero. Now choose a differentiable real-valued function g defined on the domain of $\mathbf{F}$ such that the vector field $\mathbf{G} = g\mathbf{F}$ is *not* a gradient field (this solver chooses $g(x,y,z) = y$ and observes that $\operatorname{curl}\mathbf{G} = (0,0,-1)$, so that by part (a), $\mathbf{G} = (y,0,0)$ is not a gradient field). The left side of the identity given in Exercise 16 is therefore the statement "$\mathbf{G} \cdot \operatorname{curl}\mathbf{G}$ is identically zero".

(c) Let $\mathbf{F} = (z,x,y)$. Then the only nonzero partials are $\partial F_1/\partial z = 1$, $\partial F_2 \partial x = 1$ and $\partial F_3/\partial y = 1$. The formula for $\operatorname{curl}\mathbf{F}$ gives $\operatorname{curl}\mathbf{F} = \mathbf{i} + \mathbf{j} + \mathbf{k}$. Hence,

$$
\mathbf{F} \cdot \operatorname{curl}\mathbf{F} = (z,x,y) \cdot (1,1,1) = x + y + z,
$$

which is not always zero.

19. (a) Let $\mathbf{F}$ be a 3-dimensional vector field that is twice continuously differentiable. If the coordinate functions of $\mathbf{F}$ are F_1, F_2 and F_3 then

$$
\begin{aligned}
\operatorname{div}(\operatorname{curl}\mathbf{F}) &= \operatorname{div}\left[\left(\frac{\partial F_3}{\partial y} - \frac{\partial F_2}{\partial z}\right)\mathbf{i} + \left(\frac{\partial F_1}{\partial z} - \frac{\partial F_3}{\partial x}\right)\mathbf{j} + \left(\frac{\partial F_2}{\partial x} - \frac{\partial F_1}{\partial y}\right)\mathbf{k}\right] \\
&= \frac{\partial}{\partial x}\left(\frac{\partial F_3}{\partial y} - \frac{\partial F_2}{\partial z}\right) + \frac{\partial}{\partial y}\left(\frac{\partial F_1}{\partial z} - \frac{\partial F_3}{\partial x}\right) + \frac{\partial}{\partial z}\left(\frac{\partial F_2}{\partial x} - \frac{\partial F_1}{\partial y}\right) \\
&= \frac{\partial^2 F_3}{\partial x \partial y} - \frac{\partial^2 F_2}{\partial x \partial z} + \frac{\partial^2 F_1}{\partial y \partial z} - \frac{\partial^2 F_3}{\partial y \partial x} + \frac{\partial^2 F_2}{\partial z \partial x} - \frac{\partial^2 F_1}{\partial z \partial y} \\
&= \left(\frac{\partial^2 F_3}{\partial x \partial y} - \frac{\partial^2 F_3}{\partial y \partial x}\right) + \left(\frac{\partial^2 F_2}{\partial z \partial x} - \frac{\partial^2 F_2}{\partial x \partial z}\right) + \left(\frac{\partial^2 F_1}{\partial y \partial z} - \frac{\partial^2 F_1}{\partial z \partial y}\right).
\end{aligned}
$$

Since $\mathbf{F}$ is twice continuously differentiable, the mixed partials of F_3 with respect to x and y are equal, the mixed partials of F_2 with respect to x and z are equal, and the mixed partials of F_1 with respect to y and z are equal. That is, each of the parenthetical terms of the last member above is zero. Hence, $\operatorname{div}(\operatorname{curl}\mathbf{F})$ is identically zero.

(b) With curl**F** replacing **F** in the continuity equation (Equation 4.3 in the text), div(curl**F**) = 0 implies that there is no change in local density during motion along flow lines of curl**F**. That is, neither expansion nor contraction occurs and volume is therefore preserved.

21. Let $f : \mathcal{R}^2 \longrightarrow \mathcal{R}$ be a twice continuously differentiable function. The definition of Δf the Laplacian of f is the same as the definition of the divergence of the gradient field $\mathbf{F} = \nabla f$. Thus, if $\mathrm{div}\nabla f > 0$ in a region R_0 then the effect of flow with velocity ∇f is to move the mass in R_0 to a new region R_1 with an area larger than that of R_0. The total mass is the same but it is spread out over more area and therefore its density per unit area has decreased. At the other extreme, if $\mathrm{div}\nabla f < 0$ in a region R_0 then the mass in R_0 is moved to a new region R_1 with an area smaller than that of R_0. Again, the total mass is the same but is compressed into a smaller area and therefore the density per unit area on R_1 has increased. But if f is a harmonic function then, by definition, $\Delta f = 0$ so that regions are moved to regions of the same area and density is preserved.

Chapter Review

(pgs. 395-399)

1. Let s be the square with corners at $(0,0)$, $(1,0)$, $(1,1)$ and $(0,1)$. Denoting the four sides of the square by s_1 (bottom side), s_2 (right side), s_3 (top side), and s_4 (left side), we see that $dy = 0$, $y = 0$ and x goes from 0 to 1 on s_1; $dx = 0$, $x = 1$ and y goes from 0 to 1 on s_2; $dy = 0$, $y = 1$ and x goes from 1 to 0 on s_3; and $dx = 0$, $x = 0$ and y goes from 1 to 0 on s_4. Thus,

$$\int_s xy\, dx + (x^2 + y^2)\, dy = \int_{s_1} xy\, dx + (x^2 + y^2)\, dy + \int_{s_2} xy\, dx + (x^2 + y^2)\, dy$$

$$+ \int_{s_3} xy\, dx + (x^2 + y^2)\, dy + \int_{s_4} xy\, dx + (x^2 + y^2)\, dy$$

$$= \int_0^1 0\, dx + \int_0^1 (1 + y^2)\, dy + \int_1^0 x\, dx + \int_1^0 y^2\, dy$$

$$= \int_0^1 (1 + y^2)\, dy - \int_0^1 x\, dx - \int_0^1 y^2 dy$$

$$= \int_0^1 1\, dy - \int_0^1 x\, dx = \left[y\right]_0^1 - \left[\frac{1}{2}x^2\right]_0^1 = \frac{1}{2}.$$

3. Let $\mathbf{F}(x,y,z) = (y^2 z^3, 2xyz^3, 3xy^2 z^2)$ and note that $\mathbf{F} = \nabla f$, where $f(x,y,z) = xy^2 z^3$. With $\mathbf{a} = (1,1,1)$ and $\mathbf{b} = (2,2,2)$, Theorem 1.3 in this chapter gives

$$\int_{\mathbf{a}}^{\mathbf{b}} y^2 z^3 dx + 2xyz^3 dy + 3xy^2 z^2 dz = \int_{(1,1,1)}^{(2,2,2)} (y^2 z^3, 2xyz^3, 3xy^2 z^2) \cdot (dx, dy, dz)$$

$$= \int_{(1,1,1)}^{(2,2,2)} \mathbf{F} \cdot d\mathbf{x} = \int_{(1,1,1)}^{(2,2,2)} \nabla f \cdot d\mathbf{x} = f(2,2,2) - f(1,1,1) = 63.$$

5. The counterclockwise circular path c is parametrized by $g(t) = (\cos t, \sin t)$, $0 \le t \le 2\pi$. So, $x = \cos t$, $y = \sin t$ and $dx = -\sin t\, dt$. The given integral then becomes

$$\int_c y\, dx = \int_0^{2\pi} (\sin t)(-\sin t\, dt) = \int_0^{2\pi} -\sin^2 t\, dt = \int_0^{2\pi} \frac{1}{2}(\cos 2t - 1)\, dt$$

$$= \frac{1}{2}\left[\frac{1}{2}\sin 2t - t\right]_0^{2\pi} = -\pi.$$

NOTE: In order to answer the questions posed in Exercise 9, one must first work each of the Exercises 6, 7 and 8. Therefore, a general solution will be given to Exercises 6, 7 and 8, and the conclusions asked for in Exercise 9 will reduce to special cases of the overriding principle.

General solution to Exercises 6 through 8: For real numbers $0 < n < m$, let γ be the curve parametrized by

$$g(t) = (t^n, t^m), \quad 0 \le t \le 1,$$

and note that regardless of the choice of n and m, γ starts at the point $(0,0)$ and ends at $(1,1)$. We have $x = t^n$, $y = t^m$, $dy = mt^{m-1}dt$ and $x^2y = t^{2n}t^m = t^{2n+m}$. Hence,

$$\int_\gamma x^2 y \, dy = \int_0^1 (t^{2n+m})(mt^{m-1}dt) = m \int_0^1 t^{2n+2m-1}dt = m\left[\frac{t^{2n+2m}}{2n+2m}\right]_0^1 = \frac{m}{2n+2m}.$$

In particular, denoting the integrals in Exercises 6, 7 and 8 by $I(\gamma_1)$, $I(\gamma_2)$ and $I(\gamma_3)$, respectively, we have

$$I(\gamma_1) = \frac{3}{2(2+3)} = \frac{3}{10}, \qquad I(\gamma_2) = \frac{2}{2(1+2)} = \frac{1}{3}, \qquad I(\gamma_3) = \frac{4}{2(2+4)} = \frac{1}{3}.$$

9. The general solution given above shows quite clearly what the criterion is for two integrals of the given field to be equal. Specifically, if $g_1(t) = (t^{n_1}, t^{m_1})$ and $g_2(t) = (t^{n_2}, t^{m_2})$ then the associated integrals are equal if, and only if,

$$\frac{m_1}{2n_1 + 2m_1} = \frac{m_2}{2n_2 + 2m_2}.$$

This can be rearranged to get the ratio $n_2/n_1 = m_2/m_1$. This condition holds in Exercises 7 and 8, where the common ratio is 2. No such relationship exists between the values of the pair $(2,3)$ in Exercise 6 and the pairs $(1,2)$ and $(2,4)$ given in Exercises 7 and 8. This criterion for the equality of two such integrals is more than an algebraic one. Indeed, using the variable name "u" in g_2 instead of "t", we note that the function

$$\phi(u) = u^{n_2/n_1}$$

satisfies $\phi(0) = 0$, $\phi(1) = 1$, $\phi'(u) = (n_2/n_1)u^{n_2/n_1-1} > 0$ for $0 < u < 1$ and

$$g_1(\phi(u)) = g_1(u^{n_2/n_1}) = \left((u^{n_2/n_1})^{n_1}, (u^{n_2/n_1})^{m_1}\right) = (u^{n_2}, u^{m_2}) = g_2(u),$$

where the fourth member follows because $(n_2/n_1)m_1 = (m_2/m_1)m_1 = m_2$. In other words, $I(\gamma_2) = I(\gamma_3)$ because the parametrizations used are equivalent via the function $\phi(u) = u^2$. But $I(\gamma_1)$ has a different value because the parametrization used is not equivalent to either of those used in Exercises 7 and 8.
(Note: The fact that the image curves are the same in Exercises 7 and 8 but different from that in Exercise 6 is in no way indicative of the equality or nonequality of the integrals in question.)
11. Let $\mathbf{F}(x,y,z) = (x, 2y, z)$ be a force field and suppose a particle of unit mass (say) travels in a straight line from the point $(1,0,0)$ to the point $(-1,2,1)$. The associated directed line segment can be written as

$$t(-1,2,1) + (1-t)(1,0,0) = (1-2t, 2t, t), \quad 0 \le t \le 1.$$

Thus, $g(t) = (1-2t, 2t, t)$, $0 \le t \le 1$ is a parametrization for the line segment. We also have $\mathbf{F}(g(t)) = (1-2t, 4t, t)$ and $g'(t) = (-2,2,1)$. The work done by $\mathbf{F}$ on the particle is therefore

$$W = \int_0^1 \mathbf{F}(g(t)) \cdot g'(t) \, dt = \int_0^1 (1-2t, 4t, t) \cdot (-2,2,1) \, dt$$

$$= \int_0^1 (-2 + 13t) \, dt = \left[-2t + \frac{13}{2}t^2\right]_0^1 = \frac{9}{2}.$$

13. Let γ be a smooth curve parametrized by $\mathbf{x}(t)$, $t_0 \leq t \leq t_1$. To say that a vector field $\mathbf{F}$ assigns to each $\mathbf{x}$ on γ the unit tangent vector to the curve in the direction of traversal means that $\mathbf{F}\big(\mathbf{x}(t)\big) = \dot{\mathbf{x}}(t)/|\dot{\mathbf{x}}(t)|$, for $t_0 \leq t \leq t_1$. Hence,

$$\int_\gamma \mathbf{F} \cdot d\mathbf{x} = \int_{t_0}^{t_1} \mathbf{F}\big(\mathbf{x}(t)\big) \cdot \dot{\mathbf{x}}(t)\, dt = \int_{t_0}^{t_1} \frac{\dot{\mathbf{x}}(t)}{|\dot{\mathbf{x}}(t)|} \cdot \dot{\mathbf{x}}(t)\, dt = \int_{t_0}^{t_1} \frac{|\dot{\mathbf{x}}(t)|^2}{|\dot{\mathbf{x}}(t)|}\, dt = \int_{t_0}^{t_1} |\dot{\mathbf{x}}(t)|\, dt.$$

By definition, the integral on the far right is the arc length of γ.

15. (a) Let $\mathbf{F}(t, x, y)$ be the time-dependent 2-dimensional vector field

$$\mathbf{F}(t, x, y) = \big((1 - t)x - ty, tx + (1 - t)y\big).$$

Let γ be the unit circle and suppose the motion of a particle of unit mass (say) traveling on γ is given by $g(t) = (\cos t, \sin t)$, $0 \leq t \leq 2\pi$. Then the work done by the field on the particle is

$$\begin{aligned}
W &= \int_\gamma \mathbf{F}\big(t, g(t)\big) \cdot g'(t)\, dt \\
&= \int_0^{2\pi} \big((1 - t)\cos t - t\sin t, t\cos t + (1 - t)\sin t\big) \cdot (-\sin t, \cos t)\, dt \\
&= \int_0^{2\pi} \big(-(1 - t)\cos t \sin t + t\sin^2 t + t\cos^2 t + (1 - t)\sin t \cos t\big)\, dt \\
&= \int_0^{2\pi} t\, dt = \left[\frac{1}{2}t^2\right]_0^{2\pi} = 2\pi^2.
\end{aligned}$$

(b) Instead of having the vector field $\mathbf{F}$ given in part (a) vary with time, we keep $t = t_0$ fixed in the formula for $\mathbf{F}$ and write the vector field as

$$\mathbf{F}(t_0, x, y) = \mathbf{F}_{t_0}(x, y) = \big((1 - t_0)x - t_0 y, t_0 x + (1 - t_0)y\big).$$

In examining the role that the variable t in $\mathbf{F}(t, x, y)$ played in part (a) in the evaluation of the integral, we see that the only change is that the integrand of the final integral is the constant t_0 rather than the variable t. As a consequence, the work done under the present conditions is given by

$$W = \int_0^{2\pi} t_0\, dt = \big[t_0 t\big]_0^{2\pi} = 2\pi t_0.$$

If $t_0 = 0$ then $\mathbf{F}_0(x, y) = (x, y)$ and the vector field is perpendicular to the path traced by $g(t)$, so that the line integral defining the work is zero. If $t_0 = 1$ then $\mathbf{F}_1(x, y) = (-y, x)$ and the field vectors point in the same direction as the velocity vector $g'(t)$. Noting that $g(t)$ is an arc length parametrization of γ, we conclude that the work integral must evaluate to the arc length 2π of the unit circle.

17. The base of the sculpture is a curve γ parametrized by $g(t) = (t^3 - 3t, 3t^2)$, for $1 \leq t \leq 2$, where each coordinate is measured in meters. Using the "HINT" given in the text, we think of each point on γ as being weighted (literally) by the material above it. Since the height at the point $(x, y) = (t^3 - 3t, 3t^2)$ is $y = 3t^2$, it follows that the mass $\mu\big(g(t)\big)$ of the material above the point $g(t)$ is $\mu\big(g(t)\big) = 3\rho t^2 = 90t^2$, where $\rho = 30$ kilograms per square meter is the constant density of the material. With

$$|g'(t)| = |(3t^2 - 3, 6t)| = \sqrt{(3t^2 - 3)^2 + 36t^2} = \sqrt{9t^4 + 18t^2 + 9} = \sqrt{(3t^2 + 3)^2} = 3t^2 + 3,$$

the formula in the text for the mass of a weighted curve gives

$$M = \int_1^2 \mu(g(t))|g'(t)|\,dt$$

$$= \int_1^2 (90t^2)(3t^2 + 3)\,dt = 270 \int_1^2 (t^4 + t^2)\,dt = 270 \left[\frac{1}{5}t^5 + \frac{1}{3}t^3\right]_1^2 = 2304 \text{ kilograms.}$$

19. Since curvature depends solely on the geometry of the curve and not on the orientation of the curve with respect to a coordinate system in $\mathcal{R}^2$, we can suppose that the parabola we are concerned with is the graph of an equation of the form $f(x) = ax^2$, where a is a positive constant. We have $f'(x) = 2ax$ and $f''(x) = 2a$. To simplify matters, we will use the formula for curvature given in Exercise 13 in Section 3 of this chapter; namely,

$$\kappa(x) = \frac{|f''(x)|}{\left[1 + \left(f'(x)\right)^2\right]^{3/2}} = \frac{2a}{(1 + 4a^2x^2)^{3/2}}.$$

Clearly, the fraction on the right is maximized when the denominator is minimal, which occurs at $x = 0$. This corresponds to the point $(0, f(0)) = (0,0)$, the vertex of the parabola.

21. Let $\mathbf{x}(t) = (a\cos t, a\sin t, bt)$, where a and b are nonnegative constants. If $a = b = 0$ then the only point on the curve is $(0,0,0)$, in which case the concept of curvature is meaningless. Thus, assuming at least one of a or b is nonzero, the curvature of the curve is a constant

$$\kappa = \frac{a}{a^2 + b^2}.$$

which was derived in Example 2 in Section 3 of this chapter.

(a) For a given $b > 0$, we view the curvature as a function of a. Differentiating κ twice with respect to a gives

$$\frac{d\kappa}{da} = \frac{-a^2 + b^2}{(a^2 + b^2)^2} \qquad \text{and} \qquad \frac{d^2\kappa}{da^2} = \frac{3a^3 - 6ab^2}{(a^2 + b^2)^3}.$$

The critical points are those for which the numerator on the left is zero; namely, $a = b$. Since the second derivative on the right above is negative when $a = b$, the condition $a = b$ gives the maximum curvature. The minimum curvature occurs at $a = 0$, where $\kappa = 0$. (Note: the case $a = 0$ gives a vertical line.)

(b) For a given $a > 0$, we observe that the fraction defining κ is maximal when $b = 0$. This corresponds to a circle of radius a. However, the curve has no minimum curvature since κ can never be zero but can be made arbitrarily close to zero by taking b large enough.

23. Let $\mathbf{x} = \mathbf{x}(t)$ be a smooth curve in $\mathcal{R}^3$. If $x = x(t)$, $y = y(t)$, $z = z(t)$ are the coordinate functions of $\mathbf{x}$ then the expression under the radical in Equation 3.5 in Section 3 of this chapter can be written as

$$(\dot{x}^2 + \dot{y}^2 + \dot{z}^2)(\ddot{x}^2 + \ddot{y}^2 + \ddot{z}^2) - (\dot{x}\ddot{x} + \dot{y}\ddot{y} + \dot{z}\ddot{z})^2.$$

When the first term is multiplied out there are nine terms, three of which are $\dot{x}^2\ddot{x}^2$, $\dot{y}^2\ddot{y}^2$ and $\dot{z}^2\ddot{z}^2$. When the second term above is expanded there are also nine terms, three of which are $\dot{x}^2\ddot{x}^2$, $\dot{y}^2\ddot{y}^2$ and $\dot{z}^2\ddot{z}^2$. Thus, in expanding both terms above, these three indicated terms cancel each other. The six remaining terms in the first expansion are

$$\dot{x}^2\ddot{y}^2 + \dot{x}^2\ddot{z}^2 + \dot{y}^2\ddot{x}^2 + \dot{y}^2\ddot{z}^2 + \dot{z}^2\ddot{x}^2 + \dot{z}^2\ddot{y}^2$$

(*)
$$= (\dot{y}^2\ddot{z}^2 + \dot{z}^2\ddot{y}^2) + (\dot{z}^2\ddot{x}^2 + \dot{x}^2\ddot{z}^2) + (\dot{x}^2\ddot{y}^2 + \dot{y}^2\ddot{x}^2),$$

where we have judiciously grouped them for future use. Similarly, the six remaining terms of the second expansion can be paired up to get

(∗∗)
$$-2\dot{y}\ddot{y}\dot{z}\ddot{z} - 2\dot{x}\ddot{x}\dot{z}\ddot{z} - 2\dot{x}\ddot{x}\dot{y}\ddot{y}.$$

We now observe that the sum of the first parenthetical term of (∗) and the first term of (∗∗) is a perfect square $(\dot{y}\ddot{z} - \dot{z}\ddot{y})^2$; the sum of the second parenthetical term of (∗) and the second term of (∗∗) is a perfect square $(\dot{z}\ddot{x} - \dot{x}\ddot{z})^2$; and the sum of the third parenthetical term of (∗) and the third term of (∗∗) is a perfect square $(\dot{x}\ddot{y} - \dot{y}\ddot{x})^2$. So, adding (∗) and (∗∗) gives

$$
\begin{aligned}
(\dot{y}\ddot{z} - \dot{z}\ddot{y})^2 + (\dot{z}\ddot{x} - \dot{x}\ddot{z})^2 + (\dot{x}\ddot{y} - \dot{y}\ddot{x})^2 &= |(\dot{y}\ddot{z} - \dot{z}\ddot{y}, \dot{z}\ddot{x} - \dot{x}\ddot{z}, \dot{x}\ddot{y} - \dot{y}\ddot{x})|^2 \\
&= |(\dot{x}, \dot{y}, \dot{z}) \times (\ddot{x}, \ddot{y}, \ddot{z})|^2 \\
&= |\dot{\mathbf{x}} \times \ddot{\mathbf{x}}|^2.
\end{aligned}
$$

Hence, Equation 3.5 can be written as

$$\kappa = \frac{\sqrt{|\dot{\mathbf{x}} \times \ddot{\mathbf{x}}|^2}}{|\dot{\mathbf{x}}|^3} = \frac{|\dot{\mathbf{x}} \times \ddot{\mathbf{x}}|}{|\dot{\mathbf{x}}|^3},$$

as was to be shown.

Chapter 9: Vector Field Theory

Section 1: GREEN'S THEOREM

Exercise Set 1ABC (pgs. 408-409)

Note: In Exercises 1 and 3, we let $F(x,y) = y$ and $G(x,y) = x^2$ and note that F and G are continuously differentiable everywhere on $\mathcal{R}^2$. We have

$$\int_D \left(\frac{\partial G}{\partial x} - \frac{\partial F}{\partial y} \right) dx \, dy = \int_D (2x - 1) \, dx \, dy,$$

where D is a plane region of finite area whose boundary is a piecewise smooth curve γ.

1. Since the given circle γ is the counterclockwise oriented unit circle, the region D is the unit disk, which is a simple region. By Green's Theorem,

$$\oint_\gamma y \, dx + x^2 \, dy = \int_D (2x - 1) \, dx \, dy = \int_{-1}^1 dy \int_{-\sqrt{1-y^2}}^{\sqrt{1-y^2}} (2x - 1) \, dx$$

$$= \int_{-1}^1 \left[x^2 - x \right]_{-\sqrt{1-y^2}}^{\sqrt{1-y^2}} dy = \int_{-1}^1 \left(-2\sqrt{1-y^2} \right) dy = -2 \int_{-1}^1 \sqrt{1-y^2} \, dy = -\pi,$$

where we recognize the last integral as one-half the area of the unit circle.

3. The region D whose boundary is the counterclockwise oriented square γ is defined by $0 \le x \le 1$, $0 \le y \le 1$, and is a simple region. By Green's Theorem

$$\oint_\gamma y \, dx + x^2 \, dy = \int_D (2x - 1) \, dx \, dy = \int_0^1 dy \int_0^1 (2x - 1) \, dx = \left[\frac{1}{2} y \right]_0^1 \left[x^2 - x \right]_0^1 = 0.$$

5. Let $F(x,y) = y$ and $G(x,y) = x$ and let γ_1 be parametrized by $g(t) = (t, t^2)$, $0 \le t \le 1$. Direct computation of the line integral of the vector field (y, x) along γ_1 gives

$$\int_{\gamma_1} y \, dx + x \, dy = \int_{\gamma_1} (y, x) \cdot (dx, dy) = \int_0^1 (t^2, t) \cdot (1, 2t) \, dt = \int_0^1 3t^2 \, dt = \left[t^3 \right]_0^1 = 1.$$

Now let γ_2 be parametrized by $g_2(t) = (t, t)$, $0 \le t \le 1$, which has the same initial and terminal points as γ_1. Direct computation of the line integral of (y, x) along γ_2 gives

$$\int_{\gamma_2} y \, dx + x \, dy = \int_{\gamma_2} (y, x) \cdot (dx, dy) = \int_0^1 (t, t) \cdot (1, 1) \, dt = \int_0^1 2t \, dt = \left[t^2 \right]_0^1 = 1.$$

Finally, let γ_3 be the curve consisting of the horizontal line segment $(t, 0)$, for $0 \le t \le 1$, and the vertical line segment $(1, t)$, for $0 \le t \le 1$, and note that the initial and terminal points of γ_3 are the same as γ_1 and γ_2 above. The line integral of (y, x) along γ_3 is then given by

$$\int_{\gamma_3} y \, dx + x \, dy = \int_{\gamma_1} (y, x) \cdot (dx, dy) = \int_0^1 (0, t) \cdot (1, 0) \, dt + \int_0^1 (t, 1) \cdot (0, 1) \, dt$$

$$= \int_0^1 0 \, dt + \int_0^1 1 \, dt = 0 + \left[t \right]_0^1 = 1.$$

7. The given triangle γ with vertices $(0,0)$, $(1,0)$ and $(1,1)$ is traced counterclockwise, and the functions $F(x,y) = x - y$ and $G(x,y) = x + y$ are continuously differentiable on the triangular simple region D with boundary γ. By Green's Theorem,

$$\oint_\gamma (x - y)\,dx + (x + y)\,dy = \int_D \left(\frac{\partial(x + y)}{\partial x} - \frac{\partial(x - y)}{\partial y} \right) dx\,dy$$

$$= \int_D \left(1 - (-1)\right) dx\,dy = 2 \int_D dx\,dy = 2A(D) = 1,$$

where $A(D) = 1/2$ is the area of D.

9. Let c be the *clockwise* oriented unit circle and let $F(x,y) = x^2 - y^2$, $G(x,y) = x^2 + y^2$. Then F and G are continuously differentiable on the simple region D (closed unit disk) with boundary c. Since c^- is traced counterclockwise, Theorem 1.1 can be used with Green's Theorem to get

$$\oint_c (x^2 - y^2)\,dx + (x^2 + y^2)\,dy = -\oint_{c^-} (x^2 - y^2)\,dx + (x^2 + y^2)\,dy$$

$$= -\int_D \left(\frac{\partial(x^2 + y^2)}{\partial x} - \frac{\partial(x^2 - y^2)}{\partial y} \right) dx\,dy = -\int_D (2x - (-2y))\,dx\,dy$$

$$= -2 \int_D (x + y)\,dx\,dy = -2 \int_{-1}^1 dy \int_{-\sqrt{1-y^2}}^{\sqrt{1-y^2}} (x + y)\,dx$$

$$= -2 \int_{-1}^1 \left[\frac{1}{2}x^2 + xy \right]_{-\sqrt{1-y^2}}^{\sqrt{1-y^2}} dy = -2 \int_{-1}^1 2y\sqrt{1 - y^2}\,dy = 0,$$

where the last equality follows because the integrand is an odd function being integrated over a symmetric interval.

11. Let D be a simple region bounded by a piecewise smooth counterclockwise oriented curve γ, and consider the continuous field $\mathbf{F} = (F, G) = (-y, x)$. By Green's Theorem

$$\oint_\gamma (-y\,dx + x\,dy) = \int_D \left(\frac{\partial(x)}{\partial x} - \frac{\partial(-y)}{\partial y} \right) dx\,dy = \int_D \left(1 - (-1)\right) dx\,dy = 2 \int_D dx\,dy.$$

By definition, the rightmost integral is the area of the region D. Hence, dividing the above equation by 2 gives

$$A(D) = \frac{1}{2} \oint_\gamma (-y\,dx + x\,dy).$$

13. (a) Let $f(x,y) = \arctan(y/x)$ for $x > 0$. Then

$$\nabla f = \left(\frac{\partial\big(\arctan(y/x)\big)}{\partial x}, \frac{\partial\big(\arctan(y/x)\big)}{\partial y} \right)$$

$$= \left(\frac{1}{1 + (y/x)^2}\left(-\frac{y}{x^2}\right), \frac{1}{1 + (y/x)^2}\frac{1}{x} \right) = \left(-\frac{y}{x^2 + y^2}, \frac{x}{x^2 + y^2} \right).$$

(b) Let $F(x,y) = -y/(x^2 + y^2)$ and $G(x,y) = x/(x^2 + y^2)$ and define the vector field $\mathbf{F} = \big(F(x,y), G(x,y)\big)$ for $(x,y) \neq (0,0)$. Since x, y and $x^2 + y^2$ are continuous on $\mathcal{R}^2$, so are the quotients defining F and G as long as the denominator $x^2 + y^2 \neq 0$; i.e., as long as $(x, y) \neq (0, 0)$. So, $\mathbf{F}(x, y)$ is continuous for $(x, y) \neq (0, 0)$.

(c) We want to show that the vector field $\mathbf{F}$ defined in part (b) is not a gradient field. To do this, we consider two paths from $(1, 0)$ to $(-1, 0)$. The first path γ_1 is the counterclockwise oriented semicircle $h_1(t) = (\cos t, \sin t)$, $0 \leq t \leq \pi$; and the second path γ_2 is the clockwise oriented semicircle $h_2(t) = (\cos t, -\sin t)$, $0 \leq t \leq \pi$. On γ_1,

$$\int_{\gamma_1} \mathbf{F} \cdot d\mathbf{x} = \int_0^\pi \mathbf{F}\big(h_1(t)\big) \cdot h_1'(t)\, dt = \int_0^\pi (-\sin t, \cos t) \cdot (-\sin t, \cos t)\, dt = \int_0^\pi 1\, dt = \pi.$$

But on γ_2,

$$\int_{\gamma_2} \mathbf{F} \cdot d\mathbf{x} = \int_0^\pi \mathbf{F}\big(h_2(t)\big) \cdot h_2'(t)\, dt = \int_0^\pi (\sin t, \cos t) \cdot (-\sin t, -\cos t)\, dt = \int_0^\pi -1\, dt = -\pi.$$

If $\mathbf{F}$ were a gradient field then these line integrals would be equal. Since they are not equal, there is no function g such that $\nabla g(x, y) = \mathbf{F}(x, y)$ for all $(x, y) \neq 0$.

(NOTE: By part (a), $\mathbf{F}$ restricted to the right half of the xy-plane ($x > 0$) is a gradient field. Therefore, in order to come up with the proof in part (c), it was essential that at least one of the paths touched or crossed the y-axis where $x > 0$ does not hold.)

15. Let $\mathbf{F}$ be a gradient field and let f be such that $\nabla f = \mathbf{F}$, so that $\mathbf{F} = (\partial f / \partial x, \partial f / \partial y)$. Let γ be a piecewise smooth counterclockwise oriented circuit and let D be the region bounded by γ. If $\mathbf{n}$ is the unit normal to γ that points away from the region D then Gauss's Theorem in the plane gives

$$\oint_\gamma \nabla f \cdot \mathbf{n}\, ds = \oint_\gamma \mathbf{F} \cdot \mathbf{n}\, ds = \int_D \text{div}\mathbf{F}\, dA$$

$$= \int_D \left[\frac{\partial}{\partial x}\left(\frac{\partial f}{\partial x} \right) + \frac{\partial}{\partial y}\left(\frac{\partial f}{\partial y} \right) \right] dA = \int_D \left(\frac{\partial^2 f}{\partial x^2} + \frac{\partial^2 f}{\partial y^2} \right) dA = \int_D \Delta f\, dA,$$

where $\Delta f = (\partial^2 f / \partial x^2) + (\partial^2 f / \partial y^2)$ is the Laplacian of f.

17. Consider the vector field $\mathbf{F}(x, y) = \big(-y/(x^2 + y^2)\big)\mathbf{i} + \big(x/(x^2 + y^2)\big)\mathbf{j}$, for $(x, y) \neq (0, 0)$.

(a) With $F(x, y) = -y/(x^2 + y^2)$ and $G(x, y) = x/(x^2 + y^2)$, the definition of $\text{div}\mathbf{F}$ gives

$$\text{div}\mathbf{F} = \frac{\partial F}{\partial x} + \frac{\partial G}{\partial y} = \frac{\partial}{\partial x}\left(\frac{-y}{x^2 + y^2} \right) + \frac{\partial}{\partial y}\left(\frac{x}{x^2 + y^2} \right)$$

$$= \frac{(x^2 + y^2)(0) - (-y)(2x)}{(x^2 + y^2)^2} + \frac{(x^2 + y^2)(0) - x(2y)}{(x^2 + y^2)^2}$$

$$= \frac{2xy}{(x^2 + y^2)^2} + \frac{-2xy}{(x^2 + y^2)^2} = 0.$$

Interpreting $\mathbf{F}$ as the velocity field of a fluid flow, the above result says that a given region not containing the origin always flows into a region of the same area.

(b) With $F(x, y) = -y/(x^2 + y^2)$ and $G(x, y) = x/(x^2 + y^2)$, the definition of $\text{curl}\mathbf{F}$ gives

$$\text{curl}\mathbf{F} = \frac{\partial G}{\partial x} - \frac{\partial F}{\partial y} = \frac{\partial}{\partial x}\left(\frac{x}{x^2 + y^2} \right) + \frac{\partial}{\partial y}\left(\frac{-y}{x^2 + y^2} \right)$$

$$= \frac{(x^2 + y^2)(1) - x(2x)}{(x^2 + y^2)^2} - \frac{(x^2 + y^2)(-1) - (-y)(2y)}{(x^2 + y^2)^2}$$

$$= \frac{-x^2 + y^2}{(x^2 + y^2)^2} - \frac{-x^2 + y^2}{(x^2 + y^2)^2} = 0.$$

As in part (a), we interpret $\mathbf{F}$ as the velocity field of a fluid flow. If γ is any piecewise smooth counterclockwise oriented circuit that does not go around the origin, D is the region enclosed by γ, and $\mathbf{t}$ is the unit tangent to γ in the direction of traversal then Stokes's Theorem in the plane gives

$$\oint_\gamma \mathbf{F} \cdot \mathbf{t}\, ds = \int_D \operatorname{curl}\mathbf{F}\, dA = \int_D 0\, dA = 0.$$

Since the leftmost member of this equation is the circulation of $\mathbf{F}$ around γ, the circulation is zero.

(c) For $a > 0$, let γ_a be the curve parametrized by $g(t) = (a\cos t, a\sin t)$, $0 \le t \le 2\pi$. Then γ_a is the counterclockwise oriented circle of radius a centered at the origin. We have

$$\mathbf{F}\big(g(t)\big) = \left(\frac{-a\sin t}{a^2\cos^2 t + a^2\sin^2 t}, \frac{a\cos t}{a^2\cos^2 t + a^2\sin^2 t} \right) = \left(-\frac{1}{a}\sin t, \frac{1}{a}\cos t \right),$$

so that

$$\oint_\gamma \mathbf{F}\big(g(t)\big) \cdot g'(t)\, dt = \int_0^{2\pi} \left(-\frac{1}{a}\sin t, \frac{1}{a}\cos t \right) \cdot (-a\sin t, a\cos t)\, dt = \int_0^{2\pi} 1\, dt = 2\pi.$$

Since Green's Theorem applies only to vector fields that are continuous on a region D, and since Stokes's Theorem in the plane is a variation of Green's Theorem, the result obtained here in part (c) cannot be obtained by using Stokes's Theorem in the plane as was done in part (b). In other words, Stokes's Theorem in the plane does not apply here and therefore parts (b) and (c) do not contradict each other.

19. With $i^2 = -1$ and $i^3 = -i$, the vector-valued function $f(x,y) = (x+iy)^3$ can be written as

$$f(x,y) = (x+iy)^3 = x^3 + 3x^2 iy + 3xi^2 y^2 + i^3 y^3 = x^3 + i3x^2 y - 3xy^2 - iy^3$$
$$= (x^3 - 3xy^2) + i(3x^2 y - y^3) = u(x,y) + iv(x,y),$$

where $u(x,y) = x^3 - 3xy^2$ and $v(x,y) = 3x^2 y - y^3$ are the real and imaginary parts (resp.) of f. Letting $\mathbf{F} = \big(u(x,y), -v(x,y)\big) = (x^3 - 3xy^2, -3x^2 y + y^3)$, we have

$$\operatorname{div}\mathbf{F} = \frac{\partial(x^3 - 3xy^2)}{\partial x} + \frac{\partial(-3x^2 y + y^3)}{\partial y} = (3x^2 - 3y^2) + (-3x^2 + 3y^2) = 0,$$

$$\operatorname{curl}\mathbf{F} = \frac{\partial(-3x^2 y + y^3)}{\partial x} - \frac{\partial(x^3 - 3xy^2)}{\partial y} = (-6xy) - (-6xy) = 0.$$

So $\mathbf{F}$ is irrotational and incompressible.

21. With $f(x,y) = \frac{1}{2}\ln(x^2 + y^2) + i\arctan(y/x)$, the real and imaginary parts of f are $u(x,y) = \frac{1}{2}\ln(x^2 + y^2)$ and $v(x,y) = \arctan(y/x)$, respectively. Letting

$$\mathbf{F}(x,y) = \big(u(x,y), -v(x,y)\big) = \big((1/2)\ln(x^2 + y^2), -\arctan(y/x)\big),$$

we have

$$\operatorname{div}\mathbf{F} = \frac{\partial\big((1/2)\ln(x^2 + y^2)\big)}{\partial x} + \frac{\partial\big(-\arctan(y/x)\big)}{\partial y}$$
$$= \frac{1}{2}\frac{1}{x^2 + y^2}(2x) + \left(-\frac{1}{1 + (y/x)^2}\frac{1}{x} \right) = \frac{x}{x^2 + y^2} + \left(-\frac{x}{x^2 + y^2} \right) = 0,$$

$$\operatorname{curl}\mathbf{F} = \frac{\partial\big(-\arctan(y/x)\big)}{\partial x} - \frac{\partial\big((1/2)\ln(x^2 + y^2)\big)}{\partial y}$$
$$= -\frac{1}{1 + (y/x)^2}\left(-\frac{y}{x^2} \right) - \frac{1}{2}\frac{1}{x^2 + y^2}(2y) = \frac{y}{x^2 + y^2} - \frac{y}{x^2 + y^2} = 0.$$

So $\mathbf{F}$ is irrotational and incompressible.

Section 2: CONSERVATIVE VECTOR FIELDS

Exercise Set 2A-D (pgs. 418-419)

1. Consider the vector field $\mathbf{F}(x, y, z) = (0, 0, -g)$ acting on a particle of mass 1 (here, g is the gravitational constant).

(a) We first observe that the partial derivatives of the coordinate functions of $\mathbf{F}$ are zero, so that the consistency condition holds and $\mathbf{F}$ is a gradient field. Therefore the potential energy function $U(x, y, z)$ for $\mathbf{F}$ exists. By inspection, the function $-gz$ is a field potential of $\mathbf{F}$ because $\nabla(-gz) = (0, 0, -g) = \mathbf{F}$, and therefore every field potential of $\mathbf{F}$ has the form $-gz + C$, where C is a real constant. Since $U(x, y, z)$ is the negative of the field potential function, U must satisfy

$$U(x, y, z) = gz - C, \quad \text{where } U(0, 0, 0) = 0.$$

So, $0 = U(0, 0, 0) = g(0) - C = -C$ implies $C = 0$ and the desired potential energy function is $U(x, y, z) = gz$.

(b) Let $\mathbf{x}(t)$ be the position of the particle at time $t \geq 0$. Since $\mathbf{F}$ is the only force acting on the particle, Newton's Second Law of Motion says that $\mathbf{F}(\mathbf{x}(t)) = m\ddot{\mathbf{x}}(t) = \ddot{\mathbf{x}}(t)$, where $m = 1$ is the mass of the particle. Using the initial condition $\dot{\mathbf{x}}(0) = (v_1, v_2, v_3)$, we use the fundamental theorem of calculus to integrate $\ddot{\mathbf{x}}$ from 0 to t and get

$$\dot{\mathbf{x}}(t) = \dot{\mathbf{x}}(0) + \int_0^t \ddot{\mathbf{x}}(u) \, du$$

$$= (v_1, v_2, v_3) + \int_0^t (0, 0, -g) \, du = (v_1, v_2, v_3) + (0, 0, -gt) = (v_1, v_2, -gt + v_3).$$

Using the condition $\mathbf{x}(0) = (0, 0, 0)$, we integrate again from 0 to t to get

$$\mathbf{x}(t) = \mathbf{x}(0) + \int_0^t \dot{\mathbf{x}}(u) \, du = (0, 0, 0) + \int_0^t (v_1, v_2, -gu + v_3) \, du = \left(v_1 t, v_2 t, -\frac{1}{2}gt^2 + v_3 t\right),$$

which holds for $t \geq 0$.

(c) The kinetic energy K of the particle at time t is by definition $K(\mathbf{x}(t)) = \frac{m}{2}|\dot{\mathbf{x}}(t)|^2$, where m is its mass. In this case,

$$K(\mathbf{x}(t)) = \frac{1}{2}|\dot{\mathbf{x}}(t)|^2 = \frac{1}{2}|(v_1, v_2, -gt + v_3)|^2$$

$$= \frac{1}{2}\left(v_1^2 + v_2^2 + (-gt + v_3)^2\right) = \frac{1}{2}(v_1^2 + v_2^2 + v_3^2) + \frac{1}{2}g^2 t^2 - gv_3 t.$$

If $x = x(t)$, $y = y(t)$ and $z = z(t)$ are the coordinate functions of $\mathbf{x}(t)$ then, from part (a) and part (b), the potential energy at time t is

$$U(\mathbf{x}(t)) = gz(t) = g\left(-\frac{1}{2}gt^2 + v_3 t\right) = -\frac{1}{2}g^2 t^2 + gv_3 t.$$

Hence, the total energy E is

$$E = U(\mathbf{x}(t)) + K(\mathbf{x}(t))$$

$$= -\frac{1}{2}g^2 t^2 + gv_3 t + \frac{1}{2}(v_1^2 + v_2^2 + v_3^2) + \frac{1}{2}g^2 t^2 - gv_3 t = \frac{1}{2}(v_1^2 + v_2^2 + v_3^2),$$

which is a constant.

3. The vector field $\mathbf{F}(x,y) = (x - y, x + y)$, for (x,y) in $\mathcal{R}^2$, is continuous. Moreover, the circuit γ defined by $g(t) = (\cos t, \sin t)$, $0 \leq t \leq 2\pi$, is smooth. However,

$$\oint_\gamma \mathbf{F} \cdot d\mathbf{x} = \oint_\gamma \mathbf{F}\big(g(t)\big) \cdot g'(t)\, dt = \int_0^{2\pi} (\cos t - \sin t, \cos t + \sin t) \cdot (-\sin t, \cos t)\, dt$$

$$= \int_0^{2\pi} (-\cos t \sin t + \sin^2 t + \cos^2 t + \sin t \cos t)\, dt = \int_0^{2\pi} 1\, dt = 2\pi \neq 0.$$

By the equivalence of parts (b) and (c) of Theorem 2.4, we conclude that $\mathbf{F}$ is not a gradient field on $\mathcal{R}^2$.

5. The vector field $\mathbf{H}(x,y) = (-y/(x^2 + y^2), x/(x^2 + y^2))$, for $(x,y) \neq 0$, is continuous and its region of definition is a polygonally connected open subset of $\mathcal{R}^2$. If γ is the unit circle parametrized by $g(t) = (\cos t \sin t)$, $0 \leq t \leq 2\pi$, we have

$$\oint_\gamma \mathbf{H} \cdot d\mathbf{x} = \oint_\gamma \mathbf{H}\big(g(t)\big) \cdot g'(t)\, dt$$

$$= \int_0^{2\pi} \left(\frac{-\sin t}{\cos^2 t + \sin^2 t}, \frac{\cos t}{\cos^2 t + \sin^2 t} \right) \cdot (-\sin t, \cos t)\, dt$$

$$= \int_0^{2\pi} (\sin^2 t + \cos^2 t)\, dt = \int_0^{2\pi} 1\, dt = 2\pi \neq 0.$$

By the equivalence of parts (b) and (c) of Theorem 2.4, we conclude that $\mathbf{H}$ is not a gradient field for $(x,y) \neq 0$.

7. The vector fields $\mathbf{F}(x,y) = (x - y, x + y)$ and $\mathbf{G}(x,y,z) = (y, z, x)$ given in Exercises 3 and 4 are continuously differentiable on $\mathcal{R}^2$ and $\mathcal{R}^3$, respectively. Letting $F_1(x,y) = x - y$, $F_2(x,y) = x + y$, $G_1(x,y,z) = y$, $G_2(x,y,z) = z$ and $G_3(x,y,z) = x$, the Jacobian matrices of $\mathbf{F}$ and $\mathbf{G}$ are given by

$$\mathbf{F}'(x,y) = \begin{pmatrix} (F_1)_x & (F_1)_y \\ (F_2)_x & F_2)_y \end{pmatrix} = \begin{pmatrix} 1 & -1 \\ 1 & 1 \end{pmatrix},$$

$$\mathbf{G}'(x,y,z) = \begin{pmatrix} (G_1)_x & (G_1)_y & (G_1)_z \\ (G_2)_x & (G_2)_y & (G_2)_z \\ (G_3)_x & (G_3)_y & (G_3)_z \end{pmatrix} = \begin{pmatrix} 0 & 1 & 0 \\ 0 & 0 & 1 \\ 1 & 0 & 0 \end{pmatrix},$$

neither of which is symmetric with respect to its main diagonal. By Theorem 2.5, $\mathbf{F}$ is not a gradient field on $\mathcal{R}^2$ and $\mathbf{G}$ is not a gradient field on $\mathcal{R}^3$.

9. The given vector field and its domain of definition D can be described as

$$\mathbf{F}(x,y,z) = \left(\frac{-y}{x^2 + y^2}, \frac{x}{x^2 + y^2}, 0 \right), \quad (x,y,z) \neq (0,0,z), \quad \text{for } -\infty < z < \infty.$$

We claim that $\mathbf{F}$ is not a gradient field on D. To see this, note that D is a polygonally connected open subset of $\mathcal{R}^3$ and that $\mathbf{F}$ is continuous on D. Now consider the curve γ defined by $g(t) = (\cos t, \sin t, 0)$, $0 \leq t \leq 2\pi$, and note that γ is the unit circle on the xy-plane and is therefore contained in D. Referring to the computations in Exercise 5 above, we have

$$\oint_\gamma \mathbf{F} \cdot d\mathbf{x} = \oint_\gamma \mathbf{F}\big(g(t)\big) \cdot g'(t)\, dt$$

$$= \int_0^{2\pi} (-\sin t, \cos t, 0) \cdot (-\sin t, \cos t, 0)\, dt = \int_0^{2\pi} 1\, dt = 2\pi \neq 0.$$

By the equivalence of parts (b) and (c) of Theorem 2.4, we conclude that $\mathbf{F}$ is not a gradient field on D.

11. Let $\mathbf{G}(x,y) = (y\cos xy, x\cos xy)$. If $g(x,y)$ is a potential for $\mathbf{G}$ (i.e., if $\mathbf{G} = \nabla g$), then the equations $g_x(x,y) = y\cos xy$ and $g_y(x,y) = x\cos xy$ must simultaneously hold. Holding y fixed, we integrate $g_x(x,y)$ with respect to x to get

$$g(x,y) = \int g_x(x,y)\,dx = \int y\cos xy\,dx = \sin xy + C(y),$$

where $C(y)$ is a differentiable function of y and is independent of x. Differentiating the outer members of the above equation with respect to y gives

$$g_y(x,y) = x\cos xy + C'(y).$$

Comparing this with $g_y(x,y) = x\cos xy$, we see that $C'(y) = 0$, so that $C(y) = c$, a real constant. Hence, the most general field potential for $\mathbf{G}$ is

$$g(x,y) = \sin xy + c.$$

Any specific choice for c gives a valid potential.

13. Let $\mathbf{K}(x,y) = \left(x/(x^2 + y^2), y/(x^2 + y^2)\right)$, $(x,y) \neq 0$. If $k(x,y)$ is a potential for $\mathbf{K}$ (i.e., if $\mathbf{K} = \nabla k$) then the equations $k_x(x,y) = x/(x^2 + y^2)$ and $k_y(x,y) = y/(x^2 + y^2)$ must simultaneously hold. Holding y fixed, we integrate $k_x(x,y)$ with respect to x to get

$$k(x,y) = \int k_x(x,y)\,dx = \int \frac{x}{x^2 + y^2}\,dx = \frac{1}{2}\int \frac{1}{u}\,du = \frac{1}{2}\ln u + C(y) = \frac{1}{2}\ln(x^2 + y^2) + C(y),$$

where $C(y)$ is a differentiable function of y that is independent of x, and the third equality follows via the substitution $u = x^2 + y^2$. Differentiating the outer members of the above equation with respect to y gives

$$k_y(x,y) = \frac{\partial}{\partial y}\left(\frac{1}{2}\ln(x^2 + y^2) + C(y)\right) = \frac{1}{2}\frac{1}{x^2 + y^2}(2y) + C'(y) = \frac{y}{x^2 + y^2} + C'(y).$$

Comparing this with $k_y(x,y) = y/(x^2 + y^2)$ shows that $C'(y) = 0$, so that $C(y) = c$, a real constant. Hence, the most general field potential for $\mathbf{K}$ is

$$k(x,y) = \frac{1}{2}\ln(x^2 + y^2) + c.$$

Any specific choice for c gives a valid potential.

15. Let $\mathbf{F}$ be a continuous vector field defined in a region D of $\mathcal{R}^n$. We wish to show that $\mathbf{F}$ satisfies Relation 2.2 in the text if, and only if, it satisfies Relation 2.3.

Assume $\mathbf{F}$ satisfies Relation 2.2 and let γ be a piecewise smooth closed curve lying in D. Let $\mathbf{x}_1$ and $\mathbf{x}_2$ be any two distinct points on γ and let γ_1 be that part of γ that is traced out from $\mathbf{x}_1$ to $\mathbf{x}_2$ and let γ_2 be that part of γ that is traced out from $\mathbf{x}_2$ to $\mathbf{x}_1$. Then γ_1 and γ_2^- are two paths from $\mathbf{x}_1$ to $\mathbf{x}_2$. By assumption,

$$\int_{\gamma_1} \mathbf{F} \cdot d\mathbf{x} = \int_{\gamma_2^-} \mathbf{F} \cdot d\mathbf{x}.$$

321

Hence,

$$\oint_\gamma \mathbf{F} \cdot d\mathbf{x} = \oint_{\gamma_1 \cup \gamma_2} \mathbf{F} \cdot d\mathbf{x} = \int_{\gamma_1} \mathbf{F} \cdot d\mathbf{x} + \int_{\gamma_2} \mathbf{F} \cdot d\mathbf{x} = \int_{\gamma_1} \mathbf{F} \cdot d\mathbf{x} - \int_{\gamma_2^-} \mathbf{F} \cdot d\mathbf{x} = 0.$$

Since γ was an arbitrary piecewise smooth closed curve in D, $\mathbf{F}$ satisfies Relation 2.3.

Conversely, assume $\mathbf{F}$ satisfies Relation 2.3. Let $\mathbf{x}_1$ and $\mathbf{x}_2$ be any two points in D and let γ_1 and γ_2 be two piecewise smooth curves lying in D with initial point $\mathbf{x}_1$ and terminal point $\mathbf{x}_2$. Let γ be the curve that first traces out γ_1 and then traces out γ_2 in the reverse direction. That is, let $\gamma = \gamma_1 \cup \gamma_2^-$ and note that γ is a piecewise smooth circuit lying in D. We then have

$$\int_{\gamma_1} \mathbf{F} \cdot d\mathbf{x} - \int_{\gamma_2} \mathbf{F} \cdot d\mathbf{x} = \int_{\gamma_1} \mathbf{F} \cdot d\mathbf{x} + \int_{\gamma_2^-} \mathbf{F} \cdot d\mathbf{x} = \int_{\gamma_1 \cup \gamma_2^-} \mathbf{F} \cdot d\mathbf{x} = \int_\gamma \mathbf{F} \cdot d\mathbf{x}.$$

Since the rightmost member of the above equation is zero (by assumption), so is the leftmost member. That is,

$$\int_{\gamma_1} \mathbf{F} \cdot d\mathbf{x} = \int_{\gamma_2} \mathbf{F} \cdot d\mathbf{x}.$$

Since $\mathbf{x}_1$ and $\mathbf{x}_2$ were arbitrary points in D, and since γ_1 and γ_2 were arbitrary piecewise smooth paths in D with initial point $\mathbf{x}_1$ and terminal point $\mathbf{x}_2$, $\mathbf{F}$ satisfies Relation 2.2.

17. The vector field $\mathbf{F}(x,y) = \left(2x/(x^2+y^2), 2y/(x^2+y^2)\right)$, for $(x,y) \neq (0,0)$, is continuously differentiable on its domain of definition D. We assume that $\mathbf{F}$ is a gradient field and that $f : D \longrightarrow \mathcal{R}$ is a continuously differentiable function such that $\mathbf{F}(x,y) = \nabla f(x,y)$ for all $(x,y) \in D$. Then $f_x(x,y) = 2x/(x^2+y^2)$ and $f_y(x,y) = 2y/(x^2+y^2)$. Integrating $f_x(x,y)$ with respect to x while holding y fixed gives

$$f(x,y) = \int f_x(x,y)\,dx = \int \frac{2x}{x^2+y^2}\,dx = \int \frac{1}{u}\,du = \ln u + C(y) = \ln(x^2+y^2) + C(y),$$

where the third equality is obtained via the substitution $u = x^2 + y^2$, and $C(y)$ is a differentiable function of y that is independent of x. Differentiating the outer members of the above equation with respect to y gives

$$f_y(x,y) = \frac{1}{x^2+y^2}(2y) + C'(y) = \frac{2y}{x^2+y^2} + C'(y).$$

Comparing this with $f_y(x,y) = 2y/(x^2+y^2)$ shows that $C'(y) = 0$, so that $C(y) = c$, a real constant. Thus, the most general potential for $\mathbf{F}$ is

$$f(x,y) = \ln(x^2+y^2) + c.$$

Any specific value for c gives a valid potential.

19. If $\nabla f(x,y) = (e^y, xe^y)$ then $f_x(x,y) = e^y$ and $f_y(x,y) = xe^y$. Integrating $f_x(x,y)$ with respect to x while holding y fixed gives

$$f(x,y) = \int f_x(x,y)\,dx = \int e^y\,dx = xe^y + C(y),$$

where $C(y)$ is a differentiable function of y that is independent of x. Differentiating the outer members of the above equation with respect to y gives

$$f_y(x,y) = xe^y + C'(y).$$

Comparing this with $f_y(x,y) = xe^y$ shows that $C'(y) = 0$, so that $C(y) = c$, a real constant. Hence, the most general solution f of the equation $\nabla f(x,y) = (e^y, xe^y)$ is

$$f(x,y) = xe^y + c.$$

21. If $\nabla f(x,y,z) = (y+z, z+x, x+y)$ then $f_x(x,y,z) = y+z$, $f_y(x,y,z) = z+x$ and $f_z(x,y,z) = x+y$. Integrating $f_x(x,y,z)$ with respect to x while holding y and z fixed gives

$$f(x,y,z) = \int f_x(x,y,z)\, dx = \int (y+z)\, dx = xy + xz + C(y,z),$$

where $C(y,z)$ is differentiable function of y and z that is independent of x. Differentiating the outer members of the above equation with respect to y gives

$$f_y(x,y,z) = x + C_y(y,z).$$

Comparing this with $f_y(x,y,z) = z + x$ shows that $C_y(y,z) = z$, so that integration with respect to y while holding z fixed gives $C(y,z) = yz + D(z)$, where $D(z)$ is a differentiable function of z that is independent of y. So,

$$f(x,y,z) = xy + xz + yz + D(z).$$

Differentiating this equation with respect to z gives $f_z(x,y,z) = x + y + D'(z)$. Comparing this with $f_z(x,y,z) = x + y$ shows that $D'(z) = 0$, so that $D(z) = c$, a real constant. Hence, the most general solution f of the equation $\nabla f(x,y,z) = (y+z, z+x, x+y)$ is

$$f(x,y,z) = xy + xz + yz + c.$$

23. Let $\nabla f(x,y) = (y^2 + 2xy, 2xy + x^2)$. Since ∇f is continuous, we can deduce from Theorem 1.3 of Chapter 8 that, for each $\mathbf{x} = (x,y)$ in $\mathcal{R}^2$,

$$f(x,y) = f(0,0) + \int_{(0,0)}^{(x,y)} \nabla f \cdot d\mathbf{y} = f(0,0) + \int_{(0,0)}^{(x,y)} (v^2 + 2uv, 2uv + u^2) \cdot (du, dv)$$

$$= f(0,0) + \int_{(0,0)}^{(x,y)} (v^2 + 2uv)\, du + (2uv + u^2)\, dv,$$

where $\mathbf{y} = (u,v)$ is the dummy variable we will use in the line integral. Since the value of the integral depends only on the point (x,y) and not on the path form $(0,0)$ to (x,y), we can let the path from $(0,0)$ to (x,y) consist of the horizontal line segment from $(0,0)$ to $(x,0)$ followed by the vertical line segment from $(x,0)$ to (x,y). On the first piece, $v = 0$ is constant, so that $dv = 0$. Thus, the line integral along this piece is zero. On the second piece, $u = x$ is constant so that $du = 0$. In this case, the line integral reduces to $\int_0^y (2xv + x^2)\, dv$. Hence,

$$f(x,y) = f(0,0) + \int_0^y (2xv + x^2)\, dv = f(0,0) + \left[xv^2 + x^2 v \right]_0^y = f(0,0) + xy^2 + x^2 y.$$

Here, the value of $f(0,0)$ can be chosen to be any real constant. Once this choice is made, f is uniquely determined. Otherwise, we can simply decide to leave $f(0,0)$ arbitrary, write $f(0,0) = c$, and give the most general potential as

$$f(x,y) = xy^2 + x^2 y + c.$$

25. Consider the vector field $\mathbf{F}(x,y) = 2x/(x^2 + y^2)\mathbf{i} = 2y/(x^2 + y^2)\mathbf{j}$. Let γ be any piecewise smooth circuit that does not go around the origin and let D be the region bounded by γ. It was shown in Exercise 17 above that $\mathbf{F}$ is a gradient field on its domain of definition and therefore $\mathbf{F}$ is continuously differentiable for $(x, y) \neq (0, 0)$. In particular, $\mathbf{F}$ is continuously differentiable on D. Letting $F_1(x, y) = 2x/(x^2 + y^2)$ and $F_2(x, y) = 2y/(x^2 + y^2)$, Green's Theorem gives

$$\oint_\gamma \mathbf{F} \cdot d\mathbf{x} = \int_D \left(\frac{\partial F_2}{\partial x} - \frac{\partial F_1}{\partial y} \right) dx\, dy = \int_D \left[\frac{\partial}{\partial x} \left(\frac{2y}{x^2 + y^2} \right) - \frac{\partial}{\partial y} \left(\frac{2x}{x^2 + y^2} \right) \right] dx\, dy$$

$$= \int_D \left(\frac{-4xy}{(x^2 + y^2)^2} - \frac{-4xy}{(x^2 + y^2)^2} \right) dx\, dy = \int_D 0\, dx\, dy = 0.$$

Since γ was an arbitrary piecewise smooth circuit that does not go around the origin, the above holds for all such curves.

27. A particle of unit mass moves with constant angular velocity ω on a circle of radius a about the origin in $\mathcal{R}^2$. The centripetal force field $\mathbf{F}(\mathbf{x}) = -k\omega^2 \mathbf{x}$, where $k > 0$ is a constant, is acting throughout $\mathcal{R}^2$ and tends to constrain the particle to remain on its circular path.

(a) The function $f(\mathbf{x}) = -\frac{1}{2}k\omega^2 |\mathbf{x}|^2$, for $\mathbf{x} = (x, y)$ in $\mathcal{R}^2$, can be written as $f(x, y) = -\frac{1}{2}k\omega^2 (x^2 + y^2)$. Hence,

$$\nabla f(x, y) = \left(-\frac{1}{2}k\omega^2 (2x), -\frac{1}{2}k\omega^2 (2y) \right) = (-k\omega^2 x, -k\omega^2 y) = -k\omega^2 (x, y) = \mathbf{F}(x, y).$$

That is, f is a potential for $\mathbf{F}$ and therefore $\mathbf{F}$ is a gradient field.

(b) Since $\mathbf{F}$ is a gradient field, the equivalence of (b) and (c) of Theorem 2.4 implies that the line integral of $\mathbf{F}$ around the circle is zero. This fact can also be deduced from Theorem 1.3 in Chapter 8. Perhaps a more direct approach is to simply note that the direction of traversal along the circle is perpendicular to the direction of the field and therefore, not only is the total work done by the field during one complete traversal of the circle equal to zero, but the total work done by the field during the traversal of *any* part of the circle is equal to zero.

(c) Suppose the particle leaves the circular path of radius a at the point $\mathbf{x}_0$ and follows a smooth path to a point $\mathbf{x}_1$, where it then proceeds to orbit in a circular path of radius b. The work done by the field during this "orbital transition" is therefore

$$W = \int_{\mathbf{x}_0}^{\mathbf{x}_1} \mathbf{F} \cdot d\mathbf{x} = \int_{\mathbf{x}_0}^{\mathbf{x}_1} \nabla f \cdot d\mathbf{x} = f(\mathbf{x}_1) - f(\mathbf{x}_0) = -\frac{1}{2}k\omega^2 |\mathbf{x}_1|^2 + \frac{1}{2}k\omega^2 |\mathbf{x}_0|^2$$

$$= -\frac{1}{2}k\omega^2 b^2 + \frac{1}{2}k\omega^2 a^2 = -\frac{1}{2}k\omega^2 (b^2 - a^2),$$

where the fifth equality holds because we have assumed that $\mathbf{x}_0$ is a units from the origin (so that $|\mathbf{x}_0| = a$) and that $\mathbf{x}_1$ is b units from the origin (so that $|\mathbf{x}_1| = b$). Notice that the work done by the field is negative. This is consistent with the overall definition of work; namely, *force over distance in the direction of motion*. Since the particle is, in general, moving away from the origin during the orbital transition, while the field is acting directly toward the origin, we should expect the total work done by the field during the transition to be negative.

Section 3: SURFACE INTEGRALS

Exercise Set 3A-D (pgs. 429-431)

1. (a) The region R in $\mathcal{R}^2$ defined by $u \geq 0$, $v \geq 0$, $u + v \leq 1$ is the triangle in the first quadrant bounded by the coordinate axes and the line $u + v = 1$. Its image $g(R) = T$ under the transformation $g(u, v) = (2u + v, v, 3u + v)$ is the triangle in $\mathcal{R}^3$ shown on the right. Note that the y-axis has been scaled to give a better perspective of T.

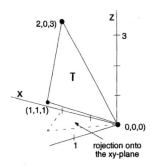

(b) To find the area of T, we first compute $g_u(u, v) = (2, 0, 3)$, $g_v(u, v) = (1, 1, 1)$ and

$$g_u(u, v) \times g_v(u, v) = \left(\begin{vmatrix} 0 & 1 \\ 3 & 1 \end{vmatrix}, \begin{vmatrix} 3 & 1 \\ 2 & 1 \end{vmatrix}, \begin{vmatrix} 2 & 1 \\ 0 & 1 \end{vmatrix} \right) = (-3, 1, 2).$$

Hence, the area differential is $d\sigma = |(-3, 1, 2)| \, du \, dv = \sqrt{14} \, du \, dv$ and the area of T is then

$$\sigma(T) = \int_R \sqrt{14} \, du \, dv = \int_0^1 dv \int_0^{-v+1} \sqrt{14} \, du$$

$$= \sqrt{14} \int_0^1 (-v + 1) \, dv = \sqrt{14} \left[-\frac{1}{2} v^2 + v \right]_0^1 = \frac{\sqrt{14}}{2}.$$

3. (a) Let D be the closed unit disk in the xy-plane. The points on the graph of $z = y^2 - x^2$ that lie on or above D are of the form $(x, y, y^2 - x^2)$, where $|y| \geq |x|$ and $x^2 + y^2 \leq 1$. Let S be this part of the surface. The projection of S onto the xy-plane therefore consists of two sectors of D: the first is the part of D (call it D^+) that lies above the x-axis and between the lines $y = x$ and $y = -x$; and the second is the part of D (call it D^-) lying below the x-axis and between the lines $y = x$ and $y = -x$. The details are shown in the figure on the right.

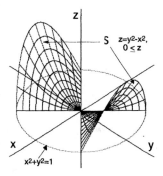

(b) The remarks made in part (a), show a number of things. First, a parametrization of S is given by

$$g(x, y) = (x, y, y^2 - x^2), \quad |y| > |x|, \; x^2 + y^2 \leq 1.$$

Hence, with $f(x, y) = y^2 - x^2$ in Equation 3.3 in the text, the area differential is given by

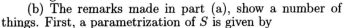

$$d\sigma = \sqrt{|\nabla(y^2 - x^2)|^2 + 1} \, dx \, dy = \sqrt{|(-2x, 2y)|^2 + 1} \, dx \, dy = \sqrt{4x^2 + 4y^2 + 1} \, dx \, dy.$$

Changing to polar coordinates, this becomes $d\sigma = r\sqrt{4r^2 + 1} \, dr \, d\theta$. Moreover, the description of D^+ given in part (a) becomes $0 \leq r \leq 1$, $\pi/4 \leq \theta \leq 3\pi/4$. Notice that S is symmetric with respect to the xz-plane. This fact allows us to compute only the area of the part of S that lies above D^+ and then multiply the result by 2. We therefore have

$$\sigma(S) = 2 \int_{D^+} d\sigma = 2 \int_{\pi/4}^{3\pi/4} d\theta \int_0^1 r\sqrt{4r^2 + 1} \, dr$$

$$= 2 \left[\theta \right]_{\pi/4}^{3\pi/4} \left[\frac{1}{12} (4r^2 + 1)^{3/2} \right]_0^1 = \frac{\pi}{12} (5\sqrt{5} - 1).$$

5. Let R be the square $0 \leq x \leq 1$, $0 \leq y \leq 1$ in the xy-plane, and let P be the portion of the graph of $z = x^2 + y$ that lies on or above R. Observe that no part of the surface defined by $z = x^2 + y$ is below the square.

(a) We use the parametrization $g(x, y) = (x, y, x^2 + y)$, for (x, y) in R. We have $g_x(x, y) = (1, 0, 2x)$ and $g_y(x, y) = (0, 1, 1)$, so that $g_x(x, y) \times g_y(x, y) = (-2x, -1, 1)$. If the weighted density of P is $\mu(x, y, z) = x$ then its mass is given by

$$M(P) = \int_R \mu(x, y, z)|g_x(x, y) \times g_y(x, y)| \, dx \, dy = \int_R x|(-2x, -1, 1)| \, dx \, dy$$

$$= \int_R x\sqrt{4x^2 + 2} \, dx \, dy = \int_0^1 dy \int_0^1 x\sqrt{4x^2 + 2} \, dx$$

$$= \int_0^1 dy \int_2^6 \frac{1}{8} u^{1/2} \, du = [y]_0^1 \left[\frac{1}{12} u^{3/2} \right]_2^6 = \frac{1}{12}(6^{3/2} - 2^{3/2}),$$

where the inside integral in the first iterated integral in the third line follows via the substitution $u = 4x^2 + 2$.

(b) For the vector field $\mathbf{F}(x, y, z) = (-x, y, z)$, we have $\mathbf{F}(g(x, y)) = \mathbf{F}(x, y, x^2 + y) = (-x, y, x^2 + y^2)$. Hence, the flux of $\mathbf{F}$ across P is

$$\Phi(\mathbf{F}, P) = \int_R \mathbf{F}(g(x, y)) \cdot (g_x(x, y) \times g_y(x, y)) \, dx \, dy$$

$$= \int_R (-x, y, x^2 + y) \cdot (-2x, -1, 1) \, dx \, dy = \int_R 3x^2 \, dx \, dy$$

$$= \int_0^1 dy \int_0^1 3x^2 \, dx = [y]_0^1 [x^3]_0^1 = 1.$$

7. Let S_a be the sphere of radius a centered at the origin in $\mathcal{R}^3$. In order to find the flux of the repelling electric field $\mathbf{E}(\mathbf{x}) = |\mathbf{x}|^{-3} \mathbf{x}$ across S_a, we will find the flux of $\mathbf{E}$ across the top hemisphere H_a of S_a and then multiply the result by 2 (since the flux across the top hemisphere is the same as the flux across the bottom hemisphere). Also, we will use the parametrization $g(x, y) = (x, y, \sqrt{a^2 - x^2 - y^2})$, where (x, y) are points on the closed disk D_a of radius a with center at the origin in the xy-plane. We have

$$g_x(x, y) \times g_y(x, y) = (1, 0, -x/\sqrt{a^2 - x^2 - y^2}) \times (0, 1, -y/\sqrt{a^2 - x^2 - y^2})$$

$$= (x/\sqrt{a^2 - x^2 - y^2}, y/\sqrt{a^2 - x^2 - y^2}, 1).$$

We now make the simplifying observation that $|g(x, y)| = a$, so that the repelling force at each point of H_a is $\mathbf{E}(g(x, y)) = |g(x, y)|^{-3} g(x, y) = a^{-3}(x, y, \sqrt{a^2 - x^2 - y^2})$. Hence, the total flux across S_a is

$$\Phi(\mathbf{E}, S_a) = 2\Phi(\mathbf{E}, H_a) = 2\int_{D_a} \mathbf{E}(g(x, y)) \cdot (g_x(x, y) \times g_y(x, y)) \, dx \, dy$$

$$= 2\int_{D_a} a^{-3}(x, y, \sqrt{a^2 - x^2 - y^2}) \cdot (x/\sqrt{a^2 - x^2 - y^2}, y/\sqrt{a^2 - x^2 - y^2}, 1) \, dx \, dy$$

$$= 2\int_{D_a} a^{-3} \left(\frac{x^2}{\sqrt{a^2 - x^2 - y^2}} + \frac{y^2}{\sqrt{a^2 - x^2 - y^2}} + \sqrt{a^2 - x^2 - y^2} \right) dx \, dy$$

$$= \frac{2}{a} \int_{D_a} \frac{1}{\sqrt{a^2 - x^2 - y^2}} \, dx \, dy.$$

The above integral is improper since the integrand becomes unbounded near the boundary of D_a. Hence, we will perform the integration over the disk D_b of radius $0 < b < a$ and then let $b \to a$ from below. Changing to polar coordinates gives

$$\Phi(\mathbf{E}, S_a) = \frac{2}{a} \int_{D_a} \frac{r}{\sqrt{a^2 - r^2}} \, dr \, d\theta = \lim_{b \to a^-} \frac{2}{a} \int_{D_b} \frac{r}{\sqrt{a^2 - r^2}} \, dr \, d\theta$$

$$= \lim_{b \to a^-} \frac{2}{a} \int_0^{2\pi} d\theta \int_0^b \frac{r}{\sqrt{a^2 - r^2}} \, dr = \lim_{b \to a^-} \frac{2}{a} [\theta]_0^{2\pi} \left[-\sqrt{a^2 - r^2} \right]_0^b$$

$$= \lim_{b \to a^-} \frac{4\pi}{a} \left(-\sqrt{a^2 - b^2} + a \right) = 4\pi.$$

9. (a) Let R be the rectangle $0 \le u \le 1$, $0 \le v \le 2$ in $\mathcal{R}^2$ and let S be the surface parametrized by $g(u, v) = (u - v, u + v, uv)$, for (u, v) in R. We have

$$g_u(u, v) \times g_v(u, v) = (1, 1, v) \times (-1, 1, u) = \left(\begin{vmatrix} 1 & 1 \\ v & u \end{vmatrix}, \begin{vmatrix} v & u \\ 1 & -1 \end{vmatrix}, \begin{vmatrix} 1 & -1 \\ 1 & 1 \end{vmatrix} \right)$$

$$= (u - v, -u - v, 2).$$

On S, the field $\mathbf{F}(x, y, z) = (x, y, z)$ satisfies $\mathbf{F}(g(u, v)) = g(u, v) = (u - v, u + v, uv)$. Hence, the surface integral of $\mathbf{F}$ over S is

$$\int_S \mathbf{F} \cdot d\mathbf{S} = \int_R \mathbf{F}(g(u, v)) \cdot (g_u(u, v) \times g_v(u, v)) \, du \, dv$$

$$= \int_R (u - v, u + v, uv) \cdot (u - v, -u - v, 2) \, du \, dv = \int_R (-2uv) \, du \, dv$$

$$= -2 \int_0^2 v \, dv \int_0^1 u \, du = -2 \left[\frac{1}{2} v^2 \right]_0^2 \left[\frac{1}{2} u^2 \right]_0^1 = -2.$$

(b) Let R be the rectangle $0 \le u \le 1$, $0 \le v \le 2\pi$ in $\mathcal{R}^2$ and let S be the surface parametrized by $g(u, v) = (u \cos v, u \sin v, v)$, for (u, v) in R. We have

$$g_u(u, v) \times g_v(u, v) = (\cos v, \sin v, 0) \times (-u \sin v, u \cos v, 1)$$

$$= \left(\begin{vmatrix} \sin v & u \cos v \\ 0 & 1 \end{vmatrix}, \begin{vmatrix} 0 & 1 \\ \cos v & -u \sin v \end{vmatrix}, \begin{vmatrix} \cos v & -u \sin v \\ \sin v & u \cos v \end{vmatrix} \right)$$

$$= (\sin v, -\cos v, u).$$

On S, the field $\mathbf{F}(x, y, z) = (x^2, 0, 0)$ satisfies $\mathbf{F}(g(u, v)) = (u^2 \cos^2 v, 0, 0)$. Hence, the surface integral of $\mathbf{F}$ over S is

$$\int_S \mathbf{F} \cdot d\mathbf{S} = \int_R \mathbf{F}(g(u, v)) \cdot (g_u(u, v) \times g_v(u, v)) \, du \, dv$$

$$= \int_R (u^2 \cos^2 v, 0, 0) \cdot (\sin v, -\cos v, u) \, du, dv = \int_R u^2 \cos^2 v \sin v \, du \, dv$$

$$= \int_0^{2\pi} \cos^2 v \sin v \, dv \int_0^1 u^2 \, du = \left[-\frac{1}{3} \cos^3 v \right]_0^{2\pi} \left[\frac{1}{3} u^3 \right]_0^1 = 0.$$

327

11. (a) Let $\mathbf{x} = g(u,v)$, for (u,v) in D, and $\mathbf{x} = h(s,t)$, for (s,t) in B, be equivalent parametrizations for the piece of smooth surface S in $\mathcal{R}^3$. Then there exists a continuously differentiable one-to-one transformation $T : D \longrightarrow B$ with $T(u,v) = (s,t)$ such that $h\big(T(u,v)\big) = g(u,v)$. That is, $s = s(u,v)$ and $t = t(u,v)$ are continuously differentiable functions of u and v. By the chain rule, the partial derivatives of g satisfy

$$g_u = h_s s_u + h_t t_u \qquad \text{and} \qquad g_v = h_s s_v + h_t t_v.$$

Hence, using the fact that the partial derivatives of s and t are scalar functions (and therefore can be moved past the cross product), we get

$$
\begin{aligned}
g_u \times g_v &= \big[h_s s_u + h_t t_u\big] \times \big[h_s s_v + h_t t_v\big] \\
&= \big[h_s s_u \times h_s s_v\big] + \big[h_s s_u \times h_t t_v\big] + \big[h_t t_u \times h_s s_v\big] + \big[h_t t_u \times h_t t_v\big] \\
&= \big[s_u s_v(h_s \times h_s)\big] + \big[s_u t_v(h_s \times h_t)\big] + \big[-t_u s_v(h_s \times h_t)\big] + \big[t_u t_v(h_t \times h_t)\big] \\
&= (0,0,0) + (h_s \times h_t)(s_u t_v - t_u s_v) + (0,0,0) \\
&= (h_s \times h_t)(s_u t_v - t_u s_v),
\end{aligned}
$$

where the anti-commutativity property of the cross product was used to obtain the third bracketed term in the third row, and the two zero terms in the fourth row result from the fact that h_s is parallel to h_s and h_t is parallel to h_t, so that their respective cross products vanish. Moreover,

$$T'(u,v) = \begin{pmatrix} s_u & s_v \\ t_u & t_v \end{pmatrix} \qquad \text{implies} \qquad \det T'(u,v) = s_u t_v - t_u s_v.$$

Using this in the last member above, the magnitudes of the outer members are related by

$$(*) \qquad |g_u(u,v) \times g_v(u,v)| = |h_s \times h_t|\big(T(u,v)\big)\,|\det T'(u,v)|,$$

where we have written $|h_s \times h_t|$ composed with T. Since T is one-to-one on D and $\det T' \neq 0$ in the interior of D, Jacobi's Theorem says that

$$\int_B |h_s \times h_t|(s,t)\,ds\,dt = \int_{T(D)} |h_s \times h_t|(s,t)\,ds\,dt = \int_D |h_s \times h_t|\big(T(u,v)\big)\,|\det T'(u,v)|\,du\,dv.$$

Hence, the integrand of the rightmost integral can be replaced with the left member of $(*)$ to get

$$\int_B |h_s \times h_t|\,ds\,dt = \int_D |g_u \times g_v|\,du\,dv.$$

By definition, the left side is the surface area of S using the parametrization h and the right member is the surface area of S using the parametrization g. That is, equivalent parametrizations of the same piece of smooth surface give the same surface area.

(b) Let $\mathbf{F}$ be a continuous vector field defined on $\bar{D}$. With the same conditions on g, h and T as in part (a), we can write

$$\int_D \mathbf{F}\big(g(u,v)\big) \cdot \big(g_u(u,v) \times g_v(u,v)\big)\,du\,dv$$

$$= \int_D \mathbf{F}\Big(h\big(T(u,v)\big)\Big) \cdot (h_s \times h_t)\big(T(u,v)\big)\,\det T'(u,v)\,du\,dv$$

$$= \int_D [(\mathbf{F} \circ h) \cdot (h_s \times h_t)]\big(T(u,v)\big)\,|\det T'(u,v)|\,du\,dv,$$

328

where the first integral is the surface integral of $\mathbf{F}$ over S with respect to the parametrization g, the second integral is obtained by using $(*)$ (without the absolute value bars) in part (a), and the third integral is obtained by using the fact that the Jacobian determinant of T is positive on D. Notice that the dot product function $(\mathbf{F} \circ h) \cdot (h_s \times h_t)$ is a real-valued function being evaluated at $T(u, v)$ and therefore can be viewed as a function of the parameter pair (u, v). Moreover, this dot product function is continuous on D. We now apply Jacobi's Theorem to the last integral to get

$$\int_D [(\mathbf{F} \circ h) \cdot (h_s \times h_t)] \left(T(u, v) \right) |\det T'(u, v)| \, du \, dv,$$

$$= \int_{T(D)} [(\mathbf{F} \circ h) \cdot (h_s \times h_t)] (s, t) \, ds \, dt = \int_B \mathbf{F}\bigl(h(s, t)\bigr) \cdot \bigl(h_s(s, t) \times h_t(s, t)\bigr) \, ds \, dt,$$

where the last integral is, by definition, the surface integral of $\mathbf{F}$ over S with respect to the parametrization h. In other words, for each continuous vector field defined on a given piece of smooth surface S, the surface integrals of $\mathbf{F}$ over S have the same value with respect to any two equivalent parametrizations of S.

13. The gradient of the Newtonian potential $(x^2 + y^2 + z^2)^{-1/2}$ is the attractive force field

$$\mathbf{F}(x, y, z) = \left(\frac{-x}{(x^2 + y^2 + z^2)^{3/2}}, \frac{-y}{(x^2 + y^2 + z^2)^{3/2}}, \frac{-z}{(x^2 + y^2 + z^2)^{3/2}} \right),$$

defined for $(x, y, z) \neq (0, 0, 0)$. We want to show that the flux of $\mathbf{F}$ across a sphere of radius a centered at the origin in $\mathcal{R}^3$ is independent of a. To do this, it will be sufficient to show that the flux across the top hemisphere H_a of such a sphere is independent of a. Thus, let D_a be the region in the xy-plane defined by $x^2 + y^2 \leq a^2$ and use the parametrization

$$g(x, y) = \bigl(x, y, (a^2 - x^2 - y^2)^{1/2}\bigr), \quad (x, y) \in D_a.$$

It was shown in Exercise 7 above that

$$g_x(x, y) \times g_y(x, y) = \left(\frac{x}{(a^2 - x^2 - y^2)^{1/2}}, \frac{y}{(a^2 - x^2 - y^2)^{1/2}}, 1 \right)$$

$$= \frac{1}{(a^2 - x^2 - y^2)^{1/2}} \bigl(x, y, (a^2 - x^2 - y^2)^{1/2}\bigr) = \frac{g(x, y)}{(a^2 - x^2 - y^2)^{1/2}}.$$

Moreover, on H_a we have $x^2 + y^2 + z^2 = a^2$, so that

$$\mathbf{F}\bigl(g(x, y)\bigr) = \left(-\frac{x}{a^3}, -\frac{y}{a^3}, -\frac{(a^2 - x^2 - y^2)^{1/2}}{a^3} \right) = -\frac{1}{a^3} \bigl(x, y, (a^2 - x^2 - y^2)^{1/2}\bigr) = -\frac{g(x, y)}{a^3}.$$

Hence, since $|g(x, y)| = a$,

$$\mathbf{F}\bigl(g(x, y)\bigr) \cdot \bigl(g_x(x, y) \times g_y(x, y)\bigr) = \left(-\frac{g(x, y)}{a^3} \right) \cdot \left(\frac{g(x, y)}{(a^2 - x^2 - y^2)^{1/2}} \right)$$

$$= \frac{-g(x, y) \cdot g(x, y)}{a^3 (a^2 - x^2 - y^2)^{1/2}} = \frac{-|g(x, y)|^2}{a^3 (a^2 - x^2 - y^2)^{1/2}} = \frac{-1}{a(a^2 - y^2 - y^2)^{1/2}}.$$

The flux of $\mathbf{F}$ across H_a is therefore

$$\Phi(\mathbf{F}, H_a) = -\int_{D_a} \frac{1}{a\sqrt{a^2 - x^2 - y^2}}\, dx\, dy.$$

Again, it was shown in Exercise 7 above that this integral evaluates to 2π. So, in this case, $\Phi(\mathbf{F}, H_a) = -2\pi$, which is independent of a.

15. (a) Suppose $G(x, y, z)$ is a continuously differentiable real-valued function that implicitly determines a piece of smooth surface S on which $\partial G/\partial z \neq 0$. If S lies over a region D of the xy-plane such that only one point of S lies over each point in D, then S is the graph of a function $z = g(x, y)$ implicitly determined by G such that $G\big(x, y, g(x,y)\big) = 0$. By the implicit function theorem, g is continuously differentiable on D. With $\mathbf{x} = (x, y)$, and $\mathbf{y} = z$ in Theorem 3.3 of Chapter 6 we get

$$g'(x, y) = -G_z^{-1} G_{(x, y)} = -\frac{\partial G}{\partial z}^{-1}\left(\frac{\partial G}{\partial x}, \frac{\partial G}{\partial y}\right) = \left(-\frac{\partial G}{\partial z}^{-1}\frac{\partial G}{\partial x}, -\frac{\partial G}{\partial z}^{-1}\frac{\partial G}{\partial y}\right),$$

where the matrix G_z is a 1-by-1 matrix whose single entry is $\partial G/\partial z$ and therefore the "exponent" -1 can simultaneously be viewed as both a symbol for the inverse matrix of G_z or as the reciprocal of the real nonzero number $\partial G/\partial z$. On the other hand, $g'(x, y) = (g_x, g_y)$, so that

$$g_x(x, y) = -\frac{\partial G}{\partial z}^{-1}\frac{\partial G}{\partial x} \qquad \text{and} \qquad g_y(x, y) = -\frac{\partial G}{\partial z}^{-1}\frac{\partial G}{\partial y}.$$

Using the result of Exercise 14(a) in this section, we get

$$\sigma(S) = \int_D \sqrt{1 + (g_x)^2 + (g_y)^2}\, dx\, dy = \int_D \sqrt{1 + \left(-\frac{\partial G}{\partial z}^{-1}\frac{\partial G}{\partial x}\right)^2 + \left(-\frac{\partial G}{\partial z}^{-1}\frac{\partial G}{\partial y}\right)^2}\, dx\, dy$$

$$= \int_D \sqrt{\left(\frac{\partial G}{\partial x}\right)^2 + \left(\frac{\partial G}{\partial y}\right)^2 + \left(\frac{\partial G}{\partial z}\right)^2}\, \left|\frac{\partial G}{\partial z}\right|^{-1}\, dx\, dy$$

(b) Let $a > 0$ be a constant, let D be the circle of radius a centered at the origin in the xy-plane, and Let S be the *open* hemisphere $x^2 + y^2 + z^2 = a^2$, $z > 0$. The function $G(x, y, z) = x^2 + y^2 + z^2 - a^2$ is continuously differentiable and implicitly determines S as a piece of smooth surface and $G_z = 2z \neq 0$ on S. Since $G_x = 2x$ and $G_y = 2y$, the formula proven in part (a) becomes

$$\sigma(S) = \int_D \sqrt{(2x)^2 + (2y)^2 + (2z)^2}\, \frac{1}{2z}\, dx dy = \int_D \frac{\sqrt{x^2 + y^2 + z^2}}{z}\, dx dy.$$

Since $G(x, y, z) = 0$ on S, we can solve for z to get $z = \sqrt{a^2 - x^2 - y^2}$. Inserting this expression for z into the above integrand and converting to polar coordinates gives

$$\sigma(S) = \int_D \frac{a}{\sqrt{a^2 - x^2 - y^2}}\, dx dy = \int_D \frac{ar}{\sqrt{a^2 - r^2}}\, dr d\theta = a\int_0^{2\pi} d\theta \int_0^a \frac{r}{\sqrt{a^2 - r^2}}\, dr$$

$$= a\big[\theta\big]_0^{2\pi}\big[-\sqrt{a^2 - r^2}\big]_0^a = 2\pi a^2.$$

330

17. The surface S of the can is comprised of two pieces. The first piece S_1 is the cylindrical frustum defined by $x^2 + y^2 = 1$, $0 \leq z \leq 1$, and the second piece is the closed unit disk in the xy-plane. Consider the obvious parametrizations of S_1 and S_2 given respectively by

$$g_1(u, v) = (\cos u, \sin u, v), \quad \text{for } 0 \leq u \leq 2\pi, \ 0 \leq v \leq 1,$$

$$g_2(u, v) = (v \cos u, v \sin u, 0), \quad \text{for } 0 \leq u \leq 2\pi, \ 0 \leq v \leq 1.$$

Let D_1 and D_2 be the above shown rectangular domains of g_1 and g_2, respectively. Note that S_1 and S_2 have a common border; namely, the unit circle in the xy-plane. On S_1, this border is traced *counterclockwise* as the boundary of D_1 is traced counterclockwise from $(0,0)$ to $(2\pi, 0)$. On S_2, this border is traced *clockwise* as the boundary of D_2 is traced counterclockwise from $(2\pi, 1)$ to $(0, 1)$. Thus, the two parametrizations are coherent along this common border and S is therefore oriented. Moreover,

$$\frac{\partial g_1}{\partial u} \times \frac{\partial g_1}{\partial v} = (-\sin u, \cos u, 0) \times (0, 0, 1) = (\cos u, \sin u, 0),$$

so that the unit normal to S at the point $(1, 0, 0)$ is the vector $(1, 0, 0)$, which points outward, away from the can. Because S is oriented, all normal unit vectors defined by the cross product of the partials of g_1 and g_2 are pointing outward.

19. The trough S is comprised of two pieces. The first piece S_1 is a rectangle defined by $z = y$, $0 \leq x \leq 1$, $0 \leq z \leq 1$, and the second piece is a rectangle defined by $z = -y$, $0 \leq x \leq 1$, $0 \leq z \leq 1$. They share a common border; namely, the piece of the x-axis from $(0, 0, 0)$ to $(1, 0, 0)$. Consider the two parametrizations

$$g_1(u, v) = (v, u, u), \quad \text{for } 0 \leq u \leq 1, \ 0 \leq v \leq 1,$$

$$g_2(u, v) = (v, u - 1, 1 - u), \quad \text{for } 0 \leq u \leq 1, \ 0 \leq v \leq 1.$$

For the purpose of identifying orientations, we distinguish the domain rectangles of g_1 and g_2 by the names D_1 and D_2, respectively. On S_1, the border on the x-axis is traced from $(1, 0, 0)$ to $(0, 0, 0)$ as the boundary of D_1 is traced counterclockwise from $(0, 1)$ to $(0, 0)$. On S_2, the border on the x-axis is traced from $(0, 0, 0)$ to $(1, 0, 0)$ as the boundary of D_2 is traced counterclockwise from $(1, 0)$ to $(1, 1)$. Thus, the two parametrizations are coherent along this common border and S is therefore oriented. Moreover, it is easy to check that

$$\frac{\partial g_1}{\partial u} \times \frac{\partial g_1}{\partial v} = (0, 1, -1) \quad \text{and} \quad \frac{\partial g_2}{\partial u} \times \frac{\partial g_2}{\partial v} = (0, -1, -1).$$

The first is a vector pointing outward on S_1, away from the "inside" of the trough, and the second is a vector pointing outward on S_2, away from the "inside" of the trough.

21. The oriented surface S in Exercise 17 is the union of two pieces of smooth surface S_1 and S_2, defined by parametrically by g_1 and g_2, respectively. The surface integral of the field $\mathbf{F}(x, y, z) = (x, y, 2z - x - y)$ over S can therefore be written as

$$\int_S \mathbf{F} \cdot d\mathbf{S} = \int_{S_1} \mathbf{F} \cdot d\mathbf{S} + \int_{S_2} \mathbf{F} \cdot d\mathbf{S}$$

$$= \int_{D_1} \mathbf{F}(g_1(u, v)) \cdot \left(\frac{\partial g_1}{\partial u} \times \frac{\partial g_1}{\partial v} \right) du \, dv + \int_{D_2} \mathbf{F}(g_2(u, v)) \cdot \left(\frac{\partial g_2}{\partial u} \times \frac{\partial g_2}{\partial v} \right) du \, dv,$$

where D_1 and D_2 are the parameter domains given for g_1 and g_2 (resp.) in Exercise 17. The two cross products are

$$\frac{\partial g_1}{\partial u} \times \frac{\partial g_1}{\partial v} = (\cos u, \sin u, 0) \quad \text{and} \quad \frac{\partial g_2}{\partial u} \times \frac{\partial g_2}{\partial v} = (0, 0, -v).$$

331

Thus, the two integrands are

$$\mathbf{F}\big(g_1(u,v)\big) \cdot \left(\frac{\partial g_1}{\partial u} \times \frac{\partial g_1}{\partial v}\right) = (\cos u, \sin u, 2v - \cos u - \sin u) \cdot (\cos u, \sin u, 0) = 1,$$

$$\mathbf{F}\big(g_2(u,v)\big) \cdot \left(\frac{\partial g_2}{\partial u} \times \frac{\partial g_2}{\partial v}\right) = (v\cos u, v\sin u, -v\cos u - v\sin u) \cdot (0,0,-v)$$

$$= v^2(\cos u + \sin u).$$

Hence,

$$\int_S \mathbf{F} \cdot d\mathbf{S} = \int_{D_1} 1\, du\, dv + \int_{D_2} v^2(\cos u + \sin u)\, du\, dv$$

$$= \int_0^1 dv \int_0^{2\pi} dv + \int_0^1 v^2 dv \int_0^{2\pi} (\cos u + \sin u)\, du = \big[v\big]_0^1 \big[u\big]_0^{2\pi} + 0 = 2\pi,$$

where the integrand $\cos u + \sin u$ in the second line forces the associated integral to be zero.
23. (Refer to Exercises 17 and 21 in this section). It was shown in the solution to Exercise 21 that the normals point downward on S_2. Specifically, $\partial g_2/\partial u \times \partial g_2/\partial v = (0,0,-v)$. So, we keep the parametrization g_2. It follows that the surface integral of $\mathbf{F}$ (where $\mathbf{F}$ is given in Exercise 21) over S_2 will not change. That is, it's value is zero, as computed in Exercises 21. To change the direction of the normals to S_1 from outward to inward, it is only necessary to interchange the variables u and v and define the new parametrization of S_1 by

$$f(u,v) = (\cos v, \sin v, u), \quad \text{for } 0 \le u \le 1,\ 0 \le v \le 2\pi.$$

The integrand of the surface integral of $\mathbf{F}$ over S_1 is therefore

$$\mathbf{F}\big(f(u,v)\big) \cdot \left(\frac{\partial f}{\partial u} \times \frac{\partial f}{\partial v}\right) = (\cos v, \sin v, 2u - \cos v - \sin v) \cdot (-\cos v, -\sin v, 0) = -1,$$

which shows that the surface integral of $\mathbf{F}$ over S_1 is the negative of what it was when it was computed with respect to the parametrization g_1. Thus, using the parametrizations f and g_2 instead of g_1 and g_2, we get

$$\int_S \mathbf{F} \cdot d\mathbf{S} = \int_{S_1} \mathbf{F} \cdot d\mathbf{S} + \int_{S_2} \mathbf{F} \cdot d\mathbf{S} = -2\pi.$$

25. Let $\mathbf{F}$, S and M be as given and, choosing an orientation for S, let $\mathbf{n} = \mathbf{n}(\mathbf{x})$ be the associated unit normal to S at the point $\mathbf{x}$. It was shown in Exercise 2 in Section 3 of Chapter 7 that the absolute value of an integral of a real-valued function f of several variables is never greater than the integral of $|f|$. Also, the Cauchy-Schwarz inequality and the definition of M gives

$$|\mathbf{F}(\mathbf{x}) \cdot \mathbf{n}(\mathbf{x})| \le |\mathbf{F}(\mathbf{x})|\,|\mathbf{n}(\mathbf{x})| = |\mathbf{F}(\mathbf{x})| \le M,$$

where we have used the fact that $\mathbf{n}(\mathbf{x})$ is a unit vector for all $\mathbf{x}$ in S. Hence, two applications of Theorem 3.5 in Chapter 7 gives

$$\left| \int_S \mathbf{F} \cdot d\mathbf{S} \right| = \left| \int_S \mathbf{F} \cdot \mathbf{n}\, d\sigma \right| \le \int_S |\mathbf{F} \cdot \mathbf{n}|\, d\sigma \le \int_S |\mathbf{F}|\,|\mathbf{n}|\, d\sigma \le \int_S M\, d\sigma = M \int_S d\sigma = M\sigma(S).$$

27. Let f be a real-valued continuously differentiable function of one real variable x and suppose $f(x) \geq 0$ for $a \leq x \leq b$. Let S be the surface generated by rotating the graph of f about the x-axis.

(a) Let (x_0, y_0, z_0) be any point on S. The last two coordinates will lie on a circle of radius $f(x_0)$ in the plane $x = x_0$ and can therefore be associated with an angle θ_0 such that $y_0 = f(x_0) \cos \theta_0$ and $z_0 = f(x_0) \sin \theta_0$. That is, the numbers x_0 and θ_0 can be used as parameter values to describe the point (x_0, y_0, z_0) on S. It follows that a valid parametrization of S is given by

$$g(x, \theta) = \big(x, f(x) \cos \theta, f(x) \sin \theta\big), \quad \text{for } a \leq x \leq b,\ 0 \leq \theta \leq 2\pi.$$

(b) Using the parametrization found in part (a), we have

$$g_x(x, \theta) \times g_\theta(x, \theta) = \big(1, f'(x) \cos \theta, f'(x) \sin \theta\big) \times \big(0, -f(x) \sin \theta, f(x) \cos \theta\big)$$
$$= \big(f'(x) f(x), -f(x) \cos \theta, -f(x) \sin \theta\big).$$

Hence, letting D be the domain of g defined in part (a),

$$\sigma(S) = \int_D |g_x(x, \theta) \times g_\theta(x, \theta)|\, dx\, d\theta$$
$$= \int_D \sqrt{(f'(x))^2 f^2(x) + f^2(x) \cos^2 \theta + f^2(x) \sin^2 \theta}\, dx\, d\theta$$
$$= \int_D f(x) \sqrt{1 + \big(f'(x)\big)^2}\, dx\, d\theta = \int_0^{2\pi} d\theta \int_a^b f(x) \sqrt{1 + \big(f'(x)\big)^2}\, dx$$
$$= [\theta]_0^{2\pi} \int_a^b f(x) \sqrt{1 + \big(f'(x)\big)^2}\, dx = 2\pi \int_a^b f(x) \sqrt{1 + \big(f'(x)\big)^2}\, dx,$$

where the absolute value is not needed on $f(x)$ in the first integral in the third line because $f(x)$ is assumed to be nonnegative for $a \leq x \leq b$.

29. Let $S = S_1 \cup S_2$ be the oriented two-piece surface described in Example 6 of the text. With the orientation as in the example, the normals on S_1 and S_2 are $\mathbf{n}_1 = (0, 0, 1)$ and $\mathbf{n}_2 = (0, 1, 0)$, respectively. Consider the field $\mathbf{G}(x, y, z) = (0, y, z)$. Since S_1 is in the xy-plane, $z = 0$ for all points on S_1. That is, the flux of $\mathbf{G}$ at the point $(x, y, 0)$ on S_1 is,

$$\mathbf{G}(x, y, 0) \cdot \mathbf{n}_1 = (0, y, 0) \cdot (0, 0, 1) = 0.$$

Similarly, since S_2 is in the xz-plane, $y = 0$ for all points on S_2 and therefore, the flux of $\mathbf{G}$ at the point $(x, 0, z)$ on S_2 is

$$\mathbf{G}(x, 0, z) \cdot \mathbf{n}_2 = (0, 0, z) \cdot (0, 1, 0) = 0.$$

It follows that the integrals defining the total flux across S_1 and S_2 are also zero, so that $\Phi(\mathbf{G}, S) = \Phi(\mathbf{G}, S_1) + \Phi(\mathbf{G}, S_2) = 0$.

31. Let S be a piece of smooth surface in $\mathcal{R}^3$ and let $f(x, y, z)$ be a continuously differentiable function defined on S. Let (x_0, y_0, z_0) be a point on S and let $f(x_0, y_0, z_0) = k$. This defines a level surface S_k in $\mathcal{R}^3$ that intersects S at (x_0, y_0, z_0). Since f is continuously differentiable on S, its gradient exists at (x_0, y_0, z_0) and is continuous. Moreover, $\nabla f(x_0, y_0, z_0)$ is normal to S_k. Given that the unit normal to S at (x_0, y_0, z_0) (which we denote by $\mathbf{n}(x_0, y_0, z_0)$) is perpendicular to S_k, it follows that $\nabla f(x_0, y_0, z_0)$ and $\mathbf{n}(x_0, y_0, z_0)$ are perpendicular vectors and therefore $\nabla f(x_0, y_0, z_0) \cdot \mathbf{n}(x_0, y_0, z_0) = 0$. Since (x_0, y_0, z_0) was an arbitrary point on S, we must have $\nabla f(x, y, z) \cdot \mathbf{n}(x, y, z) = 0$ for all (x, y, z) in S. Hence,

$$\int_S \nabla f \cdot d\mathbf{S} = \int_S \nabla f \cdot \mathbf{n}\, d\sigma = \int_S \nabla f(x, y, z) \cdot \mathbf{n}(x, y, z)\, d\sigma = \int_S 0\, d\sigma = 0.$$

Section 4: GAUSS'S THEOREM

Exercise Set 4AB (pgs. 437-438)

1. With $\mathbf{F}(x, y, z) = (x^2, y^2, z^2)$, we have

$$\text{div}\mathbf{F}(x, y, z) = \frac{\partial(x^2)}{\partial x} + \frac{\partial(y^2)}{\partial y} + \frac{\partial(z^2)}{\partial z} = 2x + 2y + 2z.$$

3. With $\mathbf{F}(x, y, z) = (y, z, x)$, we have

$$\text{div}\mathbf{F}(x, y, z) = \frac{\partial(y)}{\partial x} + \frac{\partial(z)}{\partial y} + \frac{\partial(x)}{\partial z} = 0 + 0 + 0 = 0.$$

NOTE: *The regions R for Exercises 5 and 7 below are the same; namely, R is a solid cylinder of radius 1 and height 1, with the unit disk in the xy-plane as its base. The surface S that encloses R is composed of three pieces, which we denote by T (top) C (cylindrical side) and B (bottom). In both exercises, we parametrize these pieces by f, g and h, respectively, which are defined by*

$$f(u, v) = (u \cos v, u \sin v, 1), \quad 0 \le u \le 1, \ 0 \le v \le 2\pi;$$
$$g(u, v) = (\cos u, \sin u, v), \quad 0 \le u \le 2\pi, \ 0 \le v \le 1;$$
$$h(u, v) = (v \cos u, v \sin u, 0), \quad 0 \le u \le 2\pi, \ 0 \le v \le 1.$$

The cross products of their respective partial derivatives are found to be

$$f_u \times f_v = (0, 0, u), \qquad g_u \times g_v = (\cos u, \sin u, 0), \qquad h_u \times h_v = (0, 0, -v),$$

where we observe that the normal vectors for T point straight up, the normal vectors for C point outward from the origin, and the normal vectors for B point straight down. That is, all normal vectors are outward-pointing, the surface is positively oriented, and S can therefore be denoted by ∂R. The region R is shown above on the right along with a few of its outward-pointing normal vectors.

5. Let $\mathbf{F}(x, y, z) = (x^2, y^2, z^2)$. The surface integral of $\mathbf{F}$ over T evaluates to

$$\int_T \mathbf{F} \cdot d\mathbf{S} = \int_T \mathbf{F}\big(f(u, v)\big) \cdot (f_u \times f_v) \, du \, dv = \int_T (u^2 \cos^2 v, u^2 \sin^2 v, 1) \cdot (0, 0, u) \, du \, dv$$

$$= \int_T u \, du \, dv = \int_0^{2\pi} dv \int_0^1 u \, du = \big[v\big]_0^{2\pi} \left[\frac{1}{2}u^2\right]_0^1 = \pi.$$

The surface integral of $\mathbf{F}$ over C evaluates to

$$\int_C \mathbf{F} \cdot d\mathbf{S} = \int_C \mathbf{F}\big(g(u, v)\big) \cdot (g_u \times g_v) \, du \, dv = \int_C (\cos^2 u, \sin^2 u, v^2) \cdot (\cos u, \sin u, 0) \, du \, dv$$

$$= \int_C (\cos^3 u + \sin^3 u) \, du \, dv = \int_0^1 dv \int_0^{2\pi} (\cos^3 u + \sin^3 u) \, du = 0,$$

334

where the last equality is justified because both $\sin^3 u$ and $\cos^3 u$ are symmetric with respect to the point $(u, v) = (\pi, 0)$ (the midpoint of the interval of integration) and therefore their integrals over this interval are zero.

Since, the field vectors are parallel with the xy-plane on B, they are perpendicular to the normals on B, so that the surface integral of $\mathbf{F}$ over B is zero. The total flux of $\mathbf{F}$ across ∂R is therefore

$$\int_{\partial R} \mathbf{F} \cdot d\mathbf{S} = \int_T \mathbf{F} \cdot d\mathbf{S} + \int_C \mathbf{F} \cdot d\mathbf{S} + \int_B \mathbf{F} \cdot d\mathbf{S} = \pi.$$

The field $\mathbf{F}$ is the same as that given in Exercise 1 above where its divergence was found to be $\text{div}\mathbf{F} = 2x + 2y + 2z$. Therefore, the other side of Gauss's formula evaluates to

$$\int_R \text{div}\mathbf{F} \, dV = \int_R (2x + 2y + 2z) \, dV = \int_R (2r\cos\theta + 2r\sin\theta + 2z) \, r \, dr \, d\theta \, dz$$

$$= \int_0^1 dz \int_0^{2\pi} d\theta \int_0^1 (2r^2 \cos\theta + 2r^2 \sin\theta + 2rz) \, dr$$

$$= \int_0^1 dz \int_0^{2\pi} \left[\frac{2}{3} r^3 \cos\theta + \frac{2}{3} r^3 \sin\theta + r^2 z \right]_0^1 d\theta$$

$$= \int_0^1 dz \int_0^{2\pi} \left(\frac{2}{3} (\cos\theta + \sin\theta) + z \right) d\theta$$

$$= \int_0^1 z \, dz \int_0^{2\pi} d\theta = \left[\frac{1}{2} z^2 \right]_0^1 \left[\theta \right]_0^{2\pi} = \pi.$$

So, for the region R, Gauss's formula is verified for this particular vector field $\mathbf{F}$.

7. Let $\mathbf{F}(x, y, z) = (0, 0, z)$. First note that the field vectors are perpendicular to the normal vectors on C, so that the surface integral of $\mathbf{F}$ over C is zero. Second, the field vectors are zero on the xy-plane, so that the surface integral over B is also zero. However, on T we have

$$\int_T \mathbf{F} \cdot d\mathbf{S} = \int_T \mathbf{F}\big(f(u,v)\big) \cdot (f_u \times f_v) \, du \, dv = \int_T (0, 0, 1) \cdot (0, 0, u) \, du \, dv$$

$$= \int_0^{2\pi} dv \int_0^1 u \, du = \left[v \right]_0^{2\pi} \left[\frac{1}{2} u^2 \right]_0^1 = \pi.$$

The total flux of $\mathbf{F}$ across ∂R is therefore

$$\int_{\partial R} \mathbf{F} \cdot d\mathbf{S} = \int_T \mathbf{F} \cdot d\mathbf{S} + \int_C \mathbf{F} \cdot d\mathbf{S} + \int_B \mathbf{F} \cdot d\mathbf{S} = \pi.$$

By inspection, we see that $\text{div}\mathbf{F}$ is identically equal to 1. So, the other side of Gauss's formula evaluates to

$$\int_R \text{div}\mathbf{F} \, dV = \int_R 1 \, dV = V(R) = \pi,$$

which verifies Gauss's formula for this particular vector field $\mathbf{F}$.

9. Let $\mathbf{F}(x, y, z) = (x, y, z)$. The given surface S is a sphere of radius 2 centered at the origin in $\mathcal{R}^3$ and is shown on the next page. Assuming S is positively oriented, we can denote it by $S = \partial R$, where R is the solid sphere enclosed by S. By inspection, one can see that $\text{div}\mathbf{F}(x, y, z) = 3$. Hence, by Gauss's Theorem

$$\int_S \mathbf{F} \cdot d\mathbf{S} = \int_{\partial R} \mathbf{F} \cdot d\mathbf{S} = \int_R \text{div}\mathbf{F} \, dV = \int_R 3 \, dV = 3V(R) = 32\pi,$$

where we have used the fact that the volume of a sphere of radius 2 is $\frac{4}{3}\pi(2)^3 = 32\pi/3$.

11. Let $\mathbf{F}(x, y, z) = (xz, -yz, xy)$. The given surface S is an ellipsoid with semi-axes of lengths 1, $1/\sqrt{2}$ and $1/\sqrt{3}$ and is shown below on the right. Assuming S is positively oriented, we can denote it by $S = \partial R$, where R is the solid ellipsoid enclosed by S. By Gauss's Theorem,

$$\int_S \mathbf{F} \cdot d\mathbf{S} = \int_{\partial R} \mathbf{F} \cdot d\mathbf{S} = \int_R \operatorname{div}\mathbf{F} \, dV = \int_R \left(\frac{\partial(xz)}{\partial x} + \frac{\partial(-yz)}{\partial y} + \frac{\partial(xy)}{\partial z} \right) dV$$

$$= \int_R (z - z + 0) \, dV = \int_R 0 \, dV = 0.$$

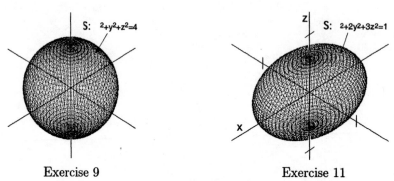

Exercise 9	Exercise 11

13. Let $\mathbf{F}(x, y, z) = \big(F_1(x, y, z), F_2(x, y, z), F_3(x, y, z)\big)$ be a twice continuously vector field. Then

$$\operatorname{div}(\operatorname{curl}\mathbf{F})(\mathbf{x}) = \operatorname{div}\left(\frac{\partial F_3}{\partial y} - \frac{\partial F_2}{\partial z}, \frac{\partial F_1}{\partial z} - \frac{\partial F_3}{\partial x}, \frac{\partial F_2}{\partial x} - \frac{\partial F_1}{\partial y} \right)$$

$$= \frac{\partial}{\partial x}\left(\frac{\partial F_3}{\partial y} - \frac{\partial F_2}{\partial z} \right) + \frac{\partial}{\partial y}\left(\frac{\partial F_1}{\partial z} - \frac{\partial F_3}{\partial x} \right) + \frac{\partial}{\partial z}\left(\frac{\partial F_2}{\partial x} - \frac{\partial F_1}{\partial y} \right)$$

$$= \frac{\partial^2 F_3}{\partial x \partial y} - \frac{\partial^2 F_2}{\partial x \partial z} + \frac{\partial^2 F_1}{\partial y \partial z} - \frac{\partial^2 F_3}{\partial y \partial x} + \frac{\partial^2 F_2}{\partial z \partial x} - \frac{\partial^2 F_1}{\partial z \partial y}$$

$$= \left(\frac{\partial^2 F_3}{\partial x \partial y} - \frac{\partial^2 F_3}{\partial y \partial x} \right) + \left(\frac{\partial^2 F_1}{\partial y \partial z} - \frac{\partial^2 F_1}{\partial z \partial y} \right) + \left(\frac{\partial^2 F_2}{\partial z \partial x} - \frac{\partial^2 F_2}{\partial x \partial z} \right),$$

where the partials in the third line have been judiciously grouped in the last line. Since $\mathbf{F}$ is twice continuously differentiable, F_1, F_2 and F_3 are all twice continuously differentiable. Hence, the two mixed partials in each of the parenthetical terms in the last line above are equal (Clairaut's Theorem extended to three variables) and therefore each parenthetical term is identically zero. That is,

$$\operatorname{div}(\operatorname{curl}\mathbf{F})(\mathbf{x}) \equiv 0.$$

15. (a) Let $f(x, y, z) = (x^2 + y^2 + z^2)^{-1/2}$. Then, for $\mathbf{x} \neq \mathbf{0}$,

$$\operatorname{div}(\nabla f)(\mathbf{x}) = \operatorname{div}\big(-x(x^2 + y^2 + z^2)^{-3/2}, -y(x^2 + y^2 + z^2)^{-3/2}, -z(x^2 + y^2 + z^2)^{-3/2}\big)$$

$$= \frac{\partial}{\partial x}\left(\frac{-x}{(x^2 + y^2 + z^2)^{3/2}} \right) + \frac{\partial}{\partial y}\left(\frac{-y}{(x^2 + y^2 + z^2)^{3/2}} \right) + \frac{\partial}{\partial z}\left(\frac{-z}{(x^2 + y^2 + z^2)^{3/2}} \right)$$

$$= \frac{2x^2 - y^2 - z^2}{(x^2 + y^2 + z^2)^{5/2}} + \frac{-x^2 + 2y^2 - z^2}{(x^2 + y^2 + z^2)^{5/2}} + \frac{-x^2 - y^2 + 2z^2}{(x^2 + y^2 + z^2)^{5/2}} = 0.$$

(b) Consider the function $f(x, y, z) = x^2$, which clearly has continuous derivatives of all orders. However, for all $\mathbf{x}$ in $\mathcal{R}^3$,

$$\operatorname{div}(\nabla f)(\mathbf{x}) = \operatorname{div}(2x, 0, 0) = \frac{\partial(2x)}{\partial x} + \frac{\partial(0)}{\partial y} + \frac{\partial(0)}{\partial z} = 2 \neq 0.$$

(c) For twice continuously differentiable real-valued functions $f : \mathcal{R}^n \longrightarrow \mathcal{R}$, define the operator Δ by $\Delta f = \operatorname{div}(\nabla f)$. We have

$$\Delta f = \operatorname{div}(\nabla f) = \operatorname{div}\left(\frac{\partial f}{\partial x_1}, \ldots, \frac{\partial f}{\partial x_n}\right) = \frac{\partial}{\partial x_1}\left(\frac{\partial f}{\partial x_1}\right) + \cdots + \frac{\partial}{\partial x_n}\left(\frac{\partial f}{\partial x_n}\right) = \frac{\partial^2 f}{\partial x_1^2} + \cdots + \frac{\partial^2 f}{\partial x_n^2}.$$

17. Let S be the sphere of radius 1 centered at the origin in $\mathcal{R}^3$ and let R be the region enclosed by S. Assuming S is positively oriented, we can denote the boundary of R as $S = \partial R$. For the field $\mathbf{F}(x, y, z) = (x^2, y^2, z^2)$, Gauss's Theorem gives

$$\int_S \mathbf{F} \cdot d\mathbf{S} = \int_{\partial R} \mathbf{F} \cdot d\mathbf{S} = \int_R \operatorname{div}\mathbf{F} \, dV = \int_R (2x + 2y + 2z) \, dx \, dy \, dz$$

$$= \int_R (2r\cos\theta\sin\phi + 2r\sin\theta\sin\phi + 2r\cos\phi) \, r^2\sin\phi \, dr \, d\phi \, d\theta$$

$$= \int_R \left(2r^3(\cos\theta + \sin\theta)\sin^2\phi + 2r^3\cos\phi\sin\phi\right) dr \, d\phi \, d\theta$$

$$= \int_0^{2\pi} (\cos\theta + \sin\theta) \, d\theta \int_0^\pi \sin^2\phi \, d\phi \int_0^1 2r^3 dr + \int_0^{2\pi} d\theta \int_0^\pi \sin 2\phi \, d\phi \int_0^1 r^3 dr = 0,$$

where the last equality holds because the integral of $\cos\theta + \sin\theta$ over an interval of length 2π is zero and the integral of $\sin 2\phi$ over an integral of length π is zero.

19. Let R be a region to which Gauss's Theorem applies and let ∂R be the positively oriented boundary of R. If $\mathbf{F}(x, y, z) = \left(F_1(x, y, z), F_2(x, y, z), F_3(x, y, z)\right)$ is a continuously differentiable vector field on R then, in terms of the coordinate functions of $\mathbf{F}$, Gauss's formula reads

$$\int_R \left(\frac{\partial F_1}{\partial x} + \frac{\partial F_2}{\partial y} + \frac{\partial F_3}{\partial z}\right) dV = \int_{\partial R} F_1 dy \, dz + F_2 dz \, dx + F_3 dx \, dy.$$

In particular, for the vector field $F(x, y, z) = (x, y, z)$, the above formula reduces to

$$\int_R 3 \, dV = \int_{\partial R} x \, dy \, dz + y \, dz \, dx + z \, dx \, dy.$$

Dividing both members by 3 and noting that the resulting left member is $\int_R dV$, which by definition is the volume of R, we get

$$V(R) = \frac{1}{3}\int_{\partial R} x \, dy \, dz + y \, dz \, dx + z \, dx \, dy.$$

21. Let S be the ellipsoid $x^2/a^2 + y^2/b^2 + z^2/c^2 = 1$, where a, b and c are positive constants, and let $D(\mathbf{x})$ be the distance from the origin to the tangent plane to S at $\mathbf{x}$.

(a) Let $\mathbf{F}(\mathbf{x}) = (x/a^2, y/b^2, z/c^2)$. Note that S is a level surface of the function $f(\mathbf{x}) = x^2/a^2 + y^2/b^2 + z^2/c^2$ and that

$$\nabla f(\mathbf{x}) = \left(2\frac{x}{a^2}, 2\frac{y}{b^2}, 2\frac{z}{c^2}\right) = 2\mathbf{F}(\mathbf{x}).$$

Hence, since $\nabla f(\mathbf{x})$ is normal to S for all $\mathbf{x}$ on S, the outward-pointing unit normal to S at a fixed point $\mathbf{x}_0$ on S is given by

$$\mathbf{n}(\mathbf{x}_0) = \frac{\nabla f(\mathbf{x}_0)}{|\nabla f(\mathbf{x}_0)|} = \frac{2\mathbf{F}(\mathbf{x}_0)}{|2\mathbf{F}(\mathbf{x}_0)|} = \frac{\mathbf{F}(\mathbf{x}_0)}{|\mathbf{F}(\mathbf{x}_0)|}.$$

Multiplying throughout by $|\mathbf{F}(\mathbf{x}_0)|$, dotting both sides by $\mathbf{n}(\mathbf{x}_0)$ and using the fact that $\mathbf{n} \cdot \mathbf{n} = 1$ gives

(*) $$\mathbf{F}(\mathbf{x}_0) \cdot \mathbf{n}(\mathbf{x}_0) = |\mathbf{F}(\mathbf{x}_0)|.$$

On the other hand, if P_0 is the tangent plane to S at $\mathbf{x}_0$ then its normalized vector equation is given by

$$\mathbf{n}(\mathbf{x}_0) \cdot (\mathbf{x} - \mathbf{x}_0) = 0, \qquad \mathbf{x} \text{ in } P_0.$$

Since the origin is on the side of P_0 opposite toward which $\mathbf{n}(\mathbf{x}_0)$ points, the distance from the origin to P_0 is

(**) $$D(\mathbf{x}_0) = -\mathbf{n}(\mathbf{x}_0) \cdot (0 - \mathbf{x}_0) = \mathbf{n}(\mathbf{x}_0) \cdot \mathbf{x}_0 = \frac{\mathbf{F}(\mathbf{x}_0)}{|\mathbf{F}(\mathbf{x}_0)|} \cdot \mathbf{x}_0 = \frac{\mathbf{F}(\mathbf{x}_0) \cdot \mathbf{x}_0}{|\mathbf{F}(\mathbf{x}_0)|} = \frac{1}{|\mathbf{F}(\mathbf{x}_0)|},$$

where the last member follows because $\mathbf{x}_0 = (x_0, y_0, z_0)$ is on S and therefore

$$\mathbf{F}(\mathbf{x}_0) \cdot \mathbf{x}_0 = \left(\frac{x_0}{a^2}, \frac{y_0}{b^2}, \frac{z_0}{c^2}\right) \cdot (x_0, y_0, z_0) = \frac{x_0^2}{a^2} + \frac{y_0^2}{b^2} + \frac{z_0^2}{c^2} = 1.$$

Since $\mathbf{x}_0$ was an arbitrary point on S, comparison of equations (*) and (**) shows that

$$\mathbf{F}(\mathbf{x}) \cdot \mathbf{n}(\mathbf{x}) = D^{-1}(\mathbf{x}) \qquad \text{for all } \mathbf{x} \text{ on } S.$$

(b) It was shown in part (a) that S was positively oriented, so that we can denote S by ∂R. By Gauss's Theorem, we can use the main result from part (a) to get

$$\int_S D^{-1} d\sigma = \int_S \mathbf{F} \cdot \mathbf{n} \, d\sigma = \int_S \mathbf{F} \cdot d\mathbf{S} = \int_{\partial R} \mathbf{F} \cdot d\mathbf{S} = \int_R \operatorname{div}\mathbf{F} \, dV$$

$$= \int_R \left(\frac{\partial(x/a^2)}{\partial x} + \frac{\partial(y/b^2)}{\partial y} + \frac{\partial(z/c^2)}{\partial z}\right) dV = \int_R \left(\frac{1}{a^2} + \frac{1}{b^2} + \frac{1}{c^2}\right) dV$$

$$= \left(\frac{1}{a^2} + \frac{1}{b^2} + \frac{1}{c^2}\right) \int_R dV = \left(\frac{1}{a^2} + \frac{1}{b^2} + \frac{1}{c^2}\right) V(R)$$

$$= \left(\frac{1}{a^2} + \frac{1}{b^2} + \frac{1}{c^2}\right) \frac{4}{3}\pi abc = \frac{4\pi}{3}\left(\frac{bc}{a} + \frac{ac}{b} + \frac{ab}{c}\right),$$

where the first expression in the fourth line follows from the fact that the volume of the given ellipsoid is $4\pi(abc)/3$.

23. Let R be a region in $\mathcal{R}^3$ whose boundary consists of finitely many smooth surfaces. Assume that $u(x, y, z)$ is twice continuously differentiable in R and that $u_{xx} + u_{yy} + u_{zz} = 0$ in R (i.e., u is harmonic in R). By Gauss's Theorem, the flux across ∂R is

$$\Phi(\nabla u, \partial R) = \int_{\partial R} \nabla u \cdot d\mathbf{S} = \int_R \operatorname{div}(\nabla u)\, dV = \int_R \operatorname{div}(u_x, u_y, u_z)\, dV$$

$$= \int_R (u_{xx} + u_{yy} + u_{zz})\, dV = \int_R 0\, dV = 0.$$

(NOTE: See Exercise 15(c) of this section.)

25. Let $\mathbf{F}$ be a divergence-free field and let S be a smooth closed surface, which we assume is positively oriented and therefore can be denoted by ∂R, where R is the region enclosed by S. By Gauss's Theorem, the flux of $\mathbf{F}$ across S is

$$\Phi(\mathbf{F}, S) = \int_S \mathbf{F} \cdot d\mathbf{S} = \int_{\partial R} \mathbf{F} \cdot d\mathbf{S} = \int_R \operatorname{div}\mathbf{F}\, dV = \int_R 0\, dV = 0.$$

27. Let $\mu(\mathbf{y})$ be an integrable mass density function defined on a region R and consider the gravitational field $\mathbf{F}$ defined by

$$\mathbf{F}(\mathbf{x}) = G \int_R \frac{\mu(\mathbf{y})(\mathbf{y} - \mathbf{x})}{|\mathbf{y} - \mathbf{x}|^3}\, dV_{\mathbf{y}}, \quad \mathbf{x} \neq \mathbf{y}.$$

(a) Let S be a smooth closed surface with no points of R inside or on S. The flux across S is therefore

$$\int_S \mathbf{F} \cdot d\mathbf{S} = \int_S \left[G \int_R \frac{\mu(\mathbf{y})(\mathbf{y} - \mathbf{x})}{|\mathbf{y} - \mathbf{x}|^3}\, dV_{\mathbf{y}} \right] \cdot d\mathbf{S}.$$

Interchanging the order of integration on the surface and volume integrals amounts to using the homogeneity of the dot product to pull the scalar differential $dV_{\mathbf{y}}$ to the outside of the dot product to get

$$(*) \qquad \int_S \mathbf{F} \cdot d\mathbf{S} = G \int_R \mu(\mathbf{y}) \left[\int_S \frac{\mathbf{y} - \mathbf{x}}{|\mathbf{y} - \mathbf{x}|^3} \cdot d\mathbf{S} \right] dV_{\mathbf{y}},$$

where we have pulled everything that depends only on the variable $\mathbf{y}$ outside of the surface integral. Note that, for each fixed $\mathbf{y} = (x_0, y_0, z_0)$ in R, the integrand of the bracketed inner integral is a function $\mathbf{H}_{\mathbf{y}}$ of $\mathbf{x} = (x, y, z)$. Moreover, since S and R do not intersect, this function is a continuously differentiable vector field at all points not in R. In terms of coordinate functions, $\mathbf{H}_{\mathbf{y}}(\mathbf{x})$ is given by

$$\mathbf{H}_{\mathbf{y}}(\mathbf{x}) = \frac{\mathbf{y} - \mathbf{x}}{|\mathbf{y} - \mathbf{x}|^3} = \left(\frac{x_0 - x}{\left[(x_0 - x)^2 + (y_0 - y)^2 + (z_0 - z)^2 \right]^{3/2}}, \right.$$

$$\frac{y_0 - y}{\left[(x_0 - x)^2 + (y_0 - y)^2 + (z_0 - z)^2 \right]^{3/2}},$$

$$\left. \frac{z_0 - z}{\left[(x_0 - x)^2 + (y_0 - y)^2 + (z_0 - z)^2 \right]^{3/2}} \right).$$

Hence, leaving the details to the reader, we have

$$\text{div}\mathbf{H_y}(\mathbf{x}) = \frac{2(x_0 - x)^2 - (y_0 - y)^2 - (z_0 - z)^2}{\left[(x_0 - x)^2 + (y_0 - y)^2 + (z_0 - z)^2\right]^{5/2}}$$

$$+ \frac{2(y_0 - y)^2 - (x_0 - x)^2 - (z_0 - z)^2}{\left[(x_0 - x)^2 + (y_0 - y)^2 + (z_0 - z)^2\right]^{5/2}}$$

$$+ \frac{2(z_0 - z)^2 - (x_0 - x)^2 - (y_0 - y)^2}{\left[(x_0 - x)^2 + (y_0 - y)^2 + (z_0 - z)^2\right]^{5/2}} = 0.$$

Therefore, assuming S is positively oriented and D is the region enclosed by S, Gauss's Theorem can be applied to the field $\mathbf{H_y}$ across S, so that $(*)$ becomes

$$\int_S \mathbf{F} \cdot d\mathbf{S} = G \int_R \mu(\mathbf{y}) \left[\int_R \text{div}\mathbf{H_y}\, dV_\mathbf{x}\right] dV_\mathbf{y} = G \int_R \mu(\mathbf{y}) \left[\int_R 0\, dV_\mathbf{x}\right] dV_\mathbf{y} = 0.$$

(b) Let S be a smooth closed surface that contains all points of R in its interior and, for each fixed $\mathbf{y}$ in R, let $\mathbf{H_y}(\mathbf{x})$ be as in part (a). Now fix $\mathbf{y} = (x_0, y_0, z_0)$ in R and let S_a be a sphere of radius a with center $\mathbf{y}$, where a is chosen large enough so that S_a and S do not intersect, and consider the surface $S \cup S_a$. If this surface is oriented so that normals on S point inward and normals on S_a point outward then the region D with boundary $\partial D = S \cup S_a$ contains no points of R and therefore, by part (a), $\text{div}\mathbf{H_y}(\mathbf{x}) = 0$ for $\mathbf{x}$ in D. By the Surface Independence Principle, the flux of $\mathbf{H_y}$ across S is equal to the flux of $\mathbf{H_y}$ across S_a. Hence, since equation $(*)$ in part (a) holds, it can be written as

$$(**) \qquad \int_S \mathbf{F} \cdot d\mathbf{S} = G \int_R \mu(\mathbf{y}) \left[\int_{S_a} \mathbf{H_y} \cdot d\mathbf{S}\right] dV_\mathbf{y}.$$

We can orient S_a with its normals pointing outward by using the parametrization

$$g(\phi, \theta) = \begin{pmatrix} x_0 + a\cos\theta\sin\phi \\ y_0 + a\sin\theta\sin\phi \\ z_0 + a\cos\phi \end{pmatrix}, \quad 0 \leq \theta \leq 2\pi, \ 0 \leq \phi \leq \pi.$$

Since $g_\phi \times g_\theta = a^2 \sin\phi(\cos\theta\sin\phi, \sin\theta\sin\phi, \cos\phi)$ are vectors pointing outward, g is a parametrization under which our conclusions are valid. Furthermore, for $\mathbf{x} = g(\phi, \theta)$, we have $\mathbf{y} - g(\phi, \theta) = -a(\cos\theta\sin\phi, \sin\theta\sin\phi, \cos\phi)$ and $|\mathbf{y} - g(\phi, \theta)| = a$. So,

$$\mathbf{H_y}\big(g(\phi, \theta)\big) = \frac{\mathbf{y} - g(\phi, \theta)}{|\mathbf{y} - g(\phi, \theta)|^3} = -\frac{1}{a^2}(\cos\theta\sin\phi, \sin\theta\sin\phi, \cos\phi), \quad \text{for } \mathbf{x} \text{ in } S_a.$$

Therefore, by inspection, $\mathbf{H_y}\big(g(\phi, \theta)\big) \cdot (g_\phi \times g_\theta) = -\sin\phi$ and we have

$$\int_{S_a} \mathbf{H_y} \cdot d\mathbf{S} = \int_{S_a} \mathbf{H_y}\big(g(\phi, \theta)\big) \cdot (g_\phi \times g_\theta)\, d\phi\, d\theta$$

$$= -\int_{S_a} \sin\phi\, d\phi\, d\theta = -\int_0^{2\pi} d\theta \int_0^\pi \sin\phi\, d\phi = -\big[\theta\big]_0^{2\pi}\big[-\cos\phi\big]_0^\pi = -4\pi.$$

Since this is independent of $\mathbf{y}$, it can be pulled outside the integral over R in $(**)$. When this is done and the variable $\mathbf{y}$ is suppressed, the result is

$$\int_S \mathbf{F} \cdot d\mathbf{S} = -4\pi G \int_R \mu\, dV.$$

Section 5: STOKES'S THEOREM

Exercise Set 5ABC (pgs. 447-449)

1. With $\mathbf{F}(x,y,z) = (y - z^2, z - x^2, x - y^2)$,

$$\text{curl}\mathbf{F} = \left(\frac{\partial(x - y^2)}{\partial y} - \frac{\partial(z - x^2)}{\partial z}, \frac{\partial(y - z^2)}{\partial z} - \frac{\partial(x - y^2)}{\partial x}, \frac{\partial(z - x^2)}{\partial x} - \frac{\partial(y - z^2)}{\partial y} \right)$$

$$= (-2y - 1, -2z - 1, -2x - 1).$$

3. With $\mathbf{F}(x,y,z) = (x - y, z - x, y - z)$,

$$\text{curl}\mathbf{F} = \left(\frac{\partial(y - z)}{\partial y} - \frac{\partial(z - x)}{\partial z}, \frac{\partial(x - y)}{\partial z} - \frac{\partial(y - z)}{\partial x}, \frac{\partial(z - x)}{\partial x} - \frac{\partial(x - y)}{\partial y} \right)$$

$$= \left(1 - 1, 0 - 0, -1 - (-1) \right) = (0, 0, 0).$$

5. Let D be the closed disk of radius 1 centered at the origin in $\mathcal{R}^2$, and let $g(u,v) = (u, v, \sqrt{1 - u^2 - v^2})$, for (u,v) in D. The surface S parametrized by g is the top hemisphere of the unit sphere in $\mathcal{R}^3$. The border ∂S of S is the image under g of the boundary of D and is parametrized by $h(t) = (\cos t, \sin t)$, $0 \le t \le 2\pi$. That is,

$$\partial S: \quad g(h(t)) = (\cos t, \sin t, 0), \quad 0 \le t \le 2\pi,$$

so that ∂S is the circle of radius 1 centered at the origin in $\mathcal{R}^3$ and lying on the plane $z = 0$. Note that h orients the boundary of D in a counterclockwise direction, so that the inherited orientation of ∂S is positive. A sketch of S and the orientation of its border is shown on the right.

Let $\mathbf{F}(x,y,z) = (x, y, z)$. On one hand, the direction of traversal around ∂S is perpendicular to the field arrows (which are parallel to the xy-plane and point away from the origin). Hence, the line integral of $\mathbf{F}$ around ∂S is zero. On the other hand, $\mathbf{F}$ is the gradient of the potential $f(x,y,z) = \frac{1}{2}x^2 + \frac{1}{2}y^2 + \frac{1}{2}z^2$, so that $\text{curl}\mathbf{F} = \text{curl}(\nabla f)$ is identically zero. Hence, the surface integral of $\text{curl}\mathbf{F}$ over S is zero. In other words,

$$\oint_{\partial S} \mathbf{F} \cdot d\mathbf{x} = \int_S \text{curl}\mathbf{F} \cdot d\mathbf{S} = 0$$

and Stokes's formula is verified in this particular case.

7. Let D be the closed disk of radius 2 centered at the origin in $\mathcal{R}^2$ and let $g(u,v) = (u, v, u^2 + v^2)$, for (u,v) in D. The surface S parametrized by g is a four unit high circular paraboloid opening upward with vertex at the origin in $\mathcal{R}^3$. The border ∂S of S is the image under g of the boundary of D parametrized by $h(t) = (2 \cos t, 2 \sin t)$, $0 \le t \le 2\pi$. That is,

$$\partial S: \quad g(h(t)) = (2 \cos t, 2 \sin t, 4), \quad 0 \le t \le 2\pi,$$

so that ∂S is the circle of radius 2 centered at $(0,0,4)$ and lying on the plane $z = 4$. Note that h orients the boundary of D in a counterclockwise direction, so that the inherited orientation of ∂S is positive. A sketch of S and the orientation of its border is shown on the right.

Let $\mathbf{F}(x, y, z) = (x, y, 0)$, On one hand, the direction of traversal of ∂S is perpendicular to the field arrows (which are all parallel to the xy-plane and point away from the z-axis). Hence, the line integral of $\mathbf{F}$ around ∂S is zero. On the other hand, $\mathbf{F}$ is the gradient of the potential $f(x, y, z) = \frac{1}{2}x^2 + \frac{1}{2}y^2$, so that $\mathrm{curl}\mathbf{F} = \mathrm{curl}(\nabla f)$ is identically zero. In other words,

$$\oint_{\partial S} \mathbf{F} \cdot d\mathbf{x} = \int_S \mathrm{curl}\mathbf{F} \cdot d\mathbf{S} = 0$$

and Stokes's formula is verified in this particular case.

9. Let D be the closed unit disk in $\mathcal{R}^2$ and let $g(u, v) = (u, v, \sqrt{1 - u^2 - v^2})$, for (u, v) in D. The surface S parametrized by g is the top hemisphere of the sphere of radius 1 centered at the origin in $\mathcal{R}^3$. In order to properly orient S so that Stokes's Theorem can be applied, we parametrize the boundary of D by $h(t) = (\cos t, \sin t)$, $0 \leq t \leq 2\pi$. Since h traces the boundary of D in a counterclockwise direction, the composite function $f = g \circ h$ traces the border ∂S of S in a positive direction. That is,

$$\partial S: \quad f(t) = g(h(t)) = (\cos t, \sin t, 0), \quad 0 \leq t \leq 2\pi.$$

For the field $\mathbf{F}(x, y, z) = (y, z, x)$, we have $\mathbf{F}(f(t)) = (\sin t, 0, \cos t)$. By Stokes's Theorem

$$\int_S \mathrm{curl}\mathbf{F} \cdot d\mathbf{S} = \oint_{\partial S} \mathbf{F} \cdot d\mathbf{x} = \oint_{\partial S} \mathbf{F}(f(t)) \cdot f'(t) \, dt = \oint_{\partial S} (\sin t, 0, \cos t) \cdot (-\sin t, \cos t, 0) \, dt$$

$$= \int_0^{2\pi} -\sin^2 t \, dt = \int_0^{2\pi} -\frac{1}{2}(1 - \cos 2t) \, dt = \left[-\frac{1}{2}t + \frac{1}{4}\sin 2t \right]_0^{2\pi} = -\pi.$$

11. (a) Let S be a planar surface in the xy-plane such that its border is a piecewise smooth curve. Since S is planar, we can parametrize S by $g(x, y) = (x, y, 0)$, for (x, y) in D, where D is the name we give to the set of points in S whenever we refer to the parameter domain of g. Now let $h(t) = (x(t), y(t))$, for $a \leq t \leq b$, be a parametrization of the boundary γ of D that traces γ counterclockwise. Then the composite function $f(t) = g(h(t)) = (x(t), y(t), 0)$, $a \leq t \leq b$, is a positive orientation of the border ∂S of S. Therefore, as a surface, S is properly oriented to apply Stokes's Theorem. Observe that $g_x \times g_y = (1, 0, 0) \times (0, 1, 0) = (0, 0, 1)$.

Now, suppose $\mathbf{F}$ is a field of the form $\mathbf{F}(x, y, z) = (F_1(x, y, z), F_2(x, y, z), 0)$, where F_1 and F_2 are independent of z. We then have

$$\mathrm{curl}\mathbf{F} = \left(\frac{\partial(0)}{\partial y} - \frac{\partial F_2}{\partial z}, \frac{\partial F_1}{\partial z} - \frac{\partial(0)}{\partial x}, \frac{\partial F_2}{\partial x} - \frac{\partial F_1}{\partial y} \right) = \left(0, 0, \frac{\partial F_2}{\partial x} - \frac{\partial F_1}{\partial y} \right),$$

so that, on one hand, the surface integral of $\mathrm{curl}\mathbf{F}$ over S in Stokes's formula becomes

$$\int_S \mathrm{curl}\mathbf{F} \cdot d\mathbf{S} = \int_S \left(0, 0, \frac{\partial F_2}{\partial x} - \frac{\partial F_1}{\partial y} \right) \cdot (0, 0, 1) \, dx \, dy = \int_S \left(\frac{\partial F_2}{\partial x} - \frac{\partial F_1}{\partial y} \right) dx \, dy,$$

On the other hand, the line integral of $\mathbf{F}$ around ∂S in Stokes's formula is just

$$\oint_{\partial S} \mathbf{F} \cdot d\mathbf{x} = \oint_{\partial S} (F_1(x, y, z), F_2(x, y, z), 0) \cdot (dx, dy, dz) = \oint_{\partial S} F_1(x, y, z) \, dx + F_2(x, y, z) \, dy,$$

342

The two sides of Stokes's formula are in terms of functions F_1 and F_2 with domain in $\mathcal{R}^3$, so that their equality by Stokes's Theorem is not quite the same thing as Green's formula. However, since F_1 and F_2 are independent of z, we can let $F(x,y) = F_1(x,y.z)$ and $G(x,y) = F_2(x,y,z)$. This implies that $\partial F/\partial y(x,y) = \partial F_1/\partial y(x,y,z)$ and $\partial G/\partial x(x,y) = \partial F_2/\partial x(x,y,z)$. Hence, we can replace F_1 and F_2 with F and G (resp.) in the two sides of Stokes's formula and still retain a true equation. Doing this, gives

$$\int_D \left(\frac{\partial G}{\partial x} - \frac{\partial F}{\partial y} \right) dx\,dy = \oint_\gamma F\,dx + G\,dy,$$

where the limits of integration D and γ reflect the fact that we are viewing the situation as a planar parameter domain D in $\mathcal{R}^2$ bounded by a piecewise smooth curve γ (instead of a surface S in $\mathcal{R}^3$ with border ∂S). This is Green's formula.

(b) Let $g(u,v) = (u\cos v, u\sin v, 0)$, for $1 \le u \le 2$, $0 \le v \le 2\pi$, and let R be the rectangle described by these parameter limits. Setting $x = u\cos v$, $y = u\sin v$ and $z = 0$ shows that $x^2 + y^2 = u^2$, where $1 \le u \le 2$. Thus, the surface S parametrized by g is the annulus $1 \le x^2 + y^2 \le 4$ in the xy-plane. To precisely describe the border ∂S of S, we consider a counterclockwise path γ around R. Clearly, γ is comprised of four pieces γ_1 (bottom), γ_2 (right side), γ_3 (top), and γ_4 (left side).

On γ_1, v is constant at $v = 0$ and the part of ∂S that corresponds to γ_1 under g is parametrized by $g(u,0) = (u,0,0)$, with u varying from $u = 1$ to $u = 2$. This traces the line segment in the xy-plane from the point $(1,0,0)$ to the point $(2,0,0)$.

On γ_2, u is constant at $u = 2$ and the part of ∂S that corresponds to γ_2 under g is parametrized by $g(2,v) = (2\cos v, 2\sin v, 0)$, with v varying from $v = 0$ to $v = 2\pi$. This traces the circle of radius 2 centered at the origin in $\mathcal{R}^3$ in a counterclockwise direction.

On γ_3, v is constant at $v = 2\pi$ and the part of ∂S that corresponds to γ_3 under g is parametrized by $g(u,2\pi) = (u,0,0)$, with u varying from $u = 2$ to $u = 1$. This traces the same line line segment as for γ_1 except it is traced in the opposite direction; i.e., from the point $(2,0,0)$ to the point $(1,0,0)$.

On γ_4, u is constant at $u = 1$ and the part of ∂S that corresponds to γ_4 under g is parametrized by $g(1,v) = (\cos v, \sin u, 0)$, with v varying from $v = 2\pi$ to $v = 0$. This traces the unit circle in the xy-plane in a *clockwise* direction.

(c) The vector field $\mathbf{F}(x,y,z) = (x,x,0)$ is of the type described in part (a). The surface S is explicitly described in part (b). By Stokes's Theorem, the line integral of $\mathbf{F}$ over the positively oriented border ∂S of S is equal to the surface integral of $\mathrm{curl}\mathbf{F}$ over S. Therefore, with D being the annulus $1 \le x^2 + y^2 \le 4$, we use the result of part (a) to obtain

$$\oint_{\partial S} \mathbf{F} \cdot d\mathbf{x} = \int_D \left(\frac{\partial(x)}{\partial x} - \frac{\partial(x)}{\partial y} \right) dx\,dy = \int_D 1\,dx\,dy = A(D) = \pi(2)^2 - \pi(1)^2 = 3\pi,$$

where the area of D is computed by subtracting the area of a circle of radius 1 from the area of a circle of radius 2.

13. Let $\mathbf{F}$ be a continuously differentiable vector field on a disk of radius r centered at $\mathbf{x}_0$. Since $\mathrm{curl}\mathbf{F}$ is then a continuous vector field, the result of Exercise 25 in Section 3 says

$$\lim_{r \to 0} \frac{1}{\sigma(D_r)} \int_{D_r} \mathrm{curl}\mathbf{F} \cdot d\mathbf{S} = \mathrm{curl}\mathbf{F}(\mathbf{x}_0) \cdot \mathbf{n}_0,$$

where $\mathbf{n}_0 = \mathbf{n}(\mathbf{x}_0)$ is the unit normal to the surface D_r at the point $\mathbf{x}_0$. However, by Exercise 12 above, Stokes's formula can be written as

$$\int_{D_r} \mathrm{curl}\mathbf{F} \cdot d\mathbf{S} = \oint_c \mathbf{F} \cdot \mathbf{t}\,ds,$$

where c is the border of D_r, $\mathbf{t}$ is the unit tangent vector in the direction of traversal of c and ds is the arc length differential. Also, note that $A(D_r) = \sigma(D_r)$. Hence, the above limit equation can be written as

$$\lim_{r \to 0} \frac{1}{A(D_r)} \oint_c \mathbf{F} \cdot \mathbf{t} \, ds = \mathrm{curl} \mathbf{F}(\mathbf{x}_0) \cdot \mathbf{n}_0.$$

15. Let $\mathbf{F}(\mathbf{x}) = \big(F(\mathbf{x}), G(\mathbf{x}), H(\mathbf{x})\big)$ be a differentiable vector field defined on an open subset B of $\mathcal{R}^3$. The derivative matrix of F and its transpose are given by

$$\mathbf{F}'(\mathbf{x}) = \begin{pmatrix} F_x & F_y & F_z \\ G_x & G_y & G_z \\ H_x & H_y & H_z \end{pmatrix}, \qquad \big[\mathbf{F}'(\mathbf{x})\big]^t = \begin{pmatrix} F_x & G_x & H_x \\ F_y & G_y & H_y \\ F_z & G_z & H_z \end{pmatrix}.$$

Hence, for any vector $\mathbf{y} = (u, v, w)$ in $\mathcal{R}^3$,

$$\big(\mathbf{F}'(\mathbf{x}) - [\mathbf{F}'(\mathbf{x})]^t\big)\mathbf{y} = \begin{pmatrix} 0 & F_y - G_x & F_z - H_x \\ G_x - F_y & 0 & G_z - H_y \\ H_x - F_z & H_y - G_z & 0 \end{pmatrix} \begin{pmatrix} u \\ v \\ w \end{pmatrix}$$

$$= \begin{pmatrix} (F_y - G_x)v + (F_z - H_x)w \\ (G_x - F_y)u + (G_z - H_y)w \\ (H_x - F_z)u + (H_y - G_z)v \end{pmatrix}.$$

On the other hand, using the same $\mathbf{y}$ as above,

$$\mathrm{curl} \mathbf{F}(\mathbf{x}) \times \mathbf{y} = \begin{pmatrix} H_y - G_z \\ F_z - H_x \\ G_x - F_y \end{pmatrix} \times \begin{pmatrix} u \\ v \\ w \end{pmatrix} = \begin{pmatrix} (F_z - H_x)w - (G_x - F_y)v \\ (G_x - F_y)u - (H_y - G_z)w \\ (H_y - G_z)v - (F_z - H_x)u \end{pmatrix}$$

$$= \begin{pmatrix} (F_y - G_x)v + (F_z - H_x)w \\ (G_x - F_y)u + (G_z - H_y)w \\ (H_x - F_z)u + (H_y - G_z)v \end{pmatrix}.$$

Comparing the above results shows that

$$(*) \qquad \big(\mathbf{F}'(\mathbf{x}) - [\mathbf{F}'(\mathbf{x})]^t\big)\mathbf{y} = \mathrm{curl} \mathbf{F}(\mathbf{x}) \times \mathbf{y}, \quad \text{for all } \mathbf{y} \text{ in } \mathcal{R}^3.$$

We now decompose the square matrix $\mathbf{F}'(\mathbf{x})$ into

$$\mathbf{F}'(\mathbf{x}) = \frac{1}{2}\big(\mathbf{F}'(\mathbf{x}) + [\mathbf{F}'(\mathbf{x})]^t\big) + \frac{1}{2}\big(\mathbf{F}'(\mathbf{x}) - [\mathbf{F}'(\mathbf{x})]^t\big) = S(\mathbf{x}) + \frac{1}{2}\big(\mathbf{F}'(\mathbf{x}) - [\mathbf{F}'(\mathbf{x})]^t\big),$$

where $S(\mathbf{x})$ is one half the sum of a square matrix and its transpose, which is clearly a symmetric matrix. Multiplying each term above on the right by the column vector $\mathbf{y}$ and using the identity $(*)$ gives

$$\mathbf{F}'(\mathbf{x})\mathbf{y} = S(\mathbf{x})\mathbf{y} + \frac{1}{2}\mathrm{curl} \mathbf{F}(\mathbf{x}) \times \mathbf{y}.$$

344

17. The first term on the left side of Equation (4) in the text shows a derivative of a product of two functions. Thus, the product rule is done first to get

$$\frac{\partial}{\partial u}\left(F_1 \circ g \frac{\partial g_1}{\partial v}\right) = F_1 \circ g \frac{\partial}{\partial u}\left(\frac{\partial g_1}{\partial v}\right) + \frac{\partial(F_1 \circ g)}{\partial u}\frac{\partial g_1}{\partial v} = F_1 \circ g \frac{\partial^2 g_1}{\partial u \partial v} + \frac{\partial(F_1 \circ g)}{\partial u}\frac{\partial g_1}{\partial v}.$$

The second term of the rightmost member above now displays a derivative of a composite function. The chain rule then gives

$$(*) \qquad \frac{\partial}{\partial u}\left(F_1 \circ g \frac{\partial g_1}{\partial v}\right) = F_1 \circ g \frac{\partial^2 g_1}{\partial u \partial v} + \left(\frac{\partial F_1}{\partial x}\frac{\partial g_1}{\partial u} + \frac{\partial F_1}{\partial y}\frac{\partial g_2}{\partial u} + \frac{\partial F_1}{\partial z}\frac{\partial g_3}{\partial u}\right)\frac{\partial g_1}{\partial v}.$$

In exactly an analogous sequence of steps, the second term of the left side of Equation (4) can be expanded to get

$$(**) \qquad \frac{\partial}{\partial v}\left(F_1 \circ g \frac{\partial g_1}{\partial u}\right) = F_1 \circ g \frac{\partial^2 g_1}{\partial v \partial u} + \left(\frac{\partial F_1}{\partial x}\frac{\partial g_1}{\partial v} + \frac{\partial F_1}{\partial y}\frac{\partial g_2}{\partial v} + \frac{\partial F_1}{\partial z}\frac{\partial g_3}{\partial v}\right)\frac{\partial g_1}{\partial u}.$$

When the left member of $(**)$ is subtracted from the left member of $(*)$ to get the left member of Equation (4), the corresponding first terms of the right members of $(*)$ and $(**)$ cancel out because of the equality of the mixed partials of g_1 with respect to u and v. This leaves the second term of the right member of $(**)$ subtracted from the second term of the right member of $(*)$. Moreover, when the remaining expression is multiplied out and regrouped with respect to the various derivatives of F_1, the terms involving $\partial F_1/\partial x$ cancel out, leaving two groups of terms: one multiplied by $\partial F_1/\partial y$ and one multiplied by $\partial F_1/\partial z$. Specifically,

$$\left(\frac{\partial F_1}{\partial x}\frac{\partial g_1}{\partial u} + \frac{\partial F_1}{\partial y}\frac{\partial g_2}{\partial u} + \frac{\partial F_1}{\partial z}\frac{\partial g_3}{\partial u}\right)\frac{\partial g_1}{\partial v} - \left(\frac{\partial F_1}{\partial x}\frac{\partial g_1}{\partial v} + \frac{\partial F_1}{\partial y}\frac{\partial g_2}{\partial v} + \frac{\partial F_1}{\partial z}\frac{\partial g_3}{\partial v}\right)\frac{\partial g_1}{\partial u}$$

$$= -\frac{\partial F_1}{\partial y}\left(\frac{\partial g_1}{\partial u}\frac{\partial g_2}{\partial v} - \frac{\partial g_1}{\partial v}\frac{\partial g_2}{\partial u}\right) + \frac{\partial F_1}{\partial z}\left(\frac{\partial g_3}{\partial u}\frac{\partial g_1}{\partial v} - \frac{\partial g_3}{\partial v}\frac{\partial g_1}{\partial u}\right)$$

$$= -\frac{\partial F_1}{\partial y}\begin{vmatrix}\frac{\partial g_1}{\partial u} & \frac{\partial g_1}{\partial v} \\ \frac{\partial g_2}{\partial u} & \frac{\partial g_2}{\partial v}\end{vmatrix} + \frac{\partial F_1}{\partial z}\begin{vmatrix}\frac{\partial g_3}{\partial u} & \frac{\partial g_3}{\partial v} \\ \frac{\partial g_1}{\partial u} & \frac{\partial g_1}{\partial v}\end{vmatrix}$$

$$= -\frac{\partial F_1}{\partial y}\frac{\partial(g_1, g_2)}{\partial(u, v)} + \frac{\partial F_1}{\partial z}\frac{\partial(g_3, g_1)}{\partial(u, v)}.$$

In other words, the last member above is equal to the left side of Equation (4). This establishes Equation (4).

19. If $\mathbf{F}(x, y, z) = (2y^2, x^2, 3z^2)$ then $\text{curl}\mathbf{F} = (0, 0, 2x - 4y)$. For points (x, y, z) on the side of the can, this vector is perpendicular to the normal shown in the figure. Hence, the surface integral of curl$\mathbf{F}$ over this part of the can is zero. On the other hand, this vector is parallel to the normals on the bottom of the can. We shall assume that the bottom surface of the can, which we denote by S_0, is parametrized by $g(x, y) = (x, y, 0)$, for $x^2 + y^2 \leq 1$. This gives, $g_x \times g_y = (0, 0, 1)$. Therefore, if S_1 denotes the side of the can then by Stokes's Theorem

$$\oint_{\partial C} \mathbf{F} \cdot d\mathbf{x} = \int_S \text{curl}\mathbf{F} \cdot d\mathbf{S} = \int_{S_1} \text{curl}\mathbf{F} \cdot d\mathbf{S} + \int_{S_0} \text{curl}\mathbf{F} \cdot d\mathbf{S} = \int_{S_0} \text{curl}\mathbf{F} \cdot d\mathbf{S}$$

Since, $\mathbf{F}(x, y, z) = (2y^2, x^2, 0)$ on S_0, $\mathbf{F}$ is of the form given in Exercise 11(a), so that the surface integral above can be written as the planar surface integral in Green's formula; namely,

$$\int_{S_0} \text{curl} \mathbf{F} \cdot d\mathbf{S} = \int_{S_0} (2x - 4y)\, dx\, dy = \int_0^{2\pi} d\theta \int_0^1 (2r\cos\theta - 4r\sin\theta)\, r\, dr$$

$$= \int_0^{2\pi} (2\cos\theta - 4\sin\theta)\, d\theta \int_0^1 r^2\, dr = 0$$

where the last equality follows because the integrals of the sine and cosine functions over an interval of length 2π is zero.

21. Let R be $\mathcal{R}^2$ with the origin deleted. Consider the vector field

$$\mathbf{F}(x, y) = \left(\frac{-y}{x^2 + y^2}, \frac{x}{x^2 + y^2} \right), \quad (x, y) \in R.$$

It is continuously differentiable on R and its scalar curl is given by

$$\text{curl} \mathbf{F} = \frac{\partial}{\partial x}\left(\frac{x}{x^2 + y^2} \right) - \frac{\partial}{\partial y}\left(\frac{-y}{x^2 + y^2} \right) = \frac{y^2 - x^2}{(x^2 + y^2)^2} - \frac{y^2 - x^2}{(x^2 + y^2)^2} = 0.$$

However, when the unit circle is parametrized by $g(t) = (\cos t, \sin t)$, $0 \le t \le 2\pi$, the line integral of $\mathbf{F}$ around γ is computed to be

$$\oint_\gamma \mathbf{F}(g(t)) \cdot g'(t)\, dt = \oint_\gamma (-\sin t, \cos t) \cdot (-\sin t, \cos t)\, dt = \int_0^{2\pi} 1\, dt = 2\pi.$$

Since this is nonzero, and since R is polygonally connected, the equivalence of (b) and (c) in Theorem 2.4 of this chapter shows that $\mathbf{F}$ can't be a gradient field on R. By Theorem 5.4, we conclude that condition (a) cannot hold on R. That is, R is not simply connected.

23. (a) If $\mathbf{F} = (F_1, F_2, F_3)$ is a continuously differentiable three-dimensional vector field then the statement "$\mathbf{G} = (G_1, G_2, G_3)$ is a vector field such that $\text{curl} \mathbf{G} = \mathbf{F}$" is equivalent to saying that the three differentiable functions "G_1, G_2 and G_3 are such that

$$\text{curl} \mathbf{G} = \left(\frac{\partial G_2}{\partial y} - \frac{\partial G_3}{\partial z}, \frac{\partial G_1}{\partial z} - \frac{\partial G_3}{\partial x}, \frac{\partial G_2}{\partial x} - \frac{\partial G_1}{\partial y} \right) = (F_1, F_2, F_3) = \mathbf{F}."$$

But this holds if, and only if, corresponding coordinates are equal; i.e., if and only if, they satisfy the system of equations

$$(*) \qquad \frac{\partial G_3}{\partial y} - \frac{\partial G_2}{\partial z} = F_1, \qquad \frac{\partial G_1}{\partial z} - \frac{\partial G_3}{\partial x} = F_2, \qquad \frac{\partial G_2}{\partial x} - \frac{\partial G_1}{\partial y} = F_3.$$

(b) We are told that there is a solution of the system $(*)$ in part (a) where $G_3 =$ constant and G_1, G_2 are given by

$$G_1(x, y, z) = \int_0^z F_2(x, y, t)\, dt - \int_0^y F_3(x, t, 0)\, dt, \qquad G_2(x, y, z) = -\int_0^z F_1(x, y, t)\, dt.$$

346

To verify this, we first notice that if G_3 is constant then $(*)$ simplifies to

$$(**) \qquad -\frac{\partial G_2}{\partial z} = F_1, \qquad \frac{\partial G_1}{\partial z} = F_2, \qquad \frac{\partial G_2}{\partial x} - \frac{\partial G_1}{\partial y} = F_3.$$

With G_1 and G_2 as proposed, we first have

$$-\frac{\partial G_2}{\partial z} = -\frac{\partial}{\partial z}\left[-\int_0^z F_1(x,y,t)\,dt\right] = \frac{\partial}{\partial z}\int_0^z F_1(x,y,t)\,dt = F_1(x,y,z),$$

where the last equation follows from the fundamental theorem of calculus. Thus, the first equation of $(**)$ is satisfied. Next, we have

$$\frac{\partial G_1}{\partial z} = \frac{\partial}{\partial z}\left[\int_0^z F_2(x,y,t)\,dt - \int_0^y F_3(x,t,0)\,dt\right]$$
$$= \frac{\partial}{\partial z}\int_0^z F_2(x,y,t)\,dt - \frac{\partial}{\partial z}\int_0^y F_3(x,t,0)\,dt = F_2(x,y,z),$$

where the first integral in the second line is computed by the fundamental theorem of calculus and the second integral in the second line is zero because it's independent of z. Thus, the second equation of $(**)$ is satisfied. And finally, we have

$$\frac{\partial G_2}{\partial x} - \frac{\partial G_1}{\partial y} = \frac{\partial}{\partial x}\left[-\int_0^z F_1(x,y,t)\,dt\right] - \frac{\partial}{\partial y}\left[\int_0^z F_2(x,y,t)\,dt - \int_0^y F_3(x,t,z)\,dt\right]$$
$$= -\left(\frac{\partial}{\partial xs}\int_0^z F_1(x,y,t)\,dt + \frac{\partial}{\partial y}\int_0^z F_2(x,y,t)\,dt\right) + \frac{\partial}{\partial y}\int_0^y F_3(x,t,0)\,dt$$
$$= -\left(\int_0^z \frac{\partial F_1}{\partial x}(x,y,t)\,dt + \int_0^z \frac{\partial F_2}{\partial y}(x,y,t)\,dt\right) + F_3(x,y,0)$$
$$= \int_0^z \left(-\frac{\partial F_1}{\partial x} - \frac{\partial F_2}{\partial y}\right)(x,y,t)\,dt + F_3(x,y,0),$$

where the Leibnitz rule is used twice inside the parenthetical expression in the second line and the fundamental theorem of calculus is used to compute the last integral in the second line. Since $\mathbf{F}$ is necessarily divergence-free,

$$\frac{\partial F_1}{\partial x} + \frac{\partial F_2}{\partial y} + \frac{\partial F_3}{\partial z} = 0, \quad \text{so that} \quad -\frac{\partial F_1}{\partial x} - \frac{\partial F_2}{\partial y} = \frac{\partial F_3}{\partial z}.$$

Therefore, the last integral above can be written as

$$\int_0^z \left(-\frac{\partial F_1}{\partial x} - \frac{\partial F_2}{\partial x}\right)(x,y,t)\,dt = \int_0^z \frac{\partial F_3}{\partial z}(x,y,t)\,dt = F_3(x,y,z) - F_3(x,y,0),$$

where the last equality is the result of applying the fundamental theorem of calculus. Thus,

$$\frac{\partial G_2}{\partial x} - \frac{\partial G_1}{\partial y} = \big(F_3(x,y,z) - F_3(x,y,0)\big) + F_3(x,y,0) = F_3(x,y,z)$$

347

and the third equation of (∗∗) is satisfied. That is, the proposed solutions given by (∗) are indeed solutions and therefore the function $\mathbf{G} = (G_1, G_2, G_3)$ satisfies $\text{curl}\mathbf{G} = \mathbf{F}$.

25. Let $\mathbf{F}(x, y, z) = (y, z, x)$, with coordinate functions $F_1(x, y, z) = y$, $F_2(x, y, z) = z$ and $F_3(x, y, z) = x$. With $G_1(x, y, z)$ and $G_2(x, y, z)$ as defined in part (b) of Exercise 23(b) of this section, we get

$$G_1(x, y, z) = \int_0^z F_2(x, y, t)\, dt - \int_0^y F_3(x, t, 0)\, dt = \int_0^z t\, dt - \int_0^y x\, dt$$

$$= \left[\frac{1}{2}t^2\right]_0^z - [xt]_0^y = \frac{1}{2}z^2 - xy.$$

$$G_2(x, y, z) = -\int_0^z F_1(x, y, t)\, dt = -\int_0^z y\, dt = -[yt]_0^z = -yz.$$

Now let $G_3(x, y, z) = c$, a real constant, and define the vector field $\mathbf{G}$ by

$$\mathbf{G}(x, y, z) = \big(G_1(x, y, z), G_2(x, y, z), G_3(x, y, z)\big) = (z^2/2 - xy, -yz, c).$$

To show that $\text{curl}\mathbf{G} = \mathbf{F}$, we compute

$$\text{curl}\mathbf{G} = \left(\frac{\partial(c)}{\partial y} - \frac{\partial(-yz)}{\partial z}, \frac{\partial(z^2/2 - xy)}{\partial z} - \frac{\partial(c)}{\partial x}, \frac{\partial(-yz)}{\partial x} - \frac{\partial(z^2/2 - xy)}{\partial y}\right)$$

$$= \big(0 - (-y), z - 0, 0 - (-x)\big) = (y, z, x) = \mathbf{F}(x, y, z).$$

27. Let $\mathbf{F}(x, y, z) = (x, -y, 3x)$, with coordinate functions $F_1(x, y, z) = x$, $F_2(x, y, z) = -y$ and $F_3(x, y, z) = 3x$. With $G_1(x, y, z)$ and $G_2(x, y, z)$ as defined in part (b) of Exercise 23(b) of this section, we get

$$G_1(x, y, z) = \int_0^z F_2(x, y, t)\, dt - \int_0^y F_3(x, t, 0)\, dt = \int_0^z -y\, dt - \int_0^y 3x\, dt$$

$$= \big[-yt\big]_0^z - [3xt]_0^y = -yz - 3xy,$$

$$G_2(x, y, z) = -\int_0^z F_1(x, y, t)\, dt = -\int_0^z x\, dt = -[xt]_0^z = -xz.$$

Now let $G_3(x, y, z) = c$, a real constant, and define the vector field $\mathbf{G}$ by

$$\mathbf{G}(x, y, z) = \big(G_1(x, y, z), G_2(x, y, z), G_3, (x, y, z)\big) = (-yz - 3xy, -xz, c).$$

To show that $\text{curl}\mathbf{G} = \mathbf{F}$, we compute

$$\text{curl}\mathbf{G} = \left(\frac{\partial(c)}{\partial y} - \frac{\partial(-xz)}{\partial z}, \frac{\partial(-yz - 3xy)}{\partial z} - \frac{\partial(c)}{\partial x}, \frac{\partial(-xz)}{\partial x} - \frac{\partial(-yz - 3xy)}{\partial y}\right)$$

$$= \big(0 - (-x), -y - 0, -z - (-z - 3x)\big) = (x, -y, 3x) = \mathbf{F}(x, y, z).$$

29. If $\mathbf{G}$ and $\mathbf{H}$ are continuously differentiable vector fields on $\mathcal{R}^3$ such that $\text{curl}\mathbf{G}(\mathbf{x}) = \text{curl}\mathbf{H}(\mathbf{x})$, for all $\mathbf{x}$ in $\mathcal{R}^3$, then

$$\text{curl}\mathbf{G}(\mathbf{x}) - \text{curl}\mathbf{H}(\mathbf{x})$$

$$= \left(\frac{\partial G_3}{\partial y} - \frac{\partial G_3}{\partial z}, \frac{\partial G_1}{\partial z} - \frac{\partial G_3}{\partial x}, \frac{\partial G_2}{\partial x} - \frac{\partial G_1}{\partial y}\right) - \left(\frac{\partial H_3}{\partial y} - \frac{\partial H_3}{\partial z}, \frac{\partial H_1}{\partial z} - \frac{\partial H_3}{\partial x}, \frac{\partial H_2}{\partial x} - \frac{\partial H_1}{\partial y}\right)$$

$$= \left(\frac{\partial(G_3 - H_3)}{\partial y} - \frac{\partial(G_2 - H_2)}{\partial z}, \frac{\partial(G_1 - H_1)}{\partial z} - \frac{\partial(G_3 - H_3)}{\partial x}, \frac{\partial(G_2 - H_2)}{\partial x} - \frac{\partial(G_1 - H_1)}{\partial y}\right)$$

$$= \text{curl}(\mathbf{G} - \mathbf{H})(\mathbf{x}) = \mathbf{0},$$

for all **x** in $\mathcal{R}^3$. Since $\mathcal{R}^3$ is connected, Theorem 5.4 in this chapter says that $\mathbf{G} - \mathbf{H}$ is a gradient field on $\mathcal{R}^3$. That is, $\mathbf{G} - \mathbf{H} = \nabla f$, for some $f : \mathcal{R}^3 \longrightarrow \mathcal{R}$.

31. The given vector field $\mathbf{F}(x, y, z) = (y, z, x)$ satisfies $\mathrm{div}\mathbf{F}(\mathbf{x}) = 0$, for all $\mathbf{x}$ in $\mathcal{R}^3$. The computation of the integral defining $\mathbf{G}$ as given in the preamble to this exercise is as follows

$$
\mathbf{G}(\mathbf{x}) = \int_0^1 \left[\mathbf{F}(t\mathbf{x}) \times t\mathbf{x} \right] dt = \int_0^1 (ty, tz, tx) \times t(x, y, z)\, dt = \int_0^1 (y, z, x) \times (x, y, z)\, t^2 dt
$$

$$
= \int_0^1 \left(\begin{vmatrix} z & y \\ x & z \end{vmatrix}, \begin{vmatrix} x & z \\ y & x \end{vmatrix}, \begin{vmatrix} y & x \\ z & y \end{vmatrix} \right) t^2 dt = \int_0^1 (z^2 - xy, x^2 - yz, y^2 - x^z)\, t^2 dt
$$

$$
= \left[\frac{1}{3}(z^2 - xy, x^2 - yz, y^2 - xz)t^3 \right]_0^1 = \frac{1}{3}(z^2 - xy, x^2 - yz, y^2 - xz).
$$

As a check that $\mathrm{curl}\mathbf{G} = \mathbf{F}$, we compute

$$
\mathrm{curl}\mathbf{G}(x, y, z)
$$

$$
= \frac{1}{3} \left(\frac{\partial(y^2 - xz)}{\partial y} - \frac{\partial(x^2 - yz)}{\partial z}, \frac{\partial(z^2 - xy)}{\partial z} - \frac{\partial(y^2 - xz)}{\partial x}, \frac{\partial(x^2 - yz)}{\partial x} - \frac{\partial(z^2 - xy)}{\partial y} \right)
$$

$$
= \frac{1}{3} \left(2y - (-y), 2z - (-z), 2x - (-x) \right) = (y, z, x) = \mathbf{F}(x, y, z),
$$

which holds for all (x, y, z) in $\mathcal{R}^3$.

33. The given vector field $\mathbf{F}(x, y, z) = (x, -y, 3x)$ satisfies $\mathrm{div}\mathbf{F}(\mathbf{x}) = 0$, for all $\mathbf{x}$ in $\mathcal{R}^3$. The computation of the integral defining $\mathbf{G}$ as given in the preamble to this exercise is as follows:

$$
\mathbf{G}(\mathbf{x}) = \int_0^1 \left[\mathbf{F}(t\mathbf{x}) \times t\mathbf{x} \right] dt = \int_0^1 (tx, -ty, 3tx) \times t(x, y, z)\, dt
$$

$$
= \int_0^1 (x, -y, 3x) \times (x, y, z)\, t^2 dt = \int_0^1 \left(\begin{vmatrix} -y & y \\ 3x & z \end{vmatrix}, \begin{vmatrix} 3x & z \\ x & x \end{vmatrix}, \begin{vmatrix} x & x \\ -y & y \end{vmatrix} \right) t^2 dt
$$

$$
= \int_0^1 (-yz - 3xy, 3x^2 - xz, 2xy)t^2 dt = \left[\frac{1}{3}(-yz - 3xy, 3x^2 - xz, 2xy)t^3 \right]_0^1
$$

$$
= \frac{1}{3}(-yz - 3xy, 3x^2 - xz, 2xy),
$$

As a check that $\mathrm{curl}\mathbf{G} = \mathbf{F}$, we compute

$$
\mathrm{curl}\mathbf{G}(x, y, z)
$$

$$
= \frac{1}{3} \left(\frac{\partial(2xy)}{\partial y} - \frac{\partial(3x^2 - xz)}{\partial z}, \frac{\partial(-yz - 3xy)}{\partial z} - \frac{\partial(2xy)}{\partial x}, \frac{\partial(3x^2 - xz)}{\partial x} - \frac{\partial(-yz - 3xy)}{\partial y} \right)
$$

$$
= \frac{1}{3} \left(2x - (-x), -y - 2y, 6x - z - (-z - 3x) \right) = (x, -y, 3x) = \mathbf{F}(x, y, z),
$$

which holds for all (x, y, z) in $\mathcal{R}^3$.

Section 6: THE OPERATORS ∇, $\nabla\times$ AND $\nabla\cdot$

Exercise Set 6ABC (pgs. 456-457)

1. Let f and g be real-valued differentiable functions defined on a common domain R of $\mathcal{R}^3$, and let a and b be real constants, so that $af + bg$ is a real-valued differentiable function on R. Using the definition of ∇ and the linearity of the partial derivative, $af + bg$ satisfies

$$\nabla(af + bg) = \left(\frac{\partial(af + bg)}{\partial x}, \frac{\partial(af + bg)}{\partial y}, \frac{\partial(af + bg)}{\partial z} \right)$$

$$= \left(a\frac{\partial f}{\partial x} + b\frac{\partial g}{\partial x}, a\frac{\partial f}{\partial y} + b\frac{\partial g}{\partial y}, a\frac{\partial f}{\partial z} + b\frac{\partial g}{\partial z} \right)$$

$$= a\left(\frac{\partial f}{\partial x}, \frac{\partial f}{\partial y}, \frac{\partial f}{\partial z} \right) + b\left(\frac{\partial g}{\partial x}, \frac{\partial g}{\partial y}, \frac{\partial g}{\partial z} \right) = a\nabla f + b\nabla g,$$

which holds on all of R. This establishes identity (1) in the text.

3. Let $\mathbf{F} = (F_1, F_2, F_3)$ and $\mathbf{G} = (G_1, G_2, G_3)$ be differentiable vector fields defined on a common domain R of $\mathcal{R}^3$, and let a and b be real constants. Then the function $a\mathbf{F} + b\mathbf{G} = (aF_1 + bG_1, aF_2 + bG_2, aF_3 + bG_3)$ is a differentiable vector field on R. Using the definition of $\nabla\times$ and the linearity of the partial derivative we can show that each coordinate of $\nabla \times (a\mathbf{F} + b\mathbf{G})$ is the sum of the corresponding coordinates of $a\nabla \times \mathbf{F}$ and $b\nabla \times \mathbf{G}$. For example, the first coordinate of $\nabla \times (a\mathbf{F} + b\mathbf{G})$ can be written as

$$\frac{\partial(aF_3 + bG_3)}{\partial y} - \frac{\partial(aF_2 + bG_2)}{\partial z} = a\left(\frac{\partial F_3}{\partial y} - \frac{\partial F_2}{\partial z} \right) + b\left(\frac{\partial G_3}{\partial y} - \frac{\partial G_2}{\partial z} \right),$$

where the grouping on the right shows that the first coordinate of $\nabla \times (a\mathbf{F} + b\mathbf{G})$ is a times the first coordinate of $\nabla \times \mathbf{F}$ plus b times the first coordinate of $\nabla \times \mathbf{G}$. A similar result holds for the other coordinates of $\nabla \times (a\mathbf{F} + b\mathbf{G})$. It follows that $\nabla \times (a\mathbf{F} + b\mathbf{G})$ must itself be the sum of a times $\nabla \times \mathbf{F}$ and b times $\nabla \times \mathbf{G}$. That is,

$$\nabla \times (a\mathbf{F} + b\mathbf{G}) = a\nabla \times \mathbf{F} + b\nabla \times \mathbf{G} \quad \text{on } R..$$

This establishes identity (3) in the text.

5. Let $\mathbf{F} = (F_1, F_2, F_3)$ and $\mathbf{G} = (G_1, G_2, G_3)$ be differentiable vector fields defined on a common domain R of $\mathcal{R}^3$, and let a and b be real constants. Then the function $a\mathbf{F} + b\mathbf{G} = (aF_1 + bG_1, aF_2 + bG_2, aF_3 + bG_3)$ is a differentiable vector field on R. Thus, on R, the definition of $\nabla\cdot$ and the linearity of the partial derivative allows us to write

$$\nabla \cdot (a\mathbf{F} + b\mathbf{G}) = \frac{\partial(aF_1 + bG_1)}{\partial x} + \frac{\partial(aF_2 + bG_2)}{\partial y} + \frac{\partial(aF_3 + bG_3)}{\partial z}$$

$$= a\frac{\partial F_1}{\partial x} + b\frac{\partial G_1}{\partial x} + a\frac{\partial F_2}{\partial y} + b\frac{\partial G_2}{\partial y} + a\frac{\partial F_3}{\partial z} + b\frac{\partial G_3}{\partial z}$$

$$= a\left(\frac{\partial F_1}{\partial x} + \frac{\partial F_2}{\partial y} + \frac{\partial F_3}{\partial z} \right) + b\left(\frac{\partial G_1}{\partial x} + \frac{\partial G_2}{\partial y} + \frac{\partial G_3}{\partial z} \right)$$

$$= a\nabla \cdot \mathbf{F} + b\nabla \cdot \mathbf{G}.$$

This establishes identity (5) in the text.

7. If $\mathbf{F} = (F_1, F_2, F_3)$ and $\mathbf{G} = (G_1, G_2, G_3)$ are differentiable vector fields defined on a common domain R of $\mathcal{R}^3$ then $\mathbf{F} \times \mathbf{G}$ is also a differentiable vector field on R. Since

$$\mathbf{F} \times \mathbf{G} = \left(\begin{vmatrix} F_2 & G_2 \\ F_3 & G_3 \end{vmatrix}, \begin{vmatrix} F_3 & G_3 \\ F_1 & G_1 \end{vmatrix}, \begin{vmatrix} F_1 & G_1 \\ F_2 & G_2 \end{vmatrix} \right) = (F_2 G_3 - F_3 G_2, F_3 G_2 - F_1 G_3, F_1 G_2 - F_2 G_1),$$

the definition of $\nabla \cdot$ and the product rule for partial derivatives gives

$$\nabla \cdot (\mathbf{F} \times \mathbf{G}) = \frac{\partial (F_2 G_3 - F_3 G_2)}{\partial x} + \frac{\partial (F_3 G_2 - F_1 G_3)}{\partial y} + \frac{\partial (F_1 G_2 - F_2 G_1)}{\partial z}$$

$$= F_2 \frac{\partial G_3}{\partial x} + G_3 \frac{\partial F_2}{\partial x} - F_3 \frac{\partial G_2}{\partial x} - G_2 \frac{\partial F_3}{\partial x}$$

$$+ F_3 \frac{\partial G_1}{\partial y} + G_1 \frac{\partial F_3}{\partial y} - F_1 \frac{\partial G_3}{\partial y} - G_3 \frac{\partial F_1}{\partial y}$$

$$+ F_1 \frac{\partial G_2}{\partial z} + G_2 \frac{\partial F_1}{\partial z} - F_2 \frac{\partial G_1}{\partial z} - G_1 \frac{\partial F_2}{\partial z}$$

$$= G_1 \left(\frac{\partial F_3}{\partial y} - \frac{\partial F_2}{\partial z} \right) + G_2 \left(\frac{\partial F_1}{\partial z} - \frac{\partial F_3}{\partial x} \right) + G_3 \left(\frac{\partial F_2}{\partial x} - \frac{\partial F_1}{\partial y} \right)$$

$$- F_1 \left(\frac{\partial G_3}{\partial y} - \frac{\partial G_2}{\partial z} \right) - F_2 \left(\frac{\partial G_1}{\partial z} - \frac{\partial G_3}{\partial x} \right) - F_3 \left(\frac{\partial G_2}{\partial x} - \frac{\partial G_1}{\partial y} \right)$$

$$= \mathbf{G} \cdot (\nabla \times \mathbf{F}) - \mathbf{F} \cdot (\nabla \times \mathbf{G}),$$

which holds on all of R. This establishes identity (7) in the text.

9. Let f be a twice continuously differentiable real-valued function defined on an open set R in $\mathcal{R}^3$. Then

$$\nabla \times (\nabla f) = \nabla \times \left(\frac{\partial f}{\partial x}, \frac{\partial f}{\partial y}, \frac{\partial f}{\partial z} \right)$$

$$= \left(\frac{\partial}{\partial y} \left(\frac{\partial f}{\partial z} \right) - \frac{\partial}{\partial z} \left(\frac{\partial f}{\partial y} \right), \frac{\partial}{\partial z} \left(\frac{\partial f}{\partial x} \right) - \frac{\partial}{\partial x} \left(\frac{\partial f}{\partial z} \right), \frac{\partial}{\partial x} \left(\frac{\partial f}{\partial y} \right) - \frac{\partial}{\partial y} \left(\frac{\partial f}{\partial x} \right) \right)$$

$$= \left(\frac{\partial^2 f}{\partial y \partial z} - \frac{\partial^2 f}{\partial z \partial y}, \frac{\partial^2 f}{\partial z \partial x} - \frac{\partial^2 f}{\partial x \partial z}, \frac{\partial^2 f}{\partial x \partial y} - \frac{\partial^2 f}{\partial y \partial x} \right),$$

which holds on all of R. Since f is twice continuously differentiable, the two mixed partials in any given coordinate are equal and therefore cancel out. That is,

$$\nabla \times (\nabla f) = \mathbf{0} \quad \text{on } R.$$

This establishes identity (9) in the text.

11. Let $\mathbf{v} = (a, b, c)$ be a constant vector and let $\mathbf{F}(\mathbf{x})$ be the field defined by

$$\mathbf{F}(\mathbf{x}) = \mathbf{v} \times \mathbf{x} = (a, b, c) \times (x, y, z) = \left(\begin{vmatrix} b & y \\ c & z \end{vmatrix}, \begin{vmatrix} c & z \\ a & x \end{vmatrix}, \begin{vmatrix} a & x \\ b & y \end{vmatrix} \right)$$

$$= (bz - cy, cx - az, ay - bx).$$

In identity (4) in the text, let $f(\mathbf{x}) = 1/|\mathbf{x}|$ and $\mathbf{F} = \mathbf{v} \times \mathbf{x}$ to get

$$(*) \qquad \nabla \times \frac{\mathbf{v} \times \mathbf{x}}{|\mathbf{x}|} = \frac{1}{|\mathbf{x}|} \nabla \times (\mathbf{v} \times \mathbf{x}) + \nabla \left(\frac{1}{|\mathbf{x}|} \right) \times (\mathbf{v} \times \mathbf{x}).$$

Examination of the two terms on the right suggests we compute $\nabla \times (\mathbf{v} \times \mathbf{x})$ and $\nabla(1/|\mathbf{x}|)$. The reader can verify that

$$\nabla \times (\mathbf{v} \times \mathbf{x}) = 2\mathbf{v} \qquad \text{and} \qquad \nabla\left(\frac{1}{|\mathbf{x}|}\right) = -\frac{\mathbf{x}}{|\mathbf{x}|^3}.$$

Equation $(*)$ now becomes

$$(**)\qquad\qquad \nabla \times \frac{\mathbf{v} \times \mathbf{x}}{|\mathbf{x}|} = 2\frac{\mathbf{v}}{|\mathbf{x}|} - \frac{1}{|\mathbf{x}|^3}[\mathbf{x} \times (\mathbf{v} \times \mathbf{x})].$$

Now observe that

$$\begin{aligned}
\mathbf{x} \times (\mathbf{v} \times \mathbf{x}) &= (x,y,z) \times (bz - cy, cx - az, ay - bx) \\
&= \left(\begin{vmatrix} y & cx - az \\ z & ay - bx \end{vmatrix}, \begin{vmatrix} z & ay - bx \\ x & bz - cy \end{vmatrix}, \begin{vmatrix} x & bz - cy \\ y & cx - az \end{vmatrix}\right) \\
&= (ay^2 - bxy - cxz + az^2, bz^2 - cyz - axy + bx^2, cx^2 - axz - byz + cy^2) \\
&= (x^2 + y^2 + z^2)(a,b,c) - (ax + by + cz)(x,y,z) = |\mathbf{x}|^2\mathbf{v} - (\mathbf{v} \cdot \mathbf{x})\mathbf{x}.
\end{aligned}$$

Hence, $(**)$ becomes

$$\nabla \times \frac{\mathbf{v} \times \mathbf{x}}{|\mathbf{x}|} = 2\frac{\mathbf{v}}{|\mathbf{x}|} - \frac{1}{|\mathbf{x}|^3}\left(|\mathbf{x}|^2\mathbf{v} - (\mathbf{v} \cdot \mathbf{x})\mathbf{x}\right) = \frac{\mathbf{v}}{|\mathbf{x}|} + \frac{\mathbf{v} \cdot \mathbf{x}}{|\mathbf{x}|^3}\mathbf{x}.$$

13. Let $f(\mathbf{x}) = 1/|\mathbf{x}|$, with $\mathbf{x} \neq \mathbf{0}$ in $\mathcal{R}^3$. Since $\nabla f = -\mathbf{x}/|\mathbf{x}|^3$ (see Exercise 12 in this section), we use identity (10) in the text to get

$$\begin{aligned}
\nabla^2 f = \nabla \cdot \nabla f &= \nabla \cdot (-\mathbf{x}/|\mathbf{x}|^3) \\
&= \nabla \cdot \left(-x(x^2 + y^2 + z^2)^{-3/2}, -y(x^2 + y^2 + z^2)^{-3/2}, -z(x^2 + y^2 + z^2)^{-3/2}\right) \\
&= \frac{\partial}{\partial x}\left(-x(x^2 + y^2 + z^2)^{-3/2}\right) + \frac{\partial}{\partial y}\left(-y(x^2 + y^2 + z^2)^{-3/2}\right) \\
&\qquad\qquad + \frac{\partial}{\partial z}\left(-z(x^2 + y^2 + z^2)^{-3/2}\right) \\
&= (2x^2 - y^2 - z^2)(x^2 + y^2 + z^2)^{-3/2} + (-x^2 + 2y^2 - z^2)(x^2 + y^2 + z^2)^{-5/2} \\
&\qquad\qquad + (-x^2 - y^2 + 2z^2)(x^2 + y^2 + z^2)^{-5/2} = 0.
\end{aligned}$$

15. Let R be an open set in $\mathcal{R}^3$ and let $T(\mathbf{x})$ be a steady-state temperature function that is twice continuously differentiable on R. We show that $\nabla^2 T(\mathbf{x}) = 0$ for all $\mathbf{x}$ in R. To do this, assume $\mathbf{x}_0$ is a point in R such that $\nabla^2 T(\mathbf{x}_0) > 0$. Since the second derivatives of T are continuous on R, so is $\nabla^2 T$. Because of this continuity, there is some open ball B in R centered at $\mathbf{x}_0$ on which $\nabla^2 T$ is positive and bounded away from zero, say $0 < \delta \le \nabla^2 T(\mathbf{x})$, for all $\mathbf{x}$ in B. Thus,

$$(*)\qquad\qquad \int_B \nabla^2 T \, dV \ge \int_B \delta \, dV = \delta V(B) > 0.$$

By Equation (12) in the text,

$$\int_B \nabla^2 T \, dV = \int_S \frac{\partial T}{\partial \mathbf{n}} \, d\sigma,$$

where S is the spherical surface bounding B. The integrand of the right side is the directional derivative of T in the direction of the unit normal $\mathbf{n}$ and is the dot product of the gradient of T and the vector $\mathbf{n}$. So,

$$\int_B \nabla^2 T \, dV = \int_S \nabla T \cdot \mathbf{n} \, d\sigma = \int_S \nabla T \cdot d\mathbf{S}.$$

The last integral is the flux of the temperature gradient across the smooth surface S which, by hypothesis, is zero. This contradicts equation ($*$). We conclude that $\nabla^2 T$ is identically zero on R, so that T is harmonic on R.

17. Let R be $\mathcal{R}^3$ with the origin deleted and consider the Newtonian potential function $N(\mathbf{x}) = |\mathbf{x}|^{-1}$. By Theorem 1.3 in Chapter 8, if $\mathbf{x}_0$ is a fixed but arbitrary point of R then, for $\mathbf{y}$ in R, the line integral of ∇N along a smooth path from $\mathbf{y}$ to $\mathbf{x}_0$ has the value

$$\int_{\mathbf{y}}^{\mathbf{x}_0} \nabla N \cdot d\mathbf{x} = N(\mathbf{x}_0) - N(\mathbf{y}) \qquad \text{or} \qquad \int_{\mathbf{y}}^{\mathbf{x}_0} \nabla N \cdot d\mathbf{x} = N(\mathbf{x}_0) - \frac{1}{|\mathbf{y}|}.$$

Note that the value on the left is also the work done in moving a particle from $\mathbf{y}$ to $\mathbf{x}_0$. Now, recall that for a continuous vector variable $\mathbf{y}$, the notation "$\mathbf{y} \to \infty$" implies "$|\mathbf{y}| \to \infty$". Thus, we let $\mathbf{y} \to \infty$ on both sides of the righthand equation. The constant $N(\mathbf{x}_0)$ on the right is unaffected and the limit of the second term on the right can be interpreted as $\lim_{|\mathbf{y}| \to \infty} |\mathbf{y}|^{-1} = 0$. We therefore have

$$\lim_{\mathbf{y} \to \infty} \int_{\mathbf{y}}^{\mathbf{x}_0} \nabla N \cdot d\mathbf{x} = N(\mathbf{x}_0).$$

Thus, $N(\mathbf{x}_0)$ is the limiting value of the work done in moving a particle from $\mathbf{y}$ to $\mathbf{x}_0$ as $\mathbf{y}$ gets farther and farther away from $\mathbf{x}_0$. It seems reasonable to write

$$\int_{\infty}^{\mathbf{x}_0} \nabla N \cdot d\mathbf{x} = N(\mathbf{x}_0),$$

and interpret this to be the work done in moving a particle from ∞ to $\mathbf{x}_0$ along some smooth path through the field ∇N.

19. Let $f(x, y, z)$ be twice differentiable and suppose that f can be written as

$$f(x, y, z) = \overline{f}\big((x^2 + y^2 + z^2)^{1/2}\big).$$

Letting $r = r(x, y, z) = (x^2 + y^2 + z^2)^{1/2}$, we have $f(x, y, z) = \overline{f}(r)$. By the chain rule, $\partial f / \partial x$ can be written as

$$\frac{\partial f}{\partial x} = \frac{\partial \overline{f}}{\partial r} \frac{\partial r}{\partial x} = \frac{\partial \overline{f}}{\partial r} \frac{\partial\big((x^2 + y^2 + z^2)^{1/2}\big)}{\partial x} = \frac{\partial \overline{f}}{\partial r} \left(\frac{x}{(x^2 + y^2 + z^2)^{1/2}} \right) = \frac{x}{r} \frac{\partial \overline{f}}{\partial r}.$$

Similarly,

$$\frac{\partial f}{\partial y} = \frac{y}{r}\frac{\partial \overline{f}}{\partial r} \qquad \text{and} \qquad \frac{\partial f}{\partial z} = \frac{z}{r}\frac{\partial \overline{f}}{\partial r}.$$

Therefore,

$$(*) \quad \nabla^2 f = \nabla \cdot (\nabla f) = \nabla \cdot \left(\frac{\partial f}{\partial x}, \frac{\partial f}{\partial y}, \frac{\partial f}{\partial z}\right) = \frac{\partial}{\partial x}\left(\frac{x}{r}\frac{\partial \overline{f}}{\partial r}\right) + \frac{\partial}{\partial y}\left(\frac{y}{r}\frac{\partial \overline{f}}{\partial r}\right) + \frac{\partial}{\partial z}\left(\frac{z}{r}\frac{\partial \overline{f}}{\partial r}\right).$$

The partials are taken using the "derivative of a product of three functions" rule. For example, the first term is the derivative of a product of the three functions: $1/r$, x and $\partial \overline{f}/\partial r$. Clearly, $\partial x/\partial x = 1$, while the derivatives of $1/r$ and $\partial \overline{f}/\partial r$ with respect to x are

$$\frac{\partial}{\partial x}\left(\frac{1}{r}\right) = \frac{\partial}{\partial x}\left(\frac{1}{(x^2+y^2+z^2)^{1/2}}\right) = -\frac{x}{(x^2+y^2+z^2)^{3/2}} = -\frac{x}{r^3}$$

$$\frac{\partial}{\partial x}\left(\frac{\partial \overline{f}}{\partial r}\right) = \frac{\partial^2 \overline{f}}{\partial r^2}\frac{\partial r}{\partial x} = \frac{\partial^2 \overline{f}}{\partial r^2}\frac{x}{r}.$$

We therefore have,

$$\frac{\partial}{\partial x}\left(\frac{x}{r}\frac{\partial \overline{f}}{\partial r}\right) = \frac{\partial}{\partial x}\left(\frac{1}{r}\right)x\frac{\partial \overline{f}}{\partial r} + \frac{1}{r}\frac{\partial x}{\partial x}\frac{\partial \overline{f}}{\partial r} + \frac{1}{r}x\frac{\partial}{\partial x}\left(\frac{\partial \overline{f}}{\partial r}\right) = -\frac{x^2}{r^3}\frac{\partial \overline{f}}{\partial r} + \frac{1}{r}\frac{\partial \overline{f}}{\partial r} + \frac{x^2}{r^2}\frac{\partial^2 \overline{f}}{\partial r^2}.$$

The terms on the right of $(*)$ involving the partials with respect to y and z are derived in the same manner, with identical results except that the variables y and z replace the variable x in the their corresponding terms. Hence, $(*)$ can be written as

$$\nabla^2 f = -\frac{x^2}{r^3}\frac{\partial \overline{f}}{\partial r} + \frac{1}{r}\frac{\partial \overline{f}}{\partial r} + \frac{x^2}{r^2}\frac{\partial^2 \overline{f}}{\partial r^2} - \frac{y^2}{r^3}\frac{\partial \overline{f}}{\partial r} + \frac{1}{r}\frac{\partial \overline{f}}{\partial r} + \frac{y^2}{r^2}\frac{\partial^2 \overline{f}}{\partial r^2} - \frac{z^2}{r^3}\frac{\partial \overline{f}}{\partial r} + \frac{1}{r}\frac{\partial \overline{f}}{\partial r} + \frac{z^2}{r^2}\frac{\partial^2 \overline{f}}{\partial r^2}$$

$$= -\left(\frac{x^2+y^2+z^2}{r^3}\right)\frac{\partial \overline{f}}{\partial r} + \frac{3}{r}\frac{\partial \overline{f}}{\partial r} + \left(\frac{x^2+y^2+z^2}{r^2}\right)\frac{\partial^2 \overline{f}}{\partial r^2}$$

$$= -\frac{r^2}{r^3}\frac{\partial \overline{f}}{\partial r} + \frac{3}{r}\frac{\partial \overline{f}}{\partial r} + \frac{r^2}{r^2}\frac{\partial^2 \overline{f}}{\partial r^2} = \frac{2}{r}\frac{\partial \overline{f}}{\partial r} + \frac{\partial^2 \overline{f}}{\partial r^2}.$$

21. The four terms in the expression for $\partial^2 u/\partial x^2$ derived in text Example 4 can be further expanded by applying the product rule to each term. Of the eight terms resulting from this expansion, seven are nonzero. When these seven terms are combined according to the type of partial derivative involved, the result is five terms, one of which results from using the fact that the mixed partials of $\overline{u}$ are equal. The final compact form is

$$\frac{\partial^2 u}{\partial x^2} = \cos^2\theta\frac{\partial^2 \overline{u}}{\partial r^2} - \frac{2}{r}\sin\theta\cos\theta\frac{\partial^2 \overline{u}}{\partial r\partial\theta} + \frac{2}{r^2}\cos\theta\sin\theta\frac{\partial \overline{u}}{\partial\theta} + \frac{1}{r}\sin^2\theta\frac{\partial \overline{u}}{\partial r} + \frac{1}{r^2}\sin^2\theta\frac{\partial^2 \overline{u}}{\partial\theta^2}.$$

A similar expansion of the four terms in the expression for $\partial^2 u/\partial y^2$ in text Example 4 also leads to seven nonzero terms which can be combined to give

$$\frac{\partial^2 u}{\partial y^2} = \sin^2\theta\frac{\partial^2 \overline{u}}{\partial r^2} + \frac{2}{r}\sin\theta\cos\theta\frac{\partial^2 \overline{u}}{\partial r\partial\theta} - \frac{2}{r^2}\cos\theta\sin\theta\frac{\partial \overline{u}}{\partial\theta} + \frac{1}{r}\cos^2\theta\frac{\partial \overline{u}}{\partial r} + \frac{1}{r^2}\cos^2\theta\frac{\partial^2 \overline{u}}{\partial\theta^2}.$$

Adding the two equations above member by member we first observe that the second and third terms of each of the right members cancel each other directly. The resulting sum therefore ostensibly has six terms, which can be grouped as follows

$$\frac{\partial^2 u}{\partial x^2} + \frac{\partial^2 u}{\partial y^2} = (\cos^2\theta + \sin^2\theta)\frac{\partial^2 \overline{u}}{\partial r^2} + \frac{1}{r}(\sin^2\theta + \cos^2\theta)\frac{\partial \overline{u}}{\partial r} + \frac{1}{r^2}(\sin^2\theta + \cos^2\theta)\frac{\partial^2 \overline{u}}{\partial\theta^2}$$

$$= \frac{\partial^2 \overline{u}}{\partial r^2} + \frac{1}{r}\frac{\partial \overline{u}}{\partial r} + \frac{1}{r^2}\frac{\partial^2 \overline{u}}{\partial\theta^2}.$$

(pgs. 457-459)

1. (a) Let f satisfy $\nabla f(x,y) = (3x^2y, x^3 + 3y^2)$. Then

$$f_x(x,y) = 3x^2y \qquad \text{and} \qquad f_y(x,y) = x^3 + 3y^2.$$

Holding y constant for the moment, we integrate the first equation with respect to x to get

$$f(x,y) = x^3y + C(y),$$

where $C(y)$ is the "constant" of integration and represents a differentiable function of y that is independent of x. Differentiating this equation with respect to y and setting the result equal to the above expression for $f_y(x,y)$ gives

$$\frac{\partial\left(x^3y + C(y)\right)}{\partial y} = x^3 + C_y(y) = x^3 + 3y^2.$$

It follows that $C_y(y) = 3y^2$ and therefore $C(y) = y^3 + c$, where c is a real constant. Choosing $c = 0$, we get

$$f(x,y) = x^3y + y^3.$$

(b) Let γ be a piecewise smooth path from $(1,1)$ to $(1,2)$. Then, with $f(x,y) = x^3y + y^3$ (as found in part (a)),

$$\int_\gamma 3x^2y\,dx + (x^3 + 3y^2)\,dy = \int_\gamma (3x^2y, x^3 + 3y^2) \cdot d\mathbf{x} = \int_{(1,1)}^{(1,2)} \nabla(x^3y + y^3) \cdot d\mathbf{x}$$

$$= \int_{(1,1)}^{(1,2)} \nabla f \cdot d\mathbf{x} = f(1,2) - f(1,1) = 8.$$

3. Consider the curve γ parametrized by $g(t) = (1 + t^2, t - t^2)$, $0 \le t \le 1$. Since $t - t^2 \ge 0$ when $0 \le t \le 1$, γ is traced out above the x-axis starting at the point $(1,0)$ and ending at $(2,0)$. Let R be the region above the x-axis and below γ. Let γ_0 be the oriented line segment from $(2,0)$ to $(1,0)$, so that the closed path $\lambda = \gamma \cup \gamma_0$ is the boundary of R and is oriented in a *clockwise* direction. Letting λ^- be the path λ traced in the opposite direction (i.e., counterclockwise), we use Green's Theorem with $F(x,y) = 0$ and $G(x,y) = x$ to get

$$A(R) = \int_R dx\,dy = \int_R \left(\frac{\partial(x)}{\partial x} - \frac{\partial(0)}{\partial y}\right) dx\,dy = \oint_{\lambda^-} 0\,dx + x\,dy = \oint_{\lambda^-} x\,dy$$

$$= -\oint_\lambda x\,dy = -\int_\gamma x\,dy - \int_{\gamma_0} x\,dy = -\int_\gamma x\,dy,$$

where $dy = 0$ along γ_0 forces $-\int_{\gamma_0} x\,dy = 0$. On γ, $x = 1 + t^2$ and $dy = (1 - 2t)\,dt$, so that

$$A(R) = -\int_\gamma x\,dy = -\int_0^1 (1 + t^2)(1 - 2t)\,dt = -\left[t - t^2 + \frac{1}{3}t^3 - \frac{1}{2}t^4\right]_0^1 = \frac{1}{6}.$$

5. Let S be the closed surface bounding the solid region R inside the cylinder $x^2 + y^2 = 4$ from $z = 0$ to $z = 2$. Assume that S is positively oriented and that S and R are in the flow field $\mathbf{F}(x, y, z) = (x^3, y^3 + x, xy)$. By Gauss's Theorem, the flux of $\mathbf{F}$ across S is

$$\Phi(\mathbf{F}, S) = \int_S \mathbf{F} \cdot d\mathbf{S} = \int_R \operatorname{div}\mathbf{F} \, dV = \int_R \left(\frac{\partial(x^3)}{\partial x} + \frac{\partial(y^3 + x)}{\partial y} + \frac{\partial(xy)}{\partial z} \right) dx \, dy \, dz$$

$$= \int_R (3x^2 + 3y^2) \, dx \, dy \, dz = \int_0^{2\pi} d\theta \int_0^2 dz \int_0^2 (3r^2 \cos^2 \theta + 3r^2 \sin^2 \theta) \, r \, dr$$

$$= \int_0^{2\pi} d\theta \int_0^2 dz \int_0^2 3r^3 dr = [\theta]_0^{2\pi} [z]_0^2 \left[\frac{3}{4} r^4 \right]_0^2 = 48\pi.$$

7. Consider the vector field $\mathbf{F}(x, y) = (-x - y)$ relative to the circle $\gamma : x^2 + y^2 = 1$.

(a) The circulation of $\mathbf{F}$ around γ is the value of the line integral of $\mathbf{F}$ in the direction of counterclockwise traversal around γ. By Stokes's Theorem for the plane, this is equal to the value of the surface integral of the scalar curl of $\mathbf{F}$ over the unit disk D. That is,

$$\text{circulation} = \oint_\gamma \mathbf{F} \cdot d\mathbf{x} = \int_D \left(\frac{\partial(-y)}{\partial x} - \frac{\partial(-x)}{\partial y} \right) dx \, dy = \int_D 0 \, dx \, dy = 0.$$

(b) The total flux across γ is the value of the line integral of $\mathbf{F}$ in the direction of outward-pointing unit normals around γ. By Gauss's Theorem in the plane, this is equal to the value of the surface integral of $\operatorname{div}\mathbf{F}$ over the unit disk D. That is,

$$\text{total flux} = \oint_\gamma \mathbf{F} \cdot \mathbf{n} \, ds = \int_D \operatorname{div}\mathbf{F} \, dA = \int_D \left(\frac{\partial(-x)}{\partial x} + \frac{\partial(-y)}{\partial y} \right) dA$$

$$= \int_D (-2) \, dA = -2A(D) = -2\pi.$$

9. For each positive constant α, define the two dimensional vector field

$$\mathbf{F}_\alpha(x, y) = \frac{-y}{(x^2 + y^2)^\alpha} \mathbf{i} + \frac{x}{(x^2 + y^2)^\alpha} \mathbf{j}, \quad (x, y) \neq 0.$$

(a) The scalar curl of $\mathbf{F}_\alpha(x, y)$ is

$$\operatorname{curl}\mathbf{F} = \frac{\partial}{\partial x} \left(\frac{x}{(x^2 + y^2)^\alpha} \right) - \frac{\partial}{\partial y} \left(\frac{-y}{(x^2 + y^2)^\alpha} \right)$$

$$= \frac{(x^2 + y^2) - 2\alpha x^2}{(x^2 + y^2)^{\alpha+1}} - \frac{-(x^2 + y^2) + 2\alpha y^2}{(x^2 + y^2)^{\alpha+1}} = \frac{2(1 - \alpha)}{(x^2 + y^2)^\alpha}.$$

Hence, the scalar curl of $\mathbf{F}_\alpha(x, y)$ is zero if, and only if, $\alpha = 1$, and in this case it is identically zero.

(b) Let γ be a smooth closed curve that doesn't contain the origin and let R be the region enclosed by γ. Then $\mathbf{F}_\alpha$ is continuously differentiable on R and γ. If γ is traversed once in a counterclockwise direction then Stokes's Theorem in the plane gives

$$\oint_\gamma \mathbf{F}_\alpha \cdot d\mathbf{x} = \int_R \operatorname{curl}\mathbf{F} \, dx \, dy = \int_R \frac{2(1 - \alpha)}{(x^2 + y^2)^\alpha} \, dx \, dy = 2(1 - \alpha) \int_R (x^2 + y^2)^{-\alpha} dx \, dy,$$

where the leftmost line integral is, by definition, the circulation of $\mathbf{F}_\alpha$ around γ in a counterclockwise direction. Because the integrand of the rightmost integral is positive, so is the value of the integral. Therefore, for a given region R (i.e., a given curve γ) the circulation of $\mathbf{F}_\alpha$ around γ is positive for $\alpha < 1$ (the fluid tends to flow in a counterclockwise direction around γ), is zero for $\alpha = 1$, and is negative for $\alpha > 1$ (the fluid tends to flow in a clockwise direction around γ).

(c) Let γ be a circle of radius $a > 0$ centered at the origin. In this case, $\mathbf{F}_\alpha$ is not continuously differentiable on the disk enclosed by γ, so that Stokes's Theorem in the plane can't be applied. We must therefore compute the circulation directly as a line integral. To do this, we first parametrize γ by $g(t) = (a\cos t, a\sin t)$, $0 \le t \le 2\pi$, and observe that γ is traced exactly once in a counterclockwise direction. Since $x^2 + y^2 = a^2$ on γ, we also observe that

$$\mathbf{F}_\alpha\big(g(t)\big) = \left(\frac{-a\sin t}{a^{2\alpha}}, \frac{a\cos t}{a^{2\alpha}}\right) = a^{-2\alpha}(-a\sin t, a\cos t).$$

Hence,

$$\int_\gamma \mathbf{F}_\alpha \cdot d\mathbf{x} = \int_0^{2\pi} \mathbf{F}_\alpha\big(g(t)\big) \cdot g'(t)\, dt = \int_0^{2\pi} a^{-2\alpha}(-a\sin t, a\cos t) \cdot (-a\sin t, a\cos t)\, dt$$

$$= \int_0^{2\pi} a^{2-2\alpha} dt = \left[a^{2-2\alpha}t\right]_0^{2\pi} = 2\pi a^{2-2\alpha}.$$

This shows that circulation around circles of radius a centered at the origin is always positive. For the special case $\alpha = 1$, the circulation is independent of the radius of the circle and is constant at 2π. However, by part (a), curl$\mathbf{F}_1$ is identically zero, so that the surface integral of curl$\mathbf{F}_1$ over the region enclosed by the circle is zero.

11. As functions on $\mathcal{R}^2$, x and y are continuously differentiable on the given region R and its counterclockwise oriented boundary ∂R. Hence, by Green's Theorem

$$\int_{\partial R} x\, dy = \int_{\partial R} 0\, dx + x\, dy = \int_R \left(\frac{\partial(x)}{\partial x} - \frac{\partial(0)}{\partial y}\right) dx\, dy = \int_R 1\, dx\, dy = A(R),$$

$$\int_{\partial R} -y\, dx = \int_{\partial R} -y\, dx + 0\, dy = \int_R \left(\frac{\partial(0)}{\partial x} - \frac{\partial(-y)}{\partial y}\right) dx\, dy = \int_R 1\, dx\, dy = A(R),$$

$$\int_{\partial R} x\, dx = \int_{\partial R} x\, dx + 0\, dy = \int_R \left(\frac{\partial(0)}{\partial x} - \frac{\partial(x)}{\partial y}\right) dx\, dy = \int_R 0\, dx\, dy = 0,$$

$$\int_{\partial R} y\, dy = \int_{\partial R} 0\, dx + y\, dy = \int_R \left(\frac{\partial(y)}{\partial x} - \frac{\partial(0)}{\partial y}\right) dx\, dy = \int_R 0\, dx\, dy = 0.$$

That is, $\int_{\partial R} x\, dy = -\int_{\partial R} y\, dx = A(R)$ and $\int_{\partial R} x\, dx = \int_R y\, dy = 0$.

13. Let S be the closed surface consisting of a conical piece C defined by $z = 2\sqrt{x^2 + y^2}$, $0 \le z \le 2$, and a hemispherical top H defined by $z = 2 + \sqrt{1 - x^2 - y^2}$, $x^2 + y^2 \le 1$.

(a) We will parametrize both pieces of S using x and y as parameters. Specifically,

$$g(x,y) = \big(x, y, 2 + \sqrt{1 - x^2 - y^2}\big) \quad \text{(for H)}, \qquad h(x,y) = \big(y, x\sqrt{x^2 + y^2}\big), \quad \text{(for C)},$$

where the parameter domain for both functions is the closed unit disk D. We have

$$g_x \times g_u = \left(\frac{x}{\sqrt{1 - x^2 - y^2}}, \frac{y}{\sqrt{1 - x^2 - y^2}}, 1\right),$$

$$h_x \times h_y = \left(\frac{2y}{\sqrt{x^2 + y^2}}, \frac{2x}{\sqrt{x^2 + y^2}}, -1\right)$$

Note that the z coordinate of $g_x \times g_y$ is positive, so that normals point in an upward and therefore outward direction. In the same way, the third coordinate of $h_x \times h_y$ is negative, so that the normals point in a downward and therefore (again) outward direction.

(b) To find the total surface area, we do each piece separately. On C we have

$$\sigma(C) = \int_D |h_x \times h_y| \, dx \, dy = \int_D \sqrt{\frac{4y^2}{x^2+y^2} + \frac{4x^2}{x^2+y^2} + 1} \, dx \, dy$$

$$= \int_D \sqrt{5} \, dx \, dy = \sqrt{5} \, A(D) = \pi\sqrt{5}$$

and on H,

$$\sigma(H) = \int_D |g_x \times g_y| \, dx \, dy = \int_D \sqrt{\frac{x^2}{1-x^2-y^2} + \frac{y^2}{1-x^2-y^2} + 1} \, dx \, dy$$

$$= \int_D \frac{1}{\sqrt{1-x^2-y^2}} \, dx \, dy = \int_D \frac{1}{\sqrt{1-r^2}} r \, dr \, d\theta. = \lim_{\delta \to 1^-} \int_0^{2\pi} d\theta \int_0^\delta \frac{r}{\sqrt{1-r^2}} \, dr$$

$$= \lim_{\delta \to 1^-} [\theta]_0^{2\pi} \left[-\sqrt{1-r^2} \right]_0^\delta = \lim_{\delta \to 1^-} 2\pi \left(1 - \sqrt{1-\delta^2} \right) = 2\pi.$$

The total surface area is therefore

$$\sigma(S) = \sigma(C) + \sigma(H) = \pi\sqrt{5} + 2\pi.$$

These answers are consistent with the answers obtained using well-known formulas for surface area of common geometric figures. Rather than compute the volume of the region enclosed by S, which we denote by D, we shall simply observe that the volume of D is the volume of the hemispherical top plus the volume of the cone. From well-known formulas for volumes of common geometric figures, the volume of the cone is $\frac{1}{3}\pi R^2 h = \frac{2}{3}\pi$, where $R = 1$ is the radius of the base and $h = 2$ is the altitude, and the volume of the hemispherical top is $\frac{2}{3}\pi R^3 = \frac{2}{3}\pi$, where $R = 1$ is the radius. Thus, the volume of D is

$$V(D) = \frac{2}{3}\pi + \frac{2}{3}\pi = \frac{4}{3}\pi.$$

(c) Let $\mathbf{F}(x,y,z) = (x,y,z)$ and observe that $\operatorname{div}\mathbf{F} = \frac{\partial(x)}{\partial x} + \frac{\partial(y)}{\partial y} + \frac{\partial(z)}{\partial z} = 3$. Thus, with D as defined in part (b) and with $V(D)$ as computed in part (b), Gauss's Theorem gives the flux across S as

$$\Phi(\mathbf{F}, S) = \int_{\partial D} \mathbf{F} \cdot d\mathbf{S} = \int_D \operatorname{div}\mathbf{F} \, dV = \int_D 3 \, dV = 3V(D) = 3\left(\frac{4}{3}\pi\right) = 4\pi.$$

15. Let f and g be real-valued functions that are twice continuously differentiable on an open subset R of $\mathcal{R}^3$.

(a) Replacing $\mathbf{F}$ by ∇g in identity (4) of Section 6, we have

$$\operatorname{curl}(f\nabla g) = \nabla \times (f\nabla g) = f\nabla \times \nabla g + \nabla f \times \nabla g = f\left(\nabla \times (\nabla g)\right) + (\nabla f) \times (\nabla g).$$

358

By identity (9) of Section 6, $\nabla \times (\nabla g) = 0$. Hence, the first term of the rightmost member vanishes and therefore equality of the two outer members can be written as

$$\text{curl}(f\nabla g) = (\nabla f) \times (\nabla g).$$

(b) Let S be the hemisphere $z = \sqrt{1 - x^2 - y^2}$, let $\mathbf{n}$ be the upward directed unit normal to S, and let D be the closed unit disk in $\mathcal{R}^2$. Notice that the boundary of D (the unit circle) is traced counterclockwise by $(\cos t, \sin t)$, $0 \le t \le 2\pi$, and the border ∂S (the unit circle) is traced by $h(t) = (\cos t, \sin t, 0)$, $0 \le t \le 2\pi$.

Now let $f(x, y, z) = x + y + z$ and $g(x, y, z) = x^2 + y^2 - z^2$. Since the form of $g\nabla f$ is simpler than the form of $f\nabla g$, we find that $g\nabla f(x, y, z) = (x^2 + y^2 + z^2)(1, 1, 1)$, so that the field $g\nabla f$ along ∂S is given by

$$g\nabla f\big(h(t)\big) = (\cos^2 + \sin^2 + 0^2)(1, 1, 1) = (1, 1, 1).$$

By part (a) (with f and g interchanged) and Stokes's Theorem,

$$\int_S [(\nabla f) \times (\nabla g)] \cdot \mathbf{n} \, dS = -\int_S [(\nabla g) \times (\nabla f)] \cdot \mathbf{n} \, d\sigma = -\int_S \text{curl}(g\nabla f) \cdot d\mathbf{S}$$

$$= -\int_{\partial S} (g\nabla f) \cdot d\mathbf{x} = -\int_0^{2\pi} (g\nabla f)\big(h(t)\big) \cdot h'(t) \, dt$$

$$= -\int_0^{2\pi} (1, 1, 1) \cdot (-\sin t, \cos t, 0) \, dt = -\int_0^{2\pi} (-\sin t + \cos t) \, dt = 0.$$

17. Let γ be the counterclockwise path consisting of the part of the circle $x^2 + y^2 = 1$ lying in the first quadrant together with the line segments $0 \le x \le 1$ and $0 \le y \le 1$ on the x- and y-axis, respectively. If R is the region bounded by γ then Green's Theorem gives

$$\int_\gamma xy \, dx + y \, dy = \int_R \left(\frac{\partial(y)}{\partial x} - \frac{\partial(xy)}{\partial y} \right) dx \, dy = -\int_R x \, dx \, dy = -\int_R r \cos\theta \, r \, dr \, d\theta$$

$$= -\int_0^{\pi/2} \cos\theta \, d\theta \int_0^1 r^2 dr = -\big[\sin\theta\big]_0^{\pi/2} \left[\frac{1}{3} r^3\right]_0^1 = -\frac{1}{3}.$$

19. Let S be the piece of smooth surface defined by $g(u, v) = (u+v, 2u+v, u-v)$, $0 \le u \le 1$, $0 \le v \le 1$, and let ∂S denote the border of S with the positive orientation. Since we are seeking a surface integral, we will need

$$g_u \times g_v = (1, 2, 1) \times (1, 1, -1) = \left(\begin{vmatrix} 2 & 1 \\ 1 & -1 \end{vmatrix}, \begin{vmatrix} 1 & -1 \\ 1 & 1 \end{vmatrix}, \begin{vmatrix} 1 & 1 \\ 2 & 1 \end{vmatrix} \right) = (-3, 2, -1).$$

Also, since we will be using Stokes's Theorem, we will need the curl of the given field $\mathbf{F}(x, y, z) = (y, z, x)$:

$$\text{curl}\mathbf{F} = \left(\frac{\partial(x)}{\partial y} - \frac{\partial(z)}{\partial z}, \frac{\partial(y)}{\partial z} - \frac{\partial(x)}{\partial x}, \frac{\partial(z)}{\partial x} - \frac{\partial(y)}{\partial y} \right) = (0 - 1, 0 - 1, 0 - 1) = (-1, -1, -1).$$

Therefore, by Stokes's Theorem,

$$\int_{\partial S} \mathbf{F} \cdot d\mathbf{x} = \int_S \text{curl}\mathbf{F} \cdot d\mathbf{S} = \int_S (-1, -1, -1) \cdot (-3, 2, -1) \, du \, dv = \int_0^1 \int_0^1 2 \, du \, dv = 2.$$

21. Let B be a region in $\mathcal{R}^3$ with a piecewise smooth surface boundary ∂B.
(a) Let $\mathbf{F}(\mathbf{x}) = \mathbf{x}$. We have

$$V(B) = \int_B dV = \frac{1}{3}\int_B 3\,dV = \frac{1}{3}\int_B \left(\frac{\partial(x)}{\partial x} + \frac{\partial(y)}{\partial y} + \frac{\partial(z)}{\partial z}\right)dV = \frac{1}{3}\int_B \mathrm{div}\mathbf{F}\,dV$$

$$= \frac{1}{3}\int_{\partial B}\mathbf{F}\cdot d\mathbf{S} = \frac{1}{3}\int_{\partial B}\mathbf{x}\cdot d\mathbf{S},$$

where Gauss's Theorem was used to obtain the first integral in the second line.

(b) Let $\mathbf{F}_1(x,y,z) = (x,0,0)$, $\mathbf{F}_2(x,y,z) = (0,y,0)$ and $\mathbf{F}_3(x,y,z) = (0,0,z)$. Three applications of Gauss's Theorem gives

$$\int_{\partial B} x\,dy\,dz = \int_{\partial B}(x,0,0)\cdot d\mathbf{S} = \int_B \mathrm{div}\mathbf{F}_1 dV = \int_B 1\,dV = V(B),$$

$$\int_{\partial B} y\,dx\,dz = \int_{\partial B}(0,y,0)\cdot d\mathbf{S} = \int_B \mathrm{div}\mathbf{F}_2 dV = \int_B 1\,dV = V(B),$$

$$\int_{\partial B} z\,dx\,dy = \int_{\partial B}(0,0,z)\cdot d\mathbf{S} = \int_B \mathrm{div}\mathbf{F}_3 dV = \int_B 1\,dV = V(B),$$

23. Let $\mathbf{F}$ be a continuously differentiable vector field defined on a simply connected region B in $\mathcal{R}^3$ and suppose $\mathrm{div}\mathbf{F} = 0$ on B. Let S_1 and S_2 be smooth surfaces in B with the same border. If the two surfaces intersect only on their common border then $S_1 \cup S_2$ encloses a region R in B. If, relative to R, the normals to S_1 are inward-pointing and the normals to S_2 are outward-pointing (or vice-versa) then the Surface Independence Principle (SIP) says that the flux across S_1 is equal to the flux across S_2.

However, if S_1 and S_2 intersect at places other than their common border, there may be no region bounded by the two surfaces to which Gauss's Theorem can be applied. This means that SIP cannot be applied directly. To remedy this, let S be a third surface with the same common border, but one which intersects S_1 and S_2 only on the common border. Then the surfaces $S_1 \cup S$ and $S_2 \cup S$ bound two regions R_1 and R_2 to which Gauss's Theorem can be applied. Note that the normals to S_1 are outward-pointing with respect to R_1 if, and only if, the normals to S_2 are outward-pointing with respect to R_2. So, choose a parametrization for S with normals pointing in a direction opposite to that of S_1 and S_2. By SIP, the flux across S_1 equals the flux across S and the flux across S_2 equals the flux across S. It follows that the flux across S_1 equals the flux across S_2.

25. Let S be a sphere of radius a centered at the point (x_0, y_0, z_0) and let R be the solid ball bounded by S. For purposes of this solution, view R as having a mass described by the density function $\mu(x,y,z) \equiv 1$, so that the mass M of the sphere is numerically equal to the volume of the sphere; i.e., $M = 4\pi a^3/3$. With a positively oriented S, we apply Gauss's Theorem to get

$$\frac{1}{M}\int_S x^2 dy\,dz + y^2 dx\,dz + z^2 dx\,dy = \frac{1}{M}\int_S (x^2, y^2, z^2)\cdot d\mathbf{S}$$

$$= \frac{1}{M}\int_R \mathrm{div}(x^2, y^2, z^2)\,dV = \frac{1}{M}\int_R (2x + 2y + 2y)\,dV$$

$$= \frac{2}{M}\int_R x\,dV + \frac{2}{M}\int_R y\,dV + \frac{2}{M}\int_R z\,dV = 2x_0 + 2y_0 + 2z_0,$$

where we used the fact that (x_0, y_0, z_0) is the centroid of R and applied the definition of a coordinate of the centroid of R three times to obtain the last expression. Hence,

$$\int_S x^2 dy\,dz + y^2 dx\,dz + z^2 dx\,dy = 2M(x_0 + y_0 + z_0) = \frac{8}{3}\pi a^3(x_0 + y_0 + z_0).$$

Chapter 10: First-Order Differential Equations

Section 1: DIRECTION FIELDS

Exercise Set 1A (pg. 466)

1. If $y = Ce^x - 1$, with C constant, then $y' = Ce^x = (Ce^x - 1) + 1 = y + 1$, so that the given function is a solution of $y' = y + 1$ for any constant C. To find the solution satisfying $y(0) = 2$, set $x = 0$, $y = 2$ in $y = Ce^x - 1$ and get $2 = C - 1$. So $C = 3$ and

$$y(x) = 3e^x - 1, \quad -\infty < x < \infty.$$

3. If $y = Ke^{-x}$, with K constant, then $y' = -Ke^{-x} = -y$, so that $y' + y = 0$ and the given function is a solution of $y' + y = 0$ for any constant K. To find the solution satisfying $y(5) = 6$, set $x = 5$, $y = 6$ in $y = Ke^{-x}$ and get $6 = Ke^{-5}$. So $K = 6e^5$ and

$$y(x) = 6e^5 e^{-x} = 6e^{5-x}, \quad -\infty < x < \infty.$$

5. If $y = (C - x)^{-1}$, with C constant, then $y' = (C - x)^{-2} = y^2$, so that the given function is a solution of $y' = y^2$ for all constants C. To find the solution satisfying $y(3) = 2$, set $x = 3$, $y = 2$ in $y = (C - x)^{-1}$ and get $2 = (C - 3)^{-1}$. Hence, $C = 7/2$ and

$$y(x) = \left(\frac{7}{2} - x\right)^{-1}, \quad x < \frac{7}{2},$$

where the domain is the largest open interval containing $x = 3$ on which $y(x)$ is continuous.

7. $y' = y/x$, $(x_0, y_0) = (1, 2)$ **9.** $dy/dx = y + x$, $(x_0, y_0) = (1, -1)$

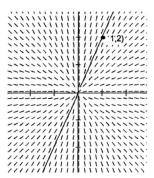
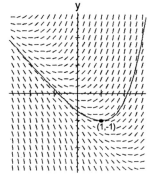

11. An isocline associated with the differential equation $y' = -y/x$ has an equation of the form $m = -y/x$, or $y = -mx$. This is a straight line through the origin with slope $-m$. The direction field and the isoclines for $m = 1/2$, $m = -1$ and $m = -2$ are shown in the left figure on the following page.

13. An isocline associated with the differential equation $y' = x^2 + y^2$ has an equation of the form $m = x^2 + y^2$. For $m > 0$, this is a circle of radius $\sqrt{m}$ centered at the origin. The direction field and the isoclines for $m = 0.64$, $m = 1.69$ and $m = 3.24$ are shown in the middle figure on the following page.

15. An isocline associated with the differential equation $y' = (1 - y)/x$ has an equation of the form $m = (1 - y)/x$, or $y = -mx + 1$. This is a straight line passing through the point $(0, 1)$ with slope $-m$. The direction field and the isoclines for $m = 1$, $m = -1$ and $m = 1/3$ are shown in the right figure on the following page.

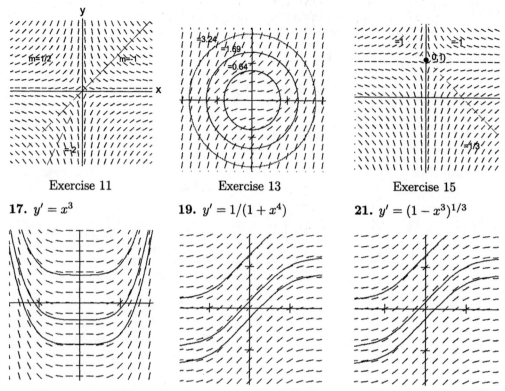

Exercise 11   Exercise 13   Exercise 15

17. $y' = x^3$ **19.** $y' = 1/(1 + x^4)$ **21.** $y' = (1 - x^3)^{1/3}$

23. (a) Suppose $F(x)$ is continuous on an interval I containing x_0. Define $y = y(x)$ by

$$y(x) = y_0 + \int_{x_0}^{x} F(t)\,dt, \quad x \in I.$$

Since F is continuous, the integral on the right exists. Moreover, by the fundamental theorem of calculus, the integral is differentiable on I and has derivative $F(x)$. It follows that $y(x)$ is differentiable on I and, by taking the derivative with respect to x of both sides,

$$y'(x) = F(x), \quad \text{for } x \in I.$$

So, y is a solution (on I) of the D.E. $y' = F(x)$.

    (b) Suppose $y_1(x)$ and $y_2(x)$ are two solutions on I of $y' = F(x)$ such that $y_1(x_0) = y_0$ and $y_2(x_0) = y_0$. Let $y = y_1 - y_2$ and observe that

$$y' = (y_1 - y_1)' = y_1' - y_2' = F(x) - F(x) = 0, \quad \text{for all } x \in I.$$

Hence, $y(x)$ is constant on I. That is, there exists a constant c such that $y_1(x) - y_2(x) = c$, for all $x \in I$. But since x_0 is in I and $y_1(x_0) - y_2(x_0) = y_0 - y_0 = 0$, it follows that $c = 0$. So, $y_1(x) = y_2(x)$ for all $x \in I$. Therefore, the solution given in part (a) is uniquely determined by the condition $y(x_0) = y_0$.

362

25. Consider the differential equation

$$y' = \sqrt{1 - y^2}.$$

For each real number a such that $|a| \geq \pi/2$, define the function $y_a(x)$ by

$$y_a(x) = \begin{cases} 1, & x \geq -a + \pi/2; \\ \sin(x + a), & -a - \pi/2 < x < -a + \pi/2; \\ -1, & x \leq -a - \pi/2. \end{cases}$$

Clearly, each piece of y_a is continuous on its respective interval of definition; and since $\sin(x+a) \to 1$ as $x \to -a+\pi/2$ and $\sin(x+a) \to -1$ as $x \to -a-\pi/2$, y_a is also continuous at $-a+\pi/2$ and $-a-\pi/2$. Each piece of y_a is also differentiable on its interval of definition, with the derivative of the two "outer" pieces being identically zero. As the derivative of the middle piece is $\cos(x+a)$, it tends to 0 as $x \to -a+\pi/2$ and as $x \to -a-\pi/2$. Thus, each y_a is differentiable for all x. Since each piece is a solution of $y' = \sqrt{1 - y^2}$ on its respective interval of definition, y_a itself is a solution of the D.E. on all of $\mathcal{R}$. Typical solution graphs are shown for $a \geq \pi/2$ and $a \leq -\pi/2$.

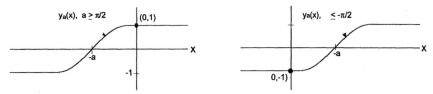

If $a \geq \pi/2$ then all nonnegative values of x satisfy $x \geq -a + \pi/2$ and therefore $y_a(0) = 1$ (as in the left diagram). Similarly, if $a \leq -\pi/2$ then all nonpositive values of x satisfy $x \leq -a - \pi/2$ and therefore $y_a(0) = -1$ (as in the right diagram). Thus, each of the infinitely many functions $y_a(x)$ with $a \geq \pi/2$ is a solution of the initial-value problem $y' = \sqrt{1 - y^2}$, $y_a(0) = 1$; and each of the infinitely many functions $y_a(x)$ with $a \leq -\pi/2$ is a solution of the initial-value problem $y' = \sqrt{1 - y^2}$, $y_a(0) = -1$. But since none of these solutions satisfies $|y_a(x)| > 1$ for any x, none are continuous on an *open* rectangle containing $(0, 1)$ or $(0, -1)$. Therefore, the uniqueness part of Theorem 1.1 is not contradicted by this example.

Exercise Set 1B (pg. 469)

1. $y' = \sin(x - y)$

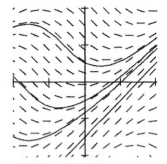

3. $y' = \sqrt{9 - y^3}$

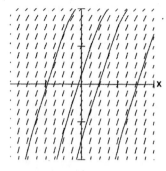

5. $y' = \sin(x^2 + y^2)$

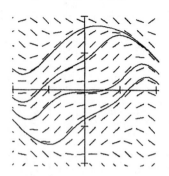

7. $y' = \cos(x^2 + y^2)$

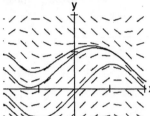

9. $y' = (1 + x^4)^{-1}$

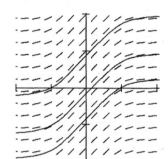

11. $y' = e^{-y^2}$

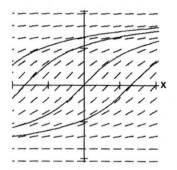

Section 2: APPLIED INTEGRATION

Exercise Set 2AB (pgs. 478-480)

1. With $y' = x(1-x)$, we integrate both sides with respect to x to get

$$y(x) = \int x(1-x)\,dx + C = \int (x - x^2)\,dx + C = \frac{1}{2}x^2 - \frac{1}{3}x^3 + C.$$

To find C, the condition $y(0) = 1$ allows us to set $x = 0$, $y = 1$ in the above expression for $y(x)$. This immediately gives $C = 1$. The particular solution is therefore

$$y(x) = \frac{1}{2}x^2 - \frac{1}{3}x^3 + 1, \quad -\infty < x < \infty.$$

3. With $y' = x/(1-x^2)$, we integrate both sides with respect to x to get

$$y(x) = \int \frac{x}{1-x^2}\,dx + C = \int \frac{-1/2}{u}\,du + C = -\frac{1}{2}\ln|u| + C = -\frac{1}{2}\ln|1-x^2| + C,$$

where the second integral is obtained by the substitution $u = 1 - x^2$. Since $y(0) = 1$, $x = 0$ is in the domain of $y(x)$. The largest interval containing this value of x on which $y(x)$ is continuous is $-1 < x < 1$. This is the domain of the solution we seek. It also allows us to remove the absolute value sign on $1 - x^2$. Using the condition $y(0) = 1$, set $x = 0$, $y = 1$ in the above equation for $y(x)$ and immediately get $C = 1$. The particular solution is therefore

$$y(x) = -\frac{1}{2}\ln(1-x^2) + 1, \quad |x| < 1.$$

5. With $y'' = \sin x$, we integrate both sides with respect to x to get $y'(x) = -\cos x + C_1$. The condition $y'(0) = 1$ then gives $1 = y'(0) = -\cos 0 + C_1 = -1 + C_1$, from which $C_1 = 2$ and $y'(x) = -\cos x + 2$. Integrating a second time gives

$$y(x) = \int (-\cos x + 2)\,dx + C_2 = -\sin x + 2x + C_2.$$

The condition $y(0) = 1$ immediately implies $C_2 = 1$. The particular solution is therefore

$$y(x) = -\sin x + 2x + 1, \quad -\infty < x < \infty.$$

7. With $dz/dt = te^t$, we integrate both sides with respect to t to get

$$z(t) = \int te^t dt = te^t - e^t + C,$$

where the integral is done using integration by parts. The condition $z(0) = 1$ allows us to set $t = 0$, $z = 1$ in the above expression for $z(t)$ to get $1 = z(0) = -1 + C$, so that $C = 2$. The particular solution is therefore

$$z(t) = te^t - e^t + 2, \quad -\infty < t < \infty.$$

9. With $dx^2/dt^2 = e^t$, we integrate with respect to t twice in succession to get

$$\frac{dx}{dt} = e^t + C_1 \qquad \text{and} \qquad x(t) = e^t + C_1 t + C_2.$$

The conditions $dx/dt(0) = 1$ and $x(0) = 1$ allows us to set $t = 0$ in both equations, $dx/dt = 1$ in the first equation and $x = 1$ in the second equation to get the two equations $1 = 1 + C_1$ and $1 = 1 + C_2$. These easily give $C_1 = C_2 = 0$. Hence, the particular solution is

$$x(t) = e^t, \quad -\infty < t < \infty.$$

11. Let $y(t)$ be the projectile's distance above the ground at time $t \geq 0$. Since the only force acting on the projectile is the constant acceleration of gravity $g = 32$ ft/sec^2, the acceleration of the projectile is $y''(t) = -32$. Given that the initial velocity is 5000 feet per second, the initial-value problem describing the motion of the projectile is

$$y''(t) = -32, \quad y'(0) = 5000, \ y(0) = 0.$$

Integrating this D.E. with respect to t gives $y'(t) = -32t + C_1$. The condition $y'(0) = 5000$ immediately gives $C_1 = 5000$, so that

$$y'(t) = -32t + 5000.$$

Integrating again with respect to t gives $y(t) = -16t^2 + 5000t + C_2$. The condition $y(0) = 0$ immediately gives $C_2 = 0$, so that

$$y(t) = -16t^2 + 5000t.$$

The maximum altitude y_{max} is attained when the velocity is zero. Setting $y'(t) = 0$ in the above expression for y' gives $t = 5000/32 = 156.25$ seconds. The maximum altitude attained is therefore

$$y_{max} = -16(156.25)^2 + 5000(156.25) = 390625 \text{ ft} \approx 74 \text{ miles.}$$

13. Assume that the projectile and the weight described in Exercises 11 and 12 are released at the same time and are aimed directly at each other.

(a) Let $y_1(t)$ be the projectile's distance above the ground at time $t \geq 0$, and let $y_2(t)$ be the weight's distance above the ground at time $t \geq 0$. From Exercises 11 and 12 we have

$$y_1(t) = -16t^2 + 5000t \quad \text{and} \quad y_2(t) = -16t^2 + 5000,$$

where both equations hold for the same value of t for all t such that both the projectile and the weight are in the air. Since the two objects will meet when their altitudes are the same, we want the value of t for which $y_1(t) = y_2(t)$. We have $-16t^2 + 5000t = -16t^2 + 5000$, which implies $t = 1$ second. Thus, they will meet at an altitude of $y_1(1) = -16 + 5000 = 4984$ feet.

(b) In order to determine the initial velocity v_0 of the projectile in Exercise 11 such that the weight and the projectile meet 2500 feet above the ground, we need to solve a new initial-value problem for the projectile's motion that leaves the initial velocity undetermined. Thus, we let $y_3(t)$ be the projectiles distance above the ground with respect to the initial conditions $y'(0) = v_0$, $y(0) = 0$. The relevant initial-value problem in therefore

$$y_3''(t) = -32, \quad y_3'(0) = v_0, \ y_3(0) = 0.$$

Going through exactly the same steps to find $y_3(t)$ as we did to find $y(t)$ in Exercise 11, we find that
$$y_3(t) = -16t^2 + v_0 t.$$

With $y_2(t)$ as given in part (a), the weight will be at 2500 feet when $2500 = -16t^2 + 5000$, or $t = 12.5$ seconds. In order for the weight to meet the projectile at this time, the projectile must be at 2500 feet when $t = 12.5$ seconds. That is, v_0 must satisfy

$$2500 = y_3(12.5) = -16(12.5)^2 + v_0(12.5).$$

366

Solving for v_0 gives

$$v_0 = \frac{2500 + 16(12.5)^2}{12.5} = 400 \text{ ft/sec}.$$

15. Let $y(t)$ be the weight's distance above the ground at time $t \geq 0$. We assume that the acceleration of gravity is $g = 32$ ft/sec^2. Since gravity is the only force acting on the weight, its acceleration is given by $y''(t) = -32$. Letting v_0 be its initial velocity, the relevant initial-value problem governing its motion is

$$y''(t) = -32, \quad y'(0) = v_0, \ y(0) = 5000.$$

Integrating this D.E. with respect to t gives $y'(t) = -32t + C_1$. The condition $y'(0) = v_0$ immediately gives $C_1 = v_0$, so that

$$y'(t) = -32t + v_0.$$

Integrating again with respect to t gives $y(t) = -16t^2 + v_0 t + C_2$. The condition $y(0) = 5000$ immediately gives $C_2 = 5000$, so that

$$y(t) = -16t^2 + v_0 t + 5000.$$

Since the weight reaches the ground in 10 seconds, v_0 must be such that $y(10) = 0$. Using this in the above equation gives

$$0 = -16(10)^2 + 10v_0 + 5000 \quad \text{so that} \quad v_0 = \frac{-5000 + 16(10)^2}{10} = -340 \text{ ft/sec},$$

where the negative sign indicates that the weight is thrown *downward*.

17. Consider the differential equation $dy/dt = 2y$. Observe at the outset that $y(t) = 0$, for all t, is a solution. For $y \neq 0$, we write the D.E. in separated variable form:

$$\frac{1}{y} \, dy = 2 \, dt.$$

Integrating both sides gives $\ln|y| = 2t + C$, and exponentiating gives $|y| = e^C e^{2t}$. Removing the absolute value sign, we formally obtain

$$y(t) = C_1 e^{2t},$$

where C_1 can be any nonzero constant. However, since the zero function is a solution, we can allow $C_1 = 0$.

Given the initial condition $y(0) = 2$, set $t = 0$, $y = 2$ in the above solution formula to get $2 = y(0) = C_1$. Hence, the particular solution is

$$y(t) = 2e^{2t}, \quad -\infty < t < \infty.$$

As a check, we have $y'(t) = 4e^{2t} = 2(2e^{2t}) = 2y(t)$, or $dy/dt = 2y$.

19. Consider the differential equation $y' = x/y^2$. With $dy/dx = y'$, write the D.E. in separated form:

$$y^2 dy = x \, dx.$$

367

Integrating both sides gives

$$\frac{1}{3}y^3 = \frac{1}{2}x^2 + C,$$

where C can be any real constant. Note that this solution formula can be solved explicitly for y and that each of the resulting solutions in continuous everywhere. Note too that positive values of C correspond to solutions where $y = 0$ does not occur, as the given D.E. suggests. However, if $C \leq 0$ then $y = 0$ does occur and the graphs of these solutions have vertical tangent lines at points where $y = 0$, just as the given D.E. suggests.

Given the initial condition $y(1) = 2$, set $x = 1$, $y = 2$ in the above solution formula to get $8/3 = 1/2 + C$, so that $C = 13/6$ and $y^3/3 = x^2/2 + 13/6$. Solving this for y gives the particular solution

$$y(x) = \left(\frac{3}{2}x^2 + \frac{13}{2}\right)^{1/3}, \quad -\infty < x < \infty.$$

As a check, we have

$$\frac{dy}{dx} = \frac{1}{3}\left(\frac{3}{2}x^2 + \frac{13}{2}\right)^{-2/3} 3x = \frac{x}{\left(\frac{3}{2}x^2 + \frac{13}{2}\right)^{2/3}} = \frac{x}{y^2}.$$

That is, $y(x) = \left((3/2)x^2 + (13/2)\right)^{1/3}$ is is a solution.

21. (a) Let $V = V(t)$ be the volume of the spherical ball of dry ice for $t \geq 0$. Since the rate of evaporation dV/dt is proportional to the radius r, there exists a constant c such that

$$\frac{dV}{dt} = cr.$$

Since we are talking about evaporation, it follows that the volume is decreasing with time. As $r > 0$, we conclude that $c < 0$. On the other hand, $V = \frac{4}{3}\pi r^3$, so that implicit differentiation of this equation with respect to t gives

$$\frac{dV}{dt} = 4\pi r^2 \frac{dr}{dt}.$$

Equating the right members of the two equations displayed above gives $4\pi r^2 \frac{dr}{dt} = cr$. Solving for dr/dt gives

$$\frac{dr}{dt} = \frac{cr}{4\pi r^2} = \frac{c/(4\pi)}{r} = \frac{k}{r},$$

where $k = c/(4\pi)$ is a negative constant.

(b) Separating the variables of the D.E. derived in part (a) gives $r\, dr = kt\, dt$. Integrating both sides gives

$$\frac{1}{2}r^2 = \frac{1}{2}kt^2 + C_1 \qquad \text{or equivalently} \qquad r^2 = kt^2 + C,$$

where $C = 2C_1$. Assuming r is in inches and t is in hours, we have two initial conditions: $r(0) = 1$ and $r(1) = 1/2$. Inserting these into the right equation above gives the two equations

$$1 = C, \qquad \frac{1}{2} = k + C.$$

368

Obviously, $C = 1$ and, from the second equation, $k = -1/2$. Hence, the radius of the ball of dry ice at time t satisfies

$$r^2(t) = -\frac{1}{2}t^2 + 1 \quad \text{or, since } r \geq 0, \qquad r(t) = \sqrt{1 - \frac{1}{2}t^2}.$$

(c) The ball will be completely evaporated when $r(t) = 0$. By part (b), this means that t must satisfy

$$0 = \sqrt{1 - \frac{1}{2}t^2}.$$

By inspection $t = \sqrt{2}$. So the ice will completely evaporate in approximately 1.414 hours (about 1 hr, 25 minutes).

23. Let $S(t)$ be the number of pounds of salt in the tank at time t with $S(0) = 150$. To find $S(t)$, we use

$$\frac{dS}{dt} = \{\text{inflow rate}\} - \{\text{outflow rate}\}.$$

The inflow rate is 2 gallons per minute at a concentration of 0 pounds per gallon (pure water is being added); i.e., inflow rate = 0 pounds per minute. The outflow rate is 2 gallons per minute at a concentration of S pounds per 100 gallons; i.e., outflow rate = $2S/100$ pounds per minute. Hence,

$$\frac{dS}{dt} = -\frac{2S}{100}.$$

Separating variables leads to $dS/S = -dt/50$. Integrating gives $\ln|S| = -t/50 + C$, exponentiating gives $|S| = e^C e^{-t/50}$, and removing the absolute value sign gives

$$S(t) = Ke^{-t/50},$$

where K is a positive constant (the amount of salt can't be negative or constantly zero). The initial condition $S(0) = 150$ immediately gives $K = 150$, so that

$$S(t) = 150e^{-t/50}, \quad t \geq 0.$$

Clearly, S is decreasing and $\lim_{t \to \infty} S(t) = 0$.

(b) If the process described in part (a) is modified as described in the statement of this part of this exercise, the equation

$$\frac{dS}{dt} = \{\text{inflow rate}\} - \{\text{outflow rate}\},$$

still holds. The inflow rate is 2 gallons per minute at a concentration of 1 pound per gallon; i.e., inflow rate = 1 pound per minute. The outflow rate is still 2 gallons per minute but only one gallon of it has salt in it (one gallon is drawn off and one gallon is boiled away as steam) at a concentration of $S(t)$ pound per 100 gallons; i.e., outflow rate = $S/100$ pounds per minute. Hence,

$$\frac{dS}{dt} = 1 - \frac{S}{100} = \frac{100 - S}{100}.$$

Separating variables leads to $dS/(S - 100) = -dt/100$. Integrating gives $\ln|S - 100| = -t/100 + C$, exponentiating gives $|S - 100| = e^C e^{-t/100}$, and removing the absolute value sign gives

$$S - 100 = Ke^{-t/100} \qquad \text{or} \qquad S(t) = 100 + Ke^{-t/100},$$

where K is a nonzero constant. The initial condition $S(0) = 150$ gives $150 = 100 + K$, so that $K = 50$. Hence,

$$S(t) = 100 + 50e^{-t/100}, \quad t \geq 0.$$

As in part (a), $S(t)$ is decreasing. However, here, $\lim_{t \to \infty} S(t) = 100$.

25. Given the function $u(t) = 10 - 5e^{-kt}$ and the condition $u(2) = 5$, we have $5 = u(2) = 10 - 5e^{-2k}$. Hence,

$$5e^{-2k} = 10 - 5 \quad \Rightarrow \quad e^{-2k} = 1 \quad \Rightarrow \quad -2k = 0 \quad \Rightarrow \quad k = 0.$$

27. Consider the two differential equations

$$\frac{dy}{dx} = \frac{y}{x} \quad \text{and} \quad \frac{dy}{dx} = -\frac{x}{y}.$$

These are both defined at all points (x, y) that are not on either coordinate axis. Let (x_0, y_0) be such a point. Then the first D.E. has a solution y_1 whose graph passes through (x_0, y_0), and the second D.E. has a solution y_2 whose graph passes through (x_0, y_0). That is, the graphs of y_1 and y_2 intersect at (x_0, y_0). The slope of the tangent line to the graph of y_1 at $x = x_0$ is the number $dy_1/dx(x_0) = y_0/x_0$, and the slope of the graph of y_2 is the number $dy_2/dx(x_0) = -x_0/y_0$. Since the product of these two slopes is manifestly equal to -1, it follows that the graphs of y_1 and y_2 are perpendicular at (x_0, y_0). This is equivalent to saying that the direction fields of the two D.E. are perpendicular to (x_0, y_0). Since (x_0, y_0) was an arbitrary point at which both D.E. are defined, the conclusion holds at every point where both D.E. are defined. The direction fields are shown below

(b) Both D.E. given in part (a) can be solved separating variables. We have

$$\frac{1}{y} \, dy = \frac{1}{x} \, dx \quad \text{and} \quad y \, dy = -x \, dx.$$

Integrating the first equation gives $\ln |y| = \ln |x| + K_1$. Exponentiating gives $|y| = e^{K_1}|x|$. And removing the absolute value signs gives

$$y(x) = C_1 x, \quad x \neq 0.$$

Integrating the second equation gives $\frac{1}{2}y^2 = -\frac{1}{2}x^2 + K_2$ or,

$$x^2 + y^2 = C_2, \quad y \neq 0,$$

where $C_2 = 2K_2$ is evidently a positive constant.

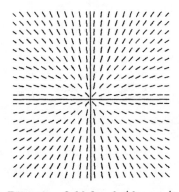

Direction field for $dy/dx = y/x$

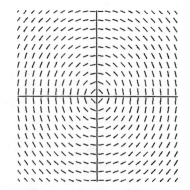

Direction field for $dy/dx = -x/y$.

29. (a) We assume the satellite is in free-fall so that its acceleration is $d^2r/dt^2 = -g = -0.006$ mi/sec^2. The D.E. given in Example 7 in the text then becomes

$$-0.006 = -\frac{GM}{r^2} \quad \text{or} \quad GM = (0.006)r^2.$$

Setting $r = 4000$ miles gives $GM = 96000$ mi^3/sec^2.

(b) From Example 7 in the text, the "escape velocity" formula can be rewritten in terms of the distance d (in miles) that an object is from the center of mass of the earth. In this case, we let $r_0 = 4000 + d$ and $GM = 96000$ (from part (a)) to get

$$v_0 = \sqrt{\frac{2GM}{r_0}} = \sqrt{\frac{192000}{4000+d}},$$

where v_0 is the initial velocity required for the object to "escape" earth's gravity. In particular, if the object is a projectile fired from the surface of the earth then $d = 0$ and

$$v_0 = \sqrt{\frac{192000}{4000}} = \sqrt{48} \approx 6.93 \text{ mi/sec} \approx 24900 \text{ mi/hr}.$$

(c) If the projectile is 1000 miles above the surface of the earth, set $d = 1000$ in the formula derived in part (b) to get

$$v_0 = \sqrt{\frac{192000}{5000}} = \sqrt{38.4} \approx 6.20 \text{ mi/sec} \approx 22300 \text{ mi/hr}.$$

31. (a) Let g be a positive constant representing the acceleration of gravity near the earth's surface. Let $s = s(t) = \frac{1}{2}gt^2$, $t \geq 0$, be the distance that an object falls in time t after being dropped near the earth's surface. Then $ds/dt = gt$. From the formula $s = \frac{1}{2}gt^2$ we get $2sg = g^2t^2$, so that $\sqrt{2gs} = gt$. Thus,

$$\frac{ds}{dt} = \sqrt{2sg}.$$

(b) Consider the one-parameter family of functions given by

$$s = \left(\sqrt{\frac{g}{2}}\,t + c\right)^2 = \frac{1}{2}gt^2 + \sqrt{2g}\,ct + c^2,$$

where c is the parameter. Differentiation with respect to t of each member of this family gives

$$\frac{ds}{dt} = gt + c\sqrt{2g} = \sqrt{2g}\left(\sqrt{\frac{g}{2}}\,t + c\right) = \sqrt{2g\left(\sqrt{\frac{g}{2}}\,t + c\right)^2} = \sqrt{2gs}.$$

So, each member of the family is a solution of $ds/dt = \sqrt{2gs}$.

(c) Let $s = s(t)$ be a function as given in part (b). The result of part (b) says that s satisfies the D.E. given in part (a). So, with $s(0) = s_0$ and $ds/dt(0) = v_0$, we can evaluate the D.E. at $t = 0$ to get

$$v_0 = \frac{ds}{dt}(0) = \sqrt{2gs_0}.$$

Squaring both sides gives $v_0^2 = 2gs_0$.

33. (a) Consider the function $h(t) = (bt + c)^2$, $t \geq 0$, where b and c are constants to be chosen (if possible) such that $h = h(t)$ satisfies the D.E. $dh/dt = -(a/A)\sqrt{2gh}$.

First of all, irrespective of whether $h(t)$ is a solution of the given D.E. equation, we can set $t = 0$ to get $h(0) = c^2$, so that $c = \sqrt{h(0)}$ or $c = -\sqrt{h(0)}$. We will therefore assume that $c = \sqrt{h(0)}$, derive the proper expression for b (if possible), and then conclude that the constants $-b$, $-c$ can also be chosen (since $(bt + c)^2 = (-bt - c)^2$ for all t).

Next, we see that $dh/dt = 2b(bt + c)$, so that if $h(t)$ is a solution of the D.E. then

$$(*) \qquad 2b(bt + c) = -\frac{a}{A}\sqrt{2g(bt + c)^2} = -\frac{a}{A}\sqrt{2g}|bt + c|.$$

If $bt + c < 0$ for some $t \geq 0$ then $|bt + c| = -(bt + c)$ and $(*)$ simplifies to $2b = (a/A)\sqrt{2g}$, so that $b > 0$. This is impossible because b, t and c can't all be nonnegative while $bt + c < 0$. We must therefore have $bt + c \geq 0$ for all $t \geq 0$. Now, $(*)$ clearly holds when $bt + c = 0$. For $bt + c > 0$, $|bt + c| = bt + c$ and $(*)$ simplifies to $2b = -(a/A)\sqrt{2g}$. Solving for b gives

$$b = -\frac{a}{A}\sqrt{g/2}.$$

As already noted, the constants $c = -\sqrt{(h(0))}$, $b = (a/A)\sqrt{g/2}$ also work.

(b) Letting $h(0) = h_0$, the solution found in part (a) can be written as

$$h(t) = \left(-\frac{a}{A}\sqrt{g/2}\,t + \sqrt{h_0}\right)^2, \quad 0 \leq t \leq t_e,$$

where t_e is the time when the tank first becomes empty and $h(t)$ no longer models the physical situation. Hence, since $h(t)$ is the height of the fluid in the tank at time t,

$$0 = h(t_e) = \left(-\frac{a}{A}\sqrt{g/2}\,t_e + \sqrt{h_0}\right)^2 \quad \text{which implies} \quad 0 = -\frac{a}{A}\sqrt{g/2}\,t_e + \sqrt{h_0}.$$

Solving for t_e gives

$$t_e = \frac{-\sqrt{h_0}}{-(a/A)\sqrt{g/2}} = \frac{A}{a}\sqrt{2h_0/g}.$$

In particular, if $A = 25\pi$ sq/ft (tank is 10 feet in diameter), $a = \pi/16$ sq/ft (outlet hole is 6 inches in diameter), $h_0 = 20$ feet, and $g = 32.2$ ft/sec^2 then

$$t_e = \frac{25\pi}{\pi/16}\sqrt{40/32.2} \approx 446 \text{ sec} \approx 7.4 \text{ minutes}.$$

Section 3: LINEAR EQUATIONS

Exercise Set 3AB (pgs. 487-488)

1. In the expression $y' + 2y$, we assume that x is the independent variable. Since this expression has the proper form of the left member of a normalized linear equation, we see that the coefficient of y is 2 and compute the exponential multiplier $M(x)$ as follows:

$$M(x) = e^{\int 2\,dx} = e^{2x}.$$

3. The given expression has the proper form of the left member of a normalized linear equation. We therefore identify the coefficient of y to be $2/x$ and compute the exponential multiplier $M(x)$ as follows:

$$M(x) = e^{\int (2/x)\,dx} = e^{2\ln x} = x^2.$$

5. The differential equation $ds/dt + ts = t$ is linear and in normalized form. Thus, an exponential integrating factor is $e^{\int t\,dt} = e^{t^2/2}$. Multiplying the normalized D.E. throughout by this quantity gives

$$e^{t^2/2}\frac{ds}{dt} + te^{t^2/2}s = te^{t^2/2}, \qquad \text{or equivalently,} \qquad \frac{d}{dt}\left(e^{t^2/2}s\right) = te^{t^2/2}.$$

Integrating with respect to t gives

$$e^{t^2/2}s = e^{t^2/2} + C.$$

Multiplying throughout by $e^{-t^2/2}$ gives the general solution

$$s(t) = 1 + Ce^{-t^2/2}.$$

The initial condition $s(0) = 0$ gives $0 = s(0) = 1 + C$, so that $C = -1$. The particular solution is therefore

$$s(t) = 1 - e^{-t^2/2}, \quad -\infty < t < \infty.$$

7. When the given D.E. is written in normalized form, the result is

$$\frac{dy}{dt} - \frac{x}{2}y = 0.$$

Multiplying this equation by the integrating factor $e^{\int (-x/2)\,dx} = e^{-x^2/4}$ gives

$$e^{-x^2/4}\frac{dy}{dx} - \frac{x}{2}e^{-x^2/4}y = 0, \qquad \text{or equivalently,} \qquad \frac{d}{dt}\left(e^{-x^2/4}y\right) = 0.$$

Integrating with respect to x and multiplying the result by $e^{x^2/4}$ gives the general solution

$$y(x) = Ce^{x^2/4}.$$

The initial condition $y(1) = 0$ immediately gives $0 = C$, so that the particular solution is the zero solution; i.e.,

$$y(x) = 0, \quad -\infty < x < \infty.$$

9. Let $S(t)$ be the number of pounds of salt in the tank at time $t \geq 0$. Since the tank is initially full of pure water, $S(0) = 0$. The two sources supply the tank with a total of 5 gallons of solution each minute (2 gallons from one source and 3 gallons from the other) and the volume of solution leaving the tank is 5 gallons per minute. Therefore, the total volume of solution in the tank remains constant at 100 gallons. The inflow rate (pounds of salt per minute) is the sum of two quantities; namely,

inflow rate = $(2 \text{ gal/min})(1 \text{ lb/gal}) + (3 \text{ gal/min})(2e^{-2t} \text{ lb/gal}) = (2 + 6e^{-2t}) \text{ lb/gal}.$

The outflow rate is just

outflow rate = $(5 \text{ gal/min})\big(S(t)/100 \text{ lb/gal}\big) = \dfrac{S(t)}{20} \text{ lb/min}.$

The pertinent D.E. is then

$$\frac{dS}{dt} = 2 + 6e^{-2t} - \frac{S}{20}, \quad S(0) = 0.$$

This is linear and can be written in normalized form as

$$\frac{dS}{dt} - \frac{1}{20}S = 2 + 6e^{-2t}.$$

Multiplying throughout by the integrating factor $M(t) = e^{\int(-1/20)dt} = e^{-t/20}$ and then writing the resulting left member as the derivative of $M(t)S$ gives

$$\frac{d}{dt}\big(e^{-t/20}S\big) = 2e^{-t/20} + 6e^{-2t-t/20} = 2e^{-t/20} + 6e^{-41t/20}.$$

Integrating with respect to t gives

$$e^{-t/20}S = -40e^{-t/20} - \frac{120}{41}e^{-41t/20} + C,$$

and multiplying throughout by $e^{t/20}$ gives

$$S(t) = -40 - \frac{120}{41}e^{-2t} + Ce^{t/20}.$$

The condition $S(0) = 0$ gives $0 = -40 - 120/41 + C$, so that $C = 1760/41$. Hence,

$$S(t) = -40 - \frac{120}{41}e^{-2t} + \frac{1760}{41}e^{t/20}, \quad t \geq 0.$$

11. Consider the differential equation $mdv/dt = mg - kv$, where m, g and k are positive constants. Recognizing this as linear, we first write it in normalized form:

$$\frac{dv}{dt} + \frac{k}{m}v = g.$$

374

Multiplying throughout by the integrating factor $M(t) = e^{\int (k/m)dt} = e^{kt/m}$ and writing the result as the derivative of $M(t)v$ gives

$$\frac{d}{dt}\left(e^{kt/m}v\right) = ge^{kt/m}.$$

Integrating with respect to t gives

$$e^{kt/m}v = \frac{mg}{k}e^{kt/m} + C,$$

and multiplying throughout by $e^{-kt/m}$ gives

$$v(t) = \frac{mg}{k} + Ce^{-kt/m}.$$

Setting $t = 0$, we obtain $v(0) = mg/k + C$, so that $C = v(0) - mg/k$. Hence, the solution can be written in terms of the initial velocity $v(0)$ as

$$v(t) = \frac{mg}{k} + \left(v(0) - \frac{mg}{k}\right)e^{-kt/m},$$

which is what we were asked to show.

13. Let k be a positive constant and let $f(t) = e^{-2t}$, for $t \geq 0$. Then, in normalized form, the differential equation describing Newton's law of cooling is given by

$$\frac{du}{dt} + ku = ke^{-2t}, \quad t \geq 0.$$

Multiplying throughout by the integrating factor $e^{\int k\,dt} = e^{kt}$ gives

$$e^{kt}\frac{du}{dt} + ke^{kt}u = ke^{kt-2t}, \quad \text{or equivalently,} \quad \frac{d}{dt}\left(e^{kt}u\right) = ke^{(k-2)t}.$$

In order to integrate this equation with respect to t, we must consider two cases: (i) $k = 2$ and (ii) $k \neq 2$. Doing the integration for these two cases gives

$$e^{kt}u = \begin{cases} 2t + C, & \text{if } k = 2; \\ \left(k/(k-2)\right)e^{(k-2)t} + C, & \text{if } k \neq 2. \end{cases}$$

Multiplying throughout by e^{-kt} gives

$$u(t) = \begin{cases} 2te^{-2t} + Ce^{-2t}, & \text{if } k = 2; \\ \left(k/(k-2)\right)e^{-2t} + Ce^{-kt}, & \text{if } k \neq 2. \end{cases}$$

If $k = 2$, the initial condition $u(0) = 10$ leads to $10 = u(0) = C$, so that

$$u(t) = 2te^{-2t} + 10e^{-2t}, \quad t \geq 0.$$

If $k \neq 2$ then the same initial condition leads to $10 = k/(k-2)+C$, so that $C = 10-k/(k-2)$ and

$$u(t) = \frac{k}{k-2}e^{-2t} + \left(10 - \frac{k}{k-2}\right)e^{-kt}, \quad t \geq 0.$$

15. (a) Let $u = u(t)$ be the temperature of the milk (in degrees Fahrenheit) for $t \geq 0$, with $u(0) = 70$. Since the ambient temperature is a constant $30°$ F (temperature of the ice and brine), the initial-value problem describing the temperature of the milk at time $t \geq 0$ is

$$\frac{du}{dt} = k(30 - u), \quad u(0) = 70.$$

By Example 5 in the text, this has the solution

$$u(t) = 30 + 40e^{-kt}.$$

To determine k, we use the given side condition $u(15) = 40$ to get $40 = 30 + 40e^{-15t}$. Solving for k gives

$$k = \frac{\ln 4}{15} \approx 0.0924.$$

(b) Setting $u(t) = 35$ in the solution formula given in part (a) leads to $35 = 30 + 40e^{-kt}$, or $1/8 = e^{-kt}$. Solving this for t and then using the value of k found in part (a), gives

$$t = \frac{1}{k} \ln 8 = \frac{15}{\ln 4} \ln 8 = 15 \frac{3 \ln 2}{2 \ln 2} = 22.5 \text{ minutes.}$$

That is, the milk will have cooled to a temperature of $35°$ F in 22.5 minutes.

17. If $y_1 = y_1(x)$ is a solution of $y' + gy = 0$ and c is a real number then

$$(cy_1)' + g(c_1 y) = c(y_1' + gy_1) = 0,$$

where the second equality holds because y_1 is a solution of $y' + gy = 0$. So, cy_1 is a solution of $y' + gy = 0$.

19. Let $S = S(t)$ be the number of pounds of salt in the tank at time $t \geq 0$, with $S(0) = 10$. Since the same volume of solution is being added per unit time as is being drawn off, the volume of solution in the tank remains constant at 100 cubic-feet.

(a) For the first hour (i.e., for $0 \leq t \leq 1$), the concentration $C(t)$ of the inflow rate decreases at the constant rate of 1 pound per cubic-foot per hour and is zero at the end of the first hour. It follows that $C(t) = 1 - t$ pounds per cubic-foot. The inflow rate is therefore the product of 1 cubic-foot per hour and $1 - t$ pounds per cubic-foot, or inflow $= 1 - t$ pounds per hour. The outflow rate is 1 cubic-foot per hour at a concentration of $S/100$ pounds per cubic-foot, or outflow $= S/100$ pounds per hour. The relevant differential equation for the first hour is therefore

$$\frac{dS}{dt} = 1 - t - \frac{S}{100}, \quad S(0) = 10, \ 0 \leq t \leq 1.$$

The normalized form $dS/dt + (1/100)S = 1 - t$ has the integrating factor $e^{\int (1/100)\, dt} = e^{t/100}$. Multiplying throughout by this factor gives

$$e^{t/100}\frac{dS}{dt} + \frac{1}{100}e^{t/100}S = (1 - t)e^{t/100}, \quad \text{or equivalently,} \quad \frac{d}{dt}\left(e^{t/100}S\right) = (1 - t)e^{t/100}.$$

Integrating gives $e^{t/100}S = 100(1 - t)e^{t/100} + 10000e^{t/100} + C$, so that

$$S(t) = 100(1 - t) + 10000 + Ce^{-t/100} = 10100 - 100t + Ce^{-t/100}.$$

376

The condition $S(0) = 10$ leads to $10 = 10100 + C$, so that $C = -10090$ and

$$S(t) = 10100 - 100t - 10090e^{-t/100}, \quad 0 \le t \le 1.$$

In particular, at the the end of the first hour we have

$$S(1) = 10000 - 10090e^{-1/100} \approx 10.397 \text{ pounds}.$$

(b). After the first hour, we can re-initialize the process by letting $S_1 = S_1(t)$ be the number of pounds of salt in the tank at time $t \ge 0$, where the time variable t is actually $t = 1$ with respect to part (a). The initial amount of salt in the tank is therefore $S(1) = 10000 - 10090e^{-1/100}$ pounds (exactly). Since pure water is now being added, the inflow rate is zero. The outflow rate is now 1 cubic-foot per hour at a concentration of $S_1/100$ pounds per cubic-foot, or outflow= $S_1/100$ pounds per hour. The relevant D.E. is now

$$\frac{dS_1}{dt} = -\frac{S_1}{100}, \quad S_1(0) = 10000 - 10090e^{-1/100}.$$

Separating variables gives $(1/S_1)\,dS_1 = -(1/100)\,dt$, integrating gives $\ln S_1 = -t/100 + C_1$ and exponentiating gives $S_1(t) = Ce^{-t/100}$. The initial condition $S_1(0) = S(1)$ immediately gives $10000 - 10090e^{-1/100} = C$, so that

$$S_1(t) = (10000 - 10090e^{-1/100})e^{-t/100}, \quad t \ge 0.$$

To find out when the tank has 5 pounds of salt in it, set $S_1(t) = 5$ and solve for t. This gives

$$t = 100\ln(2000 - 2018e^{-1/100}) \approx 73.2 \text{ hours}.$$

21. Let $S = S(t)$ be the number of pounds of salt in the tank at time $t \ge 0$, with $S(0) = 0$. First note that the total volume of solution in the tank remains constant at 100 gallons. The inflow rate is the product of 1 gallon per minute and the concentration of the incoming solution. Since this concentration is 1 pound per gallon at $t = 0$ and increases at a constant rate to 2 pounds per gallon at $t = 10$, it follows that the concentration is given by $1 + t/10$ pounds per gallon. The inflow rate is therefore $1 + t/10$ pounds per minute. The outflow rate is 1 gallon per minute at a concentration of $S/100$ pounds per gallon, or outflow rate $= S/100$ pounds per minutes. The relevant differential equation is therefore

$$\frac{dS}{dt} = 1 + \frac{t}{10} - \frac{S}{100}, \quad S(0) = 0.$$

Its normalized form is $dS/dt + \frac{1}{100}S = 1 + t/10$, from which we deduce the integrating factor $e^{t/100}$. Multiplying throughout by this factor gives

$$e^{t/100}\frac{dS}{dt} + \frac{1}{100}e^{t/100}S = \left(1 + \frac{t}{10}\right)e^{t/10}, \quad \text{or} \quad \frac{d}{dt}\left(e^{t/100}S\right) = \left(1 + \frac{t}{10}\right)e^{t/100}.$$

Integrating gives $e^{t/100}S = -900e^{t/100} + 10te^{t/100} + C$, or $S(t) = -900 + 10t + Ce^{-t/100}$. The condition $S(0) = 0$ leads to the equation $0 = -900 + C$, so that $C = 900$ and

$$S(t) = -900 + 10t + 900e^{-t/100}.$$

At $t = 10$ we have
$$S(10) = -900 + 100 + 900e^{-1/10} \approx 14.35 \text{ pounds.}$$

23. (a) Let $S = S(t)$ be the number of pounds of salt in the tank at time $t \geq 0$, with $S(0) = 10$. First note that the tank gains 1 gallon of solution every minute. Since it initially contains 50 gallons of solution and since the tank is a 100 gallon tank, it follows that the volume of solution in the tank at time t is $50 + t$ and that it will begin to overflow at time $t = 50$. The inflow rate is 2 gallons per minute at a concentration of 2 pounds per gallon, or inflow rate = 4 pounds per minute; and the outflow rate is 1 gallon per minute at a concentration of $S/(50+t)$ pounds per gallon, or outflow rate = $S/(50+t)$ pounds per minute. The relevant differential equation is therefore

$$\frac{dS}{dt} = 4 - \frac{S}{50 + t}, \quad S(0) = 10, \ 0 \leq t \leq 50.$$

Its normalized form is $dS/dt + \frac{1}{50+t}S = 4$, from which we obtain the integrating factor $e^{\int \left(1/(50+t)\right) dt} = e^{\ln(50+t)} = 50 + t$. Multiplying throughout by this factor gives

$$(50 + t)\frac{dS}{dt} + S = 200 + 4t, \quad \text{or} \quad \frac{d}{dt}\big((50 + t)S\big) = 200 + 4t.$$

Integrating gives $(50+t)S = 200t + 2t^2 + C$. The condition $S(0) = 10$ directly gives $C = 500$, so that solving for $S(t)$ gives

$$S(t) = \frac{500 + 200t + 2t^2}{50 + t}, \quad 0 \leq t \leq 50.$$

When the tank begins to overflow (i.e., at $t = 50$), we have

$$S(50) = \frac{500 + 200(50) + 2(50)^2}{100} = \frac{15500}{100} = 155 \text{ pounds.}$$

(b) If the process described above is allowed to continue after overflow is attained but with an additional drain off of 1 gallon per minute then 2 gallons of solution are being added each minute and 2 gallons are being drawn off each minute, so that the total volume of solution in the tank remains constant from then on at 100 gallons. We therefore re-initialize the process by letting $S_1 = S_1(t)$ be the number of pounds of salt in the tank at time $t \geq 0$, with $S_1(0) = 155$ (from part (a)). The inflow rate is the same as before; namely, 4 pounds per minute. The outflow rate is now 2 gallons per minute at a concentration of $S_1/100$ pounds per gallon, or outflow rate = $2S_1/100 = S_1/50$ pounds per minute. The new D.E. is therefore

$$\frac{dS_1}{dt} = 4 - \frac{S_1}{50}, \quad S_1(0) = 155, \ t \geq 0.$$

Its normalized form is $dS_1/dt + \frac{1}{50}S_1 = 4$, from which we obtain the integrating factor $e^{t/50}$. Multiplying throughout by this factor gives

$$e^{t/50}\frac{dS_1}{dt} + \frac{1}{50}e^{t/50}S_1 = 4e^{t/50}, \quad \text{or} \quad \frac{d}{dt}\big(e^{t/50}S_1\big) = 4e^{t/50}.$$

Integrating gives $e^{t/50}S_1 = 200e^{t/50} + C$, or $S_1(t) = 200 + Ce^{-t/50}$. The initial condition $S_1(0) = 155$ leads to the equation $155 = 200 + C$, so that $C = -45$ and

$$S_1(t) = 200 - 45e^{-t/50}, \quad t \geq 0.$$

Note that since $e^{-t/50}$ is always positive $S_1(t)$ is always less than 200. However, since $\lim_{t \to \infty} e^{-t/50} = 0$, the upper limit for the total amount of salt in the tank is 200 pounds. Moreover, $S_1(t) = 175$ for t satisfying $175 = 200 - 45e^{-t/50}$. Solving for t gives

$$t = 50\ln(9/5) \approx 29.4 \text{ minutes.}$$

Chapter Review

(pgs. 488-489)

1. The given differential equation $x(dy/dx) + y - x = 0$ is linear and its normalized form is $dy/dx + (1/x)y = 1$, $x \neq 0$. However, viewing y as a function of x, we recognize the expression $x(dy/dx) + y$ as the derivative of xy, so that the given D.E. can be written as

$$\frac{d}{dx}(xy) = x.$$

Integrating with respect to x gives $xy = x^2/2 + C$. Solving for y gives

$$y(x) = \frac{x}{2} + \frac{C}{x}.$$

Note that $x \neq 0$ if $C \neq 0$, but that x can be any value if $C = 0$. In the latter case, $y = x/2$, which we can see by inspection is a solution of the D.E.

3. The differential equation is linear and its normalized form is

$$\frac{dx}{dt} - tx = e^t.$$

Using the integrating factor $e^{-t^2/2}$ gives

$$e^{-t^2/2}\frac{dx}{dt} - te^{-t^2/2} = e^{t-t^2/2}, \qquad \text{or} \qquad \frac{d}{dt}\left(e^{-t^2/2}y\right) = e^{t-t^2/2}.$$

Integrating with respect to t and multiplying the result by $e^{t^2/2}$ gives

$$y(t) = e^{t^2/2}\int e^{t-t^2/2}dt + Ce^{t^2/2}.$$

It is convenient to choose the antiderivative of e^{t-t^2} whose value at $t = 0$ is zero and write the above as a definite integral

$$y(x) = e^{t^2/2}\int_0^t e^{u-u^2/2}du + Ce^{t^2/2}.$$

5. The given differential equation can be solved by separating variables. So, with $y' = dy/dx$, write it in its separated form

$$\frac{y^3}{y^4 + 1}dy = e^x dx.$$

Integrating gives

$$\frac{1}{4}\ln(y^4 + 1) = e^x + C.$$

Because of the economy of the notation, it is perhaps best to leave it in this implicit form.

7. For $x \neq 0$, the given differential equation can be written in its normalized form

$$y' + \left(2 - \frac{3}{x}\right) y = x^3.$$

An integrating factor is

$$e^{\int (2-3/x)\, dx} = e^{2x-3\ln|x|} = e^{2x} e^{\ln(|x|^{-3})} = |x|^{-3} e^{2x}.$$

multilying the normalized form by this factor allows us to write the result as

$$\frac{d}{dx}\left(|x|^{-3} e^{2x} y\right) = \frac{x^3}{|x|^3} e^{2x} = \pm e^{2x},$$

where the sign on the right is chosen to coincide with the sign of x. Integrating with respect to x gives

$$|x|^{-3} e^{2x} y = \pm\frac{1}{2} e^{2x} + C, \qquad \text{or} \qquad y(x) = \pm\frac{|x|^3}{2} + C|x|^3 e^{-2x}.$$

Notice that if $x > 0$ then $|x| = x$, the plus sign is chosen and the first term on the right becomes $x^3/2$; and if $x < 0$ then $|x| = -x$, the minus sign is chosen and the first term on the right is again $x^3/2$. Thus, we can remove both the absolute value on the first term and the symbol "$\pm$". Moreover, because the arbitrary constant C multiplies $|x|^3$ in the second term, we can remove the absolute value there as well. In other words, the general solution can be written as

$$y(x) = \frac{x^3}{2} + Cx^3 e^{-2x}.$$

Finally, recall that this solution formula was derived under the condition that $x \neq 0$. But observe that $y'(x)$ is defined for all x, including $x = 0$ and that $y(0) = 0$ regardless of the value of C. By inspection, we can see that the original D.E. therefore holds with $x = 0$. In other words, the above solution formula holds for all x.

9. For $t \neq 0$, write the given differential equation in its normalized form

$$\frac{dx}{dt} + \frac{2}{t} x = t^2.$$

Using the integrating factor $e^{\int (2/t)\, dt} = e^{2\ln|t|} = e^{\ln(t^2)} = t^2$, results in

$$\frac{d}{dt}\left(t^2 x\right) = t^4.$$

Integrating with respect to t gives $t^2 x = t^5/5 + C$, and solving for $x = x(t)$ gives

$$x(t) = \frac{1}{5} t^3 + \frac{C}{t^2}.$$

The condition $x(2) = 1$ leads to the equation $1 = 8/5 + C/4$, so that $C = -12/5$ and the particular solution is therefore

$$x(t) = \frac{1}{5} t^3 - \frac{12}{5t^2}, \quad t > 0,$$

where the domain of the solution is the largest open interval containing $t = 2$ on which $x(t)$ is continuous.

380

11. First note that the zero function $x(t) \equiv 0$ is a solution of $dx/dt = -3x^2$. For $x \neq 0$, the variables can be separated to get

$$-x^{-2}dx = 3\,dt.$$

Integration gives $1/x = 3t + C$, or $x(t) = 1/(3t + C)$. The system of solutions is therefore given by

$$x(t) = \frac{1}{3t + C} \quad \text{and} \quad x(t) \equiv 0.$$

13. Let $x + t = y$. Implicit differentiation with respect to t gives $dx/dt + 1 = dy/dt$, or $dx/dt = dy/dt - 1$. The given differential equation then becomes $dy/dt - 1 = y^2$, whose variables can be separated to get

$$\frac{1}{1 + y^2}\,dy = dt.$$

Integrating gives $\arctan y = t + C$, or $y = \tan(t + C)$. Resubstituting gives $x + t = \tan(t + C)$, or

$$x(t) = t + \tan(t + C).$$

15. Consider the differential equation $dy/dx = e^{x-y}$

(a) Since the exponential is always positive, the D.E. shows that all solutions are increasing over the entire xy-plane.

(b) Implicit differentiation with respect to x of both sides of the given D.E. gives

$$(*) \qquad\qquad \frac{d^2y}{dx^2} = \left(1 - \frac{dy}{dx}\right)e^{x-y}.$$

Since a positive second derivative implies upward concavity, all solutions are concave up when $dy/dx < 1$. From the original D.E., this happens only if $e^{x-y} < 1$. And this happens if, and only if, $x < y$. This is the region above the line $y = x$ in the xy-plane.

(c) Yes, since both sides of the given D.E. are identically equal to 1 for $y = x$.

(d) Yes, isoclines are lines on which the derivative is constant. In this case, $y = x$ implies dy/dx is the constant 1.

(e) Separating variables gives $e^y dy = e^x dx$, and integration gives $e^y = e^x + C$, or

$$y(x) = \ln(e^x + C) = x + \ln(1 + Ce^{-x}),$$

where we will use both forms of the solution family to attempt to answer parts (a)-(d).

For part (a), since $e^x + C$ is strictly increasing for any constant C, and since the log function is strictly increasing, it follows that the composite function $\ln(e^x + C)$ is strictly increasing for any constant C. The first form of the solution family then shows that all solutions are strictly increasing over the entire xy-plane.

The answer to part (b) is not easy to obtain from the general solution formula. This solver sees no way to do so except to formally compute d^2y/dx^2 from the solution formula and write the result as in $(*)$ above. But then nothing has been gained in solving the D.E. in the first place.

For part (c), we use the second form with $C = 0$ to get $y(x) = x$. So, the line $y = x$ is a solution graph.

For part (d), we use the fact that $y = x$ is a solution and that its derivative has constant value 1; i.e., $dy/dx = 1$. So, the line $y = x$ is an isocline.

The above analysis suggests that the answers to parts (a), (c) and (d) are just as easily obtained from the solution formula as from the D.E., but that nothing is gained in attempting to do this for part (b).

17. Let a and c be constants with $a \neq 0$ and consider the family $y' + ay = c$ of linear differential equations

(a) Choosing and fixing a and c selects the family member $y' + ay = c$. An isocline of the direction field of this differential equation satisfies $dy/dx = k$ for some constant k. The D.E. can then be written as $k + ay = c$. Since $a \neq 0$, we can solve for y to find that the isocline has the equation $y = (c - k)/a$, which is a horizontal line.

Conversely, let L be a horizontal line with constant y-coordinate k. Without loss of generality, we can suppose that x is the independent variable, so that a point is on L if, and only if, it has the form (x, k) for some real number x. Let $y = y(x)$ be any solution of our chosen D.E. that crosses L. Then $y = y(x_0) = k$ for some x_0. Evaluating the D.E. at (x_0, k) gives $y'(x_0) + ak = c$, or $y'(x_0) = c - ak$. As a, c and k are independent of which solution is chosen, it follows that the derivative of any solution crossing L has value $c - ak$ at the point of its intersection with L. This is precisely what is meant for L to be an isocline.

(b) Below is the direction field associated with the differential equation $dy/dx + 2y = 1$. Note that the derivatives of the solutions along any horizontal line are constant.

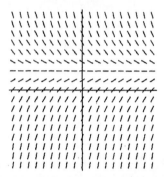

19. Let $S = S(t)$ be the number of pounds of salt in the vat at time $t \geq 0$, with $S(0) = S_0$. Note that 2 gallons of solution are added to the vat each minute and 2 gallons are leaving the vat each minute (1 gallon by evaporation and 1 gallon by overflow), so that the total volume of solution in the vat is constant at 100 gallons. The inflow rate is 2 gallons per minute at a concentration of 1 pounds per gallon, or intake rate = 2 pounds per minute. The outflow rate is 2 gallons per minute but only one gallon of it has salt in it at a concentration of $S/100$ pounds per gallon. The outflow rate is therefore $S/100$ pounds per gallon. The relevant D.E. is then

$$\frac{dS}{dt} = 2 - \frac{S}{100}, \quad S(0) = S_0.$$

When its normalized form $dS/dt + (1/100)S = 2$ is multiplied by the integrating factor $e^{t/100}$, the result can be written as

$$\frac{d}{dt}\left(e^{t/100}S\right) = 2e^{t/100}.$$

Integrating gives $e^{t/100}S = 200e^{t/100} + C$, so that $S(t) = 200 + Ce^{-t/100}$. The condition $S(0) = S_0$ leads to the equation $S_0 = 200 + C$, so that $C = S_0 - 200$ and

$$S(t) = 200 + (S_0 - 200)e^{-t/100}, \quad t \geq 0.$$

Letting S_0 take on the two values $S_0 = 0$ and $S_0 = 50$ gives the associated solutions

$$S(t) = 200 - 200e^{-t/100} \quad \text{and} \quad S(t) = 200 - 150e^{-t/100},$$

both of which hold for $t \geq 0$.

Chapter 11: Second-Order Equations

Section 1: DIFFERENTIAL OPERATORS

Exercise Set 1AB (pgs. 498-499)

1. $(D+1)e^{-2x} = De^{-2x} + e^{-2x} = -2e^{-2x} + e^{-2x} = -e^{-2x}$.

3. $D^3 e^{3x} = D(D(De^{3x})) = D(D3e^{3x}) = D9e^{3x} = 27e^{3x}$.

5.
$$(D^2+1)x\cos x = D(Dx\cos x) + x\cos x$$
$$= D(-x\sin x + \cos x) + x\cos x$$
$$= D(-x\sin x) + D\cos x + x\cos x$$
$$= -x\cos x - \sin x - \sin x + x\cos x$$
$$= -2\sin x.$$

7. Char. eq.: $r^2 + r - 6 = 0$; roots: $r_1 = 2$, $r_2 = -3$; general solution:
$$y(x) = c_1 e^{2x} + c_2 e^{-3x}.$$

Using this and its derivative $y'(x) = 2c_1 e^{2x} - 3c_2 e^{-3x}$, the initial conditions $y(0) = 2$, $y'(0) = 2$ lead to the equations $2 = c_1 + c_2$ and $2 = 2c_1 - 3c_2$, with solution $c_1 = 8/5$, $c_2 = 2/5$. The particular solution is therefore
$$y(x) = \frac{8}{5}e^{2x} + \frac{2}{5}e^{-3x}.$$

9. Char. eq.: $r^2 + 2r + 1 = 0$; roots: $r_1 = r_2 = -1$; general solution:
$$y(x) = c_1 e^{-x} + c_2 x e^{-x}.$$

Using this and its derivative $y'(x) = (c_2 - c_1)e^{-x} - c_2 x e^{-x}$, the initial conditions $y(0) = 1$, $y'(0) = 2$ lead to the equations $1 = c_1$ and $2 = c_2 - c_1$, with solution $c_1 = 1$, $c_2 = 3$. The particular solution is therefore
$$y(x) = e^{-x} + 3x e^{-x}.$$

11. Char. eq.: $r^2 - r = 0$; roots: $r_1 = 0$, $r_2 = 1$; general solution
$$y(x) = c_1 + c_2 e^{x}.$$

The boundary conditions $y(0) = 1$, $y(1) = 0$ then lead to the equations $1 = c_1 + c_2$ and $0 = c_1 + c_2 e$, with solution $c_1 = e/(e-1)$, $c_2 = -1/(e-1)$. The particular solution is therefore
$$y(x) = \frac{e}{e-1} - \frac{1}{e-1}e^{x}.$$

13. Char.eq.: $2r^2 - 3r + 1 = 0$; roots: $r_1 = 1$, $r_2 = 1/2$; general solution:
$$y(x) = c_1 e^{x} + c_2 e^{x/2}.$$

Using this and its derivative $y'(x) = c_1 e^e + \frac{1}{2}c_2 e^{x/2}$, the initial conditions $y(0) = 0$, $y'(0) = 0$ lead to the equations $0 = c_1 + c_2$ and $0 = c_1 + c_2/2$, with solution $c_1 = c_2 = 0$. The particular solution is therefore the identically zero solution
$$y(x) \equiv 0.$$

15. In order that $y = xe^{-x}$ be a solution of a differential equation of the given form, the roots of the associated characteristic equation must be equal (because of the coefficient x in xe^{-x}), and the value of the double root must be -1 (the coefficient of x in the exponent on e^{-x}). So, the characteristic equation must be $(r - (-1))^2 = 0$, or $r^2 + 2r + 1 = 0$. The corresponding D.E. is therefore

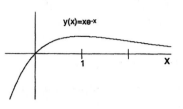

$$y'' + 2y' + y = 0, \quad \text{with general solution} \quad y(x) = c_1 e^{-x} + c_2 x e^{-x}.$$

The given function is the special case with $c_1 = 0$, $c_2 = 1$.

17. Let $y = 1 + x$. Analysis will be simpler if we write this as $y = e^{0x} + xe^{0x}$. In order that this be a solution of a differential equation of the given form, the roots of the characteristic equation must be equal (because of the coefficient x in xe^{0x}), and the value of the double root must be 0 (the coefficient of x in the exponent e^{0x}). So, the characteristic equation must be $(r - 0)^2 = 0$, or $r^2 = 0$. The corresponding D.E. is therefore

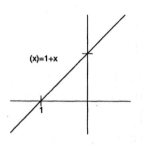

$$y'' = 0, \quad \text{with general solution} \quad y(x) = c_1 + c_2 x.$$

The given function is a special case with $c_1 = c_2 = 1$.

19. The justification that $y = xe^{-x} - e^{-x}$ is a solution of the differential equation

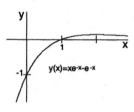

$$y'' + 2y' + y = 0 \quad \text{with general solution} \quad y(x) = c_1 e^{-x} + c_2 x e^{-x}$$

is exactly the same as in the solution of Exercise 15 above. Clearly, the given function is a special case with $c_1 = -1$, $c_2 = 1$.

21. Operator form: $(D^2 + 2D + 1)y = 0$; factored operator: $D^2 + 2D + 1 = (D+1)(D+1)$.

23. Operator form: $(2D^2 - 1)y = 0$; factored operator: $2D^2 - 1 = 2(D - 1/\sqrt{2})(D + 1/\sqrt{2})$.

25. Operator form: $D^2 y = 0$; factored operator: $D^2 = DD$.

27. Write the given differential equation in the form $(D - 0)(D - 3)y = 0$ and let $r_1 = 0$, $r_2 = 3$. If we let $(D - 3)y = z$ then the given D.E. becomes a first-order equation in the dependent variable z; namely, $(D - 0)z = 0$, or $z' = 0$. Direct integration gives $z = C_1$. As $(D - 3)y = z$, we now solve the first-order equation $(D - 3)y = C_1$, or $y' - 3y = C_1$. The integrating factor e^{-3x} leads to the equation $(d/dx)(e^{-3x}y) = C_1 e^{-3x}$, so that one integration gives $e^{-3x}y = -(C_1/3)e^{-3x} + C_2$, or $y(x) = -(C_1/3) + C_2 e^{3x}$. Setting $c_1 = -C_1/3$ and $c_2 = C_2$, we get the general solution of the given D.E.:

$$y(x) = c_1 + c_2 e^{3x}.$$

29. The operator form of the given differential equation is $(D^2 - 1)y = 1$, or in factored form, $(D - 1)(D + 1)y = 1$. With $r_1 = 1$ and $r_2 = -1$, setting $(D + 1)y = z$ reduces the D.E. to a first-order linear equation in the dependent variable z; namely, $(D - 1)z = 1$, or $z' - z = 1$. The integrating factor e^{-x} leads to the equation $(d/dx)(e^{-x}z) = e^{-x}$, and one integration gives $e^{-x}z = -e^{-x} + C_1$, or $z = -1 + C_1 e^{x}$. As $(D + 1)y = z$, we now solve the first-order equation $(D + 1)y = -1 + C_1 e^{x}$, or $y' + y = -1 + C_1 e^{x}$. The integrating factor e^{x} leads to the equation $(d/dx)(e^{x}y) = -e^{x} + C_2 e^{2x}$, and integration with respect to

x gives $e^x y = -e^x + (C_1/2)e^{2x} + C_2$, or $y(x) = -1 + (C_1/2)e^x + C_2e^{-x}$. Setting $c_1 = C_1/2$ and $c_2 = C_2$, we get the general solution of the given D.E.:

$$y(x) = -1 + c_1 e^x + c_2 e^{-x}.$$

31. (a) Let y be any twice differentiable function of x. Then

$$D(D + 1/x)y = D\big((D + 1/x)y\big) = D\big(Dy + (1/x)y\big) = D\big(y' + (1/x)y\big)$$
$$= Dy' + D\big((1/x)y\big) = y'' + (1/x)y' + (-1/x^2)y.$$

In particular, if $y = y(x)$ satisfies the given differential equation then $D(D + 1/x)y = 0$.

(b) Let $(D + 1/x)y = z$, so that $D(D + 1/x)y = Dz = 0$. Solve, $Dz = 0$, or equivalently, solve $z' = 0$. By integration, $z = C_1$, where C_1 is a constant. Now solve $(D + 1/x)y = z = C_1$, or equivalently, solve $y' + (1/x)y = C_1$. Use the integrating factor x to get $(d/dx)(xy) = C_1 x$, and integrate to get $xy = (C_1/2)x^2 + C_2$, or $y(x) = (C_1/2)x + C_2/x$, where C_2 is a constant. Replacing $C_1/2$ and C_2 with c_1 and c_2, respectively, gives the solution
$$y(x) = c_1 x + c_2/x.$$

(c) In order that $D(D + 1/x) = (D + 1/x)D$, it must be true that if y is a twice differentiable function of x then $D(D + 1/x)y = (D + 1/x)Dy$. In particular, it must be true for any nonzero constant function $y(x) = c$. However, $(D + 1/x)Dc = (D + 1/x)0 = 0$, while

$$D(D + 1/x)c = D\big((D + 1/x)c\big) = D\big(Dc + (1/x)c\big) = D(c/x) = -c/x^2 \neq 0,$$

Hence, the two operators are not the same.

(d) Let $Dy = z$ and solve $(D + 1/x)Dy = (D + 1/x)z = 0$, or equivalently, solve $z' + (1/x)z = 0$. The integrating factor x leads to $z = c_1/x$, where c_1 is a constant. Now solve $Dy = z = c_1/x$, or equivalently, solve $y' = c_1/x$. By integration, $y(x) = c_1 \ln |x| + c_2$, where c_2 is a constant.

33. (a) The roots of the characteristic equation $Ar^2 + Br + C = 0$ are

$$r_{1,2} = \frac{1}{2A}\big(-B \pm \sqrt{B^2 - 4AC}\big).$$

These roots are real if, and only if, the expression under the radical is nonnegative; that is, if, and only if, $B^2 \geq 4AC$.

(b) If $B^2 > 4AC$ then the roots from part (a) are real and unequal. The general solution of $Ay'' + By' + Cy = 0$ is therefore

$$y(x) = c_1 e^{\frac{-B + \sqrt{B^2 - 4AC}}{2A}x} + c_2 e^{\frac{-B - \sqrt{B^2 - 4AC}}{2A}x}$$
$$= e^{-\frac{B}{2A}x}\left(c_1 e^{\frac{\sqrt{B^2 - 4AC}}{2A}x} + c_2 e^{-\frac{\sqrt{B^2 - 4AC}}{2A}x}\right) = e^{\alpha x}\big(c_1 e^{\beta x} + c_2 e^{-\beta x}\big),$$

where $\alpha = -B/(2A)$ and $\beta = (\sqrt{B^2 - 4AC})/(2A)$. The result now follows form part (a) of Exercise 32 in this section.

35. The characteristic equation is $r^2 - 2r + 1 = (r - 1)^2 = 0$, with the double root $r = 1$. The general solution is therefore $y(x) = c_1 e^x + c_2 x e^x$. Since $1 = y(0) = c_1$, a solution passes through the point $(0,1)$ if, and only if, it has the form $y(x) = e^x + c_2 x e^x$, for some constant c_2. The derivatives of the solutions in this subfamily are given by

$$y'(x) = e^x + c_2 x e^x + c_2 e^x.$$

Using the additional conditions $y'(0) = -1$, $y'(0) = 0$ and $y'(0) = 1$ gives the three equations $-1 = 1 + c_2$, $0 = 1 + c$ and $1 = 1 + c_2$, respectively. Hence, $c_2 = -2$, $c_2 = -1$, and $c_2 = 0$, respectively. So, the solutions passing through $(0,1)$ with these three slopes are

$$y(x) = e^x - 2x e^x, \qquad y(x) = e^x - x e^x, \qquad y(x) = e^x.$$

Their graphs are shown below.

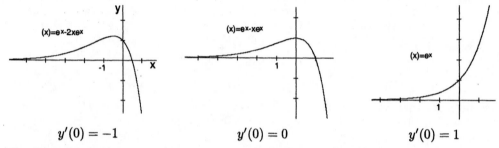

$$y'(0) = -1 \qquad\qquad y'(0) = 0 \qquad\qquad y'(0) = 1$$

37. (a) Let $m = l\delta$ be the mass of the chain and let $a = d^2 y/dt^t$ be the acceleration of the chain. By Newton"s Law, the force F acting on the chain is given by

$$F = ma = (l\delta)\frac{d^2 y}{dt^2}.$$

F is also the acceleration of gravity (g) times the mass of that portion of the chain hanging over the side $(y\delta)$. That is, $F = gy\delta$. So, $l\delta(d^2 y/dt^2) = gy\delta$, or

$$\frac{d^2 y}{dt^2} = (g/l)y, \quad 0 \le y \le l.$$

(b) Letting $t = t_1$ be the time at which the last link goes over the side, $y(t_1) = l$. By part (a),

$$y''(t_1) = \frac{d^2}{dt^2}(t_1) = (g/l)y(t_1) = g.$$

(c) Write the D.E for $y(t)$ as $y'' - (g/l)y = 0$ and, with $k = \sqrt{g/l}$, use the result of Exercise 32 in this section to get

$$y(t) = d_1 \cosh\sqrt{g/l}\,t + d_2 \sinh\sqrt{g/l}\,t.$$

Since $y' = d_1\sqrt{g/l}\sinh\sqrt{g/l}\,t + d_2\sqrt{g/l}\cosh\sqrt{g/l}\,t$, the initial-conditions $y_0 = y(0)$ and $v_0 = y'(0)$ lead to the equations $y_0 = d_1$ and $v_0 = d_2\sqrt{g/l}$, from which $d_1 = y_0$ and $d_2 = v_0\sqrt{l/g}$. Hence,

$$y(t) = y_0 \cosh\sqrt{g/l}\,t + v_0\sqrt{l/g}\sinh\sqrt{g/l}\,t, \quad 0 \le y \le l.$$

(d) The chain starting from rest implies $y'(0) = v_0 = 0$, and the definition of t_1 requires $y(t_1) = l$. From part (c), $l = y(t_1) = y_0 \cosh \sqrt{g/l}\, t_1$, or

$$\cosh \sqrt{g/l}\, t_1 = l/y_0.$$

Since $l/y_0 \geq 1$, we can use the inverse hyperbolic cosine function $\cosh^{-1} x = \ln(x + \sqrt{x^2 - 1})$, $x \geq 1$, to solve for t_1. This gives

$$t_1 = \sqrt{l/g} \cosh^{-1}(l/y_0) = \sqrt{l/g} \ln\big((l/y_0) + \sqrt{(l/y_0)^2 - 1}\big) = \sqrt{l/g} \ln\left(\frac{l + \sqrt{l^2 - y_0^2}}{y_0}\right).$$

39. Let r and h be real constants.

(a) When the linear operator $(D - r)(D - (r + h)) = D^2 - (2r + h)D + r(r + h)$ is applied to a twice differentiable function y, the result is

$$\big(D^2 - (2r + h)D + r(r + h)\big)y = D^2 y - (2r + h)Dy + r(r + h)y = y'' - (2r + h)y' + r(r + h).$$

In particular, if y is a solution then the above expressions are all equal to zero.

(b) The characteristic equation of the differential equation $y'' - (2r + h)y' + r(r + h)y = 0$ is

$$\lambda^2 - (2r + h)\lambda + r(r + h) = (\lambda - r)(\lambda - (r + h)) = 0,$$

with roots $\lambda_1 = r$ and $\lambda_2 = r + h$. If $h \neq 0$, the roots are real and unequal, so that the general solution is

$$y_h(x) = c_1 e^{(r+h)x} + c_2 e^{rx}.$$

(c) Set $c_1 = 1/h$, $c_2 = -1/h$ in the solution formula given in part (b) to get

$$y_h(x) = \frac{1}{h}e^{(r+h)x} - \frac{1}{h}e^{rx} = \frac{e^{(r+h)x} - e^{rx}}{h}.$$

Hence, for each x, one application of l'Hôpital's rule gives

$$\lim_{h \to 0} y_h(x) = \lim_{h \to 0} \frac{e^{(r+h)x} - e^{rx}}{h} = \lim_{h \to 0} \frac{x e^{(r+h)x} - 0}{1} = x e^{rx}.$$

(d) With x fixed, let $f(r) = e^{rx}$. The formal definition of $f'(r)$ is

$$\frac{df}{dr} = f'(r) = \lim_{h \to 0} \frac{f(r+h) - f(r)}{h} = \lim_{h \to 0} \frac{e^{(r+h)x} - e^{rx}}{h}.$$

This is the same limit evaluated in part (c).

41. (a) Let r and s be constants and let I be an open x-interval. Suppose there exists a constant k such that $e^{rx} = ke^{sx}$ for $x \in I$. Then $e^{(r-s)x} = k$ for all $x \in I$. Since the exponential function is either strictly increasing on I (if $r > s$)) or strictly decreasing on I (if $r < s$), it follows that $r = s$. We conclude that if $r \neq s$, then no such constant k can exist. That is, e^{rx} is not a constant multiple of e^{sx}. The same argument with the roles of r and s reversed produces the same result. Thus, e^{rx} and e^{sx} are linearly independent on I whenever $r \neq s$. As I was an arbitrary open interval, the conclusion holds on every open interval.

(b) With r and I as in part (a), suppose there exists a constant k such that either of the following equations hold on I:

$$e^{rx} = kxe^{rx} \qquad \text{or} \qquad ke^{rx} = xe^{rx}.$$

Then e^{rx} can be canceled in both equations to get either $1 = kx$ or $k = x$. The first equation says that $x = 1/k$ is constant on I and the second equation says $x = k$ is constant on I. As I is an open interval, there is more than one value of x in I, so that neither of these possibilities can occur. That is, neither of the functions e^{rx} nor xe^{rx} is a constant multiple of the other on I and they are therefore linearly independent on I. Again, since I was an arbitrary open interval, the conclusion holds on every open interval.

Section 2: COMPLEX SOLUTIONS

Exercise Set 2A (pgs. 506-507)

1. $|i| = |0 + i \cdot 1| = \sqrt{0^2 + 1^2} = 1$; $x = \pi/2$

3. $\left| \frac{1}{\sqrt{2}} + i\frac{-1}{\sqrt{2}} \right| = \sqrt{(1/\sqrt{2})^2 + (-1/\sqrt{2})^2} = 1$; $x = -\pi/4$

NOTE: The following computations are used in answering Exercises 5 and 7.

$$
\begin{aligned}
c_1 e^{(\alpha+i\beta)x} + c_2 e^{(\alpha-i\beta)x} &= e^{\alpha x}\left(c_1 e^{i\beta x} + c_2 e^{-i\beta x}\right) \\
&= e^{\alpha x}\left(c_1(\cos\beta x + i\sin\beta x) + c_2(\cos\beta x - i\sin\beta x)\right) \\
&= e^{\alpha x}\left((c_1 + c_2)\cos\beta x + i(c_1 - c_2)\sin\beta x\right).
\end{aligned}
$$

Set $d_1 = c_1 + c_2$, $d_2 = i(c_1 - c_2)$ and solve for c_1, c_2 to get $c_1 = \frac{1}{2}(d_1 - id_2)$ and $c_2 = \frac{1}{2}(d_1 + id_2)$. c_1 and c_2 can now be found once d_1 and d_2 are given.

5. $c_1 = \frac{1}{2}(1 - i \cdot 0) = 1/2$, $c_2 = \frac{1}{2}(1 + i \cdot 0) = 1/2$.

7. $c_1 = \frac{1}{2}(0 - i\pi) = -i\pi/2$, $c_2 = \frac{1}{2}(0 + i\pi) = i\pi/2$.

9. (a) If k is an integer then $e^{i(x+2k\pi)} = \cos(x + 2k\pi) + i\sin(x + 2k\pi) = \cos x + i\sin x = e^{ix}$ for all x.

(b) p is a period of $e^{i\beta x}$ only if $1 = e^{i\beta \cdot 0} = e^{i\beta(0+p)} = e^{i\beta p} = \cos\beta p + i\sin\beta p$. Equating real and imaginary parts gives $1 = \cos\beta p$ and $0 = \sin\beta p$. These are simultaneously true if, and only if, $\beta p = 2\pi n$ for some integer n, or equivalently, $p = 2\pi n/\beta$. Since p is desired to be the smallest positive period, choose $n = 1$ so that $p = 2\pi/\beta$. That $2\pi/\beta$ is actually a period follows from $e^{i\beta(x+2\pi/\beta)} = e^{i\beta x} \cdot e^{i2\pi} = e^{i\beta x}(\cos 2\pi + i\sin 2\pi) = e^{i\beta x}$ for all x.

11. In operator form, the given differential equation is $(D^2 + 1)y = 1$, or factoring the operator, $(D - i)(D + i)y = 1$. Let $(D + i)y = z$ and solve $(D - i)z = 1$, or equivalently, solve $z' - iz = 1$. An integrating factor is e^{-ix}, so that

$$
\frac{d}{dx}\left(e^{-ix}z\right) = e^{-ix}.
$$

Integrate both sides to get $e^{-ix}z = (-1/i)e^{-ix} + C_1 = ie^{-ix} + C_1$, where C_1 is a real or complex constant. Hence, $z = i + C_1 e^{ix}$. Now solve $(D + i)y = z = i + C_1 e^{ix}$, or equivalently, solve $y' + iy = i + C_1 e^{ix}$. An integrating factor is e^{ix}, so that

$$
\frac{d}{dx}\left(e^{ix}y\right) = ie^{ix} + C_1 e^{2ix}.
$$

Integrate both sides to get $e^{ix}y = e^{ix} + (1/(2i))C_1 e^{2ix} + C_2$, where C_2 is a real or complex constant. Hence,

$$
y(x) = 1 + \frac{1}{2i}C_1 e^{ix} + C_2 e^{-ix} = 1 + c_1 e^{ix} + c_2 e^{-ix},
$$

where $c_1 = 1/(2i)$ and $c_2 = C_2$ are real or complex constants. Therefore, the real solutions are given by
$$
y(x) = 1 + d_1 \cos x + d_2 \sin x,
$$
where d_1 and d_2 are real constants.

13. In operator form, the given differential equation is $(D^2 + 2)y = 0$, or factoring the operator, $(D - i\sqrt{2})(D + i\sqrt{2})y = 0$. Let $(D + i\sqrt{2})y = z$ and solve $(D - i\sqrt{2})z = 0$, or equivalently, solve $z' - i\sqrt{2}\,z = 0$. An integrating factor is $e^{-i\sqrt{2}\,x}$, so that

$$\frac{d}{dx}\left(e^{-i\sqrt{2}\,x}z\right) = 0.$$

Integrate both sides to get $e^{-i\sqrt{2}\,x}z = C_1$, or $z = C_1 e^{i\sqrt{2}\,x}$, where C_1 is a real or complex constant. Now solve $(D + i\sqrt{2})y = z = C_1 e^{i\sqrt{s}\,x}$, or equivalently, solve $y' + i\sqrt{2}\,y = C_1 e^{i\sqrt{2}\,x}$. An integrating factor is $e^{i\sqrt{2}\,x}$, so that

$$\frac{d}{dx}\left(e^{i\sqrt{2}\,x}y\right) = C_1 e^{2i\sqrt{2}\,x}.$$

Integrate both sides to get $e^{i\sqrt{2}\,x}y = (1/(2i\sqrt{2}))C_1 e^{2i\sqrt{2}\,x} + C_2$, where C_2 is a real or complex constant. Hence,

$$y(x) = \frac{1}{2i\sqrt{2}}C_1 e^{i\sqrt{2}\,x} + C_2 e^{-i\sqrt{2}\,x} = c_1 e^{i\sqrt{2}\,x} + c_2 e^{-i\sqrt{2}\,x},$$

where $c_1 = 1/(2i\sqrt{2})C_1$ and $c_2 = C_2$ are real or complex constants. Therefore, the real solutions are given by

$$y(x) = d_1 \cos \sqrt{2}\,x + d_2 \sin \sqrt{2}\,x,$$

where d_1 and d_2 are real constants.

15. Char. eq.: $r^2 + 2 = 0$; roots: $r_1 = i\sqrt{2}$, $r_2 = -i\sqrt{2}$; general solution:

$$y(x) = c_1 \cos \sqrt{2}\,x + c_2 \sin \sqrt{2}\,x.$$

Using this along with its derivative $y'(x) = -\sqrt{2}\,c_1 \sin \sqrt{2}\,x + \sqrt{2}\,c_2 \cos \sqrt{2}\,x$, the initial conditions $y(0) = 0$, $y'(0) = 1$ lead to the equations $0 = c_1$ and $1 = \sqrt{2}\,c_2$, with solution $c_1 = 0$, $c_2 = 1/\sqrt{2}$. The particular solution is therefore

$$y(x) = \frac{1}{\sqrt{2}} \sin \sqrt{2}\,x.$$

17. Char. eq.: $r^2 - 2r + 2 = 0$; roots: $r_{1,2} = 1 \pm i$; general solution:

$$y(x) = e^x(c_1 \cos x + c_2 \sin x).$$

Since the zero function is a solution ($c_1 = c_2 = 0$), and since the zero function satisfies the initial conditions $y(\pi) = 0$, $y'(\pi) = 0$, it follows that the particular solution we seek is $y(x) \equiv 0$.

19. Char. eq.: $2r^2 + r - 1 = 0$; roots: $r_1 = 1/2$, $r_2 = -1$; general solution:

$$y(x) = c_1 e^{x/2} + c_2 e^{-x}.$$

Using this along with its derivative $y'(x) = (1/2)c_1 e^{x/2} - c_2 e^{-x}$, the initial conditions $y(0) = 0$, $y'(0) = 2$ lead to the equations $0 = c_1 + c_2$ and $2 = (1/2)c_1 - c_2$, with solution $c_1 = 4/3$, $c_2 = -4/3$. The particular solution is therefore

$$y(x) = \frac{4}{3}e^{x/2} - \frac{4}{3}e^{-x}.$$

21. Char. eq.: $2r^2 + r + 1 = 0$; roots: $r_{1,2} = -1/4 \pm i\sqrt{7}/4$; general solution

$$y(x) = e^{-x/4}\left(c_1 \cos \frac{\sqrt{7}}{4}x + c_2 \sin \frac{\sqrt{7}}{4}x\right).$$

Since the zero function is a solution ($c_1 = c_2 = 0$), and since the zero function satisfies the initial conditions $y(0) = 0$, $y'(0) = 0$, it follows that the particular solution we seek is $y(x) \equiv 0$.

23. The presence of a trigonometric function indicates that the roots of the characteristic equation are complex conjugates $r_{1,2} = \alpha + i\beta$, where $\alpha = 0$ because the exponential that accompanies such solutions has exponent zero, and $\beta = 2$ is the coefficient of x in the expression $\sin 2x$. So, the characteristic equation is

$$(r - r_1)(r - r_2) = (r - 2i)(r + 2i) = r^2 + 4 = 0.$$

The associated second-order differential equation is therefore

$$y'' + 4y = 0.$$

25. The presence of a trigonometric function indicates that the roots of the characteristic equation are complex conjugates $r_{1,2} = \alpha + i\beta$, where $\alpha = 1$ is the coefficient of x on the exponential and $\beta = 2$ is the coefficient of x in the expression $\cos 2x$. So, the characteristic equation is

$$(r - r_1)(r - r_2) = (r - 1 - 2i)(r - 1 + 2i) = r^2 - 2r + 5 = 0.$$

The associated second-order differential equation is therefore

$$y'' - 2y' + 5y = 0.$$

27. The presence of a trigonometric function indicates that the roots of the characteristic equation are complex conjugates $r_{1,2} = \alpha + i\beta$, where $\alpha = 0$ because the exponential that accompanies such solutions has exponent zero, and $\beta = 1/2$ is the coefficient of x in the expression $\cos(x/2)$. So, the characteristic equation is

$$(r - r_1)(r - r_2) = (r - i/2)(r + i/2) = r^2 + 1/4 = 0, \qquad \text{or} \qquad 4r^2 + 1 = 0.$$

The associated second-order differential equation is therefore

$$4y'' + y = 0.$$

29. The presence of a trigonometric function indicates that the roots of the characteristic equation are complex conjugates $r_{1,2} = \alpha + i\beta$, where $\alpha = 0$ because the exponential that accompanies such solutions has exponent zero, and $\beta = 3$ is the coefficient of x in the expression $\sin 3x - \cos 3x$. So, the characteristic equation is

$$(r - r_1)(r - r_2) = (r - 3i)(r + 3i) = r^2 + 9 = 0.$$

The associated second-order differential equation is therefore

$$y'' + 9y = 0.$$

31. Consider the boundary-value problem $y'' + y = 0$, $y(0) = 1$, $y(\pi) = 1$. The characteristic equation is $r^2 + 1 = 0$ with roots $r_{1,2} = \pm i$. The general solution is therefore

$$y(x) = c_1 \cos c + c_2 \sin x.$$

The boundary conditions lead to the equations $1 = c_1$ and $1 = -c_1$, which are inconsistent. So, the given boundary-value problem has no solution.

33. Let r_1 and r_2 be the roots of the characteristic equation associated with the constant-coefficient equation $y'' + ay' + by = 0$. Consider the problem of determining solutions that satisfy the boundary conditions $y(x_1) = y_1$, $y(x_2) = y_2$, where $x_1 \neq x_2$. There are two cases to check: (i) $r_1 \neq r_2$; (ii) $r_1 = r_2 = r$.

If $r_1 \neq r_2$, the general solution has the form $y(x) = c_1 e^{r_1 x} + c_2 e^{r_2 x}$. The boundary conditions lead to the equations $y_1 = c_1 e^{r_1 x_1} + c_2 e^{r_2 x_1}$ and $y_2 = c_1 e^{r_1 x_2} + c_2 e^{r_2 x_2}$. Multiply the first equation by $e^{-r_2 x_1}$, the second equation by $e^{-r_2 x_2}$, solve each equation for c_2 and obtain

$$c_2 = y_1 e^{-r_2 x_1} - c_1 e^{(r_1 - r_2)x_1} = y_2 e^{-r_2 x_2} - c_1 e^{(r_1 - r_2)x_2}.$$

The two right members can be written as

$$y_1 e^{-r_2 x_1} - y_2 e^{-r_2 x_2} = c_1 \left(e^{(r_1 - r_2)x_1} - e^{(r_1 - r_2)x_2} \right).$$

Since, $r_1 \neq r_2$ and $x_1 \neq x_2$, this equation can be solved uniquely for c_1. In view of the above expressions for c_2, there corresponds a unique value of c_2.

If $r_1 = r_2 = r$, the general solution has the form $y(x) = c_1 e^{rx} + c_2 x e^{rx} = e^{rx}(c_1 + c_2 x)$. The boundary conditions give $y_1 = e^{r x_1}(c_1 + c_2 x_1)$ and $y_2 = e^{r x_2}(c_1 + c_2 x_2)$. Multiply the first equation by $e^{-r x_1}$, the second equation by $e^{-r x_2}$, solve each equation for c_1, and obtain

$$c_1 = y_1 e^{-r x_1} - c_2 x_1 = y_2 e^{-r x_2} - c_2 x_2.$$

The two right members can be written as $y_1 e^{-r x_1} - y_2 e^{-r x_2} = c_2(x_1 - x_2)$. Since $x_1 \neq x_2$, this equation can be solved uniquely for c_2. In view of the above expressions for c_1, there corresponds a unique value for c_1.

35. (a) From trigonometry, if (c_2, c_1) is any nonzero point in the xy-plane then there exists a unique angle $0 \leq \theta < 2\pi$ such that $c_2 = \sqrt{c_1^2 + c_2^2} \cos \theta$ and $c_1 = \sqrt{c_1^2 + c_2^2} \sin \theta$ (there is no such angle if (c_2, c_1) is the origin). Let $A = \sqrt{c_1^2 + c_2^2}$ and note that $A \neq 0$ (since (c_2, c_1) is a nonzero point). Hence, $\cos \theta = c_2/A$ and $\sin \theta = c_1/A$. Now use the identity $\sin(a + b) = \sin a \cos b + \cos a \sin b$ to get

$$A \sin(\beta x + \theta) = A(\sin \beta x \cos \theta + \cos \beta x \sin \theta) = A \left(\frac{c_1}{A} \cos \beta x + \frac{c_2}{A} \sin \beta x \right)$$

$$= c_1 \cos \beta x + c_2 \sin \beta x.$$

(b) Let $y(x) = \cos 2x + \sqrt{3} \sin 2x$. With $c_1 = 1$, $c_2 = \sqrt{3}$, let $A = \sqrt{c_1^2 + c_2^2} = 2$. From the equations $\cos \theta = c_2/A = \sqrt{3}/2$ and $\sin \theta = c_1/A = 1/2$, we conclude that $\theta = \pi/6$. With $\beta = 2$, we have $y(x) = 2 \sin(2x + \pi/6)$.

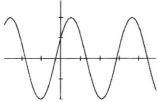

(x)=cos 2x+(3)$^{1/2}$sin 2x = 2sin(2x+π/6)

(c) Use the cofunction identity $\sin(\pi/2-a) = \cos a$, and the fact that the cosine function is an even function (i.e., $\cos a = \cos(-a)$) to get

$$A\sin(\beta x + \theta) = A\sin\big(\pi/2 - (\pi/2 - \beta x - \theta)\big) = A\cos(\pi/2 - \beta x - \theta)$$
$$= A\cos\big(\beta x - (\pi/2 - \theta)\big) = A\cos(\beta x - \phi),$$

where $\phi = \pi/2 - \theta$.

37. Since $e^{(\alpha+i\beta)x} = e^{\alpha x} \cdot e^{i\beta x} = e^{\alpha x}(\cos\beta x + i\sin\beta x) = e^{\alpha x}\cos\beta x + ie^{\alpha x}\sin\beta x$, use the linearity of the integral to write

$$\int e^{(\alpha+i\beta)x}\,dx = \int e^{\alpha x}(\cos\beta x + i\sin\beta x)\,dx = \int e^{\alpha x}\cos\beta x\,dx + i\int e^{\alpha x}\sin\beta x\,dx.$$

The integrals above have real integrands and can therefore be evaluated using standard integral tables. When this is done, the result can be rearranged to get the expression

$$e^{\alpha x}\left(\frac{\alpha - i\beta}{\alpha^2 + \beta^2}\cos\beta x + \frac{\beta + i\alpha}{\alpha^2 + \beta^2}\sin\beta x\right).$$

Since $-1 = i^2$, $\beta + i\alpha = -i^2\beta + i\alpha = i(\alpha - i\beta)$. Use this and factor out $(\alpha - i\beta)/(\alpha^2 + \beta^2)$ from each term in the parentheses to get

$$\frac{\alpha - i\beta}{\alpha^2 + \beta^2}e^{\alpha x}(\cos\beta x + i\sin\beta x) = \frac{\alpha - i\beta}{\alpha^2 + \beta^2}e^{(\alpha+i\beta)x}.$$

Finally, note that

$$\frac{1}{\alpha + i\beta} = \frac{1}{\alpha + i\beta} \cdot \frac{\alpha - i\beta}{\alpha - i\beta} = \frac{\alpha - i\beta}{\alpha^2 + \beta^2}.$$

Putting all of this together gives

$$\int e^{(\alpha+i\beta)x}\,dx = \frac{1}{\alpha + i\beta}e^{(\alpha+i\beta)x},$$

which is Equation 2.2 in the text.

39. Write the given differential equation in its operator form $(D^2 + i)y = 0$. It is easy to verify that the squares of the complex numbers $\pm(1 + i)/\sqrt{2}$ are both equal to i, so that the operator form can be written as $\big(D - (1 + i)/\sqrt{2}\big)\big(D + (1 + i)/\sqrt{2}\big)y = 0$. Therefore, let $\big(D + (1 + i)/\sqrt{2}\big)y = z$ and solve $\big(D - (1 + i)/\sqrt{2}\big)z = 0$, or equivalently, solve

$$z' - \frac{1+i}{\sqrt{2}}z = 0.$$

Multiplying by the integrating factor $e^{-(1+i)x/\sqrt{2}}$ leads to the solution

$$z = C_1 e^{(1+i)x/\sqrt{2}},$$

where C_1 is a real or complex constant. Now solve $\big(D + (1 + i)/\sqrt{2}\big)y = z = C_1 e^{(1+i)x/\sqrt{2}}$, or equivalently, solve

$$y' + \frac{1+i}{\sqrt{2}}y = C_1 e^{(1+i)x/\sqrt{2}}.$$

392

Multiplying by the integrating factor $e^{(1+i)x/\sqrt{2}}$ leads to the equation

$$\frac{d}{dx}\left(e^{(1+i)x/\sqrt{2}}y\right) = C_1 e^{2(1+i)x/\sqrt{2}}.$$

Using Equation 2.2 in the text, integrate both sides to get

$$e^{(1+i)x/\sqrt{2}}y = \frac{\sqrt{2}}{2(1+i)}C_1 e^{2(1+i)x/\sqrt{2}} + C_2,$$

and multiply throughout by $e^{-(1+i)x/\sqrt{2}}$ to obtain

$$y(x) = \frac{\sqrt{2}}{2(1+i)}C_1 e^{(1+i)x/\sqrt{2}} + C_2 e^{-(1+i)x/\sqrt{2}} = c_1 e^{(1+i)x/\sqrt{2}} + c_2 e^{-(1+i)x/\sqrt{2}},$$

where $c_1 = C_1\sqrt{2}/(2(1+i))$ and $c_2 = C_2$ are real or complex constants.
(NOTE: It can be shown that the method used to derive the above solution formula gives *all* solutions, both real and complex.)

41. $(f+g)' = (u+iv+s+it)' = \left((u+s)+i(v+t)\right)' = (u+s)' + i(v+t)'$
$\qquad = u' + s' + i(v'+t') = u' + iv' + s' + it' = (u+iv)' + (s+it)'$
$\qquad = f' + g'.$

43. $(fg)' = \left((u+iv)(s+it)\right)' = \left((us-vt)+i(ut+vs)\right)' = (us-vt)' + i(ut+vs)'$
$\qquad = (us)' - (vt)' + i\left((ut)' + (vs)'\right) = us' + u's - vt' - v't + i(ut' + u't + vs' + v's)$
$\qquad = (us' + iut') + (ivs' - vt') + (u's + iu't) + (iv's - v't)$
$\qquad = u(s' + it') + iv(s' + it') + u'(s+it) + iv'(s+it)$
$\qquad = (u+iv)(s'+it') + (u'+iv')(s+it) = (u+iv)(s+it)' + (u+iv)'(s+it)$
$\qquad = fg' + f'g.$

45. (a) Suppose $y = y(t)$ is a solution of $y'' + ay' + by = 0$, so that $y''(t) + ay'(t) + by(t) = 0$ for each real number t. Specifically, suppose

$(*)$ $\qquad\qquad \dfrac{d^2 y}{dt^2}(t) + a\dfrac{dy}{dt}(t) + by(t) = 0 \quad$ for all t.

Let c be a fixed constant and define the new variable x by the equation $x = t - c$. Then, $y_c(x) = y(x+c) = y(t)$ defines a function y_c of x for each real number x. Moreover, the chain rule implies that

$$\frac{dy_c}{dx}(x) = \frac{dy}{dt}(t) \quad\text{and}\quad \frac{d^2 y_c}{dx^2}(x) = \frac{d^2 y}{dt^2}(t) \quad\text{both hold for all } x.$$

Thus, equation $(*)$ can be written as

$$\frac{d^2 y_c}{dx^2}(x) + a\frac{dy_c}{dx}(x) + by_c(x) = 0 \quad\text{for all } x.$$

That is, with respect to the variable x, y_c is a solution of $y'' + ay' + by = 0$.

(b) If $y = y(t)$ is a solution of the given n-th order constant-coefficient equation then, specifically,

$(**)$ $\qquad \dfrac{d^n y}{dt^n}(t) + a_{n-1}\dfrac{d^{n-1}y}{dt^{n-1}}(t) + \cdots + a_1\dfrac{dy}{dt}(t) + a_0 y(t) = 0 \quad$ for all t.

Let c be a fixed constant and define the new variable x by the equation $x = t - c$. Then $y_c(x) = y(x + c) = y(t)$ defines a function y_c of x for each real number x. As in part (a), the chain rule implies that

$$\frac{d^k y_c}{dx^k}(x) = \frac{d^k y}{dt^k}(t) \quad \text{for all } x \text{ and } k = 0, \ldots, n.$$

Thus, equation (**) can be written as

$$\frac{d^n y_c}{dx^n}(x) + a_{n-1}\frac{d^{n-1} y_c}{dx^{n-1}}(x) + \cdots + a_1 \frac{dy_c}{dx}(x) + a_0 y_c(x) = 0 \quad \text{for all } x.$$

That is, with respect to the variable x, y_c is a solution of $y^{(n)} + a_{n-1}y^{(n-1)} + \cdots + a_0 y = 0$.

(c) Suppose $y = y(t)$ is a solution of $y'' = F(y, y')$ on the interval $a < t < b$, so that $y''(t) = F\big(y(t), y'(t)\big)$ for each t satisfying $a < t < b$. Specifically, suppose

$$(***) \qquad\qquad \frac{d^2 y}{dt^2}(t) = F\left(y(t), \frac{dy}{dt}(t)\right) \quad \text{for } a < t < b.$$

Let c be a fixed constant and define the new variable x by the equation $x = t - c$. Then $a < t < b$ implies $a < x + c < b$, which is equivalent to $a - c < x < b - c$. Thus, $y_c(x) = y(x + c) = y(t)$ defines a function y_c of x for $a - c < x < b - c$. As in parts (a) and (b), the chain rule implies that

$$\frac{d^2 y_c}{dx^2}(x) = \frac{d^2 y}{dt^2}(t) \quad \text{and} \quad \frac{dy_c}{dx}(x) = \frac{dy}{dt}(x) \quad \text{both hold for } a - c < x < b - c.$$

Thus, equation (***) can be written as

$$\frac{d^2 y_c}{dx^2}(x) = F\left(y_c(x), \frac{dy_c}{dt}(t)\right) \quad \text{for } a - c < x < b - c.$$

That is, with respect to the variable x, y_c is a solution of $y'' = F(y, y')$ for $a - c < x < b - c$.

47. Let α and β be real numbers. For $\beta \neq 0$, assume $e^{\alpha x} \cos \beta x$ and $e^{\alpha x} \sin \beta x$ are *not* linearly independent on some open interval I and that c is a real constant such that $e^{\alpha x} \sin \beta x = c e^{\alpha x} \cos \beta x$ for all x in I. Since $e^{\alpha x}$ is never zero, this is equivalent to

$$\sin \beta x = c \cos \beta x \quad \text{for } x \in I.$$

Assuming $\beta \neq 0$, $\cos \beta x = 0$ only for x of the form $x = (2n + 1)\pi/(2\beta)$, where n is an integer. It follows that there is an open subinterval I_0 of I on which $\cos \beta x \neq 0$. Thus, for x in I_0, we can divide both sides of the above equation by $\cos \beta x$ to get

$$\tan \beta x = c \quad \text{for all } x \in I_0.$$

But this is impossible because the left member is an increasing function of x while the right member is constant. A similar contradiction is obtained if we assume $e^{\alpha x} \cos \beta x$ is a constant multiple of $e^{\alpha x} \sin \beta x$. It follows that $e^{\alpha x} \cos \beta x$ and $e^{\alpha x} \sin \beta x$ are linearly independent on every open interval I.

If $\beta = 0$ then $e^{\alpha x} \sin \beta x$ is the zero function, which is a constant multiple (the zero multiple) of $e^{\alpha x} \cos \beta x$. In this case, $e^{\alpha x} \sin \beta x$ and $e^{\alpha x} \cos \beta x$ are not linearly independent on any open interval I.

Exercise Set 2BC (pgs. 512-513)

1. Char. eq.: $r^3 + 1 = (r+1)(r^2 - r + 1) = 0$; roots: $r_1 = -1$, $r_{2,3} = (1 \pm \sqrt{3})/2$; general solution:

$$y(x) = c_1 e^{-x} + c_2 e^{x/2} \cos \frac{\sqrt{3}}{2} x + c_3 e^{x/2} \sin \frac{\sqrt{3}}{2} x.$$

3. Char. eq.: $r^3 - 2r = r(r - \sqrt{2})(r + \sqrt{2}) = 0$; roots: $r_1 = 0$, $r_{2,3} = \pm\sqrt{2}$; general solution:

$$y(x) = c_1 + c_2 e^{\sqrt{2}x} + c_3 e^{-\sqrt{2}x}.$$

5. Char. eq.: $r^3 - 16r = r(r-4)(r+4) = 0$; roots: $r_1 = 0$, $r_{2,3} = \pm 4$; general solution:

$$y(x) = c_1 + c_2 e^{4x} + c_3 e^{-4x}.$$

7. Char. eq.: $(r^2 + 4)(r^2 - 1) = (r^2 + 4)(r-1)(r+1) = 0$; roots: $r_{1,2} = \pm 1$, $r_{3,4} = \pm 2i$; general solution:

$$y(x) = c_1 e^x + c_2 e^{-x} + c_3 \cos 2x + c_4 \sin 2x.$$

9. Char. eq.: $r^3 - 2r^2 = r^2(r-2) = 0$; roots: $r_{1,2} = 0$, $r_3 = 2$; general solution:

$$y(x) = c_1 + c_2 x + c_3 e^{2x}.$$

11. Every constant-coefficient linear D.E. having e^{5x} as a solution has a characteristic equation with $r_1 = 5$ as a root. Since $r - 5$ is the polynomial of least degree with this property, the D.E. of minimal order with e^{5x} as a solution is the associated D.E.

$$y' - 5y = 0.$$

13. If a constant-coefficient linear D.E. has x^2 as a solution then every function of the form $y(x) = c_1 + c_2 x + c_3 x^2$ is also a solution. Its characteristic equation then has the three roots $r_1 = 0$ (multiplicity 3). Since, r^3 is the polynomial of least degree with this property, the D.E. of minimal order with x^2 as a solution is the associated D.E.

$$y''' = 0.$$

15. If a constant-coefficient linear D.E. has $x + e^x$ as a solution then every function of the form $y(x) = c_1 + c_2 x + c_3 e^x$ is also a solution. Its characteristic equation then has the three roots $r_1 = 0$ (multiplicity 2) and $r_2 = 1$. Since $r^2(r-1) = r^3 - r^2$ is the polynomial of least degree with this property, the D.E. of minimal order having $x + e^x$ as a solution is the associated D.E.

$$y''' - y'' = 0.$$

17. If a constant-coefficient linear D.E. has $x^5 + x^2 e^x$ as a solution then every function of the form $y(x) = c_1 + c_2 x + c_3 x^2 + c_4 x^3 + c_5 x^4 + c_6 x^5 + c_7 e^x + c_8 x e^x + c_9 x^2 e^x$ is also a solution. Its characteristic equation then has the nine roots $r_1 = 0$ (multiplicity 6) and $r_2 = 1$ (multiplicity 3). Since $r^6(r-1)^3 = r^9 - 3r^8 + 3r^7 - r^6$ is the polynomial of least degree with this property, the D.E. of minimal order having $x^5 + x^2 e^x$ as a solution is the associated D.E.

$$y^{(9)} - 3y^{(8)} + 3y^{(7)} - y^{(6)} = 0.$$

19. If a constant-coefficient linear D.E. has $\cos 4x$ as a solution then every function of the form $y(x) = c_1 \cos 4x + c_2 \sin 4x$ is also a solution. Its characteristic equation then has the two roots $r_1 = 4i$ and $r_2 = -4i$. Since $r^2 + 16$ is the polynomial of least degree with this property, the D.E. of minimal order having $\cos 4x$ as a solution is the associated D.E.

$$y'' + 16y = 0.$$

21. If a constant-coefficient linear D.E. has $x^2 \cos 4x$ as a solution then every function of the form $y(x) = c_1 \cos 4x + c_2 \sin 4x + c_3 x \cos 4x + c_4 x \sin x + c_5 x^2 \cos 4x + c_6 x^2 \sin 4x$ is also a solution. Its characteristic equation then has the six roots $r_1 = 4i$ (multiplicity 3) and $r_2 = -4i$ (multiplicity 3). Since $(r^2 + 16)^3 = r^6 + 48r^4 + 768r^2 + 4096$ is the polynomial of least degree with this property, the D.E. of minimal order having $x^2 \cos 4x$ as a solution is the associated D.E.

$$y(x) = y^{(6)} + 48y^{(4)} + 768y'' + 4096y = 0.$$

A more compact form of this answer is $(D^2 + 16)^3 y = 0$.

23. If a constant-coefficient linear D.E. has $xe^x \sin x$ as a solution then every function of the form $y(x) = c_1 e^x \cos x + c_2 e^x \sin x + c_3 x e^x \cos x + c_4 x e^x \sin x$ is also a solution. Its characteristic equation then has the four roots $r_1 = 1 + i$ (multiplicity 2) and $r_2 = 1 - i$ (multiplicity 2). Since

$$\big(r - (1 + i)\big)^2 \big(r - (1 - i)\big)^2 = (r^2 - 2r + 2)^2 = r^4 - 4r^3 + 8r^2 - 8r + 4$$

is the polynomial of least degree with this property, the D.E. of minimal order having $xe^x \sin x$ as a solution is the associated D.E.

$$y^{(4)} - 4y''' + 8y'' - 8y' + 4y = 0.$$

25. If a constant-coefficient linear D.E. has x^5 as a solution then every function of the form $y(x) = c_1 + c_2 x + c_3 x^2 + c_4 x^3 + c_5 x^4 + c_6 x^5$ is also a solution. Its characteristic equation then has the six roots $r_1 = 0$ (multiplicity 6). Since r^6 is the polynomial of least degree with this property, the D.E. of minimal order having x^5 as a solution is the associated D.E.

$$y^{(6)} = 0.$$

27. If a constant-coefficient linear D.E. has $e^{-x} \cos x$ as a solution then every function of the form $y(x) = c_1 e^{-x} \cos x + c_2 e^{-x} \sin x$ is also a solution. Its characteristic equation then has the two roots $r_1 = -1 + i$ and $r_2 = -1 - i$. Since

$$\big(r - (-1 + i)\big)\big(r - (-1 - i)\big) = r^2 + 2r + 2$$

is the polynomial of least degree with this property, the D.E. of minimal order having $e^{-x} \cos x$ as a solution is the associated D.E.

$$y'' + 2y' + 2y = 0.$$

29. There are three characteristic roots associated with the given family: $r_1 = 1$ and $r_2 = 0$ (multiplicity 2). Its characteristic equation is therefore $r^2(r - 1) = r^3 - r^2 = 0$ and the D.E. of minimal order having the given family as solutions is the associated D.E.

$$y''' - y'' = 0.$$

31. There are three characteristic roots associated with the given family: $r_1 = 1$, $r_2 = i$ and $r_3 = -i$. Its characteristic equation is therefore $(r - 1)(r^2 + 1) = r^3 - r^2 + r - 1 = 0$ and the D.E. of minimal order having the given family as solutions is the associated D.E.

$$y''' - y'' + y' - y = 0.$$

33. There are three characteristic roots associated with the given family: $r_1 = 2i$, $r_2 = -2i$ and $r_3 = -1$. Its characteristic equation is therefore $(r + 1)(r^2 + 4) = r^3 + r^2 + 4r + 4 = 0$ and the D.E. of minimal order having the given family as solutions is the associated D.E.

$$y''' + y'' + 4y' + 4y = 0.$$

35. There are three characteristic roots associated with the given family: $r_1 = 3i$, $r_2 = -3i$ and $r_3 = 0$. Its characteristic equation is therefore $r(r^2 + 9) = r^3 + 9r = 0$ and the D.E. of minimal order having the given family as solutions is the associated D.E.

$$y''' + 9y = 0.$$

37. There are three characteristic roots associated with the given family: $r_1 = 1 + 2i$, $r_2 = 1 - 2i$ and $r_3 = 0$. Its characteristic equation is therefore $r(r - 1 - i)(r - 1 + i) = r^3 - 2r^2 + 5r = 0$ and the D.E. of minimal order having the given family as solutions is the associated D.E.

$$y''' - 2y'' + 5y' = 0.$$

39. With $y(x) = c_1 + c_2 x + c_3 e^x$, we have $y'(x) = c_2 + c_3 e^x$ and $y''(x) = c_3 e^x$. The initial conditions lead to the three equations

$$1 = y(0) = c_1 + c_3, \qquad 2 = y'(0) = c_2 + c_3, \qquad 1 = y''(0) = c_3.$$

By inspection, $c_1 = 0$, $c_2 = c_3 = 1$.

41. With $y(x) = c_1 \cos x + c_2 \sin x + c_3 e^x$, we have $y'(x) = -c_1 \sin x + c_2 \cos x + c_3 e^x$ and $y''(x) = -c_1 \cos x - c_2 \sin x + c_3 e^x$. The initial conditions lead to the three equations

$$2 = y(0) = c_1 + c_3, \qquad -3 = y'(0) = c_2 + c_3, \qquad -3 = y''(0) = -c_1 + c_3.$$

The method of elimination gives $c_1 = 5/2$, $c_2 = -5/2$, $c_3 = -1/2$.

43. Char. eq.: $r^4 - 1 = (r^2 - 1)(r^2 + 1) = 0$; roots: $r_{1,2} = \pm 1$, $r_{3,4} = \pm i$; general solution:

$$y(x) = c_1 e^x + c_2 e^{-x} + c_3 \cos x + c_4 \sin x.$$

45. Char.eq.: $r^4 - 2r = r(r^3 - 2) = 0$; roots: $r_1 = 0$ and, by De Moivre's Theorem from trigonometry, $r_2 = 2^{1/3}$, $r_3 = 2^{1/3} e^{2\pi i/3} = 2^{1/3}\left(\cos(2\pi/3) + i\sin(2\pi/3)\right) = 2^{1/3}\left(-\frac{1}{2} + i\frac{\sqrt{3}}{2}\right)$, $r_4 = 2^{1/3} e^{4\pi i/3} = 2^{1/3}\left(\cos(4\pi/3) + i\sin(4\pi/3)\right) = 2^{1/3}\left(-\frac{1}{2} - i\frac{\sqrt{3}}{2}\right)$; general solution:

$$y(x) = c_1 + c_2 e^{2^{1/3}x} + e^{-2^{-2/3}x}\left(c_3 \cos(2^{-2/3}\sqrt{3}\,x) + c_4 \sin(2^{-2/3}\sqrt{3}\,x)\right).$$

47. Char. eq.: $(r^2 - 4)(r^2 - 1) = 0$; roots: $r_1 = 2$, $r_2 = -2$, $r_3 = 1$, $r_4 = -1$; general solution:

$$y(x) = c_1 e^{2x} + c_2 e^{-2x} + c_3 e^x + c_4 e^{-x}.$$

49. Any constant-coefficient linear equation having $y(x) = \cos x + \sin 2x$ as a solution must have a characteristic equation having $\pm i$ and $\pm 2i$ as roots (complex roots must come in pairs). Since a characteristic equation with this property has degree 4 or more, the associated differential equation must be of order four or more. Therefore, the given function can't be a solution of a second-order equation of the form $y'' + ay' + by = 0$.

However, note that the polynomial equation $(r^2 + 1)(r^2 + 4) = r^4 + 5r^2 + 4 = 0$ has the required roots. Moreover, no polynomial equation of degree less than four has these roots. It follows that the associated D.E.

$$y^{(4)} + 5y'' + 4y = 0$$

is the D.E. of minimal order having $\cos x + \sin 2x$ as a solution.

51. (a) The boundary-value problem to be solved is

$$y^{(4)} = -P, \qquad y(0) = y(L) = 0, \ y'(0) = 0, \ y''(L) = 0.$$

Four successive integrations gives the general solution

$$y(x) = c_1 + c_2 x + c_3 x^2 + c_4 x^3 - \frac{1}{24} P x^4, \quad 0 \le x \le L.$$

With $y'(x) = c_2 + 2c_3 x + 3c_4 x^2 - \frac{1}{6} P x^3$ and $y''(x) = 2c_3 + 6c_4 x - \frac{1}{2} P x^2$, the boundary conditions lead to the four equations

$$0 = y(0) = c_1, \quad 0 = y(L) = c_1 + c_2 L + c_3 L^2 + c_4 L^3 - \frac{1}{24} P L^4,$$

$$0 = y'(0) = c_2, \quad 0 = y''(L) = 2c_3 + 6c_4 L - \frac{1}{2} P L^2.$$

The two equations in the left column manifestly give $c_1 = c_2 = 0$. With these values in the two equations in the right column, one finds that $c_3 = -PL^2/16$ and $c_4 = 5PL/48$. The particular solution whose graph on $0 \le x \le L$ describes the shape of the beam is therefore

$$y(x) = -\frac{1}{16} P L^2 x^2 + \frac{5}{48} P L x^3 - \frac{1}{24} P x^4 = -\frac{1}{48} P(3L^2 x^2 - 5Lx^3 + 2x^4), \quad 0 \le x \le L.$$

(b) The first two derivatives of $y(x)$ are given by

$$y'(x) = -\frac{1}{48} P(6L^2 x - 15Lx^2 + 8x^3) \qquad \text{and} \qquad y''(x) = -\frac{1}{48} P(6L^2 - 30Lx + 24x^2).$$

Inflection points on the graph of $y(x)$ can occur only at those values of x such that $y''(x) = 0$. Since $P \ne 0$, these correspond to the roots of the quadratic equation $6L^2 - 30Lx + 24x^2 = 0$. That is, at $x = L$ and $x = L/4$. Noting that $x = L/4$ is in the open x-interval $(0, L)$, we check that it is indeed an inflection point. To do this, we observe that the graph of $y''(x)$ is a parabola opening downward and that the root $x = L/4$ is to the left of the root $x = L$. It follows that the concavity of the graph of $y(x)$ changes from negative to positive at $x = L/4$. That is, the graph is concave down on $(0, L/4)$ and concave up on $(L/4, L)$, so that $x = L/4$ is an inflection point on the desired interval.

The maximum downward vertical deflection from level 0 on $0 < x < L$ occurs either at the end(s) of the beam or at a critical point of $y(x)$. But since $y(0) = y(L) = 0$ (by hypothesis), maximum deflection does not occur at the end(s) of the beam. To find the critical points of $y(x)$, set $y'(x) = 0$ and solve

$$-\frac{1}{48}P(6L^2x - 15Lx^2 + 8x^3) = 0, \quad \text{or equivalently, solve} \quad 6L^2x - 15Lx^2 + 8x^3 = 0.$$

Writing the equation on the right as $x(8x^2 - 15Lx + 6L^2) = 0$, we see that 0 is a root and the quadratic formula gives the other roots as $\left(\frac{15\pm\sqrt{33}}{16}\right)L$. Only the root $x_0 = \left(\frac{15-\sqrt{33}}{16}\right)L$ ($\approx 0.58L$) is in the open x-interval $(0, L)$. Moreover, it is straightforward (but tedious) to check that $y''(x_0) > 0$, so that $y(x_0)$ is a minimum value of $y(x)$ on $(0, L)$. That is, the maximum downward deflection of the beam is

$$y(x_0) = y\left(\frac{15-\sqrt{33}}{16}L\right)$$

$$= -\frac{1}{48}P\left[3L^2\left(\frac{15-\sqrt{33}}{16}L\right)^2 - 5L\left(\frac{15-\sqrt{33}}{16}L\right)^3 + 2\left(\frac{15-\sqrt{33}}{16}L\right)^4\right]$$

$$\approx -0.005416122PL^4.$$

(c) We arbitrarily set $P = 0.02$ and $L = 10$ and use the solution formula found in part (a) to get the particular solution

$$y(x) = -0.000\overline{3}x^4 + 0.020\overline{3}x^3 - 0.125x^2.$$

The results of part (b) give an inflection point at approximately $(2.5, -0.48)$, and the maximum downward deflection point at approximately $(5.8, 1.08)$. These important features are indicated in the graph of the solution shown below.

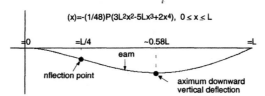

53. Let $y_1(x), \ldots, y_n(x)$ ($n \geq 2$) be functions defined on a common open interval I, and suppose $c_1, \ldots, c_n$ are constants such that $c_1y_1 + \cdots + c_ny_n = 0$ on I. If $y_1, \ldots, y_n$ are not linearly independent then some c_i is nonzero. Without loss of generality, we can suppose $c_1 \neq 0$. This allows us to solve for y_1 to get

$$\left(-\frac{c_2}{c_1}\right)y_2 + \cdots + \left(-\frac{c_n}{c_1}\right)y_n = y_1 \quad \text{on } I.$$

Thus, y_1 is a linear combination of the remaining y_is.

Conversely, if some y_i is a linear combination of the others, say $d_2y_2 + \cdots + d_ny_n = y_1$, where $d_2, \ldots, d_n$ are constants, then $(-1)y_1 + d_2y_2 + \cdots + d_ny_n = 0$ on I and not all of the coefficients are zero. Thus, $y_1, \ldots, y_n$ are not linearly independent on I.

Section 3: NONHOMOGENEOUS EQUATIONS

Exercise Set 3AB (pgs. 520-521)

1. The homogeneous equation $y'' - y = 0$ has characteristic equation $r^2 - 1 = 0$, with roots $r_{1,2} = \pm 1$. The homogeneous solution is therefore

$$y_h = c_1 e^x + c_2 e^{-x}.$$

Since the nonhomogeneous part e^{2x} is not a homogeneous solution, there is a particular solution of the nonhomogeneous equation of the form $y_p = Ae^{2x}$, for some constant A. Substitute this and $y_p'' = 4Ae^{2x}$ into the given D.E. to get

$$y_p'' - y_p = 4Ae^{2x} - Ae^{2x} = 3Ae^{2x} = e^{2a}.$$

Comparing coefficients gives $3A = 1$, or $A = 1/3$. Thus, $y_p = \frac{1}{3}e^{2x}$ is a particular solution. The general solution is therefore of the form

$$y(x) = y_h(x) + y_p(x) = c_1 e^x + c_2 e^{-x} + \frac{1}{3}e^{2x}.$$

Using this and its derivative $y'(x) = .c_1 e^x - c_2 e^{-x} + (2/3)e^{2x}$, apply the initial conditions to obtain the two equations

$$0 = y(0) = c_1 + c_2 + \frac{1}{3} \qquad \text{and} \qquad 1 = y'(0) = c_1 - c_2 + \frac{2}{3}.$$

The elimination method gives $c_1 = 0$, $c_2 = -1/3$, so that the solution satisfying the given initial conditions is

$$y(x) = -\frac{1}{3}e^{-x} + \frac{1}{3}e^{2x}.$$

3. The homogeneous equation has characteristic equation $r^2 + 2r + 1 = (r+1)^2 = 0$, with the double root $r_1 = r_2 = -1$. The homogeneous solution is therefore.

$$y_h(x) = c_1 e^{-x} + c_2 x e^{-x}.$$

Since the nonhomogeneous part e^x is not a homogeneous solution, there is a a particular solution of the nonhomogeneous equation of the form $y_p(x) = Ae^x$, for some constant A. Substitute this and $y_p' = y_p'' = Ae^x$ into the given nonhomogeneous equation to get

$$y_p'' + 2y_p' + y_p = Ae^x + 2Ae^x + Ae^x = 4Ae^x = e^x.$$

Comparing coefficients gives $4A = 1$, or $A = 1/4$. Thus, $y_p(x) = \frac{1}{4}e^x$ is a particular solution. The general solution is therefore of the form

$$y(x) = y_h(x) + y_p(x) = c_1 e^{-x} + c_2 x e^{-x} + \frac{1}{4}e^x.$$

400

Using this and its derivative $y'(x) = -c_1e^{-x} - c_2xe^{-x} + c_2e^{-x} + \frac{1}{4}e^x$, apply the initial conditions to obtain the two equations

$$0 = y'(0) = c_1 + \frac{1}{4} \qquad \text{and} \qquad 1 = y'(0) = -c_1 + c_2 + \frac{1}{4}.$$

By inspection, $c_1 = -1/4$, $c_2 = 1/2$, so that the solution satisfying the given initial conditions is

$$y(x) = -\frac{1}{4}e^{-x} + \frac{1}{2}xe^{-x} + \frac{1}{4}e^x.$$

5. The homogeneous equation $y'' - y = 0$ has characteristic equation $r^2 - 1 = 0$, with roots $r_{1,2} = \pm 1$. The homogeneous solution is therefore

$$y_h = c_1e^x + c_2e^{-x}.$$

Because the nonhomogeneous part $e^x + x$ is the sum of two functions with different derivative sets, we will use the principle of superposition to construct a particular solution of the nonhomogeneous equation. That is, we will find particular solutions to each of the nonhomogeneous equations

$$(*) \qquad\qquad y'' - y = e^x \qquad \text{and} \qquad y'' - y = x.$$

For the first equation, we note that e^x is a homogeneous solution and that $\{e^x\}$ is its derivative set. Thus, we multiply e^x by x (because xe^x is *not* a homogeneous solution) and seek a particular solution of the form $y_{p1}(x) = Axe^x$, where A is a constant. Substituting this and $y_{p1}'' = Axe^x + 2Ae^x$ into the first equation of $(*)$ gives

$$y_{p1}'' - y_{p1} = Axe^x + 2Ae^x - Axe^x = 2Ae^x = e^x.$$

Comparing coefficients gives $2A = 1$, or $A = 1/2$. Thus, $y_{p1}(x) = \frac{1}{2}xe^x$ is a particular solution of the first equation of $(*)$.

For the second equation in $(*)$, note that x is not a homogeneous solution and that $\{1, x\}$ is its derivatives set. So, it has a particular solution of the form $y_{p2}(x) = A + Bx$ for some constants A and B. Substituting this and $y_{p2}'' = 0$ into the second equation of $(*)$ gives

$$y_{p2}'' - y_{p2} = 0 - (A + Bx) = -A - Bx = x.$$

Comparing coefficients of like powers of x gives $A = 0$, $B = -1$. Thus, $y_{p2}(x) = -x$ is a particular solution of the second equation of $(*)$.

By the principle of superposition, $y_p(x) = y_{p1}(x) + y_{p2}(x)$ is a particular solution of the original nonhomogeneous equation. That is, the general solution of the original D.E. is

$$y(x) = y_h(x) + y_p(x) = y_h(x) + y_{p1}(x) + y_{p2}(x) = c_1e^x + c_2e^{-x} + \frac{1}{2}xe^x - x.$$

Using this and its derivative $y'(x) = c_1e^x - c_2e^{-x} + \frac{1}{2}xe^x + \frac{1}{2}e^x - 1$, apply the initial conditions to obtain the two equations

$$0 = y(0) = c_1 + c_2 \qquad \text{and} \qquad 1 = y'(0) = c_1 - c_2 + \frac{1}{2} - 1.$$

By the elimination method, $c_1 = 3/4$, $c_2 = -3/4$, so that the solution satisfying the given initial conditions is

$$y(x) = \frac{3}{4}e^x - \frac{3}{4}e^{-x} + \frac{1}{2}xe^x - x.$$

(NOTE: It was really not necessary to use the method of undetermined coefficients to find a particular solution of the second equation of $(*)$. By inspection, alert readers should be able to see that $-x$ is a solution. However, not everyone is this proficient at simply "seeing" the answer. For this reason, this solver derived the solution in a more formal way.)

7. The homogeneous equation $y'' + y = 0$ has characteristic equation $r^2 + 1 = 0$, with roots $r_{1,2} = \pm i$. The homogeneous solution is therefore

$$y_h(x) = c_1 \cos x + c_2 \sin x.$$

Since the nonhomogeneous term $\cos x$ is a homogeneous solution and its derivative set is $\{\cos x, \sin x\}$, multiply each member of the derivative set by x (because neither $x \cos x$ nor $x \sin x$ is a homogeneous solution) and deduce that there is a particular solution of the nonhomogeneous equation of the form $y_p(x) = Ax \cos x + Bx \sin x$. Substitute this and $y_p'' = -Ax \cos x - 2A \sin x - Bx \sin x + 2B \cos x$ into the nonhomogeneous equation to get

$$y_p'' + y_p = (-Ax \cos x - 2A \sin x - Bx \sin x + 2B \cos x) + (Ax \cos x + Bx \sin x)$$
$$= -2A \sin x + 2B \cos x = \cos x.$$

Comparing coefficients of like functions gives $-2A = 0$ and $2B = 1$, from which $A = 0$, $B = 1/2$. Thus, $y_p(x) = \frac{1}{2}x \sin x$ is a particular solution of the nonhomogeneous equation. The general solution of the given D.E. is then

$$y(x) = y_h(x) + y_p(x) = c_1 \cos x + c_2 \sin x + \frac{1}{2}x \sin x.$$

Using this and $y'(x) = -c_1 \sin x + c_2 \cos x + \frac{1}{2}x \cos x + \frac{1}{2}\sin x$, apply the initial conditions to obtain the two equations $0 = y(0) = c_1$ and $1 = y'(0) = c_2$. The solution satisfying the given initial conditions is therefore

$$y(x) = \sin x + \frac{1}{2}x \sin x.$$

9. The homogeneous equation $y'' + y = 0$ has characteristic equation $r^2 + 1 = 0$, with roots $r_{1,2} = \pm i$. The homogeneous solution is therefore

$$y_h(x) = c_1 \cos x + c_2 \sin x.$$

Since $\{\cos x, \sin x, x \cos x, x \sin x\}$ is the derivative set of the nonhomogeneous expression $x \cos x$, and since it contains homogeneous solutions, multiply each member of the set by x (because none of the resulting members will then be a homogeneous solution) and form the linear combination

$$y_p(x) = Ax \cos x + Bx \sin x + Cx^2 \cos x + Dx^2 \sin x,$$

where A, B, C and D are constants to be determined. Computing y_p'' gives

$$y_p'' = (2B + 2C) \cos x + (-2A + 2D) \sin x + (-A + 4D)x \cos x$$
$$+ (-B - 4C)x \sin x - Cx^2 \cos x - Dx^2 \sin x.$$

Substituting this and y_p into the nonhomogeneous equation gives

$$y_p'' + y_p = (2B + 2C)\cos x + (-2A + 2D)\sin x + 4Dx\cos x - 4Cx\sin x = x\cos x.$$

Equating coefficients of like functions leads to the system of four equations

$$2B + 2C = 0, \quad -2A + 2D = 0, \quad 4D = 1, \quad -4C = 0.$$

The third and fourth equations immediately give $D = 1/4$ and $C = 0$. When these are inserted into each of the first two equations, we find that $A = 1/4$ and $B = 0$. Thus, $y_p(x) = \frac{1}{4}x\cos x + \frac{1}{4}x^2\sin x$. The general solution of the given D.E. is then

$$y(x) = y_h(x) + y_p(x) = c_1\cos x + c_2\sin x + \frac{1}{4}x\cos x + \frac{1}{4}x^2\sin x.$$

Using this and its derivative $y'(x) = -c_1\sin x + c_2\cos x + \frac{1}{4}x\sin x + \frac{1}{4}\cos x + \frac{1}{4}x^2\cos x$, apply the initial conditions and obtain the two equations

$$0 = y(0) = c_1 \qquad \text{and} \qquad 1 = y'(0) = c_2 + \frac{1}{4},$$

from which $c_1 = 0$, $c_2 = 3/4$. Therefore, the solution satisfying the given initial conditions is

$$y(x) = \frac{3}{4}\sin x + \frac{1}{4}x\cos x + \frac{1}{4}x^2\sin x.$$

11. Write the given differential equation in its operator form $(D^2 + D - 2)y = e^x$ and then factor the operator to get

$$(D - 1)(D + 2)y = e^x.$$

Let $(D + 2)y = z$ and solve $(D - 1)z = e^x$, or $Dz - z = e^x$. Multiplication by e^{-x} gives

$$e^{-x}Dz - e^{-x}z = 1, \qquad \text{or equivalently,} \qquad D(e^{-x}z) = 1.$$

Integrate to get $e^{-x}z = x + C_1$, or $z = xe^x + C_1 e^x$. Now solve $(D + 2)y = z = xe^x + C_1 e^x$, or $Dy + 2y = xe^x + C_1 e^x$. Multiplication by e^{2x} gives

$$e^{2x}Dy + 2e^{2x}y = xe^{3x} + C_1 e^{3x}, \qquad \text{or equivalently,} \qquad D(e^{2x}y) = xe^{3x} + C_1 e^{3x}.$$

Integrate to get $e^x y = \frac{1}{3}xe^{3x} - \frac{1}{9}e^{3x} + (C_1/3)e^{3x} + C_2$. Multiplying by e^{-2x} gives the general solution

$$y(x) = \frac{1}{3}xe^x - \frac{1}{9}e^x + (C_1/3)e^x + C_2 e^{-2x} = \frac{1}{3}xe^x + c_1 e^x + c_2 e^{-2x},$$

where $c_1 = -1/9 + C_1/3$ and $c_2 = C_2$.

13. Write the given differential equation in its operator form $(D^2 + 1)y = e^{ix}$ and then factor the operator to get

$$(D - i)(D + i)y = e^{ix}.$$

Let $(D + i)y = z$ and solve $(D - i)z = e^{ix}$, or $Dz - iz = e^{ix}$. Multiplication by e^{-ix} gives

$$e^{-ix}Dz - ie^{-ix}z = 1, \qquad \text{or equivalently,} \qquad D(e^{-ix}z) = 1.$$

Integrate to get $e^{-ix}z = x + C_1$, or $z = xe^{ix} + C_1e^{ix}$, where C_1 is a real or complex constant. Now solve $(D + i)y = z = xe^{ix} + C_1e^{ix}$, or $Dy + iy = xe^{ix} + C_1e^{ix}$. Multiplication by e^{ix} gives

$$e^{ix}Dy + ie^{ix}y = xe^{2ix} + C_1e^{2ix}, \quad \text{or equivalently,} \quad D(e^{ix}y) = xe^{2ix} + C_1e^{2ix}.$$

Integrate to get

$$
\begin{aligned}
e^{ix}y &= \int \left(xe^{2ix} + C_1e^{2ix}\right)dx = \int xe^{2ix}dx + C_1\int e^{2ix}dx \\
&= \frac{1}{2i}xe^{2ix} - \frac{1}{2i}\int e^{2ix}dx + \frac{C_1}{2i}e^{2ix} \\
&= -\frac{1}{2}ixe^{2ix} - \frac{1}{(2i)^2}e^{2ix} - \frac{C_1}{2}ie^{2ix} + C_2 \\
&= -\frac{1}{2}ixe^{2ix} - \frac{1}{-2}e^{2ix} - \frac{C_1}{2}ie^{2ix} + C_2 \\
&= -\frac{1}{2}ixe^{2ix} + \left(\frac{1}{2} - \frac{C_1}{2}i\right)e^{2ix} + C_2 \\
&= -\frac{1}{2}ixe^{2ix} + c_1e^{2ix} + c_2,
\end{aligned}
$$

where $c_1 = 1/2 - C_1i/2$ and $c_2 = C_2$ are real or complex constants, and the first two terms in the second line are obtained using integration by parts. Multiplying by e^{-ix} gives the general solution (including all complex-valued solutions)

$$y(x) = -\frac{1}{2}ie^{ix} + c_1e^{ix} + c_2e^{-ix}.$$

15. The function $e^x + 2e^{2x}$ is a solution of all constant-coefficient homogeneous differential equations whose general solution includes functions of the form

$$y(x) = c_1e^x + c_2e^{2x}.$$

Such a D.E. must have a characteristic equation with the two real roots $r_1 = 1$, $r_2 = 2$. Since the characteristic equation of lowest degree with these roots is $(r-1)(r-2) = r^2 - 3r + 2 = 0$, it follows that the D.E. of lowest order with the given function as a solution is

$$y'' - 3y' + 2y = 0.$$

17. The function $x + 1$ is a solution of all constant-coefficient homogeneous differential equations whose general solution includes functions of the form

$$y(x) = c_1 + c_2x.$$

Such a D.E. must have a characteristic equation with the double root $r_1 = r_2 = 0$. Since the characteristic equation of lowest degree with these roots is $r^2 = 0$, it follows that the D.E. of lowest order having the given function as a solution is

$$y'' = 0.$$

19. The function $x \sin 3x$ is a solution of all constant-coefficient homogeneous differential equations whose general solution includes functions of the form

$$y(x) = c_1 \cos 3x + c_2 \sin 3x + c_3 x \cos 3x + c_4 x \sin 3x.$$

Such a D.E. must have a characteristic equation with the double pair of conjugate roots $r_1 = r_2 = 3i$, $r_3 = r_4 = -3i$. Since the characteristic equation of lowest degree with these roots is $(r - 3i)^2 (r + 3i)^2 = (r^2 + 9)^2 = r^4 + 18r^2 + 81 = 0$, it follows that the D.E. of lowest order with the given function as a solution is

$$y^{(4)} + 18y'' + 81y = 0.$$

21. The function $xe^x \sin x$ is a solution of all constant-coefficient homogeneous differential equations whose general solution includes functions of the form

$$y(x) = e^x(c_1 \cos c + c_2 \sin x) + xe^x(c_3 \cos x + c_4 \sin x).$$

Such a D.E. must have a characteristic equation with the double pair of conjugate roots $r_1 = r_2 = 1 + i$, $r_3 = r_4 = 1 - i$. Since the characteristic equation of lowest degree with these roots is $(r - 1 - i)^2 (r - 1 + i)^2 = (r^2 - 2r + 2)^2 = r^4 - 4r^3 + 8r^2 - 8r + 4 = 0$, it follows that the D.E. of lowest order with the given function as a solution is

$$y^{(4)} - 4y''' + 8y'' - 8y' + 4y = 0.$$

23. The characteristic equation of the homogeneous equation $y'' - y = 0$ is $r^2 - 1 = 0$, with roots $r_{1,2} = \pm 1$. The homogeneous solution is therefore of the form $y_h(x) = c_1 e^x + c_2 e^{-x}$. The forcing function $\cos x$ of the given nonhomogeneous equation has the derivative set $\{\cos x, \sin x\}$, none of whose members is a homogeneous solution. Thus, the appropriate form for a trial particular solution of the nonhomogeneous equation is

$$y_p(x) = A \cos x + B \sin x.$$

25. The characteristic equation of the homogeneous equation $y'' - y = 0$ is $r^2 - 1 = 0$, with roots $r_{1,2} = \pm 1$. The homogeneous solution is therefore of the form $y_h(x) = c_1 e^x + c_2 e^{-x}$. The forcing function e^x of the given nonhomogeneous equation has the derivative set $\{e^x\}$, whose member is a homogeneous solution. Since xe^x is *not* a homogeneous solution, the appropriate form for a trial particular solution of the nonhomogeneous equation is

$$y_p(x) = Axe^x.$$

27. The characteristic equation of the homogeneous equation $y'' - 2y' + y = 0$ is $r^2 - 2r + 1 = (r - 1)^2 = 0$, with the double root $r_1 = r_2 = 1$. The homogeneous solution is therefore of the form $y_h(x) = c_1 e^x + c_2 xe^x$. The forcing function xe^x of the given nonhomogeneous equation has the derivative set $\{e^x, xe^x\}$, which contains homogeneous solutions. Since x times each of these members still results in one of them being a homogeneous solution, while x^2 times each member results in neither function being a homogeneous solution, the appropriate form for a trial particular solution of the nonhomogeneous equation is

$$y_p(x) = Ax^2 e^x + Bx^3 e^x.$$

29. The characteristic equation $r^2 + 4r + 4 = (r + 2)^2 = 0$ has the double root $r_1 = r_2 = -2$, so that the homogeneous solution is

$$y_h(x) = c_1 e^{-2x} + c_2 xe^{-2x}.$$

The derivative set of the forcing function $3x$ of the given nonhomogeneous equation is $\{1, x\}$ and does not contain any homogeneous solutions. Thus, the appropriate particular solution is $y_p(x) = A + Bx$. Substituting this and its derivatives $y_p' = B$, $y_p'' = 0$ into the nonhomogeneous equation gives

$$y_p'' + 4y_p' + 4y_p = 0 + 4B + 4(A + Bx) = (4B + 4A) + 4Bx = 3x.$$

Equating coefficients of like powers of x gives $4B + 4A = 0$ and $4B = 3$. Hence, $A = -3/4$, $B = 3/4$ and $y_p(x) = -\frac{3}{4} + \frac{3}{4}x$. The general solution of the nonhomogeneous equation is therefore

$$y(x) = c_1 e^{-2x} + c_2 x e^{-2x} - \frac{3}{4} + \frac{3}{4}x.$$

Using this and its derivative $y'(x) = (-2c_1 + c_2)e^{-2x} - 2c_2 x e^{-2x} + \frac{3}{4}$, apply the initial conditions to get

$$0 = y(0) = c_1 - \frac{3}{4} \qquad \text{and} \qquad 1 = y'(0) = -2c_1 + c_2 + \frac{3}{4}.$$

By inspection, $c_1 = 3/4$ and $c_2 = 7/4$. The solution satisfying the given initial conditions is then

$$y(x) = \frac{3}{4}e^{-2x} + \frac{7}{4}x e^{-2x} - \frac{3}{4} + \frac{3}{4}x.$$

The graph of this solution is at the top left of the next page.

31. The characteristic equation $r^2 + 2r + 2 = 0$ has roots $r_{1,2} = -1 \pm i$, so that the homogeneous solution is

$$y_h(x) = c_1 e^{-x} \cos x + c_2 e^{-x} \sin x.$$

The derivative set of the forcing function e^x of the given nonhomogeneous equation is $\{e^x\}$ and does not contain a homogeneous solution. Thus, the appropriate form for the particular solution is $y_p(x) = Ae^x$. Substitute this and its derivatives $y_p' = y_p'' = Ae^x$ into the nonhomogeneous equation to get

$$y_p'' + 2y_p' + 2y_p = Ae^x + 2Ae^x + 2Ae^x = 5Ae^x = e^x.$$

Equating coefficients gives $5A = 1$, or $A = 1/5$. So, $y_p(x) = \frac{1}{5}e^x$ and the general solution of the nonhomogeneous equation is

$$y(x) = c_1 e^{-x} \cos x + c_2 e^{-x} \sin x + \frac{1}{5}e^x.$$

Using this and $y'(x) = (-c_1 + c_2)e^{-x} \cos x + (-c_1 - c_2)e^{-x} \sin x + \frac{1}{5}e^x$, apply the initial conditions to obtain the two equations

$$0 = y(0) = c_1 + \frac{1}{5} \qquad \text{and} \qquad 1 = y'(0) = -c_1 + c_2 + \frac{1}{5}.$$

By inspection, $c_1 = -1/5$ and $c_2 = 3/5$. The solution satisfying the given initial conditions is then

$$y(x) = -\frac{1}{5}e^{-x} \cos x + \frac{3}{5}e^{-x} \sin x + \frac{1}{5}e^x.$$

The graph of this solution is at the top right of the next page

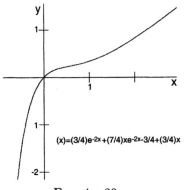

$(x)=(3/4)e^{-2x}+(7/4)xe^{-2x}-3/4+(3/4)x$

Exercise 29

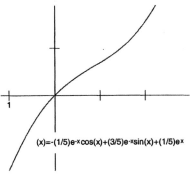

$(x)=-(1/5)e^{-x}\cos(x)+(3/5)e^{-x}\sin(x)+(1/5)e^{x}$

Exercise 31

Exercises 33, 35, 37, 39 and 41 are all worked in the same way. Each exercise deals with a nonhomogeneous D.E. with a forcing function f of the form $f = f_1 + \cdots + f_n$, where each f_i is a constant multiple of a basic solution of a constant-coefficient homogeneous equation. For each f_i, k_i is the least nonnegative integer with the property that if each member of the derivative set of f_i is multiplied by x^{k_i} then none of the resulting functions is a homogeneous solution. The trial particular solution is then the linear combination of all functions resulting from this process.

33. The characteristic equation $r^2 - 4 = 0$ has roots $r_{1,2} = \pm 2$, so that the basic homogeneous solutions are e^{2x} and e^{-2x}. Here, $f = xe^{2x} + e^{2x}$, so that $f_1 = xe^{2x}$ and $f_2 = e^{2x}$. Moreover, $k_1 = k_2 = 1$. Thus,

$$y_p(x) = Ax^2e^{2x} + Bxe^{2x}.$$

35. The characteristic equation $r^2 - 5r + 6 = (r - 2)(r - 3) = 0$ has roots $r_1 = 2$, $r_2 = 3$, so that the basic homogeneous solutions are e^{2x} and e^{3x}. Here, $f = xe^{2x} + e^{3x}$, so that $f_1 = xe^{2x}$ and $f_2 = e^{3x}$. Moreover, $k_1 = k_2 = 1$. Thus,

$$y_p(x) = Axe^{2x} + Bx^2e^{2x} + Cxe^{3x}.$$

37. The characteristic equation $r^2 - 4 = 0$ has roots $r_{1,2} = \pm 2$, so that the basic homogeneous solutions are e^{2x} and e^{-2x}. Here, $f = e^{2x} + 5\cos x$, so that $f_1 = e^{2x}$ and $f_2 = 5\cos x$. Moreover, $k_1 = 1$ and $k_2 = 0$. Thus,

$$y_p(x) = Axe^{2x} + B\cos x + C\sin x.$$

39. The characteristic equation $r^2 - r = r(r - 1) = 0$ has roots $r_1 = 0$, $r_2 = 1$, so that the basic homogeneous solutions are 1 and e^x. Here, $f = x^2 + 2e^x$, so that $f_1 = x^2$ and $f_2 = 2e^x$. Moreover, $k_1 = k_2 = 1$. Thus,

$$y_p(x) = Ax + Bx^2 + Cx^3 + Dxe^x.$$

41. The characteristic equation $r^3 = 0$ has the triple root $r_1 = r_2 = r_3 = 0$, so that the basic homogeneous solutions are 1, x and x^2. Here, $f = 1 + x + x^3$, so that $f_1 = 1$, $f_2 = x$ and $f_3 = x^3$. Moreover, $k_1 = k_2 = k_3 = 3$. Thus,

$$y_p(x) = Ax^3 + Bx^4 + Cx^5 + Dx^6.$$

43. (a) The characteristic equation is $r^2 + (k/m)r = r(r + k/m) = 0$, with roots $r_1 = 0$, $r_2 = -k/m$. The homogeneous solution is therefore $y_h(t) = c_1 + c_2 e^{-kt/m}$. The derivative set of the forcing function g of the given nonhomogeneous equation is the singleton set $\{1\}$, which contains a homogeneous solution. Multiplying the sole member of this set by t results in a function that is not a homogeneous solution. Thus, $y_p(t) = At$ is a trial particular solution of the nonhomogeneous equation. When $dy_p/dt = A$ and $d^2y_p/dt^2 = 0$ are substituted into the nonhomogeneous equation, we get

$$\frac{d^2 y_p}{dt^2} + \frac{k}{m}\frac{dy_p}{dt} = 0 + \frac{k}{m}A = \frac{k}{m}A = g.$$

Solving for A gives $A = mg/k$. So, $y_p(t) = \frac{mg}{k}t$ and the general solution of the given nonhomogeneous equation is

$$y(t) = y_h(t) + y_p(t) = c_1 + c_2 e^{-kt/m} + \frac{mg}{k}t.$$

(b) From the general solution derived in part (a), $y'(t) = -\frac{k}{m}c_2 e^{-kt/m} + \frac{mg}{k}$. Applying the initial conditions leads to the two equations

$$y_0 = y(0) = c_1 + c_2 \qquad \text{and} \qquad z_0 = y'(0) = -\frac{k}{m}c_2 + \frac{mg}{k}.$$

Solving the second equation for c_2 gives $c_2 = (z_0 - mg/k)/(-k/m) = (m/k)(mg/k - z_0)$. The first equation then gives $c_1 = y_0 - c_2 = y_0 - (m/k)(mg/k) - z_0$.

(c) When the expression for c_2 found in part (b) is inserted into the equation for $y'(t)$, the result is

$$y'(t) = -\frac{k}{m}\left[\frac{m}{k}\left(\frac{mg}{k} - z_0\right)\right]e^{0kt/m} + \frac{mg}{k} = \left(z_0 - \frac{mg}{k}\right)e^{-kt/m} + \frac{mg}{k}.$$

Since k and m are positive constants, the exponential $e^{-kt/m}$ tends to zero as t increases without bound. Thus, regardless of the initial conditions, $\lim_{t\to\infty} y'(t) = mg/k$.

(d) Using the expression for $y'(t)$ in part (c), set $y'(t) = 0$ and solve for $e^{kt/m}$ to get

$$e^{kt/m} = -\left[z_0 - (mg/k)\right]/(mg/k) = 1 - \frac{z_0 k}{mg}.$$

Since m, g and k are all positive, the expression on the right is positive whenever $z_0 < 0$. In this case, we can take logarithms of each member and solve for t. This gives

$$t = \frac{m}{k}\ln\left(1 - \frac{z_0 k}{mg}\right).$$

45. (a) Differentiating the general solution derived in Exercise 43(a) in this section gives

$$\dot{y}(t) = -\frac{k}{m}c_2 e^{-kt/m} + \frac{mg}{k}.$$

The condition $\dot{y}(0) = v_0$ gives $v_0 = -(k/m)c_2 + mg/k$, or $c_2 = (m^2 g - mkv_0)/k^2$. Inserting this expression for c_2 in the above formula for $\dot{y}(t)$ gives

$$\dot{y}(t) = -\frac{k}{m}\left(\frac{m^2 g - mkv_0}{k^2}\right)e^{-kt/m} + \frac{mg}{k} = \left(\frac{kv_0 - mg}{k}\right)e^{-kt/m} + \frac{mg}{k}$$

$$= \left(v_0 - \frac{mg}{k}\right)e^{-kt/m} + \frac{mg}{k}.$$

(b) Write the formula for $\dot{y}(t)$ found in part (a) as

$$\dot{y}(t) = v_0 e^{-kt/m} + mg\left(\frac{1 - e^{-kt/m}}{k}\right).$$

Letting $k \to 0$, the first term on the right goes to v_0. The second term requires the use of L'Hôpital's rule. We have

$$\lim_{k \to 0+} \frac{1 - e^{-kt/m}}{k} = \lim_{k \to 0+} \frac{(t/m)}{1} = \frac{t}{m}.$$

Hence,

$$\lim_{k \to 0+} \dot{y}(t) = v_0 + mg\left(\frac{t}{m}\right) = v_0 + gt.$$

47. Assume f is integrable on finite intervals and consider the initial-value problem

$$(D - r_1)(D - r_2)y = f(x), \quad y(x_0) = y'(x_0) = 0.$$

Let $(D - r_2)y = z$ and solve $(D - r_1)z = f(x)$, or equivalently, solve $Dz - r_1 z = f(x)$. Since this is in normalized form, we can multiply by the exponential multiplier $e^{-r_1 x}$ to get

$$e^{-r_1 x} Dz - r_1 e^{-r_1 x} z = e^{-r_1 x} f(x), \quad \text{or equivalently,} \quad D\left(e^{-r_1 x} z\right) = e^{-r_1 x} f(x).$$

Integrating from x_0 to x and choosing the solution $z = z(x)$ such that $z(x_0) = 0$, we get

$$e^{-r_1 x} z(x) - e^{-r_1 x_0} z(x_0) = \int_0^x e^{-r_1 s} f(s)\, ds \quad \text{or} \quad z(x) = e^{r_1 x} \int_{x_0}^x e^{-r_1 s} f(s)\, ds.$$

If we now try to solve $(D - r_2)y = z$ by the same method and choose the solution $y = y(x)$ such that $y(x_0) = 0$, we will get the same formula pattern as was obtained above for $z(x)$ but with z replaced by y, f replaced by z, r_1 replaced by r_2 and the dummy variable s replaced by the dummy variable t. That is,

$$y(x) = e^{r_2 x} \int_{x_0}^x e^{-r_2 t} z(t)\, dt = e^{r_2 x} \int_{x_0}^x e^{-r_2 t} \left[e^{r_1 t} \int_{x_0}^t e^{-r_1 s} f(s)\, ds\right] dt$$

$$= e^{r_2 x} \int_{x_0}^x e^{(r_1 - r_2)t} \left[\int_{x_0}^t e^{-r_1 s} f(s)\, ds\right] dt.$$

Because we have chosen the solution satisfying $y(x_0) = 0$, the first initial condition is automatically satisfied. For the other initial condition, we write $(D - r_2)y = z$ in one of its equivalent forms $z(x) = y'(x) - r_2 y(x)$ and evaluate at $x = x_0$ to get

$$z(x_0) = y'(x_0) - r_2 y(x_0) = y'(x_0).$$

Since $z(x_0) = 0$ by assumption, $y'(x_0) = 0$ and the second initial condition is satisfied. So $y(x)$ satisfies the given initial-value problem.

Exercise Set 3CD (pgs. 528-530)

1. The homogeneous equation has characteristic equation $r^2 - 4r + 4 = (r-2)^2 = 0$, with the double root $r = 2$. Thus, we choose the solution $y_1 = e^{2x}$ as our homogeneous solution and let $y(x) = y_1(x)u(x) = e^{2x}u(x)$ be a solution of the given nonhomogeneous equation, where $u = u(x)$ is to be determined. We have

$$y' = u'e^{2x} + 2ue^{2x} \quad \text{and} \quad y'' = u''e^{2x} + 4u'e^{2x} + 4ue^{2x}.$$

Substituting into the given D.E. results in

$$y'' - 4y' + 4y = (u''e^{2x} + 4u'e^{2x} + 4ue^{2x}) - 4(u'e^{2x} + 2ue^{2x}) + 4ue^{2x} = u''e^{2x} = e^x.$$

The last equality implies $u'' = e^{-x}$. Two successive integrations gives $u = e^{-x} + c_1 x + c_2$. Hence,

$$y(x) = e^{2x}u(x) = e^{2x}(e^{-x} + c_1 x + c_2) = e^x + c_1 x e^{2x} + c_2 e^{2x}.$$

3. Assume there exists a number n such that x^n is a solution of $x^2 y'' - 3xy' + 3y = 0$. Let $y(x) = x^n u(x)$ be a solution of the given nonhomogeneous equation, where $u = u(x)$ and n need to be determined. We compute

$$y' = x^n u' + nx^{n-1}u \quad \text{and} \quad y'' = x^n u'' + 2nx^{n-1}u' + n(n-1)x^{n-2}u.$$

Substituting into the nonhomogeneous equation gives

$$\begin{aligned}
x^2 y'' - 3xy' + 3y &= x^2\left(x^n u'' + 2nx^{n-1}u' + n(n-1)x^{n-2}u\right) - 3x(x^n u' + nx^{n-1}u) + 3x^n u \\
&= x^{n+2}u'' + (2n-3)x^{n+1}u' + \left(n(n-1) - 3n + 3\right)x^n u \\
&= x^n\left(x^2 u'' + (2n-3)xu' + (n^2 - 4n + 3)u\right) = x^4,
\end{aligned}$$

or dividing by x^n $(x > 0)$,

$$x^2 u'' + (2n-3)xu' + (n^3 - 4n + 3)u = x^{4-n}.$$

Observe that if n is chosen so that $n^2 - 4n + 3 = 0$, then the above reduces to a first order linear equation in u', which can be solved using the exponential multiplier method. Since $n^2 - 4n + 3 = (n-1)(n-3)$, we choose $n = 1$. With this choice, the above equation becomes $x^2 u'' - xu' = x^3$, or in normalized form,

$$u'' - \frac{1}{x}u' = x.$$

Multiplying this equation by the integrating factor $e^{\int(-1/x)\,dx} = e^{-\ln x} = 1/x$ leads to

$$\frac{d}{dx}((1/x)u') = 1,$$

and integrating gives $(1/x)u' = x + C_1$, or $u' = x^2 + C_1 x$. A second Integration gives $u = \frac{1}{3}x^3 + \frac{1}{2}C_1 x^2 + C_2$. Replacing $\frac{1}{2}C_1$ and C_2 with c_1 and c_2, respectively, we get

$$y(x) = x^n u(x) = x\left(\frac{1}{3}x^3 + c_1 x^2 + c_2\right) = \frac{1}{3}x^4 + c_1 x^3 + c_2 x.$$

5. The associated homogeneous equation $y'' + y' - 2y = 0$ has characteristic equation $r^2 + r - 2 = (r-1)(r+2) = 0$, with roots $r_1 = 1$, $r_2 = -2$. The homogeneous solution is then $y_h(x) = c_1 e^x + c_2 e^{-2x}$. Noting that the given D.E. equation is in normalized form, we let $y_1 = e^x$, $y_1' = e^x$, $y_2 = e^{-2x}$, $y_2' = -2e^{-2x}$ and $f(x) = e^{2x}$ in Equations 3.2 to get

$$e^x u_1' + e^{-2x} u_2' = 0,$$
$$e^x u_1' - 2e^{-2x} u_2' = e^{2x},$$

where $u_1 = u_1(x)$ and $u_2 = u_2(x)$ are to be determined. Subtracting the second equation from the first equation gives $3e^{-2x} u_2' = -e^{2x}$, or $u_2' = -\frac{1}{3} e^{4x}$. Inserting this into the first equation and solving for u_1' gives $u_1' = \frac{1}{3} e^x$. Now integrating the equations just derived for u_1' and u_2' to get $u_1 = \frac{1}{3} e^x$ and $u_2 = -\frac{1}{12} e^{4x}$. A particular solution of the nonhomogeneous equation is then

$$y_p(x) = y_1(x) u_1(x) + y_2(x) u_2(x) = e^x \left(\frac{1}{3} e^x \right) + e^{-2x} \left(-\frac{1}{12} e^{4x} \right) = \frac{1}{3} e^{2x} - \frac{1}{12} e^{2x} = \frac{1}{4} e^{2x}.$$

The complete solution of the given nonhomogeneous equation is therefore

$$y(x) = y_h(x) + y_p(x) = c_1 e^x + c_2 e^{-2x} + \frac{1}{4} e^{2x}.$$

7. The associated homogeneous equation $y'' + y = 0$ has solution $y_h(x) = c_1 \cos x + c_2 \sin x$. Noting that the given D.E. equation is in normalized form, we let $y_1 = \cos x$, $y_1' = -\sin x$, $y_2 = \sin x$, $y_2' = \cos x$ and $f(x) = \sec x$ in Equations 3.2 to get

$$(\cos x) u_1' + (\sin x) u_2' = 0,$$
$$(-\sin x) u_1' + (\cos x) u_2' = \sec x,$$

where $u_1 = u_1(x)$ and $u_2 = u_2(x)$ are to be determined for $|x| < \pi/2$. Multiplying the first equation by $\sin x$, the second equation by $\cos x$, and adding the resulting equations gives $u_2' = 1$. Substituting this into the first equation gives $(\cos x) u_1' + \sin x = 0$. Since $\cos x \neq 0$ for $|x| < \pi/2$, we can solve for u_1' to get $u_1' = -\sin x / \cos x = -\tan x$. Now integrate the equations just derived for u_1' and u_2' to get $u_1 = \ln |\cos x|$ and $u_2 = x$. A particular solution of the nonhomogeneous equation is then

$$y_p(x) = y_1(x) u_1(x) + y_2(x) u_2(x) = \cos x \ln |\cos x| + x \sin x = \cos x \ln(\cos x) + x \sin x,$$

where the absolute value sign is not needed because $\cos x > 0$ for $|x| < \pi/2$. The complete solution of the given nonhomogeneous equation is therefore

$$y(x) = y_h(x) + y_p(x) = c_1 \cos x + c_2 \sin x + x \sin x + \cos x \ln(\cos x), \quad -\pi/2 < x < \pi/2.$$

9. The solution of the homogeneous equation $y'' = 0$ is $y_h(x) = c_1 + c_2 x$. Noting that the given D.E. equation is in normalized form, we let $y_1 = 1$, $y_1' = 0$, $y_2 = x$, $y_2' = 1$ and $f(x) = x^2 e^x$ in Equations 3.2 to get

$$u_1' + x u_2' = 0,$$
$$u_2' = x^2 e^x.$$

Inserting the expression $u_2' = x^2 e^x$ from the second equation into the first equation gives $u_1' = -x^3 e^e$. Integrating u_1' and u_2' by parts gives

$$u_1(x) = -x^3 e^x + 3x^2 e^x - 6xe^x + 6e^x \qquad \text{and} \qquad u_2(x) = x^2 e^x - 2xe^x + 2e^x.$$

A particular solution of the nonhomogeneous equation is then

$$y_p(x) = y_1(x)u_1(x) + y_2(x)u_2(x)$$
$$= -x^3 e^x + 3x^2 e^x - 6xe^x + 6e^x + x(x^2 e^x - 2xe^x + 2e^x) = x^2 e^x - 4xe^x + 6e^x.$$

The complete solution of the given nonhomogeneous equation is therefore

$$y(x) = y_h(x) + y_p(x) = c_1 + c_2 x + x^2 e^x - 4xe^x + 6e^x.$$

11. The associated characteristic equation is $r^2 - 2r + 1 = (r-1)^2 = 0$, with the double root $r_{1,2} = 1$. The basic homogeneous solutions are therefore $y_1(x) = e^x$ and $y_2(x) = xe^x$, with Wronskian

$$w(x) = y_1(x)y_2'(x) - y_2(x)y_1'(x) = e^x(xe^x - e^x) - xe^x(e^x) = xe^{2x} - e^{2x} - xe^{2x} = -e^{2x}.$$

Since the given nonhomogeneous equation is normalized, the forcing function to be used in Equation 3.3 is $f(x) = e^x$. Hence, apply Equation 3.3 to get the particular solution

$$y_p(x) = e^x \int \frac{-xe^x\, e^x}{-e^{2x}}\, dx + xe^x \int \frac{e^x\, e^x}{-e^{2x}}\, dx = e^x \int x\, dx + xe^x \int (-1)\, dx$$
$$= \frac{1}{2}x^2 e^x - x^2 e^x = -\frac{1}{2}x^2 e^x.$$

13. The associated characteristic equation is $r^2 + 3r + 2 = (r+1)(r+2) = 0$, with roots $r_1 = -1$, $r_2 = -2$. The basic homogeneous solutions are therefore $y_1(x) = e^{-x}$ and $y_2(x) = e^{-2x}$, with Wronskian

$$w(x) = y_1(x)y_2'(x) - y_2(x)y_1'(x) = e^{-x}(-2e^{-2x}) - e^{-2x}(-e^{-x}) = -2e^{-3x} + e^{-3x} = -e^{-3x}.$$

Since the given nonhomogeneous equation is normalized, the forcing function to be used in Equation 3.3 is $f(x) = (a + e^x)^{-1}$. Hence, apply Equation 3.3 to get the particular solution

$$y_p(x) = e^{-x} \int \frac{-e^{-2x}(a+e^x)^{-1}}{-e^{-3x}}\, dx + e^{-2x} \int \frac{e^{-x}(a+e^x)^{-1}}{-e^{-3x}}\, dx$$
$$= e^{-x} \int \frac{e^x}{a + e^x}\, dx - e^{-2x} \int \frac{e^{2x}}{a + e^x}\, dx$$
$$= e^{-x} \int \frac{1}{u}\, du - e^{-2x} \int \frac{u - a}{u}\, du \qquad \text{(substitution } u = a + e^x)$$
$$= e^{-x} \ln|u| - e^{-2x}(u - a\ln|u|)$$
$$= e^{-x} \ln|a + e^x| - e^{-2x}(a + e^x - a\ln|a + e^x|)$$
$$= e^{-2x}(a + e^x)(-1 + \ln|a + e^x|).$$

15. The associated characteristic equation is $r^2 + 3r + 2 = (r + 1)(r + 2) = 0$, with roots $r_1 = -1$, $r_2 = -2$. The basic homogeneous solutions are therefore $y_1(x) = e^{-x}$ and $y_2(x) = e^{-2x}$. Since their Wronskian is $w(x) = -e^{-2x}$ (see Exercise 15 above), the associated Green's function is

$$G(x,t) = \frac{1}{-1-(-2)}(e^{-(x-t)} - e^{-2(x-t)}) = e^{-(x-t)} - e^{-2(x-t)}.$$

Since the given D.E. is already normalized, the forcing function $f(x)$ we will use in the computation of Green's integral is

$$f(x) = \begin{cases} 0, & x < 1, \\ 1, & 1 \le x. \end{cases}$$

More specifically, as the given initial conditions are at $x_0 = 0$, we must compute

$$y_p(x) = \int_0^x G(x,t)f(t)\,dt = \int_0^x (e^{-(x-t)} - e^{-2(x-t)})f(t)\,dt.$$

For $x < 1$, all values of t in the above integral are less than 1, so that $f(t) = 0$ and therefore $y_p(x) = 0$ for $x < 1$. For $x \ge 1$, we have

$$y_p(x) = \int_0^x (e^{-(x-t)} - e^{-2(x-t)})f(t)\,dt$$

$$= \int_0^1 (e^{-(x-t)} - e^{-2(x-t)})(0)\,dt + \int_1^x (e^{-(x-t)} - e^{-2(x-t)})(1)\,dt$$

$$= e^{-x}\int_1^x e^t\,dt - e^{-2x}\int_1^x e^{2t}\,dt = e^{-x}\left[e^t\right]_1^x - e^{-2x}\left[\frac{1}{2}e^{2t}\right]_1^x$$

$$= \frac{1}{2} - e^{1-x} + \frac{1}{2}e^{2(1-x)} = \frac{1}{2}(1 - e^{1-x})^2.$$

Hence, the particular solution with $y_p(0) = y'_p(0) = 0$ is

$$y_p(x) = \begin{cases} 0, & x < 1, \\ \frac{1}{2}(1 - e^{1-x})^2, & 1 \le x. \end{cases}$$

Since the homogeneous solution is $c_1 e^{-x} + c_2 e^{-2x}$, the general solution of the given nonhomogeneous equation is

$$y(x) = c_1 e^{-x} + c_2 e^{-2x} + y_p(x).$$

Using this and its derivative $y'(x) = -c_1 e^{-x} - 2c_2 e^{-2x} + y'_p(x)$, we use the fact that $y_p(0) = y'_p(0) = 0$ (by construction) and apply the given initial conditions to get the two equations

$$1 = y(0) = c_1 + c_2 \qquad \text{and} \qquad 2 = y'(0) = -c_1 - 2c_2.$$

The elimination method gives $c_1 = 4$ and $c_2 = -3$. The solution of the given initial-value problem is therefore

$$y(x) = 4e^{-x} - 3e^{-2x} + \begin{cases} 0, & x < 1, \\ \frac{1}{2}(1 - e^{1-x}), & 1 \le x. \end{cases}$$

17. The associated homogeneous equation $y'' + (1/x)y' = 0$ is a first-order linear equation in y' and can be solved using an exponential multiplier, namely, $e^{\int (1/x)\,dx} = e^{\ln|x|} = |x| = x$, where the absolute value sign is not needed because the interval over which we are interested is $x > 0$ (the initial conditions tell us this). Multiplication by x gives

$$xy'' + y' = 0, \qquad \text{or equivalently,} \qquad \frac{d}{dx}(xy') = 0.$$

Integrating gives $xy' = c_1$, or $y' = c_1/x$. Integrating again gives $y(x) = c_1 \ln x + c_2$. Hence, $y_1(x) = 1$ and $y_2(x) = \ln x$ are linearly independent solution of the homogeneous equation. Since, their Wronskian is

$$w(x) = y_1(x)y_2'(x) - y_2(x)y_1'(x) = 1(1/x) - (\ln x)(0) = 1/x,$$

the associated Green's function (given by Equation 3.5 in the text) is

$$G(x,t) = \frac{y_1(t)y_2(x) - y_2(t)y_1(x)}{w(t)} = \frac{(1)(\ln x) - (\ln t)(1)}{1/t} = t\ln x - t\ln t.$$

The initial conditions are at $x_0 = 1$. So, with $f(x) = 1/x$, the particular solution of the given nonhomogeneous equation that satisfies $y_p(1) = y_p'(1) = 0$ is

$$y_p(x) = \int_1^x G(x,t)f(t)\,dt = \int_1^x (t\ln x - t\ln t)\frac{1}{t}\,dt$$

$$= \ln x \int_1^x dt - \int_1^x \ln t\,dt = \ln x \,[t]_1^x - [t\ln t - t]_1^x = -\ln x + x - 1$$

The general solution of the given nonhomogeneous equation is therefore

$$y(x) = c_1 \ln x + c_2 + y_p(x).$$

Using this and its derivative $y'(x) = c_1/x + y_p'(x)$, we use the fact that $y_p(1) = y_p'(1) = 0$ (by construction) and apply the given initial conditions to get the two equations

$$0 = y(1) = c_2 \qquad \text{and} \qquad 2 = y'(1) = c_1.$$

The solution of the given initial-value problem is therefore

$$y(x) = 2\ln x - \ln x + x - 1 = \ln x + x - 1.$$

19. Write the Green's function Formula 3.4 in the text as

$$y_p(x) = y_1(x) \int_{x_0}^x \frac{-y_2(t)}{w(t)}f(t)\,dt + y_2(x) \int_{x_0}^x \frac{y_1(t)}{w(t)}f(t)\,dt.$$

The first step in differentiating both sides of this equation with respect to x is to write

$$y_p'(x) = \frac{d}{dx}\left[y_1(x) \int_{x_0}^x \frac{-y_2(t)}{w(t)}f(t)\,dt + y_2(x) \int_{x_0}^x \frac{y_1(t)}{w(t)}f(t)\,dt\right]$$

$$= \frac{d}{dx}\left[y_1(x) \int_{x_0}^x \frac{-y_2(t)}{w(t)}f(t)\,dt\right] + \frac{d}{dx}\left[y_2(x) \int_{x_0}^x \frac{y_1(t)}{w(t)}f(t)\,dt\right].$$

414

In carrying out the differentiation of each term on the right, apply the product rule and the fundamental theorem of calculus. For example, the first term on the right is equivalent to

$$y_1(x)\frac{-y_2(x)}{w(x)}f(x) + y_1'(x)\int_{x_0}^{x}\frac{-y_2(t)}{w(t)}f(t)\,dt = -\frac{y_1(x)y_2(x)f(x)}{w(x)} + \int_{x_0}^{x}\frac{-y_2(t)y_1'(x)}{w(t)}f(t)\,dt.$$

where $y_1'(x)$ can be brought inside the integral because it is independent of t and behaves as a constant. Similarly, the second term is equivalent to

$$y_2(x)\frac{y_1(x)}{w(x)}f(x) + y_2'(x)\int_{x_0}^{x}\frac{y_1(t)}{w(t)}f(t)\,dt = \frac{y_1(x)y_2(x)f(x)}{w(x)} + \int_{x_0}^{x}\frac{y_1(t)y_2'(x)}{w(t)}f(t)\,dt.$$

When the two pieces are added together, the first terms of each piece cancel each other and the two integrals can be written as a single integral. The result is

$$y_p'(t) = \int_{x_0}^{x}\left[\frac{y_1(t)y_2'(x)}{w(t)}f(t) + \frac{-y_2(t)y_1'(x)}{w(t)}f(t)\right]dt = \int_{x_0}^{x}\frac{y_1(t)y_2'(x) - y_2(t)y_1'(x)}{w(t)}f(t)\,dt.$$

Manifestly, $y_p'(x_0) = 0$.

21. Let r_1, r_2 be the roots of the differential equation $y'' + ay' + by = 0$.

(i) If r_1 and r_1 are real and unequal then $y_1(x) = e^{r_1 x}$ and $y_2(x) = e^{r_2 x}$ are independent solutions with Wronskian

$$w(x) = y_1(x)y_2'(x) - y_2(x)y_1'(x) = e^{r_1 x}(r_2 e^{r_2 x}) - e^{r_2 x}(r_1 e^{r_1 x}) = (r_2 - r_1)e^{(r_1 + r_2)x}.$$

Therefore, by Equation 3.5 in the text, the corresponding Green's function is

$$G(x,t) = \frac{y_1(t)y_2(x) - y_2(t)y_1(x)}{w(t)} = \frac{e^{r_1 t}e^{r_2 x} - e^{r_2 t}e^{r_1 x}}{(r_2 - r_1)e^{(r_1 + r_2)t}}$$

$$= \frac{1}{r_2 - r_1}(e^{r_1 t}e^{r_2 x}e^{-(r_1 + r_2)t} - e^{r_2 t}e^{r_1 x}e^{-(r_1 + r_2)t})$$

$$= \frac{1}{r_2 - r_1}(e^{r_2 x - r_2 t} - e^{r_1 x - r_1 t}) = \frac{1}{r_1 - r_2}(e^{r_1(x-t)} - e^{r_2(x-t)}).$$

(ii) If $r_1 = r_2 = r$ then $y_1(x) = e^{rx}$ and $y_2(x) = xe^{rx}$ are independent solutions with Wronskian

$$w(x) = y_1(x)y_2'(x) - y_2(x)y_1'(x) = e^{rx}(rxe^{rx} + e^{rx}) - xe^{rx}(re^{rx}) = e^{2rx}.$$

Therefore, by Equation 3.5, the corresponding Green's function is

$$G(x,t) = \frac{y_1(t)y_2(x) - y_2(t)y_1(x)}{w(t)} = \frac{e^{rt}xe^{rx} - te^{rt}e^{rx}}{e^{2rt}} = (x-t)e^{rt+rx}e^{-2rt} = (x-t)e^{r(x-t)}.$$

(iii) If $r_{1,2} = \alpha + i\beta$ then $y_1(x) = e^{\alpha x}\cos\beta x$ and $y_2(x) = e^{\alpha x}\sin\beta x$ are independent solutions with Wronskian

$$w(x) = y_1(x)y_2'(x) - y_2(x)y_1'(x)$$

$$= e^{\alpha x}\cos\beta x(\beta e^{\alpha x}\cos\beta x + \alpha e^{\alpha x}\sin\beta x) - e^{\alpha x}\sin\beta x(-\beta e^{\alpha x}\sin\beta x + \alpha e^{\alpha x}\cos\beta x)$$

$$= e^{2\alpha x}(\beta\cos^2\beta x + \alpha\cos\beta x\sin\beta x + \beta\sin^2\beta x - \alpha\sin\beta x\cos\beta x)$$

$$= \beta e^{2\alpha x}.$$

Therefore, by Equation 3.5, the corresponding Green's function is

$$G(x,t) = \frac{y_1(t)y_2(x) - y_2(t)y_1(x)}{w(t)} = \frac{e^{\alpha t}\cos\beta t(e^{\alpha x}\sin\beta x) - e^{\alpha t}\sin\beta t(e^{\alpha x}\cos\beta x)}{\beta e^{2\alpha t}}$$

$$= \frac{1}{\beta}e^{-\alpha t+\alpha x}(\cos\beta t\sin\beta x - \sin\beta t\cos\beta x) = \frac{1}{\beta}e^{\alpha(x-t)}\sin\beta(x-t),$$

where the last member follows from the identity $\sin(a - b) = \sin a\cos b - \cos a\sin b$.

415

23. The Green's function integral given in Example 11 in the text is restricted to $x_0 \neq 0$. In computing the case for $x_0 > 0$, x must be restricted to $x > 0$ because the interval of integration cannot contain $t = 0$. Similarly, the case for $x_0 < 0$ requires the restriction $x < 0$. Since $f(x) = x$ is the same in both cases, we have

$$y_p(x) = \int_{x_0}^{x} \left(\frac{x^2}{t} - x\right) t\, dt = x^2 \int_{x_0}^{x} dt - x \int_{x_0}^{x} t\, dt$$

$$= x^2 \left[t\right]_{x_0}^{x} - x \left[\frac{1}{2}t^2\right]_{x_0}^{x} = x^2(x - x_0) - \frac{1}{2}x(x^2 - x_0^2)$$

$$= -\frac{1}{2}x_0^2 x - x_0 x^2 + \frac{1}{2}x^3 = \frac{1}{2}x(x - x_0)^2, \quad x/x_0 > 0.$$

25. For an arbitrarily chosen initial point x_0, we set $f(x) = e^x$ in the differential equation given in Example 10 in the text and compute the Green's function integral:

$$y_p(x) = \int_{x_0}^{x} (e^{2(x-t)} - e^{(x-t)}) e^t\, dt = e^{2x} \int_{x_0}^{x} e^{-t}\, dt - e^x \int_{x_0}^{x} dt = e^{2x}\left[-e^{-t}\right]_{x_0}^{x} - e^x\left[t\right]_{x_0}^{x}$$

$$= -e^{2x}(e^{-x} - e^{-x_0}) - e^x(x - x_0) = (x_0 - 1)e^x + (e^{-x_0})e^{2x} - xe^x.$$

27. We are given that the homogeneous equation $(x - 1)y'' - xy' + y = 0$ has x and e^x as independent solutions and, by inspection, $y_p(x) = 1$ is a particular solution of the given nonhomogeneous equation. the general solution of the given D.E. is therefore

$$y(x) = c_1 x + c_2 e^x + 1.$$

Consider the attempt to determine a solution that satisfies the initial conditions $y(1) = y_0$ and $y'(1) = z_0$. With $y'(x) = c_1 + c_2 e^x$, we formally apply these conditions and obtain the equations

$$y_0 = y(1) = c_1 + c_2 e + 1 \qquad \text{and} \qquad z_0 = y'(1) = c_1 + c_2 e.$$

Subtracting the first equation from the second gives $z_0 - y_0 = -1$, or $z_0 = y_0 - 1$. On the other hand, if y_0 and z_0 are not related in this way then the above two equations are inconsistent and therefore no such solution exists. Otherwise, they are equivalent equations from which we can deduce that c_1 and c_2 must be related by $c_1 = y_0 - 1 - c_2 e$. The solutions satisfying these conditions are therefore given by the infinitely many members of the one-parameter family

$$y(x) = (y_0 - 1 - c_2 e)x + c_2 e^x + 1 = 1 + (y_0 - 1)x - c_2 e(x - e^{x-1}) = 1 + (y_0 - 1)x + c(x - e^{x-1}),$$

where $c = -c_2 e$ is arbitrary.

29. By inspection, the constant coefficients of y' and y in the given Euler equation are $a = 1$ and $b = -1$, respectively, and the associated indicial equation is $\mu^2 - 1 = 0$, with roots $r_{1,2} = \pm 1$. By part (b) of Exercise 28 in this section, x and x^{-1} are independent solutions for $x > 0$. Therefore, since the operator $x^2 D + xD - 1$ is linear, the general solution of the given Euler equation is the two-parameter family

$$y(x) = c_1 x + c_2 x^{-1}, \quad x > 0.$$

31. By inspection, the constant coefficients of y' and y in the given Euler equation are $a = 3$ and $b = 1$, respectively, and the associated indicial equation is $\mu^2 + 2\mu + 1 = (\mu + 1)^2 = 0$, with the double root $\mu_1 = -1$. By part (d) of Exercise 28 in this section, x^{-1} and $x^{-1}\ln x$ are independent solutions for $x > 0$. Therefore, since the operator $x^2 D^2 + 3xD + 1$ is linear, the general solution of the given Euler equation is the two-parameter family

$$y(x) = c_1 x^{-1} + c_2 x^{-1} \ln x, \quad x > 0.$$

33. (a) The given 3×3 determinant can be evaluated by expansion about the first column. This gives

$$0 = \begin{vmatrix} y & f & g \\ y' & f' & g' \\ y'' & f'' & g'' \end{vmatrix} = y'' \begin{vmatrix} f & g \\ f' & g' \end{vmatrix} - y' \begin{vmatrix} f & g \\ f'' & g'' \end{vmatrix} + y \begin{vmatrix} f' & g' \\ f'' & g'' \end{vmatrix}$$

$$= y''(fg' - f'g) - y'(fg'' - f''g) + y(f'g'' - f''g'),$$

where the coefficients of y, y' and y'' are functions of x only. In particular, the coefficient of y'' is $f(x)g'(x) - f'(x)g(x)$. By hypothesis, $f(x)g'(x) - f'(x)g(x) \neq 0$ on $a < x < b$. This shows that the given determinant equation is a *second-order* homogeneous linear differential equation on the interval $a < x < b$. To see that $f = f(x)$ (e.g.) is a solution of this differential equation, replace y, y' and y'' in the 3×3 determinant with f, f' and f'', respectively, and recall that an $n \times n$ determinant with two identical columns has value 0 (this holds for any integer $n \geq 2$). Similarly, $g = g(x)$ is a solution (for the same reason).

(b) With $f(x) = \sin x$ and $g(x) = x \sin x$, we have $f'(x) = \cos x$, $f''(x) = -\sin x$, $g'(x) = x\cos x + \sin x$ and $g''(x) = -x\sin x + 2\cos x$. A straightforward computation of the coefficients gives

$$fg' - f'g = \sin^2 x, \quad fg'' - f''g = 2\sin x \cos x, \quad f'g'' - f''g' = 2\cos^2 x + \sin^2 x.$$

The result of part (a) then gives

$$(\sin^2 x)y'' - (2\sin x \cos x)y' + (2\cos^2 x + \sin^2 x)y = 0.$$

Since the interval over which these solutions are to hold is $0 < x < \pi$, and since $\sin^2 x \neq 0$ on this interval, the above D.E. can be divided throughout by $\sin^2 x$ without changing the solutions. This leads to the somewhat more compact form

$$y'' - (2\cot x)y' + (2\cot^2 x + 1)y = 0, \quad 0 < x < \pi.$$

(c) With $f(x) = x$ and $g(x) = e^x$, we have $f'(x) = 1$, $f''(x) = 0$, $g'(x) = g''(x) = e^x$. A straightforward computation of the coefficients gives

$$fg' - f'g = xe^x - e^x, \quad fg'' - f''g = xe^x, \quad f'g'' - f''g' = e^x.$$

The result of part (a) then gives $(xe^x - e^x)y'' - (xe^x)y' + e^x y = 0$, or equivalently, (since e^x is never 0),

$$(x - 1)y'' - xy' + y = 0.$$

Section 4: OSCILLATIONS

Exercise Set 4ABC (pgs. 531-540)

1. (a) $m = 1$, $k = 2$, $h = 1$ and $k^2 - 4mh = 0$. The D.E. is critically damped.
   (b) Char. eq.: $r^2 + 2r + 1 = 0$; roots: $r_1 = r_2 = -1$; general solution:

$$x(t) = c_1 e^{-t} + c_2 t e^{-t}.$$

3. (a) $m = 1$, $k = 0$, $h = 9$ and $k^2 - 4mh = -36$. Since $k = 0$ the oscillations are harmonic.
   (b) Char.eq.: $r^2 + 9 = 0$; roots: $r_{1,2} = \pm 3i$; general solution: $x(t) = c_1 \cos 3t + c_2 \sin 3t$.

5. (a) $m = 1$, $k = a > 0$, $h = a^2$ and $k^2 - 4mh = -3a^2 < 0$. Since $k \neq 0$, the D.E. is underdamped
   (b) Char. eq.: $r^2 + ar + a^2 = 0$; roots: $r_{1,2} = \frac{1}{2}(-a \pm ia\sqrt{3})$; general solution:

$$x(t) = e^{-at/2}\left(c_1 \cos \frac{a\sqrt{3}}{2}t + c_2 \sin \frac{a\sqrt{3}}{2}t \right).$$

7. With $\dot{x}(t) = -2c_1 \sin 2t + 2c_2 \cos 2t$, the initial conditions give the two equations

$$0 = x(0) = c_1 \quad \text{and} \quad 1 = \dot{x}(0) = 2c_2.$$

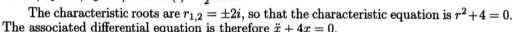

Hence, $c_1 = 0$, $c_2 = 1/2$ and $x(t) = \frac{1}{2}\sin 2t$.

   The characteristic roots are $r_{1,2} = \pm 2i$, so that the characteristic equation is $r^2 + 4 = 0$. The associated differential equation is therefore $\ddot{x} + 4x = 0$.

9. With $\dot{x}(t) = -2c_1 t e^{-2t} + (c_1 - 2c_2)e^{-2t}$, the initial conditions give the two equations

$$0 = x(0) = c_2 \quad \text{and} \quad -1 = \dot{x}(0) = c_1 - 2c_2.$$

Hence, $c_1 = -1$, $c_2 = 0$ and $x(t) = -te^{-2t}$.

   The characteristic roots are $r_1 = r_2 = -2$, so that the characteristic equation is $(r + 2)^2 = r^2 + 4r + 4 = 0$. The associated differential equation is therefore $\ddot{x} + 4\dot{x} + 4x = 0$.

11. With $\dot{x}(t) = -2c_1 e^{-2t} - 4c_2 e^{-4t}$, the initial conditions give the two equations

$$-1 = c_1 + c_2 \quad \text{and} \quad -1 = \dot{x}(0) = -2c_1 - 4c_2.$$

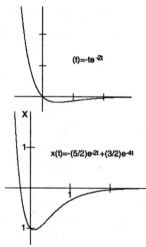

Hence, $c_1 = -5/2$, $c_2 = 3/2$ and $x(t) = -\frac{5}{2}e^{-2t} + \frac{3}{2}e^{-4t}$.

   The characteristic roots are $r_1 = -2$, $r_2 = -4$, so that the characteristic equation is $(r + 2)(r + 4) = r^2 + 6r + 8 = 0$. The associated differential equation is then $\ddot{x} + 6\dot{x} + 8x = 0$.

13. The characteristic equation is $r^2 + 3r + 2 = 0$, with roots $r_1 = -1$, $r_2 = -2$. The homogeneous solution (transient solution) is therefore

$$x_h(t) = c_1 e^{-t} + c_2 e^{-2t}.$$

Since neither the forcing function $\cos t$ nor its derivative $\sin t$ are homogeneous solutions, $x_p(t) = A\cos t + B\sin t$ is a particular solution of the nonhomogeneous equation for some constants A and B. With $\dot{x}_p(t) = -A\sin t + B\cos t$, $\ddot{x}_p(t) = -A\cos t - B\sin t$, the nonhomogeneous D.E. gives

$$\cos t = \ddot{x}_p + 3\dot{x}_p + 2x_p$$
$$= -A\cos t - B\sin t + 3(-A\sin t + B\cos t) + 2(A\cos t + B\sin t)$$
$$= (A + 3B)\cos t + (-3A + B)\sin t.$$

Equating coefficients of like functions leads to the equations $1 = A + 3B$ and $0 = -3A + B$. Hence, $A = 1/10$, $B = 3/10$ and the steady-state solution is

$$x_p(t) = \frac{1}{10}\cos t + \frac{3}{10}\sin t.$$

The general solution of the given homogeneous equation is therefore

$$x(t) = c_1 e^{-t} + c_2 e^{-2t} + \frac{1}{10}\cos t + \frac{3}{10}\sin t.$$

With $\dot{x}(t) = -c_1 e^{-t} - 2c_2 e^{-2t} - \frac{1}{10}\sin t + \frac{3}{10}\cos t$, the initial conditions lead to the two equations

$$1 = x(0) = c_1 + c_2 + \frac{1}{10} \qquad \text{and} \qquad 0 = \dot{x}(0) = -c_1 - 2c_2 + \frac{3}{10},$$

which has the solution $c_1 = 3/2$, $c_2 = -3/5$. The solution satisfying the given initial conditions is therefore

$$x(t) = \frac{3}{2}e^{-t} - \frac{3}{5}e^{-2t} + \frac{1}{10}\cos t + \frac{3}{10}\sin t,$$

where $x_h(t) = \frac{3}{2}e^{-t} - \frac{3}{5}e^{-2t}$ is the transient solution.

To find the earliest time $t = t_0$ beyond which $|x_h(t)| < 0.01$, we differentiate the transient solution $x_h(t) = \frac{3}{2}e^{-t} - \frac{3}{5}e^{-2t}$ and judiciously write the result as

$$\dot{x}_h(t) = -\frac{3}{2}e^{-t} + \frac{6}{5}e^{-2t} = \frac{3}{2}e^{-2t}\left(\frac{4}{5} - e^t\right).$$

By inspection, $\dot{x}_h(t) = 0$ implies $e^t = 4/5$, so that $t = \ln(4/5)$ (≈ -0.22). Also, $\dot{x}_h(t)$ is positive for $t < \ln(4/5)$ and negative for $t > \ln(4/5)$. So the graph of $x_h(t)$ increases on $(-\infty, \ln(4/5))$, attains a global maximum at $t = \ln(4/5)$ and decreases on $(\ln(4/5), \infty)$. Moreover, using the formula for $x_h(t)$ and our knowledge of the various limiting behaviors of exponential functions, we have $\lim_{t\to\infty} x_h(t) = 0$. It follows that t_0 is the positive solution of the equation $x_h(t) = 0.01$. Therefore, we find the positive solution of

$$\frac{2}{3}e^{-t} - \frac{3}{5}e^{-2t} = \frac{1}{100}, \qquad \text{which can be written as} \qquad e^{2t} - 150e^t + 60 = 0.$$

The equation on the right is a quadratic equation in e^t. The quadratic formula gives the two possibilities $e^t = 75 \pm \sqrt{5565}$. Solving these for t gives the two t-solutions $\ln(75 \pm \sqrt{5565})$. The smaller of these two t values is negative and is therefore discarded. Hence,

$$t_0 = \ln(75 + \sqrt{5565}) = 5.007957897\ldots.$$

419

15. (a) $h = \dfrac{\text{additional weight}}{\text{additional extension}} = \dfrac{8-5}{1-1/2} = 6.$

(b) The units for h in part (a) are lbs/ft (pounds per feet). In terms of kg/m (kilograms per meter)

$$h = 6\frac{1 \text{ lbs}}{1 \text{ ft}} = 6\frac{(1/2.2) \text{ kg}}{(1/3.28) \text{ m}} = 6\frac{3.28}{2.2} \text{ kg/m} \approx 8.9 \text{ kg/m}.$$

(c) Let W be the larger weight. Then the additional weight is $W - 20$. Hence,

$$120 = h = \frac{\text{additional weight}}{\text{additional extension}} = \frac{W - 20}{1/2}.$$

Solving for W gives $W = 80$ lbs.

(d) From the first two measurements,

$$\frac{\text{additional weight}}{\text{additional compression}} = \frac{6 \text{ kg} - 5 \text{ kg}}{20 \text{ cm} - 10 \text{ cm}} = \frac{1}{10} \text{ kg/cm};$$

while from the last two measurements,

$$\frac{\text{additional weight}}{\text{additional compression}} = \frac{7 \text{ kg} - 6 \text{ kg}}{10 \text{ cm} - 5 \text{ cm}} = \frac{1}{5} \text{ kg/cm};$$

Since the ratios are not the same, Hooke's Law is not valid over this range of compression.

17. Nontrivial solutions of the unforced equation $m\ddot{x} + k\dot{x} + hx = 0$ will oscillate if, and only if, the discriminant $k^2 - 4mh$ of the characteristic equation is negative.

(a) Set $m = 2$ in $k^2 - 4mh < 0$ and conclude that $k^2 < 8h$.

(b) Set $h = k = 1$ in $k^2 - 4mk < 0$ and conclude that $m > 1/4$.

(c) If $r_{1,2} = \alpha \pm i\beta$ are the two complex roots of the characteristic equation then $\beta = \frac{1}{2m}\sqrt{4mh - k^2}$ is the associated circular frequency. Setting $\beta = 1/2$ and $m = h = 1$, gives the equation $\frac{1}{2} = \frac{1}{2}\sqrt{4 - k^2}$, from which $k^2 = 3$. Since $k > 0$, $k = \sqrt{3}$.

(d) Set $m = h = 1$ in $k^2 - 4mh < 0$ and conclude that $k^2 < 4$. Since $k > 0$, $0 < k < 2$.

19. Since $\sqrt{2^2 + 3^2} = \sqrt{13}$, there is a unique angle $0 \le \phi < \pi/2$ such that $\cos\phi = 2/\sqrt{13}$ and $\sin\phi = 3/\sqrt{13}$. Hence,

$$2\cos t + 3\sin t = \sqrt{13}\left(\frac{2}{\sqrt{13}}\cos t + \frac{3}{\sqrt{13}}\sin t\right)$$
$$= \sqrt{13}(\cos\phi\cos t + \sin\phi\sin t) = \sqrt{13}\cos(t - \phi).$$

The coefficient of $\cos(t - \phi)$ is the amplitude $A = \sqrt{13}$ and the coefficient of t divided by 2π is the frequency $f = 1/(2\pi)$. Further, the definitions of $\cos\phi$ and $\sin\phi$ leads to $\tan\phi = 3/2$, so that $\phi = \arctan(3/2) \approx 0.9828$ radians.

21. Since $\sqrt{2^2 + 1^2} = \sqrt{5}$, there is a unique angle $0 \le \phi < \pi/2$ such that $\cos\phi = 2/\sqrt{5}$ and $\sin\phi = 1/\sqrt{5}$. Hence,

$$2\cos\pi t + \sin\pi t = \sqrt{5}\left(\frac{2}{\sqrt{5}}\cos\pi t + \frac{1}{\sqrt{5}}\sin\pi t\right)$$
$$= \sqrt{5}(\cos\phi\cos\pi t + \sin\phi\sin\pi t) = \sqrt{5}\cos(\pi t - \phi),$$

The coefficient of $\cos(\pi t - \phi)$ is the amplitude $A = \sqrt{5}$ and the coefficient of t divided by 2π is the frequency $f = \pi/(2\pi) = 1/2$. Further, the definitions of $\cos\phi$ and $\sin\phi$ leads to $\tan\phi = 1/2$, so that $\phi = \arctan(1/2) \approx 0.4636$ radians.

23. Since $\frac{\sqrt{3}}{2}\cos t + \frac{1}{2}\sin t = \cos(t - \pi/6)$, this function is out of phase with $\cos t$ by $\pi/6$ radians.

25. First note that $\sin t = \cos(t - \pi/2)$ (cofunction identity). And since

$$\frac{1}{\sqrt{2}}\cos t + \frac{1}{\sqrt{2}}\sin t = \cos(t - \pi/4),$$

the two functions are out of phase by $|\pi/2 - \pi/4| = \pi/4$ radians.

27. Let $x = x(t)$ be the displacement of the weight from equilibrium at time t. By Newton's Second Law, the force F acting to move the mass is $F = md^2x/dt^2 = \ddot{x}$, where we have used the fact that $m = 1$. By Hooke's Law, the force F_1 (due the the spring with spring constant h_1) acting to move the mass is $F_1 = -h_1 x$. Similarly, the force F_2 (due to the spring with spring constant h_2) acting to move the mass is $F_2 = -h_2 x$. Since the total force F tending to move the mass is also given by the sum of the forces F_1 and F_2, it follows that

$$\ddot{x} = F = F_1 + F_1 = -h_1 x - h_2 x = -(h_1 + h_2)x.$$

29. Let $x = x(t)$ be the positive distance the artillery piece has recoiled at time $t \geq 0$. The conditions described in the statement of the problem lead to the initial-value problem

$$m\ddot{x} + k\dot{x} + hx = 0, \quad x(0) = 0, \ \dot{x}(0) = V_0 > 0.$$

This is the D.E that will be used to work parts (a) and (b) below.

(a) The gun barrel undergoes critical damping if, and only if, the discriminant satisfies $k^2 - 4mh = 0$. In this case, there is a double characteristic root $r = -k/(2m)$ and the general solution of the above D.E. is $x(t) = (c_1 + c_2 t)e^{-kt/(2m)}$. The condition $x(0) = 0$ gives $c_1 = 0$, so that $x(t) = c_2 t e^{-kt/(2m)}$. With $\dot{x}(t) = c_2(1 - kt/(2m))e^{-kt/(2m)}$, the condition $\dot{x}(0) = V_0$ gives $c_2 = V_0$. Hence, the solution of the above initial-value problem is

$$x(t) = V_0 t e^{-kt/(2m)}, \quad t \geq 0.$$

The condition that a fixed maximum recoil distance E is always attained implies that there is a time $t = t_1$ such that $x(t_1) = E$ and $\dot{x}(t_1) = 0$. Since $\dot{x}(t) = V_0(1 - kt/(2m))e^{-kt/(2m)} = 0$ has the unique solution $t = 2m/k$, it follows that $t_1 = 2m/k$. Hence,

$$x(t_1) = x(2m/k) = V_0(2m/k)e^{-1} = E, \quad \text{which implies} \quad k = 2mV_0/(eE).$$

Using this expression for k in the equation $k^2 - 4mh = 0$ gives

$$h = k^2/(4m) = (2mV_0/(eE))^2/(4m) = mV_0^2/(eE)^2.$$

(b) Dividing the D.E. throughout my m gives $\ddot{x} + (k/m)\dot{x} + (h/m)x = 0$. Since the expressions for k and h derived in part (b) are equivalent to $k/m = 2V_0/(eE)$ and $h/m = v_0^2/(eE)^2$, respectively, the D.E. can be expressed as

$$\ddot{x} + \left(\frac{2V_0}{eE}\right)\dot{x} + \left(\frac{V_0}{eE}\right)^2 x = 0.$$

31. (a) Set $k = 0$ and $\omega = \omega_0 = \sqrt{h/m}$ in the given D.E. to get $\ddot{x} + \omega_0^2 x = (a_0/m)\sin\omega_0 t$. The homogeneous solution is $x_h(t) = c_1 \cos\omega_0 t + c_1 \sin\omega_0 t$, so that the input function $(a_0/m)\sin\omega_0 t$ is a homogeneous solution. Therefore, there is a particular solution y_p of the given D.E. of the form $x_p(t) = At\cos\omega_0 t + Bt\sin\omega_0 t$, where at least one of the coefficients A or B is nonzero. Every solution of the D.E. is therefore of the form

$$x(t) = At\cos\omega_0 t + Bt\sin\omega_0 t + c_1\cos\omega_0 t + c_2\sin\omega_0 t.$$

If $A \neq 0$ then, for $n = 1, 2, 3, \ldots$, and any solution $x(t)$

$$x\left(\frac{2n\pi}{\omega_0}\right) = A\left(\frac{2n\pi}{\omega_0}\right)\cos(2n\pi) + B\left(\frac{2n\pi}{\omega_0}\right)\sin(2n\pi)$$

$$+ c_1\cos(2n\pi) + c_2\sin(2n\pi) = \frac{2n\pi A}{\omega_0} + c_1.$$

Regardless of the actual value of A or the choice of the constant c_1, the absolute value of the rightmost member above can be made arbitrarily large by choosing n sufficiently large. Thus, $x(t)$ has arbitrarily large deviations from the equilibrium position $x = 0$. A similar result holds if $B \neq 0$ (in this case, let $t = (4n+1)\pi/(2\omega_0)$ with $n = 1, 2, 3, \ldots$).

(b) The forcing function in Example 5 is $a_0\cos\omega t$ and the forcing function here is $a_0\sin\omega t$. However, with $k > 0$ and fixed, both steady-state responses have circular frequency ω and therefore they have the same amplitude

$$\frac{|a_0|}{\sqrt{(h - \omega^2 m)^2 + \omega^2 k^2}}.$$

Since k and ω are fixed, the above amplitude is maximum when h and m are chosen such that $(h - \omega^2 m)^2$ is minimal; i.e., when $(h - \omega^2 m) = 0$. This happens if, and only if, h and m are related by $\sqrt{h/m} = \omega$. Letting h and m be related in this way (so that $(h - \omega^2 m)^2 = 0$), the above amplitude becomes $|a_0|/\sqrt{\omega^2 k^2} = |a_0|(\omega k)^{-1}$.

(c) The form of the expression for the steady-state response amplitude (as given in part (b)) shows that the larger the quantity $(h - \omega^2 m)^2$, the smaller the amplitude. Since $(h - \omega^2 m)^2 = |h - \omega^2 m|^2$, the result follows from the fact that $|h - \omega^2 m|$ increases as $|h - \omega^2 m|^2$ increases.

33. (a) The characteristic equation is $mr^2 + kr + h = 0$. Since, m, k and h are all positive, either both roots are real and negative or are complex conjugates. Let r_1, r_2 be the two roots. In the first case, the solutions of the associated homogeneous equation are of the form

$$x_h(t) = c_1 e^{r_1 t} + c_2 e^{r_2 t} \quad (r_1 \neq r_2) \qquad \text{or} \qquad x_h(t) = (c_1 + c_2 t)e^{r_1 t} \quad (r_1 = r_2).$$

Since r_1, r_2 are both negative, each of the above solutions tend to zero as $t \to \infty$. In the second case, the nonzero solutions are oscillatory and of the form

$$x_h(t) = e^{\alpha t}(c_1\cos\beta t + c_2\sin\beta t),$$

where $\alpha = -k/(2m) < 0$. Again, the exponential tends to zero (and therefore the solutions themselves tend to zero) as $t \to \infty$.

(b) Write the proposed solution as

$$x_p(t) = \left(\frac{-b_0 k\omega}{k^2\omega^2 + (h - m\omega^2)^2}\right)\cos\omega t + \left(\frac{b_0(h - m\omega^2)}{k^2\omega^2 + (h - m\omega^2)^2}\right)\sin\omega t.$$

422

Straightforward differentiation gives

$$\dot{x}_p(t) = \left(\frac{b_0\omega(h - m\omega^2)}{k^2\omega^2 + (h - m\omega^2)^2} \right) \cos\omega t + \left(\frac{b_0 k\omega^2}{k^2\omega^2 + (h - m\omega^2)^2} \right) \sin\omega t,$$

$$\ddot{x}_p(t) = \left(\frac{b_0 k\omega^3}{k^2\omega^2 + (h - m\omega^2)^2} \right) \cos\omega t + \left(\frac{-b_0\omega^2(h - m\omega^2)}{k^2\omega^2 + (h - m\omega^2)^2} \right) \sin\omega t.$$

Multiplying $x_p(t)$ by h, $\dot{x}_p(t)$ by k, $\ddot{x}_p(t)$ by m and adding the numerators of the resulting coefficients of the corresponding cosine functions, gives

$$-b_0 hk\omega + b_0 k\omega(h - m\omega^2) + b_0 mk\omega^3 = -b_0 hk\omega + b_0 k\omega h - b_0 km\omega^3 + b_0 mk\omega^3 = 0;$$

and adding the numerators of the resulting coefficients of the corresponding sine functions gives

$$b_0 h(h - m\omega^2) + b_0 k^2\omega^2 - b_0 m\omega^2(h - m\omega^2) = b_0 k^2\omega^2 + b_0(h - m\omega^2)(h - m\omega^2)$$
$$= b_0\left(k^2\omega^2 + (h - m\omega^2)^2\right).$$

It follows that when $x_p(t)$ and its derivatives are substituted into the given nonhomogeneous equation, the result is

$$m\ddot{x}_p + k\dot{x}_p + hx_p = \left(\frac{0}{k^2\omega^2 + (h - m\omega^2)^2} \right) \cos\omega t + \left(\frac{b_0\left(k^2\omega^2 + (h - m\omega^2)^2\right)}{k^2\omega^2 + (h - m\omega^2)^2} \right) \sin\omega t$$

$$= b_0 \sin\omega t.$$

Hence, the proposed function is indeed a solution of the given equation.

Next, letting $K = \sqrt{k^2\omega^2 + (h - m\omega^2)^2}$, we rewrite $x_p(t)$ as

$$x_p(t) = \frac{b_0}{K}\left(\frac{-k\omega}{K}\cos\omega t + \frac{h - m\omega^2}{K}\sin\omega t \right)$$

$$= \frac{b_0}{K}(\cos\phi\cos\omega t + \sin\phi\sin\omega t) = \frac{b_0}{K}\cos(\omega t - \phi),$$

where ϕ satisfies $\cos\phi = -k\omega/K$ and $\sin\phi = (h - m\omega^2)/K$. As $|\cos(\omega t - \phi)| \le 1$ for all t,

$$|x_p(t)| = \frac{|b_0|}{K}|\cos(\omega t - \phi)| \le \frac{|b_0|}{K}, \quad \text{for all } t.$$

Replacing K with $\sqrt{k^2\omega^2 + (h - m\omega^2)^2}$ gives the desired result.

(c) With $x_h(t)$ as in part (a) and $x_p(t)$ as in part (b), the general solution of the given nonhomogeneous D.E. is $x(t) = x_h(t) + x_p(t)$. Thus, for each t in the domain of x, the triangle inequality gives

$$|x(t)| = |x_h(t) + x_p(t)| \le |x_h(t)| + |x_p(t)|.$$

By parts (a) and (b), $x_h(t)$ and $x_p(t)$ are both bounded. Hence, so is $x(t)$.

35. (a) For $n \geq 0$, let $x_p(t) = \sum_{k=0}^{n} \frac{a_k}{2 - k^2} \cos kt$. Substituting this and its second deriva-

tive $\ddot{x}_p(t) = \sum_{k=0}^{n} \frac{-a_k k^2}{2 - k^2} \cos kt$ into the expression $\ddot{x} + 2x$ gives

$$
\begin{aligned}
\ddot{x}_p + 2x_p &= \sum_{k=0}^{n} \frac{-a_k k^2}{2 - k^2} \cos kt + 2\left(\sum_{k=0}^{n} \frac{a_k}{2 - k^2} \cos kt \right) \\
&= \sum_{k=0}^{n} \frac{-a_k k^2}{2 - k^2} \cos kt + \sum_{k=2}^{n} \frac{2a_k}{2 - k^2} \cos kt = \sum_{k=0}^{n} \left(\frac{-a_k k^2}{2 - k^2} + \frac{2a_k}{2 - k^2} \right) \cos kt \\
&= \sum_{k=0}^{n} \frac{a_k(-k^2 + 2)}{2 - k^2} \cos kt = \sum_{k=0}^{n} a_k \cos kt.
\end{aligned}
$$

So $x_p(t)$ is a solution of $\ddot{x} + 2x = \sum_{k=0}^{n} a_k \cos kt$.

(b) We seek to find one solution $x_p(t)$ of $\ddot{x} + 4x = \sum_{k=0}^{n} a_k \cos kt$ by using the method of undetermined coefficients. First, by inspection, $x_h(t) = c_1 \cos 2t + c_2 \sin 2t$ is the homogeneous solution. Next, if $k \neq 2$ then neither $\cos kt$ nor $\sin kt$ are homogeneous solutions and therefore $A_k \cos kt + B_k \sin kt$ should be included in the linear combination defining $x_p(t)$. However, if $n \geq 2$ and $k = 2$ then $\cos 2t$ *is* a homogeneous solution and therefore $A_2 t \cos 2t + B_2 t \sin 2t$ should be included in the linear combination defining $x_p(t)$ while $A_2 \cos 2t + B_2 \sin 2t$ should not. Thus, for $n \geq 0$, there exist constants $A_0, \ldots, A_n$, $B_0, \ldots, B_n$ such that

$$
x_p(t) = A_2 t \cos 2t + B_2 t \sin 2t + \sum_{k=0, k \neq 2}^{n} A_k \cos kt + \sum_{k=0, k \neq 2}^{n} B_k \sin kt,
$$

where the sum $A_2 t \cos 2t + B_2 t \sin 2t$ must vanish when $n = 0$ and $n = 1$. Also,

$$
\ddot{x}_p(t) = (4B_2 - 4A_2 t) \cos 2t + (-4A_2 - 4B_2 t) \sin 2t
$$
$$
+ \sum_{k=0, k \neq 2}^{n} (-k^2 A_k \cos kt) + \sum_{k=0, k \neq 2}^{n} (-k^2 B_k \sin kt),
$$

so that substitution into the D.E. results in

$$
\sum_{k=0}^{n} a_k \cos kt = \ddot{x}_p + 4x_p
$$
$$
= 4B_2 \cos 2t - 4A_2 \sin 2t + \sum_{k=0, k \neq 2}^{n} (4 - k^2) A_k \cos kt + \sum_{k=0, k \neq 2}^{n} (4 - k^2) B_k \sin kt.
$$

By inspection, equating coefficients of like functions leads to $A_k = a_k/(4 - k^2)$ (for $k \neq 2$), $A_2 = 0$, and $B_k = 0$ (for $k \neq 2$), $B_2 = a_2/4$ (with $a_2 = 0$ for $n = 0, 1$). Hence,

$$
x_p(t) = \frac{a_2}{4} t \sin 2t + \sum_{k=0, k \neq 2}^{n} \frac{a_k}{4 - k^2} \cos kt,
$$

where $a_2 = 0$ when $n = 0$ and $n = 1$.

37. (a) Let r be the time shift required to bring the two functions into phase. That is, r is a real number such that the phase difference between $\cos(\omega(t+r)-\alpha)$ and $\cos(\omega t - \beta)$ is zero. As $\cos(\omega(t+r)-\alpha) = \cos(\omega t - (\alpha - \omega r))$, the phase difference between this function and $\cos(\omega t - \beta)$ is $(\alpha - \omega r) - \beta$. It follows that r must satisfy $(\alpha - \omega r) - \beta = 0$. Solving for r gives $r = (\alpha - \beta)/\omega$.

(b) Use the cofunction identity to write $\sin(\omega t - \beta) = \cos(\omega t - \beta - \pi/2)$, and observe that the phase difference between $\cos(\omega t - \alpha)$ and $\cos(\omega t - (\beta + \pi/2))$ is $\alpha - (\beta + \pi/2) = \alpha - \beta - \pi/2$.

39. The D.E. has the form $m\ddot{x} + 3\dot{x} + 2x = 0$. The nontrivial solutions will oscillate if, and only if, the discriminant $9 - 8m$ is negative. This condition is equivalent to $m > 9/8$. Any mass $m \le 9/8 \, (= m_0)$ will not oscillate. In particular, no oscillation occurs if $m = 9/8$ (critical damping).

41. The form of the solution $x(t) = e^{-t}\sin 5t$ implies that the roots of the characteristic equation are the complex conjugates $r_{1,2} = -1 \pm 5i$. Since the equation

$$\left(r - (-1 + 5i)\right)\left(r - (-1 - 5i)\right) = r^2 + 2r + 26 = 0$$

is the characteristic equation for the D.E. $\ddot{x} + 2\dot{x} + 26x = 0$, it follows that $x(t) = e^{-t}\sin 5t$ is a solution of $\ddot{x} + 2\dot{x} + 26x = 0$. The condition $h = 1$ says that the coefficient of the x term in the D.E. must equal 1. We therefore divide the D.E. throughout by 26 to get $(1/26)\ddot{x} + (1/13)\dot{x} + x = 0$. This is now in the proper form to simply read off the remaining coefficients $m = 1/26$ and $k = 1/13$.

43. The conditions $k = 3$, $h = 2$ tell us that the given D.E. $m\ddot{x} + k\dot{x} + hx = 0$ having $x(t) = e^{-4t}\sin 4t$ as a solution must be of the form $m\ddot{x} + 3\dot{x} + 2x = 0$. Therefore, on one hand, its characteristic equation $mr^2 + 3r + 2 = 0$ has the complex roots $\frac{1}{2m}(-3 \pm i\sqrt{8m - 9})$, while on the other hand, the form of the given solution $x(t) = e^{-4t}\sin 4t$ implies that the roots of the characteristic equation are $-4 \pm 4i$. So $\frac{1}{2m}(-3 \pm i\sqrt{8m - 9}) = -4 \pm 4i$. Equating imaginary parts gives $4 = \frac{1}{2m}\sqrt{8m - 9}$. Multiplying by $2m$ and squaring the result gives an equation that can be written as $64m^2 - 8m + 9 = 0$, so that m is a root of the quadratic polynomial $64\mu^2 - 8\mu + 9 = 0$. But m must be real while the roots of $64\mu^2 - 8\mu + 9 = 0$ are complex conjugates. This is a contradiction. We conclude there is no real number m such that $e^{-4t}\sin 4t$ is a solution of $m\ddot{x} + 3\dot{x} + 2x = 0$.

Exercise Set 5 (pgs. 547-548)

1. For each $s > a$ we have

$$\mathcal{L}[e^{at}](s) = \int_0^\infty e^{-st}e^{at}\,dt = \lim_{b\to\infty}\int_0^b e^{-(s-a)t}\,dt = \lim_{b\to\infty}\left[-\frac{1}{s-a}e^{-(s-a)t}\right]_0^b$$

$$= \frac{1}{s-a}\lim_{b\to\infty}\left(1 - e^{-(s-a)b}\right) = \frac{1}{s-a},$$

where the fact that $s - a > 0$ implies that the exponential $e^{-(s-a)b}$ tends to 0 as $b \to \infty$.

3. Let $\mathcal{L}[y(t)](s) = Y(s)$. Then one integration by parts gives

$$\int_0^\infty e^{-st}y'(t)\,dt = \lim_{b\to\infty}\int_0^b e^{-st}y'(t)\,dt = \lim_{b\to\infty}\left(\left[e^{-st}y(t)\right]_0^b + s\int_0^b e^{-st}y(t)\,dt\right)$$

$$= \lim_{b\to\infty}\left(e^{-sb}y(b) - y(0)\right) + s\lim_{b\to\infty}\int_0^b e^{-st}y(t)\,dt$$

$$= \lim_{b\to\infty}e^{-sb}y(b) - y(0) + s\int_0^\infty e^{-st}y(t)\,dt$$

$$= -y(0) + sY(s) = -1 + sY(s),$$

where $\lim_{b\to\infty}e^{-sb}y(b) = 0$ and $y(0) = 1$ justifies the last line.

5. With $b = 2$ in formula 9 in Table 1, we get

$$\mathcal{L}[t\sin 2t](s) = \frac{4s}{(s^2+4)^2}.$$

7. Use formulas 1, 2 and 3 (with $n = 2$) in Table 1, along with the linearity of the Laplace transform operator, to get

$$\mathcal{L}[t^2 + 2t - 1](s) = \mathcal{L}[t^2](s) + 2\mathcal{L}[t](s) - \mathcal{L}[1](s) = \frac{2!}{s^3} + 2\frac{1}{s^2} - \frac{1}{s} = \frac{2}{s^3} + \frac{2}{s^2} - \frac{1}{s}.$$

9. First, write $(2t+1)e^{3t} = 2te^{3t} + e^{3t}$. By formulas 4 and 5 in Table 1, $\mathcal{L}[e^{3t}](s) = 1/(s-3)$ and $\mathcal{L}[te^{3t}](s) = 1/(s-3)^2$. Using the linearity of the Laplace transform operator then gives

$$\mathcal{L}[(2t+1)e^{3t}](s) = 2\mathcal{L}[te^{3t}](s) + \mathcal{L}[e^{3t}](s) = \frac{2}{(s-3)^2} + \frac{1}{s-3}.$$

11. The partial fraction decomposition of the given function is

$$\frac{1}{s^2-1} = \frac{1}{2}\left(\frac{1}{s-1} - \frac{1}{s+1}\right).$$

The linearity of the inverse Laplace transform operator $\mathcal{L}^{-1}$, together with Formula 4 (with $a = 1$ and $a = -1$) in Table 1, gives

$$\mathcal{L}^{-1}\left[\frac{1}{s^2-1}\right](t) = \mathcal{L}^{-1}\left[\frac{1}{2}\left(\frac{1}{s-1} - \frac{1}{s+1}\right)\right](t)$$

$$= \frac{1}{2}\mathcal{L}^{-1}\left[\frac{1}{s-1}\right](t) - \frac{1}{2}\mathcal{L}^{-1}\left[\frac{1}{s+1}\right](t) = \frac{1}{2}e^t - \frac{1}{2}e^{-t}.$$

13. With $a = 2$ and $b = 3$ in formula 11 in Table 1, use the linearity of $\mathcal{L}^{-1}$ to obtain

$$\mathcal{L}^{-1}\left[\frac{1}{(s-2)^2+9}\right](t) = \mathcal{L}^{-1}\left[\frac{1}{3}\frac{3}{(s-2)^2+3^2}\right](t)$$

$$= \frac{1}{3}\mathcal{L}^{-1}\left[\frac{3}{(s-2)^2+3^2}\right](t) = \frac{1}{3}e^{2t}\sin 3t.$$

15. A direct application of formula 9 (with $b = 2$) in Table 1 gives

$$\mathcal{L}^{-1}\left[\frac{4s}{(s^2+4)^2}\right](t) = \mathcal{L}^{-1}\left[\frac{2(2)s}{(s^2+2^2)^2}\right] = t\sin 2t.$$

17. Let $\mathcal{L}[y](s) = Y(s)$. Applying the Laplace transform to the left side of the given differential equation, along with Equation 5.2 and the initial condition $y(0) = 2$, we obtain

$$\mathcal{L}[y'-y](s) = \mathcal{L}[y'](s) - \mathcal{L}[y](s) = -y(0) + sY(s) - Y(s) = -2 + (s-1)Y(s).$$

By formula 2 in Table 1, the Laplace transform of the right side of the given D.E. is $\mathcal{L}[t] = 1/s^2$. It follows that

$$-2 + (s-1)Y(s) = \frac{1}{s^2}, \qquad \text{or solving for } Y(s), \qquad Y(s) = \frac{2}{s-1} + \frac{1}{s^2(s-1)}.$$

Partial fraction decomposition of the second term of the right member of the equation on the right gives

$$\frac{1}{s^2(s-1)} = -\frac{1}{s} - \frac{1}{s^2} + \frac{1}{s-1}.$$

Hence, $Y(s)$ can be written as

$$Y(s) = -\frac{1}{s} - \frac{1}{s^2} + \frac{3}{s-1}.$$

Applying $\mathcal{L}^{-1}$ to both sides and using formulas 1, 2 and 4 in Table 1, gives

$$y(t) = -1 - t + 3e^t.$$

As a check, $y' = -1 + 3e^t$ so that $y' - y = (-1 + 3e^t) - (-1 - t + 3e^t) = t$.

19. Let $\mathcal{L}[y](s) = Y(s)$. Applying the Laplace transform to the left side of the given differential equation, along with Equation 5.2 and the initial condition $y(0) = 0$, we obtain

$$\mathcal{L}[y'+3y](s) = \mathcal{L}[y'](s) + 3\mathcal{L}[y](s) = -y(0) + sY(s) + 3Y(s) = (s+3)Y(s).$$

By formula 8 in Table 1 the Laplace transform of the right side of the given differential equation is $\mathcal{L}[\cos 2t](s) = s/(s^2+4)$. It follows that

$$(s+3)Y(s) = \frac{s}{s^2+4}, \qquad \text{or} \qquad Y(s) = \frac{s}{(s^2+4)(s+3)}.$$

Partial fraction decomposition gives

$$Y(s) = \frac{(3/13)s + (4/13)}{s^2 + 4} + \frac{-3/13}{s+3} = \frac{3}{13}\left(\frac{s}{s^2+4}\right) + \frac{2}{13}\left(\frac{2}{s^2+4}\right) - \frac{3}{13}\left(\frac{1}{s+3}\right),$$

where each fraction has been judiciously rewritten so that Table 1 can be used to more easily compute $\mathcal{L}^{-1}[Y(s)](t)$. By formulas 4, 7 and 8 in Table 1, we apply $\mathcal{L}^{-1}$ to get

$$y(t) = \frac{3}{13}\cos 2t + \frac{2}{13}\sin 2t - \frac{3}{13}e^{-3t}.$$

As a check, $y' = -(6/13)\sin 2t + (4/13)\cos 2t + (9/13)e^{-3t}$ and we have

$$y' + 3y = \left(-\frac{6}{13}\sin 2t + \frac{4}{13}\cos 2t + \frac{9}{13}e^{-3t}\right) + 3\left(\frac{2}{13}\cos 2t + \frac{2}{13}\sin 2t - \frac{3}{13}e^{-3t}\right)$$

$$= \left(\frac{4}{13} + \frac{9}{13}\right)\cos 2t + \left(-\frac{6}{13} + \frac{6}{13}\right)\sin 2t + \left(\frac{9}{13} - \frac{9}{13}\right)e^{-3t} = \cos 2t.$$

21. Let $\mathcal{L}[y](s) = Y(s)$. Applying the Laplace transform to the given differential equation gives

$$2\left(-y'(0) - sy(0) + s^2Y(s)\right) - \left(-y(0) + sY(s)\right) = \frac{2s}{s^2 + 9}.$$

Setting $y(0) = 0$, $y'(0) = 2$ gives

$$-4 + 2s^2Y(s) - sY(s) = \frac{2s}{s^2 + 9}, \quad \text{so that} \quad Y(s) = \frac{2}{(2s-1)(s^2+9)} + \frac{4}{s(2s-1)}.$$

Partial fraction decomposition allows us to write the expression for $Y(s)$ as

$$Y(s) = -\frac{4}{37}\left(\frac{s}{s^2+9}\right) - \frac{2}{111}\left(\frac{3}{s^2+9}\right) - 4\left(\frac{1}{s}\right) + \frac{152}{37}\left(\frac{1}{s-1/2}\right).$$

Applying $\mathcal{L}^{-1}$ gives

$$y(t) = -\frac{4}{37}\cos 3t - \frac{2}{111}\sin 3t - 4 + \frac{152}{37}e^{t/2}.$$

As a check, we compute

$$y'(t) = \frac{12}{37}\sin 3t - \frac{6}{111}\cos 3t + \frac{76}{37}e^{t/2},$$

$$y''(t) = \frac{36}{37}\cos 3t + \frac{18}{111}\sin 3t + \frac{38}{37}e^{t/2},$$

and substitute into the given D.E. to get

$$2y'' - y' = 2\left(\frac{36}{37}\cos 3t + \frac{18}{111}\sin 3t + \frac{38}{37}e^{t/2}\right) - \left(\frac{12}{37}\sin 3t - \frac{6}{111}\cos 3t + \frac{76}{37}e^{t/2}\right)$$

$$= \left(\frac{72}{37} + \frac{6}{111}\right)\cos 3t + \left(\frac{36}{111} - \frac{12}{27}\right)\sin 3t + \left(\frac{76}{37} - \frac{76}{37}\right)e^{t/2} = 2\cos 3t.$$

23. (a) If $a \leq 0$ then $t - a \geq 0$ for all $t \geq 0$. Hence, $H(t - a) = 1$ for all $t \geq 0$. By definition of the Laplace transform,

$$\mathcal{L}[H(t-a)](s) = \int_0^\infty e^{-st} H(t-a)\,dt = \int_0^\infty e^{-st}(1)\,dt = \mathcal{L}[1](s) = \frac{1}{s}.$$

If $a > 0$ then $t - a \geq 0$ only for $t \geq a$. Hence, $H(t-a) = 0$ for $0 \leq t < a$, and $H(t-a) = 1$ for $t \geq a$. The definition of the Laplace transform then gives

$$\mathcal{L}[H(t-a)](s) = \int_0^\infty e^{-st} H(t-a)\,dt = \int_0^a e^{-st}(0)\,dt + \int_a^\infty e^{-st}(1)\,dt = \int_a^\infty e^{-st}\,dt$$

$$= \lim_{b \to \infty} \int_a^b e^{-st}\,dt = \lim_{\to \infty} \left[-\frac{1}{s}e^{-st} \right]_a^b = \frac{1}{s} \lim_{b \to \infty} (e^{-at} - e^{-bt}) = \frac{1}{s}e^{-at}.$$

(b) If $a \geq 0$ and $g(t) = H(t-a)f(t)$ for $t \geq 0$ then

$$\mathcal{L}[g](s) = \mathcal{L}[H(t-a)f(t)](s) = \int_0^\infty e^{-st} H(t-a)f(t)\,dt$$

$$= \int_0^a e^{-st}(0)f(t)\,dt + \int_a^\infty e^{-st}(1)f(t)\,dt = \int_a^\infty e^{-st}f(t)\,dt$$

$$= \int_0^\infty e^{-s(u+a)}f(u+a)\,du = e^{-as}\int_0^\infty e^{-su}f(u+a)\,du = e^{-as}\mathcal{L}[f(t+a)](s),$$

where the first integral in the last line is obtained by means of the substitution $u = t - a$.

(c) The function $H(t) - H(t-1)$ satisfies

$$H(t) - H(t-1) = \begin{cases} 0, & t < 0, \\ 1, & 0 \leq t < 1, \\ 0, & 1 \leq t. \end{cases}$$

H(t)-H(t-1)

Its graph is shown on the right.

(d) For $a > 0$, we can use the result of part (a) to apply the Laplace transform to both sides of $y'' = H(t-a)$ to get

$$-y'(0) - sy(0) + s^2 Y(s) = \frac{1}{s}e^{-as},$$

where $\mathcal{L}[y](s) = Y(s)$. Setting $y(0) = 1$, $y'(0) = 0$ and solving for $Y(s)$ gives

$$Y(s) = \frac{1}{s} + \frac{1}{s^3}e^{-as}.$$

If $f(t) = \frac{1}{2}(t-a)^2$ then $\mathcal{L}[f(t+a)](s) = \mathcal{L}[(1/2)t^2](s) = \frac{1}{2}\mathcal{L}[t^2](s) = 1/s^3$. Hence, part (b) can be used to write $Y(s)$ as

$$Y(s) = \frac{1}{s} + e^{-as}\mathcal{L}[f(t+a)](s) = \frac{1}{s} + \mathcal{L}[H(t-a)f(t)](s).$$

429

Applying $\mathcal{L}^{-1}$ gives

$$y(t) = 1 + H(t-a)f(t) = 1 + \frac{1}{2}(t-a)^2 H(t-a).$$

25. (a) If $f(0+) = \lim_{t \to 0+} f(t)$ then

$$
\begin{aligned}
\mathcal{L}[f'](s) &= \int_0^\infty e^{-st} f'(t)\, dt = \lim_{\epsilon \to 0+} \int_\epsilon^\infty e^{-st} f'(t)\, dt \\
&= \lim_{\epsilon \to 0+} \left\{ \left[e^{-st} f(t) \right]_\epsilon^\infty + s \int_\epsilon^\infty e^{-st} f(t)\, dt \right\} \\
&= \lim_{\epsilon \to 0+} \left\{ -e^{-s\epsilon} f(\epsilon) + s \int_\epsilon^\infty e^{-st} f(t)\, dt \right\} \\
&= -f(0+) + s \int_0^\infty e^{-st} f(t)\, dt = -f(0+) + s\mathcal{L}[f](s).
\end{aligned}
$$

(b) Part (a) established the case $n = 1$ for a proof by induction on n, the order of the derivative of f. Thus, suppose Formula 5.3 has been generalized for all orders of derivatives up to $n = n_0$. That is, for our induction hypothesis, we suppose

$$\mathcal{L}[f^{(n_0)}](s) = -f^{(n_0-1)}(0+) - \cdots - s^{n_0-1} f(0+) + s^{n_0} \mathcal{L}[f](s).$$

Assume $f^{(n_0)}(0+) = \lim_{t \to 0+} f^{(n_0)}(t)$ and let $g = f^{(n_0)}$, so that $g(0+) = \lim_{t \to 0+} g(t)$ and $g' = f^{(n_0+1)}$. Applying part (a) to the function g gives

$$
\begin{aligned}
\mathcal{L}[g'](s) &= -g(0+) + s\mathcal{L}[g](s) = -f^{(n_0)}(0+) + s\mathcal{L}[f^{(n_0)}](s) \\
&= -f^{(n_0)}(0+) + s\left(-f^{(n_0-1)}(0+) - \cdots - s^{n_0-1} f(0+) + s^{n_0} \mathcal{L}[f](s) \right) \\
&= -f^{(n_0)}(0+) - s f^{(n_0-1)}(0+) - \cdots - s^{n_0} f(0+) + s^{n_0+1} \mathcal{L}[f](s),
\end{aligned}
$$

where we have used the induction hypothesis to obtain the second line. Since the leftmost member is $\mathcal{L}[f^{(n_0+1)}](s)$, the generalization of Formula 5.3 holds for $n = n_0 + 1$ whenever it holds for $n = n_0$. By mathematical induction, it holds for all $n \geq 1$.

Section 6: CONVOLUTION

Exercise Set 6 (pgs. 552-553)

1. With $f(t) = t$ and $g(t) = e^{-t}$, we have

$$f * g(t) = \int_0^t u e^{-(t-u)} \, du = e^{-t} \int_0^t u e^u \, du = e^{-t} \left[u e^u - e^u \right]_0^t = e^{-t} \left(t e^t - e^t + 1 \right) = t - 1 + e^{-t}.$$

3. With $f(t) = 1$ and $g(t) = 1$, we have

$$f * g(t) = \int_0^t (1)(1) \, du = [u]_0^t = t.$$

5. Since $\mathcal{L}[t](s) = 1/s^2$ and $\mathcal{L}[e^{-t}](s) = 1/(s+1)$, Theorem 6.1 says

$$\mathcal{L}[t * e^{-t}](s) = \mathcal{L}[t](s)\mathcal{L}[e^{-t}](s) = \frac{1}{s^2(s+1)}, \quad \text{so that} \quad t * e^{-t} = \mathcal{L}^{-1}\left[\frac{1}{s^2(s+1)} \right],$$

where the equation on the right is obtained by applying $\mathcal{L}^{-1}$ to both sides of the equation on the left and then using the fact that the Laplace operator is 1-to-1. From Exercise 1 above, $t * e^{-t} = -1 + t + e^{-t}$. It follows that

$$\mathcal{L}^{-1}\left[\frac{1}{s^2(s+1)} \right](t) = -1 + t + e^{-t},$$

7. Since $\mathcal{L}[t](s) = 1/s^2$ and $\mathcal{L}[H(t-2)](s) = e^{-2s}/s$ (from Exercise 23 in the last section), Theorem 6.1 says

$$\mathcal{L}[H(t-2) * t](s) = \mathcal{L}[H(t-2)](s)\mathcal{L}[t](s) = \frac{e^{-2s}}{s^3}.$$

Next, we compute

$$H(t-2) * t = \int_0^t H(u-2)(t-u) \, du.$$

This integral is zero for $0 \leq t < 2$ because $H(u-2) = 0$ for $0 \leq u \leq t < 2$. If $t \geq 2$ then $H(u-s) = 0$ for $0 \leq u < 2$ and $H(u-2) = 1$ for $2 \leq u \leq t$. In this case, we have

$$\int_0^t H(u-2)(t-u) \, du = \int_2^t (t-u) \, du = \left[tu - \frac{1}{2}u^2 \right]_2^t = \frac{1}{2}t^2 - 2t + 2 = \frac{1}{2}(t-2)^2.$$

Thus, for all $t \geq 0$,

$$\int_0^t H(u-2)(t-u) \, du = \frac{1}{2}(t-2)^2 H(t-2).$$

Therefore,

$$\mathcal{L}^{-1}\left[\frac{e^{-2s}}{s^3} \right](t) = \frac{1}{2}(t-2)^2 H(t-2).$$

9. By Partial fraction decomposition we have

$$\frac{1}{s(s+3)^2} = \frac{1}{9}\left(\frac{1}{s}\right) - \frac{1}{9}\left(\frac{1}{s+3}\right) - \frac{1}{3}\left(\frac{1}{(s+3)^2}\right).$$

Applying $\mathcal{L}^{-1}$ gives

$$\mathcal{L}^{-1}\left[\frac{1}{s(s+3)^2}\right](t) = \frac{1}{9} - \frac{1}{9}e^{-3t} - \frac{1}{3}te^{-3t}.$$

11. Since $s^2 + 2s + 2 = (s+1)^2 + 1$,

$$\mathcal{L}^{-1}\left[\frac{1}{s^2+2s+2}\right](t) = \mathcal{L}^{-1}\left[\frac{1}{(s+1)^2+1}\right](t) = e^{-t}\sin t.$$

13. Since $(e^{-s}+1)/s = e^{-s}/s + 1/s$, we have

$$\mathcal{L}^{-1}\left[\frac{e^{-s}+1}{s}\right](t) = \mathcal{L}^{-1}\left[\frac{e^{-s}}{s}\right](t) + \mathcal{L}^{-1}\left[\frac{1}{s}\right](t) = H(t-1) + 1.$$

15. With $\mathcal{L}[y](s) = Y(s)$, and the initial conditions $y(0) = 1$, $y'(0) = -1$, take the Laplace transform of the given differential equation and obtain

$$s^2Y(s) - s + 1 - Y(s) = \frac{2}{s^2+4} + \frac{1}{s}.$$

Solving for $Y(s)$ and using partial fraction decomposition gives

$$Y(s) = \frac{s^4 - s^3 + 5s^2 - 2s + 4}{s(s-1)(s+1)(s^2+4)} = -\frac{1}{s} + \frac{7}{10}\left(\frac{1}{s-1}\right) + \frac{13}{10}\left(\frac{1}{s+1}\right) - \frac{1}{5}\left(\frac{2}{s^2+4}\right).$$

Applying $\mathcal{L}^{-1}$ gives

$$y(t) = -1 + \frac{7}{10}e^t + \frac{13}{10}e^{-t} - \frac{1}{5}\sin 2t.$$

As a check, first note that $y(0) = 1$, so that the first initial condition holds. Then compute

$$y'(t) = \frac{7}{10}e^t - \frac{13}{10}e^{-t} - \frac{2}{5}\cos 2t, \qquad y''(t) = \frac{7}{10}e^t + \frac{13}{10}e^{-t} + \frac{4}{5}\sin 2t.$$

Next, observe that $y'(0) = -1$, so that the second initial condition holds. Next, substitute $y(t)$ and $y''(t)$ back into the given differential equation to get

$$y'' - y = \left(\frac{7}{10}e^t + \frac{13}{10}e^{-t} + \frac{4}{5}\sin 2t\right) - \left(-1 + \frac{7}{10}e^t + \frac{13}{10}e^{-t} - \frac{1}{5}\sin 2t\right) = 1 + \sin 2t.$$

So, $y(t)$ as given above is indeed the solution of the given initial-value problem.

17. With $\mathcal{L}[y](s) = Y(s)$, and the initial conditions $y(0) = 2$, $y'(0) = 1$, take the Laplace transform of the given differential equation and obtain

$$s^2Y(s) - 2s - 1 + sY(s) - 2 = \frac{1}{s^2} + \frac{1}{s+1}.$$

Solving for $Y(s)$ and using partial fraction decomposition gives

$$Y(s) = \frac{2s^4 + 5s^3 + 4s^2 + s + 1}{s^3(s+1)^2} = 5\left(\frac{1}{s}\right) - \frac{1}{s^2} + \frac{1}{2}\left(\frac{2}{s^3}\right) - 3\left(\frac{1}{s+1}\right) - \frac{1}{(s+1)^2}.$$

Applying $\mathcal{L}^{-1}$ gives

$$y(t) = 5 - t + \frac{1}{2}t^2 - 3e^{-t} - te^{-t}.$$

As a check, observe that $y(0) = 2$, so that the first initial condition holds. Then compute

$$y'(t) = -1 + t + 2e^{-t} + te^{-t}, \qquad y''(t) = 1 - e^{-t} - te^{-t}.$$

Next, observe that $y'(0) = 1$, so that the second initial condition holds. Next, substitute $y''(t)$ and $y'(t)$ back into the given differential equation to get

$$y'' + y' = \left(1 - e^{-t} - te^{-t}\right) + \left(-1 + t + 2e^{-t} + te^{-t}\right) = t + e^{-t}.$$

So, $y(t)$ as given above is indeed the solution of the given initial-value problem.

19. With $\mathcal{L}[y](s) = Y(s)$, and the initial condition $y(0) = 1$, use formula 4 in Table 2 and take the Laplace transform of the given equation to obtain

$$sY(s) - 1 + Y(s) = \frac{1}{s}Y(s) + \frac{1}{s^2}.$$

Solving for $Y(s)$ and using partial fraction decomposition give

$$Y(s) = \frac{s^2 + 1}{s(s^2 + s - 1)} = -\frac{1}{s} + \frac{1}{s - \frac{-1+\sqrt{5}}{2}} + \frac{1}{s - \frac{-1-\sqrt{5}}{2}}.$$

Applying $\mathcal{L}^{-1}$ gives

$$y(t) = -1 + e^{\frac{-1+\sqrt{5}}{2}t} + e^{\frac{-1-\sqrt{5}}{2}t} = -1 + 2e^{-t/2}\frac{e^{\frac{\sqrt{5}}{2}t} + e^{-\frac{\sqrt{5}}{2}t}}{2} = -1 + 2e^{-t/2}\cosh\frac{\sqrt{5}}{2}t.$$

21. (a) Let $F(s) = \mathcal{L}[f](s)$, and for $n \geq 0$ define $f_n(t)$ by

$$f_n(t) = \int_0^{t_0}\left(\int_0^{t_1}\left(\cdots\int_0^{t_n} f(t_{n+1})\,dt_{n+1}\right)\cdots dt_2\right)dt_1.$$

We will use mathematical induction to show that $\mathcal{L}[f_n](s) = \frac{1}{s^{n+1}}F(s)$ for all $n \geq 0$.
For $n = 0$, formula 4 in Table 2 directly gives

$$\mathcal{L}[f_0](s) = \mathcal{L}\left[\int_0^{t_0} f(t_1)\,dt_1\right](s) = \frac{1}{s}F(s).$$

So the formula holds for $n = 0$.

Next, assume $\mathcal{L}[f_{n_0}](s) = \frac{1}{s^{n_0+1}}F(s)$ for some $n_0 \geq 0$. Then formula 4 in Table 2, along with the induction hypothesis gives

$$\mathcal{L}[f_{n_0+1}](s) = \mathcal{L}\left[\int_0^{t_0}\left(\int_0^{t_1}\left(\cdots\int_0^{t_{n_0+1}} f(t_{n_0+2})\,dt_{n_0+2}\right)\cdots dt_2\right)dt_1\right](s)$$

$$= \frac{1}{s}\mathcal{L}\left[\int_0^{t_1}\left(\cdots\int_0^{t_{n_0+1}} f(t_{n_0+2})\,dt_{n_0+2}\right)\cdots dt_2\right](s)$$

$$= \frac{1}{s}\mathcal{L}[f_{n_0}](s) = \frac{1}{s}\left(\frac{1}{s^{n_0+1}}F(s)\right) = \frac{1}{s^{n_0+2}}F(s).$$

Hence, the formula holds for $n = n_0 + 1$ whenever it holds for $n = n_0$. By induction, the formula holds for all $n \geq 0$.

(b) From Table 1, $\mathcal{L}[t^n](s) = n!/s^{n+1}$ for all $n \geq 0$. So Theorem 6.1 gives

$$\frac{1}{s^{n+1}}F(s) = \frac{1}{n!}\left(\frac{n!}{s^{n+1}}\right)F(s) = \frac{1}{n!}\mathcal{L}[t^n * f(t)](s) = \frac{1}{n!}\mathcal{L}\left[\int_0^t (t-u)^n f(u)\,du\right](s)$$

$$= \mathcal{L}\left[\frac{1}{n!}\int_0^t (t-u)^n f(u)\,du\right](s).$$

Taking the inverse Laplace transform of both sides and using the formula established in part (a), we get

$$f_n(t) = \frac{1}{n!}\int_0^t (t-u)^n f(u)\,du,$$

which holds for all $n \geq 0$. Because of how $f_n(t)$ was defined in part (a), we are done.

23. (a) Let $f(t) = e^{t+e^t}\sin e^{e^t}$. Using the formal substitution $u = e^{e^t}$, we have

$$\mathcal{L}[f](s) = \int_0^\infty e^{-st}f(t)\,dt = \int_0^\infty e^{-st}e^{t+e^t}\sin e^{e^t}\,dt = \int_e^\infty \frac{\sin u}{(\ln u)^s}\,du.$$

To show that this is finite, write the integral on the right as

$$\int_e^\infty \frac{\sin u}{(\ln u)^s}\,du = \int_e^\pi \frac{\sin u}{(\ln u)^s}\,du + \int_\pi^{2\pi} \frac{\sin u}{(\ln u)^s}\,du + \cdots + \int_{n\pi}^{(n+1)\pi} \frac{\sin u}{(\ln u)^s}\,du + \cdots$$

$$= \int_e^\pi \frac{\sin u}{(\ln u)^s}\,du + \sum_{n=1}^\infty \int_{n\pi}^{(n+1)\pi} \frac{\sin u}{(\ln u)^s}\,du.$$

Since $\sin u$ is positive when n is even and negative when n is odd, the n-th term of the infinite series is positive for even n and negative for odd n. Thus, the series is an alternating series. Moreover, for $s > 0$ the sequence of absolute values of the terms of the series is a decreasing sequence with limit 0, which means that the series converges to a finite limit for each $s > 0$. Since the integral with lower limit e and upper limit π is obviously finite for each s, $\mathcal{L}[f](s)$ is finite for $s > 0$.

(b) Replacing $f(t)$ with $|f(t)|$ in the derivations obtained in part (a), one first gets the analogous equation

$$\mathcal{L}[|f|](s) = \int_0^\infty e^{-st}|f(t)|\,dt = \int_e^\infty \frac{|\sin u|}{(\ln u)^s}\,du.$$

434

Then, proceeding as we did in part (a), one arrives at the equation

$$\int_0^\infty e^{-st}|f(t)|dt = \int_e^\pi \frac{|\sin u|}{(\ln u)^s}du + \sum_{n=1}^\infty \int_{n\pi}^{(n+1)\pi} \frac{|\sin u|}{(\ln u)^s}du$$

$$= \int_e^\pi \frac{|\sin u|}{(\ln u)^s}du + \sum_{n=2}^\infty \int_{(n-1)\pi}^{n\pi} \frac{\sin u}{(\ln u)^s}du,$$

where we have shifted the index of summation for ease of computation. By the Cauchy condensation test, the above positive term series converges or diverges with the series

$$\sum_{k=1}^\infty 2^{k-1} \int_{(2^k-1)\pi}^{2^k\pi} \frac{|\sin u|}{(\ln u)^s}du \geq \sum_{k=1}^\infty 2^{k-1} \int_{(2^k-1)\pi}^{2^k\pi} \frac{|\sin u|}{\left(\ln(2^k\pi)\right)^s}du$$

$$= \sum_{k=1}^\infty \frac{2^{k-1}}{\left(\ln(2^k\pi)\right)^s} \int_{(2^k-1)\pi}^{2^k\pi} |\sin u|du$$

$$= \sum_{k=1}^\infty \frac{2^k}{(k\ln 2 + \ln \pi)^s},$$

where we have used the fact that the area under the curve of $|\sin u|$ equals 2 on intervals of length π. The numerator 2^k in the last series is an exponential function of k, which increases faster than the denominator, a power of a linear function of k. It follows that the terms in the last series do not tend to 0. and therefore the series diverges, regardless of the value of s. We conclude that $\mathcal{L}[|f|](s)$ is infinite for every s.

NOTE: See *Principles of Mathematical Analysis* by Walter Rudin (McGraw-Hill) for a clear presentation of the Cauchy condensation test.

(c) Recall that a basic result of Theorem 3.5 in Chapter 7 is that if a function $g(t)$ is integrable over every finite t-interval of the form $I = (0, b)$ then the absolute value of the integral of g over I is no greater than the integral of $|g|$ over I. Hence, if $g(t) = e^{-st}f(t)$, where f is integrable over I for each sufficiently large s then

$$\left|\int_0^b e^{-st}f(t)\,dt\right| \leq \int_0^b |e^{-st}f(t)|\,dt = \int_0^b e^{-st}|f(t)|\,dt.$$

Letting $b \to \infty$, we obtain

$$\left|\mathcal{L}[f](s)\right| = \left|\int_0^\infty e^{-st}f(t)\,dt\right| \leq \int_0^\infty e^{-st}|f(t)|\,dt = \mathcal{L}[|f|](s),$$

which holds for each sufficiently large s. Thus, $\left|\mathcal{L}[f](s)\right|$ is finite for all s for which $\mathcal{L}[|f|](s)$ is finite. Hence, if $\mathcal{L}[|f|](s)$ is finite then so is $\mathcal{L}[f](s)$.

Section 7: NONLINEAR EQUATIONS

Exercise Set 7AB (pgs. 557-558)

1. Let $z = \dot{y}$, $\dot{z} = \ddot{y}$. The given initial-value problem then leads to the first-order system

$$t\dot{z} + z = 0, \quad z(1) = 1,$$
$$\dot{y} = z, \quad y(1) = 0.$$

The left side of the first equation is the derivative of tz, so that the equation can be written as

$$\frac{d}{dt}(tz) = 0.$$

Integrating gives $tz = c_1$. The condition $z(1) = 1$ implies $c_1 = 1$, so that $tz = 1$, or equivalently, $z(t) = 1/t$. Note that $t_0 = 1$ is in the domain of z. Therefore, the largest interval containing $t_0 = 1$ on which $z = \dot{y}(t)$ is continuous is $0 < t < \infty$.

The second equation can now be written as $\dot{y} = 1/t$. One integration gives $y = \ln t + c_2$, where the absolute value sign is not needed because $t > 0$. The condition $y(1) = 0$ implies $c_2 = 0$, so that

$$y(t) = \ln t, \quad t > 0.$$

3. Let $z = \dot{y}$, $\dot{z} = \ddot{y}$. The given initial-value problem then leads to the first-order system

$$\dot{z} + z^2 = 0, \quad z(0) = 1,$$
$$\dot{y} = z, \quad y(0) = 0.$$

Assuming $z \neq 0$, the first equation can be written in separable form as

$$-z^{-2}dz = dt.$$

Hence, $z^{-1} = t + c_1$. The condition $z(0) = 1$ gives $c_1 = 1$, so that $z^{-1} = t + 1$, or equivalently

$$z(t) = \frac{1}{t+1}.$$

Note that $t_0 = 0$ is in the domain of z. Thus, the largest interval on which $z = \dot{y}(t)$ is continuous is $-1 < t < \infty$.

The second equation above can now be written as $\dot{y} = 1/(t+1)$, which can be solved by integration with respect to t. This gives $y = \ln(t+1) + c_2$, where the absolute value sign is not needed because $t + 1 > 0$. The condition $y(0) = 0$ gives $c_2 = 0$, so that

$$y(t) = \ln(t+1), \quad t > -1.$$

5. Let $z = \dot{y}$, $\dot{z} = \ddot{y}$. The given initial-value problem then leads to the first-order system

$$t\dot{z} + z = t^3, \quad z(1) = 1,$$
$$\dot{y} = z, \quad y(1) = 1.$$

The left side of the first equation is the derivative of tz, so that the equation can be written as

$$\frac{d}{dt}(tz) = t^3.$$

Integration leads to $tz = \frac{1}{4}t^4 + c_1$, and the condition $z(1) = 1$ gives $c_1 = 3/4$. So,

$$z = \frac{1}{4}t^3 + \frac{3}{4t}.$$

Note that $t_0 = 1$ is in the domain of z. Thus, the largest interval on which $z = \dot{y}(t)$ is continuous is $0 < t < \infty$.

The second equation above can now be written as $\dot{y} = t^3/4 + 3/(4t)$. One integration gives $y = t^4/(16) + (3/4) \ln t + c_2$, where the absolute value sign on $\ln |t|$ is not needed because $t > 0$. The condition $y(1) = 1$ leads to $c_2 = 15/16$ and therefore

$$y(t) = \frac{1}{16}t^4 + \frac{3}{4} \ln t + \frac{15}{16}, \quad t > 0.$$

7. Let $z = dy/dt = \dot{y}$ and $z\,dz/dy = \ddot{y}$. The given initial-value problem then leads to the first-order system

$$yz\frac{dz}{dy} - z^2 = 0, \quad z(0) = 1,$$

$$\frac{dy}{dt} = z, \quad y(0) = 1.$$

If neither z or y is 0, then the the variables can be separated in the first equation to get $z^{-1}dz = y^{-1}dy$. Integrating gives $\ln |z| = \ln |y| + c_1$. Evaluating this equation at $t_0 = 0$ by setting $y = z = 1$, we find that $c_1 = 0$. So, $\ln |z| = \ln |y|$. Note that since $t_0 = 0$ makes both y and z positive, no value of t is permitted that makes either y or z negative. Hence, $y(t) > 0$ and $z(t) > 0$ for all t in the domain of y. Thus, we can remove the absolute value signs and conclude that $z = y$. The second equation above then becomes $dy/dt = y$, so that $y = c_2 e^t$. The condition $y(0) = 1$ gives $c_2 = 1$ and therefore

$$y(t) = e^t, \quad -\infty < t < \infty.$$

9. Let $z = dy/dt = \dot{y}$ and $z\,dz/dy = \ddot{y}$. The given initial-value problem then leads to the first-order system

$$z\frac{dz}{dy} - z^3 = 0, \quad z(0) = 1,$$

$$\frac{dy}{dt} = z, \quad y(0) = 1.$$

For $z \neq 0$, the variables in the first equation can be separated to get $z^{-2}dz = dy$. Hence, $-z^{-1} = y + c_1$, or equivalently, $z = -1/(y + c_1)$. Evaluating this equation at $t_0 = 0$, the initial conditions lead to the equation $1 = -1/(1 + c_1)$. Thus, $c_1 = -2$ and

$$z = \frac{1}{2 - y}.$$

Note that z is discontinuous at $y = 2$, so that $y(0) = 1$ implies that $y(t) < 2$ for all t in the domain of y.

The second equation above can now be written as

$$\frac{dy}{dt} = \frac{1}{2 - y}, \quad \text{or in separated form,} \quad (2 - y)\,dy = dt.$$

437

Integrating gives $2y - \frac{1}{2}y^2 = t + c_2$. The condition $y(0) = 1$ implies the equation $2 - \frac{1}{2} = c_2$, so that $c_2 = 3/2$. Hence, $2y - \frac{1}{2}y^2 = t + \frac{3}{2}$, or

$$-\frac{1}{2}y^2 + 2y - \frac{3}{2} = t.$$

Note that since t is being viewed as the independent variable, the graph of this relationship in a ty-coordinate system is a parabola opening to the left with vertex at $(t, y) = (1/2, 2)$. As $y(t) < 2$ must hold for all t in the domain of y, it follows that the graph of the solution we seek is the set of points on the lower half of the parabola. Therefore, we can explicitly solve for y by using the quadratic formula. Taking care to choose the correct sign in front of the radical, we get

$$y = 2 - \sqrt{1 - 2t}, \quad t < \frac{1}{2}.$$

11. Let $z = dy/dt = \dot{y}$ and $z\,dz/dy = \ddot{y}$. The given initial-value problem then leads to the first-order system

$$z\frac{dz}{dy} + z^2 = 1, \quad z(0) = 0,$$

$$\frac{dy}{dt} = z, \quad y(0) = 0.$$

For $|z| \neq 1$, the variables in the first equation can be separated as

$$\frac{-2z}{1 - z^2}\,dz = -2\,dy, \quad \text{or equivalently,} \quad \left(\frac{1}{z-1} + \frac{1}{z+1}\right)dz = -2\,dy,$$

where partial fraction decomposition was used to obtain the equation on the right. Integrating and combining the resulting log functions gives $\ln|z^2 - 1| = -2y + c_1$. Evaluating this equation at $t_0 = 0$ by setting $z = y = 0$, we find that $c_1 = 0$, so that $\ln|z^2 - 1| = -2y$. Since $z(0) = 0$, we can remove the absolute value sign and write

$$\ln(1 - z^2) = -2y, \quad |z| < 1.$$

This implies $-2y \leq 0$, so that $y(t) \geq 0$ for all t in the solution we seek. Solving this equation for z, we see that we have two possibilities

$$z = \pm\sqrt{1 - e^{-2y}}, \quad y \geq 0.$$

The second equation above can then be separated to get

$$\frac{dy}{\sqrt{1 - e^{-2y}}} = \pm dt.$$

In order to evaluate $\int(1 - e^{-2y})^{-1/2}dy$, multiply the numerator and denominator of the integrand by e^y and use the substitution $u = \sqrt{e^{2y} - 1}$. This results in the integrand $1/\sqrt{u^2 + 1}$. Using integral tables and converting back to the variable y gives the result

$$\ln(e^y + \sqrt{e^{2y} - 1}) = \pm t + c_2.$$

438

The condition $y(0) = 0$ leads to $c_2 = 0$, so that $\ln(e^y + \sqrt{e^{2y} - 1}) = \pm t$. Exponentiating both sides, subtracting e^y and squaring, gives an equation that can be written as

$$e^y = \frac{1}{2}(e^{\pm t} + e^{\mp t}) = \cosh t.$$

Finally, taking logarithms gives

$$y(t) = \ln(\cosh t), \quad -\infty < t < \infty.$$

13. Consider the initial-value problem $t^2 \ddot{y} + \dot{y}^2 = 0$, $y(0) = \dot{y}(0) = 0$. First note that the zero function is a solution. For a nontrivial solution, set $z = \dot{y}$, $\dot{z} = \ddot{y}$ and write the equation as $t^2 \dot{z} + z^2 = 0$. If neither t nor z is 0, separate variables to get $-z^{-2} dz = t^{-2} dt$. Integrating gives $z^{-1} = -t^{-1} + C$, or equivalently,

$$z(t) = \frac{t}{Ct - 1}.$$

Observing that the initial condition $z(0) = \dot{y}(0) = 0$ holds for every value of C, set $C = 0$ (to make the computations easier) to get $z = -t$. As $z = \dot{y}$, $\dot{y} = -t$. Integration then gives $y = -\frac{1}{2}t^2 + K$. The condition $y(0) = 0$ then forces $K = 0$ and a second solution of the initial-value problem is $y(t) = -\frac{1}{2}t^2$. (NOTE: Although it was necessary to initially assume that neither t nor z were zero in order to separate variables, it is straightforward to check that this solution satisfies the differential equation for all values of t.)

This does not contradict the uniqueness condition in Theorem 7.1. Here's why. When $\ddot{y}$ is isolated on one side of the given D.E., the result is

$$\ddot{y} = -t^{-2}\dot{y}^2.$$

Therefore we identify the function f in Theorem 7.1 as $f(t, y, \dot{y}) = -t^{-2}\dot{y}^2$, and observe that it not continuous at $t = 0$. We cannot therefore expect uniqueness to hold.

15. (a) The equations $\ddot{y} = -y$ and $\ddot{y} = y$ are constant-coefficient linear homogeneous equations. The corresponding characteristic equations are $r^2 + 1 = 0$ and $r^2 - 1 = 0$, respectively, with roots $r_{1,2} = \pm i$ and $r_{1,2} = \pm 1$, respectively. The corresponding general solutions are therefore

$$y_1(t) = c_1 \cos t + c_2 \sin t \quad \text{and} \quad y_2(t) = c_1 e^t + c_2 e^{-t}.$$

With $\dot{y}_1(t) = -c_1 \sin t + c_2 \cos t$, the initial conditions give the two equations $0 = y_1(0) = c_1$ and $1 = \dot{y}_1(0) = c_2$. Hence, $y_1(t) = \sin t$. And with $\dot{y}_2(t) = c_1 e^t - c_2 e^{-t}$, the initial conditions give the equations $0 = y_2(0) = c_1 + c_2$ and $1 = \dot{y}_2(0) = c_1 - c_2$. Hence, $c_1 = 1/2$, $c_2 = -1/2$ and $y_2(t) = \frac{1}{2}e^t - \frac{1}{2}e^{-t} = \sinh t$.

(b) The graphs of $y_1(t) = \sin t$ and $y_2(t) = \sinh t$ are shown on the right on the same set of axes.

(c) We have

$$y_2(it) = \sinh it = \frac{e^{it} - e^{-it}}{2} = \frac{(\cos t + i \sin t) - (\cos t - i \sin t)}{2} = i \sin t = i y_1(t).$$

17. (a) If $z = dy/dt = \dot{y}$ then $z\,dz/dy = \ddot{y}$ and the given initial-value problem can be written as a first-order initial-value problem in terms of y and z:

$$z\frac{dz}{dy} = -\sin y, \quad \text{where } y = y_0, \; z = z_0 \text{ when } t = 0.$$

Separating variables gives $z\,dz = -\sin y\,dy$, and integrating gives $\frac{1}{2}z^2 = \cos y + C$. Evaluating this equation at $t = 0$ by setting $y = y_0$ and $z = z_0$ gives $\frac{1}{2}z_0^2 = \cos y_0 + C$. Hence, $C = -\cos y_0 + \frac{1}{2}z_0^2$ and we have

$$\frac{1}{2}\dot{y}^2 = \cos y - \cos y_0 + \frac{1}{2}z_0^2,$$

where z has been replaced with $\dot{y}$.

NOTE: For parts (b) and (c) below, let $C_0 = -\cos y_0 + \frac{1}{2}z_0^2$.

(b) If $C_0 > 1$ then the equation established in part (a) satisfies

$$\frac{1}{2}\dot{y}^2 = \cos y + C_0 > \cos y + 1.$$

Since $\cos\alpha \geq -1$ for all α, it follows that $\cos y(t) + 1 \geq 0$ for all t. So, the quantity $\frac{1}{2}\dot{y}^2(t)$ is always positive. This means that the angular velocity $\dot{y}(t)$ is never zero, so that the pendulum never comes to rest and begins motion in the opposite direction. That is, the pendulum repeatedly goes "over the top".

(c) The result of part (b) shows that oscillatory motion does *not* occur if $C_0 > 1$. Thus, to show that oscillatory motion implies $C_0 < 1$, it is only necessary to show that oscillatory motion cannot occur if $C_0 = 1$. To this end, suppose oscillatory motion occurs and that $C_0 = 1$. By part (a), we have

$$\frac{1}{2}\dot{y}^2(t) = \cos y(t) + 1, \quad \text{for all } t.$$

But since oscillatory motion occurs, the pendulum must come to a stop at some angle y lying strictly between $-\pi$ and π. So, for such an angle, say $y = y_1$, we have $\dot{y} = 0$. Hence, $\cos y_1 + 1 = 0$, or equivalently, $\cos y_1 = -1$. But this is a contradiction since there is no angle lying strictly between $-\pi$ and π whose cosine is equal to -1. We conclude that oscillatory motion is inconsistent with the assumption that $C_0 = 1$. As already noted, this is sufficient to conclude that oscillatory motion implies $C_0 = -\cos y_0 + \frac{1}{2}z_0^2 < 1$.

440

Exercise Set 7C (pg. 561)

1. The given differential equation is a constant-coefficient linear nonhomogeneous equation. The associated homogeneous equation has solutions $y_h(y) = c_1 \cos t + c_2 \sin t$. By inspection, a particular solution of the given nonhomogeneous equation is $y_p(t) = 1$. The general solution of the given equation is then $y(t) = 1 + c_1 \cos t + c_2 \sin t$. The initial condition $y(0) = 2$ gives the equation $2 = 1 + c_1$, so that $c_1 = 1$ and $y(t) = 1 + \cos t + c_2 \sin t$. With $\dot{y} = -\sin t + c_2 \cos t$, the condition $\dot{y}(0) = 0$ implies $c_2 = 0$. Hence,

$$y(t) = 1 + \cos t \quad \text{and} \quad z(t) = \dot{y}(t) = -\sin t.$$

Writing the first equation as $y - 1 = \cos t$, we have

$$(y-1)^2 + z^2 = \cos^2 t + \sin^2 t = 1.$$

The phase curve of the solution is therefore a circle of radius 1 centered at $(y, z) = (1, 0)$.

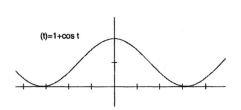
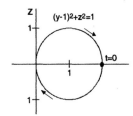

3. The differential equation can be solved by two successive integrations. The first integration gives $\dot{y} = t + c_1$ and the condition $\dot{y}(0) = 0$ forces $c_1 = 0$, so that $\dot{y} = t$. The second integration gives $y(t) = \frac{1}{2}t^2 + c_2$ and the condition $y(0) = 0$ forces $c_2 = 0$. Hence,

$$y(t) = \frac{1}{2}t^2 \quad \text{and} \quad z(t) = \dot{y}(t) = t.$$

Eliminating t from these equations gives $y = \frac{1}{2}z^2$. The phase curve of the solution is therefore a parabola opening to the right with vertex at the origin.

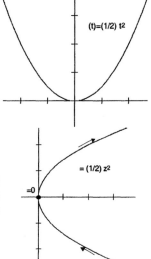

5. The differential equation is a constant-coefficient linear homogeneous equation with solutions $y(t) = c_1 + c_2 e^{-t}$. With $\dot{y}(t) = -c_2 e^{-t}$, the initial conditions lead to the equations $2 = c_1 + c_2$, $0 = -c_2$. So, $c_1 = 2$, $c_2 = 0$ and

$$y(t) = 2 \quad \text{and} \quad z(t) = \dot{y}(t) = 0.$$

The phase curve of the solution is therefore the single point $(y, z) = (2, 0)$.

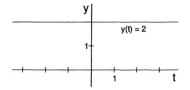
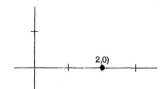

441

7. Let $z = dy/dt = \dot{y}$, so that $z\,dz/dy = \ddot{y}$. The given differential equation can then be written as

$$z\frac{dz}{dy} + y = 1, \quad \text{or in separated form,} \quad z\,dz = (1-y)\,dy.$$

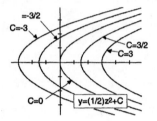

Integration gives $\frac{1}{2}z^2 = y - \frac{1}{2}y^2 + C_1$. Multiplying throughout by 2 and adding $y^2 - 2y + 1 = (y-1)^2$ to both sides of the resulting equation gives

$$(y-1)^2 + z^2 = C, \quad C \geq 0,$$

where $C = 2C_1 + 1$. Except for the equilibrium point $(1,0)$, all phase curves are circles centered at $(1,0)$. Hence, all nonequilibrium solutions are periodic.

9. Let $z = dy/dt = \dot{y}$, so that $z\,dz/dy = \ddot{y}$. The given differential equation can then be written as

$$z\frac{dz}{dy} = 1, \quad \text{or in separated form,} \quad z\,dz = dy.$$

Integrating gives $\frac{1}{2}z^2 = y + C_1$, or equivalently.

$$y = \frac{1}{2}z^2 + C,$$

where $C = -C_1$. All phase curves are parabolas opening to the right with their vertices on the y-axis. There are no equilibrium points or periodic solutions.

11. Let $z = dy/dt = \dot{y}$, so that $z\,dz/dy = \ddot{y}$. The given differential equation can then be written as

$$z\frac{dz}{dy} + y = 0, \quad \text{or in separated form,} \quad z\,dz = -y\,dy.$$

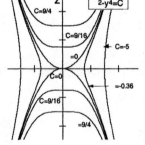

Integrating gives $\frac{1}{2}z^2 = -\frac{1}{2}y^2 + C_1$. Multiplying throughout by 2 and rearranging the result gives

$$y^2 + z^2 = C, \quad C \geq 0,$$

where $C = 2C_1$. Except for the equilibrium point $(0,0)$, all phase curves are circles centered at $(0,0)$. As in Exercise 7 above, all nonequilibrium solutions are periodic.

13. Let $z = dy/dt = \dot{y}$, so that $z\,dz/dy = \ddot{y}$. The given differential equation can then be written as

$$z\frac{dz}{dy} - 2y^3 = 0, \quad \text{or in separated form,} \quad z\,dz = 2y^3\,dy.$$

Integrating gives $\frac{1}{2}z^2 = \frac{1}{2}y^4 + C_1$. Multiplying throughout by 2 and rearranging the result gives

$$z^2 - y^4 = C,$$

where $C = 2C_1$. The accompanying phase portrait shows that there are no periodic solutions. However, $(0,0)$ is an equilibrium point.

442

15. (a) The equation $\ddot{y} - \dot{y} = 0$ has the characteristic equation $r^2 - r = r(r-1) = 0$, with roots $r_1 = 0$, $r_2 = 1$. The general solution is then $y = c_1 + c_2 e^t$. Hence, $z = \dot{y} = c_2 e^t$ and we have $y = c_1 + z$, or equivalently,

$$z = y + c,$$

where $c = -c_1$ is a constant uniquely determined by initial conditions $y(t_0) = y_0$, $\dot{y}(t_0) = z_0$. For fixed c, the line L defined by the above equation is a straight line in phase space with slope 1 and z-intercept at $(0, c)$. Specifically, if $y = y(t)$ is the solution satisfying the initial conditions and whose trace lies on L then $c = z_0 - y_0$.

If y is a constant solution then $\dot{y}(t) = z(t) = 0$ for all t. It follows that $z_0 = 0$ and the point $(y_0, 0) = (-c, 0)$ on L is an equilibrium point.

On the other hand, If y is not a constant solution then $z(t) \neq 0$ for some t. In particular, for our solution on L, suppose that t_0 is such a value of t and that $z_0 > 0$. Then the point (y_0, z_0) is above the y-axis. This means that $y(t)$ is increasing at t_0 and therefore the trace of this solution moves to the right. That is, it moves up the line L (and therefore with increasing velocity), away from the equilibrium point. It follows that if $\dot{y}(t) > 0$ for some t then it remains positive for all time. Similarly, if t_0 is such that $z_0 < 0$ then the point

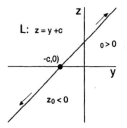

(y_0, z_0) is below the y-axis and $y(t)$ is decreasing at t_0 and therefore the trace of this solution moves to the left. That is, it moves down the line L (and therefore with increasing absolute velocity), away from the equilibrium point. Again, it follows that if $\dot{y}(t) < 0$ for any t then it remains negative for all time. The details are shown in the accompanying drawing.

(b) Since the analysis made in part (a) holds for any line $z = y + c$ in phase space, it follows that the equilibrium points of the given differential equation are precisely the points in yz-space that lie on the y-axis. Further, regardless of the size of $|z_0|$ (i.e., the proximity of an initial point (y_0, z_0) to the y-axis) a nonequilibrium solution moves away from the equilibrium point and does so with increasing absolute velocity. Thus, all equilibrium points are unstable.

Section 8: NUMERICAL METHODS

Exercise Set 8AB (pgs. 566-568)

1. The improved Euler method was used on $\ddot{y} = -ty$ with initial conditions $y_0 = y(0) = 0$, $z_0 = \dot{y}(0) = a$ (for various values of a) and step size $h = 0.001$. Of the 12,000 calculations, every tenth one was printed out and the approximations below correspond to the closest printed value.

(a) For $a = 2$, there are nine t-values on $0 \le t \le 12$ such that $y(t) = 0$. They are approximately

$$t_1 = 0.00, \quad t_4 = 5.75, \quad t_7 = 9.20,$$
$$t_2 = 2.68, \quad t_5 = 6.99, \quad t_8 = 10.22,$$
$$t_3 = 4.35, \quad t_6 = 8.14, \quad t_9 = 11.17.$$

Replacing $\dot{y}(0) = 2$ with $\dot{y}(0) = a$ for various values of $a \ne 0$ does not alter the above zeros of y. However, $a = 0$ gives the zero solution.

(b) The given D.E. shows that, for $t < 0$, $y(t)$ and $\ddot{y}(t)$ have the same sign. That is, the graphs of nontrivial solutions are concave up if $y(t) > 0$ and concave down if $y(t) < 0$. Suppose $y(t_0) = 0$ for some $t_0 < 0$ and let I be the interval $(t_0, 0)$. Since $y(0) = 0$, continuity of $y(t)$ implies that there exist $t_1 \in I$ such that $|y(t_1)| \ge |y(t)|$ for all $t \in I$. In other words, if $y(t) > 0$ on I then $y(t)$ takes on a maximum somewhere on I, and if $y(t) < 0$ then $y(t)$ takes on a minimum somewhere on I. But if $y(t) > 0$ on I then the graph must be concave down at t_1, which is inconsistent with $\ddot{y}(t_1) > 0$. Similarly, if $y(t) < 0$ on I then the graph must be concave up at t_1, which is inconsistent with $\ddot{y}(t_1) < 0$. This contradiction leads to the conclusion that if $y(t)$ is a nontrivial solution then $y(t) = 0$ is impossible for $t < 0$.

3. The equation $\ddot{y} = -y$ has the general solution $y(t) = c_1 \cos t + c_2 \sin t$. The initial conditions $y(0) = 0$, $\dot{y}(0) = 1$ lead to $c_1 = 0$, $c_2 = 1$ and the solution $y(t) = \sin t$. The zeros of this solution are $t = k\pi$ where k is an integer. In particular, consecutive zeros differ by $\pi \approx 3.14$ seconds. Also, maximum angular deviation from the downward vertical position is 1 radian and maximum speed attained (at odd integer multiples of $\pi/2$ seconds) is 1 rad/sec.

Using the improved Euler method with $h = 0.001$, one finds that the solution to the nonlinear equation $\ddot{y} = -\sin t$ (same initial conditions) has its zeros at $t \approx 3.37k$, where k is a nonnegative integer. The maximum angular deviation from the downward vertical position is about 1.0472 radians and the maximum speed attained is 1 rad/sec. As should be expected, the result of the inequality $\sin y < y$ is a wider range of motion and a longer period for solutions of the nonlinear equation $\ddot{y} = -\sin y$.

5. (a) The appropriate D.E. is $y'' + (\sin x)y' + (\cos x)y = 0$, or $y'' = -(\sin x)y' - (\cos x)y$. 6,284 values of the desired solution $y(x)$ on the interval $0 \le x \le 2\pi$ were computed using the improved Euler method with the initial conditions $y_0 = 0$, $z_0 = 1$, $x_0 = 0$ and step size $h = 0.001$. The following table lists 26 equally spaced values of $y(x)$.

x	$y(x)$	x	$y(x)$	x	$y(x)$	x	$y(x)$
0.00	0.000000	1.75	0.894722	3.50	1.93317	5.25	12.5157
0.25	0.244888	2.00	0.928126	3.75	2.43701	5.50	15.51
0.50	0.461284	2.25	0.976694	4.00	3.15283	5.75	18.3345
0.75	0.630306	2.50	1.05199	4.25	4.15872	6.00	20.5036
1.00	0.747292	2.75	1.16638	4.50	5.54429	6.25	21.5787
1.25	0.82011	3.00	1.3354	4.75	7.39153		
1.50	0.863721	3.25	1.58067	5.00	9.7368		

444

One can change the "PRINT" command in the program to a "PLOT" command and get a nice graphical representation of this solution on the interval $0 \leq x \leq 2\pi$. Similar tables and graphs are obtained by using "PRINT Y,Z" and "PLOT Y,Z". This results in a table of points in phase space and the relevant portion of the phase curve of the solution (resp.).

(b) The appropriate D.E. is $y'' + e^{-x/2}y' + e^{-x/3}y = 0$, or $y'' = -e^{-x/2}y' - e^{-x/3}y$. 1,000 values of the desired solution $y(x)$ on the interval $0 \leq x \leq 1$ were computed using the improved Euler method with the initial conditions $y_0 = 0$, $z_0 = 1$, $x_0 = 0$ and step size $h = 0.001$. The following table lists 21 equally spaced values of $y(x)$.

x	$y(x)$	x	$y(x)$	x	$y(x)$	x	$y(x)$
0.00	0.000000	0.30	0.25722	0.60	0.437481	0.90	0.552986
0.05	0.048760	0.35	0.292266	0.65	0.460914	0.95	0.566741
0.10	0.095083	0.40	0.325235	0.70	0.482604	1.00	0.579058
0.15	0.139029	0.45	0.356184	0.75	0.502605		
0.20	0.180661	0.50	0.385171	0.80	0.520968		
0.25	0.220038	0.55	0.412252	0.85	0.537745		

7. (a) With the parameter values as indicated, the appropriate D.E. is $\ddot{y} = -k\dot{y}^{1.5} + 32.2$, where k takes on the four values 0, 0.1, 0.5 and 1. Using the improved Euler method (once for each value of k) with the initial conditions $y_0 = 0$, $z_0 = 0$, $t_0 = 0$ and step size $h = 0.001$, 20,000 values of $y(t)$ were generated for equally spaced values of t on the interval $0 \leq t \leq 20$. The following tables lists 20 values of $y(t)$ (at 1 time unit intervals), where y_1, y_2, y_3, y_4 are the solutions corresponding to $k = 0, 0.1, 0.5, 1$, respectively. (NOTE: The graphs below are only for $0 \leq t \leq 8$ because an appropriate scale for all four graphs was extremely awkward to depict for $t > 8$.)

t	$y_1(t)$	$y_2(t)$	$y_3(t)$	$y_4(t)$
1	16.1	14.2821	10.0937	7.63053
2	64.4	47.7528	25.8119	17.727
3	144.9	89.636	41.8611	27.8481
4	257.6	134.757	57.9269	37.9694
5	402.5	181.066	73.9934	48.0908
6	579.6	227.805	90.06	58.2121
7	788.9	274.699	106.127	68.3334
8	1030.4	321.647	122.193	78.4547
9	1304.1	368.615	138.26	88.576
10	1610	415.59	154.326	98.6974
11	1948.1	462.568	170.393	108.819
12	2318.4	509.546	186.46	118.94
13	2720.9	556.525	202.526	129.061
14	3155.6	603.504	218.593	139.183
15	3622.5	650.483	234.659	149.304
16	4121.6	697.462	250.726	159.425
17	4652.9	744.441	266.793	169.547
18	5216.4	791.42	282.859	179.668
19	5812.1	838.399	208.926	189.789
20	6440	885.378	314.992	199.911

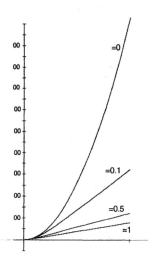

(b) Using the table constructed in part (a), one finds that that the desired value of k lies between $k = 0.5$ and $k = 1$. Linear interpolation at $t = 5$, 10 and 15 yield the three k values 0.6543, 0.6547 and 0.6566, respectively. Averaging these values gives $k = 0.6554$. Using the improved Euler method shows that this value of k gives y values that are too low (specifically, $y(5) = 62.653$, $y(10) = 129.723$ and $y(15) = 196.794$). To get slightly larger distances, k must be slightly smaller that 0.6554. A guess of $k = 0.6$ gives $y(5) = 66.1673$, $y(10) = 137.306$ and $y(15) = 208.445$. Rounded to the nearest unit, these agree with the three desired approximate values.

9. The wire has the shape of the graph of $y = f(x)$, where $f(x) = -x^3 + 4x^2 - 3x$, so that $f'(x) = -3x^2 + 8x - 3$ and $f''(x) = -6x + 8$. With $g = 32$, the differential equation of motion for x is then

$$\ddot{x} = \frac{-\left[32 + (-6x + 8)\dot{x}^2\right](-3x^2 + 8x - 3)}{1 + (-3x^2 + 8x - 3)^2}.$$

Experimenting with a second-order D.E. solver such as the applet 2ORD using $x(0) = 0$ and gradually increasing small values of $\dot{x}(0) > 0$ shows that $x(t)$ oscillates in the approximate range $-0.4 \le x \le 2.2$ until $\dot{x}(0)$ reaches 3.677051 (the exact value of the upper bound for this range of oscillation can be analytically derived by simply setting $f'(x) = 0$ and solving for the larger root to get $x = \frac{1}{3}(4 + \sqrt{7}) \approx 2.2$). A little beyond this value, $x(t)$ increases without bound, as the bead goes over the top of the hump. In particular, the value $\dot{x}(0) = 3.677052$ leads to unbounded $x(t)$ with no oscillation.

11. When the given parameter values are inserted into the given damped pendulum equation, the result is

$$\ddot{\theta} = -1.61 \sin\theta - 0.006\dot{\theta}.$$

The improved Euler method was used with initial time $t_0 = 0$ and step size $h = 0.001$. The initial conditions $\theta(0) = 0$, $\dot{\theta}(0) = 0.2$ were used for parts (a) and (b); and $\theta(0) = 0$, $\dot{\theta}(0) = 2$ were used for part (c).

(a) There are three local maximums for the solution on the interval $0 \le t \le 15$. They are approximately $\theta(1.24) = 0.1572$, $\theta(6.2) = 0.154875$ and $\theta(11.16) = 0.152584$.

(b) There are seven values of t on the interval $0 \le t \le 15$ such that $\theta(t) = 0$. They are approximately $t = 0$, 2.48, 4.96, 7.44, 9.92, 12.40 and 14.88.

(c) There are five values of t on the interval $0 \le t \le 15$ such that $\theta(t) = 0$. They are approximately $t = 0$, 3.1, 6.18, 9.25 and 12.3.

13. (a) With $k(t) = 0.2(1 - e^{-0.1t})$, $h = 5$, and $m = 1$, the unforced oscillator equation can be written as

$$\ddot{x} = -0.2(1 - e^{-0.1t})\dot{x} - 5x.$$

Applying the improved Euler method with initial conditions $x(0) = 0$, $\dot{x}(0) = 5$, $t_0 = 0$ and step size $h = 0.001$, 20,000 values of the solution $x(t)$ on the interval $0 \le t \le 20$ were computed. These values were then plotted in the tx-plane. Note that

$$\lim_{t \to \infty} k(t) = 0.2.$$

This implies that the long-term behavior of the solution of the given initial-value problem should approximate the long-term behavior of the solution of the initial-value problem $\ddot{x} + 0.2\dot{x} + 5x = 0$, $x(0) = 0$, $\dot{x}(0) = 5$. Indeed, this is suggested by the solution graph shown on the right.

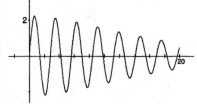

(b) With $k = 0$, $h = 5(1 - e^{-0.2t})$ and $m = 1$, the unforced oscillator equation can be written as

$$\ddot{x} = -5(1 - e^{-0.2t})x.$$

As in part (a), the improved Euler method was applied with initial conditions $x(0) = 0$, $\dot{x}(0) = 5$, $t_0 = 0$ and step size $h = 0.001$. The 20,000 corresponding values of $x(t)$ on the interval $0 \le t \le 20$ were computed and plotted in the tx-plane. Note that

$$\lim_{t \to \infty} h(t) = 5.$$

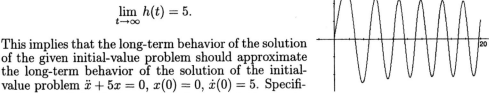

This implies that the long-term behavior of the solution of the given initial-value problem should approximate the long-term behavior of the solution of the initial-value problem $\ddot{x} + 5x = 0$, $x(0) = 0$, $\dot{x}(0) = 5$. Specifically, for sufficiently large t, the desired solution is virtually indistinguishable from a pure harmonic oscillation. This suggested by the solution graph shown on the right.

15. The accompanying graphs are for $k = -0.5$ and were generated by applying the improved Euler method with step size $h = 0.001$ to the given **hard spring oscillator equation** with initial conditions $y(0) = 0$, $\dot{y}(0) = 3$ on the interval $0 \le t \le 40$. Experimentation of these graphs suggest that the solution does approach periodic behavior. This is especially evident in the phase plot (indeed, this is one of the benefits in using phase plots; i.e., one is not required to look at solution graphs over t-intervals that are so large as to be impractical to reproduce). Experimentation shows that as k becomes more negative, long-term approach to periodic behavior becomes more rapid.

However, the long-term behavior of the solution corresponding to $k = 0$ does have some nice regularity — a feature that is evident by using the initial conditions $y(0) = \dot{y}(0) = 0$ and constructing the corresponding phase curve on the interval $0 \le t \le 200$.

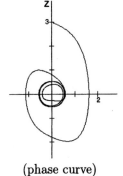

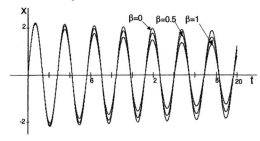

      (solution graph)   (phase curve)

$$\ddot{y} = -y - y^3 - \tfrac{1}{2}\dot{y} + \tfrac{3}{10}\cos t, \quad y(0) = 0, \ \dot{y}(0) = 3, \text{ for } 0 \le t \le 40.$$

17. With the given parameter values, the D.E. becomes $\ddot{x} + (0.02)\dot{x}|\dot{x}|^\beta + 5x = 0$. On the interval $0 \le t \le 20$, the improved Euler method was applied to the equations corresponding to $\beta = 0$, $\beta = 1/2$ and $\beta = 1$. In each case, the initial conditions were $x(0) = 0$, $\dot{x}(0) = 5$ and the step size was 0.001. Of the resulting 20,000 computed values of $x(t)$, the nonnegative integer values of t are listed in the table on the following page.

By changing the "PRINT T,Y" command to "PLOT T,Y", the same program was used to plot the three corresponding solution graphs below. They have been drawn on the same set of axes in order to more clearly see their differences as well as their common features.

447

Table for solutions of the initial-value problem $\ddot{x} + (0.02)\dot{x}|\dot{x}|^\beta + 5x = 0$, $x(0) = 0$, $\dot{x}(0) = 5$, for $\beta = 0, 0.5, 1$ and for nonnegative integer values of t on the interval $0 \le t \le 20$.

t	$x(t)$ $(\beta = 0)$	$x(t)$ $(\beta = 0.5)$	$x(t)$ $(\beta = 1)$
0	0.0000	0.0000	0.0000
1	1.7417	1.7242	1.6908
2	−2.1288	−2.0851	−2.0004
3	0.8946	0.8673	0.8164
4	0.9932	0.9546	0.8853
5	−2.0908	−1.9897	−1.8147
6	1.5819	1.4902	1.3365
7	0.1159	0.1086	0.0966
8	−1.6923	−1.5672	−1.3709
9	1.9548	1.7927	1.5463
10	−0.7303	−0.6636	−0.5654
11	−1.0234	−0.9237	−0.7804
12	1.9667	1.7593	1.4699
13	−1.4006	−1.2416	−1.0261
14	−0.2158	−0.1905	−0.1567
15	1.6367	1.4301	1.1637
16	−1.7889	−1.5494	−1.2491
17	0.5821	0.4999	0.3996
18	1.0420	0.8904	0.7086
19	−1.8441	−1.5633	−1.2349
20	1.2326	1.0365	0.8128

Chapter Review

(pgs. 568-570)

1. The characteristic equation is $r^2 + 2r + 1 = (r+1)^2 = 0$, with roots $r_1 = r_2 = -1$. The two linearly independent basic homogeneous solutions are therefore e^{-x} and xe^{-x}. Since the first term of the forcing function $e^{-x} + 3e^x$ is a homogeneous solution (as is xe^{-x}), and since the derivative set of the second term of the forcing function does not contain a homogeneous solution, there must exist a nonhomogeneous solution of the form $y_p(x) = Ae^x + Bx^2 e^{-x}$. Substituting this and

$$y_p'(x) = Ae^x + (-Bx^2 + 2Bx)e^{-x} \quad \text{and} \quad y_p''(x) = Ae^x + (Bx^2 - 4Bx + 2B)e^{-x}$$

into the nonhomogeneous equation gives

$$y_p'' + 2y_p' + y_p$$
$$= Ae^x + (Bx^2 - 4Bx + 2B)e^{-x} + 2\big(Ae^x + (-Bx^2 + 2Bx)e^{-x}\big) + Ae^x + Bx^2 e^{-x}$$
$$= 4Ae^x + 2Be^{-x} = e^{-x} + 3e^x.$$

Equating coefficients of like functions gives $4A = 3$ and $2B = 1$. Hence, $A = 3/4$, $B = 1/2$ and $y_p = \frac{3}{4}e^x + \frac{1}{2}x^2 e^{-x}$. The general solution of the given nonhomogeneous equation is therefore

$$y(x) = c_1 e^{-x} + c_2 x e^{-x} + \frac{1}{2}x^2 e^{-x} + \frac{3}{4}e^x.$$

3. The characteristic equation is $r^2 - 1 = (r-1)(r+1) = 0$, with roots $r_1 = 1$, $r_2 = -1$. The two linearly independent basic homogeneous solutions are therefore e^x and e^{-x}. Since no member of the derivative set of the forcing function $\sin x$ is a homogeneous solution, there is a nonhomogeneous solution of the form $y_p(x) = A\cos x + B\sin x$. Substituting this and $y_p''(x) = -A\cos x - B\sin x$ into the nonhomogeneous equation gives

$$y_p'' - y_p = -A\cos x - B\sin x - (A\cos x + B\sin x) = -2A\cos x - 2B\sin x = \sin x.$$

Equating coefficients of like functions gives $-2A = 0$ and $-2B = 1$. Hence, $A = 0$, $B = -1/2$ and $y_p(x) = -\frac{1}{2}\sin x$. The general solution of the given nonhomogeneous equation is therefore

$$y(x) = c_1 e^x + c_2 e^{-x} - \frac{1}{2}\sin x.$$

5. The characteristic equation is $r^2 + 2r + 3 = 0$, with complex roots $r_{1,2} = -1 \pm i\sqrt{2}$. The two linearly independent basic homogeneous solutions are therefore $e^{-x}\cos(\sqrt{2}\,t)$ and $e^{-x}\sin(\sqrt{2}\,t)$. Since the forcing function is a constant (which is not a homogeneous solution), there must be a constant solution $y_p(x) = A$ of the nonhomogeneous equation. Since $y_p'(x)$ and $y_p''(x)$ are both identically zero, the given D.E. implies $3y_p(x) = 1$. So, $y_p(x) = 1/3$ is a solution of the nonhomogeneous equation and the general solution of the nonhomogeneous equation is therefore

$$y(x) = c_1 e^{-x}\cos(\sqrt{2}\,t) + c_2 e^{-x}\sin(\sqrt{2}\,t) + \frac{1}{3}.$$

449

7. Three successive integrations of the given equation results in

$$y'' = \frac{1}{2}x^2 + c_1, \qquad y' = \frac{1}{6}x^3 + c_1 x + c_2, \qquad y = \frac{1}{24}x^4 + \frac{1}{2}c_2 x^2 + c_2 x + c_3.$$

Replacing $\frac{1}{2}c_1$ with c_1' gives the general solution as

$$y(x) = \frac{1}{24}x^4 + c_1' x^2 + c_2 x + c_3.$$

9. The characteristic equation $r^2 + 9 = 0$ has the characteristic roots $r_{1,2} = \pm 3i$. The two basic homogeneous solutions are therefore $\cos 3x$ and $\sin 3x$, which comprise the members of the derivative set of the forcing function $\sin 3x$. It follows that there is a nonhomogeneous solution of the form $y_p(x) = Ax \cos 3x + Bx \sin 3x$. Substitute this and

$$y_p''(x) = -9Ax \cos 3x - 6A \sin 3x - 9Bx \sin 3x + 6B \cos 3x$$

into the nonhomogeneous equation to get

$$y_p'' + 9y_p = -9Ax \cos 3x - 6A \sin 3x - 9Bx \sin 3x + 6B \cos 3x + 9(Ax \cos 3x + Bx \sin 3x)$$
$$= -6A \sin 3x + 6B \cos 3x = \sin 3x.$$

Equating coefficients of like functions gives $-6A = 1$ and $6B = 0$. Hence, $A = -1/6$, $B = 0$ and $y_p(x) = -\frac{1}{6}x \cos 3x$. The general solution is therefore

$$y(x) = c_1 \cos 3x + c_2 \sin 3x - \frac{1}{6}x \cos 3x.$$

The condition $y(0) = 1$ immediately gives $c_1 = 1$, so that $y(x) = \cos 3x - \frac{1}{6}x \cos 3x + c_2 \sin 3x$. Since

$$y'(x) = -3 \sin 3x + \frac{1}{2}x \sin 3x - \frac{1}{6} \cos 3x + 3c_2 \cos 3x,$$

the condition $y'(0) = 0$ implies $0 = -1/6 + 3c_2$. Hence, $c_2 = 1/18$ and the solution of the given initial-value problem is

$$y(x) = \cos 3x - \frac{1}{6}x \cos 3x + \frac{1}{18} \sin 3x.$$

11. The simplest free harmonic oscillator equation $y'' + y = 0$ has the general solution $y(x) = c_1 \cos x + c_2 \sin x$. The two boundary conditions lead to $-1 = y(0) = c_1$ and $1 = y(\pi) = -c_1$, which are equivalent. Hence, $c_1 = -1$ and c_2 remains arbitrary. The solution of the given boundary-value problem is therefore $y(x) = -\cos x + c_2 \sin x$.

13. The cubic characteristic equation $r^3 + ar^2 + br + c = 0$ has three roots r_1, r_2, r_3, one of which, say r_1, is always real. If r_2 and r_3 are complex conjugates, say $r_{1,2} = \alpha \pm i\beta$ ($\beta \neq 0$), then the basic solution set is

$$\{e^{r_1 x}, e^{\alpha x} \cos \beta x, e^{\alpha x} \sin \beta x\}.$$

Otherwise, r_2 and r_3 are real. In this case, there are three possibilities:

$$\{e^{r_1 x}, e^{r_2 x}, e^{r_3 x}\}, \quad \text{where } r_1,\ r_2,\ r_3 \text{ are all different;}$$
$$\{e^{r_1 x}, x e^{r_1 x}, e^{r_3 x}\}, \quad \text{where } r_1 = r_2 \neq r_3;$$
$$\{e^{r_1 x}, x e^{r_1 x}, x^2 e^{r_1 x}\}, \quad \text{where } r_1 = r_2 = r_3.$$

15. (a) Multiplying the given equation by $\dot{y}$ gives $\dot{y}\ddot{y} + \dot{y}y = 0$. It follows that

$$\int \dot{y}\ddot{y}\, dt + \int \dot{y}y\, dt = C_1,$$

where C_1 is a constant. Since the integrand of the first integral is the derivative of $\frac{1}{2}\dot{y}^2$ with respect to t and the integrand of the second integral is the derivative of $\frac{1}{2}y^2$ with respect to t, the above equation is equivalent to

$$\frac{1}{2}\dot{y}^2 + \frac{1}{2}y^2 = C_1.$$

(b) Evaluating the equation derived in part (a) at $t = 0$ gives $\frac{1}{2}(1)^2 + \frac{1}{2}(0)^2 = C_1$, so that $C_1 = 1/2$ and $\dot{y}^2 + y^2 = 1$. Solving for $\dot{y}$ gives $\dot{y}(t) = \pm\sqrt{1 - y^2}$. But since $\dot{y}(0) = 1$, the minus sign must be rejected in favor of $\dot{y} = \sqrt{1 - y^2}$. For $|y| < 1$, the variables can be separated to get

$$\frac{dy}{\sqrt{1 - y^2}} = dt.$$

Integration gives $\arcsin y = t + C_2 = t$, where the constant C_2 is chosen so that $y(0) = 0$. So, $y(t) = \sin t$. In order to derive this conclusion, we needed to assume that $-1 < y < 1$, which restricted the domain of the solution to $-\pi/2 < t < \pi/2$. However, since $\ddot{y} = -\sin t$ for all t, the equation $\ddot{y} + y = 0$ is satisfied by $y = \sin t$ for all t..

17. Let r and s be different complex numbers. Suppose c_1 and c_2 are real or complex constants such that the equation

$$c_1 e^{rx} + c_2 e^{sx} = 0$$

holds on an open interval I. If c_1 and c_2 are both nonzero then the above equation can be written as

$$e^{(r-s)x} = -\frac{c_2}{c_1},$$

which holds on I. Since this can only happen if $r = s$, it follows that one of c_1 or c_2 is zero. Without loss of generality, we can suppose $c_1 = 0$. Then the equation reduces to $c_2 e^{2x} = 0$. Since e^{sx} is never zero, we must have $c_2 = 0$. So, e^{rx} and e^{sx} are linearly independent on I. Since I was arbitrary, they are linearly independent on every open interval.

19. Let $I(t) = A\sin(t - \alpha)$, where $A > 0$ and $-\pi \le \alpha < \pi$, be a solution of the given differential equation. Substitute this, $I'(t) = A\cos(t - \alpha)$ and $I''(t) = -A\sin(t - \alpha)$ into the given D.E to get

$$-A\sin(t - \alpha) + RA\cos(t - \alpha) + A\sin(t - \alpha) = \sin t, \qquad \text{or} \qquad RA\cos(t - \alpha) = \sin t.$$

Since $\cos(t - \alpha) = \cos t \cos \alpha + \sin t \sin \alpha$, the equation on the right can be rewritten as

$$(RA\cos\alpha)\cos t = (1 - RA\sin\alpha)\sin t.$$

Equating coefficients of like functions leads to the equations $RA\cos\alpha = 0$ and $RA\sin\alpha = 1$. Since $RA > 0$, the first equation says $\cos\alpha = 0$ and the second equation says $\sin\alpha > 0$. The only value of α in the required range is $\alpha = \pi/2$. The second equation then becomes $RA = 1$, or equivalently, $A = 1/R$.

21. (a) Suppose that for all x in an open interval I and some constants c_1, c_2,

$$c_1 e^x + c_2 e^{-x} = 0.$$

Let $x_0 \in I$ and let $\epsilon > 0$ be such that $x_0 + \epsilon \in I$. We then have the two equations

$$c_1 e^{x_0} + c_2 e^{-x_0} = 0 \quad \text{and} \quad c_1 e^{x_0+\epsilon} + c_2 e^{-x_0-\epsilon} = 0.$$

When the second equation is subtracted from e^ϵ times the first equation, the result is

$$c_2 e^{-x_0+\epsilon} - c_2 e^{-x_0-\epsilon} = 2 c_2 e^{-x_0}\left(\frac{e^\epsilon - e^{-\epsilon}}{2}\right) = 2 c_2 e^{-x_0} \sinh \epsilon = 0.$$

Since $\epsilon \neq 0$, $\sinh \epsilon \neq 0$. Therefore, since $e^{x_0} \neq 0$, the last equality above shows that $c_2 = 0$. The original equation then becomes $c_1 e^x = 0$, which implies $c_1 = 0$. It follows that e^x and e^{-x} are linearly independent on any open interval.

(b) Given the equation $2e^x - 3e^{-x} = 0$, multiply through by e^x, add 3 to both sides, divide both sides by 2, take logarithms and divide by 2 to obtain $x = \frac{1}{2}\ln(3/2)$. This is the only solution.

On the other hand, since the exponential function is positive for all real x, the expression $2e^x + 3e^{-x}$ is positive for all real x. Therefore, $2e^x + 3e^{-x} = 0$ has no real solutions.

(c) Let $x = a + ib$ be a complex solution of the equation $2e^x + 3e^{-x} = 0$. Then

$$2e^{a+ib} + 3e^{-a-ib} = 2e^a(\cos b + i \sin b) + 3e^{-a}(\cos b - i \sin b)$$
$$= (2e^a + 3e^{-a})\cos b + i(2e^a - 3e^{-a})\sin b = 0.$$

It follows that the real and imaginary parts of the left side are both zero. That is,

$$(2e^a + 3e^{-a})\cos b = 0 \quad \text{and} \quad (2e^a - 3e^{-a})\sin b = 0.$$

The second result of part (b) says that $2e^a + 3e^{-a} \neq 0$. It therefore follows from the first equation that $\cos b = 0$. Hence, $b = (n + 1/2)\pi$, with n an integer. As a result of this, $\sin b \neq 0$, so that the second equation implies $2e^a - 3e^{-a} = 0$. By part (b), the only possibility is $a = \frac{1}{2}\ln(3/2)$. Thus, the complex solutions of $2e^x + 3e^{-x} = 0$ are necessarily of the form

$$x = \frac{1}{2}\ln(3/2) + i\left(n + \frac{1}{2}\right)\pi, \quad n = 0, \pm 1, \pm 2, \ldots.$$

That these are all solutions can be shown by selecting a candidate at random and computing

$$2e^{\frac{1}{2}\ln(3/2)+i(n+1/2)\pi} = 2\sqrt{3/2}\, e^{in\pi} e^{i\pi/2} = i(-1)^n\sqrt{6}$$
$$3e^{-\frac{1}{2}\ln(3/2)-i(n+1/2)\pi} = 3\sqrt{2/3}\, e^{-in\pi} e^{-i\pi/2} = -i(-1)^n\sqrt{6}.$$

Hence, $2e^x + 3e^{-x} = 0$ and the given form for x is a solution for each n.

23. Pick $x = x_0$ and fix it. Given an n-tuple of initial values $(z_0, z_1, \ldots, z_{n-1})$, Theorem 2.5 asserts the existence of a unique solution $y(x)$ such the $y^{(k)}(x_0) = z_k$, for $k = 0, \ldots, n-1$. On the other hand, given any solution $y(x)$, the n-tuple $(y(x_0), \ldots, y^{(n-1)}(x_0))$ of real numbers is unique with respect to the given solution. Thus, the map sending n-tuples of real numbers to solutions of $L(y) = 0$ is a bijection for each fixed x_0.

452

25. (a) With the linear differential operator $Dy = y'$, we have

$$(D - r)\big(D - (r + h)\big)y = \big(D^2 - (2r + h)D + r(r + h)\big)y$$
$$= D^2 y - (2r + h)Dy - r(r + h)y = y'' - (2r + h)y' + r(r + h)y = 0.$$

(b) The factored form of the characteristic equation is $(\lambda - r)\big(\lambda - (r + h)\big) = 0$ with roots $\lambda_1 = r$, $\lambda_2 = r + h$. If $h \neq 0$ then the roots are different real numbers and the general solution is

$$y(x) = c_1 e^{(r+h)x} + c_2 e^{rx}.$$

(c) Consider the solution $y(x) = \frac{1}{h} e^{(r+h)x} - \frac{1}{h} e^{rx}$. For each fixed x we have

$$\lim_{h \to 0} y(x) = \lim_{h \to 0} \frac{e^{(r+h)x} - e^{rx}}{h} = \frac{d(e^{rx})}{dr} = x e^{rx},$$

where we recognize the second limit expression as the definition of the derivative of e^{rx} *with respect to r.*

27. Let $z = c_1 y_1 + c_2 y_2 + c_3 y_3$ and require that $L(z) = w_1 + 2w_2 - 4w_3$. The linearity of L together with $L(y_1) = w_1$, $L(y_2) = w_2$, $L(y_3) = w_3$, gives

$$L(z) = L(c_1 y_1 + c_2 y_2 + c_3 y_3) = c_1 L(y_1) + c_2 L(y_2) + c_3 L(y_3) = c_1 w_1 + c_2 w_2 + c_3 w_3.$$

It follows that $c_1 w_1 + c_2 w_2 + c_3 w_3 = w_2 + 2w_2 - 4w_3$. We therefore choose $c_1 = 1$, $c_2 = 2$ and $c_3 = -4$ and get $z = y_1 + 2y_2 - 4y_3$.

29. Let $z = c_1 y_1 + c_2 y_2 + c_3 y_3$ and require that $L(z) = 0$. The linearity of L together with $L(y_1) = w_1$, $L(y_2) = w_2$, $L(y_3) = w_3$, gives

$$L(z) = L(c_1 y_1 + c_2 y_2 + c_3 y_3) = c_1 L(y_1) + c_2 L(y_2) + c_3 L(y_3) = c_1 w_1 + c_2 w_2 + c_3 w_3.$$

It follows that $c_1 w_1 + c_2 w_2 + c_3 w_3 = 0$. We therefore choose $c_1 = c_2 = c_3 = 0$ and get $z = 0$.

31. Let $f(t) = t \ln t$ for $t > 0$. Since $\lim_{t \to 0^+} f(t) = 0$, extend the definition of f by defining $f(0) = 0$. This makes f continuous on $[0, \infty)$ and bounded on every closed interval of the form $[a, b]$, so that $\int_a^b e^{-st} t \ln t \, dt$ exists for any constant s and any a and b satisfying $0 < a < b$. Also, the fact that $0 < \ln t < t$ for $t > 1$ implies $0 < e^{-st} f(t) < e^{-st} t^2$ for $t > 1$. Hence, for $b > 1$, $s > 0$,

$$\int_1^b e^{-st} f(t) \, dt \leq \int_1^b e^{-st} t^2 dt < \int_0^\infty e^{-st} t^2 \, dt = \mathcal{L}[t^2](s) = \frac{2}{s^3}.$$

For each $s > 0$, the integral on the left defines a function $F_s(b)$ of b, where the integrand is positive on $(1, b)$. Therefore, $F_s(b)$ is an increasing function on $(1, \infty)$ that is also bounded above by $2/s^3$. So, $\lim_{b \to \infty} F_s(b)$ exists as a function of s for $s > 0$. Moreover,

$$\lim_{b \to \infty} F_s(b) = \lim_{b \to \infty} \int_1^b e^{-st} f(t) \, dt = \int_1^\infty e^{-st} f(t) \, dt = \int_0^\infty e^{-st} f(t) \, dt - \int_0^1 e^{-st} f(t) \, dt.$$

Since $\int_0^1 e^{-st} f(t) \, dt$ is a real number for each $s \geq 0$, it defines a function of s for $s > 0$. Thus, equality of the two outer members in the last displayed equation shows that $\int_0^\infty e^{-st} f(t) \, dt$ defines function of s for $s > 0$ and is therefore, by definition, the Laplace transform of $t \ln t$.

33. For each real number s, the condition $\lim_{t\to\infty} e^{-st} f(t) = 0$ is necessary for the existence of the integral defining the Laplace transform of f. In this case, we let $f(t) = e^{t^2}$ and look at the expression $e^{-st} e^{t^2} = e^{-st+t^2}$. Observe that the quadratic exponent $-st + t^2$ is an increasing positive function of t for $t > \max\{0, s\}$. This means that $e^{-st} e^{t^2}$ is an increasing positive function of t for $t > \max\{0, s\}$. Hence, $\lim_{t\to\infty} e^{-st} e^{t^2} \neq 0$ and the Laplace transform of e^{t^2} does not exist for any value of s.

35. Letting $u = \ln t$ in the integral $\int_0^\infty e^{-st} \sin(\ln t)\, dt$ results in

$$\int_0^\infty e^{-st} \sin(\ln t)\, dt = \int_{-\infty}^\infty e^{-se^u + u} \sin u\, du$$

$$= \int_{-\infty}^0 e^{-se^u + u} \sin u\, du + \int_0^\infty e^{-se^u + u} \sin u\, du$$

$$= -\int_0^\infty e^{-se^{-u} - u} \sin u\, du + \int_0^\infty e^{-se^u + u} \sin u\, du,$$

where the first integral in the bottom row is obtained by the substitution $v = -u$ and then writing the result using the dummy variable name "u".

For each $s > 0$, the exponent $-se^{-u} - u$ on the exponential in the first integral is less than $-u$ for all u, and the exponent $-se^u + u$ on the exponential in the second integral is less than $-u$ when $e^u > 2u/s$. Thus, the absolute value of both integrands are dominated by $e^{-u} |\sin u|$ for all sufficiently large u. Since $\int_0^\infty e^{-u} |\sin u|\, du$ exists, so do both of the above displayed integrals, and therefore so does the Laplace transform of $\sin(\ln t)$.

37. For each real number s, $e^{-st} > 1 - st$, and for each $t \leq 1/s$, $1 - st \geq 0$. Hence,

$$\int_0^{1/s} e^{-st} t^{-1} dt = \lim_{\epsilon \to 0^+} \int_\epsilon^{1/s} e^{-st} t^{-1} dt \geq \lim_{\epsilon \to 0^+} \int_\epsilon^{1/s} (1 - st) t^{-1} dt$$

$$= \lim_{\epsilon \to 0^+} \int_\epsilon^{1/s} (t^{-1} - s) dt = \lim_{\epsilon \to 0^+} (\ln|1/s| - \ln|\epsilon| - 1 + \epsilon) = +\infty.$$

That is, $\int_0^{1/s} e^{-st} t^{-1} dt$ diverges to $+\infty$ for every s. Therefore, since $e^{-st} t^{-1}$ is positive for $t > 0$, we can write

$$\int_0^\infty e^{-st} t^{-1} dt = \int_0^{1/s} e^{-st} t^{-1} dt + \int_{1/s}^\infty e^{-st} t^{-1} dt \geq \int_0^{1/s} e^{-st} t^{-1} dt = +\infty.$$

So $\int_0^\infty e^{-st} t^{-1} dt$ does not converge for any s. That is, the Laplace transform of t^{-1} does not exist for any s.

Chapter 12: Introduction to Systems

Section 1: VECTOR FIELDS

Exercise Set 1ABC (pgs. 578-580)

1. The first equation of the given system can be written as $\dot{x} - x = 1$, so that the integrating factor e^{-t} leads to

$$\frac{d}{dt}\left(e^{-t}x\right) = e^{-t} \quad \text{and therefore} \quad e^{-t}x = -e^{-t} + c_1.$$

The condition $x(0) = 1$ gives $c_1 = 2$, so that $x(t) = -1 + 2e^t$.
 The second equation of the given system can be written as the homogeneous equation $\dot{y} - y = 0$. The root of its linear characteristic equation $r - 1 = 0$ is $r_1 = 1$, so that $y = c_2e^t$. The condition $y(0) = 2$ then gives $c_2 = 2$ and therefore $y(t) = 2e^t$. The solution of the system is then

$$\left(x(t), y(t)\right) = (-1 + 2e^t, 2e^t), \quad -\infty < t < \infty.$$

3. The system can be written as the three homogeneous equations $\dot{x} - x = 0$, $\dot{y} - \frac{1}{2}y = 0$ and $\dot{z} - \frac{1}{3}z = 0$. Their respective characteristic equations are $r - 1 = 0$, $r - 1/2 = 0$ and $r - 1/3 = 0$, with the respective roots $r_1 = 1$, $r_2 = 1/2$ and $r_3 = 1/3$. The general solution of the system is then

$$\left(x(t), y(t), z(t)\right) = (c_1e^t, c_2e^{t/2}, c_3e^{t/3}).$$

The initial conditions then gives $(0, 1, -1) = \left(x(0), y(0), z(0)\right) = (c_1, c_2, c_3)$. So, $c_1 = 0$, $c_2 = 1$, $c_3 = -1$ and the desired solution is

$$\left(x(t), y(t), z(t)\right) = (0, e^{t/2}, -e^{t/3}), \quad -\infty < t < \infty.$$

5. (a) If $\mathbf{F}(t, \mathbf{x})$, with $\mathbf{x} = (x, y)$, is such that $d\mathbf{x}/dt = \mathbf{F}(t, \mathbf{x})$ is the vector form of the system given in Exercise 2 above then

$$F(t, \mathbf{x}) = \frac{d\mathbf{x}}{dt} = \left(\frac{dx}{dt}, \frac{dy}{dt}\right) = (t, y).$$

(b) The speed of the trajectory is the magnitude of $d\mathbf{x}/dt$. By part (a), this is the magnitude of $\mathbf{F}(t, \mathbf{x}) = (t, y)$, or $\sqrt{t^2 + y^2}$. To write this as a function of t, solve the second equation of the system in Exercise 2 for $y = y(t)$. The form of the equation itself shows that the solution is a constant multiple of an exponential function, which means that it can take on the value 0 if, and only if, it is identically zero. The given condition $y(1) = 0$ then implies that $y = 0$ identically. So, the speed of the trajectory is

$$\left|\frac{d\mathbf{x}}{dy}\right| = \sqrt{t^2 + 0^2} = |t|.$$

455

7. The vector field $\mathbf{F}(x, y) = (x + 1, y)$

9. The vector field $\mathbf{F}(x, y, z) = (x, y/2, z/3)$.

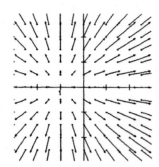
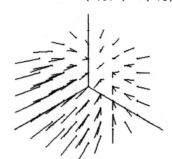

11. The vector field shown on the right describes the motion of vector functions $\mathbf{x} = \mathbf{x}(t) = \big(x(t), y(t)\big)$ that satisfy the system $\dot{\mathbf{x}} = \mathbf{F}(\mathbf{x}) = (-y, x)$. That is, since $\dot{\mathbf{x}} = (\dot{x}, \dot{y})$, it depicts the general trajectory motion of solutions of the system

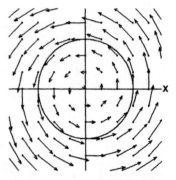

$$\dot{x} = -y,$$
$$\dot{y} = x.$$

Differentiating the first equation gives $\ddot{x} = -\dot{y}$, so that the second equation forces $\ddot{x} = -x$, or equivalently, $\ddot{x} + x = 0$. The solutions of this equation are $x(t) = c_1 \cos t + c_2 \sin t$, with derivative $\dot{x}(t) = -c_1 \sin t + c_2 \cos t$. On the other hand, $\dot{x} = -y$, so that $y(t) = -c_2 \cos t + c_1 \sin t$. Squaring the equations for x and y and adding the results gives $x^2(t) + y^2(t) = c_1^2 + c_2^2$. That is, the trajectories are circles of radius $\sqrt{c_1^2 + c_2^2}$ centered at the origin. In particular, the trajectory starting at $(x, y) = (1, 0)$ (corresponding to the initial conditions $x(0) = 1$, $y(0) = 0$) is described by a counterclockwise tracing of the curve $x^2(t) + y^2(t) = 1$, which is also shown in the sketch.

13. If $\dot{y} = z$ then $\ddot{y} = \dot{z}$ and the given differential equation becomes $\dot{z} + z + y = 0$, or $\dot{z} = -y - z$. The initial condition $\dot{y}(0) = 1$ translates to $z(0) = 1$. The resulting system is therefore

$$\dot{y} = z,$$
$$\dot{z} = -y - z, \quad y(0) = 1, \ z(0) = 1.$$

To solve the system, replace z and $\dot{z}$ in the second equation with $\dot{y}$ and $\ddot{y}$, respectively, to get $\ddot{y} = -y - \dot{y}$, or equivalently, $\ddot{y} + \dot{y} + y = 0$. Since the characteristic equation $r^2 + r + 1 = 0$ has the complex roots $r_{1,2} = -\frac{1}{2} \pm i\frac{\sqrt{3}}{2}$, the general solution for $y = y(t)$ is of the form

$$y(t) = c_1 e^{-t/2} \cos \frac{\sqrt{3}}{2} t + c_2 e^{-t/2} \sin \frac{\sqrt{3}}{2} t.$$

Since $z = \dot{y}$,

$$z(t) = \dot{y}(t) = \left(-\frac{1}{2}c_1 + \frac{\sqrt{3}}{2}c_2 \right) e^{-t/2} \cos \frac{\sqrt{3}}{2} t + \left(-\frac{\sqrt{3}}{2}c_1 - \frac{1}{2}c_2 \right) e^{-t/2} \sin \frac{\sqrt{3}}{2} t.$$

The initial conditions conditions $y(0) = 1$, $z(0) = 1$ imply the equations

$$1 = y(0) = c_1 \qquad \text{and} \qquad 1 = z(0) = -\frac{1}{2}c_1 + \frac{\sqrt{3}}{2}c_2.$$

456

So $c_1 = 1$, $c_2 = \sqrt{3}$ and the solution of the above system is

$$y(t) = e^{-t/2}\left(\cos\frac{\sqrt{3}}{2}t + \sqrt{3}\sin\frac{\sqrt{3}}{2}\right), \qquad z(t) = e^{-t/2}\left(\cos\frac{\sqrt{3}}{2}t - \sqrt{3}\sin\frac{\sqrt{3}}{2}\right).$$

15. Let $dy/dt = z$. Then $d^2y/dt^2 = dz/dt$ and the given differential equation becomes $dz/dt = z^2 + y^2 = e^t$. The equivalent two-dimensional first-order system is therefore

$$\frac{dy}{dt} = z,$$
$$\frac{dz}{dt} = -y^2 - z^2 + e^t.$$

17. Let $dy/dt = z$ and $d^2y/dt^2 = dz/dt = w$. Then $d^3y/dt^3 = d^2z/dt^2 = dw/dt$ and the given differential equation can be written as $dw/dt = w^2 - yz - t$. The equivalent three-dimensional first-order system is therefore

$$\frac{dy}{dt} = z, \qquad \frac{dz}{dt} = w, \qquad \frac{dw}{dt} = w^2 - yz - t.$$

19. Adding and subtracting the given equations gives $2dx/dt = t + y$ and $2dy/dx = t - y$. Hence, the equivalent normal system is

$$\frac{dx}{dt} = \frac{1}{2}(t + y),$$
$$\frac{dy}{dt} = \frac{1}{2}(t - y).$$

21. Subtract the second equation of the given system from the first to get $dx/dt - x + 3y = t$, or $dx/dt = x - 3y + t$. Substitute this expression for dx/dt into the given second equation and get $x - 3y + t + dy/dt + 2x + 2y = 0$, or $dy/dt = -3x + y - t$. Hence, the equivalent normal system is

$$\frac{dx}{dt} = x - 3y + t,$$
$$\frac{dy}{dt} = -3x + y - t.$$

23. Let $\dot{y} = z$, so that $\ddot{y} = z\,dz/dy$. The given differential equation becomes $z\,dz/dy = z$. For $z \neq 0$, the equation is equivalent to the separated equation $dz = dy$. Integrating gives $z = y + C$. Note that once C is chosen, y is restricted so that $y + C$ is always the same sign (because $z \neq 0$). Thus, a solution $y = y(t)$ with a positive initial velocity is strictly increasing for all t and therefore the associated phase curve is the part of the line $z = y + C$ lying above the y-axis. Similarly, a solution with a negative initial velocity is strictly decreasing for all t and therefore the associated phase curve is the part of the line $z = y + C$ lying below the y-axis. A solution with initial velocity $z(0) = 0$ is a constant solution $y(t) = -C$ and the associated phase curve is the equilibrium point $(-C, 0)$. The details are shown on the right for $C = -1, 0, 1$.

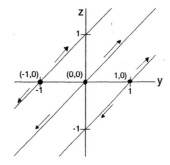

25. (a) If $x = z + w$ and $y = 2z - 2w$ then $\dot{x} = \dot{z} + \dot{w}$ and $\dot{y} = 2\dot{z} - 2\dot{w}$. Using these equations, replace each occurrence of x, y, $\dot{x}$ and $\dot{y}$ in the given coupled system with their corresponding expressions in terms of z, w, $\dot{z}$ and $\dot{w}$ and then simplify to get

$$\dot{z} + \dot{w} = 3z - w,$$
$$\dot{z} - \dot{w} = 3z + w.$$

Adding and subtracting these equations gives $2\dot{z} = 6z$ and $2\dot{w} = -2w$, respectively. Dividing each of these equations by 2 produces an uncoupled system of the form

$$\dot{z} = 3z, \qquad \dot{w} = -w.$$

(b) The solutions of $\dot{z} = 3z$ are the constant multiples of e^{3t} and the solutions of $\dot{w} = -w$ are the constant multiples of e^{-t}. It follows that

$$x = z + w = c_1 e^{3t} + c_2 e^{-t} \qquad \text{and} \qquad y = 2x - 2y = 2c_1 e^{3t} - 2c_2 e^{-t}.$$

As a check, we observe that $x + y = c_1 e^{3t} + c_2 e^{-t} + 2c_1 e^{3t} - 2c_2 e^{-t} = 3c_1 e^{3t} - c_2 e^{-t} = \dot{x}$ and $4x + y = 4(c_1 e^{3t} + c_2 e^{-t}) + 2c_1 e^{3t} - 2c_2 e^{-t} = 6c_1 e^{3t} + 2c_2 e^{-t} = \dot{y}$, so that the pair $x(t)$, $y(t)$ as given above is a solution of the given system.

27. With $dx/dt = e^{2y}$ and $dy/dt = e^{x+y}$, the chain rule allows us to form the quotient to get

$$\frac{dy}{dx} = \frac{dy/dt}{dx/dt} = \frac{e^{x+y}}{e^{2y}} = e^{x-y}.$$

This can be separated to get $e^y dy = e^x dy$. Integration gives $e^y = e^x + C$, where C is an arbitrary constant. This can be solved explicitly for either x or y. For example,

$$y = \ln(e^x + C).$$

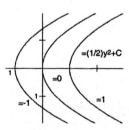

29. With $dx/dt = xy + y^2$ and $dy/dt = x + y$, the chain rule allows us to form the quotient to get

$$\frac{dy}{dx} = \frac{dy/dt}{dy/dt} = \frac{x+y}{xy+y^2} = \frac{x+y}{(x+y)y} = \frac{1}{y}.$$

The variables can be separated to get $y\, dy = dx$. Integration gives $\frac{1}{2}y^2 = x + C_1$, or equivalently,

$$x = \frac{1}{2}y^2 + C,$$

where $C = -C_1$ is an arbitrary constant.

31. (a) Write the first equation of the given system as $\ddot{x} + k\dot{x} = 0$ and note that its characteristic equation $r^2 + kr = 0$ has the roots $r_1 = 0$, $r_2 = -k$. Since $k > 0$, these roots are different. Hence, the general solution is $x(t) = c_1 + c_2 e^{-kt}$. With $\dot{x}(t) = -c_2 k e^{-kt}$, the initial conditions lead to $0 = x(0) = c_1 + c_2$ and $z_0 = \dot{x}(0) = -c_2 k$. So $c_1 = z_0/k$, $c_2 = -z_0/k$ and

$$x(t) = \frac{z_0}{k}(1 - e^{-kt}).$$

Now write the second equation as $\ddot{y}+k\dot{y} = -g$ and note that its characteristic equation is the same as that for the first equation of the system. Hence, the homogeneous solution of the second equation is $y_h(t) = c_1+c_2 e^{-kt}$. The forcing function $-g$ is a homogeneous solution, so that instead of $y_p = A$ as a trial particular solution, we use instead $y_p(t) = At$. Substituting $\dot{y}(t) = A$ and $\ddot{y}(t) = 0$ into the second equation of the system gives $0 + kA = -g$, so that $A = -g/k$ and $y_p(t) = -gt/k$. The general solution of the second equation is therefore $y(t) = c_1 + c_2 e^{-kt} - gt/k$. With $\dot{y}(t) = -c_2 k e^{-kt} - g/k$, the initial conditions lead to the equations $0 = y(0) = c_1 + c_2$ and $w_0 = \dot{y}(0) = -c_2 k - g/k$. Solving for c_2 from the second of these equations gives $c_2 = -(kw_0 + g)/k^2$. The first equation then gives $c_1 = (kw_0 + g)/k^2$. Hence,

$$y(t) = \frac{kw_o + g}{k^2}(1 - e^{-kt}) - \frac{g}{k}t.$$

(b) Since the vertical position of the projectile is given by $y(t)$ as found in part (a), first compute

$$\dot{y}(t) = \frac{kw_0 + g}{k}e^{-kt} - \frac{g}{k}.$$

Setting this equal to 0 and solving for $t = t_{\max}$ gives

$$t_{\max} = \frac{1}{k}\ln\left(\frac{kw_0}{g} + 1\right).$$

By inspection, we see that $\ddot{y}(t) < 0$ for all t. Hence, the critical point found must give a maximum value for $y(t)$. This justifies the subscript "max" on $t_{\max}$.

(c) All that needs to be done here is to evaluate $x_{\max} = x(t_{max})$ and $y_{max} = y(t_{max})$, where t_{max} is from part (b). However, it is perhaps easiest to note that the equation $\dot{y}(t_{max}) = 0$ implies $e^{-kt_{max}} = g/(kw_0 + g)$, so that

$$1 - e^{-kt_{\max}} = 1 - \frac{g}{kw_0 + g} = \frac{kw_0 + g - g}{kw_0 + g} = \frac{kw_0}{kw_0 + g}.$$

Hence, the coordinates of the position of maximum height are

$$x_{\max} = x(t_{\max}) = \frac{z_0}{k}(1 - e^{-kt_{\max}}) = \frac{z_0 w_0}{kw_0 + g}$$

$$y_{\max} = y(t_{max}) = \frac{kw_0 + g}{k^2}(1 - e^{kt_{max}}) - \frac{g}{k}t_{\max} = \frac{w_0}{k} - \frac{g}{k^2}\ln\left(\frac{kw_0}{g} + 1\right).$$

(d) Using $y_{\max}$ found in part (c), rewrite it in a form amenable for applying l'Hôpital's rule as $k \to 0^+$. This gives

$$\lim_{k\to 0^+}\frac{kw_0 - g\ln(kw_0/g + 1)}{k^2} = \lim_{k\to 0^+}\frac{w_0 - \frac{w_0}{kw_0/g+1}}{2k} = \lim_{k\to 0^+}\frac{w_0^2}{2g(kw_0/g + 1)} = \frac{w_0^2}{2g}.$$

33. (a) First, observe that

$$q\frac{du}{dt} + p\frac{dv}{dt} = -qp(u - v) + pq(u - v) = 0.$$

Therefore, since p and q are constants, integration with respect to t gives

$$qu + pv = c_0, \quad \text{for some constant } c_0.$$

459

Using the initial conditions $u(0) = u_0$ and $v(0) = v_0$, evaluate the above equation at $t = 0$ to get $qu_0 + pv_0 = c_0$.

(b) Since p is nonzero, the equation $qu + pv = c_0$ derived in part (a) can be solved for v to get $v = (c_0 - qu)/p$. The given differential equation involving du/dt ($= \dot{u}$) can then be written in terms of u and $\dot{u}$ as

$$\dot{u} = -p(u - v) = -p\left(u - \frac{c_0 - qu}{p}\right) = -pu + (c_0 - qu) = -(p + q)u + c_0.$$

Write this as $\dot{u} + (p + q)u = c_0$, multiply by $e^{(p+q)t}$ and derive the equivalent equation $d/dt\big(e^{(p+q)t}u\big) = c_0 e^{(p+q)t}$. Integrating gives

$$e^{(p+q)t}u = \frac{c_0}{p + q}e^{(p+q)t} + C, \qquad \text{or} \qquad u = \frac{c_0}{p + q} + Ce^{-(p+q)t}.$$

The condition $u(0) = u_0$ gives $u_0 = c_0/(p + q) + C$, so that $C = u_0 - c_0/(p + q)$. Hence,

$$u(t) = \frac{c_0}{p + q} + \left(u_0 - \frac{c_0}{p + q}\right)e^{-(p+q)t} = \frac{1}{p + q}\Big(c_0 + \big(u_0(p + q) - c_0\big)e^{-(p+q)t}\Big).$$

From part (a), $c_0 = qu_0 + pv_0$, so that $u_0(p + q) - c_0 = pu_0 + qu_0 - (qu_0 + pv_0) = p(u_0 - v_0)$. The above equation for $u(t)$ can therefore be written as

$$u(t) = \frac{1}{p + q}\Big(c_0 + p(u_0 - v_0)e^{-(p+q)t}\Big).$$

(c) The equation for $v = v(t)$ can be found by solving $qu + pv = c_0$ for v and using the result of part (b) to get

$$\begin{aligned}
v &= \frac{1}{p}(c_0 - qu) = \frac{1}{p}\left(c_0 - q\left[\frac{1}{p + q}\Big(c_0 + p(u_0 - v_0)e^{-(p+q)t}\Big)\right]\right)\\
&= \frac{1}{p(p + q)}\Big(c_0(p + q) - q\big(c_0 + p(u_0 - v_0)e^{-(p+q)t}\big)\Big)\\
&= \frac{1}{p(p + q)}\Big(c_0 p - qp(u_0 - v_0)e^{-(p+q)t}\Big)\\
&= \frac{1}{p + q}\Big(c_0 - q(u_0 - v_0)e^{-(p+q)t}\Big).
\end{aligned}$$

(d) Since $u = u(t)$ is the temperature of the warmer body, the initial temperature difference between the two bodies is $u(0) - v(0) = u_0 - v_0 > 0$. We therefore want the time $t = t_0$ for which $u(t) - v(t) = \frac{1}{2}(u_0 - v_0)$. Using the formulas for $u(t)$ and $v(t)$ derived in parts (b) and (c), we find that t_0 must satisfy

$$\begin{aligned}
\frac{1}{2}(u_0 - v_0) &= \frac{1}{p + q}\Big(c_0 + p(u_0 - v_0)e^{-(p+q)t_0}\Big) - \frac{1}{p + q}\Big(c_0 - q(u_0 - v_0)e^{-(p+q)t_0}\Big)\\
&= \frac{1}{p + q}\Big(p(u_0 - v_0) + q(u_0 - v_0)\Big)e^{-(p+q)t_0}\\
&= (u_0 - v_0)e^{-(p+q)t_0}.
\end{aligned}$$

If $u_0 - v_0 \neq 0$ (i.e., if the initial temperatures of the two bodies are different) then $u_0 - v_0$ can be canceled from both sides to get $1/2 = e^{-(p+q)t_0}$. Solving for t_0 gives

$$t_0 = \frac{\ln 2}{p + q}.$$

Exercise Set 1D (pgs. 584-585)

1. With $\dot{x} = dx/dt$, the variables in $\dot{x} = ax^2$ can be separated to get $-x^{-2}dx = -a\,dt$. Integration gives $x^{-1} = -at + C$. The condition $x(0) = 1$ forces $C = 1$, so that $x^{-1} = 1 - at$, or

$$x(t) = \frac{1}{1 - at}.$$

The domain of this solution is the largest interval containing $t = 0$ on which $x(t)$ is differentiable. Since $x(t)$ is differentiable everywhere it's defined, it follows that its domain is also the largest interval containing $t = 0$ on which $x(t)$ is defined. Since $x(0) = 1 > 0$, $x(t)$ must remain positive on its entire domain. So the domain is $t < 1/a$.

3. Suppose $\mathbf{F}(\mathbf{x})$ is a continuously differentiable vector field and that B is a region of positive area. The transformation $T_t(\mathbf{x}) = \mathbf{y}(t)$, where $\mathbf{y}(0) = \mathbf{x}$, has Jacobian determinant J_t, which is nonzero on the image of B under T_t. By Jacobi's Theorem,

$$\int_{T_t(B)} dA_{\mathbf{y}} = \int_B |J_t(\mathbf{x})|\, dA_{\mathbf{x}}.$$

For each $t \geq 0$, the left side is $A\big(T_t(B)\big)$, the area of the set of points $T_t(\mathbf{x})$ as $\mathbf{x}$ ranges over B. The right side is nonzero for each t because the Jacobian determinant is nonzero on B and therefore the integral has a positive value for every t. It follows that $A\big(T_t(B)\big) > 0$ for all t. That is, a region of positive area can never be sent by $\mathbf{F}$ to a region of zero area in time t for any t.

5. (a) Let $\mathbf{F}(x, y) = (\partial H/\partial y, -\partial H/\partial x) = (\dot{x}, \dot{y})$ be the vector field for the given Hamiltonian system. Since $H = H(x, y)$ is assumed to be twice continuously differentiable, $\mathbf{F}$ is continuously differentiable and the mixed second-order partial derivatives of H are equal. Hence,

$$\text{div}\,\mathbf{F} = \frac{\partial}{\partial x}\left(\frac{\partial H}{\partial y}\right) + \frac{\partial}{\partial y}\left(-\frac{\partial H}{\partial x}\right) = \frac{\partial^2 H}{\partial x \partial y} - \frac{\partial^2 H}{\partial y \partial x} = 0.$$

By Theorem 1.6, $T_t(\mathbf{x})$ is volume preserving on B for any region B in $\mathcal{R}^2$.

(b) In order to show that the system is Hamiltonian, we need only find a Hamiltonian function $H(x, y)$ for which $\dot{x} = \partial H/\partial y$ and $\dot{y} = -\partial H/\partial x$. We have

$$\dot{x} = \frac{\partial H}{\partial y} \quad \Leftrightarrow \quad -y = \frac{\partial H}{\partial y} \quad \Leftrightarrow \quad H(x, y) = -\frac{1}{2}y^2 + f(x).$$

$$\dot{y} = -\frac{\partial H}{\partial x} \quad \Leftrightarrow \quad x = -\frac{\partial H}{\partial x} = -f'(x) \quad \Leftrightarrow \quad f(x) = -\frac{1}{2}x^2 + C,$$

where $f(x)$ in the first line is a differentiable function of x and is independent of y and C is a real constant. Hence, we can choose $C = 0$ and take

$$H(x, y) = -\frac{1}{2}(x^2 + y^2)$$

as a Hamiltonian for the system.

(c) Let $x = x(t)$, $y = y(t)$ be a solution of a two-dimensional Hamiltonian system. Along the trajectory of the solution, the chain rule gives

$$\frac{dH}{dt} = \frac{\partial H}{\partial x}\dot{x} + \frac{\partial H}{\partial y}\dot{y} = \frac{\partial H}{\partial x}\frac{\partial H}{\partial y} - \frac{\partial H}{\partial y}\frac{\partial H}{\partial x} = 0.$$

It follows the $H(x, y) = C$ is constant on this trajectory (flow line). This is equivalent to saying that the flow line follows the level curve of H (specifically, level C).

7. Let $U = U(x, y)$ be a potential function that is twice continuously differentiable and let $\mathbf{F} = (U_x, U_y)$ be the associated gradient system. Then

$$\text{div}\mathbf{F} = \frac{\partial U_x}{\partial x} + \frac{\partial U_y}{\partial y} = U_{xx} + U_{yy}.$$

By Theorem 1.6, the flow of the system preserves areas if, and only if, $\text{div}\mathbf{F} = 0$. So, the flow of the system preserves areas if, and only if, $U_{xx} + U_{yy} = 0$ (i.e., if, and only if, $U(x, y)$ is a harmonic function).

9. (a) Let $A(t) = \begin{pmatrix} a_{11}(t) & a_{12}(t) \\ a_{21}(t) & a_{22}(t) \end{pmatrix}$, and let $A^{(1)}(t)$ and $A^{(2)}(t)$ be the matrices whose second and first rows are, respectively, the second and first rows of $A(t)$ and whose first and second rows are, respectively, the derivatives with respect to t of the first and second rows of $A(t)$. Then

$$\frac{d}{dt} \begin{vmatrix} a_{11}(t) & a_{12}(t) \\ a_{21}(t) & a_{22}(t) \end{vmatrix} = \frac{d}{dt} \Big(a_{11}(t)a_{22}(t) - a_{12}(t)a_{21}(t) \Big)$$

$$= a_{11}(t)a'_{22}(t) + a'_{11}(t)a_{22}(t) - a_{12}(t)a'_{21}(t) - a'_{12}(t)a_{21}(t)$$

$$= \Big(a'_{11}(t)a_{22}(t) - a'_{12}(t)a_{21}(t) \Big) + \Big(a_{11}(t)a'_{22}(t) - a_{12}(t)a'_{21}(t) \Big)$$

$$= \begin{vmatrix} a'_{11}(t) & a'_{12}(t) \\ a_{21}(t) & a_{22}(t) \end{vmatrix} + \begin{vmatrix} a_{11}(t) & a_{12}(t) \\ a'_{21}(t) & a'_{22}(t) \end{vmatrix} = \det A^{(1)}(t) + \det A^{(2)}(t).$$

(b) Part (a) establishes the case $n = 2$ as a basis for a proof by induction on n, the dimension of the matrix $A(t)$. Suppose that the claim has been proven for some $n = n_0 \geq 2$. That is, assume that if $A(t) = \big(a_{ij}(t) \big)$ is an $n_0 \times n_0$ matrix where each entry $a_{ij}(t)$ is a differentiable function of t, then

$$\frac{d}{dt} \det A(t) = \det A_1(t) + \cdots + \det A_{n_0}(t),$$

where $A_k(t)$ $(k = 1, \ldots, n_0)$ denotes the matrix $A(t)$ with its kth row replaced by the derivatives $a'_{k1}(t), \cdots, a'_{kn_0}(t)$. Let $B(t) = \big(a_{ij}(t) \big)$ be a square matrix of dimension $n_0 + 1$, where each entry is a differentiable function of t. Expanding by the first row, $\det B(t)$ becomes

$$\det B(t) = a_{11} \det B_{11} - a_{12} \det B_{12} + \cdots + (-1)^{n_0+1} a_{1(n_0+1)} \det B_{1(n_0+1)},$$

where B_{1j} is the matrix B with the first row and jth column deleted. Consider the effect of differentiating the above equation. the product rule must be applied to each term. When this is done to the jth term the result is

$$(-1)^{j+1} a_{1j} \frac{d}{dt} \det B_{1j} + (-1)^{j+1} a'_{1j} \det B_{ij}$$

$$= (-1)^{j+1} a_{1j} \sum_{k=1}^{n_0} \det B_{1j}^{(k)} + (-1)^{j+1} a'_{1j} \det B_{ij},$$

where the induction hypothesis is applied to the $n_0 \times n_0$ matrix B_{1j} to obtain the summation on the right, and $B_{1j}^{(k)}$ ($k = 1, \ldots, n_0$) is the matrix B_{1j} with its kth row differentiated with respect to t. Now sum the expression on the right from $j = 1$ to $j = n_0 + 1$ to get

$$\frac{d}{dt} \det B(t) = \sum_{j=1}^{n_0+1} \left((-1)^{j+1} a_{1j} \sum_{k=1}^{n_0} \det B_{1j}^{(k)} + (-1)^{j+1} a_{1j}' \det B_{ij} \right)$$

$$= \sum_{j=1}^{n_0+1} \left((-1)^{j+1} a_{1j} \sum_{k=1}^{n_0} \det B_{1j}^{(k)} \right) + \sum_{j=1}^{n_0+1} (-1)^{j+1} a_{1j}' \det B_{ij}$$

$$= \sum_{k=1}^{n_0} \left(\sum_{j=1}^{n_0+1} (-1)^{j+1} a_{1j} \det B_{1j}^{(k)} \right) + \sum_{j=1}^{n_0+1} (-1)^{j+1} a_{1j}' \det B_{ij},$$

where the order of summation has been interchanged. Note that the the last summation is just $\det B^{(1)}(t)$ expanded about the first row. Moreover, with respect to the double sum in the last line, the kth term of the outside summation is the determinant of the matrix $B^{(k+1)}(t)$ expanded about the first row. In other words, the above equation can be written more succinctly as

$$\frac{d}{dt} \det B(t) = \det B^{(1)}(t) + \cdots + \det B^{(n_0+1)}(t).$$

Hence, the claim holds for $(n_0 + 1) \times (n_0 + 1)$ matrices if it holds for all $n_0 \times n_0$ matrices. By induction, the claim holds for all $n \times n$ matrices with $n \geq 2$.

11. (a) If $\text{div}\mathbf{F}$ is constant then $\int_0^t \text{div}\mathbf{F}\, du = \text{div}\mathbf{F} \int_0^t du = t\, \text{div}\mathbf{F}$.

(b) Let $\mathbf{x}$ be fixed. The transformation $T_t(\mathbf{x})$ is defined to be the point $\mathbf{y}(t)$ on the flow line that is the trajectory of the solution $\mathbf{y}(t)$ of the system $\mathbf{F} = \dot{\mathbf{y}}$ with initial condition $\mathbf{y}(0) = \mathbf{x}$. Therefore, as t varies from $t = 0$, the flow line (trajectory) of $\mathbf{y}(t)$ is traced out starting at $\mathbf{x}$. Thus, for a given t, $\text{div}\mathbf{F}\big(T_t(\mathbf{x})\big)$ is the divergence of the field $\mathbf{F}$ along the flow line from the point $\mathbf{x}$ to the point $T_t(\mathbf{x})$. Therefore, the time-average of the divergence along this flow line is given by the integral

$$\frac{1}{t} \int_0^t \text{div}\mathbf{F}\big(T_u(\mathbf{x})\big)\, du.$$

The result follows by noting that multiplication by t gives the exponent $\int_0^t \text{div}\mathbf{F}\big(T_u(\mathbf{x})\big)\, du$.

Section 2: LINEAR SYSTEMS

Exercise Set 2AB (pgs. 592-594)

1. The presence of the x^2 term makes the given system nonlinear.

3. With $\mathbf{x} = \begin{pmatrix} x \\ y \end{pmatrix}$, $A(t) = \begin{pmatrix} t^2 & 1 \\ 0 & 0 \end{pmatrix}$ and $\mathbf{b}(t) = \begin{pmatrix} e^t \\ 1 \end{pmatrix}$, the given system can be written as

$$\frac{d\mathbf{x}}{dt} = A(t)\mathbf{x} + \mathbf{b}(t).$$

So the system is linear.

5. Write the given system in differential operator form as

$$(D - 6)x - 8y = 0,$$
$$4x + (D + 6)y = 0.$$

Multiply the first equation by -4, operate on the second equation with $D - 6$ and add the results to get $(D^2 - 36)y + 32y = 0$, or equivalently, $(D^2 - 4)y = 0$. So the characteristic roots are ± 2 and the general solution for $y = y(t)$ is $y(t) = c_1 e^{2t} + c_2 e^{-2t}$. The condition $y(0) = 0$ leads to the equation $0 = c_1 + c_2$, or $c_2 = -c_1$. The solution for y can then be written as

$$y(t) = c_1(e^{2t} - e^{-2t}) = 2c_1 \sinh 2t.$$

Solving the second equation for x, we can now compute

$$x = -\frac{1}{4}\frac{dy}{dt} - \frac{3}{2}y = -\frac{1}{4}(4c_1 \cosh 2t) - \frac{3}{2}(2c_1 \sinh 2t) = -c_1 \cosh 2t - 3c_1 \sinh 2t.$$

The condition $x(0) = 1$ directly gives $c_1 = -1$. The complete solution is therefore

$$x(t) = \cosh 2t + 3 \sinh 2t,$$
$$y(t) = -2 \sinh 2t.$$

7. Using the *Hint* in the text, let $\mathbf{x}_p(t) = \begin{pmatrix} x_p(t) \\ y_p(t) \end{pmatrix} = \begin{pmatrix} at + b \\ ct + d \end{pmatrix}$ be a trial particular solution. Substitute this and $\begin{pmatrix} dx_p/dt \\ dy_p/dt \end{pmatrix} = \begin{pmatrix} a \\ c \end{pmatrix}$ into the given system to get

$$\begin{pmatrix} a \\ c \end{pmatrix} = \begin{pmatrix} 6 & 8 \\ -4 & -6 \end{pmatrix}\begin{pmatrix} at + b \\ ct + d \end{pmatrix} + \begin{pmatrix} 1 \\ t \end{pmatrix} = \begin{pmatrix} 6(at + b) + 8(ct + d) + 1 \\ -4(at + b) - 6(ct + d) + t \end{pmatrix}$$
$$= \begin{pmatrix} (6a + 8c)t + (6b + 8d + 1) \\ (-4a - 6c + 1)t + (-4b - 6d) \end{pmatrix}.$$

Hence, $a = (6a + 8c)t + (6b + 8d + 1)$ and $c = (-4a - 6c + 1)t + (-4b - 6d)$. Equating coefficients of like powers of t leads to the four equations

$$6a + 8c = 0, \qquad 6b + 8d + 1 = a, \qquad -4a - 6c + 1 = 0, \qquad -4b - 6d = c.$$

This has the unique solution $a = -2$, $b = -3/2$, $c = 3/2$, $d = 3/4$, so that a particular solution of the given nonhomogeneous system is

$$\mathbf{x}_p(t) = \begin{pmatrix} x_p(t) \\ y_p(t) \end{pmatrix} = \begin{pmatrix} -2t - 3/2 \\ 3t/2 + 3/4 \end{pmatrix}.$$

From Exercise 5 above, the homogeneous solution is

$$\mathbf{x}_h(t) = \begin{pmatrix} \cosh 2t + 3\sinh 2t \\ -2\sinh 2t \end{pmatrix}.$$

The general solution of the given nonhomogeneous system is therefore

$$\begin{aligned} \mathbf{x}(t) &= \mathbf{x}_h(t) + \mathbf{x}_p(t) \\ &= \begin{pmatrix} \cosh 2t + 3\sinh 2t \\ -2\sinh 2t \end{pmatrix} + \begin{pmatrix} -2t - 3/2 \\ 3t/2 + 3/4 \end{pmatrix} = \begin{pmatrix} \cosh 2t + 3\sinh 2t - 2t - 3/2 \\ -2\sinh 2t + 3t/2 + 3/4 \end{pmatrix}. \end{aligned}$$

9. The third equation of the given system is independent of both x and y and can therefore be solved directly for $z = z(t)$ to get $z(t) = c_1 e^{-t}$. The condition $z(0) = 2$ gives $c_1 = 2$, so that

$$z(t) = 2e^{-t}.$$

Inserting this into the first equation of the given system results in

$$\frac{dx}{dt} = x + 2e^{-t}, \qquad \text{or in normalized form,} \qquad \frac{dx}{dt} - x = 2e^{-t}.$$

The integrating factor e^{-t} leads to the equivalent equation $d/dt(e^{-t}x) = 2e^{-2t}$, and integration gives $e^{-t}x = -e^{-2t} + c_2$, or equivalently, $x = -e^{-t} + c_2 e^t$. The condition $x(0) = 1$ gives $1 = -1 + c_2$. So $c_2 = 2$ and

$$x(t) = -e^{-t} + 2e^t.$$

Inserting this into the second equation of the given system results in

$$\frac{dy}{dt} = -e^{-t} + 2e^t + 2y, \qquad \text{or in normalized form,} \qquad \frac{dy}{dt} - 2y = -e^{-t} + 2e^t.$$

The integrating factor e^{-2t} leads to the equivalent equation $d/dt(e^{-2t}y) = -e^{-3t} + 2e^{-t}$, and integration gives $e^{-2t}y = \frac{1}{3}e^{-3t} - 2e^{-t} + c_3$, or equivalently, $y = \frac{1}{3}e^{-t} - 2e^t + c_3 e^{2t}$. The condition $y(0) = -1$ gives $-1 = \frac{1}{3} - 2 + c_3$. So, $c_3 = 2/3$ and

$$y(t) = \frac{1}{3}e^{-t} - 2e^t + \frac{2}{3}e^{2t}.$$

11. Subtract the second equation from the first equation to eliminate dx/dt. This gives

$$3\frac{dy}{dt} = ty - t^2, \qquad \text{or equivalently,} \qquad \frac{dy}{dt} = (t/3)y - t^2/3.$$

Using this result, the first equation of the given system can be solved for dx/dt to get

$$\frac{dx}{dt} = x + ty - \frac{dy}{dt} = x + ty - \left[(t/3)y - t^2/3\right] = x + (2t/3)y + t^2/3.$$

Setting $\mathbf{x} = \begin{pmatrix} x \\ y \end{pmatrix}$, $A(t) = \begin{pmatrix} 1 & 2t/3 \\ 0 & t/3 \end{pmatrix}$ and $\mathbf{b}(t) = \begin{pmatrix} t^2/3 \\ -t^2/3 \end{pmatrix}$, the above system can be written in the desired standard vector form $d\mathbf{x}/dt = A(t)\mathbf{x} + \mathbf{b}(t)$.

13. (a) Let $\mathbf{x}_1$ and $\mathbf{x}_2$ be solutions of the system $d\mathbf{x}/dt = A(t)\mathbf{x}$, where $A(t)$ is an $n \times n$ matrix. For constants c_1 and c_2, consider the vector function $c_1\mathbf{x}_1 + c_2\mathbf{x}_2$. We have

$$\frac{d(c_1\mathbf{x}_1 + c_2\mathbf{x}_2)}{dt} = c_1\frac{d\mathbf{x}_1}{dt} + c_2\frac{d\mathbf{x}_2}{dt}$$
$$= c_1 A(t)\mathbf{x}_1 + c_2 A(t)\mathbf{x}_2 = A(t)\big(c_1\mathbf{x}_1 + c_2\mathbf{x}_2\big),$$

where the first member in the second line follows from the fact that $d\mathbf{x}_1/dt = A(t)\mathbf{x}_1$ and $d\mathbf{x}_2/dt = A(t)\mathbf{x}_2$ (i.e., $\mathbf{x}_1$ and $\mathbf{x}_2$ are solutions of the given system). Equality of the first and last members implies that $c_1\mathbf{x}_1(t) + c_2\mathbf{x}_2(t)$ is a solution of the given system.

(b) Given that A is a matrix with n columns, $A\mathbf{x}$ is defined for vectors $\mathbf{x}$ only if $\mathbf{x}$ has dimension n. Moreover, since $\mathbf{x}$ and its derivative have the same dimension, $d\mathbf{x}/dt$ has dimension n. Thus, if $\mathbf{x}$ is a solution of $d\mathbf{x}/dt = A\mathbf{x}$ then $A\mathbf{x}$ has dimension n. That is, A has n rows and is therefore a square matrix.

(c) Let $\mathbf{x}_1$ and $\mathbf{x}_2$ be solutions of the system $d\mathbf{x}/dt = A(t)\mathbf{x} + \mathbf{b}(t)$, where $A(t)$ is an $n \times n$ matrix and $\mathbf{b}(t)$ is an n-dimensional vector-valued function of t. For constants c_1 and c_2, consider the vector function $c_1\mathbf{x}_1 + c_2\mathbf{x}_2$. We have

$$\frac{d(c_1\mathbf{x}_1 + c_2\mathbf{x}_2)}{dt} = c_1\frac{d\mathbf{x}_1}{dt} + c_2\frac{d\mathbf{x}_2}{dt}$$
$$= c_1 A(t)\mathbf{x}_I + \mathbf{b}(t) + c_2 A(t)\mathbf{x}_2 + \mathbf{b}(t) = A(t)\big(c_1\mathbf{x}_1 + c_2\mathbf{x}_2\big) + 2\mathbf{b}(t),$$

where the first member in the second line follows from the fact that $\mathbf{x}_1$ and $\mathbf{x}_2$ are solutions of the given system. Equality of the first and last members implies that $c_1\mathbf{x}_1 + c_2\mathbf{x}_2$ is a solution of the system $d\mathbf{x}/dt = A(t)\mathbf{x} + 2\mathbf{b}(t)$. If it were also a solution of the given system then it must be the case that $2\mathbf{b}(t) = \mathbf{b}(t)$ for all t, or equivalently, $\mathbf{b}(t) = \mathbf{0}$ for all t. So, the conclusion of part (a) holds for the system $d\mathbf{x}/dt = A(t)\mathbf{x} + \mathbf{b}(t)$ only if $\mathbf{b}(t) = \mathbf{0}$.

15. The system is nonlinear because of the term y^2 in the second equation of the system.

17. The system is nonlinear because of the term x^2 in the first equation of the system.

19. Write the given system as

$$dx/dt + y = -t, \qquad \text{or in operator form,} \qquad Dx + y = -t,$$
$$-x + dy/dt = t, \qquad\qquad\qquad\qquad\qquad -x + Dy = t.$$

Applying D to the second equation of the system on the right gives $-Dx + D^2y = 1$. Adding this to the first equation of the system eliminates x and results in $D^2y + y = 1 - t$, or

$$(D^2 + 1)y = 1 - t.$$

The homogeneous solution of this equation is $y_h(t) = c_1\cos t + c_2\sin t$. Using $y_p(t) = At + B$ as a trial particular solution of the nonhomogeneous equation, substitute into the D.E. to get

$$1 - t = (D^2 + 1)y_p = (D^2 + 1)(At + B) = D^2(At + B) + At + B = At + B.$$

Equating coefficients of like powers of t directly gives $A = -1$, $B = 1$, so that $y_p(t) = -t + 1$. The general solution of the above nonhomogeneous equation for $y = y(t)$ is therefore

$$y(t) = c_1\cos t + c_2\sin t + 1 - t.$$

Since the second equation of the system is equivalent to $x = dy/dt - t$, the general solution for $x = x(t)$ is

$$x(t) = \frac{dy}{dt} - t = -c_1\sin t + c_2\cos t - 1 - t.$$

466

21. Subtracting the first equation of the given system from the second gives $dy/dt = x - y$. Inserting this expression for dy/dt into the first equation gives $dx/dt + x - y = y$, or $dx/dt = -x + 2y$. The standard form for the given system is therefore

$$\frac{dx}{dt} = -x + 2y,$$
$$\frac{dy}{dt} = x - y.$$

23. Adding and subtracting the equations of the given system gives $2dx/dt = \sin t + \cos t$ and $2dy/dt = \sin t - \cos t$, respectively. The standard form for the given system is therefore

$$\frac{dx}{dt} = \frac{1}{2}(\sin t + \cos t),$$
$$\frac{dy}{dt} = \frac{1}{2}(\sin t - \cos t).$$

25. First, write the given system in operator form:

$$D^2 x - Dy = 0,$$
$$Dx + D^2 y = 0.$$

Apply D to the first equation and add the result to the second equation to get $D^3 x + Dx = 0$, or equivalently, $D(D^2 + 1)x = 0$. Since the characteristic roots are $r_1 = 0$, $r_{2,3} = \pm i$, the general solution for $x = x(t)$ is

$$x(t) = c_1 + c_2 \cos t + c_3 \sin t.$$

Since $D^2 x = -c_2 \cos t - c_3 \sin t$, the first equation of the system is equivalent to

$$Dy = -c_2 \cos t - c_3 \sin t.$$

The homogeneous solution is $y_h(t) = c_4$. Moreover, we can take $y_p(t) = A \cos t + B \sin t$ as a trial particular solution of the nonhomogeneous equation. With $Dy_p = -A \sin t + B \cos t$, substituting into the nonhomogeneous equation gives $-c_2 \cos t - c_3 \sin t = -A \sin t + B \cos t$, from which we immediately get $A = c_3$ and $B = -c_2$. So $y_p(t) = c_3 \cos t - c_2 \sin t$ and the general solution for $y = y(t)$ is

$$y(t) = c_4 + c_3 \cos t - c_2 \sin t.$$

With $\dot{x}(t) = -c_2 \sin t + c_3 \cos t$ and $\dot{y}(t) = -c_3 \sin t - c_2 \cos t$, the initial conditions lead to the system

$$1 = x(0) = c_1 + c_2,$$
$$0 = \dot{x}(0) = c_3,$$
$$0 = y(0) = c_4 + c_3,$$
$$0 = \dot{y}(0) = -c_2.$$

The unique solution is found to be $c_1 = 1$, $c_2 = c_3 = c_4 = 0$. Hence, the complete solution of the given system is

$$x(t) = 1, \qquad y(t) = 0.$$

27. First, write the given system in operator form:

$$D^2x - y = e^t,$$
$$x + D^2y = 0.$$

Apply D^2 to the first equation and add the result to the second equation to get $D^4x + x = e^t$, or equivalently,

$$(D^4 + 1)x = e^t.$$

The roots of the characteristic equation $r^4 + 1 = 0$ are the four fourth roots of -1; namely,

$$r_{1,2} = \frac{\sqrt{2}}{2} \pm i\frac{\sqrt{2}}{2}, \quad r_{3,4} = -\frac{\sqrt{2}}{2} \pm i\frac{\sqrt{2}}{2}.$$

Letting $\omega = \sqrt{2}/2$ (to reduce clutter), the homogeneous solution has the form

$$x_h(t) = e^{\omega t}(c_1 \cos \omega t + c_2 \sin \omega t) + e^{-\omega t}(c_3 \cos \omega t + c_4 \sin \omega t).$$

The method of undetermined coefficients leads to the particular solution $x_p(t) = \frac{1}{2}e^t$. The general solution for $x = x(t)$ is therefore

$$x(t) = \frac{1}{2}e^t + e^{\omega t}(c_1 \cos \omega t + c_2 \sin \omega t) + e^{-\omega t}(c_3 \cos \omega t + c_4 \sin \omega t).$$

Rewrite the first equation of the given system as $y = D^2x - e^t$. Computing $D^2x = \ddot{x}$, the general solution for $y = y(t)$ is found to be

$$y(t) = -\frac{1}{2}e^t + 2\omega^2 e^{\omega t}(c_2 \cos \omega t - c_1 \sin \omega t) + 2\omega^2 e^{-\omega t}(-c_4 \cos \omega t + c_3 \sin \omega t).$$

With

$$\dot{x}(t) = \frac{1}{2}e^t + \omega e^{\omega t}\Big((c_1 + c_2)\cos \omega t + (-c_1 + c_2)\sin \omega t\Big)$$
$$+ \omega e^{-\omega t}\Big((-c_3 + c_4)\cos \omega t + (-c_3 - c_4)\sin \omega t\Big)$$

and

$$\dot{y}(t) = -\frac{1}{2}e^t + 2\omega^3 e^{\omega t}\Big((-c_1 + c_2)\cos \omega t + (-c_1 - c_2)\sin \omega t\Big)$$
$$+ 2\omega^3 e^{-\omega t}\Big((c_3 + c_4)\cos \omega t + (-c_3 + c_4)\sin \omega t\Big),$$

the initial conditions gives a system of four equations that can be written as

$$c_1 + c_2 - c_3 + c_4 = -1/(2\omega),$$
$$-c_1 + c_2 + c_3 + c_4 = 1/(4\omega^3),$$
$$c_1 + c_3 = -1/2,$$
$$c_2 - c_4 = 1/(4\omega^2).$$

Without showing the details, the elimination method gives the unique solution

$$c_1 = -\frac{1}{4}(1 + \sqrt{2}), \quad c_2 = \frac{1}{4}, \quad c_3 = -\frac{1}{4}(1 - \sqrt{2}), \quad c_4 = -\frac{1}{4}.$$

Hence,

$$x(t) = \frac{1}{2}e^t - \frac{1}{4}e^{\frac{\sqrt{2}}{2}t}\left((1+\sqrt{2})\cos\frac{\sqrt{2}}{2}t - \sin\frac{\sqrt{2}}{2}t \right)$$

$$- \frac{1}{4}e^{-\frac{\sqrt{2}}{2}t}\left((1-\sqrt{2})\cos\frac{\sqrt{2}}{2}t + \sin\frac{\sqrt{2}}{2}t \right),$$

$$y(t) = -\frac{1}{2}e^t + \frac{1}{4}e^{\frac{\sqrt{2}}{2}t}\left(\cos\frac{\sqrt{2}}{2}t + (1+\sqrt{2})\sin\frac{\sqrt{2}}{2}t \right)$$

$$+ \frac{1}{4}e^{-\frac{\sqrt{2}}{2}t}\left(\cos\frac{\sqrt{2}}{2}t + (-1+\sqrt{2})\sin\frac{\sqrt{2}}{2}t \right).$$

29. If $\dot{x} = u$ and $\dot{y} = v$ then $\ddot{x} = \dot{u}$ and $\ddot{y} = \dot{v}$. Using these in the given system results in the two equations $\dot{u} - v = 0$ and $u + \dot{v} = 0$. The equivalent 4-dimensional first-order normal form for the given system is therefore

$$\dot{x} = u,$$
$$\dot{y} = v,$$
$$\dot{u} = v,$$
$$\dot{v} = -u.$$

31. If $\dot{x} = u$ and $\dot{y} = v$ then $\ddot{x} = \dot{u}$ and $\ddot{y} = \dot{v}$. Using these in the given system results in the two equations $\dot{u} - y = e^t$ and $\dot{v} + x = 0$. The equivalent 4-dimensional first-order normal form for the given system is therefore

$$\dot{x} = u,$$
$$\dot{y} = v,$$
$$\dot{u} = y + e^t,$$
$$\dot{v} = -x.$$

33. Subtracting the two equations gives $x(t) = 0$. The first equation (e.g.) then reduces to $dy/dt = 0$, so that $y(t) = c$. The system therefore has the one-parameter family of solutions

$$x(t) = 0,$$
$$y(t) = c.$$

35. Subtracting the two equations gives $x(t) = y(t)$, so that $dx/dt = dy/dt$. This allows us to write the second equation in terms of x and dx/dt only to get $dx/dt = \frac{1}{2}x$, and the first equation in terms of y and dy/dt only to get $dy/dt = \frac{1}{2}y$. These two equations have the solutions $x(t) = c_1 e^{t/2}$ and $y(t) = c_2 e^{t/2}$, respectively. However, since $x(t) = y(t)$, we must have $c_1 = c_2$. Hence, dispensing with the subscripts on the constants, the given system has the one-parameter family of solutions

$$x(t) = ce^{t/2}, \qquad y(t) = ce^{t/2}.$$

37. Differentiating the first two equations of the given system gives $\ddot{x} = \dot{z} + \dot{w}$ and $\ddot{y} = \dot{z} - \dot{w}$. The variables z and w can then be eliminated by using the third and fourth equations of the given system to get

$$\ddot{x} = \dot{z} + \dot{w} = (x - y) + (x + y) = 2x \qquad \text{and} \qquad \ddot{y} = \dot{z} - \dot{w} = (x - y) - (x + y) = -2y.$$

This produces the following uncoupled second-order 2-dimensional system satisfied by x and y:

$$\ddot{x} = 2x,$$
$$\ddot{y} = -2y.$$

The associated characteristic equations are $r^2 - 2 = 0$ and $r^2 + 2 = 0$ (resp.), with roots $r_{1,2} = \pm\sqrt{2}$ and $r_{1,2} = \pm i\sqrt{2}$, respectively. The solutions for $x = x(t)$ and $y = y(t)$ of this second-order system are then

$$x(t) = c_1 e^{\sqrt{2}t} + c_2 e^{-\sqrt{2}t},$$
$$y(t) = c_3 \cos \sqrt{2}t + c_4 \sin \sqrt{2}t.$$

Adding the first two equations of the original system eliminates w and allows us to solve for $z = z(t)$ to get

$$z = \frac{1}{2}(\dot{x} + \dot{y}) = \frac{1}{2}\left(c_1\sqrt{2}e^{\sqrt{2}t} - c_2\sqrt{2}e^{-\sqrt{2}t} - c_3\sqrt{2}\sin\sqrt{2}t + c_4\sqrt{2}\cos\sqrt{2}t\right)$$
$$= \frac{\sqrt{2}}{2}\left(c_1 e^{\sqrt{2}t} - c_2 e^{-\sqrt{2}t} - c_3 \sin\sqrt{2}t + c_4 \cos\sqrt{2}t\right).$$

Similarly, subtracting the second equation of the original system from the first eliminates z and allows us to solve for $w = w(t)$ to get

$$w(t) = \frac{1}{2}(\dot{x} - \dot{y}) = \frac{1}{2}\left(c_1\sqrt{2}e^{\sqrt{2}t} - c_2\sqrt{2}e^{-\sqrt{2}t} + c_3\sqrt{2}\sin\sqrt{2}t - c_4\sqrt{2}\cos\sqrt{2}t\right)$$
$$= \frac{\sqrt{2}}{2}\left(c_1 e^{\sqrt{2}t} - c_2 e^{-\sqrt{2}t} + c_3 \sin\sqrt{2}t - c_4 \cos\sqrt{2}t\right).$$

(b) Differentiating the second two equations of the given system gives $\ddot{z} = \dot{x} - \dot{y}$ and $\ddot{w} = \dot{x} + \dot{y}$. The variables x and y can then be eliminated by using the first and second equations of the given system to get

$$\ddot{z} = \dot{x} - \dot{y} = (z + w) - (z - w) = 2w \qquad \text{and} \qquad \ddot{w} = \dot{x} + \dot{y} = (z + w) + (z - w) = 2z.$$

This produces the second-order 2-dimensional system satisfied by z and w:

$$\ddot{z} = 2w,$$
$$\ddot{w} = 2z.$$

Section 3: APPLICATIONS

Exercise Set 3 (pgs. 601-606)

1. (a) The volume of solution in each tank is fixed at 100 gallons, therefore $y/100$ and $z/100$ are the respective concentrations of salt in the tanks. For tank Y, four gallons of solution are entering each minute at a concentration of $z/100$ pounds per gallon, and four gallons are leaving each minute at a concentration of $y/100$ pounds per minute. That is, each minute, $4z/100$ pounds of salt are entering and $4y/100$ pounds are leaving. For the Z tank, one gallon of solution is entering each minute at a concentration of $y/100$ pounds per gallon, and four gallons are leaving each minute at a concentration of $z/100$ pounds per gallon. That is, each minute, $y/100$ pounds of salt are entering and $4z/100$ pounds are leaving. The first-order system governing the flow of salt in the two tanks is therefore

$$\frac{dy}{dt} = \frac{4}{100}z - \frac{4}{100}y, \qquad\qquad \left(D + \frac{4}{100}\right)y - \frac{4}{100}z = 0,$$

$$\frac{dz}{dt} = \frac{1}{100}y - \frac{4}{100}z, \quad \text{or in operator form,} \quad -\frac{1}{100}y + \left(D + \frac{4}{100}\right)z = 0.$$

(b) Operating on the first equation with $D + 4/100$, multiplying the second equation by $4/100$, and adding the resulting equations eliminates z to get

$$\left(D + \frac{4}{100}\right)^2 y - \frac{4}{(100)^2}y = \left(D^2 + \frac{8}{100}D + \frac{12}{(100)^2}\right)y = \left(D + \frac{2}{100}\right)\left(D + \frac{6}{100}\right)y = 0.$$

The characteristic roots are therefore $r_1 = -2/100 = -1/50$ and $r_2 = -6/100 = -3/50$, and the general solution for $y = y(t)$ is

$$y(t) = c_1 e^{-t/50} + c_2 e^{-3t/50}.$$

Since the first equation of the system on the above left is equivalent $z = 25\dot{y} + y$, the general solution for $z = z(t)$ is

$$z(t) = 25\dot{y} + y = 25\left(-\frac{1}{50}c_1 e^{-t/50} - \frac{3}{50}c_2 e^{-3t/50}\right) + c_1 e^{-t/50} + c_2 e^{-3t/50}$$

$$= \frac{1}{2}(c_1 e^{-t/50} - c_2 e^{-3t/50}).$$

The initial conditions $y(0) = 10$, $z(0) = 20$ lead to the equations $10 = y(0) = c_1 + c_2$ and $20 = \frac{1}{2}c_1 - \frac{1}{2}c_2$, which has the unique solution $c_1 = 25$, $c_2 = -15$. So, the particular solution satisfying the given initial conditions is

$$y(t) = 25e^{-t/50} - 15e^{-3t/50},$$
$$z(t) = 12.5e^{-t/50} + 7.5e^{-3t/50}.$$

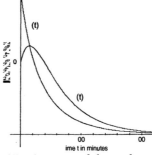

(c) The graph on the right shows that the amount of salt in tank Y is at a maximum of about 12 pounds at about 15 minutes and from then on tends to zero as t increases without bound; while the amount of salt in tank Z is decreasing for $t > 0$ and tends to zero as t increases without bound. The limiting behavior for each tank is confirmed by the solution equations themselves, since they show that the amount of salt in each tank tends to zero as $t \to \infty$.

471

3. (a) With

$$\dot{x}(t) = -c_1 \sin t + c_2 \cos t + c_3 \sqrt{3} \sin \sqrt{3}t - c_4 \sqrt{3} \cos \sqrt{3}t,$$
$$\dot{y}(t) = -c_1 \sin t + c_2 \cos t - c_3 \sqrt{3} \sin \sqrt{3}t + c_4 \sqrt{3} \cos \sqrt{3}t,$$

the initial conditions lead to the four equations

$$0 = x(0) = c_1 - c_3, \quad 1 = \dot{x}(0) = c_2 - c_4 \sqrt{3},$$
$$1 = y(0) = c_1 + c_3, \quad 0 = \dot{y}(0) = c_2 + c_4 \sqrt{3}.$$

The two equations in the first column has the solution $c_1 = c_3 = 1/2$, and the two equations in the right column has the solution $c_2 = 1/2$, $c_3 = -1/(2\sqrt{3})$.

(b) With $\dot{x}(t)$ and $\dot{y}(t)$ as given in part (a), the initial conditions lead to the four equations

$$x_0 = c_1 - c_3, \quad u_0 = c_2 - c_4 \sqrt{3},$$
$$y_0 = c_1 + c_3, \quad v_0 = c_2 + c_4 \sqrt{3}.$$

Writing these in matrix form yields

$$\begin{pmatrix} 1 & -1 \\ 1 & 1 \end{pmatrix} \begin{pmatrix} c_1 \\ c_3 \end{pmatrix} = \begin{pmatrix} x_0 \\ y_0 \end{pmatrix} \quad \text{and} \quad \begin{pmatrix} 1 & -\sqrt{3} \\ 1 & \sqrt{3} \end{pmatrix} \begin{pmatrix} c_2 \\ c_4 \end{pmatrix} = \begin{pmatrix} u_0 \\ v_0 \end{pmatrix}.$$

The two coefficient matrices have nonzero determinants, so that the inverses of both of these matrices exist. The above matrix equations can therefore be solved for $\begin{pmatrix} c_1 \\ c_3 \end{pmatrix}$ and $\begin{pmatrix} c_2 \\ c_4 \end{pmatrix}$ by multiplying on the left by the appropriate inverse. So, c_1, c_2, c_3 and c_4 can always be chosen to satisfy any set of initial conditions.

5. (a) Newton's equations of planetary motion are given in three dimensions by Equations 4.3 in the text, where the origin of the coordinate system is maintained at the center of mass of one of the bodies. However, since the motion takes place in a plane, we can assume that all motion acts in the xy-plane, so that $z = z(t)$ is identically zero and the third equation involving $\ddot{z}$ vanishes. Furthermore, if $y = y(t)$ is assumed to be identically zero then the equation involving $\ddot{y}$ also vanishes, all motion takes place on a line connecting the two bodies and we are left with the single equation

$$\ddot{x} = \frac{-kx}{(x^2)^{3/2}} = -\frac{k}{x^2},$$

where $k = GM$, G is the gravitational constant and M is the sum of the masses of the two bodies.

(b) For purposes of visualisation, let the two bodies be a star and a planet, with the origin of the coordinate system at the center of mass of the star. The differential equation obtained in part (a) shows that acceleration is always negative, implying that the velocity of the planet is always decreasing. Since the initial velocity is $\dot{x}(0) = 0$, we must have $dx/dt < 0$ for all $t > 0$. Therefore, velocity is increasing in absolute value; i.e., the planet's speed is increasing with time. This implies that the planet is being pulled toward the star and will therefore eventually collide with it.

7. (a) Since 2 gallons of solution are flowing into tank X each minute but only 1 gallon is flowing out each minute, it follows that the number of gallons of solution in tank X is increasing by 1 gallon per minute. Since it initially contains 50 gallons of solution and its capacity is 100 gallons, it will begin to overflow at time $t_1 = 50$ minutes. Also, salt enters the system (via tank Y) at a rate of 1 pound per minute. Since the system initially contains no salt, the total amount of salt in both tanks at time t, for $0 \le t \le 50$ is given by

$$x(t) + y(t) = t, \quad 0 \le t \le 50,$$

where $x(t)$ and $y(t)$ are the number of pounds of salt in tanks X and Y, respectively, at time t.

(b) For $0 \le t \le 50$, part (a) shows that tank X contains $50 + t$ gallons of solution, so that the concentration of salt in tank X is $x/(50+t)$. Hence, salt enters tank X at a rate of $2y/100$ pounds per minute and leaves at a rate of $x/(50+t)$ pounds per minute. This gives the equation shown below for dx/dt. The equation for dy/dt is found by differentiating $x(t) + y(t) = t$ (from part (a)) to get $dy/dt = 1 - dx/dt$ and then using the equation for dx/dt already derived. The appropriate system of differential equations for $0 \le t \le 50$ is therefore

$$\frac{dx}{dt} = \frac{2y}{100} - \frac{x}{50+t}, \quad x(0) = 0,$$

$$\frac{dy}{dt} = 1 + \frac{x}{50+t} - \frac{2y}{100}, \quad y(0) = 0, \qquad 0 \le t \le 50.$$

(c) For $t \ge 50$, the number of gallons in each tank is constant at 100 gallons. The concentrations are therefore $x/100$ pounds per gallon and $y/100$ pounds per gallon for tanks X and Y, respectively. Thus, salt enters tank X at a rate of $2y/100$ pounds per minute and leaves at a rate of $2x/100$ pounds per minute; while salt enters tank Y at a rate of $1 + x/100$ pounds per minute and leaves at a rate of $2y/100$ pounds per minute. The appropriate system of differential equations for $t \ge 50$ is therefore

$$\frac{dx}{dt} = \frac{2y}{100} - \frac{2x}{100},$$

$$\frac{dy}{dt} = 1 + \frac{x}{100} - \frac{2y}{100}, \qquad t \ge 50.$$

9. Let $x(t)$ be the number of pounds of salt in the 100 gallons tank at time t (hereafter referred to as "tank X"), and let $y(t)$ be the number of pounds of salt in the 200 gallon tank at time t (hereafter referred to as "tank Y").

(a) Tank X has 2 gallons of solution flowing into it each minute, one gallon per minute from tank Y and 1 gallon per minute of pure water from a source outside the system. Thus, the total amount of solution in tank X at time t is constant at 50 gallons (since it is initially half full). Tank Y gains 2 gallons of solution each minute from tank X but loses only 1 gallon per minute. Therefore tank Y gains 1 gallon of solution each minute and the total amount of solution in tank Y at time t is $100 + t$ (since it initially contains 100 gallon). Since tank Y has a capacity of 200 gallons, it follows that it will begin to overflow at $t = 100$ minutes and therefore the process will stop at time $t = 100$ minutes..

(b) The analysis done in part (a) shows that the concentration of salt in tank X is $x/50$ pounds per gallon and the concentration in tank Y is $y/(100+t)$ pounds per gallon. For tank X, $y/(100+t)$ pounds of salt are gained each minute and $2x/50$ pounds are lost each minute. For tank Y, $2x/50$ pounds are gained each minute and $y/(100+t)$ pounds are lost each minute. The system of differential equations and initial conditions that describes the process as a function of time is therefore

$$\frac{dx}{dt} = \frac{y}{100+t} - \frac{2x}{50}, \quad x(0) = 0,$$

$$\frac{dy}{dt} = \frac{2x}{50} - \frac{y}{100+t}, \quad y(0) = 10, \qquad 0 \le t \le 100.$$

473

As a check, note that $dx/dt+dy/dt$ is identically zero. Integration then gives $x(t)+y(t) = C$, a constant. Evaluating this at $t = 0$ and using the initial conditions gives $C = 10$, from which $x(t) + y(t) = 10$.

(c) The check alluded to in part (b) shows that $y = x - 10$. Using this in the first equation derived in part (b) gives (after a little algebra and judicious rearranging)

$$\frac{dx}{dt} + \left(\frac{1}{25} + \frac{1}{100+t}\right)x = \frac{10}{100+t}, \quad x(0) = 0, \quad 0 \le t \le 100.$$

(d) The equation found in part (c) can be solved by using the integrating factor

$$e^{\int \left(\frac{1}{25} + \frac{1}{100+t}\right)dt} = e^{t/25 + \ln(100+t)} = e^{t/25}(100 + t).$$

Hence, the equation is equivalent to $d/dt\left(e^{t/25}(100+t)x\right) = 10e^{t/25}$. Integrating and solving for $x = x(t)$ gives the general solution $x(t) = 250(100 + t)^{-1} + C(100 + t)^{-1}e^{-t/25}$. The initial condition $x(0) = 0$ leads to $C = -250$ and the particular solution is therefore

$$x(t) = \frac{250}{100+t}(1 - e^{-t/25}), \quad 0 \le t \le 100.$$

Since $y = 10 - x$,

$$y(t) = 10 - \frac{250}{100+t}(1 - e^{-t/25}), \quad 0 \le t \le 100.$$

Evaluating both of these functions at $t = 100$ gives the amount of salt in each tank when the process stops. Specifically,

$$x(100) \approx 1.227 \text{ pounds} \quad \text{and} \quad y(100) \approx 8.773 \text{ pounds}.$$

11. Set $m_1 = m_2 = 1$, $k_1 = k_2 = 1$, $k_3 = 2$ in the system derived in Example 2 in the text to get

$$\ddot{x} = -2x + y, \quad \text{or in operator form,} \quad (D^2 + 2)x - y = 0,$$
$$\ddot{y} = x - 3y, \quad \quad\quad\quad\quad\quad\quad\quad\quad -x + (D^2 + 3)y = 0.$$

Applying D^2+3 to the first equation and adding the resulting equations gives a single fourth-order equation in x; namely, $(D^4 + 5D^2 + 5)x = 0$. Hence, the characteristic equation is $r^4 + 5r^2 + 5 = 0$. Viewing this as a quadratic equation in r^2, the quadratic formula gives $r^2 = \frac{1}{2}(-5 \pm \sqrt{5})$, both of which are negative. Thus, the roots are pure imaginary and are given by

$$r_{1,2} = \pm i\sqrt{\frac{5 - \sqrt{5}}{2}} \quad \text{and} \quad r_{3,4} = \pm i\sqrt{\frac{5 + \sqrt{5}}{2}}.$$

So the circular frequencies associated with the system are

$$\mu_1 = \sqrt{\frac{5 - \sqrt{5}}{2}} \approx 1.1756 \quad \text{and} \quad \mu_2 = \sqrt{\frac{5 + \sqrt{5}}{2}} \approx 1.9021.$$

13. Set $m_1 = 1$, $m_2 = 2$, $k_1 = 2$, $k_2 = k_3 = 3$ in the system derived in Example 2 in the text to get

$$\ddot{x} = -5x + 3y, \quad \text{or in operator form,} \quad (D^2 + 5)x - 3y = 0,$$
$$\ddot{y} = \frac{3}{2}x - 3y, \quad \quad\quad\quad\quad\quad\quad\quad\quad -\frac{3}{2}x + (D^2 + 3)y = 0.$$

474

Applying $D^2 + 3$ to the first equation, multiplying the second equation by 3, and adding the resulting equations gives a single fourth-order equation in x; namely, $(D^4 + 8D^2 + 31/2)x = 0$. Hence, the characteristic equation is $r^4 + 8r^2 + 31/2 = 0$. Viewing this a quadratic equation in r^2, the quadratic formula gives $r^2 = \frac{1}{2}(-8 \pm \sqrt{2})$, both of which are negative. Thus, the roots are pure imaginary and are given by

$$r_{1,2} = \pm i\sqrt{\frac{8 - \sqrt{2}}{2}} \quad \text{and} \quad r_{3,4} = \pm i\sqrt{\frac{8 + \sqrt{2}}{2}}.$$

So the circular frequencies associated with the system are

$$\mu_1 = \sqrt{\frac{8 - \sqrt{2}}{2}} \approx 1.8146 \quad \text{and} \quad \mu_2 = \sqrt{\frac{8 + \sqrt{2}}{2}} \approx 2.1696.$$

15. (a) For the given two-body system, we seek a potential function $U(x, y)$ such that

$$\frac{\partial U(x, y)}{\partial x} = (k_1 + k_2)x - k_2 y \quad \text{and} \quad \frac{\partial U(x, y)}{\partial y} = -k_2 x + (k_2 + k_3)y.$$

Integrating the first equation with respect to x gives

$$U(x, y) = \frac{1}{2}(k_1 + k_2)x^2 - k_2 xy + C(y),$$

where $C(y)$ is a differentiable function of y that is independent of x. Differentiating this equation with respect to y and setting the result equal to the expression for $\partial U(x, y)/\partial y$ given above results in

$$-k_2 x + C'(y) = \frac{\partial U(x, y)}{\partial y} = -k_2 x + (k_2 + k_3)y.$$

Hence, $C'(y) = (k_2 + k_3)y$, so that $C(y) = \frac{1}{2}(k_2 + k_3)y^2 + K$, where K is an arbitrary real constant. For simplicity, we choose $K = 0$ to get the potential

$$U(x, y) = \frac{1}{2}(k_1 + k_2)x^2 - k_2 xy + \frac{1}{2}(k_2 + k_3)y^2.$$

(b) Let $U(x, y)$ be any potential function of a two-dimensional conservative system of the type considered here. Then the system can be written as

$$m_1 \ddot{x} = -\frac{\partial U(x, y)}{\partial x},$$
$$m_2 \ddot{y} = -\frac{\partial U(x, y)}{\partial y}.$$

Multiply the first equation by $\dot{x}$, the second by $\dot{y}$, and add the resulting equations to get

$$m_1 \dot{x}\ddot{x} + m_2 \dot{y}\ddot{y} = -\frac{\partial U(x, y)}{\partial x}\dot{x} - \frac{\partial U(x, y)}{\partial y}\dot{y} = -\frac{d}{dt}\big(U(x, y)\big),$$

475

where the last equality is a result of the chain rule. We can now integrate both sides with respect to t to get

$$\frac{1}{2}m_1\dot{x}^2 + \frac{1}{2}m_2\dot{y}^2 = -U(x,y) + C,$$

where C is a constant. The left side is equal to $T(x,y)$, the kinetic energy, so that the above can be written as

$$T(x,y) + U(x,y) = C.$$

That is, the total energy $T + U$ is constant.

17. A homogeneous spherical body of mass M has its center of mass located at its geometric center, so that the surface of the body is R units from its center of mass, where R is the radius of the body. If an object of mass m is on the surface of this body then the inverse square law says that magnitude of the mutually attractive force between them is $F = GMm/R^2$, where G is the gravitational constant in appropriate units of measurement. If g is the acceleration of gravity at the surface of the body then, by Newton's second force law, the magnitude of the force of gravity is also $F = mg$. Hence,

$$\frac{GMm}{R^2} = mg, \quad \text{or solving for } g, \quad g = \frac{GM}{R^2}.$$

19. In general, the escape speed is given by

$$v_e = \sqrt{\frac{2G(m_1 + m_2)}{r}},$$

where $G = 6.673 \times 10^{-11}$, $r = 6.500 \times 10^6$ meters, $m_1 = 5.976 \times 10^{24}$ kg. If $m_2 = 100$ kg or $m_2 = 1000$ kg then $m_1 + m_2 \approx m_1$. In these cases,

$$v_e = \sqrt{\frac{2(6.673 \times 10^{-11})(5.976 \times 10^{24})}{6.500 \times 10^6}} \approx 1.1077 \times 10^4 \text{ m/s}.$$

If $m_2 = 10^{22}$ kg then $v_e \approx 1.1086 \times 10^4$ m/s.

21. Integrating $\ddot{x} = -g$ gives $\dot{x}(0) = -gt + C_1$. The condition $\dot{x}(0) = v_0$ then gives $C_1 = v_0$, so that $\dot{x}(t) = -gt + v_0$. A second integration gives $x(t) = -\frac{1}{2}gt^2 + v_0 t + C_2$, and the condition $x(0) = 0$ gives $C_2 = x_0$. So,

$$x(t) = -\frac{1}{2}gt^2 + v_0 t + x_0.$$

Since $g > 0$, this is an equation of a downward opening parabola in a tx-coordinate system. This means that $x(t)$ has a global finite maximum at the vertex. This is true regardless of the sizes of x_0 and v_0. Specifically, this maximum value occurs at time $t = v_0/g$ and is equal to $x(v_0/g) = v_0^2/g + x_0$.

23. (a) If v and x_0 are positive constants then the orbit $\mathbf{x}(t) = x_0\big(\cos(vt/x_0), \sin(vt/x_0)\big)$ satisfies

$$\dot{\mathbf{x}}(t) = v\big(-\sin(vt/x_0), \cos(vt/x_0)\big),$$

$$\ddot{\mathbf{x}}(t) = -\frac{v^2}{x_0}\big(\cos(vt/x_0), \sin(vt/x_0)\big) = -\frac{v^2}{x_0^2}\mathbf{x}(t).$$

Note that the acceleration vector points toward the origin. Also, $\big(\cos(vt/x_0), \sin(vt/x_0)\big)$ and $\big(-\sin(vt/x_0), \cos(vt/x_0)\big)$ are unit vectors, so that

$$|\mathbf{x}(t)| = x_0, \qquad |\dot{\mathbf{x}}(t)| = v, \qquad |\ddot{\mathbf{x}}(t)| = \frac{v^2}{x_0}.$$

So the orbit is circular of radius x_0, the satellite has constant orbital speed v, and the magnitude of the centripetal acceleration is v^2/x_0.

(b) Since the magnitude of the gravitational acceleration must equal the magnitude of the centripetal acceleration, we must have

$$\frac{G(m_1 + m_2)}{x_0^2} = \frac{v^2}{x_0},$$

where the centripetal acceleration on the right is from part (a). Solving this for $v = v_1$, the required uniform orbital speed is

$$v_1 = \sqrt{\frac{G(m_1 + m_2)}{x_0}}.$$

(c) Since escape speed is given by $v_e = \sqrt{2G(m_1 + m_2)/x_0}$, the uniform orbital speed found in part (b) is related to escape speed by

$$v_e = \sqrt{\frac{2G(m_1 + m_2)}{x_0}} = \sqrt{2}\sqrt{\frac{G(m_1 + m_2)}{x_0}} = \sqrt{2}\,v_1.$$

(d) A lunar month P is the time it takes for the moon to complete one orbit around the earth. Assuming that this orbit is circular of radius x_0 and uniform orbital speed v_1, we have $P = 2\pi x_0/v_1$ (circumference of the orbit divided by the speed of traversal). With $x_0 = 3.844 \times 10^8$ meters, $G = 6.673 \times 10^{-11}$ (gravitational constant), $m_1 = 5.976 \times 10^{24}$ kg (mass of the earth), and $m_2 = 7.176 \times 10^{22}$ kg (mass of the moon), we can use the expression for v_1 found in part (b) to get

$$P = \frac{2\pi x_0}{v_1} = 2\pi\sqrt{\frac{x_0^3}{G(m_1 + m_2)}} = 2.357 \times 10^6 \text{ seconds.}$$

Since there are 86,400 seconds in one day,

$$P = \frac{2.357 \times 10^6}{8.640 \times 10^4} \approx 27.28 \text{ days.}$$

25. (a) If $x(t) = a\cos\omega t$, $y(t) = a\sin\omega t$ is a solution of the given system, where a and ω are positive constants, then the orbit is circular of radius $a = \sqrt{x^2 + y^2}$ and the system can be written as

$$\ddot{x} = -\frac{k}{a^3}\, x, \qquad \ddot{y} = -\frac{k}{a^3}\, y.$$

Substituting $x = x(t)$, $y = y(t)$, $\ddot{x} = \ddot{x}(t) = -a\omega^2 \cos\omega t$ and $\ddot{y} = \ddot{y}(t) = -a\omega^2 \sin\omega t$ into the system gives

$$-a\omega^2 \cos\omega t = -\frac{k}{a^3}\, a\cos\omega t \qquad \text{and} \qquad -a\omega^2 \sin\omega t = -\frac{k}{a^3}\, a\sin\omega t.$$

Canceling $-a\cos\omega t$ from both sides of the first equation and $-a\sin\omega t$ from both sides of the second equation results is two identical equations; namely, $\omega^2 = k/a^3$. This is the desired relationship.

(b) Since we are concerned here with a circular orbit, $a\omega$ is the orbital speed. The square of the time it takes for one complete revolution is then

$$T^2 = \left(\frac{2\pi a}{a\omega}\right)^2 = \frac{4\pi^2}{\omega^2} = \frac{4\pi^2 a^3}{k},$$

where $\omega^2 = k/a^3$ is used form part (b) to obtain the last equality. Since $k = G(m_1 + m_2)$, we have

$$T^2 = \frac{4\pi^3 a^3}{G(m_1 + m_2)},$$

which is Kepler's third law for the case of circular orbits.

(c) In general, if $r = f(\theta)$ is in polar coordinates, where $\theta = \theta(t)$ is a function of t, then the area bounded by the graph of f over the time period $[0, t]$, where $\theta(0) = 0$, is

$$A(t) = \int_0^t \frac{1}{2} r^2 \frac{d\theta}{du} \, du.$$

Differentiating with respect to t gives $dA/dt = \frac{1}{2}r^2 d\theta/dt$. In particular, if $\theta(t) = \omega t$, and $r(t) = a$, where a and ω are constants (i.e., if the orbit is circular and given by $x(t) = a\cos\omega t$, $y(t) = a\sin\omega t$), then

$$\frac{dA}{dt} = \frac{1}{2} a^2 \omega.$$

So $A(t) = \alpha t$, where $\alpha = a^2\omega/2$. Therefore, over any fixed time interval of length Δt, the area swept out over this time interval is

$$A(t + \Delta t) - A(t) = \alpha(t + \Delta t) - \alpha t = \alpha \Delta t,$$

a constant. This is Kepler's second law for the case of circular orbits.

27. (a) Each t determines a plane P_t with normal vector $\mathbf{x}(t) \times \dot{\mathbf{x}}(t)$ that contains the two bodies. Since the mass m is a nonzero constant $\mathbf{x}(t) \times m\dot{\mathbf{x}}(t)$ is also normal to P_t. Using Equation 1.6 in Chapter 4, the time derivative of this normal vector is

$$\frac{d}{dt}\left(\mathbf{x}(t) \times m\dot{\mathbf{x}}(t)\right) = \dot{\mathbf{x}}(t) \times m\dot{\mathbf{x}}(t) + \mathbf{x}(t) \times m\ddot{\mathbf{x}}(t) = \mathbf{x}(t) \times m\ddot{\mathbf{x}}(t),$$

where we have used the fact that the cross product of two parallel vectors is zero. Since the path of motion obeys the inverse-square law $m\ddot{\mathbf{x}} = -(k/|\mathbf{x}|^3)\mathbf{x}$, the above can be written as

$$\frac{d}{dt}\left(\mathbf{x}(t) \times m\dot{\mathbf{x}}(t)\right) = \mathbf{x}(t) \times \left(-\frac{k}{|\mathbf{x}(t)|^3}\right)\mathbf{x}(t) = 0,$$

where we have again used the fact that the cross product of two parallel vectors is zero. This equation shows that the normal is independent of t, so that all P_t are parallel to each other. Since each P_t contains the two bodies, it follows that all P_t are the same. In other words, the entire orbit lies in a fixed plane containing both bodies.

(b) If motion is governed by an equation of the form $\ddot{\mathbf{x}} = G\mathbf{x}$, where G is a real-valued function, set $m = 1$ in the argument given in part (a) and replace the scalar $-k/|\mathbf{x}|^3$ with the more general scalar G. The identical argument still holds and, at each step, for the same reasons.

29. (a) The vector equation $\ddot{\mathbf{x}} = G\mathbf{x}$ is equivalent to the system of two scalar equations

$$\ddot{x} = G(x,y)x,$$
$$\ddot{y} = G(x,y)y,$$

where $\mathbf{x} = (x,y)$ and $\ddot{\mathbf{x}} = (\ddot{x}, \ddot{y})$. Hence, for motion governed by a central force law,

$$x\ddot{y} - y\ddot{x} = x\big(G(x,y)y\big) - y\big(G(x,y)x\big) = 0.$$

(b) By the product rule for differentiation, we have

$$\frac{d}{dt}(x\dot{y} - y\dot{x}) = x\ddot{y} + \dot{x}\dot{y} - (y\ddot{x} + \dot{y}\dot{x}) = x\ddot{y} - y\ddot{x} = 0,$$

where the last equality follows from part (a). Hence, $x\dot{y} - y\dot{x} = h$ is constant.

(c) Let $x = r\cos\theta$, $y = r\sin\theta$, where r and θ are functions of t. Then,

$$\begin{aligned}
x\dot{y} - y\dot{x} &= (r\cos\theta)(r\dot{\theta}\cos\theta + \dot{r}\sin\theta) - (r\sin\theta)(-r\dot{\theta}\sin\theta + \dot{r}\cos\theta)\\
&= r^2\dot{\theta}\cos^2\theta + r\dot{r}\cos\theta\sin\theta + r^2\dot{\theta}\sin^2\theta - r\dot{r}\sin\theta\cos\theta\\
&= r^2\dot{\theta}(\cos^2\theta + \sin^2\theta) = r^2\dot{\theta}.
\end{aligned}$$

But by part (a), $x\dot{y} - y\dot{x} = h$. Hence, $r^2\dot{\theta} = h$ for some constant h.

(d) The formula for area in polar coordinates allows us to write the area swept out by the line joining the two bodied over the time interval $[t_0, t]$ as

$$A(t) = \int_{\theta(t_0)}^{\theta(t)} t\frac{1}{2}r^2\theta\, d\theta = \int_{t_0}^{t} r^2\dot{\theta}\, du$$

or, using the result of part (c),

$$A(t) = \int_{t_0}^{t} \frac{1}{2}h\, du = \left[\frac{1}{2}hu\right]_{t_0}^{t} = \frac{1}{2}ht - \frac{1}{2} = \frac{1}{2}ht + c,$$

where $c = -ht_0/2$.

To explain why this proves Kepler's second law, fix a positive real number τ. The function $A_\tau(t)$, defined by $A_\tau(t) = A(t+\tau) - A(t)$ for all t, is the area swept out over the time interval $[t, t+\tau]$. Thus,

$$A_\tau(t) = \left(\frac{1}{2}h(t+\tau) + c\right) - \left(\frac{1}{2}ht + c\right) = \frac{1}{2}h\tau.$$

That is, over any time interval of length τ, the area swept out is the constant $h\tau/2$, and Kepler's second law is therefore proved for time intervals of length τ. But since τ is arbitrary, the conclusion holds in general.

(e) Designate a time $t = 0$ and, for each $t \geq 0$, let D_t be the region swept out over the time interval $[0, t]$. The boundary of D_t consists of three pieces: the straight line segments γ_1 and γ_2 joining the origin to the points $\mathbf{x}_0 = \big(x(0), y(0)\big)$ and $\mathbf{x}(t) = \big(x(t), y(t)\big)$ (resp.) and the part of the orbit curve staring at $\mathbf{x}_0$ and ending at $\mathbf{x}(t)$, call it $\gamma(t)$.

Now consider the vector field $\mathbf{F}(x, y) = (-y, x)$. From previous work, we've seen that the flow lines of $\mathbf{F}$ are circles centered at the origin. Thus, a line integral of $\mathbf{F}$ over any line segment emanating from the origin is zero. It follows that if γ denotes the boundary of D_t then the line integral around γ reduces to the line integral over $\gamma(t)$, where the interval of integration can be taken to be the interval $[0, t]$. Thus,

$$\oint_\gamma \mathbf{F} \cdot d\mathbf{x} = \oint_\gamma (-y, x) \cdot (dx, dy) = \oint_\gamma -y\, dx + x\, dy = \int_{\gamma(t)} \left(-y\frac{dx}{du} + x\frac{dy}{du} \right) du = \int_0^t (x\dot{y} - y\dot{x})\, du.$$

By part (b), $x\dot{y} - y\dot{x} = h$, a constant, so that the value of the last integral above is evidently ht. Hence,

$$(*) \qquad\qquad\qquad \oint_\gamma -y\, dx + x\, dy = ht.$$

On the other hand, with $F(x, y) = -y$ and $G(x, y) = x$,

$$\int_{D_t} \left(\frac{\partial G}{\partial x} - \frac{\partial F}{\partial y} \right) dx\, dy = \int_{D_t} 2\, dx\, dy = 2A(D_t).$$

By Green's Theorem, the left side of this equation is equal to the left side of $(*)$. Hence,

$$2A(D_t) = ht_0, \quad \text{or equivalently,} \quad A(D_t) = \frac{1}{2}ht.$$

By the explanation requested in part (c), this is sufficient to conclude the truth of Kepler's second law.

31. (a) Using Equation 1.6 in Chapter 4, we see that

$$\frac{d}{dt}(\dot{\mathbf{x}} \cdot \dot{\mathbf{x}}) = \dot{\mathbf{x}} \cdot \ddot{\mathbf{x}} + \ddot{\mathbf{x}} \cdot \dot{\mathbf{x}} = 2(\dot{\mathbf{x}} \cdot \ddot{\mathbf{x}}).$$

Hence, if $v = \sqrt{\dot{\mathbf{x}} \cdot \dot{\mathbf{x}}}$ then

$$\dot{\mathbf{x}} \cdot \ddot{\mathbf{x}} = \frac{1}{2}(2(\dot{\mathbf{x}} \cdot \ddot{\mathbf{x}})) = \frac{1}{2}\frac{d}{dt}(\dot{\mathbf{x}} \cdot \dot{\mathbf{x}}) = \frac{1}{2}\frac{d}{dt}(v^2) = \frac{d}{dt}\left(\frac{v^2}{2} \right).$$

(b) Since $|\mathbf{x}| = (x^2 + y^2 + z^2)^{1/2}$, we have

$$\nabla \frac{1}{|\mathbf{x}|} = \nabla (x^2 + y^2 + z^2)^{-1/2} = \left(\frac{x}{(x^2 + y^2 + z^2)^{3/2}}, \frac{y}{(x^2 + y^2 + z^2)^{3/2}}, \frac{z}{(x^2 + y^2 + z^2)^{3/2}} \right)$$

$$= -\frac{1}{(x^2 + y^2 + z^2)^{3/2}}(x, y, z) = -\frac{1}{|\mathbf{x}|^3}\mathbf{x}.$$

Therefore,

$$-k\frac{\mathbf{x} \cdot \dot{\mathbf{x}}}{|\mathbf{x}|^3} = k\left(-\frac{\mathbf{x}}{|\mathbf{x}|^3} \right) \cdot \dot{\mathbf{x}} = k\left(\nabla \frac{1}{|\mathbf{x}|} \right) \cdot \dot{\mathbf{x}}.$$

(c) If $f(x, y, z)$ is a real-valued function defined on $\mathcal{R}^3$ then, by the chain rule,

$$\frac{df}{dt} = \frac{\partial f}{\partial x}\dot{x} + \frac{\partial f}{\partial y}\dot{y} + \frac{\partial f}{\partial z}\dot{z} = \nabla f \cdot \dot{\mathbf{x}}.$$

In particular, if $f(x, y, z) = 1/|\mathbf{x}|$ then

$$\frac{d}{dt}\left(\frac{1}{|\mathbf{x}|}\right) = \left(\nabla \frac{1}{|\mathbf{x}|}\right) \cdot \dot{\mathbf{x}}.$$

Using the result of part (b) then gives

$$-k\frac{\mathbf{x} \cdot \dot{\mathbf{x}}}{|\mathbf{x}|^3} = k\frac{d}{dt}\left(\frac{1}{|\mathbf{x}|}\right) = k\frac{d}{dt}\left(\frac{1}{r}\right),$$

where $r = |\mathbf{x}|$. Therefore, the result of part (a) allows us to conclude that the given scalar equation is equivalent to

$$\frac{d}{dt}\left(\frac{v^2}{2}\right) = k\frac{d}{dt}\left(\frac{1}{r}\right).$$

(d) Integrating both sides of the equation derived in part (c) from 0 to t gives

$$\int_0^t \frac{d}{du}\left(\frac{v^2}{2}\right)du = k\int_0^t \frac{d}{dt}\left(\frac{1}{r}\right)dt,$$

or by the fundamental theorem of calculus,

$$\frac{v^2}{2} - \frac{v_0^2}{2} = \frac{k}{r} - \frac{k}{r_0},$$

where $v_0 = v(0)$ and $r_0 = r(0)$.

33. By the chain rule,

$$\begin{aligned}
\frac{d}{dt}\left[H(x, y, t)\right] &= H_x(x, y, t)\frac{dx}{dt} + H_y(x, y, t)\frac{dy}{dt} + H_t(x, y, t)\frac{dt}{dt} \\
&= H_x(x, y, t)\frac{dx}{dt} + H_y(x, y, t)\frac{dy}{dt} + H_t(x, y, t).
\end{aligned}$$

Since $(x(t), y(t))$ satisfies the given Hamiltonian system, we can replace dy/dt and dx/dt in the above equation with $-H_x(x, y, t)$ and $H_y(x, y, t)$, respectively. This causes the first two terms on the right to cancel each other, leaving only $H_t(x, y, t)$. That is,

$$\frac{d}{dt}\left[H(x, y, t)\right] = H_t(x, y, t).$$

Setting $x = x(t)$ and $x = y(t)$ then gives

$$\frac{d}{dt}\left[H(x(t), y(t), t)\right] = H_t(x(t), y(t), t).$$

481

Section 4: NUMERICAL METHODS

Exercise Set 4AB (pgs. 612-615)

1. (b) Given the functions $x(t) = c_1 e^t + c_2 e^{-t}$ and $y(t) = c_1 e^t - c_2 e^{-t}$, we have

$$\dot{x}(t) = c_1 e^t - c_2 e^{-t} = y(t) \qquad \text{and} \qquad \dot{y}(t) = c_1 e^t + c_2 e^{-t} = x(t).$$

That is, $(x, y) = \big(x(t), y(t)\big)$ satisfies $\dot{x} = y$ and $\dot{y} = x$.

(b) The given conditions lead to the two equations $1 = c_1 + c_2$ and $2 = c_1 - c_2$, with the unique solution $c_1 = 3/2$, $c_2 = -1/2$. Hence, the desired particular solution is

$$x(t) = \frac{3}{2}e^t - \frac{1}{2}e^{-t}, \qquad y(t) = \frac{3}{2}e^t + \frac{1}{2}e^{-t}.$$

(c) In the table below, the columns headed "IE x" and "IE y" are values obtained by using the improved Euler method with step size $h = 0.1$. The columns headed "actual x" and "actual y" were computed from the solution formulas given in part (b).

t	IE x	actual x	IE y	actual y
0.0	1.00000	1.00000	2.00000	2.00000
0.1	1.20500	1.20533	2.11000	2.11017
0.2	1.42202	1.42273	2.24105	2.24146
0.3	1.65324	1.65437	2.39446	2.39519
0.4	1.90095	1.90257	2.57175	2.57289
0.5	2.16763	2.16981	2.77471	2.77634

3. (a) Let $\dot{y} = x$, so that $\ddot{y} = \dot{x}$. The given second-order equation becomes $\dot{x} - 2x + y = t$, or equivalently, $\dot{x} = 2x - y + t$. Also, the condition $\dot{y}(0) = 2$ becomes $x(0) = 2$. The equivalent first-order system is therefore

$$\dot{x} = 2x - y + t, \quad x(0) = 2,$$
$$\dot{y} = x, \qquad\qquad y(0) = 1.$$

(b) The improved Euler method with step size $h = .01$ was used to approximate a numerical solution of the system found in part (a) for $0 \le t \le 1$. Of the 100 values computed, every 20th one is shown in the table on the left below. The table on the right was computed directly from the solution formula derived in part (c).

t	IE x	IE y	t	actual y	actual $\dot{y}$
0.0	2.00000	1.00000	0.0	1.00000	2.00000
0.2	2.70993	1.46714	0.2	1.46715	2.70996
0.4	3.68521	2.10158	0.4	2.10163	3.68528
0.6	5.00851	2.96431	0.6	2.96442	5.00866
0.8	6.78615	4.13513	0.8	4.13532	6.78640
1.0	9.15444	5.71797	1.0	5.71828	9.15484

(c) The characteristic equation of the second-order differential equation given in part (a) is $r^2 - 2r + 1 = (r-1)^2 = 0$, with the double root $r_1 = r_2 = 1$. The homogeneous solution is therefore $y_h(t) = c_1 e^t + c_2 t e^t$. With $y_p(t) = A + Bt$ as the trial particular solution of the nonhomogeneous equation, we have $\dot{y}_p = B$, and $\ddot{y}_p = 0$. Substituting into the nonhomogeneous equation gives

$$t = \ddot{y}_p - 2\dot{y}_p + y_p = 0 - 2B + A + Bt = -2B + A + At.$$

Equating coefficients of like powers of t gives $-2B + A = 0$ and $A = 1$, form which $A = 1$, $B = 1/2$ and $y_p(t) = \frac{1}{2} + t$. The general solution of the nonhomogeneous equation is therefore

$$y(t) = c_1 e^t + c_2 t e^t + \frac{1}{2} + t.$$

With $\dot{y}(t) = (c_1 + c_2) e^t + c_2 t e^t + 1$, the initial conditions give the equations $1 = c_1 + \frac{1}{2}$ and $2 = c_1 + c_2 + 1$, with solution $c_1 = c_2 = 1/2$. The solution of the given initial-value problem is then

$$y(t) = \frac{1}{2} e^t + \frac{1}{2} t e^t + \frac{1}{2} + t.$$

This and $\dot{y}(t) = 1 + e^t + \frac{1}{2} t e^t$ were used to construct the table shown on the right at the bottom of the previous page.

5. The improved Euler method was used with step size $h = 0.001$ to compute a numerical approximation of the solution to the first-order, one-dimensional, initial-value problem $dx/dt = 1 - x^{1/3}$, $x(0) = \frac{1}{2}$, for $0 \le t \le 5$. Of the 5,000 values generated, the left-hand table below records every 500th value.

7. The improved Euler method was used with step size $h = 0.001$ to compute a numerical approximation of the solution to the first-order, one-dimensional, initial-value problem $dx/dt = \sqrt{1 + x^4}$, $x(0) = 0$, for $0 \le t \le 1.5$. Of the 1500 values generated, the right-hand table below records every 150th value.

t	$x(t)$		t	$x(t)$
0.0	0.500000		0.00	0.000000
0.5	0.591083		0.15	0.150008
1.0	0.662880		0.30	0.300243
1.5	0.720503		0.45	0.451852
2.0	0.767312		0.60	0.607861
2.5	0.805669		0.75	0.774373
3.0	0.837303		0.90	0.962467
3.5	0.863521		1.05	1.19204
4.0	0.885335		1.20	1.50098
4.5	0.903540		1.35	1.97104
5.0	0.918770		1.50	2.81982

Table for Exercise 5   Table for Exercise 7

9. The diagram on the right shows the paths of the four bugs. The improved Euler method was used with step size $h = 0.001$ to plot the solution of the initial-value problem

$$\dot{x} = -\frac{v(x+y)}{\sqrt{2(x^2+y^2)}}, \quad x(0) = 1,$$

$$\dot{y} = \frac{v(x-y)}{\sqrt{2(x^2+y^2)}}, \quad y(0) = 1.$$

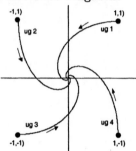

With the program on "PLOT X,Y" mode instead of "PRINT X,Y", the above initial conditions produced the path for bug 1. Since the paths of bugs 2, 3 and 4 are congruent to the path of bug 1 (only rotated clockwise about the origin by $\pi/2$ radians, π radians and $3\pi/2$ radians, respectively), the other three paths were generated using the same program but with the initial conditions $x(0) = -1$, $y(0) = 1$ (for bug 2), $x(0) = -1$, $y(0) = -1$ (for bug 3), and $x(0) = 1$, $y(0) = -1$ (for bug 4). In all four cases, the value used for their constant speed was $v = 1$. It is interesting to note that the paths of the bugs is independent of the value of v. However, the larger the value of v, the faster the paths are traced. The mathematical reason for this is apparent when one views y as a function of x and forms the quotient of the two equations of the system:

$$\frac{dy}{dx} = \frac{dy/dt}{dx/dt} = \frac{\dot{x}}{\dot{y}} = -\frac{x-y}{x+y}.$$

Observe that v does not appear in this first-order differential equation. Depending on which bug is being discussed, the solution graph (over the proper range for x) is a restricted portion of the path of that bug.

11. Let $f(x,y) = x^2 + \frac{1}{2}y^2$.

(a) The level curves of f are the trajectories of the system

$$\dot{x} = -f_y(x,y) = -y,$$
$$\dot{y} = f_x(x,y) = 2x,$$

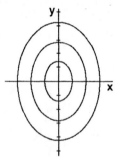

and are shown in the top sketch on the right. The improved Euler method was used with step size $h = 0.001$.

(b) Curves perpendicular to the level sets of f are the trajectories of the system

$$\dot{x} = f_x(x,y) = 2x,$$
$$\dot{y} = f_y(x,y) = y,$$

and are shown in the bottom sketch on the right.

(c) The system shown in part (b) is uncoupled. By inspection, the general solutions are seen to be $x(t) = c_1 e^{2t}$ and $y(t) = c_2 e^t$. If initial conditions are given by $x(t_0) = x_0$, $y(t_0) = y_0$ then we are led to the specific solution

$$x(t) = x_0 e^{2(t-t_0)} \qquad \text{and} \qquad y(t) = y_0 e^{t-t_0}.$$

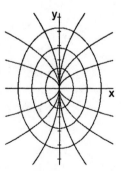

If $x_0 = y_0 = 0$ then the trajectory is the single point $(0,0)$,

484

which corresponds to the identically zero solution. If $x_0 = 0$ and $y_0 \neq 0$ then the trajectory consists of all positive multiples of y_0, which is the positive y axis if $y_0 > 0$ and the negative y-axis if $y_0 < 0$. In a similar fashion, If $x_0 \neq 0$ and $y_0 = 0$ then the solution trajectory is the positive x-axis if $x_0 > 0$ and the negative x-axis if $x_0 < 0$.

If $x_0 \neq 0$ and $y_0 \neq 0$ then note first that the corresponding trajectory stays in the same quadrant for all t. Second, the solution equations can be written as $(1/x_0)x = e^{2(t-t_0)}$ and $(1/y_0)y = e^{t-t_0}$, so that $(1/y_0)^2 y^2 = e^{2(t-t_0)} = (1/x_0)x$, or equivalently,

$$x = Cy^2,$$

where $C = x_0/y_0^2$. If $C > 0$, (i.e., if $x_0 > 0$)), this is a rightward opening parabola with vertex at the origin (but not including the origin). The associated trajectory is the top half of the parabola if $y_0 > 0$ and the bottom half if $y_0 < 0$. Similarly, if $C < 0$ (i.e., if $x_0 < 0$), this is a leftward opening parabola with vertex at the origin (but, again, not including the origin). The associated trajectory is the top half of the parabola if $y_0 > 0$ and the bottom half if $y_0 < 0$.

13. Let $x = x(t)$ be the number of pounds of salt in tank 1 at time $t \geq 0$, and let $y = y(t)$ be the number of pounds of salt in tank 2 at time $t \geq 0$.

(a) Tank 1 gains 8 gallons of solution per minute and loses 7 gallons of solution per minute. Since it initially has 50 gallons of solution, the volume of solution in tank 1 at time t is $50 + t$ gallons. Therefore, the concentration of salt in tank 1 is $x/(50 + t)$ pounds per gallon. Tank 2 gains 5 gallons per minute and loses 4 gallons per minute. Since it initially has 100 gallons of solution, the volume of solution in tank 2 at time t is $100 + t$ gallons. Therefore, the concentration of salt in tank 2 is $y/(100 + t)$ pounds per gallon. Moreover, since tank 1 has a capacity of 100 gallons, it will fill up in 50 minutes; and since tank 2 has a capacity of 200 gallons, it will fill up in 100 minutes. Therefore, the process stops when tank 1 first becomes full at time $t = 50$ minutes.

Tank 1 gains 0 pounds of salt per minute from the external source and $3y/(100 + t)$ pounds per minute from tank 2, while it loses a total of $7x/(50 + t)$ pounds of salt per minute. Tank 2 gains 2 pounds of salt per minute from an external source and $3x/(50 + t)$ pounds of salt per minute from tank 1, while it loses a total of $4y/(100 + t)$ pounds of salt per minute. Thus, if the initial conditions are given by $x(0) = x_0$ and $y(0) = y_0$ then the system of differential equations and initial conditions that describes the process is

$$\frac{dx}{dt} = \frac{3y}{100 + t} - \frac{7x}{50 + t}, \qquad x(0) = x_0,$$

$$\frac{dy}{dt} = 2 + \frac{3x}{50 + t} - \frac{4y}{100 + t}, \qquad y(0) = y_0, \quad 0 \leq t \leq 50.$$

(b) The improved Euler method with step size $h = 0.01$ was used to plot the solution graphs of the system derived in part (a) with $x_0 = 10$, $y_0 = 20$ and $0 \leq t \leq 50$. Once the graphs were plotted, the output line was changed from the "PLOT" command to the "PRINT" command and the step size was increased to $h = 0.1$. A list of numerical values was then generated and the results were examined in order to estimate the minimum amount of salt in tank 1 and the corresponding time at which this value is attained. It was found that

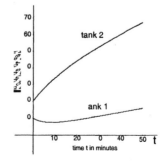

$$x(8.50) \approx x_{\min} \approx 7.45 \text{ pounds},$$

These values are consistent with the graphs shown on the right.

485

15.

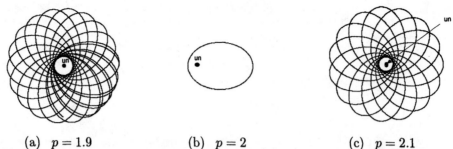

(a) $p = 1.9$ (b) $p = 2$ (c) $p = 2.1$

The improved Euler method with step size $h = 0.001$ was used to plot the orbits shown above. This solver attempted to plot the orbits with $h = 0.01$ but the resulting trajectories were wrong. The reason is that there was too much accumulated error inherent in the Euler method to produce accurate results.

The orbit shown on the left with $p = 1.9$ is for $0 \leq t \leq 60$ and is a result of the planet almost returning to its initial position after each revolution around the sun, but reaching apogee (its farthest point from the sun) about 20° counterclockwise from its previous apogee. As can be seen, this resulted in about 18 almost elliptical orbits (the petals on the overall picture). The planet did not return to its original position and initial velocity at the completion of these 18 "almost elliptical orbits". It is not clear whether the planet eventually returns to its initial condition and closes its orbit. However, it *is* clear that some value of p allows a perfect 18 petal closed orbit.

The orbit shown on the right with $p = 2.1$ is also for $0 \leq t \leq 60$. Here, with each revolution about the sun, the planet returns to apogee at about 24° *clockwise* from its previous apogee. However, unlike the case $p = 1.9$, the planet appears to return to its initial conditions after 15 revolutions about the sun and therefore its orbit appears to be closed. It is unknown if this is actually the case.

The orbit in the the center with $p = 2$ is, of course, representative of the inverse-square law and is perfectly elliptical.

17. (a) With $y = \dot{x}$ and $\dot{y} = \ddot{x}$, the van der Pol equation becomes $\dot{y} - \alpha(1 - x^2)y + x = 0$. The desired first-order nonlinear system is therefore

$$\dot{x} = y,$$
$$\dot{y} = \alpha(1 - x^2)y - x.$$

(b) The improved Euler method with step size $h = 0.001$ was used to generate each of the following three limit cycles of the system shown in part (a) (these are actually phase curves of the associated van der Pol equation). Note that as $\alpha \to 0$ the given system tends to $\dot{x} = y$, $\dot{y} = -x$, whose solution trajectories are circles of radius $\sqrt{x^2(0) + y^2(0)}$ centered at the origin. Thus, it is reasonable that small values of α should give limit cycles that are nearly circular, while larger values of α give limit cycles that are "deformed circles"

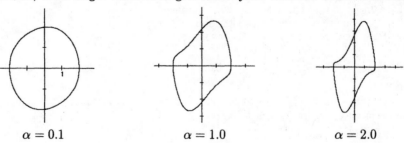

$\alpha = 0.1$ $\alpha = 1.0$ $\alpha = 2.0$

(c) The following six phase curves were generated by the improved Euler method with step size $h = 0.001$. Notice that the phase curve tends to the limit cycle in all six cases: from the inside for the initial point $(1, 1)$, and from the outside for the initial point $(3, 0)$.

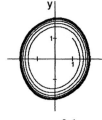
$\alpha = 0.1$

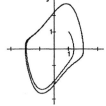
$\alpha = 1.0$

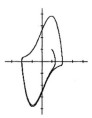
$\alpha = 2.0$

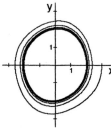
$\alpha = 0.1$

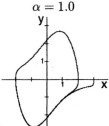
$\alpha = 1.0$

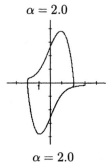
$\alpha = 2.0$

19. The improved Euler method with step size $h = 0.005$ was used to plot five phase curves of the soft spring oscillator equation $\ddot{y} = -\gamma y^3 + \delta y$ for various pairs (γ, δ). There are three equilibrium solutions; namely, $y = 0$, $y = \sqrt{\gamma/\delta}$ and $y = -\sqrt{\gamma/\delta}$, whose phase curves are the three points $(0, 0)$, $(\sqrt{\gamma/\delta}, 0)$ and $(-\sqrt{\gamma/\delta}, 0)$. The results are shown below.

(a) $\gamma = \delta = 1$

(b) $\gamma = 1$, $\delta = 2$

(c) $\gamma = 2$, $\delta = 1$

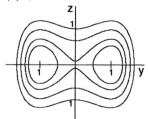

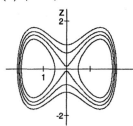

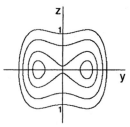

In Exercises 21, 23, and 25, the improved Euler method with step size $h = 0.001$ was used to plot the solution $y = y(t)$ of the oscillator $\ddot{y} + k(t)\dot{y} + h(t)y = \sin t$, $y(0) = 0$, $\dot{y}(0) = 1$, where the damping factor $k(t)$ and the spring stiffness $h(t)$ are time dependent. The results are shown below.

21.

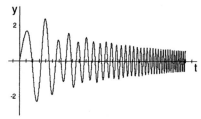

23.

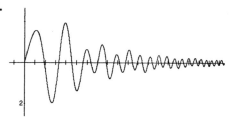

$\ddot{y} + e^{t/2}y = \sin t$, $y(0) = 0$, $\dot{y}(0) = 1$

$\ddot{y} + (1/10)\dot{y} + e^{t/2}y = \sin t$, $y(0) = 0$, $\dot{y}(0) = 0$

487

25.

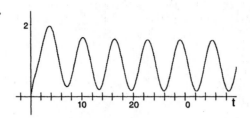

$$\ddot{y} + \dot{y} + \left(1/(1+t^2)\right)y = \sin t, \; y(0) = 0, \; \dot{y}(0) = 1.$$

27. Consider the Lotka-Volterra system $\dot{H} = (3 - 2P)H$, $\dot{P} = (\frac{1}{2}H - 1)P$.

(a) Using the improved Euler method with step size $h = 0.001$, various values of t_1 for the interval $0 \le t \le t_1$ were used to sketch a phase curve of the solution corresponding to the initial conditions $H(0) = 3$, $P(0) = \frac{3}{2}$. It was found that the corresponding phase curve almost closed with $t_1 = 3.6$ and was closed when $t_1 = 3.7$. The "PLOT H,P" command was then changed to "PRINT T,H,P" on the interval $0 \le t \le 3.7$. The numerical evidence suggested that the orbit time was approximately 3.666 time units (truncated to three places).

(b) The result of part (a) shows that the graphs of $H(t)$ and $P(t)$ are periodic of period ≈ 3.666 time units. Using the same program as was used in part (a), the "PRINT T,H,P" command was suppressed and the commands "PLOT T,H" and "PLOT T,P" were simultaneously activated to plot one period of the graphs of H and P on the same set of axes. The result is shown below in the figure on the left.

(c) Here, the "PLOT H,P" command was used to plot the trajectory in the HP-plane of the solution found in part (a). The result is shown below in the figure on the right. The arrows indicate the direction of the trajectory.

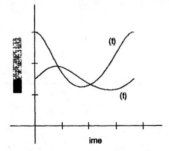
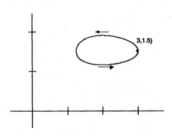

29. Set $a = 3$, $b = d = 2$, $c = 1$, $L = 4$ and $M = 3$ in the given refinement of the Lotka-Volterra equations. The conditions $H(0) = 3$, $P(0) = \frac{1}{2}$ then determine the unique solution of the initial-value problem

$$\dot{H} = (3 - 2P)H(4 - H), \quad H(0) = 3, \; P(0) = 1/2;$$
$$\dot{P} = (H - 2)P(3 - P).$$

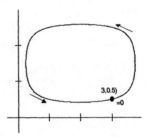

The plot of the trajectory of the solution is shown on the right. The arrows indicate the direction of the trajectory.

488

31. Set $a = 3$, $b = d = 2$, $c = 1$, $L = 4$ and $M = 3$ in the given refinement of the Lotka-Volterra equations. The conditions $H(0) = 3$, $P(0) = 2$ then determine the unique solution of the initial-value problem

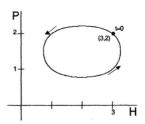

$$\dot{H} = (3 - 2P)H(4 - H), \quad H(0) = 3, \ P(0) = 2;$$
$$\dot{P} = (H - 2)P(3 - P).$$

The plot of the trajectory of the solution is shown on the right. The arrows indicate the direction of the trajectory.

33. First observe that $H(0) = 3$ for each of the initial-value problems suggested by Exercises 29, 30, 31, 32 in this section. Using the improved Euler method with step size $h = 0.001$, various values of t_1 for the interval $0 \le t \le t_1$ were used to sketch a phase curve of the four solutions corresponding to the initial conditions $H(0) = 3$, $P(0) = P_0$, where $P_0 = \frac{1}{2}, \frac{3}{2}, 2, \frac{5}{2}$. Visually comparing the curves for various values of t_1 allowed us to hone in on an estimate t_0 of the actual orbit time t_* such that $t_0 - 0.01 < t_* < t_0$. Once, t_0 was determined for a given P_0, the "PLOT H,P" command was suppressed and the "PRINT T,H,P" command was used on the time interval $0 \le t \le t_0$. Toward the end of this printout the columns for $H(t)$ and $P(t)$ contained values close to $H(t) = 3$ and $P(t) = P_0$. The corresponding values of t were therefore close to t_*. Using this method, the values of t_* for the four orbits were estimated to be

$$t_* \approx \begin{cases} 1.911, & P_0 = 1/2; \\ 1.562, & P_0 = 3/2; \\ 1.630, & P_0 = 2; \\ 1.911, & P_0 = 5/2. \end{cases}$$

489

Chapter Review

(pgs. 615-616)

1. The system is uncoupled. For the x-equation, separate variables to get $(x^2+1)^{-1}dx = dt$. Integration gives $\arctan x = t + C$, where "$\arctan x$" is taken to be the principal branch, so that $|t + \pi/4| < \pi/2$. Thus, applying the initial condition $x(0) = 1$ to get $\arctan 1 = C$ forces us to conclude that $C = \pi/4$. So $\arctan x = t + \pi/4$, or equivalently,

$$x(t) = \tan(t + \pi/4), \quad -3\pi/4 < t < \pi/4.$$

The general solution of the y-equation is $y(t) = c_2 e^t$. The condition $y(0) = 2$ gives $c_2 = 2$, so that $y(t) = 2e^t$. Since the domain of $y(t)$ is all real t, while the domain of $x(t)$ is $-3\pi/4 < t < pi/4$, the desired solution $\mathbf{x}(t) = \big(x(t), y(t)\big)$ is

$$\mathbf{x}(t) = \big(x(t), y(t)\big) = \big(\tan(t + \pi/4), 2e^t\big), \quad -3\pi/4 < t < \pi/4.$$

3. Differentiate the second equation of the system to get $d^2y/dt^2 = dx/dt$, and substitute into the first equation to get $d^2y/dt^2 = y + 1$, or equivalently, $d^2y/dt^2 - y = 1$. Since the characteristic equation $r^2 - 1 = 0$ has roots $r_{1,2} = \pm 1$, the homogeneous solution is $y_h(t) = c_1 e^t + c_2 e^{-t}$. By inspection, a particular solution is the constant $y_p(t) = -1$. The general solution for $y = y(t)$ is therefore

$$y(t) = c_1 e^t + c_2 e^{-t} - 1.$$

The derivative of this function can now be used in the second equation to get

$$x(t) = c_1 e^t - c_2 e^{-t}.$$

The initial conditions lead to the equations $1 = x(0) = c_1 - c_2$ and $2 = y(0) = c_1 + c_2 - 1$, from which $c_1 = 2$, $c_2 = 1$. So the desired solution $\mathbf{x}(t) = \big(x(t), y(t)\big)$ is

$$\mathbf{x}(t) = (2e^t - e^{-t}, 2e^t + e^{-t} - 1), \quad -\infty < t < \infty.$$

5. Write the given system in operator form:

$$(D - 3)x + 4y = 0,$$
$$-4x + (D + 7)y = 0.$$

Multiplying the first equation by 4, operating on the second equation with $D - 3$, and adding the results gives $(D - 3)(D + 7)y + 16y = 0$, or equivalently $(D - 1)(D + 5)y = 0$. Since the characteristic roots are $r_1 = 1$ and $r_2 = -5$, the general solution for $y = y(t)$ is

$$y(t) = c_1 e^t + c_2 e^{-5t}.$$

Insert this and its derivative $\dot{y}(t) = c_1 e^t - 5c_2 e^{-5t}$ into the second equation of the original system and solve for $x = x(t)$ to get the general formula for $x = x(t)$:

$$x(t) = \frac{1}{4}(c_1 e^t - 5c_2 e^{-5t} + 7c_1 e^t + 7c_2 e^{-5t}) = 2c_1 e^t + \frac{1}{2}c_2 e^{-5t}.$$

The initial conditions lead to the equations $1 = x(0) = 2c_1 + \frac{1}{2}c_2$ and $2 = y(0) = c_1 + c_2$, from which $c_1 = 0$, $c_2 = 2$. The desired solution $\mathbf{x}(t) = \big(x(t), y(t)\big)$ of the given system is therefore

$$\mathbf{x}(t) = \big(e^{-5t}, 2e^{-5t}\big), \quad -\infty < t < \infty.$$

7. Differentiate the second equation of the given system to get $\ddot{y} = 4\dot{x}$, and then use the first equation to eliminate x:

$$\ddot{y} = 4(y + t) = 4y + 4t, \quad \text{or equivalently,} \quad \ddot{y} - 4y = 4t.$$

The characteristic equation is $r^2 - 4 = 0$, with roots $r_{1,2} = \pm 2$. The homogeneous solution is therefore $y_h(t) = c_1 e^{2t} + c_2 e^{-2t}$. By inspection, a particular solution is $y_p(t) = -t$. So the general solution for $y = y(t)$ is

$$y(t) = c_1 e^{2t} + c_2 e^{-2t} - t.$$

Insert $\dot{y} = 2c_1 e^{2t} - 2c_2 e^{-2t} - 1$ into the second equation of the system and solve for x to get the general solution for $x = x(t)$:

$$x(t) = \frac{1}{2}c_1 e^{2t} - \frac{1}{2}c_2 e^{-2t}.$$

The initial conditions lead to the equations $1 = x(0) = \frac{1}{2}c_1 - \frac{1}{2}c_2$ and $2 = y(0) = c_1 + c_2$, from which $c_1 = 2$, $c_2 = 0$. The desired solution $\mathbf{x}(t) = \big(x(t), y(t)\big)$ is therefore

$$\mathbf{x}(t) = \big(e^{2t}, 2e^{2t} - t\big), \quad -\infty < t < \infty.$$

9. Write the given system in operator form:

$$(D^2 + 3)x - 2y = 0,$$
$$-2x + (D^2 + 2)y = 0.$$

Multiply the second equation by 2, apply $D^2 + 2$ to the first equation and add the results to get $(D^2 + 2)(D^2 + 3)x - 4x = 0$, or equivalently, $(D^4 + 5D^2 + 2)x = 0$. Since the characteristic equation $r^4 + 5r^2 + 2 = 0$ is quadratic in r^2, the quadratic formula gives $r^2 = \frac{1}{2}(-5 \pm \sqrt{17})$. Since both squares are negative, the four roots of the characteristic equation are pure imaginary:

$$r_{1,2} = \pm i\sqrt{\frac{5 - \sqrt{17}}{2}}, \qquad r_{3,4} = \pm i\sqrt{\frac{5 + \sqrt{17}}{2}}.$$

For ease of computation, let $\omega_1 = \sqrt{(5 - \sqrt{17})/2}$ and $\omega_2 = \sqrt{(5 + \sqrt{17})/2}$, so that the general solution for $x = x(t)$ is

$$x(t) = c_1 \cos \omega_1 t + c_2 \sin \omega_1 t + c_3 \cos \omega_2 t + c_4 \sin \omega_2 t.$$

Insert this and $\ddot{x}(t) = -c_1 \omega_1^2 \cos \omega_1 t - c_2 \omega_1^2 \sin \omega_1 t - c_3 \omega_2^2 \cos \omega_2 t - c_4 \omega_2^2 \sin \omega_2 t$ into the first equation of the system and solve for y to obtain the general solution for $y = y(t)$:

$$y(t) = \frac{1}{2}(3 - \omega_1^2)(c_1 \cos \omega_1 t + c_2 \sin \omega_1 t) + \frac{1}{2}(3 - \omega_2^2)(c_3 \cos \omega_2 t + c_4 \sin \omega_2 t).$$

Computing $\dot{x}(t)$ and $\dot{y}(t)$ gives

$$\dot{x}(t) = -c_1\omega_1\sin\omega_1 t + c_2\omega_1\cos\omega_1 t - c_3\omega_2\sin\omega_2 t + c_4\omega_2\cos\omega_2 t,$$
$$\dot{y}(t) = \frac{1}{2}(3-\omega_1^2)\omega_1(-c_1\sin\omega_1 t + c_2\cos\omega_1 t) + \frac{1}{2}(3-\omega_2^2)\omega_2(-c_3\sin\omega_2 t + c_4\cos\omega_2 t).$$

The initial conditions lead to the system

$$3 = x(0) = c_1 + c_3,$$
$$0 = \dot{x}(0) = c_2\omega_1 + c_4\omega_2,$$
$$3 = y(0) = \frac{1}{2}(3-\omega_1^2)c_1 + \frac{1}{2}(3-\omega_2^2)c_3,$$
$$0 = \dot{y}(0) = \frac{1}{2}(3-\omega_1^2)\omega_1 c_2 + \frac{1}{2}(3-\omega_2^2)\omega_2 c_4,$$

with solution $c_1 = -3\left(\dfrac{1-\omega_2^2}{\omega_2^2-\omega_1^2}\right)$, $c_3 = 3\left(\dfrac{1-\omega_1^2}{\omega_2^2-\omega_1^2}\right)$, $c_2 = c_4 = 0$. Thus, the desired solution is

$$x(t) = -3\left(\frac{1-\omega_2^2}{\omega_2^2-\omega_1^2}\right)\cos\omega_1 t + 3\left(\frac{1-\omega_1^2}{\omega_2^2-\omega_1^2}\right)\cos\omega_2 t,$$
$$y(t) = -\frac{3}{2}\left(\frac{(3-\omega_1^2)(1-\omega_2^2)}{\omega_2^2-\omega_1^2}\right)\cos\omega_1 t + \frac{3}{2}\left(\frac{(3-\omega_2^2)(1-\omega_1^2)}{\omega_2^2-\omega_1^2}\right)\cos\omega_2 t.$$

When the values for ω_1 and ω_2 are inserted into the coefficients of the terms on the right, the resulting expressions can be written as

$$x(t) = \frac{3}{2}\left(1+\frac{7}{\sqrt{17}}\right)\cos\omega_1 t + \frac{3}{2}\left(1-\frac{7}{\sqrt{17}}\right)\cos\omega_2 t,$$
$$y(t) = \frac{3}{2}\left(1+\frac{15}{\sqrt{17}}\right)\cos\omega_1 t + \frac{3}{2}\left(1-\frac{15}{\sqrt{17}}\right)\cos\omega_2 t.$$

In this form, one can see by inspection that these functions satisfy the four initial conditions.
11. By inspection, the general solution for $z = z(t)$ comes directly from the third equation of the given system; namely, $z(t) = c_1 e^t$. The initial condition $z(0) = -1$ then gives $z(t) = -e^t$.

Now differentiate the second equation to get $d^2y/dt^2 = dx/dt$, and then use the first equation of the given system to conclude that $d^2y/dt^2 = -y$, or equivalently, $\ddot{y} + y = 0$. This has the general solution

$$y(t) = c_1\cos t + c_2\sin t.$$

Since $\dot{y}(t) = -c_1\sin t + c_2\cos t$, the second equation of the given system provides the general solution for $x = x(t)$:

$$x(t) = -c_1\sin t + c_2\cos t.$$

The initial conditions $x(0) = 0$, $y(0) = 1$ immediately give $0 = c_2$ and $1 = c_1$, so that $x(t) = -\sin t$ and $y(t) = \cos t$. The general solution $\mathbf{x}(t) = (x(t), y(t), z(t))$ is therefore

$$\mathbf{x}(t) = (-\sin t, \cos t, -e^t).$$

492

13. For each solution $\mathbf{x}_* = \mathbf{x}_*(t)$ of the autonomous system $\dot{\mathbf{x}} = F(\mathbf{x})$, define the vector function $\mathbf{y} = \mathbf{y}(t)$ by

$$\mathbf{y}(t) = \mathbf{x}_*(-t),$$

where t can take on all values such that $-t$ is in the domain of $\mathbf{x}_*$. Differentiating the above equation with respect to t and then using the relation $\dot{\mathbf{x}}_*(-t) = F(\mathbf{x}_*(-t))$ gives

$$\dot{\mathbf{y}}(t) = -\dot{\mathbf{x}}_*(-t) = -F(\mathbf{x}_*(-t)) = -F(\mathbf{y}(t)).$$

Suppressing the variable t gives

$$\dot{\mathbf{y}} = -F(\mathbf{y}).$$

So, $\mathbf{y} = \mathbf{y}(t)$ is a solution of the system $\dot{\mathbf{x}} = -F(\mathbf{x})$. Moreover, given that $\mathbf{x}_*(0) = \mathbf{x}_0$ and $\mathbf{x}_*(1) = \mathbf{x}_1$, it follows that we have $\mathbf{y}(0) = \mathbf{x}_0$ and $\mathbf{y}(-1) = \mathbf{x}_1$.

15. If $f(x, y)$ is a continuously differentiable function then its gradient is perpendicular to the surface $z = f(x, y)$. In particular, its gradient is perpendicular to each level curve $f(x, y) = c$ of f. Thus, if $\mathbf{x} = (x, y)$ is a solution of the associated gradient system $\dot{\mathbf{x}} = \nabla f$ then $(\dot{x}, \dot{y})$ is both the gradient of f and a tangent vector to the trajectory. In other words, the solution trajectory is perpendicular to each level curve of f.

17. Recall that the integral of a gradient field ∇f along a piecewise smooth curve starting at $\mathbf{x}_0$ and ending at $\mathbf{x}_1$ can be evaluated by means of the formula

$$\int_{\mathbf{x}_0}^{\mathbf{x}_1} \nabla f \cdot d\mathbf{x} = f(\mathbf{x}_1) - f(\mathbf{x}_0).$$

In particular if $\mathbf{x}_0$ and $\mathbf{x}_1$ lie on the same level set of f then $f(\mathbf{x}_0) = f(\mathbf{x}_1)$ and the right side of the above integral equation is zero.

Now suppose $\mathbf{x}(t)$ is a solution of the gradient system $\dot{\mathbf{x}} = \nabla f$ with $\mathbf{x}(t_0) = \mathbf{x}_0$ and $\mathbf{x}(t_1) = \mathbf{x}_1$, then the above integral can be written in terms of t; namely,

$$\int_{\mathbf{x}_0}^{\mathbf{x}_1} \nabla f \cdot d\mathbf{x} = \int_{\mathbf{x}_0}^{\mathbf{x}_1} \nabla f(\mathbf{x}) \cdot \dot{\mathbf{x}} \, dt = \int_{t_0}^{t_1} \dot{\mathbf{x}} \cdot \dot{\mathbf{x}} \, dt = \int_{t_0}^{t_1} |\dot{\mathbf{x}}(t)|^2 dt.$$

It follows that if $\mathbf{x}_0$ and $\mathbf{x}_1$ lie on the same level set of f then

$$\int_{t_0}^{t_1} |\dot{\mathbf{x}}(t)|^2 dt = 0.$$

Since the integrand is continuous and nonnegative on $[t_0, t_1]$, the above equation can hold only if the integrand is identically zero on $[t_0, t_1]$. So the velocity vector $\dot{\mathbf{x}}(t)$ in the zero vector on $[t_0, t_1]$ and therefore $\mathbf{x}(t) = \mathbf{c}$ is constant on $[t_0, t_1]$. Since this argument holds for any t_0 and t_1 in the domain of $\mathbf{x}(t)$, it follows that $\mathbf{x}(t) = \mathbf{c}$ for all t in the domain of $\mathbf{x}(t)$.

19. Let $H(x, y) = x^2 - y^2$ and consider the corresponding Hamiltonian system

$$\dot{x} = \frac{\partial H}{\partial y} = -2y, \qquad \dot{y} = -\frac{\partial H}{\partial x} = -2x.$$

Differentiating the first equation gives $\ddot{x} = -2\dot{y}$, and the second equation then implies $\ddot{x} = 4x$, or equivalently, $\ddot{x} - 4x = 0$. The characteristic equation is $r^2 - 4 = 0$, with roots $r_{1,2} = \pm 2$. The general solution for $x = x(t)$ is then

$$x(t) = c_1 e^{2t} + c_2 e^{-2t}.$$

Writing the first equation of the system as $y = -\frac{1}{2}\dot{x}$, compute the general solution for $y = y(t)$:

$$y(t) = -\frac{1}{2}(2c_1e^{2t} - 2c_2e^{-2t}) = -c_1e^{2t} + c_2e^{-2t}.$$

Hence,

$$x + y = 2c_2e^{-2t} \qquad \text{and} \qquad x - y = 2c_1e^{2t}.$$

Multiplying these equations member by member gives

$$x^2 - y^2 = 4c_1c_2.$$

So, $H(x, y) = 4c_1c_2$ is constant on the trajectory and the trajectory is therefore a level curve of H.

Chapter 13: Matrix Methods

Section 1: EIGENVALUES & EIGENVECTORS

Exercise Set 1AB (pgs. 624-625)

1. Let $A = \begin{pmatrix} 8 & -3 \\ 10 & -3 \end{pmatrix}$. If I is the 2×2 identity matrix then λ is an eigenvalue of A if λ satisfies

$$\det(A - \lambda I) = \begin{vmatrix} 8 - \lambda & -3 \\ 10 & -3 - \lambda \end{vmatrix} = (8 - \lambda)(-3 - \lambda) + 30 = \lambda^2 - 5\lambda + 6 = (\lambda - 2)(\lambda - 3) = 0.$$

Hence, $\lambda_1 = 2$ and $\lambda_2 = 3$ are the eigenvalues of A.

For $\lambda_1 = 2$, we want a nonzero vector $\mathbf{u} = \begin{pmatrix} u \\ v \end{pmatrix}$ such that

$$(A - 2I)\mathbf{u} = \begin{pmatrix} 6 & -3 \\ 10 & -5 \end{pmatrix} \begin{pmatrix} u \\ v \end{pmatrix} = \begin{pmatrix} 6u - 3v \\ 10u - 5v \end{pmatrix} = (2u - v) \begin{pmatrix} 3 \\ 5 \end{pmatrix} = \begin{pmatrix} 0 \\ 0 \end{pmatrix}.$$

This implies $2u - v = 0$. Choosing $u = 1$, $v = 2$ gives the eigenvector $\mathbf{u} = \begin{pmatrix} 1 \\ 2 \end{pmatrix}$.

For $\lambda_2 = 3$, we want $\mathbf{v} = \begin{pmatrix} u \\ v \end{pmatrix}$ to satisfy

$$(A - 3I)\mathbf{v} = \begin{pmatrix} 5 & -3 \\ 10 & -6 \end{pmatrix} \begin{pmatrix} u \\ v \end{pmatrix} = \begin{pmatrix} 5u - 3v \\ 10u - 6v \end{pmatrix} = (5u - 3v) \begin{pmatrix} 1 \\ 2 \end{pmatrix} = \begin{pmatrix} 0 \\ 0 \end{pmatrix}.$$

This implies $5u - 3v = 0$. Choosing $u = 3$, $v = 5$ gives the eigenvector $\mathbf{v} = \begin{pmatrix} 3 \\ 5 \end{pmatrix}$.

3. Let $A = \begin{pmatrix} -1 & 3 \\ -2 & -4 \end{pmatrix}$. Every eigenvalue λ of A satisfies

$$\det(A - \lambda I) = \begin{vmatrix} -1 - \lambda & 3 \\ -2 & -4 - \lambda \end{vmatrix} = (-1 - \lambda)(-4 - \lambda) + 6 = \lambda^2 + 5\lambda + 10 = 0.$$

Here, the eigenvalues are complex: $\lambda_{1,2} = -\frac{5}{2} \pm i\frac{\sqrt{15}}{2}$.

For $\lambda_1 = -\frac{5}{2} + i\frac{\sqrt{15}}{2}$, we want $\mathbf{u} = \begin{pmatrix} u \\ v \end{pmatrix}$ to satisfy

$$(A - \lambda_1 I)\mathbf{u} = \begin{pmatrix} \frac{3}{2} - i\frac{\sqrt{15}}{2} & 3 \\ -2 & -\frac{3}{2} - i\frac{\sqrt{15}}{2} \end{pmatrix} \begin{pmatrix} u \\ v \end{pmatrix} = \begin{pmatrix} \left(\frac{3}{2} - i\frac{\sqrt{15}}{2}\right)u + 3v \\ -2u + \left(-\frac{3}{2} - i\frac{\sqrt{15}}{2}\right)v \end{pmatrix} = \begin{pmatrix} 0 \\ 0 \end{pmatrix}.$$

The two scalar equations implied by the above equation are equivalent to the single equation $(3 - i\sqrt{15})u + 6v = 0$. Choosing $u = -6$, $v = 3 - i\sqrt{15}$ gives the eigenvector $\mathbf{u} = \begin{pmatrix} -6 \\ 3 - i\sqrt{15} \end{pmatrix}$.

For $\lambda_2 = -\frac{5}{2} - i\frac{\sqrt{15}}{2}$, we want $\mathbf{v} = \begin{pmatrix} u \\ v \end{pmatrix}$ to satisfy

$$(A - \lambda_2 I)\mathbf{v} = \begin{pmatrix} \frac{3}{2} + i\frac{\sqrt{15}}{2} & 3 \\ -2 & -\frac{3}{2} + i\frac{\sqrt{15}}{2} \end{pmatrix} \begin{pmatrix} u \\ v \end{pmatrix} = \begin{pmatrix} \left(\frac{3}{2} + i\frac{\sqrt{15}}{2}\right)u + 3v \\ -2u + \left(-\frac{3}{2} + i\frac{\sqrt{15}}{2}\right)v \end{pmatrix} = \begin{pmatrix} 0 \\ 0 \end{pmatrix}.$$

The two scalar equations implied by the above equation are equivalent to the single equation $(3 + i\sqrt{15})u + 6v = 0$. Choosing $u = -6$, $v = 3 + i\sqrt{15}$ gives the eigenvector $\mathbf{v} = \begin{pmatrix} -6 \\ 3 + i\sqrt{15} \end{pmatrix}$.

5. Let $A = \begin{pmatrix} 1 & 1 & 1 \\ 0 & 1 & 0 \\ 0 & 0 & 2 \end{pmatrix}$. If I is the 3×3 identity matrix then λ is an eigenvalue of A if λ satisfies

$$\det(A - \lambda I) = \begin{vmatrix} 1 - \lambda & 1 & 1 \\ 0 & 1 - \lambda & 0 \\ 0 & 0 & 2 - \lambda \end{vmatrix} = (1 - \lambda)(1 - \lambda)(2 - \lambda) = 0.$$

Thus, $\lambda_1 = 1$ and $\lambda_2 = 2$ are the only eigenvalues of A.

For $\lambda_1 = 1$, we want a nonzero vector $\mathbf{u} = \begin{pmatrix} u \\ v \\ w \end{pmatrix}$ to satisfy

$$(A - I)\mathbf{u} = \begin{pmatrix} 0 & 1 & 1 \\ 0 & 0 & 0 \\ 0 & 0 & 1 \end{pmatrix} \begin{pmatrix} u \\ v \\ w \end{pmatrix} = \begin{pmatrix} v + w \\ 0 \\ w \end{pmatrix} = \begin{pmatrix} 0 \\ 0 \\ 0 \end{pmatrix}.$$

This implies the two nontrivial scalar equations $v + w = 0$ and $w = 0$, from which $v = w = 0$. As u is arbitrary, choose $u = 1$ to get the eigenvector $\mathbf{u} = \begin{pmatrix} 1 \\ 0 \\ 0 \end{pmatrix}$.

For $\lambda_2 = 2$, we want $\mathbf{v} = \begin{pmatrix} u \\ v \\ w \end{pmatrix}$ to satisfy

$$(A - 2I)\mathbf{v} = \begin{pmatrix} -1 & 1 & 1 \\ 0 & -1 & 0 \\ 0 & 0 & 0 \end{pmatrix} \begin{pmatrix} u \\ v \\ w \end{pmatrix} = \begin{pmatrix} -u + v + w \\ -v \\ 0 \end{pmatrix} = \begin{pmatrix} 0 \\ 0 \\ 0 \end{pmatrix}.$$

This implies the two nontrivial scalar equations $-u + v + w = 0$ and $-v = 0$. The second equation gives $v = 0$, and substitution into the first equation gives $-u + w = 0$. Choosing $u = w = 1$ gives the eigenvector $\mathbf{v} = \begin{pmatrix} 1 \\ 0 \\ 1 \end{pmatrix}$.

7. First, identify the constant matrix A to be $A = \begin{pmatrix} -3 & 2 \\ -4 & 3 \end{pmatrix}$. The eigenvalues λ of A satisfy

$$\det(A - \lambda I) = \begin{vmatrix} -3 - \lambda & 2 \\ -4 & 3 - \lambda \end{vmatrix} = (-3 - \lambda)(3 - \lambda) + 8 = (\lambda - 1)(\lambda + 1) = 0.$$

Hence, $\lambda_1 = 1$, $\lambda_2 = -1$ are the eigenvalues of A.

For $\lambda_1 = 1$, we want $\mathbf{u} = \begin{pmatrix} u \\ v \end{pmatrix}$ to satisfy

$$(A - I)\mathbf{u} = \begin{pmatrix} -4 & 2 \\ -4 & 2 \end{pmatrix} \begin{pmatrix} u \\ v \end{pmatrix} = \begin{pmatrix} -4u + 2v \\ -4u + 2v \end{pmatrix} = (-4u + 2v) \begin{pmatrix} 1 \\ 1 \end{pmatrix} = \begin{pmatrix} 0 \\ 0 \end{pmatrix}.$$

This implies $-4u + 2v = 0$, or equivalently, $2u = v$. Choosing $u = 1$, $v = 2$ gives the eigenvector $\mathbf{u} = \begin{pmatrix} 1 \\ 2 \end{pmatrix}$.

For $\lambda_2 = -1$, we want $\mathbf{v} = \begin{pmatrix} u \\ v \end{pmatrix}$ to satisfy

$$(A + I)\mathbf{v} = \begin{pmatrix} -2 & 2 \\ -4 & 4 \end{pmatrix} \begin{pmatrix} u \\ v \end{pmatrix} = \begin{pmatrix} -2u + 2v \\ -4u + 4v \end{pmatrix} = (-2u + 2v) \begin{pmatrix} 1 \\ 2 \end{pmatrix} = \begin{pmatrix} 0 \\ 0 \end{pmatrix}.$$

This implies $-2u + 2v = 0$, or equivalently, $u = v$. Choosing $u = v = 1$ gives the eigenvector $\mathbf{v} = \begin{pmatrix} 1 \\ 1 \end{pmatrix}$. Therefore, the general solution $\mathbf{x}(t) = \big(x(t), y(t)\big)$ is

$$\mathbf{x}(t) = c_1 e^{\lambda_1 t} \mathbf{u} + c_2 e^{\lambda_2 t} \mathbf{v} = c_1 e^t \begin{pmatrix} 1 \\ 2 \end{pmatrix} + c_2 e^{-t} \begin{pmatrix} 1 \\ 1 \end{pmatrix}.$$

Setting $t = 0$ in this solution, the initial condition gives

$$\begin{pmatrix} 1 \\ 0 \end{pmatrix} = \mathbf{x}(0) = c_1 \begin{pmatrix} 1 \\ 2 \end{pmatrix} + c_2 \begin{pmatrix} 1 \\ 1 \end{pmatrix} = \begin{pmatrix} c_1 + c_2 \\ 2c_1 + c_2 \end{pmatrix}.$$

This implies the equations $1 = c_1 + c_2$ and $0 = 2c_1 + c_2$, with solution $c_1 = -1$, $c_2 = 2$. So the solution satisfying the given initial condition is

$$\mathbf{x}(t) = -e^t \begin{pmatrix} 1 \\ 2 \end{pmatrix} + 2e^{-t} \begin{pmatrix} 1 \\ 1 \end{pmatrix} = \begin{pmatrix} -e^t + 2e^{-t} \\ -2e^t + 2e^{-t} \end{pmatrix}.$$

9. The given system can be written as

$$\begin{pmatrix} dx/dt \\ dy/dt \end{pmatrix} = \begin{pmatrix} 1 & 4 \\ 0 & 5 \end{pmatrix} \begin{pmatrix} x \\ y \end{pmatrix}, \qquad \begin{pmatrix} x(0) \\ y(0) \end{pmatrix} = \begin{pmatrix} 1 \\ 1 \end{pmatrix},$$

where we identify the matrix A to be $A = \begin{pmatrix} 1 & 4 \\ 0 & 5 \end{pmatrix}$. The eigenvalues λ of A satisfy

$$\det(A - \lambda I) = \begin{vmatrix} 1 - \lambda & 4 \\ 0 & 5 - \lambda \end{vmatrix} = (1 - \lambda)(5 - \lambda) = 0.$$

So $\lambda_1 = 1$, $\lambda_2 = 5$ are the eigenvalues of A.

For $\lambda_1 = 1$, we want $\mathbf{u} = \begin{pmatrix} u \\ v \end{pmatrix}$ to satisfy

$$(A - I)\mathbf{u} = \begin{pmatrix} 0 & 4 \\ 0 & 4 \end{pmatrix} \begin{pmatrix} u \\ v \end{pmatrix} = \begin{pmatrix} 4v \\ 4v \end{pmatrix} = 4v \begin{pmatrix} 1 \\ 1 \end{pmatrix} = \begin{pmatrix} 0 \\ 0 \end{pmatrix}.$$

This implies $4v = 0$, so that $v = 0$. As u is arbitrary, choose $u = 1$ to get the eigenvector $\mathbf{u} = \begin{pmatrix} 1 \\ 0 \end{pmatrix}$.

For $\lambda_2 = 5$, we want $\mathbf{v} = \begin{pmatrix} u \\ v \end{pmatrix}$ to satisfy

$$(A - 5I)\mathbf{v} = \begin{pmatrix} -4 & 4 \\ 0 & 0 \end{pmatrix} \begin{pmatrix} u \\ v \end{pmatrix} = \begin{pmatrix} -4u + 4v \\ 0 \end{pmatrix} = \begin{pmatrix} 0 \\ 0 \end{pmatrix}.$$

This implies the single nontrivial scalar equation $-4u + 4v = 0$. Since this is equivalent to $u = v$, we choose $u = v = 1$ to get the eigenvector $\mathbf{v} = \begin{pmatrix} 1 \\ 1 \end{pmatrix}$. Therefore, the general solution $\mathbf{x}(t) = \big(x(t), y(t)\big)$ of the given system is

$$\mathbf{x}(t) = c_1 e^{\lambda_1 t}\mathbf{u} + c_2 e^{\lambda_2 t}\mathbf{v} = c_1 e^t \begin{pmatrix} 1 \\ 0 \end{pmatrix} + c_2 e^{5t} \begin{pmatrix} 1 \\ 1 \end{pmatrix}.$$

Setting $t = 0$ in this solution, the initial condition gives

$$\begin{pmatrix} 1 \\ 1 \end{pmatrix} = c_1 \begin{pmatrix} 1 \\ 0 \end{pmatrix} + c_2 \begin{pmatrix} 1 \\ 1 \end{pmatrix} = \begin{pmatrix} c_1 + c_2 \\ c_2 \end{pmatrix}.$$

By inspection, $c_1 = 0$ and $c_2 = 1$. So the solution satisfying the given initial condition is

$$\mathbf{x}(t) = e^{5t} \begin{pmatrix} 1 \\ 1 \end{pmatrix} = \begin{pmatrix} e^{5t} \\ e^{5t} \end{pmatrix}.$$

11. (a) Let $A = \begin{pmatrix} 1 & -1 & 4 \\ 3 & 2 & -1 \\ 2 & 1 & -1 \end{pmatrix}$. The eigenvalues λ of A satisfy

$$\det(A - \lambda I) = \begin{vmatrix} 1 - \lambda & -1 & 4 \\ 3 & 2 - \lambda & -1 \\ 2 & 1 & -1 - \lambda \end{vmatrix} = -\lambda^3 + 2\lambda^2 + 5\lambda - 6 = -(\lambda + 2)(\lambda - 1)(\lambda - 3).$$

Hence, $\lambda_1 = -2$, $\lambda_2 = 1$, $\lambda_3 = 3$ are the eigenvalues of A.

For $\lambda_1 = -2$, we want $\mathbf{u} = \begin{pmatrix} u \\ v \\ w \end{pmatrix}$ to satisfy

$$(A + 2I)\mathbf{u} = \begin{pmatrix} 3 & -1 & 4 \\ 3 & 4 & -1 \\ 2 & 1 & 1 \end{pmatrix} \begin{pmatrix} u \\ v \\ w \end{pmatrix} = \begin{pmatrix} 3u - v + 4w \\ 3u + 4v - w \\ 2u + v + w \end{pmatrix} = \begin{pmatrix} 0 \\ 0 \\ 0 \end{pmatrix}.$$

This leads to the scalar equations $3u - v + 4w = 0$, $3u + 4v - w = 0$, $2u + v + w = 0$. Using the elimination method we find that each coordinate can be written in terms of the arbitrary parameter u; namely, $u = u$, $v = -u$ and $w = -u$. Choosing $u = 1$, $v = w = -1$ gives the eigenvector $\mathbf{u} = \begin{pmatrix} 1 \\ -1 \\ -1 \end{pmatrix}$.

For $\lambda_2 = 1$, we want $\mathbf{v} = \begin{pmatrix} u \\ v \\ w \end{pmatrix}$ to satisfy

$$(A - I)\mathbf{v} = \begin{pmatrix} 0 & -1 & 4 \\ 3 & 1 & -1 \\ 2 & 1 & -2 \end{pmatrix} \begin{pmatrix} u \\ v \\ w \end{pmatrix} = \begin{pmatrix} -v + 4w \\ 3u + v - w \\ 2u + v - 2w \end{pmatrix} = \begin{pmatrix} 0 \\ 0 \\ 0 \end{pmatrix}.$$

This leads to the scalar equations $-v + 4w = 0$, $3u + v - w = 0$, $2u + v - 2w = 0$. Using the elimination method we find that each coordinate can be written in terms of the arbitrary parameter w; namely, $u = -w$ and $v = 4w$ and $w = w$. Choosing $u = -1$, $v = 4$, $w = 1$ gives the eigenvector $\mathbf{v} = \begin{pmatrix} -1 \\ 4 \\ 1 \end{pmatrix}$.

For $\lambda_3 = 3$, we want $\mathbf{w} = \begin{pmatrix} u \\ v \\ w \end{pmatrix}$ to satisfy

$$(A - 3I)\mathbf{w} = \begin{pmatrix} -2 & -1 & 4 \\ 3 & -1 & -1 \\ 2 & 1 & -4 \end{pmatrix} \begin{pmatrix} u \\ v \\ w \end{pmatrix} = \begin{pmatrix} -2u - v + 4w \\ 3u - v - w \\ 2u + v - 4w \end{pmatrix} = \begin{pmatrix} 0 \\ 0 \\ 0 \end{pmatrix}.$$

This leads to the scalar equations $-2u - v + 4w = 0$, $3u - v - w = 0$, $2u + v - 4w = 0$. Using the elimination method we find that each coordinate can be written in terms of the arbitrary parameter u; namely, $u = u$, $v = 2u$ and $w = u$. Choosing $u = 1$, $v = 2$, $w = 1$ gives the eigenvector $\mathbf{w} = \begin{pmatrix} 1 \\ 2 \\ 1 \end{pmatrix}$.

The general homogeneous solution of the given system can therefore be written as

$$\mathbf{x}_h(t) = c_1 e^{-2t} \begin{pmatrix} 1 \\ -1 \\ -1 \end{pmatrix} + c_2 e^t \begin{pmatrix} -1 \\ 4 \\ 1 \end{pmatrix} + c_3 e^{3t} \begin{pmatrix} 1 \\ 2 \\ 1 \end{pmatrix}.$$

(b) Since the nonhomogeneous part of the given system is a constant vector $\mathbf{c} = \begin{pmatrix} 1 \\ 0 \\ 2 \end{pmatrix}$, it is reasonable that the nonhomogeneous system has a nonzero constant solution $\mathbf{x}_p(t)$. With this and its derivative (the zero vector) substituted into the given nonhomogeneous system, the result is $\mathbf{0} = A\mathbf{x}_p + \mathbf{c}$. Since $\det A = -6 \neq 0$, A^{-1} exists, so we can left-multiply each member of this equation by A^{-1} to get $\mathbf{0} = \mathbf{x}_p + A^{-1}\mathbf{c}$. Computing A^{-1} (which we leave to the reader), we can solve for $\mathbf{x}_p$ to get

$$\mathbf{x}_p = -A^{-1}\mathbf{c} = -\frac{1}{6} \begin{pmatrix} 1 & -3 & 7 \\ -1 & 9 & -13 \\ 1 & 3 & -5 \end{pmatrix} \begin{pmatrix} 2 \\ 0 \\ 1 \end{pmatrix} = -\frac{1}{6} \begin{pmatrix} 9 \\ -15 \\ -3 \end{pmatrix} = \begin{pmatrix} -3/2 \\ 5/2 \\ 1/2 \end{pmatrix}.$$

(c) Using the results of parts (a) and (b), the general solution of the given system is

$$\mathbf{x}(t) = c_1 e^{-2t} \begin{pmatrix} 1 \\ -1 \\ -1 \end{pmatrix} + c_2 e^t \begin{pmatrix} -1 \\ 4 \\ 1 \end{pmatrix} + c_3 e^{3t} \begin{pmatrix} 1 \\ 2 \\ 1 \end{pmatrix} + \begin{pmatrix} -3/2 \\ 5/2 \\ 1/2 \end{pmatrix}.$$

Setting $t = 0$ and using the condition $\mathbf{x}(0) = \begin{pmatrix} 1 \\ 1 \\ 2 \end{pmatrix}$ gives

$$\begin{pmatrix} 1 \\ 1 \\ 2 \end{pmatrix} = c_1 \begin{pmatrix} 1 \\ -1 \\ -1 \end{pmatrix} + c_2 \begin{pmatrix} -1 \\ 4 \\ 1 \end{pmatrix} + c_3 \begin{pmatrix} 1 \\ 2 \\ 1 \end{pmatrix} + \begin{pmatrix} -3/2 \\ 5/2 \\ 1/2 \end{pmatrix} = \begin{pmatrix} c_1 - c_2 + c_3 - 3/2 \\ -c_1 + 4c_2 + 2c_3 + 5/2 \\ -c_1 + c_2 + c_3 + 1/2 \end{pmatrix}.$$

This gives three scalar equations, which can be written as

$$c_1 - c_2 + c_3 = \frac{5}{2}, \qquad -c_1 + 4c_2 + 2c_3 = -\frac{3}{2}, \qquad -c_1 + c_2 + c_3 = \frac{3}{2}.$$

This has the solution $c_1 = -7/6$, $c_2 - 5/3$, $c_3 = 2$. So the solution satisfying the given condition is

$$\mathbf{x}(t) = -\frac{7}{6} e^{-2t} \begin{pmatrix} 1 \\ -1 \\ -1 \end{pmatrix} - \frac{5}{3} e^t \begin{pmatrix} -1 \\ 4 \\ 1 \end{pmatrix} + 2e^{3t} \begin{pmatrix} 1 \\ 2 \\ 1 \end{pmatrix} + \begin{pmatrix} -3/2 \\ 5/2 \\ 1/2 \end{pmatrix}.$$

13. (a) The second equation of the given system can be directly solved to get

$$y(t) = c_1 e^{-t}.$$

The first equation then becomes $dx/dt = -x + c_1 e^{-t}$, or equivalently, $dx/dt + x = c_1 e^{-t}$. The integrating factor e^t leads to $d/dt(e^t x) = c_1$, and integration leads to $e^t x = c_1 t + c_2$. Solving for $x = x(t)$ gives

$$x(t) = c_1 t e^{-t} + c_2 e^{-t}.$$

(b) Denoting the given matrix by A, we see that the D.E. given in part (a) can be written in vector form as $\dot{\mathbf{x}} = A\mathbf{x}$. We can therefore identify the eigenvalues λ of A with the coefficients of the exponents in the exponentials appearing in the general solution derived in part (a); namely, $\lambda = -1$ is a double eigenvalue. It follows that the eigenvectors of -1 (and therefore the eigenvectors of A) are precisely the nonzero vectors $\mathbf{u} = \begin{pmatrix} u \\ v \end{pmatrix}$ such that

$$(A + I)\mathbf{u} = \begin{pmatrix} 0 & 1 \\ 0 & 0 \end{pmatrix} \begin{pmatrix} u \\ v \end{pmatrix} = \begin{pmatrix} v \\ 0 \end{pmatrix} = v \begin{pmatrix} 1 \\ 0 \end{pmatrix} = \begin{pmatrix} 0 \\ 0 \end{pmatrix}.$$

This implies $v = 0$. Since $u \neq 0$ is arbitrary, the eigenvectors of A are those vectors of the form $\mathbf{u} = \begin{pmatrix} u \\ 0 \end{pmatrix} = u \begin{pmatrix} 1 \\ 0 \end{pmatrix}$, $u \neq 0$.

15. Let $A = \begin{pmatrix} -a & -b \\ 1 & 0 \end{pmatrix}$. The eigenvalues of A are the solutions of $\det(A - \lambda I) = 0$, so that

$$\det(A - \lambda I) = \det \begin{pmatrix} -a - \lambda & -b \\ 1 & -\lambda \end{pmatrix} = (-a - \lambda)(-\lambda) + b = \lambda^2 + a\lambda + b = 0.$$

Now simply observe that the characteristic equation associated with the given second-order D.E. is $r^2 + ar + b = 0$. Since the polynomials are the same, the eigenvalues of A are the roots of the characteristic equation.

17. (a) The kth n-dimensional basis vector $\mathbf{e}_k$ has the property that if M is an $n \times n$ matrix then $M\mathbf{e}_k$ is the kth column of M. In particular, since $\mathbf{u}_k$ is the kth column of U, and since the kth column of $D\mathbf{e}_k$ is λ_k times $\mathbf{e}_k$, we have

$$U\mathbf{e}_k = \mathbf{u}_k, \qquad \text{and} \qquad D\mathbf{e}_k = \lambda_k \mathbf{e}_k,$$

both of which hold for $k = 1, \ldots, n$. The first equation implies $A(U\mathbf{e}_k) = A\mathbf{u}_k$. Multiplying the equation $A = UDU^{-1}$ on the right by $U\mathbf{e}_k$ gives

$$A\mathbf{u}_k = UDU^{-1}(U\mathbf{e}_k) = UD(I_n\mathbf{e}_k) = U(D\mathbf{e}_k) = U(\lambda_k\mathbf{e}_k) = \lambda_k(U\mathbf{e}_k) = \lambda_k\mathbf{u}_k,$$

where I_n is the $n \times n$ identity matrix.

(b) Let $U = \begin{pmatrix} 2 & 1 \\ 3 & 1 \end{pmatrix}$, so that the given vectors $\mathbf{u}_1$ and $\mathbf{u}_2$ are the first and second columns of U, respectively. Then the inverse of U exists and is given by $U^{-1} = \begin{pmatrix} -1 & 1 \\ 3 & -2 \end{pmatrix}$.

Letting $D = \begin{pmatrix} 3 & 0 \\ 0 & 1 \end{pmatrix}$, define the matrix A by

$$A = UDU^{-1} = \begin{pmatrix} 2 & 1 \\ 3 & 1 \end{pmatrix}\begin{pmatrix} 3 & 0 \\ 0 & 1 \end{pmatrix}\begin{pmatrix} -1 & 1 \\ 3 & -2 \end{pmatrix} = \begin{pmatrix} -3 & 4 \\ -6 & 7 \end{pmatrix}.$$

By part (a), $A\mathbf{u}_k = \lambda_k\mathbf{u}_k$ for $k = 1, 2$. By the definitions of *eigenvalue* and *eigenvector*, λ_1 and λ_2 are the eigenvalues of A and $\mathbf{u}_1$, $\mathbf{u}_2$ are corresponding eigenvectors.

(c) For $k = 1, \ldots, n$, let $\mathbf{x}_k(t) = e^{\lambda_k t}\mathbf{u}_k$. Then $\dot{\mathbf{x}}_k(t) = \lambda_k e^{\lambda_k t}\mathbf{u}_k$ for $k = 1, \ldots, n$. By part (a), $\lambda_k\mathbf{u}_k = A\mathbf{u}_k$ for $k = 1, \ldots, n$. Scalar multiplication of this equation by $e^{\lambda_k t}$ results in an equation that can be written as $\lambda_k e^{\lambda_k t}\mathbf{u}_k = A(e^{\lambda_k t}\mathbf{u}_k)$, which is equivalent to $\dot{\mathbf{x}}_k = A\mathbf{x}_k$. That is, $\mathbf{x}_k(t) = e^{\lambda_k t}\mathbf{u}_k$ is a solution of the system $\dot{\mathbf{x}} = A\mathbf{x}$ for $k = 1, \ldots, n$.

(d) Let $\mathbf{u}_1$, $\mathbf{u}_2$, λ_1, λ_2 and A be as in part (b). Then part (c) shows that

$$e^{\lambda_1 t}\mathbf{u}_1 = e^{3t}\begin{pmatrix} 2 \\ 3 \end{pmatrix} \qquad \text{and} \qquad e^{\lambda_2 t}\mathbf{u}_2 = e^t\begin{pmatrix} 1 \\ 1 \end{pmatrix}$$

are both solutions of the linear system $\dot{\mathbf{x}} = A\mathbf{x}$. Since these solutions are linearly independent, the general solution of the system is

$$\mathbf{x}(t) = c_1 e^{3t}\begin{pmatrix} 2 \\ 3 \end{pmatrix} + c_2 e^t\begin{pmatrix} 1 \\ 1 \end{pmatrix}.$$

When $\mathbf{x}(t) = \begin{pmatrix} x(t) \\ y(t) \end{pmatrix} = \begin{pmatrix} x \\ y \end{pmatrix}$ and $A = \begin{pmatrix} -3 & 4 \\ -6 & 7 \end{pmatrix}$ are substituted into $\dot{\mathbf{x}} = A\mathbf{x}$, the system becomes

$$\begin{pmatrix} \dot{x} \\ \dot{y} \end{pmatrix} = \begin{pmatrix} -3 & 4 \\ -6 & 7 \end{pmatrix}\begin{pmatrix} x \\ y \end{pmatrix} = \begin{pmatrix} -3x + 4y \\ -6x + 7y \end{pmatrix}.$$

Section 2: MATRIX EXPONENTIALS

Exercise Set 2ABC (pgs. 630-631)

1. $A = \begin{pmatrix} -1 & 0 \\ 0 & 1 \end{pmatrix}$, $A^2 = \begin{pmatrix} 1 & 0 \\ 0 & 1 \end{pmatrix}$, ..., $A^k = \begin{pmatrix} (-1)^k & 0 \\ 0 & 1 \end{pmatrix}$, ...

$$e^{tA} = \sum_{k=0}^{\infty} \frac{t^k}{k!} A^k = \sum_{k=0}^{\infty} \frac{t^k}{k!} \begin{pmatrix} (-1)^k & 0 \\ 0 & 1 \end{pmatrix} = \begin{pmatrix} e^{-t} & 0 \\ 0 & e^t \end{pmatrix}.$$

3. $A = \begin{pmatrix} 0 & 1 \\ 0 & 1 \end{pmatrix}$, $A^2 = \begin{pmatrix} 0 & 1 \\ 0 & 1 \end{pmatrix}$, ..., $A^k = \begin{pmatrix} 0 & 1 \\ 0 & 1 \end{pmatrix}$, ...

$$e^{tA} = \sum_{k=0}^{\infty} \frac{t^k}{k!} A^k = \sum_{k=0}^{\infty} \frac{t^k}{k!} \begin{pmatrix} 0 & 1 \\ 0 & 1 \end{pmatrix} = e^t \begin{pmatrix} 0 & 1 \\ 0 & 1 \end{pmatrix} = \begin{pmatrix} 0 & e^t \\ 0 & e^t \end{pmatrix}.$$

5. $A = \begin{pmatrix} 1 & 0 & 1 \\ 0 & 1 & 0 \\ 0 & 0 & 1 \end{pmatrix}$, $A^2 = \begin{pmatrix} 1 & 0 & 2 \\ 0 & 1 & 0 \\ 0 & 0 & 1 \end{pmatrix}$, ..., $A^k = \begin{pmatrix} 1 & 0 & k \\ 0 & 1 & 0 \\ 0 & 0 & 1 \end{pmatrix}$, ...

$$
e^{tA} = \sum_{k=0}^{\infty} \frac{t^k}{k!} A^k = \sum_{k=0}^{\infty} \frac{t^k}{k!} \begin{pmatrix} 1 & 0 & k \\ 0 & 1 & 0 \\ 0 & 0 & 1 \end{pmatrix} = \begin{pmatrix} e^t & 0 & \sum_0^{\infty} \frac{kt^k}{k!} \\ 0 & e^t & 0 \\ 0 & 0 & e^t \end{pmatrix}
$$

$$
= \begin{pmatrix} e^t & 0 & \sum_1^{\infty} \frac{t^k}{(k-1)!} \\ 0 & e^t & 0 \\ 0 & 0 & e^t \end{pmatrix} = \begin{pmatrix} e^t & 0 & \sum_0^{\infty} \frac{t^{k+1}}{k!} \\ 0 & e^t & 0 \\ 0 & 0 & e^t \end{pmatrix} = \begin{pmatrix} e^t & 0 & t\sum_0^{\infty} \frac{t^k}{k!} \\ 0 & e^t & 0 \\ 0 & 0 & e^t \end{pmatrix}
$$

$$
= \begin{pmatrix} e^t & 0 & te^t \\ 0 & e^t & 0 \\ 0 & 0 & e^t \end{pmatrix}.
$$

7. (a) The equation

$$\exp\left(t \begin{pmatrix} 1 & 1 \\ 0 & 1 \end{pmatrix} \right) = \begin{pmatrix} e^t & te^t \\ 0 & e^t \end{pmatrix}$$

holds for all real numbers t. It follows that the equation obtained by replacing t with $-t$ also holds for all real t. That is

$$\exp\left(-t \begin{pmatrix} 1 & 1 \\ 0 & 1 \end{pmatrix} \right) = \begin{pmatrix} e^{-t} & -te^{-t} \\ 0 & e^{-t} \end{pmatrix}.$$

Multiplying the two equations displayed above (member by member) gives

$$\exp\left(t \begin{pmatrix} 1 & 1 \\ 0 & 1 \end{pmatrix} \right) \exp\left(-t \begin{pmatrix} 1 & 1 \\ 0 & 1 \end{pmatrix} \right) = \begin{pmatrix} e^t & te^t \\ 0 & e^t \end{pmatrix} \begin{pmatrix} e^{-t} & -te^{-t} \\ 0 & e^{-t} \end{pmatrix} = \begin{pmatrix} 1 & 0 \\ 0 & 1 \end{pmatrix} = I.$$

Hence, $\exp\left(t \begin{pmatrix} 1 & 1 \\ 0 & 1 \end{pmatrix} \right)$ and $\exp\left(-t \begin{pmatrix} 1 & 1 \\ 0 & 1 \end{pmatrix} \right)$ are inverses of each other.

(b) For each pair of real numbers t and s,

$$\exp\left(t\begin{pmatrix}1 & 1\\ 0 & 1\end{pmatrix}\right)\exp\left(s\begin{pmatrix}1 & 1\\ 0 & 1\end{pmatrix}\right)$$

$$=\begin{pmatrix}e^t & te^t\\ 0 & e^t\end{pmatrix}\begin{pmatrix}e^s & se^s\\ 0 & e^s\end{pmatrix}=\begin{pmatrix}e^te^s & e^t(se^s)+(te^t)e^s\\ 0 & e^te^s\end{pmatrix}$$

$$=\begin{pmatrix}e^{t+s} & (t+s)e^{t+s}\\ 0 & e^{t+s}\end{pmatrix}=\exp\left((t+s)\begin{pmatrix}1 & 1\\ 0 & 1\end{pmatrix}\right).$$

(c) On one hand,

$$\frac{d}{dt}\exp\left(t\begin{pmatrix}1 & 1\\ 0 & 1\end{pmatrix}\right)=\frac{d}{dt}\begin{pmatrix}e^t & te^t\\ 0 & e^t\end{pmatrix}=\begin{pmatrix}\frac{d}{dt}(e^t) & \frac{d}{dt}(te^t)\\ \frac{d}{dt}(0) & \frac{d}{dt}(e^t)\end{pmatrix}=\begin{pmatrix}e^t & te^t+e^t\\ 0 & e^t\end{pmatrix};$$

and on the other hand,

$$\begin{pmatrix}1 & 1\\ 0 & 1\end{pmatrix}\exp\left(t\begin{pmatrix}1 & 1\\ 0 & 1\end{pmatrix}\right)=\begin{pmatrix}1 & 1\\ 0 & 1\end{pmatrix}\begin{pmatrix}e^t & te^t\\ 0 & e^t\end{pmatrix}=\begin{pmatrix}e^t & te^t+e^t\\ 0 & e^t\end{pmatrix}.$$

Hence,

$$\frac{d}{dt}\exp\left(t\begin{pmatrix}1 & 1\\ 0 & 1\end{pmatrix}\right)=\begin{pmatrix}1 & 1\\ 0 & 1\end{pmatrix}\exp\left(t\begin{pmatrix}1 & 1\\ 0 & 1\end{pmatrix}\right).$$

(d) It was shown in the text that solution $\mathbf{x}=\mathbf{x}(t)$ of the system $\dot{\mathbf{x}}=A\mathbf{x}$ that satisfies $\mathbf{x}(0)=\mathbf{x}_0$ is

$$\mathbf{x}(t)=e^{tA}\mathbf{x}_0.$$

In particular, with $A=\begin{pmatrix}1 & 1\\ 0 & 1\end{pmatrix}$, $\mathbf{x}(t)=(x(t),y(t))$ and $\mathbf{x}_0=(-1,2)$, we see that the solution $\mathbf{x}=\mathbf{x}(t)$ satisfying $\mathbf{x}_0=(-1,2)$ is

$$\begin{pmatrix}x(t)\\ y(t)\end{pmatrix}=\exp\left(t\begin{pmatrix}1 & 1\\ 0 & 1\end{pmatrix}\right)\begin{pmatrix}-1\\ 2\end{pmatrix}=\begin{pmatrix}e^t & te^t\\ 0 & e^t\end{pmatrix}\begin{pmatrix}-1\\ 2\end{pmatrix}=\begin{pmatrix}-e^t+2te^t\\ 2e^t\end{pmatrix}.$$

9. With $A=2I$, the definition of e^{tA} and the fact that $I^k=I$ for every identity matrix I gives

$$e^{tA}=e^{2tI}=\sum_{k=0}^{\infty}\frac{(2tI)^k}{k!}=\sum_{k=0}^{\infty}\frac{(2t)^k}{k!}I^k=\sum_{k=0}^{\infty}\frac{(2t)^k}{k!}I=e^{2t}I.$$

11. Given $A=\begin{pmatrix}-3 & 2\\ -4 & 3\end{pmatrix}$, the computation of e^{tA} using Equation 2.2 in the text requires an eigenvector matrix Λ_t and an invertible eigenvalue matrix U.

First, the eigenvalues λ of A satisfy

$$\det(A-\lambda I)=\begin{vmatrix}-3-\lambda & 2\\ -4 & 3-\lambda\end{vmatrix}=(-3-\lambda)(3-\lambda)+8=\lambda^2-1=0.$$

Hence, $\lambda_1=1$, $\lambda_2=-1$ and an eigenvalue matrix is

$$\Lambda_t=\begin{pmatrix}e^t & 0\\ 0 & e^{-t}\end{pmatrix}.$$

An eigenvector $\mathbf{u} = \begin{pmatrix} u \\ v \end{pmatrix}$ for $\lambda_1 = 1$ satisfies

$$(A - I)\mathbf{u} = \begin{pmatrix} -4 & 2 \\ -4 & 2 \end{pmatrix}\begin{pmatrix} u \\ v \end{pmatrix} = \begin{pmatrix} -4u + 2v \\ -4u + 2v \end{pmatrix} = (-4u + 2v)\begin{pmatrix} 1 \\ 1 \end{pmatrix} = \begin{pmatrix} 0 \\ 0 \end{pmatrix}.$$

This implies $-4u + 2v = 0$, or equivalently, $v = 2u$. Thus, with $u = 1$, $v = 2$ we get the eigenvector $\mathbf{u} = \begin{pmatrix} 1 \\ 2 \end{pmatrix}$.

An eigenvector $\mathbf{v} = \begin{pmatrix} u \\ v \end{pmatrix}$ for $\lambda_2 = -1$ satisfies

$$(A + I)\mathbf{v} = \begin{pmatrix} -2 & 2 \\ -4 & 4 \end{pmatrix}\begin{pmatrix} u \\ v \end{pmatrix} = \begin{pmatrix} -2u + 2v \\ -4u + 4v \end{pmatrix} = (-2u + 2v)\begin{pmatrix} 1 \\ 2 \end{pmatrix} = \begin{pmatrix} 0 \\ 0 \end{pmatrix}.$$

This implies $-2u + 2v = 0$, or equivalently, $u = v$. Thus, with $u = v = 1$ we get the eigenvector $\mathbf{v} = \begin{pmatrix} 1 \\ 1 \end{pmatrix}$.

Now let $\mathbf{u}$ and $\mathbf{v}$ be the columns of the eigenvector matrix U, so that

$$U = \begin{pmatrix} 1 & 1 \\ 2 & 1 \end{pmatrix}.$$

With $U^{-1} = \begin{pmatrix} -1 & 1 \\ 2 & -1 \end{pmatrix}$, we can now use Equation 2.2 to get

$$e^{tA} = U\Lambda_t U^{-1} = \begin{pmatrix} 1 & 1 \\ 2 & 1 \end{pmatrix}\begin{pmatrix} e^t & 0 \\ 0 & e^{-t} \end{pmatrix}\begin{pmatrix} -1 & 1 \\ 2 & -1 \end{pmatrix} = \begin{pmatrix} -e^t + 2e^{-t} & e^t - e^{-t} \\ -2e^t + 2e^{-t} & 2e^t - e^{-t} \end{pmatrix}.$$

The inverse matrix is found by replacing t with $-t$ in the above formula. Specifically,

$$e^{-tA} = \begin{pmatrix} -e^{-t} + 2e^t & e^{-t} - e^t \\ -2e^{-t} + 2e^t & 2e^{-t} - e^t \end{pmatrix}.$$

As a check on our original computation, note that both sides of the formula found for e^{tA} is the identity matrix when $t = 0$. Also note that

$$\frac{d}{dt}\begin{pmatrix} -e^t + 2e^{-t} & e^t - e^{-t} \\ -2e^t + 2e^{-t} & 2e^t - e^{-t} \end{pmatrix} = \begin{pmatrix} -e^t - 2e^{-t} & e^t + e^{-t} \\ -2e^t - 2e^{-t} & 2e^t + e^{-t} \end{pmatrix},$$

so that evaluation at $t = 0$ gives the original matrix $A = \begin{pmatrix} -3 & 2 \\ -4 & 3 \end{pmatrix}$.

13. Given $A = \begin{pmatrix} 2 & -1 \\ 1 & 2 \end{pmatrix}$, the computation of e^{tA} using Equation 2.2 in the text requires an eigenvector matrix Λ_t and an invertible eigenvalue matrix U.

First, the eigenvalues λ of A satisfy

$$\det(A - \lambda I) = \begin{vmatrix} 2 - \lambda & -1 \\ 1 & 2 - \lambda \end{vmatrix} = (2 - \lambda)^2 + 1 = 0.$$

504

Hence, $\lambda_1 = 2 + i$, $\lambda_2 = 2 - i$ and an eigenvalue matrix is

$$\Lambda_t = \begin{pmatrix} e^{(2+i)t} & 0 \\ 0 & e^{(2-i)t} \end{pmatrix} = e^{2t} \begin{pmatrix} e^{it} & 0 \\ 0 & e^{-it} \end{pmatrix}.$$

An eigenvector $\mathbf{u} = \begin{pmatrix} u \\ v \end{pmatrix}$ for $\lambda_1 = 2 + i$ satisfies

$$(A - (2+i)I)\mathbf{u} = \begin{pmatrix} -i & -1 \\ 1 & -i \end{pmatrix} \begin{pmatrix} u \\ v \end{pmatrix} = \begin{pmatrix} -iu - v \\ u - iv \end{pmatrix} = (u - iv) \begin{pmatrix} i \\ 1 \end{pmatrix} = \begin{pmatrix} 0 \\ 0 \end{pmatrix}.$$

This implies $u - iv = 0$. Thus, with $u = i$, $v = 1$ we get the eigenvector $\mathbf{u} = \begin{pmatrix} i \\ 1 \end{pmatrix}$.

An eigenvector $\mathbf{v} = \begin{pmatrix} u \\ v \end{pmatrix}$ for $\lambda_2 = 2 - i$ satisfies

$$(A - (2-i)I)\mathbf{v} = \begin{pmatrix} i & -1 \\ 1 & i \end{pmatrix} \begin{pmatrix} u \\ v \end{pmatrix} = \begin{pmatrix} iu - v \\ u + iv \end{pmatrix} = (u + iv) \begin{pmatrix} i \\ 1 \end{pmatrix} = \begin{pmatrix} 0 \\ 0 \end{pmatrix}.$$

This implies $u + iv = 0$. Thus, with $u = 1$, $v = i$ we get the eigenvector $\mathbf{v} = \begin{pmatrix} 1 \\ i \end{pmatrix}$.

Now let $\mathbf{u}$ and $\mathbf{v}$ be the columns of the eigenvector matrix U, so that

$$U = \begin{pmatrix} i & 1 \\ 1 & i \end{pmatrix}.$$

With $U^{-1} = \frac{1}{2} \begin{pmatrix} -i & 1 \\ 1 & -i \end{pmatrix}$, we can now use Equation 2.2 to get

$$e^{tA} = U\Lambda_t U^{-1} = \frac{1}{2} e^{2t} \begin{pmatrix} i & 1 \\ 1 & i \end{pmatrix} \begin{pmatrix} e^{it} & 0 \\ 0 & e^{-it} \end{pmatrix} \begin{pmatrix} -i & 1 \\ 1 & -i \end{pmatrix} = \begin{pmatrix} e^{2t}\cos t & -e^{2t}\sin t \\ e^{2t}\sin t & e^{2t}\cos t \end{pmatrix}.$$

The inverse matrix e^{-tA} is found by replacing t with $-t$ in the above formula. Specifically, use the fact that the cosine function is even and the sine function is odd to get

$$e^{-tA} = \begin{pmatrix} e^{-2t}\cos t & e^{-2t}\sin t \\ -e^{-2t}\sin t & e^{-2t}\cos t \end{pmatrix}.$$

As a check on our original computation, note that both sides of the formula found for e^{tA} is the identity matrix when $t = 0$. Also note that

$$\frac{d}{dt} \begin{pmatrix} e^{2t}\cos t & -e^{2t}\sin t \\ e^{2t}\sin t & e^{2t}\cos t \end{pmatrix} = e^{2t} \begin{pmatrix} 2\cos t - \sin t & -\cos t - \sin t \\ \cos t + 2\sin t & 2\cos t - \sin t \end{pmatrix},$$

so that evaluation at $t = 0$ gives the original matrix $A = \begin{pmatrix} 2 & -1 \\ 1 & 2 \end{pmatrix}$.

15. Given $A = \begin{pmatrix} 4 & 5 \\ 3 & 4 \end{pmatrix}$, the computation of e^{tA} using Equation 2.2 in the text requires an eigenvalue matrix Λ_t and an invertible eigenvector matrix U.

First, the eigenvalues λ of A satisfy

$$\det(A - \lambda I) = \begin{vmatrix} 4 - \lambda & 5 \\ 3 & 4 - \lambda \end{vmatrix} = (4 - \lambda)^2 - 15 = 0.$$

Hence, $\lambda_1 = 4 + \sqrt{15}$, $\lambda_2 = 4 - \sqrt{15}$ and an eigenvalue matrix is

$$\Lambda_t = \begin{pmatrix} e^{(4+\sqrt{15})t} & 0 \\ 0 & e^{(4-\sqrt{15})t} \end{pmatrix} = e^{4t} \begin{pmatrix} e^{\sqrt{15}\,t} & 0 \\ 0 & e^{-\sqrt{15}\,t} \end{pmatrix}.$$

An eigenvector $\mathbf{u} = \begin{pmatrix} u \\ v \end{pmatrix}$ for $\lambda_1 = 4 + \sqrt{15}$ satisfies

$$(A - (4 + \sqrt{15})I)\mathbf{u} = \begin{pmatrix} -\sqrt{15} & 5 \\ 3 & -\sqrt{15} \end{pmatrix} \begin{pmatrix} u \\ v \end{pmatrix} = \begin{pmatrix} -\sqrt{15}\,u + 5v \\ 3u - \sqrt{15}\,v \end{pmatrix} = \begin{pmatrix} 0 \\ 0 \end{pmatrix}.$$

This gives the equivalent equations $-\sqrt{15}\,u + 5v = 0$ and $3u - \sqrt{15}\,v = 0$. Writing the first of these as $5v = \sqrt{15}\,u$, we choose $u = \sqrt{15}$, $v = 3$ to get the eigenvector $\mathbf{u} = \begin{pmatrix} \sqrt{15} \\ 3 \end{pmatrix}$.

An eigenvector $\mathbf{v} = \begin{pmatrix} u \\ v \end{pmatrix}$ for $\lambda_2 = 4 - \sqrt{15}$ satisfies

$$(A - (4 - \sqrt{15})I)\mathbf{v} = \begin{pmatrix} \sqrt{15} & 5 \\ 3 & \sqrt{15} \end{pmatrix} \begin{pmatrix} u \\ v \end{pmatrix} = \begin{pmatrix} \sqrt{15}\,u + 5v \\ 3u + \sqrt{15}\,v \end{pmatrix} = \begin{pmatrix} 0 \\ 0 \end{pmatrix}.$$

This gives the equivalent equations $\sqrt{15}\,u + 5v = 0$ and $3u + \sqrt{15}\,v = 0$. Writing the first of these as $\sqrt{15}\,u = -5v$, we choose $u = -5$, $v = \sqrt{15}$ to get the eigenvector $\mathbf{v} = \begin{pmatrix} -5 \\ \sqrt{15} \end{pmatrix}$.

Now let $\mathbf{u}$ and $\mathbf{v}$ be the columns of the eigenvector matrix U, so that

$$U = \begin{pmatrix} \sqrt{15} & -5 \\ 3 & \sqrt{15} \end{pmatrix}.$$

With $U^{-1} = \frac{1}{30} \begin{pmatrix} \sqrt{15} & 5 \\ -3 & \sqrt{15} \end{pmatrix}$, we can now use Equation 2.2 to get

$$e^{tA} = U\Lambda_t U^{-1} = \frac{1}{30} e^{4t} \begin{pmatrix} \sqrt{15} & -5 \\ 3 & \sqrt{15} \end{pmatrix} \begin{pmatrix} e^{\sqrt{15}\,t} & 0 \\ 0 & e^{-\sqrt{14}\,t} \end{pmatrix} \begin{pmatrix} \sqrt{15} & 5 \\ -3 & \sqrt{15} \end{pmatrix}$$

$$= \frac{1}{30} e^{4t} \begin{pmatrix} 15(e^{\sqrt{15}\,t} + e^{-\sqrt{15}\,t}) & 5\sqrt{15}(e^{\sqrt{15}\,t} - e^{-\sqrt{15}\,t}) \\ 3\sqrt{15}(e^{\sqrt{15}\,t} - e^{-\sqrt{15}\,t}) & 15(e^{\sqrt{15}\,t} + e^{-\sqrt{15}\,t}) \end{pmatrix}$$

$$= \frac{1}{30} e^{4t} \begin{pmatrix} 30\cosh\sqrt{15}\,t & 10\sqrt{15}\sinh\sqrt{15}\,t \\ 6\sqrt{15}\sinh\sqrt{15}\,t & 30\cosh\sqrt{15}\,t \end{pmatrix}$$

$$= \frac{1}{15} \begin{pmatrix} 15e^{4t}\cosh\sqrt{15}\,t & 5\sqrt{15}e^{4t}\sinh\sqrt{15}\,t \\ 3\sqrt{15}e^{4t}\sinh\sqrt{15}\,t & 15e^{4t}\cosh\sqrt{15}\,t \end{pmatrix},$$

where the fraction $\frac{1}{15}$ is left outside of the last matrix to avoid displaying any fractional entries.

The inverse of e^{-tA} is found by simply replacing t with $-t$ in the formula above. Specifically, use the fact that the hyperbolic sine function is odd and the hyperbolic cosine is even to obtain

$$e^{-tA} = \frac{1}{15} \begin{pmatrix} 15e^{-4t}\cosh\sqrt{15}\,t & -5\sqrt{15}e^{-4t}\sinh\sqrt{15}\,t \\ -3\sqrt{15}e^{-4t}\sinh\sqrt{15}\,t & 15e^{-4t}\cosh\sqrt{15}\,t \end{pmatrix}.$$

As a check on our original computation, note that both sides of the formula found for e^{tA} is the identity matrix when $t = 0$. Also note that

$$\frac{d}{dt}\left[\frac{1}{30}e^{4t} \begin{pmatrix} 30\cosh\sqrt{15}\,t & 10\sqrt{15}\sinh\sqrt{15}\,t \\ 6\sqrt{15}\sinh\sqrt{15}\,t & 30\cosh\sqrt{15}\,t \end{pmatrix} \right]$$
$$= \frac{1}{15}e^{4t} \begin{pmatrix} 15\sqrt{15}\sinh\sqrt{15}\,t + 60\cosh\sqrt{15}\,t & 20\sqrt{15}\sinh\sqrt{15}\,t + 75\cosh\sqrt{15}\,t \\ 12\sqrt{15}\sinh\sqrt{15}\,t + 45\cosh\sqrt{15}\,t & 15\sqrt{15}\sinh\sqrt{15}\,t + 60\cosh\sqrt{15}\,t \end{pmatrix},$$

so that evaluation at $t = 0$ gives the original matrix $A = \begin{pmatrix} 4 & 5 \\ 3 & 4 \end{pmatrix}$.

17. Given $A = \begin{pmatrix} 0 & 1 \\ -6 & 5 \end{pmatrix}$, the computation of e^{tA} requires an eigenvalue matrix Λ_t and an invertible eigenvector matrix U.

An eigenvalue λ of A satisfies

$$\det(A - \lambda I) = \begin{vmatrix} -\lambda & 1 \\ -6 & 5 - \lambda \end{vmatrix} = -\lambda(5 - \lambda) + 6 = (\lambda - 2)(\lambda - 3) = 0.$$

So the eigenvalues of A are $\lambda_1 = 2$, $\lambda_2 = 3$ and an eigenvalue matrix is

$$\Lambda_t = \begin{pmatrix} e^{2t} & 0 \\ 0 & e^{3t} \end{pmatrix}.$$

An eigenvector $\mathbf{u} = \begin{pmatrix} u \\ v \end{pmatrix}$ for $\lambda_1 = 2$ satisfies

$$(A - 2I)\mathbf{u} = \begin{pmatrix} -2 & 1 \\ -6 & 3 \end{pmatrix} \begin{pmatrix} u \\ v \end{pmatrix} = \begin{pmatrix} -2u + v \\ -6u + 3v \end{pmatrix} = (-2u + v)\begin{pmatrix} 1 \\ 3 \end{pmatrix} = \begin{pmatrix} 0 \\ 0 \end{pmatrix}.$$

This implies $-2u + v = 0$, or equivalently, $v = 2u$. Thus, with $u = 1$, $v = 2$ we get the eigenvector $\mathbf{u} = \begin{pmatrix} 1 \\ 2 \end{pmatrix}$.

An eigenvector $\mathbf{v} = \begin{pmatrix} u \\ v \end{pmatrix}$ for $\lambda_2 = 3$ satisfies

$$(A - 3I)\mathbf{v} = \begin{pmatrix} -3 & 1 \\ -6 & 2 \end{pmatrix} \begin{pmatrix} u \\ v \end{pmatrix} = \begin{pmatrix} -3u + v \\ -6u + 2v \end{pmatrix} = (-3u + v)\begin{pmatrix} 1 \\ 2 \end{pmatrix} = \begin{pmatrix} 0 \\ 0 \end{pmatrix}.$$

This implies $-3u + v = 0$, or equivalently, $v = 3u$. Thus, with $u = 1$, $v = 3$ we get the eigenvector $\mathbf{v} = \begin{pmatrix} 1 \\ 3 \end{pmatrix}$.

Now let $\mathbf{u}$ and $\mathbf{v}$ be the columns of the eigenvector matrix U, so that

$$U = \begin{pmatrix} 1 & 1 \\ 2 & 3 \end{pmatrix}.$$

With $U^{-1} = \begin{pmatrix} 3 & -1 \\ -2 & 1 \end{pmatrix}$, we can now use Equation 2.2 to get

$$e^{tA} = U\Lambda_t U^{-1} = \begin{pmatrix} 1 & 1 \\ 2 & 3 \end{pmatrix} \begin{pmatrix} e^{2t} & 0 \\ 0 & e^{3t} \end{pmatrix} \begin{pmatrix} 3 & -1 \\ -2 & 1 \end{pmatrix} = \begin{pmatrix} 3e^{2t} - 2e^{3t} & -e^{2t} + e^{3t} \\ 6e^{2t} - 6e^{3t} & -2e^{2t} + 3e^{3t} \end{pmatrix}.$$

The inverse matrix e^{-tA} is found by replacing t with $-t$ in the above formula. Specifically,

$$e^{-tA} = \begin{pmatrix} 3e^{-2t} - 2e^{-3t} & -e^{-2t} + e^{-3t} \\ 6e^{-2t} - 6e^{-3t} & -2e^{-2t} + 3e^{-3t} \end{pmatrix}.$$

As a check on our original computation, note that both sides of the formula found for e^{tA} is the identity matrix when $t = 0$. Also note that

$$\frac{d}{dt} \begin{pmatrix} 3e^{2t} - 2e^{3t} & -e^{2t} + e^{3t} \\ 6e^{2t} - 6e^{3t} & -2e^{2t} + 3e^{3t} \end{pmatrix} = \begin{pmatrix} 6e^{2t} - 6e^{3t} & -2e^{2t} + 3e^{3t} \\ 12e^{2t} - 16e^{3t} & -4e^{2t} + 9e^{3t} \end{pmatrix},$$

so that evaluation at $t = 0$ gives the original matrix $A = \begin{pmatrix} 0 & 1 \\ -6 & 5 \end{pmatrix}$.

19. Given $A = \begin{pmatrix} -1 & 0 & 0 \\ 0 & \frac{3}{2} & -\frac{1}{2} \\ 0 & -\frac{1}{2} & \frac{3}{2} \end{pmatrix}$, the computation of e^{tA} requires an eigenvalue matrix Λ_t and an invertible eigenvector matrix U.

First, the eigenvalues λ of A satisfy

$$\det(A - \lambda I) = \begin{vmatrix} -1 - \lambda & 0 & 0 \\ 0 & \frac{3}{2} - \lambda & -\frac{1}{2} \\ 0 & -\frac{1}{2} & \frac{3}{2} - \lambda \end{vmatrix} = (-1 - \lambda) \begin{vmatrix} \frac{3}{2} - \lambda & -\frac{1}{2} \\ -\frac{1}{2} & \frac{3}{2} - \lambda \end{vmatrix}$$

$$= (-1 - \lambda) \left[\left(\frac{3}{2} - \lambda \right)^2 - \frac{1}{4} \right] = (-1 - \lambda)(1 - \lambda)(2 - \lambda) = 0.$$

So the eigenvalues of A are $\lambda_1 = 1$, $\lambda_2 = -1$ and $\lambda_3 = 2$, and an eigenvalue matrix is

$$\Lambda_t = \begin{pmatrix} e^t & 0 & 0 \\ 0 & e^{-t} & 0 \\ 0 & 0 & e^{2t} \end{pmatrix}.$$

An eigenvector $\mathbf{u} = \begin{pmatrix} u \\ v \\ w \end{pmatrix}$ for $\lambda_1 = 1$ satisfies

$$(A - I)\mathbf{u} = \begin{pmatrix} -2 & 0 & 0 \\ 0 & \frac{1}{2} & -\frac{1}{2} \\ 0 & -\frac{1}{2} & \frac{1}{2} \end{pmatrix} \begin{pmatrix} u \\ v \\ w \end{pmatrix} = \begin{pmatrix} -2u \\ \frac{1}{2}v - \frac{1}{2}w \\ -\frac{1}{2}v + \frac{1}{2}w \end{pmatrix} = \begin{pmatrix} 0 \\ 0 \\ 0 \end{pmatrix}.$$

This gives the equations $-2u = 0$, $\frac{1}{2}v - \frac{1}{2}w = 0$ and $-\frac{1}{2}v + \frac{1}{2}w = 0$. The first of these gives $u = 0$. The last two are both equivalent to the single relation $v = w$. Thus, choosing $v = w = 1$ gives the eigenvector $\mathbf{u} = \begin{pmatrix} 0 \\ 1 \\ 1 \end{pmatrix}$.

An eigenvector $\mathbf{v} = \begin{pmatrix} u \\ v \\ w \end{pmatrix}$ for $\lambda_2 = -1$ satisfies

$$(A + I)\mathbf{v} = \begin{pmatrix} 0 & 0 & 0 \\ 0 & \frac{5}{2} & -\frac{1}{2} \\ 0 & -\frac{1}{2} & \frac{5}{2} \end{pmatrix} \begin{pmatrix} u \\ v \\ w \end{pmatrix} = \begin{pmatrix} 0 \\ \frac{5}{2}v - \frac{1}{2}w \\ -\frac{1}{2}v + \frac{5}{2}w \end{pmatrix} = \begin{pmatrix} 0 \\ 0 \\ 0 \end{pmatrix}.$$

This gives the two independent equations $\frac{5}{2}v - \frac{1}{2}w = 0$ and $-\frac{1}{2}v + \frac{5}{2}w = 0$, which are equivalent to $v = 5w$ and $w = 5v$. These have the unique solution $v = w = 0$. Since u is arbitrary, we choose $u = 1$ to get the eigenvector $\mathbf{v} = \begin{pmatrix} 1 \\ 0 \\ 0 \end{pmatrix}$.

An eigenvector $\mathbf{w} = \begin{pmatrix} u \\ v \\ w \end{pmatrix}$ for $\lambda_3 = 2$ satisfies

$$(A - 2I)\mathbf{w} = \begin{pmatrix} -3 & 0 & 0 \\ 0 & -\frac{1}{2} & -\frac{1}{2} \\ 0 & -\frac{1}{2} & -\frac{1}{2} \end{pmatrix} \begin{pmatrix} u \\ v \\ w \end{pmatrix} = \begin{pmatrix} -3u \\ -\frac{1}{2}v - \frac{1}{2}w \\ -\frac{1}{2}v - \frac{1}{2}w \end{pmatrix} = \begin{pmatrix} 0 \\ 0 \\ 0 \end{pmatrix}.$$

This gives the two independent equations $-3u = 0$ and $-\frac{1}{2}v - \frac{1}{2}w = 0$. The first implies $u = 0$ and the second implies $v = -w$. Thus, choosing $v = 1$, $w = -1$ gives the eigenvector $\mathbf{w} = \begin{pmatrix} 0 \\ 1 \\ -1 \end{pmatrix}$.

Now let $\mathbf{u}$, $\mathbf{v}$ and $\mathbf{w}$ be the columns of the eigenvector matrix U, so that

$$U = \begin{pmatrix} 0 & 1 & 0 \\ 1 & 0 & 1 \\ 1 & 0 & -1 \end{pmatrix}.$$

With $U^{-1} = \begin{pmatrix} 0 & \frac{1}{2} & \frac{1}{2} \\ 1 & 0 & 0 \\ 0 & \frac{1}{2} & -\frac{1}{2} \end{pmatrix}$, we can now use Equation 2.2 to get

$$e^{tA} = U\Lambda_t U^{-1} = \begin{pmatrix} 0 & 1 & 0 \\ 1 & 0 & 1 \\ 1 & 0 & -1 \end{pmatrix} \begin{pmatrix} e^t & 0 & 0 \\ 0 & e^{-t} & 0 \\ 0 & 0 & e^{2t} \end{pmatrix} \begin{pmatrix} 0 & \frac{1}{2} & \frac{1}{2} \\ 1 & 0 & 0 \\ 0 & \frac{1}{2} & -\frac{1}{2} \end{pmatrix}$$

$$= \frac{1}{2} \begin{pmatrix} 2e^{-t} & 0 & 0 \\ 0 & e^t + e^{2t} & e^t - e^{2t} \\ 0 & e^t - e^{2t} & e^t + e^{2t} \end{pmatrix},$$

509

where $\frac{1}{2}$ is left outside the final matrix to avoid fractional entries. The inverse of e^{tA} is found by replacing t with $-t$ in the above matrix. Specifically,

$$e^{-tA} = \frac{1}{2} \begin{pmatrix} 2e^t & 0 & 0 \\ 0 & e^{-t} + e^{-2t} & e^{-t} - e^{-2t} \\ 0 & e^{-t} - e^{-2t} & e^{-t} + e^{-2t} \end{pmatrix}.$$

As a check on our original computation, note that both sides of the formula found for e^{tA} is the identity matrix when $t = 0$. Also note that

$$\frac{d}{dt}\left[\frac{1}{2} \begin{pmatrix} 2e^{-t} & 0 & 0 \\ 0 & e^t + e^{2t} & e^t - e^{2t} \\ 0 & e^t - e^{2t} & e^t + e^{2t} \end{pmatrix} \right] = \frac{1}{2} \begin{pmatrix} -2e^{-t} & 0 & 0 \\ 0 & e^t + 2e^{2t} & e^t - 2e^{2t} \\ 0 & e^t - 2e^{2t} & e^t + 2e^{2t} \end{pmatrix},$$

so that evaluation at $t = 0$ gives the original matrix A.

21. The real and imaginary parts of the series $\sum_{k=0}^{\infty}(i)^k \frac{t^k}{k!} A^k$ correspond to the sum of the even terms and the sum of the odd terms, respectively. Hence,

$$\cos tA = \sum_{j=0}^{\infty}(-1)^j \frac{t^{2j}}{(2j)!} A^{2j} \qquad \text{and} \qquad \sin tA = \sum_{j=0}^{\infty}(-1)^j \frac{t^{2j+1}}{(2j+1)!} A^{2j+1}.$$

(a) Using the series for $\cos tA$ and $\sin tA$ given in part (a), we have

$$\cos(-tA) = \cos((-t)A) = \sum_{j=0}^{\infty}(-1)^j \frac{(-t)^{2j}}{(2j)!} A^{2j} = \sum_{j=0}^{\infty}(-1)^j \frac{t^{2j}}{(2j)!} A^{2j} = \cos tA,$$

$$\sin(-tA) = \sin((-t)A) = \sum_{j=0}^{\infty}(-1)^j \frac{(-t)^{2j+1}}{(2j+1)!} A^{2j+1} = -\sum_{j=0}^{\infty}(-1)^j \frac{t^{2j+1}}{(2j+1)!} A^{2j+1}$$

$$= -\sin tA,$$

where $(-1)^{2j} = 1$ for all j and $(-1)^{2j+1} = -1$ for all j is used along with the fact that -1 can be factored out of each term of the second series.
(b) By part (a),

$$e^{-itA} = \cos(-tA) + i\sin(-tA) = \cos tA - i\sin tA.$$

Adding this and its negative to the equation $e^{itA} = \cos tA + i\sin tA$ allows us to solve for $\cos tA$ and $\sin tA$:

$$\cos tA = \frac{1}{2}(e^{itA} + e^{-itA}) \qquad \text{and} \qquad \sin tA = -\frac{i}{2}(e^{itA} - e^{-itA}).$$

Using part (c) of Theorem 2.1, we can differentiate the right sides of these equations. The results are

$$\frac{d}{dt}\cos tA = \frac{1}{2}(iAe^{itA} - iAe^{-itA}) = -A\left(-\frac{i}{2}(e^{itA} - e^{-itA})\right) = -A\sin tA,$$

$$\frac{d}{dt}\sin tA = -\frac{i}{2}(iAe^{itA} - (-iA)\sin tA) = A\left(\frac{1}{2}(e^{itA} + e^{-itA})\right) = A\cos tA.$$

510

(c) Using the formulas for $\cos tA$ and $\sin tA$ derived in part (b), we can directly compute the sums of the squares of these matrix functions. In doing so, we use the fact all powers of e^{itA} commute with each other as well as with every power of e^{-itA}, and the fact that $e^{itA}e^{-itA} = I$, where I is the identity matrix of the proper dimension. With this in mind,

$$(\cos tA)^2 + (\sin tA)^2 = \left(\frac{1}{2}(e^{itA} + e^{-itA})\right)^2 + \left(-\frac{i}{2}(e^{itA} - e^{-itA})\right)^2$$

$$= \frac{1}{4}(e^{2itA} + 2I + e^{-2itA}) - \frac{1}{4}(e^{2itA} - 2I + e^{-2itA}) = \frac{1}{2}I + \frac{1}{2}I = I.$$

23. Let A be an n-by-n matrix and consider the family of vector-valued function

$$\mathbf{x}(t) = (\cos tA)\mathbf{c}_1 + (\sin tA)\mathbf{c}_2,$$

where $\mathbf{c}_1$, $\mathbf{c}_2$ range over all pairs of constant n-dimensional column vectors. By part (b) of Exercise 21 in this section, $\dot{\mathbf{x}}(t)$ and $\ddot{\mathbf{x}}(t)$ both exist for all t and, using the formulas for their derivatives, we have

$$\frac{d^2\mathbf{x}(t)}{dt^2} = \frac{d}{dt}\left(\frac{d\mathbf{x}(t)}{dt}\right) = \frac{d}{dt}\left((-A\sin tA)\mathbf{c}_1 + (A\cos tA)\mathbf{c}_2\right)$$

$$= (-A^2\cos tA)\mathbf{c}_1 + (-A^2\sin tA)\mathbf{c}_2 = -A^2\left((\cos tA)\mathbf{c}_1 + (\sin tA)\mathbf{c}_2\right)$$

$$= -A^2\mathbf{x}(t).$$

Equality of the outermost members is equivalent to $\ddot{\mathbf{x}}(t) + A^2\mathbf{x}(t) = \mathbf{0}$. So $\mathbf{x} = \mathbf{x}(t)$ satisfies the system $\ddot{\mathbf{x}} + A^2\mathbf{x} = \mathbf{0}$. Moreover, this is always the most general solution.

25. Let $A = \begin{pmatrix} 1 & 0 \\ 1 & 0 \end{pmatrix}$ and $B = \begin{pmatrix} 0 & 1 \\ 0 & 1 \end{pmatrix}$. A simple calculation shows that $AB = B$ and $BA = A$, so that A and B do not commute. Furthermore, since $A^2 = A$ and $B^2 = B$, it inductively follows that $A^k = A$ for $k \geq 1$ and $B^k = B$ for $k \geq 1$. The definitions of e^A and e^B then give

$$e^B = \sum_{k=0}^{\infty} \frac{A^k}{k!} = I + \sum_{k=1}^{\infty} \frac{1}{k!}A = I + (e-1)A = \begin{pmatrix} e & 0 \\ e-1 & 1 \end{pmatrix},$$

$$e^B = \sum_{k=0}^{\infty} \frac{B^k}{k!} = I + \sum_{k=1}^{\infty} \frac{1}{k!}B = I + (e-1)B = \begin{pmatrix} 1 & e-1 \\ 0 & e \end{pmatrix}.$$

Hence,

$$e^A e^B = \begin{pmatrix} e & 0 \\ e-1 & 1 \end{pmatrix}\begin{pmatrix} 1 & e-1 \\ 0 & e \end{pmatrix} = \begin{pmatrix} e & e^2-e \\ e-1 & e^2-e+1 \end{pmatrix}.$$

On the other hand, $A + B = \begin{pmatrix} 1 & 1 \\ 1 & 1 \end{pmatrix}$. By part (a) of Exercise 24 in this section, $(A+B)^k = 2^{k-1}(A+B)$ for $k \geq 1$. Hence,

$$e^{A+B} = \sum_{k=0}^{\infty} \frac{(A+B)^k}{k!} = I + \sum_{k=1}^{\infty} \frac{2^{k-1}}{k!}(A+B) = I + \frac{1}{2}\sum_{k=1}^{\infty} \frac{2^k}{k!}(A+B)$$

$$= I + \frac{1}{2}(e^2-1)(A+B) = \frac{1}{2}\begin{pmatrix} e^2+1 & e^2-1 \\ e^2-1 & e^2+1 \end{pmatrix}.$$

Manifestly, $e^A e^B \neq e^{A+B}$.

Exercise Set 2D (pgs. 636-637)

1. Let $A = \begin{pmatrix} 8 & -3 \\ 10 & -3 \end{pmatrix}$. Solving $\det(A - \lambda I) = \lambda^2 - 5\lambda + 6 = 0$ for λ gives the eigenvalues $\lambda_1 = 2$ and $\lambda_2 = 3$. With $n = 2$, Theorem 2.4 says that e^{2t} and e^{3t} can be written as

$$e^{2t} = b_0 + 2b_1, \qquad e^{3t} = b_0 + 3b_1,$$

for some functions $b_0 = b_0(t)$, $b_1 = b_1(t)$. Solving these equations for b_0 and b_1 gives

$$b_0 = 3e^{2t} - 2e^{3t}, \qquad b_1 = -e^{2t} + e^{3t}.$$

Since the pair b_0, b_1 also satisfies $e^{tA} = b_0 I + b_1 A$, we have

$$e^{tA} = b_0 I + b_1 A = (3e^{2t} - 2e^{3t}) \begin{pmatrix} 1 & 0 \\ 0 & 1 \end{pmatrix} + (-e^{2t} + e^{3t}) \begin{pmatrix} 8 & -3 \\ 10 & -3 \end{pmatrix}$$

$$= \begin{pmatrix} -5e^{2t} + 6e^{3t} & 3e^{2t} - 3e^{3t} \\ -10e^{2t} + 10e^{3t} & 6e^{2t} - 5e^{3t} \end{pmatrix}.$$

The solution of $\dot{\mathbf{x}} = A\mathbf{x}$ satisfying the initial condition $\mathbf{x}(0) = \begin{pmatrix} 2 \\ -3 \end{pmatrix}$ is then

$$\mathbf{x}(t) = e^{tA}\mathbf{x}(0) = \begin{pmatrix} -5e^{2t} + 6e^{3t} & 3e^{2t} - 3e^{3t} \\ -10e^{2t} + 10e^{3t} & 6e^{2t} - 5e^{3t} \end{pmatrix} \begin{pmatrix} 2 \\ -3 \end{pmatrix} = \begin{pmatrix} -19e^{2t} + 21e^{3t} \\ -38e^{2t} + 35e^{3t} \end{pmatrix}.$$

3. Let $A = \begin{pmatrix} 1 & 1 & 2 \\ 0 & 1 & -1 \\ 0 & 0 & 2 \end{pmatrix}$. By inspection, $\det(A - \lambda I) = (1 - \lambda)^2(2 - \lambda)$, so that the eigenvalues of A are $\lambda_1 = 1$ and $\lambda_2 = 2$. With $n = 3$, Theorem, 2.4 immediately gives the two equations

$$e^t = b_0 + b_1 + b_2, \qquad e^{2t} = b_0 + 2b_1 + 4b_2,$$

for some functions $b_0 = b_0(t)$, $b_1 = b_1(t)$, $b_2 = b_2(t)$. Since $\lambda_1 = 1$ is a double eigenvalue, Theorem 2.4 also provides a third equation. Specifically, the equation obtained by differentiating $e^{t\lambda} = b_0 + b_1\lambda + b_2\lambda^2$ with respect to λ and then evaluating the result at $\lambda = \lambda_1 = 1$. When this is done, one obtains $te^t = b_1 + 2b_2$. The resulting three-equation system can then be solved for b_0, b_1, b_2 by elimination to compute

$$b_0 = e^{2t} - 2te^t, \qquad b_1 = 2e^t - 2e^{2t} + 3te^t, \qquad b_2 = -e^t + e^{2t} - te^t.$$

With $A^2 = \begin{pmatrix} 1 & 2 & 5 \\ 0 & 1 & -3 \\ 0 & 0 & 4 \end{pmatrix}$, we can use the equation $e^{tA} = b_0 I + b_1 A + b_2 A^2$ to get

$$e^{tA} = b_0 I + b_1 A + b_2 A^2 = \begin{pmatrix} e^t & te^t & -e^t + e^{2t} + te^t \\ 0 & e^t & e^t - e^{2t} - 2te^t \\ 0 & 0 & e^{2t} \end{pmatrix}.$$

The solution of $\dot{\mathbf{x}} = A\mathbf{x}$ satisfying the initial condition $\mathbf{x}(0) = \begin{pmatrix} 0 \\ 1 \\ 0 \end{pmatrix}$ is then

$$\mathbf{x}(t) = e^{tA}\mathbf{x}(0) = \begin{pmatrix} e^t & te^t & -e^t + e^{2t} + te^t \\ 0 & e^t & e^t - e^{2t} - 2te^t \\ 0 & 0 & e^{2t} \end{pmatrix} \begin{pmatrix} 0 \\ 1 \\ 0 \end{pmatrix} = \begin{pmatrix} te^t \\ e^t \\ 0 \end{pmatrix}.$$

5. For the given matrix A, we compute

$$\det(A - \lambda I) = \begin{vmatrix} 1 - \lambda & 1 & 1 & 0 \\ 0 & -\lambda & -1 & 0 \\ 1 & 1 & 2 - \lambda & 0 \\ 1 & 0 & -1 & 1 - \lambda \end{vmatrix} = -\lambda(1 - \lambda)^2(2 - \lambda).$$

The eigenvalues of A are $\lambda = 0, 1, 2$. Theorem 2.4 asserts that there exist coefficient functions $b_0 = b_0(t)$, $b_1 = b_1(t)$, $b_2 = b_2(t)$, $b_3 = b_3(t)$ such that

$$e^{\lambda t} = b_0 + b_1\lambda + b_2\lambda^2 + b_3\lambda^3, \quad \text{for } \lambda = 0,\ 1,\ 2.$$

Since $\lambda = 1$ is a double eigenvalue, Theorem 2.4 also asserts that these coefficient functions also satisfy a fourth equation, found by differentiating the above equation with respect to λ and then evaluating the result at $\lambda = 1$. The four equations obtained are

$$1 = b_0,$$
$$e^t = b_0 + b_1 + b_2 + b_3,$$
$$e^{2t} = b_0 + 2b_1 + 4b_2 + 8b_3,$$
$$te^t = b_1 + 2b_2 + 3b_3.$$

When the b_i are solved for in terms of e^t, e^{2t}, te^t, the result is

$$b_0(t) = 1, \quad b_1(t) = 2e^t + \frac{1}{2}e^{2t} - 2te^t - \frac{5}{2}, \quad b_2 = -e^t - e^{2t} + 3te^t + 2, \quad b_3 = \frac{1}{2}e^{2t} - te^t - \frac{1}{2}.$$

We also have

$$A^2 = \begin{pmatrix} 2 & 2 & 2 & 0 \\ -1 & -1 & -2 & 0 \\ 3 & 3 & 4 & 0 \\ 1 & 0 & -2 & 1 \end{pmatrix} \quad \text{and} \quad A^3 = \begin{pmatrix} 4 & 4 & 4 & 0 \\ -3 & -3 & -4 & 0 \\ 7 & 7 & 8 & 0 \\ 0 & -1 & -4 & 1 \end{pmatrix}.$$

The computation of $e^{tA} = b_0 I + b_1 A + b_2 A^2 + b_3 A^3$ is straightforward but algebraically tedious. So, without showing the details,

$$e^{tA} = \begin{pmatrix} \frac{1}{2}(e^{2t} + 1) & \frac{1}{2}(e^{2t} - 1) & \frac{1}{2}(e^{2t} - 1) & 0 \\ e^t - \frac{1}{2}(e^{2t} + 1) & e^t - \frac{1}{2}(e^{2t} + 1) & -\frac{1}{2}(e^{2t} - 1) & 0 \\ -e^t + e^{2t} & -e^t + e^{2t} & e^{2t} & 0 \\ e^t + te^t - \frac{1}{2}(e^{2t} + 1) & te^t - \frac{1}{2}(e^{2t} - 1) & -\frac{1}{2}(e^{2t} - 1) & e^t + te^t - \frac{1}{2}(e^{2t} - 1) \end{pmatrix}.$$

Using $\frac{1}{2}(e^{2t}+1) = \frac{1}{2}e^t(e^t+e^{-t}) = e^t\cosh t$ and $\frac{1}{2}(e^{2t}-1) = \frac{1}{2}e^t(e^t-e^{-t}) = e^t\sinh t$, the matrix for e^{tA} can be simplified to

$$e^{tA} = e^t \begin{pmatrix} \cosh t & \sinh t & \sinh t & 0 \\ 1-\cosh t & 1-\sinh t & -\sinh t & 0 \\ -1+e^t & -1+e^t & e^t & 0 \\ 1+t-\cosh t & t-\sinh t & -\sinh t & 1+t-\sinh t \end{pmatrix}.$$

Due to the large amount of algebra required to derive this exponential matrix, the reader should check to see if the derivative of the right side is equal to A when $t=0$.

7. The given system in matrix form is

$$\dot{\mathbf{x}} = \begin{pmatrix} 2 & 0 & 1 \\ -1 & 3 & 1 \\ -1 & 0 & 4 \end{pmatrix} \mathbf{x}.$$

Defining A to be the above coefficient matrix, we see that $\det(A-\lambda I) = (3-\lambda)^3$ and $\lambda=3$ is a triple eigenvalue of A. Theorem 2.4 says that the coefficient functions b_0, b_1, b_2 satisfying $e^{tA} = b_0I + b_1A + b_2A^2$ also satisfy the eigenvalue equation $e^{\lambda t} = b_0 + b_1\lambda + b_2\lambda^2$ whenever λ is an eigenvalue of A. If λ happens to be a triple eigenvalue, it also satisfies $te^{\lambda t} = b_1 + 2b_2\lambda$ and $t^2e^{\lambda t} = 2b_2$. In particular, if $\lambda=3$ then we obtain the three equations

$$e^{3t} = b_0 + 3b_1 + 9b_2, \qquad te^{3t} = b_1 + 6b_2, \qquad t^2e^{3t} = 2b_2.$$

Solving these for b_0, b_1 and b_2 gives

$$b_0(t) = \frac{1}{2}(2 - 6t + 9t^2)e^{3t}, \qquad b_1(t) = (t - 3t^2)e^{3t}, \qquad b_2(t) = \frac{1}{2}t^2e^{3t}.$$

Since $A^2 = \begin{pmatrix} 3 & 0 & 6 \\ -6 & 9 & 6 \\ -6 & 0 & 15 \end{pmatrix}$, we obtain (without showing the tedious details)

$$e^{3t} = b_0I + b_1A + b_2A^2 = e^{3t}\begin{pmatrix} 1-t & 0 & t \\ -t & 1 & t \\ -t & 0 & 1+t \end{pmatrix}.$$

The vector general solution $\mathbf{x} = \mathbf{x}(t)$ is therefore given by $\mathbf{x}(t) = e^{tA}\mathbf{c}$, where $\mathbf{c} = \begin{pmatrix} c_1 \\ c_2 \\ c_3 \end{pmatrix}$ is a constant vector in $\mathcal{R}^3$. Setting $\mathbf{x}(t) = \begin{pmatrix} x \\ y \\ z \end{pmatrix}$, we have

$$\begin{pmatrix} x \\ y \\ z \end{pmatrix} = e^{3t}\begin{pmatrix} 1-t & 0 & t \\ -t & 1 & t \\ -t & 0 & 1+t \end{pmatrix}\begin{pmatrix} c_1 \\ c_2 \\ c_3 \end{pmatrix} = e^{3t}\begin{pmatrix} c_1 + (-c_1 + c_3)t \\ c_2 + (-c_1 + c_3)t \\ c_3 + (-c_1 + c_3)t \end{pmatrix},$$

or in scalar form,

$$x(t) = \Big(c_1 + (-c_1 + c_3)t\Big)e^{3t}, \quad y(t) = \Big(c_2 + (-c_1 + c_3)t\Big)e^{3t}, \quad z(t) = \Big(c_3 + (-c_1 + c_3)t\Big)e^{3t}.$$

9. If $A = \begin{pmatrix} \alpha & 1 \\ 0 & \beta \end{pmatrix}$ then

$$\det(A - \lambda I) = \begin{vmatrix} \alpha - \lambda & 1 \\ 0 & \beta - \lambda \end{vmatrix} = (\alpha - \lambda)(\beta - \lambda),$$

so that α and β are the eigenvalues of A.

(a) Assume $\alpha \neq \beta$ and that $\mathbf{u}$ and $\mathbf{v}$ are eigenvectors for α and β, respectively. If $\mathbf{u}$ and $\mathbf{v}$ are not linearly independent then $\mathbf{u} = k\mathbf{v}$ for some nonzero scalar k. Hence,

$$(A - \alpha I)\mathbf{u} = \big(A - (\beta + \alpha - \beta)I\big)(k\mathbf{v}) = k(A - \beta I)\mathbf{v} + k(\beta - \alpha)I\mathbf{v} = k(A - \beta I)\mathbf{v} + k(\beta - \alpha)\mathbf{v}.$$

Since $(A - \alpha I)\mathbf{u} = \mathbf{0}$ and $(A - \beta I)\mathbf{v} = \mathbf{0}$, equality of the two outer members reduces to $\mathbf{0} = k(\beta - \alpha)\mathbf{v}$. This is impossible because neither of the scalars k or $\beta - \alpha$ is zero and $\mathbf{v}$ is not the zero vector (because it is an eigenvector). This contradiction shows that $\mathbf{u}$ and $\mathbf{v}$ are linearly independent.

(b) If $\alpha = \beta$ then α is the only eigenvalue of $A = \begin{pmatrix} \alpha & 1 \\ 0 & \alpha \end{pmatrix}$. Therefore, the eigenvectors of A are precisely the eigenvectors for α. Every eigenvector $\mathbf{u} = \begin{pmatrix} u \\ v \end{pmatrix}$ of α satisfies

$$(A - \alpha I)\mathbf{u} = \begin{pmatrix} 0 & 1 \\ 0 & 0 \end{pmatrix}\begin{pmatrix} u \\ v \end{pmatrix} = \begin{pmatrix} v \\ 0 \end{pmatrix} = v\begin{pmatrix} 1 \\ 0 \end{pmatrix} = \begin{pmatrix} 0 \\ 0 \end{pmatrix}.$$

This implies $v = 0$. So all eigenvectors of α are of the form $\mathbf{u} = \begin{pmatrix} u \\ 0 \end{pmatrix}$, with $u \neq 0$. By our previous remarks, these are the eigenvectors of A.

(c) Let A be as in part (a). Then each eigenvalue has multiplicity 1. By Theorem 2.4, the coefficient functions b_0, b_1 satisfying $e^{tA} = b_0 I + b_1 A$ also satisfy $e^{\lambda t} = b_0 + b_1 \lambda$ for each eigenvalue λ. In this case, $e^{\alpha t} = b_0 + \alpha b_1$ and $e^{\beta t} = b_0 + \beta b_1$. Solving for b_0 and b_1 gives

$$b_0(t) = \frac{\beta e^{\alpha t} - \alpha e^{\beta t}}{\beta - \alpha}, \qquad b_1(t) = \frac{e^{\beta t} - e^{\alpha t}}{\beta - \alpha}.$$

Hence, using $e^{tA} = b_0 I + b_1 A$,

$$e^{tA} = \frac{\beta e^{\alpha t} - \alpha e^{\beta t}}{\beta - \alpha}\begin{pmatrix} 1 & 0 \\ 0 & 1 \end{pmatrix} + \frac{e^{\beta t} - e^{\alpha t}}{\beta - \alpha}\begin{pmatrix} \alpha & 1 \\ 0 & \beta \end{pmatrix} = \begin{pmatrix} e^{\alpha t} & (e^{\beta t} - e^{\alpha t})/(\beta - \alpha) \\ 0 & e^{\beta t} \end{pmatrix}.$$

Now let A be as in part (b). Then $\lambda = \alpha$ is an eigenvalue of A with multiplicity 2. By Theorem 2.4, the coefficient functions b_0, b_1 satisfying $e^{tA} = b_0 I + b_1 A$ also satisfy $e^{\lambda t} = b_0 + b_1 \lambda$ and $te^{\lambda t} = b_1$ for $\lambda = \alpha$. That is, $e^{\alpha t} = b_0 + \alpha b_1$ and $te^{\alpha t} = b_1$. Solving for b_0 and b_1 gives
$$b_0(t) = e^{\alpha t} - \alpha t e^{\alpha t}, \qquad b_1(t) = te^{\alpha t}.$$

Hence, using $e^{tA} = b_0 I + b_1 A$,

$$e^{tA} = (e^{\alpha t} - \alpha t e^{\alpha t})\begin{pmatrix} 1 & 0 \\ 0 & 1 \end{pmatrix} + te^{\alpha t}\begin{pmatrix} \alpha & 1 \\ 0 & \alpha \end{pmatrix} = \begin{pmatrix} e^{\alpha t} & te^{\alpha t} \\ 0 & e^{\alpha t} \end{pmatrix}.$$

11. Let I be the n-dimensional identity matrix. Then $I^k = I$ for all integers k. Using the power series definition of e^{tA} with $A = I$ gives

$$e^{tI} = \sum_{k=0}^{\infty} \frac{(tI)^k}{k!} = \sum_{k=0}^{\infty} \frac{t^k}{k!} I = \begin{pmatrix} \sum_0^{\infty}(t^k/k!) & \cdots & 0 \\ \vdots & \ddots & \vdots \\ 0 & \cdots & \sum_0^{\infty}(t^k/k!) \end{pmatrix} = \begin{pmatrix} e^t & \cdots & 0 \\ \vdots & \ddots & \vdots \\ 0 & \cdots & e^t \end{pmatrix} = e^t I.$$

13. The characteristic polynomial of the matrix $A = \begin{pmatrix} 1 & 2 \\ 3 & 4 \end{pmatrix}$ is

$$P(\lambda) = \det(A - \lambda I) = \begin{vmatrix} 1-\lambda & 2 \\ 3 & 4-\lambda \end{vmatrix} = (1-\lambda)(4-\lambda) - 6 = \lambda^2 - 5\lambda - 2.$$

By the Cayley-Hamilton Theorem, $P(A) = A^2 - 5A - 2I = 0$. Solving for A^2 gives

$$A^2 = 5A + 2I = 5\begin{pmatrix} 1 & 2 \\ 3 & 4 \end{pmatrix} + 2\begin{pmatrix} 1 & 0 \\ 0 & 1 \end{pmatrix} = \begin{pmatrix} 7 & 10 \\ 15 & 22 \end{pmatrix}.$$

Multiplying $A^2 = 5A + 5I$ by A gives $A^3 = 5A^2 + 2A$. Hence,

$$A^3 = 5A^2 + 2A = 5\begin{pmatrix} 7 & 10 \\ 15 & 22 \end{pmatrix} + 2\begin{pmatrix} 1 & 2 \\ 3 & 4 \end{pmatrix} = \begin{pmatrix} 37 & 54 \\ 81 & 118 \end{pmatrix}.$$

15. If A is the given matrix then expansion about the first row gives $\det A = 1$, so that A^{-1} exists. Moreover, The characteristic polynomial of A is

$$P(\lambda) = \det(A - \lambda I) = \begin{vmatrix} 1-\lambda & 0 & 0 \\ 3 & 1-\lambda & 5 \\ -2 & 0 & 1-\lambda \end{vmatrix} = (1-\lambda)\begin{vmatrix} 1-\lambda & 5 \\ 0 & 1-\lambda \end{vmatrix}$$

$$= (1-\lambda)(1-\lambda)^2 = -\lambda^3 + 3\lambda^2 - 3\lambda + 1.$$

By the Cayley-Hamilton Theorem, $P(A) = 0$, so that $-A^3 + 3A^2 - 3A + I = 0$. Multiply by A^{-1} to get $-A^2 + 3A - 3I + A^{-1} = 0$, and then solve for A^{-1} to get $A^{-1} = A^2 - 3A + 3I$. Using the fact that $A^2 = \begin{pmatrix} 1 & 0 & 0 \\ -4 & 1 & 10 \\ -4 & 0 & 1 \end{pmatrix}$, we have

$$A^{-1} = \begin{pmatrix} 1 & 0 & 0 \\ -4 & 1 & 10 \\ -4 & 0 & 1 \end{pmatrix} - 3\begin{pmatrix} 1 & 0 & 0 \\ 3 & 1 & 5 \\ -2 & 0 & 1 \end{pmatrix} + 3\begin{pmatrix} 1 & 0 & 0 \\ 0 & 1 & 0 \\ 0 & 0 & 1 \end{pmatrix} = \begin{pmatrix} 1 & 0 & 0 \\ -13 & 1 & -5 \\ 2 & 0 & 1 \end{pmatrix}.$$

17. If A is the given matrix then expansion about the second row gives $\det A = 4$, so that A^{-1} exists. Moreover, the characteristic polynomial of A is

$$P(\lambda) = \det(A - \lambda I) = \begin{vmatrix} 2-\lambda & 4 & 8 \\ 1 & -\lambda & 0 \\ 1 & -3 & -7-\lambda \end{vmatrix} = -\begin{vmatrix} 4 & 8 \\ -3 & -7-\lambda \end{vmatrix} - \lambda\begin{vmatrix} 2-\lambda & 8 \\ 1 & -7-\lambda \end{vmatrix}$$

$$= -\lambda^3 - 5\lambda^2 + 26\lambda + 4.$$

By the Cayley-Hamilton Theorem $P(A) = 0$, so that $-A^3 - 5A^2 + 26A + 4I = 0$. Multiply by A^{-1} to get $-A^2 - 5A + 26I + 4A^{-1} = 0$, and then solve for A^{-1} to get $A^{-1} = \frac{1}{4}(A^2 + 5A - 26I)$.

With $A^2 = \begin{pmatrix} 16 & -16 & -40 \\ 2 & 4 & 8 \\ -8 & 25 & 57 \end{pmatrix}$, $A^{-1} = \frac{1}{4}(A^2 + 5A - 26I)$ gives

$$A^{-1} = \frac{1}{4}(A^2 + 5A - 26I) = \begin{pmatrix} 0 & 1 & 0 \\ 7/4 & -11/2 & 2 \\ -3/4 & 5/2 & -1 \end{pmatrix}.$$

19. The given matrix, call it A, is a triangular matrix and, as such, its determinant is the product of its diagonal entries. Hence, $\det A = (1)(0)(3) = 0$, so that A^{-1} does not exist.

21. The given matrix, call it A, is a triangular matrix and, as such, its determinant is the product of its diagonal entries. So, by inspection, $\det A = 8$. Moreover, since $A - \lambda I$ is also a triangular matrix, its determinant is also the product of its diagonal entries. The characteristic polynomial of A is therefore

$$P(\lambda) = \det(A - \lambda I) = (1 - \lambda)^2(2 - \lambda)(4 - \lambda) = \lambda^4 - 8\lambda^3 + 21\lambda^2 - 22\lambda + 8.$$

By the Cayley-Hamilton Theorem, $P(A) = 0$, so that $A^4 - 8A^3 + 21A^2 - 22A + 8I = 0$. Multiply by A^{-1} to get $A^3 - 8A^2 + 21A - 22I + 8A^{-1} = 0$, and solve for A^{-1} to get $A^{-1} = \frac{1}{8}(-A^3 + 8A^2 - 21A + 22I)$. With

$$A^2 = \begin{pmatrix} 1 & 6 & -2 & 16 \\ 0 & 4 & 0 & 6 \\ 0 & 0 & 1 & 5 \\ 0 & 0 & 0 & 16 \end{pmatrix} \quad \text{and} \quad A^3 = \begin{pmatrix} 1 & 14 & -3 & 71 \\ 0 & 8 & 0 & 28 \\ 0 & 0 & 1 & 21 \\ 0 & 0 & 0 & 64 \end{pmatrix},$$

$A^{-1} = \frac{1}{8}(-A^3 + 8A^2 - 21A + 22I)$ gives

$$A^{-1} = \frac{1}{8}(-A^3 + 8A^2 - 21A + 22I) = \begin{pmatrix} 1 & -1 & 1 & -3/4 \\ 0 & 1/2 & 0 & -1/8 \\ 0 & 0 & 1 & -1/4 \\ 0 & 0 & 0 & 1/4 \end{pmatrix}.$$

23. If A is the given matrix then the fact that A is a triangular matrix means that its determinant is the product of its diagonal entries. That is, $\det A = e^{2t}$. Since this is never zero, A^{-1} exists for all real t. Moreover, since $A - \lambda I$ is also triangular, its determinant is also the product of its diagonal entries. Hence, the characteristic polynomial of A is

$$P(\lambda) = \det(A - \lambda I) = (1 - \lambda)(e^t - \lambda)^2 = -\lambda^3 + (1 + 2e^t)\lambda^2 - (2e^t + e^{2t})\lambda + e^{2t}.$$

By the Cayley-Hamilton Theorem, $P(A) = 0$, so that

$$-A^3 + (1 + 2e^t)A^2 - (2e^t + e^{2t})A + e^{2t}I = 0.$$

Multiply by A^{-1} to get $-A^2 + (1 + 2e^t)A - (2e^t + e^{2t})I + e^{2t}A^{-1} = 0$, and solve for A^{-1} to get $A^{-1} = e^{-2t}\left(A^2 - (1 + 2e^t)A + (2e^t + e^{2t}I)\right)$. With $A^2 = \begin{pmatrix} 1 & 0 & 0 \\ 0 & e^{2t} & 2te^{2t} \\ 0 & 0 & e^{2t} \end{pmatrix}$, we obtain

$$A^{-1} = e^{-2t}\left(A^2 - (1 + 2e^t)A + (2e^t + e^{2t}I)\right) = \begin{pmatrix} 1 & 0 & 0 \\ 0 & e^{-t} & -te^{-t} \\ 0 & 0 & e^{-t} \end{pmatrix}.$$

Exercise Set 2E (pgs. 639-640)

1. Let $e^{tA} = \begin{pmatrix} 1 & 0 \\ 0 & e^t \end{pmatrix}$. Differentiation gives $Ae^{tA} = \begin{pmatrix} 0 & 0 \\ 0 & e^t \end{pmatrix}$, and setting $t = 0$ gives

$$A = \begin{pmatrix} 0 & 0 \\ 0 & 1 \end{pmatrix}.$$

The vector function $\mathbf{x}(t) = e^{tA}\mathbf{c}$, where $\mathbf{c} = \begin{pmatrix} c_1 \\ c_2 \end{pmatrix}$ is a constant vector, can be written as

$$\mathbf{x}(t) = \begin{pmatrix} 1 & 0 \\ 0 & e^t \end{pmatrix}\begin{pmatrix} c_1 \\ c_2 \end{pmatrix} = \begin{pmatrix} c_1 \\ c_2 e^t \end{pmatrix} = c_1\begin{pmatrix} 1 \\ 0 \end{pmatrix} + c_2\begin{pmatrix} 0 \\ e^t \end{pmatrix}.$$

Denote the first and second columns of e^{tA} by $\mathbf{x}_1 = x_1(t)$ and $\mathbf{x}_2 = \mathbf{x}_2(t)$, respectively, so that $\dot{\mathbf{x}}_1(t) = \begin{pmatrix} 0 \\ 0 \end{pmatrix}$ and $\dot{\mathbf{x}}_2(t) = \begin{pmatrix} 0 \\ e^t \end{pmatrix}$. We then have

$$A\mathbf{x}_1 = \begin{pmatrix} 0 & 0 \\ 0 & 1 \end{pmatrix}\begin{pmatrix} 1 \\ 0 \end{pmatrix} = \begin{pmatrix} 0 \\ 0 \end{pmatrix} = \dot{\mathbf{x}}_1 \quad \text{and} \quad A\mathbf{x}_2 = \begin{pmatrix} 0 & 0 \\ 0 & 1 \end{pmatrix}\begin{pmatrix} 0 \\ e^t \end{pmatrix} = \begin{pmatrix} 0 \\ e^t \end{pmatrix} = \dot{\mathbf{x}}_2.$$

So $\mathbf{x}_1(t)$ and $\mathbf{x}_2(t)$ are solutions of the system $\dot{\mathbf{x}} = A\mathbf{x}$.

3. Let $e^{tA} = \begin{pmatrix} e^t & 0 & 0 \\ 0 & e^{2t} & 0 \\ 0 & 0 & e^{3t} \end{pmatrix}$. Differentiation gives $Ae^{tA} = \begin{pmatrix} e^t & 0 & 0 \\ 0 & 2e^{2t} & 0 \\ 0 & 0 & 3e^{3t} \end{pmatrix}$, and setting $t = 0$ gives

$$A = \begin{pmatrix} 1 & 0 & 0 \\ 0 & 2 & 0 \\ 0 & 0 & 3 \end{pmatrix}.$$

The vector function $\mathbf{x}(t) = e^{tA}\mathbf{c}$, where $\mathbf{c} = \begin{pmatrix} c_1 \\ c_2 \\ c_3 \end{pmatrix}$ is a constant vector, can be written as

$$\mathbf{x}(t) = \begin{pmatrix} e^t & 0 & 0 \\ 0 & e^{2t} & 0 \\ 0 & 0 & e^{3t} \end{pmatrix}\begin{pmatrix} c_1 \\ c_2 \\ c_3 \end{pmatrix} = \begin{pmatrix} c_1 e^t \\ c_2 e^{2t} \\ c_3 e^{3t} \end{pmatrix} = c_1\begin{pmatrix} e^t \\ 0 \\ 0 \end{pmatrix} + c_2\begin{pmatrix} 0 \\ e^{2t} \\ 0 \end{pmatrix} + c_3\begin{pmatrix} 0 \\ 0 \\ e^{3t} \end{pmatrix}.$$

Denote the first, second and third columns of e^{tA} by $\mathbf{x}_1 = \mathbf{x}_1(t)$, $\mathbf{x}_2 = \mathbf{x}_2(t)$ and $\mathbf{x}_3 = \mathbf{x}_3(t)$, respectively, so that $\dot{\mathbf{x}}_1 = \begin{pmatrix} e^t \\ 0 \\ 0 \end{pmatrix}$, $\dot{\mathbf{x}}_2 = \begin{pmatrix} 0 \\ 2e^{2t} \\ 0 \end{pmatrix}$ and $\dot{\mathbf{x}}_3 = \begin{pmatrix} 0 \\ 0 \\ 3e^{3t} \end{pmatrix}$. We then have

$$A\mathbf{x}_1 = \begin{pmatrix} 1 & 0 & 0 \\ 0 & 2 & 0 \\ 0 & 0 & 3 \end{pmatrix}\begin{pmatrix} e^t \\ 0 \\ 0 \end{pmatrix} = \begin{pmatrix} e^t \\ 0 \\ 0 \end{pmatrix} = \dot{\mathbf{x}}_1,$$

$$A\mathbf{x}_2 = \begin{pmatrix} 1 & 0 & 0 \\ 0 & 2 & 0 \\ 0 & 0 & 3 \end{pmatrix}\begin{pmatrix} 0 \\ e^{2t} \\ 0 \end{pmatrix} = \begin{pmatrix} 0 \\ 2e^{2t} \\ 0 \end{pmatrix} = \dot{\mathbf{x}}_2,$$

$$A\mathbf{x}_3 = \begin{pmatrix} 1 & 0 & 0 \\ 0 & 2 & 0 \\ 0 & 0 & 3 \end{pmatrix}\begin{pmatrix} 0 \\ 0 \\ e^{3t} \end{pmatrix} = \begin{pmatrix} 0 \\ 0 \\ 3e^{3t} \end{pmatrix} = \dot{\mathbf{x}}_3.$$

So $\mathbf{x}_1(t)$, $\mathbf{x}_2(t)$ and $\mathbf{x}_3(t)$ are solutions of the system $\dot{\mathbf{x}} = A\mathbf{x}$.

5. Let $M_t = \begin{pmatrix} e^t & 1 \\ 0 & e^t \end{pmatrix}$. As vector functions, neither column of M_t is a constant multiple

of the other for any value of t. So the columns of M_t are linearly independent.

Next, suppose M_t is an exponential matrix, so that $M_t = e^{tA}$ for some 2-by-2 matrix A. Then, by Theorem 2.1, part (b), $M_t M_{-t} = I$ for all t. In particular, $M_1 M_{-1} = I$. But by direct calculation,

$$M_1 M_{-1} = \begin{pmatrix} e & 1 \\ 0 & e \end{pmatrix} \begin{pmatrix} e^{-1} & 1 \\ 0 & e^{-1} \end{pmatrix} = \begin{pmatrix} 1 & e + e^{-1} \\ 0 & 1 \end{pmatrix} \neq I.$$

a contradiction. It follows that M_t is not an exponential matrix.

7. Let $M_t = \begin{pmatrix} e^t & te^{2t} \\ 0 & e^t \end{pmatrix}$. As vector functions, neither column of M_t is a constant multiple

of the other for any value of t. So the columns of M_t are linearly independent.

Next, suppose M_t is an exponential matrix, so that $M_t = e^{tA}$ for some 2-by-2 matrix A. Then, by Theorem 2.1, part (b), $M_t M_{-t} = I$ for all t. In particular, $M_1 M_{-1} = I$. But by direct calculation,

$$M_1 M_{-1} = \begin{pmatrix} e & e^2 \\ 0 & e \end{pmatrix} \begin{pmatrix} e^{-1} & -e^{-2} \\ 0 & e^{-1} \end{pmatrix} = \begin{pmatrix} 1 & e - e^{-1} \\ 0 & 1 \end{pmatrix} \neq I.$$

a contradiction. It follows that M_t is not an exponential matrix.

9. Let $A = A(t)$ be an $n \times n$ invertible matrix for all t in some open interval J, and let the columns of A be the vector functions $\mathbf{x}_1 = \mathbf{x}_1(t), \ldots, \mathbf{x}_n = \mathbf{x}_n(t)$. Suppose $c_1, \ldots, c_n$ are scalars such that

$$c_1 \mathbf{x}_1 + \cdots + c_n \mathbf{x}_n = \mathbf{0} \quad \text{for } t \in J.$$

By assumption, $A^{-1} = A^{-1}(t)$ exists for all $t \in J$, so we can left-multiply both sides of this equation by A^{-1} to obtain

$$c_1 A^{-1} \mathbf{x}_1 + \cdots + c_n A^{-1} \mathbf{x}_n = \mathbf{0} \quad \text{for } t \in J.$$

Since $A^{-1} A = I$, $A^{-1} \mathbf{x}_k = \mathbf{e}_k$, the kth column of the identity matrix. The above equation is therefore equivalent to

$$c_1 \mathbf{e}_1 + \ldots + c_n \mathbf{e}_n = \mathbf{0}.$$

Since $\mathbf{e}_1, \ldots, \mathbf{e}_n$ are linearly independent vectors, $c_k = 0$ for each k. Hence, the columns of $A(t)$ are linearly independent vector functions of t on J.

11. Let $\mathbf{x}_1(t), \ldots, \mathbf{x}_m(t)$ be vector-valued functions of the same dimension. First, suppose that one of these functions is a linear combination of the others on a t-interval I. Without loss of generality, we can suppose that $\mathbf{x}_1(t)$ is a linear combination of the others, so that

$$\mathbf{x}_1(t) = c_2 \mathbf{x}_2(t) + \cdots + c_m \mathbf{x}_m(t), \quad \text{for } t \in I.$$

Hence,

$$(-1)\mathbf{x}_1(t) + c_2 \mathbf{x}_2 + \cdots + c_m \mathbf{x}_m(t) = 0, \quad \text{for } t \in I.$$

Thus, some linear combination of the $\mathbf{x}_k(t)$ is the zero vector on I and therefore the $\mathbf{x}_k(t)$ are not linearly independent on I.

Conversely, if the $\mathbf{x}_k(t)$ are not linearly independent on I then, on I,

$$c_1 \mathbf{x}_1(t) + \cdots + c_m \mathbf{x}_m(t) = 0$$

for some scalars $c_1, \ldots, c_m$, not all zero. Without loss of generality, we can suppose that $c_1 \neq 0$. We can then divide throughout by c_1 and rearrange the result to get

$$\mathbf{x}_1(t) = (-c_2/c_1)\mathbf{x}_2(t) + \cdots + (-c_m/c_1)\mathbf{x}_m(t), \quad \text{for } t \in I.$$

That is, $\mathbf{x}_1(t)$ is a linear combination of the others on I. So, the $\mathbf{x}_k(t)$ are not linearly independent on I.

Section 3: NONHOMOGENEOUS SYSTEMS

Exercise Set 3ABC (pgs. 644-645)

1. Let $A = \begin{pmatrix} 3 & 0 \\ 0 & 2 \end{pmatrix}$, $\mathbf{b}(t) = \begin{pmatrix} e^t - 1 \\ e^{-t} \end{pmatrix}$ and $\mathbf{x} = \begin{pmatrix} x \\ y \end{pmatrix}$, so that the given system can be written as $\dot{\mathbf{x}} = A\mathbf{x} + \mathbf{b}(t)$. The exponential matrix e^{tA} is most easily found by noting that $A^k = \begin{pmatrix} 3^k & 0 \\ 0 & 2^k \end{pmatrix}$ and then using the series definition of e^{tA} to get $e^{tA} = \begin{pmatrix} e^{3t} & 0 \\ 0 & e^{2t} \end{pmatrix}$. The general solution of the homogeneous system $\dot{\mathbf{x}} = A\mathbf{x}$ is therefore

$$\mathbf{x}_h(t) = e^{tA}\mathbf{c} = \begin{pmatrix} e^{3t} & 0 \\ 0 & e^{2t} \end{pmatrix}\begin{pmatrix} c_1 \\ c_2 \end{pmatrix} = \begin{pmatrix} c_1 e^{3t} \\ c_2 e^{2t} \end{pmatrix}.$$

Further, Equation 3.2 provides a particular solution of the given nonhomogeneous system:

$$\mathbf{x}_p(t) = e^{tA}\int e^{-tA}\mathbf{b}(t)\,dt = \begin{pmatrix} e^{3t} & 0 \\ 0 & e^{2t} \end{pmatrix}\int\begin{pmatrix} e^{-3t} & 0 \\ 0 & e^{-2t} \end{pmatrix}\begin{pmatrix} e^t - 1 \\ e^{-t} \end{pmatrix}dt$$

$$= \begin{pmatrix} e^{3t} & 0 \\ 0 & e^{2t} \end{pmatrix}\int\begin{pmatrix} e^{-2t} - e^{-3t} \\ e^{-3t} \end{pmatrix}dt = \begin{pmatrix} e^{3t} & 0 \\ 0 & e^{2t} \end{pmatrix}\begin{pmatrix} -\frac{1}{2}e^{-2t} + \frac{1}{3}e^{-3t} \\ -\frac{1}{3}e^{-3t} \end{pmatrix}$$

$$= \begin{pmatrix} -\frac{1}{2}e^t + \frac{1}{3} \\ -\frac{1}{3}e^{-t} \end{pmatrix}.$$

The general solution $\mathbf{x}(t)$ of the given system is therefore

$$\mathbf{x}(t) = \mathbf{x}_h(t) + \mathbf{x}_p(t) = \begin{pmatrix} c_1 e^{3t} \\ c_2 e^{2t} \end{pmatrix} + \begin{pmatrix} -\frac{1}{2}e^t + \frac{1}{3} \\ -\frac{1}{3}e^{-t} \end{pmatrix} = \begin{pmatrix} c_1 e^{3t} - \frac{1}{2}e^t + \frac{1}{3} \\ c_2 e^{2t} - \frac{1}{3}e^{-t} \end{pmatrix}.$$

The initial condition $\mathbf{x}(0) = \begin{pmatrix} -1 \\ -1 \end{pmatrix}$ gives the equation $\begin{pmatrix} -1 \\ -1 \end{pmatrix} = \begin{pmatrix} -\frac{1}{2} + \frac{1}{3} + c_1 \\ -\frac{1}{3} + c_2 \end{pmatrix}$. By inspection, $c_1 = -\frac{5}{6}$ and $c_2 = -\frac{2}{3}$. Hence, the solution of the given initial-value problem is

$$\mathbf{x}(t) = \begin{pmatrix} -\frac{5}{6}e^{3t} - \frac{1}{2}e^t + \frac{1}{3} \\ -\frac{2}{3}e^{2t} - \frac{1}{3}e^{-t} \end{pmatrix}.$$

3. Let $A = \begin{pmatrix} 1 & 4 \\ 0 & 5 \end{pmatrix}$, $\mathbf{b}(t) = \begin{pmatrix} 1 \\ e^t \end{pmatrix}$ and $\mathbf{x} = \begin{pmatrix} x \\ y \end{pmatrix}$, so that the given system can be written as $\dot{\mathbf{x}} = A\mathbf{x} + \mathbf{b}(t)$. From Exercise 12 (Sec.2ABC, Ch.13) we have $e^{tA} = \begin{pmatrix} e^t & e^{5t} - e^t \\ 0 & e^{5t} \end{pmatrix}$. The general solution of the homogeneous system $\dot{\mathbf{x}} = A\mathbf{x}$ is therefore

$$\mathbf{x}_h(t) = e^{tA}\mathbf{c} = \begin{pmatrix} e^t & e^{5t} - e^t \\ 0 & e^{5t} \end{pmatrix}\begin{pmatrix} c_1 \\ c_2 \end{pmatrix} = \begin{pmatrix} c_1 e^t + c_2(e^{5t} - e^t) \\ c_2 e^{5t} \end{pmatrix}.$$

Further, Equation 3.2 provides a particular solution of the given nonhomogeneous system:

$$\mathbf{x}_p(t) = e^{tA}\int e^{-tA}\mathbf{b}(t)\,dt = \begin{pmatrix} e^t & e^{5t} - e^t \\ 0 & e^{5t} \end{pmatrix}\int\begin{pmatrix} e^{-t} & e^{-5t} - e^{-t} \\ 0 & e^{-5t} \end{pmatrix}\begin{pmatrix} 1 \\ e^t \end{pmatrix}dt$$

$$= \begin{pmatrix} e^t & e^{5t} - e^t \\ 0 & e^{5t} \end{pmatrix}\int\begin{pmatrix} e^{-t} + e^{-4t} - 1 \\ e^{-4t} \end{pmatrix}dt = \begin{pmatrix} e^t & e^{5t} - e^t \\ 0 & e^{5t} \end{pmatrix}\begin{pmatrix} -e^{-t} - \frac{1}{4}e^{-4t} - t \\ -\frac{1}{4}e^{-4t} \end{pmatrix}$$

$$= \begin{pmatrix} -1 - \frac{1}{4}e^t - te^t \\ -\frac{1}{4}e^t \end{pmatrix}.$$

The general solution $\mathbf{x}(t)$ of the given system is therefore

$$\mathbf{x}(t) = \mathbf{x}_h(t) + \mathbf{x}_p(t) = \begin{pmatrix} c_1 e^t + c_2(e^{5t} - e^t) \\ c_2 e^{5t} \end{pmatrix} + \begin{pmatrix} -1 - \frac{1}{4}e^t - te^t \\ -\frac{1}{4}e^t \end{pmatrix}$$

$$= \begin{pmatrix} c_1 e^t + c_2(e^{5t} - e^t) - 1 - \frac{1}{4}e^t - te^t \\ c_2 e^{5t} - \frac{1}{4}e^t \end{pmatrix}.$$

The initial condition $\mathbf{x}(0) = \begin{pmatrix} 0 \\ 1 \end{pmatrix}$ gives the equation $\begin{pmatrix} 0 \\ 1 \end{pmatrix} = \begin{pmatrix} c_1 - 1 - \frac{1}{4} \\ c_2 - \frac{1}{4} \end{pmatrix}$. By inspection, $c_1 = c_2 = \frac{5}{4}$. Hence, the solution of the given initial-value problem is

$$\mathbf{x}(t) = \begin{pmatrix} -1 - \frac{1}{4}e^t - te^t + \frac{5}{4}e^{5t} \\ \frac{5}{4}e^{5t} - \frac{1}{4}e^t \end{pmatrix}.$$

5. (a) Consider a solution of the form $\mathbf{x}(t) = e^{tA}\mathbf{c}$ and suppose $\mathbf{x}(t_0) = \mathbf{x}_0$. Setting $t = t_0$ gives $\mathbf{x}_0 = e^{t_0 A}\mathbf{c}$. Since $e^{-t_0 A}$ is the inverse of $e^{t_0 A}$, we can left-multiply both sides of this equation by $e^{-t_0 A}$ to get $\mathbf{c} = e^{-t_0 A}\mathbf{x}_0$.

(b) Let $X(t)$ be any $n \times n$ matrix whose columns are linearly independent vector functions on an open interval J containing t_0. Then, using the result from Exercise 19 in this section, $X^{-1}(t)$ exists for all $t \in J$. In particular, $X^{-1}(t_0)$ exists. Assuming that the function $\mathbf{x}(t) = X(t)\mathbf{c}$ satisfies $\mathbf{x}(t_0) = \mathbf{x}_0$, we can set $t = t_0$ to get $\mathbf{x}_0 = X(t_0)\mathbf{c}$. Left-multiplying this equation by $X^{-1}(t_0)$ gives $\mathbf{c} = X^{-1}(t_0)\mathbf{x}_0$.

7. In scalar form, the system given in Exercise 1 is

$$\dot{x} = 3x + e^t - 1,$$
$$\dot{y} = 2y + e^{-t};$$

and the scalar form of the homogeneous solution consists of the two equations $x_h(t) = c_1 e^{3t}$ and $y_h(t) = c_2 e^{2t}$, where $\mathbf{x}_h(t) = \begin{pmatrix} x_h(t) \\ y_h(t) \end{pmatrix}$. Note that the system is uncoupled. This means that we can use the method of undetermined coefficients on each equation separately. Since none of the terms in the nonhomogeneous parts of the above system are homogeneous solutions, we can therefore use $x_p(t) = A + Be^t$ and $y_p(t) = Ce^{-t}$ as trial particular solutions. Substitution into the above system gives

$$\dot{x}_p - 3x_p = Be^t - 3(A + Be^t) = -3A - 2Be^t = e^t - 1,$$
$$\dot{y}_p - 2y_p = -Ce^{-t} - 2Ce^{-t} = -3Ce^{-t} = e^{-t}.$$

Equating coefficients of like functions gives the three equations $-3A = -1$, $-2B = 1$ and $-3C = 1$, from which we get $A = 1/3$, $B = -1/2$ and $C = -1/3$. Therefore, the particular solutions are

$$x_p(t) = \frac{1}{3} - \frac{1}{2}e^t \quad \text{and} \quad y_p(t) = -\frac{1}{3}e^{-t},$$

which are the same solutions found in Exercise 1.

9. In scalar form, the system given in Exercise 3 is

$$\dot{x} = x + 4y + 1,$$
$$\dot{y} = 5y + e^t;$$

521

and the scalar form of the homogeneous solution consists of the two equations

$$x_h(t) = (c_1 - c_2)e^t + c_2 e^{5t} \qquad \text{and} \qquad y_h(t) = c_2 e^{5t},$$

where $\mathbf{x}_h(t) = \begin{pmatrix} x_h(t) \\ y_h(t) \end{pmatrix}$.

The second equation of the system is independent of x and the derivative set of the nonhomogeneous term e^t does not contain a homogeneous solution. Therefore, the trial solution we will use for y is $y_p(t) = Ae^t$. Substituting this into the second equation of the above system gives

$$\dot{y}_p - 5y_p = Ae^t - 5Ae^t = -4Ae^t = e^t.$$

Equating coefficients gives $-4A = 1$, so that $A = -1/4$ and $y_p(t) = -\frac{1}{4}e^t$.

Using this in the first equation of the system gives

$$\dot{x} = x + 4y_p + 1 = x - e^t + 1, \qquad \text{or equivalently,} \qquad \dot{x} - x = -e^t + 1,$$

where e^t (a solution of the homogeneous equation $\dot{x} - x = 0$) is now a nonhomogeneous term. Therefore, the proper form for a particular solution of this equation is $x_p(t) = A + Bte^t$. Substituting into the equation gives

$$\dot{x}_p - x_p = Bte^t + Be^t - (A + Bte^t) = -A + Be^t = -e^t + 1.$$

Equating coefficients provides the solutions $A = -1$, $B = -1$. So $x_p(t) = -1 - te^t$.

11. Let $\mathbf{x}_1 = \mathbf{x}_1(t) = \begin{pmatrix} t \\ 1 \end{pmatrix}$ and $\mathbf{x}_2 = \mathbf{x}_2(t) = \begin{pmatrix} -t^2 \\ t \end{pmatrix}$ and note that if $t = 0$ then $\mathbf{x}_2$ is the zero vector, which is linearly dependent with every vector. But if $t \neq 0$ then neither vector is a constant multiple of the other. That is, for each $t \neq 0$, $\mathbf{x}_1(t)$ and $\mathbf{x}_2(t)$ are linearly independent *vectors*. Therefore, as vector functions of t, $\mathbf{x}_1$ and $\mathbf{x}_2$ are linearly independent on every open interval.

Next, we form the fundamental matrix $X(t) = (\mathbf{x}_1(t), \mathbf{x}_2(t)) = \begin{pmatrix} t & -t^2 \\ 1 & t \end{pmatrix}$ and, for $t \neq 0$, compute its inverse $X^{-1}(t) = \begin{pmatrix} 1/(2t) & 1/2 \\ -1/(2t^2) & 1/(2t) \end{pmatrix}$. Then, with $\mathbf{b}(t) = \begin{pmatrix} t^3 \\ 2t^2 \end{pmatrix}$, Equation 3.3 gives the particular solution

$$\mathbf{x}_p(t) = X(t) \int X^{-1}(t)\mathbf{b}(t)\,dt = \begin{pmatrix} t & -t^2 \\ 1 & t \end{pmatrix} \int \begin{pmatrix} 1/(2t) & 1/2 \\ -1/(2t^2) & 1/(2t) \end{pmatrix} \begin{pmatrix} t^3 \\ 2t^2 \end{pmatrix} dt$$

$$= \begin{pmatrix} t & -t^2 \\ 1 & t \end{pmatrix} \int \begin{pmatrix} 3t^2/2 \\ t/2 \end{pmatrix} dt = \begin{pmatrix} t & -t^2 \\ 1 & t \end{pmatrix} \begin{pmatrix} t^3/2 \\ t^2/4 \end{pmatrix} = \begin{pmatrix} t^4/4 \\ 3t^3/4 \end{pmatrix}, \qquad t \neq 0.$$

13. (a) Let $X(t)$ be the $n \times n$ matrix with columns $\mathbf{x}_1(t), \ldots, \mathbf{x}_n(t)$. The differentiability of the columns together with their linear independence for each t implies that the matrix $A(t) = X'(t)X^{-1}(t)$ is defined for all t. Since $X^{-1}(t)\mathbf{x}_k = \mathbf{e}_k$ is the kth column of the identity matrix, we have

$$A(t)\mathbf{x}_k(t) = X'(t)X^{-1}(t)\mathbf{x}_k(t) = X'(t)\mathbf{e}_k = \dot{\mathbf{x}}_k(t) \qquad \text{for } k = 1, \ldots, n.$$

So each $\mathbf{x}_k(t)$ is a solution of the system $\dot{\mathbf{x}} = A(t)\mathbf{x}$ and $X(t)$ is a fundamental matrix.

(b) First of all, by inspection, neither of the given vectors $\mathbf{x}_1(t)$, $\mathbf{x}_2(t)$ is a multiple of the other for any t, so that $\mathbf{x}_1(t)$ and $\mathbf{x}_2(t)$ are linearly independent vectors for each t. Moreover, they are evidently differentiable. Thus, let $X(t) = (\,\mathbf{x}_1(t) \quad \mathbf{x}_2(t)\,) = \begin{pmatrix} e^t & 1 \\ 2e^{2t} & e^t \end{pmatrix}$ and compute

$$X'(t) = \begin{pmatrix} e^t & 0 \\ 4e^{2t} & e^t \end{pmatrix} \qquad \text{and} \qquad X^{-1}(t) = \begin{pmatrix} -e^{-t} & e^{-2t} \\ 2 & -e^{-t} \end{pmatrix}.$$

By part (a), if $A(t)$ is defined by

$$A(t) = X'(t)X^{-1}(t) = \begin{pmatrix} e^t & 0 \\ 4e^{2t} & e^t \end{pmatrix}\begin{pmatrix} -e^{-t} & e^{-2t} \\ 2 & -e^{-t} \end{pmatrix} = \begin{pmatrix} -1 & e^{-t} \\ -2e^t & 3 \end{pmatrix}$$

then $X(t)$ is a fundamental matrix for the system $\dot{\mathbf{x}} = A(t)\mathbf{x}$. Therefore, $\mathbf{x}_1(t)$ and $\mathbf{x}_2(t)$ are solutions of $\dot{\mathbf{x}} = A(t)\mathbf{x}$. With $\mathbf{x} = \begin{pmatrix} x \\ y \end{pmatrix}$, the scalar form of this system is

$$\dot{x} = -x + e^{-t}y, \qquad \dot{y} = -2e^t x + 3y.$$

15. If A is an n-by-n constant matrix then e^{tA} exists for all t. Moreover, if $\mathbf{x}_1(t), \ldots, \mathbf{x}_n(t)$ are the columns of e^{tA} then the general solution $\mathbf{x}(t)$ of $\dot{\mathbf{x}} = A\mathbf{x}$ can be written as

$$\mathbf{x}(t) = e^{tA}\mathbf{c} = (\,\mathbf{x}_1(t) \quad \cdots \quad \mathbf{x}_n(t)\,)\begin{pmatrix} c_1 \\ \vdots \\ c_n \end{pmatrix} = c_1\mathbf{x}_1(t) + \cdots + c_n\mathbf{x}_n(t),$$

where $\mathbf{c} = \begin{pmatrix} c_1 \\ \vdots \\ c_n \end{pmatrix}$ is an arbitrary constant vector in $\mathcal{R}^n$. In particular, the cases where $\mathbf{c}$ is the standard basis vector $\mathbf{e}_k$, all coefficients are zero except $c_k = 1$ and we get the solution $\mathbf{x}(t) = \mathbf{x}_k(t)$. That is, since this holds for $k = 1, \ldots, n$, each column of e^{tA} is a solution of $\dot{\mathbf{x}} = A\mathbf{x}$. Also, since e^{tA} is invertible, its columns are linearly independent. Therefore, by definition, $X(t) = e^{tA}$ is a fundamental matrix for the system $\dot{\mathbf{x}} = A\mathbf{x}$. Moreover, setting $t = 0$ gives $X(0) = I$.

17. Let $A = A(t)$ be a square matrix with entries that are real-valued functions of t. The series definition of the exponential matrix and the linearity of the derivative gives

$$\frac{d}{dt}e^A = \frac{d}{dt}\sum_{k=0}^{\infty}\frac{A^k}{k!} = \frac{d}{dt}\left(I + \sum_{k=1}^{\infty}\frac{A^k}{k!}\right) = \frac{dI}{dt} + \frac{d}{dt}\sum_{k=1}^{\infty}\frac{A^k}{k!} = 0 + \sum_{k=1}^{\infty}\frac{1}{k!}\frac{d}{dt}A^k = \sum_{k=1}^{\infty}\frac{1}{k!}\frac{d}{dt}A^k.$$

If $A(t)$ and $dA(t)/dt$ commute, the result of part (b) of Exercise 16 can be used to get

$$\sum_{k=1}^{\infty}\frac{1}{k!}\frac{d}{dt}A^k = \sum_{k=1}^{\infty}\frac{1}{k!}kA^{k-1}\frac{dA}{dt} = \sum_{k=1}^{\infty}\frac{1}{(k-1)!}A^{k-1}\frac{dA}{dt} = \sum_{k=0}^{\infty}\frac{A^k}{k!}\frac{dA}{dt} = e^A\frac{dA}{dt}.$$

Hence, $\dfrac{de^{A(t)}}{dt} = e^{A(t)}\dfrac{dA(t)}{dt}$.

19. Let $A(t)$ be an $n \times n$ matrix with entries that are all continuous on some common open interval J; let $\mathbf{x}_1(t), \ldots, \mathbf{x}_n(t)$ be linearly independent solutions of $\dot{\mathbf{x}} = A(t)\mathbf{x}$ on J; and let $X(t)$ be the fundamental matrix whose kth column is $\mathbf{x}_k(t)$ for $k = 1, \ldots, n$.

(a) Suppose $X(t_0)$ is not invertible for some $t_0 \in J$. Because of how it is computed, its characteristic polynomial $\det\big(X(t_0) - \lambda I\big)$ is a polynomial in λ of degree n. Letting $a \neq 0$ be the coefficient of λ^n, if $\lambda_1, \ldots, \lambda_n$ are the solutions of the corresponding characteristic equation $\det\big(X(t_0) - \lambda I\big) = 0$ then the characteristic polynomial can be factored as

$$\det\big(X(t_0) - \lambda I\big) = a(\lambda_1 - \lambda) \cdots (\lambda_n - \lambda).$$

Setting $\lambda = 0$ gives $\det X(t_0) = (-1)^n a(\lambda_1 \cdots \lambda_n)$. However, as $X(t_0)$ is assumed to be noninvertible, we must have $\det X(t_0) = 0$. So $(-1)^n a(\lambda_1 \cdots \lambda_n) = 0$. Since $a \neq 0$, $\lambda_k = 0$ for some $1 \leq k \leq n$. Thus, if $\mathbf{c}$ is an eigenvector for $\lambda_k = 0$ then $\mathbf{c} \neq \mathbf{0}$ and

$$X(t_0)\mathbf{c} = \lambda_k \mathbf{c} = \mathbf{0}.$$

(b) By definition, a fundamental matrix of the system $\dot{\mathbf{x}} = A(t)\mathbf{x}$ is a matrix solution of the system. So,

$$\frac{d}{dt}X(t) = A(t)X(t), \quad t \in J.$$

With $\mathbf{c}$ as found in part (a), the function $\mathbf{x}(t) = X(t)\mathbf{c}$ satisfies

$$\dot{\mathbf{x}}(t) = \frac{d}{dt}(X(t)\mathbf{c}) = \left(\frac{d}{dt}X(t)\right)\mathbf{c} + X(t)\frac{d\mathbf{c}}{dt} = \left(\frac{d}{dt}X(t)\right)\mathbf{c} + \mathbf{0}$$
$$= \big(A(t)X(t)\big)\mathbf{c} = A(t)\big(X(t)\mathbf{c}\big) = A(t)\mathbf{x}(t).$$

So $\mathbf{x}(t)$ is a solution of the system on J. Moreover, setting $t = t_0$ and using the fact that $X(t_0)\mathbf{c} = \mathbf{0}$ gives $\mathbf{x}(t_0) = 0$, so that $\mathbf{x}(t)$ is a solution of the initial-value problem $\dot{\mathbf{x}} = A(t)\mathbf{x}$, $\mathbf{x}(t_0) = 0$. Observe that this initial-value problem also has the zero vector as a solution.

(c) Since $A(t)$ has continuous entries, Theorem 1.1 in Chapter 12, Section 1D, guarantees a unique solution on some open subinterval J_0 of J containing t_0. By part (b), this solution must be the zero solution, so that $X(t)\mathbf{c} = \mathbf{0}$ identically on J_0. This is equivalent to saying that if $\mathbf{c} = \begin{pmatrix} c_1 \\ \vdots \\ c_n \end{pmatrix}$ then

$$X(t)\mathbf{c} = \big(\, \mathbf{x}_1(t) \quad \cdots \quad \mathbf{x}_n(t)\,\big) \begin{pmatrix} c_1 \\ \vdots \\ c_n \end{pmatrix} = c_1\mathbf{x}_1(t) + \cdots + c_n\mathbf{x}_n(t) = \mathbf{0}, \quad \text{for all } t \in J_0.$$

Since $\mathbf{c} \neq \mathbf{0}$, some c_k are nonzero. Thus, the above equation implies that the columns of $X(t)$ are not linearly independent, contrary to hypothesis. We conclude that our original assumption of the noninvertibility of $X(t)$ at some point in J was false. So, $X^{-1}(t)$ exists for all $t \in J$.

Section 4: EQUILIBRIUM & STABILITY

Exercise Set 4A (pgs. 652-653)

1. The coefficient matrix of the given system is $A = \begin{pmatrix} -3 & 2 \\ -4 & 3 \end{pmatrix}$. The eigenvalues of A are the roots of

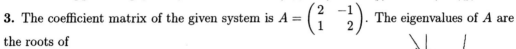

$$\det(A - \lambda I) = \begin{vmatrix} -3 - \lambda & 2 \\ -4 & 3 - \lambda \end{vmatrix}$$
$$= (-3 - \lambda)(3 - \lambda) + 8 = \lambda^2 - 1 = 0.$$

So the eigenvalues are $\lambda_1 = -1$ and $\lambda_2 = 1$. Since these are real and of opposite sign, the system is a saddle (Type II).

3. The coefficient matrix of the given system is $A = \begin{pmatrix} 2 & -1 \\ 1 & 2 \end{pmatrix}$. The eigenvalues of A are the roots of

$$\det(A - \lambda I) = \begin{vmatrix} 2 - \lambda & -1 \\ 1 & 2 - \lambda \end{vmatrix}$$
$$= (2 - \lambda)^2 + 1 = \lambda^2 - 4\lambda + 5 = 0.$$

The eigenvalues are therefore $\lambda_{1,2} = 2 \pm i$. Since the roots are complex with positive real part, the system is an unstable spiral (Type IV).

5. The coefficient matrix of the given system is $A = \begin{pmatrix} 2 & -3 \\ -2 & -2 \end{pmatrix}$. The eigenvalues of A are the roots of

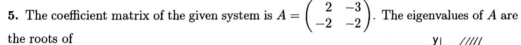

$$\det(A - \lambda I) = \begin{vmatrix} 2 - \lambda & -3 \\ -2 & -2 - \lambda \end{vmatrix}$$
$$= (2 - \lambda)(-2 - \lambda) - 6 = \lambda^2 - 10 = 0.$$

So the eigenvalues are $\lambda_1 = -\sqrt{10}$ and $\lambda_2 = \sqrt{10}$, which are real and opposite in sign. The system is a saddle (Type II).

7. The coefficient matrix of the given system is $A = \begin{pmatrix} 1 & 1 \\ 2 & 0 \end{pmatrix}$. The eigenvalues of A are the roots of

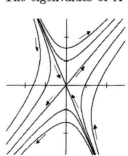

$$\det(A - \lambda I) = \begin{vmatrix} 1 - \lambda & 1 \\ 2 & -\lambda \end{vmatrix}$$
$$= -\lambda(1 - \lambda) - 2 = (\lambda - 2)(\lambda + 1) = 0.$$

The eigenvalues are therefore $\lambda_1 = -1$ and $\lambda_2 = 2$. Since the roots are real and of opposite sign, the system is a saddle (Type II).

9. The characteristic equation of the given second-order equation is $r^2 - 1 = 0$, with roots $r_1 = 1$ and $r_2 = -1$. These are the same as the eigenvalues of the first-order 2-dimensional system obtained by letting $dx/dt = y$. Since the roots are real and opposite in sign, the equation (and system) is of saddle type (Type II). Typical phase plots of solutions of the given equation are shown below on the left.

11. The characteristic equation of the given second-order equation is $r^2 + r + 1 = 0$, with roots $r_{1,2} = -\frac{1}{2} \pm i\frac{\sqrt{3}}{2}$. These are the same as the eigenvalues of the first-order 2-dimensional system obtained by letting $dx/dt = y$. Since the roots are complex with negative real part, the equation (and system) is an asymptotically stable spiral (Type VI). Typical phase plots of the solutions of the given equation are shown below on the right.

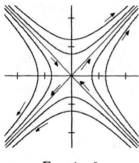
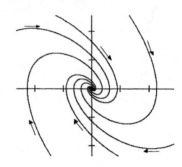

Exercise 9   Exercise 11

13. The characteristic equation of the given second-order equation is $r^2 + kr + 1 = 0$, with roots

$$r_1 = \frac{-k + \sqrt{k^2 - 4}}{2} \quad \text{and} \quad r_2 = \frac{-k - \sqrt{k^2 - 4}}{2}.$$

For $k = 0$, the roots are pure imaginary, which corresponds to a stable center (Type V).

For $0 < k < 2$, the roots are complex conjugates with negative real part $-k/2$, which corresponds to an asymptotically stable spiral (Type VI).

For $k = 2$, the roots are both equal to -1, which corresponds to an asymptotically stable star (Type VIII).

For $k > 2$, the roots are unequal and negative, which corresponds to an asymptotically stable node (Type III).

15. The characteristic equation of the given second-order equation is $2r^2 + kr + 1 = 0$, with roots

$$r_1 = \frac{-k + \sqrt{k^2 - 8}}{4} \quad \text{and} \quad r_2 = \frac{-k - \sqrt{k^2 - 8}}{4}.$$

For $k = 0$, the roots are pure imaginary, which corresponds to a stable center (Type V).

For $0 < k < \sqrt{8}$, the roots are complex conjugates with negative real part $-k/4$, which corresponds to an asymptotically stable spiral (Type VI).

For $k = \sqrt{8}$, the roots are both equal to $-\sqrt{2}/2$, which corresponds to an asymptotically stable star (Type VIII).

For $k > \sqrt{8}$, the roots are unequal and negative, which corresponds to an asymptotically stable node (Type III).

17. If $ad - bc = 0$ then the characteristic equation for $A = \begin{pmatrix} a & b \\ c & d \end{pmatrix}$ has the form

$$\det(A - \lambda I) = \begin{vmatrix} a - \lambda & b \\ c & d - \lambda \end{vmatrix} = (a - \lambda)(d - \lambda) - bc$$
$$= \lambda^2 - (a + d)\lambda + ad - bc = \lambda^2 - (a + d)\lambda = \lambda(\lambda - a - d) = 0,$$

so that $\lambda = 0$ and $\lambda = a + d$ are the eigenvalues of the system.

If $a + d \neq 0$ then the eigenvalues are different and the general solution for $x = x(t)$ is

$$x(t) = c_1 + c_2 e^{(a+d)t}.$$

The first equation of the system can then be written as

$$by = \dot{x} - ax = (a + d)c_2 e^{(a+d)t} - ac_1 - ac_2 e^{(a+d)t} = -ac_1 + dc_2 e^{(a+d)t}.$$

If $b = 0$, this becomes $0 = -ac_1 + dc_2 e^{(a+d)t}$, which must hold for every pair of constants c_1, c_2. But this can happen only if $a = d = 0$, contrary to the assumption that $a + d \neq 0$. It follows that $b \neq 0$ and we can solve the above equation for $y = y(t)$ to get

$$y(t) = -\frac{a}{b}c_1 + \frac{d}{b}c_2 e^{(a+d)t}.$$

Choosing $c_2 = 0$ shows that $(x(t), y(t)) = (c_1, -ac_1/b) = (1, -a/b)c_1$ is an equilibrium solution for every constant c_1.

If $a + d = 0$ then $\lambda = 0$ is a double eigenvalue and the general solution for $x = x(t)$ is

$$x(t) = c_1 + c_2 t.$$

The first equation of the system can then be written as

$$by = c_2 - ac_1 - ac_2 t = -ac_1 + (1 - at)c_2.$$

If $b = 0$, this becomes $0 = -ac_1 + (1 - at)c_2$, which must hold for every pair of constants c_1, c_2. In particular, the pair $c_1 = -1$, $c_2 = 0$ directly gives $a = 0$. Hence, the equation $0 = -ac_1 + (1 - at)c_2$ reduces to $0 = c_2$, which must hold for all constants c_2. Since this is absurd, $b \neq 0$. We can therefore solve the above equation for $y = y(t)$ to get

$$y(t) = -\frac{a}{b}c_1 + \frac{1 - at}{b}c_2.$$

Choosing $c_2 = 0$ shows that $(x(t), y(t)) = (c_1, -ac_1/b) = (1, -a/b)c_1$ is an equilibrium solution for every constant c_1. In either case, $ad - bc = 0$ implies that there are infinitely many equilibrium points.

Conversely, if $ad - bc \neq 0$ then $\det A = ad - bc \neq 0$ and therefore A^{-1} exists. Hence, if $\mathbf{x}$ is an arbitrarily chosen equilibrium solution then the system $\dot{\mathbf{x}} = A\mathbf{x}$ becomes $\mathbf{0} = A\mathbf{x}$. Left-multiplying both sides of this equation by A^{-1} gives $\mathbf{0} = \mathbf{x}$. That is, the "arbitrarily chosen" equilibrium solution must be the zero solution. It follows that this is the only equilibrium solution. The contrapositive of what we have just proven is "if there are infinitely many equilibrium solutions then $ad - bc = 0$".

19. In both cases, the trajectories in the xy-plane satisfy the differential equation

$$\frac{dy}{dt} = \frac{g(x,y)}{f(x,y)}.$$

Hence, the portraits are identical except for the direction of the arrow — they reverse for the system when minus signs are introduced. This corresponds to time reversal.

21. (a) Differentiating the first equation gives $\ddot{x} = \dot{y}$. The second equation of the system can then be written as $\ddot{x} = -x - \dot{x}$, or equivalently, $\ddot{x} + \dot{x} + x = 0$. The characteristic equation $r^2 + r + 1 = 0$ has the roots $r_{1,2} = \frac{1}{2}(-1 \pm i\sqrt{3})$. Thus, the solution for $x = x(t)$ is

$$x(t) = e^{-t/2}\left[c_1 \cos\frac{\sqrt{3}}{2}t + c_2 \sin\frac{\sqrt{3}}{2}t\right].$$

The general solution for $y = y(t)$ can then be obtained from the first equation of the system:

$$y(t) = \dot{x}(t) = \frac{1}{2}e^{-t/2}\left[(\sqrt{3}c_2 - c_1)\cos\frac{\sqrt{3}}{2}t - (\sqrt{3}c_1 + c_2)\sin\frac{\sqrt{3}}{2}t\right].$$

The third equation of the system can be written as the nonhomogeneous equation

$$\dot{z} + z = e^{-t/2}\left[c_1 \cos\frac{\sqrt{3}}{2}t + c_2 \sin\frac{\sqrt{3}}{2}t\right].$$

The homogeneous equation $\dot{z} + z = 0$ has the solution $z_h(t) = c_3 e^{-t}$. Using the method of undetermined coefficients, we find that a particular solution of the nonhomogeneous equation is

$$z_p(t) = \frac{1}{2}e^{-t/2}\left[(c_1 - \sqrt{3}c_2)\cos\frac{\sqrt{3}}{2}t + (\sqrt{3}c_1 + c_2)\sin\frac{\sqrt{3}}{2}t\right].$$

The general solution for $z = z(t)$ is therefore

$$z(t) = c_3 e^{-t} + \frac{1}{2}e^{-t/2}\left[(c_1 - \sqrt{3}c_2)\cos\frac{\sqrt{3}}{2}t + (\sqrt{3}c_1 + c_2)\sin\frac{\sqrt{3}}{2}t\right].$$

(b) The general solutions $x(t)$, $y(t)$, $z(t)$ found in part (a) all exponentially decay to zero as $t \to \infty$. Hence, the trajectory $(x(t), y(t), z(t))$ of every solution tends to the equilibrium point $(0,0,0)$ as $t \to \infty$. So, $(0,0,0)$ is asymptotically stable.

(c) The general solutions for $x(t)$ and $y(t)$ found in part (a) remain unchanged if the last equation of the system is replaced by $\dot{z} = x + z$. However, the homogeneous equation $\dot{z} - z = 0$ corresponding to the changed third equation has the solution $z_h(t) = c_3 e^t$, which is unbounded for $c_3 \neq 0$. The particular solution of the nonhomogeneous equation, obtained by using the method of undetermined coefficients, is still $e^{-t/2}$ times a linear combination of $\cos\frac{\sqrt{3}}{2}t$ and $\sin\frac{\sqrt{3}}{2}t$, and therefore still tends to zero as $t \to \infty$. Thus, the general solution for $z = z(t)$ is unbounded for $c_3 \neq 0$. It follows that the trajectories of such solutions recede farther and farther from the origin as $t \to \infty$. So, the equilibrium point $(0,0,0)$ is unstable.

Exercise Set 4B (pgs. 657-658)

1. Since A is a 2×2 nonzero matrix with $\det A = 0$, we can let $A = \begin{pmatrix} a & b \\ c & d \end{pmatrix}$, where a, b, c, d are constants, not all zero, such that $ad - bc = 0$. The associated characteristic equation is then

$$\det(A - \lambda I) = \begin{vmatrix} a - \lambda & b \\ c & d - \lambda \end{vmatrix} = (a - \lambda)(d - \lambda) - bc$$
$$= \lambda^2 - (a + d)\lambda + ad - bc = \lambda^2 - (a + d)\lambda = \lambda(\lambda - a - d) = 0.$$

So $\lambda = 0$ is an eigenvalue of the system.

Setting $\mathbf{x} = \begin{pmatrix} x \\ y \end{pmatrix}$, the scalar form of the system is $\dot{x} = ax + by$, $\dot{y} = cx + dy$, so that every equilibrium solution satisfies the algebraic system

$$0 = ax + by,$$
$$0 = cx + dy.$$

Since A is not the 0 matrix, at least one of a, b, c, d is nonzero. Therefore, assume $a \neq 0$. The first equation can then be solved for x to get $x = -(b/a)y$. Inserting this into the second equation gives $0 = c(-(b/a)y) + dy$, or equivalently, $0 = (ad - bc)y$. Since $ad - bc = 0$, this equation leaves y arbitrary. The equilibrium solutions are therefore of the form

$$\begin{pmatrix} x \\ y \end{pmatrix} = \begin{pmatrix} -(b/a)y \\ y \end{pmatrix} = \begin{pmatrix} -b/a \\ 1 \end{pmatrix} y, \quad \text{with } y \text{ arbitrary.}$$

which is the equation of a line in $\mathcal{R}^2$. Similar results are obtained for the three cases $b \neq 0$, $c \neq 0$ and $d \neq 0$.

3. By inspection, it can be seen that $(0, 0, 0)$ is an equilibrium point of the Lorentz system. To find the remaining equilibrium points when $\rho > 1$, we solve the algebraic system

$$\sigma(y - x) = 0,$$
$$\rho x - y - xz = 0,$$
$$-\beta z + xy = 0.$$

Since $\sigma > 0$, the first equation gives $x = y$, so that the remaining equations are equivalent to

$$x(\rho - 1 - z) = 0,$$
$$-\beta z + x^2 = 0.$$

The first of these equations shows that $x = 0$ or $z = \rho - 1$. If $x = 0$ then the last equation becomes $-\beta z = 0$, so that $\beta > 0$ forces $z = 0$. This, along with $x = y$ gives the equilibrium solution $(0, 0, 0)$, which we already have. Thus, we suppose that $z = \rho - 1$. Inserting this into the last equation gives an equation that can be written as $x^2 = \beta(\rho - 1)$. By hypothesis, $\beta > 0$ and $\rho > 1$, so that this equation has two solutions for x; namely, $x = \pm\sqrt{\beta(\rho - 1)}$. The two remaining equilibrium points are therefore

$$(x, y, z) = (\pm\sqrt{\beta(\rho - 1)}, \pm\sqrt{\beta(\rho - 1)}, \rho - 1).$$

5. (a) To find the equilibrium points of the given system, we solve the algebraic system

$$0 = y + \alpha x(x^2 + y^2),$$
$$0 = -x + \alpha y(x^2 + y^2).$$

If $\alpha = 0$, then the equations directly give $x = y = 0$, so that $(0,0)$ is the only equilibrium point. If $\alpha \neq 0$, multiply the first equation by x, the second equation by y, and add the results to get $0 = \alpha x^2(x^2 + y^2) + \alpha y^2(x^2 + y^2) = \alpha(x^2 + y^2)^2$. Hence, $x^2 + y^2 = 0$, from which we conclude that $x = y = 0$. Again, $(0,0)$ is the only equilibrium point. In any case, $(0,0)$ is the only equilibrium point, regardless of the value of α.

(b) With $\mathbf{F}(x,y) = \begin{pmatrix} y + \alpha x(x^2 + y^2) \\ -x + \alpha y(x^2 + y^2) \end{pmatrix}$, we first compute its derivative matrix to get

$$\mathbf{F}'(x,y) = \begin{pmatrix} 3\alpha x^2 + \alpha y^2 & 1 + 2\alpha xy \\ -1 + 2\alpha xy & \alpha x^2 + 3\alpha y^2 \end{pmatrix}.$$

Evaluating this at the equilibrium point $(0,0)$ gives $\mathbf{F}'(0,0) = \begin{pmatrix} 0 & 1 \\ -1 & 0 \end{pmatrix}$. The linearized system associated with $(0,0)$ is therefore

$$\dot{\mathbf{x}} = \mathbf{F}'(0,0)\mathbf{x} = \begin{pmatrix} 0 & 1 \\ -1 & 0 \end{pmatrix}\begin{pmatrix} x \\ y \end{pmatrix} = \begin{pmatrix} y \\ -x \end{pmatrix},$$

from which we obtain the scalar system $\dot{x} = y$, $\dot{y} = -x$. The eigenvalues λ of this system satisfy

$$\det(\mathbf{F}'(0,0) - \lambda I) = \begin{vmatrix} -\lambda & 1 \\ -1 & -\lambda \end{vmatrix} = \lambda^2 + 1 = 0.$$

So, the eigenvalues are $\lambda_{1,2} = \pm i$. We conclude that the equilibrium point $(0,0)$ for the linearized system is a stable center.

(c) With $x = r\cos\theta$ and $y = r\sin\theta$, the chain rule gives

$$\frac{dx}{dt} = \frac{dx}{dr}\frac{dr}{dt} + \frac{dx}{d\theta}\frac{d\theta}{dt} = \dot{r}\cos\theta - \dot{\theta}r\sin\theta,$$
$$\frac{dy}{dt} = \frac{dy}{dr}\frac{dr}{dt} + \frac{dy}{d\theta}\frac{d\theta}{dt} = \dot{r}\sin\theta + \dot{\theta}r\cos\theta.$$

Thus, when the given system is converted to polar coordinates, the result is

$$\dot{r}\cos\theta - \dot{\theta}r\sin\theta = r\sin\theta + \alpha r^3\cos\theta,$$
$$\dot{r}\sin\theta + \dot{\theta}r\cos\theta = -r\cos\theta + \alpha r^3\sin\theta.$$

Multiplying the first equation by $\cos\theta$, the second equation by $\sin\theta$, and then adding the results gives $\dot{r} = \alpha r^3$. When this expression for $\dot{r}$ is inserted into the above polar form of the given system, the two resulting equations reduce to $-\dot{\theta}r\sin\theta = r\sin\theta$ and $\dot{\theta}r\cos\theta = -r\cos\theta$. If $r \neq 0$ (i.e., if $(x,y) \neq (0,0)$) then these equations together imply the relation $\dot{\theta} = -1$. Thus, the polar form of the given system is

$$\dot{r} = \alpha r^3,$$
$$\dot{\theta} = -1.$$

(d) The system derived in part (c) is uncoupled. The second equation directly gives $\theta = -t + c_1$. Separating the variables in the first equation gives $r^{-3}dr = \alpha dt$, where one integration results in $-\frac{1}{2}r^{-2} = \alpha t + c_2$. Setting $t = 0$ and letting $r(0) = r_0$ (where we observe that $r_0 \neq 0$), we obtain $-\frac{1}{2}r_0^{-2} = c_2$. Hence, $-\frac{1}{2}r^{-2} = \alpha t - \frac{1}{2}r_0^{-2}$, or solving for r^2,

$$r^2 = \frac{r_0^2}{1 - 2\alpha t r_0^2}.$$

If $\alpha > 0$ then $r^2 \to \infty$ as $t \to 1/(2\alpha r_0^2)$, so that all nonequilibrium solutions are unbounded. Thus, $(0,0)$ is unstable. On the other hand, if $\alpha < 0$ then $r^2 \to 0$ as $t \to \infty$, so that all nonequilibrium solutions tend to $(0,0)$. Hence, $(0,0)$ is stable.

7. (a) Assume that (x,y) is an equilibrium solution. The left sides of the equations in the given system are then identically zero and the system can be written as

$$A - x(B + 1 - xy) = 0,$$
$$x(B - xy) = 0.$$

The second equation shows that either $x = 0$ or $B = xy$. However, if $x = 0$ then the first equation reduces to $A = 0$, contrary to hypothesis. It follows that $B = xy$. The first equation then shows that $x = A$. The equation $B = xy$ then becomes $B = Ay$, which can be solved for y (because $A \neq 0$) to get $y = B/A$. Thus, there is only one equilibrium solution; namely,

$$\big(x(t), y(t)\big) = (A, B/A).$$

(b) The phase portrait displayed in the figure on the right is for the case $A = 1$, $B = 3$. The associated system is

$$\frac{dx}{dt} = 1 - 3x - x + x^2 y,$$
$$\frac{dy}{dt} = 3x - x^2 y.$$

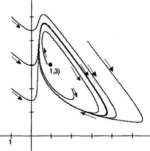

The equilibrium point is $(1,3)$. As the portrait shows, trajectories starting near $(1,3)$ exhibit limit cycle behavior.

9. The equilibrium points of the given system are the solutions of the algebraic system

$$0 = -y(1 - x^2 - y^2),$$
$$0 = x(1 - x^2 - y^2).$$

Clearly, the points on the unit circle $1 - x^2 - y^2 = 0$ are all equilibrium points. For points off this circle, the above two equations are satisfied only for $x = y = 0$. Thus, the equilibrium points of the system are all points on the circle $x^2 + y^2 = 1$, together with the origin $(0,0)$.

To show that all other trajectories are circles, we form the quotient

$$\frac{dy}{dx} = \frac{dy/dt}{dx/dt} = -\frac{x(1 - x^2 - y^2)}{y(1 - x^2 - y^2)} = -\frac{x}{y},$$

which must be satisfied for all portions of trajectories that are not on the unit circle or the x-axis (i.e., $1 - x^2 - y^2 \neq 0$ and $y \neq 0$). This first-order equation can be solved by separating the variables to get $y\,dy = -x\,dx$, then integrating to get $\frac{1}{2}y^2 = -\frac{1}{2}x^2 + c$, or equivalently,

$$x^2 + y^2 = C^2,$$

531

where $C = \sqrt{2c}$ is a positive constant different from 1. Thus, all portions of trajectories that are not on the unit circle or the x-axis are portions of nondegenerate circles of radius $C \neq 1$. Since a complete trajectory must be a connected curve, it follows that every trajectory is either a circle of positive radius different from 1 or an equilibrium point.

11. The equilibrium points of the given system are the solutions of the algebraic system

$$0 = -x(x^2 + y^2 - 1),$$
$$0 = -y(x^2 + y^2 + 1).$$

Since $x^2 + y^2 + 1 \neq 0$ for any pair x, y, the second equation is satisfied only for $y = 0$. The first equation then reduces to $0 = -x(x^2 - 1)$, which is satisfied for $x = -1, 0, 1$. Thus, there are three equilibrium solutions: $(-1, 0)$, $(0, 0)$ and $(1, 0)$.

Now let $\mathbf{F}(x, y) = \begin{pmatrix} -x(x^2 + y^2 + 1) \\ -y(x^2 + y^2 + 1) \end{pmatrix}$ and compute its derivative matrix:

$$\mathbf{F}'(x, y) = \begin{pmatrix} -3x^2 - y^2 + 1 & -2xy \\ -2xy & -x^2 - 3y^2 - 1 \end{pmatrix}.$$

Evaluating this at the three equilibrium points gives the coefficient matrix of the linearization of the given system at each of the equilibrium points. Specifically,

$$\mathbf{F}'(-1, 0) = \begin{pmatrix} -2 & 0 \\ 0 & -2 \end{pmatrix}, \quad \mathbf{F}'(0, 0) = \begin{pmatrix} 1 & 0 \\ 0 & -1 \end{pmatrix}, \quad \mathbf{F}'(1, 0) = \begin{pmatrix} -2 & 0 \\ 0 & -2 \end{pmatrix}.$$

By inspection, the first and third matrices have -2 as a double eigenvalue. By Theorem 4.2, the equilibrium points $(\pm 1, 0)$ are stable. However, the second matrix has the eigenvalues ± 1, so that Theorem 4.2 implies $(0, 0)$ is unstable.

13. (a) To show that the given system has circular trajectories of radius 1, it is only necessary to verify that the pair of functions $x = x(t) = \cos(t + \alpha)$, $y = y(t) = \sin(t + \alpha)$ is a solution of the system for each real number $0 \leq \alpha < 2\pi$. To do this, note that $x^2 + y^2 = 1$ for each such pair, and that $\dot{x} = -y$ and $\dot{y} = x$ for each such pair. Hence,

$$\dot{x} = x \cdot 0 - y = x(1 - x^2 + y^2) - y \quad \text{and} \quad \dot{y} = y \cdot 0 + x = y(1 - x^2 - y^2) + x,$$

which shows that each such pair is a solution of the given system.

(b) If $x = x(t)$, $y = y(t)$ is a nonconstant solution pair of the given system then, for all t such that $x(t) \neq 0$,

$$\frac{d}{dt}(y/x) = \frac{x\dot{y} - y\dot{x}}{x^2} = \frac{x\left(x + y(1 - x^2 - y^2)\right) - y\left(y(1 - x^2 - y^2) - y\right)}{x^2}$$
$$= \frac{x^2 + y^2}{x^2} = 1 + (y/x)^2.$$

(c) Set $v = y/x$ in the differential equation derived in part (b) and separate variables to get $(1 + v^2)^{-1}dv = dt$. Integrating gives $\arctan v = t + C$, or converting back to the variable x and y, $\arctan(y/x) = t + C$. Thus, for a given constant C, the polar coordinate form $x = r\cos\theta$, $y = r\sin\theta$ implies $y/x = \tan\theta$, so that the polar angle θ of a point on a given nonconstant trajectory satisfies $\theta = \arctan(y/x) = t + C$. The fact that the arctangent function is multiple-valued, allows is to conclude that since $t + C$ is continuous for all t, so is the polar angle function $\theta = \theta(t)$. Hence, each nonconstant solution winds around the origin infinitely often as $t \to \infty$.

(d) Using the given system, we compute
$$x\dot{x} + y\dot{y} = \left(x^2(1 - x^2 - y^2) - xy\right) + \left(yx + y^2(1 - x^2 + y^2)\right) = (x^2 + y^2)(1 - x^2 - y^2).$$

(e) Let $x = r\cos\theta$, $y = r\sin\theta$. On one hand, the right side of the equation derived in part (d) is $r^2(1 - r^2)$. On the other hand, the chain rule gives $\dot{x} = -\dot{\theta}r\sin\theta + \dot{r}\cos\theta$ and $\dot{y} = \dot{\theta}r\cos\theta + \dot{r}\sin\theta$, so that
$$x\dot{x} + y\dot{y} = r\cos\theta(-\dot{\theta}r\sin\theta + \dot{r}\cos\theta) + r\sin\theta(\dot{\theta}r\cos\theta + \dot{r}\sin\theta) = r\dot{r}.$$

Hence, $r\dot{r} = r^2(1 - r^2)$, or since $r \neq 0$, $\dot{r} = r(1 - r^2)$. Separating variables, partial fraction decomposition gives the equation
$$\left[\frac{1}{r} + \frac{1}{2}\left(\frac{1}{1-r} - \frac{1}{1+r}\right)\right]dr = dt.$$

integrating gives $\ln r - \frac{1}{2}(\ln|1 - r| + \ln|1 + r|) = t + C$. Multiplying through by 2, combining log functions and exponentiating gives $r^2/|1 - r^2| = k^2 e^{2t}$, where $k = e^C$. This can be written without an absolute value sign as $r^2/(1 - r^2) = \pm k^2 e^{2t}$, where the plus sign is chosen if $0 < r < 1$ and the minus sign is chosen if $r > 1$. Solving for r gives

$$r = \frac{ke^t}{\sqrt{k^2 e^{2t} \pm 1}}, \quad \text{which can be written as} \quad r = \frac{1}{\sqrt{1 \pm k^{-2}e^{-2t}}}.$$

The expression for $r = r(t)$ on the right shows that $r(t)$ can be made arbitrarily close to 1 by taking t sufficiently large. This means that every nonconstant trajectory approaches arbitrarily close to the circular trajectory $x^2 + y^2 = 1$ and therefore has this trajectory as a limit.

15. If $f(x, y) = y$ and $g(x, y) = -x - \alpha(x^2 - 1)y$ then
$$f_x(x, y) = 0, \quad f_y(x, y) = 1, \quad g_x(x, y) = -1 - 2\alpha xy, \quad g_y(x, y) = -\alpha(x^2 - 1),$$

and the linearization of the given system at he point $(x_0, y_0) = (0, 0)$ is given by
$$\dot{x} = f_x(0, 0)(x - 0) + f_y(0, 0)(y - 0) = y,$$
$$\dot{y} = g_x(0, 0)(x - 0) + g_y(0, 0)(y - 0) = -x + \alpha y.$$

(b) Let A be the coefficient matrix $\begin{pmatrix} 0 & 1 \\ -1 & \alpha \end{pmatrix}$ of the linearization found in part (a). Its eigenvalues are the solutions of
$$\det(A - \lambda I) = \begin{vmatrix} -\lambda & 1 \\ -1 & \alpha - \lambda \end{vmatrix} = -\lambda(\alpha - \lambda) + 1 = \lambda^2 - \alpha\lambda + 1 = 0.$$

Hence, the eigenvalues of the system are
$$\lambda_1 = \frac{\alpha - \sqrt{\alpha^2 - 4}}{2} \quad \text{and} \quad \lambda_2 = \frac{\alpha + \sqrt{\alpha^2 - 4}}{2}.$$

For $\alpha > 0$, there are three distinct kinds of behavior of solutions near $(0, 0)$. Specifically, (i) if $0 < \alpha < 2$ then λ_1 and λ_2 are complex conjugates with positive real part $\alpha/2$ and the solutions are unstable spirals; (ii) if $\alpha = 2$ then $\lambda_1 = \lambda_2 = 1 > 0$ and solutions exhibit unstable star behavior; and (iii) if $\alpha > 2$ then $0 < \lambda_1 < \lambda_2$ and $(0, 0)$ is an unstable node.

For $\alpha \leq 0$, there are four distinct kinds of behavior of solutions near $(0, 0)$: (i) if $\alpha = 0$ then $\lambda_{1,2} = \pm i$ and $(0, 0)$ is a stable center; (ii) if $-2 < \alpha < 0$ then λ_1 and λ_2 are complex conjugates with negative real part $\alpha/2$ and the solutions are asymptotically stable spirals; (iii) if $\alpha = -2$ then $\lambda_1 = \lambda_2 = -1 < 0$ and solutions exhibit asymptotically stable star behavior; and (iv) if $\alpha < -2$ then $\lambda_1 < \lambda_2 < 0$ and $(0, 0)$ is an asymptotically stable node.

Chapter Review

(pgs. 658-659)

1. Write the given system as the equivalent initial-value problem,

$$\dot{x} = 2x - 3y + 1,$$
$$\dot{y} = x - 2y + e^t, \quad x(0) = 1, \, y(0) = 2.$$

Then use the elimination method to derive the second-order equation in y: $\ddot{y} - y = 1 - e^t$. The characteristic roots are $r_{1,1} = \pm 1$, the homogeneous solution is $y_h(t) = c_1 e^t + c_2 e^{-t}$, and the method of undetermined coefficients gives the particular nonhomogeneous solution $y_p(t) = -1 - \frac{1}{2}te^t$. The general solution for $y = y(t)$ is therefore

$$y(t) = c_1 e^t + c_2 e^{-t} - \frac{1}{2}te^t - 1.$$

Now use the second equation of the above system to obtain the general solution for $x = x(t)$:

$$x(t) = \frac{3}{2}(2c_1 - 1)e^t + c_2 e^{-t} - \frac{3}{2}te^t - 2.$$

The initial conditions lead to $c_1 = 3/4$ and $c_2 = 9/4$. The solution of the given initial-value problem is therefore

$$x(t) = \frac{3}{4}e^t + \frac{9}{4}e^{-t} - \frac{3}{2}te^t - 2,$$
$$y(t) = \frac{3}{4}e^t + \frac{9}{4}e^{-t} - \frac{1}{2}te^t - 1.$$

3. Write the given system as the equivalent initial-value problem

$$\dot{x} = x - y + e^t,$$
$$\dot{y} = -y + e^{2t}, \quad x(0) = 1, \, y(0) = 0.$$

Use the integrating factor e^t on the second equation to derive the general solution for $y = y(t)$: $y(t) = c_1 e^{-t} + \frac{1}{3}e^{2t}$. The initial condition $y(0) = 0$ leads to $c_1 = -1/3$, so that

$$y(t) = -\frac{1}{3}e^{-t} + \frac{1}{3}e^{2t}.$$

Insert this into the first equation and use the integrating factor e^{-t} to obtain the general solution for $x = x(t)$: $x(t) = c_2 e^t + te^t - \frac{1}{6}e^{-t} - \frac{1}{3}e^{2t}$. The initial condition $x(0) = 1$ leads to $c_2 = 3/2$ and

$$x(t) = \frac{3}{2}e^t + te^t - \frac{1}{6}e^{-t} - \frac{1}{3}e^{2t}.$$

5. Write the given system as the equivalent initial-value problem

$$\dot{x} = 2x - y + 1,$$
$$\dot{y} = x + 5y + e^t, \quad x(0) = 1, \, y(0) = -1.$$

534

The elimination method gives the second-order equation in y: $\ddot{y} - 7\dot{y} + 11y = 1 - e^t$. The characteristic roots $r_1 = \frac{1}{2}(7 + \sqrt{5})$, $r_2 = \frac{1}{2}(7 - \sqrt{5})$ give the homogeneous solution $y_h(t) = c_1 e^{\frac{1}{2}(7+\sqrt{5})t} + c_2 e^{\frac{1}{2}(7-\sqrt{5})t}$, and the method of undetermined coefficients gives the particular nonhomogeneous solution $y_p(t) = \frac{1}{11} - \frac{1}{5}e^t$. The general solution for $y = y(t)$ is then

$$y(t) = c_1 e^{\frac{1}{2}(7+\sqrt{5})t} + c_2 e^{\frac{1}{2}(7-\sqrt{5})t} + \frac{1}{11} - \frac{1}{5}e^t.$$

Use this in the second equation to obtain the general solution for $x = x(t)$:

$$x(t) = \frac{1}{2}(-3 + \sqrt{5})c_1 e^{\frac{1}{2}(7+\sqrt{5})t} - \frac{1}{2}(3 + \sqrt{5})c_2 e^{\frac{1}{2}(7-\sqrt{5})t} - \frac{5}{11} - \frac{1}{5}e^t.$$

Setting $t = 0$ in the above formulas for $y(t)$ and $x(t)$ and using the initial conditions gives a two-equation system in c_1, c_2 with solution $c_1 = \frac{7}{110}(-7 + \sqrt{5})$, $c_2 = -\frac{7}{110}(7 + \sqrt{5})$. The complete solution is therefore

$$y(t) = \frac{7}{110}(-7 + \sqrt{5})e^{\frac{1}{2}(7+\sqrt{5})t} - \frac{7}{110}(7 + \sqrt{5})e^{\frac{1}{2}(7-\sqrt{5})t} + \frac{1}{11} - \frac{1}{5}e^t,$$

$$x(t) = \frac{7}{110}(13 - 5\sqrt{5})e^{\frac{1}{2}(7+\sqrt{5})t} + \frac{7}{110}(13 + 5\sqrt{5})e^{\frac{1}{2}(7-\sqrt{5})t} - \frac{5}{11} - \frac{1}{5}e^t.$$

7. Each eigenvalue of the coefficient matrix $A = \begin{pmatrix} 1 & -1 & 1 \\ 0 & 0 & 1 \\ 0 & -1 & 2 \end{pmatrix}$ satisfies

$$\det(A - \lambda I) = \begin{vmatrix} 1 - \lambda & -1 & 1 \\ 0 & -\lambda & 1 \\ 0 & -1 & 2 - \lambda \end{vmatrix} = (1 - \lambda)\begin{vmatrix} -\lambda & 1 \\ -1 & 2 - \lambda \end{vmatrix}$$

$$= (1 - \lambda)\big[- \lambda(2 - \lambda) + 1 \big] = (1 - \lambda)(\lambda^2 - 2\lambda + 1) = (1 - \lambda)^3 = 0.$$

Hence, $\lambda = 1$ is as a triple eigenvalue. By Theorem 2.4, the coefficient functions b_0, b_1, b_2 satisfying $e^{tA} = b_0 I + b_1 A + b_2 A^2$ also satisfy the three equations

$$e^t = b_0 + b_1 + b_2, \qquad te^t = b_1 + 2b_2, \qquad t^2 e^t = 2b_2.$$

Solving for b_0, b_1 and b_2 gives $b_0 = e^t - te^t + \frac{1}{2}t^2 e^t$, $b_1 = te^t - t^2 e^t$ and $b_2 = \frac{1}{2}t^2 e^t$. With $A^2 = \begin{pmatrix} 1 & -2 & 2 \\ 0 & -1 & 2 \\ 0 & -2 & 3 \end{pmatrix}$, we have

$$e^{tA} = b_0 I + b_1 A + b_2 A^2 = \begin{pmatrix} e^t & -te^t & te^t \\ 0 & e^t - te^t & te^t \\ 0 & -te^t & e^t + te^t \end{pmatrix}.$$

So the general solution can be written as $\mathbf{x}(t) = e^{tA}\mathbf{c}$. Setting $t = 0$ and applying the initial condition $\mathbf{x}(0) = \begin{pmatrix} 1 \\ 2 \\ 0 \end{pmatrix}$ immediately gives $\mathbf{c} = \begin{pmatrix} 1 \\ 2 \\ 0 \end{pmatrix}$. The desired solution is therefore

$$\mathbf{x}(t) = \begin{pmatrix} e^t & -te^t & te^t \\ 0 & e^t - te^t & te^t \\ 0 & -te^t & e^t + te^t \end{pmatrix} \begin{pmatrix} 1 \\ 2 \\ 0 \end{pmatrix} = \begin{pmatrix} e^t - 2te^t \\ 2e^t - 2te^t \\ -2te^t \end{pmatrix}.$$

9. Each eigenvalue of the coefficient matrix $A = \begin{pmatrix} 3 & -1 & -1 \\ 1 & 1 & -1 \\ 1 & -1 & 1 \end{pmatrix}$ satisfies

$$\det(A - \lambda I) = \begin{vmatrix} 3-\lambda & -1 & -1 \\ 1 & 1-\lambda & -1 \\ 1 & -1 & 1-\lambda \end{vmatrix}$$

$$= (3-\lambda)\begin{vmatrix} 1-\lambda & -1 \\ -1 & 1-\lambda \end{vmatrix} - \begin{vmatrix} -1 & -1 \\ -1 & 1-\lambda \end{vmatrix} + \begin{vmatrix} -1 & -1 \\ 1-\lambda & -1 \end{vmatrix} = (1-\lambda)(2-\lambda)^2.$$

Hence, the eigenvalues of A are 1 and 2, where 2 is a double eigenvalue. By Theorem 2.4, the coefficient functions b_0, b_1, b_2 satisfying $e^{tA} = b_0 I + b_1 A + b_2 A^2$ also satisfy the three equations

$$e^t = b_0 + b_1 + b_2, \qquad e^{2t} = b_0 + 2b_1 + 4b_2, \qquad te^{2t} = b_1 + 4b_2.$$

Solving for b_0, b_1 and b_2 gives

$$b_0 = 4e^t - 3e^{2t} + 2te^{2t}, \qquad b_1 = -4e^t + 4e^{2t} - 3te^{2t}, \qquad b_2 = e^t - e^{2t} + te^{2t}.$$

With $A^2 = \begin{pmatrix} 7 & -3 & -3 \\ 3 & 1 & -3 \\ 3 & -3 & 1 \end{pmatrix}$, we have

$$e^{tA} = b_0 I + b_1 A + b_2 A^2 = \begin{pmatrix} -e^t + 2e^{2t} & e^t - e^{2t} & e^t - e^{2t} \\ -e^t + e^{2t} & e^t & e^t - e^{2t} \\ -e^t + e^{2t} & e^t - e^{2t} & e^t \end{pmatrix}.$$

So the general solution can be written as $\mathbf{x}(t) = e^{tA}\mathbf{c}$. Setting $t = 0$ and applying the initial condition $\mathbf{x}(0) = \begin{pmatrix} 1 \\ 2 \\ 0 \end{pmatrix}$ immediately gives $\mathbf{c} = \begin{pmatrix} 1 \\ 2 \\ 0 \end{pmatrix}$. The desired solution is therefore

$$\mathbf{x}(t) = \begin{pmatrix} -e^t + 2e^{2t} & e^t - e^{2t} & e^t - e^{2t} \\ -e^t + e^{2t} & e^t & e^t - e^{2t} \\ -e^t + e^{2t} & e^t - e^{2t} & e^t \end{pmatrix}\begin{pmatrix} 1 \\ 2 \\ 0 \end{pmatrix} = \begin{pmatrix} e^t \\ e^t + 2e^{2t} \\ e^t - 2e^{2t} \end{pmatrix}.$$

11. Write the given system as the equivalent initial-value problem

$$\dot{x} = x + w + 1,$$
$$\dot{y} = 2y,$$
$$\dot{z} = 2z,$$
$$\dot{w} = w + e^t, \quad x(0) = 1,\ y(0) = 2,\ z(0) = 0,\ w(0) = 1.$$

The second and third equations immediately give $y(t) = c_1 e^{2t}$ and $z(t) = c_2 e^{2t}$. The initial conditions on y and z give $c_1 = 2$ and $c_2 = 0$, so that

$$y(t) = 2e^{2t} \qquad \text{and} \qquad z(t) \equiv 0.$$

536

The fourth equation can be solved using the integrating factor e^{-t} to get $w(t) = te^t + c_3e^t$. The initial condition on w gives $c_3 = 1$, so that

$$w(t) = te^t + e^t.$$

The first equation can then be written as $\dot{x} - x = te^t + e^t + 1$, where the integrating factor e^{-t} leads to the general solution $x(t) = \frac{1}{2}t^2e^t + te^t - 1 + c_4e^t$. The initial condition on x then gives $c_4 = 2$, so that

$$x(t) = \frac{1}{2}t^2e^t + te^t + 2e^t - 1.$$

13. Let $\dot{x} = z$ and $\dot{y} = w$, so that $\ddot{x} = \dot{z}$ and $\ddot{y} = \dot{w}$. Inserting these expressions for $\ddot{x}$ and $\ddot{y}$ into the given system yields $\dot{z} = x + y$ and $\dot{w} = x - y$. The equivalent first-order system of dimension 4 is then

$$\dot{x} = z,$$
$$\dot{y} = w,$$
$$\dot{z} = x + y,$$
$$\dot{w} = x - y,$$

or in matrix form,

$$\begin{pmatrix} \dot{x} \\ \dot{y} \\ \dot{z} \\ \dot{w} \end{pmatrix} = \begin{pmatrix} 0 & 0 & 1 & 0 \\ 0 & 0 & 0 & 1 \\ 1 & 1 & 0 & 0 \\ 1 & -1 & 0 & 0 \end{pmatrix} \begin{pmatrix} x \\ y \\ z \\ w \end{pmatrix}.$$

The characteristic equation of the system is found to be $\lambda^4 - 2 = 0$, so that the eigenvalues of the system are the four fourth-roots of 2. In order to simplify the computations, we let $\omega = 2^{1/4}$ an write the eigenvalues as $\lambda_1 = \omega$, $\lambda_2 = -\omega$, $\lambda_3 = i\omega$ and $\lambda_4 = -i\omega$. Letting A be the above displayed coefficient matrix, we use Theorem 2.4 in this chapter to conclude that the coefficient functions b_0, b_1, b_2, b_3 satisfying $e^{tA} = b_0I + b_1A + b_2A^2 + b_3A^3$ also satisfy the three equations

$$e^{\omega t} = b_0 + \omega b_1 + \omega^2 b_2 + \omega^3 b_3,$$
$$e^{-\omega t} = b_0 - \omega b_1 + \omega^2 b_2 - \omega^3 b_3,$$
$$e^{i\omega t} = b_0 + i\omega b_1 - \omega^2 b_2 - i\omega^3 b_3.$$

Using the fact that the b_k are real-valued, write both sides of the third equation in standard form for complex numbers and equate real and imaginary parts to obtain the two real equations

$$(*) \qquad \cos \omega t = b_0 - \omega^2 b_2 \qquad \text{and} \qquad \sin \omega t = \omega b_1 - \omega^3 b_3.$$

Adding and subtracting the first two equations of the three equation system gives two equations which, when divided by 2, can be written as

$$(**) \qquad \cosh \omega t = b_0 + \omega^2 b_2 \qquad \text{and} \qquad \sinh \omega t = \omega b_1 + \omega^3 b_3.$$

The left equation of $(*)$ and the left equation of $(**)$ can now be used to solve for b_0 and b_2. Similarly, the right equation of $(*)$ and the right equation of $(**)$ can be used to solve for b_1 and b_3. The results are

$$b_0 = \frac{1}{2}(\cos \omega t + \cosh \omega t),$$
$$b_1 = \frac{1}{2\omega}(\sin \omega t + \sinh \omega t),$$
$$b_2 = -\frac{1}{2\omega^2}(\cos \omega t - \cosh \omega t),$$
$$b_3 = -\frac{1}{2\omega^3}(\sin \omega t - \sinh \omega t).$$

Now use the equation $e^{tA} = b_0 I + b_1 A + b_2 A^2 + b_3 A^3$ along with the fact that

$$A^2 = \begin{pmatrix} 1 & 1 & 0 & 0 \\ 1 & -1 & 0 & 0 \\ 0 & 0 & 1 & 1 \\ 0 & 0 & 1 & -1 \end{pmatrix} \quad \text{and} \quad A^3 = \begin{pmatrix} 0 & 0 & 1 & 1 \\ 0 & 0 & 1 & -1 \\ 2 & 0 & 0 & 0 \\ 0 & 2 & 0 & 0 \end{pmatrix}$$

to obtain

$$e^{tA} = \begin{pmatrix} b_0 + b_2 & b_2 & b_1 + b_3 & b_3 \\ b_2 & b_0 - b_2 & b_3 & b_1 - b_3 \\ b_1 + 2b_3 & b_1 & b_0 + b_2 & b_2 \\ b_1 & -b_1 + 2b_3 & b_2 & b_0 - b_2 \end{pmatrix}.$$

Thus, the general solution of the above matrix system is

$$\mathbf{x}(t) = e^{tA}\mathbf{c} = \begin{pmatrix} b_0 + b_2 & b_2 & b_1 + b_3 & b_3 \\ b_2 & b_0 - b_2 & b_3 & b_1 - b_3 \\ b_1 + 2b_3 & b_1 & b_0 + b_2 & b_2 \\ b_1 & -b_1 + 2b_3 & b_2 & b_0 - b_2 \end{pmatrix} \begin{pmatrix} c_1 \\ c_2 \\ c_3 \\ c_4 \end{pmatrix}$$

$$= \begin{pmatrix} c_1(b_0 + b_2) + c_2 b_2 + c_3(b_1 + b_3) + c_4 b_3 \\ c_1 b_2 + c_2(b_0 - b_2) + c_3 b_3 + c_4(b_1 - b_3) \\ c_1(b_1 + 2b_3) + c_2 b_1 + c_3(b_0 + b_2) + c_4 b_2 \\ c_1 b_1 + c_2(-b_1 + 2b_3) + c_3 b_2 + c_4(b_0 - b_2) \end{pmatrix}.$$

In scalar form, this becomes

$$x(t) = c_1(b_0 + b_2) + c_2 b_2 + c_3(b_1 + b_3) + c_4 b_3,$$
$$y(t) = c_1 b_2 + c_2(b_0 - b_2) + c_3 b_3 + c_4(b_1 - b_3),$$
$$z(t) = c_1(b_1 + 2b_3) + c_2 b_1 + c_3(b_0 + b_2) + c_4 b_2,$$
$$w(t) = c_1 b_1 + c_2(-b_1 + 2b_3) + c_3 b_2 + c_4(b_0 - b_2),$$

where b_0, b_1, b_2 and b_3 are given above. The general solution of the given second-order system consists of the above equations for $x(t)$ and $y(t)$. We leave the explicit expansion of these equations to the reader.

15. (a) Let $\mathbf{x} = (x_1, \ldots, x_n)$ be a solution of the n-dimensional second-order initial-value problem $\ddot{\mathbf{x}} = -\mathbf{x}$, $\mathbf{x}(0) = \mathbf{x}_0$, $\dot{\mathbf{x}}(0) = \mathbf{z}_0$. Letting $\mathbf{x}_0 = (a_1, \ldots, a_n)$ and $\mathbf{z}_0 = (b_1, \ldots, b_n)$, the scalar form of the system is

$$\ddot{x}_1 = -x_1, \quad x_1(0) = a_1, \; \dot{x}_1(0) = b_1,$$

$$\vdots$$

$$\ddot{x}_n = -x_n, \quad x_n(0) = a_n, \; \dot{x}_n(0) = b_n.$$

For $k = 1, \ldots, n$, let $y_k = \dot{x}_k$. Then the kth equation of the above system becomes $\dot{y}_k = -x_k$. The pair of equations $\dot{x}_k = y_k$, $\dot{y}_k = -x_k$, along with the initial conditions $x_k(0) = a_k$, $y_k(0) = b_k$, is then a coupled 2-dimensional first-order initial-value problem. It follows that the first-order system of dimension $2n$

$$\begin{array}{llll} \dot{x}_1 = & y_1, & x_1(0) = a_1, \\ \dot{y}_1 = -x_1, & y_1(0) = b_1, \end{array} \quad \cdots \quad \begin{array}{ll} \dot{x}_n = & y_n, & x_n(0) = a_n, \\ \dot{y}_n = -x_n, & y_n(0) = b_n. \end{array}$$

538

is an uncoupled sequence of coupled 2-dimensional first-order initial-value problems. It is also a $2n$-dimensional first-order initial-value problem that is equivalent to the given n-dimensional second-order initial-value problem.

(b) Let $n = 1$ so that the given second-order system is $\ddot{x}_1 = -x_1$, or equivalently, $\ddot{x}_1 + x_1 = 0$. The characteristic equation $r^2 + 1 = 0$ gives the roots $\pm i$, so that the general solution is $x_1(t) = c_1 \cos t + c_2 \sin t$. Since $\dot{x}_1 = c_2 \cos t - c_1 \sin t$, we have $a_1 = x_1(0) = c_1$ and $b_1 = \dot{x}_1(0) = c_2$. Hence, the solution of the second-order initial-value problem is

$$x_1(t) = a_1 \cos t + b_1 \sin t.$$

Now let $n > 1$. The given initial-value problem can be solved by solving the equivalent $2n$-dimensional first-order initial-value problem that was derived in part (a). This amounts to independently solving each of the 2-dimensional first-order initial-value problems of which it is comprised. The above result for the case $n = 1$ implies that the kth 2-dimensional system has the solution

$$x_k(t) = a_k \cos t + b_k \sin t,$$
$$y_k(t) = - a_k \sin t + b_k \cos t,$$

and therefore the solution $\mathbf{x} = \mathbf{x}(t)$ of the given n-dimensional second-order system is

$$\mathbf{x}(t) = \begin{pmatrix} x_1(t) \\ \vdots \\ x_n(t) \end{pmatrix} = \begin{pmatrix} a_1 \cos t + b_1 \sin t \\ \vdots \\ a_n \cos t + b_n \sin t \end{pmatrix}.$$

(c) For each k, the 2-dimensional first-order initial-value problem $\dot{x}_k = y_k$, $\dot{y}_k = -x_k$, $x_k(0) = a_k$, $y_k(0) = b_k$, in matrix form is

$$\begin{pmatrix} \dot{x}_k \\ \dot{y}_k \end{pmatrix} = \begin{pmatrix} 0 & 1 \\ -1 & 0 \end{pmatrix} \begin{pmatrix} x_k \\ y_k \end{pmatrix}, \qquad \begin{pmatrix} x_k(0) \\ y_k(0) \end{pmatrix} = \begin{pmatrix} a_k \\ b_k \end{pmatrix}.$$

Let $A = \begin{pmatrix} 0 & 1 \\ -1 & 0 \end{pmatrix}$ and observe that the solution of this 2-dimensional initial-value problem can be written as

$$\begin{pmatrix} x_k(t) \\ y_k(t) \end{pmatrix} = \begin{pmatrix} a_k \cos t + b_k \sin t \\ -a_k \sin t + b_k \cos t \end{pmatrix} = \begin{pmatrix} \cos t & \sin t \\ -\sin t & \cos t \end{pmatrix} \begin{pmatrix} a_k \\ b_k \end{pmatrix},$$

where $e^{tA} = \begin{pmatrix} \cos t & \sin t \\ -\sin t & \cos t \end{pmatrix}$. For the general case, we first write the $2n$-dimensional system derived in part (a) in matrix form:

$$\begin{pmatrix} \dot{x}_1 \\ \dot{y}_1 \\ \dot{x}_2 \\ \dot{y}_2 \\ \vdots \\ \dot{x}_n \\ \dot{y}_n \end{pmatrix} = \begin{pmatrix} 0 & 1 & 0 & 0 & \cdots & 0 & 0 \\ -1 & 0 & 0 & 0 & \cdots & 0 & 0 \\ 0 & 0 & 0 & 1 & \cdots & 0 & 0 \\ 0 & 0 & -1 & 0 & \cdots & 0 & 0 \\ \vdots & \vdots & \vdots & \vdots & \ddots & \vdots & \vdots \\ 0 & 0 & 0 & 0 & \cdots & 0 & 1 \\ 0 & 0 & 0 & 0 & \cdots & -1 & 0 \end{pmatrix} \begin{pmatrix} x_1 \\ y_1 \\ x_2 \\ y_2 \\ \vdots \\ x_n \\ y_n \end{pmatrix}.$$

Note that the above coefficient matrix, call it $\mathcal{A}$, has n copies of the 2×2 matrix A "down its diagonal" and zeros elsewhere. It follows that $e^{t\mathcal{A}}$ has n copies of the 2×2 exponential matrix e^{tA} "down its diagonal" and zeros elsewhere. That is,

$$
e^{t\mathcal{A}} = \begin{pmatrix}
\cos t & \sin t & 0 & 0 & \cdots & 0 & 0 \\
-\sin t & \cos t & 0 & 0 & \cdots & 0 & 0 \\
0 & 0 & \cos t & \sin t & \cdots & 0 & 0 \\
0 & 0 & -\sin t & \cos t & \cdots & 0 & 0 \\
\vdots & \vdots & \vdots & \vdots & \ddots & \vdots & \vdots \\
0 & 0 & 0 & 0 & \cdots & \cos t & \sin t \\
0 & 0 & 0 & 0 & \cdots & -\sin t & \cos t
\end{pmatrix}.
$$

By how $\mathcal{A}$ was constructed, this is fundamental matrix for the $2n$-dimensional system derived in part (a). Moreover, evaluating this matrix at $t = 0$ gives the $2n$-dimensional identity matrix, and evaluation of its derivative at $t = 0$ gives the matrix $\mathcal{A}$. This is good evidence that it is an exponential matrix and therefore must be the exponential matrix $e^{t\mathcal{A}}$.

Chapter 14: Infinite Series

Section 1: EXAMPLES AND DEFINITIONS

Exercise Set 1 (pgs. 662-664)

1. Let $a_k = \frac{1}{k} - \frac{1}{k+1}$ and set $m = 1$, $n = 5$ in the definition $\sum_{k=m}^{n} a_k = a_m + \cdots + a_n$ to get

$$\sum_{k=1}^{5} \left(\frac{1}{k} - \frac{1}{k+1} \right) = \left(\frac{1}{1} - \frac{1}{2} \right) + \left(\frac{1}{2} - \frac{1}{3} \right) + \left(\frac{1}{3} - \frac{1}{4} \right) + \left(\frac{1}{4} - \frac{1}{5} \right) + \left(\frac{1}{5} - \frac{1}{6} \right) = 1 - \frac{1}{6} = \frac{5}{6}.$$

3. Let $a_k = 2^{-k}$ and set $m = 0$, $n = 4$ in the definition $\sum_{k=m}^{n} a_k = a_m + \cdots + a_n$ to get

$$\sum_{k=0}^{4} 2^{-k} = 1 + 2^{-1} + 2^{-2} + 2^{-3} + 2^{-4} = \frac{1 - 2^{-5}}{1 - 2^{-1}} = \frac{1 - \frac{1}{32}}{\frac{1}{2}} = \frac{32 - 1}{16} = \frac{31}{16}.$$

5. $a_k = \dfrac{1}{k}$ **7.** $a_k = \dfrac{1}{k(k+2)}$ **9.** $a_k = \dfrac{k+3}{(k+1)3^{k+1}}$

11. Using Example 3 in the text,

$$\sum_{k=1}^{n} \frac{1}{2k(2k+2)} = \sum_{k=1}^{n} \frac{1}{4k(k+1)} = \frac{1}{4} \sum_{k=1}^{n} \frac{1}{k(k+1)}$$

$$= \frac{1}{4} \sum_{k=1}^{n} \left(\frac{1}{k} - \frac{1}{k+1} \right) = \frac{1}{4} \left(1 - \frac{1}{n+1} \right) = \frac{n}{4(n+1)}.$$

Hence,

$$\sum_{k=1}^{\infty} \frac{1}{2k(2k+2)} = \lim_{n \to \infty} \frac{n}{4(n+1)} = \frac{1}{4} \lim_{n \to \infty} \frac{1}{1 + 1/n} = \frac{1}{4}.$$

13. By inspection, the formula holds for $n = 1$. Now let n_0 be a positive integer such that $\sum_{k=1}^{n_0} 2k = n_0(n_0 + 1)$. Then

$$\sum_{k=1}^{n_0+1} 2k = \sum_{k=1}^{n_0} 2k + 2(n_0 + 1) = n_0(n_0 + 1) + 2(n_0 + 1) = (n_0 + 1)(n_0 + 2).$$

Hence, the formula holds for $n = n_0 + 1$ whenever it holds for $n = n_0$. By mathematical induction, it holds for all $n \geq 1$. However, since $\lim_{n \to \infty} n(n + 1) = \infty$, the given series diverges to ∞.

15. The common ratio is $\frac{1}{6}$, with absolute value less than 1. By Example 1 in the text,

$$\sum_{k=0}^{\infty} \left(\frac{1}{6} \right)^k = \frac{1}{1 - 1/6} = \frac{6}{5}.$$

17. The common ratio is $1/(1+\pi)$, with absolute value less than 1. However, the first term begins with $k=1$ instead of $k=0$, as exemplified in Example 1 in the text. It follows that the given sum equals the sum as computed in Example 1, with the first term 1 subtracted off. That is,

$$\sum_{k=1}^{\infty}\left(\frac{1}{1+\pi}\right)^k = -1 + \sum_{k=0}^{\infty}\left(\frac{1}{1+\pi}\right)^k = -1 + \frac{1}{1-\frac{1}{1+\pi}} = -1 + \frac{1+\pi}{\pi} = \frac{1}{\pi}.$$

19. The common ratio is e^{-2}, with absolute value less than 1. By Example 1 in the text,

$$\sum_{k=0}^{\infty}e^{-2k} = \frac{1}{1-e^{-2}} = \frac{e^2}{e^2-1}.$$

21. We have

$$0.888\overline{8} = \frac{8}{10} + \frac{8}{100} + \frac{8}{1000} + \cdots = \frac{8}{10} + \frac{8}{10^2} + \frac{8}{10^3} + \cdots = \sum_{k=1}^{\infty}\frac{8}{10^k} = 8\sum_{k=1}^{\infty}\left(\frac{1}{10}\right)^k$$

$$= 8\left[-1 + \sum_{k=0}^{\infty}\left(\frac{1}{10}\right)^k\right] = 8\left[-1 + \frac{1}{1-1/10}\right] = 8\left[-1 + \frac{10}{9}\right] = \frac{8}{9}.$$

23. We have

$$0.123123\overline{123} = \frac{123}{1000} + \frac{123}{1\,000\,000} + \frac{123}{1\,000\,000\,000} + \cdots = \frac{123}{1000} + \frac{123}{1000^2} + \frac{123}{1000^3} + \cdots$$

$$= \sum_{k=1}^{\infty}\frac{123}{1000^k} = 123\sum_{k=1}^{\infty}\left(\frac{1}{1000}\right)^k = 123\left[-1 + \sum_{k=0}^{\infty}\left(\frac{1}{1000}\right)^k\right]$$

$$= 123\left[-1 + \frac{1}{1-1/1000}\right] = 123\left[-1 + \frac{1000}{999}\right] = \frac{123}{999}.$$

25. To simplify matters, we shall say that a real number has the "**repeating property**" if it has an infinite decimal expansion that repeats periodically from some point on. Let ω have the repeating property. If ω is an integer then ω is rational and there is nothing to show. If $|\omega| > 1$ and is not an integer then there exists a positive integer n and a real number $0 < \omega' < 1$ such that $\omega = n + \omega'$ (if $\omega > 1$) or $\omega = -(n+\omega')$ (if $\omega < -1$). In either case, ω' has the repeating property and ω is rational if ω' is rational. It follows that we can restrict our attention to those ω with the repeating property that satisfy $0 < \omega < 1$. For such ω we can write

$$\omega = 0.a_1 \cdots a_r \overline{a_{r+1} \cdots a_{r+s}},$$

where $r \geq 1$, $s \geq 1$, and each a_k is an integer satisfying $0 \leq a_k \leq 9$. We then have the following string of equivalent equations:

$$10^r\omega = a_1 \cdots a_r.\overline{a_{r+1} \cdots a_{r+s}},$$

$$10^r\omega - (a_1 \cdots a_r) = 0.\overline{a_{r+1} \cdots a_{r+s}},$$

$$10^s\left[10^r\omega - (a_1 \cdots a_r)\right] = a_{r+1} \cdots a_{r+s}.\overline{a_{r+s+1} \cdots a_{r+2s}},$$

$$10^s\left[10^r\omega - (a_1 \cdots a_r)\right] - (a_{r+1} \cdots a_{r+s}) = 0.\overline{a_{r+s+1} \cdots a_{r+2s}},$$

$$10^s\left[10^r\omega - (a_1 \cdots a_r)\right] - (a_{r+1} \cdots a_{r+s}) = 10^r\omega - (a_1 \cdots a_r).$$

Solving for ω gives

$$\omega = \frac{1}{10^r} \left[\frac{a_{r+1} \cdots a_{r+s}}{10^s - 1} + (a_1 \cdots a_r) \right].$$

Since $a_1 \cdots a_r$, $a_{r+1} \cdots a_{r+s}$, $10^s - 1$ and 10^r are all integers, ω is rational. By what we have shown above, every real number with the repeating property is rational.

27. The general term of the given series can be written as

$$\frac{1}{2 \cdot 4 \cdots (2k)} = \frac{1}{2^k (1 \cdot 2 \cdots k)} = \frac{1}{2^k k!}.$$

The given series can therefore be written as

$$\sum_{k=1}^{\infty} \frac{1}{2^k k!} = \frac{1}{2 \cdot 1!} + \frac{1}{2^2 \cdot 2!} + \frac{1}{2^3 \cdot 3!} + \cdots = \frac{1}{2} + \frac{1}{8} + \frac{1}{48} + \cdots.$$

29. The general term of the given series can be written as

$$\frac{1}{k(k+1) \cdots (2k-1)(2k)} = \frac{1 \cdot 2 \cdots (k-1)}{1 \cdot 2 \cdots (k-1)} \cdot \frac{1}{k(k+1) \cdots (2k-1)(2k)} = \frac{(k-1)!}{(2k)!}.$$

The given series can therefore be written as

$$\sum_{k=1}^{\infty} \frac{(k-1)!}{(2k)!} = \frac{0!}{2!} + \frac{1!}{4!} + \frac{2!}{6!} + \cdots = \frac{1}{2} + \frac{1}{24} + \frac{1}{360} + \cdots.$$

31. The general term of the given series can be written as

$$\frac{(-1)^k}{1 \cdot 3 \cdots (2k+1)} = \frac{2 \cdot 4 \cdots (2k)}{2 \cdot 4 \cdots (2k)} \cdot \frac{(-1)^k}{1 \cdot 3 \cdot (2k+1)} = (-1)^k \frac{2^k (1 \cdot 2 \cdots k)}{1 \cdot 2 \cdot 3 \cdot 4 \cdots (2k)(2k+1)}$$

$$= (-1)^k \frac{2^k k!}{(2k+1)!}.$$

The given series can therefore be written as

$$\sum_{k=0}^{\infty} (-1)^k \frac{2^k k!}{(2k+1)!} = \frac{2^0 \cdot 0!}{1!} - \frac{2 \cdot 1!}{3!} + \frac{2^2 \cdot 2!}{5!} + \cdots = 1 - \frac{1}{3} + \frac{1}{15} - \cdots.$$

33. Setting $n = 1, 2, 3$ in the given expression for s_n gives $s_1 = 2 - \frac{1}{1} = 1$, $s_2 = 2 - \frac{1}{2} = \frac{3}{2}$ and $s_3 = 2 - \frac{1}{3} = \frac{5}{3}$. Also,

$$\lim_{n \to \infty} s_n = \lim_{n \to \infty} \left(2 - \frac{1}{n} \right) = 2.$$

35. Setting $n = 1, 2, 3$ in the given expression for s_n gives $s_1 = \frac{2}{3^2} = \frac{2}{9}$, $s_2 = \frac{2^2}{3^3} = \frac{4}{27}$ and $s_3 = \frac{2^3}{3^4} = \frac{8}{81}$. Also,

$$\lim_{n \to \infty} s_n = \lim_{n \to \infty} \frac{2^n}{3^{n+1}} = \frac{1}{3} \lim_{n \to \infty} \left(\frac{2}{3} \right)^n = \frac{1}{3} \cdot 0 = 0.$$

37. Setting $n = 1, 2, 3$ in the given expression for s_n gives

$$s_1 = \frac{1^{10} + 1}{1(1^9 + 1)} = 1, \qquad s_2 = \frac{2^{10} + 1}{2(2^{10} + 1)} = \frac{1025}{1026}, \qquad s_3 = \frac{3^{10} + 1}{3(3^9 + 1)} = \frac{29525}{29526}.$$

Also,

$$\lim_{n \to \infty} s_n = \lim_{n \to \infty} \frac{n^{10} + 1}{n(n^9 + 1)} = \lim_{n \to \infty} \frac{1 + n^{-10}}{1 + n^{-9}} = 1.$$

39. The given series is a geometric series with common ratio $\frac{1}{3}$ but starts at $k = 2$ instead of $k = 0$. However, we have

$$\sum_{k=2}^{\infty} \frac{1}{3^k} = \sum_{k=0}^{\infty} \frac{1}{3^{k+2}} = \frac{1}{9} \sum_{k=0}^{\infty} \frac{1}{3^k} = \frac{1}{9} \left(\frac{1}{1 - 1/3} \right) = \frac{1}{6}.$$

41. The given series can be written as the telescoping series $\sum_{k=3}^{\infty} \left(\frac{1}{k} - \frac{1}{k+1} \right)$. The nth partial sum is then $\sum_{k=3}^{n} (\frac{1}{k} - \frac{1}{k+1}) = \frac{1}{3} - \frac{1}{n+1}$. Letting $n \to \infty$ gives the desired result.

43. So far, the only convergent series we know of whose terms alternate in sign is the geometric series $\sum_{k=0}^{\infty} x^k$ with $-1 < x < 0$. Since the sum of such a series is $1/(1-x)$, it follows that all such series have sums in the open interval $(\frac{1}{2}, 1)$. Let r be any real number in this interval and solve the equation $1/(1-x) = r$ to get $x = 1 - 1/r$. Then consecutive integer powers of x have opposite signs and

$$\sum_{k=0}^{\infty} \left(1 - \frac{1}{r} \right)^k = r \qquad \text{implies} \qquad \sum_{k=0}^{\infty} \frac{3}{r} \left(1 - \frac{1}{r} \right)^k = \frac{3}{r} \sum_{k=0}^{\infty} \left(1 - \frac{1}{r} \right)^k = \frac{3}{r} \cdot r = 3.$$

This provides the reader with infinitely many "alternating series" that converge to 3.

45. (a) First of all, the number of subintervals remaining after the nth step is twice the number remaining after the $(n-1)$st step. Moreover, it is fairly easy to see that the subintervals remaining after the nth step are all of the same length. Specifically, we have

After the 1st step, there are two subintervals of length $1/3$;

After the 2nd step, there are four subintervals of length $1/9$;

After the 3rd step, there are eight subintervals of length $1/27$;

$$\vdots$$

Thus, we recognize the general pattern to be

After the nth step, there are 2^n subintervals of length $\dfrac{1}{3^n}$.

It follows that the sum c_n of the lengths of the intervals remaining after n steps is

$$c_n = 2^n \cdot \frac{1}{3^n} = \left(\frac{2}{3} \right)^n, \quad n = 1, 2, 3, \ldots.$$

(b) Let $\epsilon > 0$ be given. Now let n be any positive integer satisfying $n > \ln \epsilon / \ln(2/3)$. Multiply both sides by the negative number $\ln(2/3)$ (thereby reversing the inequality sign) and exponentiate the result to get $0 < (2/3)^n < \epsilon$, or equivalently, $0 < c_n < \epsilon$. Thus, c_n is within ϵ units of zero for all sufficiently large n. Since ϵ is arbitrary, $\lim_{n \to \infty} c_n = 0$.

Section 2: TAYLOR SERIES

Exercise Set 2AB (pgs. 669-670)

1. With $f(x) = 1 + x + x^2$, we have $f'(x) = 1 + 2x$, $f''(x) = 2$ and $f^{(k)}(x) = 0$ for $k \geq 3$. Letting $x = 2$ gives $f(2) = 7$, $f'(2) = 5$, $f''(2) = 2$ and $f^{(k)}(2) = 0$ for $k \geq 3$. By Theorem 2.1,

$$f(x) = f(2) + f'(2)(x - 2) + \frac{f''(2)}{2}(x - 2)^2 = 7 + 5(x - 2) + (x - 2)^2.$$

3. With $f(x) = 1 + x^2$, we have $f'(x) = 2x$, $f''(x) = 2$ and $f^{(k)}(x) = 0$ for $k \geq 3$. Letting $x = -1$ gives $f(-1) = 2$, $f'(-1) = -2$, $f''(-1) = 2$ and $f^{(k)}(-1) = 0$ for $k \geq 0$. By Theorem 2.1,

$$f(x) = f(-1) + f'(-1)(x + 1) + \frac{f''(-1)}{2}(x + 1)^2 = 2 - 2(x + 1) + (x + 1)^2.$$

5. The function $f(x) = x^{1/3}$ has derivatives of all orders for $x > 0$. The first three are

$$f'(x) = \frac{1}{3}x^{-2/3}, \quad f''(x) = -\frac{2}{9}x^{-5/3}, \quad f'''(x) = \frac{10}{27}x^{-8/3}.$$

Setting $x = 1$ gives $f(1) = 1$, $f'(1) = 1/3$ and $f''(1) = -2/9$. Moreover, for each $x > 0$ there is a number c between x and 1 such that the remainder $R_2(x)$ associated with the 2nd degree Taylor polynomial of $f(x)$ at $x = 1$ is given by

$$R_2(x) = \frac{f'''(c)}{3!}(x - 1)^3 = \frac{5}{81}c^{-8/3}(x - 1)^3.$$

Hence, by Theorem 2.2,

$$f(x) = \sum_{k=0}^{2} \frac{1}{k!}f^{(k)}(1)(x - 1)^k + R_2(x) = 1 + \frac{1}{3}(x - 1) - \frac{1}{9}(x - 1)^2 + \frac{5}{81}c^{-8/3}(x - 1)^3.$$

7. The function $f(x) = x^{-1}$ has derivatives of all orders for $x < 0$. The first three are

$$f'(x) = -x^{-2}, \quad f''(x) = 2x^{-3}, \quad f'''(x) = -6x^{-4}.$$

Setting $x = -1$ gives $f(-1) = -1$, $f'(-1) = -1$, $f''(-1) = -2$. Moreover, for each $x < 0$ there is a number c between x and -1 such that the remainder $R_2(x)$ associated with the 2nd degree Taylor polynomial of $f(x)$ at $x = -1$ is given by

$$R_2(x) = \frac{f'''(c)}{3!}(x + 1)^3 = -c^{-4}(x + 1)^3.$$

Hence, by Theorem 2.2,

$$f(x) = \sum_{k=0}^{2} \frac{1}{k!}f^{(k)}(-1)(x + 1)^k + R_2(x) = -1 - (x + 1) - (x + 1)^2 - c^{-4}(x + 1)^3.$$

9. Since the infinite series given by Formula 2.4(a) in the text converges to e^x for all x, replace x with $-x$ to get

$$e^{-x} = \sum_{k=0}^{\infty} \frac{(-x)^k}{k!} = \sum_{k=0}^{\infty} \frac{(-1)^k}{k!} x^k,$$

which holds for $-\infty < x < \infty$.

11. Since the infinite series given by Formula 2.4(c) in the text converges to $\sin x$ for all x, replace x with $x/2$ to get

$$\sin(x/2) = \sum_{k=0}^{\infty} \frac{(-1)^k}{(2k+1)!} \left(\frac{x}{2}\right)^{2k+1} = \sum_{k=0}^{\infty} \frac{(-1)^k}{2^{2k+1}(2k+1)!} x^{2k+1},$$

which converges for $-\infty < x < \infty$.

13. Since the infinite series given by Formula 2.4(a) in the text converges to e^x for all x, replace x with x^2 to get

$$e^{x^2} = \sum_{k=0}^{\infty} \frac{1}{k!} (x^2)k = \sum_{k=0}^{\infty} \frac{1}{k!} x^{2k},$$

which holds for all x.

15. We have

$$\cosh x = \frac{e^x + e^{-x}}{2} = \frac{1}{2} e^x + \frac{1}{2} e^{-x} = \frac{1}{2} \sum_{k=0}^{\infty} \frac{1}{k!} x^k + \frac{1}{2} \sum_{k=0}^{\infty} \frac{1}{k!} (-x)^k$$

$$= \frac{1}{2} \sum_{k=0}^{\infty} \frac{1}{k!} x^k + \frac{1}{2} \sum_{k=0}^{\infty} (-1)^k \frac{1}{k!} x^k$$

$$= \frac{1}{2} \sum_{k=0}^{\infty} \left(\frac{1}{k!} + (-1)^k \frac{1}{k!} \right) x^k.$$

The coefficient of x^k is 0 when k is odd, and is $2/k!$ when k is even. Thus, we can replace k with $2k$ to obtain

$$\cosh x = \frac{1}{2} \sum_{k=0}^{\infty} \frac{2}{(2k)!} x^{2k} = \sum_{k=0}^{\infty} \frac{x^{2k}}{(2k)!}.$$

17. Let $f(x) = (x+a)^n$. Taking successive derivatives of f, one finds that the k th derivative has the form

$$f^{(k)}(x) = \begin{cases} (x+a)^n, & \text{for } k = 0; \\ [n(n-1)\cdots(n-k+1)](x+a)^{n-k}, & \text{for } 1 \leq k \leq n; \\ 0, & \text{for } k > n. \end{cases}$$

Thus, evaluation at $x = 0$ gives

$$f^{(k)}(0) = \begin{cases} a^n, & \text{for } k = 0; \\ [n(n-1)\cdots(n-k+1)]a^{n-k}, & \text{for } 0 \leq k \leq n; \\ 0, & \text{for } k > n. \end{cases}$$

By Theorem 2.1, when the polynomial of degree n is written as powers $(x - 0)^k = x^k$, the coefficient of x^k is

$$\frac{f^{(k)}(0)}{k!} = \frac{[n(n-1)\cdots(n-k+1)]a^{n-k}}{k!} = \frac{n(n-1)\cdots(n-k+1)}{k!}a^{n-k}$$

$$= \frac{n(n-1)\cdots(n-k+1)}{k!} \cdot \frac{(n-k)!}{(n-k)!}a^{n-k} = \frac{n!}{k!(n-k)!}a^{n-k} = \binom{n}{k}a^{n-k}.$$

That is,

$$(x + a)^n = \sum_{k=0}^{n} \binom{n}{k} a^{n-k} x^k,$$

which is the binomial theorem.

19. Let $f(x) = \cos x$. Given x, there exists a number c between x and 0 such that the remainder $R_4(x)$ associated with the 4th degree Taylor approximation of $f(x)$ is

$$R_4(x) = \frac{f^{(5)}(c)}{5!}x^5 = -\frac{1}{120}(\sin c)x^5.$$

Using the fact that $|\sin c| \leq 1$ for *any* real number c, on the interval $-\pi \leq x \leq \pi$, the above formula for $R_4(x)$ satisfies

$$|R_4(x)| = \frac{1}{120}|\sin c||x|^5 \leq \frac{\pi^5}{120}.$$

The graph of $f(x)$ and the approximation $T_4(x) = 1 - \frac{1}{2}x^2 + \frac{1}{24}x^4$ of f for $\pi \leq x \leq \pi$ are shown on the same set of axes (below left).

21. Let $f(x) = \sin x$. Given x, there exists a number c between x and 0 such that the remainder $R_5(x)$ associated with the 5th degree Taylor approximation of $f(x)$ about $x = 0$ is

$$R_5(x) = \frac{f^{(6)}(c)}{6!}x^6 = -\frac{1}{720}(\sin c)x^6.$$

Using the fact that $|\sin c| \leq 1$ for *any* real number c, on the interval $-\pi \leq x \leq \pi$, the above formula for $R_5(x)$ satisfies

$$|R_5(x)| = \frac{1}{720}|\sin c||x|^6 \leq \frac{\pi^6}{720}.$$

The graph of $f(x)$ and the approximation $T_5(x) = x - \frac{1}{6}x^3 + \frac{1}{120}x^5$ of f for $\pi \leq x \leq \pi$ are shown on the same set of axes (below right).

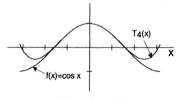

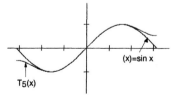

Exercise 19   Exercise 21

547

23. (a) Let $f(x) = \cos x$. We have $f'(x) = -\sin x$, $f''(x) = -\cos x$, $f'''(x) = \sin x$ and $f^{(4)}(x) = \cos x = f(x)$. Hence, the four-term finite sequence $\cos x$, $-\sin x$, $-\cos x$, $\sin x$ repeats infinitely often. Specifically, for any integer $k \geq 0$, we have

$$f^{(4k)}(x) = \cos x, \quad f^{(4k+1)}(x) = -\sin x, \quad f^{4k+2}(x) = -\cos x, \quad f^{(4k+3)}(x) = \sin x.$$

Evaluation at $x = 0$ shows that the odd-order derivatives (i.e., $n = 4k+1$ and $n = 4k+3$) are zero, and the even-order derivatives (i.e., $n = 4k$ and $n = 4k+2$) alternate between $+1$ and -1. Since $n/2$ is even when $n = 4k$ and odd when $n = 4k+2$, it follows that $f^{(n)}(0) = (-1)^{n/2}$ whenever n is even. That is, the Taylor coefficients of f about $x = 0$ are

$$\frac{f^{(n)}(0)}{n!} = \begin{cases} 0, & n \text{ odd}, \\ (-1)^{n/2}/n!, & n \text{ even}. \end{cases}$$

(b) By part (a), the Taylor remainder theorem implies that, for each $n \geq 0$,

$$(*) \qquad \cos x = \sum_{k=0}^{n} \frac{(-1)^k}{(2k)!} x^{2k} + R_n(x).$$

For each fixed x, the remainder $R_n(x)$ satisfies

$$|R_n(x)| = \left| \frac{f^{(n+1)}(c)}{(n+1)!} \right| |x|^{n+1} = |f^{(n+1)}(c)| \left| \frac{x}{(n+1)!} \right|^{n+1}$$

for some c between x and 0. Since $|f^{(n+1)}(c)|$ is either $|\cos c|$ or $|\sin c|$ (depending on the parity of n), and since both of these are bounded above by 1, it follows that

$$|R_n(x)| \leq \left| \frac{x}{(n+1)!} \right|^{n+1}.$$

By Example 4 in the text, for all x, the expression on the right tends to zero as $n \to \infty$, so that $R_n(x) \to 0$. Letting $n \to \infty$ on both sides of $(*)$ and using the definition of an infinite series as the limit of partial sums establishes the formula

$$\cos x = \sum_{k=0}^{\infty} \frac{(-1)^k}{(2k)!} x^{2k}, \quad -\infty < x < \infty.$$

25. Let $f(x) = \ln(1+x)$. Taking successive derivatives, one finds that the general formula for the kth derivatives of f is

$$f^{(k)}(x) = (-1)^{k+1}(k-1)!(1+x)^{-k}, \quad k = 1, 2, \ldots.$$

Setting $x = 0$ gives $f^{(k)}(0) = (-1)^{k+1}(k-1)!$, so that the Taylor coefficients of the expansion of f about $x = 0$ are

$$\frac{f^{(k)}(0)}{k!} = \frac{(-1)^{k+1}(k-1)!}{k!} = \frac{(-1)^{k+1}}{k}, \quad k = 1, 2, \ldots.$$

Using the fact that $f(0) = 0$, the Taylor remainder theorem gives

$$(*)\quad \ln(1+x) = \sum_{k=1}^{n} \frac{f^{(k)}(0)}{k!} x^k + R_n(x) = \sum_{k=1}^{n} \frac{(-1)^{k+1}}{k} x^k + \frac{(-1)^{n+2}(1+c)^{-n-1}}{n+1} x^{n+1}$$

$$= \sum_{k=1}^{n} \frac{(-1)^{k+1}}{k} x^k + \frac{(-1)^n}{n+1} \left(\frac{x}{1+c}\right)^{n+1},$$

where c is a number between x and 0. We now show that the remainder $R_n(x)$ tends to zero as $n \to \infty$.

For $x > 0$ we have $0 < c < x$, so that $1 + c > 1$. Thus, $0 < x/(1+c) < 1$ when $x < 1$, so that $(x/(1+c))^{n+1} \to 0$ as $n \to \infty$. For $x < 0$ we have $x < c < 0$, so that $x - 1 < -1 < c$ when $x > -1$. This is equivalent to $x < 1 + c$ for $-1 < x < 0$. Again, $(x/(1+c))^{n+1} \to 0$ as $n \to \infty$. Since the factor $(-1)^n/(n+1)$ also tends to zero as $n \to \infty$, it follows that

$$\lim_{n \to \infty} R_n(x) = \lim_{n \to \infty} \left[\frac{(-1)^n}{n+1} \left(\frac{x}{1+c}\right)^{n+1}\right] = 0, \quad \text{for } -1 < x < 1.$$

Letting $n \to \infty$ in $(*)$ and using the definition of the sum of an infinite series as the limit of the sequence of its partial sums, we obtain

$$\ln(1+x) = \sum_{k=1}^{\infty} \frac{(-1)^{k+1}}{k} x^k, \quad -1 < x < 1.$$

27. Setting $x = 1$ and $x = -1$ in Formula 2.4(a) in the text gives

$$e = \sum_{k=0}^{\infty} \frac{1}{k!} \quad \text{and} \quad e^{-1} = \sum_{k=0}^{\infty} \frac{(-1)^k}{k!},$$

respectively. It follows that

$$\left(\sum_{k=0}^{\infty} \frac{(-1)^k}{k!}\right) \left(\sum_{k=0}^{\infty} \frac{1}{k!}\right) = e^{-1}e = 1.$$

29. (a) Let $f(x)$ be an even function, so that $f(x) = f(-x)$ for all x. Differentiating both sides of this equation n times gives

$$f^{(n)}(x) = (-1)^n f^{(n)}(-x).$$

In particular, if n is odd then $f^{(n)}(x) = -f^{(n)}(-x)$. Setting $x = 0$ gives $f^{(n)}(0) = -f^{(n)}(0)$, or equivalently, $2f^{(n)}(0) = 0$. Hence, $f^{(n)}(0) = 0$ when n is odd.

(b) Let $f(x)$ be an odd function, so that $f(-x) = -f(x)$ for all x. Differentiating both sides of this equation n times gives

$$(-1)^n f^{(n)}(-x) = -f^{(n)}(x).$$

In particular, if n is even then $f^{(n)}(-x) = -f^{(n)}(x)$. Setting $x = 0$ gives $f^{(n)}(0) = -f^{(n)}(0)$, or equivalently, $2f^{(n)}(0) = 0$. Hence, $f^{(n)}(0) = 0$ when n is even.

(c) Since the coefficient of x^n in the Taylor expansion about $x = 0$ is a constant multiple of $f^{(n)}(0)$, part (a) shows that the coefficients associated with odd powers of x are all zero whenever f is an even function, and part (b) shows that the coefficients associated with even powers of x are all zero whenever f is an odd function. Thus, the Taylor expansions are of the form

$$f(x) = \sum_{k=0}^{\infty} a_k x^{2k} \quad (f \text{ even}) \quad \text{and} \quad f(x) = \sum_{k=0}^{\infty} a_k x^{2k+1} \quad (f \text{ odd}).$$

Section 3: CONVERGENCE CRITERIA

Exercise Set 3AB (pgs. 673-674)

1. We have

$$\lim_{n \to \infty} s_n = \lim_{n \to \infty} \left(1 + \frac{1}{n^2} \right) = 1 + \lim_{n \to \infty} \frac{1}{n^2} = 1,$$

where the last equality follows from equation 3.2(a) in the text with $\alpha = 2$.

3. Since

$$s_n = \frac{1 + n^2}{n} = \frac{1}{n} + n > n,$$

it follows that $\lim_{n \to \infty} s_n \geq \lim_{n \to \infty} n = +\infty$. So the given sequence does not converge.

5. Using Equation 3.2(b) in the text with $b = 3$ and $b = 3/2$, we obtain

$$\lim_{n \to \infty} s_n = \lim_{n \to \infty} \frac{1 + 2^n}{3^n} = \lim_{n \to \infty} \left(\frac{1}{3^n} + \frac{2^n}{3^n} \right) = \lim_{n \to \infty} \left(\frac{1}{3^n} + \frac{1}{(3/2)^n} \right) = 0.$$

7. Since $\cos n\pi = (-1)^n$, the terms of the given sequence can be written as $s_n = (-1)^n / n$. By Equation 3.2(a) in the text with $\alpha = 1$, $\lim_{n \to \infty} |s_n| = \lim_{n \to \infty} 1/n = 0$. It follows that

$$\lim_{n \to \infty} s_n = 0.$$

9. Assuming $s_1 = \frac{1}{2}$, we observe that each of the given terms has a integer denominator that is one more than its integer numerator. Moreover, the denominator of each term is the numerator of the next term. Given these observations, the infinite sequence $\{s_n\}$ suggested by the given four terms is

$$s_n = \frac{n}{n+1}, \quad n = 1, 2, \ldots.$$

Since $n/(n+1) = 1/(1 + 1/n)$, we have

$$\lim_{n \to \infty} s_n = \lim_{n \to \infty} \frac{1}{1 + 1/n} = \frac{\lim_{n \to \infty} 1}{\lim_{n \to \infty} (1 + 1/n)} = \frac{1}{1 + \lim_{n \to \infty} (1/n)} = \frac{1}{1 + 0} = 1,$$

where the fourth equality follows from Equation 3.2(a) in the text with $\alpha = 1$.

11. Assuming $s_1 = \frac{1}{5}$, we observe that the numerators and denominators of the given terms are all of the same form; namely, each is an integer added to the square of the next integer. Moreover, the denominator of each term is the numerator of the next term. Given these observations, the infinite sequence $\{s_n\}$ suggested by the given four terms is

$$s_n = \frac{n - 1 + n^2}{n + (n+1)^2} = \frac{n^2 + n - 1}{n^2 + 3n + 1}, \quad n = 1, 2, \ldots.$$

Hence,

$$\lim_{n \to \infty} s_n = \lim_{n \to \infty} \frac{n^2 + n - 1}{n^2 + 3n + 1} = \lim_{n \to \infty} \frac{1 + 1/n - 1/n^2}{1 + 3/n + 1/n^2} = \frac{\lim_{n \to \infty} (1 + 1/n - 1/n^2)}{\lim_{n \to \infty} (1 + 3/n + 1/n^2)} = \frac{1}{1} = 1,$$

where the fourth equality follows from Equation 3.2(a) in the text with $\alpha = 1$ and $\alpha = 2$.

13. (a) Let $s_1, s_2, \ldots$, be an arbitrary infinite sequence of real numbers. Define the infinite sequence $a_1, a_2, \ldots$ by the rule $a_k = s_k - s_{k-1}$, for $n \geq 2$, and $a_1 = s_1$. Then, for $n \geq 2$,

$$\sum_{k=1}^{n} a_k = a_1 + \sum_{k=2}^{n} a_k = s_1 + \sum_{k=2}^{n} (s_k - s_{k-1})$$

$$= s_1 + (s_2 - s_1) + (s_3 - s_2) + \cdots + (s_n - s_{n-1}) = s_n.$$

Noting that $\sum_{k=1}^{n} a_k = s_n$ also holds for $n = 1$, the equation holds for all $n \geq 1$.

(b) Given the infinite sequence $\{1 + 1/k\}_{k=1}^{\infty}$, define the infinite sequence $a_1, a_2, \ldots$

$$a_1 = 2, \qquad a_k = \left(1 + \frac{1}{k}\right) - \left(1 + \frac{1}{k-1}\right) = \frac{1}{k} - \frac{1}{k-1} = -\frac{1}{k(k-1)}, \qquad \text{for } k = 2, 3, \ldots.$$

By part (a), the nth partial sum of the infinite series with the terms $a_1, a_2, \ldots$ is $s_n = 1 + 1/n$. Since $\lim_{n \to \infty} s_n = \lim_{n \to \infty} (1 + 1/n) = 1$, we deduce $\sum_{k=1}^{\infty} a_k = 1$.

15. Consider the function $f(x) = (x+1)^x / x^x$, where $x > 0$ is a continuous variable. We compute $\lim_{x \to \infty} f(x)$ by using l'Hôpital's rule to compute $\lim_{x \to \infty} \ln f(x)$. We have

$$\lim_{x \to \infty} \ln f(x) = \lim_{x \to \infty} \ln\left(\frac{(x+1)^x}{x^x}\right) = \lim_{x \to \infty} \ln\left(\frac{x+1}{x}\right)^x = \lim_{x \to \infty} \left[x \ln\left(1 + \frac{1}{x}\right)\right]$$

$$= \lim_{x \to \infty} \left[\frac{\ln(1 + 1/x)}{1/x}\right] = \lim_{x \to \infty} \left(\frac{\frac{1}{1+(1/x)} \cdot (-1/x^2)}{-1/x^2}\right) = \lim_{x \to \infty} \frac{1}{1 + 1/x} = 1,$$

By continuity of the log function, $\ln\left(\lim_{x \to \infty} f(x)\right) = 1$, so that exponentiation gives

$$\lim_{x \to \infty} f(x) = \lim_{x \to \infty} \frac{(x+1)^x}{x^x} = e.$$

Since $f(n) = (n+1)^n / n^n$ when n is a positive integer, it follows that

$$\lim_{n \to \infty} \frac{(n+1)^n}{n^n} = e.$$

17. Using Theorem 3.3, we have

$$\sum_{k=1}^{\infty} 2^{-k} + \sum_{k=1}^{\infty} 2^{-k+1} = \sum_{k=1}^{\infty} (2^{-k} + 2^{-k+1}) = \sum_{k=1}^{\infty} (1 + 2) 2^{-k} = \sum_{k=1}^{\infty} 3 \cdot 2^{-k}.$$

19. Using Theorem 3.3, we have

$$\sum_{k=0}^{\infty} (2/3)^k - \sum_{k=0}^{\infty} (2/3)^{k+1} = \sum_{k=0}^{\infty} \left((2/3)^k - (2/3)^{k+1}\right) = \sum_{k=0}^{\infty} (1 - 2/3)(2/3)^k$$

$$= \sum_{k=0}^{\infty} (1/3)(2/3)^k = \sum_{k=0}^{\infty} \frac{2^k}{3^{k+1}}.$$

21. Let $\epsilon > 0$ be given. Fix an integer n_0 satisfying $n_0 > \epsilon^{-1/\alpha}$, so that $n > \epsilon^{-1/\alpha}$ for all $n \geq n_0$. Taking reciprocals gives $n^{-1} < \epsilon^{1/\alpha}$ for all $n \geq n_0$, and raising both sides to the α (> 0) power gives $n^{-\alpha} < \epsilon$ for all $n \geq n_0$. This is equivalent to

$$\lim_{n \to \infty} n^{-\alpha} = 0.$$

Exercise Set 3CDE (pgs. 681-682)

1. Since

$$\lim_{k \to \infty} \frac{k^2}{k^2 + 1} = \lim_{k \to \infty} \frac{1}{1 + 1/k^2} = 1 \neq 0,$$

the series $\sum_{k=1}^{\infty} k^2/(k^2 + 1)$ diverges by the term test.

3. Letting $f(x) = xe^{-x}$, we have

$$\int_1^{\infty} f(x)\,dx = \int_1^{\infty} xe^{-x}dx = \lim_{b \to \infty} \int_1^b xe^{-x}dx = \lim_{b \to \infty} \left[-xe^{-x} - e^{-x} \right]_1^b$$

$$= \lim_{b \to \infty} \left[-be^{-b} - e^{-b} + 2e^{-1} \right] = 2e^{-1},$$

i.e., $\int_1^{\infty} f(x)\,dx$ converges. So $\sum_{k=1}^{\infty} f(k) = \sum_{k=1}^{\infty} ke^{-k}$ converges by the integral test.

5. Letting $f(x) = 1/(x^2 + 1)$, we have

$$\int_1^{\infty} f(x)\,dx = \int_1^{\infty} \frac{1}{x^2 + 1}dx = \lim_{b \to \infty} \int_1^b \frac{1}{x^2 + 1}dx = \lim_{b \to \infty} \left[\arctan x \right]_1^b$$

$$= \lim_{b \to \infty} \left[\arctan b - \arctan 1 \right] = \frac{\pi}{2} - \frac{\pi}{4} = \pi/4,$$

i.e., $\int_1^{\infty} f(x)\,dx$ converges. So $\sum_{k=1}^{\infty} f(k) = \sum_{k=1}^{\infty} 1/(k^2 + 1)$ converges by the integral test.

7. For each $k \geq 1$ we have

$$\frac{2^k}{3^{k+1}} = \frac{2^k}{3^k}\frac{1}{3} < \frac{2^k}{3^k} = \left(\frac{2}{3} \right)^k.$$

Since the geometric series $\sum_{k=1}^{\infty} (2/3)^k$ converges, the series $\sum_{k=1}^{\infty} 2^k/3^{k+1}$ converges by the comparison test.

An alternate method is to note that if a_k denotes the kth term then $a_{k+1}/a_k = 2/3$ for all k. Hence, $\lim_{k \to \infty} a_{k+1}/a_k = 2/3 < 1$. By the ratio test, the given series converges.

9. Note that if $n \geq 3$ then $\ln n > 1$, so that

$$\frac{1}{n^2 \ln n} < \frac{1}{n^2}, \quad \text{for } n \geq 3.$$

Since $\sum_{n=1}^{\infty} 1/n^2$ is a convergent p-series (with $p = 2$), the series $\sum_{n=3}^{\infty} 1/n^2$ is also convergent. By the comparison test, the series $\sum_{n=3}^{\infty} 1/(n^2 \ln n)$ converges. Therefore, the given series $\sum_{n=2}^{\infty} 1/(n^2 \ln n)$ converges.

11. For $j \geq 1$ we have

$$\frac{j}{j^3 + 1} < \frac{j}{j^3} = \frac{1}{j^2}.$$

Since $\sum_{j=1}^{\infty} 1/j^2$ is a convergent p-series (with $p = 2$), the given series $\sum_{j=1}^{\infty} j/(j^3 + 1)$ converges by the comparison test.

13. The absolute values of the terms of the given series are $1/(k^2 \ln k)$, $k = 1, 2, \ldots$. Noting that $\ln k > 1$ for $k \geq 3$, we have $1/(k^2 \ln k) < 1/k^2$ for $k \geq 3$. Since $\sum_{k=1}^{\infty} 1/k^2$ is a convergent p-series (with $p = 2$), the series $\sum_{k=3}^{\infty} 1/k^2$ converges. By the comparison test, the series $\sum_{k=3}^{\infty} 1/(k^2 \ln k)$ converges, so that the given alternating series (with the extra term $4 \ln 2$) converges absolutely.

15. The absolute values of the terms of the given alternating series are $1/\sqrt{j}$, $j = 1, 2, \dots$. But since $\sum_{j=1}^{\infty} 1/\sqrt{j}$ is a divergent p-series (with $p = 1/2$), the given alternating series $\sum_{j=1}^{\infty} (-1)^j/\sqrt{j}$ does not converge absolutely.

However, the infinite sequence $\{1/\sqrt{j}\}$ is a decreasing sequence with limit 0, so that conditions (i) and (ii) of the Leibniz test hold. So the given alternating series converges.

17. The absolute values of the terms of the given series are $1/(m^2 + 1)$, $m = 0, 1, \dots$. Noting that $1/(m^2 + 1) < 1/m^2$ for $m \geq 1$, and that $\sum_{m=1}^{\infty} 1/m^2$ is a convergent p-series (with $p = 2$), the comparison test shows that $\sum_{m=1}^{\infty} 1/(m^2 + 1)$ converges. Hence, so does $\sum_{m=0}^{\infty} 1/(m^2 + 1)$. That is, the given alternating series $\sum_{m=0}^{\infty} (-1)^m/(m^2 + 1)$ converges absolutely.

19. Write the given alternating series as the alternating series

$$\sum_{k=2}^{\infty} \frac{(-1)^k}{\ln(1/k)} = \sum_{k=2}^{\infty} \frac{(-1)^k}{-\ln k} = \sum_{k=2}^{\infty} \frac{(-1)^{k+1}}{\ln k}.$$

Since the sequence $\{1/\ln k\}$ is a decreasing sequence with limit 0, the given alternating series converges by the Leibniz test.

However, the fact that $\ln k < k$ for $k \geq 2$ implies $1/k < 1/\ln k$. Since $\sum_{k=1}^{\infty} 1/k$ is the divergent harmonic series, the series $\sum_{k=2}^{\infty} 1/k$ also diverges. By the comparison test, the series $\sum_{k=2}^{\infty} 1/\ln k$ diverges. So the given alternating series does not converge absolutely.

21. Since $(-2)^k = (-1)^k 2^k$, the given series is an alternating series. We claim that the given series diverges because the sequence of its terms does not converge to 0.

To prove the above claim, consider the function $2^x/(x^2 + 1)$, where $x > 0$ is a continuous variable. We have

$$\lim_{x \to \infty} \frac{2^x}{x^2 + 1} = \lim_{x \to \infty} \frac{2^x \ln 2}{2x} = \lim_{x \to \infty} \frac{2^x (\ln 2)^2}{2} = +\infty,$$

where the first and second equalities follow by two applications of L'hôpital's rule. It follows that $\lim_{k \to \infty} 2^k/(k^2 + 1) = +\infty$ and, from this, that $\lim_{k \to \infty} (-1)^k 2^k/(x^2 + 1)$ does not exist. Therefore the terms of the given series do not converge to 0 and, as claimed, the given series diverges by the term test.

(NOTE: The fact that the term test is presented in Section 3C, which deals with nonnegative term series, might suggest that the term test is valid only for nonnegative term series. However, examination of the proof given in the text for the validity of the term test shows that the signs of the terms of the series is irrelevant. That is, the term test is valid for alternating series as well.)

23. The given series is a positive term series. Moreover,

$$\left(1 + \frac{1}{k}\right) 2^{-k} \leq 2 \cdot 2^{-k} = 2^{-(k-1)}, \quad \text{for } k \geq 1.$$

Since $\sum_{k=1}^{\infty} 2^{-(k-1)} = \sum_{k=0}^{\infty} 2^{-k}$ is a convergent geometric series (with common ratio $\frac{1}{2}$), the given series $\sum_{k=1}^{\infty} (1 + 1/k) 2^{-k}$ converges absolutely by the comparison test.

25. Let $a_k = k^2/3^k$ and form the quotients a_{k+1}/a_k. We have

$$\lim_{k \to \infty} \frac{(k+1)^2/3^{k+1}}{k^2/3^k} = \lim_{k \to \infty} \frac{(k+1)^2}{k^2} \frac{1}{3} = \lim_{k \to \infty} \left(1 + \frac{1}{k}\right)^2 \frac{1}{3} = \frac{1}{3} < 1,$$

where the last equality follows because $\lim_{k\to\infty}(1+1/k)^2 = 1$. By the ratio test, the given series $\sum_{k=1}^{\infty} k^2/3^k$ is absolutely convergent. Since absolute convergence implies convergence, the given series converges.

27. Let $a_k = k!/k^k$ and form the quotient a_{k+1}/a_k. We have

$$\lim_{k\to\infty} \frac{(k+1)!/(k+1)^{k+1}}{k!/k^k} = \lim_{k\to\infty} \left[\frac{(k+1)!}{k!} \frac{k^k}{(k+1)^{k+1}} \right]$$

$$= \lim_{k\to\infty} \left[(k+1)\frac{k^k}{(k+1)(k+1)^k} \right] = \lim_{k\to\infty} \left(\frac{k}{k+1} \right)^k.$$

It was shown in Exercise 15 of the last exercise section that $\lim_{k\to}(1+1/k)^k = e$. Since $k/(k+1) = 1/(1+1/k)$, it follows that the last limit in the above display exists and equals $1/e < 1$. By the ratio test, the given series $\sum_{k=1}^{\infty} k!/k^k$ is absolutely convergent. Since absolute convergence implies convergence, the given series converges.

29. Since $|\sin k| \le 1$ for $k \ge 1$, we have

$$\frac{|\sin k|}{k^2} \le \frac{1}{k^2}.$$

Since $\sum_{k=1}^{\infty} 1/k^2$ is a convergent p-series (with $p = 2$), the series $\sum_{k=1}^{\infty} |\sin k|/k^2$ converges. That is, the given series $\sum_{k=1} \sin k/k^2$ is absolutely convergent. Since absolute convergence implies convergence, the given series converges.

31. (a) Letting $a_k = kx^k$ (where each a_k is a function of x), we form the quotients a_{k+1}/a_k and compute

$$\lim_{k\to\infty} \left| \frac{a_{k+1}}{a_k} \right| = \lim_{k\to\infty} \frac{(k+1)|x|^{k+1}}{k|x|^k} = \lim_{k\to\infty} \left(1 + \frac{1}{k}\right)|x| = |x| \lim_{k\to\infty} \left(1 + \frac{1}{k}\right) = |x|,$$

where the last equality follows from the fact that $\lim_{k\to\infty}(1+1/k) = 1$. By the ratio test, the series $\sum_{k=1}^{\infty} kx^k$ converges absolutely for $|x| < 1$.

(b) For $n \ge 1$, we formally multiply the nth partial sum $\sum_{k=1}^{n} kx^k$ by $1 - x$ to get

$$(1 - x)\sum_{k=1}^{n} kx^k = (1 - x)(x + 2x^2 + 3x^3 + \cdots + nx^n)$$

$$= (x + 2x^2 + 3x^3 + \cdots + nx^n) - (x^2 + 2x^3 + \cdots + (n-1)x^n + nx^{n+1})$$

$$= x + (2x^2 - x^2) + (3x^3 - 2x^3) + \cdots + (nx^n - (n-1)x^n) - nx^{n+1}$$

$$= x + x^2 + x^3 + \cdots + x^n - nx^{n+1} = \sum_{k=1}^{n} x^k - nx^{n+1}.$$

Letting $n \to \infty$ results in the equation

$$(1 - x)\sum_{k=1}^{\infty} kx^k = \sum_{k=1}^{\infty} x^k - \lim_{n\to\infty} nx^{n+1}.$$

By part (a), the series on the left converges for $|x| < 1$. The term test then implies that $\lim_{n\to\infty} nx^n = 0$. But this limit equation can be multiplied through by x to get $\lim_{n\to\infty} nx^{n+1} = 0$. So,

$$(1 - x)\sum_{k=1}^{\infty} kx^k = \sum_{k=1}^{\infty} x^k.$$

33. (a) For $n \geq 1$, we formally multiply the nth partial sum $\sum_{k=1}^{n} kx^k$ by $(1-x)^2$ to get

$$(1-x)^2 \sum_{k=1}^{n} kx^k = (1 - 2x + x^2)(x + 2x^2 + 3x^3 + \cdots + nx^n)$$

$$= (x + 2x^2 + 3x^3 + \cdots + nx^n)$$
$$- \left(2x^2 + 4x^3 + 6x^4 + \cdots + 2(n-1)x^n + 2nx^{n+1}\right)$$
$$+ \left(x^3 + 2x^4 + 3x^5 + \cdots + (n-2)x^n + (n-1)x^{n+1} + nx^{n+2}\right)$$
$$= x - (n+1)x^{n+1} + nx^{n+2} = x - nx^n(x - x^2) - x^{n+1},$$

where the last expression is written to facilitate what follows next. Now refer to the proof of Exercise 31(a) above, where the ratio test is used to show that $\sum_{k=1}^{\infty} kx^k$ converges for $|x| < 1$. The term test then implies that, for such x, $kx^k \to 0$ as $k \to \infty$. Thus, for $|x| < 1$, both $nx^n(x - x^2)$ and x^{n+1} tend to 0 as $n \to \infty$. It follows that letting $n \to \infty$ in the above displayed equation gives

$$(1-x)^2 \sum_{k=1}^{\infty} kx^k = x, \quad \text{for } |x| < 1.$$

The result now follows by dividing throughout by $(1-x)^2$.

(b) By part (a), we can set $x = \frac{1}{2}$ in both sides of $\sum_{k=1}^{\infty} kx^k = x/(1-x)^2$ to get

$$\sum_{k=1}^{\infty} k2^{-k} = \frac{1/2}{(1-1/2)^2} = \frac{1/2}{1/4} = 2.$$

35. Let $a_k = x^k/k^2$ and form the ratio a_{k+1}/a_k. We have

$$\lim_{k \to \infty} \left| \frac{a_{k+1}}{a_k} \right| = \lim_{k \to \infty} \frac{|x|^{k+1}/(k+1)^2}{|x|^k/k^2} = \lim_{k \to \infty} \left(\frac{k}{k+1} \right)^2 |x| = |x| \lim_{k \to \infty} \left(\frac{1}{1+1/x} \right)^2 = |x|.$$

Thus, by the ratio test, the series $\sum_{k=1}^{\infty} x^k/k^2$ converges absolutely (and therefore converges) for $|x| < 1$, and diverges for $|x| > 1$. The points where $|x| = 1$ must be tested separately. Setting $x = 1$ and $x = -1$ in the given series yields

$$\sum_{k=1}^{\infty} \frac{1}{k^2} \quad \text{and} \quad \sum_{k=1}^{\infty} \frac{(-1)^k}{k^2},$$

respectively. The first series is a convergent p-series with $p = 2$. The second series is an alternating series that converges by the Leibniz test. Thus, the given series converges if, and only if, $|x| \leq 1$.

37. When the given series is written as

$$\sum_{k=1}^{\infty} \frac{1}{2^k}(x-1)^k = \sum_{k=1}^{\infty} \left(\frac{x-1}{2} \right)^k,$$

we recognize it as a geometric series with common ratio $(x-1)/2$. Thus, the given series converges if, and only if,

$$-1 < \frac{x-1}{2} < 1, \quad \text{or equivalently,} \quad -1 < x < 3.$$

39. Let $a_j = \left[j/(j^2+1) \right] x^j$ and form the quotient a_{j+1}/a_j. We have

$$\lim_{j\to\infty} \left| \frac{a_{j+1}}{a_j} \right| = \lim_{j\to\infty} \frac{\left[(j+1)/((j+1)^2+1) \right] |x|^{j+1}}{[j/(j^2+1)]|x|^j} = \lim_{j\to\infty} \frac{(j+1)(j^2+1)}{j((j+1)^2+1)}|x|$$

$$= |x| \lim_{j\to\infty} \frac{1+1/j+1/j^2+1/j^3}{1+2/j+2/j^2} = |x|.$$

Thus, by the ratio test, the series $\sum_{j=1}^{\infty} \left[j/(j^2+1) \right] x^j$ converges absolutely (and therefore converges) for $|x| < 1$, and diverges for $|x| > 1$. The points where $|x| = 1$ must be tested separately.

Setting $x = 1$ in the given series yields the series $\sum_{j=1}^{\infty} j/(j^2+1)$. For $j \geq 1$

$$\frac{1}{2j} = \frac{j}{2j^2} = \frac{j}{j^2+j^2} \leq \frac{j}{j^2+1}.$$

Since $\sum_{j=1}^{\infty} 1/(2j) = \frac{1}{2} \sum_{j=1}^{\infty} 1/j$ is a constant multiple of the divergent harmonic series, it follows that $\sum_{j=1}^{\infty} j/(j^2+1)$ diverges by the comparison test. That is, the given series diverges for $x = 1$.

Setting $x = -1$ in the given series yields the alternating series $\sum_{j=1}^{\infty} (-1)^j j/(j^2+1)$. Since $\{j/(j^2+1)\}$ is a decreasing sequence with limit 0, the series converges by the Leibniz test. That is, the given series converges for $x = -1$.

To summarize, the given series converges if, and only if, $-1 \leq x < 1$.

41. The given series is a geometric series with common ratio $3/(1+x^2)$. It therefore converges if, and only if,

$$-1 < \frac{3}{1+x^2} < 1.$$

The middle member of this inequality is always positive with a denominator that is always greater than 1. Thus, the inequality will hold if, and only if, the denominator is greater than 3. That is, x must satisfy $1 + x^2 > 3$, or equivalently, $x^2 > 2$. It follows that the given series will converge if, and only if, $|x| > \sqrt{2}$.

43. (a) For $x > 0$, let $x^x = e^{x \ln x}$. Since the exponential function is continuous

$$\lim_{x\to 0^+} e^{x \ln x} = e^{\lim_{x\to 0^+} x \ln x}.$$

Applying L'hôpital's rule gives

$$\lim_{x\to 0^+} x \ln x = \lim_{x\to 0^+} \frac{\ln x}{1/x} = \lim_{x\to 0^+} \frac{1/x}{-1/x^2} = \lim_{x\to 0^+} (-x) = 0.$$

Hence,

$$\lim_{x\to 0^+} x^x = e^0 = 1.$$

Moreover, since $x \to 0^+$ if, and only if, $1/x \to +\infty$, the two limit equations

$$\lim_{x \to 0^+} x^x = 1 \qquad \text{and} \qquad \lim_{x \to +\infty} (1/x)^{1/x} = 1$$

are equivalent. Replacing x with k in the right limit equation gives the desired result.

(b) By part (a), the terms of the series $\sum_{k=1}^{\infty} (1/k)^{1/k}$ tend to the nonzero limit 1. So the series diverges by the term test.

(c) Since $k > 1/k$ for $k \geq 2$, we have $k^{1/k} > (1/k)^{1/k}$ for $k \geq 2$. Therefore, using the result from part (a),

$$\lim_{k \to \infty} k^{1/k} \geq \lim_{k \to \infty} (1/k)^{1/k} = 1 > 0.$$

So the series $\sum_{k=1}^{\infty} k^{1/k}$ diverges by the term test.

45. Let a be a constant and consider the continuous function $1/[x(\ln x)^a]$, for $x \geq 2$. Assuming for the moment that $a \neq 1$, let $b > 3$ and compute the finite integral

$$\int_2^b \frac{1}{x(\ln x)^a}\,dx = \int_{\ln 2}^{\ln b} \frac{1}{u^a}\,du = \frac{1}{1-a}\left[\frac{1}{u^{a-1}}\right]_{\ln 2}^{\ln b} = \frac{1}{1-a}\left[\frac{1}{(\ln b)^{a-1}} - \frac{1}{(\ln 2)^{a-1}}\right].$$

Hence,

$$\int_2^{\infty} \frac{1}{x(\ln x)^a}\,dx = \lim_{b \to \infty}\int_2^b \frac{1}{x(\ln x)^a}\,dx = \lim_{b \to \infty}\frac{1}{1-a}\left[\frac{1}{(\ln b)^{a-1}} - \frac{1}{(\ln 2)^{a-1}}\right]$$

$$= \begin{cases} (\ln 2)^{a-1}/(a-1), & a > 1; \\ +\infty, & a < 1. \end{cases}$$

That is, $\int_2^{\infty} \left(1/[x(\ln x)^a]\right)dx$ converges if $a > 1$ and diverges if $a < 1$.

For $a = 1$, we have

$$\int_2^{\infty} \frac{1}{x(\ln x)^a}\,dx = \lim_{b \to \infty}\int_2^b \frac{1}{x\ln x}\,dx = \lim_{b \to \infty}\int_{\ln 2}^{\ln b}\frac{1}{u}\,du$$

$$= \lim_{b \to \infty}\left[\ln u\right]_{\ln 2}^{\ln b} = \lim_{b \to \infty}[\ln(\ln b) - \ln(\ln 2)] = \infty,$$

so that $\int_2^{\infty} \left(1/[x(\ln x)^a]\right)dx$ diverges for $a = 1$.

Combining the above results shows that $\int_2^{\infty} \left(1/[x(\ln x)^a]\right)dx$ converges if, and only if, $a > 1$. Since $1/[x(\ln x)^a]$ is positive for $x \geq 2$, the integral test allows us to conclude that

$$\sum_{k=2}^{\infty} \frac{1}{k(\ln k)^a} \qquad \text{converges if, and only if, } a < 1.$$

Section 4: UNIFORM CONVERGENCE

Exercise Set 4 (pgs. 687-688)

1. If $0 < d < 1$ then (i) $|x|^k \leq d^k$ for all $-d \leq x \leq d$; and (ii) $\sum_{k=0}^{\infty} d^k$ is a convergent geometric series. By the Weierstrass test, $\sum_{k=0}^{\infty} x^k$ converges uniformly on $[-d, d]$.

3. Let $S = [-\pi, \pi]$. Since the given series converges uniformly on S, and since the terms of the series are continuous on S, Corollary 4.3 says that the series defines a continuous function $f(x)$ on S. So we can write

$$f(x) = \frac{a_0}{2} + \sum_{k=1}^{\infty} (a_k \cos kx + b_k \sin kx), \quad \text{for } x \in S.$$

The definition of uniform convergence then implies that, given $\epsilon > 0$, there is an integer K such that for all $x \in S$ and all $N > K$,

$$\left| \frac{a_0}{2} + \sum_{k=1}^{N} (a_k \cos kx + a_k \sin kx) - f(x) \right| < \epsilon.$$

Now, for each real number x there exists a unique integer m_x such that $-\pi < x + 2m_x\pi \leq \pi$. Thus, we can extend f to a periodic function (which we also denote by f) defined on all of $x \in \mathcal{R}$ by

$$f(x) = f(x + 2m_x\pi) \quad \text{for } x \in \mathcal{R}.$$

Let ϵ and K be given as above. Since 2π is a period of each term of the series, so is $2m_x\pi$. Hence, for all $N > K$ and all x,

$$\left| \frac{a_0}{2} + \sum_{k=1}^{\infty} (a_k \cos kx_0 + a_k \sin kx_0) - f(x_0) \right|$$
$$= \left| \frac{a_0}{2} + \sum_{k=1}^{\infty} (a_k \cos k(x_0 + 2m\pi) + a_k \sin k(x_0 + 2m\pi)) - f(x_0 + 2m\pi) \right| < \epsilon.$$

So the given series converges uniformly on $\mathcal{R}$.

5. We consider the trigonometric series

$$\sum_{k=1}^{\infty} (a_k \cos kx + b_k \sin kx).$$

Since A and B are constants such that $|a_k| \leq A/k^2$ and $|b_k| \leq B/k^2$ for all $k \geq 1$, we use the facts that $|\cos kx| \leq 1$ and $|\sin kx| \leq 1$ for all k and all x to get

$$|a_k \cos kx + b_k \sin kx| \leq |a_k||\cos kx| + |b_k||\sin kx| \leq \frac{A}{k^2} + \frac{B}{k^2} = \frac{A+B}{k^2}.$$

Since $\sum_{k=1}^{\infty} 1/k^2$ converges (p-series with $p = 2$), $A + B$ times this series also converges. By the Weierstrass test, the above trigonometric series converges uniformly for all x.

7. We first consider the series $\sum_{k=1}^{\infty}(-1)^k(1-x)x^k$. Note that each term is zero when $x = 1$, so that the series converges to 0 when $x = 1$. Now write the series as $(1-x)\sum_{k=1}^{\infty}(-x)^k$ and observe that the series is a geometric series that converges for $-1 < -x < 1$, or equivalently, for $-1 < x < 1$. In particular, it converges for $0 \leq x < 1$. Therefore the series converges for x in $[0, 1]$.

Next, for each $x \in [0, 1]$, the Nth partial sum $s_N(x)$ differs from the sum $s(x)$ of the series by an amount that is no greater than the absolute value of the $(N+1)$st term (see Theorem 3.10, Sec.3E, Ch.14). That is,

$$|s_N(x) - s(x)| \leq |(-1)^{N+1}(1-x)x^{N+1}| = (1-x)x^{N+1} \quad \text{for } x \text{ in } [0, 1].$$

Let $f_N(x) = (1-x)x^{N+1}$. Then $f_N'(x) = \big((N+1) - (N+2)x\big)x^{N+1} = 0$ for $x = 0$ and for $x = (N+1)/(N+2)$. Let $x_N = (N+1)/(N+2)$. The first derivative test then shows that

$$f_N(x_N) = (1 - x_N)x_N^{N+1} = \left(1 - \frac{N+1}{N+2}\right)\left(\frac{N+1}{N+2}\right)^{N+1} = \frac{1}{N+2}\left(\frac{N+1}{N+2}\right)^{N+1}$$

is the absolute maximum value of $f_N(x)$ on $[0, 1]$. That is,

$$|s_N(x) - s(x)| \leq f_N(x) \leq f_N(x_N) = \frac{1}{N+2}\left(\frac{N+1}{N+2}\right)^{N+1}, \quad \text{for all } N \text{ and all } x \in [0, 1].$$

Of the the two factors comprising the expression for $f_N(x_N)$, the first one tends to 0 as $N \to \infty$ while the second one remains bounded in the interval $(0, 1)$. Thus, $f_N(x_N) \to 0$ as $N \to \infty$. This means that given $\epsilon > 0$ there exist K such that $|s_N(x) - s(x)| < \epsilon$ for all x in $[0, 1]$ and all $N \geq K$. That is, the series is uniformly convergent on $[0, 1]$.

We now consider the series $\sum_{k=1}^{\infty}(1-x)x^k$. In this case, the Nth partial sum $s_N(x)$ is given by

$$s_N(x) = \sum_{k=1}^{N}(1-x)x^k = (1-x)\sum_{k=1}^{N}x^k = x(1-x)\sum_{k=1}^{N}x^{k-1} = x(x-1)\sum_{k=0}^{N-1}x^k$$

$$= x(1-x)\frac{1-x^N}{1-x} = x(1-x^N).$$

Note that $s_N(1) = 0$ for all N, and for $0 \leq x < 1$, $s_N(x) \to x$ as $N \to \infty$. That is, the series converges pointwise to the function

$$f(x) = \begin{cases} x, & 0 \leq x < 1; \\ 0, & x = 1. \end{cases}$$

Hence,

$$|s_N(x) - f(x)| = \begin{cases} x^{N+1}, & 0 \leq x < 1; \\ 0, & x = 1, \end{cases}$$

For each N, x^{N+1} can be made arbitrarily close to 1 by choosing x sufficiently close to 1. Hence, if $0 < \epsilon < 1$ is chosen then x can be chosen so that $\epsilon < |s_N(x) - f(x)| < 1$. So the series can't be uniformly convergent on $[0, 1]$.

Section 5: POWER SERIES

Exercise Set 5A-D (pgs. 695-696)

1. The Taylor expansion for $\ln(1+x)$ about $x=0$ is given by Equation 2.4(d) in Section 2B of this chapter and holds for $-1 < x < 1$. Hence, the formula holds for $-1 < x < 1$ when x is replaced with $-x$. This gives

$$\ln(1-x) = \sum_{k=1}^{\infty} \frac{(-1)^{k+1}}{k}(-x)^k = \sum_{k=1}^{\infty} \frac{(-1)^{2k+1}}{k} x^k = -\sum_{k=1}^{\infty} \frac{1}{k} x^k.$$

So the given series converges to $-\ln(1-x)$ for $-1 < x < 1$. If $x = -1$ the given series is the alternating series $\sum_{k=1}^{\infty}(-1)^k/k$, which converges by the Leibniz test; and if $x = 1$ the given series is the divergent harmonic series. The interval of convergence is therefore $-1 \le x < 1$.

3. The kth term of the series is $a_k = x^{2k}/(2k)!$. We form the quotient a_{k+1}/a_k and compute

$$\lim_{k \to \infty} \left| \frac{a_{k+1}}{a_k} \right| = \lim_{k \to \infty} \left| \frac{x^{2k+2}/(2k+2)!}{x^{2k}/(2k)!} \right| = \lim_{k \to \infty} \frac{x^2}{(2k+1)(2k+2)} = 0,$$

which holds for all x. By the ratio test, the given series converges absolutely for all x. The interval of convergence is therefore $-\infty < x < \infty$.

5. The kth term of the series is $a_k = \big(k/(k+1)\big)(x+2)^k$. We form the quotient a_{k+1}/a_k and compute

$$\lim_{k \to \infty} \left| \frac{a_{k+1}}{a_k} \right| = \lim_{k \to \infty} \left| \frac{\big((k+1)/(k+2)\big)(x+2)^{k+1}}{\big(k/(k+1)\big)(x+2)^k} \right| = \lim_{k \to \infty} \frac{(k+1)^2}{k(k+2)} |x+2|$$

$$= |x+2| \lim_{k \to \infty} \frac{k^2 + 2k + 1}{k^2 + 2k} = |x+2|.$$

By the ratio test, the series converges for $|x+2| < 1$ and diverges for $|x+2| > 1$. The endpoints of the interval defined by $|x+2| < 1$ are $x = -3$ and $x = -1$. When these values of x are inserted into the given series we get

$$\sum_{k=1}^{\infty} \frac{k}{k+1}(-1)^k \quad (x = -3) \qquad \text{and} \qquad \sum_{k=1}^{\infty} \frac{k}{k+1} \quad (x = -1).$$

Since the sequence of terms of these series do not converge to 0, the term test shows that neither of these series converge. The interval of convergence is therefore $-3 < x < -1$.

7. The kth term of the series is $a_k = \big(1/\sqrt{k}\big)(x+3)^{2k+1}$. We form the quotient a_{k+1}/a_k and compute

$$\lim_{k \to \infty} \left| \frac{a_{k+1}}{a_k} \right| = \lim_{k \to \infty} \left| \frac{(1/\sqrt{k+1})(x+3)^{2k+3}}{(1/\sqrt{k})(x+3)^{2k+1}} \right| = \lim_{k \to \infty} (x+3)^2 \sqrt{\frac{k}{k+1}} = (x+3)^2.$$

By the ratio test, the given series converges absolutely for $(x+3)^2 < 1$ and diverges for $(x+3)^2 > 1$. This is equivalent to saying that the series converges absolutely for $|x+3| < 1$

and diverges for $|x + 3| > 1$. The inequality $|x + 3| < 1$ can be written as $-4 < x < -2$, so that the endpoints correspond to $x = -4$ and $x = -2$. Inserting these values of x into the given series gives the two series

$$\sum_{k=1}^{\infty} \frac{(-1)^{2k+1}}{\sqrt{k}} = -\sum_{k=1}^{\infty} \frac{1}{\sqrt{k}} \quad \text{and} \quad \sum_{k=1}^{\infty} \frac{1}{\sqrt{k}}.$$

Since these are divergent p-series (with $p = 1/2$), the given series diverges for $x = -4$ and $x = -2$. The interval of convergence is therefore $-4 < x < -2$.

9. Replace x with $-x$ in the given Taylor expansion to get

$$e^{-x} = \sum_{k=0}^{\infty} \frac{1}{k!}(-x)^k = \sum_{k=0}^{\infty} \frac{(-1)^k}{k!}x^k, \quad -\infty < x < \infty.$$

11. Use the given Taylor expansion along with the the Taylor expansion derived above in Exercise 9 to get

$$\sinh x = \frac{1}{2}(e^x - e^{-x}) = \frac{1}{2}\left(\sum_{k=0}^{\infty} \frac{1}{k!}x^k - \sum_{k=0}^{\infty} \frac{(-1)^k}{k!}x^k\right) = \frac{1}{2}\sum_{k=0}^{\infty} \frac{1 - (-1)^k}{k!}x^k.$$

Since $1 - (-1)^k$ is 0 when k is even and 2 when k is odd, we can replace k with $2k + 1$ in in all of the odd terms to get

$$\sinh x = \frac{1}{2}\sum_{k=0}^{\infty} \frac{2}{(2k + 1)!}x^{2k+1} = \sum_{k=0}^{\infty} \frac{1}{(2k + 1)!}x^{2k+1}, \quad -\infty < x < \infty.$$

13. Replace x with x^2 in the given Taylor expansion to get

$$e^{x^2} = \sum_{k=1}^{\infty} \frac{1}{k!}x^{2k}, \quad -\infty < x < \infty.$$

15. Replace x with $-x$ in the given Taylor expansion to get

$$\ln(1 + x) = -\sum_{k=1}^{\infty} \frac{1}{k}(-x)^k = \sum_{k=1}^{\infty} \frac{(-1)^{k+1}}{k}x^k, \quad -1 < x < 1.$$

17. Replace x with $-x^2$ in the given Taylor expansion to get

$$\ln(1 + x^2) = -\sum_{k=1}^{\infty} \frac{1}{k}(-x^2)^k = \sum_{k=1}^{\infty} \frac{(-1)^{k+1}}{k}x^{2k}, \quad -1 < x < 1.$$

19. Replace x with $-x^2$ in the given Taylor expansion to get

$$\frac{1}{1 + x^2} = \sum_{k=0}^{\infty} (-x^2)^k = \sum_{k=0}^{\infty} (-1)^k x^{2k}, \quad |x| < 1.$$

21. Replacing x with $1-x$ in the given Taylor expansion gives

$$\frac{1}{1-(1-x)} = \sum_{k=0}^{\infty}(1-x)^k, \quad \text{or equivalently,} \quad \frac{1}{x} = \sum_{k=0}^{\infty}(-1)^k(x-1)^k,$$

which is valid for $|1-x| < 1$, or equivalently, for $0 < x < 2$. Note that the resulting series is a power series about the point $x = 1$ and converges to $1/x$ for $0 < x < 2$. It follows that the series is the Taylor expansion of the function $1/x$ about the point $x = 1$, valid for $0 < x < 2$.

23. The relations $d(\arctan x)/dx = 1/(1+x^2)$ and $\arctan x = \int_0^x(1+t^2)^{-1}dt$ are equivalent via the fundamental theorem of calculus. Therefore, using the fact that the result of Exercise 19 above is valid for $|x| < 1$, Theorem 5.1 implies

$$\arctan x = \int_0^x \frac{1}{1+t^2}dt = \int_0^x\left[\sum_{k=0}^{\infty}(-1)^k t^{2k}\right]dt = \sum_{k=0}^{\infty}\int_0^x(-1)^k t^{2k}dt$$

$$= \sum_{k=0}^{\infty}\left[(-1)^k\frac{1}{2k+1}t^{2k+1}\right]_0^x = \sum_{k=0}^{\infty}\frac{(-1)^k}{2k+1}x^{2k+1},$$

provided $|x| < 1$. Since the last series is a power series about $x = 0$ that converges to $\arctan x$ for $|x| < 1$, it follows that it must be the Taylor expansion of $\arctan x$ about $x = 0$ and is valid for $|x| < 1$.

25. The Taylor expansion $(1-x)^{-1} = \sum_{k=0}^{\infty}x^k$, valid for $|x| < 1$, can be differentiated term by term to obtain the Taylor expansion of $d(1-x)^{-1}/dx = (1-x)^{-2}$, which can then be differentiated term by term to obtain the Taylor expansion of $d(1-x)^{-2}/dx = 2(1-x)^{-3}$. The results are

$$\frac{1}{(1-x)^2} = \frac{d}{dx}\left(\frac{1}{1-x}\right) = \frac{d}{dx}\sum_{k=0}^{\infty}x^k = \sum_{k=1}^{\infty}kx^{k-1} = \sum_{k=0}^{\infty}(k+1)x^k$$

$$\frac{2}{(1-x)^3} = \frac{d}{dx}\left(\frac{1}{(1-x)^2}\right) = \frac{d}{dx}\sum_{k=0}^{\infty}(k+1)x^k = \sum_{k=1}^{\infty}k(k+1)x^{k-1} = \sum_{k=0}^{\infty}(k+1)(k+2)x^k,$$

both of which are valid for $|x| < 1$. By partial fraction decomposition, we also have

$$\frac{x(x-1)}{(1-x)^3} = \frac{1}{1-x} - \frac{3}{(1-x)^2} + \frac{2}{(1-x)^3}.$$

In this expression, replace $1/(1-x)$, $1/(1-x)^2$ and $2/(1-x)^3$ with their respective Taylor expansions derived above to get

$$\frac{x(x-1)}{(1-x)^3} = \sum_{k=0}^{\infty}x^k - 3\sum_{k=0}^{\infty}(k+1)x^k + \sum_{k=0}^{\infty}(k+1)(k+2)x^k$$

$$= \sum_{k=0}^{\infty}\Big(1 - 3(k+1) + (k+1)(k+2)\Big)x^k = \sum_{k=0}^{\infty}k^2x^k,$$

which is valid for $|x| < 1$. Since the last series is a power series about $x = 0$ that converges to $x(x+1)/(1-x)^2$ for $|x| < 1$, it follows that it must be the Taylor expansion of $x(x+1)/(1-x)^2$ about $x = 0$ and is valid for $|x| < 1$.

27. (a) Let $f(x) = (1+x)^\alpha$, where α is a real number. Then

$$f^{(k)}(x) = \frac{d^k}{dx^k}(1+x)^\alpha = \alpha(\alpha-1)\cdots(\alpha-k+1)(1+x)^{\alpha-k}, \quad \text{for } k = 1, 2, 3, \ldots.$$

So $f^{(k)}(0) = \alpha(\alpha-1)\cdots(\alpha-k+1)$, for $k \geq 1$, and the Taylor expansion of $(1+x)^\alpha$ about $x = 0$ is

$$(1+x)^\alpha = \sum_{k=0}^{\infty} \frac{f^{(k)}(0)}{k!} x^k = 1 + \sum_{k=1}^{\infty} \frac{\alpha(\alpha-1)\cdots(\alpha-k+1)}{k!} x^k.$$

If α is not a nonnegative integer then the ratio test shows that the series converges absolutely if $|x| < 1$ and diverges if $|x| > 1$. But if α is a nonnegative integer then only finitely many coefficients are nonzero, so that the series is a finite sum and therefore converges for all x.

(b) Setting $\alpha = 3$, $\alpha = -3$ and $\alpha = \frac{1}{2}$ in the series in part (a) gives

$$(1+x)^3 = 1 + \sum_{k=1}^{\infty} \frac{3(3-1)\cdots(3-k+1)}{k!} x^k = 1 + \frac{3}{1!}x + \frac{3(2)}{2!}x^2 + \frac{3(2)(1)}{3!}x^3$$

$$= 1 + 3x + 3x^2 + x^3,$$

$$(1+x)^{-3} = 1 + \sum_{k=1}^{\infty} \frac{(-3)(-3-1)\cdots(-3-k+1)}{k!} x^k$$

$$= 1 + \frac{(-3)}{1!}x + \frac{(-3)(-4)}{2!}x^2 + \frac{(-3)(-4)(-5)}{3!}x^3 + \cdots$$

$$= 1 - 3x + 6x^2 - 10x^3 + \cdots,$$

$$(1+x)^{1/2} = 1 + \sum_{k=1}^{\infty} \frac{\frac{1}{2}(\frac{1}{2}-1)\cdots(\frac{1}{2}-k+1)}{k!} x^k = 1 + \frac{\frac{1}{2}}{1!}x + \frac{\frac{1}{2}(-\frac{1}{2})}{2!}x^2 + \frac{\frac{1}{2}(-\frac{1}{2})(-\frac{3}{2})}{3!}x^3$$

$$= 1 + \frac{1}{2}x - \frac{1}{8}x^2 + \frac{1}{16}x^3 - \cdots.$$

29. Suppose $\sum_{k=0}^{\infty} a_k x^k$ converges to a continuous function $f(x)$ for $|x| < R$. Let a be a fixed real number and let $t = x + a$. Then $|x| < R$ and $|t - a| < R$ are equivalent and

$$f(x) = \sum_{k=0}^{\infty} a_k x^k, \quad |x| < R \qquad \text{and} \qquad f(t-a) = \sum_{k=0}^{\infty} a_k(t-a)^k, \quad |t-a| < R,$$

are equivalent. Now let t_1 be interior to the interval of convergence of the series on the right and set $x_1 = t_1 - a$ Since we are given that

$$\int_0^{x_1} f(x)\,dx = \int_0^{x_1} \sum_{k=0}^{\infty} a_k x^k dx = \sum_{k=0}^{\infty} a_k \frac{x_1^{k+1}}{k+1}, \quad |x_1| < R,$$

it follows that for t_1 satisfying $|t_1 - a| < R$

$$\int_a^{t_1} \sum_{k=0}^{\infty} a_k(t-a)^k dt = \int_a^{t_1} f(t-a)\,dt = \int_0^{x_1} f(x)\,dx = \sum_{k=0}^{\infty} a_k \frac{x_1^{k+1}}{k+1} = \sum_{k=0}^{\infty} a_k \frac{(t_1-a)^{k+1}}{k+1}.$$

Now replace t with x and t_1 with x_1 in the two outer members.

31. The equation $\ln(1+x) = \sum_{k=1}^{\infty}(-1)^{k+1}x^k/k$ is valid for $|x| < 1$. For $x \neq 0$, replace x with x^2 and multiply the resulting members by $1/x^2$ to get

$$\frac{\ln(1+x^2)}{x^2} = \frac{1}{x^2}\sum_{k=1}^{\infty}\frac{(-1)^{k+1}(x^2)^k}{k} = \sum_{k=1}^{\infty}\frac{(-1)^{k+1}x^{2k-2}}{k} = 1 - \frac{x^2}{2} + \frac{x^4}{3} - \cdots.$$

Letting $x \to 0$ on the left is equivalent to setting $x = 0$ on the right. That is,

$$\lim_{x \to 0}\frac{\ln(1+x^2)}{x^2} = 1.$$

33. The equation $\ln(1-x) = -\sum_{k=1}^{\infty}x^k/k$ is valid for $|x| < 1$. For $x \neq 0$, add x to both sides and multiply the resulting members by $1/x^2$ to get

$$\frac{x + \ln(1-x)}{x^2} = \frac{1}{x^2}\left(x - \sum_{k=1}^{\infty}\frac{x^k}{k}\right) = \frac{1}{x^2}\left[x - \left(x + \frac{x^2}{2} + \frac{x^3}{3} + \frac{x^4}{4}\cdots\right)\right]$$

$$= -\frac{1}{x^2}\left(\frac{x^2}{2} + \frac{x^3}{3} + \frac{x^4}{4} + \cdots\right) = -\frac{1}{2} - \frac{x}{3} - \frac{x^2}{4} - \cdots.$$

Letting $x \to 0$ on the left is equivalent to setting $x = 0$ on the right. That is,

$$\lim_{x \to 0}\frac{x + \ln(1-x)}{x^2} = -\frac{1}{2}.$$

35. Since $c \neq 0$, we can write $1/(x+c) = (1/c)/\big(1 - (-x/c)\big)$, which we recognize as the sum of the geometric series $\sum_{k=0}^{\infty}(-x/c)^k$, provided $|-x/c| < 1$, or equivalently, provided $|x| < |c|$. Hence,

$$f(x) = \frac{1}{x+c} = \frac{1}{c}\sum_{k=0}^{\infty}(-x/c)^k = \frac{1}{c}\sum_{k=0}^{\infty}\frac{(-1)^k}{c^k}x^k = \sum_{k=0}^{\infty}\frac{(-1)^k}{c^{k+1}}x^k, \quad \text{for } |x| < |c|.$$

Since this is a power series about $x = 0$ that converges to $f(x)$ for $|x| < |c|$, it must be the Taylor expansion of $f(x)$ about $x = 0$.

37. Suppose $f(x) = \sum_{k=0}^{\infty}c_k x^k$ in some interval $|x - a| < R$. Differentiating both sides twice with respect to x gives

$$f''(x) = \sum_{k=2}^{\infty}k(k-1)c_k x^{k-2} = \sum_{k=0}^{\infty}(k+2)(k+1)c_{k+2}x^k, \quad |x - a| < R.$$

Note that the first sum starts with $k = 2$. This is permissible since the terms involving $k = 0$ and $k = 1$ are 0 and can therefore be deleted. Also note that in moving from the first series to the second series the index of summation has been *decreased* by two units (from $k = 2$ to $k = 0$). This is permissible if compensation is made by *increasing* the value of k by 2 units in the general expression for the terms. The reason this second step was made is that we anticipate adding this series to the series defining $f(x)$ and, in so doing, x^k will be the power of x corresponding to the kth term in each series. Specifically, we can now easily write

$$f''(x) + f(x) = \sum_{k=0}^{\infty}(k+2)(k+1)c_{k+2}x^k + \sum_{k=0}^{\infty}c_k x^k = \sum_{k=0}^{\infty}[(k+2)(k+1)c_{k+2} + c_k]x^k,$$

which holds for $|x - a| < R$.

Section 6: DIFFERENTIAL EQUATIONS

Exercise Set 6 (pg. 698)

1. Given the initial-value problem $y' = y^2 + y$, $y(0) = 1$, we compute

$$y' = y^2 + y, \quad y'(0) = 2;$$
$$y'' = 2yy' + y', \quad y''(0) = 6.$$

Hence, the first three nonzero terms in the Taylor expansion expansion about $x = 0$ of the solution $y = y(x)$ of the given initial-value problem are

$$y(x) = y(0) + \frac{y'(0)}{1!}x + \frac{y''(0)}{2!}x^2 + \cdots = 1 + 2x + 3x^2 + \cdots.$$

3. Given the initial-value problem $y' = xy$, $y(0) = 2$, we compute

$$y' = xy, \quad y'(0) = 0;$$
$$y'' = xy' + y, \quad y''(0) = 2;$$
$$y''' = xy'' + 2y', \quad y'''(0) = 0;$$
$$y^{(4)} = xy''' + 3y'', \quad y^{(4)}(0) = 6.$$

Hence, the first three nonzero terms in the Taylor expansion expansion about $x = 0$ of the solution $y = y(x)$ of the given initial-value problem are

$$y(x) = y(0) + \frac{y'(0)}{1!}x + \frac{y''(0)}{2!}x^2 + \frac{y'''(0)}{3!}x^3 + \frac{y^{(4)}(0)}{4!}x^4 + \cdots = 2 + x^2 + \frac{1}{4}x^4 + \cdots.$$

5. Given the initial-value problem $y'' = yy'$, $y(0) = y'(0) = 1$, we compute

$$y'' = yy', \quad y''(0) = 1;$$
$$y''' = yy'' + (y')^2, \quad y'''(0) = 2.$$

Hence, the first four nonzero terms in the Taylor expansion about $x = 0$ of the solution $y = y(x)$ of the given initial-value problem are

$$y(x) = y(0) + \frac{y'(0)}{1!}x + \frac{y''(0)}{2!}x^2 + \frac{y'''(0)}{3!}x^3 + \cdots = 1 + x + \frac{1}{2}x^2 + \frac{1}{3}x^3 + \cdots.$$

7. Given the initial-value problem $y'' = x^2 y$, $y(0) = c_0$, $y'(0) = c_1$, we compute

$$y'' = x^2 y, \quad y''(0) = 0;$$
$$y''' = x^2 y' + 2xy, \quad y'''(0) = 0;$$
$$y^{(4)} = x^2 y'' + 4xy' + 2y, \quad y^{(4)} = 2c_0;$$
$$y^{(5)} = x^2 y''' + 6xy'' + 6y', \quad y^{(5)}(0) = 6c_1.$$

Hence, the solution of the given initial-value problem has the form

$$y(x) = y(0) + \frac{y'(0)}{1!}x + \frac{y''(0)}{2!}x^2 + \frac{y'''(0)}{3!}x^3 + \frac{y^{(4)}(0)}{4!}x^4 + \frac{y^{(5)}(0)}{5!}x^5 + \cdots$$

$$= c_0 + c_1 x + \frac{1}{12}c_0 x^4 + \frac{1}{20}c_1 x^5 + \cdots = c_0\left(1 + \frac{1}{12}x^4 + \cdots\right) + c_1\left(x + \frac{1}{20}x^5 + \cdots\right).$$

Section 7: POWER SERIES SOLUTIONS

Exercise Set 7 (pgs. 705-706)

1. We assume that all solutions of $y'' - y = 0$ have power series expansions about $x = 0$. That is, if $y = y(x)$ is a solution then we assume that $y = \sum_{k=0}^{\infty} c_k x^k$. Since

$$y'' = \sum_{k=2}^{\infty} k(k-1)c_k x^{k-2} = \sum_{k=0}^{\infty} (k+2)(k+1)c_{k+2} x^k,$$

where we have shifted the index of summation by 2 units to make the powers of x agree with those for the series for y, we have

$$y'' - y = \sum_{k=0}^{\infty} (k+2)(k+1)c_{k+2} x^k - \sum_{k=0}^{\infty} c_k x^k = \sum_{k=0}^{\infty} [(k+2)(k+1)c_{k+2} - c_k] x^k = 0.$$

Setting the coefficient of x^k in the sum on the right equal to 0 gives

$$(k+2)(k+1)c_{k+2} - c_k = 0, \quad \text{or equivalently,} \quad c_{k+2} = \frac{1}{(k+1)(k+2)} c_k,$$

which holds for $k = 0, 1, 2, \cdots$. Since the indices in the recurrence relation on the right differ by 2 units, we look at the even subscripts and the odd subscripts separately. We have

$$c_2 = \frac{1}{(1)(2)} c_0 = \frac{1}{2!} c_0, \qquad c_3 = \frac{1}{(2)(3)} c_1 = \frac{1}{3!} c_1,$$

$$c_4 = \frac{1}{(3)(4)} c_2 = \frac{1}{4!} c_0, \qquad c_5 = \frac{1}{(4)(5)} c_3 = \frac{1}{5!} c_1,$$

$$c_6 = \frac{1}{(5)(6)} c_4 = \frac{1}{6!} c_0, \qquad c_7 = \frac{1}{(6)(7)} c_5 = \frac{1}{7!} c_1,$$

$$\vdots \qquad\qquad\qquad \vdots$$

The general formulas are evidently of the form

$$c_{2k} = \frac{1}{(2k)!} c_0, \quad k = 0, 1, 2, \ldots \qquad \text{and} \qquad c_{2k+1} = \frac{1}{(2k+1)!} c_1, \quad k = 0, 1, 2, \ldots.$$

Setting $y(0) = c_0 = 1$, $y'(0) = c_1 = 0$ gives the particular solution $y_1 = y_1(x)$ whose power series contains only even powers of x, and setting $y(0) = c_0 = 0$, $y'(0) = c_1 = 1$ gives the particular solution $y_2 = y_2(x)$ whose power series contains only odd powers of x. That is, we have the linearly independent solutions

$$y_1(x) = \sum_{k=0}^{\infty} \frac{1}{(2k)!} x^{2k} \qquad \text{and} \qquad y_2(x) = \sum_{k=0}^{\infty} \frac{1}{(2k+1)!} x^{2k+1}.$$

Recognizing these series as the Taylor expansions of $\cosh x$ and $\sinh x$ about $x = 0$, we can write the general solution as

$$y(x) = c_0 \cosh x + c_1 \sinh x,$$

where c_0 and c_1 are arbitrary constants.

3. (a) Assume that $y = y(x) = \sum_{k=0}^{\infty} c_k x^k$ is a power series solution of $y' + 2xy = 0$. Since, $y' = \sum_{k=1}^{\infty} k c_k x^{k-1}$, we have

$$y' + 2xy = \sum_{k=1}^{\infty} k c_k x^{k-1} + 2x \sum_{k=0}^{\infty} c_k x^k = \sum_{k=1}^{\infty} k c_k x^{k-1} + \sum_{k=0}^{\infty} 2 c_k x^{k+1}$$

$$= \sum_{k=0}^{\infty} (k+1) c_{k+1} x^k + \sum_{k=1}^{\infty} 2 c_{k-1} x^k = c_1 + \sum_{k=1}^{\infty} [(k+1) c_{k+1} + 2 c_{k-1}] x^k = 0.$$

Setting each coefficient in the last series equal to 0 gives $c_1 = 0$ and $(k+1) c_{k+1} + 2 c_{k-1} = 0$, for $k = 1, 2, \ldots$. The recurrence relation defines c_{k+1} in terms of c_{k-1}, so that the condition $c_1 = 0$ implies that all coefficients with odd subscripts are 0. Thus, we can set $k = 2j - 1$ in the recurrence relation and write it as

$$c_{2j} = \frac{-1}{j} c_{2j-2}, \quad j = 1, 2, 3 \ldots.$$

Repeated application of this recurrence relation gives the general formula

$$c_{2j} = \left(\frac{-1}{j}\right)\left(\frac{-1}{j-1}\right) \cdots \left(\frac{-1}{1}\right) c_0 = \frac{(-1)^j}{j!} c_0,$$

which not only holds for $j = 1, 2, \ldots$, but also for $j = 0$. Thus,

$$y(x) = \sum_{k=0}^{\infty} c_k x^k = \sum_{j=0}^{\infty} c_{2j} x^{2j} = \sum_{j=0}^{\infty} \frac{(-1)^j}{j!} c_0 x^{2j} = c_0 \sum_{j=0}^{\infty} \frac{(-1)^j}{j!} x^{2j}.$$

(b) Multiplying the given D.E. by the integrating factor e^{x^2} leads to $d(e^{x^2} y)/dx = 0$, and one integration gives $e^{x^2} y = c$, or equivalently,

$$y(x) = c e^{-x^2}.$$

(c) The power series about $x = 0$ for the basic solution e^{-x^2} found in part (b) has the form

$$e^{-x^2} = \sum_{k=0}^{\infty} \frac{1}{k!} (-x^2)^k = \sum_{k=0}^{\infty} \frac{(-1)^k}{k!} x^{2k},$$

which holds for all x. Comparing this with the series found in part (a), one sees that they are identical (except for the variable names used for the two sets of indices). Thus, the general solutions found in parts (a) and (b) agree for all x.

5. (a) Assume $y_h = y_h(x) = \sum_{k=0}^{\infty} c_k x^k$ is a power series solution of the homogeneous equation $y'' + xy' + y = 0$. Since

$$y_h' = \sum_{k=1}^{\infty} k c_k x^{k-1} \qquad \text{and} \qquad y_h'' = \sum_{k=2}^{\infty} k(k-1) c_k x^{k-2},$$

we substitute into the D.E. to get

$$y_h'' + xy_h' + y_h = \sum_{k=2}^{\infty} k(k-1)c_k x^{k-2} + x\sum_{k=1}^{\infty} kc_k x^{k-1} + \sum_{k=0}^{\infty} c_k x^k$$

$$= \sum_{k=2}^{\infty} k(k-1)c_k x^{k-2} + \sum_{k=1}^{\infty} kc_k x^k + \sum_{k=0}^{\infty} c_k x^k$$

$$= \sum_{k=0}^{\infty} (k+2)(k+1)c_{k+2} x^k + \sum_{k=0}^{\infty} kc_k x^k + \sum_{k=0}^{\infty} c_k x^k$$

$$= \sum_{k=0}^{\infty} \left[(k+2)(k+1)c_{k+2} + (k+1)c_k\right] x^k = 0.$$

Setting each coefficient in the last series equal to 0 gives $(k+1)(k+2)c_{k+2} + (k+1)c_k = 0$, for $k = 0, 1, 2, \ldots$. After dividing out the nonzero factor $k+1$, the resulting recurrence relation can be written as

$$c_{k+2} = \frac{-1}{k+2} c_k, \quad k = 0, 1, 2, \ldots.$$

Setting $k = 2j - 2$ in the recurrence relation gives $c_{2j} = \left(-1/(2j)\right)c_{2j-2}$, for $j = 1, 2, \ldots$, so that repeatedly applying this formula produces the general formula for c_{2j} in terms of c_0:

$$c_{2j} = \left(\frac{-1}{2j}\right)\left(\frac{-1}{2j-2}\right) \cdots \left(\frac{-1}{2}\right) c_0 = \frac{(-1)^j}{2^j j!} c_0,$$

which also holds for $j = 0$. Similarly, setting $k = 2j - 1$ in the recurrence relation gives $c_{2j+1} = \left(-1/(2j+1)\right)c_{2j-1}$, for $j = 1, 2, \ldots$, so that repeatedly applying this formula produces the general formula for c_{2j+1} in terms of c_1:

$$c_{2j+1} = \left(\frac{-1}{2j+1}\right)\left(\frac{-1}{2j-1}\right) \cdots \left(\frac{-1}{3}\right) c_1 = \frac{(-1)^j}{(3)(5)\cdots(2j+1)} c_1$$

$$= \frac{(-1)^j}{(3)(5)\cdots(2j+1)} \cdot \frac{(2)(4)\cdots(2j)}{(2)(4)\cdots(2j)} c_1 = \frac{(-1)^j 2^j j!}{(2j+1)!} c_1,$$

which also holds for $j = 0$. Since each series converges absolutely on the interior of its respective interval of convergence, we can break up the series for $y_h(x)$ into its even and odd terms to get

$$y_h(x) = \sum_{k=0}^{\infty} c_k x^k = \sum_{j=0}^{\infty} c_{2j} x^{2j} + \sum_{j=0}^{\infty} c_{2j+1} x^{2j+1} = \sum_{j=0}^{\infty} \frac{(-1)^j}{2^j j!} c_0 x^{2j} + \sum_{j=0}^{\infty} \frac{(-1)^j 2^j j!}{(2j+1)!} c_1 x^{2j+1}$$

$$= c_0 \sum_{j=0}^{\infty} \frac{(-1)^j}{2^j j!} x^{2j} + c_1 \sum_{j=0}^{\infty} \frac{(-1)^j 2^j j!}{(2j+1)!} x^{2j+1} = c_0 y_0(x) + c_1 y_1(x),$$

where $y_0(x)$ and $y_1(x)$ are given by

$$y_0(x) = \sum_{j=0}^{\infty} \frac{(-1)^j}{2^j j!} x^{2j} \quad \text{and} \quad y_1(x) = \sum_{j=0}^{\infty} \frac{(-1)^j 2^j j!}{(2j+1)!} x^{2j+1}.$$

(b) The power series for $e^{-x^2/2}$ can be obtained by replacing x with $-x^2/2$ in the power series for e^x. When this is done, the result can be written as

$$e^{-x^2/2} = \sum_{k=0}^{\infty} \frac{1}{k!}\left(-\frac{x^2}{2}\right)^k = \sum_{k=0}^{\infty} \frac{(-1)^k}{2^k k!} x^{2k},$$

which is valid for all x. Now observe that, except for the variable name for the index of summation, this series is precisely the series identified as $y_0(x)$ in part (a). That is, $e^{-x^2/2}$ is a solution of the given homogeneous D.E.

(c) Assume $y_p = \sum_{k=0}^{\infty} c_k x^k$ is a particular power series solution of the nonhomogeneous equation $y'' + xy' + y = x$. When y_p, y_p' and y_p'' are inserted into this equation, we arrive at the equation

$$y_p'' + xy_p' + y_p = \sum_{k=0}^{\infty} \left[(k+2)(k+1)c_{k+2} + (k+1)c_k\right]x^k = x,$$

where the series has the same form as the series representation for $y'' + xy' + y$ that was found in part (a). The only difference here is that the series is not identically 0 but rather is the Taylor series representation for the polynomial x. Thus, all coefficients of this series are zero except the coefficient of x (corresponding to $k = 1$), which is 1. That is,

$$(k+1)(k+2)c_{k+2} + (k+1)c_k = 0 \quad \text{for } k = 0, 2, 3, \ldots \qquad \text{while} \qquad 6c_3 + 2c_1 = 1.$$

If we choose $c_0 = 0$ and $c_3 = 0$ then the recurrence relation implies that all c_k with $k \neq 1$ are zero. With $c_3 = 0$ in the equation on the right we obtain $c_1 = 1/2$. In other words, $y_p = y_p(x) = \frac{1}{2}x$ is a particular solution of the given nonhomogeneous equation.

(d) With the homogeneous power series solution y_h found in part (a) and the particular power series solution y_p of the nonhomogeneous equation found in part (c), the general series solution of the nonhomogeneous equation $y'' + xy' + y = x$ is

$$y(x) = y_p(x) + y_h(x) = \frac{x}{2} + c_0 \sum_{j=0}^{\infty} \frac{(-1)^j}{2^j j!} x^{2j} + c_1 \sum_{j=0}^{\infty} \frac{(-1)^j 2^j j!}{(2j+1)!} x^{2j+1}$$

$$= \frac{x}{2} + c_0 e^{-x^2/2} + c_1 \sum_{j=0}^{\infty} \frac{(-1)^j 2^j j!}{(2j+1)!} x^{2j+1},$$

where we have used the result of part (b) to write the final answer in a more familiar form.

7. (a) Multiply the given Bessel equation by x^2 and set $n = 0$ to get $x^2 y'' + xy' + x^2 y = 0$. Let $y = y(x)$ be a power series solution with

$$y = \sum_{k=0}^{\infty} c_k x^k, \qquad y' = \sum_{k=1}^{\infty} k c_k x^{k-1}, \qquad y'' = \sum_{k=2}^{\infty} k(k-1)c_k x^{k-2}.$$

569

Substitution into the D.E. gives

$$x^2 y'' + xy' + x^2 y = x^2 \sum_{k=2}^{\infty} k(k-1)c_k x^{k-2} + x \sum_{k=1}^{\infty} kc_k x^{k-1} + x^2 \sum_{k=0}^{\infty} c_k x^k$$

$$= \sum_{k=2}^{\infty} k(k-1)c_k x^k + \sum_{k=1}^{\infty} kc_k x^k + \sum_{k=0}^{\infty} c_k x^{k+2}$$

$$= \sum_{k=2}^{\infty} k(k-1)c_k x^k + \sum_{k=1}^{\infty} kc_k x^k + \sum_{k=2}^{\infty} c_{k-2} x^k$$

$$= c_1 x + \sum_{k=2}^{\infty} \left[k(k-1)c_k + kc_k + c_{k-2} \right] x^k = c_1 x + \sum_{k=2}^{\infty} \left[k^2 c_k + c_{k-2} \right] x^k.$$

Setting the coefficient of each power of x in the last series equal to zero gives $c_1 = 0$ and $k^2 c_k + c_{k-2} = 0$, for $k \geq 2$. Shifting the index on the recurrence relation by 2 units gives a recurrence relation that can be written as $(k+2)^2 c_{k+2} = -c_k$, for $k = 0, 1, \ldots$.

(b) With $c_1 = 0$, the recurrence relation derived in part (a) shows that all odd subscripted coefficients are zero. Thus, we need only concentrate on the even subscripted coefficients. To this end, let $k = 2j - 2$ in the recurrence relation and write it as

$$c_{2j} = \frac{-1}{(2j)^2} c_{2j-2}, \quad j = 1, 2, \cdots.$$

Repeatedly applying this recurrence relation gives the general formula for c_{2j}, for $j \geq 1$:

$$c_{2j} = \left(\frac{-1}{(2j)^2} \right) \left(\frac{-1}{(2j-2)^2} \right) \cdots \left(\frac{-1}{2^2} \right) c_0 = \frac{(-1)^j}{\left[2^j j(j-1) \cdots (2)(1) \right]^2} c_0 = \frac{(-1)^j}{2^{2j} (j!)^2} c_0.$$

Noting that this also holds for $j = 0$, we set $c_0 = 1$ to obtain the particular solution

$$y(x) = \sum_{k=0}^{\infty} c_k x^k = \sum_{j=0}^{\infty} c_{2j} x^{2j} = \sum_{j=0}^{\infty} (-1)^j 2^{-2j} (j!)^{-2} x^{2j}.$$

This is $J_0(x)$, the Bessel function of order 0. It should be noted that the ratio test shows that the series converges for all x.

(c) By part (b), $x^2 J_0'' + x J_0' + x^2 J_0 = 0$ for all x. Differentiation of the series representation for $J_0(x)$ derived in part (b) shows that $J_0'(0) = 0$. This means that $x^2 J_0'' + x J_0' + x^2 J_0 = 0$ and $x J_0'' + J_0' + x J_0 = 0$ both hold for all x. Thus, we can use the second equation without affecting the result we seek. Differentiation with respect to x gives

$$x J_0''' + 2 J_0'' + x J_0' + J_0 = 0, \quad \text{which implies} \quad x^2 J_0''' + 2x J_0'' + x^2 J_0' + x J_0 = 0.$$

The equation $x J_0'' + J_0' + x J_0 = 0$ also implies $x J_0 = -x J_0'' - J_0'$, so that the equation on the right above can written as

$$x^2 J_0''' + 2x J_0'' + x^2 J_0' - x J_0'' - J_0' = 0, \quad \text{or equivalently,} \quad x^2 (J_0')'' + x(J_0')' + (x^2 - 1)(J_0') = 0.$$

The equation on the right shows that $J_0'(x)$ is a solution of the Bessel equation of index 1. Hence every constant multiple of $J_0'(x)$ is also a solution. In particular, $-J_0'(x) = J_1(x)$ is a solution.

Section 8: FOURIER SERIES

Exercise Set 8ABC (pgs. 712-713)

1. Let $f(x) = x$, for $-\pi < x \leq \pi$. Since $x \cos kx$ is an odd function for $k \geq 0$, $a_k = 0$ for $k \geq 0$. And since $x \sin kx$ is an even function for $k \geq 1$, we integrate by parts to get

$$b_k = \frac{1}{\pi} \int_{-\pi}^{\pi} x \sin kx \, dx = \frac{2}{\pi} \int_0^{\pi} x \sin kx = \frac{2}{\pi} \left[-\frac{x \cos k\pi}{k} + \frac{\sin kx}{k^2} \right]_0^{\pi} = \frac{2(-1)^{k+1}}{k}.$$

Therefore, the Fourier series of $f(x)$ is

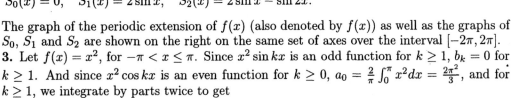

$$\sum_{k=1}^{\infty} \frac{2(-1)^{k+1}}{k} \sin kx = 2 \sin x - \sin 2x + \frac{2}{3} \sin 3x - \cdots,$$

and the first three partial sums are

$$S_0(x) = 0, \quad S_1(x) = 2 \sin x, \quad S_2(x) = 2 \sin x - \sin 2x.$$

The graph of the periodic extension of $f(x)$ (also denoted by $f(x)$) as well as the graphs of S_0, S_1 and S_2 are shown on the right on the same set of axes over the interval $[-2\pi, 2\pi]$.

3. Let $f(x) = x^2$, for $-\pi < x \leq \pi$. Since $x^2 \sin kx$ is an odd function for $k \geq 1$, $b_k = 0$ for $k \geq 1$. And since $x^2 \cos kx$ is an even function for $k \geq 0$, $a_0 = \frac{2}{\pi} \int_0^{\pi} x^2 dx = \frac{2\pi^2}{3}$, and for $k \geq 1$, we integrate by parts twice to get

$$a_k = \frac{2}{\pi} \int_0^{\pi} x^2 \cos kx \, dx = \frac{2}{\pi} \left[\frac{x^2}{k} \sin kx + \frac{2x}{k^2} \cos kx - \frac{2}{k^3} \sin kx \right]_0^{\pi} = \frac{4(-1)^k}{k^2}.$$

Therefore, the Fourier series of $f(x)$ is

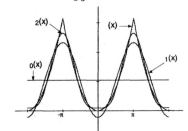

$$\frac{\pi^2}{3} + \sum_{k=1}^{\infty} \frac{4(-1)^k}{k^2} \cos kx$$

$$= \frac{\pi^2}{3} - 4 \cos x + \cos 2x - \frac{4}{9} \cos 3x + \cdots,$$

and the first three partial sums are

$$S_0(x) = \frac{\pi^2}{3}, \qquad S_1(x) = \frac{\pi^2}{3} - 4 \cos x, \qquad S_2(x) = \frac{\pi^2}{3} - 4 \cos x + \cos 2x.$$

The graph of the periodic extension of $f(x)$ (also denoted by $f(x)$) as well as the graphs of S_0, S_1 and S_2 are shown on the right on the same set of axes over the interval $[-2\pi, 2\pi]$.

5. Let $f(x) = \begin{cases} 0, & -\pi < x \leq 0; \\ 1, & 0 < x \leq \pi. \end{cases}$. Since $f(x) = 0$ for $-\pi < x \leq 0$, the integrals for the Fourier coefficients are zero on $[-\pi, 0]$. Hence,

$$a_k = \frac{1}{\pi} \int_0^{\pi} \cos kx \, dx = \begin{cases} 1, & k = 0; \\ 0, & k \geq 1, \end{cases}$$

$$b_k = \frac{1}{\pi} \int_0^{\pi} \sin kx \, dx = \frac{(-1)^{k+1} + 1}{k\pi} = \begin{cases} 0, & k \text{ even}; \\ 2/(k\pi), & k \text{ odd}. \end{cases}$$

571

Therefore the Fourier series of $f(x)$ is

$$\frac{1}{2} + \sum_{j=0}^{\infty} \frac{2}{(2j+1)\pi} \sin(2j+1)x$$

$$= \frac{1}{2} + \frac{2}{\pi} \sin x + \frac{2}{3\pi} \sin 3x + \frac{2}{5\pi} \sin 5x + \cdots,$$

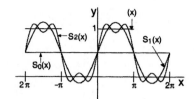

and the first three partial sums are

$$S_0(x) = \frac{1}{2}, \qquad S_1(x) = \frac{1}{2} + \frac{2}{\pi} \sin x, \qquad S_2(x) = \frac{1}{2} + \frac{2}{\pi} \sin x + \frac{2}{3\pi} \sin 3x.$$

The graph of the periodic extension of $f(x)$ (also denoted by $f(x)$) as well as the graphs of S_0, S_1 and S_2 are shown on the right on the same set of axes over the interval $[-2\pi, 2\pi]$.

7. Let $f(x) = \begin{cases} -\pi, & -\pi < x < 0; \\ \pi, & 0 \le x \le \pi. \end{cases}$ Since f is an odd function, $f(x)\cos kx$ is an odd function for $k \ge 0$ and $f(x)\sin kx$ is an even function for $k \ge 1$. Thus, integration over the symmetric interval $[\pi, \pi]$ gives $a_k = 0$, for $k \ge 0$, and

$$b_k = \frac{2}{\pi} \int_0^\pi \pi \sin kx \, dx = 2 \int_0^\pi \sin kx \, dx = 2\frac{(-1)^{k+1} + 1}{k} = \begin{cases} 0, & k \text{ even}; \\ 4/k, & k \text{ odd}. \end{cases}$$

Therefore, the Fourier series of $f(x)$ is

$$\sum_{j=0}^{\infty} \frac{4}{2j+1} \sin(2j+1)x$$

$$= 4 \sin x + \frac{4}{3} \sin 3x + \frac{4}{5} \sin 5x + \cdots,$$

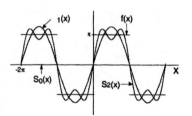

and the first three partial sums are

$$S_0(x) = 0, \quad S_1(x) = 4 \sin x, \quad S_2(x) = 4 \sin x + \frac{4}{3} \sin 3x.$$

The graph of the periodic extension of $f(x)$ (also denoted by $f(x)$) as well as the graphs of S_0, S_1 and S_2 are shown on the right on the same set of axes over the interval $[-2\pi, 2\pi]$.

9. Let $f(x) = -|x|$, for $-\pi \le x \le \pi$. Since $f(x)$ is an even function, $f(x)\cos kx$ is an even function for $k \ge 0$ and $f(x)\sin kx$ is an odd function for $k \ge 1$. Thus, integration over the symmetric interval $[-\pi, \pi]$ gives $b_k = 0$ for $k \ge 1$. For $k = 0$, $a_0 = \frac{2}{\pi}\int_0^\pi (-x)dx = -\pi$, and for $k \ge 1$, we use integration by parts to get

$$a_k = \frac{2}{\pi} \int_0^\pi (-x \cos kx) \, dx = -\frac{2}{\pi} \left[\frac{x}{k} \sin kx + \frac{1}{k^2} \cos kx \right]_0^\pi = \begin{cases} 0, & k \text{ even}; \\ 4/(k^2\pi), & k \text{ odd}. \end{cases}$$

Therefore, the Fourier series of $f(x)$ is

$$-\frac{\pi}{2} + \sum_{j=0}^{\infty} \frac{4}{(2j+1)^2\pi} \cos(2j+1)x$$

$$= -\frac{\pi}{2} + \frac{4}{\pi} \cos x + \frac{4}{9\pi} \cos 3x + \frac{4}{25\pi} \cos 5x + \cdots,$$

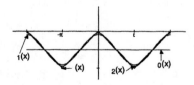

and the first three partial sums are

$$S_0(x) = -\frac{\pi}{2}, \quad S_1(x) = -\frac{\pi}{2} + \frac{4}{\pi}\cos x, \quad S_2(x) = -\frac{\pi}{2} + \frac{4}{\pi}\cos x + \frac{4}{9\pi}\cos 3x.$$

The graph of the periodic extension of $f(x)$ (also denoted by $f(x)$) as well as the graphs of S_0, S_1 and S_2 are shown on the right side of the previous page on the same set of axes over the interval $[-2\pi, 2\pi]$.

11. Let $f(x) = x$ for $-\pi < x \le \pi$ (from Exercise 1), and let $F(x)$ be the function to which the Fourier series of f converges on $[-2\pi, 2\pi]$. On this interval, the discontinuities of the periodic extension of f are at $x = \pm\pi$. By Theorem 8.6,

$$F(\pi) = \frac{1}{2}[f(\pi-) + f(-\pi+)] = \frac{1}{2}[\pi + (-\pi)] = 0.$$

Since $F(x)$ has period 2π, $F(-\pi) = 0$ as well. Thus,

$$F(x) = \begin{cases} x + 2\pi, & -2\pi \le x < -\pi; \\ x, & -\pi < x < \pi; \\ x - 2\pi, & \pi < x \le 2\pi; \\ 0, & x = \pm\pi. \end{cases}$$

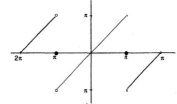

13. Let $f(x) = x^2$ for $-\pi < x \le \pi$ (from Exercise 3), and let $F(x)$ be the function to which the Fourier series of f converges on $[-2\pi, 2\pi]$. Since the periodic extension of f has no discontinuities, Theorem 8.6 implies

$$F(x) = \begin{cases} (x + 2\pi)^2, & -2\pi \le x \le -\pi; \\ x^2, & -\pi < x \le \pi; \\ (x - 2\pi)^2, & \pi < x \le 2\pi. \end{cases}$$

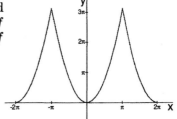

15. Let $f(x) = \begin{cases} 0, & -\pi < x \le \pi; \\ 1, & 0 < x \le \pi, \end{cases}$ (from Exercise 5), and let $F(x)$ be the function to which the Fourier series of f converges on $[-2\pi, 2\pi]$. On this interval, the discontinuities of the periodic extension of f are at $x = \pm\pi$, $x = 0$ and $x = \pm 2\pi$. By Theorem 8.6,

$$F(0) = \frac{1}{2}[f(0-) + f(0+)] = \frac{1}{2}[0 + 1] = \frac{1}{2}, \qquad F(\pi) = \frac{1}{2}[f(\pi-) + f(-\pi+)] = \frac{1}{2}[1 + 0] = \frac{1}{2}.$$

Since F has period 2π, $F(-2\pi) = \frac{1}{2}$, $F(-\pi) = \frac{1}{2}$ and $F(2\pi) = \frac{1}{2}$. Thus,

$$F(x) = \begin{cases} 1, & -2\pi < x < -\pi,\ 0 < x < \pi; \\ 0, & -\pi < x < 0,\ \pi < x < 2\pi; \\ \frac{1}{2}, & x = 0,\ \pm\pi,\ \pm 2\pi. \end{cases}$$

17. Let $f(x) = \begin{cases} -\pi, & -\pi < x < 0; \\ \pi, & 0 \le x \le \pi, \end{cases}$ (from Exercise 7), and let $F(x)$ be the function to which the Fourier series of f converges on $[-2\pi, 2\pi]$. On this interval, the discontinuities of the periodic extension of f are at $x = \pm\pi$, $x = 0$ and $x = \pm 2\pi$. By Theorem 8.6,

$$F(0) = \frac{1}{2}[f(0-) + f(0+)] = \frac{1}{2}[-\pi + \pi] = 0, \quad F(\pi) = \frac{1}{2}[f(\pi-) + f(-\pi+)] = \frac{1}{2}[\pi + (-\pi)] = 0.$$

Since F has period 2π, $F(-2\pi) = 0$, $F(-\pi) = 0$ and $F(2\pi) = 0$. Thus,

$$F(x) = \begin{cases} \cdot \ \pi, & -2\pi < x < -\pi,\ 0 < x < \pi; \\ -\pi, & -\pi < x < 0,\ \pi < x < 2\pi; \\ 0, & x = 0, \pm\pi, \pm2\pi. \end{cases}$$

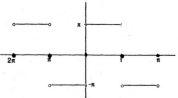

19. Let $f(x) = -|x|$ for $-\pi \le x \le \pi$ (from Exercise 9), and let $F(x)$ be the function to which the Fourier series of f converges on $[-2\pi, 2\pi]$. Since the periodic extension of f has no discontinuities, Theorem 8.6 implies

$$F(x) = \begin{cases} -|x + 2\pi|, & -2\pi \le x \le -\pi; \\ -|x|, & -\pi < x \le \pi; \\ -|x - 2\pi|, & \pi < x \le 2\pi. \end{cases}$$

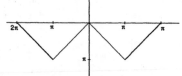

21. Let $f(x)$ and $g(x)$ have the Fourier coefficients a_k, b_k and a'_k, b'_k, respectively. Let α and β be constants and let $h(x) = \alpha f(x) + \beta g(x)$. If A_k and B_k are the Fourier coefficients of h then, by Equations 8.2 in the text

$$A_k = \frac{1}{\pi}\int_{-\pi}^{\pi} h(x)\cos kx\,dx = \frac{1}{\pi}\int_{-\pi}^{\pi} [\alpha f(x) + \beta g(x)]\cos kx\,dx$$

$$= \alpha\left[\frac{1}{\pi}\int_{-\pi}^{\pi} f(x)\cos kx\,dx\right] + \beta\left[\frac{1}{\pi}\int_{-\pi}^{\pi} g(x)\cos kx\,dx\right] = \alpha a_k + \beta a'_k,$$

and

$$B_k = \frac{1}{\pi}\int_{-\pi}^{\pi} h(x)\sin kx\,dx = \frac{1}{\pi}\int_{-\pi}^{\pi} [\alpha f(x) + \beta g(x)]\sin kx\,dx$$

$$= \alpha\left[\frac{1}{\pi}\int_{-\pi}^{\pi} f(x)\sin kx\,dx\right] + \beta\left[\frac{1}{\pi}\int_{-\pi}^{\pi} g(x)\sin kx\,dx\right] = \alpha b_k + \beta b'_k,$$

where the last member of each display follows from Equations 8.2 applied to the functions f and g, respectively.

23. The square of the half-angle identity for the cosine function is $\cos^2 x = \frac{1}{2} + \frac{1}{2}\cos 2x$, so that multiplication by $\cos x$ gives $\cos^3 x = \frac{1}{2}\cos x + \frac{1}{2}\cos x \cos 2x$. The first identity given in Exercise 28 of this section can then be used with $\alpha = 2x$ and $\beta = x$ to get

$$\cos^3 x = \frac{1}{2}\cos x + \frac{1}{2}\left(\frac{1}{2}\cos 3x + \frac{1}{2}\cos x\right) = \frac{3}{4}\cos x + \frac{1}{4}\cos 3x.$$

25. The square of the half-angle identity for the sine function is $\sin^2 x = \frac{1}{2} - \frac{1}{2}\cos 2x$, so that multiplication by $\sin x$ gives $\sin^3 x = \frac{1}{2}\sin x - \frac{1}{2}\sin x \cos 2x$. The second identity given in Exercise 28 of this section can then be used with $\alpha = x$ and $\beta = 2x$ to get

$$\sin^3 x = \frac{1}{2}\sin x - \frac{1}{2}\left(\frac{1}{2}\sin 3x + \frac{1}{2}\sin(-x)\right)$$

$$= \frac{1}{2}\sin x - \frac{1}{2}\left(\frac{1}{2}\sin 3x - \frac{1}{2}\sin x\right) = \frac{3}{4}\sin x - \frac{1}{4}\sin 3x.$$

27. The given trigonometric function is already in the form of a trigonometric polynomial, and therefore it is its own Fourier series.

29. When the identities $\cos nx = \frac{1}{2}(e^{inx} + e^{-inx})$ and $\sin nx = \frac{1}{2}(e^{inx} - e^{-inx})$ are inserted into the three integrals in Equations 8.4 in the text and the resulting integrands are multiplied out, one obtains the equivalent integrals

(1a) $\quad \dfrac{1}{\pi} \displaystyle\int_{-\pi}^{\pi} \cos kx \sin lx \, dx = \dfrac{1}{4\pi i} \displaystyle\int_{-\pi}^{\pi} \left[e^{i(k+l)x} - e^{i(k-l)x} + e^{-i(k-l)x} - e^{-i(k+l)x} \right] dx,$

(2a) $\quad \dfrac{1}{\pi} \displaystyle\int_{-\pi}^{\pi} \cos kx \cos lx \, dx = \dfrac{1}{4\pi} \displaystyle\int_{-\pi}^{\pi} \left[e^{i(k+l)x} + e^{i(k-l)x} + e^{-i(k-l)x} + e^{-i(k+l)x} \right] dx,$

(3a) $\quad \dfrac{1}{\pi} \displaystyle\int_{-\pi}^{\pi} \sin kx \sin lx \, dx = - \dfrac{1}{4\pi} \displaystyle\int_{-\pi}^{\pi} \left[e^{i(k+l)x} - e^{i(k-l)x} - e^{-i(k-l)x} + e^{-i(k+l)x} \right] dx.$

First, assume $k \neq l$. Each of the integrals on the right sides of (1a), (2a) and (3a) can be broken in a natural way into four integrals with integrands of the form $e^{i(k+l)x}$, $e^{-i(k+l)x}$, $e^{i(k-l)x}$ and $e^{-i(k-l)x}$. When any of these are integrated, the result is a constant multiple of the exponential in question. Since each such exponential has the same value at $x = \pi$ and $x = -\pi$, the value of the integral of such an exponential over the interval $[-\pi, \pi]$ is 0. It follows that the values of the integrals on the left sides of (1a), (2a) and (3a) are all 0.

If $k = l \neq 0$ then equations (1a), (2a) and (3a) become

(1b) $\qquad \dfrac{1}{\pi} \displaystyle\int_{-\pi}^{\pi} \cos kx \sin kx \, dx = \dfrac{1}{4\pi i} \displaystyle\int_{-\pi}^{\pi} \left[e^{i2kx} - e^{-i2kx} \right] dx,$

(2b) $\qquad \dfrac{1}{\pi} \displaystyle\int_{-\pi}^{\pi} \cos^2 kx \, dx = \dfrac{1}{4\pi} \displaystyle\int_{-\pi}^{\pi} \left[e^{i2kx} + 2 + e^{-i2kx} \right] dx,$

(3b) $\qquad \dfrac{1}{\pi} \displaystyle\int_{-\pi}^{\pi} \sin^2 kx \, dx = - \dfrac{1}{4\pi} \displaystyle\int_{-\pi}^{\pi} \left[e^{i2kx} - 2 + e^{-i2kx} \right] dx,$

The observations made in the case $k \neq l$ can be used here as well. Specifically, the integral on the right side of (1b) can be broken in a natural way into two integrals, both of which have value 0; and the integrals on the right sides of (2b) and (3b) can be broken in a natural way into three integrals, two of which have value zero and the third that evaluates to

$$\dfrac{1}{4\pi} \int_{-\pi}^{\pi} 2 \, dx = \dfrac{1}{4\pi} \left[2x \right]_{-\pi}^{\pi} = 1.$$

That is, the left side of (1b) equals 0 and the left sides of (2b) and (3b) equal 1. This completes the proof of the orthogonality relations given by Equations 8.4.

31. Yes. To see this, one need only look at the graph of f on $[-\pi, \pi]$:

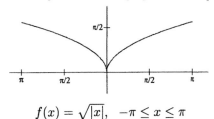

$$f(x) = \sqrt{|x|}, \quad -\pi \leq x \leq \pi$$

Besides the fact that f is bounded on $[-\pi, \pi]$, it is monotone decreasing on $[-\pi, 0]$ and monotone increasing on $[0, \pi]$. That is, f is a bounded piecewise monotone function on $[-\pi, \pi]$ and therefore satisfies the hypotheses of Theorem 8.6.

Section 9: APPLIED FOURIER EXPANSIONS

Exercise Set 9ABC (pgs. 720-721)

1. Since $f(x) = -x$, $-2 < x < 2$, is an odd function, $-x \cos \frac{k\pi x}{2}$ is an odd function for $k \geq 0$ and $-x \sin \frac{k\pi x}{2}$ is an even function for $k \geq 1$. Hence, on the symmetric interval $[-2, 2]$, Equations 9.1 in the text (with $p = 2$) give

$$a_k = \frac{1}{2} \int_{-2}^{2} -x \cos \frac{k\pi x}{2} dx = 0, \quad k \geq 0,$$

$$b_k = \frac{1}{2} \int_{-2}^{2} -x \sin \frac{k\pi x}{2} dx = \int_{0}^{2} -x \sin \frac{k\pi x}{2} dx = \left[\frac{2x}{k\pi} \cos \frac{k\pi x}{2} - \frac{4}{k^2 \pi^2} \sin \frac{k\pi x}{2} \right]_{0}^{2}$$

$$= \frac{4}{k\pi} (-1)^k, \quad k \geq 1,$$

where integration by parts was used to obtain the last expression in the second line. The Fourier series of $f(x)$ is therefore

$$\sum_{k=1}^{\infty} \frac{4(-1)^k}{k\pi} \sin \frac{k\pi x}{2} = \frac{4}{\pi} \left(-\sin \frac{\pi x}{2} + \frac{1}{2} \sin \pi x - \frac{1}{3} \sin \frac{3\pi x}{2} + \cdots \right).$$

Note that all terms are zero whenever x is an even integer. In particular, the series converges to 0 at $x = -2$ and $x = 2$.

3. Let $f(x)$ be an odd function on $[-p, p]$. Then $f(x) \cos(k\pi x/p)$ is an odd function for $k \geq 0$ and its integral over $[-p, p]$ is zero for $k \geq 0$; and $f(x) \sin(k\pi x/p)$ is an even function for $k \geq 1$ and its integral over $[-p, p]$ is twice its integral over $[0, p]$. Hence, by Equations 9.1 in the text, the Fourier coefficients a_k, b_k of f are given by

$$a_k = \frac{1}{p} \int_{-p}^{p} f(x) \cos \frac{k\pi x}{p} dx = 0, \quad k \geq 0,$$

$$b_k = \frac{1}{p} \int_{-p}^{p} f(x) \sin \frac{k\pi x}{p} dx = \frac{2}{p} \int_{0}^{p} f(x) \sin \frac{k\pi x}{p} dx, \quad k \geq 1.$$

Similarly, let $g(x)$ be an even function on $[-p, p]$. Then $g(x) \sin(k\pi x/p)$ is an odd function for $k \geq 1$ and its integral over $[-p, p]$ is zero for $k \geq 1$; and $g(x) \cos(k\pi x/p)$ is an even function for $k \geq 0$ and its integral over $[-p, p]$ is twice its integral over $[0, p]$. Again, by Equations 9.1, the Fourier coefficients a'_k, b'_k of g are given by

$$a'_k = \frac{1}{p} \int_{-p}^{p} g(x) \cos \frac{k\pi x}{p} dx = \frac{2}{p} \int_{0}^{p} g(x) \cos \frac{k\pi x}{p} dx, \quad k \geq 0,$$

$$b'_k = \frac{1}{p} \int_{-p}^{p} g(x) \sin \frac{k\pi x}{p} dx = 0, \quad k \geq 1.$$

5. The odd periodic extension $f_o(x)$ of $f(x)$ has period 2 and its graph is shown on the following page for $-2 \leq x \leq 2$. The jump discontinuities are at $x = 2n$ (n an integer), where f_o is normalized to have the value 0. The simple discontinuities are at $x = 2n + 1$, where f_o is defined to be 0. The complete description of f_o is therefore

$$f_o(x) = \begin{cases} 2n + 1 - x, & 2n < x < 2n + 2; \\ 0, & x = 2n, \end{cases}$$

where $n = 0, \pm 1, \pm 2, \ldots$. The Fourier series of $f_o(x)$ consists only of sine terms and converges to $f_o(x)$ for each x. In particular, it converges to $f(x)$ for $0 < x < 1$. By Equation 9.3 (with $p = 1$), $a_k = 0$ for $k \geq 0$, and for $k \geq 1$,

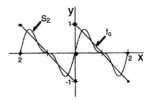

$$b_k = 2 \int_0^1 (1 - x) \sin k\pi x \, dx = 2 \left[-\frac{1 - x}{k\pi} \cos k\pi x - \frac{1}{k^2\pi^2} \sin k\pi x \right]_0^1 = \frac{2}{k\pi}.$$

The Fourier series of $f_o(x)$ is therefore

$$\sum_{k=1}^{\infty} \frac{2}{k\pi} \sin k\pi x = \frac{2}{\pi} \left(\sin \pi x + \frac{1}{2} \sin 2\pi x + \frac{1}{3} \sin 3\pi x + \cdots \right).$$

The sum of the first two nonzero terms of the series is the 2nd partial sum

$$S_2(x) = \frac{2}{\pi} \sin \pi x + \frac{1}{\pi} \sin 2\pi x,$$

whose graph is shown on the same set of axes as the graph of f_o.

7. The odd periodic extension $f_o(x)$ of $f(x)$ has period π and its graph is shown on the right for $-\pi \leq x \leq \pi$. The jump discontinuities are at the integer multiples of π, where f_o is normalized to have the value 0. The simple discontinuities are at odd integer multiples of π, where f_e is defined to 0. The complete description of $f_o(x)$ over the entire real line is therefore

$$f_o(x) = \begin{cases} \cos x, & 2n\pi < x < (2n+1)\pi; \\ -\cos x, & (2n-1)\pi < x < 2n\pi; \\ 0, & x = n\pi, \end{cases}$$

where $n = 0, \pm 1, \pm 2, \ldots$. The Fourier series of $f_o(x)$ consists only of sine terms and converges to $f_o(x)$ for each x. In particular, it converges to $f(x)$ for $0 < x < \pi/2$. By Equation 9.3 (with $p = \pi/2$), $a_k = 0$ for $k \geq 0$, and for $k \geq 1$,

$$b_k = \frac{4}{\pi} \int_0^{\pi/2} \cos x \sin 2kx \, dx = \frac{2}{\pi} \int_0^{\pi/2} \left[\sin(2k+1)x + \sin(2k-1)x \right] dx$$

$$= \frac{2}{\pi} \left[-\frac{1}{2k+1} \cos(2k+1)x - \frac{1}{2k-1} \cos(2k-1)x \right]_0^{\pi/2}$$

$$= \frac{2}{\pi} \left(\frac{1}{2k+1} + \frac{1}{2k-1} \right) = \frac{8k}{(4k^2 - 1)\pi}.$$

The Fourier series of $f_o(x)$ is therefore

$$\sum_{k=1}^{\infty} \frac{8k}{(4k^2 - 1)\pi} \sin 2kx = \frac{8}{\pi} \left(\frac{1}{3} \sin 2x + \frac{2}{15} \sin 4x + \frac{3}{35} \sin 6x + \cdots \right).$$

577

The sum of the first two nonzero terms of this series is the 2nd partial sum

$$S_2(x) = \frac{8}{3\pi} \sin 2x + \frac{16}{15\pi} \sin 4x,$$

whose graph is shown on the same set of axes as the graph of f_o.

9. The odd periodic extension $f_o(x)$ of $f(x)$ has period 4 and its graph is shown on the right for $-3 \le x \le 3$. The jump discontinuities are at the odd integer values of x, where f_o is normalized to have value $\frac{1}{2}$ at $x = 4n - 1$ and value $-\frac{1}{2}$ at $4n + 1$. There are also simple discontinuities at the even integer values of x, where f_o is defined to be 0. The complete description of $f_o(x)$ over the entire real line is given by

$$f_o(x) = \begin{cases} 0, & 4n - 1 < x < 4n + 1; \\ x - 2 - 4n, & 4n + 1 < x < 4n + 3; \\ 1/2, & x = 4n - 1; \\ -1/2, & x = 4n + 1, \end{cases}$$

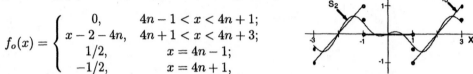

where $n = 0, \pm 1, \pm 2, \cdots$. The Fourier series of $f_o(x)$ consists only of sine terms and converges to $f_o(x)$ for each x. In particular, it converges to $f(x)$ on $0 < x \le 2$, except at $x = 1$ where it converges to its normalized value of $-\frac{1}{2}$. By Equation 9.3 (with $p = 2$), $a_k = 0$ for $k \ge 0$, and for $k \ge 1$,

$$b_k = \int_0^2 f(x) \sin \frac{k\pi x}{2} \, dx = \int_1^2 (x - 2) \sin \frac{k\pi x}{2} \, dx$$

$$= \left[-\frac{2(x-2)}{k\pi} \cos \frac{k\pi x}{2} + \frac{4}{k^2\pi^2} \sin \frac{k\pi x}{2} \right]_1^2 = -\frac{2}{k\pi} \cos \frac{k\pi}{2} - \frac{4}{k^2\pi^2} \sin \frac{k\pi}{2}.$$

The Fourier series of $f_o(x)$ is therefore

$$\sum_{k=1}^{\infty} \left(-\frac{2}{k\pi} \cos \frac{k\pi}{2} - \frac{4}{k^2\pi^2} \sin \frac{k\pi}{2} \right) \sin \frac{k\pi x}{2}.$$

The sum of the first two nonzero terms of the series is the 2nd partial sum

$$S_2(x) = -\frac{4}{\pi^2} \sin \frac{\pi x}{2} + \frac{1}{\pi} \sin \pi x,$$

whose graph is shown on the same set of axes as the graph of f_o.

11. The even periodic extension $f_e(x)$ of $f(x)$ has period 2 and its graph is shown on the following page for $-3 \le x \le 3$. The discontinuities are all simple discontinuities and occur at the integer values of x, where $f_e(x)$ is defined to be 1 when x is an even integer and 0 when x is an odd integer. So $f_e(x)$ everywhere continuous. The complete description of $f_e(x)$ is therefore

$$f_e(x) = \begin{cases} 1 - 2n + x, & 2n - 1 \le x \le 2n; \\ 1 + 2n - x, & 2n < x < 2n + 1, \end{cases}$$

where $n = 0, \pm 1, \pm 2, \ldots$. The Fourier series of $f_e(x)$ consists only of cosine terms and converges to $f_e(x)$ for each x. In particular, it converges to $f(x)$ on $0 < x < 1$. By Equation 9.4 (with $p = 1$), $b_k = 0$ for $k \ge 1$. Also, $a_0 = 2 \int_0^1 (1 - x) \, dx = 1$, and for $k \ge 1$,

$$a_k = 2 \int_0^1 (1 - x) \cos k\pi x \, dx = 2 \left[\frac{1 - x}{k\pi} \sin k\pi x - \frac{1}{k^2\pi^2} \cos k\pi x \right]_0^1$$

$$= \frac{2}{k^2\pi^2} [1 - (-1)^k] = \begin{cases} 0, & k \text{ even}; \\ 4/(k^2\pi^2), & k \text{ odd}. \end{cases}$$

578

The Fourier series for $f_e(x)$ is therefore

$$\frac{1}{2} + \sum_{j=0}^{\infty} \frac{4}{(2j+1)^2 \pi^2} \cos(2j+1)\pi x = \frac{1}{2} + \frac{4}{\pi^2}\left(\cos \pi x + \frac{1}{9}\cos 3\pi x + \frac{1}{25}\cos 5\pi x + \cdots \right).$$

The sum of the first two nonzero terms of the series is the 1st partial sum

$$S_1(x) = \frac{1}{2} + \frac{4}{\pi^2}\cos \pi x,$$

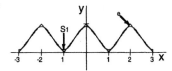

whose graph is shown on the right on the same set of axes as the graph of f_e.

13. The even periodic extension $f_e(x)$ of $f(x)$ has period π and its graph is shown on the right for $-\pi \leq x \leq \pi$. All discontinuities are simple discontinuities and occur at the integer multiples of $\pi/2$. Defining $f_e(x) = 0$ when x is an even integer multiple of $\pi/2$ and $f_e(x) = 1$ when x is an odd integer multiple of $\pi/2$ makes $f_e(x)$ everywhere continuous. The complete description of $f_e(x)$ is therefore

$$f_e(x) = \begin{cases} \sin x, & 2n\pi < x < (2n+1)\pi; \\ -\sin x, & (2n-1)\pi \leq x \leq 2n\pi, \end{cases}$$

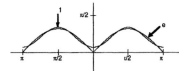

where $n = 0, \pm 1, \pm 2, \ldots$. The Fourier series of $f_e(x)$ consists only of cosine terms and converges to $f_e(x)$ for each x. In particular, it converges to $f(x)$ on $0 < x < \pi/2$. By Equation 9.4 (with $p = \pi/2$), $b_k = 0$ for $k \geq 1$. Also, $a_0 = \frac{4}{\pi}\int_0^{\pi/2} \sin x\, dx = \frac{4}{\pi}$, and for $k \geq 1$,

$$a_k = \frac{4}{\pi}\int_0^{\pi/2} \sin x \cos 2kx\, dx = \frac{2}{\pi}\int_0^{\pi/2} \big[\sin(2k+1)x - \sin(2k-1)x\big]dx$$

$$= \frac{2}{\pi}\left[-\frac{1}{2k+1}\cos(2k+1)x + \frac{1}{2k-1}\cos(2k-1)x \right]_0^{\pi/2}$$

$$= \frac{2}{\pi}\left[\frac{1}{2k+1} - \frac{1}{2k-1} \right] = -\frac{4}{(4k^2-1)\pi}.$$

The Fourier series of $f_e(x)$ is therefore

$$\frac{2}{\pi} + \sum_{k=1}^{\infty} \frac{-4}{(4k^2-1)\pi}\cos 2kx = \frac{2}{\pi} - \frac{4}{\pi}\left(\frac{1}{3}\cos 2x + \frac{1}{15\pi}\cos 4x + \frac{1}{35}\cos 6x + \cdots \right).$$

The sum of the first two nonzero terms of the series is the 1st partial sum

$$S_1(x) = \frac{2}{\pi} - \frac{4}{3\pi}\cos 2x,$$

whose graph is shown on the same set of axes as the graph of f_e.

15. The even periodic extension $f_e(x)$ of $f(x)$ has period 4 and its graph is shown on the following page for $-4 \leq x \leq 4$. The jump discontinuities are at the odd integer values of x, where f_e is normalized to have value $\frac{1}{2}$. There are also simple discontinuities at the even

579

integer values of x that are not divisible by 4, where f_e is defined to be 0. The complete description of $f_e(x)$ over the entire real line is given by

$$f_e(x) = \begin{cases} -x + 4n, & 4n - 1 < x \le 4n; \\ x - 4n, & 4n < x < 4n + 1; \\ 0, & 4n + 1 < x < 4n + 3; \\ \frac{1}{2}, & x = 2n + 1, \end{cases}$$

where $n = 0, \pm 1, \pm 2, \ldots$. The Fourier series of $f_e(x)$ consists only of cosine terms and converges to $f_e(x)$ for each x. In particular, it converges to $f(x)$ on $0 < x < 2$, except at $x = 1$ where it converges to the normalized value $\frac{1}{2}$. By Equation 9.4 (with $p = 2$), $b_k = 0$ for $k \ge 1$. Also, since $f(x) = 0$ on $[1, 2]$,

$$a_0 = \int_0^2 f(x)\, dx = \int_0^1 x\, dx = \frac{1}{2},$$

and for $k \ge 1$,

$$a_k = \int_0^2 f(x) \cos \frac{k\pi x}{2}\, dx = \int_0^1 x \cos \frac{k\pi x}{2}\, dx$$

$$= \left[\frac{2x}{k\pi} \sin \frac{k\pi x}{2} + \frac{4}{k^2 \pi^2} \cos \frac{k\pi x}{2} \right]_0^1 = \frac{2}{k\pi} \sin \frac{k\pi}{2} + \frac{4}{k^2 \pi^2} \left(\cos \frac{k\pi}{2} - 1 \right).$$

The Fourier series of f_e is therefore

$$\frac{1}{4} + \sum_{k=1}^{\infty} \left[\frac{2}{k\pi} \sin \frac{k\pi}{2} + \frac{4}{k^2 \pi^2} \left(\cos \frac{k\pi}{2} - 1 \right) \right] \cos \frac{k\pi x}{2}.$$

The sum of the first two nonzero terms of the series is the 1st partial sum

$$S_1(x) = \frac{1}{4} + \left(\frac{2}{\pi} - \frac{4}{\pi^2} \right) \cos \frac{\pi x}{2},$$

whose graph is shown on the same set of axes as the graph of f_e.

17. If $f(x)$ and $g(x)$ are even functions on the same symmetric interval $(-p, p)$, where $p = \infty$ is allowed, then the equations $f(-x) = f(x)$ and $g(-x) = g(x)$ both hold on $(-p, p)$. Hence, the product function $h(x)$, defined by $h(x) = f(x)g(x)$ for $-p < x < p$, satisfies

$$h(-x) = f(-x)g(-x) = f(x)g(x) = h(x), \quad -p < x < p.$$

That is, h is an even function on $(-p, p)$.

19. Let $f(x)$ be an even function on the symmetric interval $(-p, p)$, where $p = \infty$ is allowed, and let $g(x)$ be an odd function on the same interval. Then the equations $f(-x) = f(x)$ and $g(-x) = -g(x)$ both hold on $(-p, p)$. Hence, the product function $h(x)$, defined by $h(x) = f(x)g(x)$ for $-p < x < p$, satisfies

$$h(-x) = f(-x)g(-x) = f(x)\big(-g(x)\big) = -f(x)g(x) = -h(x), \quad -p < x < p.$$

That is, h is an odd function on $(-p, p)$.

21. Let $f(x)$ be periodic with period p, so that $f(x+p) = f(x)$ for all x. If f is also differentiable, then the definition of $f'(x)$ as the limit of a quotient gives

$$f'(x+p) = \lim_{h \to 0} \frac{f(x+p+h) - f(x+p)}{h} = \lim_{h \to 0} \frac{f(x+h) - f(x)}{h} = f'(x), \quad \text{for all } x.$$

So, f' is periodic with period p.

23. Let $f(x)$ be an even function on the symmetric interval $(-p, p)$, where $p = \infty$ is allowed, so that $f(-x) = f(x)$ for $-p < x < p$. If f is differentiable on $(-p, p)$ then, for $-p < x < p$, the definition of $f'(x)$ as the limit of a quotient satisfies

$$f'(-x) = \lim_{h \to 0} \frac{f(-x+h) - f(-x)}{h} = \lim_{h \to 0} \frac{f(x-h) - f(x)}{h} = -\lim_{h \to 0} \frac{f(x-h) - f(x)}{-h}$$

$$= -\lim_{-h \to 0} \frac{f(x-h) - f(x)}{-h} = -\lim_{\hat{h} \to 0} \frac{f(x+\hat{h}) - f(x)}{\hat{h}} = -f'(x),$$

where the first limit in the second line follows from the fact that $-h \to 0$ and $h \to 0$ are equivalent operations, and the last limit quotient is simply the previous limit quotient with $-h$ replaced by $\hat{h}$. In other words, $f'(x)$ is an odd function on $(-p, p)$.

25. Using the given identity with $\alpha = k\pi x/p$ and $\beta = l\pi x/p$, the given integral becomes

$$(*) \qquad \frac{2}{p} \int_0^p \sin \frac{k\pi x}{p} \sin \frac{l\pi x}{p} \, dx = \frac{1}{p} \int_0^p \left[\cos \frac{(k-l)\pi x}{p} - \cos \frac{(k+l)\pi x}{p} \right] dx.$$

If $k \neq l$ then the integral on the right can be integrated directly to get

$$\frac{1}{p} \int_0^p \left[\cos \frac{(k-l)\pi x}{p} - \cos \frac{(k+l)\pi x}{p} \right] dx$$

$$= \frac{1}{p} \left[\frac{p}{(k-l)\pi} \sin \frac{(k-l)\pi x}{p} - \frac{p}{(k+l)\pi} \sin \frac{(k+l)\pi x}{p} \right]_0^p = 0.$$

On the other hand, if $k = l \, (\neq 0)$ then equation $(*)$ becomes

$$\frac{2}{p} \int_0^p \sin \frac{k\pi x}{p} \sin \frac{l\pi x}{p} \, dx = \frac{1}{p} \int_0^p \left[1 - \cos \frac{2k\pi x}{p} \right] dx = \frac{1}{p} \left[x - \frac{p}{2k\pi} \sin \frac{2k\pi x}{p} \right]_0^p = 1.$$

Putting these two cases together gives the desired result.

27. The square-wave function given in Example 5 of the text was shown to have the Fourier expansion

$$f(t) = \frac{4}{\pi} \sum_{n=0}^{\infty} \frac{1}{2n+1} \sin(2n+1)\pi t,$$

with the understanding that the series converges to 0 at the jump discontinuities. Thus, in seeking a particular solution $y_p(t)$ of the differential equation $\ddot{y} - a^2 y = f(t)$, we first look for a particular solution of each of the equations

$$(*) \qquad \ddot{y} - a^2 y = \sin(2n+1)\pi t, \quad n = 0, 1, 2, \ldots.$$

581

Note that the associated characteristic equation $r^2 - a^2 = 0$ is independent of n and has the roots $r_{1,2} = \pm a$, so that $y_h(t) = c_1 e^{at} + c_2 e^{-at}$ is the homogeneous solution for all n. Thus, given $n \geq 0$, there is a solution $y_{p_n} = y_{p_n}(t)$ of $(*)$ of the form

$$y_{p_n} = A_n \cos(2n+1)\pi t + B_n \sin(2n+1)\pi t,$$

where A_n and B_n are constant coefficients to be determined. Inserting y_{p_n} and its second derivative

$$\ddot{y}_{p_n} = -(2n+1)^2\pi^2 A_n \cos(2n+1)\pi t - (2n+1)^2\pi^2 B_n \sin(2n+1)\pi t$$

into $(*)$ gives

$$\begin{aligned}
\ddot{y}_{p_n} - a^2 y_{p_n} &= -(2n+1)^2\pi^2 A_n \cos(2n+1)\pi t - (2n+1)^2\pi^2 B_n \sin(2n+1)\pi t \\
&\quad - a^2\Big(A_n \cos(2n+1)\pi t + B_n \sin(2n+1)\pi t\Big) \\
&= -\big[(2n+1)^2\pi^2 + a^2\big] A_n \cos(2n+1)\pi t - \big[(2n+1)^2\pi^2 + a^2\big] B_n \sin(2n+1)\pi t \\
&= \sin(2n+1)\pi t.
\end{aligned}$$

Equating coefficients of like functions gives the two equations

$$-\big[(2n+1)^2\pi^2 + a^2\big] A_n = 0 \qquad \text{and} \qquad -\big[(2n+1)^2\pi^2 + a^2\big] B_n = 1.$$

Since $(2n+1)^2\pi^2 + a^2$ is nonzero for all possible values of n and a, the first equation implies $A_n = 0$. The second equation implies $B_n = -1/\big[(2n+1)^2\pi^2 + a^2\big]$. Hence,

$$y_{p_n}(t) = -\frac{1}{(2n+1)^2\pi^2 + a^2} \sin(2n+1)\pi t$$

is a particular solution of $(*)$. By the principle of superposition, a particular solution of the differential equation $\ddot{y} - a^2 y = f(t)$ is given by

$$y_p(t) = \sum_{n=0}^{\infty} \frac{4}{(2n+1)\pi} y_{p_n}(t) = \frac{4}{\pi} \sum_{n=0}^{\infty} \frac{1}{(2n+1)\big[(2n+1)^2\pi^2 + a^2\big]} \sin(2n+1)\pi t.$$

Finally, note that the series has the form of a Fourier series and must therefore be the Fourier expansion of $y_p(t)$.

Section 10: HEAT AND WAVE EQUATIONS

Exercise Set 10AB (pgs. 727-728)

1. The homogeneous boundary conditions allows us to use the solution formula 10.2 in the text, where $u(x,0) = h(x) = \sin(\pi x/p)$ and $p > 0$ is unspecified. However, since $\sin(\pi x/p)$ is its own Fourier sine expansion, it follows that the Fourier coefficients b_k given by Equation 10.2 are $b_1 = 1$ and $b_k = 0$ for $k \geq 2$. Hence, the desired solution consists of the single term

$$u(x,t) = e^{-(\pi^2 a^2/p^2)t} \sin \frac{\pi x}{p}, \quad 0 \leq x \leq p, \ t \geq 0.$$

(NOTE: If the b_k are formally computed from Equation 10.2, one discovers that the formula for b_k is just the orthogonality relation already proven in Exercise 25 in the last exercise section and that the same result is obtained as was obtained above by the more direct method.)

3. The homogeneous boundary conditions allows us to use the solution formula 10.2 in the text, where $u(x,0) = h(x) = 1 - x$ and $p = 1$. The b_k are given by

$$b_k = 2 \int_0^p (1-x) \sin k\pi x \, dx = 2 \left[-\frac{1-x}{k\pi} \cos k\pi x - \frac{1}{k^2\pi^2} \sin k\pi x \right]_0^1 = \frac{2}{k\pi}, \quad k \geq 1.$$

The desired solution is therefore

$$u(x,t) = \sum_{k=1}^{\infty} \frac{2}{k\pi} e^{-k^2 a^2 \pi^2 t} \sin k\pi x, \quad 0 \leq x \leq 1, \ t \geq 0.$$

5. The homogeneous boundary conditions allows us to use the solution formula 10.2 in the text, where $u(x,0) = h(x) = \sin x + \frac{1}{2}\sin 2x$ and $p = \pi$. However, since $\sin x + \frac{1}{2}\sin 2x$ is its own Fourier sine expansion, it follows that the Fourier coefficients b_k given by Equation 10.2 are $b_1 = 1$, $b_2 = \frac{1}{2}$ and $b_k = 0$ for $k \geq 3$. Hence, the desired solution consists of the two terms

$$u(x,t) = e^{-a^2 t} \sin x + \frac{1}{2} e^{-4a^2 t} \sin 2x, \quad 0 \leq x \leq \pi, \ t \geq 0.$$

(See "NOTE" following the solution to Exercise 1 above.)

7. Set $u(x,t) = v(x) = \alpha + \beta x$, so that the given conditions can be written as $v(0) = -1$ and $v(2) = 1$. These give the two equations $-1 = \alpha$ and $1 = \alpha + 2\beta$, from which $\alpha = -1$ and $\beta = 1$. Hence,

$$v(x) = -1 + x.$$

9. Set $u(x,t) = v(x) = \alpha + \beta x$, so that the given conditions can be written as $v'(0) = 1$ and $v(1) = 2$. As $v'(x) = \beta$, the first condition immediately gives $\beta = 1$. The condition $v(1) = 2$ then implies $\alpha + 1 = 2$, or equivalently, $\alpha = 1$. Hence,

$$v(x) = 1 + x.$$

11. If $v(x) = \alpha + \beta x$ is the steady-state solution of the given heat equation then the boundary conditions become $v(0) = 1$ and $v(p) = 3$. The first condition immediately gives $\alpha = 1$, so that $v(x) = 1 + \beta x$. The second condition then gives $1 + \beta p = 3$, so that $\beta = 2/p$. Hence, $v(x) = 1 + (2/p)x$ for $0 \leq x \leq p$.

We next find the coefficients of the Fourier sine expansion of the function

$$h(x) - v(x) = \sin \frac{\pi x}{p} - 1 - \frac{2x}{p}, \quad 0 < x < p.$$

These are given by the second equation of Equations 10.3 in the text:

$$b_k = \frac{2}{p} \int_0^p \left(\sin \frac{\pi x}{p} - 1 - \frac{2x}{p} \right) \sin \frac{k\pi x}{p} \, dx$$

$$= \frac{2}{p} \int_0^p \sin \frac{\pi x}{p} \sin \frac{k\pi x}{p} \, dx - \frac{2}{p} \int_0^p \left(1 + \frac{2x}{p} \right) \sin \frac{k\pi x}{p} \, dx.$$

The first integral in the second line is the integral associated with the orthogonality relation given in Exercise 25 in the last exercise set. Its value is 1 when $k = 1$ and is 0 when $k \geq 2$. The second integral can be done by parts to get

$$\frac{2}{p} \int_0^p \left(1 + \frac{2x}{p} \right) \sin \frac{k\pi x}{p} \, dx = \frac{2}{p} \left[-\left(1 + \frac{2x}{p} \right) \frac{p}{k\pi} \cos \frac{k\pi x}{p} + \frac{2p}{k^2\pi^2} \sin \frac{k\pi x}{p} \right]_0^p$$

$$= \frac{2}{k\pi} [1 - 3(-1)^k].$$

Thus,

$$b_1 = 1 - \frac{8}{\pi} \quad \text{and} \quad b_k = \frac{2}{k\pi} [3(-1)^k - 1] \quad \text{for } k \geq 2.$$

Using the first equation of Equation 10.3, the desired solution is

$$u(x,t) = 1 + \frac{2x}{p} + \left(1 - \frac{8}{\pi} \right) e^{-(a^2\pi^2/p^2)t} \sin \frac{\pi x}{p}$$

$$+ \sum_{k=2}^{\infty} \frac{2}{k\pi} [3(-1)^k - 1] e^{-k^2(a^2\pi^2/p^2)t} \sin \frac{k\pi x}{p}, \quad 0 \leq x \leq p, \ t > 0.$$

13. If $v(x) = \alpha + \beta x$ is the steady-state solution of the given heat equation then the boundary conditions become $v(0) = 0$ and $v(1) = 1$. The first condition immediately gives $\alpha = 0$, so that $v(x) = \beta x$. The second condition then gives $\beta = 1$. Hence, $v(x) = x$ for $0 \leq x \leq 1$.

We next find the coefficients of the Fourier sine expansion of the function

$$h(x) - v(x) = 1 - x - x = 1 - 2x, \quad 0 < x < 1.$$

With $p = 1$, these are given by the second equation of Equations 10.3 in the text:

$$b_k = 2 \int_0^1 (1 - 2x) \sin k\pi x \, dx = 2 \left[-\frac{1 - 2x}{k\pi} \cos k\pi x - \frac{2}{k^2\pi^2} \sin k\pi x \right]_0^1$$

$$= \frac{2}{k\pi} [(-1)^k + 1] = \begin{cases} 4/(k\pi), & k \text{ even}; \\ 0, & k \text{ odd}. \end{cases}$$

Using the first equation of Equation 10.3, the desired solution is

$$u(x,t) = x + \sum_{j=1}^{\infty} \frac{4}{2j\pi} e^{-4j^2 a^2 \pi^2 t} \sin 2j\pi x$$

$$= x + \frac{2}{\pi} \sum_{j=1}^{\infty} \frac{1}{j} e^{-4j^2 a^2 \pi^2 t} \sin 2j\pi x, \quad 0 \leq x \leq 1, \ t \geq 0.$$

15. Let $u(x,t) = X(x)T(t)$ be a nonzero product solution of $u_{xx} + u_x = u_t$. Since

$$u_x = \frac{\partial(XT)}{\partial x} = X'T, \qquad u_{xx} = \frac{\partial^2(XT)}{\partial x^2} = X''T, \qquad u_t = \frac{\partial(XT)}{\partial t} = XT',$$

insertion into the D.E. gives $X''T + X'T = XT'$, and separating variables gives

$$\frac{X''}{X} + \frac{X'}{X} = \frac{T'}{T}.$$

Setting each side equal to the constant λ results in the two ordinary differential equations

$$X'' + X' - \lambda X = 0 \qquad \text{and} \qquad T' - \lambda T = 0.$$

The general solution of the second equation is $T(t) = c_0 e^{\lambda t}$. For the first equation, the characteristic roots are $r_{1,2} = (-1 \pm \sqrt{1+4\lambda})/2$. These will be unequal if $\lambda \neq -1/4$. In this case, the general solution is $X(x) = c_1 e^{r_1 x} + c_2 e^{r_2 x}$. Otherwise, $r_1 = r_2 = -1/2$ and the general solution is $X(x) = c_1 e^{-x/2} + c_2 x e^{-x/2}$. Thus,

$$u(x,t) = X(x)T(t) = \begin{cases} \left(C_1 e^{r_1 x} + C_2 e^{r_2 x}\right)e^{\lambda t}, & \lambda \neq -1/4; \\ \left(C_1 e^{-x/2} + C_2 x e^{-x/2}\right)e^{-t/4}, & \lambda = -1/4, \end{cases}$$

where $C_1 = c_0 c_1$, $C_2 = c_0 c_2$.

17. Let $u(x,t) = X(x)T(t)$ be a nonzero product solution of $xu_x = 2u_t$, where $x > 0$. Since $u_x = X'T$ and $u_t = XT'$, insertion into the D.E. gives $xX'T = 2XT'$, and separating variables gives

$$x\frac{X'}{X} = 2\frac{T'}{T}.$$

Setting each side equal to the constant λ results in the two ordinary differential equations

$$xX' - \lambda X = 0 \qquad \text{and} \qquad T' - \frac{\lambda}{2}T = 0.$$

The second equation has the general solution $T(t) = c_0 e^{\lambda t/2}$. Writing the first equation as $X' - (\lambda/x)X = 0$, use the integrating factor $e^{\int (-\lambda/x)dx} = x^{-\lambda}$ to obtain $d(x^{-\lambda}X)/dx = 0$. Integrating and solving for $X = X(x)$ gives $X(x) = c_1 x^\lambda$. Thus,

$$u(x,t) = X(x)T(t) = Cx^\lambda e^{\lambda t/2}, \qquad x > 0,$$

where $C = c_0 c_1$.

19. Let $u(x,t) = v(x)$ be a steady-state solution of $u_{xx} = u_t + 2$ and write the given boundary conditions as $v(0) = 1$ and $v(1) = 2$. Since such solutions are independent of t, $u_{xx}(x,t) = v''(x)$ and $u_t(x,t) = 0$, so that insertion into the D.E. results in the equation $v''(x) = 2$. Two integrations gives the general solution $v(x) = x^2 + c_1 x + c_2$. The boundary conditions therefore imply the two equations $c_2 = 1$ and $1 + c_1 + c_2 = 2$, from which $c_1 = 0$, $c_2 = 1$. Thus, the unique steady-state solution is

$$v(x) = x^2 + 1.$$

21. Let $u(x,t) = v(x)$ be a steady-state solution of $u_{xx} = u_t + x$ and write the given boundary conditions as $v(0) = 1$ and $v(1) = 2$. Since such solutions are independent of t, $u_{xx}(x,t) = v''(x)$ and $u_t(x,t) = 0$, so that insertion into the D.E. results in the equation $v''(x) = x$. Two integrations gives the general solution $v(x) = \frac{1}{6}x^3 + c_1 x + c_2$. The boundary conditions therefore implies the two equations $c_2 = 1$ and $\frac{1}{6} + c_1 + c_2 = 2$, from which $c_1 = \frac{5}{6}$, $c_2 = 1$. Thus, the unique steady-state solution is

$$v(x) = \frac{1}{6}x^3 + \frac{5}{6}x + 1.$$

23. (a) In this part of the exercise, we consider the more general boundary-value problem

$$u_{xx} = u_t, \quad u_x(0,t) = u_x(p,t) = 0, \quad 0 \le x \le p, \ t \ge 0,$$

where the initial temperature condition $u(x,0) = f(x)$ is not used. We want to show that the product solutions of this general boundary-value problem are precisely the functions $u_k(x,t)$ of the form

$$(*) \qquad u_k(x,t) = a_k e^{-k^2\pi^2 t/p^2} \cos\frac{k\pi x}{p}, \qquad \text{where } k = 0,1,2,\ldots.$$

Let $u(x,t) = X(x)T(t)$ be a product solution. Then the boundary conditions become $u_x(0,t) = X'(0)T(t) = 0$ and $u_x(p,t) = X'(p)T(t) = 0$, which must hold for all $t \ge 0$. In order to derive nonzero product solutions, we need $T(t)$ to *not* be identically zero. Hence, $X'(0) = 0$ and $X'(p) = 0$.

Next, since $u_{xx} = X''T$ and $u_t = XT'$, insertion into $u_{xx} = u_t$ gives $X''T = XT'$, or in separated form, $X''/X = T'/T$. Setting each side equal to the convenient constant $-\lambda^2$ results in the two ordinary differential equations

$$(**) \qquad\qquad X'' + \lambda^2 X = 0 \qquad \text{and} \qquad T' + \lambda^2 T = 0.$$

Because the form of the solution of the first equation depends on whether $\lambda = 0$ or $\lambda \ne 0$, we need to consider each case separately.

Setting $\lambda = 0$ in $(**)$ gives the equations $X'' = 0$ and $T' = 0$, so that the general solutions are of the form

$$X(x) = c_1 + c_2 x \qquad \text{and} \qquad T(t) = c_0,$$

where c_0, c_1, c_2 are constants. Since $X'(x) = c_2$, the condition $X'(0) = 0$ directly gives $c_2 = 0$. The corresponding family of product solutions is therefore of the form

$$u(x,t) = X(x)T(t) = c_1 c_0 = C.$$

Setting $C = a_0$ and writing $u(x,t)$ as $u_0(x,t)$, we obtain the family in $(*)$ with $k = 0$.

If $\lambda \ne 0$, the general solution of the first equation of $(**)$ is

$$X(x) = c_1 \cos\lambda x + c_2 \sin\lambda x.$$

With $X'(x) = -c_1\lambda\sin\lambda x + c_2\lambda\cos\lambda x$, the condition $X'(0) = 0$ gives $0 = c_2\lambda\cos\lambda x$. Since $\lambda \ne 0$, this can hold for all x in $0 \le x \le p$ only if $c_2 = 0$. Thus, $X'(x) = -c_1\lambda\sin\lambda x$. The other boundary condition $X'(p) = 0$ then gives $0 = -c_1\lambda\sin\lambda p$. Again, since $\lambda \ne 0$, this can happen only if $c_1 = 0$ or $\sin\lambda p = 0$. However, since $c_2 = 0$, we can have nonzero product solutions only if c_1 is allowed to be nonzero. Therefore, $\sin \lambda p = 0$, so that λ must be chosen to satisfy $\lambda p = k\pi$, where k is a nonzero integer. Hence, $\lambda = k\pi/p$ and $X(x)$ has the form

$$X(x) = c_1 \cos \frac{k\pi x}{p},$$

where the constant c_1 can be chosen differently for each k.

The general solution of the second equation of $(**)$ is $T(t) = c_0 e^{-\lambda^2 t}$. Using the expression for λ found above, we see that $T(t)$ has the form

$$T(t) = c_0 e^{-k^2 \pi^2 t/p^2},$$

where the constant c_0 can be chosen differently for each k. Hence, the corresponding product solutions are given by

$$u(x,t) = X(x)T(t) = c_0 c_1 e^{-k^2 \pi^2 t/p^2} \cos \frac{k\pi x}{p}, \quad k = \pm 1, \pm 2, \ldots.$$

Noting that the right side of this equation has the same value if k is replaced by $-k$, it follows that we can restrict k to be positive. Finally, we reflect the dependence on k by setting $c_0 c_1 = a_k$ and replacing $u(x,t)$ with $u_k(x,t)$. The resulting functions are then of the same form as the functions in $(*)$ with $k \geq 1$.

(b) Since the heat equation is linear, if two functions are a solution of the boundary-value problem in part (a) then so is their sum. It follows that the sum $\sum_{k=0}^{\infty} u_k(x,t)$ of all of the product solutions derived in part (a) is a solution whenever the series converges uniformly on the half-strip S defined by $0 \leq x \leq p$, $t \geq 0$. That is,

$$u(x,t) = u_0(x,t) + \sum_{k=1}^{\infty} u_k(x,t) = a_0 + \sum_{k=1}^{\infty} a_k e^{-k^2 \pi^2 t/p^2} \cos \frac{k\pi x}{p}$$

is a solution whenever the constants $a_0, a_1, a_2, \ldots$ are chosen so that the series is uniformly convergent on S.

Now suppose that $f(x)$ is a bounded piecewise monotone function defined in $0 \leq x \leq p$ and that $u(x,t)$ is a solution of the above form such that $u(x,0) = f(x)$ on $0 \leq x \leq p$. By uniform convergence of the series solution, we can set $t = 0$ in each term of the series to conclude that $f(x)$ must satisfy

$$f(x) = a_0 + \sum_{k=1}^{\infty} a_k \cos \frac{k\pi x}{p} = \frac{a_0'}{2} + \sum_{k=1}^{\infty} a_k \cos \frac{k\pi x}{p} \quad \text{for all } 0 \leq x \leq p,$$

where we have temporarily replaced a_0 with $a_0'/2$. By Equation 9.4 in Section 9 of this chapter, we see that the above series solution will satisfy the initial temperature condition if the constants $a_0', a_1, a_2, \ldots$ are chosen to be the coefficients of the Fourier cosine expansion of $f(x)$ on $0 \leq x \leq p$; i.e., if $a_0, a_1, a_2, \ldots$ are chosen to satisfy

$$a_0 = \frac{1}{p} \int_0^p f(x)\, dx, \qquad a_k = \frac{2}{p} \int_0^p f(x) \cos \frac{k\pi x}{p}\, dx, \quad k \geq 1.$$

(c) If $u(x,t)$ is the series solution found in part (b) then the symbol $u(x,\infty)$ for the steady-state temperature function is defined for each fixed $0 \leq x \leq p$ by

$$u(x,\infty) = \lim_{t \to \infty} u(x,t).$$

On the other hand, since each term of the series solution found in part (b) goes to zero as $t \to \infty$, uniform convergence of the series allows us to write

$$\lim_{t \to \infty} u(x,t) = a_0 + \sum_{k=1}^{\infty} \lim_{t \to \infty} e^{-k^2 \pi^2 t/p^2} \cos \frac{k \pi x}{p} = a_0.$$

Therefore, given the specific expression for a_0 derived in by part (b),

$$u(x, \infty) = \frac{1}{p} \int_0^p f(x)\, dx, \quad \text{for } 0 \le x \le p.$$

25. The result of Exercise 23(b) shows that the series solution of the insulated endpoints problem $a^2 u_{xx} = u_t$, $u_x(0,t) = u_x(1,t) = 0$, $u(x,0) = 1$, $0 < x < 1$,

$$u(x,t) = a_0 + \sum_{k=1}^{\infty} a_k e^{-k^2 a^2 \pi^2 t} \cos k \pi x, \quad 0 \le x \le 1,\ t \ge 0,$$

where the a_k are the coefficients of the Fourier cosine expansion of the constant function $f(x) = 1$ on $0 \le x \le 1$. However, since this function is its own Fourier cosine expansion, it follows that $a_0 = 1$ and $a_k = 0$ for $k \ge 1$. That is, the desired solution is

$$u(x,t) = 1, \quad 0 \le x \le 1,\ t \ge 0.$$

27. (a) Let $u(x,t) = X(x)T(t)$ be a product solution to $t u_{xx} = u_t$. Taking the requisite derivatives and inserting them into the given D.E. results in the equation $t X''T = XT'$. Separating variables gives

$$\frac{X''}{X} = \frac{T'}{tT}.$$

Setting each side equal to the convenient common constant $-\lambda^2$ gives $X''/X = -\lambda^2$ and $T'/(tT) = -\lambda^2$, or equivalently,

$$X'' + \lambda^2 X = 0 \quad \text{and} \quad T' + \lambda^2 t T = 0.$$

(b) Since the product solution arising out of the of the two ordinary equations derived in part (a) have different forms depending on whether $\lambda = 0$ or $\lambda \ne 0$, we treat each case separately.

If $\lambda = 0$ then the equations can be directly integrated to get $X(x) = c_1 + c_2 x$ and $T(t) = c_0$, so that the corresponding product solution is

$$u(x,t) = X(x)T(t) = (c_1 + c_2 x)c_0 = c_1 c_0 + c_2 c_0 x = C_1 + C_2 x,$$

where $C_1 = c_1 c_0$ and $C_2 = c_2 c_0$ are arbitrary constants.

If $\lambda \ne 0$ then the first equation has the solution $X(x) = c_1 \cos \lambda x + c_2 \sin \lambda x$, which can be written more compactly as $X(x) = A \cos \lambda(x - \alpha)$, where A and α are arbitrary constants. The second equation can be solved using the integrating factor $e^{\lambda^2 t^2/2}$ to obtain $T(t) = c_0 e^{-\lambda^2 t^2/2}$. The corresponding product solution is therefore of the form

$$u(x,t) = X(x)T(t) = A c_0 e^{-\lambda^2 t^2/2} \cos \lambda(x - \alpha) = C e^{-\lambda^2 t^2/2} \cos \lambda(x - \alpha),$$

where $C = Ac_0$.

29. If $u(x, t_0)$ is concave up for $x_0 < x < x_1$ then $u_{xx}(x, t_0) > 0$ for $x_0 < x < x_1$. This also means that $a^2 u_{xx}(x, t_0) > 0$ for $x_0 < x < x_1$. It follows that if u satisfies $a^2 u_{xx} = u_t$ then $u_t(x, t_0) > 0$ for $x_0 < x < x_1$. For each such x, $u_t(x, t)$ will remain positive over some interval (by continuity of u_t) as t increases over that interval. That is, there is a time $t(x)$ such that $u(x, t)$ increases as t increases from t_0 to $t(x)$.

If $u(x, t_0)$ is concave down for $x_0 < x < x_1$ then $u_{xx}(x, t_0) < 0$ for $x_0 < x < x_1$. As in the previous case, we conclude that $u_t(x, t_0) < 0$ for $x_0 < x < x_1$. For each such x, $u_t(x, t)$ will remain negative over some interval (by continuity of u_t) as t increases over that interval. That is, there is a time $t(x)$ such that $u(x, t)$ decreases as t increases from t_0 to $t(x)$.

31. Let $u(x, t) = X(x)T(t)$ be a nonzero product solution of the given homogeneous boundary-value problem and write the boundary conditions as $X(0) = X(p) = 0$. Taking the requisite partial derivatives and inserting them into the D.E. leads to the separated equation $a^2 X''/X = T'/T$. Setting both sides equal to the constant C gives the two ordinary equations

$$a^2 X'' - CX = 0 \qquad \text{and} \qquad T' - CT = 0.$$

The characteristic roots of the first equation are $\pm\sqrt{C}/a$, and the characteristic root of the second equation is C. Hence, the solutions are

$$X(x) = c_1 e^{(\sqrt{C}/a)x} + c_2 e^{-(\sqrt{C}/a)x} \qquad \text{and} \qquad T(t) = c_3 e^{Ct},$$

where c_1, c_2 and c_3 are real or complex constants to be chosen. The condition $X(0) = 0$ immediately gives $0 = c_1 + c_2$, or equivalently, $c_2 = -c_1$. (Note that $c_1 \neq 0$ because $u(x, t)$ is not the zero solution). The X equation can then be written as

$$X(x) = c_1 e^{(\sqrt{C}/a)x} - c_1 e^{-(\sqrt{C}/a)x} = c_1 e^{-(\sqrt{C}/a)x}\left(e^{2(\sqrt{C}/a)x} - 1\right).$$

The condition $X(p) = 0$ gives $0 = c_1 e^{-(\sqrt{C}/a)p}\left(e^{2(\sqrt{C}/a)p} - 1\right)$. Since $c_1 \neq 0$ and the exponential function is never zero, we must have $e^{2(\sqrt{C}/a)p} - 1 = 0$, or equivalently, $e^{2(\sqrt{C}/a)p} = 1$. Using the fact that $e^z = 1$ if, and only if, $z = 2k\pi i$ for some integer k, we have $2(\sqrt{C}/a)p = 2k\pi i$. Dividing by $2p$ gives $\sqrt{C}/a = k\pi i/p$, so that the X solution becomes

$$X(x) = c_1\left(e^{(k\pi i/p)x} - e^{-(k\pi i/p)x}\right) = c_1\left[\left(\cos\frac{k\pi x}{p} + i\sin\frac{k\pi x}{p}\right) - \left(\cos\frac{k\pi x}{p} - i\sin\frac{k\pi x}{p}\right)\right]$$

$$= 2ic_1 \sin\frac{k\pi x}{p}.$$

For the T equation, we square both sides of $\sqrt{C}/a = k\pi i/p$ and solve for C to get $C = -k^2 a^2 \pi^2/p^2$, so that

$$T(t) = c_3 e^{-(k^2 a^2 \pi^2/p^2)t}.$$

The product solutions are therefore of the form

$$u(x, t) = X(x)T(t) = 2ic_1 c_3 e^{-(k^2 a^2 \pi^2/p^2)t} \sin\frac{k\pi x}{p}.$$

Finally, we choose c_1 and c_3 so that $2ic_1 c_3 = 1$ (e.g., $c_1 = -i/2$, $c_3 = 1$) and replace $u(x, t)$ with $u_k(x, t)$ to indicate the dependence on k.

589

Exercise Set 10C (pgs. 731-732)

1. This exercise is a special case of Example 3 in the text. Specifically, we need only set $p = \pi$ and $A = 1$ in the solution found in Example 3 to get the solution desired here; namely,

$$u(x,t) = \cos at \sin x.$$

3. Using Equations 10.6 with $p = \pi$, $f(x) = 0$ and $g(x) = \begin{cases} 0, & 0 < x \le \pi/2; \\ 1, & \pi/2 < x < \pi, \end{cases}$ we see that $A_k = 0$ for $k \ge 1$, and

$$B_k = \frac{2}{k\pi a} \int_0^\pi g(x) \sin kx \, dx = \frac{2}{k\pi a} \int_{\pi/2}^\pi \sin kx \, dx = \frac{2}{k\pi a} \left[-\frac{1}{k} \cos kx \right]_{\pi/2}^\pi$$

$$= \frac{2}{k^2\pi a} \left[(-1)^{k+1} + \cos \frac{k\pi}{2} \right], \quad \text{for } k \ge 1.$$

The desired solution is therefore

$$u(x,t) = \sum_{k=1}^\infty \frac{2}{k^2\pi a} \left[(-1)^{k+1} + \cos \frac{k\pi}{2} \right] \sin kat \sin kx.$$

5. (a) Equilibrium solutions of $a^2 u_{xx} = u_{tt} + g$ are those that are independent of t. Let $u(x,t) = v(x)$ be such a solution. Taking the requisite partial derivatives and inserting them into the D.E. gives $a^2 v''(x) = g$, or equivalently, $v''(x) = g/a^2$. Two integrations with respect to x yields

$$v(x) = \frac{g}{2a^2} x^2 + c_1 x + c_2,$$

where c_1, c_2 are arbitrary constants.

(b) With respect to the equilibrium solutions found in part (a), the conditions $u(0,t) = u(p,t) = 0$ translate to $v(0) = v(p) = 0$. This leads to the two equations

$$0 = c_2 \qquad \text{and} \qquad 0 = \frac{g}{2a^2} x^2 + c_1 p + c_2,$$

from which $c_1 = -(gp)/(2a^2)$ and $c_2 = 0$. Hence,

$$v(x) = \frac{g}{2a^2} x^2 - \frac{gp}{2a^2} x = \frac{g}{2a^2} x(x - p).$$

(c) First solve the homogeneous wave equation $a^2 u_{xx} = u_{tt}$, $u(0,t) = u(p,t) = 0$, and let $w(x,t)$ denote the solution. Then

$$a^2 w_{xx} = w_{tt} \qquad \text{and} \qquad w(0,t) = w(p,t) = 0.$$

With $v(x)$ as found in part (b), use the boundary conditions on the right to conclude that the function $u(x,t) = w(x,t) + v(x)$ satisfies $u(0,t) = u(p,t) = 0$. Moreover, since $u_{xx}(x,t) = w_{xx}(x,t) + v''(x)$ and $u_{tt}(x,t) = w_{tt}(x,t)$, we have

$$a^2 u_{xx} = a^2 (w_{xx} + v'') = a^2 w_{xx} + a^2 v'' = w_{tt} + g = u_{tt} + g.$$

That is, $a^2 u_{xx} = u_{tt} + g$, so that $u(x,t)$ is a solution of the nonhomogeneous wave equation.

7. (a) Let $U(x)$ and $V(x)$ be twice differentiable functions that are defined for all real x, and let $u(x,t) = U(x + at)$ and $v(x,t) = V(x - at)$, where $a > 0$ is fixed. Let $t_1 > t_0$ be fixed but arbitrarily chosen times and set $b = a(t_1 - t_0) \ (> 0)$. Then, for each fixed x,

$$u(x,t_0) = U(x + at_0) = U(x + at_1 - b) = U(x - b + at_1) = u(x - b, t_1).$$

Since $x - b < x$, the two outer members of the above equation shows that as t increases from $t = t_0$ to $t = t_1$, the position of the wave changes from x to $x - b$. That is, the wave moves to the left with increasing t. Moreover, since the wave moves b units over the time interval $[t_0, t_1]$, the equation $b = a(t_1 - t_0)$ implies that $a =$ speed of the wave's motion.

In a similar fashion, for each fixed x,

$$v(x,t_0) = V(x - at_0) = V(x + b - at_1) = v(x + b, t_1),$$

which shows that as t increases from $t = t_0$ to $t = t_1$, the position of the the wave changes from x to $x + b$. That is, the wave moves to the right with increasing t. Moreover, as in the case of the first wave, $a =$ speed of the wave's motion.

(b) The identity $\cos(\alpha - \beta) = \cos\alpha\cos\beta + \sin\alpha\sin\beta$ with $\alpha = k\pi at/p$, $\beta = \theta$ implies

$$\sqrt{A_k^2 + B_k^2} \cos\left(\frac{k\pi at}{p} - \theta\right) = \sqrt{A_k^2 + B_k^2}\left[\cos\frac{k\pi at}{p}\cos\theta + \sin\frac{k\pi at}{p}\sin\theta\right]$$

$$= \left(\sqrt{A_k^2 + B_k^2}\cos\theta\right)\cos\frac{k\pi at}{p} + \left(\sqrt{A_k^2 + B_k^2}\sin\theta\right)\sin\frac{k\pi at}{p}.$$

If A_k and B_k are not both zero, we can define θ to be that unique angle $-\pi < \theta \leq \pi$ such that

$$\cos\theta = \frac{A_k}{\sqrt{A_k^2 + B_k^2}} \qquad \text{and} \qquad \sin\theta = \frac{B_k}{\sqrt{A_k^2 + B_k^2}}.$$

With this definition of θ, equality of the two outer member of the first displayed equation above implies that the general nonzero term of the solution given by Equation 10.6 in the text can be written as

$$(*) \qquad \left[A_k\cos\frac{k\pi at}{p} + B_k\sin\frac{k\pi at}{p}\right]\sin\frac{k\pi x}{p} = \sqrt{A_k^2 + B_k^2}\cos\left(\frac{k\pi at}{p} - \theta\right)\sin\frac{k\pi x}{p}.$$

(c) When the identity $\cos\beta\sin\alpha = \frac{1}{2}\sin(\alpha + \beta) + \frac{1}{2}\sin(\alpha - \beta)$, with $\alpha = k\pi x/p$ and $\beta = k\pi at/p - \theta$, is applied to the right side of equation $(*)$ in part (b), the result is

$$\sqrt{A_k^2 + B_k^2}\cos\left(\frac{k\pi at}{p} - \theta\right)\sin\frac{k\pi x}{p}$$

$$= \sqrt{A_k^2 + B_k^2}\left[\frac{1}{2}\sin\left(\frac{k\pi x}{p} + \frac{k\pi at}{p} - \theta\right) + \frac{1}{2}\sin\left(\frac{k\pi x}{p} - \frac{k\pi at}{p} + \theta\right)\right]$$

$$= \frac{1}{2}\sqrt{A_k^2 + B_k^2}\left[\sin\left(\frac{k\pi}{p}(x + at) - \theta\right) + \sin\left(\frac{k\pi}{p}(x - at) + \theta\right)\right].$$

Hence, the left side of equation $(*)$ in part (b) has the desired form. What's more, the last member above is a d'Alembert solution $u(x,t) = U(x + at) + V(x - at)$ of the wave equation with U and V defined by

$$U(x + at) = \frac{1}{2}\sqrt{A_k^2 + B_k^2}\sin\left(\frac{k\pi}{p}(x + at) - \theta\right),$$

$$V(x - at) = \frac{1}{2}\sqrt{A_k^2 + B_k^2}\sin\left(\frac{k\pi}{p}(x - at) + \theta\right).$$

9. (a) Let $G(x)$ be twice differentiable on $[0, p]$ and suppose $G'(0) = G'(p) = 0$. Extend G to the interval $[-p, 0)$ by defining $G(x) = G(-x)$ for $-p \leq x < 0$. Now extend G to $-\infty < x < \infty$ to have period $2p$. This makes $G(x)$ everywhere continuous. Further, for each x there is a unique integer n such that $(2n - 1)p < x \leq (2n + 1)p$, so that $x - 2np$ is in the interval $(-p, p]$. Hence,

$$G(x) = G(x - 2np) = G(-x + 2np) = G(-x),$$

where the first and third equality follows from the fact that G has period $2p$. Thus, G is an even function.

(b) The even periodic extension $G_e(x)$ of the function $G(x) = x^2(1 - x^2)$ is sketched on the right.

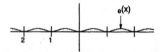

(c) Let $G(x)$ be an even function that is twice continuously differentiable on all of $\mathcal{R}$, and suppose G is periodic with period $2p$. Further suppose that $G'(0) = G'(p) = 0$. By Exercise 6(b) in this section, the function

$$u(x, t) = \frac{1}{2a}[G(x + at) - G(x - at)]$$

is a d'Alembert solution to $a^2 u_{xx} = u_t t$. To show that $u(x, t)$ satisfies the initial conditions, we first observe that $u(x, 0) = \frac{1}{2a}[G(x) - G(x)] = 0$ for all x. Moreover, with

$$u_t(x, t) = \frac{1}{2a}[aG'(x + at) - (-a)G'(x - at)] = \frac{1}{2}[G'(x + at) + G'(x - at)],$$

we find that $u_t(x, 0) = \frac{1}{2}[G'(x) + G'(x)] = G'(x)$. The boundary conditions are satisfied because

$$u(0, t) = \frac{1}{2a}[G(at) - G(-at)] = \frac{1}{2a}[G(at) - G(at)] = 0,$$

$$u(p, t) = \frac{1}{2a}[G(p + at) - G(p - at)] = \frac{1}{2a}[G(p + at) - G(-p + at)]$$

$$= \frac{1}{2a}[G(p + at) - G(p + at)] = 0,$$

where the third member in each of the first two lines follows from the fact that G is an even function, and the first member in the last line follows from the fact that G has period $2p$.

Finally, in part (a), G was asserted to be twice continuously differentiable only on $[0, p]$. The conditions $G'(0) = G'(p) = 0$ are therefore needed to insure that when G is extended to all of $\mathcal{R}$ the resulting extension will be continuously differentiable at all points of the form $x = np$, where n is an integer. This, in turn, forces the existence of $G''(np)$ for all n. In part (c), twice differentiability was assumed on all of $\mathcal{R}$, which can occur only if $G'(0) = G'(p) = 0$. In any case, these conditions are necessary to insure that $u(x, t)$ as defined above is a d'Alembert solution to the wave equation.

11. Let $u(x, t) = X(x)T(t)$ be a product solution of $t u_{xx} = u_{tt}$. Taking the requisite partial derivatives and inserting them into the D.E. gives $tX''T = XT''$. Separating variables results in $X''/X = T''/(tT)$, where $t \neq 0$. Setting each side equal to the constant λ leads to the two ordinary differential equations

$$X'' - \lambda X = 0 \qquad \text{and} \qquad T'' - \lambda t T = 0 \quad (t \neq 0).$$

Chapter Review

(pgs. 733-734)

1. The given series is a geometric series that converges for $|5-x| < 1$ and diverges otherwise. Specifically,

$$\sum_{k=0}^{\infty}(-1)^k(x-5)^k = \sum_{k=0}^{\infty}(5-x)^k = \frac{1}{1-(5-x)} = \frac{1}{x-4}, \quad 4 < x < 6.$$

3. The given series is a geometric series with the common ratio $1/(x^2+1)$. Since this ratio is positive and less than 1 for $x \neq 0$, the series converges for all nonzero x. We therefore have

$$\sum_{k=0}^{\infty}(x^2+1)^{-k} = \sum_{k=0}^{\infty}\left(\frac{1}{x^2+1}\right)^k = \frac{1}{1-[1/(x^2+1)]} = \frac{x^2+1}{x^2} = 1 + \frac{1}{x^2}, \quad x \neq 0.$$

5. The given series is of the form $e^X = \sum_{k=0}^{\infty} X^k/k!$ and converges for all real numbers X. Specifically, $X = (x+1)^2$ and we have

$$\sum_{k=0}^{\infty}\frac{(x+1)^{2k}}{k!} = \sum_{k=0}^{\infty}\frac{[(x+1)^2]^k}{k!} = e^{(x+1)^2}, \quad -\infty < x < \infty.$$

7. The given series is of the form $\cos X = \sum_{k=0}^{\infty}(-1)^k X^{2k}/(2k)!$ and converges for all real numbers X. Specifically, $X = x - 5$ and we have

$$\sum_{k=0}^{\infty}(-1)^k\frac{(x-5)^{2k}}{(2k)!} = \cos(x-5), \quad -\infty < x < \infty.$$

9. The given series is of the form $\cosh X = \sum_{k=0}^{\infty} X^{2k}/(2k)!$ and converges for all real numbers X. Specifically, $X = 1/(x^2-1)$ and we have

$$\sum_{k=0}^{\infty}\frac{(x^2-1)^{-2k}}{(2k)!} = \sum_{k=0}^{\infty}\frac{[1/(x^2-1)]^{2k}}{(2k)!} = \cosh\left(\frac{1}{x^2-1}\right), \quad |x| \neq 1.$$

11. The function $1/(1-x^3)$ has the form of the sum of a geometric series with common ratio x^3. That is,

$$\sum_{k=0}^{\infty} x^{3k} = \sum_{k=0}^{\infty}(x^3)^k = \frac{1}{1-x^3},$$

which holds for all x satisfying $-1 < x^3 < 1$, or equivalently, for $-1 < x < 1$. Now observe that the series is a power series about $x = 0$. Therefore, it must be the Taylor expansion about $x = 0$ for the function $(1-x^3)^{-1}$.

593

13. The Taylor expansion about $x = 0$ for the function $\ln(1+x)$ (given by Formula 2.4(d) in this chapter) holds for $-1 < x < 1$. When x is replaced with $-2x$, the series converges to $\ln(1-2x)$ for $-1 < -2x < 1$, or equivalently, for $-\frac{1}{2} < x < \frac{1}{2}$. That is,

$$\ln(1-2x) = \sum_{k=1}^{\infty}(-1)^{k+1}\frac{(-2x)^k}{k} = \sum_{k=1}^{\infty}(-1)^{k+1}\frac{(-1)^k 2^k x^k}{k} = \sum_{k=1}^{\infty}\frac{-2^k}{k}x^k, \quad -\frac{1}{2} < x < \frac{1}{2}.$$

This is the required Taylor expansion about $x = 0$ for $\ln(1-2x)$. The actual domain of convergence is a little larger than that shown above. To see this, note that when $x = -\frac{1}{2}$ is inserted into the above Taylor series, the result is an alternating series that converges by the Leibniz test. The domain of convergence is therefore $-\frac{1}{2} \le x < \frac{1}{2}$. Moreover, continuity of the log function implies that the series converges to $\ln 2$, a fact that does not follow directly from any of the theorems presented in the text.

15. To find the Taylor expansion about $x = 1$ for the function e^{x-1}, replace x with $x - 1$ in the Taylor expansion about $x = 0$ for the function e^x (given by Formula 2.4(a) in this chapter). This gives

$$e^{x-1} = \sum_{k=0}^{\infty}\frac{1}{k!}(x-1)^k, \quad -\infty < x < \infty.$$

Now observe that the resulting series is a power series about $x = 1$ and therefore it must be the Taylor expansion about $x = 1$ for the function e^{x-1}.

17. Replace x with $2x$ in the power series for $\cos x$ given by Formula 2.4(b) in this chapter to get

$$\cos 2x = \sum_{k=0}^{\infty}(-1)^k\frac{(2x)^{2k}}{(2k)!} = \sum_{k=0}^{\infty}\frac{(-1)^k 4^k}{(2k)!}x^{2k}, \quad -\infty < x < \infty.$$

This is the Taylor expansion about $x = 0$ for the function $\cos 2x$.

19. Since $\sin(x + \pi) = \sin x \cos \pi + \cos x \sin \pi = -\sin x$, we need only write down the negative of the power series for $\sin x$ given by Formula 2.4(c) in this chapter to get

$$\sin(x + \pi) = -\sin x = -\sum_{k=0}^{\infty}(-1)^k\frac{x^{2k+1}}{(2k+1)!} = \sum_{k=0}^{\infty}\frac{(-1)^{k+1}}{(2k+1)!}x^{2k+1}, \quad -\infty < x < \infty.$$

This is the Taylor expansion about $x = 0$ for the function $\sin(x + \pi)$.

21. Using the Taylor expansion about $x = 0$ for the function e^x given by Formula 2.4(a) in this chapter, we obtain

$$e^x - e^{2x} = \sum_{k=0}^{\infty}\frac{x^k}{k!} - \sum_{k=0}^{\infty}\frac{(2x)^k}{k!} = \sum_{k=0}^{\infty}\frac{1}{k!}x^k - \sum_{k=0}^{\infty}\frac{2^k}{k!}x^k = \sum_{k=0}^{\infty}\left(\frac{1-2^k}{k!}\right)x^k,$$

which holds for $-\infty < x < \infty$. Since this is a power series about $x = 0$ that converges to $e^x - e^{2x}$ for all x, it must be the Taylor expansion about $x = 0$ for the function $e^x - e^{2x}$.

23. Let $a_k = k^2 x^k$ for $k \ge 1$, form the quotient a_{k+1}/a_k and compute

$$\lim_{k\to\infty}\left|\frac{a_{k+1}}{a_k}\right| = \lim_{k\to\infty}\left|\frac{(k+1)^2 x^{k+1}}{k^2 x^k}\right| = |x|\lim_{k\to\infty}\left(\frac{k+1}{k}\right)^2 = |x|.$$

By the ratio test, the given series converges if $|x| < 1$ and diverges if $|x| > 1$. Testing the points $x = \pm 1$ separately, we obtain the two series $\sum_{k=1}^{\infty}k^2$ (for $x = 1$) and $\sum_{k=1}^{\infty}(-1)^k k^2$

(for $x = -1$), both of which diverge by the term test. Therefore, the series $\sum_{k=1}^{\infty} k^2 x^k$ converges if, and only if, $|x| < 1$.

25. Let $a_k = (x/k)^k$, form the quotient a_{k+1}/a_k and compute

$$\lim_{k\to\infty}\left|\frac{a_k}{a_{k+1}}\right| = \lim_{k\to\infty}\left|\frac{[x/(k+1)]^{k+1}}{(x/k)^k}\right| = \lim_{k\to\infty}\left|\frac{x}{k+1}\cdot\frac{[x/(k+1)]^k}{(x/k)^k}\right|$$

$$= |x|\lim_{k\to\infty}\left[\frac{1}{k+1}\cdot\left(\frac{k}{k+1}\right)^k\right].$$

Since $\lim_{k\to\infty} 1/(k+1) = 0$ and $\lim_{k\to\infty}\left[k/(k+1)\right]^k = 1/e$, the above limit is 0 for all x. By the ratio test, the series $\sum_{k=1}^{\infty}(x/k)^k$ is absolutely convergent for all x. Since absolute convergence implies convergence, the given series converges for all x.

27. For each integer $k \geq 0$ and each real number x we have

$$|2^{-k}\cos kx| \leq 2^{-k}.$$

Since the series $\sum_{k=1}^{\infty} 2^{-k}$ is a convergent geometric series (with ratio $r = 1/2 < 1$), the Weierstrass test shows that the series $\sum_{k=0}^{\infty} 2^{-k}\cos kx$ converges uniformly on $\mathcal{R}$. Since uniform convergence on $\mathcal{R}$ implies pointwise convergence on $\mathcal{R}$, the given series converges for all x

29. The kth term of the divergent harmonic series $\sum_{k=1}^{\infty} 1/k$ satisfies

$$\frac{1}{k} = \frac{k}{k^2} < \frac{k+1}{k^2},$$

which holds for all $k \geq 1$. By the comparison test, the series $\sum_{k=1}^{\infty}(k+1)/k^2$ diverges.

31. Two applications of L'hôpital's rule gives

$$\lim_{x\to\infty}\frac{e^x}{x^2} = \lim_{x\to\infty}\frac{e^x}{2x} = \lim_{x\to\infty}\frac{e^x}{2} = \infty.$$

Hence, $\lim_{k\to\infty}(-1)^k e^k/k^2 \neq 0$. Therefore, $\sum_{k=1}^{\infty}(-1)^k e^k/k^2$ diverges by the term test.

33. Noting that $2k \leq 2^k$ for $k \geq 1$, we have

$$\left(\frac{k}{2^k}\right)^k \leq \left(\frac{k}{2k}\right)^k = \left(\frac{1}{2}\right)^k, \quad k \geq 1.$$

Since $\sum_{k=1}^{\infty}(1/2)^k$ is a convergent geometric series, the series $\sum_{k=1}^{\infty}(k/2^k)^k$ converges by the comparison test. Moreover, since the series is a positive term series, it is absolutely convergent.

35. (a) For twice differentiable functions u, let L be the operator defined by $L(u) = u_{xx} - u_t$. Let $u = u(x,t)$ and $v = v(x,t)$ be twice differentiable functions and let a and b be constants. Then the linearity of the partial derivative operator gives

$$L(au+bv) = (au+bv)_{xx} - (au+bv)_t = (au_{xx} + bv_{xx}) - (au_t + bv_t)$$

$$= (au_{xx} - au_t) + (bv_{xx} - bv_t) = a(u_{xx} - u_t) + b(v_{xx} - v_t) = aL(u) + bL(v).$$

That is, L is a linear operator. It follows that linear combinations of solutions of the heat equation $u_{xx} - u_t = 0$ are also solutions.

(b) Let a and b be constants and let $u(x, t)$ and $v(x, t)$ be functions such that for all t

$$u(a, t) = u(b, t) = 0 \qquad \text{and} \qquad v(a, t) = v(b, t) = 0.$$

Now let c_1 and c_2 be any constants and observe that

$$c_1 u(a, t) + c_2 v(a, t) = c_1(0) + c_2(0) = 0 \quad \text{and} \quad c_1 u(b, t) + c_2 v(b, t) = c_1(0) + c_2(0) = 0.$$

Hence, linear combinations of u and v satisfy the boundary conditions and therefore the boundary conditions are linear in the sense asserted.

(c) Let a be a constant and let $u(x, t)$ and $v(x, t)$ be functions such that for all t, $u(a, t) = 1$ and $v(a, t) = 1$. If c_1 and c_2 are any constants then

$$c_1 u(a, t) + c_2 v(a, t) = c_1(1) + c_2(1) = c_1 + c_2.$$

Since $c_1 + c_2$ is not always equal to 1, a linear combination of two functions that satisfy the condition $u(a, t) = 1$ does not necessarily satisfy the same condition. That is, the given boundary condition is not linear in the sense of part (b).

(d) Let $u = u(x, t)$ and $v = v(x, t)$ have D as a common domain in $\mathcal{R}^2$ and let D_1 be that subset of $\mathcal{R}$ consisting of all x such that $(x, t) \in D$. Let $f(x)$ be defined on D_1 and let it remain undefined off D_1. Now suppose u and v both satisfy the initial condition $u(x, 0) = f(x)$ and $v(x, 0) = f(x)$ for all $x \in D_1$. Then for constants c_1, c_2 we have

$$(*) \qquad c_1 u(x, 0) + c_2 v(x, 0) = c_1 f(x) + c_2 f(x) = (c_1 + c_2) f(x) \quad \text{for all } x \in D_1.$$

On one hand, if $f(x_0) \neq 0$ for some $x_0 \in D_1$ and c_1, c_2 are chosen to satisfy $c_1 + c_2 = 0$ then $(*)$ reduces to

$$c_1 u(x_0, 0) + c_2 v(x_0, 0) = (c_1 + c_2) f(x_0) = 0 \neq f(x_0).$$

Thus, it is not true that every linear combination of $u(x, 0)$ and $v(x, 0)$ is equal to $f(x)$ for every $x \in D_1$ and therefore the initial condition is not linear in the sense of part (b).

On the other hand, if the initial condition is assumed to be linear in the sense of part (b) then the right side of $(*)$ must equal $f(x)$ for all $x \in D_1$. That is, for every pair of constants c_1, c_2, we must have

$$(c_1 + c_2) f(x) = f(x) \quad \text{for all } x \in D_1.$$

In particular, setting $c_1 = c_2 = 1$ leads to $2f(x) = f(x)$ for all $x \in D_1$. Hence, $f(x) = 0$ for all $x \in D_1$. In other words, the given initial condition is linear in the sense of part (b) if, and only if, $f(x)$ is identically zero on its domain of definition.

37. First, multiply the given Laplace equation by $r^2 \sin^2 \phi$ to get

$$(r^2 \sin^2 \phi) \frac{\partial^2 u}{\partial r^2} + (2r \sin^2 \phi) \frac{\partial u}{\partial r} + (\sin^2 \phi) \frac{\partial^2 u}{\partial \phi^2} + (\sin \phi \cos \phi) \frac{\partial u}{\partial \phi} + \frac{\partial^2 u}{\partial \theta^2} = 0.$$

Now assume there is a function of the form $u(r, \phi, \theta) = R(r) \Phi(\phi) \Theta(\theta)$ that satisfies this differential equation. Computing the requisite partial derivatives and inserting them into the above form of the D.E. gives

$$(r^2 \sin^2 \phi) R'' \Phi \Theta + (2r \sin^2 \phi) R' \Phi \Theta + (\sin^2 \phi) R \Phi'' \Theta + (\sin \phi \cos \phi) R \Phi' \Theta + R \Phi \Theta'' = 0.$$

Dividing throughout by $R\Phi\Theta$ and rearranging the result gives

$$(r^2\sin^2\phi)\frac{R''}{R} + (2r\sin^2\phi)\frac{R'}{R} + (\sin^2\phi)\frac{\Phi''}{\Phi} + (\sin\phi\cos\phi)\frac{\Phi'}{\Phi} = -\frac{\Theta''}{\Theta}.$$

Notice that the left member above is independent of θ, so that the right member is also independent of θ. Moreover, the right side is independent of both r and ϕ as well. That is, the right side is independent of all variables and must therefore be constant. Setting both members equal to the real constant λ gives the two equations

$$(r^2\sin^2\phi)\frac{R''}{R} + (2r\sin^2\phi)\frac{R'}{R} + (\sin^2\phi)\frac{\Phi''}{\Phi} + (\sin\phi\cos\phi)\frac{\Phi'}{\Phi} = \lambda \quad \text{and} \quad -\frac{\Theta''}{\Theta} = \lambda.$$

The second equation is a second-order ordinary differential equation in Θ and can be written in the standard form

(∗) $$\Theta'' + \lambda\Theta = 0.$$

The first equation can be divided throughout by $\sin^2\phi$ and then rearranged to get

$$r^2\frac{R''}{R} + 2r\frac{R'}{R} = \frac{\lambda - \sin\phi\big(\Phi''/\Phi + \cos\phi\,(\Phi'/\Phi)\big)}{\sin^2\phi}.$$

Since the right side is a function of ϕ only and the left side is a function or r only, varying either r or ϕ can change neither side of the equation. That is, both sides are independent of both r and ϕ and therefore both must be equal to the same constant. Setting both sides equal to the real constant γ gives

$$r^2\frac{R''}{R} + 2r\frac{R'}{R} = \gamma \quad \text{and} \quad \frac{\lambda - \sin\phi\big(\Phi''/\Phi + \cos\phi\,(\Phi'/\Phi)\big)}{\sin^2\phi} = \gamma,$$

which can be written in standard form as two second-order ordinary differential equations

$$r^2 R'' + 2rR' - \gamma R = 0 \quad \text{and} \quad (\sin\phi)\Phi'' + (\sin\phi\cos\phi)\Phi' + (\gamma\sin^2\phi - \lambda)\Phi = 0.$$

These two equations together with the equation given in (∗) are the desired second-order ordinary differential equations.